HIGH RESOLUTION SPECTROSCOPY

HIGH RESOLUTION SPECTROSCOPY

Second Edition

J. Michael Hollas
University of Reading, UK

JOHN WILEY & SONS
Chichester · New York · Weinheim · Brisbane · Singapore · Toronto

Baffins Lane, Chichester,
West Sussex PO19 1UD, England

National 01243 779777
International (+44) 1243 779777
e-mail (for orders and customer service enquiries): cs-books@wiley.co.uk
Visit our Home Page on http://www.wiley.co.uk
or http://www.wiley.com

Other Wiley Editorial Offices

John Wiley & Sons, Inc., 605 Third Avenue, New York,
NY 10158-0012, USA

WILEY-VCH Verlag GmbH, Pappelallee 3,
D-69469 Weinheim, Germany

Jacaranda Wiley Ltd, 33 Park Road, Milton,
Queensland 4064, Australia

John Wiley & Sons (Asia) Pte Ltd, Clementi Loop #02-01,
Jin Xing Distripark, Singapore 129809

John Wiley & Sons (Canada) Ltd, 22 Worcester Road,
Rexdale, Ontario M9W 1L1, Canada

Library of Congress Cataloging-in-Publication Data
Hollas, J. Michael (John Michael)
High resolution spectroscopy / J. Michael Hollas.—2nd ed.
p. cm.
Includes bibliographical references and index.
ISBN 0-471-97421-8 (cloth : alk. paper)
1. High resolution spectroscopy. I. Title
QC454.H618H64 1998 97-44183
543′.0858—dc21 CIP

British Library Cataloguing in Publication Data
A catalogue record for this book is available from the British Library

ISBN 0 471 97421 8

Typeset in 10/12pt Times by Dobbie Typesetting Limited, Tavistock, Devon
Printed and bound in Great Britain by Bookcraft (Bath) Ltd
This book is printed on acid-free paper responsibly manufactured from sustainable forestry, in which at least two trees are planted for each one used for paper production

CONTENTS

Preface to First Edition xiii

Preface to Second Edition xv

Fundamental Constants and Useful Conversion Factors xvii

Chapter 1 Quantization of Energy 1
1.1 Historical evolution of quantum theory 1
1.2 The Schrödinger equation 11
1.3 Some important solutions of the Schrödinger equation 13
1.3.1 Types of quantization in atoms and molecules 13
1.3.2 Solution of the Schrödinger equation for the hydrogen atom 15
1.3.3 Quantization of molecular rotational, electron spin and nuclear spin angular momenta 22
1.3.4 Quantization of vibrational energy: the harmonic oscillator 27
1.3.5 Summary of quantized quantities in atoms and molecules 31
Bibliography 31

Chapter 2 Interaction of Electromagnetic Radiation with Atoms and Molecules 32
2.1 Nature of electromagnetic radiation 32
2.2 Absorption and emission processes 33
2.3 Line widths of transitions 39
2.3.1 Natural line broadening 40
2.3.2 Doppler broadening 41
2.3.3 Pressure broadening 42
2.3.4 Wall collision broadening 43
2.3.5 Power saturation broadening 44
2.3.6 Modulation broadening 44
2.3.7 Summary 44
Bibliography 45

Chapter 3 General Experimental Methods 46
3.1 Regions of the electromagnetic spectrum 46
3.2 General features of instrumentation 47
3.3 Microwave and millimetre wave spectroscopy 49
3.4 Methods for reducing Doppler broadening 53
3.4.1 Spectroscopy in an effusive beam 53
3.4.2 Lamb dip spectroscopy 54
3.4.3 Spectroscopy in a supersonic free jet or molecular beam 55
3.5 Prisms, diffraction gratings and interferometers as dispersing elements 58

3.5.1 Resolution and resolving power 58
3.5.2 Prisms 60
3.5.3 Diffraction gratings 61
3.5.4 Interferometers 64
3.6 Far infrared spectroscopy 70
3.7 Mid and near infrared spectroscopy 74
3.8 Visible and near ultraviolet spectroscopy 78
3.8.1 Molecular samples 78
3.8.2 Atomic samples 82
3.9 Far (vacuum) ultraviolet spectroscopy 84
3.10 Low and high resolution spectroscopy 86
Bibliography 87

Chapter 4 Rotational Spectroscopy 89
4.1 Classification as linear molecules, symmetric rotors, spherical rotors or asymmetric rotors 89
4.2 Pure rotational infrared, millimetre wave and microwave spectra of diatomic and linear polyatomic molecules 94
4.2.1 Transition frequencies in the rigid rotor approximation 94
4.2.2 Intensities 100
4.2.3 Centrifugal distortion 102
4.2.4 The Stark effect 102
4.2.5 Nuclear hyperfine splitting 104
4.2.6 Vibrational satellites 105
4.2.6.1 Diatomic molecules 105
4.2.6.2 Linear polyatomic molecules 107
4.3 Pure rotational infrared, millimetre wave and microwave spectra of symmetric rotor molecules 110
4.4 Nuclear spin statistical weights and their effect on intensities 114
4.5 Pure rotational infrared, millimetre wave and microwave spectra of asymmetric rotor molecules 118
4.6 Pure rotational infrared, millimetre wave and microwave spectra of spherical rotor molecules 126
4.7 Interstellar molecules detected by their pure rotation spectra 127
4.8 Pure rotational Raman spectroscopy 130
4.8.1 Theory 130
4.8.2 Experimental techniques 135
4.9 Structure determination from rotational constants 139
4.10 Rotational spectroscopy of weakly bound complexes 141
4.10.1 Molecular beam electric resonance spectroscopy of van der Waals and hydrogen bonded complexes 143
4.10.2 Microwave spectroscopy of van der Waals and hydrogen bonded complexes 146
4.10.2.1 Microwave spectroscopy without a supersonic jet 146
4.10.2.2 Microwave spectroscopy with a supersonic jet: pulsed nozzle Fourier transform microwave spectroscopy 152
Bibliography 157

Chapter 5 Vibrational Spectroscopy 159
5.1 Diatomic molecules 159
5.1.1 Harmonic oscillator approximation 159
5.1.2 Anharmonicity 162
5.1.3 Vibration–rotation spectroscopy 166
5.1.3.1 Infrared vibration–rotation spectra 166
5.1.3.2 Raman vibration–rotation spectra 170
5.2 Polyatomic molecules 171
5.2.1 Group vibrations 171
5.2.2 Molecular symmetry 181
5.2.2.1 Elements of symmetry 181
5.2.2.1(i) *n*-Fold axis of symmetry, C_n 183
5.2.2.1(ii) Plane of symmetry, σ 183
5.2.2.1(iii) Centre of inversion (or centre of symmetry), i 185
5.2.2.1(iv) *n*-Fold rotation–reflection axis of symmetry, S_n 185
5.2.2.1(v) The identity element of symmetry, I (or E) 186
5.2.2.1(vi) Generation of elements 186
5.2.2.1(vii) Commutation and non-commutation of operations 187
5.2.2.2 Point groups 187
5.2.2.2(i) C_n point groups 187
5.2.2.2(ii) S_n point groups 188
5.2.2.2(iii) $C_{n\mathrm{v}}$ point groups 188
5.2.2.2(iv) D_n point groups 189
5.2.2.2(v) $C_{n\mathrm{h}}$ point groups 189
5.2.2.2(vi) $D_{n\mathrm{d}}$ point groups 189
5.2.2.2(vii) $D_{n\mathrm{h}}$ point groups 189
5.2.2.2(viii) T_d point group 190
5.2.2.2(ix) O_h point group 190
5.2.2.2(x) K_h point group 191
5.2.2.2(xi) I_h point group 191
5.2.2.3 Point groups as examples of groups in general 191
5.2.2.4 Point group character tables 193
5.2.2.4(i) $C_{2\mathrm{v}}$ character table 193
5.2.2.4(ii) $C_{3\mathrm{v}}$ character table 196
5.2.2.4(iii) $C_{\infty\mathrm{v}}$ character table 198
5.2.3 Determination of normal modes of vibration 201
5.2.3.1 Introduction 201
5.2.3.2 The stretching vibrations of a linear molecule ABC 204
5.2.3.3 Use of symmetry coordinates 206
5.2.3.4 Use of Wilson's FG matrix method 209
5.2.4 Vibrational selection rules 210
5.2.4.1 Electric dipole selection rules 210
5.2.4.2 Raman selection rules 215
5.2.5 Vibration–rotation spectroscopy 218
5.2.5.1 Linear molecules 219
5.2.5.1(i) Infrared spectra 220
5.2.5.1(ii) Raman spectra 223

5.2.5.2 Symmetric rotor molecules . . . 223
5.2.5.2(i) Infrared spectra . . . 223
5.2.5.2(ii) Raman spectra . . . 230
5.2.5.3 Spherical rotor molecules . . . 231
5.2.5.3(i) Infrared spectra . . . 232
5.2.5.3(ii) Raman spectra . . . 233
5.2.5.4 Asymmetric rotor molecules . . . 234
5.2.5.4(i) Infrared spectra . . . 237
5.2.5.4(ii) Raman spectra . . . 242
5.2.6 Anharmonicity . . . 244
5.2.6.1 Potential energy surfaces . . . 244
5.2.6.2 Anharmonic vibrations . . . 246
5.2.6.3 Fermi and Darling–Dennison resonances . . . 247
5.2.6.4 Local mode treatment of vibrations . . . 249
5.2.7 Vibrational potential functions with more than one minimum . . . 252
5.2.7.1 Inversion vibrations . . . 252
5.2.7.2 Ring-puckering vibrations . . . 256
5.2.7.3 Bending vibrations in non-linear triatomic molecules . . . 259
5.2.7.4 Torsional vibrations . . . 260
Bibliography . . . 266

Chapter 6 Electronic Spectroscopy . . . 267
6.1 Electronic spectroscopy of atoms . . . 267
6.1.1 The periodic table . . . 267
6.1.2 Vector representation of momenta and vector coupling approximations . . . 273
6.1.2.1 Angular momenta and magnetic moments . . . 273
6.1.2.2 Coupling of angular momenta . . . 275
6.1.2.2(i) Russell–Saunders coupling approximation . . . 276
6.1.2.2(ii) *jj* Coupling approximation . . . 281
6.1.2.2(iii) Coupling of nuclear spin to other angular momenta . . . 282
6.1.3 Spectra of the alkali metal atoms . . . 283
6.1.4 Spectrum of the hydrogen atom . . . 287
6.1.5 Spectra of the helium atom and the alkaline earth metal atoms . . . 289
6.1.6 Spectra of other polyelectronic atoms . . . 291
6.1.7 Symmetry selection rules . . . 293
6.1.8 Atoms in a magnetic field . . . 294
6.1.8.1 The Zeeman effect . . . 294
6.1.8.2 The Paschen–Back effect . . . 297
6.1.9 Atoms in an electric field . . . 298
6.2 Electronic spectroscopy of diatomic molecules . . . 298
6.2.1 Electronic structure . . . 298
6.2.1.1 Homonuclear diatomic molecules . . . 298
6.2.1.2 Heteronuclear diatomic molecules . . . 306
6.2.2 Rydberg orbitals . . . 308
6.2.3 Classification of electronic states: selection rules . . . 310
6.2.3.1 Classification of states . . . 310
6.2.3.2 Electric dipole selection rules for transitions with $\Delta S = 0$ . . . 312

6.2.3.3 Electric dipole selection rules for transitions with $\Delta S \neq 0$ 313
6.2.3.4 Magnetic dipole transitions 315
6.2.4 Vibrational coarse structure 316
6.2.4.1 Excited state potential energy curves 316
6.2.4.2 Progressions and sequences 318
6.2.4.3 Franck–Condon principle 320
6.2.4.4 Deslandres tables 324
6.2.4.5 Dissociation energies 325
6.2.4.6 Repulsive states, continuous spectra and predissociation 327
6.2.4.7 Ion-pair states 329
6.2.5 Rotational fine structure 330
6.2.5.1 $^1\Sigma$–$^1\Sigma$ electronic and vibronic transitions 330
6.2.5.2 $^1\Pi$–$^1\Sigma$ electronic and vibronic transitions 334
6.2.5.3 Hund's coupling case (a) 337
6.2.5.4 Hund's coupling case (b) 338
6.2.5.5 Hund's coupling case (c) 339
6.2.5.6 Hund's coupling case (d) 340
6.2.5.7 Hund's coupling case (e) 340
6.2.5.8 Rotational fine structure of electronic transitions involving states with non-zero multiplicity 340
6.2.5.9 Nuclear hyperfine structure 341
6.3 Electronic spectroscopy of polyatomic molecules 343
6.3.1 Orbitals, states and electronic transitions 343
6.3.1.1 AH_2 molecules 344
6.3.1.2 AB_2 and HAB molecules 353
6.3.1.3 AH_3 molecules 358
6.3.1.4 HAAH molecules 360
6.3.1.5 H_2AB molecules 362
6.3.1.6 H_2AAH_2 molecules 364
6.3.1.7 XHC—CHY molecules 367
6.3.1.8 Benzene and substituted benzenes 369
6.3.1.9 Naphthalene 371
6.3.1.10 Aza-aromatic molecules 373
6.3.1.11 Crystal field and ligand field molecular orbitals 374
6.3.1.11(i) Crystal field theory 374
6.3.1.11(ii) Ligand field theory 379
6.3.2 Selection rules and intensities for electronic transitions 380
6.3.3 Chromophores 381
6.3.4 Vibrational coarse structure 382
6.3.4.1 Herzberg–Teller intensity stealing 382
6.3.4.2 Progressions: Franck–Condon principle 387
6.3.4.2(i) Totally symmetric vibrations 387
6.3.4.2(ii) Non-totally symmetric vibrations 390
6.3.4.3 Sequences and cross-sequences 391
6.3.4.4 The Renner–Teller effect 393
6.3.4.5 The Jahn–Teller effect 397
6.3.5 Rotational fine structure 400

6.3.5.1 Linear molecules 400
6.3.5.1(i) Transitions between singlet states 400
6.3.5.1(ii) Transitions involving states with non-zero multiplicity 401
6.3.5.2 Symmetric rotor molecules 404
6.3.5.2(i) Transitions between singlet states 404
6.3.5.2(ii) Transitions involving states with non-zero multiplicity 411
6.3.5.3 Asymmetric rotor molecules 411
6.3.5.3(i) Transitions between singlet states 411
6.3.5.3(ii) Transitions involving states with non-zero multiplicity 417
6.3.5.4 Spherical rotor molecules 418
6.3.5.5 Effects of magnetic and electric fields 419
6.3.5.6 Diffuse spectra 421
Bibliography 425

Chapter 7 Photoelectron Spectroscopy 426
7.1 Introduction 426
7.2 Experimental methods 428
7.2.1 UPS spectrometers 428
7.2.2 XPS spectrometers 432
7.3 Ionization processes in photoelectron spectra 434
7.4 Koopmans' theorem 436
7.5 Photoelectron spectra and their interpretation 437
7.5.1 Ultraviolet photoelectron spectroscopy (UPS) 437
7.5.1.1 UPS of atoms 437
7.5.1.2 UPS of diatomic molecules 439
7.5.1.2(i) Hydrogen 439
7.5.1.2(ii) Nitrogen 441
7.5.1.2(iii) Oxygen, S_2, SO and nitric oxide 443
7.5.1.2(iv) The hydrogen halides 449
7.5.1.3 UPS of triatomic molecules 450
7.5.1.3(i) Linear molecules 450
7.5.1.3(ii) Bent molecules 451
7.5.1.4 UPS of larger polyatomic molecules 453
7.5.1.4(i) AH_3 molecules 453
7.5.1.4(ii) Ethylene 453
7.5.1.4(iii) Methane 454
7.5.1.4(iv) Benzene 456
7.5.1.5 Angular distribution of photoelectrons 457
7.5.2 X-ray photoelectron spectroscopy (XPS) 459
7.5.2.1 The chemical shift 459
7.5.2.2 Line widths in XPS 464
7.5.2.3 Electron shakeup and shakeoff satellites 464
Bibliography 467

Chapter 8 Lasers and Laser Spectroscopy 468
8.1 Lasers and masers 468
8.1.1 General features of lasers and their design 468

8.1.2 Properties of laser radiation . . . 470
8.1.3 Methods of obtaining population inversion . . . 471
8.1.4 Laser cavity modes . . . 472
8.1.5 *Q*-switching . . . 474
8.1.6 Mode locking . . . 476
8.1.7 Harmonic generation . . . 477
8.1.8 Examples of lasers and masers . . . 477
8.1.8.1 The ammonia maser . . . 478
8.1.8.2 The ruby and alexandrite lasers . . . 479
8.1.8.3 The titanium–sapphire laser . . . 481
8.1.8.4 The neodymium–YAG laser . . . 481
8.1.8.5 Diode (or semiconductor), spin-flip Raman and colour centre lasers . . . 483
8.1.8.6 The helium–neon laser . . . 487
8.1.8.7 The argon ion and krypton ion lasers . . . 490
8.1.8.8 The nitrogen laser . . . 491
8.1.8.9 Excimer and exciplex lasers . . . 492
8.1.8.10 The carbon dioxide laser . . . 493
8.1.8.11 The carbon monoxide laser . . . 495
8.1.8.12 The water and hydrogen cyanide lasers . . . 495
8.1.8.13 Chemical lasers . . . 496
8.1.8.14 Dye lasers . . . 497
8.1.8.15 The optical parametric oscillator . . . 501
8.2 Laser spectroscopy . . . 503
8.2.1 Resonance Raman and time-resolved resonance Raman spectroscopy . . . 504
8.2.2 Hyper Raman spectroscopy . . . 510
8.2.3 Stimulated Raman and Raman gain spectroscopy . . . 512
8.2.4 Inverse Raman spectroscopy . . . 518
8.2.5 Coherent anti-Stokes and coherent Stokes Raman scattering spectroscopy . . . 520
8.2.6 Laser magnetic resonance (or laser Zeeman) spectroscopy . . . 522
8.2.7 Laser Stark (or laser electric resonance) spectroscopy . . . 529
8.2.8 Spectroscopy with diode and colour centre lasers . . . 533
8.2.9 Spectroscopy with a difference–frequency laser system . . . 540
8.2.10 Spectroscopy with a tunable sideband laser . . . 544
8.2.11 Optothermal and vibrational predissociation spectroscopy . . . 548
8.2.12 Infrared laser induced fluorescence . . . 555
8.2.13 Photoacoustic spectroscopy . . . 557
8.2.14 Cavity ring-down absorption spectroscopy . . . 562
8.2.15 Infrared laser and microwave spectroscopy of molecules near to dissociation . . . 566
8.2.16 Rotationally resolved fluorescence excitation spectroscopy . . . 571
8.2.17 Fluorescence excitation and single vibronic level (or dispersed) fluorescence spectroscopy with vibrational resolution . . . 587
8.2.18 Multiphoton absorption and ionization spectroscopy . . . 601
8.2.19 Zero kinetic energy photoelectron spectroscopy . . . 617
8.2.20 Laser photoelectron or photodetachment spectroscopy of negative ions . . . 625
8.2.21 Double resonance spectroscopy . . . 630

8.2.21.1 Microwave–optical double resonance spectroscopy 632
8.2.21.2 Optical–optical double resonance spectroscopy 634
8.2.22 Stimulated emission pumping spectroscopy 644
8.2.23 Saturation, intermodulated fluorescence and polarization spectroscopy 655
8.2.24 Level crossing spectroscopy, including the Hanle effect 659
8.2.25 Level anticrossing spectroscopy 664
8.2.26 Quantum beat spectroscopy 668
8.2.27 Rotational coherence spectroscopy 672
8.2.28 Hydrogen atom photofragment translational spectroscopy 680
Bibliography 684

Appendix: Character Tables 686

References 699

Index of Atoms and Molecules 711

Subject Index 729

PREFACE TO FIRST EDITION

The subject matter which I have tried to cover in this book makes it suitable primarily for postgraduate students and teachers of spectroscopy. At the same time I have attempted to present the material in such a way that it will prove rewarding to an undergraduate student who wishes to read selected sections of the book in order to understand more clearly a part of spectroscopy which may be covered too superficially in an undergraduate text.

In the past 10 years or so, development of instrumentation in spectroscopy has caused what was already a vast subject to continue to expand, and at an even greater rate. What makes the increased expansion seem particularly remarkable is that it started at a time when the subject had become somewhat stagnant in the sense that developments tended to be confined to the application of standard experimental techniques to new and interesting molecules.

In this book I have confined myself mostly to what is implied by the title, namely high resolution spectroscopy. In general, this means that the sample is assumed to be in the gas or vapour phase. All the main branches of spectroscopy are covered with the exceptions of electron spin resonance and nuclear magnetic resonance. Photoelectron spectroscopy is included, not because it is basically a high resolution technique, but because the resolution is so much better than in previous attempts to obtain ionization energies of polyatomic molecules. In addition the subject is so closely related to electronic spectroscopy as to merit inclusion for this reason alone.

It has been my aim to present an overview of microwave, millimetre wave, infrared, visible, ultraviolet, Raman, photoelectron and laser spectroscopy with a bias towards the experimental aspects and to provide the reader with many illustrations of spectra. In the high resolution field such instrumental developments as continuously recording microwave spectrometers, sources for millimetre wave spectroscopy, infrared and visible interferometers, ultraviolet and X-ray photoelectron spectrometers and, particularly, lasers have created revolutionary changes in research and also in teaching. It never has been the case that the most recent advances in research are necessarily too difficult to understand for them to be included in an undergraduate course, and that is probably more true now than ever it was. One example is photoelectron spectroscopy, which has been developed almost entirely since 1963 and which seems tailor-made to drive home the validity of the concepts of atomic and molecular orbitals. Another example is provided by lasers, which we now read and hear about so much in science generally that an undergraduate should learn something of their construction and uses.

With such rapid developments in high resolution spectroscopy it has become quite common for advances in restricted parts of the field to be reported in the form of reviews, each written by a different author and published in the mushrooming numbers of review publications. Throughout this book I have referred to many such reviews which have an important part to play in the scientific literature. Complementary to these is a book, such as this one, in which a single author is able to stress the unifying aspects of all the branches of the subject. Such unification is probably more easily appreciated by today's spectroscopists than those of say 20 years ago. In those days many spectroscopists had to build their own instruments which necessitated, for example, a microwave spectroscopist having to be an electronics expert and an infrared or ultraviolet spectroscopist requiring a thorough knowledge of optics. It is not surprising that acquiring this expertise, building instruments and recording spectra in one branch of spectroscopy took so much time that those who were able to make important contributions in more than one branch were exceptional.

In a book covering such a wide range it is not possible to give a comprehensive list of references. In the first six chapters I have relied mainly on the books and reviews referred to in the Bibliographies (at the end of the chapters) for providing references. The material covered in Chapter 7 and, more especially, Chapter 8 is generally much more recently developed so that I have found it necessary to provide more references for these chapters. The list of references is at the end of the book.

I have tried as far as possible to avoid acronyms and abbreviations which abound in the literature of spectroscopy but which may often confuse the reader.

During the writing of this book I have been greatly helped by stimulating discussions with colleagues at Reading University, particularly Professor I. M. Mills, Professor G. W. Series and Dr A. G. Robiette.

I am especially grateful to Dr J. P. Maier, Dr J. W. C. Johns, Dr A. G. Robiette, Dr J. K. G. Watson and Professor D. H. Whiffen, who have each read through a chapter of the book at the manuscript stage and without whose comments, criticisms and corrections there would be more errors and omissions than those which may still be present, and also to Dr T. Ridley, who helped in checking the proofs.

Some of the figures have been made from original spectra and I wish to express my thanks to the following who supplied them:

Figure 4.13 Professor I. M. Mills
Figure 4.15 Dr A. G. Robiette
Figure 4.21 Dr R. K. Heenan and Dr A. G. Robiette
Figure 4.28 Professor I. Ozier
Figure 4.35 Dr R. J. Butcher
Figure 5.10 Professor S. Brodersen
Figure 5.52 Professor I. M. Mills and Dr P. H. Turner
Figure 5.56 Dr A. G. Robiette
Figure 5.57 Professor I. M. Mills and Dr P. H. Turner
Figure 5.59 Professor H. Wieser
Figure 5.60 Professor I. M. Mills and Dr P. H. Turner
Figure 5.61 Professor I. M. Mills and Dr P. H. Turner
Figure 6.46 Dr R. F. Barrow
Figure 6.49 Dr R. F. Barrow
Figure 6.104 Dr A. E. Douglas
Figure 6.105 Dr G. Herzberg
Figure 6.107 Professor I. M. Mills

The rest of the figures, other than those which are reproduced directly from journals or books, have been drawn by Mr H. Nichol and I am very appreciative of the way in which he has managed to interpret my intentions.

My thanks are due also to various typists who have been involved with producing the final manuscript but especially to Mrs. A. Gillett who did such an excellent job in typing about three-quarters of it.

J. Michael Hollas

PREFACE TO SECOND EDITION

The first edition of *High Resolution Spectroscopy* was published in 1982. When a book has been out of print for many years, as this has been, an author can despair of ever having the chance of bringing it up to date and also of correcting any errors and misprints. I very much appreciate the opportunity that the publishers of this second edition, John Wiley & Sons, have given me to do this.

The magnitude of the task of bringing the first edition up to date can be appreciated when realising that it contained virtually no material published after 1980. In the following 17 years the subject of high resolution spectroscopy has blossomed. For example, in microwave spectroscopy, the Fourier transform method was developed only in 1979 and Fourier transform infrared spectroscopy, although starting in 1966, was only used routinely much later when commercial instruments became available.

Although the first laser, the ruby laser, was produced in 1960, this proved to be of limited use in spectroscopy. It was not until the development of more flexible, particularly tunable, lasers, such as the dye and diode lasers in the early 1970s, that they began to make an appreciable impact. Since then, many ingenious spectroscopic techniques have been developed. In this new edition the section on Laser Spectroscopy contains 28 sub-sections, each devoted to a particular technique. Inevitably, I have not been able to cover all techniques—there are just too many of them—but I have tried to include those which have made the greatest impact.

Use of the supersonic jet, the 'free' jet or skimmed to form a molecular beam, has had an impact on high resolution spectroscopy which is almost as great as that of lasers. The first jet spectra were published in 1975 so that the first edition could give only some impression of the breakthrough that this development was to prove. The stabilization of weakly bound complexes in a supersonic jet has led to structural information, not only in the ground state but also in vibrationally and electronically excited states, which could not have been dreamt of in earlier days.

The extreme rotational, and less extreme vibrational, cooling experienced by molecules seeded into a jet has opened up high resolution to the study of very much larger molecules than was possible previously.

Much high resolution spectroscopy has been done with resolution limited by the line width imposed by the Doppler effect. In 1969, the first sub-Doppler spectra were obtained; these were in the microwave region. More recently, laser techniques such as the use of a skimmed supersonic jet and, in two-photon spectroscopy, counterpropagating laser beams, have been devised for obtaining sub-Doppler resolution in other regions of the electromagnetic spectrum. To take advantage of the sub-Doppler capability the laser band width must be small compared with the sub-Doppler line width. This is readily achieved with, for example, a ring dye or diode laser.

Unfortunately, since 1982, there has been little progress in recognizing the error of our ways in saying, for example, 'a frequency of 3000 wavenumbers' when what we should say is 'a wavenumber of 3000 centimetre-to-the-minus-one' or 'a wavenumber of 3000 reciprocal centimetres.' The main reason for this is the awkwardness of saying 'centimetre-to-the-minus-one' or 'reciprocal centimetres.' When we really are talking about frequency no such problem arises: we say, for example, 'a frequency of 200 hertz.' The reason that there is no problem is that the 'second-to-the-minus-one' or 'reciprocal second' has been conveniently called the hertz (Hz).

What we need is a corresponding name for the 'centimetre-to-the-minus-one' or 'reciprocal centimetre.' The name 'kaiser' had a brief life but, even then, only among predominantly analytical spectroscopists. It would be an appropriate tribute to someone who has contributed so much to spectroscopy if it were to be called the 'herzberg,' for which the symbol would be Hg. In this way we would say 'a wavenumber of 3000 herzbergs,' meaning 3000 Hg, and the problem would be solved.

Finally, I would like to thank Professor B. van der Veken for Figure 5.9 and all those authors of papers and review articles for the use of many of the figures which have been reproduced here. Making choices of which figures to use and, indeed, which examples to use of all the wonderful work that has been done in the field of high resolution spectroscopy is invidious. I can only hope that those which have been chosen give a useful impression of the present state of the art.

J. Michael Hollas
e-mail: michaelhollas@compuserv.com

FUNDAMENTAL CONSTANTS AND USEFUL CONVERSION FACTORS

FUNDAMENTAL CONSTANTS

Quantity	Symbol	Value and units[†]
Speed of light (*in vacuo*)	c	$2.997\,924\,58 \times 10^{8}$ m s^{-1} (exactly)
Vacuum permeability	μ_0	$4\pi \times 10^{-7}$ H m^{-1} (exactly)
Vacuum permittivity	$\varepsilon_0\,(= \mu_0{}^{-1}\, c^{-2})$	$8.854\,187\,816 \times 10^{-12}$ F m^{-1}
Charge on proton	e	$1.602\,177\,33\,(49) \times 10^{-19}$ C
Planck constant	h	$6.626\,075\,5\,(40) \times 10^{-34}$ J s
Molar gas constant	R	$8.314\,510\,(70)$ J mol^{-1} K^{-1}
Avogadro constant	N_A, L	$6.022\,136\,7\,(36) \times 10^{23}$ mol^{-1}
Boltzmann constant	$k\,(= RN_A{}^{-1})$	$1.380\,658\,(12) \times 10^{-23}$ J K^{-1}
Atomic mass unit	$u\,(= 10^{-3}$ kg mol$^{-1}\, N_A{}^{-1})$	$1.660\,540\,2\,(10) \times 10^{-27}$ kg
Rest mass of electron	m_e	$9.109\,389\,7\,(54) \times 10^{-31}$ kg
Rest mass of proton	m_p	$1.672\,623\,1\,(10) \times 10^{-27}$ kg
Rydberg constant	R_∞	$1.097\,373\,153\,4\,(13) \times 10^{7}$ m^{-1}
Bohr radius	a_0	$5.291\,772\,49\,(24) \times 10^{-11}$ m
Bohr magneton	$\mu_B\,[= e\hbar(2m_e)^{-1}]$	$9.274\,015\,4\,(31) \times 10^{-24}$ J T^{-1}
Nuclear magneton	μ_N	$5.050\,786\,6\,(17) \times 10^{-27}$ J T^{-1}
Electron magnetic moment	μ_e	$9.284\,770\,1\,(31) \times 10^{-24}$ J T^{-1}
g-Factor for free electron	$g_e(= 2\mu_e\mu_B{}^{-1})$	$2.002\,319\,304\,386\,(20)$

[†]Values taken from International Union of Pure and Applied Chemistry, *Quantities, Units and Symbols in Physical Chemistry*, 2nd Edition, Blackwell Science, Oxford (1993). The uncertainties in the final digits are given in parentheses.

USEFUL CONVERSION FACTORS

Unit	cm^{-1}	MHz	kJ	eV	kJ mol^{-1}
1 cm^{-1}	1	29 979.25	$1.986\,447 \times 10^{-26}$	$1.239\,842 \times 10^{-4}$	$11.962\,66 \times 10^{-3}$
1 MHz	$3.335\,64 \times 10^{-5}$	1	$6.626\,076 \times 10^{-31}$	$4.135\,669 \times 10^{-9}$	$3.990\,313 \times 10^{-7}$
1 kJ	$5.034\,11 \times 10^{25}$	$1.509\,189 \times 10^{30}$	1	$6.241\,506 \times 10^{21}$	$6.022\,137 \times 10^{23}$
1 eV	8065.54	$2.417\,988 \times 10^{8}$	$1.602\,177 \times 10^{-22}$	1	96.485 3
1 kJ mol^{-1}	83.593 5	$2.506\,069 \times 10^{6}$	$1.660\,540 \times 10^{-24}$	$1.036\,427 \times 10^{-2}$	1

1 QUANTIZATION OF ENERGY

1.1 HISTORICAL EVOLUTION OF QUANTUM THEORY

During the late nineteenth century, much attention was focused on phenomena which defied explanation in terms of the newtonian laws of classical mechanics. One such phenomenon was the emission spectrum of atomic hydrogen which had been observed to consist of discrete wavelengths rather than the continuous range of wavelengths which classical mechanics predicted. Balmer, in 1885, was able to fit the wavelengths λ, observed in the visible region of the spectrum and comprising what we now call the Balmer series, to the empirical formula

$$\lambda = n'^2 G/(n'^2 - 4) \tag{1.1}$$

where G is a constant and n' is an integer which can take only the values 3, 4, 5,... For electromagnetic radiation, of which visible light is a part, the frequency ν is related to the wavelength by

$$c = \nu\lambda \tag{1.2}$$

where c is the speed of light. Equation (1.1) can be written more usefully, in terms of frequency, in the form

$$\nu = R_H[(1/2^2) - (1/n'^2)] \tag{1.3}$$

R_H is the Rydberg constant for hydrogen and is named after Rydberg who, in 1890, proposed that expressions of the form of equation (1.3) should be applicable to the spectra of not only hydrogen but also other elements. In 1908, Ritz realized the importance of the fact implied by equation (1.3) that the frequencies in an atomic spectrum can be expressed as the difference between two terms—in this case the terms are $R_H/2^2$ and R_H/n'^2.

Another phenomenon which could not be interpreted classically was the frequency distribution of radiation from a black body. The distribution of the energy over a range of frequencies and also the way in which the distribution changes with temperature is illustrated in Figure 1.1. The attempts by Rayleigh and Jeans in 1900 to derive an equation to reproduce the observations were successful at low frequencies but were unable to predict the maximum in the energy distribution and the decrease at high frequencies. On the other hand, in 1894 Wien had been able to predict how the frequency at which there is a maximum in the energy distribution changes with temperature.

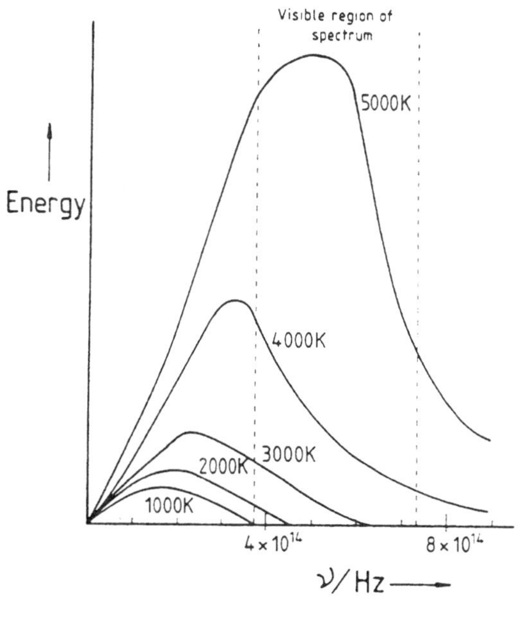

Figure 1.1 Frequency distribution of black body radiation and its dependence on temperature

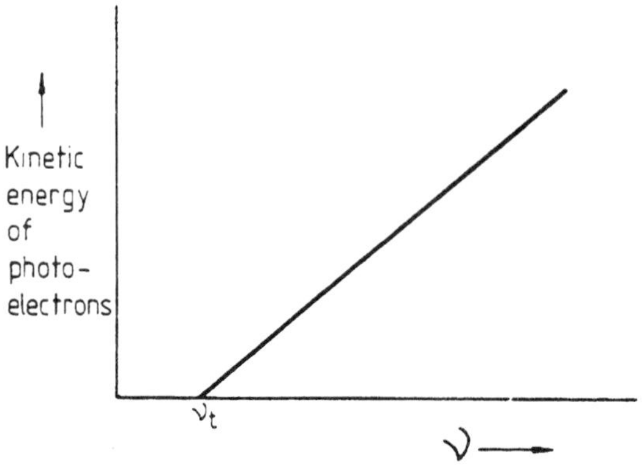

Figure 1.2 Dependence of the kinetic energy of photoelectrons on the frequency of incident radiation

The photoelectric effect also defies classical laws. The effect is observed when radiation falls on the surface of a metal. If the frequency of the radiation is continuously increased nothing happens until a frequency ν_t, the threshold frequency, is reached when electrons, so-called photoelectrons, are emitted from the metal surface. As the frequency is increased, further electron emission is instantaneous and the photoelectrons have kinetic energy which is proportional to the frequency. This behaviour is shown in Figure 1.2. All metals have their own characteristic threshold frequency. What was expected from classical physics was that the electrons in the metal would require a certain specific amount of energy, known as the work function,† to release them from the surface. If the radiation was of low frequency, i.e. of low energy, then it would take a considerable time, which would be dependent on the quantity of radiation falling on the surface, for the metal to absorb sufficient energy to release an electron: for radiation of high frequency the time for electron release would be shorter. However, the experimental observations are clearly a contradiction of these expectations.

†The work function is really a solid-state ionization energy.

A further difficulty, inexplicable using classical laws, was the temperature dependence of the molar heat capacity, C_v, of a solid at low temperatures. In 1819, Dulong and Petit had derived the expression

$$C_v = 3R \tag{1.4}$$

where R is the ideal gas constant, which requires that C_v is independent of temperature; however, this is far from the case at low temperatures, as shown by Figure 1.3.

The breakthrough which led to satisfactory explanations of the hydrogen atom spectrum, black body radiation, the photoelectric effect and heat capacities of solids was made by Planck in 1900. His theory of black body radiation requires that the oscillators of which a black body is made up and which are responsible for the emission of energy can oscillate only in such a way that the energy E emitted is given by

$$E = nh\nu \tag{1.5}$$

where n is an integer, ν is the fundamental frequency of the oscillator and h is a constant which we now know as the Planck constant. The energy is said to be quantized in discrete packets, or quanta, each of magnitude $h\nu$. The expression which Planck obtained for $\rho(\nu)$, the density of radiation of frequency ν for a black body at an absolute temperature T, is

$$\rho(\nu) = 8\pi h\nu^3\{c^3[\exp(h\nu/kT) - 1]\}^{-1} \tag{1.6}$$

where c is the speed of light.

The fact that the idea of quantization of energy was not formulated until 1900 was due undoubtedly to its appearing, at first sight, to be at variance with our experience of the macroscopic world. For example, in the case of a simple pendulum swinging at a constant frequency we can give the pendulum *any* energy we choose by starting it swinging with a larger or smaller amplitude. Similarly, if we construct

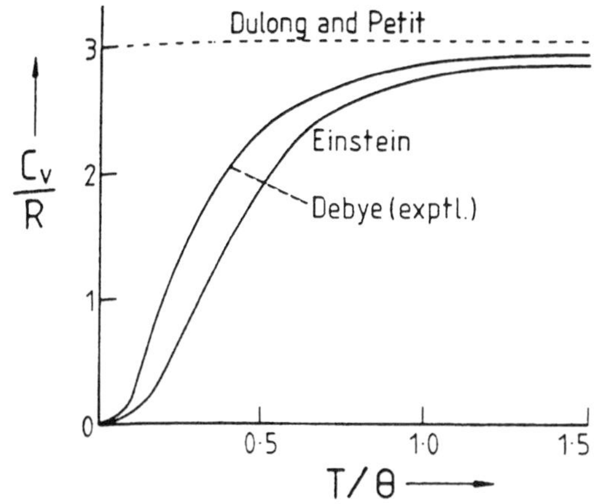

Figure 1.3 Temperature dependence of the molar heat capacity, C_v, of a solid at low temperatures. The quantity θ is the Einstein characteristic temperature where $\theta = h\nu/k$

a ball and spring model to simulate an oscillator in a black body we can, by choosing the initial impetus which we give to it, choose the energy with which it oscillates.

Part of the answer to this apparent anomaly in the behaviour of microscopic systems, regarded here as those on an atomic or molecular scale, and macroscopic systems is the very small value of the Planck constant: the currently accepted value is

$$h=(6.626\,075\,5 \pm 0.000\,004\,0)\times 10^{-34}\ \text{J s} \quad (1.7)$$

Therefore, the magnitude of energy quanta of a simple pendulum oscillating with a period of, say, 1 s is only $6.626\,075\,5\times 10^{-34}$ J. It is hardly surprising that quantization escaped notice in macroscopic systems!

In 1906 Einstein explained the photoelectric effect in terms of quantization of the energy, in the form of visible or ultraviolet light, falling on the metal surface. Only when the quanta of the radiation have sufficient energy $h\nu$ to overcome the forces binding the electron to the metal, the work function, will a photoelectron be ejected. As the energy and frequency of the radiation increase, the excess energy of $h\nu$ over the work function will appear as kinetic energy of the photoelectron.

In the same year, Einstein applied the new quantum theory to the problem of heat capacities of solids at low temperatures and obtained the following expression:

$$C_v/N_\text{A}k = [3(\theta_\text{E}/T)^2 \exp(\theta_\text{E}/T)]/[\exp(\theta_\text{E}/T) - 1]^2 \quad (1.8)$$

where N_A is the Avogadro constant,† k is the Boltzmann constant and T is the absolute temperature. θ_E is the Einstein characteristic temperature and is given by

$$\theta_\text{E} = h\nu_0/k \quad (1.9)$$

Einstein's incorporation of quantum theory into the theory of heat capacity involved the assumption that *all* the $3N$ vibrations of the N particles in the solid lattice have the frequency ν_0. This is only very approximately true and although, as shown in Figure 1.3, equation (1.8) is a good approximation to the experimental specific heat variation at low temperatures, agreement is not perfect. Debye was able to improve on Einstein's theory by taking into account the fact that there is a distribution of frequencies among the oscillators, and agreement with experiment was obtained.

In a somewhat empirical way, Bohr, in 1913, amalgamated classical mechanics and quantum theory in order to explain the observations of Balmer, Rydberg and Ritz regarding the spectrum of atomic hydrogen. Bohr assumed that the electron in the hydrogen atom can move only in specific circular orbits around the nucleus. He proposed that the orbits are those for which the electron has angular momentum, p_θ, given by

†N_A strictly should not be called the Avogadro number, as it frequently is, since it is *not* dimensionless and therefore *not* a pure number: it usually has units of mol^{-1}.

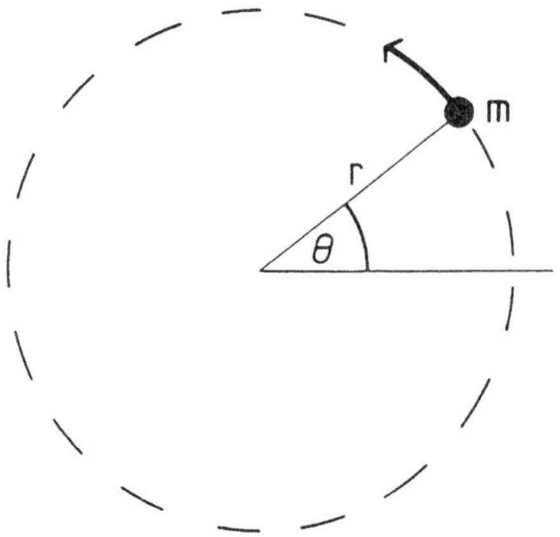

Figure 1.4 Angular momentum due to a ball tracing a circular path at the end of a string

$$p_\theta = nh/2\pi \tag{1.10}$$

where n is any non-zero integer.

In understanding angular momentum, we recall that the linear momentum p of a body of mass m moving with linear velocity v is given by

$$p = mv \tag{1.11}$$

If we consider a ball of mass m at the end of a string of length r, as shown in Figure 1.4, rotating with an angular velocity ω, then the angular momentum p_θ is given by

$$p_\theta = I\omega \tag{1.12}$$

where I is the moment of inertia, which is mr^2 in this example; ω is related to the angle θ, which is swept out, by

$$\omega = \dot{\theta} = \mathrm{d}\theta/\mathrm{d}t \tag{1.13}$$

where θ is measured from zero at time $t = 0$.

Bohr assumed that, in the hydrogen atom, energy is emitted or absorbed only when the electron transfers from one orbit to another and this energy ΔE is related to the frequency of the radiation emitted or absorbed by

$$\Delta E = h\nu \tag{1.14}$$

Calculation of the energy E_n of the electron in the nth orbit using these quantum postulates together with classical mechanics gave

$$E_n = -(\mu e^4/8h^2\varepsilon_0^2)(1/n^2) \tag{1.15}$$

where μ [$= m_e m_p/(m_e + m_p)$] is the reduced mass of the system of electron of mass m_e plus proton of mass m_p; e is the charge on the electron and ε_0 is the permittivity of free space. The energy levels are shown in Figure 1.5. It is important to remember that, conventionally, the zero of energy is taken to be that of the $n = \infty$ level which corresponds to ionization of the atom† in which the electron is removed from the influence of the nucleus. Below the $n = \infty$ level the energy levels are discrete and, above it, continuous.

From equation (1.15), it follows that for a transition of the electron from a lower, n'', to an upper, n', orbit‡

$$\Delta E = (\mu e^4/8h^2\,\varepsilon_0^2)\,[(1/n''^2) - (1/n'^2)] \tag{1.16}$$

From equations (1.14) and (1.16) we obtain, for the frequency ν of the radiation emitted or absorbed,

$$\nu = (\mu e^4/8h^3\,\varepsilon_0^2)\,[(1/n''^2) - (1/n'^2)] \tag{1.17}$$

Comparing this with equation (1.3), obtained empirically 18 years earlier, we can see that $n'' = 2$ for the Balmer series. This series, and also the Lyman ($n'' = 1$), Paschen ($n'' = 3$), Brackett ($n'' = 4$) and Pfund ($n'' = 5$) series, is indicated in Figure 1.5. There is in fact an infinite number of series, with $n'' = 1-\infty$, but only those with $n'' = 1$–5 have their discoverer's

†It is incorrect to speak, as is sometimes done, of 'the ionization of an electron.' In the process, involving an atom A, in which $A \rightarrow A^+ + e$, it is clearly A which is ionized, i.e. converted into an ion, and not the electron.

‡It is a general spectroscopic convention, and one which will be used throughout this book, to use double primes (″) and a single prime (′) to indicate the lower and upper states, respectively, of a transition.

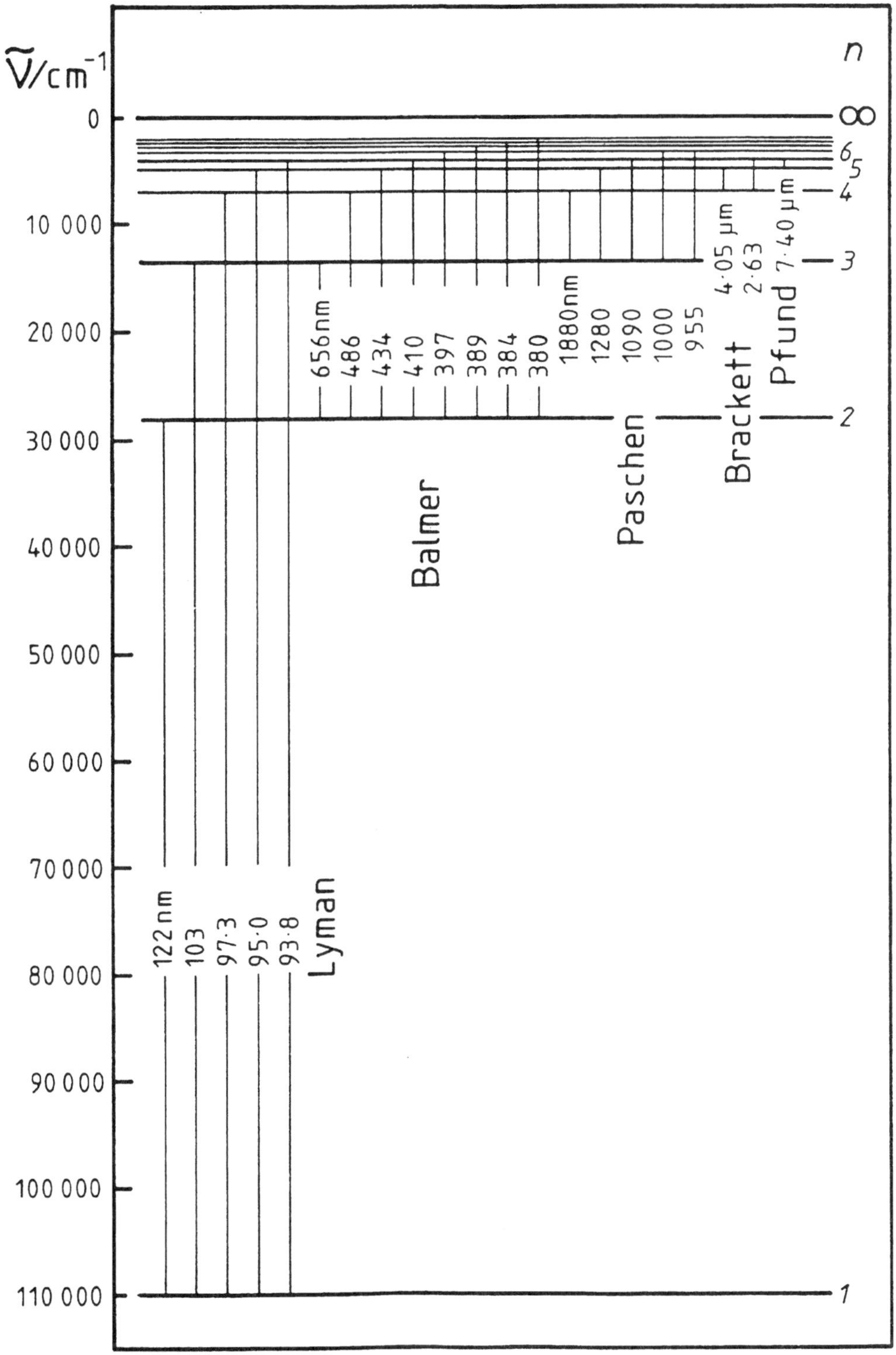

Figure 1.5 Energy levels of the hydrogen atom, according to the Bohr theory

names attached. Many members of series with high values of n'' have been observed, by techniques of radioastronomy, in the interstellar medium where there is a large amount of atomic hydrogen. For example, the $n' = 167 - n'' = 166$ transition† has been observed with $\nu = 1.425$ GHz ($\lambda = 21.04$ cm).

We can see also from equations (1.17) and (1.3) that the Rydberg constant with the dimensions of frequency is given by $\mu e^4/8h^3\,\varepsilon_0^{\,2}$. However, this constant is usually used and quoted with the dimensions of wavenumber, where the wavenumber $\tilde{\nu}$ is related to the frequency by‡

$$\nu = c\tilde{\nu} \tag{1.18}$$

†It is a general spectroscopic convention, and one which will be used throughout this book, to indicate a transition between two states by putting the *upper state first* and the *lower state second.*

‡Equation (1.18), relating frequency ν and wavenumber $\tilde{\nu}$, demonstrates that they have different dimensions. The dimensions of ν are (time)$^{-1}$ and those of $\tilde{\nu}$ are (length)$^{-1}$. Provided that the dimensions are correct there is a choice of units for either of these physical quantities: commonly used are s^{-1} (or Hz) for frequency and cm^{-1} for wavenumber. Therefore, we should speak of, for example, 'a frequency of 20 Hz' and 'a wavenumber of 1300 cm^{-1}'. While all this is undeniably correct, what commonly happens in practice is that, although frequency is treated properly, people speak of, for example: 'a frequency of 1300 wavenumbers' instead of 'a wavenumber of 1300 cm^{-1}.' The use of wavenumber as the unit instead of cm^{-1} probably stems from the difficulty of saying 'centimetre-to-the-minus-one' or 'reciprocal centimetre.' What is needed is a new, acceptable name for the cm^{-1} to parallel the Hertz for s^{-1}. Indeed, the kaiser has sometimes been used but it has not been generally accepted and, in any case, the abbreviation K can be confused with the unit of temperature. The use of frequency instead of wavenumber for the physical quantities then arises because, otherwise, we should be speaking of, for example, 'a wavenumber of 1300 wavenumbers.' (See Preface to Second Edition for suggested alternative of the herzberg (Hg) for cm^{-1}.)

Generally, in this book, I have fallen in with general usage but, in respect of wavenumber, where general usage is so clearly incorrect, I feel that I cannot. The reader, therefore, will have to get used to such expressions as 'a wavenumber of 1300 cm^{-1} and, when we come to molecular vibrations, 'a vibration wavenumber of 3200 cm^{-1}, whereas, unfortunately, he or she is more likely to encounter elsewhere 'a frequency of 1300 wavenumbers' and 'a vibration frequency of 3200 wavenumbers.'

We use the symbol $\tilde{R}_H$ for the Rydberg constant with dimensions of wavenumber and it is given by

$$\tilde{R}_H = \mu e^4/8h^3\,\varepsilon_0^{\,2}c \tag{1.19}$$

A rather more realistic picture of the motion of the electron in the hydrogen atom is due to Sommerfeld. Still trying to marry Planck's quantum theory to classical mechanics, he suggested, in 1916, that the electron moves in elliptical, as well as circular, orbits with the nucleus at one focus of the ellipse. Each orbit has a characteristic value of p_ϕ, the total angular momentum of the orbit, where ϕ is the angle defining the position of the electron in the orbit. This angular momentum is given by

$$p_\phi = kh/2\pi \tag{1.20}$$

where $k = 1, 2, 3, \ldots, n$ and is called the azimuthal quantum number. When $k = n$ the orbit is the circular one obtained in the Bohr theory but, when $k < n$, the length a of the major axis of the elliptical orbit is always equal to the diameter of the circular orbit. The length b of the minor axis is related to a by

$$a/b = n/k \tag{1.21}$$

The possibility that $k = 0$ was rejected because an orbit with $b = 0$ would involve the electron travelling backwards and forwards along a straight line through the nucleus!

The three possible orbits for $n = 3$ are shown in Figure 1.6.

It is an important result of the Sommerfeld theory that, when relativity is neglected, the total energy of the hydrogen atom is still the same as that given in equation (1.15) and therefore independent of k. When relativity is introduced into the theory there is a very slight energy dependence on k but the lack of success of relativity in accounting for the experimental observation that many lines in the spectrum of atomic hydrogen are split into a number of very closely spaced components—the so-called fine

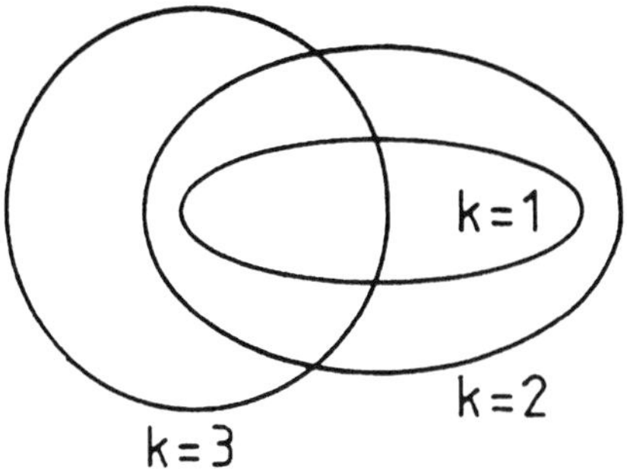

Figure 1.6 Sommerfeld orbits for an electron with $n = 3$ in the hydrogen atom

structure—led to the eventual demise of the Sommerfeld theory.

In the absence of any electric or magnetic field, the Bohr–Sommerfeld orbits, such as those in Figure 1.6, may take up any orientation in space. If we introduce directionality in the form of an electric or magnetic field along what we conventionally take to be the z-axis, the vector $\boldsymbol{p}$ representing the magnitude $(kh/2\pi)$ and direction of the angular momentum can take up only certain orientations relative to the z-axis. These orientations are those which produce a z-component of $\boldsymbol{p}$ of magnitude

$$p_z = mh/2\pi \tag{1.22}$$

where m is known as the magnetic quantum number and can take the values ± 1, ± 2, $\pm 3, \ldots, \pm k$. The effect on the orbits is called space quantization. Figure 1.7 shows the direction of the vector $\boldsymbol{p}$ perpendicular to the plane of the orbit and Figure 1.8 shows the effect of space quantization on the $n = 3$, $k = 3$ orbit of Figure 1.6.

The successes of Planck's quantum theory in going a long way towards explaining the spectrum of the hydrogen atom, the radiation of a black body, the photoelectric effect and the heat capacities of solids were considerable, but there were still many difficulties unresolved. One of these concerned the photoelectric effect in which the radiation falling on the metal surface behaves as if it consists of particles each with energy $h\nu$ (later, in 1924, Lewis called them photons), whereas the phenomena of interference and diffraction had been interpreted by assuming that light behaves as if it consists of waves. This apparent dual wave–particle nature of light, and indeed of any particle or radiation, was resolved in 1924 by the equation proposed by de Broglie:

$$p = h/\lambda \tag{1.23}$$

which relates the momentum p, in the particle picture of radiation, with the wavelength λ, in the wave picture. The importance of this

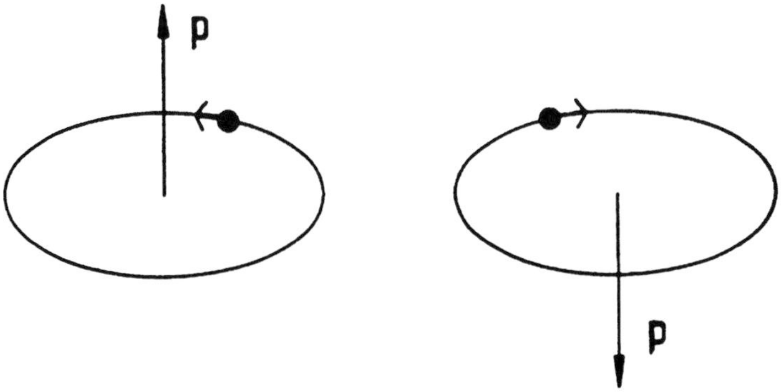

Figure 1.7 Direction of the angular momentum vector $\boldsymbol{p}$ for an electron in an orbit

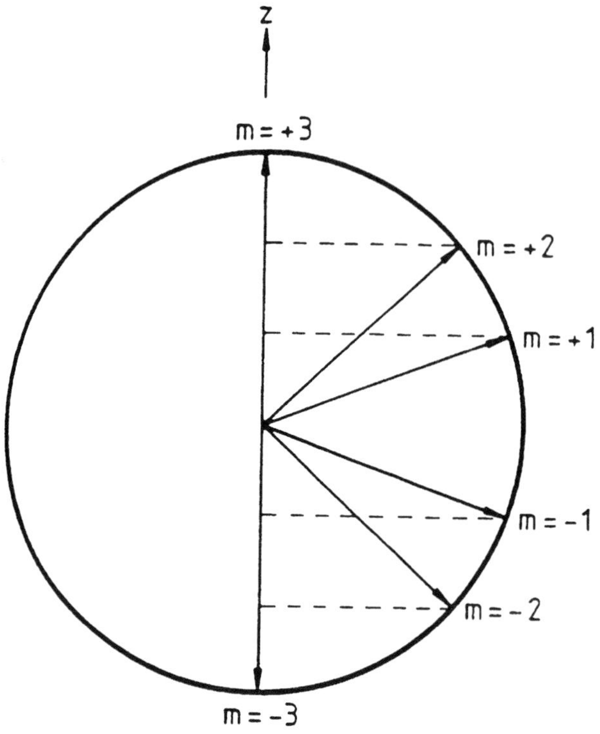

Figure 1.8 Space quantization for an $n = 3$, $k = 3$ orbit

beautifully simple equation cannot be too strongly emphasized. It brings together and rationalizes the behaviour of microscopic particles, with which we would usually associate a wavelength, and macroscopic particles, with which we would usually associate a momentum. For example, each photon of violet light with a wavelength of 4.5×10^{-7} m, or 450 nm,† has a momentum of 1.472×10^{-27} kg m s^{-1}—so small that there is little wonder that we tend not to think of radiation as consisting of particles. At the other extreme, a ball of mass 100 g travelling with a velocity of 10 m s^{-1} has a momentum of 1 kg m s^{-1} and a wavelength of 6.6×10^{-34} m—so small that we would not expect to observe any effects of its wave nature.

Not only did the de Broglie relation reconcile the wave and particle nature of phenomena in which one or other viewpoint had previously been the obvious one to take, it also suggested that some microscopic particles, which had previously been thought of only as particles, might behave in an observable way as waves. One of the most important of such particles is the electron.

The kinetic energy of an electron of mass m_e moving with velocity v is $m_e v^2/2$. If this energy is due to the electron moving in a potential

†The nanometre (nm) is the most convenient SI unit of wavelength of light but the ångstrom (1 Å = 10^{-10} m) is often encountered.

difference V then, if e is the charge on the electron,

$$eV = m_e v^2/2$$
$$\therefore m_e v = (2eVm_e)^{\frac{1}{2}} \quad (1.24)$$
$$\therefore \lambda = h/(2em_e)^{\frac{1}{2}} V^{\frac{1}{2}}$$

$$= (1.227 \times 10^{-9}\ \mathrm{J\,s\,C^{-\frac{1}{2}}\,kg^{-\frac{1}{2}}})/V^{\frac{1}{2}} \quad (1.25)$$

For a potential difference of, for example, 100 V, then λ is 1.227×10^{-10} m (122.7 pm or 1.227 Å).[†] This wavelength is of the same order as the separation of atoms in a molecule or the separation of atoms, ions or molecules in a crystal. Therefore, diffraction of electrons by molecules and crystals was anticipated. In 1925, Davisson and Germer showed experimentally that diffraction of electrons does occur. Figure 1.9 shows a diffraction pattern similar to one obtained by Mark and Wierl in 1930. In this experiment, a beam of electrons with a wavelength of 0.0645 Å is directed at a piece of silver foil perpendicular to the beam. The resulting diffraction pattern is detected by blackening of a photographic plate and is in the form of rings centred on the point where the undiffracted electron beam falls on the plate.

After the introduction of the de Broglie equation, it became clear why the electron in the hydrogen atom may be only in specific Bohr orbits with quantized values of the angular momentum given by equation (1.10). In the wave picture of the electron, the possible orbits are restricted to those whose total length $2\pi r$, where r is the radius of the orbit, contains an integral number of wavelengths, i.e.

$$n\lambda = 2\pi r \quad (1.26)$$

where $n = 1, 2, 3, \ldots, \infty$. Under these conditions, a standing wave is set up. Figure 1.10

Figure 1.9 Diffraction pattern caused by an electron beam falling normally on silver foil and passing through it

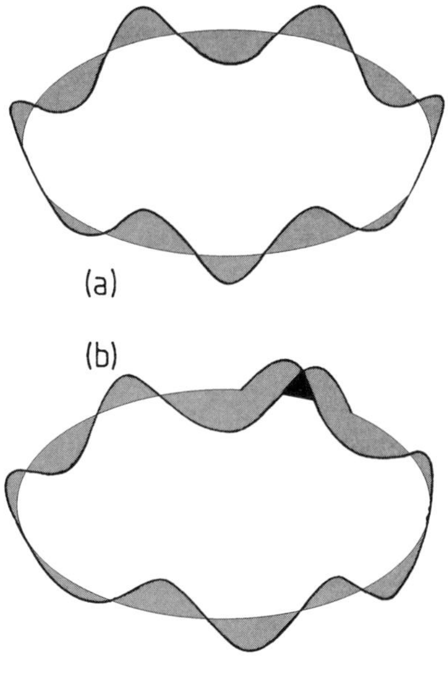

Figure 1.10 (a) A standing wave for an electron in an orbit with $n = 6$. (b) A travelling wave which results when n is not an integer

[†]Although the picometre is a convenient SI unit to replace the ångstrom when speaking of atomic and molecular dimensions, I have the impression that the ångstrom in this context is not being replaced and I shall continue to use it.

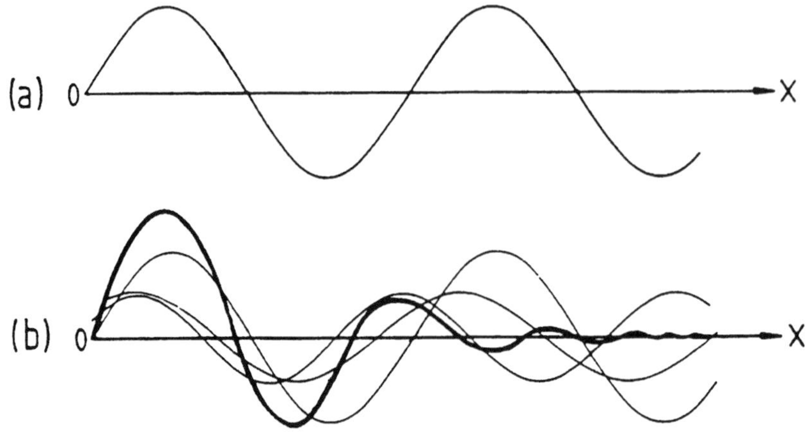

Figure 1.11 (a) The wave due to an electron travelling in the x-direction. (b) Superposition of waves of different wavelengths and which tend to reinforce each other near to $x = 0$

illustrates a standing wave for $n = 6$ and also shows how, if n is not an integer, the wave interferes on successive circuits of the orbit and is destroyed. Combination of equations (1.26) and (1.23) gives equation (1.10), which was Bohr's fundamental assumption in explaining the hydrogen atom spectrum.

It is appropriate at this stage, though historically slightly out of context, to mention the extremely important uncertainty principle introduced by Heisenberg in 1927.

The picture of the electron in an orbit being represented by a standing wave of a specific wavelength and therefore of specific energy goes a long way towards explaining the spectrum of atomic hydrogen but it does pose an important question which, in a classical treatment of the electron as a particle, does not arise. The question is, 'where is the electron?' We shall consider the answer to this question in the more general case of an electron travelling in a particular direction x with constant velocity. The de Broglie picture of this is of a wave, with a specific wavelength, travelling in the x-direction as illustrated in Figure 1.11(a). As in the case of an electron in a Bohr orbit, to ask where the electron is creates an apparent difficulty.

Let us go to the other extreme and consider the electron as a localized particle whose position we can specify, for example when it impinges on a phosphorescent screen and produces an observable scintillation. How can we reconcile the particle picture with the de Broglie picture of the electron? Figure 1.11(b) shows how a superposition of several waves of different wavelengths travelling in the x-direction can reinforce each other in the region of a particular value of x to produce a large amplitude whereas elsewhere they tend to cancel each other. If there were an infinite number of wavelengths, the resulting sum would produce a large amplitude at a specific value of x, say x_s, and zero amplitude elsewhere. This superposition of waves produces what is called a wave packet at x_s and we can say that the electron is behaving as if it were a particle at x_s.

In the picture of the electron as a travelling wave of a single wavelength λ we are, by the de Broglie relationship, specifying the momentum p_x. However, the position x is completely undefined or, in other words, is infinitely uncertain. In the wave packet picture we know that the particle is at x_s but the momentum is infinitely uncertain since the wave packet is made up of an infinite number of wavelengths and, therefore, momenta. But what happens to these uncertainties as the picture of the electron changes smoothly from that of a particle to that of a wave or vice versa? It was Heisenberg who answered this question by his uncertainty principle in which he proposed that the uncertainty Δx in a linear coordinate x and the uncertainty Δp_x in the corresponding linear momentum are related by[†]

$$\Delta p_x \Delta x \geqslant \hbar \tag{1.27}$$

Here the symbol $\hbar$ is used for $h/2\pi$, a quantity which occurs often in quantum mechanics and spectroscopy. We can see from equation (1.27) that, in the extreme case of $\Delta x = 0$, we have $\Delta p_x = \infty$, corresponding to the particle picture; also, when $\Delta p_x = 0$ we have $\Delta x = \infty$, corresponding to the wave picture. Similarly, for Δp_y and Δp_z,

$$\Delta p_y \Delta y \geqslant \hbar; \quad \Delta p_z \Delta z \geqslant \hbar \tag{1.28}$$

There is also an important form of the uncertainty principle relating the uncertainties in time t and energy E:

$$\Delta t \Delta E \geqslant \hbar \tag{1.29}$$

It follows from equation (1.29) that for states with accurately known energy, i.e. $\Delta E = 0$, we have $\Delta t = \infty$. This means that such states do not change in any way with time; such states are called stationary states.

The arguments which have been reviewed here regarding the resolution of the wave–particle picture of the electron can be applied to any other small corpuscle such as a positron, a proton or a neutron. These arguments all closely parallel similar arguments which were subsequent to the experiment by Thomas Young in 1807 demonstrating interference fringes caused by light emerging from two close pinholes. These arguments were applied to the nature of light and concerned whether it should be regarded as consisting of waves or particles (photons); the wave picture is necessary in explaining such phenomena as interference and diffraction, whereas problems in geometrical optics can be solved using the particle picture.

[†]The relationship is written alternatively as $\Delta p_x \Delta x \geqslant h$ or $\Delta p_x \Delta x \geqslant \frac{1}{2}\hbar$ and it is not clear which is more valid. There is even some uncertainty about the statement of the uncertainty principle!

1.2 THE SCHRÖDINGER EQUATION

The justification of the Schrödinger equation is not easy and part of the difficulty is due to the fact that the major steps in its development resulted from postulates rather than firm proof. The validity of the postulates was shown by the correctness of the results obtained by applying them.

The analogy between the wave nature of light and of the electron formed an important part of the background to the Schrödinger equation. Just as travelling light can be represented by a function expressing its amplitude at a particular position in space and at a particular time, so it was proposed that there is a wave function $\Psi(x, y, z, t)$, a function of position and time, which describes the amplitude of a de Broglie electron wave. In 1926, Born related the wave and particle views by saying that we must speak not of a particle being at a particular point at a particular time (as we do in the particle view) but of the *probability* of finding the particle there. This probability is given by $\Psi^*\Psi$, where Ψ^* is the complex conjugate of Ψ and is obtained by replacing i $(= \sqrt{-1})$ in Ψ by $-$i. It follows that

$$\int \Psi^*\Psi \mathrm{d}\tau = 1 \tag{1.30}$$

because the probability of finding the particle anywhere in space is unity. In the integral, $\mathrm{d}\tau$ is a small element of volume which is given by $\mathrm{d}x\mathrm{d}y\mathrm{d}z$ in cartesian coordinates. It seemed reasonable also that the total probability of finding a particle anywhere in space is independent of time, giving

$$\partial\left(\int \Psi^*\Psi \mathrm{d}\tau\right)/\partial t = 0 \tag{1.31}$$

However, when the theory of relativity is taken into account, as it was by Dirac in 1928, this is not strictly true since particles can be created and destroyed. However, the non-relativistic quantum mechanics of Schrödinger, developed in 1926, is adequate for many purposes and it is this with which we shall be concerned.

The form postulated for the wave function is

$$\Psi = b\exp(\mathrm{i}A/\hbar) \tag{1.32}$$

where b is a constant and A is called the action. Action is related to the total energy, given by the sum of the kinetic energy T and the potential energy V, by

$$-\partial A/\partial t = T + V = H \tag{1.33}$$

where H is known as the hamiltonian in classical mechanics. Therefore, ∂A is the negative of the energy given to the system during a period of time ∂t.

From equations (1.32) and (1.33), it follows that

$$H\Psi = \mathrm{i}\hbar\partial\Psi/\partial t \tag{1.34}$$

Schrödinger postulated that the form of the hamiltonian appropriate to quantum mechanics is

$$H = -(\hbar^2/2m)\nabla^2 + V \tag{1.35}$$

where, in cartesian coordinates,

$$\nabla^2 = (\partial^2/\partial x^2) + (\partial^2/\partial y^2) + (\partial^2/\partial z^2) \tag{1.36}$$

and the symbol ∇ is called 'del.' In other coordinates ∇^2, which is the laplacian, will have a different form.

Equations (1.33) and (1.35) show that the hamiltonian in quantum mechanics is obtained from that in classical mechanics by replacing the kinetic energy by $-(\hbar^2/2m)\nabla^2$ but leaving the potential energy in the same form. From equations (1.34) and (1.35) we obtain

$$-(\hbar^2/2m)\nabla^2\Psi + V\Psi = \mathrm{i}\hbar(\partial\Psi/\partial t) \tag{1.37}$$

which is the time-dependent Schrödinger equation.

For most of the uses to which we shall be putting the Schrödinger equation we shall be dealing with standing waves, such as that illustrated in Figure 1.10(a), so we shall require to use only the part of the equation which is independent of time.

We consider, for ease of manipulation, the application of the time-dependent Schrödinger equation to a one-dimensional system involving the coordinate x and assume that $\Psi(x, t)$ can be factorized, giving

$$\Psi(x, t) = \psi(x)\theta(t) \tag{1.38}$$

where θ is the time-dependent and ψ the time-independent part of the total wave function. Combination of equation (1.37), for a one-dimensional system, and equation (1.38) gives

$$-[\hbar^2/2m\psi(x)][\partial^2\psi(x)/\partial x^2] + V(x) = [\mathrm{i}\hbar/\theta(t)][\partial\theta(t)/\partial t] \tag{1.39}$$

For the left-hand, x-dependent, side of the equation to be equal to the right-hand, t-dependent, side they must both be constant. This constant must have the same dimensions as $V(x)$, i.e. those of energy, and so we give to the constant the symbol E and obtain

$$-(\hbar^2/2m)[\partial^2\psi(x)/\partial x^2] + V(x)\psi(x) = E\psi(x) \tag{1.40}$$

and

$$\mathrm{i}\hbar[\partial\theta(t)\partial t] = E\theta(t) \tag{1.41}$$

Equation (1.40) is the one-dimensional time-independent Schrödinger equation, often called simply the wave equation, and describes the behaviour of standing waves.

The form of $\theta(t)$ which satisfies equation (1.41) is given by

$$\theta(t) = (\text{constant})\exp(-\mathrm{i}Et/\hbar) \qquad (1.42)$$

so that if the constant is incorporated into $\psi(x)$ we have, from equations (1.38) and (1.42),

$$\Psi(x, t) = \psi(x)\exp(-\mathrm{i}Et/\hbar) \qquad (1.43)$$

It is an important result, which follows from the form of $\theta(t)$, that

$$\Psi^*\Psi = \psi^*\psi \qquad (1.44)$$

so that the probability of finding a particle somewhere in space is independent of time in non-relativistic quantum mechanics.

Equation (1.40) can be written in the general form

$$H\psi = E\psi \qquad (1.45)$$

This form of the Schrödinger equation appears deceptively simple and requires further explanation. H is the quantum mechanical hamiltonian and we can see from equation (1.40) that, for a one-dimensional system, it is given by

$$H = -(\hbar^2/2m)(\partial^2/\partial x^2) + V(x) \qquad (1.46)$$

Comparing this and the three-dimensional equation (1.35) with the classical hamiltonian of equation (1.33), we see that the classical kinetic energy has been replaced by an operator.† Therefore, equation (1.45) is true for wave functions ψ which, when operated upon by H, give the result of ψ being multiplied by a quantity E. A simple example will serve to illustrate this point. If H is the operator $\mathrm{d}/\mathrm{d}x$ and ψ is the function $\exp(2x)$, then

$$(\mathrm{d}/\mathrm{d}x)\exp(2x) = 2\exp(2x) \qquad (1.47)$$

which is of the same form as equation (1.45).

In the parts of quantum mechanics which we require in this book we shall be concerned almost entirely with stationary states, or eigenstates, which are independent of time and therefore described by the time-independent wave equation (1.45). We require to find the functions ψ, the eigenfunctions, and the energy values E, the eigenvalues, for a particular system. The details of the methods used for solving the Schrödinger equation for ψ and E do not concern us here but are described in several books on quantum mechanics, some of which are listed in the Bibliography at the end of this chapter.

Before we go on to look at some solutions to the Schrödinger equation, it is worthwhile pointing out that in 1926, at the same time as Schrödinger was developing his quantum mechanics, Heisenberg was obtaining the same results by matrix mechanics. Matrix algebra is the main mathematical tool used in matrix mechanics, whereas operator algebra is used in quantum mechanics. Shortly afterwards, Schrödinger and Eckart showed that the two approaches are equivalent.

1.3 SOME IMPORTANT SOLUTIONS OF THE SCHRÖDINGER EQUATION

1.3.1 Types of quantization in atoms and molecules

In the semi-classical Bohr–Sommerfeld treatment of the hydrogen atom we have already encountered quantization of the total energy associated with the electron orbiting the nucleus

†An operator operates on a function to convert it into another function; for example, the operator $\mathrm{d}/\mathrm{d}x$ operates on x^2 to convert it into $2x$, i.e. $(\mathrm{d}/\mathrm{d}x)(x^2) = 2x$. Most of the operators in quantum mechanics involve differentials.

in specific orbits, and of the angular momentum which the electron may have when it is in one of these orbits.[†] In a quantum mechanical treatment the total energy is quantized in an identical way and the orbital angular momentum in a similar way but, in addition, there are two other kinds of quantized angular momentum, the electron spin and nuclear spin angular momenta. These names tend to conjure up a picture of the electron and proton of the hydrogen atom resembling that of the moon rotating on its own axis (spinning electron) orbiting around the earth which is also rotating on its own axis (spinning nucleus). Indeed the names 'electron spin' and 'nuclear spin' were given to these two angular momenta before the days of quantum mechanics and when analogies with the earth and moon rotating on their own axes did not seem unreasonable. Unfortunately, as we shall see in Section 1.3.3, both electron and nuclear spin are phenomena which arise only in quantum mechanics and there are no analogues in classical mechanics. In the case of electron spin, this lack of a classical equivalent seems more reasonable if we try to imagine what we mean by 'electron spin' when we think of the electron in its orbit not as a particle but as a standing wave.

Just as there is space quantization of the orbital angular momentum (see equation 1.22) due, in the Bohr–Sommerfeld treatment, to the orbits being able to take only certain orientations in space in the presence of an electric or magnetic field, so there is space quantization of both electron and nuclear spin angular momenta.

In a polyelectronic atom the orbital and spin angular momenta of the electrons are quantized. The nuclear spin angular momentum is also quantized but, unlike the nucleus of the hydrogen atom, the angular momentum may be zero, depending on the number of protons and neutrons contained in the nucleus—we shall return to this in Section 1.3.3.

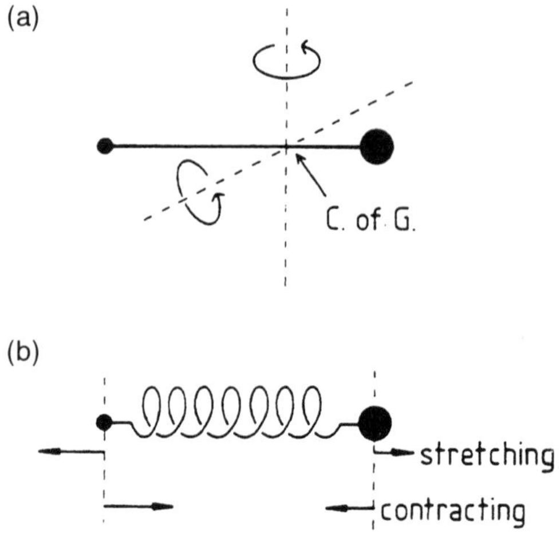

Figure 1.12 (a) End-over-end rotation of a heteronuclear diatomic molecule about axes perpendicular to the bond and through the centre of gravity. (b) Vibration of a heteronuclear diatomic molecule

In a diatomic molecule there is quantization associated with the orbital angular momentum of the electrons but, because of the fact that the electrons are moving in the electrostatic field of the two nuclei, this is analogous to the space quantization of electronic orbital angular momentum in atoms. The electrons have quantized spin angular momentum and, as in the corresponding atoms, the nuclear spin angular momentum is quantized but may in some cases be zero.

In addition, there are two further kinds of quantized energy which contribute to the total energy of a diatomic molecule and which are not to be found in atoms. The first of these is the energy associated with end-over-end rotation of the molecule about any two mutually perpendicular axes through the centre of gravity and perpendicular to the line joining the nuclei, as shown in Figure 1.12(a). This motion provides a further quantized angular momentum which also shows space quantization just like the other angular momenta we have encountered.

[†]Quantization refers, strictly, only to energy but it is convenient to extend it to angular momentum.

The second quantized energy in a diatomic molecule which is not to be found in atoms is that due to vibration of the nuclei in the form of a periodic lengthening and shortening of the internuclear distance. This vibration can be visualized by considering a ball-and-spring model of the molecule in which the spring represents the flexible bond between the nuclei and can stretch and contract as the nuclei move away from and then towards each other. This motion is illustrated in Figure 1.12(b) for a heteronuclear diatomic molecule, such as HF, in which the heavier atom moves through a smaller distance than the lighter one.

In linear polyatomic molecules, such as HCN and acetylene (HC≡CH), quantization of electronic, nuclear and overall rotational energy is analogous to that in diatomic molecules. The situation is somewhat different with regard to vibrational energy since, in general, an N-atomic linear molecule can vibrate in $(3N-5)$ ways. For the case of a diatomic molecule $N = 2$ and the molecule can vibrate in only one way, as we have assumed already.

The number $(3N-5)$ is easily justified when we remember that a linear system of N particles (nuclei) has $3N$ degrees of freedom, which can be regarded as being described by an x-, y- and z-coordinate for each particle. Of the $3N$ degrees of freedom, three represent translations of the set of particles as a whole along the x-, y- or z-axis, and two represent rotation of the set of particles as a whole about the x- or y-axis. (If we take the line of the N nuclei to be the z-axis, rotation about this axis does not represent a degree of freedom since, taking the nuclei to be point masses, the moment of inertia for the rotation about this axis is zero and therefore there is no energy associated with it.) The rest of the degrees of freedom, $(3N-5)$, represent motions in which nuclei move relative to each other, namely vibrational degrees of freedom. As for the single vibrational mode of a diatomic, the energy associated with *each* of the $(3N-5)$ modes of vibration of a linear polyatomic molecule is quantized.

In non-linear polyatomic molecules, such as iodomethane (CH_3I) or fluorobenzene (C_6H_5F), electron spin and nuclear spin behave rather as in diatomic and linear polyatomic molecules. The energy associated with all the vibrational modes is quantized but there are $(3N-6)$ of these since a non-linear set of N particles has three translational and three rotational (as opposed to two rotational in a linear molecule) degrees of freedom. There is no quantization associated with the electron orbital motion in the non-linear electrostatic field of the nuclei. Quantization of rotational angular momentum is complete only for the particular groups of molecules called symmetric rotors (or symmetric tops) and spherical rotors (or spherical tops). Symmetric rotors are molecules having two equal principal moments of inertia and spherical rotors have three equal principal moments of inertia. For molecules having three unequal principal moments of inertia, the asymmetric rotors, quantization of rotational angular momentum is only partial and insufficient to describe completely the rotational energy levels.

This brief summary of the various types of quantization which we shall meet in atoms and molecules serves to show how the situation changes from the simplest atom, in which all types of energy are quantized, to a nonlinear polyatomic molecule in which only a few types are quantized.

1.3.2 Solution of the Schrödinger equation for the hydrogen atom

The treatment of the hydrogen atom by the methods of Schrödinger occupies a very important position in quantum mechanics since the Schrödinger equation [equation (1.45)] can be solved exactly for electronic motion in atoms or molecules only for the case of one electron, for example in H, He^+, Li^{2+} (hydrogen-like atoms), $H_2{}^+$, etc.

For the hydrogen atom the hamiltonian of equation (1.46) becomes, in three dimensions,

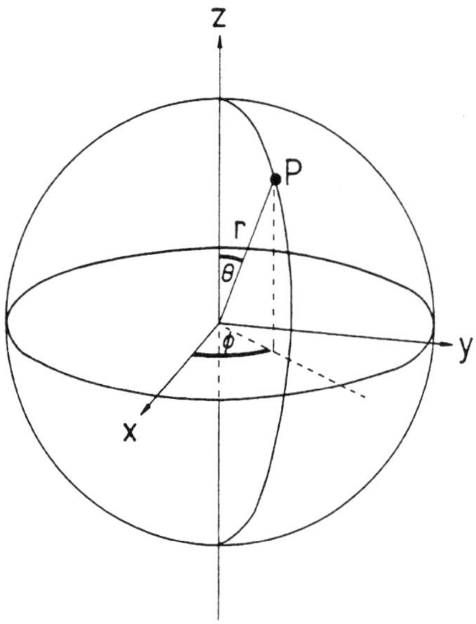

Figure 1.13 Illustration of the spherical polar coordinates r, θ and ϕ

$$H = -(\hbar^2\nabla^2/2\mu) - (e^2/4\pi\varepsilon_0 r) \qquad (1.48)$$

The first term on the right-hand side is the quantum mechanical equivalent of the classical kinetic energy. In this term μ is the reduced mass of the system [$= m_e m_p/(m_e + m_p)$ as in equation (1.15)] and takes account of the fact that the electron and the nucleus move around a common centre, rather than the electron moving around the stationary nucleus. ∇^2 is the laplacian operator of equation (1.36). However, it is much more convenient in quantum mechanics to use the spherical polar coordinates r, θ and ϕ instead of the cartesian coordinates x, y and z to locate a point in space. Figure 1.13 illustrates the meaning of r, the distance of the point P from the origin, θ the co-latitude, and ϕ the azimuth. In the r, θ, ϕ coordinates the laplacian becomes

$$\nabla^2 = \frac{1}{r^2 \sin\theta}\left[\sin\theta\frac{\partial}{\partial r}\left(r^2\frac{\partial}{\partial r}\right) + \frac{\partial}{\partial\theta}\left(\sin\theta\frac{\partial}{\partial\theta}\right) + \frac{1}{\sin\phi}\frac{\partial^2}{\partial\phi^2}\right] \qquad (1.49)$$

Although this is a much more clumsy-looking expression than equation (1.36), the use of spherical polar coordinates has the great virtue of allowing the factorization of the wave function ψ into two parts:

$$\psi(r, \theta, \phi) = R_{nl}(r)Y_{lm_l}(\theta, \phi) \qquad (1.50)$$

In this equation $Y_{lm_l}(\theta, \phi)$ is called the angular wave function. Being a function of θ and ϕ only, it represents the distribution of a hydrogen atom wave function over the surface of a sphere of a particular radius r. For this reason, the Y_{lm_l} wave functions are sometimes known as spherical harmonics and must satisfy the boundary conditions appropriate to a particle moving on the surface of a sphere. The values that the quantum numbers l and m_l can take follow from these conditions. The quantum number l is called the azimuthal quantum number (as was k in the Bohr–Sommerfeld theory) and can take the values

$$l = 0, 1, 2, \ldots, (n - 1) \qquad (1.51)$$

where n is the principal quantum number and has a very similar significance to n in the Bohr theory. The electron orbitals, which are described by the wave functions of equation (1.50) and which are the quantum mechanical analogues of the 'orbits' of the Bohr–Sommerfeld theory, are labelled by the values of n and l. The symbols s, p, d, f, g..., are used to indicate values of $l = 0, 1, 2, 3, 4, \ldots$, respectively. The symbols originated from early observations of atomic spectra and they will be referred to again in Section 6.1.3. Thus, we speak of 1s, 2p, 3p, 3d, etc., orbitals, where the 1, 2, 3, etc., refer to the value of n.

The quantum number l is analogous to Sommerfeld's k quantum number of equation (1.20) but differs from it in that l can be zero whereas k cannot.

The quantity m_l is the magnetic quantum number. It can take the values

$$m_l = 0, \pm 1, \pm 2, \ldots, \pm l \qquad (1.52)$$

and, as in the Bohr–Sommerfeld treatment, is associated with the space quantization of orbital angular momentum.

Y_{lm_l} can be factorized further to give

$$Y_{lm_l}(\theta, \phi) = \Theta_{lm_l}(\theta)\Phi_{m_l}(\phi) \tag{1.53}$$

Φ_{m_l} is, as is clear from Figure 1.13, the angular distribution of the wave function about the z-axis and is given by

$$\Phi_{m_l}(\phi) = \frac{1}{(2\pi)^{\frac{1}{2}}} \exp(\mathrm{i}m_l\phi) \tag{1.54}$$

Since Φ_{m_l} must have the same values for $\phi = 0, 2\pi, 4\pi, \ldots$, it follows that m_l can take only the values $0, \pm 1, \pm 2, \ldots$.

Y_{lm_l} is now given by

$$Y_{lm_l} = \frac{1}{(2\pi)^{\frac{1}{2}}}[\Theta_{lm_l}(\theta) \exp(\mathrm{i}m_l\phi)] \tag{1.55}$$

The Θ_{lm_l} functions are known as the associated Legendre polynomials, of which a few are given in Table 1.1. Unlike the R_{nl} wave functions, the Θ_{lm_l} and therefore the Y_{lm_l} wave functions are independent of Z, the nuclear charge number, and are the same for hydrogen and hydrogen-like atoms, such as He^+ and Li^{2+}.

R_{nl} is called the radial wave function, since it is a function of r but not of θ or ϕ, and its form depends on the values of the quantum numbers n and l. The method of solving the Schrödinger equation for the R_{nl} wave functions follows the procedure, well known in mathematics, for dealing with a particular kind of differential equation whose solutions are known as the associated Laguerre functions. Some of the R_{nl} wave functions are given in Table 1.2 for hydrogen and hydrogen-like atoms with a nuclear charge of $+Ze$. The frequently recurring quantity a_0 is the radius of the orbit in the Bohr theory with $n = 1$ and is given by

$$a_0 = \hbar^2 4\pi\varepsilon_0/m_e e^2 \tag{1.56}$$

and its value is 0.529 Å. The quantity ρ is used for convenience and is related to r by

Table 1.1 Some Θ_{lm_l} wave functions for hydrogen and hydrogen-like atoms

l	m_l	$\Theta_{lm_l}(\theta)$
0	0	$1/2^{\frac{1}{2}}$
1	0	$(6^{\frac{1}{2}}/2)\cos\theta$
1	±1	$(3^{\frac{1}{2}}/2)\sin\theta$
2	0	$(10^{\frac{1}{2}}/4)(3\cos^2\theta - 1)$
2	±1	$(15^{\frac{1}{2}}/2)\sin\theta\cos\theta$
2	±2	$(15^{\frac{1}{2}}/4)\sin^2\theta$

Table 1.2 Some radial wave functions R_{nl} for hydrogen and hydrogen-like atoms

n	l	$R_{nl}(r)$
1	0	$(Z/a_0)^{\frac{3}{2}}2\exp(-\rho)$
2	0	$(Z/a_0)^{\frac{3}{2}}(1/2^{\frac{1}{2}})(1-\rho/2)\exp(-\rho/2)$
2	1	$(Z/a_0)^{\frac{3}{2}}(1/2)(1/6^{\frac{1}{2}})\rho\exp(-\rho/2)$

$$\rho = Zr/a_0 \tag{1.57}$$

There are three useful ways of representing graphically the radial part of the wave function:

(1) Plot R_{nl} against ρ (or r). This is done for R_{10}, R_{20} and R_{21} in Figure 1.14(a). It can be seen that, although R_{10} and R_{21} are always positive, R_{20} changes from positive to negative and, at one particular value of ρ, $R_{20} = 0$.
(2) Plot R_{nl}^2 against ρ (or r). Since R_{nl} is a real function, $R_{nl} = R^*_{nl}$ and $R_{nl}^2\,\mathrm{d}r$ is the probability of finding the electron between r and $r+\mathrm{d}r$. A plot of R_{nl}^2 against ρ represents the radial probability distribution of the electron. Such distributions are shown in Figure 1.14(b).
(3) Plot $4\pi r^2 R_{nl}^2$ against ρ (or r). The quantity $4\pi r^2 R_{nl}^2$ is the probability of finding an electron within an element of volume

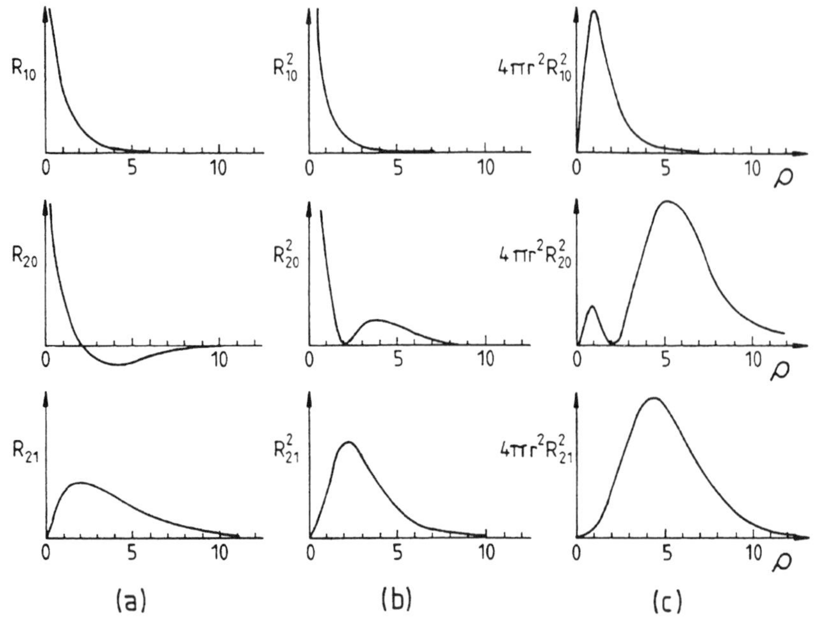

Figure 1.14 Plots of (a) the radial wave function R_{nl}, (b) the radial probability distribution function R_{nl}^{2} and (c) the radial charge density function $4\pi r^2\ R_{nl}^{2}$ against ρ $(= Zr/a_0)$

consisting of a spherical shell of thickness dr and with a volume of $4\pi r^2\,\mathrm{d}r$. This probability is $4\pi r^2\ R_{nl}^{2}\,\mathrm{d}r$. Plots of $4\pi r^2\ R_{nl}^{2}$, called the radial charge density, against ρ are shown in Figure 1.14(c).

Before we can make diagrammatic representations of the Y_{lm_l} functions of equation (1.55), we must convert them from imaginary into real functions. This does not apply to s $(l = 0)$ functions since $m_l = 0$ and the functions are real.

In the absence of an electric or magnetic field, all the Y_{lm_l} functions with $l \neq 0$ are $(2l + 1)$-fold degenerate; this means that there are $(2l + 1)$ functions, each having one of the $(2l + 1)$ possible values of m_l, which have the same energy. It is a property of degenerate functions that linear combinations of them are also solutions of the Schrödinger equation. For example, just as $\psi_{2p,1}$ and $\psi_{2p,-1}$ are solutions of the Schrödinger equation, so are the following linear combinations, ψ_{2p_x} and ψ_{2p_y}, where

$$\begin{aligned} \psi_{2p_x} &= (\psi_{2p,1} + \psi_{2p,-1})/2^{\frac{1}{2}} \\ \psi_{2p_y} &= -\mathrm{i}(\psi_{2p,1} - \psi_{2p,-1})/2^{\frac{1}{2}} \end{aligned} \tag{1.58}$$

From equations (1.58), (1.55) and (1.50), together with the Θ_{lm_l} wave functions in Table 1.1, it follows that

$$\begin{aligned} \psi_{2p_x} &= [R_{21}(r)3^{\frac{1}{2}} \sin\theta\,(\exp(\mathrm{i}\phi) \\ &\qquad + \exp(-\mathrm{i}\phi))]/2(4\pi)^{\frac{1}{2}} \\ \psi_{2p_y} &= [\mathrm{i}R_{21}(r)3^{\frac{1}{2}} \sin\theta\,(\exp(\mathrm{i}\phi) \\ &\qquad - \exp(-\mathrm{i}\phi))]/2(4\pi)^{\frac{1}{2}} \end{aligned} \tag{1.59}$$

But, since

$$\begin{aligned} \exp(\mathrm{i}\phi) + \exp(-\mathrm{i}\phi) &= 2\cos\phi \\ \exp(\mathrm{i}\phi) - \exp(-\mathrm{i}\phi) &= 2\mathrm{i}\sin\phi \end{aligned} \tag{1.60}$$

equations (1.59) become

$$\psi_{2p_x} = R_{21}(r)3^{\frac{1}{2}} \sin\theta \cos\phi/(4\pi)^{\frac{1}{2}} \tag{1.61}$$

$$\psi_{2p_y} = R_{21}(r)3^{\frac{1}{2}} \sin\theta \sin\phi/(4\pi)^{\frac{1}{2}} \tag{1.62}$$

and, for completeness, the third degenerate $\psi_{2p,0}$ wave function with $m_l = 0$ is always real, since $\exp(im_l\phi) = 1$, and is given by

$$\psi_{2p_z} = R_{21}(r)3^{\frac{1}{2}} \cos\theta/(4\pi)^{\frac{1}{2}} \tag{1.63}$$

The wave functions ψ_{np}, with $n > 2$, differ from those of equations (1.61)–(1.63) only in the form of $R_{nl}(r)$.

All the ψ_{ns} wave functions are real since $m_l = 0$ and so are the ψ_{nd}, ψ_{nf}, etc., wave functions for $m_l = 0$. However, for ψ_{nd} with $m_l = \pm 1$, it is necessary to form linear combinations of the imaginary wave functions $\psi_{nd,1}$ and $\psi_{nd,-1}$ in order to obtain real ones. A similar treatment is necessary to obtain real ψ_{nd} wave functions for $m_l = \pm 2$ from the imaginary functions $\psi_{nd,2}$ and $\psi_{nd,-2}$. The ψ_{nd} orbitals for any $n > 2$ are distinguished by subscripts as follows: nd_{z^2} ($m_l = 0$), nd_{xz} and nd_{yz} ($m_l = \pm 1$), and nd_{xy} and $nd_{x^2-y^2}$ ($m_l = \pm 2$). There are seven nf orbitals for any $n > 3$ but we shall not consider them here.

In Figure 1.15(a) the real Y_{lm_l} wave functions for the $1s$, the three $2p$ and the five $3d$ orbitals are plotted in the form of polar diagrams. The construction of these polar diagrams may be illustrated by the simple case of a $2p_z$ orbital. According to the wave function ψ_{2p_z} in equation (1.63), the angular part is a function only of the angle θ (see Figure 1.13) and is independent of the angle ϕ. More specifically, ψ_{2p_z} is proportional to $\cos\theta$. The polar diagram consists of points on a surface obtained by marking off, on lines drawn radially outwards from the nucleus in all directions, distances proportional to $|\cos\theta|$. The resulting surface consists of the two touching spheres shown in Figure 1.15(a).

Similar surfaces may be obtained for $2p_x$ and $2p_y$ by marking off distances proportional to $|\sin\theta\cos\phi|$ and $|\sin\theta\sin\phi|$, as indicated by equations (1.61) and (1.62), respectively. The $3d$ radial wave functions can be represented in the same way and all are shown in Figure 1.15(a).

An alternative method of representing the angular distribution is by marking off distances from the nucleus proportional to the square of the angular wave function, for example $(\cos\theta)^2$, $(\sin\theta\cos\phi)^2$ and $(\sin\theta\sin\phi)^2$ for $2p_z$, $2p_x$ and $2p_y$, respectively. These plots represent the angular probability distribution. They appear qualitatively similar to those in Figure 1.15(a) but there are quantitative differences. One such difference is the shape of the probability distribution for p orbitals. Figure 1.15(b) shows, for $2p_x$, that this is like two eggs touching at the nucleus rather than the two spheres in Figure 1.15(a).

It is worth noting at this stage that the apparently unique behaviour of the z-axis compared with the x- and y-axes, as for example in the $3d_{z^2}$ orbital in Figure 1.15(a), is not incompatible with the fact that an atom, in which orbitals with a particular value of n are filled or empty, is spherically symmetrical. It must be remembered that we arbitrarily chose to treat the z-axis uniquely in defining the angle θ in Figure 1.13; we could equally well have chosen the x-axis to give $3d_{x^2}$, $3d_{z^2-y^2}$, etc., or the y-axis to give $3d_{y^2}$, $3d_{x^2-z^2}$, etc., orbitals.

A more difficult point is again concerned with spherical symmetry. How can we reconcile with spherical symmetry the picture of an atom having, for example, three $2p$ orbitals projecting with their axes mutually at right angles? The answer lies in the fact that the orbitals were obtained for the particular set of cartesian axes with the orientation in space shown in Figure 1.13. But we could have chosen *any* orientation of the axes and obtained a set of orbitals bearing the same relation to those axes as do the orbitals in Figure 1.15(a) to the axis orientation we happened to choose.

The non-spherically symmetrical form of some of the Y_{lm_l} wave functions shown in Figure 1.15(a) prompts us to look back at the R_{nl} wave functions plotted in Figure 1.14(a) and

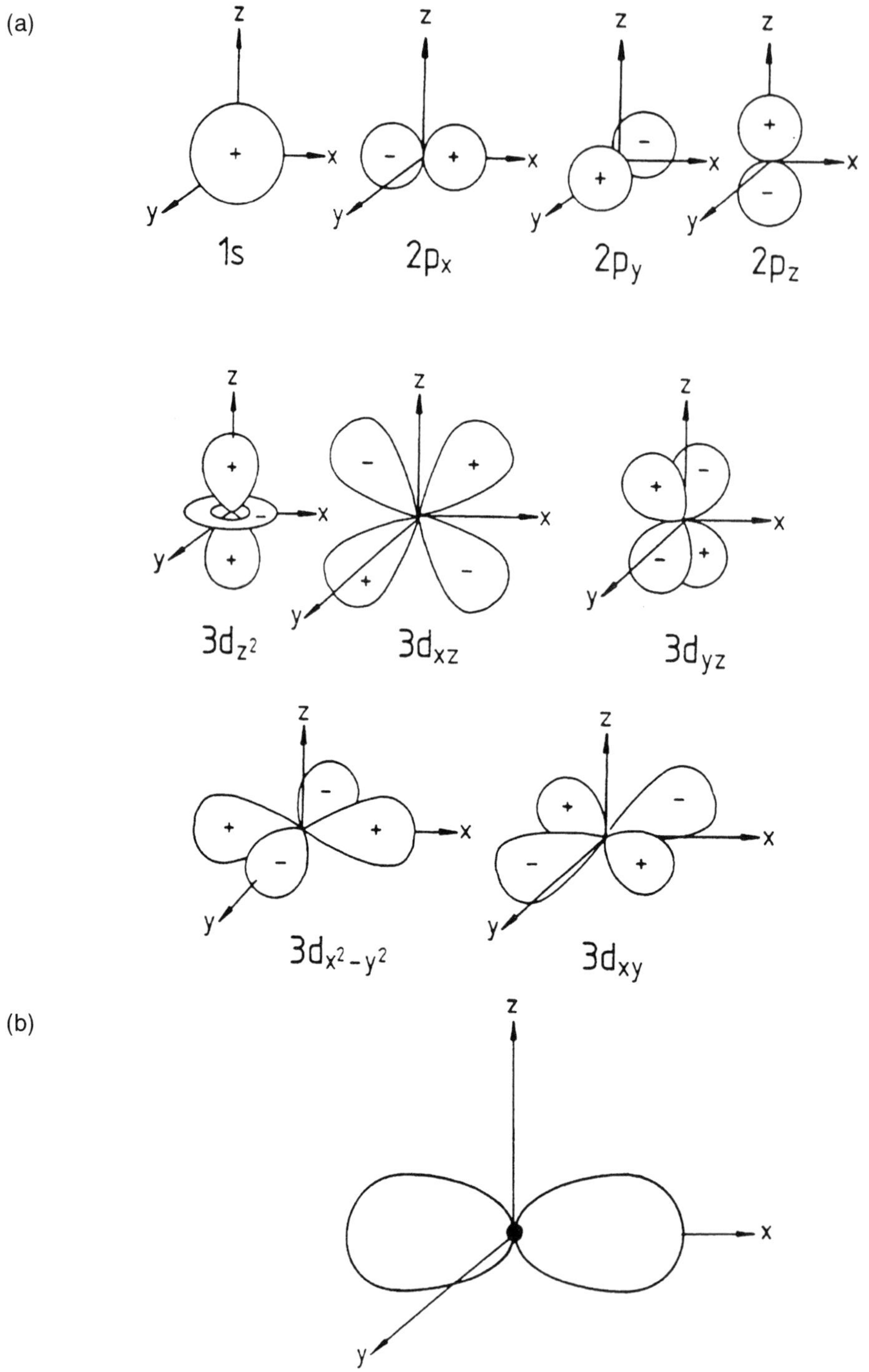

Figure 1.15 (a) Polar diagrams for 1*s*, 2*p* and 3*d* atomic orbitals showing the distribution of the angular wave functions. (b) Polar diagram for the $2p_x$ orbital showing the angular probability distribution

(b) and consider which directions from the nucleus the R_{nl} plots refer to. Since the 1s and 2s orbitals are spherically symmetrical, the R_{10}, R_{20} and R_{10}^2, R_{20}^2 functions are the same in whatever direction from the nucleus ρ is measured. The R_{21} and R_{21}^2 functions are plotted for ρ measured along the z-axis for the $2p_z$ orbital [$\cos\theta = 1$ in equation (1.63)] and along the x- and y-axes for the $2p_x$ and $2p_y$ orbitals [$\sin\theta = 1$ in equations (1.61) and (1.62)]. For the d orbitals, plots of R_{nl} and R_{nl}^2 are still more directionally dependent and therefore less useful.

For all orbitals except 1s there are regions in space where $\psi(r, \theta, \phi) = 0$ because either $Y_{lm_l} = 0$ or $R_{nl} = 0$: these regions are called nodal surfaces or, simply, nodes. Figure 1.15(a) shows that 2p orbitals each have one nodal plane, for example the xy-plane of $2p_z$. The 3d orbitals each have two nodal planes and, in general, there are l such angular nodes where $Y_{lm_l} = 0$. Figure 1.14 shows that the 1s orbital has no radial nodes but that the 2s orbital has one radial node forming a sphere of zero wave function around the nucleus separating a region of positive from one of negative wave function. In fact there are $(n - 1)$ radial nodes in an ns orbital (or n nodes if we include the one at infinity).

Solution of the Schrödinger equation in order to determine the energy gives the result

$$E_n = -[(\mu e^4)/8h^2 \varepsilon_0^2](Z^2/n^2) \tag{1.64}$$

This is exactly the same expression as that produced by the Bohr–Sommerfeld theory [equation (1.15)] and, again, the energy is dependent only on n, and not l or m_l. The dependence of the energy only on n is unique for hydrogen and hydrogen-like atoms in which the only contribution to the potential energy is the coulombic attraction between the nucleus and the electron. Whenever there is more than one electron, inter-electron repulsions contribute to the potential energy and the energy becomes strongly l-dependent. Because the energy of the hydrogen atom is dependent only on one quantum number n, all the orbitals, except 1s, are multiply degenerate. Since l can take n values, 0, 1, 2, ..., $(n - 1)$, and m_l can take $(2l + 1)$ values, 0, ± 1, $\pm 2, \ldots, \pm l$, the degree of degeneracy, which is the number of orbitals having the same energy but different wave functions, is given by

$$\sum_{l=0}^{l=n-1} (2l + 1) = n^2 \tag{1.65}$$

Although the expression for the energy levels given by equation (1.64) is the same as that given by the Bohr theory (indeed, it must be since the Bohr theory for hydrogen and hydrogen-like atoms agrees exactly with experiment, except for the fine structure of the spectrum which the Schrödinger treatment also does not explain), there are very important differences resulting from the Schrödinger treatment. Perhaps the most important and far-reaching of these is embodied in Figures 1.14 and 1.15. These portray the electron as being smeared out in various possible patterns of probability which, according to the R_{nl} functions in Figure 1.14, always approaches zero asymptotically as r tends to infinity. The probability distribution also contains regions (nodes) where there is zero probability of finding the electron.

In the process of solving the Schrödinger equation for the R_{nl} wave functions if, instead, we solve for a quantity rR_{nl}, which we shall call P_{nl}, then the equation to be solved is

$$\frac{d^2 P_{nl}}{dr^2} + \frac{2\mu}{\hbar^2}\left\{E - \left[\frac{l(l+1)\hbar^2}{2\mu r^2} - \frac{e^2}{4\pi\varepsilon_0 r}\right]\right\}P_{nl} = 0 \tag{1.66}$$

The quantity P_{nl}^2 is proportional to the radial charge density $(4\pi r^2 R_{nl}^2)$ and comparison with equation (1.40) shows that equation (1.66) is a one-dimensional Schrödinger equation for a particle of mass μ and in which the effective potential energy, V_{eff}, is given by

$$V_{\text{eff}} = [l(l+1)\hbar^2/2\mu r^2] - (e^2/4\pi\varepsilon_0 r) \quad (1.67)$$

In the Bohr–Sommerfeld treatment of the hydrogen atom the potential energy V due to an orbiting electron having an angular momentum p_ϕ is given by

$$V = (p_\phi{}^2/2\mu r^2) - (e^2/4\pi\varepsilon_0 r) \quad (1.68)$$

where, according to equation (1.20),

$$p_\phi = k\hbar \quad (1.69)$$

It is apparent from equations (1.68) and (1.67) that we have replaced the classical quantity $p_\phi{}^2$, the square of the angular momentum, by $l(l+1)\hbar^2$ and therefore we deduce that, in a quantum mechanical treatment, the magnitude of the orbital angular momentum, for which we shall use the symbol P_l in this case, is given by

$$P_l = [l(l+1)]^{\frac{1}{2}}\hbar \quad (1.70)$$

There are two important differences between equations (1.69) and (1.70): one is the quantitative difference between $[l(l+1)]^{\frac{1}{2}}$ and k, where both l and k can only be integers, and the other is that l can be zero whereas k cannot.

In classical mechanics $k = 0$ implies zero angular momentum and therefore no centrifugal force to counteract the coulombic attraction, a state of affairs which would result in the electron being drawn into the nucleus and which could not be tolerated. In quantum mechanics, however, when $l = 0$ we have

$$V_{\text{eff}} = -e^2/4\pi\varepsilon_0 r \quad (1.71)$$

In Figure 1.16, V_{eff} is plotted against r and this shows that V_{eff} decreases rapidly as r approaches zero. However, the steepness of the curve in that region implies a rapid increase in kinetic energy and the balance between potential and kinetic energy results in the maximum value of the radial charge density for a 1s orbital being some distance from the nucleus—see Figure 1.14(c).

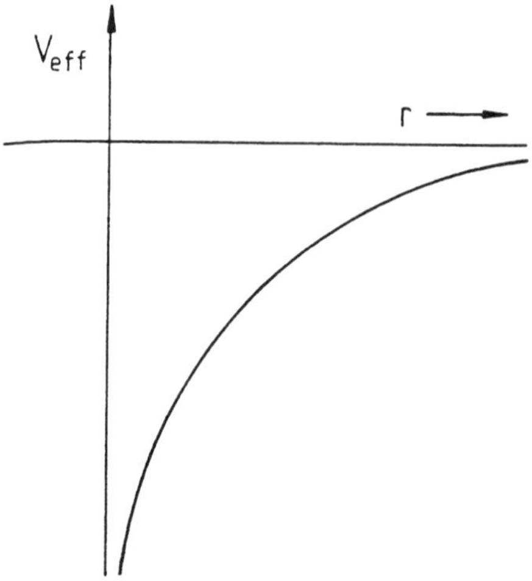

Figure 1.16 Variation of the effective potential energy V_{eff}, given by equation (1.67), with r

1.3.3 Quantization of molecular rotational, electron spin and nuclear spin angular momenta

In the previous section the quantum mechanical treatment of the hydrogen atom has been described in some detail because of the very important position which it occupies in the application of quantum mechanics to atomic and molecular systems. There we saw, in equation (1.70), that the orbital angular momentum in the hydrogen atom is quantized. In this section we look much more briefly at the results of quantum mechanics with regard to the other kinds of angular momenta which have been mentioned in Section 1.3.1.

The quantization of angular momentum in general is represented by the equation

$$P_R = [R(R+1)]^{\frac{1}{2}}\hbar \quad (1.72)$$

where P_R is the magnitude of the angular momentum and R is a quantum number† which may be zero, integral or half-integral depending on the type of angular momentum with which we are concerned. There is also space quantization of the angular momentum so that, in the presence of an electric or magnetic field along the z-azis, the component of angular momentum with respect to that axis is given by

$$(P_R)_z = M_R\hbar \qquad (1.73)$$

where M_R can take $(2R+1)$ possible values given by

$$M_R = R, R-1, R-2, \ldots, -(R-1), -R \qquad (1.74)$$

In the case of the hydrogen atom we have already encountered the magnitude of the orbital angular momentum of the electron, $[l(l+1)]^{\frac{1}{2}}\hbar$, and the magnitude of the component along the z-axis, $m_l\hbar$. Figure 1.17 is a vector diagram illustrating space quantization of orbital angular momentum when there is a magnetic field along the z-axis. This figure should be contrasted with Figure 1.8, which resulted from the Bohr–Sommerfeld treatment. In the quantum mechanical treatment m_l can be zero and the angular momentum vector can never lie along the z-axis.

Although equation (1.73) tells us the values that the component of angular momentum along the z-axis can take, it tells us nothing of the components along the x- and y-axes. In fact, $(P_R)_x$ and $(P_R)_y$ remain undefined which means that the positions of the ends of the vectors drawn in Figure 1.17 with $m_l \neq 0$ are free to rotate around the circles drawn in Figure 1.18(a). This rotation is called precession. Precessional motion occurs in classical systems also. An example is the spinning top, illustrated in Figure 1.18(b), whose axis precesses around the direction of the earth's gravitational field.

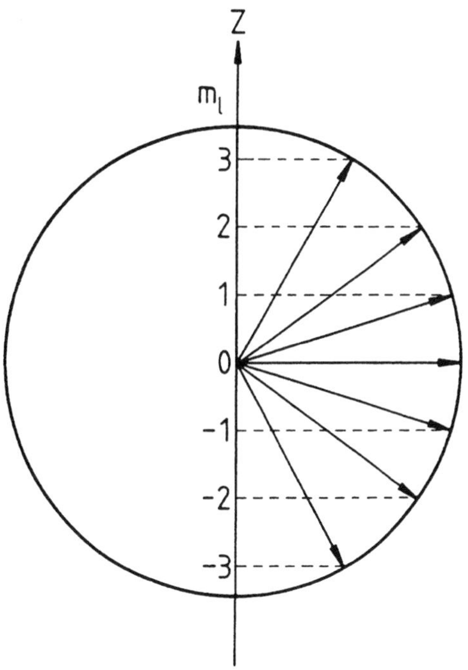

Figure 1.17 Space quantization of orbital angular momentum for $l = 3$

Before we can consider rotational angular momentum in a diatomic molecule we have to justify the separate treatment of electronic, vibrational and rotational motion.

The hamiltonian H for a diatomic molecule is the sum of the kinetic energy T, or its quantum mechanical equivalent, and the potential energy V, as in equation (1.33). The kinetic energy consists of contributions from the motions of the electrons, T_e, and the nuclei, T_n. The potential energy consists of two terms, V_{ee} and V_{nn}, due to coulombic repulsions between the electrons and the nuclei, respectively, and a third term, V_{en}, due to attractive forces between electrons and nuclei. Therefore, the hamiltonian is given by

$$H = T_e + T_n + V_{en} + V_{ee} + V_{nn} \qquad (1.75)$$

†Capital letters are used for quantum numbers referring to a system containing *many* electrons and lower-case letters for a system containing *one* electron.

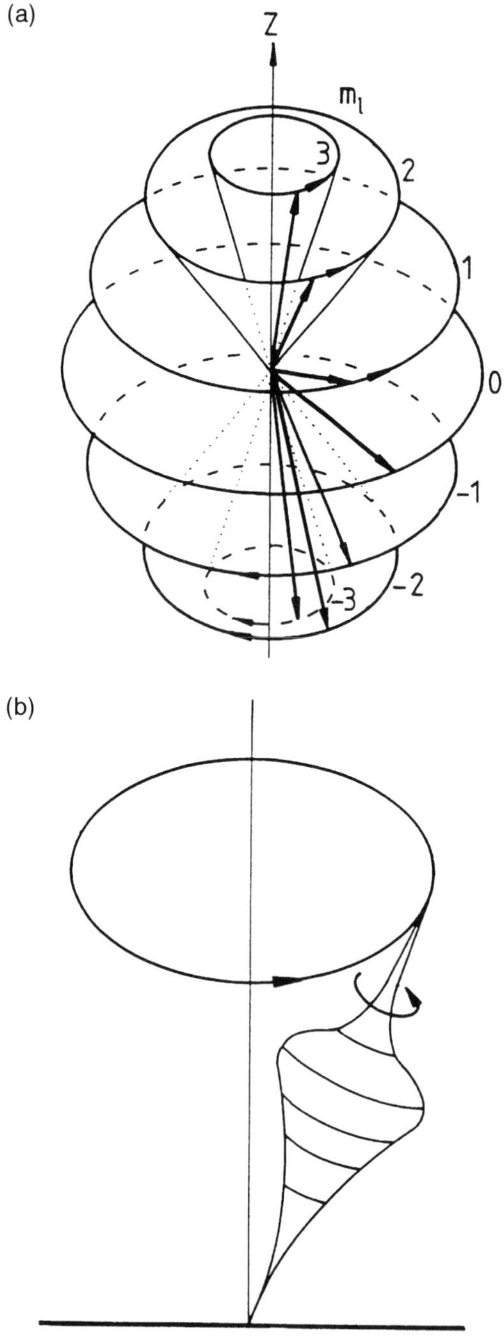

Figure 1.18 (a) Precession of the space quantized angular momentum vectors for $l = 3$ and (b) precession of a spinning top

For fixed nuclei $T_n = 0$ and V_{nn} is constant. Under these conditions there is a set of electronic wave functions ψ_e which satisfy

$$H_e\psi_e = E_e\psi_e \tag{1.76}$$

where

$$H_e = T_e + V_{en} + V_{ee} \tag{1.77}$$

Since H_e depends on nuclear coordinates, because of the V_{en} term, so do ψ_e and E_e. However, in the Born–Oppenheimer approximation,[1] of 1927, it is assumed that the vibrating nuclei move so slowly relative to the electrons that ψ_e and E_e involve the nuclear coordinates as parameters only. As a result, E_e can be calculated for all values of the nuclear coordinates. The result for a diatomic molecule, in which the nuclear coordinate is the internuclear distance r (or its displacement from equilibrium), is that a potential energy curve such as those shown later in Figures 1.20 and 1.22 can be drawn for a particular electronic state in which T_e and V_{ee} are constant.

In so far as the Born–Oppenheimer approximation is valid, the electrons adjust instantaneously to any change of the internuclear distance: they are said to follow the nuclei. For this reason, E_e can be treated as part of the potential field in which the nuclei move, giving

$$H_n = T_n + V_{nn} + E_e \tag{1.78}$$

and the Schrödinger equation is

$$H_n\psi_n = E_n\psi_n \tag{1.79}$$

It follows from what we have said about the consequences of the Born–Oppenheimer approximation that the total wave function ψ can be factorized:

$$\psi = \psi_e(q, Q)\psi_n(Q) \tag{1.80}$$

where ψ_e is a function of nuclear coordinates Q and electron coordinates q. Then the total energy E is given by

$$E = E_e + E_n \tag{1.81}$$

In Section 1.3.4 we shall look more closely at potential energy curves for diatomic molecules which can be drawn only in the context of the Born–Oppenheimer approximation and, in Chapter 6, we shall encounter situations where the approximation breaks down.

The method of solution of the Schrödinger equation (1.79) to obtain ψ_n and E_n is analogous to that for the motions of the nucleus and the electron in the hydrogen atom, the main difference being the form of the potential energy. In the same way that the hydrogen atom wave function $\psi(r, \theta, \phi)$ can be factorized into $R_{nl}(r)$ and $Y_{lm_l}(\theta, \phi)$, as in equation (1.50), the wave function ψ_n can be factorized:

$$\psi_n = \psi_v \psi_r \tag{1.82}$$

where ψ_v and ψ_r are the vibrational and rotational wave functions, respectively. The corresponding energies are again additive:

$$E_n = E_v + E_r \tag{1.83}$$

From equations (1.80) and (1.82) it follows that

$$\psi = \psi_e \psi_v \psi_r \tag{1.84}$$

and, from equations (1.81) and (1.83),

$$E = E_e + E_v + E_r \tag{1.85}$$

In an approximate treatment of the end-over-end rotation of a diatomic molecule about an axis through the centre of gravity, we assume the bond between the nuclei to be rigid and therefore of fixed length: this is the rigid rotor approximation and the rigid rotor model is illustrated in Figure 1.12(a). The magnitude of the angular momentum for rotation is given by

$$P_J = [J(J+1)]^{\frac{1}{2}}\hbar \tag{1.86}$$

The rotational quantum number J can take any integral value 0, 1, 2, 3, ... Generally, J is associated with the *total* angular momentum excluding nuclear spin, i.e. rotational plus orbital plus electron spin, but, in the numerous cases where there is no orbital or electron spin angular momentum, J is associated with just the rotation.

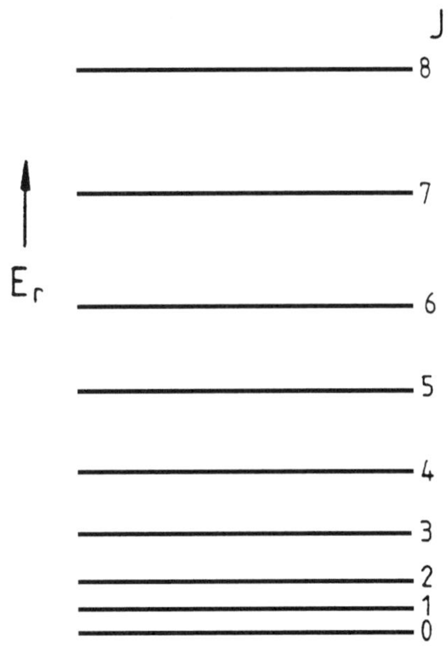

Figure 1.19 Rotational energy levels, E_r, for end-over-end rotation of a diatomic molecule

In deducing E_r for a diatomic molecule we use the fact that the classical energy E associated with an angular momentum p_θ, as for example in the system illustrated in Figure 1.4, is given by

$$E = \tfrac{1}{2} I\omega^2 = p_\theta^{\ 2}/2I \tag{1.87}$$

where ω is the angular velocity and I the moment of inertia. Then, replacing p_θ in equation (1.87) by the quantum mechanical expression in equation (1.86) gives

$$E_r = J(J+1)\hbar^2/2\mu r^2 \tag{1.88}$$

where $\mu = M_1 M_2/(M_1 + M_2)$, for nuclear masses M_1 and M_2, and r is the internuclear distance. This gives a set of diverging energy levels such as those shown in Figure 1.19.

Table 1.3 Some values of the nuclear spin quantum number I

Nucleus	I	Nucleus	I
^{1}H	$\frac{1}{2}$	^{16}O	0
^{2}H	1	^{19}F	$\frac{1}{2}$
^{10}B	3	^{28}Si	0
^{11}B	$\frac{3}{2}$	^{29}Si	$\frac{1}{2}$
^{12}C	0	^{30}Si	0
^{13}C	$\frac{1}{2}$	^{31}P	$\frac{1}{2}$
^{14}N	1	^{35}Cl	$\frac{3}{2}$
^{15}N	$\frac{1}{2}$	^{37}Cl	$\frac{3}{2}$

Space quantization of the rotational angular momentum in a diatomic molecule is expressed by

$$(P_J)_z = M_J\hbar \tag{1.89}$$

where $M_J = J, J-1, \ldots, -J$. Therefore, every rotational energy level is $(2J+1)$-fold degenerate in the absence of an electric or magnetic field.

The magnitude of the angular momentum due to the spin of one electron, as in the hydrogen atom, is given by

$$P_s = [s(s+1)]^{\frac{1}{2}}\hbar \tag{1.90}$$

but the quantum number s can take the value $\frac{1}{2}$ *only*. The expression in equation (1.90) cannot be derived from the Schrödinger equation but it can be obtained using Dirac's relativistic quantum mechanics (although electron spin is not concerned directly with relativity). Space quantization results in

$$(P_s)_z = m_s\hbar \tag{1.91}$$

where $m_s = \pm\frac{1}{2}$ only.

The magnitude of the angular momentum due to nuclear spin is given by

$$P_I = [I(I+1)]^{\frac{1}{2}}\hbar \tag{1.92}$$

where the value of the quantum number I depends on the particular nucleus but may be zero, integer or half-integer. Values of I for some of the more common nuclei are given in Table 1.3. The protons and neutrons contained in a nucleus each have spin angular momentum with $I = \frac{1}{2}$ and the way in which they couple together to give the resultant nuclear spin determines the value of I for a particular nucleus. There are three useful rules regarding the resultant value of I:

(1) I is half-integral for nuclei with an odd mass number, e.g. ^{13}C has $I = \frac{1}{2}$.
(2) I is integral for nuclei with an even mass number but an odd charge number, e.g. ^{14}N, for which $Z = 7$, has $I = 1$.
(3) I is zero for nuclei with an even mass number and an even charge number, e.g. ^{12}C, for which $Z = 6$, has $I = 0$.

Space quantization of nuclear spin angular momentum is expressed by

$$(P_I)_z = M_I\hbar \tag{1.93}$$

where $M_I = I, I-1, \ldots, -I$.

As comparison of equations (1.67) and (1.68) shows, there is a classical angular momentum which is analogous to the quantum mechanical orbital angular momentum of the electron in the hydrogen atom. There is also a classical angular momentum which is analogous to the quantum mechanical angular momentum, given in equation (1.86), for rotation of a diatomic molecule. On the other hand, there is *no* classical angular momentum analogous to either electron spin [equation (1.90)] or nuclear spin [equation (1.92)] angular momenta. Whether or not there is a classical analogue can be demonstrated by allowing h to become zero, as it is in classical mechanics, in the quantum mechanical expressions. If, as $h \to 0$, the angular momentum also tends to zero then there is no analogous classical angular momentum. From equation (1.90) and the fact that s can have only the value $\frac{1}{2}$, it is clear that $P_s \rightarrow 0$ as $h \rightarrow 0$ and electron spin angular momentum has no classical analogue. Similarly, equation (1.92) shows that $P_I \rightarrow 0$ as $h \rightarrow 0$, since I is fixed for a particular nucleus, and therefore nuclear spin has no classical analogue.

These conclusions illustrate the dangers of taking too literally the terms 'electron spin' and 'nuclear spin', which date from the days of semi-classical interpretations.

The quantum number l in equation (1.70) can take any positive value from 0 to ∞ and therefore, as $h \rightarrow 0$, the orbital angular momentum need not become zero. Similarly, J in equation (1.86) is unrestricted in the positive values it can take and so, as $h \rightarrow 0$, the rotational angular momentum need not become zero. These observations regarding orbital and rotational angular momenta illustrate the general principle that quantum mechanical behaviour approaches classical behaviour as the corresponding quantum number increases; this is known as the correspondence principle.

1.3.4 Quantization of vibrational energy: the harmonic oscillator

As a result of the Born–Oppenheimer approximation and the separability of vibrational and rotational motion, we can treat separately the problem of solving the Schrödinger equation for the vibration of a diatomic molecule.

The ball-and-spring model of vibration, shown in Figure 1.12(b), suggests that the stretching and contracting of the bond might obey Hooke's law, for *small* changes in bond length, as for any other harmonic motion. If this is so, then

$$\text{restoring force} = -\mathrm{d}V(x)/\mathrm{d}x = -kx \quad (1.94)$$

where V is the potential energy, k is the force constant[†] (large for a strong bond, small for a weak bond) and x is the displacement from equilibrium, namely $r - r_e$ where r_e is the equilibrium bond length. Integrating equation (1.94) gives

$$V(x) = kx^2/2 \quad (1.95)$$

The parabolic variation of V with x is illustrated in Figure 1.20 and gives the potential energy curve. For a diatomic molecule this curve cannot be valid either for large positive displacements, because we know that the bond weakens and eventually dissociation into atoms occurs, or for large negative displacements, because there is an increasing resistance to contraction of the bond which is due partly to nuclear repulsion. We shall return to these shortcomings of the harmonic oscillator approximation in Chapter 5.

The quantum mechanical hamiltonian for the one-dimensional harmonic oscillator is given by

$$H = -(\hbar^2/2\mu)(\mathrm{d}^2/\mathrm{d}x^2) + kx^2/2 \quad (1.96)$$

where $\mu = M_1M_2/(M_1 + M_2)$ and M_1 and M_2 are the masses of the two nuclei, and the Schrödinger equation (equation 1.40) becomes

$$(\mathrm{d}^2\psi_v/\mathrm{d}x^2)+[(2\mu E_v/\hbar^2)-(\mu kx^2/\hbar^2)]\psi_v=0 \quad (1.97)$$

The method for solving this equation will be found in most of the texts on quantum mechanics given in the Bibliography at the end of this chapter. The resulting values for the vibrational energy E_v are given by

$$E_v = h\nu(v + \tfrac{1}{2}) \quad (1.98)$$

or

$$E_v = hc\tilde{\nu}(v + \tfrac{1}{2}) \quad (1.99)$$

[†]Some authors use f for the force constant.

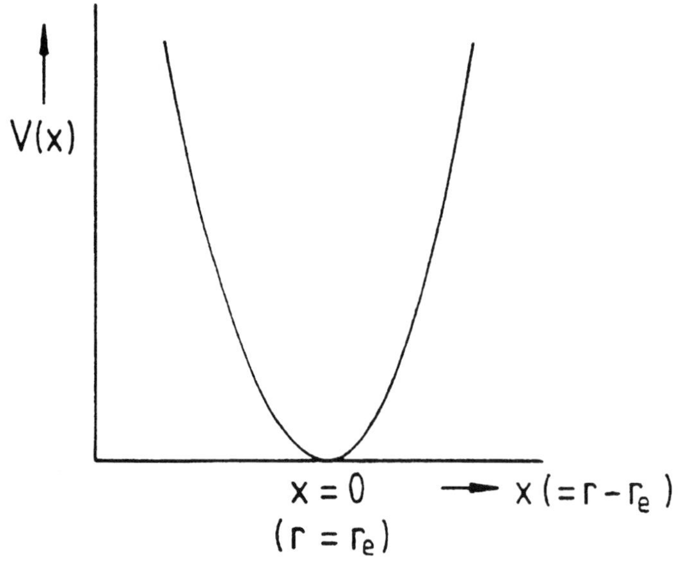

Figure 1.20 Potential energy curve for vibration of a diatomic molecule treated as a harmonic oscillator

where ν is the classical vibration frequency for the oscillator and is given by

$$\nu = (k/\mu)^{\frac{1}{2}}/2\pi \qquad (1.100)$$

As we would expect, the frequency increases with increase in k (with the stiffness of the spring or the bond) and decreases with μ. In equation (1.98), υ is the vibrational quantum number and can take the values

$$\upsilon = 0, 1, 2, 3, \ldots \qquad (1.101)$$

In equation (1.99), $\tilde{\nu}$ is the classical vibration wavenumber[†] and c is the speed of light.

The vibrational energy is sometimes expressed in the form

$$E_\upsilon = \hbar\omega(\upsilon + \tfrac{1}{2}) \qquad (1.102)$$

[†]See footnote '‡' on p. 6.

where ω is the classical circular vibration frequency and is related to ν by

$$\omega = 2\pi\nu \qquad (1.103)$$

This use of ω arises because of the analogy with the standard mathematical example of harmonic motion illustrated in Figure 1.21. The point P, starting from A at time $t = 0$, is travelling with a constant angular frequency ω around a circular path. The projection Q of the point P on the diameter perpendicular to OA executes simple harmonic motion which also has a frequency ω. Like the ω in equation (1.102) it has units of, for example, rad s^{-1}, whereas the corresponding unit of ν in equation (1.98) is s^{-1} (this unit of ν used to be called the 'cycle per second', which did have the virtue of distinguishing it clearly from the 'radian per second'). However, we shall not be using subsequently the analogy between the harmonic motion of a molecular vibration and that of the point Q in Figure 1.21 and therefore shall not be using ω

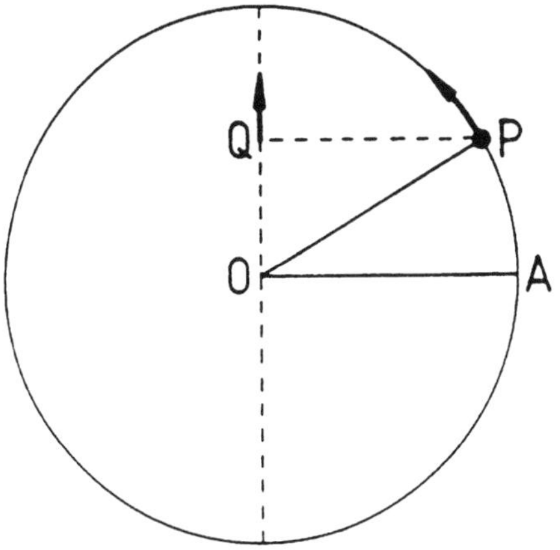

Figure 1.21 Illustration of harmonic motion and the relation between vibration frequency and circular vibration frequency

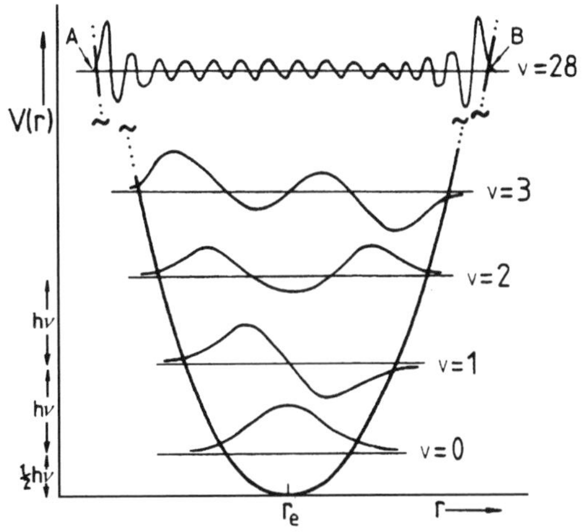

Figure 1.22 Harmonic oscillator potential energy curve showing a few energy levels and vibrational wave functions

again in this sense. Instead, when we treat molecular vibration more rigorously in Chapter 5, we shall use ω in what, for spectroscopists, is its more useful and more usual way.

Equation (1.98) shows that adjacent vibration energy levels for the harmonic oscillator model are separated by $h\nu$, The $v = 0$ level, which has an energy of $\frac{1}{2}h\nu$, is called the zero-point level and $\frac{1}{2}h\nu$ the zero-point energy. This is the minimum vibrational energy the molecule may have, even at the absolute zero of temperature, and is a consequence of the uncertainty principle. If we use this principle in the form of equation (1.27), connecting the uncertainty of momentum to that of position, we can see that, if vibrational energy and hence the momentum were zero, the uncertainty in the displacement of the nuclei would become infinite; but this would not be consistent with the fact that the motion is confined to that illustrated in the potential energy curve of Figure 1.20.

Figure 1.22 shows the harmonic oscillator potential energy curve with the energy levels indicated by a horizontal line for each value of v. Where this line intersects the curve represents a classical turning point of a vibration. At this turning point the velocity of each nucleus is zero; therefore, the kinetic energy is zero and all the energy is in the form of potential energy. At the mid-point of a vibrational energy level all the energy of the nuclei is kinetic.

The vibrational wave functions ψ_v which can be derived from equation (1.97) are given by

$$\psi_v = (2^v v! \pi^{\frac{1}{2}})^{-\frac{1}{2}} H_v(y) \exp(-y^2/2) \qquad (1.104)$$

The $H_v(y)$ are known as Hermite polynomials. These are well known in mathematics and are functions of a quantity y where, for the harmonic oscillator,

$$y = (4\pi^2 \nu\mu/h)^{\frac{1}{2}}(r - r_e) \qquad (1.105)$$

The Hermite polynomials for $v = 0$–6 are given in Table 1.4. Some of the vibrational wave functions ψ_v are plotted in Figure 1.22 with the corresponding horizontal line representing the energy level as base-line. There are several

Table 1.4 Some Hermite polynomials $H_\upsilon(y)$ occurring in harmonic oscillator wave functions

υ	$H_\upsilon(y)$
0	1
1	$2y$
2	$4y^2 - 2$
3	$8y^3 - 12y$
4	$16y^4 - 48y^2 + 12$
5	$32y^5 - 160y^3 + 120y$
6	$64y^6 - 480y^4 + 720y^2 - 120$

important features to note regarding these wave functions:

(1) They penetrate regions which lie outside the parabola and which would be forbidden in a classical system.
(2) As υ increases the two points where ψ_υ^2, the vibrational probability, has a maximum value occur closer to the classical turning points. This is illustrated by the $\upsilon = 28$ wave function in Figure 1.22, where A and B are the classical turning points. At these points the nuclei, treated classically, have zero velocity and therefore spend more time at the corresponding values of r than at any other values. This tendency towards classical behaviour as υ increases is in accord with the correspondence principle.
(3) As υ decreases the maximum value of ψ_υ^2 lies further away from the classical turning points and departure from classical behaviour increases; this is illustrated by comparing the $\upsilon = 0$–3 wave functions in Figure 1.22 with that for $\upsilon = 28$.
(4) The wavelength of the ripples in ψ_υ away from the classical turning points is less than that near the turning points. This effect increases with υ and is evident in Figure 1.22 for $\upsilon = 28$.

Table 1.5 Summary of quantized quantities in hydrogen and hydrogen-like atoms, and in diatomic molecules, discussed in Chapter 1

Physical quantity	Quantized value	Values of quantum number
(i) *H and H-like atoms*		
Total energy	$-(\mu e^4/8h^2\varepsilon_0^2)(Z^2/n^2)$	$n = 1, 2, 3, \ldots, \infty$
Orbital angular momentum (OAM)	$[l(l+1)]^{\frac{1}{2}}\hbar$	$l = 0, 1, 2, \ldots, (n-1)$
z-Component of OAM	$m_l\hbar$	$m_l = 0, \pm 1, \pm 2, \ldots, \pm l$
Electron spin angular momentum (ESAM)	$[s(s+1)]^{\frac{1}{2}}\hbar$	$s = \frac{1}{2}$
z-Component of ESAM	$m_s\hbar$	$m_s = \pm\frac{1}{2}$
Nuclear spin angular momentum (NSAM)	$[I(I+1)]^{\frac{1}{2}}\hbar$	$I = \frac{1}{2}$ (for H) but zero, integral or half-integral for other atoms
z-Component of NSAM	$M_I\hbar$	$M_I = \pm\frac{1}{2}$(for $I = \frac{1}{2}$) $M_I = 0, \pm 1, \ldots, \pm I$ (for I integral) $M_I = \pm\frac{1}{2}, \pm\frac{3}{2}, \ldots, \pm I$ (for I half-integral)
(ii) *Diatomic molecules*		
Rotational angular momentum	$[J(J+1)]^{\frac{1}{2}}\hbar$ (in rigid rotor approximation)	$J = 0, 1, 2, \ldots, \infty$
Vibrational energy	$hc\tilde{\nu}(\upsilon + \frac{1}{2})$ (in harmonic oscillator approximation)	$\upsilon = 0, 1, 2, \ldots, \infty$

1.3.5 Summary of quantized quantities in atoms and molecules

Table 1.5 summarizes the physical quantities which are quantized and which we have encountered so far in hydrogen and hydrogen-like atoms and also in diatomic molecules.

BIBLIOGRAPHY

ATKINS, P. W. and FRIEDMAN, R. S. (1997). *Molecular Quantum Mechanics*. Oxford University Press, Oxford

ATKINS, P. W. (1991). *Quanta*. Oxford University Press, Oxford

BOCKHOFF, F. J. (1969). *Elements of Quantum Theory*. Addison-Wesley, Reading, MA

FEYNMAN, R. P., LEIGHTON, R. B. and SANDS, M. (1965). *The Feynman Lectures on Physics*. Addison-Wesley, Reading, MA

KAUZMANN, W. (1957). *Quantum Chemistry*. Academic Press, New York

LANDAU, L. D. and LIFSHITZ, E. M. (1959). *Quantum Mechanics*. Pergamon Press, Oxford

PAULING, L. and WILSON, E. B. (1935). *Introduction to Quantum Mechanics*. McGraw-Hill, New York

SCHUTTE, C. J. H. (1968). *The Wave Mechanics of Atoms, Molecules and Ions*. Arnold, London

2 INTERACTION OF ELECTROMAGNETIC RADIATION WITH ATOMS AND MOLECULES

2.1 NATURE OF ELECTROMAGNETIC RADIATION

It was Maxwell, in 1855, who first concluded that visible radiation can be regarded as an electromagnetic disturbance in the form of waves, and we know that all electromagnetic radiation can be regarded in this way. It may also be considered as consisting of particles (photons) but, in this book, it will be primarily the wave picture with which we shall be concerned.

The kind of electromagnetic radiation which perhaps best illustrates its nature is that which is plane-polarized.[†] As the word electromagnetic implies, all electromagnetic radiation is of a dual character: it has an electric and a magnetic component. The electric component is in the form of an oscillating electric field, whose magnitude and direction is specified by the vector $\boldsymbol{E}$, the electric field strength. The magnetic component is in the form of an oscillating magnetic field, whose magnitude and direction are specified by the vector $\boldsymbol{H}$, the magnetic field strength. In plane-polarized radiation these two vectors oscillate in mutually perpendicular directions. If the directions in which the electric and magnetic vectors oscillate are y and z, respectively, then

$$E_y = A \sin(\omega t - kx) \tag{2.1}$$

and

$$H_z = A \sin(\omega t - kx) \tag{2.2}$$

Equations (2.1) and (2.2) are the equations of motion for plane-polarized radiation travelling along the x-axis and show that both fields oscillate sinusoidally with a radial frequency ω. The quantity k determines the *phase* of the waves and the fact that it is the same for E_y and H_z means that the two fields always oscillate in phase with each other; for example, $E_y = H_z = A$ when $t = 0$ at $x = 0$. Equations (2.1) and (2.2) show also that the fields have the same amplitude and frequency. Figure 2.1 illustrates all these properties of plane-polarized radiation.

Although the dual electric–magnetic character of electromagnetic radiation is of great fundamental importance, we shall find that, in its interaction with atoms and molecules, it is the oscillating electric field which is more often involved than the magnetic field. For example, if radiation falls on a photographic plate it may cause blackening; if it falls on a fluorescent screen it may cause fluorescence of the material of which the screen is made; and if it falls on the retina of the eye it may cause a signal to be transmitted to the brain. All these effects are due to interaction of the detector with the electric vector only.

Because the electric vector is generally the more important, the plane of polarization of

[†]Also known as linearly polarized.

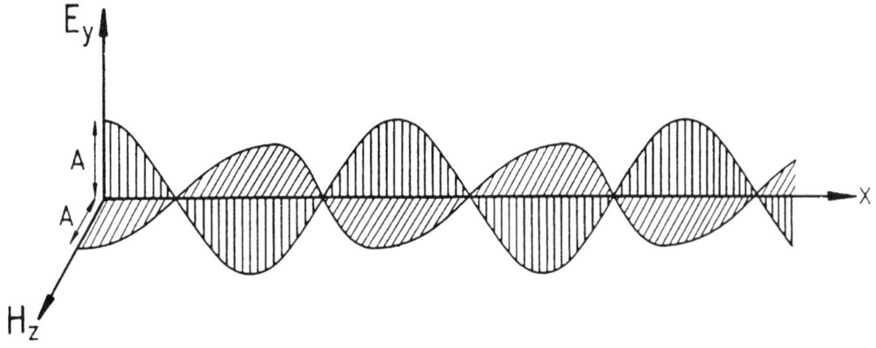

Figure 2.1 The electric and magnetic components of plane-polarized electromagnetic radiation travelling along the x-axis

plane-polarized radiation is defined conventionally as the plane containing the direction of propagation and of the electric vector: in Figure 2.1 this is the xy-plane.

2.2 ABSORPTION AND EMISSION PROCESSES

When electromagnetic radiation falls on a gaseous, liquid or solid material, which may be atomic or molecular in nature, the radiation may be (i) *transmitted* (for example, visible light is transmitted by water); (ii) *scattered* (for example, small particles suspended in water scatter blue light preferentially); or (iii) *absorbed* (for example, the red part of visible light is absorbed by copper sulphate solution, causing the remaining transmitted light, which is white light minus red light, to appear blue).

Just as a wine glass may be caused to vibrate, or resonate, in sympathy with a musical instrument playing a note containing the resonance frequency of the glass, so resonance may occur if an atom or molecule is placed in the path of electromagnetic radiation containing certain frequencies. The first process which the glass, or the atom or molecule, undergoes is to absorb energy, at the resonant frequency, from the radiation. Subsequently the energy is lost by emission or perhaps transferred to another system.

The process of absorption by an atom or molecule is illustrated by (a) in Figure 2.2, where m and n are stationary states. The following discussion applies whether the states are electronic, vibrational, rotational or whatever.

Although states m and n are stationary states, and therefore independent of time, the absorption process is not instantaneous: it takes a period of time t_1 given by

$$t_1 = \nu^{-1} = (c\tilde{\nu})^{-1} \tag{2.3}$$

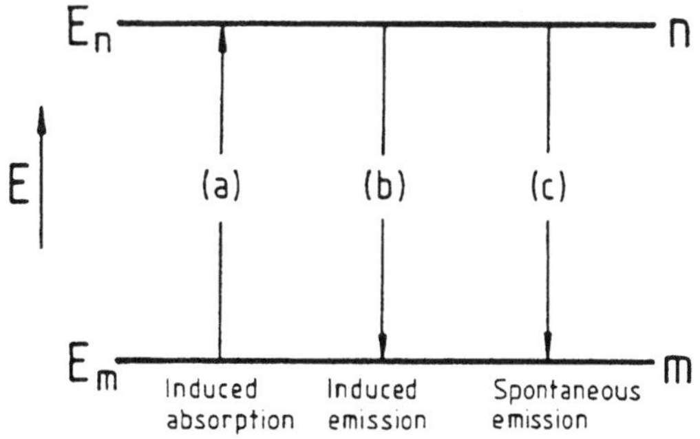

Figure 2.2 Illustration of the absorption and emission processes which may take place between two stationary states m and n

where the frequency ν, or wavenumber $\tilde{\nu}$, is related to ΔE, the energy separation of the states, by

$$\Delta E = E_n - E_m = h\nu = hc\tilde{\nu} \tag{2.4}$$

For example, if an absorption process occurs at a wavelength of 600 nm, in the green region of the visible spectrum, then

$$t_1 = (c\tilde{\nu})^{-1} = \lambda/c = 2.00 \times 10^{-15}\ \text{s} \tag{2.5}$$

This time is clearly very short, but not negligible, and we might ask what happens to the wave function describing the system during this time. What happens is self-evidently time-dependent and the quantum mechanical description involves the use of equation (1.37), the time-dependent Schrödinger equation. We shall not pursue this any further except to say that, during the time t_1, the stystem evolves from one which is described by the stationary state wave function ψ_m, through intermediate states involving mixtures of wave functions, to one described by the stationary state wave function ψ_n.

At equilibrium the populations N_n and N_m of the upper and lower states, respectively, are related, according to the Boltzmann distribution law, by

$$N_n/N_m = \exp(-\Delta E/kT) \tag{2.6}$$

where ΔE is given by equation (2.4), k is the Boltzmann constant and T is the temperature in kelvins. In a system having many stationary states, equation (2.6) gives the ratio of the populations of *any* pair of states m and n. The probability of finding an atom or molecule in state n in a multistate system is given by

$$N_n/\sum_i N_i = \exp(-E_n/kT)/\sum_i \exp(-E_i/kT) \tag{2.7}$$

The denominator of the right-hand side of this equation is the partition function—the sum of the Boltzmann factors of all the states in the system.

An alternative way of discussing populations is in terms of energy levels rather than states. This affects the discussion when degenerate states are involved, i.e. states having different wave functions but the same energy. Thus, if an energy level is g-fold degenerate then the ratio of N_n, the number of atoms or molecules with energy E_n, to N_m, the number with energy E_m, is given by

$$N_n/N_m = (g_n/g_m)\exp(-\Delta E/kT) \tag{2.8}$$

and equation (2.7) becomes

$$N_n/\sum_i N_i = g_n \exp(-E_n/kT)/\sum_i g_i \exp(-E_i/kT) \tag{2.9}$$

Note that the partition function, $\Sigma_i\, g_i \exp(-E_i/kT)$, of equation (2.9) is equal to that of equation (2.7).

From equations (2.6)–(2.9) it is clear that the equilibrium populations of states decrease rapidly (exponentially) as the energy increases.

In certain circumstances, for a particular pair of states in a particular atom or molecule, it is possible to achieve a population inversion in which $N_n > N_m$. This leads to the possibility of the production of lasers, which will be discussed in Chapter 8.

If the population distribution in the two-level system in Figure 2.2 is initially the equilibrium one given by equation (2.6) the absorption process (a) disturbs this equilibrium by increasing N_n and decreasing N_m. When the source of radiation which excites the atoms or molecules from m to n is removed equilibrium will be restored over a period of time by loss of energy ΔE. This loss may occur in various ways, which will be discussed in Section 2.3, but the one which concerns us here is the reverse of absorption, namely the emission of radiation of energy ΔE.

Emission may take place by two different processes. The first of these is known as induced or stimulated emission in which the system emits energy in the presence of radiation of the same energy. The requirement of the presence of this radiation in order to induce emission to occur may seem rather curious when the phenomenon is first encountered but an analogy can be drawn between this and the seeding of a rain cloud. The cloud is capable of spontaneously emitting a shower of rain (see later for a discussion of spontaneous emission), but conditions may not be suitable for it to do so. But if the cloud is seeded with suitable crystals it may be induced to emit a shower of rain when otherwise this would not happen.

An atom or molecule in the state n may similarly be induced to emit radiation by the presence of suitable photons, analogous to the crystals in seeding the rain cloud. The induced emission process, (b) in Figure 2.2, in an atom or molecule M can be represented by the equation

$$\mathrm{M}^* + hc\tilde{\nu} \rightarrow \mathrm{M} + 2hc\tilde{\nu} \tag{2.10}$$

where M* is an excited atom or molecule in state n and $hc\tilde{\nu}$ is a photon of radiation having a wavenumber related to the energy levels by equation (2.4).

Just as we have called the process of equation (2.10) an induced process, because it is induced by the radiation, so strictly we should call process (a), in Figure 2.2, induced absorption since, of course, this also requires the presence of radiation. Induced absorption is the process

$$\mathrm{M} + hc\tilde{\nu} \rightarrow \mathrm{M}^* \tag{2.11}$$

The second emission process, (c) in Figure 2.2, is called spontaneous emission and is the process

$$\mathrm{M}^* \rightarrow \mathrm{M} + hc\tilde{\nu} \tag{2.12}$$

In this case M* spontaneously, without the presence of radiation, emits a photon.

The rate of change of population of the upper state, $\mathrm{d}N_n/\mathrm{d}t$, due to induced absorption is proportional to the population N_m of the lower state and to the radiation density, $\rho(\tilde{\nu})$, which is the density of radiation of wavenumber $\tilde{\nu}$ falling on M; therefore, we have

$$\mathrm{d}N_n/\mathrm{d}t = N_m B_{mn} \rho(\tilde{\nu}) \tag{2.13}$$

where B_{mn} is the Einstein coefficient for induced absorption. Similarly, the change of population due to induced emission is

$$\mathrm{d}N_n/\mathrm{d}t = -N_n B_{nm} \rho(\tilde{\nu}) \tag{2.14}$$

and that due to spontaneous emission is

$$\mathrm{d}N_n/\mathrm{d}t = -N_n A_{nm} \tag{2.15}$$

where B_{nm} and A_{nm} are the Einstein coefficients for induced and spontaneous emission, respectively. The coefficients were introduced in 1917 by Einstein, who was also the first to propose that spontaneous emission may occur, in addition to induced emission, and to show that

$$B_{mn} = B_{nm} \tag{2.16}$$

In the presence of radiation of wavenumber $\tilde{\nu}$ all three processes are going on at once and the overall rate of change of the population of the upper state is

$$\mathrm{d}N_n/\mathrm{d}t = (N_m - N_n) B_{nm} \rho(\tilde{\nu}) - N_n A_{nm} \tag{2.17}$$

If the source of radiation is removed then $\rho(\tilde{\nu}) = 0$, the rate becomes $(-N_n A_{nm})$ and only spontaneous emission occurs.

When the states n and m have their equilibrium Boltzmann populations, $\mathrm{d}N_n/\mathrm{d}t = 0$ and

$$(N_m - N_n) B_{nm} \rho(\tilde{\nu}) - N_n A_{nm} = 0 \tag{2.18}$$

N_m and N_n are related, according to equation (2.6), and the radiation density of a black body has been shown, by applying Planck's quantum theory to the oscillators emitting the radiation, to be given by

$$\rho(\tilde{\nu}) = 8\pi hc\tilde{\nu}^3 [\exp(hc\tilde{\nu}/kT) - 1]^{-1} \tag{2.19}$$

From equations (2.6), (2.18) and (2.19), it follows that

$$A_{nm} = 8\pi hc\tilde{\nu}^3 B_{nm} \tag{2.20}$$

This equation illustrates the important point that spontaneous emission increases rapidly in importance relative to induced emission as the wavenumber (or frequency) increases because of the $\tilde{\nu}^3$ dependence of A_{nm}.

The magnitudes of the Einstein coefficients are dependent on the wave functions of the two combining states and, in particular, on what is called the transition moment for the transition between the state m and n. For absorption or emission involving the electric component of the radiation, which is most often the case, the transition moment $\boldsymbol{R}^{nm}$, a vector quantity having magnitude and direction, is given by

$$\boldsymbol{R}^{nm} = \int \psi_n^* \boldsymbol{\mu} \psi_m \mathrm{d}\tau \tag{2.21}$$

where $\boldsymbol{\mu}$ is also a vector† and is the electric dipole moment operator. It is given by

$$\boldsymbol{\mu} = \sum_i q_i \boldsymbol{r}_i \tag{2.22}$$

where q_i and $\boldsymbol{r}_i$ are the charge and position vector, respectively, of the ith particle. The transition moment can be thought of as the oscillating electric dipole moment caused by a transition from state m to state n.

As an illustration of this, consider a transition between two electronic states m and n in ethylene in which an electron is transferred from the π-orbital in Figure 2.3(a) to the π^*-orbital in Figure 2.3(b). Clearly there is charge movement during the transition, an oscillating electric dipole moment is set up during the transition from state m to state n, and the integral in equation (2.21) is non-zero. However, note that it is not necessary for the molecule to have a *permanent* dipole moment (this is given by equation 2.22 referred to a particular axis) for the integral to be non-zero. Ethylene has no dipole moment in either of the states m or n, owing to the symmetry of the electronic charge in both the π- and π^*-orbitals, but the transition moment is non-zero. The square of the magnitude of the transition moment vector, $|\boldsymbol{R}^{nm}|^2$, is the transition probability and is related to the Einstein coefficient B_{nm} by

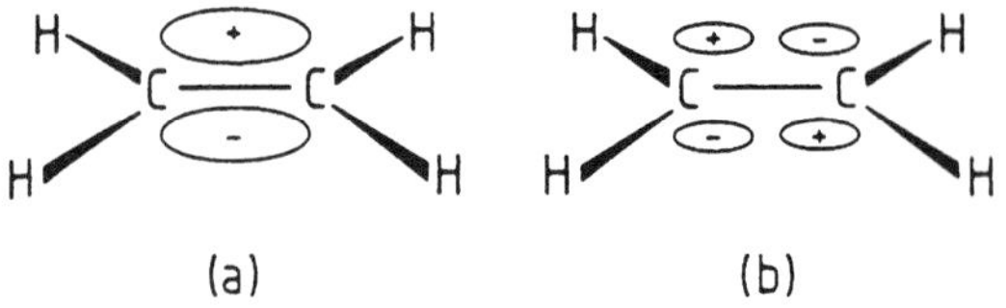

Figure 2.3 (a) A π and (b) a π^* orbital of ethylene

$$B_{nm} = [8\pi^3/(4\pi\varepsilon_0)3h^2]|\boldsymbol{R}^{nm}|^2 \tag{2.23}$$

Quantum mechanical calculation may be used to estimate the transition probability for a particular pair of states; then B_{nm} follows from equation (2.23) and A_{nm} from equation (2.20).

But how are experimental intensity measurements related to Einstein coefficients?

Consider a simple absorption experiment, shown diagrammatically in Figure 2.4(a), in which radiation of intensity I_0, and a continuous range of wavelengths, passes through a cell of length l containing a sample in the liquid phase with a concentration c and which absorbs the incident radiation over a range of wavenumbers $\tilde{\nu}_1$ to $\tilde{\nu}_2$. If the intensity of radiation of wavenumber $\tilde{\nu}$ emerging from the cell is I, then the quantity $\log_{10}(I_0/I)$ is called the absorbance, A, which is a function of $\tilde{\nu}$. A plot of A against $\tilde{\nu}$, an example of which is shown in Figure 2.4(b), is one way of presenting the absorption spectrum. In a liquid-phase experiment the absorption in, say, the visible region of the spectrum using visible radiation is typically broad, extending over a large wavenumber

†The symbol $\boldsymbol{p}$ may be used instead of $\boldsymbol{\mu}$.

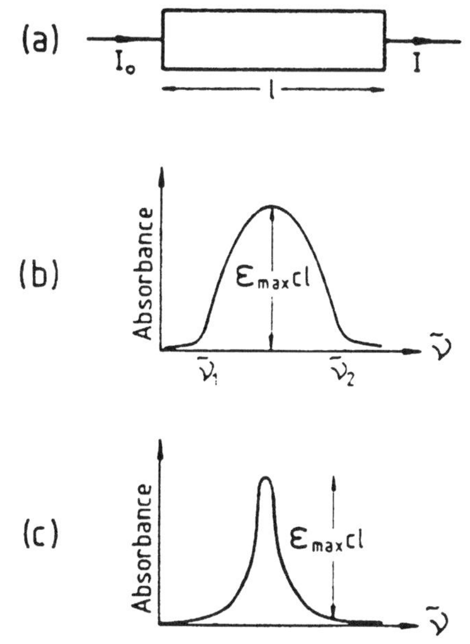

Figure 2.4 (a) A simple absorption experiment. (b) A broad absorption band. (c) A narrow absorption band with the same ε_{max} as in (b)

range. According to the Beer–Lambert law, the absorbance is proportional to the length of the cell and to the concentration of the absorbing sample, as expressed in the equation

$$A = \log_{10}(I_0/I) = \varepsilon(\tilde{\nu})cl \qquad (2.24)$$

where ε is the molar absorption coefficient† or molar absorptivity and is a function of wavenumber. Since absorbance is dimensionless, ε has dimensions of (concentration × length)$^{-1}$ and the units are very often mol^{-1} dm^3 cm^{-1}. To the maximum value of the absorbance in Figure 2.4(b) there corresponds a maximum value of the molar absorption coefficient, ε_{max}, which is sometimes used as an approximate measure of the total absorption intensity. However, this is only approximate because it ignores the width of the absorption and would give the same intensity for the broad absorption in Figure 2.4(b) as for the narrow absorption in Figure 2.4(c). A more accurate measure of intensity is the area under the absorption curve, i.e. $\int_{\tilde{\nu}_1}^{\tilde{\nu}_2} \varepsilon(\tilde{\nu})\,d\tilde{\nu}$. If the states m and n are separated by an energy difference which is large enough for N_n to be very much less than N_m, then induced emission will be negligible compared with absorption; under these conditions the area under the absorption curve is related to B_{nm} by

$$\int_{\tilde{\nu}_1}^{\tilde{\nu}_2} \varepsilon(\tilde{\nu})\,d\tilde{\nu} = N_A h\tilde{\nu}_{nm}B_{nm}/2.303 \qquad (2.25)$$

where $\tilde{\nu}_{nm}$ is the average wavenumber of the transition and N_A is the Avogadro constant.‡

If the absorption is due to a transition between *electronic* states, a quantity f_{nm}, the oscillator strength of the transition, is often used. It is related to the area under the curve by

$$f_{nm} = (4\varepsilon_0 m_e c^2 \ln 10/N_A e^2) \int_{\tilde{\nu}_1}^{\tilde{\nu}_2} \varepsilon(\tilde{\nu})\,d\tilde{\nu} \qquad (2.26)$$

where m_e is the mass and e the charge of the electron; f_{nm} is a dimensionless quantity and is the ratio of the strength of the observed transition to that of an electric dipole transition between the two states of an electron oscillating in three dimensions in a harmonic way: f_{nm} can be calculated from ψ_n and ψ_m and its maximum value is usually 1.

Transition probabilities can be obtained from B_{nm} using equation (2.23). They are concerned with the very important topic in spectroscopy of selection rules. These rules tell us whether a transition between two states m and n is allowed or forbidden and, in terms of transition probabilities, this means that

†Until recently ε was called the molar extinction coefficient and sometimes the word 'decadic' was inserted to indicate the use of the base of 10 rather than e for the logarithm in equation (2.24).

‡This equation is often given in a slightly different form to take account of the fact that the usual units of $\varepsilon(\tilde{\nu})$ are mol^{-1} dm^3 cm^{-1}; but equation (2.25) is always valid provided a *self-consistent* set of units is used for the various physical quantities.

$$\begin{aligned} |\boldsymbol{R}^{nm}|^2 &= 0 \text{ forbidden} \\ |\boldsymbol{R}^{nm}|^2 &\neq 0 \text{ allowed} \end{aligned} \tag{2.27}$$

Strictly we must speak of electric dipole allowed or electric dipole forbidden transitions, if we are concerned with interaction of the electric vector of the radiation, and magnetic dipole allowed or forbidden transitions for interaction of the magnetic vector. There are also, as we shall see later, the important Raman selection rules and the less important electric quadrupole selection rules. Normally, unless we are concerned with Raman or spin resonance spectroscopy, it can be assumed that when we speak of allowed or forbidden transitions we are referring to electric dipole selection rules unless it is stated otherwise.

The electric dipole moment is a vector $\boldsymbol{\mu}$ and has components along the cartesian axes

$$\begin{aligned} \mu_x &= \sum_i q_i x_i \\ \mu_y &= \sum_i q_i y_i \\ \mu_z &= \sum_i q_i z_i \end{aligned} \tag{2.28}$$

where q_i is the charge on the ith particle (electron or nucleus) and x_i, y_i and z_i are its coordinates. Therefore, it follows that the transition moment can be resolved into three components:

$$\begin{aligned} R_x^{nm} &= \int \psi_n^* \mu_x \psi_m \, \mathrm{d}x \\ R_y^{nm} &= \int \psi_n^* \mu_y \psi_m \, \mathrm{d}y \\ R_z^{nm} &= \int \psi_n^* \mu_z \psi_m \, \mathrm{d}z \end{aligned} \tag{2.29}$$

and the transition probability is related to these components by

$$|\boldsymbol{R}^{nm}|^2 = (R_x^{nm})^2 + (R_y^{nm})^2 + (R_z^{nm})^2 \tag{2.30}$$

In a spherically symmetrical system, such as an atom, $R_x = R_y = R_z$ and for an allowed transition in such a system the transition moment vector is not polarized in any particular direction.

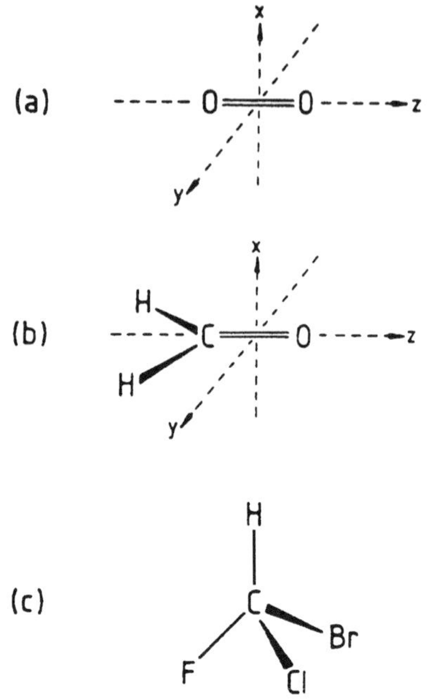

Figure 2.5 (a) Polarization directions of transition moments in the oxygen molecule. (b) Polarization directions in the formaldehyde molecule. (c) The polarization directions are not restricted by symmetry in the bromochlorofluoromethane molecule

A diatomic molecule, such as oxygen, has lower symmetry than an atom. If we take the z-axis to be the line of the nuclei, as shown in Figure 2.5(a), then for an allowed transition either $R_z \neq 0$ and $R_x = R_y = 0$ and we say that the transition moment is polarized along the z-axis, or $R_z = 0$ and $R_x = R_y \neq 0$, in which case the transition moment is polarized in the xy-plane: the symmetry of the molecule results in the x- and y-axes being indistinguishable.

A molecule with still lower symmetry is formaldehyde, which is planar and is shown in Figure 2.5(b). In this molecule the directions of the cartesian axes are defined by symmetry. The C_2 axis is conventionally taken as the z-axis. If, in general, a molecule has an n-fold axis (C_n)

this means that rotation of the molecule by $2\pi/n$ about that axis produces a configuration which is indistinguishable from the original one. In the case of the z-axis of formaldehyde, $n = 2$. Formaldehyde also has two planes of symmetry (σ) which means that reflection of the nuclei through either of these planes produces a configuration indistinguishable from the original one: these are the xz- and yz-planes in Figure 2.5(b) and they define the directions of the other cartesian axes. Conventionally the axis perpendicular to the plane of the nuclei is taken to be the x-axis. The origin of the axis system is the centre of gravity of the molecule. In such a molecule if $|\boldsymbol{R}^{nm}|^2 \neq 0$, then either $R_x^{nm} \neq 0$ or $R_y^{nm} \neq 0$ or $R_z^{nm} \neq 0$ and the polarization of the transition moment is along one of the axes and can never lie between them.

In a molecule with no symmetry at all, such as CHBrClF in Figure 2.5(c), none of the cartesian axes is defined by symmetry and therefore the direction of the transition moment is not confined at all by symmetry.

These examples of the possible restrictions of the directions of transition moments serve to illustrate one aspect of the great importance of symmetry in such problems. Molecular symmetry will be treated at length in Section 5.2.2.

2.3 LINE WIDTHS OF TRANSITIONS

Even under the most favourable conditions for observation, a transition between two states, such as m and n in Figure 2.2, is not seen as an infinitely sharp line[†] but there is a definite, reproducible line shape.

We shall see in Chapter 3 that the instrumentation used for observing a spectrum may itself be the limiting factor in the observed line shape. For example, a slit of a spectrometer may be too wide or a diffraction grating may not be of sufficiently high quality. In this section we shall assume that the experimental method used to obtain the spectrum does not impose any limitations on the line shape.

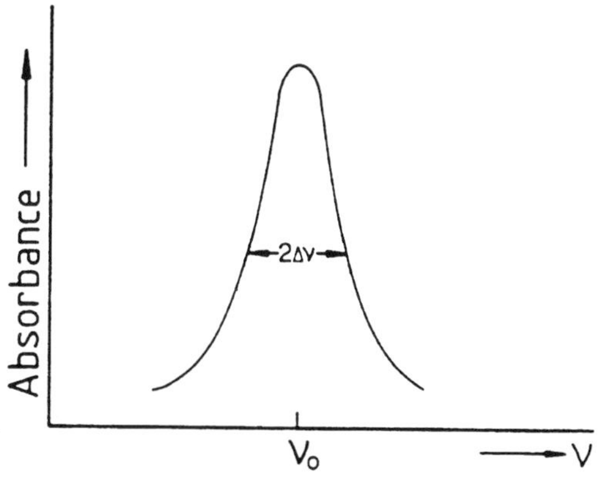

Figure 2.6 Typical line shape for a transition observed in absorption

Figure 2.6 illustrates a typical line shape for a transition observed in absorption: absorbance (equation 2.24) is plotted against frequency (or wavenumber). In referring to a parameter which is related to the width of the line we use the half intensity line width, which is the width of the line at half the maximum absorbance or, alternatively, the half width at half maximum, abbreviated to HWHM and equal to $\Delta\nu$ in the figure.

There are several factors, other than instrumental ones, which may contribute to the observed line shapes and the most important are

(1) natural line broadening;
(2) Doppler broadening;
(3) pressure broadening;
(4) wall collision broadening;
(5) power saturation broadening;
(6) modulation broadening.

[†]It is a trivial point, but worth noting here, that we often refer to a 'line' in a spectrum irrespective of whether the method of observation causes it to appear as a line, as it does when we use a spectroscope and the eye as detector, or as a peak in a plot of intensity versus wavelength.

Within these categories there is a useful distinction between contributions to the line shape which are the *same* for every atom or molecule of the sample—homogeneous line broadening—and those which result from a statistical average of a property which is not the same for all atoms or molecules—inhomogeneous line broadening.

2.3.1 Natural line broadening

Excited states which are populated in excess of their Boltzmann equilibrium population decay as a function of time. If m in Figure 2.2 is the ground state of an atom or molecule M then decay of the excited species M* to its equilibrium concentration is commonly a first-order process. This is analogous to any first-order chemical reaction. The rate of disappearance of M* is proportional to its concentration, giving

$$-\frac{\mathrm{d}N_n}{\mathrm{d}t} = kN_n \tag{2.31}$$

where k is the first-order rate constant (and not to be confused with the Boltzmann constant). Integrating from time zero to t gives

$$\ln[N_n(t)/N_n(0)] = -kt \tag{2.32}$$

or

$$N_n(t) = N_n(0)\exp(-kt) \tag{2.33}$$

For a first-order reaction, particularly a radioactive decay process, we sometimes refer to the half-life, $t_{\frac{1}{2}}$, which is the time after which the concentration has decreased to half of its initial value. From equation (2.32) it follows that

$$t_{\frac{1}{2}} = \frac{\ln 2}{k} = \frac{0.693}{k} \tag{2.34}$$

In spectroscopy we more often use the quantity τ, which is the reciprocal of the first-order rate constant:

$$\tau = \frac{1}{k} \tag{2.35}$$

and is therefore larger than the half-life. We refer to τ as the relaxation time or simply the lifetime of M* and it is the time taken for the concentration of M* to fall to 1/e times its initial value (e is the base of natural logarithms here).

There are various ways in which M* can decay to M by a first-order process. Often the most important is spontaneous emission for which the rate constant is given by

$$k = A_{nm} \tag{2.36}$$

where A_{nm} is the Einstein coefficient for spontaneous emission. Another important mechanism of decay is by transfer of energy in collision processes, which will be discussed in Sections 2.3.3 and 2.3.4. If M* is an excited electronic state then it may decay to a lower excited state, between n and m, or to the ground state m by a radiationless transition in which no energy is radiated but the excess electronic energy is converted into vibrational energy. The species M* may also undergo photochemical reaction

$$\mathrm{M}^* \rightarrow \text{products} \tag{2.37}$$

Collisional decay is second order but the other processes are first order, so we now have

$$k = A_{nm} + k_{\mathrm{rt}} + k_{\mathrm{photo}} + \ldots + k_{\mathrm{coll}}[\mathrm{M}] + \ldots \tag{2.38}$$

where k_{rt}, k_{photo} and k_{coll} are the rate constants for decay by radiationless transitions, photochemical reaction and collisions, respectively. [M] is the concentration of M, or M*, and $k_{\mathrm{coll}}[\mathrm{M}]$ is a pseudo-first-order rate constant.

The form of the Heisenberg uncertainty principle given in equation (1.29) gives the uncertainty ΔE of the energy of a state with a lifetime τ as

$$\tau\Delta E \geqslant \hbar \tag{2.39}$$

It follows that the energy of the state would be exactly defined, and the state would be a truly stationary state, only if the lifetime were infinite. In reality, τ is not infinite and the state is represented by energies smeared over a range ΔE. If we take into account the contribution to τ due only to spontaneous emission, then this is called the natural lifetime of the state and the resulting line broadening gives the natural line width.

For an excited electronic state, a typical natural lifetime is of the order of 10 ns. From equation (2.39) we obtain $\Delta E \geqslant 5.3 \times 10^{-27}$ J: converting to wavenumber gives $\Delta\tilde{\nu} \geqslant 0.000\,53\,\text{cm}^{-1}$ and, to frequency, $\Delta\nu \geqslant 16$ MHz.

At the other extreme, rotational states have long lifetimes and a typical natural line width of a rotational transition is in the range 10^{-4}–10^{-5} Hz.

From equations (2.35) and (2.36) it follows that the natural lifetime τ is related to the Einstein coefficient for spontaneous emission, A_{nm}, by

$$\tau = \frac{1}{A_{nm}} \tag{2.40}$$

Since A_{nm} is given by

$$A_{nm} = \frac{64\pi^4\nu^3}{(4\pi\varepsilon_0)3hc^3}|\boldsymbol{R}^{nm}|^2 \tag{2.41}$$

where $\boldsymbol{R}^{nm}$ is the transition moment given in equation (2.21), the frequency spread $\Delta\nu$ due to the uncertainty principle expressed in equation (2.39) is given by

$$\Delta\nu \geqslant \frac{32\pi^3\nu^3}{(4\pi\varepsilon_0)3hc^3}|\boldsymbol{R}^{nm}|^2 \tag{2.42}$$

The ν^3 dependence of $\Delta\nu$ results in much larger natural line widths in excited vibrational and, particularly, electronic states. Whereas $\Delta\nu$ is in the range 10^{-4}–10^{-5} Hz for a rotational energy level, it is increased to the order of 20 kHz in an excited vibrational state and 30 MHz in an excited electronic state.

Line broadening due to the natural line width is small relative to most other contributions and only in so-called sub-Doppler spectroscopy, which is capable of the highest precision, does it limit the accuracy of measurement of a transition frequency.

The natural line width is contributed to in an identical way by each atom or molecule and is an example of homogeneous broadening.

2.3.2 Doppler broadening

The Doppler effect is the familiar effect which causes the frequency of a train's whistle to increase as the train travels with constant velocity towards an observer and to decrease as it leaves the observer. In a similar way, the frequency of the radiation absorbed during a transition in an atom or molecule differs according to the direction of motion relative to the source of radiation. If the atom or molecule is moving with a velocity υ_a, measured positively *away* from the source, then the frequency, ν_a, at which the transition appears to the observer is related to the actual (resonance) transition frequency ν_{res} of a stationary atom or molecule by

$$\nu_a = \nu_{res}/(1 - \upsilon_a/c) \tag{2.43}$$

which c is the speed of light.

The distribution of velocities υ_a in a gas has been deduced by Maxwell and results in a broadening of a spectral line to a half width at half maximum given by

$$\Delta\nu = \frac{\nu}{c}\left(\frac{2kT\ln 2}{m}\right)^{\frac{1}{2}} \tag{2.44}$$

where k is the Boltzmann constant and m the mass of the atom or molecule. This equation shows that the Doppler effect is smaller for heavier molecules and can be overcome, to some extent, by lowering the temperature. However, a reduction of $\Delta\nu$ to about half the room temperature value is as much as is practicable.

An important way of removing, or partially removing, the Doppler line broadening is by producing the sample in an atomic or molecular beam in which all the particles are travelling in the same direction with a restricted range of velocities. Such a beam can be produced by heating the sample in a furnace and allowing the atoms or molecules to escape from it through a narrow slit. The observation of absorption or emission must be in a direction perpendicular to the beam for the Doppler broadening to be reduced.

A second way of overcoming Doppler broadening is by the technique of Lamb dip spectroscopy (see Sections 3.4.2 and Chapter 8), in which transitions are observed only in atoms or molecules having zero velocity component in the direction of propagation of the radiation.

Because atoms or molecules with different velocities absorb or emit radiation at different frequencies, Doppler broadening is an example of inhomogeneous line broadening.

The line shape, due only to the Doppler effect, is expressed by the equation

$$\frac{a}{a_{\max}} = \frac{p}{p_c}\left(\frac{\nu}{\nu_0}\right)^2 \exp\left[-\ln 2\left(\frac{\nu-\nu_0}{\Delta\nu}\right)^2\right] \tag{2.45}$$

where a is the absorption coefficient given by (see equation 2.24)

$$a = \frac{A}{l} = \varepsilon c \tag{2.46}$$

and is, therefore, the absorbance per unit length. The quantity p is the pressure and p_c is the pressure at which pressure broadening (Section 2.3.3) becomes significant. Equation (2.44) defines $\Delta\nu$, and ν_0 is the frequency corresponding to the maximum absorption coefficient $a_{\max}$.

The line shape defined by equation (2.45) is known as gaussian and is typical of all inhomogeneous contributions to the line shape.

2.3.3 Pressure broadening

If τ is the mean time between collisions in a gaseous sample and each collision results in a transition between two states m and n, there is a line broadening $\Delta\nu$ where, from equation (2.39),

$$\Delta\nu = (2\pi\tau)^{-1} \tag{2.47}$$

In predicting the line shape due to collisions, Lorentz assumed that, on collision, the oscillation in the atom or molecule is halted and, after collision, starts again with a phase completely unrelated to that before collision. On the other hand, Debye predicted the line shape assuming that the phase of oscillation is *not* random after collision.

Van Vleck and Weisskopf[2] integrated the approaches of Lorentz and Debye. They assumed that there is some phase relation of oscillations before and after collision and obtained the expression

$$a = \frac{8\pi^2 Nf}{(4\pi\varepsilon_0)3ckT}|\boldsymbol{R}^{nm}|^2\nu^2\left[\frac{\Delta\nu}{(\nu-\nu_0)^2+(\Delta\nu)^2} + \frac{\Delta\nu}{(\nu+\nu_0)^2+(\Delta\nu)^2}\right] \tag{2.48}$$

where a is the absorption coefficient given by equation (2.46), N is the number of atoms or molecules per unit volume, f is the fraction in the lower state m, ν_0 is the frequency at maximum absorbance and $\Delta\nu$ is given by equation (2.47).

At low pressure τ is large, $\Delta\nu \ll \nu_0$ and the first term of equation (2.48) is dominant so that

$$a \approx \frac{8\pi^2 Nf}{(4\pi\varepsilon_0)3ckT}|\boldsymbol{R}^{nm}|^2\nu^2\left[\frac{\Delta\nu}{(\nu-\nu_0)^2+(\Delta\nu)^2}\right] \tag{2.49}$$

This line shape is similar to that obtained by Lorentz and has become known as the lorentzian line shape.

The first term in equation (2.48) may be dominant, even at high pressure, provided that

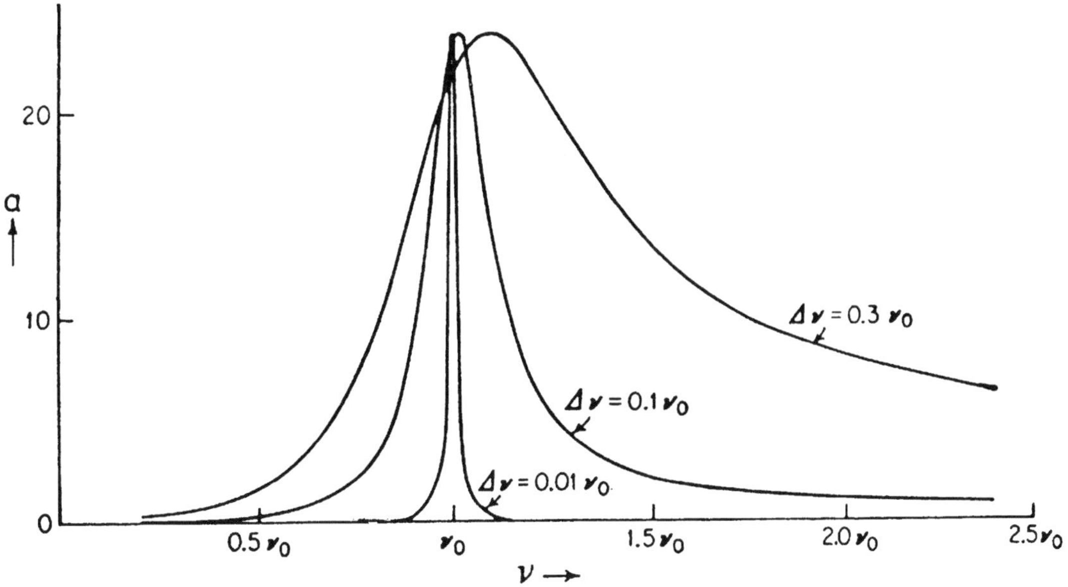

Figure 2.7 Line shape asymmetry of pressure-broadened lines is increasingly apparent with increasing $\Delta\nu/\nu_0$. (From *Microwave Spectroscopy* by C. H. Townes and A. L. Schawlow, p. 345. McGraw-Hill, New York, 1955. Reproduced by permission)

$\Delta\nu \ll \nu_0$. This is always the case in the optical regions, where ν is large, but not in the low frequency regions, such as the microwave and millimetre wave regions (see Section 3.1). In the limit the absorption coefficient at high frequency, a_∞, is given by

$$a_\infty = \int \frac{a}{\nu^2}\,\mathrm{d}\nu = \frac{8\pi N f}{(4\pi\varepsilon_0)3ckT\tau}\,|\boldsymbol{R}^{nm}|^2 \qquad (2.50)$$

and is, therefore, constant and non-zero on the high frequency side of ν_0. The resulting asymmetry of the line shape is illustrated for various $\Delta\nu/\nu_0$ in Figure 2.7.

At pressures of the order of 1 atm this high frequency tail on microwave transitions is readily observed. Quantitatively, it cannot be reproduced by the Van Vleck–Weisskopf theory which neglects the effect of collisions between more than two atoms or molecules.

2.3.4 Wall collision broadening

Collisions between the atoms or molecules of the sample and the walls of the cell in which it is contained broaden a transition in the same way as inter-particle collisions discussed in Section 2.3.3. The main difference lies in the possibility of eliminating the effect.

Wall collision broadening can be removed by making the sample cell so large that such collisions are infrequent. In high frequency regions of the spectrum this is not difficult and, in any case, collisional broadening is relatively unimportant. It is in the low frequency regions, such as the microwave region, where wall collision broadening may be appreciable and where it may be difficult to eliminate. If Stark modulation is used (Section 3.3), the metal Stark electrodes, placed inside the sample cell, cannot be too far apart and collisions with these may be important.

2.3.5 Power saturation broadening

The Boltzmann distribution law, given in equation (2.6), relates, for conditions of equilibrium, the populations N_m and N_n of the lower and upper states m and n to their energy separation ΔE. When ΔE is small, as in microwave and millimetre wave spectroscopy, N_n is not much less than N_m and absorption of radiation can easily promote a sufficient number of molecules to state n, so that $N_n \approx N_m$. The effect of the populations becoming nearly equal is known as saturation and is more likely to occur when the power of the incident radiation is high.

Under conditions of saturation the absorption coefficient a is dependent on the intensity I of incident radiation. It is given by

$$a = a_0\left(1 - \frac{2IakTt}{N_m h\nu}\right) \tag{2.51}$$

where a_0 is the absorption coefficient at low I and $t = \frac{1}{2} t_{nm}$, where $(t_{nm})^{-1}$ is the probability per unit time that a molecule will transfer from state n to m by collisions. When I is large $a \ll a_0$ and

$$\frac{2IakTt}{N_m h\nu} \approx 1 \tag{2.52}$$

and all the energy absorbed is removed only by collisions.

In microwave spectroscopy, power saturation broadening can usually be avoided by keeping the source power below about 100 μW, but this depends on the power per unit area and the magnitude of the transition moment.

Until the advent of lasers saturation effects were limited to low frequency spectroscopy but laser power can be so high that, even though $N_n \approx 0$ at equilibrium, saturation may be achieved (see Section 8.2.23).

2.3.6 Modulation broadening

As we shall see in Section 3.3, in microwave and millimetre wave spectroscopy an electric field is very often applied across the sample, normal to the incident radiation. It is usual to apply a modulation to the field so that it varies periodically with time. If the modulation frequency f is comparable to the collision frequency, the transitions between energy levels cannot exactly follow the changes in electric field strength. If these two frequencies are similar and the modulation is sinusoidal, satellites are observed separated from the main line by the modulation frequency. At lower modulation frequencies there is a line broadening to a half width at half maximum $\Delta\nu'$ given by

$$\Delta\nu' = \Delta\nu\left[1 + \left(\frac{f}{2\Delta\nu}\right)^2\right] \tag{2.53}$$

where $f \ll \Delta\nu$. In order to avoid this type of broadening, the maximum modulation frequency which can be used is about 100 kHz.

2.3.7 Summary

In most gas phase spectroscopy, line broadening is due predominantly to Doppler broadening (inhomogeneous) and pressure broadening (homogeneous), but only at low frequencies is it difficult to reduce pressure broadening to such an extent that Doppler broadening dominates. For example, in microwave and millimetre wave spectroscopy a typical half width at half maximum intensity is 1–10 kHz due to pressure broadening at a pressure of only 1 mTorr compared to about 10 kHz due to Doppler broadening.

Under these circumstances of comparable inhomogeneous and homogeneous line broadening, the line shape is a combination of gaussian and lorentzian and is called a Voigt line shape.[3]

In all kinds of spectroscopy, provided that the conditions are such that the line width would normally be Doppler limited, transitions with a line width which is sub-Doppler and dominated by the natural line width may beobserved either by having the sample in the using the Lamb dip technique. These, and other methods, will be discussed in Section 3.4 and various sections of Chapter 8 describing techniques of laser spectroscopy.

BIBLIOGRAPHY

Further discussion can be found in several of the books in the Bibliography of Chapters 3 and 4 but, particularly, in Townes, C. H. and Schawlow, A. L. (1955). *Microwave Spectroscopy*. McGraw-Hill, New York

3 GENERAL EXPERIMENTAL METHODS

3.1 REGIONS OF THE ELECTROMAGNETIC SPECTRUM

The dual electric and magnetic character of electromagnetic radiation has been discussed in Section 2.1. All electromagnetic radiation has this dual character and also travels, in a vacuum, with the same speed, c, which is referred to as the speed of light. Its value is defined exactly as

$$c = 2.997\,924\,58 \times 10^8 \text{ m s}^{-1} \qquad (3.1)$$

The radiation can be characterized by its wavelength λ *in vacuo* or frequency ν, where

$$\nu = c/\lambda \qquad (3.2)$$

or by its wavenumber $\tilde{\nu}$ *in vacuo*, where

$$\tilde{\nu} = 1/\lambda \qquad (3.3)$$

The use of wavenumber rather than frequency originated at a time when wavelengths could be measured to a higher accuracy than that to which the speed of light was known and bypassed the necessity for an accurate value of c.

Figure 3.1 shows that the range of electromagnetic radiation, or the electromagnetic spectrum as it is called, is considerable. It shows that the visible region, extending from red, with a wavelength of about 770 nm, through yellow, green, blue, indigo to violet, with a wavelength of about 390 nm, is small compared with the whole electromagnetic spectrum. Although we tend to think of the visible region as unique, because the eye is uniquely sensitive to it, we must regard it in

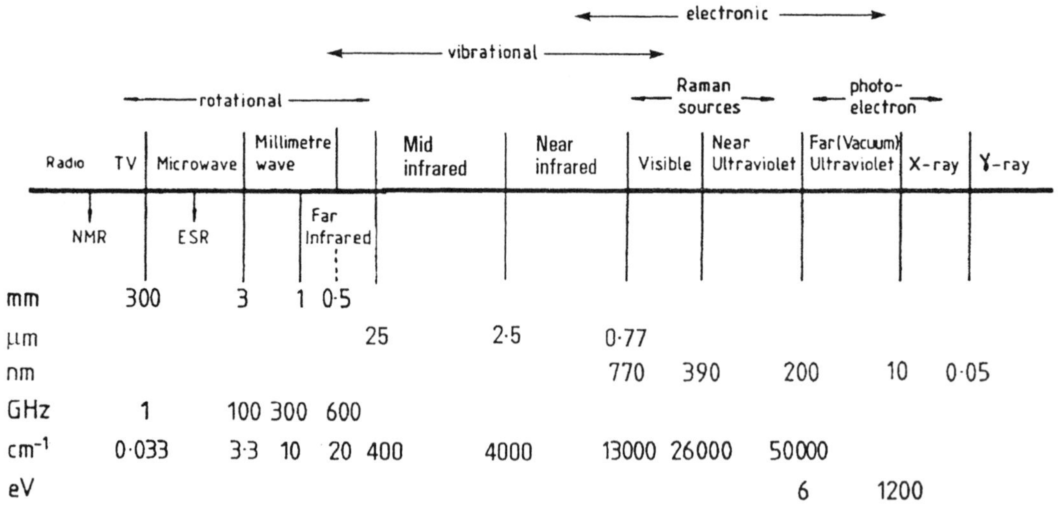

Figure 3.1 Regions of the electromagnetic spectrum

sepctroscopic terms as equal in importance to the other regions.

The division of the electromagnetic spectrum into the radio/television, microwave, millimetre wave, far infrared, mid infrared, near infrared, visible, near ultraviolet, far ultraviolet, X-ray and γ-ray regions, indicated in Figure 3.1, is artificial in the sense that there is nothing fundamental which distinguishes radiation in these regions. The regions are distinguished mainly by the different experimental techniques which are used; these are differences particularly of generating and detecting the radiation.

In Figure 3.1, the approximate limits of each region are indicated by the wavelengths in units of nm, μm or mm, wavenumbers in units of cm^{-1}, or frequencies in units of GHz ($1\,GHz = 10^9\,s^{-1}$), but in practice the divisions between regions are not precise and there is some overlap.

From the extreme of radio waves to that of γ-rays there is a very large increase in wavenumber or frequency and therefore, from equation (1.14), of energy. It follows that more and more drastic results accrue from the interaction of radiation with matter as the energy of the radiation increases. This is illustrated by the fact that radio waves, which are constantly surrounding us, have no effect on the tissues of the body whereas γ-rays are extremely dangerous. Another illustration of the energy differences across the spectrum is the effect of radiation on molecules. Generally, the energy difference between a pair of rotational energy levels is much smaller than between a pair of vibrational levels, which in turn is much smaller than between a pair of electronic levels. Therefore, rotational transitions occur in the low energy microwave, millimetre wave and far infrared regions, vibrational transitions in the far, mid and near infrared, and electronic transitions in the visible, near and far ultraviolet regions although, again, there are no rigid energy boundaries for these processes.

If we take a typical ionization energy for the removal of a valence electron from an atom or molecule as 10 eV,† we can see from Figure 3.1 that far ultraviolet radiation may have sufficient energy to cause ionization.

In this chapter we shall look at the experimental methods used to observe absorption and emission spectra from the microwave to the far ultraviolet region. These methods have many principles in common although the instrumentation used varies considerably. Spectroscopic techniques which differ in principle from those generally applicable in the microwave to far ultraviolet range will be discussed in later chapters along with the relevant theory. These techniques include Raman spectroscopy, which involves neither an absorption nor an emission but a scattering process, photoelectron spectroscopy, in which the analysis is of the velocity of electrons ejected in an ionization process rather than of the frequency or wavenumber of electromagnetic radiation, and laser spectroscopy, in which the unusual nature of the laser radiation necessitates the use of rather different experimental techniques.

3.2 GENERAL FEATURES OF INSTRUMENTATION

Only for transitions between electronic states is it relatively easy to build up a concentration of atoms or molecules in excited states which is sufficiently high for emission of radiation to be observed. The reason for this is that the transition moment of equation (2.21) is generally much larger for an electronic transition, involving considerable shift of electronic charge, than for vibrational or rotational transitions. So in what follows we shall mention techniques used in emission spectroscopy only when discussing electronic spectroscopy—the study of transitions between electronic states.

†The unit electronvolt (eV) is commonly used for expressing ionization energies. Its relation to cm^{-1} and s^{-1} units is given by $1\,eV = 8065.54\,cm^{-1}$ or $2.417988 \times 10^{14}\,s^{-1}$.

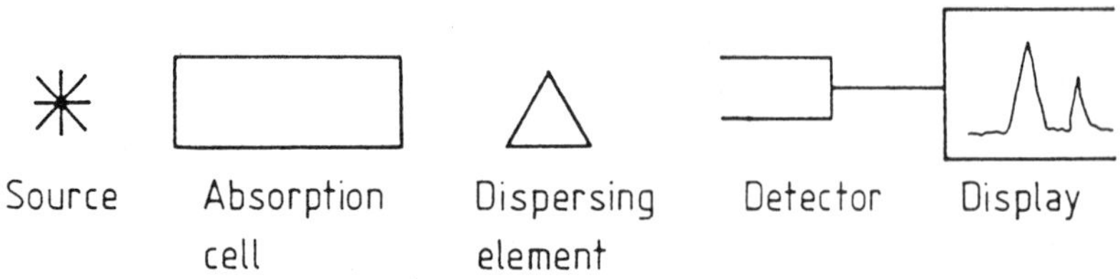

Figure 3.2 The main elements of an experimental arrangement for observing an absorption spectrum

The general experimental arrangement for an absorption experiment is shown schematically in Figure 3.2.

The source is usually a continuum, which means that it emits radiation with a continuous range of wavenumbers with, ideally, uniform intensity in the region of the spectrum being studied.

The absorption cell contains the absorbing sample and must be constructed with windows on each end which transmit the radiation from the source. The cell must also be long enough for the absorbance [equation (2.24)] to be sufficient to be detected.

The choice of phase (gas, liquid or solid) of the sample is very important. As was mentioned in Section 2.3.3, if collisions occur at a rate which is appreciable compared with the emission lifetime, the effect is to broaden the excited state n and therefore to broaden the transition. If we are trying to achieve high resolution (see Section 3.5.1 for a discussion of resolution), we must try to use conditions in which the transitions are as sharp as possible. This usually means working with the sample in the gas phase at low pressure to minimize the number of collisions. The requirement of low pressure often leads to the necessity for a long pathlength unless the molar absorption coefficient ε is large.

In the liquid phase, collisions are so frequent that, in most cases, rotational motion must be treated classically rather than quantum mechanically. The result is that no rotational transitions can be observed and rotational, vibrational and electronic spectra are all broadened considerably in liquids. However, this broadening may be beneficial for some purposes. For example, the measurement of the oscillator strength [equation (2.26)] of an electronic transition by measuring the area under the absorption curve is more conveniently done in the liquid phase where broadening of the rotational and vibrational transitions results in a smoother absorption curve than for the gas phase.

In the solid phase, in which the sample may be a pure crystal, a mixed crystal or a glassy non-crystalline solid, rotational motion is quenched owing to the molecules being held rigidly in position by intermolecular forces. Transitions between vibrational and between electronic states are mostly very broad at normal temperatures but increase in sharpness, sometimes dramatically, at very low temperatures, such as that of liquid helium (*ca* 4 K).

The dispersing element is perhaps the most important part of the instrumentation, except in microwave and millimetre wave spectroscopy where it is not required. This element separates the radiation into its component wavelengths as, for example, a prism separates white light into the colours of the visible spectrum. For highly precise, high resolution observations of a spectrum it is necessary for the dispersing element to be capable of separating radiation of very closely similar wavelengths, and this capability is very often the factor limiting the precision and resolution of which the instrumentation as a whole is capable.

Table 3.1 Bands in the microwave region

Band label	ν/GHz	Band label	ν/GHz
S	2.6–4.0	M	10.0–15.0
G	4.0–5.9	P	12.4–18.0
J	5.3–8.2	K	18.0–26.5
H	7.1–10.0	R	26.5–40.0
X	8.2–12.4		

The detector must be sensitive to whatever kind of radiation is falling on it; for example, a thermocouple will record infrared but not microwave radiation. The spectrum is displayed in the form of the intensity of radiation falling on the detector as a function of wavelength, frequency or wavenumber. Because frequency and wavenumber are proportional to energy they are much more useful quantities than wavelength in the interpretation of spectra. For this reason, it is desirable for the spectrum to be displayed as a linear function of frequency or wavenumber. The importance of wavelength is mostly confined to the optics of the instrumentation.

3.3 MICROWAVE AND MILLIMETRE WAVE SPECTROSCOPY

In the microwave and millimetre wave regions of the electromagnetic spectrum, the methods of generation and detection of the radiation make it more convenient and natural to use frequency to characterize the radiation rather than, say, wavenumber, which is a more natural choice for the regions extending from the far infrared to the far ultraviolet.

The microwave region covers approximately the range 1–100 GHz (Figure 3.1) and the millimetre wave region 100–600 GHz. It is because the latter range covers wavelengths of the order of 1 mm that the name 'millimetre wave' is used. As Figure 3.1 shows, the millimetre wave region overlaps the far infrared.

The experimental techniques of microwave and millimetre wave spectroscopy involve the use of electronic devices much more than those of infrared, visible and ultraviolet spectroscopy, which use mainly optical devices.

Of the two techniques discussed in this section, millimetre wave spectroscopy will be mentioned more briefly since it has much in common with microwave spectroscopy but is not so widely used.

Unlike infrared, visible and ultraviolet radiation, both microwave and millimetre wave radiation can be easily channelled both in straight lines and around corners by metal tubing or a waveguide. The waveguide is rectangular in cross-section, made of copper or brass and often coated internally with silver to prevent loss of power when the radiation travels along it. The cross-sectional dimensions of the waveguide depend on the frequency range for which it is being used, so that a change of the range may necessitate a change to waveguide of a different size. Any range for which a particular waveguide can be used is called a band and each band is denoted, as shown in Table 3.1, by a letter; for example, we may speak of the 8.2–12.4 GHz region as the X-band region. However, conventional letters other than those in Table 3.1 are sometimes used. The size of the waveguide decreases as the frequency increases; for example, the cross-section is 76.2×25.4 mm in the S-band and 7.02×3.15 mm in the R-band region.

(a)

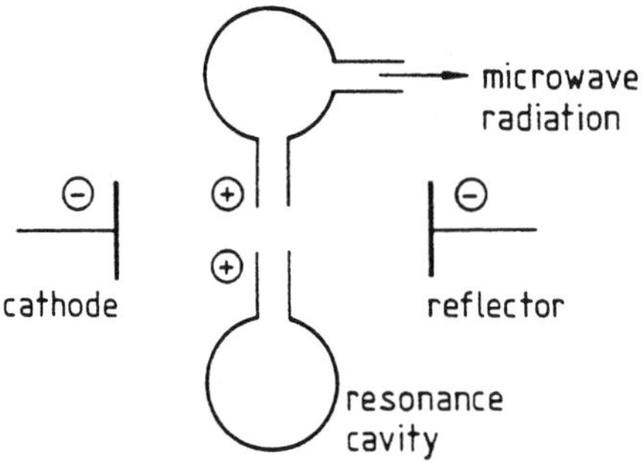

(b)

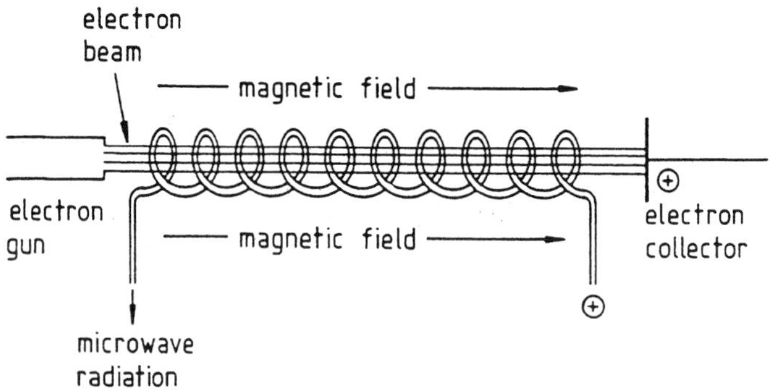

Figure 3.3 (a) A klystron microwave source and (b) a backward wave oscillator microwave source

In the early years of microwave spectroscopy, the most widely used source was the klystron, which is illustrated in Figure 3.3(a). In an evacuated container electrons are emitted from the electrically heated cathode and pass through the positively charged resonant cavity to the negatively charged grid which acts as an electron reflector. The cavity is able to retain microwave radiation emanating from the electron beam but only of a certain frequency, depending on the dimensions of the cavity. As a result, each time the electrons pass through the cavity they experience an oscillating electric field due to the microwave radiation. The result is that the electrons tend to bunch together instead of forming a homogeneous beam and microwave radiation of a particular frequency accumulates in the cavity and can be fed into a waveguide and then into the absorption cell. The frequency of the klystron can be changed slightly by electrical tuning over a range of about 30 MHz. Mechanical distortion of the cavity allows tuning over a range which is about 10% of the mean frequency. This means, for example, that it takes about four different klystrons to cover just the X-band region.

A much more convenient source, developed in the 1960s, is the backward wave oscillator illustrated in Figure 3.3(b). A beam of electrons with a specific velocity is directed from an electron gun towards a positive collector electrode. The electron beam passes through a wire helix which is also at a positive voltage but less than that of the collector. A cylindrical permanent magnet keeps the beam close to, but clear of, the helix. Bunching of the electrons is produced by voltages induced in the helix by the beam. As a bunch of electrons passes a space between turns in the helix the electric field due to the electrons passes out of the helix. At a certain frequency the oscillating electric field resonates with the electron bunches passing down the helix and a wave is generated in the opposite direction to the beam—a backward wave. The complete backward wave oscillator is contained in an evacuated tube. The frequency of the radiation is varied by changing the beam velocity and the potential of the helix. The great advantage of this type of source over the klystron is that a single backward wave oscillator can cover a complete microwave band.

The higher frequencies required for millimetre wave radiation can be obtained by frequency multipliers which generate harmonics of lower frequency microwave radiation emitted from a klystron. The unavoidable loss of power after frequency multiplication is somewhat offset by the fact that the molar absorption coefficient is proportional to the square of the absorption frequency and therefore is considerably higher for millimetre waves. Backward wave oscillators have been developed for the millimetre wave region, extending it up to a frequency of 1000 GHz (1 THz).

The microwave or millimetre wave radiation from a klystron or backward wave oscillator is highly monochromatic, which means that it comprises an extremely narrow band of frequencies. This band is very small compared with the half-intensity line width of even the sharpest absorption lines. In microwave and millimetre wave spectroscopy the line width is often limited by pressure broadening since, even at a pressure as low as 1 mTorr, the line width is about 1–10 kHz due to this effect, compared with a typical line width due to Doppler broadening of the order of 10 kHz, as discussed in Section 2.3.7.

Another possible mechanism which can produce appreciable line broadening in microwave and millimetre wave spectroscopy is saturation broadening, described in Section 2.3.5. However, saturation is not always just a problem to be overcome. It has been made use of in a technique which makes it possible to reduce the width of Doppler-limited lines to the natural line width. The technique is called Lamb dip spectroscopy and will be explained in Section 3.4.

Because the radiation from the source is monochromatic there is no need for any dispersing element in microwave or millimetre wave spectroscopy.

The waveguide conveys the radiation to the absorption cell. The cell has windows, usually made from mica, and it may be several metres in length. It is plated on the inside with copper, silver or gold to prevent troublesome absorption of, or chemical attack by, the sample gas.

In microwave and millimetre wave spectroscopy the detector is a crystal diode rectifier consisting of a tungsten whisker in point contact with a small piece of semiconductor such as silicon or germanium. However, the low sensitivity of the crystal detector can easily be a limiting factor.

The sensitivity of crystal detectors is limited by the fact that they produce random noise with a power P given by

$$P = kT\Delta\nu + CI^2\Delta\nu/\nu \qquad (3.4)$$

where C is a constant, I is the current flowing, $\Delta\nu$ is the detection band width and ν is the modulation frequency. Modulation is a periodic variation of the amplitude of the radiation reaching the crystal. The first term in equation (3.4) constitutes the so-called Johnson noise and little can be done to reduce its value. The second term, however, the crystal conversion noise, can be reduced by using as high a modulation frequency as possible and a small detection band width (which also reduces the Johnson noise somewhat).

There are several ways in which modulation can be achieved but Stark modulation is the most widely used.

If we consider the simple two energy level system shown in Figure 3.4, in which a transition between the levels occurs with a zero-field frequency ν_0, then, in the presence of a static electric field of voltage V, the energy

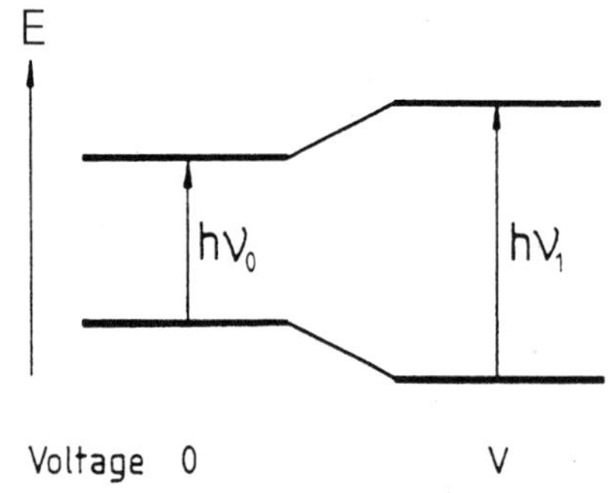

Figure 3.4 Modification of a transition frequency ν_0 to ν_1 by an electrostatic field

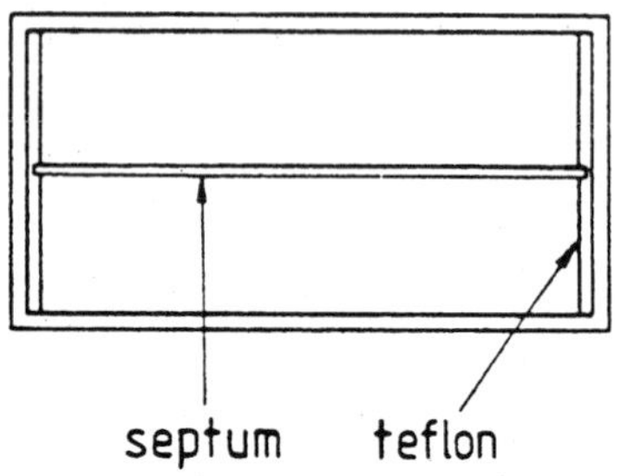

Figure 3.5 The position of the Stark electrode, or septum, in a microwave absorption cell

levels are modified and the frequency of the transition is shifted to ν_1: the effect of the applied field is known as the Stark effect. The field is applied to a metal Stark electrode, or septum, running along the length of the absorption cell parallel to, and equidistant from, the two broad cell walls, as shown in Figure 3.5. The septum is inserted into grooves in pieces of Teflon which insulate it from the walls of the cell. The Stark voltage varies periodically from 0 to V, typically hundreds of volts per centimetre, in the square wave form shown in Figure 3.6(b) and with a frequency which is usually between 50 and 100 kHz. Figure 3.6(a) shows the positions of the lines in the spectrum with the Stark voltage on and off. Figure 3.6(c) shows how the radiation detected by the crystal varies as the spectrum is scanned. In the ν_0 position it shows absorption which is 180° out-of-phase with the modulation and, in the ν_1 position, absorption which is in-phase with the modulation. Amplification and the use of a phase sensitive detector enable the 'field off' and 'field on' transitions to be recorded above and below the baseline respectively, as in Figure 3.6(d).

In addition to decreasing the noise due to the crystal detector, Stark modulation also allows measurement of the permanent dipole moment of the sample molecules. However, the effect of

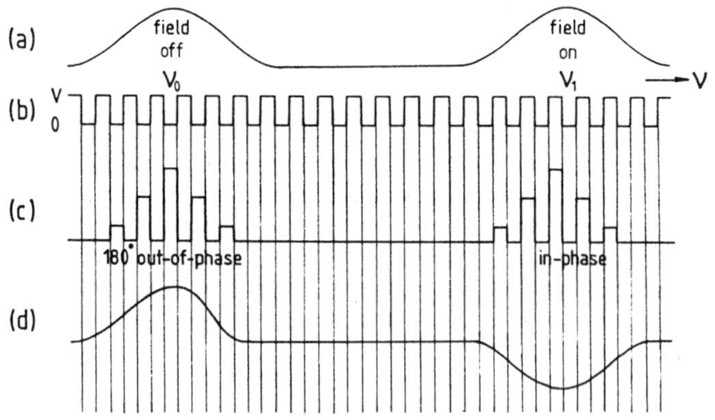

Figure 3.6 (a) Line positions with Stark field on and off. (b) Square wave variation of Stark voltage. (c) Radiation detected by the crystal. (d) Signal after amplification and phase sensitive detection

an electric field on energy levels is not usually as simple as that shown in Figure 3.4. The Stark effect often produces a complex pattern of weak lines and the pattern due to 'field off' lines may overlap that due to 'field on' lines.

Stark modulation, although the most commonly used, is not always the most suitable type. This is the case, for example, when the molecule has a small dipole moment or when the spectrum being recorded involves relatively high values of the rotational quantum number J: in both circumstances there is only a small Stark effect and, therefore, reduced sensitivity.

One alternative is to use source modulation in which a small alternating voltage is applied to the reflecting grid of the klystron, the effect of which is to cause a small alternating change of frequency. A disadvantage of source modulation is that it does not discriminate between absorption and frequency-dependent reflections of the microwaves in the waveguide system. This problem can be overcome by subtracting from the spectrum the baseline effects due to these reflections.

A second alternative is to use Zeeman modulation.[4] Whereas Stark modulation employs the Stark effect, the effect of an electric field, Zeeman modulation uses the Zeeman effect, the effect of a magnetic field (see Section 6.1.8.1). The use of the technique of Zeeman modulation is limited to molecules which have one or more unpaired electron spins.

The magnetic field is introduced transversely across the absorption cell and is rotated at a frequency ω in a plane perpendicular to the direction of microwave propagation. The selection rules for transitions in the molecule change at a frequency 2ω from $\Delta M = 0$, where M is the magnetic quantum number, when the field is parallel to the direction of the electric vector of the plane-polarized microwave radiation, to $\Delta M = \pm 1$ when it is perpendicular. Because the frequency at which absorption occurs changes with the selection rules, the absorption is modulated and may be synchronously detected at a frequency 2ω.

Zeeman modulation and, particularly, source modulation have found important applications in the microwave spectroscopy of short-lived species, both neutral molecules and positive ions.[†] The species are produced, in low concentration, in the positive column of a d.c. discharge within the absorption cell, conditions which preclude the use of Stark modulation. The cell has a large cross-section which has the effect of removing line broadening due to both wall collisions and power saturation. In addition, modulation and pressure broadening can be reduced to such an extent that line widths are almost Doppler limited (Section 2.3.2). This method was devised by Woods and first applied to obtaining the microwave spectrum of the OH radical produced in a discharge in H_2O vapour.[†]

With a klystron source, which cannot easily be tuned continuously, the microwave spectrum is usually recorded as a trace on an oscilloscope. With a continuously tunable backward wave oscillator source a chart recorder can be used.

The complete system of source, cell and detector is referred to as a microwave (or millimetre wave) spectrometer. As we shall see later in this chapter, what we call a spectrometer uses some kind of electronic means of detection, whereas a spectrograph uses photographic detection. A spectroscope uses the eye as a detector and therefore is used only in the visible region.

3.4 METHODS FOR REDUCING DOPPLER BROADENING

3.4.1 Spectroscopy in an effusive beam

An effusive beam of atoms or molecules (see Ramsey in the Bibliography) is produced by pumping them through a narrow slit, typically 20 μm wide and 1 cm long, with a pressure of

[†]See Woods, R. C. (1973). *Rev. Sci. Instrum.*, **44**, 282.

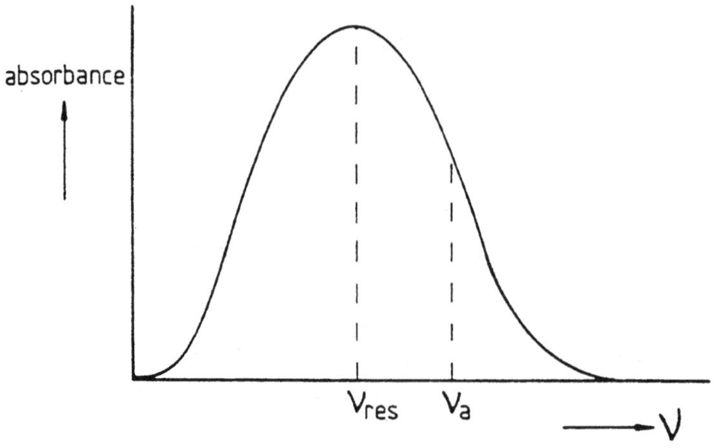

Figure 3.7 Doppler broadened line shape

only a few Torr on the source side of the slit. Because the slit width d is much less than the mean free path λ_0 for collisions between the atoms or molecules passing through it, i.e.

$$d \ll \lambda_0 \tag{3.5}$$

there are no collisions in or beyond the slit. As a result, the Maxwellian velocity distribution among the particles, and therefore their translational temperature, remains the same as it was in the reservoir of the gas forming the beam.

The effusive beam may be further collimated by placing suitable apertures along the beam.

In such a beam pressure broadening of spectral lines is very much reduced because of the necessarily low pressure of the atoms or molecules. In addition, Doppler broadening, discussed in Section 2.3.2, may also be reduced. This is because all the particles in the beam are moving in the same direction. If observation of the spectrum is made in a direction perpendicular to the beam, the velocity component in the direction of observation is zero, or at least very small. Consequently, the Doppler widths of lines in the spectrum are very much reduced, thereby increasing the accuracy of measurement.

3.4.2 Lamb dip spectroscopy

In 1969, a very elegant method was devised of eliminating the Doppler effect in favourable circumstances without the absorbing gas being in an effusive beam. We shall illustrate the technique here as it is applied in microwave and millimetre wave spectroscopy and later, in Chapter 8, discuss its importance in laser spectroscopy.

Figure 3.7 shows the shape of an absorption line which is broadened by only the Doppler effect. Figure 3.8 illustrates how the conventional source–cell–detector arrangement is modified in Lamb dip[†] spectroscopy by a reflector R which reflects the radiation back through the absorption cell to the detector. The width of a band of monochromatic microwave or millimetre wave radiation is small compared with the Doppler width, so that the source can be tuned to a frequency ν_a, say, which is, as indicated in Figure 3.7, higher than the resonance frequency. From equation (2.43) we can see that the molecules

[†]Predicted by Lamb, W. E. (1964). *Phys. Rev.*, **134**, A1429. First observed by Costain, C. C. (1969). *Can. J. Phys.*, **47**, 2431.

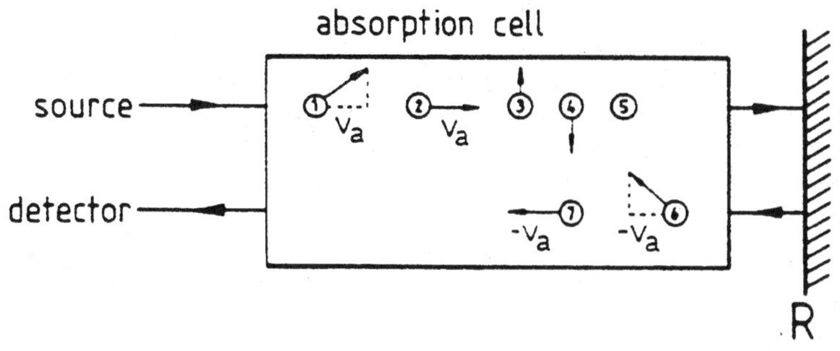

Figure 3.8 Variation of frequency of absorption by molecules travelling with different velocities in a Lamb dip experiment

absorbing at the frequency ν_a must be, like molecules 1 and 2 in Figure 3.8, moving away from the source with a velocity v_a, or a velocity component v_a, in the direction of propagation of the radiation. For this reason, the number of molecules moving with a relative velocity v_a away from the source and in the lower state of the transition is depleted compared with the usual Maxwell distribution: this is illustrated in Figure 3.9. We say that a hole has been *burnt* in the lower state velocity distribution. The small range of velocities covering the range of the hole is that which is characteristic of only the very small natural line width.

When the radiation of frequency ν_a is reflected by R a new set of molecules, such as 6 and 7 in Figure 3.8 having a velocity component $-v_a$ towards the source, are able to absorb the radiation. Therefore, twice as much absorption takes place after two passes of the cell than after one and a second hole, at $-v_a$ is burnt in the ground state velocity distribution, as shown in Figure 3.9.

On the other hand, when $\nu = \nu_{res}$, only molecules such as 3, 4 and 5 in Figure 3.8, with zero velocity relative to the source, absorb radiation. If one pass of the cell produces saturation then no more radiation is absorbed on the second pass since it is the same set of molecules which absorb on both passes. This causes a reduced absorption and a Lamb dip centred on ν_{res}. Such a dip is illustrated in Figure 3.10.

From what has been said so far about this technique we can see that some important conditions must be satisfied before Lamb dips can be observed. These are:

(1) The line width must be limited only by the Doppler effect.
(2) The two energy levels between which the transitions are occurring must be sufficiently close and the power of the source must be sufficiently high for saturation to be achieved.
(3) The radiation must be monochromatic (of a small frequency range) compared with the Doppler line width and be capable of being tuned over the frequency range of this line width.

These conditions are easily met in the microwave and millimetre wave regions but the advent of highly monochromatic and sufficiently tunable lasers has resulted in Lamb dips being observed in other regions (see Chapter 8).

3.4.3 Spectroscopy in a supersonic free jet or molecular beam

The atoms or molecules in an effusive beam, described in Section 3.4.1, do not undergo any

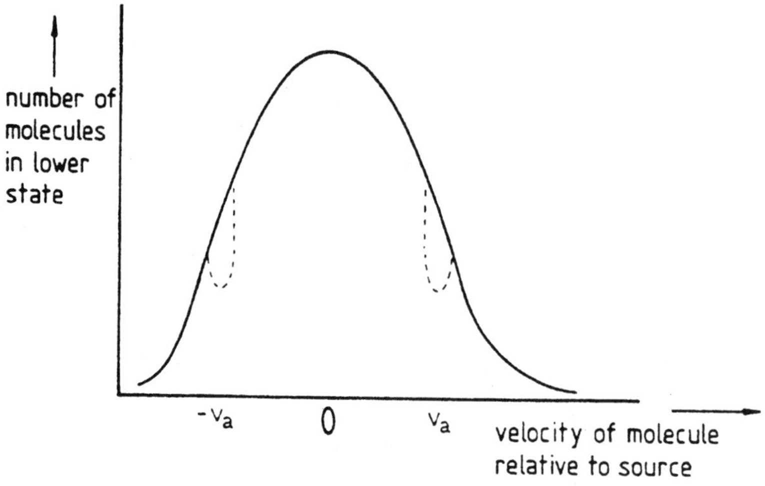

Figure 3.9 Hole burning in the velocity distribution in a Lamb dip experiment

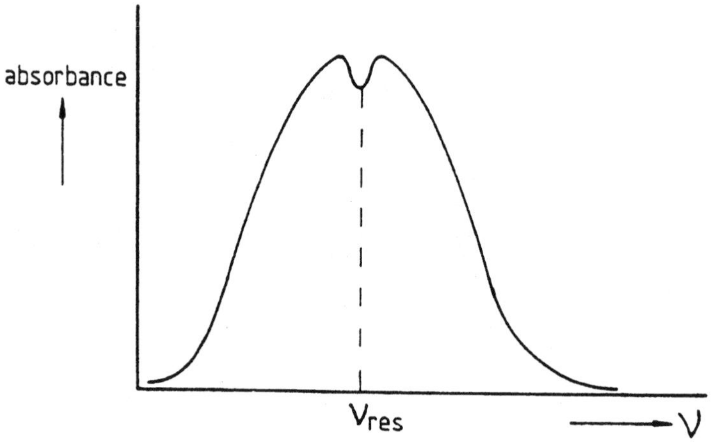

Figure 3.10 A Lamb dip in the Doppler profile

collisions in or beyond the slit: the Maxwellian velocity distribution and, therefore, the translational temperature are unaffected.

In 1951, Kantrovitz and Grey (see Bibliography) suggested the use of a wider slit and/or a higher gas pressure, known as the stagnation pressure, before the slit. Under these conditions

$$d \gg \lambda_0 \tag{3.6}$$

where d is the slit width, or diameter of a pinhole or nozzle, and λ_0 the mean free path between collisions. With d typically 40–100 μm and a stagnation pressure of several atmospheres, the numerous collisions occurring in, and immediately beyond, the pinhole will convert the random motions of the particles in the gas into mass flow in the direction of the resulting beam. Not only is the flow highly directional but also the range of velocities is very much reduced. One result of this is that the translational temperature T_{tr} of the beam produced may be extremely low, even less than 1 K, because any particle travelling in the beam has a very small velocity relative to neighbouring particles and experiences collisions only rarely. Such cooling continues in the region beyond the nozzle where collisions are still occurring. In this region the type of flow of the gas is called hydrodynamic whereas in the region beyond this, where collisions no longer occur, the flow is said to be molecular. In the region of molecular flow there is no further decrease of T_{tr} and there are decreases in pressure and Doppler line broadening similar to those produced in an effusive beam, in which the flow is always molecular.

Why are these beams, or jets, distinguished from effusive beams by their description as supersonic? In some ways this description is rather misleading, first because particles in an effusive beam may well be travelling at supersonic velocities, and second, because the name implies that something special happens when the particle velocities become supersonic whereas this is not the case. What 'supersonic' is meant to imply is that the particles may have *very* high Mach numbers (of the order of 100). The Mach number M is defined as

$$M = \frac{u}{a} \tag{3.7}$$

where u is the mass flow velocity and a is the local speed of sound, given by

$$a = \left(\frac{\gamma k T_{tr}}{m}\right)^{\frac{1}{2}} \tag{3.8}$$

where $\gamma = C_p/C_V = \frac{5}{3}$ for a monatomic gas, such as helium or argon which are commonly used in supersonic jets, and m is the mass of the gas particles. Very high Mach numbers arise not so much because u is large (although it might be twice the speed of sound in air) but because a is so small as a result of the low value of T_{tr}.

On the high pressure side of the nozzle molecules may be seeded, at a typical concentration of only a few per cent, into the helium or argon. These molecules are also cooled by the many collisions which take place.

It is necessary to remove Doppler broadening when ultrahigh resolution spectroscopy, so-called sub-Doppler spectroscopy, of the seeded molecules is being investigated. In order to do this, one or more metal skimmers are placed downstream. A skimmer is trumpet-shaped with a hole at the pointed end so that the outer regions of the jet are removed and the central core allowed to pass through. In this central core of the jet, referred to as a supersonic molecular beam, motion of the atoms and molecules is highly parallel. As a result, observations in a direction perpendicular to the beam produce spectra with minimal Doppler broadening.

For an unskimmed supersonic free jet, pumping is required to limit the background pressure in the vacuum chamber to about 10^{-5} Torr. For a skimmed supersonic molecular beam in, for example, a system using two skimmers at different downstream positions typical pressures are 10^{-4}, 10^{-6} and 10^{-7} Torr in the three vacuum chambers separated by the skimmers.

Because of the particular advantage of extreme cooling of the molecules, a supersonic jet or beam is much more commonly used in spectroscopy than an effusive beam.

A further useful modification to a supersonic jet is for it to be pulsed rather than continuous. This may have several advantages: less of the sample material may be used, the nozzle diameter may be increased, thereby increasing the molecular density in each pulse, the pumping requirements may not be so extreme and, if a pulsed laser (see Chapter 8) is used to produce the spectrum, the jet pulses can be arranged to coincide with the laser pulses.

There are various methods of pulsing the jet, one using a modified fuel injection valve. Whatever method is used, a typical pulse length is of the order of 10 μs.

The concentration of molecules seeded into a supersonic jet is typically very low. Nevertheless, there can often be a gain in sensitivity because of the extreme cooling. The molecules are concentrated into many fewer rotational and vibrational energy levels than they would be at room temperature.

In discussing the temperature of molecules seeded into a supersonic jet it is important to distinguish between translational, rotational and vibrational temperatures.

The translational temperature is the same as that of the helium or argon carrier gas and is usually of the order of 1 K.

The rotational temperature is defined as the temperature which describes the Boltzmann population distribution among rotational levels. The Boltzmann equation for the relative populations of two energy levels separated by ΔE is given by equation (2.6) and, for a diatomic molecule, the rotational energy levels are given by equation (1.88). These, and rotational energy levels for polyatomic molecules also, are closely spaced, but not as closely as the translational energy levels. For this reason, collisional energy transfer between translational and rotational levels is not totally efficient. As a result, the rotational temperature is higher and is typically of the order of 10 K. There are, however, some well established cases where the distribution of jet-cooled molecules among the rotational levels does not follow the Boltzmann equation. Then it is not possible to assign a very meaningful rotational temperature.

The vibrational temperature is defined as the temperature which describes the Boltzmann population distribution among vibrational levels. For a diatomic molecule the vibrational energy levels are given by equation (1.99) and are much more widely spaced than rotational energy levels. This applies to vibrational levels of polyatomic molecules also, and collisional energy transfer is inefficient. Consequently, a vibrational temperature of the order of 100 K is typical although, for a polyatomic molecule, this temperature may vary from one vibration to another.

One extremely important property of a supersonic jet is that the vibrational temperature is sufficiently low to stabilize complexes which are weakly bound by van der Waals or hydrogen-bonding forces. This has made possible many detailed spectroscopic studies of such species for the first time. Several examples will be found later in this book.

3.5 PRISMS, DIFFRACTION GRATINGS AND INTERFEROMETERS AS DISPERSING ELEMENTS

3.5.1 Resolution and resolving power

Figure 3.11 indicates the general principles of an optical spectrometer or spectrograph operating in the infrared, visible or ultraviolet region. Radiation emerging from an absorption cell, or from an emission source, is focused on the narrow rectangular entrance slit S; the length of the slit is perpendicular to the plane of Figure 3.11. The slit is separated from the lens L_1 by its focal length f_1 so that the lens converts the cone of rays it receives into a parallel beam. This beam falls on the dispersing element E which separates the radiation into its component wavelengths. The dispersed radiation is focused by the lens L_2 on to the detector D.

In the infrared and far ultraviolet regions the lenses are usually replaced by mirrors because of the difficulty of finding materials which transmit the radiation.

Let us consider radiation consisting of only two close-lying wavelengths λ and $\lambda + d\lambda$ falling on the slit and assume that they are focused at P

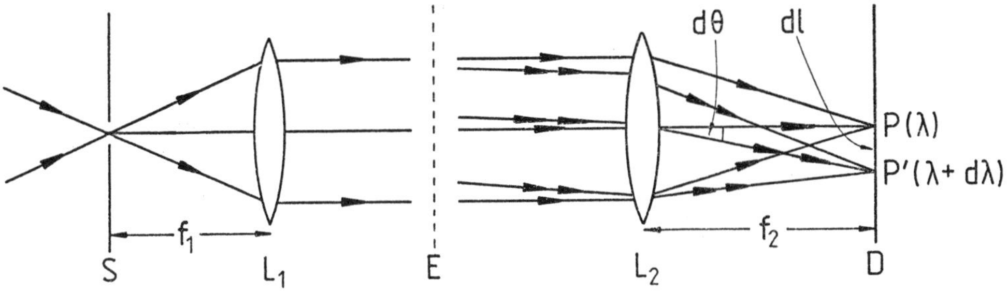

Figure 3.11 The main elements of an optical spectrometer or spectrograph

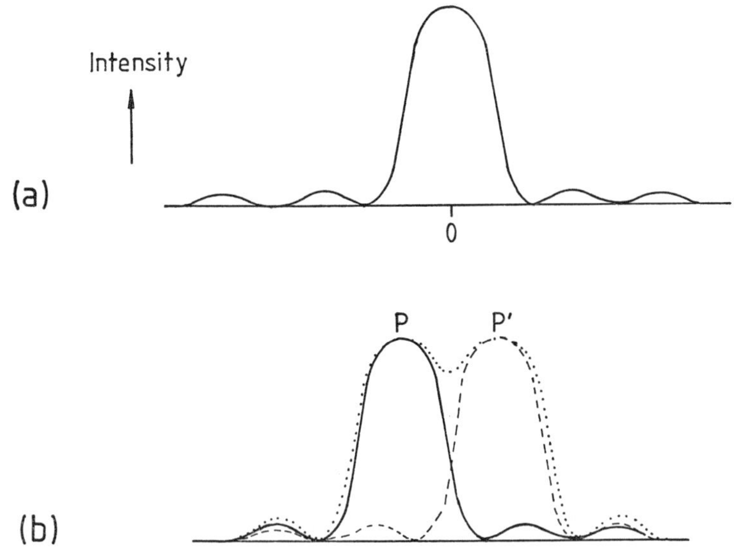

Figure 3.12 (a) Diffraction pattern produced by a narrow slit. (b) The Rayleigh criterion for resolution

and P′, respectively, shown in Figure 3.11. The ability to observe P and P′ as separated, rather than coincident, is a measure of the resolving power, R, of the dispersing element. If P and P′ can be just observably separated, then

$$R = \lambda/\mathrm{d}\lambda = \nu/\mathrm{d}\nu = \tilde{\nu}/\mathrm{d}\tilde{\nu} \qquad (3.9)$$

where $\mathrm{d}\nu$ and $\mathrm{d}\tilde{\nu}$ are the frequency and wavenumber separations, respectively, of P and P′. The quantities $\mathrm{d}\lambda$, $\mathrm{d}\nu$ and $\mathrm{d}\tilde{\nu}$ are the resolution achieved by the dispersing element. The quantity $\mathrm{d}\theta/\mathrm{d}\lambda$, where $\mathrm{d}\theta$ is the small angle shown in Figure 3.11, is the angular dispersion and $\mathrm{d}l/\mathrm{d}\lambda$, where $\mathrm{d}l$ is the distance between P and P′, is the linear dispersion.

But what do we mean when we say that P and P′ are just observably separated? We must remember that the images P and P′ are really diffraction patterns produced by the narrow slit S whose width is comparable to λ. Such a pattern is shown in Figure 3.12(a). Rayleigh suggested the convention, known as the Rayleigh criterion, that P and P′ are said to

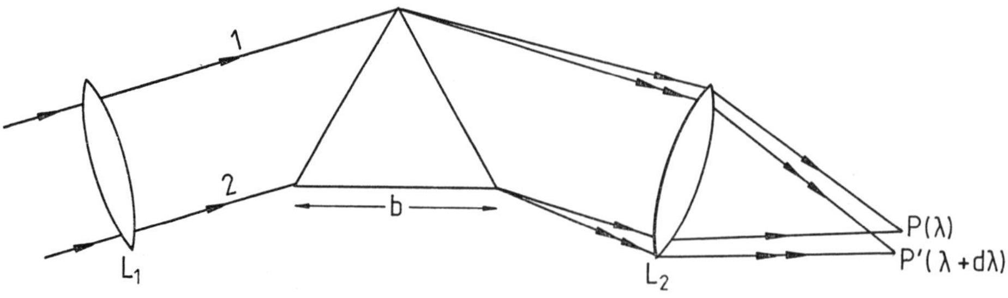

Figure 3.13 A prism as dispersing element

be resolved if the maximum intensity of the diffraction pattern due to P′ is no nearer to that of P than the first minimum in the pattern due to P. This condition is illustrated in Figure 3.12(b), but it is clear that, if the intensity at one of the wavelengths corresponding to P or P′ is much less than the other, their resolution may not be achieved.

In most spectrometers or spectrographs the quality of the dispersing element limits the resolving power of the instrument, but care has to be taken that it is not limited by the quality of the other optical components or of the detector. Another very important factor is the width of the slit. As the slit is widened the width of the main peak in the diffraction pattern increases and resolution is decreased. In some cases, particularly in visible and ultraviolet spectroscopy, it may be that the half intensity line width (Section 2.3) of the radiation entering the slit is greater than the resolution of which the instrument is capable, in which case the resolution obtained is sample-limited.

3.5.2 Prisms

As long ago as the late seventeenth century, Newton realized the capabilities of prisms as dispersing elements for visible radiation and they are still in use today but have been largely superseded by diffraction gratings and, in some regions of the spectrum and when the highest resolving power is required, by interferometers.

When a prism is used as the dispersing element E in Figure 3.11, the parallel beam from L_1 is directed at an angle to one of the prism faces, as shown in Figure 3.13. Then wavelengths λ and $\lambda + d\lambda$ are focused at points P and P′. The resolving power R of the prism is given by

$$R = \lambda/d\lambda = b(dn/d\lambda) \qquad (3.10)$$

where b is the length of the base of the prism and n is the refractive index of the prism material. From equation (3.10) we can see that, for high resolving power, b should be large and therefore the prism must be as large as possible. Also, care must be taken to fill the prism face with radiation because, strictly, b is the difference in pathlength between the extreme rays 1 and 2 in Figure 3.13 and is reduced if the face is not filled.

The factor $dn/d\lambda$ in equation (3.10) is very important in the choice of material from which the prism is made. The material must, of course, transmit the radiation being analysed but $dn/d\lambda$ typically increases rapidly as a region of the spectrum where the material absorbs is approached. Glass, for example, transmits visible radiation with wavelengths down to about 400 nm, the ultraviolet region where it becomes completely absorbing;

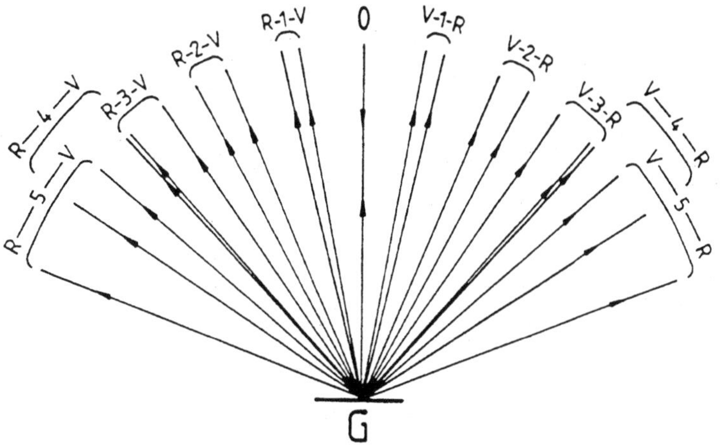

Figure 3.14 Dispersion by a reflection grating G

therefore, the resolving power of a glass prism is highest in the blue–violet region of the spectrum, close to 400 nm. On the other hand, quartz, in either crystalline or fused form, transmits visible and ultraviolet radiation down to about 185 nm where it starts to absorb. The resolving power of a quartz prism is highest, therefore, at wavelengths just higher than 185 nm, but in the visible region a quartz prism has much lower resolving power than a glass prism of the same dimensions.

3.5.3 Diffraction gratings

The theory of diffraction gratings was developed from about 1893 by Rowland. A grating consists of a series of equally separated, parallel slits, the width of each being comparable to the wavelength of the radiation to be dispersed. The slits may be supported on a material which transmits or reflects the radiation and the grating is then called a transmission or a reflection grating, respectively. Transmission gratings are very little used as they necessitate a longer light path than reflection gratings.

One method of making a reflection grating is by ruling, with a cutting diamond, a series of parallel grooves on a suitably hard material. In the early days this used to be speculum metal (33% tin, 67% copper), but it has the disadvantages of tarnishing in air and poor reflectivity in the ultraviolet. It has been largely replaced by aluminium. A glass blank is coated with chromium followed by aluminium, the chromium acting as a cement between the aluminium and the glass, and the grooves are ruled on the aluminium coating. The blank may be plane or concave depending on how the grating is to be used: a plane grating acts as a plane mirror, as well as a grating, and a concave grating acts as a concave mirror. In the near infrared a gold coating is sometimes used because of its high reflectivity and chemical inertness. In the far ultraviolet, from about 150 to 110 nm, the reflectivity of the aluminium can be enhanced by a coating of magnesium fluoride (MgF_2).

If a beam of a radiation continuum, such as white light in the visible region, falls normally on to the surface of a reflection grating G, as shown in Figure 3.14, the radiation is dispersed and reflected. In the dispersion process there is not just one spectrum produced, as with a prism, but several. There are two spectra, each extending from red (R) to violet (V), produced for each order of diffraction by the grating, one on each side of the normal. At a particular angle θ from the normal, radiation of wavelength λ will be reflected such that

$$m\lambda = d(\sin i + \sin \theta) \qquad (3.11)$$

where m is the order, d is the separation of adjacent grooves of the grating and i is the angle of incidence (0° here, giving $m\lambda = d \sin \theta$). The angular dispersion of the grating, $\mathrm{d}\theta/\mathrm{d}\lambda$, is given by

$$\mathrm{d}\theta/\mathrm{d}\lambda = m/d \cos \theta \qquad (3.12)$$

and is therefore proportional to the order, as indicated in Figure 3.14. (Note that in zero order there is no dispersion.) The resolving power R is given by

$$R = mN \qquad (3.13)$$

where N is the total number of grooves on the grating. However, this resolving power can be achieved in practice only if the beam of radiation covers all the grooves, otherwise N is effectively reduced.

Equation (3.13) indicates that, for high resolving power, we require N to be large and it is unaffected by having the grooves closely spaced on a small grating or widely spaced on a large grating. Other considerations, however, make a small grating preferable. The first is that the angular dispersion, given by equation (3.12), increases as d decreases and it is no good having a grating with high resolving power if the dispersion is so low that the detector is unable to distinguish wavelengths which the grating has resolved. The angular dispersion is related to the linear dispersion, $\mathrm{d}x/\mathrm{d}\lambda$, at the detector by

$$\mathrm{d}x/\mathrm{d}\lambda = f(\mathrm{d}\theta/\mathrm{d}\lambda) \qquad (3.14)$$

where x is the distance measured across the spectrum at the detector and f is the focal length of the lens (L_2 in Figure 3.11) or mirror which focuses the dispersed radiation on the detector. Hence a low angular dispersion can be compensated for to some extent by a large focal length, but this greatly adds to the size of the instrument. The second consideration favouring a small grating is that a large one requires large lenses, mirrors, etc., in order to fill it with radiation and this again increases the size of the instrument.

Both the dispersion (equation 3.12) and the resolving power (equation 3.13) are proportional to the order; therefore, it is desirable to work in as high an order as possible. However, Figure 3.14 illustrates how the problem of overlapping orders arises when m is large. Quantitatively, if a wavelength λ_1 in the first order is focused at one point at the detector then wavelengths λ_1/m, for *all* values of m, are focused at the same point. In the simplest cases of overlap by widely differing wavelengths, the use of suitable filters will separate orders, and the limitations of sensitivity of the detector to various wavelengths may also result in rejection of unwanted radiation.

When overlapping orders are of closely similar wavelengths the technique of predispersion is used in which the radiation from the absorption cell or source is relatively crudely dispersed and the unwanted wavelengths rejected before the radiation enters slit S (Figure 3.11). Predispersion can be achieved by a fore-prism or a fore-grating, in front of S.

Figure 3.15 shows how a fore-prism FP achieves the separation of a band of radiation, with wavelengths in the range λ_1 to λ_2, from all other radiation. This process is best understood by the principle of reverse optics, which means that we can reverse the directions of the arrows on a ray diagram and the diagram is equally valid. If we reverse the direction of the rays in Figure 3.15 then the result is very similar to Figure 3.11. Consider white light starting from the slit S and travelling to the fore-slit FS. Only wavelengths from λ_1 to λ_2 are

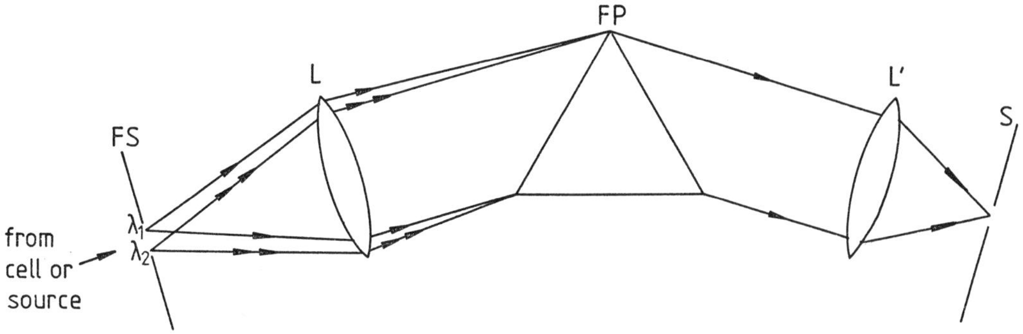

Figure 3.15 Separation of wavelengths by a fore-prism

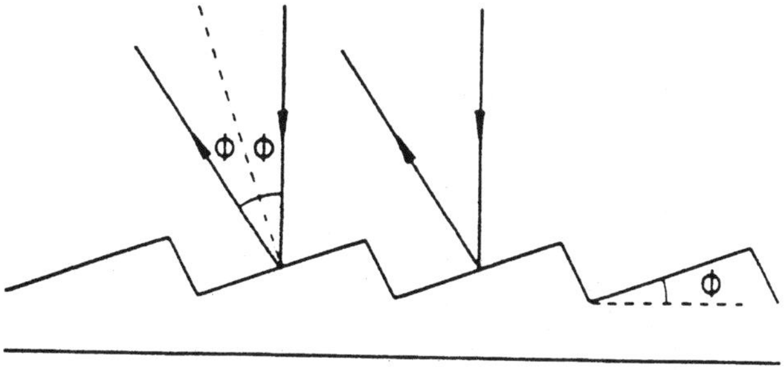

Figure 3.16 The blazing of a diffraction grating

focused within the width of FS: the rest of the wavelengths fall outside the fore-slit. Now consider what happens if white light falls on the fore-slit FS from the left of the figure and travels towards S. Because of the principle of reverse optics, only wavelengths from λ_1 to λ_2 will pass through S provided that the width of the spectrometer slit S is small compared with that of the fore-slit: the rest of the radiation will fall outside S and not enter the spectrometer. The range of wavelengths entering the spectrometer can be changed by rotating the fore-prism through an appropriate angle.

In addition to the fore-prism being replaced by a fore-grating, the lenses L and L′ may be replaced by mirrors. This is almost essential in the infrared and far ultraviolet.

Another difficulty which has been overcome in the use of gratings is that of loss of intensity. As is clear from Figure 3.14, the reflected radiation in only one order is being used in a particular observation and all the rest of the intensity is wasted. If the source of radiation is not very strong this loss of intensity could be crippling. In the early days of ruled gratings, the intensity distribution among the orders was somewhat random and unpredictable. In 1910 Wood realized the importance of the geometry of the cross-section of the ruled groove and was then able to rule gratings which concentrated up to 90% of the intensity at a particular angle. The grooves of such a grating are ruled with the kind of saw-tooth cross-section shown in Figure 3.16. This type of grating was called an echelle or echellette (from the French 'échelle', which means 'ladder'), but since most gratings are now of this type the name is not used so often.

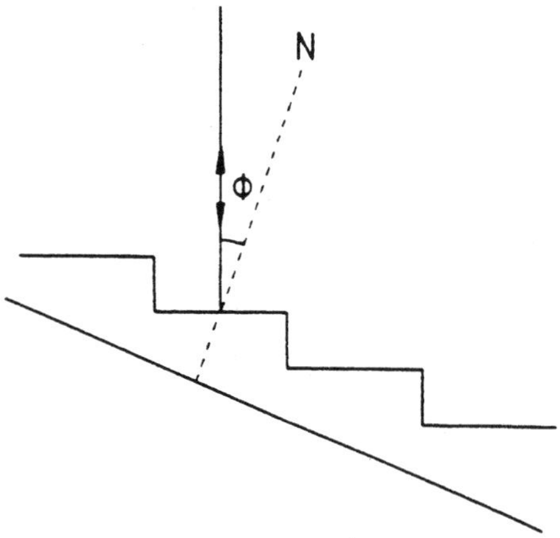

Figure 3.17 Illumination of a blazed grating for the most efficient reflection

For incident radiation directed normal to the grating, as in Figure 3.16, most of the radiation is reflected by the long edge of the saw tooth. If the length of the edge were large compared with the wavelength, the radiation reflected by the long edge would be at an angle 2ϕ to the normal, where ϕ is the angle shown in Figure 3.16 and is the blaze angle. But, of course, the purpose of the grating is to diffract the radiation and so the edge of the groove must not be long compared with the wavelength. Under these conditions the intensity of the reflected radiation is distributed among a wider range of angles but with a maximum at 2ϕ.

In many spectrographs or spectrometers utilizing an echelle grating, illumination is not normal to the grating but normal, or nearly normal, to the long edge of the groove as in Figure 3.17. For this to be achieved the grating has to be rotated so that the grating normal N makes an angle with the incident radiation which is equal, or nearly equal, to the blaze angle. The wavelength λ, the order m, the spacing of the grooves d and the angle θ, between the incident radiation and the grating normal, are related by equation (3.11) where $i = \theta$, giving

$$m\lambda = 2d \sin \theta \qquad (3.15)$$

For an echelle grating, θ is close to the blaze angle ϕ. An alternative way of making a diffraction grating is, instead of cutting the grooves with a diamond, to produce ridges on a photosensitive material by interposing two beams from the same laser. This produces an interference pattern in the form of alternate light and dark lines. When the material is developed equally spaced ridges are produced. Such holographic gratings have the advantages over ruled gratings of a lower level of scattered light and the absence of grating ghosts which arise with ruled gratings due to unavoidable periodic errors in the groove spacings. On the other hand, holographic gratings cannot be made with controllable blaze characteristics and the efficiency falls off so badly in the infrared that ruled gratings are much to be preferred in that region.

3.5.4 Interferometers

Frequently, when photographic slides, in which the film is mounted between two glass plates, are projected on to a screen small areas of coloured rings can be observed around a point where the film is touching the glass. These are Newton's rings and the fact that the white light from the projector lamp is split up into its component colours demonstrates the potential use in a dispersing element of interference involving multiple reflections.

The dispersing effect in an interference phenomenon is made use of in the Fabry–Perot interferometer, the principles of which are illustrated in Figure 3.18. Radiation from the point A, for example, in the broad radiation source S falls on two parallel plates P_1 and P_2 which have their inner surfaces coated with a material which partially transmits and partially reflects the radiation. The parallel rays emerging from P_2 interfere constructively for

$$2nd \cos \theta = m\lambda \qquad (3.16)$$

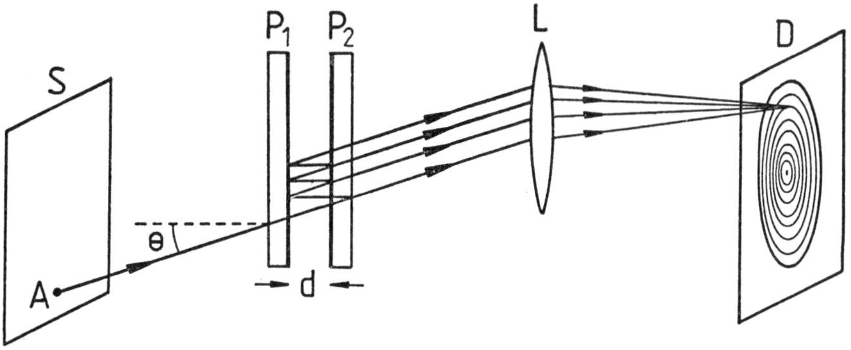

Figure 3.18 A Fabry–Perot interferometer

where d is the separation of the inner surfaces of P_1 and P_2, θ is the angle between the incident ray and the normal to P_1, m is the (integral) order, λ is the wavelength of the radiation and n is the refractive index of the material between the plates ($n = 1$ for air at 1 atm).

For a particular wavelength, order and incident angle, radiation from S constructively interferes to form a circular fringe of radiation at the detector D. For a monochromatic source S a series of concentric circular interference fringes is formed at D, each fringe being characterized by a particular value of m and a corresponding value of θ. If the source contained two wavelengths λ_1 and λ_2, each would produce a set of concentric rings with the same centre but different spacings. Radiation comprising more wavelengths would produce a more confused series of rings and some means of wavelength dispersion, such as the prism in Figure 3.19, or a diffraction grating would have to be used to separate the systems of rings for the various wavelengths. Figure 3.19 shows three fringe systems corresponding to the wavelengths λ_1, λ_2 and λ_3. If the broad source S is replaced by a slit F then what is observed at the detector D is a set of sections F_1, F_2 and F_3 of the three fringe systems and the overlapping of close systems is avoided. This kind of Fabry–Perot interferometer with fixed rather than variable plate separation is called an etalon. The resolving power of an etalon is given by

$$R = \lambda/\mathrm{d}\lambda = m\pi r^2/(1 - r^2) \qquad (3.17)$$

and so is proportional to the order m, which is equal to $2d/\lambda$. The quantity r is the reflectance of the coated surfaces of P_1 and P_2 and for high resolving power r should be close to unity.

Another method of separating the wavelengths falling on the interferometer is by pressure scanning. The presence of the refractive index n in equation (3.16) means that for a particular position at the detector, such as a fixed exit slit, the wavelength falling on it can be changed by varying the refractive index of the gas between the plates. This is achieved by placing the interferometer in an air-tight box in which the air pressure can be smoothly changed, usually in the range 0–1 atm. The wavelength interval $\Delta\lambda$ which is scanned by a pressure change ΔP is given by

$$\Delta\lambda = \lambda C \Delta P \qquad (3.18)$$

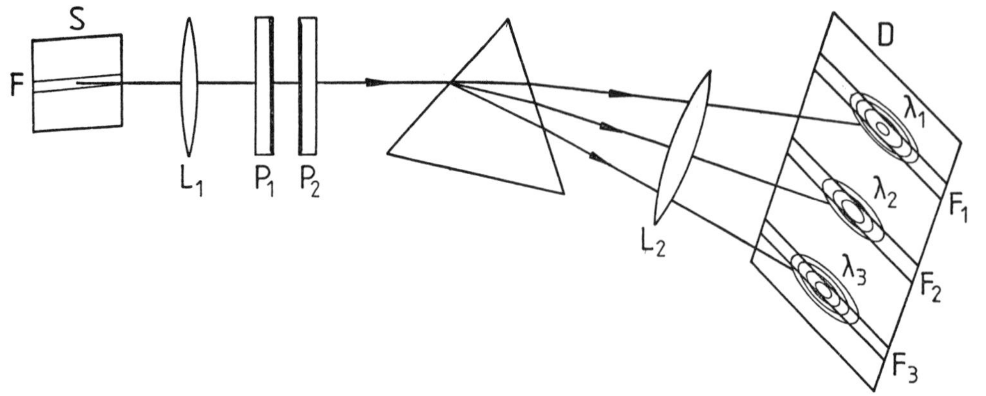

Figure 3.19 Separation of wavelengths using a Fabry–Perot interferometer

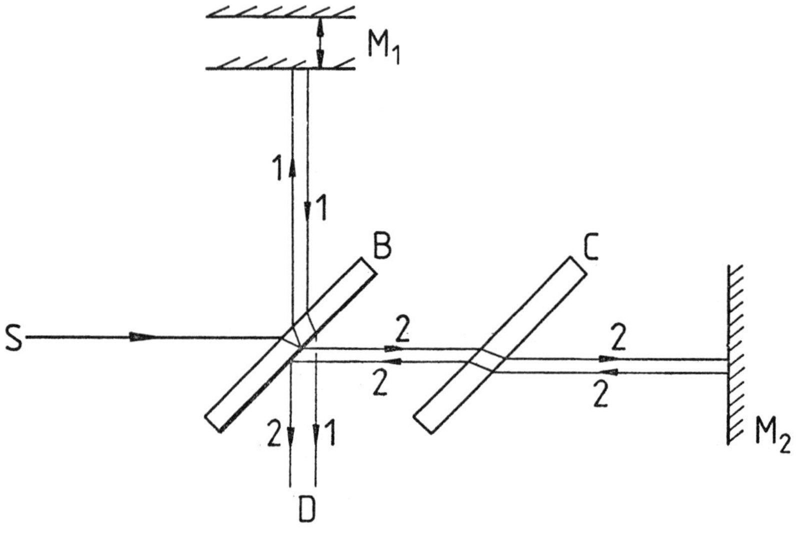

Figure 3.20 A Michelson interferometer

where C is a constant which depends on the nature of the gas between the plates. C has a value of $3 \times 10^{-4}\,\text{atm}^{-1}$ for air so the range of $\Delta\lambda$ is very small. To move to a different range the spacing d must be changed. Scanning can be achieved also by smoothly varying the separation between the plates, but this is very difficult to perform sufficiently accurately.

A different kind of interferometer used as a dispersing element is the Michelson interferometer. The principles of operation are illustrated in Figure 3.20. Radiation from a broad monochromatic source S falls on a transparent parallel-sided plate B which is coated on one side, indicated in the figure by a heavy line, with a material which makes the

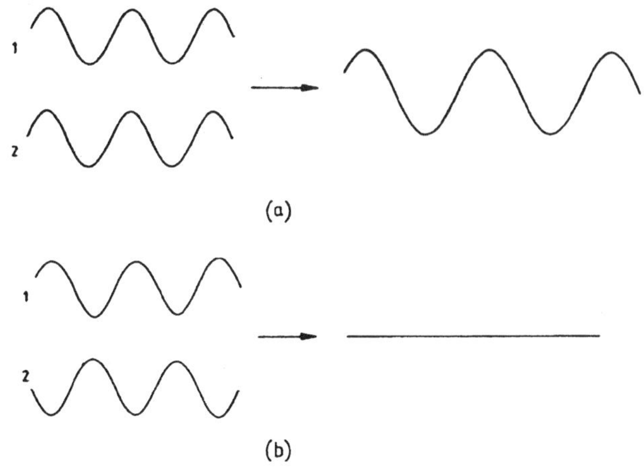

Figure 3.21 (a) Constructive and (b) destructive interference between rays 1 and 2 of monochromatic radiation

surface half transmitting and half reflecting. This has the effect of splitting the incident beam into two rays, labelled 1 and 2; for this reason B is known as a beam splitter. Ray 1 is reflected by the movable plane mirror M_1 and falls on the detector D. Ray 2 is reflected by mirror M_2 and also falls on D. C is a compensating plate of the same material and thickness as B and ensures that ray 2, like ray 1, passes twice through an identical length of the material from which the beam splitter is made. When the two rays reach D they have traversed different paths with a path difference δ. This is called the retardation and its magnitude depends on the position of M_1. Figure 3.21(a) shows that, if $\delta = 0, \lambda, 2\lambda, \ldots$, the two rays interfere constructively at the detector, while Figure 3.21(b) shows that, if $\delta = \lambda/2, \quad 3\lambda/2, \quad 5\lambda/2, \ldots$, they interfere destructively and no signal is detected. Therefore, if δ is smoothly changed from zero the detected signal intensity $I(\delta)$ changes like a cosine function as in Figure 3.22.

In a more usual emission experiment the source contains many wavelengths, the detector sees intensity due to many cosine waves of different wavelengths and the detected intensity is of the form

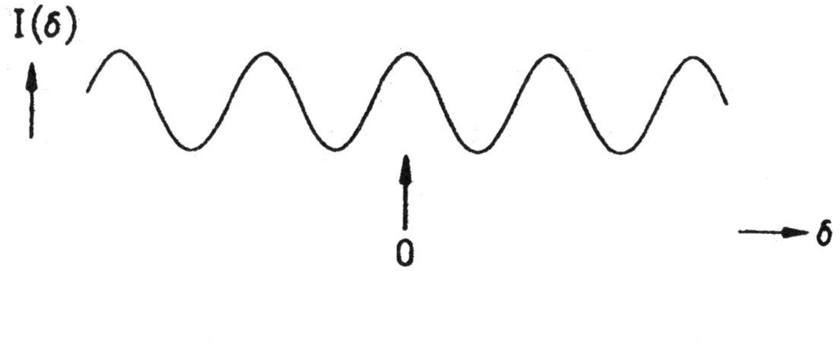

Figure 3.22 Change of signal intensity $I(\delta)$ with retardation δ

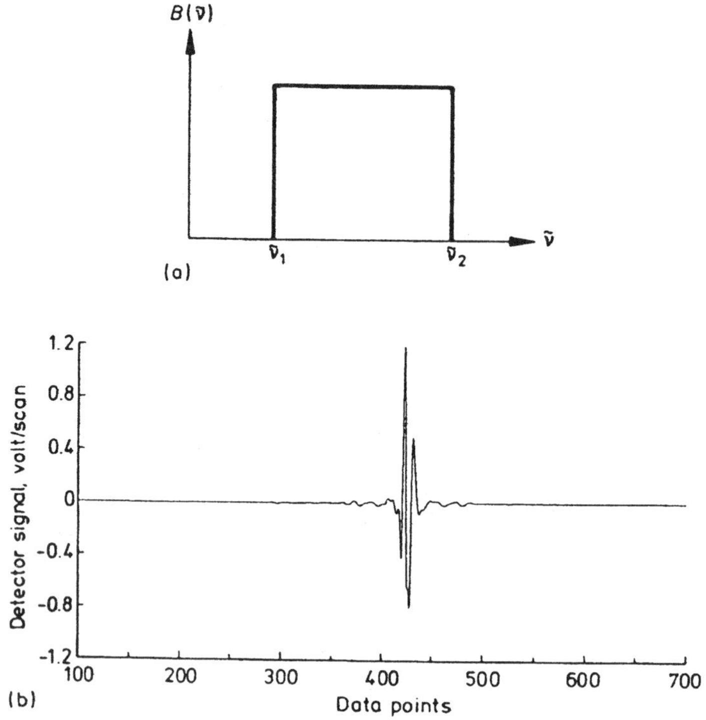

Figure 3.23 (a) Wavenumber domain spectrum of a broad band source and (b) the corresponding interferogram

$$I(\delta) = \int_0^\infty B(\tilde{\nu}) \cos 2\pi\tilde{\nu}\delta \, d\tilde{\nu} \qquad (3.19)$$

where $\tilde{\nu}$ is the wavenumber of the radiation and $B(\tilde{\nu})$ is the source intensity at that wavenumber (neglecting small corrections due to variable beamsplitter efficiency and detector response). A plot of $B(\tilde{\nu})$ against $\tilde{\nu}$ is the dispersed spectrum of the source and is obtained by the procedure of Fourier transformation, giving

$$B(\tilde{\nu}) = 2\int_0^\infty I(\delta) \cos 2\pi\tilde{\nu}\delta \, d\delta \qquad (3.20)$$

The majority of infrared spectra are obtained by an absorption rather than an emission process and, as a result, the change of signal intensity $I(\delta)$ with retardation δ appears very different from that in Figure 3.22.

Consider an infrared source emitting a wide range of wavenumbers for use in an absorption experiment. Figure 3.23(a) shows how the idealized wavenumber domain spectrum of this source, emitting continuously between $\tilde{\nu}_1$ and $\tilde{\nu}_2$, might appear. We can regard this spectrum as comprising a very large number of wavenumbers between $\tilde{\nu}_1$ and $\tilde{\nu}_2$ so that the corresponding detector signal, as a function of retardation, will be the result of adding together very many cosine waves of different wavelengths. The signal is large at $\delta = 0$ since all the waves are in-phase but elsewhere they are out-of-phase, interfere with each other

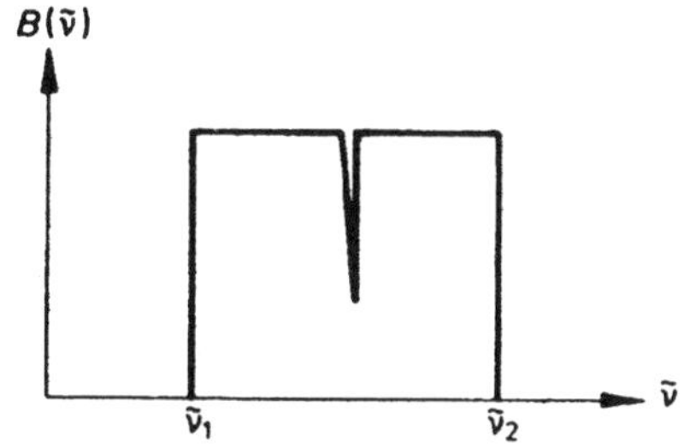

Figure 3.24 Wavenumber domain spectrum of a broad band source with a narrow absorption

and produce total cancellation of the signal. The intense signal at $\delta = 0$ is known as the centre burst and is shown in Figure 3.23(b). (Because of slight dispersion by the beamsplitter B, the waves at $\delta = 0$ are not exactly in-phase, resulting in an asymmetry of the centre burst about $\delta = 0$.)

If a single, sharp absorption occurs at a wavenumber $\tilde{\nu}_a$, as shown in the wavenumber domain spectrum in Figure 3.24, the cosine wave corresponding to $\tilde{\nu}_a$ is not cancelled out and remains in the $I(\delta)$ versus δ plot, or interferogram as it is often called. For a more complex set of absorptions the pattern of uncancelled cosine waves becomes more intense and irregular.

Figure 3.25(a) shows an interferogram resulting from the infrared absorption spectrum of air in the 400–3400 cm^{-1} region. The Fourier transformed spectrum in Figure 3.25(b) shows strong absorption bands due to CO_2 and H_2O, that of H_2O showing much fine structure because it is a lighter molecule.

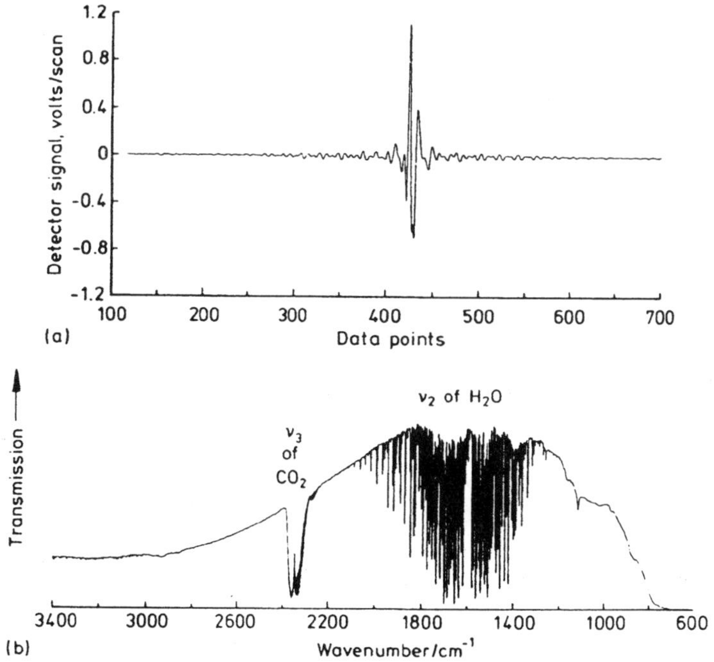

Figure 3.25 (a) Infrared interferogram of the absorption spectrum of air in the 400–3400 cm^{-1} region and (b) the Fourier transformed spectrum

The interferogram is digitized before Fourier transformation by a dedicated computer. Figure 3.25(a) shows that, in this example, there are 600 data points in the part of the interferogram shown. The computer is limited in the number of data points it can handle and the user has a choice of having the data points closer together, and neglecting the outer regions of the interferogram, which gives a wider wavenumber range but lower resolution, or having the data points further apart, which gives higher resolution but a narrower wavenumber range. The resolution $\Delta\tilde{\nu}$ is always determined by how much of the interferogram can be observed, which, in turn, depends on the maximum displacement δ of the mirror M_1. The resolution is given by

$$\Delta\tilde{\nu} = \frac{1}{\delta_{max}} \tag{3.21}$$

One of the main design problems in an FTIR spectrometer is to obtain accurate, uniform translation of M_1 over distances δ_{max} which may be as large as 1 m in a high resolution interferometer.

A great advantage of the Michelson interferometer is that radiation of *all* wavenumbers from the source falls on the detector *all* the time. This advantage is known as the multiplex, or Fellgett, advantage and any spectrometer which makes use of it is called a multiplex spectrometer.

In a more conventional scanning spectrometer in which, for example, a diffraction grating is smoothly rotated and the wavelengths falling on the exit slit in front of the detector are smoothly scanned, most of the source radiation is wasted since it falls on either side of the slit. The width of the exit slit, in addition to that of the entrance slit, determines the resolution. For example, for slits giving a resolution of 8 cm^{-1} only 0.2% of the radiation typically falls on the detector and, for 1 cm^{-1} resolution, only 0.03%.

Because of the multiplex advantage, an infrared interferometer can record a spectrum much more rapidly than a conventional spectrometer. This means that it is particularly useful for recording spectra of short-lived species and of very weak sources of radiation such as that from the night sky and from stars and planets.

A second advantage which applies to an interferometer is the étendu, or Jacquinot, advantage. This results from using a circular entrance aperture, through which the source radiation passes, which allows more efficient use of the radiation than with an entrance slit.

3.6 FAR INFRARED SPECTROSCOPY

There are two recurring problems to be overcome in far infrared instruments. The first of these is caused by the high absorbance in this region of the water vapour present in the air. This can be partially removed by continuous flushing of the whole spectrometer with dry air, but the most satisfactory solution is complete evacuation.

The second problem is that continuum sources in the far infrared are of very low intensity and for this reason far infrared spectroscopy is energy-limited. This means that the resolution which can be attained is ultimately limited by the intensity of the source. If the entrance slit to a spectrometer has to be opened wide in order to allow sufficient intensity of radiation to enter, so that the detector can distinguish signal from noise, then it can easily happen that the line width in the observed spectrum is determined by the width of the slit rather than any other effect, since an observed line is an image of the slit.

Any black body source of radiation which emits energy on heating, as illustrated in Figure 1.1, is unsuitable for the far infrared because of the steep fall-off of energy with wavenumber. The most commonly used source is a mercury arc produced in a high voltage discharge through mercury vapour. The envelope containing the discharge is made from fused quartz, which transmits the lower wavenumber far infrared radiation, generated in the plasma of

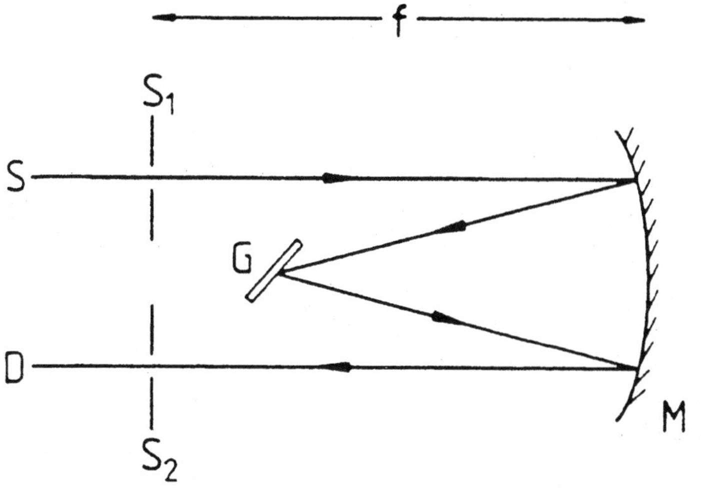

Figure 3.26 The Ebert–Fastie mounting of a diffraction grating

the discharge, fairly efficiently. The higher wavenumber far infrared radiation is emitted mostly from the hot outer surface of the quartz rather than from the plasma itself.

A suitable dispersing element for use in the far infrared is the Michelson interferometer. This interferometer, illustrated in Figure 3.20, is particularly suited to the far infrared because of the multiplex and étendu advantages, mentioned in Section 3.5.4, resulting in more efficient use of the limited source energy. The beamsplitter for the far infrared is not usually made by coating a transparent material because of the difficulty of finding a suitable material. Instead, a stretched thin film of a dielectric material such as poly(ethylene terephthalate) (Terylene) is used directly as the partially reflecting film without any supporting material.

Plane diffraction gratings are also used as dispersing elements. The possible optical arrangements for illuminating a grating and gathering the light reflected from it, not only in the far infrared but also in the mid and near infrared, visible and ultraviolet regions, are numerous and we shall mention only those more commonly used.

The Ebert system of mounting the grating was first used by Ebert in 1889, but its virtues were not fully appreciated until Fastie developed it further in 1952. The Elbert–Fastie system is shown in Figure 3.26. Radiation from the source S enters the spectrometer at the slit S_1 which is separated from the concave mirror M by its focal length f so that the grating G receives a parallel beam from M. The grating reflects and disperses the radiation, which is focused by M on to the exit slit S_2 and passes to the detector D. The required wavenumber range is scanned across S_2 by smooth rotation of the grating. An important property of the Ebert–Fastie system, not realized by Ebert, is that the optical aberrations produced by using the mirror off-axis at the first reflection are cancelled by the aberrations on the second reflection.

A useful variation on the Ebert–Fastie system is the Czerny–Turner system, first described in 1930 and shown in Figure 3.27. Here the single concave mirror of the

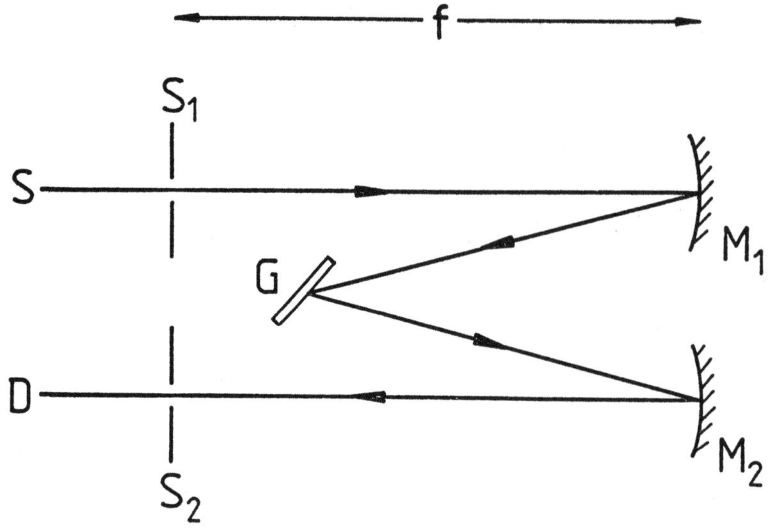

Figure 3.27 The Czerny–Turner mounting of a diffraction grating

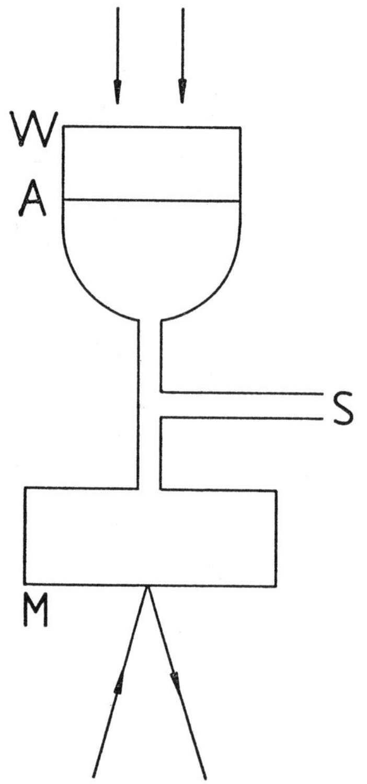

Figure 3.28 A Golay cell far infrared detector

Ebert–Fastie system is replaced by two separate mirrors M_1 and M_2 of equal focal lengths f and having the slits S_1 and S_2, respectively, at their focal points. The result of the increased flexibility produced in the Czerny–Turner system by, in effect, cutting the mirror M of the Ebert–Fastie system in half is that optical aberrations, other than those eliminated in the Ebert–Fastie system, are greatly reduced so that resolution may be limited only by grating imperfections.

Materials which transmit far infrared radiation and are therefore suitable for cell windows include polyethylene, poly(ethylene terephthalate) and polystyrene. If a window is required in front of the detector, quartz or diamond may be used.

One of the commonly used detectors in the far infrared is the Golay cell illustrated in Figure 3.28. The radiation falls on the transmitting window W and then on to a thin membrane A, consisting of aluminium deposited on collodion, which absorbs the radiation and is slightly heated as a consequence. The whole cell is filled with xenon gas, which expands on heating and distorts

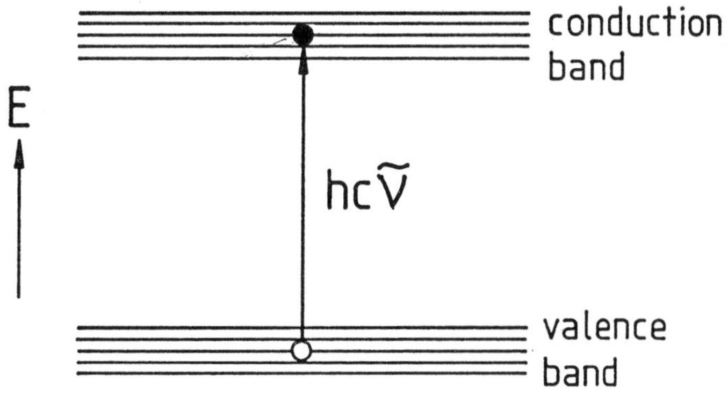

Figure 3.29 Valence and conduction bands of a semiconductor

the collodion membrane M which has a reflecting layer of antimony deposited on it. The distortion of this membrane, which is a measure of the intensity of radiation falling on W, is measured by reflecting a beam of visible light from the surface. The sidearm S on the cell connects it to a reservoir of xenon acting as a ballast.

Two other types of far infrared detectors are basically resistance thermometers which have low heat capacity so that they respond sensitively and quickly to temperature changes.

The thermocouple detector consists of a thin metal foil attached at either end to pins of positive and negative thermoelectric material. The foil is blackened in order to absorb the maximum amount of radiation falling on it. The most commonly used thermocouple is that developed by Schwartz in which the foil is blackened gold, the positive thermoelectric material is a Cu–Ag–Te–Se–S alloy (27:32:33:7:1, w/w) and the negative thermoelectric material is made from Ag_2S and Ag_2Se (1:1, w/w).

The bolometer detector uses the variation of resistance with temperature of the material from which the bolometer is made. In a thermistor bolometer, which is the most commonly used type, this consists of flakes of a sintered mixture of nickel, manganese and cobalt oxides mounted on quartz or sapphire. This type of bolometer has a greater responsivity to the radiation falling on it and a shorter time constant than a thermocouple, but it suffers, at room temperature, from a high noise level. This can be overcome by using a superconducting material such as tin or cabon very close to its transition temperature, at which the change of resistance with temperature is extremely high. The temperature required is only a few kelvin, obtained with liquid helium, and the temperature has to be held constant to within about 10^{-5} K.

Photon detectors of far infrared radiation work on a different principle. These depend on the photoconductivity of semiconductor materials. Conductivity in such materials is a measure of the number of electrons which are in the conduction band of the semiconductor and therefore relatively free. The remaining electrons are in the valence band; these are more strongly bound, as shown in Figure 3.29, and cannot conduct electricity. However, if a far infrared photon of energy $hc\tilde{\nu}$ falling on the semiconductor is able to promote an electron from the valence to the conduction band, an increase in conductivity results owing not only to an extra electron in the conduction band but also to the hole created in the valence band. Two such photoconducting materials used as far infrared detectors are germanium, doped with antimony, and indium antimonide.

All infrared detectors, including those used in the mid and near infrared regions, perform better if the incident radiation is cut off periodically, or chopped, than if it is continuous. The reasons for this are (a) that amplification of an a.c. signal is easier than that of a d.c. signal, (b) there may be a reduction in the random noise of the detector and (c) the effects on the recorded spectrum of any stray radiation which is inevitably present can be removed: this removal is achieved by comparing the detector current, when the radiation from the absorption cell is falling on it, with that when this radiation is cut off. The chopping is done either by a rotating sector or an oscillating vane which periodically interrupts the radiation from the source. The chopping frequency is usually about 10 Hz.

Many far infrared, and also mid and near infrared, visible and ultraviolet, spectrometers work with a double beam system in which the radiation from the source is divided into two beams of equal intensity. One of these goes through the sample and is called the sample beam and the other is the reference beam. As in the case of a single beam instrument, the radiation is chopped so that the detector receives radiation alternately from both beams. Phase sensitive detection is used in which the amplifier attached to the detector is able to distinguish the radiation reaching the detector via the sample and reference beams. When the two signals have been amplified they are compared electronically and a comb with wedge-shaped teeth opaque to the radiation is moved across the reference beam until the two signals are equalized. Then the distance the comb has moved is a measure of the absorption in the sample beam. This is called the optical null method. The absorption spectrum displayed is that of the sample minus that of the reference. Clearly, this technique is useful when the sample is a solute, whose spectrum is required, dissolved in a solvent. If cells of the same length, one containing the solvent only and the other containing the solution, are placed in the reference and sample beams, respectively, then the spectrum produced will be that of the solute. Nevertheless, it is always best to use a solvent which absorbs as little as possible in the region of interest. In the far infrared suitable solvents are benzene, hexane, carbon tetrachloride, chloroform and carbon disulphide.

If flushing with dry nitrogen is used to try to remove all water vapour, double beam operation can be useful in removing from the spectrum, at least partly, any absorption due to remaining traces of water.

It should be remembered, however, that double beam operation has its limitations. If absorption is strong, but not 100%, in both the sample and reference beams the detector is receiving very little radiation from either beam and is therefore insensitive to small differences. In addition, it may be difficult to arrange that the pathlength of an interfering absorber is exactly the same in both beams. If it is not, some unwanted absorption will be recorded.

The optical null method has the disadvantage of drastically reducing the intensity of the reference beam when the sample is absorbing strongly. In the infrared regions this method has been largely replaced by ratio recording in which the intensity, I_0, in the reference beam and that in the sample beam, I, are ratioed by a microprocessor to give, by equation (2.24), the absorbance A of the sample relative to that of the reference. This method is particularly advantageous when the absorbance is high.

3.7 MID AND NEAR INFRARED SPECTROSCOPY

Spectroscopy in this region is similar in principle to that in the far infrared region except that two major problems are no longer so serious. One of these is the problem of absorption by air, which does not occur and so

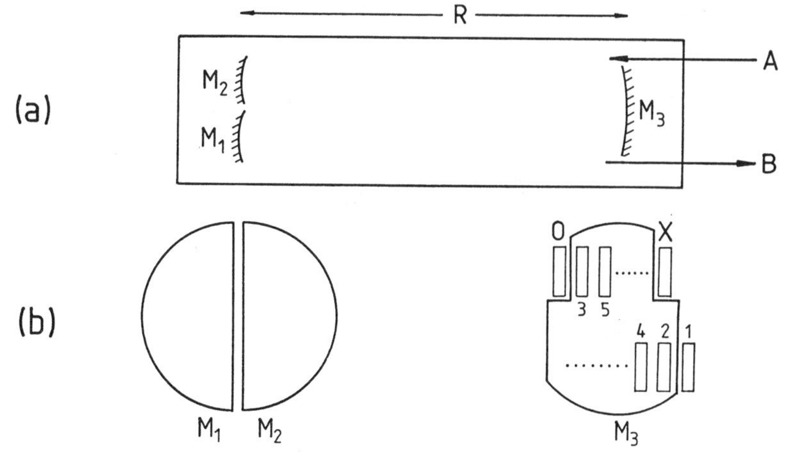

Figure 3.30 (a) Arrangement of three concave mirrors in a White multiple reflection cell. (b) Front view of the three mirrors and the images on M_3

evacuation of the spectrometer is not necessary. The second is the problem of limitation of resolution by the low intensity of the source, which is not so acute in the near infrared.

Sources of radiation are all of the black body type. One which is very commonly used is the Nernst filament, which is in the form of a cylindrical rod made from a mixture of zirconium oxide, yttrium oxide and erbium oxide in typical proportions of 90:7:3 (w/w). The rod first requires heating externally since it is an insulator at room temperature. When hot it conducts electricity which then maintains the radiation. Another common source is the globar, which is a carborundum (silicon carbide) rod. This, unlike the Nernst filament, does not require initial heating as it conducts electricity at all temperatures.

The windows of the absorption cell are commonly made from polished sodium chloride, which transmits infrared radiation down to a wavenumber of about 700 cm^{-1}. At lower wavenumbers potassium bromide, which transmits down to about 400 cm^{-1}, is used.

Because of the very high reflectivity in the mid and near infrared of some metal surfaces, such as aluminium, silver or gold, large numbers of reflections can be tolerated not only in the spectrometer but also in the absorption cell. If an absorbing sample to be investigated in the gas phase has either (i) a low vapour pressure or (ii) a low molar absorption coefficient in the region of interest, then it may be necessary to pass the source radiation many times through the gas, in a multiple reflection cell, in order to increase the absorbance. The most convenient design of a multiple reflection cell is a modification of the original design by White in 1942 made by Bernstein and Herzberg in 1948[5] and shown in Figure 3.30.

Three concave spherical mirrors, M_1, M_2 and M_3 are placed inside the absorption cell as in Figure 3.30(a). M_1 and M_2 are, as shown in Figure 3.30(b), semicircular and M_3 is essentially circular but with some parts cut away. All three mirrors have the same focal length and the distance between the M_1 and M_2 pair and M_3 is equal to the radius of curvature R (twice the focal length). Radiation enters the cell at A and is focused level with the surface of M_3 to form the image 0 in Figure 3.30(b). Radiation diverges from 0 and is directed so as to fill M_1

Table 3.2 High wavelength and the corresponding low wavenumber limits of transparency of materials used for prisms and windows in the mid and near infrared

Material	$\lambda_{max}/\mu m$	$\tilde{\nu}_{min}/cm^{-1}$
Quartz (SiO_2)	3.5	2900
Lithium fluoride (LiF)	5.5	1800
Fluorite (CaF_2)	8.5	1200
Rock salt (NaCl)	15	670
Caesium bromide (CsBr)	35	290
Caesium iodide (CsI)	50	200

and M_2. M_1 and M_2 are separately adjustable and are tilted so that they form the focused images 1 and 2, respectively (they are focused because M_3 is at the radius of curvature of M_1 and M_2). Image 1 falls off the edge of M_3 and this radiation is lost. Image 2 falls on M_3 which is tilted so as to reflect the radiation from this image on to M_1 which in turn focuses the radiation to form image 3 on M_3. These reflections continue until an image X [Figure 3.30(b)] falls off the miror M_3 and the radiation emerges from the cell at B [Figure 3.30(a)]. The number of traverses of the cell can be increased by decreasing the separation of images 1 and 2 by tilting M_2; similarly, the traverses can be decreased by increasing the separation. The total number of traverses is $2N + 2$, where N is the number of images on M_3. In the mid and near infrared it is possible to achieve more than 100 traverses of the cell.

If the sample is to be investigated in solution, then suitable solvents must be used. Apart from having to dissolve the sample they must also be transparent and not dissolve the windows. No single solvent is transparent throughout the region but a combination of solvents can be used such as carbon disulphide, which is transparent except for the region 1400–1700 cm^{-1}, and tetrachloroethylene, which is transparent in the region where carbon disulphide absorbs.

A solid sample, in the form of a finely ground powder, can be used provided that scattering of radiation is reduced by surrounding the particles with a substance of similar refractive index. For work at wavenumbers less than 1300 cm^{-1} this substance is often the liquid paraffin called nujol: the sample material is mixed with it to form a nujol mull. Another substance commonly used is potassium bromide. About 1% of the sample is finely ground with potassium bromide and the resulting powder strongly compressed under vacuum to give a thin KBr disk.

Double beam operation or ratio recording (see Section 3.6) is commonly employed. These methods are particularly important in the fairly low resolution spectrometers used for obtaining spectra of solutions, KBr disks or mulls.

In low resolution instruments, prisms are still sometimes used as dispersing elements. The material from which the prism is made must be transparent in the spectral region to be studied. All materials used in the mid and near infrared are transparent at the low wavelength end of the region but at the high wavelength end, towards the far infrared, the material has to be carefully chosen. Table 3.2 gives a list of possible materials, together with the wavelength and wavenumber of the limit of transparency towards the far infrared. These materials are also suitable for windows.

The two main problems associated with prisms in the mid and near infrared are (i) that the material may be hygroscopic, for example rock salt, and have to be kept in a moisture-free atmosphere, and (ii) that the resolving power of a prism varies with wavelength, as discussed in

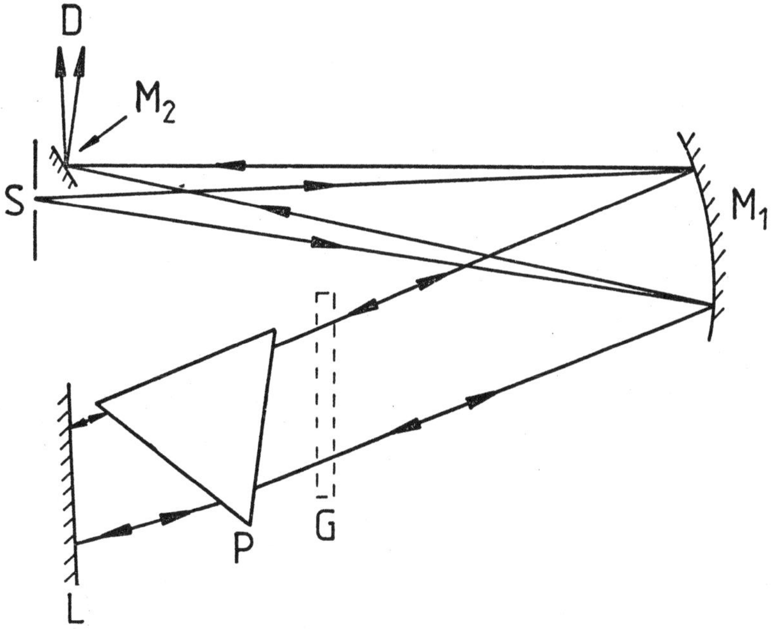

Figure 3.31 Littrow mount of prism or grating in an infrared spectrometer

Section 3.5.2. However, in an inexpensive instrument it may be cheaper to combat these problems than to use a diffraction grating.

The optical system often used in a prism spectrometer is that originally devised by Littrow in 1862 and illustrated in Figure 3.31. Radiation from the entrance slit S falls on the mirror M_1, an off-axis paraboloid to reduce aberrations, which reflects a parallel beam on to the prism P. From P the radiation falls on the plane mirror L, the so-called Littrow mirror, which reflects it back through P and, almost along the same path, to M_2, a small plane mirror which deflects the radiation towards the detector D. The wavelength reaching the detector, via an exit slit, is varied by rotating the Littrow mirror. The Littrow system does have the great virtue of passing the radiation through the same prism *twice*, thereby producing double the resolution which would result from passing it through only once.

The Littrow system can also be used with a plane diffraction grating G, as shown in Figure 3.31, replacing the prism and Littrow mirror.

Other systems employed in the near infrared and using a plane diffraction grating are the Ebert–Fastie, shown in Figure 3.26, and the Czerny–Turner, shown in Figure 3.27.

From 1966 onwards, great advances were made in high resolution, near infrared spectroscopy by the use of a Michelson type of interferometer (see Section 3.6).

One of the main problems to be overcome with a Michelson interferometer is the achievement of high accuracy in the translation of the mirror M_1 in Figure 3.20. Whereas in the far infrared an accuracy of about 1 μm was sufficient to produce an improvement on grating instruments, in the mid and near infrared an accuracy of about 1–0.1 nm was required to improve on grating instruments. Starting from the work of Connes and Connes,[6] interferometers with a resolution of 0.001 cm^{-1} have been built. They

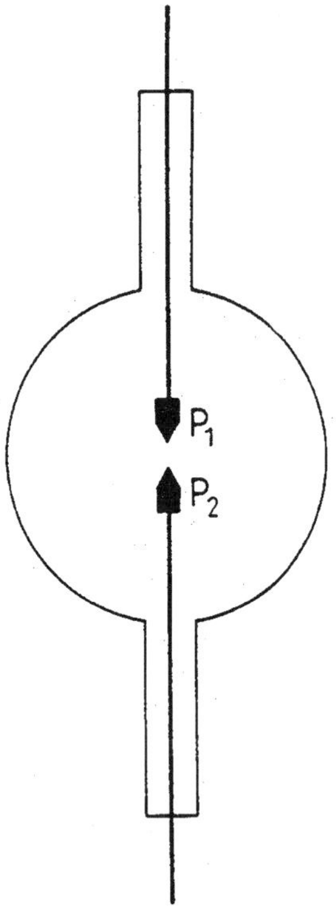

Figure 3.32 A high pressure a.c. xenon arc lamp

were developed originally for use with the low intensity extra-terrestrial sources of interest in astronomy and it is an interesting fact that mid infrared absorption spectra of carbon dioxide in the atmospheres of the planets Venus and Mars were obtained with a resolution and general quality superior to any that had been obtained from laboratory experiments on earth!

The beam splitter is in the form of a quartz or calcium fluoride plate coated with silicon or germanium and the moving mirror may be translated through a distance of up to 1–2 m. The problems of source and detector stability, of control of the path difference between the two interferometer mirrors to within about 0.6 nm and of Fourier transformation of the interferogram were formidable, but have all been overcome.

Detectors used for the mid and near infrared are the thermocouple, bolometer and Golay cell as described in Section 3.6. Photoconductive detectors employing semiconductor materials are also used but they are sensitive over only limited ranges. Indium antimonide and lead sulphide, for example, are sensitive only in the 1800 and 3000 cm^{-1} regions, respectively. The sensitivity of photoconductive detectors is generally high, compared with other types, especially when cooled to liquid nitrogen or, preferably, liquid helium temperature.

3.8 VISIBLE AND NEAR ULTRAVIOLET SPECTROSCOPY

The experimental techniques employed in this region of the spectrum are different when the sample is atomic rather than molecular. The main differences are due to the fact that vapour pressures of the elements are generally very low which means that special techniques are required to obtain an atomic absorption spectrum. Alternatively, atomic emission spectroscopy using a high temperature source is often used.

Because of these differences the techniques used for molecular and atomic samples will be discussed separately.

3.8.1 Molecular samples

A convenient continuum source of visible radiation for absorption spectroscopy is the tungsten lamp, similar to the type used in the home except that the filament is a plane metal strip instead of a coiled wire.

A more intense source is the high pressure xenon arc lamp which is illustrated in Figure 3.32. The two tungsten poles P_1 and P_2 are about 1 mm–1 cm apart and surrounded by xenon gas at a pressure of about 20 atm. When power of low voltage and high current is applied to the poles there is an intense white arc between them. The power supplied may be a.c. or d.c. but the design of the source differs slightly according to which is used. The envelope is of quartz and the lamp produces intense and continuous visible and near ultraviolet radiation at wavelengths down to about 200 nm. It also produces about 20% of its radiation in the near infrared.

A much less intense but more convenient source of near ultraviolet radiation is the high pressure hydrogen discharge lamp enclosed in a quartz envelope. Deuterium gas usually replaces hydrogen because it produces higher intensity.

The absorption cell for the visible and near ultraviolet regions may be of the straight-through or multiple reflection type (see Figure 3.30), as in the infrared. The material used for the cell windows, and also for any lenses or prisms which may be used, is usually Pyrex glass for the visible, down to a wavelength of about 350 nm, and fused quartz for the near ultraviolet, down to about 180 nm. Quartz transmits throughout the visible region also but is more expensive than glass.

Because of the larger transition moments for electronic transitions, compared with rotational and vibrational transitions, it is relatively easy to build up a concentration of molecules in excited states sufficient for the emission spectra in the visible and ultraviolet regions to be sufficiently intense to be observed.

One way of achieving this is by creating a high voltage discharge in the gas. The plasma produced in the discharge contains electrons with sufficient energy to excite the molecules, on collision with the electrons, into excited electronic states. For example, a high voltage discharge through nitrogen gas causes a deep pink glow due to two electronic transitions occurring in emission in the visible region.

Another method of obtaining emission spectra is by the use of a suitable d.c. arc struck between two poles. For example, emission spectra of CuH and AlH can be obtained from arcs between two copper or aluminium electrodes in the presence of hydrogen gas. Other examples are the emission spectra of CN and C_2 which can be observed in a carbon arc struck in air.

Most fairly low resolution visible and near ultraviolet spectrometers employ double beam operation with chopping and phase sensitive detection. This is particularly useful for solution spectra when the solvent is placed in the reference beam.

In studying the absorption spectra of short-lived or transient species, sometimes misleadingly called free radicals,† the system of the source and absorption cell must be modified considerably. The most often used technique is flash photolysis, devised in 1949 by Norrish and Porter.[7] The source and cell arrangement which they used for flash photolysis is shown in Figure 3.33.

The parent substance from which the transient species is to be made is placed in the quartz absorption cell C. The substance is usually in the gas phase but may also be solid, liquid or in solution. Parallel to the cell are flash lamps F consisting of electrodes in an inert gas contained in a quartz envelope. High capacity condensers, e.g. 100 μF charged to 10 kV, are rapidly discharged across the electrodes. The lamps emit a high intensity short pulse (commonly about 20 μs) of visible and ultraviolet radiation which passes through the quartz walls of F and C to the parent substance. If, for example, this is ammonia gas, then the ultraviolet radiation from F has sufficient energy to photolyse the ammonia:

†The name 'free radical' was first given to groups, such as CH_3, which behave chemically in many ways like an atom. Then the name came to be applied to any species with one or more unpaired electrons, such as benzene in a triplet electronic state, but then, unfortunately, this definition included such stable species as O_2 and NO! So nowadays the name 'free radical' is often used to describe a short-lived or transient species, but it is probably better not used at all.

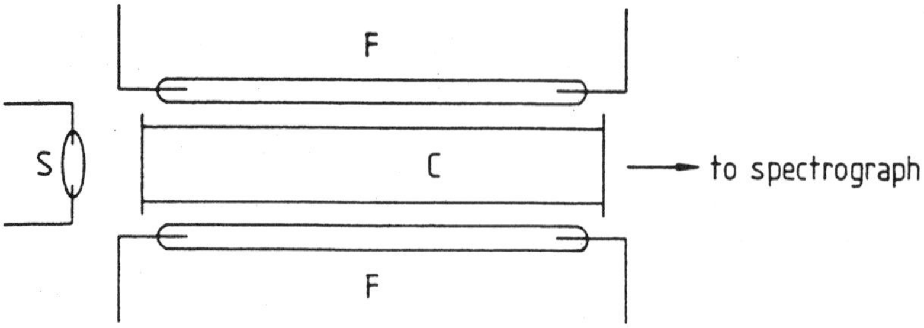

Figure 3.33 Experimental arrangement for the observation of an absorption spectrum following flash photolysis

$$NH_3 \xrightarrow{h\nu} NH_2 + H \qquad (3.22)$$

When the intensity of the flash from F begins to decrease, the concentration of the species whose absorption spectrum we wish to observe, in this case NH_2, also decreases. How long this takes depends on the half-life of the transient species in the conditions which obtain in the cell. This may be about 100 μs in the case of NH_2. In order to obtain the absorption spectrum the continuum source S is pulsed for only about 10 μs by discharging condensers, e.g. 3 μF charged to 16 kV, between the two electrodes. The timing of the source pulse after the flash lamp pulse depends on the lifetime of the transient species and is variable between about 5 and 2000 μs. The radiation from S passes through the cell to the spectrograph, where it is dispersed and recorded. Because of the low concentration of transient species, the absorption cell often contains multiple reflection mirrors of the type shown in Figure 3.30.

In addition to being used to obtain absorption spectra of transient species, these spectra being of great importance in their own right, flash photolysis has also been used in the field of kinetics to follow the changes with time of the concentrations of the transient species produced by the photolysis flash. This is achieved by changing the time interval between the photolysis and source pulses. In some cases the transient species may be a short-lived excited electronic state of a stable molecule when the time interval may be only a few nanoseconds or even picoseconds.

The dispersing element employed in visible and near ultraviolet spectroscopy may be a prism of glass (visible) or quartz (near ultraviolet). Although a quartz prism transmits visible radiation $dn/d\lambda$, where n is the refractive index, is small in this region giving poor resolving power [see equation (3.9)].

A useful and economical method of mounting a prism for use in these regions is the Littrow system shown in Figure 3.34. This is similar to that used in the infrared, in Figure 3.31, except that a glass or quartz lens can be used instead of a mirror. Also, since the prism material can itself be coated with a reflecting material such as aluminium, the 60° prism in Figure 3.31 can be replaced by a 30° prism with one reflecting surface, as in Figure 3.34.

Diffraction gratings are used almost exclusively as dispersing elements. For high resolution instruments the Czerny–Turner system, shown in Figure 3.27, employing a plane grating, is probably the best in terms of freedom from aberrations.

The Ebert system (Figure 3.26) was first used in the visible and near ultraviolet regions by King[8] in 1958, by which time the problems of

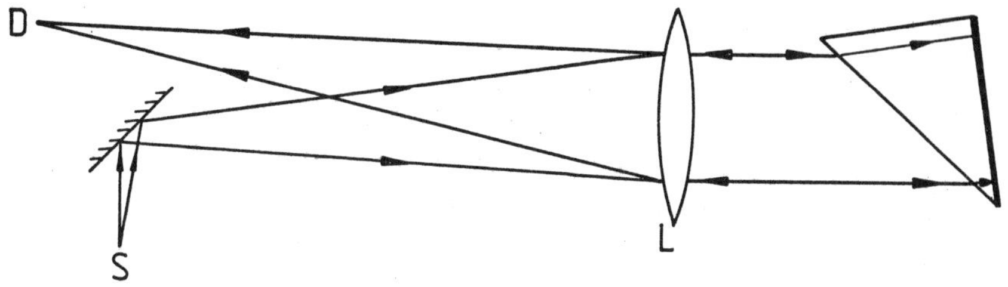

Figure 3.34 Littrow mount of a prism in a visible or near ultraviolet spectrometer

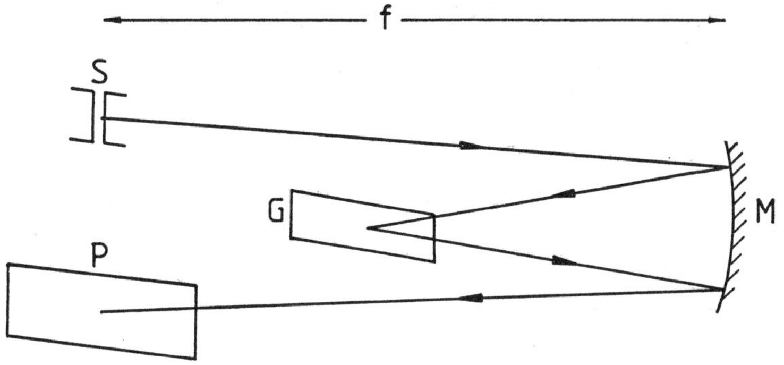

Figure 3.35 An Ebert mounting of a grating in a visible and near ultraviolet spectrograph

making a large spherical mirror free from imperfections and of the order of 40 cm in diameter had been overcome. If the spectrum is scanned across the exit slit S_2, in front of the detector, by rotating the grating then the optical arrangement is the same as for the infrared and shown in Figure 3.26. If the detection is photographic the grating is stationary and the system shown in Figure 3.35 is used. The radiation from the entrance slit S passes over the top of the grating G to the mirror M. The distance of S from M is equal to the focal length f of the mirror. A parallel beam passes from M to G. The dispersed radiation is focused by M on to the photographic plate P and passes underneath the grating.

The Fabry–Perot interferometer (see Section 3.5.4) with pressure scanning can be used as a dispersing element for visible and near ultraviolet radiation. Very high resolution has been obtained but the fact that only a small wavelength range can be scanned with a given plate separation makes it inconvenient to use. More convenient is the Michelson interferometer (see Section 3.5.4), which can be used in the visible and near ultraviolet but with lower optimum resolution (*ca* 0.05 cm^{-1}) than in the infrared.

The photographic plate detector has now been replaced almost entirely by the photomultiplier. This is used in a scanning spectrometer in which the grating is smoothly rotated and the detector remains stationary. In the photomultiplier the photons fall on to a surface such as caesium which emits one electron per photon. This electron is subjected to an accelerating potential of about 100 V and then falls on to a second surface which releases several secondary electrons. This process is repeated many times to give a very large amplification of the current which is then fed to a recorder.

A photomultiplier is convenient to use and has the additional advantage over a photographic plate of a linear response to the intensity of the radiation falling on it. A photographic emulsion responds far from linearly making quantitative intensities very difficult to obtain. On the other hand, a photographic plate has two advantges. The first is that, like the Michelson interferometer, it possesses the multiplex advantage in that it records a large wavelength range of the radiation *all* the time. The second advantage is that if the radiation falling on it is very weak the exposure time can be increased indefinitely with no decrease in the signal to noise ratio, provided that the spectrograph is free from internal stray radiation and from vibrations and provided also that the temperature and pressure are steady.

A type of detector which has, to some extent, the multiplex advantage of a photographic plate while having the advantages of a photomultiplier consists of an array of sensitive photodiodes. Each photodiode occupies a position in the focal plane of the spectrometer and the array can be thought of as replacing the photographic plate. However, whereas the resolution achieved with a photographic plate is limited, ultimately, by the grain size in the emulsion, the resolution of an array of photodiodes is limited by how closely they can be spaced and it cannot approach that of a photographic plate. Scanning of the signals from the elements of the array and accumulation over a period of time to produce the spectrum are achieved with a multichannel analyser.

A more recent, and superior, type of detector, which also benefits from the multiplex advantage, is the charge-coupled device, or CCD. The CCD used for spectroscopy has been developed from the CCD detector used in a camcorder.

A CCD is a two-dimensional array of silicon photosensors, each photosensor usually being referred to as a pixel. When radiation falls on a pixel photoelectrons are produced in numbers proportional to the intensity of the radiation. A typical wavelength range to which the CCD is sensitive is 400–1050 nm, but this may be extended down to below 1.5 nm with a phosphor which converts short wavelength into visible radiation.

Whereas a photodiode array is typically a linear array, consisting of a single row of photodiodes, the CCD is a two-dimensional array, consisting of rows and columns of pixels. In this sense it resembles a photographic plate. Each row of the CCD should, ideally, give the same spectrum (wavelength versus intensity). Computer addition of the signals from pixels in the same column then produces a spectrum with a much higher signal to noise ratio than would result from a one-dimensional array. A typical CCD may contain about 2000 columns and 800 rows of pixels, the area of each pixel being about $15 \times 15\,\mu m$.

3.8.2 Atomic samples

In studies of atomic spectra, it was not until about 1955 that absorption spectroscopy became possible for a wide range of atoms.

Atomic absorption spectroscopy is used mainly as an important quantitative analytical tool. The greatest difficulty to be overcome in its development was that of the very low vapour pressure of most atomic materials, with the exception of a few such as mercury and sodium. The technique used for producing atoms in the form of a vapour is to spray, in a very fine mist, a liquid molecular sample containing the atom concerned into a high temperature combustion flame. In the combustion process the molecules are broken up and the atom is freed. Flames consisting of air and coal (or natural) gas, air and propane, air and acetylene, or nitrous oxide and acetylene are used. The choice of flame depends on the element concerned, as illustrated in Table 3.3, but the most universally

Table 3.3 Choice of combustion flame in atomic absorption spectroscopy

Flame	Temperature/K	Suitable elements
Air–coal (or natural) gas	2100	Alkali metals, Zn, Cu, Cd, Pb
Air–propane	2200	Volatile elements (including those above) plus noble metals
Air–acetylene (non-luminous)	2600	Alkaline earth metals, and many other elements
Air–acetylene (luminous)	2600	Elements forming refractory oxides, such as Sn, Ba, Cr, Mo
N_2O–acetylene	3200	Refractory elements, such as Al, Si, V, Ti, Be

applicable is acetylene and air. A luminous air–acetylene flame provides reducing conditions for elements which would otherwise form refractory oxides, but some such elements require the higher temperature of the nitrous oxide–acetylene flame.

The source radiation which passes through the flame is not a continuum, which we would normally use in absorption spectroscopy, but a hollow cathode source of discrete radiation. The general design of such a source is shown in Figure 3.36. The anode, a tungsten wire, and the cup-shaped cathode, made from the metal being investigated, are sealed into a glass vessel with a window W sealed on to the end. The vessel contains a carrier gas such as neon at a pressure of about 5 Torr. When a high voltage is applied a coloured discharge appears and the positive column, where intense neutral atom emission occurs, is confined to the inside of the cathode by suitable choice of voltage and carrier gas pressure. The radiation emitted by this source is the emission spectrum of the atom being observed in absorption. A part of this is shown schematically in Figure 3.37, where the transition at wavelength λ_1 is from an excited to the ground electronic state and that at λ_2 is between two excited electronic states. Because of the Boltzmann factor [equation (2.6)] associated with a large ΔE, typical of electronic state separations, virtually all the atoms formed in the flame are in their ground electronic states. These atoms are able to absorb radiation of wavelength λ_1 but not λ_2, as indicated in Figure 3.37. If the dispersing element of the atomic absorption spectrometer is set so that the detector, a photoelectric cell, receives only a narrow band of wavelengths centred on λ_1, the amount of absorption by the atoms in the flame can be measured. If the system has been

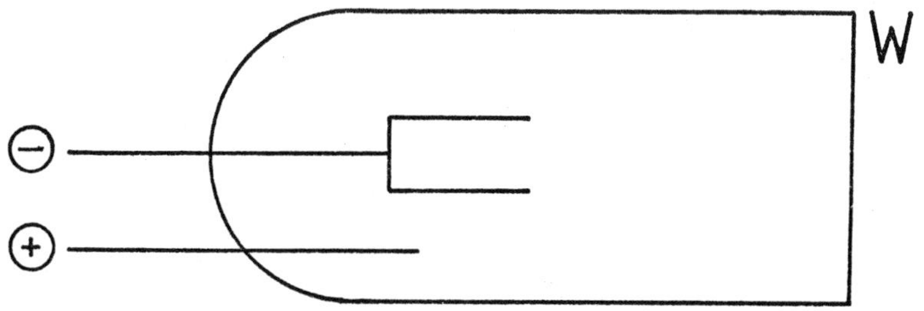

Figure 3.36 A hollow cathode lamp

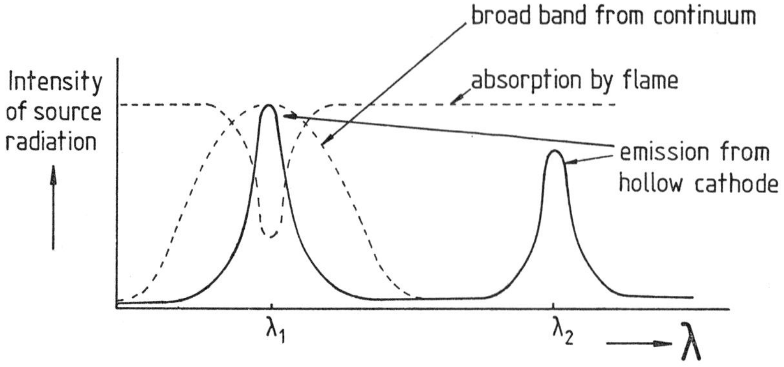

Figure 3.37 Measurement of the concentration of atoms using a hollow cathode lamp and atomic absorption spectroscopy

calibrated by spraying a mist of a solution containing a known concentration of the metal atoms, then the concentration in the sample to be analysed can be measured.

The advantage of using a hollow cathode source rather than a continuum is that a spectrometer with only low resolution is adequate. If a continuum source were used a low resolution instrument would feed to the detector a broad band of wavelengths, as shown in Figure 3.37, and the proportion absorbed in the flame would be very small and impossible to detect accurately.

Sources used in investigations of atomic emission spectra are the hollow cathode lamp (mentioned above), a low voltage d.c. arc or a high voltage spark.

If the element can be made into electrodes, as can iron or copper, then an arc can be maintained between two poles, a few millimetres apart, by a d.c. power supply of about 100 V. Arcs are usually struck in air but other gases may be used. Elements which are either non-conducting, or in short supply, or in the form of a salt, can be ground to a powder and placed in a small cup drilled into the lower electrode of a carbon arc. This method is useful for qualitative analysis of atom content but for quantitative analysis fluctuations and instability in position and temperature of the arc create problems.

For quantitative analysis a high voltage (10–50 kV) spark provided by the discharge of a condenser across two electrodes, similar to those used for arcs, is more reliable.

Dispersing elements and detectors used in atomic spectroscopy in the visible and near ultraviolet regions are of the same types as those used in molecular spectroscopy.

3.9 FAR (VACUUM) ULTRAVIOLET SPECTROSCOPY

Oxygen is opaque to ultraviolet radiation from about 185 nm to low wavelength, and spectroscopic techniques in this region require the evacuation of the complete system from source to detector.

As a source of far ultraviolet radiation, the hydrogen discharge lamp, which has been mentioned in Section 3.8.1 as a near ultraviolet source, can be used down to about 160 nm.

A high voltage repetitive spark discharge in helium or neon gas produces a continuum from

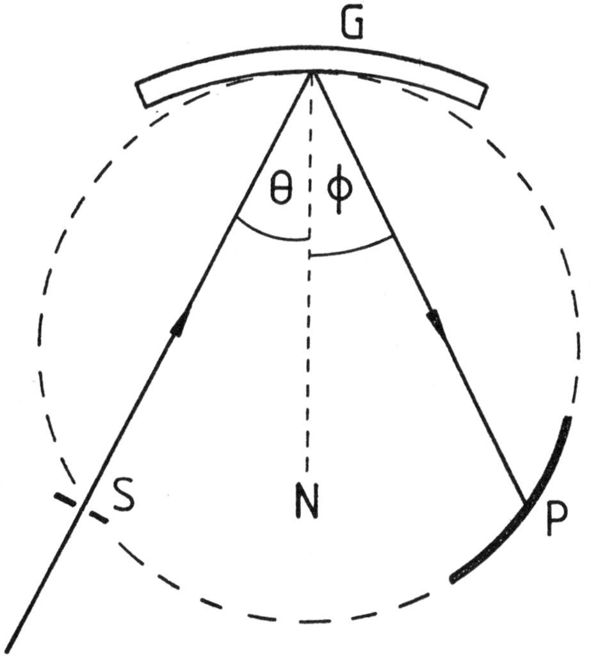

Figure 3.38 The arrangement of slit S and plate P on the Rowland circle of a grating G

100 to 60 nm or from 100 to 74 nm due to emission by the He_2 or Ne_2 molecule, respectively.

A microwave induced discharge in argon, krypton or xenon gas produces continua from 155 to 105, 180 to 125 or 200 to 148 nm, respectively. These continua also result from emission by inert gas diatomic molecules.

Perhaps the most widely used source is the Lyman continuum. This can provide a continuum extending from the visible region down to 90 nm or, with some discrete emission superimposed on the continuum, down to 30 nm. A high capacity condenser is repetitively discharged through a low pressure gas contained in a glass capillary. Surprisingly, the continuum is independent of the gas used and is thought to be emitted by species produced by the erosive action of the discharge on the glass.

Accelerating electrons emit radiation which is continuous from the visible to the X-ray region. This is made use of in the synchrotron source (see Section 7.2.1), which is particularly intense, compared to other sources, at wavelengths less than 60 nm.

Below about 180 nm, where all except some synthetic forms of quartz are opaque, the alkaline earth metal fluorides and sapphire can be used as window materials. Perhaps the most commonly used is lithium fluoride, which is transparent down to about 105 nm. In the region below this no windows can be used. This creates a problem of separation of the source, such as a Lyman discharge, from the rest of the evacuated system; this is solved by rapid pumping away of the source gas in the region where a window would normally be.

Prisms are clearly not suitable dispersing elements in the far ultraviolet. Gratings are always used and the choice of optical system in which the grating is incorporated is governed to a large extent by the low reflectivity of metal surfaces at low wavelengths. The reflectivity of aluminium is improved by coating with magnesium fluoride but it may be advantageous to avoid the three reflections which occur in a Czerny–Turner or Ebert system by using a concave reflection grating which not only reflects and disperses the radiation but also focuses it on to the detector.

The system often used is one based on the properties of the Rowland circle. This is shown in Figure 3.38 and is a circle with a diameter equal to the radius of curvature of the concave grating. The surface of the grating is tangential to the circle. An important property of the Rowland circle is that if the entrance slit S is positioned somewhere on the circle the spectrum will be focused on the circle. If a photographic plate P is used as a detector then it must be curved to form an arc of the circle.

When θ, the angle between the incident radiation and the grating normal N, is less than about 10° the grating is said to be operating under conditions of normal incidence. The Eagle system is a normal incidence system and it is used only under conditions where $\theta \approx \phi$, the angle of reflection. The system is therefore very compact, since $\theta + \phi < 20^\circ$, and the detector and slit are close together.

At wavelengths below about 30 nm the reflectivity of the grating surface is very poor. However, the reflectivity increases as the angle of incidence θ increases and the use of grazing incidence, with $\theta \geqslant 89^\circ$, is essential at these low wavelengths. A grating surface of gold or platinum is effective also in increasing the low wavelength reflectivity. Wavelengths as low as 0.7 nm have been observed using these techniques.

The photographic plate is a very useful form of detector but, as Schumann discovered in 1892, gelatin, like oxygen, is opaque at wavelengths lower than 185 nm. He therefore removed as much of it as possible and such plates are still called 'Schumann plates'. One effect of removing most of the gelatin is to make the emulsion very sensitive to abrasion. In addition to having less gelatin, the plates can also be sensitized in order to increase their sensitivity at low wavelengths.

A conventional photomultiplier detector, as used in the visible and near ultraviolet regions, can be used but it is limited to wavelengths above about 105 nm by its lithium fluoride window. A photomultiplier in combination with a fluorescent scintillator, such as sodium salicylate, can be used down to about 60 nm. A solar blind type of photomultiplier, sensitive only in the region 110–360 or 110–200 nm, depending on the photocathode material, is useful when insensitivity to visible radiation is desirable: these devices act not only as ultraviolet detectors but also as highly efficient filters.

3.10 LOW AND HIGH RESOLUTION SPECTROSCOPY

Spectroscopy in the infrared, visible and ultraviolet regions serves two main purposes. One is to provide a means of qualitative or, sometimes, quantitative analysis of the material absorbing or emitting radiation. Qualitative analysis is made possible by the fingerprint nature of spectra: they are highly characteristic of the material concerned and are mostly additive if the material is a mixture. Quantitative analysis can be carried out in cases where relative intensities are accurately measurable, as in atomic absorption spectroscopy. The second purpose of spectroscopy in these regions is as a very powerful and accurate tool for the investigation of atomic and molecular structure. This includes not only determination of molecular geometry in the ground electronic state but also how molecules behave when they vibrate and rotate and how both atoms and molecules change when they are in excited electronic states. It is these properties of atoms and molecules with which quantum mechanics is very much concerned. The real test of any quantum mechanical theory lies in its ability to reproduce observed experimental data: theory and experiment in spectroscopy are inseparable.

When we speak of low and high resolution spectroscopy, we are referring usually to the performance of the instrument used to obtain the spectra. Generally, low resolution instruments are used for analytical purposes, although this is not true of, for example, atomic emission spectroscopy, where high resolution is necessary to separate crowded spectra. On the other hand, high resolution instruments are used for obtaining more and more accurate structural information, which in turn provides impetus for theory to account for it.

The general appearance of low and high resolution instruments is usually different. The design of a low resolution instrument is orientated towards compactness, convenience and ease of use. Hence it may have double beam operation, automatic chart recording with a scale linear in wavelength (or, much more preferable, wavenumber), maximum wavelength range and compact printed circuits. In a high resolution instrument, all these desirable virtues are subordinate to the ability to achieve high resolution, so that the instrument may be bulky, of limited wavelength range and inconvenient and slow to use. However, the principles underlying the basic designs of both kinds of instruments have much in common.

Of course, the division between low and high resolution instruments is not a clear one and the choice of a particular instrument depends very much on the problem concerned. High resolution is not always a virtue and the injudicious use of a high resolution instrument could result in being unable to see the wood for the trees. For example, determination of the intensity of a transition involving integration of the area under a broad absorption curve should be carried out on a low resolution spectrum.

Microwave and millimetre wave spectroscopy are special cases in that, because of the highly monochromatic nature of the sources, they are intrinsically high resolution techniques. However, the development of continuously tunable backward wave oscillators and chart recording has made it possible, by changing the conditions of scanning the spectrum, to use the spectrometer as a relatively low resolution instrument. It is ironical that in microwave and millimetre wave spectroscopy we are now able to see the wood, whereas previously it was possible to see only the trees!

BIBLIOGRAPHY

BOUSQUET, P. (1971). *Spectroscopy and its Instrumentation.* Adam Hilger, London

CHANTRY, G. W. (1971). *Submillimetre Spectroscopy.* Academic Press, New York

DITCHBURN, R. W. (1953). *Light.* Interscience, New York

GRIFFITHS, P. R. (1975). *Chemical Infrared Fourier Transform Spectroscopy.* Wiley, New York

HARRISON, G. R., LORD, R. C. and LOOFBOUROW, J. R. (1948). *Practical Spectroscopy.* Prentice-Hall, Englewood Cliffs, NJ

HECHT, E. and ZAJAC, A. (1974). *Optics.* Addison-Wesley, Reading, MA

HOUGHTON, J. T. and SMITH, S. D. (1966). *Infrared Physics.* Oxford University Press, Oxford

JENKINS, F. A. and WHITE, H. E. (1957). *Fundamentals of Optics.* McGraw-Hill, New York

KANTROWITZ, A. and GREY, J. (1951). *Rev. Sci. Instrum.*, **22**, 328

MARTIN, A. E. (1963). 'Instrumentation and general experimental methods', Chapter 2 in *Infrared Spectroscopy and Molecular Structure* (Ed. M. Davies). Elsevier, Amsterdam

RAMSEY, N. F. (1956). *Molecular Beams.* Oxford University Press, Oxford

SAMSON, J. A. R. (1967). *Techniques of Vacuum Ultraviolet Spectroscopy.* Wiley, New York

SANDERSON, R. B. (1972). 'Fourier spectroscopy', Chapter 7.1 in *Molecular Spectroscopy: Modern Research* (Eds K. N. Rao and C. W. Mathews). Academic Press, New York

SAWYER, R. A. (1963). *Experimental Spectroscopy*. Dover, New York

SMITH, R. A., JONES, F. E. and CHASMAR, R. P. (1968). *The Detection and Measurement of Infrared Radiation*. Oxford University Press, Oxford

SUGDEN, T. M. and KENNEY, C. N. (1965). *Microwave Spectroscopy of Gases*. Van Nostrand, London

TOLANSKY, S. (1955). *An Introduction to Interferometry*. Longmans, London

TOWNES, C. H. and SCHAWLOW, A. L. (1955). *Microwave Spectroscopy*. McGraw-Hill, New York

WHIFFEN, D. H. (1966). *Spectroscopy*. Longmans, London

4 ROTATIONAL SPECTROSCOPY

4.1 CLASSIFICATION AS LINEAR MOLECULES, SYMMETRIC ROTORS, SPHERICAL ROTORS OR ASYMMETRIC ROTORS

We shall not be concerned with atoms in this or the following chapter since they have no rotational or vibrational degrees of freedom.

In respect of their rotational, vibrational and electronic spectra, molecules can be divided very usefully into classes according to their principal moments of inertia.

A molecule contains a one-, two- or three-dimensional set of nuclei, each of which is taken to be a point mass m_i with coordinates x_i, y_i, z_i with respect to any set of cartesian axes whose origin is at the centre of mass of the set of nuclei. The axes are fixed relative to the nuclei so that when the molecule rotates the axes rotate with it. These are called molecule-fixed axes, as opposed to space-fixed axes which remain fixed in space while the molecule may rotate about them. The moments of inertia about x, y and z molecule-fixed axes are given by

$$\begin{aligned} I_x &= \sum_i m_i(y_i^2 + z_i^2) \\ I_y &= \sum_i m_i(x_i^2 + z_i^2) \\ I_z &= \sum_i m_i(x_i^2 + y_i^2) \end{aligned} \qquad (4.1)$$

The products of inertia I_{xy}, I_{yz} and I_{xz} relating to the same axes are defined as

$$\begin{aligned} I_{xy} &= -\sum_i m_i x_i y_i \\ I_{yz} &= -\sum_i m_i y_i z_i \\ I_{xz} &= -\sum_i m_i x_i z_i \end{aligned} \qquad (4.2)$$

Then the three-dimensional surface given by the equation

$$I_x x^2 + I_y y^2 + I_z z^2 - 2I_{xy}xy - 2I_{yz}yz - 2I_{xz}xz = 1 \qquad (4.3)$$

represents the momental ellipsoid. Such an ellipsoid is shown in Figure 4.1. The cartesian axes in this figure represent a particular set of axes for which the products of inertia are zero and equation (4.3) reduces to

$$I_a x^2 + I_b y^2 + I_c z^2 = 1 \qquad (4.4)$$

where a and b are the major and minor axes of the ellipse in the xy-plane and b and c are the major and minor axes in the yz-plane. These axes are the principal inertial axes and I_a, I_b and I_c the principal moments of inertia. One of the principal moments of inertia, I_a in this case, is the minimum moment of inertia about any axis; another, I_c in this case, is the maximum moment of inertia and its axis must always be perpendicular to that of I_a; the third, I_b, is intermediate in magnitude and about an axis perpendicular to those of both I_a and I_c. In fact, we always maintain the convention

$$I_c \geqslant I_b \geqslant I_a \qquad (4.5)$$

and the x-, y- and z-axes corresponding to these principal moments of inertia are labelled a, b and c, respectively.

Each molecule has a characteristic momental ellipsoid which has the property that the

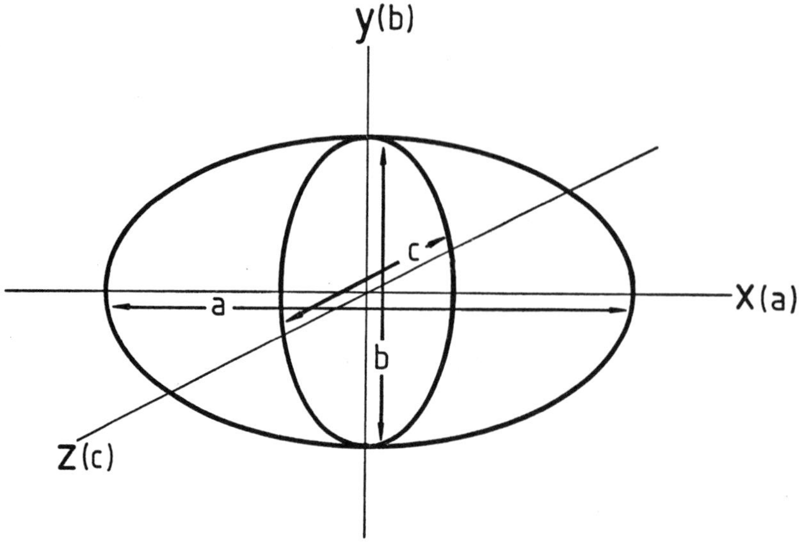

Figure 4.1 The momental ellipsoid

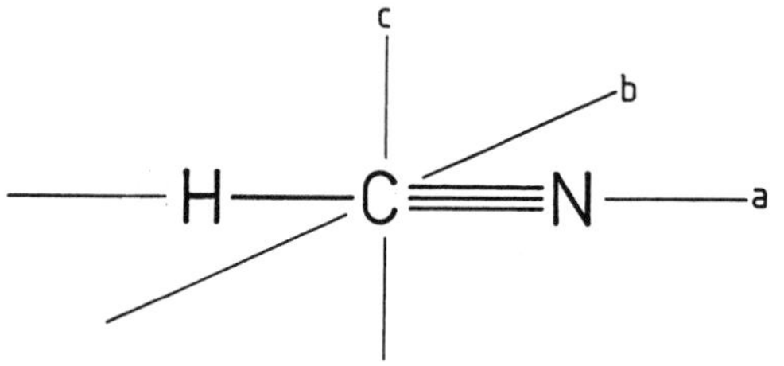

Figure 4.2 Principal axes of inertia of the HCN molecule

distance between any point on the surface and the centre of mass is $1/I^{\frac{1}{2}}$, where I is the moment of inertia of the molecule about that line.

A linear molecule, such as HCN shown in Figure 4.2, has

$$I_c = I_b > I_a = 0 \tag{4.6}$$

In this case

$$I_c = m_H r_H^2 + m_C r_C^2 + m_N r_N^2 \tag{4.7}$$

where m_X and r_X are the mass of atom X and the distance along the a-axis of atom X from the centre of mass, respectively. The cross-section of the momental ellipsoid in the bc-plane is reduced to a circle.

A symmetric rotor, or symmetric top, molecule has two equal moments of inertia and the third may be less or greater than the other two. If it is less than the other two, then

$$I_c = I_b > I_a \tag{4.8}$$

and the molecule is a prolate symmetric rotor. Methyl iodide is an example and is shown in Figure 4.3(a) As iodine is such a heavy atom it is clear that I_a is much smaller than I_b and I_c, since the iodine atom is on the a-axis and makes no contribution to I_a. It may not be immediately obvious that I_b and I_c, about *any* two perpendicular axes which are also perpendicular to the a-axis, are equal, but simple trigonometry will prove this. An example of a prolate symmetric rotor from the macroscopic world is the rugby ball shown in Figure 4.3(b). Since the cross-section perpendicular to the long a-

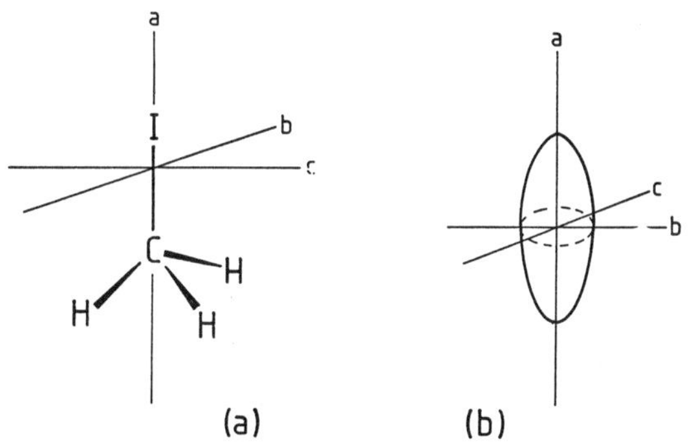

Figure 4.3 Principal axes of inertia of the prolate symmetric rotors (a) methyl iodide and (b) a rugby ball

axis is circular, it is obvious in this case that I_b and I_c are equal.

A linear molecule is a special case of a prolate symmetric rotor since the conditions in equation (4.6) are a special case of those in equation (4.8).

If the third moment of inertia is greater than the two equal ones, then

$$I_c > I_b = I_a \tag{4.9}$$

and the molecule is an oblate symmetric rotor. An example is the square planar ion $[PtCl_4]^{2-}$ shown in Figure 4.4(a). In this case it is clear that I_a and I_b are equal, but in the benzene molecule, shown in Figure 4.4(b) and also an oblate symmetric rotor, some trigonometry is required to show this equality. A macroscopic example of an oblate symmetric rotor is the discus shown in Figure 4.4(c).

It can be shown that any molecule having a C_n symmetry axis with $n > 2$ is necessarily a symmetric rotor (the definition of a C_n axis has been given on p. 38–39). The C_n axis with $n > 2$ is called the top axis or figure axis of the symmetric rotor. It can be seen from Figures 4.3 and 4.4 that the a-axis of methyl iodide is a C_3 axis, since rotation about it by $2\pi/3$ produces an indistinguishable configuration, the c-axis of $[PtCl_4]^{2-}$ is a C_4 axis and the c-axis of benzene is a C_6 axis.

While the rule that a molecule having a C_n axis with $n > 2$ is a symmetric rotor is true in all cases, this is not exactly so for the converse. The allene molecule ($H_2C{=}C{=}CH_2$), shown in Figure 4.5, provides a good illustration of this point. In allene the planes of the two CH_2 groups are mutually perpendicular and the C–C–C axis is the a-axis. It is clear from Figure 4.5 that I_b is equal to I_c and therefore the molecule is a prolate symmetric rotor in spite of

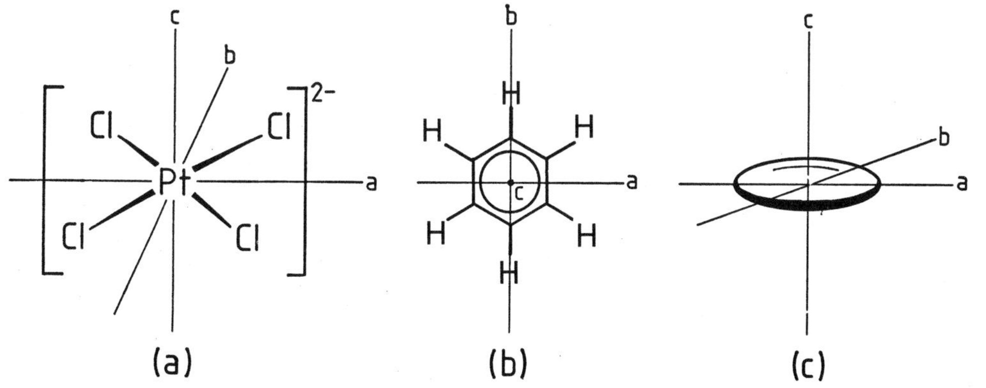

Figure 4.4 Principal axes of inertia of the oblate symmetric rotors (a) $[PtCl_4]^{2-}$, (b) the benzene molecule and (c) a discus

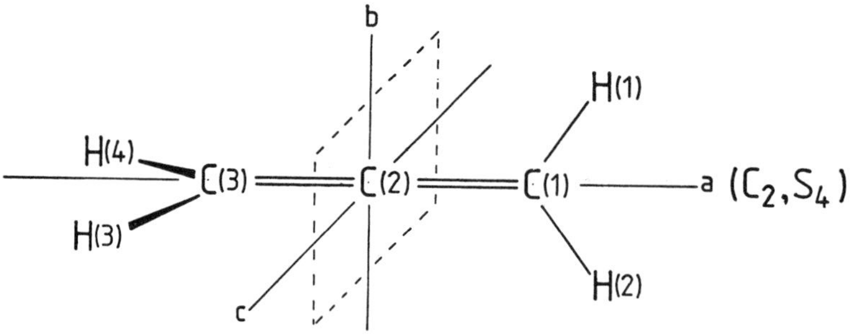

Figure 4.5 Principal axes of inertia of the allene molecule, which has an S_4 rotation–reflection axis

not having a C_n axis with $n > 3$. The fact that allene is a symmetric rotor is due to its having an S_4 rotation–reflection axis.

An S_n rotation–reflection axis is an axis such that rotation by $2\pi/n$ about it followed by reflection across a plane, which is perpendicular to the axis and passes through the centre of the molecule, produces a configuration indistinguishable from the starting one. The a-axis in allene is an S_4 axis since rotation about it by $2\pi/4$ followed by reflection across the dotted plane in Figure 4.5 has the effect of exchanging hydrogen atoms 1 and 2 for 3 and 4, and carbon atom 1 for 3.

It happens that a molecule having any S_n axis with $n > 2$, other than $n = 4$, necessarily has a C_n axis with $n > 2$. Hence we can now state the complete rule:

A molecule with either a C_n axis with $n > 2$ or an S_4 axis must be a symmetric rotor.

An asymmetric rotor molecule has no equal principal moments of inertia, i.e.

$$I_c \neq I_b \neq I_a \tag{4.10}$$

There are many examples of asymmetric rotors, including the non-planar molecule CHBrClF, shown in Figure 4.6(a), and planar H_2CO (formaldehyde), shown in Figure 4.6(b). In a molecule with such low symmetry as CHBrClF the principal inertial axes are not confined by symmetry to any particular direction or plane, but formaldehyde, and many other asymmetric rotors, have sufficiently high symmetry that the principal axes are restricted by symmetry. Any C_2 axis of symmetry in an asymmetric rotor, such as the line of the C=O bond in formaldehyde, is always a principal axis; in formaldehyde it is the a-axis. In addition, in any planar molecule the moment of inertia about the axis perpendicular to the plane is equal to the sum of the two in-plane moments of inertia. For this reason, the moment of inertia about the axis perpendicular to the plane must be the largest; therefore, the axis must be the c-axis and we have

$$I_c = I_a + I_b \tag{4.11}$$

Some molecules which are strictly asymmetric rotors have two moments of inertia which are accidentally nearly equal. If

$$I_c \approx I_b > I_a \tag{4.12}$$

the molecule is a prolate near-symmetric rotor. An example of this is *s-trans*-acrolein shown in Figure 4.7(a) (*s-trans* means that the two double bonds are in the *trans* configuration about the carbon–carbon single bond). The positions of the heavy atoms of the C=C–C=O chain are such as to make I_a small compared with I_b and I_c.

If, on the other hand,

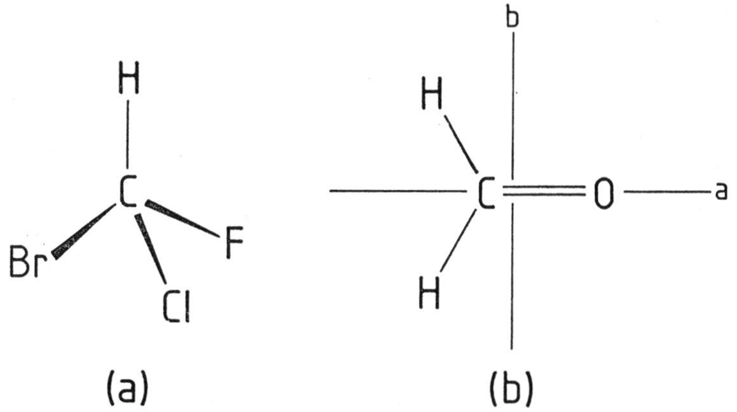

Figure 4.6 (a) A non-planar asymmetric rotor, the bromochlorofluoromethane molecule, and (b) a planar asymmetric rotor, the formaldehyde molecule

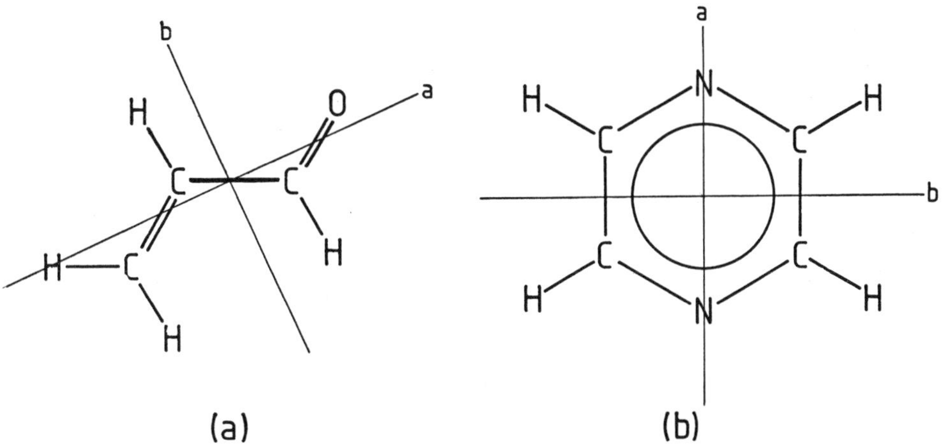

Figure 4.7 (a) A prolate near-symmetric rotor, the *s-trans*-acrolein molecule, and (b) an oblate near-symmetric rotor, the pyrazine molecule

$$I_c > I_b \approx I_a \tag{4.13}$$

the molecule is an oblate near-symmetric rotor. 1,4-Diazabenzene (pyrazine), shown in Figure 4.7(b), is an example. Because of the similarity of the masses of N and CH the principal moments of inertia are similar to those of benzene, an oblate symmetric rotor, and therefore $I_a \approx I_b$.

Finally, there are highly symmetrical molecules, the spherical rotors, for which

$$I_c = I_b = I_a \tag{4.14}$$

All regular tetrahedral molecules such as methane, shown in Figure 4.8(a), and silicon tetrafluoride are spherical rotors, although the equality of the moments on inertia may not be

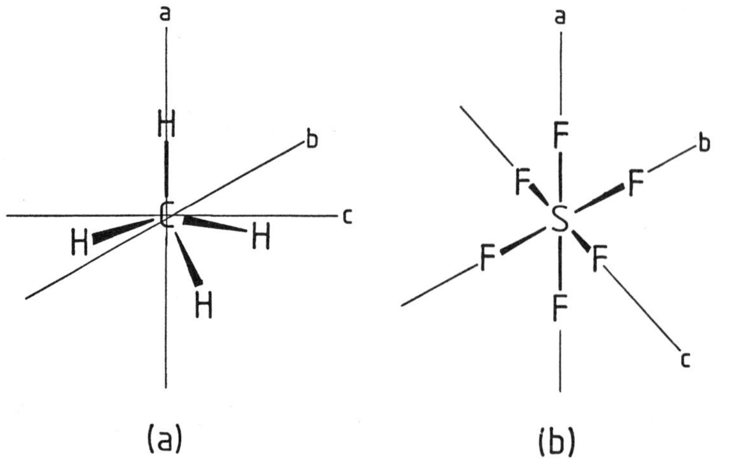

Figure 4.8 The methane (a) and sulphur hexafluoride (b) molecules are spherical rotors

obvious at first sight. Regular octahedral molecules such as sulphur hexafluoride, shown in Figure 4.8(b), and the hexacyanoferrate (III) ion, $[Fe(CN)_6]^{3-}$, are also spherical rotors in which the condition in equation (4.14) is more obviously satisfied.

Accidental near-spherical rotors are possible but rare.

4.2 PURE ROTATIONAL INFRARED, MILLIMETRE WAVE AND MICROWAVE SPECTRA OF DIATOMIC AND LINEAR POLYATOMIC MOLECULES

4.2.1 Transition frequencies in the rigid rotor approximation

All discussion here of rotational spectroscopy assumes the sample to be in collision-free conditions in the gas phase or in a supersonic jet.

The expression

$$E_r = h^2 J(J+1)/8\pi^2 I \tag{4.15}$$

for the quantized rotational energy levels E_r using the rigid rotor model of a diatomic molecule has been presented previously (equation 1.88). The quantity I is the moment of inertia: for a diatomic this is given by μr^2, where μ is the reduced mass and r the internuclear distance. Equation (4.15) applies also to the rotational energy levels of a linear polyatomic molecule such as acetylene (H–C≡C–H) or O=C=S.

As will have been apparent from the discussion of experimental methods in Chapter 3, what is measured experimentally is not the energy ΔE of a transition between energy levels, but the frequency ν or the wavenumber $\tilde{\nu}$, both of which are proportional to ΔE:

$$\Delta E = h\nu = hc\tilde{\nu} \tag{4.16}$$

For this reason, it is convenient to convert energy level expressions to what are called term value expressions which have dimensions of frequency or wavenumber rather than energy. For a diatomic or linear polyatomic molecule, the rotational term values are symbolized by $F(J)$, where

$$F(J) = E_r/h = hJ(J+1)/8\pi^2 I \tag{4.17}$$

in which $F(J)$ has the dimensions of frequency, or

$$F(J) = E_r/hc = hJ(J+1)/8\pi^2 cI \quad (4.18)$$

in which $F(J)$ has the dimensions of wavenumber. The use of the same symbol $F(J)$ for two different physical quantities with different dimensions and units is unfortunate but is so prevalent that we cannot avoid it. Equally unfortunate is the fact that the constant multiplying $J(J+1)$ in both equations (4.17) and (4.18) is symbolized by B. This quantity is the rotational constant and the rotational term value expressions become

$$F(J) = BJ(J+1) \quad (4.19)$$

where B, like $F(J)$, may have the dimensions of frequency or wavenumber[†]. Because of the nature of the experimental methods, microwave and millimetre wave spectroscopists usually use $B\ (= h/8\pi^2 I)$ with the dimensions of frequency and infrared, visible and ultraviolet spectroscopists use $B\ (= h/8\pi^2 cI)$ with the dimensions of wavenumber.

As an example of a set of rotational energy levels, or strictly term values, Figure 4.9 shows them for the molecule CO.

Pure rotational transitions may be observed in the microwave, millimetre wave or far infrared regions. They are electric dipole transitions and the transition moments $\boldsymbol{R}^{J'M_J'J''M_J''}$ are vector quantities given by

$$\boldsymbol{R}^{J'M_J'J''M_J''} = \int (\psi_r^{J'M_J'})^* \boldsymbol{\mu} (\psi_r^{J''M_J''})\, \mathrm{d}\tau \quad (4.20)$$

This is similar to equation (2.21) but applied to a transition between two rotational states, the lower having quantum numbers J'' and M_J'' (see equations 1.86 and 1.89) and the upper having quantum numbers J' and M_J'.[‡] The rotational selection rules constitute the conditions for which the transition moment is non-zero and are:

(a) the molecule must have a permanent dipole moment, i.e. $\boldsymbol{\mu}$, given by equation (2.22), must be non-zero;
(b) $\Delta J = \pm 1$;
(c) $\Delta M_J = 0, \pm 1$, a rule which is important if the molecule is in an electric or magnetic field.

It follows from the selection rule (a) that transitions are allowed in heteronuclear diatomic molecules such as CO, NO, HF and even $^1H^2H$[§] (for which $\mu = 5.9 \times 10^{-4}$ D,[¶] compared with, say, HF for which $\mu = 1.82$ D) but not in homonuclear molecules such as H_2, Cl_2 and N_2.

Similarly, those linear polyatomic molecules with no centre of symmetry, such as O=C=S, H–C≡N, H–C≡C–Cl and even ^{1}H–C≡C–^{2}H ($\mu \approx 0.012$ D) have allowed rotational transitions, whereas those with a centre of symmetry, such as S=C=S, H–C≡C–H and O=C=C=C=O (carbon suboxide) do not.

The centre of symmetry, or centre of inversion, is the fourth kind of element of symmetry we have encountered—a rotation axis, a rotation–reflection axis and a plane of

[†]It would be much simpler to use the symbols $F(J)$ and B *only* when they have the dimensions of frequency and, say, $\tilde{F}(J)$ and $\tilde{B}$ when they have the dimensions of wavenumber, giving

$$F(J) = hJ(J+1)/8\pi^2 I = BJ(J+1)$$

and

$$\tilde{F}(J) = hJ(J+1)/8\pi^2 cI = \tilde{B}J(J+1)$$

but I hesitate to use a new symbolism, however logical, when readers may encounter it only rarely elsewhere.

[‡]It is a general spectroscopic convention to indicate any quantum numbers belonging to the lower or upper state of a transition by double primes (″) or a single prime (′), respectively.

[§]We might expect that the addition of a neutron to one of the nuclei in H_2 would not produce any dipole moment. In fact, the dipole moment arises because of a breakdown of the Born–Oppenheimer approximation in that, when the molecule vibrates, the electrons do not exactly follow the motions of the nuclei and, in addition, the vibration is not exactly that of a harmonic oscillator.

[¶]1 D = 3.335 64 × 10^{-30} C m, where D is the debye unit.

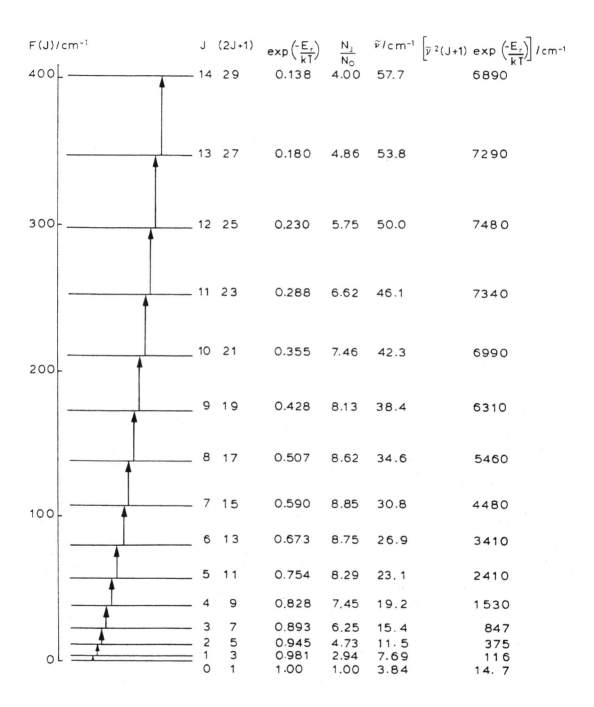

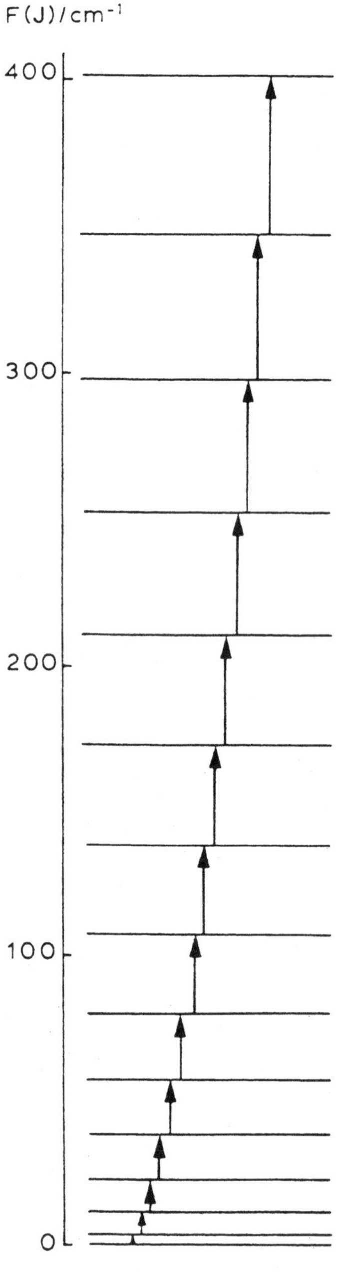

J	(2J+1)	$\exp\left(\frac{-E_r}{kT}\right)$	$\frac{N_J}{N_0}$	$\tilde{\nu}/\mathrm{cm}^{-1}$	$\left[\tilde{\nu}^2(J+1)\exp\left(\frac{-E_r}{kT}\right)\right]/\mathrm{cm}^{-1}$
14	29	0.138	4.00	57.7	6890
13	27	0.180	4.86	53.8	7290
12	25	0.230	5.75	50.0	7480
11	23	0.288	6.62	46.1	7340
10	21	0.355	7.46	42.3	6990
9	19	0.428	8.13	38.4	6310
8	17	0.507	8.62	34.6	5460
7	15	0.590	8.85	30.8	4480
6	13	0.673	8.75	26.9	3410
5	11	0.754	8.29	23.1	2410
4	9	0.828	7.45	19.2	1530
3	7	0.893	6.25	15.4	847
2	5	0.945	4.73	11.5	375
1	3	0.981	2.94	7.69	116
0	1	1.00	1.00	3.84	14.7

Figure 4.9 Rotational term values, relative populations and relative rotational transition intensities for the CO molecule

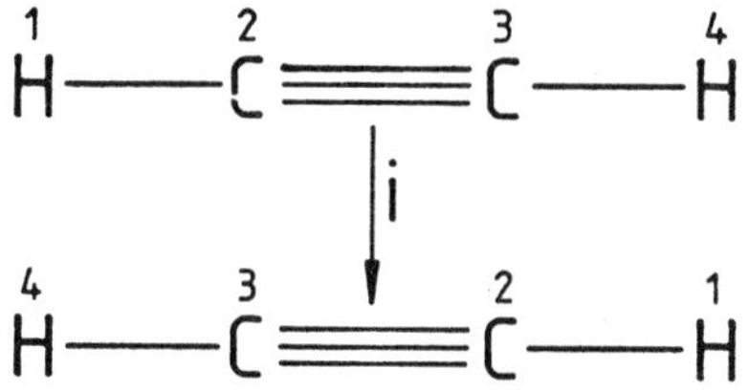

Figure 4.10 The effect of inversion on the acetylene molecule

symmetry are the other three. If the changing of the coordinates x, y, z of all the nuclei to $-x$, $-y$, $-z$ produces a configuration which is indistinguishable from the starting one, then there is a centre of symmetry; the symbol for it is i and it is at the origin of the coordinate system. As an example we will consider the effect of carrying out this operation, called *inversion*, on the acetylene molecule shown in Figure 4.10. Inversion exchanges the two hydrogen atoms and the two carbon atoms and the molecule appears the same before and after the operation. The centre of symmetry is at the centre of the C≡C bond.

Since ΔJ is taken conventionally to refer to $J' - J''$, where J' is the quantum number of the upper state and J'' that of the lower state of the transition, the selection rule (b) in J is effectively $\Delta J = +1$. The transitions have wavenumbers or frequencies given by

$$\begin{aligned} \tilde{\nu}(\text{or } \nu) &= F(J') - F(J'') \\ &= F(J+1) - F(J) \\ &= 2B(J+1) \end{aligned} \tag{4.21}$$

where, as is conventional, J is used instead of the more cumbersome J''. The transitions allowed in absorption are shown in Figure 4.9: they are, from equation (4.21), spaced $2B$ apart with the first transition, written† as $J = 1$–0, at $2B$. Whether the spectrum of a particular molecule lies in the microwave, millimetre wave or far infrared region depends on the values of B and J. The observed spectrum of CO lies partly in the millimetre wave and partly in the far infrared region.

Part of the far infrared spectrum of CO, from 15 to 40 cm^{-1}, is shown in Figure 4.11. The transitions are those with J'' in the range 3–9.

The transitions with $J'' = 0$–6 have been observed with characteristically high accuracy in the millimetre wave region. The frequencies and wavenumbers are given in Table 4.1. The transitions with $J'' = 3$–6 have been observed with both millimetre wave and far infrared techniques and, while the uncertainty in $\tilde{\nu}$ is in only the fifth decimal place for the millimetre wave measurements, it is in the third for the far infrared measurements. This superiority of the accuracy of millimetre wave measurements is typical and demonstrates how important it is to extend millimetre wave spectroscopy to higher frequencies.

The separations between adjacent transitions are given in the last column of Table 4.1. They are very nearly constant and equal to $2B_0$,‡ as predicted from the rigid rotor theory. The average separation is 115.234 9 GHz and gives a value for B_0 of 57.617 5 GHz (or 1.921 91 cm^{-1}). However, the systematic but small decrease of the separation with J is much greater than experimental error and we shall return to it in Section 4.2.3.

Linear polyatomic molecules mostly have larger moments of inertia and therefore smaller B values than diatomic molecules. Therefore, more transitions in their pure rotation spectra occur in the microwave or millimetre wave regions. For example, the $J = 1$–0 and 42–41 transitions in O=C=S occur in the microwave and millimetre wave regions at 12.162 97 GHz and 510.457 3 GHz, respectively.

†Some authors replace – by ← or → to indicate absorption or emission, respectively, and others put the lower state first and the upper state second.

‡The subscript zero refers to the zero-point ($v = 0$) vibrational level. B varies with vibrational quantum number and this variation will be discussed in Section 4.2.6.

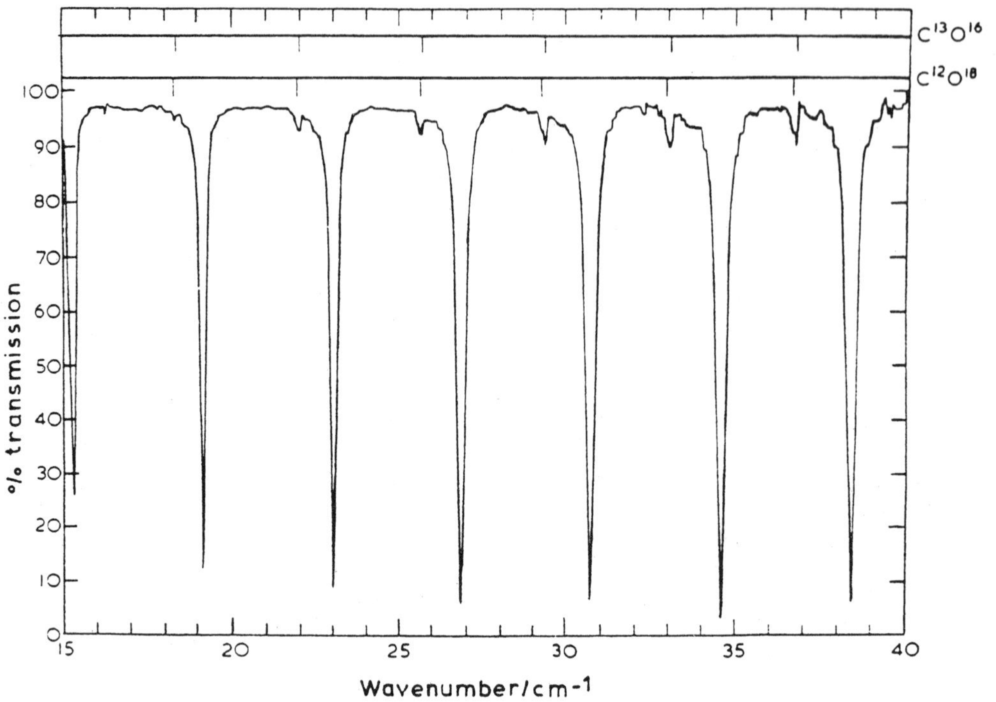

Figure 4.11 Part of the far infrared spectrum of the CO molecule showing transitions with $J'' = 3$ (at 15.38 cm^{-1}) or $J'' = 9$ (at 38.41 cm^{-1}) [Reproduced, with permission, from Fleming, J. W. and Chamberlain, J. (1974). *Infrared Phys.*, **14**, 277]

Another example of a linear polyatomic molecule is cyanodiacetylene (H–C≡C–C≡C–C≡N). This has such a small B_0 value (1331.331 MHz) that, as Figure 4.12 shows, six transitions with J'' in the range 9–14 lie in the R-band (26.5–40.0 GHz) of the microwave region. The fact that each transition is accompanied by many satellite transitions is due to some molecules being in vibrationally excited states; such satellites will be discussed further in Section 4.2.6.

Table 4.1 Frequencies and wavenumbers of pure rotational transitions of CO observed in the millimetre wave region[9,10]

$\tilde{\nu}/cm^{-1}$	J''	J'	ν/GHz[a]	$\Delta\nu_{J''}^{J''+1}$/GHz
3.845 033 19	0	1	115.271 195	115.271 195
7.689 919 07	1	2	230.537 974	115.266 779
11.534 509 6	2	3	345.795 900	115.257 926
15.378 662	3	4	461.040 68	115.244 78
19.222 223	4	5	576.267 75	115.227 07
23.065 043	5	6	691.472 60	115.204 85

[a]The uncertainties increase from 1.5×10^{-5} to 2×10^{-4} GHz with increasing J.

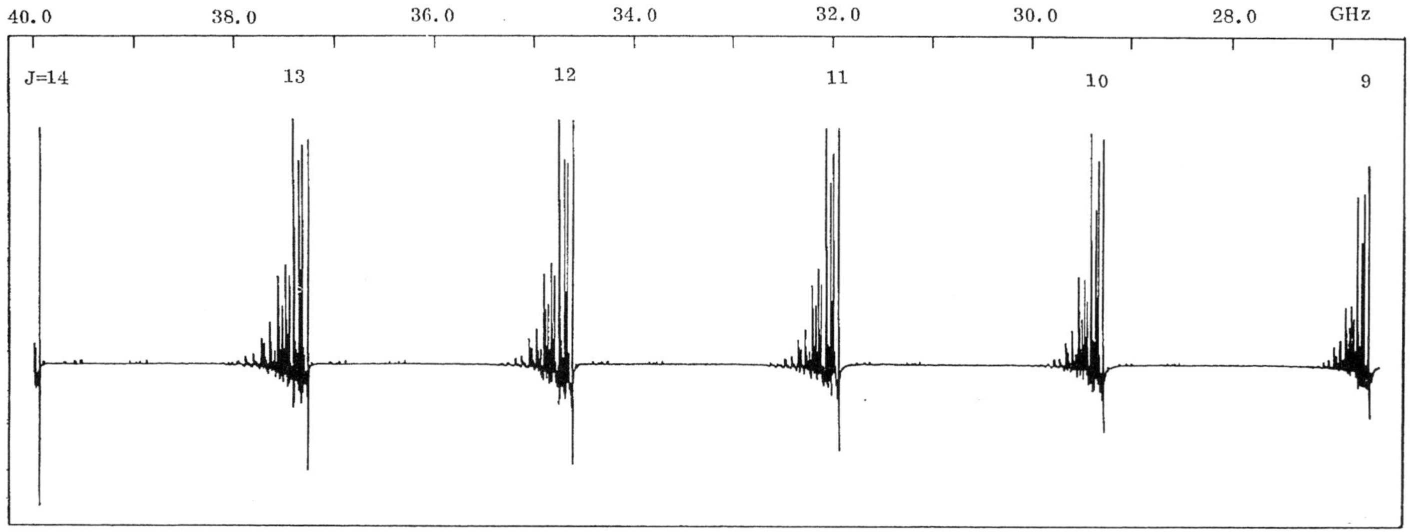

Figure 4.12 Part of the microwave spectrum of the cyanodiacetylene molecule [Reproduced, with permission, from Alexander, A. J., Kroto, H. W. and Walton, D. R. M. (1967). *J. Mol. Spectrosc.*, **62**, 175]

4.2.2 Intensities

The populations of rotational energy levels, relative to $J = 0$, are obtained from equation (2.8) and are given by

$$N_J/N_0 = (2J + 1)\exp(-E_r/kT) \tag{4.22}$$

where E_r is the rotational energy and the factor $(2J + 1)$ represents the degeneracy, in the absence of an electric or magnetic field, of each rotational energy level; this degeneracy is due to the fact that M_J [equation (1.89)] can take $(2J + 1)$ values. The values of N_J/N_0 for carbon monoxide, calculated from equations (4.15) and (4.22) assuming $B = 1.922\,\mathrm{cm}^{-1}$ and $T = 293\,\mathrm{K}$, are given in Figure 4.9: they show an increase up to $J = 7$ and a subsequent decrease. Such behaviour of rotational energy level populations is typical and is due to the fact that, at low J, the increase in N_J/N_0 due to the $(2J + 1)$ degeneracy factor in equation (4.22) dominates the decrease due to the exponential factor, but, at high J, the opposite is the case. The value of J_{max}, the J value of the energy level with the highest population, is obtained when

$$\mathrm{d}(N_J/N_0)/\mathrm{d}J = 0 \tag{4.23}$$

which gives

$$J_{max} \approx (kT/2hB)^{\frac{1}{2}} - \tfrac{1}{2} \tag{4.24}$$

for B having the dimensions of frequency. The approximation is because J_{max} must, of course, be an integer.

When considering rotational transitions accompanying either vibrational transitions (Chapter 5) or electronic transitions (Chapter 6) in an absorption spectrum, the intensity variation with J is given fairly accurately by the population variation given in equation (4.22) applied to the rotational levels of the initial state of the transition. For pure rotational transitions this is far from the case, although the population variation is an important factor.

Strictly, the intensities of pure rotational transitions are proportional to the fraction of the total number of molecules N which occupies each rotational level, rather than N_J/N_0. This fraction, N_J/N, can be obtained from equation (2.9) and is given by

$$N_J/N = [(2J + 1)\exp(-E_r/kT)]/q_r \tag{4.25}$$

where q_r is the rotational partition function given by

$$q_r = \sum_i g_i \exp[-(E_r)_i/kT] \tag{4.26}$$

and the summation is over all occupied rotational levels. At high temperatures, or for low B values and normal temperatures, many rotational levels have significant populations. Under these circumstances, the stack of rotational levels can be treated approximately as a continuum of levels and the summation can be replaced by an integral giving

$$q_r \approx \int_0^\infty [(2J + 1)\exp(-E_r/kT)]\,\mathrm{d}J$$
$$= kT/hB \tag{4.27}$$

for B having dimensions of frequency.

In the case of a set of vibrational or electronic energy levels of a molecule, the levels are often so widely separated that the summations corresponding to that in equation (4.26) are essentially equal to the first term, namely $g_0 \exp(-E_0/kT)$. If $E_0 = 0$ and $g_0 = 1$, then

$$q_v \approx 1$$

and

$$q_e \approx 1 \tag{4.28}$$

where q_v and q_e are the vibrational and electronic partition functions. Only in cases where there are low energy excited vibrational or electronic states can either of these partition functions be appreciably different from unity.

The intensity of absorption I_{abs} for a transition between a lower state m and an upper state n is the net result of stimulated absorption, stimulated emission and spontaneous emission

(Section 2.2). For the low frequency transitions typical of pure rotation spectra the population of state n is appreciable compared with that of m, so emission processes are important. However, at low frequencies spontaneous emission can be neglected and only stimulated emission is important.

The rate I_{mn} of stimulated absorption of energy is given by

$$I_{mn} = (N_m/N)B_{nm}\rho(\nu_{mn})h\nu_{mn} \tag{4.29}$$

where N_m/N is the fraction of molecules in state m and, as in Section 2.2, B_{nm} is the Einstein coefficient, $\rho(\nu_{mn})$ the radiation density of frequency ν_{mn} and $h\nu_{mn}$ the energy difference between the two states. The rate of stimulated emission I_{nm} is given by

$$I_{nm} = (N_n/N)B_{nm}\rho(\nu_{mn})h\nu_{mn} \tag{4.30}$$

so that the net absorption is

$$I_{\text{abs}} = I_{mn} - I_{nm} = (N_m - N_n)B_{nm}\rho(\nu_{mn})h\nu_{mn}/N \tag{4.31}$$

Using the Boltzmann relation between N_m and N_n, given in equation (2.8),

$$I_{\text{abs}} = N_m[1 - \exp(-h\nu_{mn}/kT)]B_{nm}\rho(\nu_{mn})h\nu_{mn}/N \tag{4.32}$$

If $\nu_{mn} \ll kT/h$, then

$$\exp(-h\nu_{mn}/kT) \approx 1 - (h\nu_{mn}/kT) \tag{4.33}$$

and equation (4.32) becomes

$$I_{\text{abs}} \approx N_m h^2 \nu_{mn}{}^2 B_{nm}\rho(\nu_{mn})/NkT \tag{4.34}$$

Since $kT/h = 6110$ GHz (and $kT/hc = 204\,\text{cm}^{-1}$) at a typical room temperature of 293 K, the approximation in equation (4.33) is a reasonable one in the microwave, millimetre wave and part of the far infrared regions at temperatures around 293 K or less.

In the rotation spectrum of a diatomic or any linear polyatomic molecule the square of the transition moment, $|\boldsymbol{R}^{nm}|^2$, varies with J according to

$$|\boldsymbol{R}^{nm}|^2 = \mu^2(J+1)/(2J+1) \tag{4.35}$$

where μ is the permanent dipole moment. Therefore, from equation (2.23),

$$B_{nm} = [8\pi^3/(4\pi\varepsilon_0)3h^2][\mu^2(J+1)/(2J+1)] \tag{4.36}$$

Combining equations (4.34) with equations (4.25), (4.27) and (4.36) gives

$$I_{\text{abs}} \approx \rho(\nu_{J''J'})[8\pi^3/3(4\pi\varepsilon_0)](hB/kT) \mu^2\nu_{J''J'}{}^2(J+1)\exp[-hBJ(J+1)/kT] \tag{4.37}$$

where $\nu_{J''J'}$ is the frequency of the transition between the rotational energy levels with quantum numbers J'' and J', and J is understood to stand for J''.

The most important difference between equation (4.22), giving population ratios of rotational energy levels, and equation (4.37) is the ν^2 factor; this has the effect of producing a maximum in the absorption intensity at a higher J value than J_{max} corresponding to the maximum population given by equation (4.24). In addition, the factor $(J+1)$ in equation (4.37) replaces $(2J+1)$ in equation (4.22). Values of $\tilde{\nu}^2(J+1)\exp[-hcBJ(J+1)/kT]$ calculated for CO ($B = 1.9314\,\text{cm}^{-1}$) are given in Figure 4.9 and show that, although the maximum population is at $J = 7$, the most intense transition is that with $J'' = 12$. This result is consistent with the observed far infrared spectrum (Figure 4.11) in which the intensity must be taken as the integrated area for each transition; since the linewidths change with wavenumber, peak intensities are not a very reliable indication of intensity.

Intensity measurements of pure rotation spectra in the far infrared have been used to determine the magnitude of the dipole moment μ. A value of 0.108 ± 0.005 D [$(3.60 \pm 0.17) \times 10^{-31}$ C m] has been obtained[11] in this way for the dipole moment of CO, which is in agreement with the value obtained using the Stark effect (Section 4.2.4) on the millimetre wave spectrum.

4.2.3 Centrifugal distortion

The separations between adjacent lines in the millimetre wave spectrum of CO, given in the last column of Table 4.1, show a steady decrease as J increases. The effect is due to centrifugal distortion.

In Chapter 1, the rigid rotor approximation for a diatomic molecule was used to obtain the rotational term values of equation (4.19). However, since the bond can stretch and contract in a vibrational motion, it cannot be entirely rigid. The effect of the bond being somewhat flexible is that the rotational motion tends to throw the nuclei outwards from the centre of mass due to centrifugal forces. Classically the effect increases with the speed of rotation and therefore, quantum mechanically, with the value of J. This centrifugal distortion of the molecule increases the internuclear distance r and, since B is proportional to I^{-1}, decreases B. The effect can be allowed for by replacing the term values of equation (4.19) by

$$F(J) = B[1 - uJ(J+1)]J(J+1) \tag{4.38}$$

where u is a constant. It is more usual, however, to write the term values in the form

$$F(J) = BJ(J+1) - DJ^2(J+1)^2 \tag{4.39}$$

D is the centrifugal distortion constant and is always positive for diatomic molecules. The transition wavenumbers or frequencies are now

$$\begin{aligned} \tilde{\nu}(\text{or } \nu) &= F(J+1) - F(J) \\ &= 2B(J+1) - 4D(J+1)^3 \end{aligned} \tag{4.40}$$

which corresponds to equation (4.21) for the rigid rotor.

From the transition frequencies of CO and allowing for centrifugal distortion, the values $B_0 = 57.635\,970\,\text{GHz}$ and $D_0 = 0.183\,90\,\text{MHz}$ have been obtained for the zero-point vibrational level.

The constant D is related to B and the vibration wavenumber $\omega^\dagger$ by

$$D = 4B^3/\omega^2 \tag{4.41}$$

We shall not derive this expression but it may be helpful in remembering it to use the fact that it must be, like all equations, dimensionally correct. Since B, D and ω all have the same dimensions, the power of B must be greater by one than that of ω.

When very accurate experimental data are used, even equation (4.39) is not satisfactory. As molecular vibration is not accurately described as that of a harmonic oscillator, but rather as that of an anharmonic oscillator (see Section 5.1.2), there are additional terms, in higher powers of $J(J+1)$, in the rotational term value expression which becomes

$$\begin{aligned} F(J) = BJ(J+1) &- DJ^2(J+1)^2 \\ &+ HJ^3(J+1)^3 + \ldots \end{aligned} \tag{4.42}$$

Subsequently we shall not be concerned with terms beyond $DJ^2(J+1)^2$.

4.2.4 The Stark effect

Space quantization of rotational angular momentum of a diatomic molecule, or of any linear polyatomic molecule, is expressed by

$$(P_J)_z = M_J\hbar \tag{4.43}$$

as already discussed in Section 1.3.3; $(P_J)_z$ is the z-component of the angular momentum P_J and M_J can take the values $J, J-1, \ldots, -J$. In the presence of an electric field $\mathcal{E}$ the $(2J+1)$-fold degeneracy is partly removed. The

†This symbol will be discussed in Chapter 5 but, in a harmonic oscillator, it is equivalent to $\tilde{\nu}$ in equation (1.99).

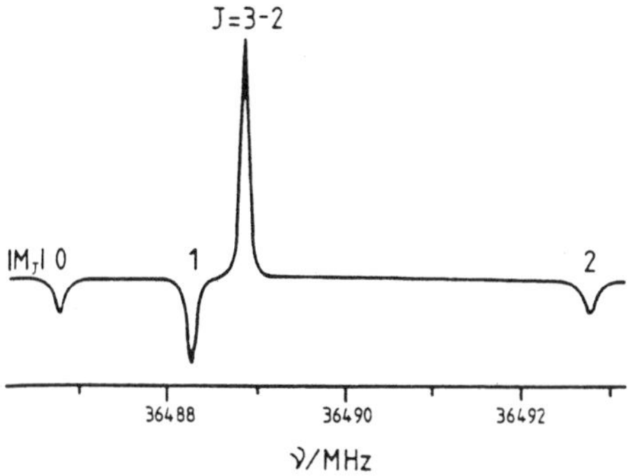

Figure 4.13 The Stark effect on the $J = 3\text{–}2$ microwave transition of the OCS molecule

rotational energy levels E_r, given in equation (4.15), are modified to $E_r + E_{\mathcal{E}}$, where

$$E_{\mathcal{E}} = \mu^2\mathcal{E}^2[J(J+1) - 3M_J^{\,2}]/[2hBJ(J+1)(2J-1)(2J+3)] \tag{4.44}$$

in which μ is the permanent electric dipole moment of the molecule. The effect of the splitting of the levels in an electric field is called the Stark effect and, because $E_{\mathcal{E}}$ is proportional to the square of the field, it is a second-order Stark effect. Because $E_{\mathcal{E}}$ [equation (4.44)] contains only an M_J^2 term involving M_J, it depends on the magnitude but not the sign of M_J. For this reason, each J level is split into only $(J+1)$ components having $|M_J| = 0, 1, 2, \ldots, J$. The selection rules governing transitions between the components are:

$\Delta M_J = 0$, for $\mathcal{E}$ polarized parallel to the oscillating electric field of the radiation;
$\Delta M_J = \pm 1$, for $\mathcal{E}$ polarized perpendicular to the oscillating electric field of the radiation.

The technique (see Section 3.3) is used mainly in microwave or millimetre wave spectroscopy in which the radiation is usually polarized, this being caused by the rectangular cross-section of the waveguide. In most spectrometers the electric field of the radiation oscillates in a direction perpendicular to the Stark electrode and therefore parallel to the Stark field. Under these conditions, transitions with $\Delta M_J = 0$ only are observed.

Figure 4.13 shows the Stark effect on the $J = 3\text{–}2$ transition of OCS. The $J = 3$ level is split into four components with $|M_J| = 0, 1, 2, 3$ and the $J = 2$ level into three components with $|M_J| = 0, 1, 2$. The $\Delta M_J = 0$ selection rule results in the three Stark components, or Stark lobes, shown in Figure 4.13. An accurate value for μ obtained in this way for OCS is $0.715\,21 \pm 0.000\,20$ D [$(2.385\,7 \pm 0.000\,7) \times 10^{-30}$ C m]. The accuracy with which any dipole moment is known from the Stark effect is often limited by the accuracy with which the field $\mathcal{E}$ is known; this in turn is limited by the difficulty of knowing the separation of the Stark electrode from the walls of the cell. For this reason, a molecule, such as OCS whose dipole moment is known accurately, may be used to determine $\mathcal{E}$.

Only the magnitude and *not the sign* of the dipole moment can be found from the Stark affect. In OCS the dipole moment is large enough for us to be confident that the fact that oxygen is more electronegative than sulphur makes oxygen the negative end. However, in CO, for which μ has the very small value of 0.112 D (3.74×10^{-31} C m), no simple argument will lead us to the correct conclusion that the carbon atom is at the negative end. Only very accurate theory is able to provide this information.

Compared with other methods, the Stark effect is the most accurate way of measuring dipole moments. This method also has the advantages of being applicable to an impure sample, of not requiring any accurate temperature or pressure measurements and of giving the dipole moment of the free molecule which is not undergoing any electrostatic interactions with neighbouring molecules as, for example, in the liquid phase. In addition, the dipole moment can be measured in excited vibrational states giving information on the way in which the vibrational motion deviates from that of a harmonic oscillator.

4.2.5 Nuclear hyperfine splitting

We have seen in Section 1.3.3 that nuclei have spin angular momentum whose magnitude P_I is given by

$$P_I = [I(I+1)]^{\frac{1}{2}}\hbar \tag{4.45}$$

For some nuclei $I = 0$, and therefore $P_I = 0$, but for others I is integral or half-integral (see Table 1.3 for some examples).

The nuclear spin angular momentum is a vector denoted by $\boldsymbol{I}$. The rotational angular momentum is also a vector, denoted by $\boldsymbol{J}$. Both these kinds of angular momenta are caused by the rotation of charged particles and therefore each creates a magnetic field. These magnetic fields may interact, causing coupling of the angular momenta to give a resultant total angular momentum (in this case rotational plus nuclear spin). The coupling can be represented by addition of the vector $\boldsymbol{I}$, of magnitude $[I(I+1)]^{\frac{1}{2}}\hbar$, and $\boldsymbol{J}$, of magnitude $[J(J+1)]^{\frac{1}{2}}\hbar$, to give the total angular momentum $\boldsymbol{F}$, of magnitude P_F, where

$$P_F = [F(F+1)]^{\frac{1}{2}}\hbar \tag{4.46}$$

and F can take the values

$$F = J+I, J+I-1, \ldots, |J-I| \tag{4.47}$$

If $J > I$ then the lowest value of F is $J - I$ but, if $J < I$, the lowest value of F is $I - J$. The series of values which F can take is called a Clebsch–Gordan series; we shall encounter several similar series in performing vector additions of various angular momenta. Figure 4.14 illustrates the vector addition of $\boldsymbol{I}$ and $\boldsymbol{J}$ and the precession of the vectors $\boldsymbol{I}$ and $\boldsymbol{J}$ around the resultant vector $\boldsymbol{F}$.

All nuclei with $I > \frac{1}{2}$ such as ^{35}Cl and ^{37}Cl, both of which have $I = \frac{3}{2}$, and ^{14}N, for which $I = 1$, have an electric quadrupole moment Q. This is due to an asymmetry of the nuclear charge caused by either an elongation of the nucleus along the axis of spin to give a shape like a cigar, or a flattening giving it a shape like

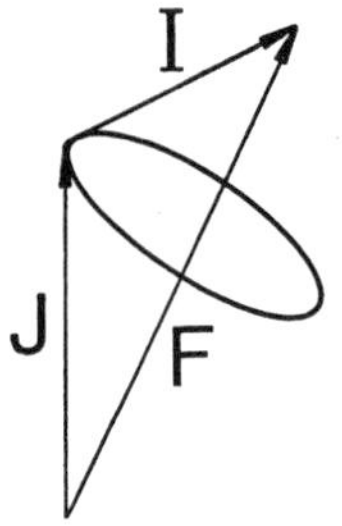

Figure 4.14 Vector addition of the rotational, $\boldsymbol{J}$, and nuclear spin, $\boldsymbol{I}$, angular momenta to give the total angular momentum, $\boldsymbol{F}$

a discus. An elongated nucleus has a positive value of Q and a flattened nucleus a negative value. The quadrupole moment interacts with the gradient of the electric field $\mathcal{E}$ surrounding it, this field being due to the electrons, especially the valence electrons, in the molecule. In a linear molecule the field gradient q along the z-axis, the axis containing the nuclei, is given by

$$q = \partial\mathcal{E}_z/\partial z \tag{4.48}$$

This field gradient causes a modification of the rotational energy levels E_r to $E_r + E_Q$, where

$$E_Q = -eQq[\tfrac{3}{4}C(C+1) - I(I+1)J(J+1)] \\ [2I(2I-1)(2J-1)(2J+3)]^{-1} \tag{4.49}$$

where e is the charge on the electron and

$$C = F(F+1) - I(I+1) - J(J+1) \tag{4.50}$$

The quantity multiplying $-eqQ$ in equation (4.49) is known as Casimir's function.

A fairly simple example of the effect on the spectrum of nuclear hyperfine splitting is shown for the $J = 3$–2 transition of $^{35}Cl\,^{12}C\,^{15}N$ in Figure 4.15. This particular molecule has been chosen because it has only one nucleus, ^{35}Cl, with a quadrupole moment (^{14}N, for which $I = 1$, has a quadrupole moment but ^{15}N, for which $I = \frac{1}{2}$, has not). Since $I = \frac{3}{2}$ for ^{35}Cl, $F = \frac{7}{2}, \frac{5}{2}, \frac{3}{2}, \frac{1}{2}$ for the $J = 2$ level and $F = \frac{5}{2}, \frac{3}{2}, \frac{1}{2}$ for $J = 1$. The selection rule governing transitions between levels split by the effects of nuclear spin is

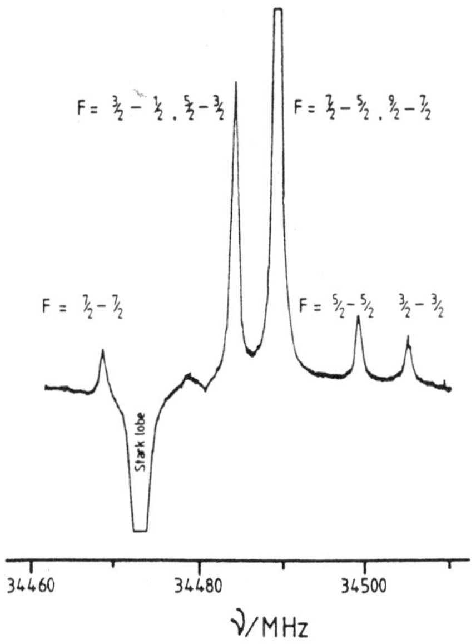

Figure 4.15 Nuclear hyperfine splitting of the $J = 3$–2 microwave transition of the $^{35}Cl\,^{12}C\,^{15}N$ molecule

$$\Delta F = 0, \ \pm 1$$

Therefore, there are nine hyperfine components of the $J = 3$–2 transition, four with $\Delta F = +1$, three with $\Delta F = 0$ and two with $\Delta F = -1$. These are indicated in Figure 4.15 except for those with $\Delta F = -1$, which are too weak to be observed. The $\Delta F = +1$ transitions are clearly the most intense and, at higher J, this effect is even more pronounced. In general, the most intense transitions are those with $\Delta F = \Delta J$.

The most important and, sometimes, troublesome consequence of nuclear hyperfine splitting is that, for any observed transition in which it occurs, the pattern has to be analysed in order to determine the frequency at which the transition would have occurred in the absence of any splitting. It is only these frequencies which can be used in equation (4.40) to obtain B and D.

The values of eqQ obtained from nuclear hyperfine splitting are of some theoretical significance since they give information regarding the field gradient in the region of the quadrupolar nucleus. In the free atom, electrons in filled $s, p, d, \ldots$ orbitals do not contribute to the field gradient since they are spherically symmetrical. For the same reason, one electron in an s orbital does not contribute. Hence the field gradient is likely to be due to p or d electrons in partially filled orbitals. The field gradient caused by these electrons is modified when the atom is contained in a molecule and the electrons are involved in bonding. The resulting value of eqQ compared with that in the free atom can be interpreted in terms of ionic or covalent character of the bond and also in terms of the degree of hybridization in the molecular orbitals.

4.2.6 Vibrational satellites

4.2.6.1 Diatomic molecules

The stack of rotational energy levels with which we have been concerned so far, such as that in Figure 4.9, is associated with the $\upsilon = 0$ vibrational level and all the rotational transitions we have discussed are within the $\upsilon = 0$ rotational stack. However, there is a similar stack associated with each vibrational energy level.

We have seen in Section 1.3.4 that the vibrational energy of a diatomic molecule is quantized and, in the harmonic oscillator approximation, is given by $hc\tilde{\nu}(\upsilon + \frac{1}{2})$ according to equation (1.99). The energy levels are illustrated in Figure 1.22. If the vibration wavenumber $\tilde{\nu}$ is sufficiently small, vibrational levels with $\upsilon > 0$ may be appreciably populated at room temperature. The population N_υ of the υth vibrational level relative to the population N_0 of the zero-point level can be obtained from equation (2.6) and is given by

$$\begin{aligned} N_\upsilon/N_0 &= \exp\{[-hc\tilde{\nu}(\upsilon + \tfrac{1}{2}) + \tfrac{1}{2}hc\tilde{\nu}]/kT\} \\ &= \exp(-hc\upsilon\tilde{\nu}/kT) \qquad (4.51) \end{aligned}$$

For example, if $N_\upsilon/N_0 = 0.10$, then $\upsilon\tilde{\nu} = 470\,\mathrm{cm}^{-1}$ at a temperature $T = 293\,\mathrm{K}$. If, in a particular molecule, this were true for the $\upsilon = 1$

level, then we would expect to observe pure rotational transitions in the $\upsilon = 1$ rotational stack with one tenth of the intensity of those in the $\upsilon = 0$ stack. In general, we can observe rotational transitions in excited vibrational states provided that the vibrational energy level is sufficiently populated. If this is not possible at room temperature then heating of the sample cell may increase the population sufficiently. However, equation (4.51) shows that if $\upsilon\tilde{\nu}$ is 940 cm^{-1} we have to double the temperature to maintain $N_\upsilon/N_0 = 0.10$. This example shows that increasing the temperature is helpful in increasing the population of excited vibrational states; however, the experimental problems associated with heating the cell, together with the fact that $\upsilon = 1$ levels may be more than 3000 cm^{-1} above $\upsilon = 0$ in some molecules, mean that the experimentalist may easily be fighting a losing battle in trying to increase appreciably the $\upsilon = 1$ population.

There is another powerful, but more limited, way of increasing vibrational level populations. This is by using an infrared laser (Chapter 8) whose intense monochromatic radiation is of a wavenumber which exactly matches the interval between the $\upsilon = 0$ level and the vibrational level whose population we wish to increase.

The values of the rotational constants B and D are both dependent on the vibrational state of the molecule so that equation (4.39) for the rotational term values should be written as

$$F_\upsilon(J) = B_\upsilon J(J+1) - D_\upsilon J^2(J+1)^2 \qquad (4.52)$$

and equation (4.40) becomes

$$\tilde{\nu}(\text{or } \upsilon) = 2B_\upsilon(J+1) - 4D_\upsilon(J+1)^3 \qquad (4.53)$$

The vibrational dependences of B and D are given by†

$$B_\upsilon = B_\mathrm{e} - \alpha(\upsilon + \tfrac{1}{2}) + [\gamma(\upsilon + \tfrac{1}{2})^2 + \ldots] \qquad (4.54)$$

and

$$D_\upsilon = D_\mathrm{e} - [\beta(\upsilon + \tfrac{1}{2}) - \ldots] \qquad (4.55)$$

B_e and D_e are the values these rotational constants have in the hypothetical equilibrium configuration of the molecule, at the bottom of the potential curve shown in Figure 1.22.

Centrifugal distortion is itself a small effect and the vibrational dependence of D_υ is a still smaller effect, but measurable in microwave and millimetre wave spectroscopy. The vibrational dependence of B_υ is a much larger effect; for example, in the short-lived molecule $^{12}C^{32}S$, $B_\mathrm{e} = 245\,84.367$ MHz, $\alpha = 177.550$ MHz and $D_\mathrm{e} = 42.9$ kHz.

It is important in observing small J-dependent effects to be able to record transitions with high values of J; for these purposes millimetre wave studies are of great importance in obtaining accurate values of D_e, β, α, γ, etc.

In subsequent discussions we shall disregard the very small terms in square brackets in equations (4.54) and (4.55).

The effect of the vibration–rotation interaction constant α is to separate pure rotational transitions with the same value of J'' but occurring in different vibrational states. Transitions are separated in far infrared and also microwave and millimetre wave spectra but, because of the lower resolution, not many cases have been recorded in the far infrared.

In order to obtain B_e and α for a particular molecule, B_υ is required in at least two, and preferably more, vibrational states; similarly D_e and β can be obtained only if at least two values of D_υ are known. From what has been said about vibrational level populations it is clear that, for diatomic molecules, B_e, α, D_e and β can be obtained from pure rotational spectroscopy alone only for relatively heavy molecules with low vibration wavenumbers. Many of these molecules, such as the alkali metal halides, have very low vapour pressures at room

†Note the *negative* sign which is used conventionally in front of the $\alpha(\upsilon + \frac{1}{2})$ term. This convention has arisen because α is then always positive.

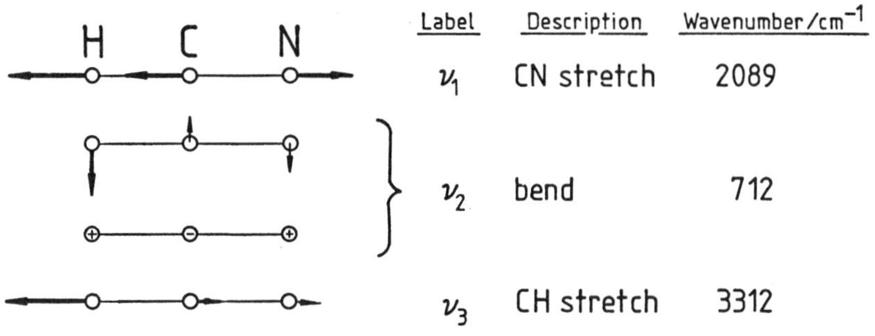

Figure 4.16 The normal modes of vibration of the HCN molecule

temperature so that the high temperatures necessary for achieving sufficiently high vapour pressure have the incidental effect of increasing populations of excited vibrational levels.

4.2.6.2 Linear polyatomic molecules

The rotational term values for linear polyatomic molecules are, like those for diatomics, dependent on the vibrational state and are also given, for any non-degenerate state, by equation (4.52). However, an N-atomic linear polyatomic molecule has $(3N - 5)$ vibrations (see Section 1.3.1) and B_υ is given by

$$B_\upsilon = B_e - \sum_i \alpha_i(\upsilon_i + d_i/2) \qquad (4.56)$$

where the summation is over all the vibrations and d_i is the degeneracy of the ith vibration.

We shall not be dealing with vibrational motion in detail until Chapter 5, but we require here to use the results of solving the equations of motion for vibration in some simple cases.

In the linear molecule HCN there are four normal modes of vibration. These are illustrated in Figure 4.16, in which the arrows attached to the nuclei are vectors representing the direction and relative magnitudes of the motions of the nuclei in the various normal modes. Both the CN stretching and CH stretching vibrations, ν_1 and ν_3, respectively, are non-degenerate [$d_i = 1$ in equation (4.56)] but the bending vibration ν_2 is doubly degenerate ($d_i = 2$). The reason for the double degeneracy is that if the bending occurs in the xz-plane, where z is the direction of the internuclear axis, then the vectors representing this motion have no components in the yz-plane. Similarly, bending in the yz-plane has no component in the xz-plane. Therefore, vibrations in the xz- and yz-planes are independent but are degenerate, which means that they have different wave functions, $\psi_\upsilon(xz)$ and $\psi_\upsilon(yz)$, but it requires the same energy to excite each of them.

Bending vibrations of all linear molecules are of particular importance in the observation of rotational transitions in excited vibrational states because, like ν_2 in HCN, they have fairly low fundamental wavenumbers. This means that at least the $\upsilon = 1$ level for such a vibration may be appreciably populated at normal temperatures. They are also important because there is an additional angular momentum associated with such a vibration.

We can see from Figure 4.16 that a rotation, clockwise or anti-clockwise, of the vectors representing the ν_2 vibrational motion of HCN converts one of the motions illustrated into the other. Therefore, when the molecule is in any excited vibrational state involving ν_2 there is a vibrational angular momentum about the internuclear axis. The magnitude of this momentum is $l\hbar$, where, in general, for a vibration ν_i

$$l = v_i, v_i - 2, v_i - 4, \ldots, -v_i$$

in which v_i is the vibrational quantum number for the normal mode ν_i. For example, in HCN:

for $v_2 = 1$; $l = \pm 1$
for $v_2 = 2$; $l = 0, \pm 2$
for $v_2 = 3$; $l = \pm 1, \pm 3$

Taking into account this additional angular momentum, the term values of equation (4.52) are modified to

$$F_v(J) = B_v[J(J+1) - l^2] - D_v[J(J+1) - l^2]^2 \tag{4.57}$$

where

$$J = |l|, |l| + 1, |l| + 2, \ldots$$

which means that the $J = 0$ level is missing for $v_2 = l = 1$, as are the $J = 0$, 1 levels for $v_2 = l = 2$, and so on.

According to equation (4.57), the term values are independent of the sign of l, but this is true only if we neglect Coriolis interaction.

If a nucleus in a molecule is moving linearly in a vibrational motion and the molecule is also rotating, then the nucleus experiences a Coriolis force f_c given by

$$f_c = 2mv_a\omega \sin \phi \tag{4.58}$$

where m is the mass of the nucleus, v_a its apparent velocity with respect to the coordinate system attached to the rotating molecule, ω the angular velocity of rotation referred to a fixed coordinate system and ϕ the angle between the axis of rotation and the direction of motion with velocity v_a. Clearly, if $\phi = 0$, there is no Coriolis force. Figure 4.17(a) shows the Coriolis forces experienced by the nuclei of HCN when the molecule is vibrating in the ν_1 normal mode and also rotating about an axis perpendicular to the plane of the figure and through the centre of gravity. The effect of the Coriolis forces is to make the molecule bend, but with a wavenumber the same as that for ν_1. If the fundamental wavenumbers of ν_1 and ν_2 were similar then ν_1 would be excited simultaneously with ν_2.

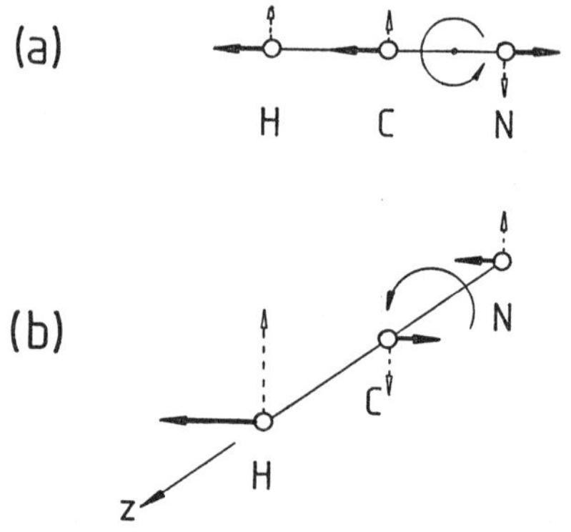

Figure 4.17 Coriolis interaction between (a) ν_1 and ν_2 and (b) the two components of ν_2 of the HCN molecule. Vibrational motion excited by Coriolis interaction is shown by broken lines

There are many cases known in which Coriolis interaction occurs between two fundamental vibrations accidentally having similar wavenumbers. The effect is to distort the rotational energy levels and transitions. In the case of HCN, Coriolis interaction between ν_1 and ν_2 is small because of their very different wavenumbers, 2089 and 712 cm^{-1}, respectively. However, the two components of ν_2 undergo a special type of Coriolis interaction. As Figure 4.17(b) shows, the rotational angular momentum about the z-axis due to excitation of ν_2 creates Coriolis forces on the nuclei which convert one component of ν_2 into the other. Since the two interacting vibrations in this case have the same wavenumber the interaction is large—a so-called first-order Coriolis interaction. The effect of this is to modify the term values of equation (4.57) to

$$F_v(J, l^{\pm}) = B_v[J(J+1) - l^2] \pm (q_i/4)(v_i + 1)J(J+1) - D_v[J(J+1) - l^2]^2 \tag{4.59}$$

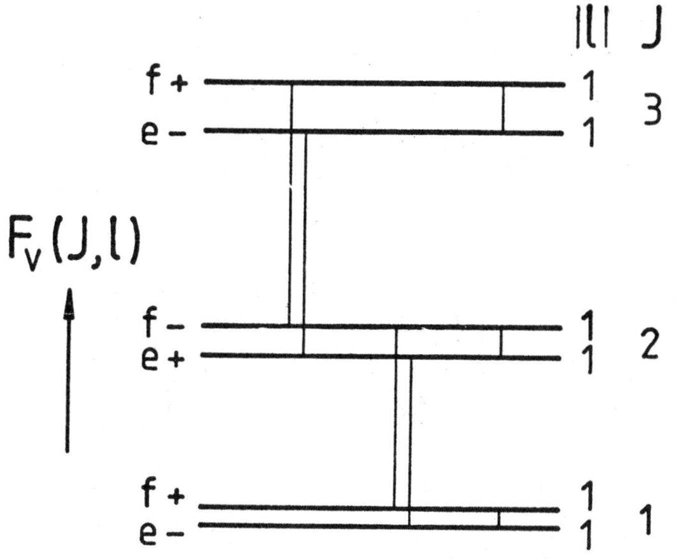

Figure 4.18 The effects of l-type doubling on rotational levels in a linear molecule. The transitions shown are allowed by electric dipole selection rules

where q_i is a parameter which depends on the coupling between the rotational and vibrational motions for vibration i (in this case $i = 2$). The effect of the term involving q_i is to split the term values which differ only in the sign of l by an amount equal to $(q_i/2)(v_i + 1)J(J + 1)$. The splitting is known as l-type doubling. It decreases with increasing $|l|$ and is appreciable only for $|l| = 1$. Some of the rotational energy levels for $|l| = 1$ are illustrated in Figure 4.18.

The components of the doubled levels are distinguished by their parity and are labelled + or −. This label refers to behaviour of the wave function ψ, which is the product $\psi_e\psi_n$ of the wave function for electronic and nuclear motions, under the operation of inversion with respect to space-fixed axes. The allowed transitions are governed by the parity selection rule:[†]

$$+ \longleftrightarrow -, \quad + \not\longleftrightarrow +, \quad - \not\longleftrightarrow -$$

[†] $\longleftrightarrow$ indicates that the transition is allowed, whichever is the upper state, and $\not\longleftrightarrow$ indicates that it is forbidden.

An alternative convention[‡] for labelling the levels is that those with parity $+(-1)^J$ are labelled e and those with parity $-(-1)^J$ are labelled f. For half-integral values of J, in molecules with non-zero electron spin, $+(-1)^{J-\frac{1}{2}}$ levels are labelled e and $-(-1)^{J-\frac{1}{2}}$ levels are labelled f. These labels are also given in Figure 4.18. The allowed transitions are then

$$e \longleftrightarrow f,\ e \not\longleftrightarrow e,\ f \not\longleftrightarrow f \quad \text{for } \Delta J = 0$$
$$e \not\longleftrightarrow f,\ e \longleftrightarrow e,\ f \longleftrightarrow f \quad \text{for } \Delta J = \pm 1$$

Whichever way the selection rules are formulated, it is clear that the allowed transitions are those shown in Figure 4.18. These include direct l-doublet transitions, between the upper and lower components, with $\Delta J = 0$. The frequency ν at which such a transition is observed is given by

$$\nu = (q_i/2)(v_i + 1)J(J + 1) \qquad (4.60)$$

[‡]For a discussion of conventions, see Brown, J. M. *et al.* (1975). *J. Mol. Spectrosc.*, **55**, 500.

The parameter q_i is small and for a linear molecule with three dissimilar atoms, such as HCN, q_2 for the bending vibration is given, for $|l| = 1$, by

$$q_2 = \frac{2B_e^2}{\omega_2}\left\{1 + 4\sum_{i \neq 2} \zeta_{2i}^2[\omega_2^2/(\omega_i^2 - \omega_2^2)]\right\} \tag{4.61}$$

where the ζ_{2i}, namely ζ_{21} and ζ_{23}, are Coriolis coupling constants. These are a measure of the Coriolis interactions between ν_2 and ν_1 and between ν_2 and ν_3, respectively. Both of these interactions usually make only a small contribution to q_2, since $(\omega_i^2 - \omega_2^2)$ is large, so q_2 can be written approximately as

$$q_2 \approx 2B_e^2/\omega_2 \tag{4.62}$$

Comparison of equations (4.61) and (4.62) shows that much the greatest contribution to q_2 is due to the first-order Coriolis interaction between the two degenerate components of ν_2. Equation (4.62) shows that q_i tends to be large for light molecules since these have large values of B_e, In HCN, for example, $q_2 = 224.47$ MHz, and transitions between the two components of the l-doubled levels lie in the microwave region for low J, but in the linear molecule HCNO, $q_4 = 23.6722$ MHz and $q_5 = 34.6391$ MHz for the bending vibrations ν_4 and ν_5 so that such transitions are of sufficiently high frequency to fall in the microwave region only for fairly high values of J. Investigations at high J have shown that small extra terms must be added to equation (4.60).

4.3 PURE ROTATIONAL INFRARED, MILLIMETRE WAVE AND MICROWAVE SPECTRA OF SYMMETRIC ROTOR MOLECULES

In a diatomic or linear polyatomic molecule, the angular momentum vector $\boldsymbol{P}$ due to the rotation of the molecule lies along the axis of rotation, as illustrated in Figure 4.19(a) (this figure is analogous to Figure 1.7 for electronic orbital angular momentum). In a prolate symmetric rotor, defined by equation (4.8) and illustrated in Figure 4.19(b) by the example of methyl iodide, the rotational angular momentum vector $\boldsymbol{P}$ need not be perpendicular to the top axis, which is the a-axis in CH_3I. The molecule rotates (or nutates) about the axis of $\boldsymbol{P}$. This axis passes through the centre of gravity of the molecule and is to be regarded as fixed in space and not to the molecule. In addition, the molecule rotates about the a-axis and the vector $\boldsymbol{P}_a$ represents the magnitude and direction of the angular momentum due to this motion. Solution of the Schrödinger equation for this system in any non-degenerate vibrational state gives the term values

$$F_v(J, K) = B_vJ(J + 1) + (A_v - B_v)K^2 \tag{4.63}$$

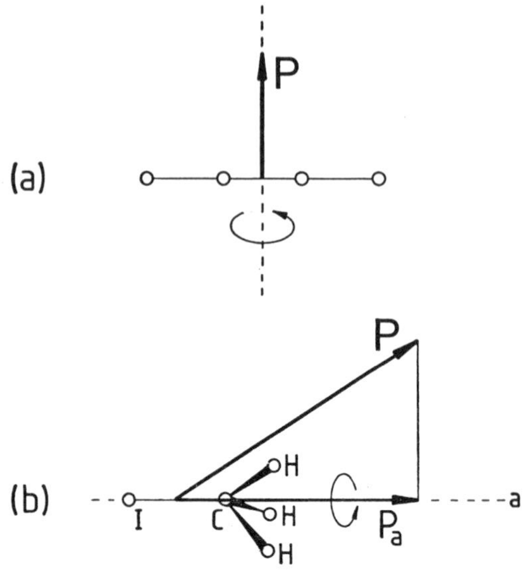

Figure 4.19 (a) The angular momentum vector $\boldsymbol{P}$ due to end-over-end rotation of a linear molecule. (b) The angular momentum vector $\boldsymbol{P}$, and the component P_a along the top axis, in the prolate symmetric rotor CH_3I

for the rigid rotor model in which centrifugal distortion has been neglected. The rotational constants A and B are related to I_a and I_b by

$$A = h/8\pi^2 I_a; \quad B = h/8\pi^2 I_b \tag{4.64}$$

if they have dimensions of frequency, and by

$$A = h/8\pi^2 cI_a; \quad B = h/8\pi^2 cI_b \tag{4.65}$$

if they have dimensions of wavenumber. In both cases $A > B = C$.

The quantum number K can take the values 0, 1, 2, 3, . . ., J. The fact that K cannot be greater than J follows from the fact that the magnitude of the vector $\boldsymbol{P}_a$ cannot be greater than that of $\boldsymbol{P}$. All rotational levels with $K > 0$ possess a double degeneracy which can be thought of, classically, as being due to clockwise or anti-clockwise rotation about the top axis resulting in the same magnitude of the angular momentum. For $K = 0$ there is no angular momentum about the top axis and therefore no K degeneracy.

An alternative convention is to replace K by a signed quantum number k which can take the values 0, ±1, ±2, ±3, . . ., ±J. In equation (4.63) it does not matter whether we use K or k since it contains only K^2, but when we discuss l-type doubling in symmetric rotors, it will be convenient to use k.[†]

There is a strong similarity between equations (4.63) and (4.57) when the centrifugal distortion term in the latter is neglected. This is because the vibrational angular momentum about the main axis of a linear molecule, taken account of in equation (4.57), is analogous to that about the top axis of a symmetric rotor due to the nuclei which do not lie on the axis. The only difference is that in a linear molecule in an excited state of a bending vibration the $A_v K^2$ term is usually incorporated into the vibrational term value.

For an oblate symmetric rotor, such as NH_3, the rotational term values for any non-degenerate vibrational state are

$$F_v(J, K) = B_v J(J+1) + (C_v - B_v)K^2 \tag{4.66}$$

where

$$C = h/8\pi^2 I_c \text{ or } h/8\pi^2 cI_c \tag{4.67}$$

and $A = B > C$.

The energy levels for a prolate and an oblate symmetric rotor are shown schematically in Figure 4.20(a) and (b), respectively. The main difference between the two sets of energy levels is that the levels for a given J and different values of K are more widely spaced with increasing K for a prolate symmetric rotor, because $A - B$ is always positive, than for an oblate symmetric rotor, for which $C - B$ is always negative.

The selection rules governing pure rotational transitions between levels in a symmetric rotor are

$$\Delta J = \pm 1; \quad \Delta K = 0$$

The result is that transition frequencies or wavenumbers for a prolate or oblate symmetric rotor are given by

$$\begin{aligned} \nu(\text{or } \tilde{\nu}) &= F_v(J+1, K) - F_v(J, K) \\ &= 2B_v(J+1) \end{aligned} \tag{4.68}$$

This equation is the same as equation (4.21) for a diatomic or linear polyatomic molecule. The result is a series of equally spaced lines separated by $2B_v$ but, since $K = 0, 1, 2, 3, \ldots, J$, each line contains $(J+1)$ components.

In practice, the symmetric rotor spectrum is not so simple. The effect of centrifugal distortion is to modify the term value equation (4.63) for a prolate symmetric rotor to

$$\begin{aligned} F_v(J, K) = {} & B_v J(J+1) + (A_v - B_v)K^2 \\ & - D_J J^2(J+1)^2 - D_{JK} J(J+1)K^2 \\ & - D_K K^4 \end{aligned} \tag{4.69}$$

[†]Some authors use the symbol K for both the signed and the unsigned quantum number.

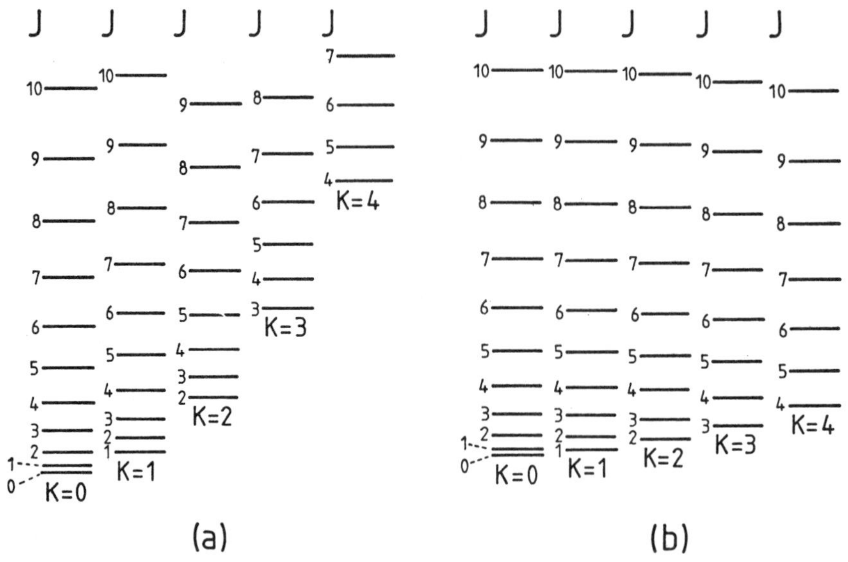

Figure 4.20 Rotational energy levels for (a) a prolate and (b) an oblate symmetric rotor

where the slight vibrational dependence of the centrifugal distortion constants D_J, D_{JK} and D_K has, for simplicity, not been indicated. Similarly, the term value equation (4.66) for an oblate symmetric rotor becomes

$$F_v(J, K) = B_v J(J+1) + (C_v - B_v)K^2 - D_J J^2(J+1)^2 - D_{JK} J(J+1)K^2 - D_K K^4 \quad (4.70)$$

Application of the selection rules $\Delta J = \pm 1$ and $\Delta K = 0$ gives the transition frequencies or wavenumbers in both prolate and oblate symmetric rotors as

$$\nu(\text{or } \tilde{\nu}) = F_v(J+1, K) - F_v(J, K) = 2(B_v - D_{JK}K^2)(J+1) - 4D_J(J+1)^3 \quad (4.71)$$

The term $-2D_{JK}K^2(J+1)$ has the effect of separating the $(J+1)$ components of each $(J+1)$–J transition. The effect is illustrated in Figure 4.21, which shows the eight components of the $J = 8$–7 transition in the vibrational ground state of silyl isothiocyanate (SiH_3–$N{=}C{=}S$), which has a linear SiNCS chain and is a prolate symmetric rotor.

Equation (4.71) does not involve either A_v (or C_v) or D_K, so that neither of these quantities can be obtained from the pure rotational spectrum.

When an electric field $\mathcal{E}$ is applied to a symmetric rotor the rigid rotor energy levels, corresponding to the term values of equations (4.63) and (4.66), are modified from E_r, in the absence of the field, to $E_r + E_{\mathcal{E}}$, where

$$E_{\mathcal{E}} = -\frac{\mu \mathcal{E} K M_J}{J(J+1)} + \frac{\mu^2 \mathcal{E}^2}{2hB}\left\{\frac{(J^2 - K^2)(J^2 - M_J^2)}{J^3(2J-1)(2J+1)} - \frac{[(J+1)^2 - K^2][(J+1)^2 - M_J^2]}{(J+1)^3(2J+1)(2J+3)}\right\} \quad (4.72)$$

in which all the symbols, apart from K, have the same significance as in equation (4.44), in which

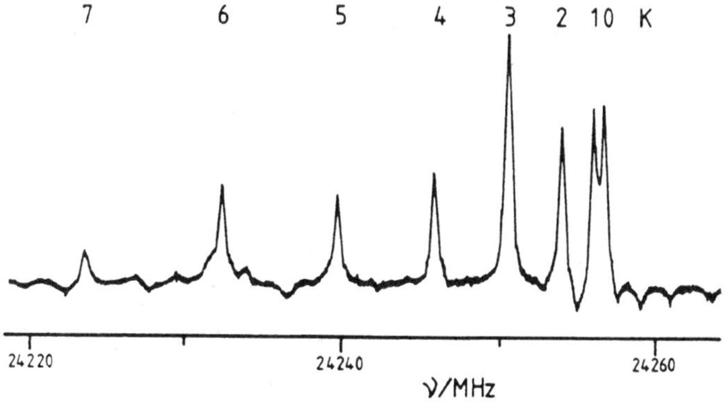

Figure 4.21 The $J = 8\text{–}7$ microwave transition in the vibrational ground state of the SiH_3NCS molecule showing the eight components split by centrifugal distortion

the expression for $E_{\mathcal{E}}$ in a linear molecule is given. By putting $K = 0$ in equation (4.72), we obtain equation (4.44). Because there is a term in equation (4.72) which is proportional to $\mathcal{E}$ and another which is proportional to $\mathcal{E}^2$, there is, for $K \neq 0$, a first- and second-order Stark effect in a symmetric rotor.

Observation of the Stark effect in a symmetric rotor spectrum makes it possible to determine the dipole moment μ; for example, a value of 1.857 ± 0.001 D [$(6.194 \pm 0.004) \times 10^{-30}$ C m] has been obtained for CH_3F from its microwave spectrum.

As with linear molecules, the rotational constants B and A (or C) change with the vibrational state of the molecule. These changes are given by the expressions

$$B_v = B_e - \sum_i \alpha_i^B \left(v_i + \frac{d_i}{2} \right) \qquad (4.73)$$

$$A_v = A_e - \sum_i \alpha_i^A \left(v_i + \frac{d_i}{2} \right) \qquad (4.74)$$

$$C_v = C_e - \sum_i \alpha_i^C \left(v_i + \frac{d_i}{2} \right) \qquad (4.75)$$

where the summation is over all the vibrations and d_i is the degeneracy.

In a non-linear N-atomic symmetric rotor molecule there are $(3N - 6)$ normal modes, of which some are doubly degenerate. In order to determine the value of B_e, B_1 has to be found for each vibration and this is possible only for small molecules having relatively low-lying $v_i = 1$ states. NF_3 is an example of such a molecule. It has six normal vibrations but there are two degenerate pairs so that four values of α_i^B have to be found. These are given in Table 4.2 for both $^{14}NF_3$ and the isotopically substituted molecule $^{15}NF_3$. The values of B_0 and B_e for these two molecules will be referred to again in Section 4.9 when we discuss structure determination.

In the case of degenerate vibrations, l-type doubling occurs. This is similar to l-type doubling in linear polyatomic molecules (Section 4.2.6.2) in that motion in a degenerate vibrational mode i gives rise to an additional angular momentum $l\hbar$, where

$$l = v_i,\ v_i - 2,\ v_i - 4,\ \ldots,\ -v_i$$

If the vibrational motion involves nuclei moving only perpendicular to the top axis then, as in a linear molecule, the vibrational angular momentum vector lies along this axis. In a symmetric rotor this need not be so and,

Table 4.2 Values of B_v and α_i^B in various vibrational states of $^{14}NF_3$ and $^{15}NF_3$ (ref. 13)

Vibrational state	$^{14}NF_3$		$^{15}NF_3$	
	B_v/MHz	α_i^B/MHz	B_v/MHz	α_i^B/MHz
Ground	10 681.02 ± 0.01	—	10 629.44 ± 0.03	—
$v_1 = 1$	10 724.47 ± 0.03	−43.45 ± 0.04	10 667.85 ± 0.04	−38.41 ± 0.07
$v_2 = 1$	10 642.38 ± 0.02	38.65 ± 0.03	10 589.30 ± 0.02	40.14 ± 0.05
$v_3 = 1$	10 602.22 ± 0.03	78.81 ± 0.04	10 553.77 ± 0.02	75.67 ± 0.05
$v_4 = 1$	10 676.54 ± 0.01	4.48 ± 0.02	10 624.80 ± 0.02	4.64 ± 0.05
$v_5 = 1$	10 671.90 ± 0.05	9.12 ± 0.06	10 620.12 ± 0.03	9.32 ± 0.06
	B_e/MHz		B_e/MHz	
	10 761.91 ± 0.2		10 710.63 ± 0.2	

in general, the component of the angular momentum along the top axis is $\zeta_i l\hbar$, where $0 \leqslant |\zeta_i| \leqslant 1$. The quantity ζ_i is the Coriolis coupling constant for vibration i and is a measure of the Coriolis interaction between the vibrational and rotational motions. If, for example, $v_i = 1$ then $l = \pm 1$ and the doubling is particularly large for transitions involving $k = \pm 1$, where $K = |k| = 1$ (see Section 4.3 for the introduction of the signed quantum number k). For transitions $(J+1) \leftarrow J$, $k \leftarrow k$ and $l \leftarrow l$, the frequencies or wavenumbers in equation (4.71) are modified to

$$\nu(\text{or } \tilde{\nu}) = 2(B_v - D_{JK}k^2)(J+1) - 4D_J(J+1)^3 + 2\eta_{iJ}(J+1)kl + \Delta P(J, k, l,) \quad (4.76)$$

where η_{iJ} is a further centrifugal distortion constant.[12] It is the ΔP term which causes the l-type doubling. For $k = l = \pm 1$ it is given by

$$\Delta P = \pm q(J+1) \quad (4.77)$$

In this equation

$$q \approx 2B_v^2/\omega \quad (4.78)$$

where ω is the frequency or wavenumber of the degenerate vibration. For $l = \pm 1$ and $k \neq l$, the doubling is given by

$$\Delta P = \pm q^2\{(J+1)[(J+1)^2 - (k \mp 1)^2]\}/\{4(k \mp 1)[(1-\zeta_i)A_v - B_v]\} \quad (4.79)$$

where the upper signs are used when k and l have the same sign and the lower signs when k and l have different signs.

4.4 NUCLEAR SPIN STATISTICAL WEIGHTS AND THEIR EFFECT ON INTENSITIES

In the $J = 8$–7 transition of SiH_3NCS shown in Figure 4.21 it can be observed that superimposed on the general decline of intensity from $K = 0$ to $K = 7$ is a marked increase of intensity of the $K = 3$ and $K = 6$ lines. This is due to the fact that the statistical weights of the rotational levels with K equal to a multiple of three are twice those for levels with K not equal to a multiple of three. This in turn is due to the fact that each of the hydrogen nuclei has a nuclear spin angular momentum for which the nuclear spin quantum number I (see Table 1.3) has the value of $\frac{1}{2}$.

When the effects of nuclear spin are included the total molecular wave function ψ of equation (1.84) becomes

$$\psi = \psi_e \psi_v \psi_r \psi_{ns} \tag{4.80}$$

where ψ_e, ψ_v and ψ_r are the electronic, vibrational and rotational wave functions and ψ_{ns} is the nuclear spin wave function. We shall not be concerned here with the detailed form of the ψ_{ns} wave functions but only with their symmetry properties.

For any molecule containing two or more identical nuclei having $I = n + \frac{1}{2}$, where n is zero or an integer, exchange of any two of them results in a change of sign of the total wave function ψ which is said to be antisymmetric to nuclear exchange. In addition, the nuclei are said to be Fermi particles (or fermions) and obey Fermi–Dirac statistics. On the other hand, if $I = n$, where n is zero or an integer, ψ is symmetric to nuclear exchange, and the nuclei are said to be Bose particles (or bosons) and obey Bose–Einstein statistics. These rules regarding the behaviour of the total wave function on exchange of identical nuclei follow from the generalized form of the Pauli principle, which states that the total wave function must be antisymmetric to the exchange of any kind of identical fermions and symmetric to the exchange of identical bosons.

We will now consider the consequences of these rules in the simple case of $^1H_2$. In this molecule both ψ_v, whatever the value of v and ψ_e, in the ground electronic state, are symmetric to nuclear exchange, so we need consider only the behaviour of $\psi_r \psi_{ns}$. Since $I = \frac{1}{2}$ for ^1H, ψ and therefore $\psi_r \psi_{ns}$ must be antisymmetric to nuclear exchange. It can be shown that, for even values of J, ψ_r is symmetric (s) to exchange and, for odd values of J, ψ_r is antisymmetric (a) to exchange, as shown in Figure 4.22

Equation (1.93) shows that, for $I = \frac{1}{2}$, space quantization of nuclear spin angular momentum results in the quantum number M_I taking the values of $\frac{1}{2}$ or $-\frac{1}{2}$. The nuclear spin wave function ψ_{ns} is usually written as α or β, corresponding to $M_I = \frac{1}{2}$ or $-\frac{1}{2}$, respectively, and both ^1H nuclei, labelled 1 and 2, can have either α or β spin wave functions. There are therefore four possible forms of ψ_{ns} for the molecule as a whole:

$$\psi_{ns} = \alpha(1)\alpha(2);\ \beta(1)\beta(2);\ \alpha(1)\beta(2);\ \text{or}\ \beta(1)\alpha(2) \tag{4.81}$$

Although the $\alpha(1)\alpha(2)$ and $\beta(1)\beta(2)$ wave functions are clearly symmetric to interchange of the 1 and 2 labels, the other two functions are neither symmetric nor antisymmetric. For this reason, it is necessary to use instead the linear combinations $2^{-\frac{1}{2}}[\alpha(1)\beta(2) + \beta(1)\alpha(2)]$ and $2^{-\frac{1}{2}}[\alpha(1)\beta(2) - \beta(1)\alpha(2)]$, where $2^{-\frac{1}{2}}$ is a normalization constant. Then three of the four nuclear spin wave functions are seen to be symmetric (s) to nuclear exchange and one is antisymmetric (a):

$$\text{(s)}\quad \psi_{ns} = \begin{cases} \alpha(1)\alpha(2) \\ 2^{-\frac{1}{2}}[\alpha(1)\beta(2) + \beta(1)\alpha(2)] \\ \beta(1)\beta(2) \end{cases} \tag{4.82}$$

$$\text{(a)}\quad \psi_{ns} = 2^{-\frac{1}{2}}[\alpha(1)\beta(2) - \beta(1)\alpha(2)] \tag{4.83}$$

In general, for a homonuclear diatomic molecule there are $(2I+1)(I+1)$ symmetric and $(2I+1)I$ antisymmetric nuclear spin wave functions; and therefore

$$\text{number of (s) functions/number of (a) functions} = (I+1)/I \tag{4.84}$$

In order that $\psi_r \psi_{ns}$ is always antisymmetric for $^1H_2$, the antisymmetric ψ_{ns} are associated with even J states and the symmetric ψ_{ns} with odd J states, as shown in Figure 4.22. Interchange between states with ψ_{ns} symmetric and antisymmetric is forbidden, so that $^1H_2$ can be regarded as consisting of two distinct forms:

(a) *para*-hydrogen with ψ_{ns} antisymmetric and with what is commonly referred to as antiparallel nuclear spins;

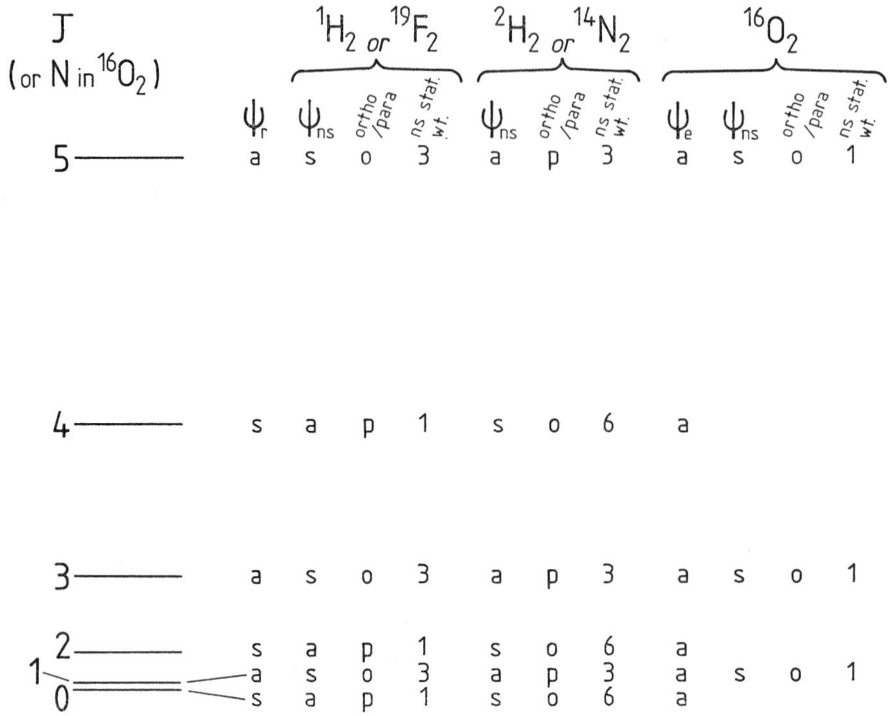

Figure 4.22 Nuclear spin statistical weights of rotational states of various diatomic molecules

(b) *ortho*-hydrogen with ψ_{ns} symmetric and parallel nuclear spins.

As indicated in Figure 4.22, *para*-$^1H_2$ can exist in only even J states and *ortho*-$^1H_2$ in odd J states. At temperatures at which there is appreciable population up to fairly high values of J there is roughly three times as much *ortho*- as there is *para*-$^1H_2$. However, at very low temperatures at which the population of all rotational levels other than $J = 0$ is small, $^1H_2$ is mostly in the *para* form.

All other homonuclear diatomic molecules with $I = \frac{1}{2}$ for each nucleus, such as $^{19}F_2$, also have *ortho* and *para* forms with odd and even J and nuclear spin statistical weights of 3 and 1, respectively, as shown in Figure 4.22.

If $I = 1$ for each nucleus, as in $^2H_2$ and $^{14}N_2$, the total wave function must be symmetric to nuclear exchange. There are nine nuclear spin wave functions, of which six are symmetric and three antisymmetric to exchange. Figure 4.22 illustrates the fact that *ortho*-$^2H_2$ (or $^{14}N_2$) can have only even J and the *para* form only odd J, and that there is roughly twice as much of the *ortho* form as there is of the *para* form at normal temperatures: at low temperatures there is a larger excess of the *ortho* form.

The effect of low temperatures affecting the *ortho*:*para* ratio is more important for light molecules, such as $^1H_2$ and $^2H_2$, than for heavy molecules, such as $^{19}F_2$ and $^{14}N_2$. The reason is that the separation of the $J = 0$ and $J = 1$ levels is smaller for a heavier molecule and a lower temperature is required before a significant deviation from the normal *ortho:para* ratio is observed.

For ^{16}O, I is zero and there are no antisymmetric nuclear spin functions in $^{16}O_2$. Since the ^{16}O nucleus is a boson the total wave function must be symmetric to nuclear exchange. In the case of $^{16}O_2$, with two electrons with unpaired spins in the ground state (see Section 6.2.1.1), ψ_e is *antisymmetric*, unlike in the other molecules we have considered. Since $I = 0$, the nuclear spin wave function $\psi_{ns} = 1$, which is symmetric to exchange of nuclei. Therefore, as Figure 4.22 shows, $^{16}O_2$ has *only* levels with the rotational quantum number having odd values. In molecules such as $^{16}O_2$ which have a resultant electron spin angular momentum due to the unpaired electrons, the quantum number used to distinguish rotational levels is N rather than J.

We usually think of all homonuclear diatomic molecules as having no permanent electric dipole moment and so not having any pure rotation spectrum. However, molecules like O_2 which have unpaired electron spins are exceptions to this rule. Owing to the unpaired electron spins, such a molecule has a magnetic dipole moment and therefore a weak pure rotation spectrum.

In the symmetric rotor SiH_3NCS, rotation about the top axis by $\pi/3$ or $2\pi/3$ has the effect of exchanging hydrogen nuclei, each of which has $I = \frac{1}{2}$ and therefore $M_I = \frac{1}{2}$ or $-\frac{1}{2}$ indicated, as previously, by an α or β nuclear spin wave function. There are eight nuclear spin wave functions for the molecule as a whole: these are

$$\begin{aligned}\psi_{ns} = {} & \alpha(1)\alpha(2)\alpha(3);\ \alpha(1)\alpha(2)\beta(3);\ \alpha(1)\beta(2)\alpha(3);\\ & \beta(1)\alpha(2)\alpha(3);\ \alpha(1)\beta(2)\beta(3);\ \beta(1)\beta(2)\alpha(3);\\ & \beta(1)\alpha(2)\beta(3);\ \beta(1)\beta(2)\beta(3)\end{aligned} \tag{4.85}$$

For the quantum number $K = 3n$, where $n = 0$, 1, 2, . . ., four of these ψ_{ns} wave functions, or linear combinations of them, combine with ψ_r to give a total wave function which is antisymmetric to nuclear exchange. For $K \neq 3n$ only two ψ_{ns} wave functions combine with ψ_r. Hence the nuclear spin statistical weights are 4, 2, 2, 4, 2, 2, 4, . . . for $K = 0, 1, 2, 3, 4, 5, 6, \ldots$. A detailed proof of this can be found in several of the books listed in the Bibliography at the end of this chapter.

It is the nuclear spin statistical weights which cause the $K = 3$ and $K = 6$ lines of SiH_3NCS in Figure 4.21 to be more intense than the neighbouring $J = 8$–7 lines. The fact that the $K = 0$ line also is not more intense is due to the fact that levels with $K = 0$ do not have the double degeneracy associated with all other values of K, as discussed in Section 4.3.

For planar XY_3 molecules, for example BF_3, with a C_3 axis perpendicular to the plane of the molecule there is, in addition to the statistical weight variation with K, a variation with J when $K = 0$. When the Y nuclei are bosons the nuclear spin statistical weight ratio for J even and odd is given by

$$\frac{\text{statistical weight for } J \text{ even}}{\text{statistical weight for } J \text{ odd}} = \frac{(2I+3)(I+1)}{(2I-1)I} \tag{4.86}$$

and, when they are fermions,

$$\frac{\text{statistical weight for } J \text{ odd}}{\text{statistical weight for } J \text{ even}} = \frac{(2I+3)(I+1)}{(2I-1)I} \tag{4.87}$$

There are also nuclear spin statistical weight alternations in symmetric rotors with a higher than threefold axis.

Finally, in addition to the nuclear spin and K degeneracy contributing to the populations of various rotational levels in symmetric rotors, there is also the $(2J + 1)$-fold degeneracy which is associated, as in diatomic molecules, with each J level.

Intensities of rotational transitions in symmetric rotors are determined by quantities analogous to those in diatomic and linear polyatomic molecules discussed in Section 4.2.2. Only the fraction of the total number of molecules occupying each rotational level [equations (4.25)–(4.27) for a linear molecule] and the square of the transition moment [equation (4.35) for a linear molecule] are modified.

In a symmetric rotor, the fraction of the total number of molecules occupying each rotational level is given by

$$N_{JK}/N = [g_K(2J+1)\exp(-E_r/kT)]/q_r \quad (4.88)$$

where q_r, the rotational partition function, is given by

$$q_r = \sum_{J=0}^{\infty}\sum_{K=0}^{J} g_k(2J+1)\exp\{-[BJ(J+1) + (A-B)K^2]hc/kT\} \quad (4.89)$$

for a prolate symmetric rotor and B and A having the dimensions of wavenumber; C replaces A for an oblate symmetric rotor. The quantity g_K in equation (4.88) takes account of the double degeneracy of all levels with $K \neq 0$, so that $g_K = 1$ for $K = 0$ and $g_K = 2$ for $K \neq 0$. For small values of B/T the partition function is given approximately by

$$q_r \approx [(kT/hc)^3(\pi/B^2 A)]^{\frac{1}{2}} \quad (4.90)$$

The square of the rotational transition moment between two states m and n is given by

$$|\boldsymbol{R}^{nm}|^2 = \mu^2[(J+1)^2 - K^2]/(J+1)(2J+1) \quad \text{for } \Delta J = +1 \text{ transitions} \quad (4.91)$$

$$|\boldsymbol{R}^{nm}|^2 = \mu^2 K^2/J(J+1) \quad \text{for } \Delta J = 0 \text{ transitions} \quad (4.92)$$

$$|\boldsymbol{R}^{nm}|^2 = \mu^2(J^2 - K^2)/J(2J+1) \quad \text{for } \Delta J = -1 \text{ transitions} \quad (4.93)$$

The expression for I_{abs}, the absorption intensity analogous to equation (4.37) for a linear molecule, becomes, for a prolate symmetric rotor,

$$I_{\text{abs}} \approx \rho(\nu_{J''K''J'K'})[8\pi^3/3(4\pi\varepsilon_0)q_r]\nu_{J''K''J'K'}{}^2|\boldsymbol{R}^{nm}|^2 \times \exp\{-[BJ(J+1) + (A-B)K^2]hc/kT\} \quad (4.94)$$

where $\rho(\nu_{J''K''J'K'})$ is the radiation density of frequency $\nu_{J''K''J'K'}$ which is the frequency of the transition, q_r is given by equation (4.90) and $|\boldsymbol{R}^{nm}|^2$ by equations (4.91)–(4.93). For an oblate symmetric rotor C replaces A.

4.5 PURE ROTATIONAL INFRARED, MILLIMETRE WAVE AND MICROWAVE SPECTRA OF ASYMMETRIC ROTOR MOLECULES

Unlike the cases of linear and symmetric rotor molecules, there are no closed formulae for the rotational term values which are generally applicable to asymmetric rotors.

Two types of term value expressions have proved very useful in the interpretation of asymmetric rotor spectra. These expressions involve an asymmetry parameter which is a measure of the degree to which a molecule deviates from being a symmetric rotor.

The first type of term value expression involves the asymmetry parameter b. For a molecule which resembles a prolate rather than an oblate symmetric rotor, a so-called prolate near-symmetric rotor for which

$$I_a < I_b \approx I_c \quad (4.95)$$

the term values can be written as

$$F(J_\tau) = \tfrac{1}{2}(B+C)J(J+1) + [A - \tfrac{1}{2}(B+C)]W(b_{\text{p}}) \quad (4.96)$$

neglecting centrifugal distortion. The function $W(b_{\text{p}})$ has the form

$$W(b_{\text{p}}) = K_a{}^2 + c_1 b_{\text{p}} + c_2 b_{\text{p}}{}^2 + \ldots \quad (4.97)$$

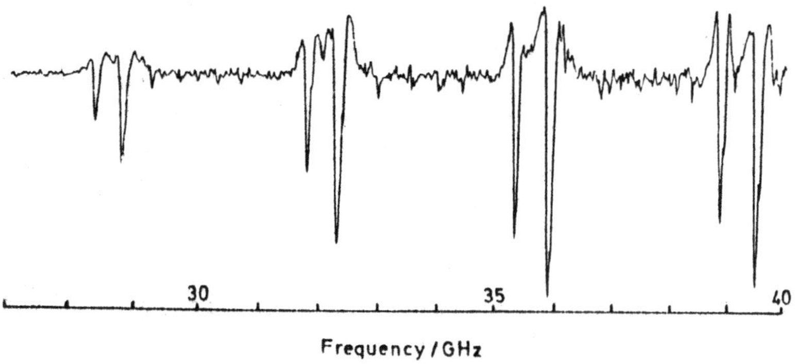

Figure 4.23 Part of the microwave spectrum of the crotonic acid molecule showing two series of transitions, the weaker series due to the *s-cis* and the stronger to the more abundant *s-trans* isomer [Reproduced, with permission, from Scharpen, L. H. and Laurie, V. W. (1972). *Anal. Chem.*, **44**, 378R]

where K_a, sometimes written as K_{-1}, is the quantum number K of the corresponding prolate symmetric rotor. For an oblate near-symmetric rotor the symbol used is K_c, or K_1, The quantity b_p is the asymmetry parameter used for a prolate near-symmetric rotor and is given by

$$b_p = (C - B)/(2A - B - C) \tag{4.98}$$

The coefficients $c_1, c_2, \ldots$ are functions of the quantum numbers and have been listed for $J = 0$–12;[†] τ is a label used to distinguish the $(2J + 1)$ rotational levels having the same value of J and can take the values

$$\tau = J, J - 1, J - 2, \ldots, -J \tag{4.99}$$

analogous to the signed quantum number k in a symmetric rotor. When $B = C$, $b_p = 0$ and the term values become those of a prolate symmetric rotor [equation (4.63)]. When $B \approx C$, $b_p \approx 0$ and the term values become

$$F(J, K_a) \approx \bar{B}J(J + 1) + (A - \bar{B})K_a^2 \tag{4.100}$$

where

$$\bar{B} = (B - C)/2 \tag{4.101}$$

The approximate term values are most accurate when $K_a \approx J$. This happens when the molecule is rotating about an axis close to the a-axis. On the other hand, the series in equation (4.97) converges too slowly for practical use when K_a becomes much less than J.

Figure 4.23 shows part of the microwave spectrum of crotonic acid, which can exist in *s-cis* and *s-trans* forms (*cis* or *trans* about the carbon–carbon single bond), as shown in Figure 4.24. Each isomer is a prolate near-symmetric rotor. Using the symmetric rotor selection rules $\Delta J = \pm 1$ and $\Delta K_a = 0$, together with equation (4.100), we can see that the spectrum should consist of a set of equally spaced lines with a separation of $2\bar{B}$, which is equal to $(B + C)$. The spectrum in Figure 4.23 shows *two* series of equally spaced lines, the stronger series being due to the more abundant *s-trans* isomer and the weaker to the *s-cis* isomer. The difference in $(B + C)$ for the two isomers is sufficiently great for the two series of lines to be separated easily and also for each series to be attributed unambiguously to a particular isomer by agreement of the observed

[†]See Townes and Schawlow in the Bibliography at the end of this chapter.

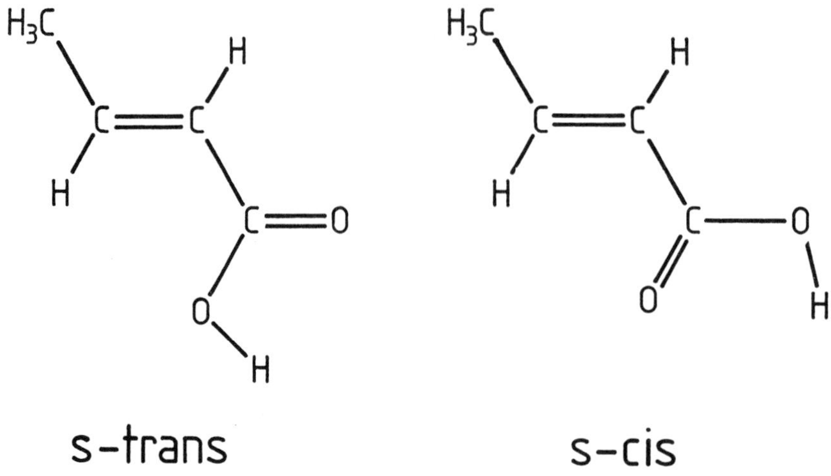

Figure 4.24 The *s-trans* and *s-cis* isomers of the crotonic acid molecule

$(B+C)$ with a value estimated from an assumed structure.

This example demonstrates the power of low resolution microwave spectroscopy in conformational analysis which is concerned with a determination of gross molecular structure, particularly with regard to possible isomers caused by rotation of a part of the molecule about a single bond.

A high resolution spectrum of crotonic acid would show that each 'line' observed at low resolution is a complex group of lines separated by the effects of asymmetry, centrifugal distortion, vibration and internal rotation of the methyl group.

For an oblate near-symmetric rotor, for which

$$I_a \approx I_b < I_c \tag{4.102}$$

the term values can be written as

$$F(J_\tau) = \tfrac{1}{2}(A+B)J(J+1) + [C - \tfrac{1}{2}(A+B)]W(b_o) \tag{4.103}$$

neglecting centrifugal distortion, where

$$W(b_o) = K_c^2 + c_1 b_o + c_2 b_o^2 + \ldots \tag{4.104}$$

and b_o, used for an oblate near-symmetric rotor, is given by

$$b_o = (A-B)/(2C-B-A) \tag{4.105}$$

When $A = B$, $b_o = 0$ and the term values become those of an oblate symmetric rotor [equation (4.66)]. When $A \approx B$, $b_o \approx 0$ and

$$F(J, K_c) \approx \bar{B}J(J+1) + (C-\bar{B})K_c^2 \tag{4.106}$$

where

$$\bar{B} = (A+B)/2 \tag{4.107}$$

Again, the approximate term values are more accurate, and the power series for $W(b_o)$ converges more quickly, when $K_c \approx J$.

The second type of useful term value expression is for a general asymmetric rotor and is of the form

Table 4.3 Asymmetric rotor $E(\kappa)$ functions for $J = 0, 1, 2, 3$

$J_{K_aK_c}$	J_τ	$E(\kappa)$	$J_{K_aK_c}$	J_τ	$E(\kappa)$
0_{00}	0_0	0	3_{30}	3_3	$5\kappa + 3 + 2(4\kappa^2 - 6\kappa + 6)^{\frac{1}{2}}$
1_{10}	1_1	$\kappa + 1$	3_{31}	3_2	$2[\kappa + (\kappa^2 + 15)^{\frac{1}{2}}]$
1_{11}	1_0	0	3_{21}	3_1	$5\kappa - 3 + 2(4\kappa^2 + 6\kappa + 6)^{\frac{1}{2}}$
1_{10}	1_{-1}	$\kappa - 1$	3_{22}	3_0	4κ
			3_{12}	3_{-1}	$5\kappa + 3 - 2(4\kappa^2 - 6\kappa + 6)^{\frac{1}{2}}$
2_{20}	2_2	$2[\kappa + (\kappa^2 + 3)^{\frac{1}{2}}]$	3_{13}	3_{-2}	$2[\kappa - (\kappa^2 + 15)^{\frac{1}{2}}]$
2_{21}	2_1	$\kappa + 3$	3_{03}	3_{-3}	$5\kappa - 3 - 2(4\kappa^2 + 6\kappa + 6)^{\frac{1}{2}}$
2_{11}	2_0	4κ			
2_{12}	2_{-1}	$\kappa - 3$			
2_{02}	2_{-2}	$2[\kappa - (\kappa^2 + 3)^{\frac{1}{2}}]$			

$$F(J_\tau) = \tfrac{1}{2}(A + C)J(J + 1) + \tfrac{1}{2}(A - C)E(\kappa) \tag{4.108}$$

where $E(\kappa)$ is a function of the asymmetry parameter κ, introduced by Ray, and given by

$$\kappa = (2B - A - C)/(A - C) \tag{4.109}$$

For a prolate symmetric rotor $B = C$ and $\kappa = -1$ and for an oblate symmetric rotor $A = B$ and $\kappa = 1$; these extreme values of κ are used as subscripts in the K_{-1} and K_1 symbols sometimes used for prolate and oblate symmetric rotors, respectively.

Instead of using the subscript τ as a label for states having the same value of J, the values of K_a and K_c are often used, as in $J_{K_aK_c}$. The quantity τ is related to K_a and K_c by

$$\tau = K_a - K_c \quad (\text{or } K_{-1} - K_1) \tag{4.110}$$

The $E(\kappa)$ functions for J values of 0–3 are given in Table 4.3 and the corresponding term values in Table 4.4.

Figure 4.25 shows how the energy levels of a prolate symmetric rotor correlate with those of an oblate symmetric rotor for the case when $A = 2C$ and B varies between A and C. This shows how, as a prolate symmetric rotor becomes slightly asymmetric, on the left-hand side of the figure, the double degeneracy of all levels with $K_a > 0$ is removed. The splitting decreases with K_a, for constant J, and increases with J for constant K_a. The same is true when an oblate symmetric rotor becomes slightly asymmetric, on the right-hand side of the figure, and we look at the splitting as a function of J and K_c. In the regions of Figure 4.25 where the energy levels are those of a more strongly asymmetric rotor they form a very irregular pattern.

Although the expressions of asymmetric rotor term values given by equations (4.96) and (4.103) and, particularly for low values of J, equation (4.108) are still useful, they have been largely superseded by the accurate computational methods. These methods involve representing the required symmetric rotor wave function as a linear combination of symmetric rotor wave functions. In order to obtain the asymmetric rotor term values, tridiagonal matrices, with symmetric rotor terms as diagonal elements, are diagonalized. Each matrix refers to a particular value of J and is a square matrix approximately $\frac{1}{2}J \times \frac{1}{2}J$. Diagonalization of such matrices, even when J is very large, is a standard operation on a large modern computer. The details of the accurate asymmetric rotor methods can be found in several of the books in the Bibliography at the end of this chapter, in particular that by Gordy and Cook.

Table 4.4 Asymmetric rotor term values $F(J_\tau)$ for $J = 0, 1, 2, 3$

$J_{K_aK_c}$	J_τ	$F(J_\tau)$
0_{00}	0_0	0
1_{10}	1_1	$A + B$
1_{11}	1_0	$A + C$
1_{01}	1_{-1}	$B + C$
2_{20}	2_2	$2A + 2B + 2C + 2[(B - C)^2 + (A - C)(A - B)]^{\frac{1}{2}}$
2_{21}	2_1	$4A + B + C$
2_{11}	2_0	$A + 4B + C$
2_{12}	2_{-1}	$A + B + 4C$
2_{02}	2_{-2}	$2A + 2B + 2C - 2[(B - C)^2 + (A - C)(A - B)]^{\frac{1}{2}}$
3_{30}	3_3	$5A + 5B + 2C + 2[4(A - B)^2 + (A - C)(B - C)]^{\frac{1}{2}}$
3_{31}	3_2	$5A + 2B + 5C + 2[4(A - C)^2 - (A - B)(B - C)]^{\frac{1}{2}}$
3_{21}	3_1	$2A + 5B + 5C + 2[4(B - C)^2 + (A - B)(A - C)]^{\frac{1}{2}}$
3_{22}	3_0	$4A + 4B + 4C$
3_{12}	3_{-1}	$5A + 5B + 2C - 2[4(A - B)^2 + (A - C)(B - C)]^{\frac{1}{2}}$
3_{13}	3_{-2}	$5A + 2B + 5C - 2[4(A - C)^2 - (A - B)(B - C)]^{\frac{1}{2}}$
3_{03}	3_{-3}	$2A + 5B + 5C - 2[4(B - C)^2 + (A - B)(A - C)]^{\frac{1}{2}}$

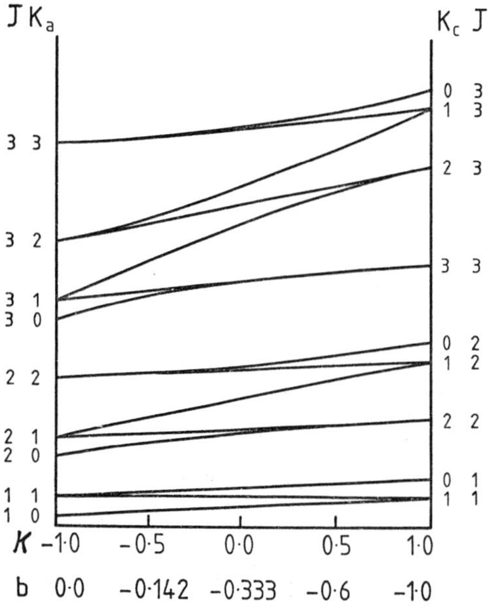

Figure 4.25 Correlation of the rotational energy levels of a prolate ($\kappa = -1.0$, $b = 0$) with those of an oblate ($\kappa = 1.0$, $b = -1.0$) symmetric rotor

The selection rule governing the changes in J in asymmetric rotors is

$$\Delta J = 0, \ \pm 1$$

Transitions with $\Delta J = 0$ are called Q-branch transitions and those with $\Delta J = +1$ and -1 are R- and P-branch transitions, respectively. In fact, $\Delta J = 0$ applies also to symmetric rotors and linear molecules in their vibrational ground states, but such transitions would have zero frequency. The splitting of the K degeneracy in asymmetric rotors results in Q-branch transitions being observed.

The selection rules replacing the $\Delta K = 0$ rule in symmetric rotors are more complex. They involve the parity, i.e. the evenness or oddness, of K_a and K_c and also the direction in the molecule of the permanent dipole moment, which must be non-zero. For example, if the dipole moment is along the a-axis, as in 1,1-difluoroethylene shown in Figure 4.26(a), the parity of either K_a or K_c must change during the transition. If we label the rotational energy

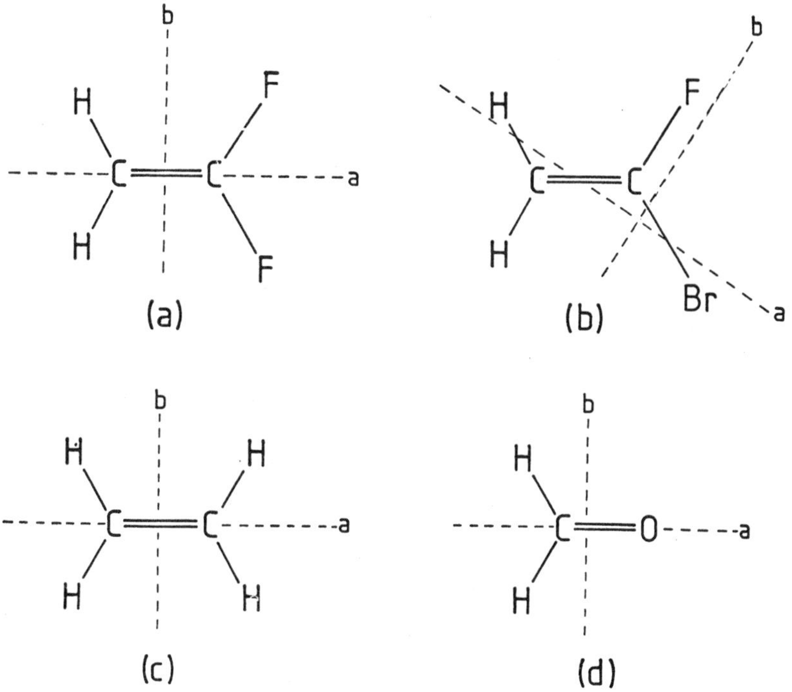

Figure 4.26 The molecules (a) 1,1-difluoroethylene, (b) 1-bromo-1-fluoroethylene, (c) ethylene and (d) formaldehyde are all asymmetric rotors. The a and b inertial axes are shown

levels as eo, implying that K_a is even and K_c is odd, oe, ee or oo the selection rules are those given in Table 4.5 for an a-axis dipole moment; the allowed transitions in this case are referred to as A-type. This table also gives the selection rules for B- and C-type transitions which occur when the dipole moment is along the b- or the c-axis, respectively.

If, as in the case of 1-bromo-1-fluoroethylene [Figure 4.26(b)], the dipole moment does not lie on a principal axis, then hybrid selection rules apply, in this case a mixture of A- and B-types.

Since the selection rules in Table 4.5 refer only to the parity of K_a and K_c, in principle ΔK_a or ΔK_c can be 0, ±2, ±4, . . . or ±1, ±3, ±5, ... In practice, if the molecule is a prolate near-symmetric rotor, only transitions with $\Delta K_a = 0$ or ± 1, the symmetric rotor selection rules, have significant intensity and, for an oblate near-symmetric rotor, only transitions with $\Delta K_c = 0$ or ± 1.

Table 4.5 Asymmetric rotor selection rules for microwave, millimetre wave and far infrared pure rotational spectra

Direction of dipole moment	Selection rule[a]	
a-Axis	ee ↔ eo oe ↔ oo	and $\Delta J = 0, \pm 1$
b-Axis	ee ↔ oo oe ↔ eo	
c-Axis	ee ↔ oe eo ↔ oo	

[a]The double arrow (↔) implies that the transition is allowed whichever of the two states involved is the upper state.

Transition moments which are necessary for the calculation of intensities can be obtained accurately only by matrix methods. For a near-

Table 4.6 Nuclear spin statistical weight factors for some asymmetric rotors

	$^{1}H_2{}^{16}O$	$^{2}H_2{}^{16}O$	$^{12}C_2{}^{1}H_2{}^{19}F_2$[a]	$^{12}C_2{}^{1}H_4$	$^{12}C_2{}^{2}H_4$	$^{13}C_2{}^{1}H_4$	$^{1}H_2{}^{12}C^{16}O$
ee	1	6	10	7	27	16	1
oe	3	3	6	3	18	12	3
eo	3	3	10	3	18	24	1
oo	1	6	6	3	18	12	3

[a] 1,1-Difluoroethylene.

symmetric rotor, and when neither of the levels involved in the transition has had the K_a (prolate) or K_c (oblate) degeneracy removed within experimental resolution, symmetric rotor transition moments are a useful approximation.

The rotational partition function q_r for an asymmetric rotor is given by

$$q_r \approx [(kT/hc)^3(\pi/ABC)]^{\frac{1}{2}} \tag{4.111}$$

for sufficiently high temperatures or sufficiently small rotational constants.

There can be nuclear spin statistical weight alternations with K_a or K_c if the a- or c-axis is a C_2 symmetry axis. For example, there is an alternation of 1:3 for K_a even:odd in formaldehyde ($^{1}H_2{}^{12}C^{16}O$), shown in Figure 4.26(d). There is an exchange of the two ^{1}H nuclei, which are fermions with $I=\frac{1}{2}$, on carrying out the C_2 operation about the a-axis. Table 4.6 shows the resulting nuclear spin statistical weight factors for the asymmetric rotor levels. In the prolate symmetric rotor limit, when the K_a degeneracy is not split, the statistical weight factors for $K_a = 0, 1, 2, 3, 4, \ldots$ are 1, 6, 2, 6, 2, . . ., where the double degeneracy for $K_a > 0$ is included.

In 1,1-difluoroethylene (1,1-$^{12}C_2{}^{1}H_2{}^{19}F_2$), shown in Figure 4.26(a), a C_2 operation about the a-axis exchanges two ^{1}H and two ^{19}F nuclei, all with $I=\frac{1}{2}$. Table 4.6 shows the nuclear spin statistical weights. The molecule as a whole shows Bose–Einstein statistics and there is an intensity alternation of 10:6 for K_a even:odd.

In the $^{1}H_2{}^{16}O$ and $^{2}H_2{}^{16}O$ molecules, the C_2 axis is the b-inertial axis. As a result, there is no intensity alternation in either K_a or K_c. For example, Table 4.6 shows that there are four nuclear spin functions for $^{1}H_2{}^{16}O$ with K_a both even and odd, and similarly for K_c.

For a molecule such as ethylene ($^{12}C_2{}^{1}H_4$) shown in Figure 4.26(c), with a- and c-axes which are both C_2 axes, there may be an intensity alternation with K_a *and* K_c. Rotation about the a- and c-axes exchanges two pairs of ^{1}H nuclei resulting in the nuclear spin statistical weights given in Table 4.6[†] and an intensity alternation of 10:6 for K_a even:odd in the prolate and of 10:6 for K_c even:odd in the oblate symmetric rotor limits. Table 4.6 also gives the nuclear spin statistical weight factors for the isotopomers $^{12}C_2{}^{2}H_4$ and $^{13}C_2{}^{1}H_4$, in which $I(^{13}C) = \frac{1}{2}$.

Ethylene, and these two isotopomers, have no permanent dipole moment and therefore have no pure rotational infrared, millimetre wave or microwave spectra. However, these statistical weight factors are important in, for example, their vibration–rotation spectra [Section 5.2.5.4(i)].

An interesting, recently discovered, molecule which shows an intensity alternation with K_c, due to nuclear spin statistical weight factors, is Si_2H_2 (disilyne), shown[15] to have the dibridged 'butterfly' structure shown in Figure 4.27(b). This contrasts with the linear structure of C_2H_2 (acetylene).

Disilyne is a prolate near-symmetric rotor and the c-axis is the C_2 symmetry axis passing through the midpoints between the hydrogen

[†] For derivation of these factors, see Wilson, E. B. (1935). *J. Chem. Phys.*, **3**, 276.

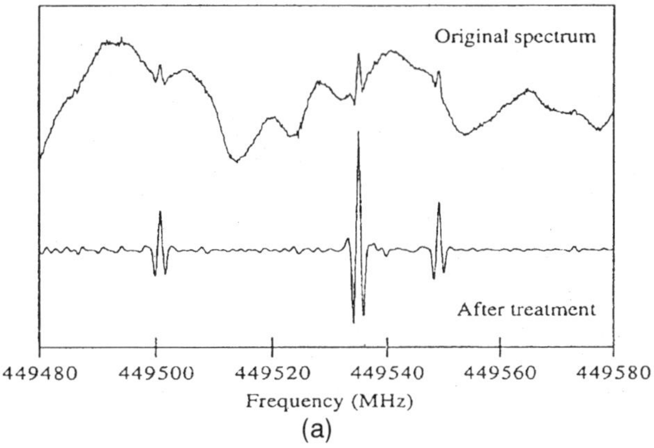

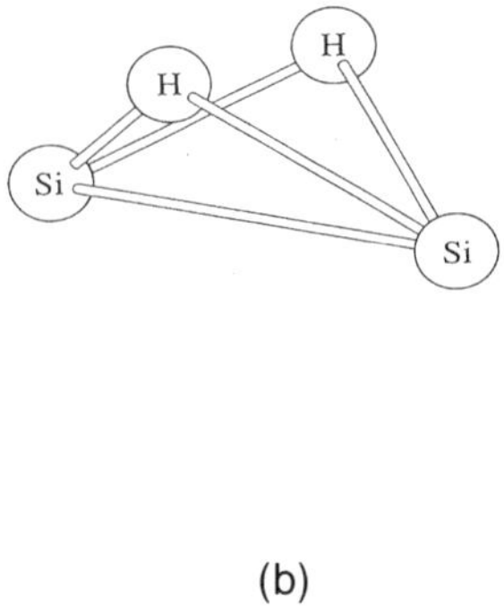

Figure 4.27 (a) Part of millimetre wave spectrum of Si_2H_2. (b) Structure of Si_2H_2. [Reproduced, with permission, from Bogey, M., Bolvin, H., Cordonnier, M., Demuynck, C., Destombes, J. L. and Császár, A. G. (1994). *J. Chem. Phys.*, **100**, 8614]

atoms and between the silicon atoms. Rotation about this axis exchanges the two equivalent hydrogen atoms resulting in a 1:3 intensity alternation for K_c even:odd.

Disilyne is an example of an unstable molecule whose pure rotational absorption spectrum, in this case in the millimetre wave region, can be studied only under special conditions which are necessary for generation of the species. Disilyne was generated[15] directly in a long (2.5 m) absorption cell by creating a plasma, excited by a d.c. glow discharge, in a mixture of silane (SiH_4) and argon. Figure 4.27(a) shows a very small part of the millimetre wave spectrum, in the 449.48–449.58 GHz region, as originally obtained and after treatment to remove the background absorption. The three lines show a characteristic second derivative shape. The lowest frequency line is assigned to the 3_{22}–3_{12} transition of disilyne. The other two transitions are due to a second, less stable, planar monobridged isomer of Si_2H_2 in which one of the hydrogen atoms is attached to only one silicon atom.

The rotational constants from four isotopomers of the dibridged isomer have been interpreted to give a substitution structure (see Section 4.9) in which r (Si–Si) = 2.2154 Å, r (Si–H) = 1.6680 Å and $\angle$ (HSiSiH) = 104.22°.

For an asymmetric rotor the vibrational dependence of A, B and C is given by

$$A_v = A_e - \sum_i \alpha_i^A (v_i + \tfrac{1}{2}) \qquad (4.112)$$

$$B_v = B_e - \sum_i \alpha_i^B (v_i + \tfrac{1}{2}) \qquad (4.113)$$

$$C_v = C_e - \sum_i \alpha_i^C (v_i + \tfrac{1}{2}) \qquad (4.114)$$

These equations are similar to equations (4.73), (4.74) and (4.75) for a symmetric rotor, except that all vibrations are non-degenerate in an asymmetric rotor: vibrational degeneracy is associated with the presence of a C_n axis with $n > 2$ or an S_4 axis, but these requirements necessarily mean that the molecule is not an asymmetric rotor.

Corresponding to the quartic centrifugal distortion constants D_J, D_{JK} and D_K of a symmetric rotor, given in equations (4.69) and (4.70), there are five quartic centrifugal distortion constants, Δ_J, Δ_{JK}, Δ_K, δ_J and δ_K, appropriate to asymmetric rotors.[14] For these constants the subscript K refers to K_a or K_c in a prolate or oblate rotor, respectively. If the molecule is a near-symmetric rotor it may be necessary to include further, sextic, centrifugal distortion constants.

Just as the term values of a rigid asymmetric rotor cannot be expressed generally in closed form, neither can those which include centrifugal distortion.

As in the case of symmetric rotors and linear molecules, the Stark effect is a useful tool in the analysis of rotational spectra of asymmetric rotors. The main difference is that, in general, the K degeneracy of a symmetric rotor is split in an asymmetric rotor. Therefore, whereas symmetric rotors show a first- *and* second-order Stark effect for $K \neq 0$ and only a second-order effect for $K = 0$ [see equation (4.72)], an asymmetric rotor shows only a second-order effect. However, a first-order effect may be observed for K_a or $K_c \neq 0$ when the K degeneracy is not split and the molecule is behaving as a near-symmetric rotor, or when two rotational levels are accidentally degenerate.

There are no closed formulae representing Stark shifts of energy levels in asymmetric rotors. These shifts have to be obtained, as for the zero-field energy levels, by matrix diagonalization.

The Stark effect is used mainly for the identification of transitions. For a Q-type transition with $\Delta J = 0$ there are J Stark components, whereas for a P- or R-type transition, with $\Delta J = -1$ or $+1$, respectively, there are $(J + 1)$ components. In addition, the magnitude of the splitting of the components can be used to determine the dipole moment.

4.6 PURE ROTATIONAL INFRARED, MILLIMETRE WAVE AND MICROWAVE SPECTRA OF SPHERICAL ROTOR MOLECULES

In a spherical rotor all three principal moments of inertia are equal [equation (4.14)]. Examples are the regular tetrahedral molecule CH_4 [Figure 4.8(a)] and the regular octahedral molecule SF_6 [Figure 4.8(b)].

The rotational term values for a spherical rotor, including the effects of centrifugal distortion, are given by

$$F_v(J) = B_v J(J + 1) - D_v J^2(J + 1)^2 + D_v{}' f(J, \kappa) \tag{4.115}$$

for any non-degenerate vibrational state. This is the same expression as for a diatomic molecule [equation (4.52)] except for the addition of the last term which involves a centrifugal distortion constant D_v^t, of similar magnitude to D_v, and a function f. This function cannot be written in closed form but values have been given by Hecht.[16]

Comparison with a symmetric rotor shows that there is a $(2J + 1)^2$-fold degeneracy associated with each J level of a spherical rotor. One $(2J + 1)$ factor arises from the $(2J + 1)$ values that M_J can take, as in all other molecules. The second $(2J + 1)$ factor arises because, as a symmetric rotor becomes a spherical rotor, all levels with the same J but different values of K become degenerate. Since K can take $(J + 1)$ values and all levels with $K > 0$ are doubly degenerate, this results in the second $(2J + 1)$ factor.

Before 1971, it was supposed that, because spherical rotors have no permanent dipole moment, they have no pure rotation spectrum. In the rigid rotor approximation this is strictly true but centrifugal distortion may produce a small dipole moment. For example, in CH_4, rotation about one of the C–H bonds, each of which is a C_3 axis, causes a centrifugal distortion in the other three bonds with a consequent asymmetry of the molecule leading to a dipole moment of the order of 10^{-5}–10^{-6} D (10^{-35}–10^{-36} C m). Such a small dipole moment leads to an extremely weak rotational spectrum.

The far infrared pure rotational spectrum of the tetrahedral molecule SiH_4, shown in Figure 4.28, was obtained with a Michelson interferometer. An absorption path of 10.6 m and a pressure of 4.03 atm were necessary, the high pressure causing considerable broadening of the transitions which obey the selection rule

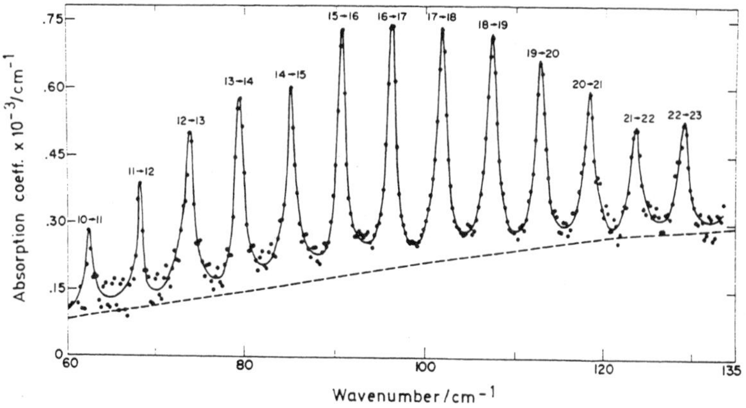

Figure 4.28 The far infrared pure rotational spectrum of the SiH_4 molecule obtained with a Michelson interferometer [Reproduced, with permission, from Rosenberg, A. and Ozier, I. (1974). *Can. J. Phys.*, **52**, 575]

$\Delta J = \pm 1$ and show an approximate spacing of $2B_0$. The dipole moment was estimated from the intensities of the rotational transitions to be 8.3×10^{-6} D (2.7×10^{-35} C m).

Not all spherical rotors possess a dipole moment due to centrifugal distortion. For example, rotation of SF_6 [Figure 4.8(b)] about a C_4 axis (any of the F–S–F directions) results in equal lengthening of each of the four S–F bonds in the plane perpendicular to the axis, but clearly no dipole moment results.

In addition some molecules which are not spherical rotors, such as BF_3 (planar) and $H_2C{=}C{=}CH_2$ (allene, with the planes of the two CH_2 groups mutually perpendicular), should have a weak pure rotation spectrum due to centrifugal distortion although, according to the rigid rotor model, they have no dipole moment.

4.7 INTERSTELLAR MOLECULES DETECTED BY THEIR PURE ROTATION SPECTRA

Radiotelescopes are used to scan the universe for radiation in the radiofrequency region (see Figure 3.1). The telescope consists of a parabolic reflecting dish. In the case of the one at Parkes in New South Wales (Australia), the dish is 64 m in diameter. The radiofrequency detector is positioned at the focus of the parabola, as shown in Figure 4.29.

Atomic hydrogen is extremely abundant, but not uniformly distributed, in the universe. It has an emission line at a wavelength of 21 cm (see Section 6.1.4) and can be detected fairly easily. For this reason, most of the early radiotelescopes were constructed so that the detector and degree of accuracy in shaping the dish were suitable for wavelengths in this region. In 1963 the first molecule was detected with such a telescope, although observations of molecular electronic spectra using optical telescopes were made much earlier. Absorption lines at wavelengths close to 18 cm, due to transitions between components of Λ-doublets (see Section 6.2.5.2) in the $v = 0$ vibrational state of OH, were observed in several sources of radiofrequency radiation in our galaxy.

The regions of space where molecules have been detected are the nebulae which are found in the Milky Way and also in galaxies other than our own. The Milky Way appears as a hazy band of light coming from millions of stars. In this band there are rifts which consist of dark clouds of interstellar dust and gas and which are associated with the luminous clouds of the nebulae. The presence of the dust in the dark clouds is indicated by the fact that visible

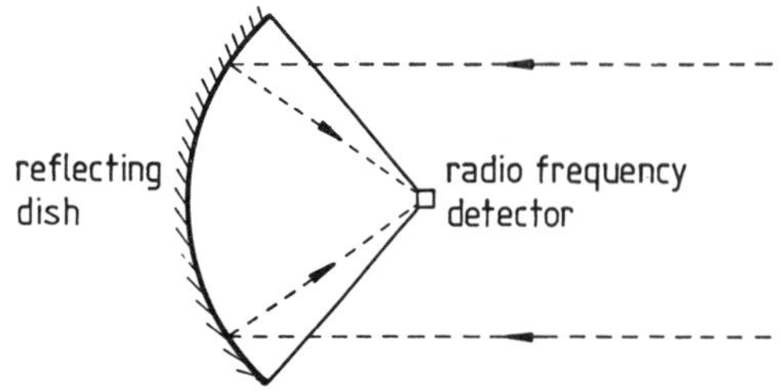

Figure 4.29 Detection of radiofrequency radiation with a radiotelescope

starlight passing through such a cloud is reddened due to preferential scattering, which is proportional to λ^{-4}, of the blue light by the dust particles. The nature of the dust particles is not known but they are about 0.2 μm in diameter. New stars in the galaxy appear to be formed by gravitational collapse in the region of nebulae which therefore must contain the raw materials from which the new stars are formed. For this reason, the detection of molecules in nebulae is of the greatest importance.

Molecules have been detected in several nebulae. The large cloud known as Sagittarius B2, situated near the centre of our galaxy, has proved to be a particularly fruitful source.

Development of radiotelescopes to work at shorter wavelengths, extending to the millimetre wave region, has enabled searches to be made for a wide range of molecules. The first polyatomic molecule was detected in 1968 using a telescope having a dish 6.3 m in diameter built at Hat Creek, California (USA) to work in the millimetre wave region. Emission lines were found in the 1.25 cm wavelength region and are due to NH_3. The transitions are not pure rotational but are between the closely spaced $v_2 = 0$ and $v_2 = 1$ levels of the inversion vibration ν_2 (see Section 5.2.7.1).

Table 4.7 lists some of the molecules which have been detected. It is interesting to note that some of them, such as C_2H, HCO^+ and N_2H^+, were found in interstellar space before they were found in the laboratory. Identification of the molecules is usually unambiguous because of the uniqueness of the highly precise transition frequencies. However, before frequencies detected in space can be compared with laboratory frequencies, they must be corrected for the Doppler effect. In Sagittarius B2 the molecules are found to be travelling fairly uniformly with a velocity of 60 km s^{-1} relative to the local standard of rest, which is taken to be certain stars close to the sun. In other clouds a wider range of molecular velocities is found.

Figure 4.30 shows the $J = 1$–0 transition of the linear molecule cyanodiacetylene (H–C≡C–C≡C–C≡N) also observed in emission in Sagittarius B2. (Figure 4.12 shows a part of the absorption spectrum of the same molecule obtained in the laboratory.) The vertical scale on the figure is a measure of the change of temperature of the antenna due to the received signal. The three hyperfine components into which the transition is split are due to interaction between the rotational angular momentum and the nuclear spin angular momentum of the ^{14}N nucleus for which $I = \frac{1}{2}$. As discussed in Section 4.2.5, for $J = 1$ and $I = 1$ F can take the values 2, 1 and 0, whereas, for $J = 0$, $F = 1$ only. The selection rule $\Delta F = 0, \pm 1$ allows the three components of the $J = 1$–0 transition shown in Figure 4.30.

The discovery of fairly large molecules in the interstellar medium came as a considerable surprise. The largest, to date, are cyanotetra-acetylene (HC_8CN), detected in Heiles' cloud 2,[17] and cyanopentaacetylene ($HC_{10}CN$). Previously it was thought that the ultraviolet radiation present throughout the galaxy would photodecompose most of the molecules, and especially the larger ones, now known to exist. It seems likely that the dust particles play an important part not only in formation of the molecules but also in preventing their decomposition.

The interdependence of interstellar and laboratory observations in the identification of interstellar molecules is nicely illustrated by the 'X-ogen' story.

Table 4.7 Interstellar molecules detected by their radiofrequency or millimetre wave spectra

Year	Molecule	Transitions[a]
1963	OH	Λ doublet (emission and absorption)
1968	NH_3	Inversion doublet, $J = 1, 2, 3, 4, 5, 6$ (emission)
1969	H_2O	$J = 6_{16} \rightarrow 5_{23}$
	H_2CO	$1_{10} \rightleftarrows 1_{11}$
		$2_{11} \leftarrow 2_{12}$
		$3_{12} \leftarrow 3_{13}$
1970	CO	$1 \rightarrow 0$
	CN	$1 \rightarrow 0$
	CH_3OH	$1_{10} \rightarrow 1_{11}$
1971	HCOOH	$1_{10} \rightarrow 1_{11}$
	CS	$3 \rightarrow 2$
	SiO	$3 \rightarrow 2$
		$2 \rightarrow 1$
	HCN	$1 \rightarrow 0$
	HNC	$1 \rightarrow 0$
	H–C≡C–CN	$1 \rightarrow 0$
	OCS	$9 \rightarrow 8$
	CH_3CN	$6 \rightarrow 5$
	NH_2CHO	$2_{11} \rightarrow 2_{12}$
	$CH_3C{\equiv}CH$	$5 \rightarrow 4$
	HNCO	$4_{04} \rightarrow 3_{03}$
		$1_{01} \rightarrow 0_{00}$
	CH_3CHO	$1_{10} \rightarrow 1_{11}$
		$2_{11} \rightarrow 2_{12}$
	H_2CS	$2_{11} \rightarrow 2_{12}$
1972	$CH_2{=}NH$	$1_{10} \rightarrow 1_{12}$
After 1972:		
Diatomics	SO, SiS, NO, NS, CH, CH^+, SiO, SiC, SiCl, SiN, SiS, NH, CP, HCl, NaCl, KCl, AlCl, AlF, PN, CO^+, SO^+	
Triatomics	H_2S, N_2H^+, SO_2, HNO, C_2H, HCO, HCO^+, HCS^+, SiCCl, SiC_2, CH_2, NH_2, C_3, C_2O, C_2S, MgNC, NaCN, N_2O	
Tetratomics	HNCS, C_2CN, H_3O^+, C_3H (linear), C_3H (cyclic), MgNCO, $HCNH^+$, H_2CN, HCCN, $HOCO^+$, C_3O, C_3S	
5-atomics	C_4H, NH_2CN, C_3H_2 (linear), C_3H_2 (cyclic), CH_2CN, CH_2CO, HC_2CN, HCCNC, HNCCC, C_4Si	
6-atomics	N_2CHO, CH_3SH, CH_3NC, HC_2CHO, HC_3NH^+, C_5H	
>6-atomics	CH_3NH_2, $CH_2{=}CHCN$, HC_4CN, C_6H, $HCOOCH_3$, CH_3C_2N, CH_3OCH_3, CH_3CH_2OH, HC_6CN, CH_3CH_2CN, CH_3C_4H, HC_8CN, $HC_{10}CN$	

[a]Absorption is indicated by ← and emission by →.

In 1970, an interstellar line at 89 189 MHz was discovered by Buhl and Snyder.[18] No laboratory-generated species had been found with an absorption at this frequency and they called this unknown molecule X-ogen. Klemperer[19] suggested that the species was the previously unknown ion HCO^+. This was later confirmed by the laboratory observation by Woods *et al.*[20] of a microwave absorption line at $89\,188.545 \pm 0.020$ MHz in a discharge in a gaseous mixture of hydrogen and carbon monoxide using the method devised by Woods and described in Section 3.3. The spectrum, in which the single line is assigned to the $J = 1$–0 transition, is shown in Figure 4.31.

As is shown in Table 4.7 most of the transitions have been observed in emission rather than in absorption. The non-Boltzmann

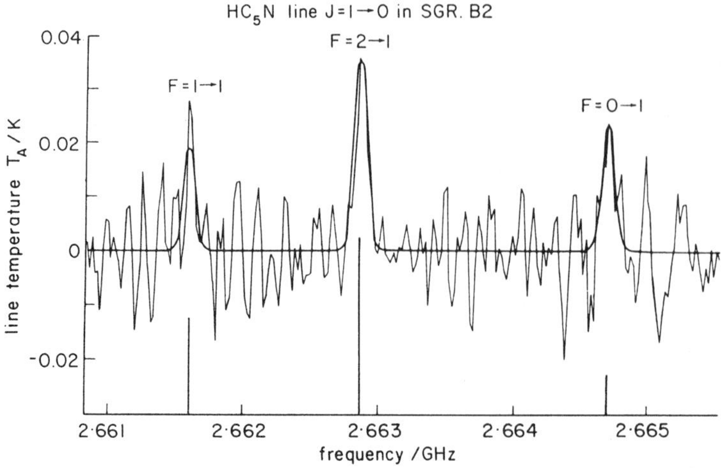

Figure 4.30 The $J = 1\text{–}0$ transition of cyanodiacetylene observed in emission in Sagittarius B2 [Reproduced, with permission, from Broton, N. W., MacLeod, J. M., Oka, T., Avery, L. W., Brooks, J. W., McGee, R. X. and Newton, L. M. (1976). *Astrophys. J.*, **209**, L143. Published by the University of Chicago Press; © 1976 The American Astronomical Society]

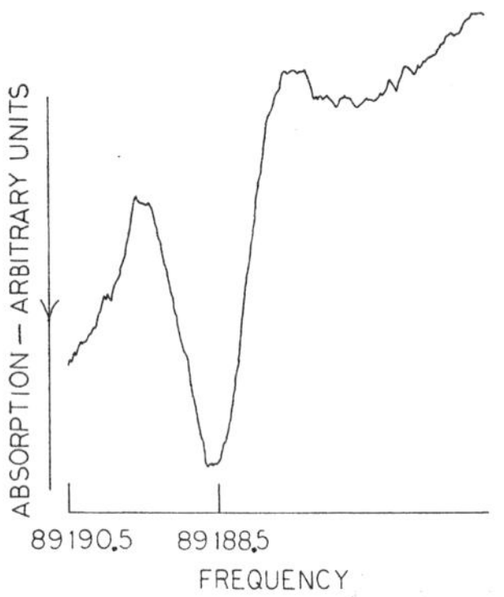

Figure 4.31 The $J = 1\text{–}0$ transition of HCO^+ observed in the laboratory. [Reproduced, with permission, from Woods, R. C., Dixon, T. A., Saykally, R. J. and Szanto, P. G. (1975). *Phys. Rev. Lett.*, **35**, 1269]

population of the levels which gives rise to this is thought to be due either to population of the upper level by collisions with the abundant H_2 molecules or to preferential population of the upper level during the formation of the molecule. The emission process is stimulated by a background continuum of radiofrequency and millimetre wave radiation corresponding to black body radiation at about 2.7 K and which is thought to be due to the 'big bang' which occurred when the universe started its life. It is this radiation which also provides the continuum for these transitions observed in absorption.

4.8 PURE ROTATIONAL RAMAN SPECTROSCOPY

4.8.1 Theory

When electromagnetic radiation falls on an atomic or molecular sample it may be absorbed if the energy of the radiation corresponds to the energy separation of two stationary states of the atoms or molecules. If it does not, the radiation will either pass straight through the sample or be scattered by it. Most of the scattered radiation is of unchanged wavelength, λ, and is called Rayleigh scattering. It was Lord Rayleigh in 1871 who showed that the intensity

I_s of scattered light is related to the wavelength λ by

$$I_s \propto \lambda^{-4} \tag{4.116}$$

This equation demonstrates that blue radiation from the sun is scattered preferentially by particles in the atmosphere; this phenomenon causes a cloudless sky to appear blue.

It was predicted in 1923 by Smekal and shown experimentally in 1928 by Raman and Krishnan[21] that a small amount of radiation scattered by a gas, liquid or solid which is transparent to the incident radiation is of increased or decreased wavelength, or wavenumber. This effect is called the Raman effect: the scattered radiation with decreased wavenumber is called the Stokes Raman scattering and that with increased wavenumber the anti-Stokes Raman scattering. The incident radiation should be highly monochromatic for the effect to be observed clearly.

The processes which cause molecular Raman scattering involve electronic, vibrational or rotational transitions accompanying the scattering process. In this chapter we shall be concerned only with rotational transitions.

The property of the sample which determines the degree of scattering is the polarizability, $\boldsymbol{\alpha}$. When the incident radiation is in the visible or ultraviolet region, as it often is in Raman spectroscopy, the polarizability is a measure of the degree to which the electrons in the molecule can be displaced relative to the nuclei. In general the polarizability of a molecule is an anisotropic property which means that, at equal distances from the centre of the molecule, $\boldsymbol{\alpha}$ is not always the same when measured in different directions. A surface drawn so that the distance from the origin to a point on the surface has a length $\alpha^{-\frac{1}{2}}$, where α is the polarizability in that direction, forms an ellipsoid, shown in Figure 4.32. The polarizability is therefore a tensor property. The tensor, $\boldsymbol{\alpha}$, can be expressed in the form of a matrix

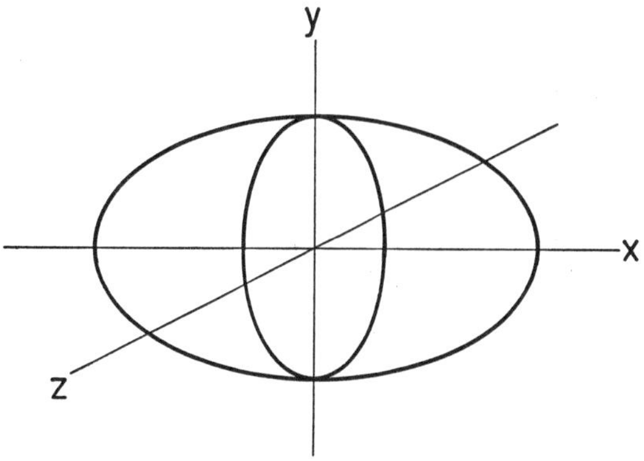

Figure 4.32 The polarizability ellipsoid

$$\boldsymbol{\alpha} = \begin{pmatrix} \alpha_{xx} & \alpha_{xy} & \alpha_{xz} \\ \alpha_{yx} & \alpha_{yy} & \alpha_{yz} \\ \alpha_{zx} & \alpha_{zy} & \alpha_{zz} \end{pmatrix} \tag{4.117}$$

where the diagonal elements α_{xx}, α_{yy} and α_{zz} are the values of $\boldsymbol{\alpha}$ along the x, y and z molecule-fixed axes. The matrix is symmetrical about the leading diagonal (top left to bottom right), i.e. $\alpha_{yx} = \alpha_{xy}$, $\alpha_{zx} = \alpha_{xz}$, and $\alpha_{zy} = \alpha_{yz}$, Therefore, in the most general case, there are six different components of $\boldsymbol{\alpha}$, namely, α_{xx}, α_{yy}, α_{zz}, α_{xy}, α_{xz} and α_{yz}. The polarizability ellipsoid is reminiscent of the momental ellipsoid in Figure 4.1; indeed, the moment of inertia is also a tensor property. Just as the momental ellipsoid becomes a sphere for a spherical rotor ($I_a = I_b = I_c$), so does the polarizability ellipsoid and $\boldsymbol{\alpha}$ is isotropic.

When monochromatic radiation falls on a molecular sample in the gas phase and is not absorbed, the oscillating electric field $\boldsymbol{E}$ of the radiation induces in the molecule an electric dipole $\boldsymbol{\mu}$ which is related to $\boldsymbol{E}$ by

$$\boldsymbol{\mu} = \boldsymbol{\alpha}\boldsymbol{E} \tag{4.118}$$

where $\boldsymbol{\mu}$ and $\boldsymbol{E}$ are vectors. Writing equation (4.118) in matrix form, we obtain

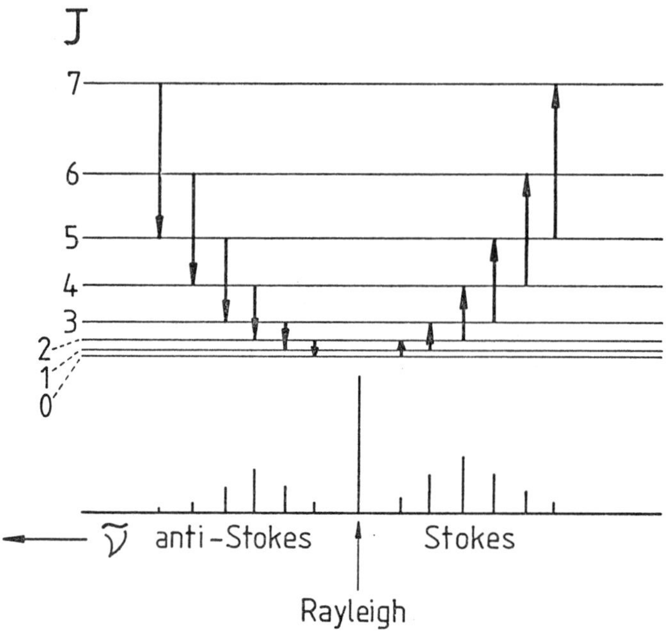

Figure 4.33 Pure rotational Raman spectrum of a diatomic or linear polyatomic molecule

$$\begin{pmatrix} \mu_x \\ \mu_y \\ \mu_z \end{pmatrix} = \begin{pmatrix} \alpha_{xx} & \alpha_{xy} & \alpha_{xz} \\ \alpha_{yx} & \alpha_{yy} & \alpha_{yz} \\ \alpha_{zx} & \alpha_{zy} & \alpha_{zz} \end{pmatrix} \begin{pmatrix} E_x \\ E_y \\ E_z \end{pmatrix} \quad (4.119)$$

where E_x, E_y, E_z refer to components of $\boldsymbol{E}$ along space-fixed axes.

The magnitude of the oscillating electric field can be represented by

$$E = A \sin 2\pi c\tilde{\nu}t \quad (4.120)$$

where A is the amplitude and $\tilde{\nu}$ is the wavenumber of the monochromatic radiation.

The magnitude α of the polarizability varies during rotation and a simple classical treatment of rotation will serve to illustrate the Raman effect.

The polarizability ellipsoid rotates with the molecule. If the frequency of rotation is ν_{rot} it is clear that α, measured in a space-fixed rather than a molecule-fixed direction, changes at *twice* the frequency of rotation since the configuration of the ellipsoid is repeated every π radians. The variation of α is given by

$$\alpha = \alpha_{0,\text{r}} + \alpha_{1,\text{r}} \sin 2\pi c(2\tilde{\nu}_{\text{rot}})t \quad (4.121)$$

where $\alpha_{0,\text{r}}$ is the average polarizability during rotation and $\alpha_{1,\text{r}}$ is the amplitude of the change of the polarizability. Substituting this expression for the magnitude of $\boldsymbol{\alpha}$, and equation (4.120) for the magnitude of $\boldsymbol{E}$, into equation (4.118) gives

$$\mu = \alpha_{0,\text{r}}A \sin 2\pi c\tilde{\nu}t - \tfrac{1}{2}\alpha_{1,\text{r}}A \cos 2\pi c(\tilde{\nu} + 2\tilde{\nu}_{\text{rot}})t + \tfrac{1}{2}\alpha_{1,\text{r}}A \cos 2\pi c(\tilde{\nu} - 2\tilde{\nu}_{\text{rot}})t \quad (4.122)$$

Therefore, classically, rotation of the molecule causes a variation of the induced electric dipole with time which can be broken down into the three terms of equation (4.122). The first term corresponds to scattered radiation of unchanged

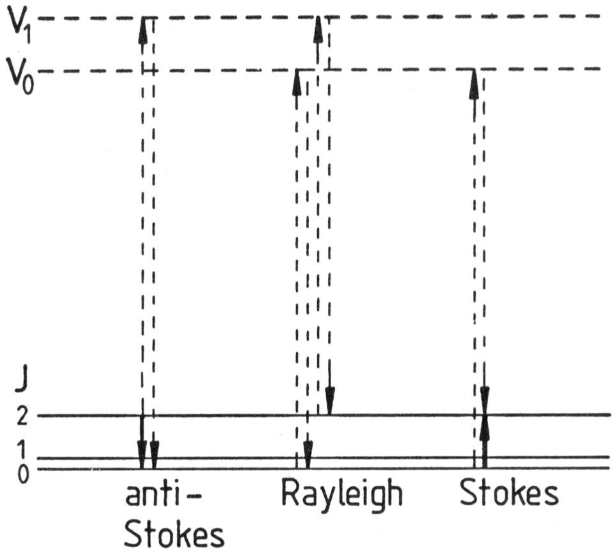

Figure 4.34 Stokes and anti-Stokes rotational Raman scattering and Rayleigh scattering processes involving virtual states V_0 and V_1

wavenumber—the Rayleigh scattering. The second and third terms correspond to scattered radiation with a wavenumber of $(\tilde{\nu} + 2\tilde{\nu}_{rot})$ and $(\tilde{\nu} - 2\tilde{\nu}_{rot})$—the anti-Stokes and Stokes Raman scattering, respectively. Although in a classical system $\tilde{\nu}_{rot}$ can take a continuous range of values, in a quantum mechanical system it can take only discrete values.

In a diatomic or linear polyatomic molecule the scattering obeys the rotational selection rule

$$\Delta J = 0, \ \pm 2$$

where $\Delta J = 0$ corresponds to the Rayleigh scattering. The resulting spectrum is illustrated in Figure 4.33. Figure 4.34 illustrates in more detail the processes involved in the first Stokes and anti-Stokes transitions and in the Rayleigh scattering. Molecules which are initially in the $J = 0$ state encounter the intense, monochromatic radiation of wavenumber $\tilde{\nu}$. Provided the energy $hc\tilde{\nu}$ does not correspond to the difference in energy between the $J = 0$ state and any other eigenstate (electronic, vibrational or rotational), it will not be absorbed but will produce an induced dipole in the molecule as expressed by equation (4.118). The molecule is said to be in a virtual state which, in the case illustrated in Figure 4.34, is V_0. When the molecule scatters the radiation it can return, according to the selection rules, to $J = 0$ (Rayleigh) or $J = 2$ (Stokes). Similarly, a molecule initially in the $J = 2$ state goes to the virtual state V_1 and returns to $J = 2$ (Rayleigh), $J = 0$ (anti-Stokes) or $J = 4$ (Stokes). The overall transitions $J = 2$–0 and 0–2 are indicated by solid lines in Figure 4.34. Figure 4.33 shows more of the overall rotational transitions. If the incident radiation is not monochromatic and covers a range of wavenumbers $\delta\tilde{\nu}$ then all the virtual states and all the rotational lines in the Raman spectrum are broadened by $\delta\tilde{\nu}$.

As is conventional we take ΔJ to be J(upper) – J(lower) so that we need consider only $\Delta J = +2$. The Raman displacements $\Delta\tilde{\nu}$ from the exciting line are given by

$$|\Delta\tilde{\nu}| = F(J+2) - F(J) \tag{4.123}$$

where $\Delta\tilde{\nu}$ is positive for anti-Stokes and negative for Stokes lines. When centrifugal distortion is neglected $F(J)$ is given by equation (4.19) and this gives, for the zero-point vibrational level,

$$\begin{aligned}|\Delta\tilde{\nu}| &= B_0(J+2)(J+3) - B_0J(J+1)\\ &= 4B_0J + 6B_0\end{aligned} \tag{4.124}$$

Therefore, the spectrum, as shown in Figure 4.33, shows two sets of equally spaced lines with a spacing of $4B_0$ and a separation between the exciting line and the first of each of the anti-Stokes and Stokes lines of $6B_0$. Since $\Delta J = 2$ for *both* the anti-Stokes and Stokes branches they are *both* called *S*-branches.†

When centrifugal distortion is included, $F(J)$ is given by equation (4.39) and we obtain

$$|\Delta\tilde{\nu}| = (4B_0 - 6D_0)(J+\tfrac{3}{2}) - 8D_0(J+\tfrac{3}{2})^3 \tag{4.125}$$

The intensities of the Raman lines reflect the populations of the rotational levels. Since the initial state of the first, second, . . . anti-Stokes line is higher in energy than that of the corresponding Stokes line, the anti-Stokes lines are correspondingly weaker, but this is a small effect except in light molecules.

Figure 4.35(a) shows the rotational Raman spectrum of $^{15}N_2$ obtained with 476.5 nm incident radiation from an argon ion laser (see Section 8.1.8.6). For the ^{15}N nucleus the nuclear spin quantum number $I = \frac{1}{2}$. As a result, there is an intensity alternation in the spectrum of 1:3 for the J value of the initial level of the transition even:odd, as explained in Section 4.4. Figure 4.35(b) shows the corresponding spectrum of $^{14}N^{15}N$ in which there is no intensity alternation as the nuclei are no longer identical.

From the spectra of $^{14}N_2$, $^{15}N_2$ and $^{14}N^{15}N$, values for B_0 of $(1.989\,506 \pm 0.000\,027)$, $(1.857\,672 \pm 0.000\,027)$ and $(1.923\,604 \pm 0.000\,020)$ cm^{-1}, respectively, have been obtained.[22] Values for r_0 of 1.100 105, 1.099 985 and 1.100 043 Å (all ±0.000 010 Å), calculated from the respective values of B_0, illustrate its small isotope dependence. On the other hand, as will be discussed in Section 4.9, the corresponding values for r_e of 1.097 651, 1.097 614 and 1.097 630 Å (all ±0.000 030 Å) are isotope independent.

In the rotational Raman spectrum of $^{16}O_2$ alternate lines are missing because alternate rotational levels have zero population (Figure 4.22).

A requirement that any molecule shall have a rotational Raman spectrum is that it must have anisotropic polarizability. This condition is satisfied in both homonuclear and heteronuclear diatomic molecules in which the polarizability ellipsoid (Figure 4.32) has two main axes equal. In fact, all molecules have anisotropic polarizability except spherical rotors. This means that molecules such as CH_4 (tetrahedral) and SF_6 (octahedral) have no pure rotational Raman spectrum unless we include the possibility that centrifugal distortion may cause a small anisotropy in the polarizability analogous to its causing a small dipole moment (Section 4.6).

The requirement of anisotropic polarizability is much less restrictive than that of a permanent

†The general convention for labelling branches of rotational lines is that, for $\Delta J = -2, -1, 0, 1, 2$, the branches are labelled *O*, *P*, *Q*, *R*, *S*, respectively. Note, however, that some authors label the anti-Stokes branch in a pure rotational Raman spectrum the *O*-branch.

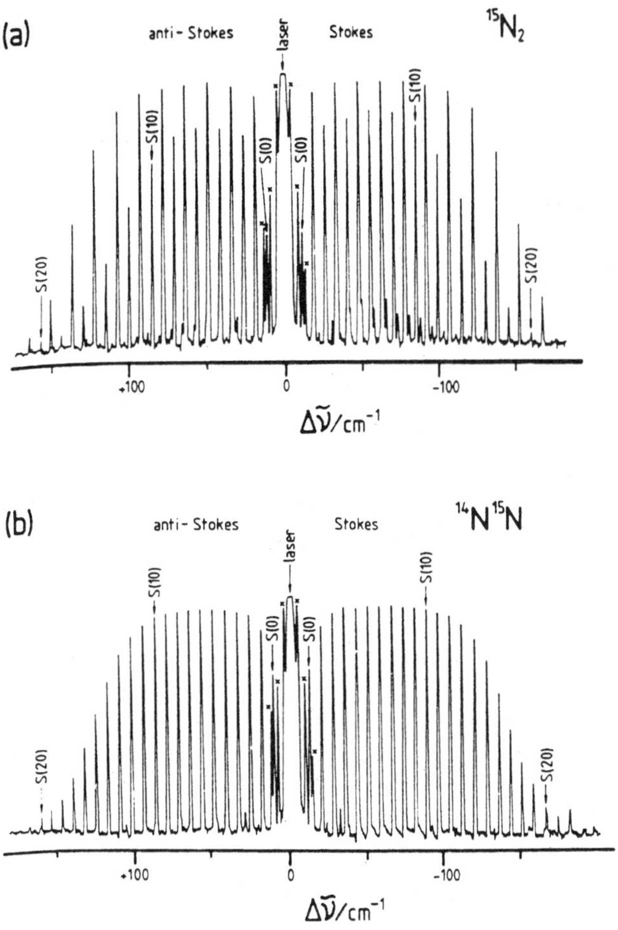

Figure 4.35 Pure rotational Raman spectra (a) $^{15}N_2$ and (b) $^{14}N^{15}N$ obtained with 476.5 nm radiation from an argon ion laser. The intensities in both spectra show maxima which are broadened owing to the effects of saturation of the photographic plates from which the traces are taken. Those lines marked with a cross are grating ghosts and are not part of the Raman spectra

dipole moment which is necessary for a microwave or far infrared pure rotational spectrum to be observed. For this reason, the technique of Raman spectroscopy is important in obtaining structural information on molecules with no permanent dipole moment.

The selection rules for pure rotational Raman spectra of symmetric, near-symmetric and asymmetric rotors are given in Table 4.8.

4.8.2 Experimental techniques

The main problems to be overcome in obtaining Raman spectra are concerned with the extreme weakness of Raman scattering and with devising an intense monochromatic source. The problem of weak scattering is worse when the sample has to be in the gas phase at low pressure, as in pure rotational Raman spectroscopy.

Table 4.8 Selection rules in pure rotational Raman spectroscopy

Type of molecule	Selection rules	
Linear	$\Delta J = 0, \pm 2$	
Symmetric rotor	$\Delta J = 0, \pm 1, \pm 2$	
	$\Delta K = 0$	
Asymmetric rotor[a]	$\Delta J = 0, \pm 1, \pm 2$ but $J' + J'' \geqslant 2$	
	and	
C_{2v}, D_{2h}, D_2	ee ↔ ee	I
	eo ↔ eo	
	oo ↔ oo	
	oe ↔ oe	
(*a*)[b] C_2, C_{2h}, C_s	ee ↔ eo	II
	oe ↔ oo	
	in addition to I	
(*b*)[b] C_2, C_{2h}, C_s	ee ↔ oo	III
	oe ↔ eo	
	in addition to I	
(*c*)[b] C_2, C_{2h}, C_s	ee ↔ oe	IV
	eo ↔ oo	
	in addition to I	
C_1, C_i	I + II + III + IV	
Near-symmetric rotor	$\Delta J = 0, \pm 1, \pm 2$ but $J' + J'' \geqslant 2$	
	and	
C_{2v}, D_{2h}, D_2	$\Delta K_a = 0, \pm 2$	
	$\Delta K_c = 0, \pm 2$	
(*a*)[b] C_2, C_{2h}, C_s	$\Delta K_a = 0, \pm 2$	
	$\Delta K_c = 0, \pm 1, \pm 2$	
(*b*)[b] C_2, C_{2h}, C_s	$\Delta K_a = 0, \pm 1, \pm 2$	
	$\Delta K_c = 0, \pm 1, \pm 2$	
(*c*)[b] C_2, C_{2h}, C_s	$\Delta K_a = 0, \pm 1, \pm 2$	
	$\Delta K_c = 0, \pm 2$	
C_1, C_i	$\Delta K_a = 0, \pm 1, \pm 2$	
	$\Delta K_c = 0, \pm 1, \pm 2$	

[a]The selection rules depend on the type of asymmetric rotor concerned and, in particular, the point group to which it belongs. Point group classification will be discussed in Section 5.2.2.
[b](*a*), (*b*), (*c*) refer to whether the unique axis of the molecule is the *a*, *b* or *c* inertial axis. The unique axis is the two-fold (C_2) rotation axis in C_2 and C_{2h} molecules and perpendicular to the plane of symmetry (σ) in C_s molecules.

In outline, the method used is to pass the monochromatic radiation through the gaseous sample and disperse and detect the scattered radiation. The scattered radiation is usually detected in directions normal to the incident radiation in order to avoid the incident radiation passing directly to the detector.

Until the advent of lasers, the most successful source of monochromatic radiation was the so-called Toronto arc. A Toronto arc is a tubular mercury discharge lamp, up to 3 m in length, operating at the vapour pressure of mercury and at high current. An operating current of 20–30 A is made possible by water cooling. The mercury discharge consists of the emission spectrum of mercury atoms. The spectrum is typical of that of an atom and consists of narrow lines each one of which is a possible

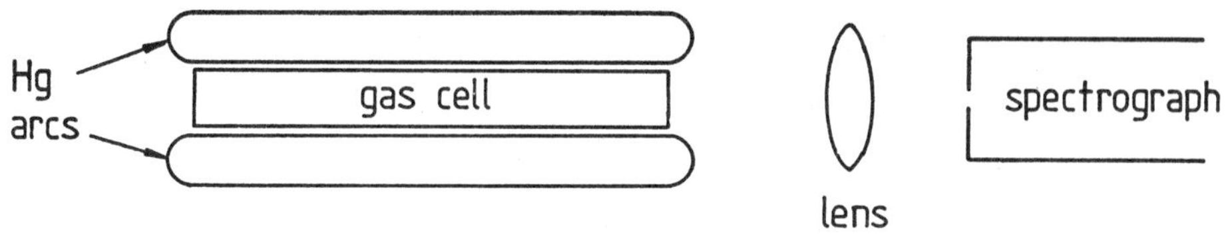

Figure 4.36 Experimental arrangement for obtaining the Raman spectrum of a gas using Toronto mercury arcs

source of monochromatic radiation. Three of the most intense lines in the mercury spectrum are at 253.7, 404.7 and 435.7 nm. Unfortunately, all three lines show nuclear hyperfine structure [see Section 6.1.2.2(iii)] and therefore are less monochromatic than they would otherwise be. The 435.7 nm line is the least broadened by this effect but even in this case the line width at half intensity of the most intense, unresolved, part of the hyperfine pattern of lines is about $0.2\,\mathrm{cm}^{-1}$, and this places a limit on the resolution which can be achieved.

One, or as many as four, Toronto arcs are placed parallel to the cell containing the gaseous sample; Figure 4.36 shows two arcs being used. Very often concave multiple reflection mirrors, similar in principle to those shown in Figure 3.26 but usually in a modified arrangement using *four* mirrors,[23] are placed inside the cell to increase the gathering power for the scattered radiation. In addition, the whole system of cell and arcs is enclosed in a reflecting container which may be coated with MgO powder to increase the amount of mercury radiation passing into the cell.

The scattered radiation is photographed with a high resolution spectrograph. Even with a sample gas pressure as high as 300 Torr, exposure times are typically several hours.

Lasers (Chaper 8) are sources of intense, highly monochromatic radiation and therefore ideal for Raman spectroscopy. Their chief advantages over Toronto arcs, which they have now entirely replaced, lie in their greater convenience, their higher intensity and the fact that they are more highly monochromatic; for example, the line width at half intensity of the 632.8 nm line from a helium–neon laser is less than $0.05\,\mathrm{cm}^{-1}$ compared with $0.2\,\mathrm{cm}^{-1}$ for the 435.7 nm mercury line. This means that the attainable resolution is increased considerably.

A laser beam is so highly parallel that multiple reflections in the sample cell can be achieved simply by using two plane mirrors M_1 and M_2, as shown in Figure 4.37. The four concave mirrors M_3–M_6 collect the scattered radiation,[23] which is analysed by a spectrograph or spectrometer.

Until the mid-1970s, only lasers operating in the visible region of the spectrum, such as the helium–neon laser (Section 8.1.8.6) at 632.8 nm and the argon ion laser (Section 8.1.8.7) at 514.5 nm, were used as sources of monochromatic radiation for Raman spectroscopy. However, many molecules are coloured and therefore absorb and fluoresce in the visible region. This fluorescence tends to mask the much weaker Raman scattering thereby excluding many molecules from investigation by the Raman effect. Vibrational Raman spectra, to be discussed in Chapter 5, are obtained, very often, with the sample in the liquid or solid phase when even a small amount of coloured impurity may produce sufficient fluorescence to interfere with the very weak Raman scattering of the main component of the sample.

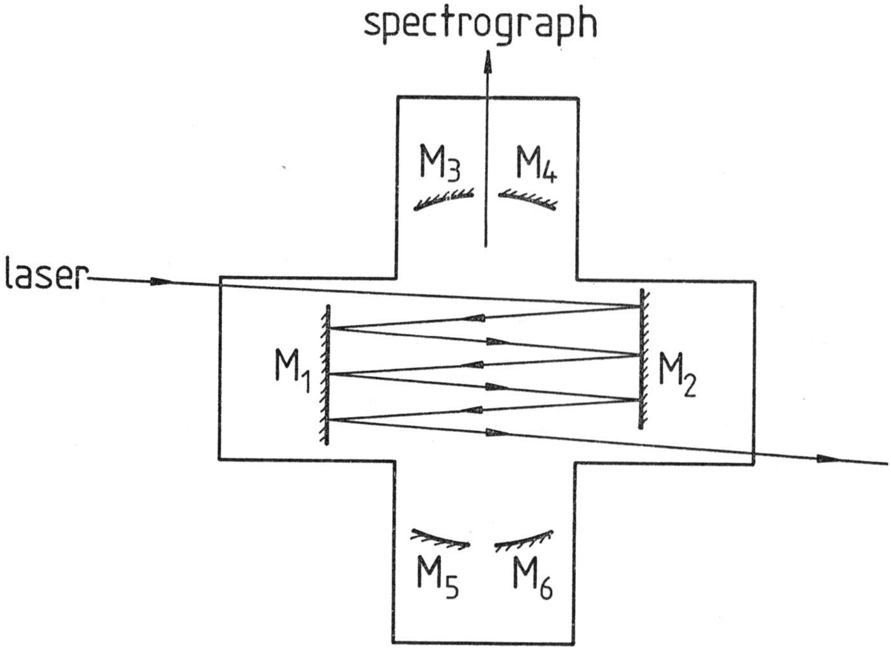

Figure 4.37 Multiple reflection sample cell for passing the laser beam through the gas many times (using mirrors M_1 and M_2) and for efficient collection of the scattered radiation (using mirrors M_3, M_4, M_5 and M_6)

Using a laser operating in the infrared overcomes the problem of fluorescence, which normally occurs following the absorption of only visible or ultraviolet radiation. However, with an ordinary spectrometer, having a diffraction grating as the dispersing element, the advantage of using an infrared laser is more than counteracted by the fact that the intensity of Raman scattering decreases as the fourth power of the wavelength, as equation (4.116) indicates, making detection extremely difficult.

It was not until the development of Fourier transform infrared (FTIR) spectrometers (see Section 3.5.4) that the possibility of using an infrared laser routinely was opened up. The intensity advantage of an infrared interferometer, with which a single spectrum can be obtained very rapidly and then many spectra co-added, coupled with the development of more sensitive Ge and InGaAs semiconductor infrared detectors, more than compensate for the loss of scattering intensity in the infrared region.

The infrared laser which is most often used in this technique of Fourier transform Raman, or FT-Raman, spectroscopy is the Nd^{3+}:YAG laser (see Section 8.1.8.4) operating at a wavelength of 1064 nm.

In FT-Raman spectroscopy the radiation emerging from the sample contains not only the Raman scattering but also the extremely intense laser radiation used to produce it. If this were allowed to contribute to the interferogram, before Fourier transformation, the corresponding cosine wave would overwhelm those due to the Raman scattering. A cut-off (interference) filter is inserted after the sample cell to remove 1064 nm (and lower wavelength) radiation but, because the cut-off is not sufficiently sharp, spectra can be recorded only to within about 50 cm^{-1} of the laser radiation. This prevents the use of FT-Raman spectroscopy to observe rotational Raman scattering.

An FT-Raman spectrometer is often simply an FTIR spectrometer adapted to accommodate the laser source, the filters to remove the laser radiation and a variety of infrared detectors.

4.9 STRUCTURE DETERMINATION FROM ROTATIONAL CONSTANTS

We have seen how the rotational constants of linear molecules, symmetric rotors, asymmetric rotors and spherical rotors can be found from their pure rotational spectra. To obtain a principal moment of inertia from the corresponding rotational constant is straightforward, using equations (4.64), (4.65) and (4.67), and the moments of inertia are obviously related to the magnitudes of bond lengths and bond angles in the molecule concerned.

Before we embark on structure determination we have to consider which rotational constants to use: the equilibrium values, the zero-point values or some other values.

The equilibrium rotational constants refer, in a case of a diatomic molecule, to the minimum in the potential energy curve, whether it is that of a harmonic oscillator (Figure 1.22) or an anharmonic oscillator (see Section 5.1.2). The potential energy curve of a harmonic or an anharmonic oscillator is independent of the isotopes in the molecule and therefore r_e is also independent; for example, r_e for $^1H^{35}Cl$ is the same as for $^2H^{35}Cl$. The fact that the bond length is independent of isotopic substitution is to be expected when we remember that bond lengths (and angles) are determined by the net effect of the interactions of the charged particles, electrons and nuclei, in the molecule. Addition or subtraction of neutrons, which have no charge, should not affect these interactions. On the other hand, if we obtain from B_0 a bond length which we call r_0 this is isotope dependent (see example of N_2 in Section 4.8.1) because of the isotope dependence of the vibrational energy levels which follows from the mass dependence of the vibration frequency [equation (1.100)]. For this reason we choose, when possible, to use equilibrium values of rotational constants for structure determination in spite of the fact that the equilibrium condition can never be realized, even at the absolute zero of temperature.

These arguments can be extended to polyatomic molecules whose equilibrium bond lengths and angles are isotope independent but whose zero-point structures are not.

As in diatomic molecules, the structure of greatest importance is the equilibrium structure. One rotational constant can give, at the most, only one structural parameter. However, in a planar molecule the out-of-plane principal moment of inertia I_c is related to the other two by

$$I_c = I_a + I_b \tag{4.126}$$

and there are only two independent rotational constants for each isotopic species. For example, in the planar asymmetric rotor formaldehyde, shown in Figure 4.26(d), determination of A_e, B_e and C_e for one isotopic species would provide only two geometrical parameters. Data from another isotopic species, such as ^2H_2CO, would be required to give the structural parameters r_e(CH), r_e(CO) and $(\angle HCH)_e$ necessary for a complete equilibrium structure.

The planar asymmetric rotor, 1,1-difluoroethylene, shown in Figure 4.26(a), has five structural parameters, r_e(CC), r_e(CH), r_e(CF), $(\angle HCH)_e$ and $(\angle FCF)_e$. It would require determination of A_e, B_e and C_e for three isotopic species, such as $F_2{}^{12}C{=}{}^{12}C^1H_2$, $F_2{}^{12}C{=}{}^{12}C^2H_2$ and $F_2{}^{13}C{=}{}^{12}C^1H_2$, to obtain a complete equilibrium structure.

Although isotopic substitution is immensely important in structure determination, it should be remembered that there is little value in substituting an atom in the same site more than once. A more serious limitation of isotopic substitution is that if an atom is close to the centre of mass, as is the central atom in the linear molecule N_2O, then the position of the atom is difficult to locate by isotopic substitution of that atom. As a result, although r_e(NN) and r_e(NO) have been determined[24] fairly accurately as $1.126\,7 \pm 0.000\,5$ Å and $1.185\,7 \pm 0.000\,5$ Å, the separation r_e(N . . . O) of the terminal atoms is known much more accurately to be $2.312\,30 \pm 0.000\,03$ Å.

The fact that the B_e values for at least two isotopic species of a linear triatomic molecule XYZ are required in order to determine r_e(XY) and r_e(YZ) is illustrated by the expression for the moment of inertia I_b:

$$I_b = m_x[r(\mathrm{XY})]^2 + m_z[r(\mathrm{YZ})]^2 - \frac{1}{M}[m_x r(\mathrm{XY}) - m_z r(\mathrm{YZ})]^2 \quad (4.127)$$

where m_i is the mass of atom i and M is the total mass of the molecule.

An example of complete structure determination for a symmetric rotor is that of NF_3 which has two structural parameters, r(NF) and ∠FNF. Combination of B_e for $^{14}NF_3$ and $^{15}NF_3$ (Table 4.2) gives $r_e(\mathrm{NF}) = 1.365 \pm 0.002$ Å and $(\angle\mathrm{FNF})_e = 102.37 \pm 0.03°$.

However, even for such a small symmetric rotor there are considerable difficulties involved in obtaining this structure. These difficulties are due to having to determine the rotational constant B_υ in at least two vibrational states, usually $\upsilon_i = 0$ and $\upsilon_i = 1$, for each of the normal modes of vibration in order to obtain all the α_i^Bs. In molecules with one or more high wavenumber fundamental vibrations, some vibrational satellites may be too weak to be observed and, even though other techniques such as vibrational and electronic spectroscopy may provide some values of α_i^B, the problem of determining eqilibrium structures of large molecules becomes insuperable. In fact, NF_3 is one of the largest for which the equilibrium structure, commonly known as the r_e structure, is known. Even in cases where rotational constants can be obtained for $\upsilon = 1$ for all the normal vibrations, there may be additional effects, such as interaction between vibrations (for example Fermi resonance—see Section 5.2.6.3), which invalidates the expressions given in equations (4.73)–(4.75) and (4.112)–(4.114) for the α_is.

For molecules in which it is not possible to determine equilibrium rotational constants, but only those for the zero-point level, the simplest way to find an approximate structure is to treat them as if they were equilibrium values, ignoring the variation of zero-point motion with isotopic substitution. The structure obtained is called the r_0 structure.

If we use the values of B_0 for $^{14}NF_3$ and $^{15}NF_3$ given in Table 4.2, we obtain an r_0 structure in which $r_0(\mathrm{NF}) = 1.371 \pm 0.002$ Å and $(\angle\mathrm{FNF})_0 = 102.17 \pm 0.03°$. The differences between this and the r_e structure given above are typical except for bond lengths between a hydrogen atom and a heavy atom in which case r_e and r_0 may differ by as much as 0.03 Å.

An improvement on the r_0 structure is the substitution structure, or r_s structure, obtained using the general method suggested by Kraitchman,[25] whose equations give, for isotopic substitution of an atom, the coordinates of that atom with respect to the principal axes of the molecule before substitution. The r_s structure is nearer to the equilibrium structure than the r_0 structure and, for a diatomic molecule, it has been shown that

$$r_s \approx \tfrac{1}{2}(r_0 + r_e) \quad (4.128)$$

One of the largest molecules for which an r_s structure has been obtained[26] is aniline, shown in Figure 4.38. The benzene ring shows small deviations from a regular haxagon in the angles but no meaningful deviations in the bond lengths. As might be expected, by comparison with the pyramidal NH_3 molecule, the plane of the NH_2 group is not coplanar with the rest of the molecule.

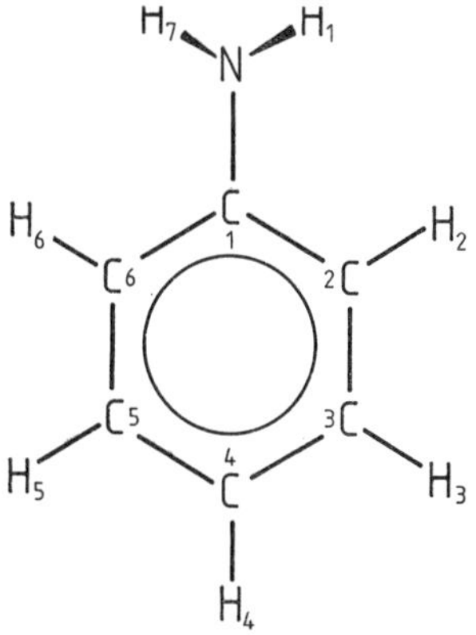

XY	r_s(XY)/Å	XYZ	$(\angle XYZ)_s/°$
NH_1	1.001 ± 0.01	H_1NH_7	113.1 ± 2
C_1N	1.402 ± 0.002	$C_6C_1C_2$	119.4 ± 0.2
C_1C_2	1.397 ± 0.003	$C_1C_2C_3$	120.1 ± 0.2
C_2C_3	1.394 ± 0.004	$H_2C_2C_3$	120.1 ± 0.2
C_3C_4	1.396 ± 0.002	$C_2C_3C_4$	120.7 ± 0.1
C_2H_2	1.082 ± 0.004	$H_3C_3C_2$	119.4 ± 0.1
C_3H_3	1.083 ± 0.002	$C_3C_4C_5$	118.9 ± 0.1
C_4H_4	1.080 ± 0.002		

Out-of-plane angle of NH_2 is $37.5 \pm 2°$

Figure 4.38 The r_s structure of the aniline molecule

4.10 ROTATIONAL SPECTROSCOPY OF WEAKLY BOUND COMPLEXES

Pure rotational transitions have been observed in a large number of complexes consisting of two, and sometimes more, weakly bound atomic or molecular species. These complexes can be usefully divided into two types, van der Waals complexes and hydrogen bonded complexes, although the binding mechanisms in both types are sufficiently similar that the distinction should not be thought of as a clear one.

Molecules and atoms resist close approach to each other because of strong repulsions between their electron clouds. The resulting potential energy V_r is inversely proportional to the separation R raised to a high power which is commonly taken to be 12, so that

$$V_r = \frac{C_{12}}{R^{12}} \tag{4.129}$$

where C_{12} is a constant for a particular complex.

At larger distances, where the V_r term becomes very small, weak attractive forces, the van der Waals forces, become important. In general, these forces may be of three kinds. If both partners in the complex have non-zero electric dipole moments, they tend to line up in such a way that the $\delta+$ end of one dipole is adjacent to the $\delta-$ end of the other. The force of attraction is then a dipole–dipole force. If one partner has no dipole moment, e.g. Ar or CO_2, then the partner which has a dipole moment will tend to induce a dipole moment in the other, the magnitude of the induced dipole moment depending on the permanent dipole moment of one partner and the polarizability (see Section 4.8.1) of the other. In this case there is a dipole–induced-dipole force of attraction. Finally, neither partner may have a permanent dipole moment. In this case, either of them may have a small charge asymmetry at any instant of time although there is no average asymmetry. The resulting small, instantaneous dipole moment can induce a small dipole moment in the other partner, and vice versa. The resulting attractive force is an induced-dipole–induced-dipole, or dispersion, force. If both molecules are polar, all three types of force may contribute to the attraction.

The potential energy due to each of the three types of van der Waals force is inversely proportional to R^6 so that the total contribution V_a of these attractive forces to the potential energy can be written as

$$V_a = -\frac{C_6}{R^6} \tag{4.130}$$

where C_6 is a constant for a particular complex. The total potential energy $V(R)$ due to repulsive and attractive forces is given by

$$V(R) = V_r + V_a = \frac{C_{12}}{R^{12}} - \frac{C_6}{R^6}$$
$$= 4\varepsilon\left[\left(\frac{R_0}{R}\right)^{12} - \left(\frac{R_0}{R}\right)^{6}\right] \tag{4.131}$$

This form of the potential energy is qualitatively, and often quantitatively, useful and is called, after its originator, the Lennard-Jones 12–6 potential.

Figure 4.39 shows the form of this potential. As $R \to \infty$, $V(R) \to 0$ and, when the partners experience maximum attraction, $R = R_e$ and the energy is decreased by ε, the depth of the potential well. This is very shallow, typically of the order of $100\,\mathrm{cm}^{-1}$ ($1.2\,\mathrm{kJ\,mol}^{-1}$). The quantity R_0 in equation (4.131) refers to the $\upsilon = 0$ vibrational level of this potential and is related to R_e by $R_e = 2^{\frac{1}{6}} R_0$.

The vibrational levels in this potential well are not equally spaced like those in Figure 1.22 but converge towards $V(R) = 0$ above which there is a continuum of levels. This behaviour resembles more closely that of the anharmonic oscillator (see Figure 5.4), to be discussed in Section 5.1.2, except that the Lennard-Jones potential curve is likely to support many fewer vibrational levels.

The potential in Figure 4.39 relates to the stretching vibration of one van der Waals partner relative to the other. In general there are two other vibrational modes introduced when complexation occurs. These are bending vibrations of one partner relative to the other. If, for example, the complex is planar, one bending vibration will be in the plane and the other perpendicular to the plane of the complex. The potentials for these two bending modes will not resemble the Lennard-Jones potential.

The hydrogen bond is also a result of weak attractive forces but which are, in general, stronger than van der Waals forces. Hydrogen bonding involves a hydrogen atom in a polar bond, such as O–H, N–H, F–H and S–H, in which the hydrogen atom is at the positive end of the bond dipole. The hydrogen atom is attracted towards an atom, such as oxygen, nitrogen or fluorine, which has lone pair electrons. This atom may be in the same molecule (intramolecular hydrogen bonding) or a neighbouring molecule (intermolecular hydrogen bonding).

Figure 4.40 shows hydrogen bonding involving O–H and N in, for example, phenol (C_6H_5OH) and pyridine (C_5H_5N). A commonly accepted, but oversimplified, picture is that the bond dipole and the dipole due to the lone pair tend to line up in the way illustrated. There are many examples of two molecules, or two parts of the same molecule, in which local dipoles could line up but the unique effectiveness of hydrogen in forming a bond is due to the very small electron density on that atom, minimizing electron repulsion between it and the lone pair.

Just as for a van der Waals complex, three new vibrational modes are introduced when two partners form a hydrogen bonded complex. Figure 4.41 illustrates these modes for a linear complex X–Y$\cdots$H–Z. The bending vibration ν_B is doubly degenerate in the linear complex but, if it is non-linear, the degeneracy is split. The stretching vibration ν_S is qualitatively similar to that in a van der Waals complex and the potential function resembles that in Figure 4.39 except that the depth of the potential well is much larger, of the order of $2000\,\mathrm{cm}^{-1}$ ($24\,\mathrm{kJ\,mol}^{-1}$).

Two experimental techniques, molecular beam electric resonance and microwave spectroscopy, have been used to observe rotational transitions in hydrogen bonded complexes and van der Waals complexes.

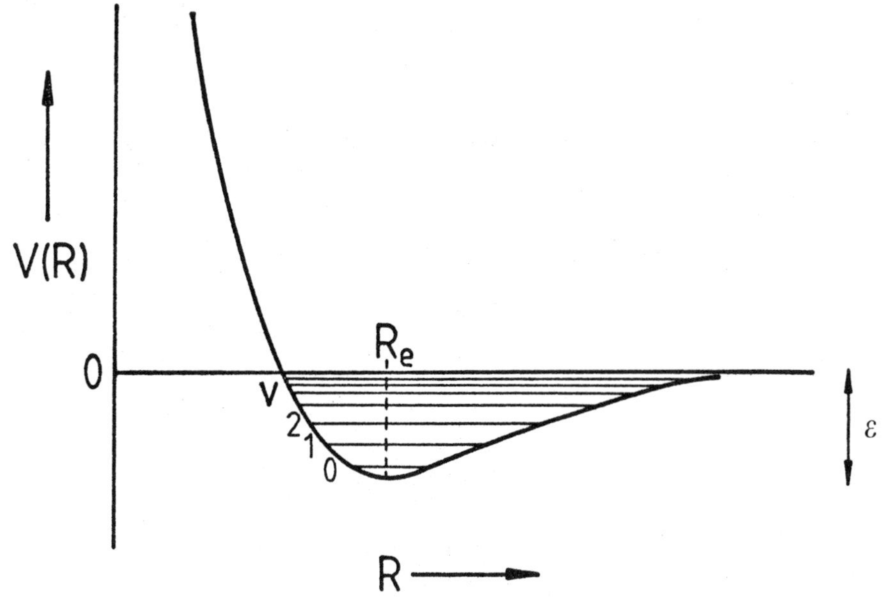

Figure 4.39 Lennard-Jones 12–6 potential curve for a van der Waals complex. The curve supports only a few vibrational levels

4.10.1 Molecular beam electric resonance spectroscopy of van der Waals and hydrogen bonded complexes

Figure 4.42 illustrates the main features of a molecular beam electric resonance spectrometer. The design is based on that originated by Rabi,† who used deflection by a magnetic rather than an electrostatic field.

The sample is in the form of a unidirectional stream of molecules. For the weakly bound van der Waals and hydrogen bonded complexes a supersonic jet (Section 3.4.3) emerging from a very small (*ca* 0.1 mm diameter) nozzle N is used exclusively. The reason for using such a supersonic jet is that the molecules undergo extreme cooling to a rotational temperature which may be of the order of 1 K and a vibrational temperature of the order of 100 K. At such vibrational temperatures, weakly bound molecules with a shallow potential well are more likely to be stable.

The molecules enter the evacuated electric resonance spectrometer in which there are two regions, A and B, in each of which there is an inhomogeneous electric field. This may be a dipole or a quadrupole electrostatic field. If it is a dipole field it is the electrical analogue of the inhomogeneous magnetic field used for the separation of atoms with different Zeeman states in the Stern–Gerlach experiment (see Section 6.1.2.1 and Figure 6.5). More commonly, in recent years, quadrupole fields have

†See Ramsay, N. F. (1956). *Molecular Beams*. Oxford University Press, Oxford.

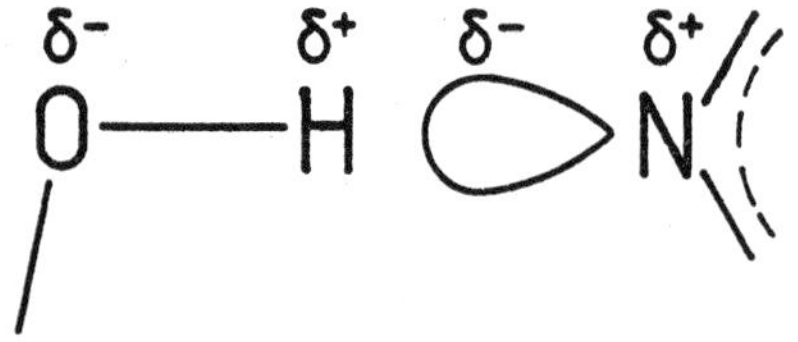

Figure 4.40 Hydrogen bonding between the phenol (C_6H_5OH) and pyridine (C_5H_5N) molecules

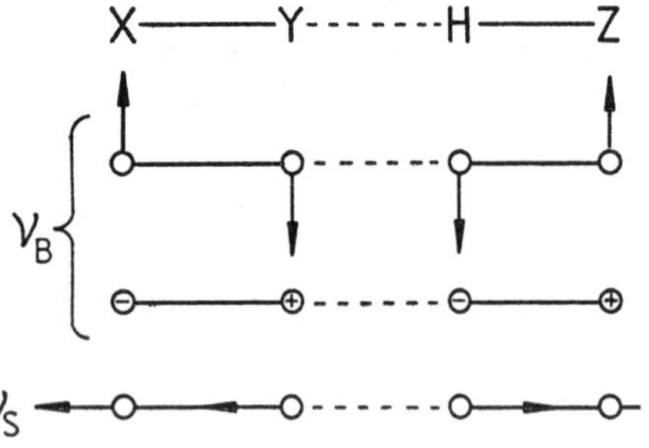

Figure 4.41 The bending and stretching vibrations, ν_B and ν_S, of a hydrogen bonded complex

been used.[27] Figure 4.42 shows how this is produced by four poles, across two of which a voltage is applied while the other two are earthed. The field increases outwards from the centre line between the poles.

We consider a molecule with just two rotational energy levels E_1 and E_2, as in Figure 4.43, and assume that the change of energy with field $\mathcal{E}$ is such that $dE_1/d\mathcal{E}$ is negative and $dE_2/d\mathcal{E}$ is positive, as shown. The effect of fields A and B is to deflect and refocus molecules in the beam but, for the energy level system in Figure 4.43, the only molecules which can be refocused are those in the upper state; those in the lower state are rejected from the beam as shown by the dotted line in Figure 4.42. The stopwire S prevents molecules which interact with the field weakly or not at all from reaching the detector D.

One type of detector which is commonly used is an electron bombardment analyser in which an electron gun is directed at the point of focus of the molecules. The resulting ions are mass analysed, as in a mass spectrometer, with a 60° sector magnet and the number of ions counted using an electron multiplier.

In the absence of component C the electric resonance spectrometer can be used for distinguishing molecules which have a permanent dipole moment from those which do not since only polar molecules can be brought to a focus at the detector. In this way it was shown, using an effusive molecular beam from a high temperature oven, that BaF_2, $BaCl_2$, $BaBr_2$, BaI_2, SrF_2, $SrCl_2$ and CaF_2 have dipole moments and must be bent molecules, in disagreement with simple valence theory, but that $CaCl_2$, $CaBr_2$, $SrBr_2$ and SrI_2 are non-polar and therfore linear.[28]

In order to detect rotational transitions in polar molecules, a resonance region C is introduced between the A and B fields. In this region molecules encounter both a homogeneous (Stark) electric field across two electrodes and radiofrequency, or microwave, radiation. If this radiation is tuned to match the separation of two energy levels in the molecules transitions between them are induced in region C but only polar molecules in one specific state will be refocused at the detector D, as shown by the solid lines in Figure 4.42. The dashed lines show how molecules in other states are brought to a focus but not at the detector.

If, for example, the radiofrequency or microwave source is tuned into resonance with the transition between the two states in Figure 4.43 and the quadrupolar fields are adjusted so that only molecules in the lower of the pair of levels are focused on the detector, there will be a *reduction* in the signal. This is called the flop out mode of operation whereas, in the flop in mode, an *increase* in signal results from detecting molecules in the upper state.

Since only molecules with occupied states with a positive Stark coefficient, like E_2 in Figure 4.43, are focused by the field we can

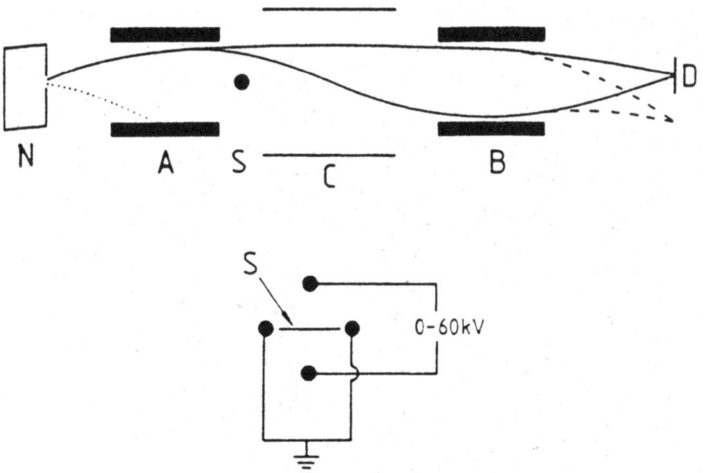

Figure 4.42 A molecular beam electric resonance spectrometer

define a polar molecule as one which has thermally occupied levels with a positive Stark coefficient.

The effect of the Stark field is to split a transition into several components which can be detected by tuning the incident radiation. If nuclear hyperfine splitting can be detected, the selection rules are:

$$\Delta J = 0, \ \pm 1; \ \Delta F = 0, \ \pm 1;$$
$$\Delta I = 0; \ \Delta M_F = 0, \ \pm 1$$

where M_F is the quantum number referring to the space quantization of the total angular momentum including nuclear spin (Section 4.2.5). When the radiation is parallel to the Stark field $\Delta M_F = 0$ and, when it is perpendicular, $\Delta M_F = \pm 1$.

Line widths are considerably reduced in this type of experiment because of the smaller Doppler line width in a unidirectional beam with a small spread of velocities. Resolution of the order of 5 kHz has been obtained.

In a supersonic jet, almost any bimolecular species held together by van der Waals forces can be observed. For example, in a mixture of Ar and HCl the species Ar_2, ArHCl and the non-polar species $(HCl)_2$ have been observed. In addition, ArNO, Ne^2HCl, XeHCl, $ArBF_3$, $KrBF_3$, $(NO)_2$, BF_3NO, BF_3CO, $(CO_2)_2$, all polar, and $(NO)_3$, $(NO)_4$, N_2O_2, NeHCl, COH_2, $(BF_3)_2$, $(BF_3)_3$ and $(BF_3)_4$, all non-polar, have been detected[29].

Rotational transitions have been observed[30] in four isotopic species of ArHCl. Of course, the transitions refer to the zero-point vibrational level and the structure determined from the rotational constants is an r_0 structure (see Section 4.9). In this level the molecule is bent and is, therefore, an asymmetric rotor, but it is nearly a prolate symmetric rotor with the asymmetry parameter κ very close to -1. With the centre of mass, G, of the H–Cl moiety as a suitable coordinate origin, R_0, the distance Ar$\cdots$G, was found to be 4.006 Å and θ_0, the angle Ar$\cdots$G$\cdots$H, about 45° in Ar^1HCl and 33° in Ar^2HCl.

The reason why this angle is so different in the two isotopic species is that, in general, van der Waals complexes have very large zero-point

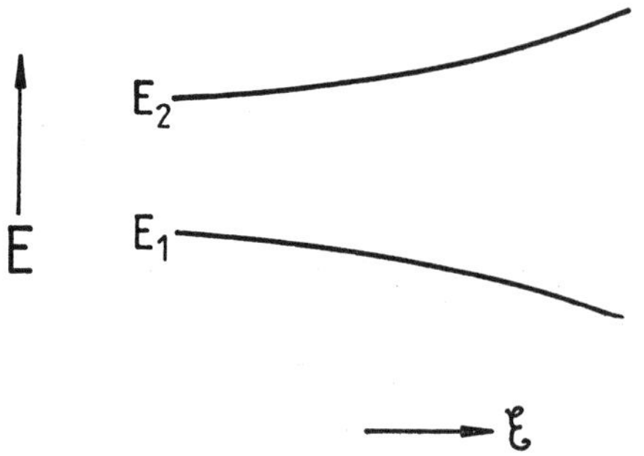

Figure 4.43 Variation of rotational energy levels E_1 and E_2 with electric field $\mathcal{E}$

motions. In the case of ArHCl the large amplitude vibrations include not only the $Ar \cdots HCl$ stretching vibration but a degenerate bending vibration (see, for example, Figure 4.41). It is the large contribution of the bending vibration to the zero-point motion which results in a bent r_0 structure.

The equilibrium, r_e, structure has been predicted [31] by accurate calculation of the potential energy as a function of R and θ. The results show that the deepest minimum corresponds to $\theta_e = 0°$, a linear molecule, and $R_e = 4.009$ Å. However, the calculations indicate a secondary minimum at $\theta_e = 180°$ with the argon attached to the chlorine atom.

From the value of the centrifugal distortion constant the Ar–HCl stretching fundamental vibration wavenumber was estimated to be $32\,cm^{-1}$, a very small value consistent with the expected shallow potential curve.

Other van der Waals molecules detected include $ArCO_2$, which is T-shaped with an $Ar \cdots C$ stretching wavenumber of $40\,cm^{-1}$, ArNa, with $R(Ar \cdots Na) = 4.991$ Å and a stretching wavenumber of $13.7\,cm^{-1}$, and KrClF, with $R(Kr \cdots Cl) = 3.388$ Å and an angle of 170°.

From observed rotational transitions in the hydrogen bonded complexes $^1HF \cdots {}^2HF$, $HF \cdots HCl$, $H_2O \cdots HOH$, the equilibrium structures shown in Figure 4.44 have been estimated.†

The structures of the complexes shown in Figure 4.40 suggest that an important factor determining them is the attraction of a hydrogen atom of one partner for a lone pair of electrons on the other. For example, this leads to a structure for $^1H–F \cdots {}^2H–F$ in which the angle shown would be 90°. Similarly, in $H_2O \cdots HOH$ the hydrogen atom of one molecule is attracted to one of the 'rabbit's ears' lone pairs of electrons on the other. Such simple arguments can give a qualitative guide to the shapes of some van der Waals complexes, but other factors are clearly important. In the T-shaped complexes $ArCO_2$ and ArC_2H_2 the argon atom appears to be attracted to the electrons in the π-orbitals of carbon dioxide and acetylene.

4.10.2 Microwave spectroscopy of van der Waals and hydrogen bonded complexes

4.10.2.1 Microwave spectroscopy without a supersonic jet

We have seen in Section 3.4.3 that van der Waals and hydrogen bonded complexes are readily formed and more easily held together in a supersonic jet. For this reason, the study of such complexes by microwave spectroscopy was, before the supersonic jet was developed, limited to a few special cases.

Hydrogen bonded complexes are more strongly bound than van der Waals complexes and therefore more likely to exist with appreciable concentration in an equilibrium mixture at vibrational temperatures accessible without the use of a supersonic jet.

The first complexes to be detected in this way were $CF_3COOH–HCOOH$ (trifluoroacetic acid–formic acid), $CF_3COOH–CH_3COOH$

†See Janda, K. C., Steed, J. M., Novick, S. E. and Klemperer, W. (1977). *J. Chem. Phys.*, **67**, 5162, and references cited therein.

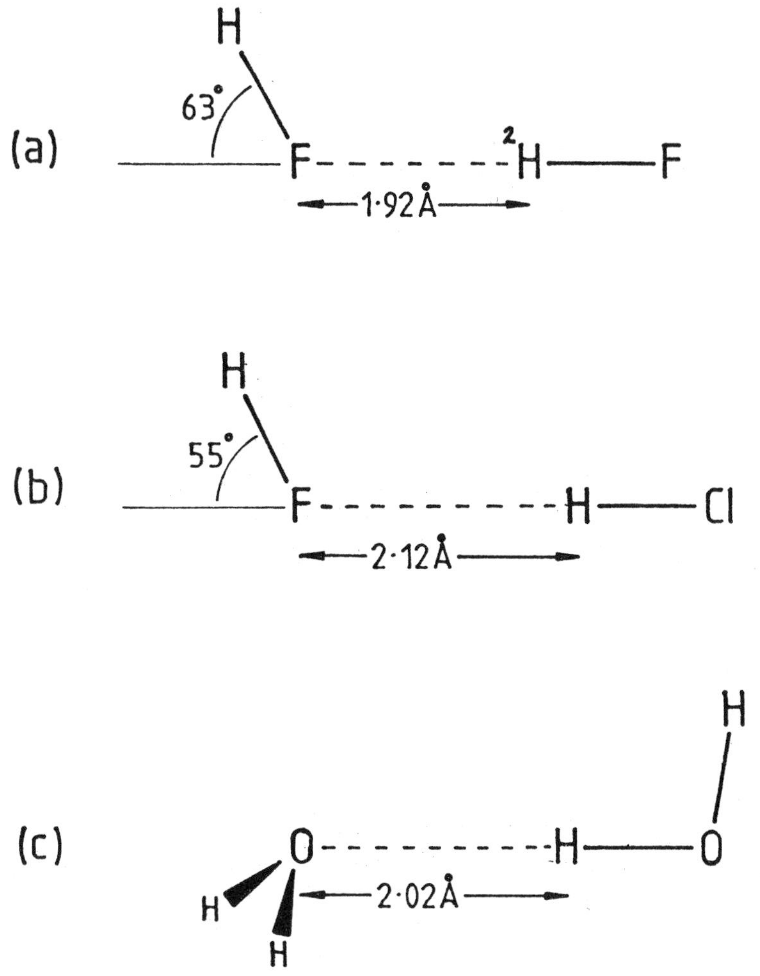

Figure 4.44 Structures of some hydrogen bonded complexes

(trifluoroacetic acid–acetic acid and CF_3COOH–CH_2FCOOH,[32] the purpose of the CF_3 groups being to give each complex a large dipole moment. These examples are particularly favourable because the molecules are joined by *two* hydrogen bonds, as shown for CF_3COOH–HCOOH in Figure 4.45, resulting in a favourable equilibrium constant. The partial pressure of the complex is expected to be about one tenth of the total pressure at 30 °C and can be increased by lowering the temperature.

The observed rotational transitions are broad; for example, the half intensity width of the $J = 16$–5 line of CF_3COOH–HCOOH is 13 MHz. The broadness is due both to the relatively high pressure of 0.1 Torr which has to be used and to the fact that transitions with $K_a = 4$ to $K_a = 15$ are not resolved (the complex is a prolate asymmetric rotor). With reasonable assumed geometries for the parent molecules and the assumption that they are little changed in the complex, the O···O

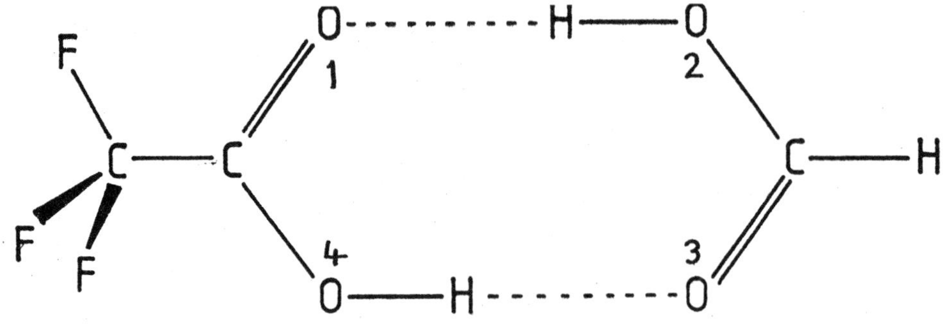

Figure 4.45 The trifluoroacetic acid–formic acid complex, held together by two hydrogen bonds

distance was estimated to be 2.69 Å. The microwave spectrum of this complex has been much more highly resolved in a supersonic jet[33] and the $O(1)\cdots O(2)$ and $O(3)\cdots O(4)$ distances were found to be 2.73 and 2.72 Å, respectively.

The microwave spectra of several more hydrogen bonded complexes, including $CH_3CN\cdots HF$, $(CH_3)_3CCN\cdots HF$, $H_2O\cdots HF$, $HCN\cdots HF$ and $HCN\cdots HCN$, have been observed.[†]

Figure 4.46 shows the spectrum of $CH_3CN\cdots HF$ in the 30 GHz region obtained at a total pressure of 0.12 Torr and a temperature of −50 °C, the low temperature favouring formation of the complex. The complex is a prolate symmetric rotor and the transitions in the figure all involve $J = 8$–7. Each transition is pressure broadened to such an extent that those with different values of K_a,

separated by the small effects of centrifugal distortion, are not resolved.

The various components in Figure 4.46 are vibrational satellites involving the two lowest wavenumber vibrations, ν_B and ν_S, of the complex. Both vibrations are characteristic of a hydrogen bonded complex and are illustrated in Figure 4.47. Vibration ν_B is a bending vibration and ν_S the hydrogen bond stretching vibration, analogous to the stretching vibration in van der Waals complexes. The wavenumbers of the $\upsilon = 1$–0 interval in ν_B and ν_S have been obtained from intensity measurements and the Boltzmann distribution law [equation (4.51)]. The intensity maximum in the ν_B series in Figure 4.46 is due to a factor of $(\nu_B + 1)$ which must be included in the Boltzmann equation because of the double degeneracy of ν_B giving

$$N_\upsilon/N_0 = (\upsilon_B + 1) \exp(-hc\tilde{\nu}/kT) \quad (4.132)$$

The resulting values of $\tilde{\nu}_B$ and $\tilde{\nu}_S$ are given in Table 4.9 for this and other hydrogen bonded complexes. Values of $\tilde{\nu}_B$ and $\tilde{\nu}_S$ have also been obtained for these and other complexes from observations of their infrared vibrational spectra.

[†]Millen, D. J. (1978). *J. Mol. Struct.*, **45**, 1. See also Legon, A. C., Millen, D. J. and Rogers, S. C. (1980). *Proc. R. Soc. (London)*, **A370**, 213, for $HCN\cdots HF$; Bevan, J. W., Legon, A. C., Millen, D. J. and Rogers, S. C. (1980). *Proc. R. Soc. (London)*, **A370**, 239 for $CH_3CN\cdots HF$; and Georgiou, A. S., Legon, A. C. and Millen, D. J. (1980). *Proc. R. Soc. (London)*, **A370**, 257 for $(CH_3)_3CCN\cdots HF$.

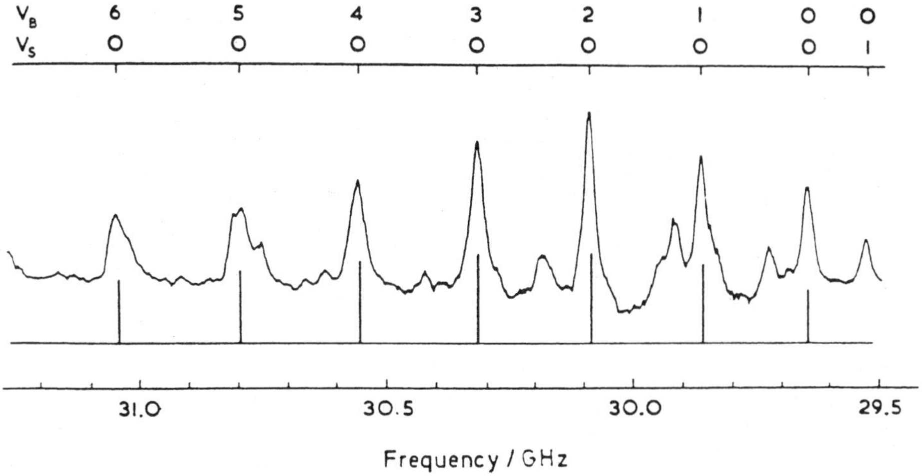

Figure 4.46 Part of the microwave spectrum of the hydrogen bonded $CH_3CN \cdots HF$ complex. Transitions shown are $J = 8–7$ and its vibrational satellites involving the hydrogen bond bending and stretching vibrations, ν_B and ν_S [Reproduced, with permission, from Millen, D. J. (1978). *J. Mol. Struct.*, **45**, 1]

Table 4.9 Hydrogen bond bending (ν_B) and stretching (ν_S) vibration wavenumbers and hydrogen bond lengths

Complex	$\tilde{\nu}_B/cm^{-1}$	$\tilde{\nu}_S/cm^{-1}$	$r(X \cdots H)/Å$
$CH_3CN \cdots HF$	45 ± 15	181 ± 20	2.74
$H_2O \cdots HF$	94 ± 20[a]	198 ± 20	2.68
	180 ± 20[b]		
$HCN \cdots HF$	91 ± 20	196 ± 15	2.81

[a]Out-of-plane bend.
[b]In-plane bend.

The $H_2O \cdots HF$ complex is a planar C_{2v} structure so that the double degeneracy of ν_B is split to give an in-plane and out-of-plane bending vibration.

Comparison of $\tilde{\nu}_S$ with the much smaller wavenumbers associated with stretching of a van der Waals bond shows how much deeper is the potential well for hydrogen bond stretching.

Table 4.9 also gives the lengths of the hydrogen bonds, which are considerably longer in these than in the complexes in Figure 4.44.

Special cases of hydrogen bonding are those in which the attractive force is between two groups in the same molecule—intramolecular hydrogen bonding. This may occur in, for example, 2-hydroxybenzaldehyde [$C_6H_4(OH)CHO$], in which there may be hydrogen bonding between the adjacent C=O and O–H groups. Of particular interest are those molecules in which, if the hydrogen involved in the bond is midway between the other atoms, the molecule has a higher degree

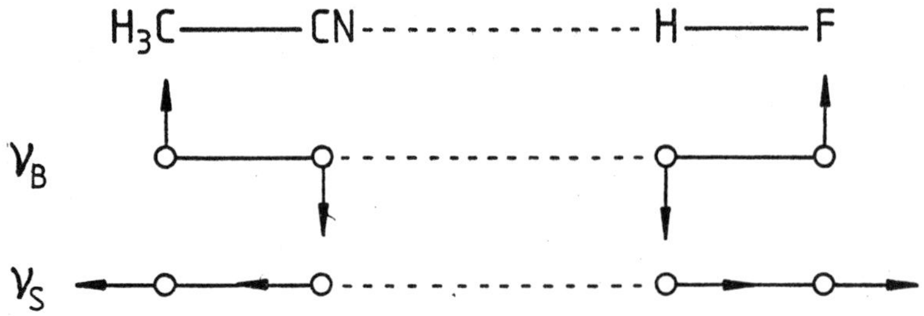

Figure 4.47 Hydrogen bond bending and stretching vibrations, ν_B and ν_S, in the $CH_3CN \cdots HF$ complex

of symmetry than it would have if the hydrogen were not in the midway position. Such examples are 6-hydroxy-2-formylfulvene (HFF), malonaldehyde and tropolone, shown in Figure 4.48(a), (b) and (c), respectively. In all three, if the hydrogen atom involved in the hydrogen bond is symmetrically disposed between the two oxygen atoms in the equilibrium configuration, each molecule has a C_2 symmetry axis, as shown in Figure 4.48(ii). If the equilibrium configuration is that shown in Figure 4.48(i), or in Figure 4.48(iii), where the configuration is indistinguishable from that in Figure 4.48(i), the C_2 axis is lost.

In internally hydrogen bonded molecules there is no longer a low wavenumber, facile bending vibration like that in Figure 4.47. The analogue of the stretching vibration remains but it involves not only the motion of the hydrogen between the two oxygen atoms but also a shift of carbon–carbon double and single bonds. The vibration associated with this complex motion we shall call ν_H.

Figure 4.49(a) illustrates the vibrational potential function for ν_H for the case where the equilibrium structure is the symmetrical one in Figure 4.48(ii). The vibrational levels are evenly spaced. Figure 4.49(b) illustrates the situation where the equilibrium configuration is (i) [or (iii)] but the potential energy barrier in going from (i) to (iii), via (ii), is not large. The effect of the barrier is to cause irregularities in the vibrational levels, bringing $v_H = 0$ and 1 close together, $v_H = 2$ or 3 close, but not as close as 0 and 1, and so on for higher levels except that above the barrier they eventually become evenly spaced. (The effect of a barrier of this kind on the vibrational levels is discussed in more detail in Section 5.2.7.1.) In an extreme case, such as that in Figure 4.49(c), the barrier is so high that the $v_H = 0$ and 1 levels have merged, as have higher pairs of levels.

Microwave spectroscopy of such molecules is able to distinguish between the three cases in Figure 4.49 using two types of evidence. The first involves nuclear spin statistical weights.

All three molecules are prolate asymmetric rotors [$\kappa = -0.34$, -0.44 and -0.28 for the molecules in Figure 4.48(a), (b) and (c), respectively] and the C_2 axis, in the symmetrical structure, is the a-axis. Therefore, if the equilibrium structure is (ii) there is, for all three molecules, a nuclear spin statistical weight alternation with the quantum number K_a of 10:6 for K_a even:odd for HFF and tropolone, since they each have two pairs of hydrogen atoms symmetrically disposed about the C_2 axis, and 1:3 for K_a even:odd for malonaldehyde, which has only one pair. Such alternations were observed in the microwave rotational spectra of HFF[34] and malonaldehyde[35] and are indicative of a potential function like that in Figure 4.49(a) or (b) in which the $v_H = 0$ and 1 levels are separated

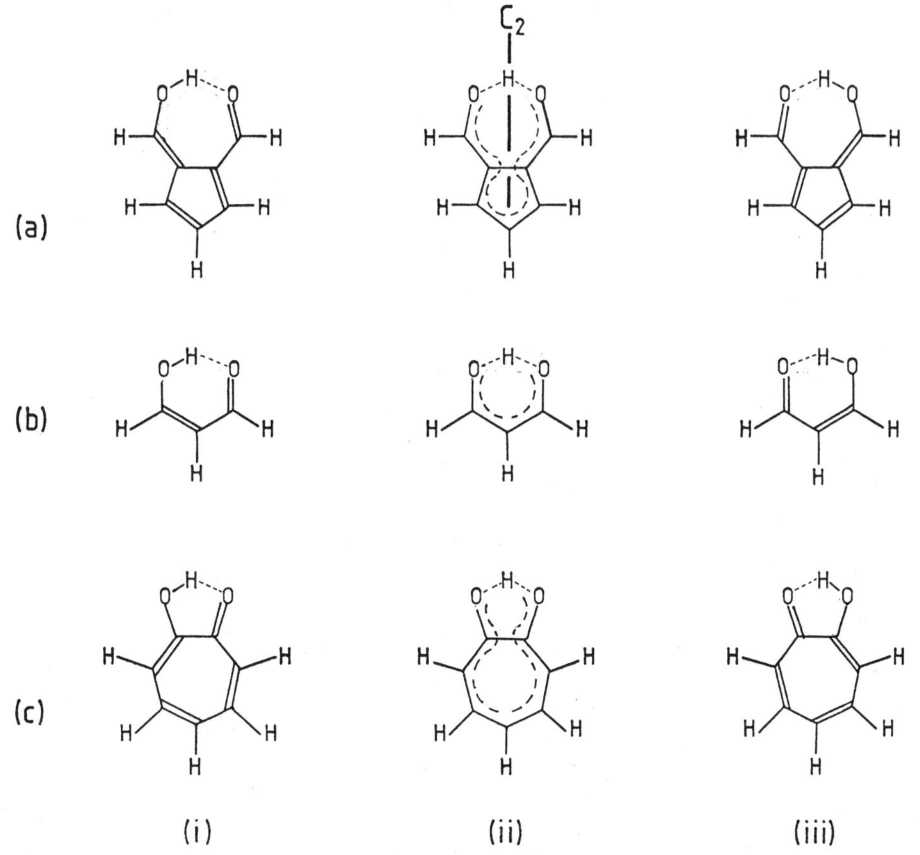

Figure 4.48 Intramolecular hydrogen bonding in the (a) 6-hydroxy-2-formylfulvene, (b) malonaldehyde and (c) tropolone molecules. In configurations (ii) the hydrogen atom is half way between the oxygen atoms and the molecules have a C_2 axis. In configuration (i) or (iii) the hydrogen atom is closer to one of the oxygen atoms and there is no longer a C_2 axis

sufficiently for rotational transitions within each of them to be resolved.

The second piece of evidence serves to distinguish a potential function such as that in Figure 4.49(a) from that in Figure 4.49(b). In the former case the $\upsilon_H = 0$ and 1 levels are relatively widely separated and any satellite rotational transitions with $\upsilon_H = 1$ are weak or unobservable whereas, in the latter case, the separation is small and satellites are observed. In addition, since the vibration ν_H is antisymmetric to rotation about the C_2 axis, the nuclear spin alternation for K_a even:odd would be reversed for $\upsilon_H = 1$ compared with $\upsilon_H = 0$.

In the microwave spectrum of HFF an intensity alternation of 10:6 is observed but there are no vibrational satellites with a 6:10 alternation. A lower limit of 150 cm^{-1} has been put on the $\upsilon_H = 1$–0 interval, which implies that the hydrogen bond is either symmetrical, or unsymmetrical with a very low barrier, and the potential function is similar to that in Figure 4.49(a).

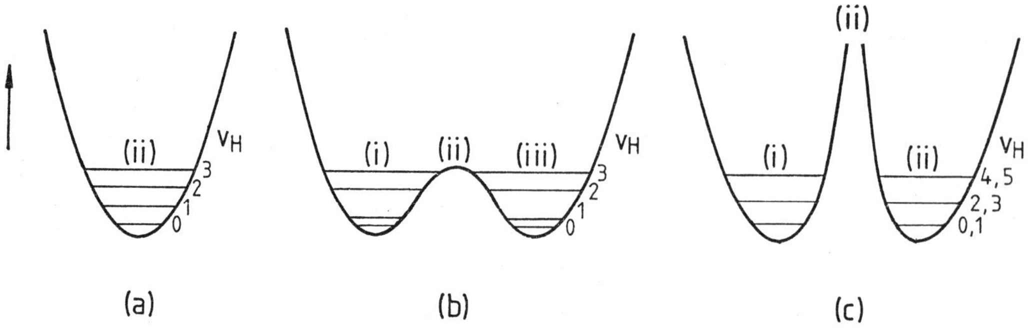

Figure 4.49 Vibrational potential functions for v_H when (a) the intramolecular hydrogen bonded molecule has a symmetrically disposed hydrogen atom, (b) the hydrogen atom is not symmetrically disposed but the barrier to the symmetrical position is not high and (c) the hydrogen atom is not symmetrically disposed but the barrier to the symmetrical position is high

In the case of malonaldehyde, the microwave spectrum shows an intensity alternation of 1:3 for K_a even:odd for rotational transitions associated with $v_H = 0$ but a very intense set of satellites with an alternation of 3:1. The potential function is, therefore, like that in Figure 4.49(b). The hydrogen bond is unsymmetrical and relative intensity measurements of rotational transitions indicate a $v_H = 1$–0 separation of $16 \pm 14\,\text{cm}^{-1}$.

The electronic spectrum of tropolone shows many pairs of intense transitions originating in the $v_H = 0$ and 1 levels of the ground electronic state.[36] Transitions from $v_H = 0$ show an intensity alternation for K_a even:odd of 10:6 and those from $v_H = 1$ show an alternation of 6:10. This observation, and the fact that the relative intensities of the two bands indicate a $v_H = 1$–0 separation of less than $50\,\text{cm}^{-1}$, show that the hydrogen bond is unsymmetrical and that the potential function is of the type shown in Figure 4.49(b). Assignment of the microwave spectrum[37] gives a value of $0.9777(11)\,\text{cm}^{-1}$ for this separation.

4.10.2.2 Microwave spectroscopy with a supersonic jet: pulsed nozzle Fourier transform microwave spectroscopy

The supersonic jet was first employed in microwave spectroscopy[38] using a molecular beam electric resonance spectrometer described in Section 4.10.1 and shown in Figure 4.42. If a microwave, rather than a radiofrequency, source is introduced in region C of the spectrometer, transitions in the microwave region of the spectrum can be observed.

A complementary method involving the use of a supersonic jet is the technique of pulsed nozzle Fourier transform microwave spectroscopy.[39] The method involves the observation of rotational transitions in the time domain, rather than the frequency domain familiar from conventional microwave spectroscopy.

The sample is subjected to a short pulse, of about $0.16\,\mu\text{s}$ duration, of microwave radiation of frequency ν. Because of the relationship

$$\Delta t \Delta E \geqslant \hbar \tag{4.133}$$

the pulsing of the radiation with a pulse length Δt produces a frequency spread $\Delta\nu(= \Delta E/hc)$ of about 1 MHz centred at the frequency ν. This range of frequencies constitutes a broad band source of microwave radiation which induces absorption by the sample at frequencies which lie within this range. After the radiation pulse has ended, emission of microwave radiation, from the rotational states populated during the pulse, takes place. Since the lifetimes of the emitting states are of the order of 100 μs the emission can be detected during this decay period, after the radiation pulse has died away and after a delay of a few further microseconds.

The microwave emission constitutes a spectrum in the time domain in which use is made of the multiplex, or Fellgett, advantage of detecting all the radiation all the time. Identical time domain spectra from many radiation pulses are collected, digitized and co-added. This spectrum then consists of a series of cosine waves of different wavelengths corresponding to the different frequencies emitted by the sample. The technique is very similar to that employed in Fourier transform NMR spectroscopy in which the pulsed source is of radiofrequency, rather than microwave, radiation.

The sample itself is in the form of a supersonic jet, pulsed to coincide with the radiation pulses. To increase the efficiency of collection of the radiation emitted by the sample the supersonic jet is directed into a Fabry–Perot cavity (see Figure 3.18), formed between two confocal concave mirrors, and perpendicular to its axis, as shown in Figure 4.50. The microwave radiation enters the cavity through one partially reflecting mirror and emerges through the other. As in the molecular beam electric resonance experiment, an advantage of the sample being in this form is that weakly bound complexes are stabilized owing to the very low vibrational temperature, and their spectra can be readily studied.

An example of a time domain spectrum is shown in Figure 4.51(a) and the corresponding frequency domain spectrum in Figure 4.51(b). The frequency domain spectrum would be expected to show just one line due to the unresolved $F = \frac{7}{2} - \frac{5}{2}$ and $\frac{5}{2} - \frac{3}{2}$ ^{35}Cl nuclear hyperfine components of the 2_{02}–1_{01} transition of the hydrogen bonded complex $H_2{}^{32}S \cdots H^{35}Cl$. The line would be at about 150 kHz and would give a time domain spectrum consisting of a single wave with a wavelength corresponding to the reciprocal of this frequency, about 6.7 μs, and dying away after about 200 μs. This wave can be seen in Figure 4.51(a) but there is clearly a 'beat frequency' superimposed due to the fact that the frequency domain spectrum shows two frequencies, ν_a and ν_b. When the corresponding two waves, of frequencies ν_a and ν_b, are added there is a resulting beating between the waves at a beat frequency ν_B given by

$$\nu_B = |\nu_a - \nu_b| \tag{4.134}$$

The doubling of the transition in Figure 4.51(b) is an artefact which is sometimes unavoidable in these experiments and is a consequence of the Doppler effect.

Using the technique of pulsed nozzle Fourier transform microwave spectroscopy, rotational spectra of the van der Waals complexes between benzene and the noble gas atoms neon,[40] argon,[41] krypton (^{84}Kr)[42] and xenon (^{129}Xe)[40] have been obtained. The complexes are all prolate symmetric rotors with the general structure shown in Figure 4.52. The noble gas atom is vertically above the centre of the benzene ring at a distance R_0. These values are given in Table 4.10 which also gives the estimated well depth ε for the van der Waals stretching mode.[43] The well depth was estimated from a vibrational force field derived from the quartic centrifugal distortion constants. The distance R_0 increases slightly with the size of the rare gas atom. The well depth increases with increasing polarizability of the noble gas atom. Xenon is the most polarizable atom, resulting in the most stable complex with a well depth of 431 cm^{-1}.

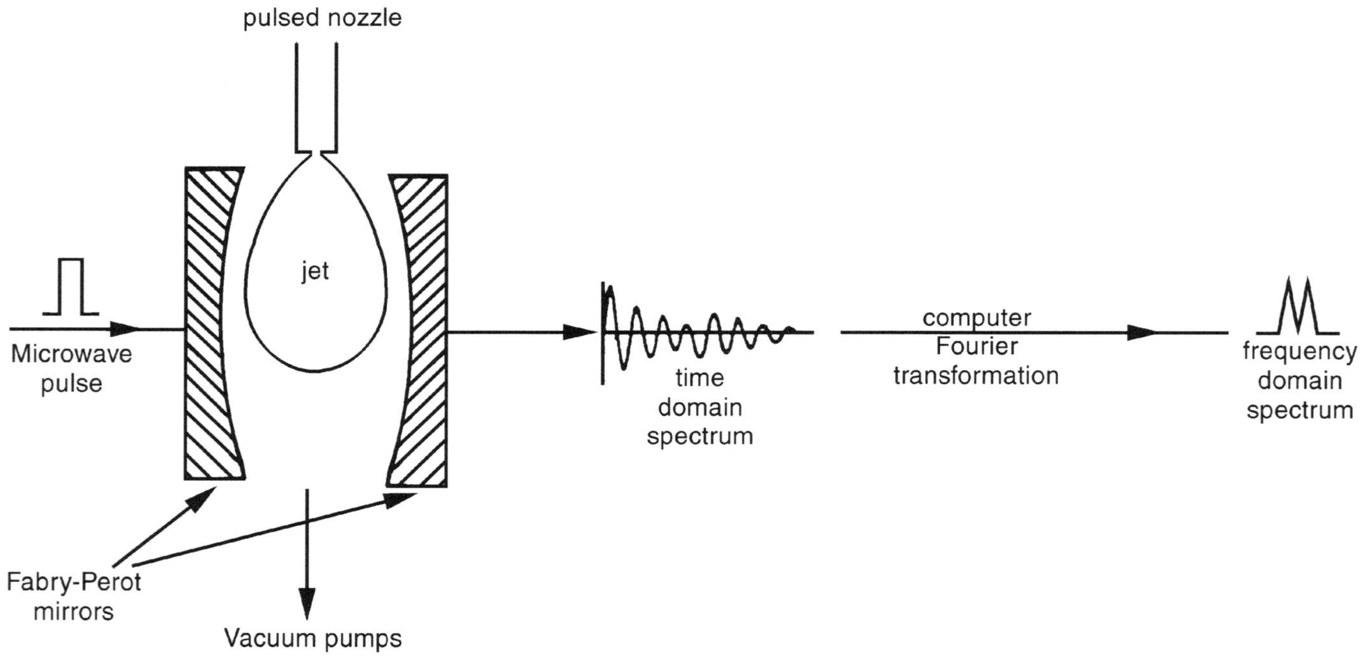

Figure 4.50 The supersonic jet in a pulsed nozzle Fourier transform microwave spectrometer

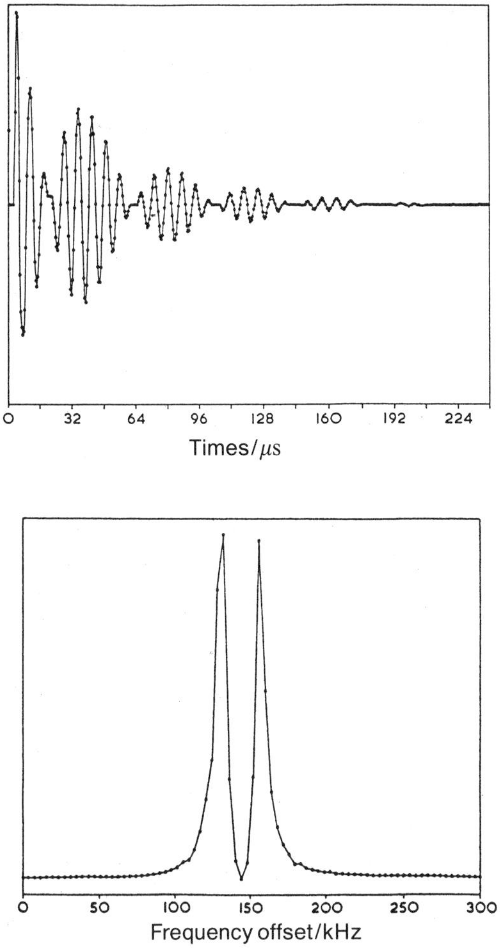

Figure 4.51 (a) Time domain spectrum showing a Doppler doubled 2_{02}–1_{01} rotational transition of $H_2{}^{32}S \cdots H^{35}Cl$. (b) The corresponding frequency domain spectrum [Reproduced, with permission, from Legon, A. C. (1983). *Annu. Rev. Phys. Chem.*, **34**, 275]

An ammonium halide, such as ammonium chloride (NH_4Cl), is normally encountered as a crystalline solid or in aqueous solution when it is completely ionized giving $NH_4{}^+$ and Cl^-. But is it ionic in the gas phase? This question, and the analogous question relating to the trimethylammonium halides [$(CH_3)_3N \cdots HX$],

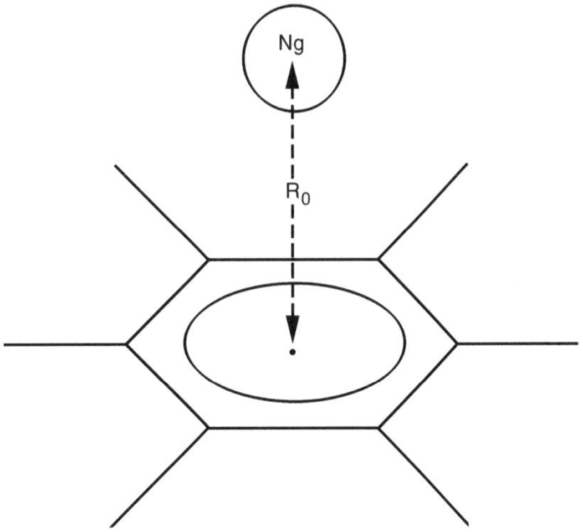

Figure 4.52 Benzene···noble gas (Ng) dimer

Table 4.10 Van der Waals bond lengths R_0 and stretching vibration well depths ε for benzene–noble gas complexes

Complex	R_0/Å	ε/cm^{-1}
Benzene···Ne	3.46	73
Benzene···Ar	3.59	251
Benzene···^{84}Kr	3.68	330
Benzene···^{129}Xe	3.83	431

has been answered using pulsed nozzle Fourier transform microwave spectroscopy.[44]

The method of formation of these species in the gas phase using a pulzed nozzle must avoid solidification. In the case of trimethylammonium iodide,[44] for example, the nozzle consisted of two concentric tubes. Down the inner one was passed a mixture of trimethylamine (*ca* 2%) in the argon carrier gas and, down the outer one, hydrogen iodide. The complex was formed only at the tip of the nozzle, thus avoiding solidification.

All the halogen atoms, except fluorine, have a nuclear spin quantum number I greater than $\frac{1}{2}$ and consequently have a nuclear quadrupole moment Q due to the asymmetry of the nuclear charge (see Section 4.2.5). The quadrupole moment interacts with the gradient of the electric field surrounding the nucleus and there is a corresponding nuclear quadrupole coupling constant $\chi(X)$ for the nucleus X. If, for example, we have a free iodide ion I^- (^{127}I has a nuclear spin quantum number of $\frac{5}{2}$), its spherical symmetry results in $\chi(I) = 0$. In solid sodium iodide, however, the spherical symmetry is slightly distorted by the nearby Na^+, resulting in the small value for $\chi(I)$ given in Table 4.11. At the opposite extreme is the covalent molecule HI, which has a large value of $\chi(I)$ in the gas phase. Comparison of this value with those for the gas phase complexes formed between phosphine and hydrogen iodide and between ammonia and hydrogen iodide indicates very little ionic character. On the other hand, the value for the gas phase complex between trimethylamine and hydrogen iodide indicates very strong ionic character and almost complete transfer of the proton from HI to the trimethylamine.

The degree of ionic character increases in the series of complexes $(CH_3)_3N \cdots HX$ with $X = F$, Cl, Br, I but there is very little ionic character in any of the $H_3N \cdots HX$ complexes. These conclusions are consistent with the inductive effect of the methyl groups giving the nitrogen atom a higher proton affinity and with a lowering of the proton affinity of X^- on going from F^- to I^-.

In all these species there is a second nuclear quadrupole coupling constant due to the nitrogen atom also having a quadrupole moment. This complicates the hyperfine splittings which are observed but is not a strong indicator of ionic character.

For the gas phase species in Table 4.11 the values for the intermolecular stretching force constant k_σ have been obtained from the centrifugal distortion constant D_J. These values, compared with that for ionic sodium iodide, again indicate a high degree of ionic character in the trimethylamine–hydrogen iodide complex.

Table 4.11 Nuclear quadrupole coupling constants $\chi(I)$ and intermolecular stretching force constants k_σ for some hydrogen bonded and ionic iodides

Species	$\chi(I)$/MHz	k_σ/N m^{-1}
HI	−1828.42	—
$H_3P \cdots HI$	−1461.022	3.409
$H_3N \cdots HI$	−1324.891	7.18
$(CH_3)_3H^+ \cdots I^-$	−341.204	66.5
$Na^+ \cdots I^-$	−262.14	77.0

The search for rotational microwave spectra of mixed noble gas dimers, e.g. Ne$\cdots$Kr, was made difficult because of the expected small dipole moment and, more importantly, the low concentration, even in a supersonic jet. Both are a consequence of the complexes being held together by only weak dispersion forces. In addition, krypton and, particularly, xenon have a number of naturally occurring isotopes. Although the resolution of the spectrum of each isotopomer is easily achieved this again contributes to the difficulties caused by very low concentration. However, spectra of Ne$\cdots$Xe, Ar$\cdots$Xe, Kr$\cdots$Xe, Ne$\cdots$Ar, Ne$\cdots$Kr and Ar$\cdots$Kr have been found[45] using pulsed nozzle Fourier transform microwave spectroscopy. Figure 4.53 shows, for example, the $J = 2$–1 transition of $^{20}Ne \cdots ^{83}Kr$. Although ^{20}Ne has a natural abundance of 90.5%, that of ^{83}Kr is only 11.5%. The ^{20}Ne nucleus has zero nuclear spin but ^{83}Kr has $I = \frac{9}{2}$. Therefore, the $J = 2$ level has $F = \frac{13}{2}, \frac{11}{2}, \frac{9}{2}, \frac{7}{2}$ and $\frac{5}{2}$ and the $J = 1$ level has $F = \frac{11}{2}, \frac{9}{2}$ and $\frac{7}{2}$. The selection rule $\Delta F = 0, \pm 1$ allows nine transitions but the $F = \frac{5}{2}$–$\frac{7}{2}$ transition is not resolved. Each transition is doubled for the same reason as that in Figure 4.51(b), a consequence of the Doppler effect.

Table 4.12 gives the dipole moment, the internuclear distance, obtained by isotopic substitution, and the stretching vibration well depth for these dimers. The dipole moments are all very small, comparable in magnitude to that

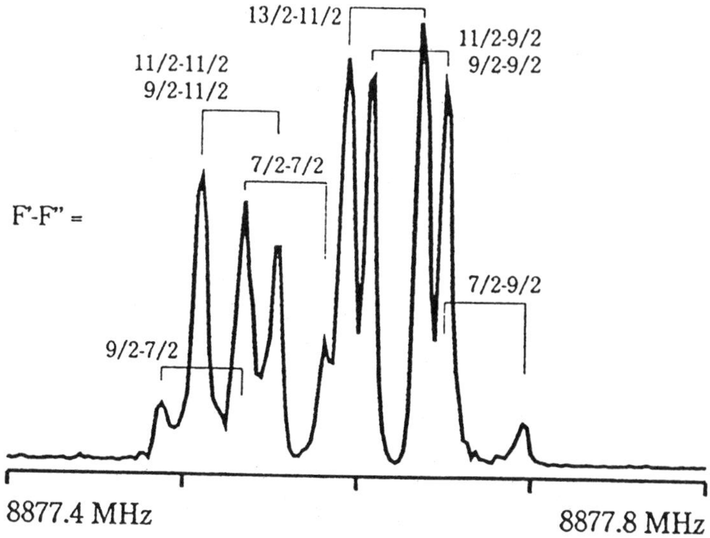

Figure 4.53 Nuclear hyperfine structure in the $J = 2$–1 transition of $^{20}Ne \cdots {}^{83}Kr$ due to the nuclear spin of $^{83}Kr(I = \frac{9}{2})$ [Reproduced, with permission, from Xu, Y., Jäger, W., Djauhari, J. and Gerry, M. C. L. (1995). *J. Chem. Phys.*, **103**, 2827]

Table 4.12 Dipole moment μ, internuclear distance R_S and well depth ε for various noble gas van der Waals dimers

Dimer	μ/D	R_S/Å	ε/cm^{-1}
Ne $\cdots$ Ar	0.0029	3.5353	45.8
Ne $\cdots$ Kr	0.011	3.6375	48.2
Ar $\cdots$ Kr	0.0034	3.8929	115.5
Ne $\cdots$ Xe	0.012	3.8810	49.0
Ar $\cdots$ Xe	0.014	4.0940	126.2
Kr $\cdots$ Xe	0.007	4.2028	155.1

due to deuterium substitution in an otherwise symmetrical molecule, such as monodeutero-acetylene. The well depths are typically small, as expected for van der Waals complexes, but the larger ones are those where one or both atoms have relatively high polarizability. Consequently, the largest well depth is for Kr $\cdots$ Xe.

BIBLIOGRAPHY

CARRINGTON, A. (1974). *Microwave Spectroscopy of Free Radicals*. Academic Press, New York
GORDY, W. and COOK, R. (1970). *Microwave Molecular Spectra*. Interscience, New York
HERZBERG, G. (1945). *Infrared and Raman Spectra*. Van Nostrand, New York
HERZBERG, G. (1950). *Spectra of Diatomic Molecules*. Van Nostrand, New York
KROTO, H. W. (1975). *Molecular Rotation Spectra*. Wiley, London; (1992) Dover, New York
LONG, D. A. (1977). *Raman Spectroscopy*. McGraw-Hill, London
SUGDEN, T. M. and KENNEY, C. N. (1965). *Microwave Spectroscopy of Gases*. Van Nostrand, London

TOWNES, C. H. and SCHAWLOW, A. L. (1955). *Microwave Spectroscopy*. McGraw-Hill, New York

WOLLRAB, J. E. (1967). *Rotational Spectra and Molecular Structure*. Academic Press, New York

Interstellar molecules

RANK, D. M., TOWNES, C. H. and WELCH, W. J. (1971). *Science*, **174**, 1083

SOMERVILLE, W. B. (1977). *Adv. At. Mol. Phys.*, **13**, 383

5 VIBRATIONAL SPECTROSCOPY

5.1 DIATOMIC MOLECULES

5.1.1 Harmonic oscillator approximation

In Section 1.3.4 we have seen how the quantum mechanical treatment of vibrational motion in a diatomic molecule, approximating to the harmonic motion of two particles of reduced mass μ connected by a spring of force constant k, leads to vibrational energy levels E_{υ}, given by

$$E_{\upsilon} = h\nu(\upsilon + \tfrac{1}{2}) \qquad (5.1)$$

where ν is the classical vibration frequency related to μ and k by

$$\nu = \frac{1}{2\pi}(k/\mu)^{\frac{1}{2}} \qquad (5.2)$$

and the vibrational quantum number υ can take any positive integral value, or zero.

The force constant can be regarded as a measure of the strength of the spring (bond) connecting the two particles. Table 5.1 gives some typical values which illustrate the increase of k with the bond order. The most favoured unit for force constants in the SI system is probably aJ Å^{-2} or N m^{-1}.[†]

Figure 1.22 illustrates the potential function, vibrational wave functions and energy levels. Vibrational term values $G(\upsilon)$, unlike rotational term values, are expressed almost invariably with the dimensions of wavenumber. It follows from equation (5.1) that

$$G(\upsilon) = \omega(\upsilon + \tfrac{1}{2}) \qquad (5.3)$$

[†]Previously the unit of k in common usage was mdyn Å^{-1} but, fortunately, 1 mdyn Å^{-1} = 1 aJ Å^{-2}. Alternative SI units of k are N m^{-1} or J m^{-2}, where 1 aJ Å^{-2} = 10^2 N m^{-1} = 10^2 J m^{-2}.

Table 5.1 Force constants, k, of some diatomic molecules

Molecule	k/aJ Å^{-2}	Molecule	k/aJ Å^{-2}
HCl	5.16	O_2	11.41
HF	9.64	NO	15.48
Cl_2	3.20	CO	18.55
F_2	4.45	N_2	22.41

where ω is the classical vibration wavenumber.[‡]

The transition moment [equation (2.21)] for a transition between lower and upper vibrational states with wave functions ψ_{υ}'' and ψ_{υ}', respectively, is given by

$$\boldsymbol{R}^{\upsilon'\upsilon''} = \int \psi_{\upsilon}'^{*}\boldsymbol{\mu}\psi_{\upsilon}''\,\mathrm{d}x \qquad (5.4)$$

where x is $(r - r_e)$, the displacement of the internuclear distance from equilibrium, and $\boldsymbol{\mu}$ is the electric dipole moment, which is zero for a homonuclear diatomic molecule. For a heteronuclear diatomic, $\boldsymbol{\mu}$ is non-zero and varies with x. This variation can be expressed as a Taylor series expansion:

$$\boldsymbol{\mu} = \boldsymbol{\mu}_e + (\mathrm{d}\boldsymbol{\mu}/\mathrm{d}x)_e x + \frac{1}{2!}(\mathrm{d}^2\boldsymbol{\mu}/\mathrm{d}x^2)_e x^2 + \ldots \qquad (5.5)$$

where the subscript e refers to the equilibrium configuration. The transition moment of equation (5.4) now becomes

[‡]ω is the symbol commonly used by spectroscopists for the classical vibration wavenumber and is to be contrasted with its equally common usage for the classical circular vibration frequency to which we referred briefly in equations (1.102) and (1.103). See also the footnote on p. 102.

$$\boldsymbol{R}^{v'v''} = \boldsymbol{\mu}_e \int \psi_v'^* \psi_v'' \, dx + (d\boldsymbol{\mu}/dx)_e \int \psi_v'^* x \psi_v'' \, dx + \ldots \quad (5.6)$$

Since ψ_v' and ψ_v'' are eigenfunctions of the same hamiltonian, namely the one given in equation (1.96), they are necessarily orthogonal which means that, when $v' \neq v''$,

$$\int \psi_v'^* \psi_v'' \, dx = 0 \quad (5.7)$$

Equation (5.6) becomes

$$\boldsymbol{R}^{v'v''} = (d\boldsymbol{\mu}/dx)_e \int \psi_v'^* x \psi_v'' \, dx + \ldots \quad (5.8)$$

The first term in equation (5.8) is non-zero only if $\Delta v = \pm 1$, which constitutes the vibrational selection rule. Since Δv refers to v(upper) − v(lower), the selection rule is effectively $\Delta v = +1$.

In the harmonic oscillator approximation all transitions obeying this selection rule are coincident at a wavenumber ω.

If the spectrum is observed in absorption, as it usually is, and at normal temperatures, intensities of the transitions decrease rapidly with increasing v'' since the population N_v of the vth vibrational level is related to N_0 by the Boltzmann factor:

$$N_v/N_0 = \exp(-E_v/kT) \quad (5.9)$$

Each vibrational transition gives rise to a band in the spectrum: the word 'line' is reserved for describing a transition between rotational levels (we shall see in Section 5.1.3 that, in the gas phase, every vibrational band has fine structure consisting of rotational lines). All bands with $v'' \neq 0$ are called hot bands because, as indicated by the Boltzmann factor in equation (5.9), the populations of the lower levels of such transitions, and therefore the intensities of the transitions, are increased at higher temperatures.

The transition intensities are dependent also on $|\boldsymbol{R}^{v'v''}|^2$ and therefore, according to equation (5.8), on $(d\boldsymbol{\mu}/dx)_e^2$. Figure 5.1 shows how μ varies† with internuclear distance in a typical heteronuclear diatomic molecule. Obviously $\mu \rightarrow 0$ when $r \rightarrow 0$ and the nuclei coalesce. For most neutral diatomic molecules $\mu \rightarrow 0$ also when $r \rightarrow \infty$ because, as $r \rightarrow \infty$, the molecule dissociates into neutral atoms. Therefore, between $r = 0$ and ∞ there must be a maximum value of μ. Figure 5.1 has been drawn with this maximum at $r < r_e$, giving a negative slope $d\mu/dr$ at r_e. The maximum may also be at $r > r_e$, giving a positive slope at r_e. If it happened that $(d\mu/dr)$ were zero at r_e then the transition moment and the intensity would be zero. In general, $(d\mu/dr)$ is sensitive to the environment of the molecule so that intensities may be somewhat different in the liquid and gas phases and, in solution, may vary according to the solvent used.

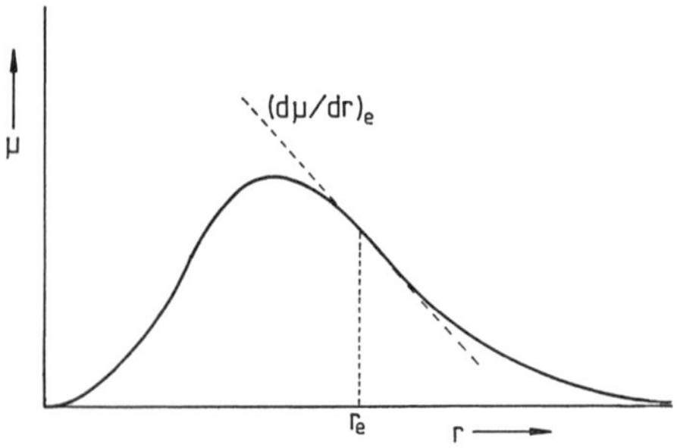

Figure 5.1 Variation of the dipole moment, μ, with internuclear distance, r, in a heteronuclear diatomic molecule

In both heteronuclear and homonuclear diatomic molecules the polarizability (see Section 4.8.1) varies during vibrational motion leading to a vibrational Raman effect. A classical treatment, analogous to that for rotation in Section 4.8.1, leads to a variation with time of the dipole moment μ, induced by irradiation of the sample with intense monochromatic radiation of wavenumber $\tilde{\nu}$, given by

†We use μ to indicate the magnitude of the vector $\boldsymbol{\mu}$.

$$\mu = \alpha_{0,\mathrm{v}} A \sin 2\pi c\tilde{\nu}t - \tfrac{1}{2}\alpha_{1,\mathrm{v}} A \cos 2\pi c(\tilde{\nu} + \tilde{\nu}_{\mathrm{vib}})t + \tfrac{1}{2}\alpha_{1,\mathrm{v}} A \cos 2\pi c(\tilde{\nu} - \tilde{\nu}_{\mathrm{vib}})t \qquad (5.10)$$

where $\alpha_{0,\mathrm{v}}$ is the average polarizability during vibration, $\alpha_{1,\mathrm{v}}$ is the amplitude of the change of polarizability due to vibration, A is the amplitude of the oscillating electric field of the incident radiation [equation (4.120)] and $\tilde{\nu}_{\mathrm{vib}}$ is the vibration wavenumber. Equation (5.10) is analogous to equation (4.122) for rotation except that the second and third terms in equation (5.10) correspond to Raman scattering with a wavenumber of $(\tilde{\nu} + \tilde{\nu}_{\mathrm{vib}})$ and $(\tilde{\nu} - \tilde{\nu}_{\mathrm{vib}})$—the anti-Stokes and Stokes Raman scattering, respectively. Whereas for rotation the polarizability changes at *twice* the rotational frequency ν_{rot}, during a complete vibrational cycle the polarizability goes through only one cycle so that the polarizability changes at the *same* frequency as the frequency of vibration, ν_{vib}; this accounts for the absence of the factor of two in equation (5.10) compared with equation (4.122).

The change of polarizability with vibrational displacement x can be represented by a Taylor series:

$$\boldsymbol{\alpha} = \boldsymbol{\alpha}_{\mathrm{e}} + (\mathrm{d}\boldsymbol{\alpha}/\mathrm{d}x)_{\mathrm{e}} x + \frac{1}{2!}(\mathrm{d}^2\boldsymbol{\alpha}/\mathrm{d}x^2)_{\mathrm{e}} x^2 + \ldots \qquad (5.11)$$

analogous to that for the variation of dipole moment given in equations (5.5). By analogy with equation (5.8), the vibrational Raman transition moment $\boldsymbol{R}^{\upsilon'\upsilon''}$ is given by

$$\boldsymbol{R}^{\upsilon'\upsilon''} = (\mathrm{d}\boldsymbol{\alpha}/\mathrm{d}x)_{\mathrm{e}} A \int \psi_{\upsilon}'^{*} x \psi_{\upsilon}'' \,\mathrm{d}x + \ldots \qquad (5.12)$$

and the first term is non-zero only if

$$\Delta\upsilon = \pm 1$$

This is the same selection rule as for infrared vibrational transitions but vibrational Raman spectroscopy has the advantage that transitions are allowed in homonuclear, as well as heteronuclear, diatomic molecules.

Intensities of Raman transitions are proportional to $|\boldsymbol{R}^{\upsilon'\upsilon''}|^2$ and therefore, taking into account only the first term of equation (5.12), to $(\mathrm{d}\boldsymbol{\alpha}/\mathrm{d}x)_{\mathrm{e}}^2$. Since $\boldsymbol{\alpha}$ is a tensor property [equation (4.117)] we cannot illustrate easily its variation with x; instead, we use the mean polarizability $\bar{\alpha}$, where

$$\bar{\alpha} = \tfrac{1}{3}(\alpha_{xx} + \alpha_{yy} + \alpha_{zz}) \qquad (5.13)$$

and is the mean value of the diagonal elements of $\boldsymbol{\alpha}$. Figure 5.2 shows typically how $\bar{\alpha}$ varies with r. $(\mathrm{d}\bar{\alpha}/\mathrm{d}r)$ is usually positive and, unlike $(\mathrm{d}\mu/\mathrm{d}r)$ in Figure 5.1, varies little with r. For this reason vibrational Raman intensities are less sensitive than infrared intensities to the environment of the molecule, such as the solvent in a solution spectrum.

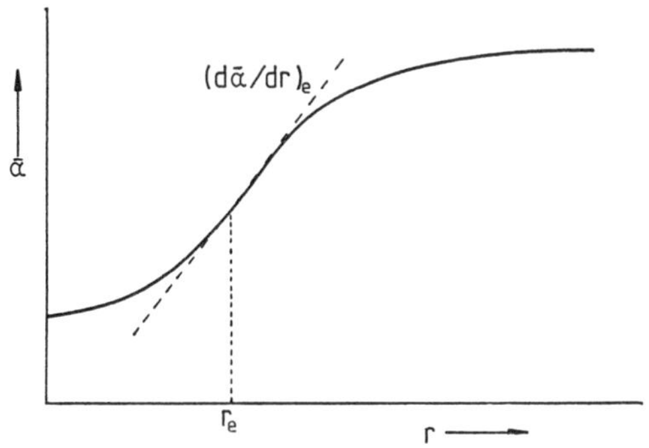

Figure 5.2 Variation of the mean polarizability, $\bar{\alpha}$, with internuclear distance, r, in a diatomic molecule

The mechanism for Stokes and anti-Stokes vibrational Raman transitions is analogous to that for rotational Raman transitions, illustrated in Figure 4.32. As shown in Figure 5.3, intense monochromatic radiation may take the molecule from the $\upsilon = 0$ state to a virtual state V_0; then it can return to $\upsilon = 0$ in a Rayleigh scattering process or to the $\upsilon = 1$ state in a Stokes Raman transition. Alternatively, it can go from the $\upsilon = 1$ state to the virtual state V_1 and return to $\upsilon = 1$ (Rayleigh) or to the $\upsilon = 0$ state in an anti-Stokes Raman transition. However, in many molecules at normal temperatures the population of the $\upsilon = 1$ state is so low that anti-Stokes transitions may be too weak to be observed.

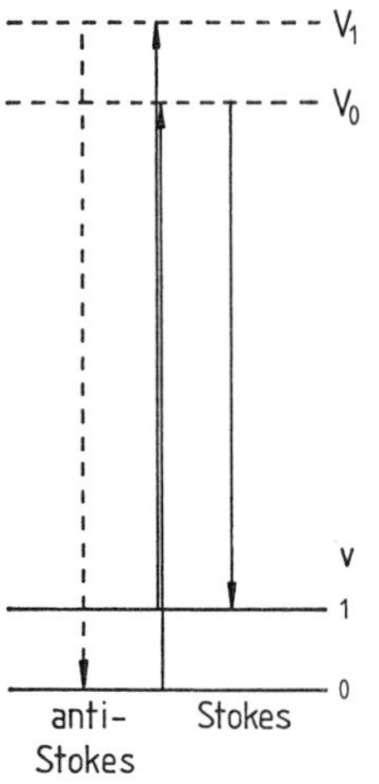

Figure 5.3 Stokes and anti-Stokes vibrational Raman scattering

5.1.2 Anharmonicity

We have already encountered one type of possible deviation from harmonic behaviour. For a harmonic oscillator the variation with x of the dipole moment μ and the polarizability α are given by equations (5.5) and (5.11), respectively. The fact that both equations contain terms in x raised to the second and higher powers represents one aspect of a phenomenon known as anharmonicity. This particular kind of anharmonicity is concerned with electrical properties, dipole moment and polarizability, of the molecule and is called electrical anharmonicity. One effect of electrical anharmonicity is to cause the vibrational selection rule $\Delta v = \pm 1$ in infrared and Raman spectroscopy to be modified to $\Delta v = \pm 1, \pm 2, \pm 3, \ldots$. Electrical anharmonicity is usually small and this is reflected in the low intensities of $\Delta v = \pm 2, \pm 3, \ldots$ transitions, called vibrational overtones, compared with $\Delta v = \pm 1$. The intensities of $\Delta v \pm 2$ transitions are dependent on $(d^2\mu/dx^2)_e$ for infrared or $(d^2\bar{\alpha}/dx^2)_e$ for Raman spectra. From what we have seen from Figures 5.1 and 5.2, it is clear that the effect of electrical anharmonicity in intensifying $\Delta v = \pm 2$ transitions is usually greater in infrared than in Raman spectra.

Just as the electrical behaviour of a real diatomic molecule is not accurately harmonic, neither is its mechanical behaviour. The potential function, vibrational energy levels and wave functions shown in Figure 1.22 were derived with the assumption that vibrational motion obeys Hooke's law, as expressed by equation (1.94). This assumption is reasonable only when r is not very different from r_e, i.e. when x is small.

At large values of r we know that the molecule dissociates: two neutral atoms are formed and, since they do not influence each other, the force constant is zero and r can then be increased to infinity with no further change of the potential energy V. Therefore, the potential function flattens out at $V = D_e$, where D_e is the dissociation energy measured relative to the equilibrium potential energy, as shown in Figure 5.4. As dissociation is approached the force constant $k \rightarrow 0$, which means that the bond gets weaker. The effect is to make the potential energy curve shallower than for a harmonic oscillator for $r > r_e$.

At small values of r the positive charges on the two nuclei cause mutual repulsion which increasingly opposes their approaching each other. This causes the potential energy curve to be steeper than for a harmonic oscillator, as in

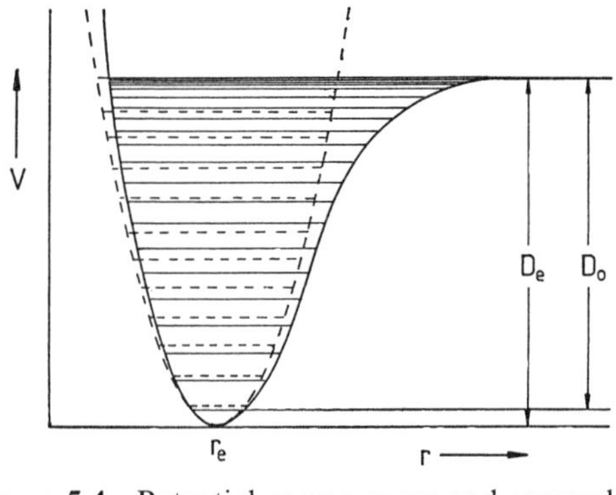

Figure 5.4 Potential energy curve and energy levels for a diatomic molecule behaving as an anharmonic oscillator compared with those for a harmonic oscillator

Figure 5.4. The deviations found in the curve for a real molecule from that of a harmonic oscillator are due to mechanical anharmonicity.

A molecule may show both electrical and mechanical anharmonicity but it is conventional to define a harmonic oscillator as one which is harmonic in the mechanical sense. It follows that a harmonic oscillator may show electrical anharmonicity.

One effect of mechanical anharmonicity is to modify the $\Delta v = \pm 1$ selection rule for a harmonic oscillator to $\Delta v = \pm 1, \pm 2, \pm 3, \ldots$, but transitions with $\Delta v = \pm 2, \pm 3, \ldots$, are usually weak compared with those with $\Delta v = \pm 1$. Since electrical anharmonicity also has this effect on the selection rule, *both* electrical *and* mechanical anharmonicity may contribute to the intensity of $\Delta v = \pm 2, \pm 3, \ldots$ transitions.

However, unlike electrical anharmonicity, mechanical anharmonicity modifies the vibrational term values and wave functions. The harmonic oscillator term values of equation (5.3) are modified to a power series in $(v + \frac{1}{2})$:

$$G(v) = \omega_e(v + \tfrac{1}{2}) - \omega_e x_e(v + \tfrac{1}{2})^2 + \omega_e y_e(v + \tfrac{1}{2})^3 + \ldots \qquad (5.14)$$

where ω_e is the vibration wavenumber which a classical harmonic oscillator would have for an infinitesimal displacement; $\omega_e x_e, \omega_e y_e, \ldots$ are anharmonic constants.† $\omega_e x_e$ has the same sign in all diatomic molecules and, in order to make it positive, the convention is to put a negative sign in front of the second term in equation (5.14); $\omega_e y_e$, etc., may be positive or negative. The $\omega_e, \omega_e x_e, \omega_e y_e, \ldots$ are of rapidly decreasing magnitude, for example for $^1H^{35}Cl$ $\omega_e = 2990.946\ \mathrm{cm}^{-1}$, $\omega_e x_e = 52.8186\ \mathrm{cm}^{-1}$, $\omega_e y_e = 0.2244\ \mathrm{cm}^{-1}$, $\omega_e z_e = -0.0122\ \mathrm{cm}^{-1}$. The effect of the positive value of $\omega_e x_e$ is to close up the vibrational term values with increasing v. The corresponding energy levels are compared with the constantly separated ones of a harmonic oscillation in Figure 5.4. The anharmonic oscillator levels converge to the dissociation limit D_e above which there is a continuum of levels.

One effect of the modified term values of equation (5.14) is that, unlike the case of the harmonic oscillator, the fundamental vibration wavenumber ω_e cannot be measured directly. The $(v+1)$–v transition‡ wavenumbers $\tilde{\nu}_{v''}^{v'}$ are given by

$$\begin{aligned}\tilde{\nu}_{v''}^{v'} &= G(v+1) - G(v)\\ &= \omega_e - \omega_e x_e(2v+2)\\ &\quad + \omega_e y_e\left(3v^2 + 6v + \frac{13}{4}\right) + \ldots \qquad (5.15)\end{aligned}$$

In order to determine ω_e, $\omega_e x_e$ and $\omega_e y_e$, at least three transition wavenumbers must be known.

The dissociation energy D_e is given approximately by

$$D_e \approx \omega_e^2/4\omega_e x_e \qquad (5.16)$$

the approximation being due to the neglect of all anharmonic constants other than $\omega_e x_e$.

Experimentally the dissociation energy is measured relative to the zero-point level and is symbolized by D_0. It is clear from Figure 5.4 that

$$D_0 = \sum_v \Delta G_{v+\frac{1}{2}} \qquad (5.17)$$

where

$$\Delta G_{v+\frac{1}{2}} = G(v+1) - G(v) \qquad (5.18)$$

If all anharmonic constants except $\omega_e x_e$ are neglected, $\Delta G_{v+\frac{1}{2}}$ is a linear function of v [equations (5.15)] and D_0 is equal to the area under a plot of $\Delta G_{v+\frac{1}{2}}$ versus v, shown by a dashed line in Figure 5.5. But in many cases only the first few ΔGs can be observed and a

†The reason why the anharmonic constants are written $\omega_e x_e, \omega_e y_e, \ldots$ rather than, say, $x_e, y_e, \ldots$ is that early authors wrote equations (5.14) in the form

$$G(v) = \omega_e[(v + \tfrac{1}{2}) - x_e(v + \tfrac{1}{2})^2 + y_e(v + \tfrac{1}{2})^3 + \ldots]$$

‡The reader is reminded that the upper state quantum number is given first and the lower state quantum number second.

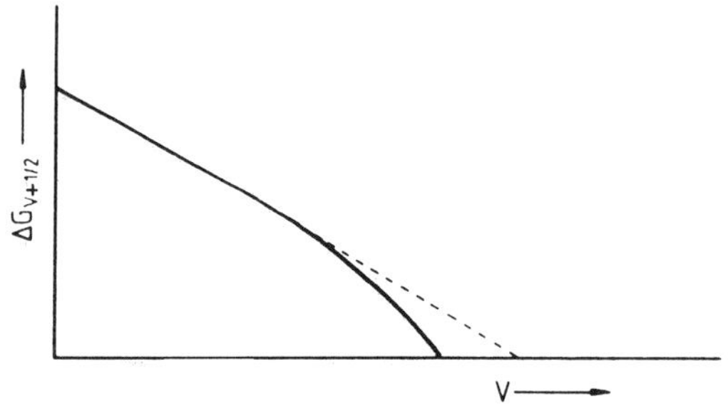

Figure 5.5 A Birge–Sponer extrapolation for determining an approximate value of the dissociation energy D_0 of a diatomic molecule. The extrapolation is shown by the dashed line whereas experimental points lie on the solid line in a typical case

linear extrapolation of the plot to $\Delta G_{v+\frac{1}{2}} = 0$ has to be made. This is a Birge–Sponer extrapolation and the area under the linearly extrapolated plot gives an approximate value for D_0. Most such plots deviate considerably from linearity at high values of v, as indicated in Figure 5.5, so that experimental values of $\Delta G_{v+\frac{1}{2}}$ are necessary at high values of v if an accurate value of D_0 is to be obtained.

If dissociation produces neutral atoms, as it usually does, the potential energy at large values of r, approaching dissociation, is dominated by van der Waals forces and there is only a *finite* number of vibrational energy levels below the dissociation limit. This number is given by the value of v for which $\Delta G_{v+\frac{1}{2}} = 0$ in a plot of $\Delta G_{v+\frac{1}{2}}$ versus v.

Should dissociation produce two oppositely charged ions, the potential energy at large values of r is dominated by coulombic attraction and there is an *infinite* number of vibrational levels below the dissociation limit. For the same reason there is an infinite number of electronic energy levels below the ionization limit of an atom, as in the case of hydrogen (Figure 1.5).

Experimental values of $\Delta G_{v+\frac{1}{2}}$ for high values of v are not normally obtainable from infrared or Raman spectroscopy because of the low intensity of $\Delta v = \pm 2, \pm 3, \ldots$ transitions and low populations of excited vibrational energy levels under normal temperature conditions. Information on vibrational levels with high values of v is obtained mostly from electronic emission spectroscopy (Chapters 6 and 8).

The dissociation energy D_e is unaffected by isotopic substitution because the potential energy curve, and therefore the force constant, is not changed by the number of neutrons in the nucleus. On the other hand the vibrational energy levels *are* changed owing to the mass dependence of the vibration wavenumber: $\omega \propto \mu^{-\frac{1}{2}}$, where μ is the reduced mass. For this reason, D_0 is isotope dependent. D_0 differs from D_e by the zero-point energy where the term value $G(0)$ corresponding to the zero-point energy is given by

$$G(0) = \tfrac{1}{2}\omega_e - \tfrac{1}{4}\omega_e x_e + \tfrac{1}{8}\omega_e y_e + \ldots \qquad (5.19)$$

For example, ω_e for $^2H_2$ is less than that for $^1H_2$, so that

$$D_0(^2H_2) > D_0(^1H_2) \qquad (5.20)$$

An important consequence of the isotope dependence of D_0 is that, if a chemical reaction involves bond dissociation in a rate-determining step, the rate of the reaction is decreased by the substitution of a heavier isotope at either end of the bond.

Owing to the effects of mechanical anharmonicity—to which we shall refer in future simply as anharmonicity, since we shall encounter electrical anharmonicity only rarely—the vibrational wave functions are also modified compared with those of a harmonic oscillator. Figure 5.6 shows some wave functions and probability density functions $\psi_v^*\psi_v$ for an anharmonic oscillator. The asymmetry in ψ_v and $\psi_v^*\psi_v$, compared with the harmonic oscillator wave functions (Figure 1.22), increases their magnitude on the shallow side of the potential curve compared to the steep side.

The experimentally observed values of $\Delta G_{v+\frac{1}{2}}$ can be used to construct the true potential curve by the Rydberg–Klein–Rees (RKR) method but this does not lead to an analytical expression for

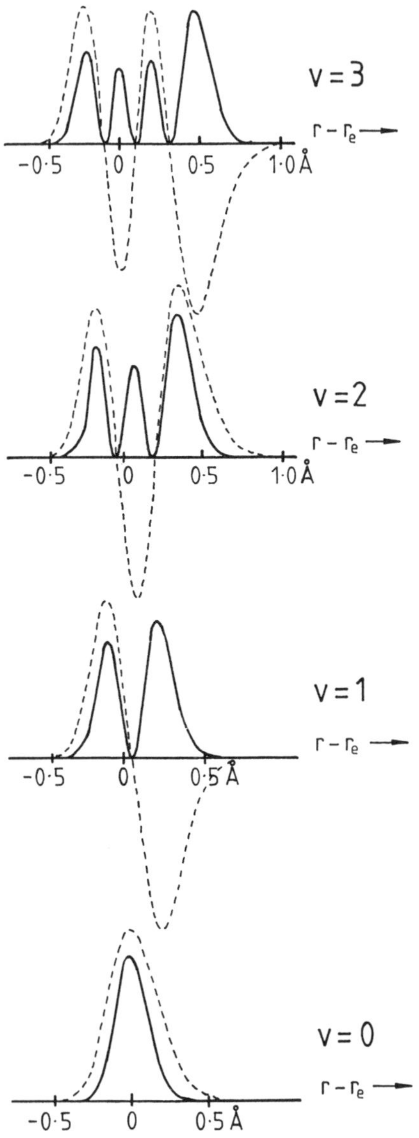

Figure 5.6 The dashed lines show some vibrational wave functions ψ_v and the solid lines some probability density functions $\psi_v^*\psi_v$ for a diatomic molecule behaving as an anharmonic oscillator

the potential function of an anharmonic oscillator.

A simple modification of the harmonic oscillator potential function of equation (1.95) to take account of anharmonicity is by the inclusion of a term gx^3 to give

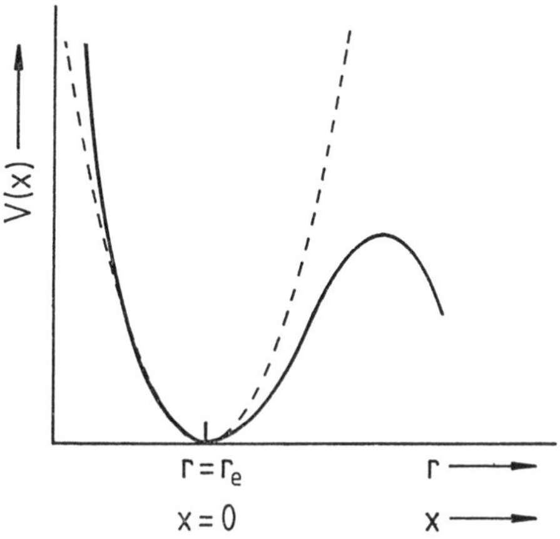

Figure 5.7 The solid line shows the effect of adding a cubic term to the harmonic oscillator potential function, shown by the dashed line

$$V(x) = \tfrac{1}{2}kx^2 + gx^3 \qquad (5.21)$$

where $x = r - r_e$ and g is negative. The effect of the additional term, as shown in Figure 5.7, is to steepen the harmonic oscillator curve for $r < r_e$ and to make it shallower for $r > r_e$. However, the use of this potential function is limited only to regions where r is not very different from r_e. The greatest anomaly which the function produces is that, as $r \to \infty$, $V(r) \to -\infty$ instead of D_e.

A more generally useful potential function is that proposed by Morse[46] in 1929, which has the form

$$V(x) = D_e[1 - \exp(-ax)]^2 \qquad (5.22)$$

where a, and also D_e, is a constant for each electronic state of each molecule. This function has the advantage over that of equation (5.21) that, when $x \to \infty$, $V(x) \to D_e$. In the region where $r \to 0$, or $x \to -r_e$, the Morse function is not well behaved in that, although $V(x)$ becomes large, it does not tend to infinity. However, this is not a serious defect as this is not an experimentally important region. Solution of the Schrödinger equation using the Morse potential gives the term value expression

$$G(\upsilon) = a(\hbar D_e/\pi c\mu)^{\frac{1}{2}}(\upsilon + \tfrac{1}{2}) - (\hbar a^2/4\pi c\mu)(\upsilon + \tfrac{1}{2})^2 \quad (5.23)$$

where μ is the reduced mass. Equating ω_e, from equation (5.14), with the coefficient of $(\upsilon + \frac{1}{2})$ in equation (5.23) gives

$$a = (\pi c\mu/D_e\hbar)^{\frac{1}{2}}\omega_e \quad (5.24)$$

Although the Morse function is convenient to use it is not quantitatively very satisfactory. An improved function was proposed by Hulburt and Hirschfelder[47] and takes the form

$$V(x) = D_e\{[1 - \exp(-ax)]^2 + ca^3x^3(1 + abx)\exp(-2ax)\} \quad (5.25)$$

where a has the same significance as in the Morse function and b and c are further constants.

The most general form for the potential function is a power series expansion in x:

$$V(x) = (k_2x^2/2!) + (k_3x^3/3!) + (k_4x^4/4!) + \ldots \quad (5.26)$$

to which equation (5.21) is an approximation, taking into account only the first two terms. The terms in equation (5.26) decrease smoothly with increasing powers of x. For HCl the values of k_2, k_3 and k_4 are 5.162 aJ Å^{-2}, −28.7 aJ Å^{-3} and 139.7 aJ Å^{-4}, respectively.

In obtaining quantitative agreement with experimentally observed vibrational term values, it is important that, if the k_3 term is to be included in $V(x)$, the k_4 term should also be included. This is particularly true if the anharmonic constant $\omega_e x_e$ is to be predicted from the potential function, since it is given by the expression

$$\omega_e x_e = (5k_3{}^2/48h^2c^2a^3\omega_e) - (k_4/16hca^2) \quad (5.27)$$

where $a = k_2^{\frac{1}{2}}\mu^{\frac{1}{2}}/\hbar$ and μ is the reduced mass, and the two terms contributing to $\omega_e x_e$ are similar in magnitude.

5.1.3 Vibration–rotation spectroscopy

5.1.3.1 Infrared vibration–rotation spectra

As we have seen in Section 4.2.1, there is a stack of rotational energy levels associated with all vibrational energy levels. In pure rotational spectroscopy we observe transitions between rotational energy levels associated with the same vibrational level. In vibration–rotation spectroscopy we observe transitions between the stacks of rotational energy levels associated with two different vibrational levels. These transitions accompany all vibrational transitions but, whereas vibrational transitions can be observed even when the sample is in a liquid or solid phase, the rotational transitions may be observed only in the gas phase at low pressure and usually in an absorption process.

When a molecule has both vibrational and rotational energy, the total term values S are given by the sum of the rotational term values $F_\upsilon(J)$, given in equation (4.52), and the vibrational term values $G(\upsilon)$, given in equation (5.14):

$$\begin{aligned} S &= G(\upsilon) + F_\upsilon(J) \\ &= \omega_e(\upsilon + \tfrac{1}{2}) - \omega_e x_e(\upsilon + \tfrac{1}{2})^2 + \ldots \\ &\quad + B_\upsilon J(J+1) - D_\upsilon J^2(J+1)^2 + \ldots \end{aligned} \quad (5.28)$$

Figure 5.8(a) illustrates the rotational energy levels associated with two vibrational levels υ' (upper) and υ'' (lower) between which a vibrational transition is allowed by the $\Delta\upsilon = \pm 1$ selection rule. The rotational selection rule governing transitions between the two stacks of levels is $\Delta J = \pm 1$, giving an R branch ($\Delta J = +1$) and a P branch ($\Delta J = -1$). Each transition is labelled $R(J)$ or $P(J)$, where J is understood to represent J'', the J-value of the lower state. The fact that $\Delta J = 0$ is forbidden means that the pure vibrational transition is not observed. The position at which it would occur is known as the band centre. Exceptions to this selection rule are molecules, such as nitric oxide, which

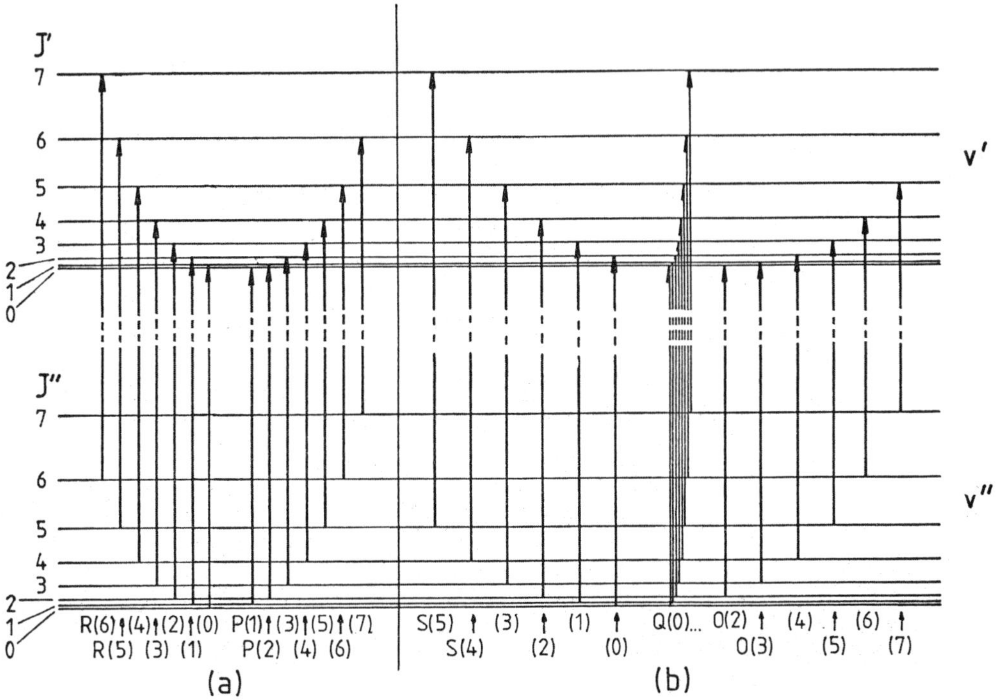

Figure 5.8 Rotational transitions accompanying a vibrational transition in (a) an infrared spectrum, for which $\Delta J = \pm 1$, and (b) a Raman spectrum, for which $\Delta J = 0, \pm 2$

have an electronic angular momentum in the ground electronic state. The rotational selection rule for such molecule is

$$\Delta J = 0, \ \pm 1$$

and the $(J' = 0)$–$(J'' = 0)$ transition, the first line of the Q-branch, marks the band centre.

More usual is the kind of vibration–rotation band shown in Figure 5.9. This spectrum was obtained with an FTIR spectrometer having a resolution of about $1\,\mathrm{cm}^{-1}$ and shows the $\upsilon = 1$–0 transition in $^{1}H^{35}Cl$ and $^{1}H^{37}Cl$. The ^{35}Cl and ^{37}Cl isotopes occur naturally with a 3:1 abundance ratio. The band due to $^{1}H^{37}Cl$ is displaced to low wavenumber relative to that due to $^{1}H^{35}Cl$ because of the larger reduced mass [see equation (5.2)].

It is clear from Figure 5.9 that the band for each isotope is fairly symmetrical about the corresponding band centre and that there is approximately equal spacing between adjacent R-branch lines and between adjacent P-branch lines, with twice this spacing between the first R- and P-branch lines, $R(0)$ and $P(1)$. This spacing between $R(0)$ and $P(1)$ is called the zero gap and it is in this region where the band centre falls.

The approximate symmetry of the band is due to the fact that $B_1 \approx B_0$, i.e. the vibration–rotation interaction constant α [equation (4.54)] is small. If we assume that $B_1 = B_0 = B$ and neglect centrifugal distortion the wavenumbers of the R-branch transitions, $\tilde{\nu}[R(J)]$, are given by

$$\begin{aligned}\tilde{\nu}[R(J)] &= \omega_0 + B(J+1)(J+2) - BJ(J+1) \\ &= \omega_0 + 2BJ + 2B \qquad (5.29)\end{aligned}$$

where ω_0 is the wavenumber of the pure vibrational transition. Similarly, the wave-

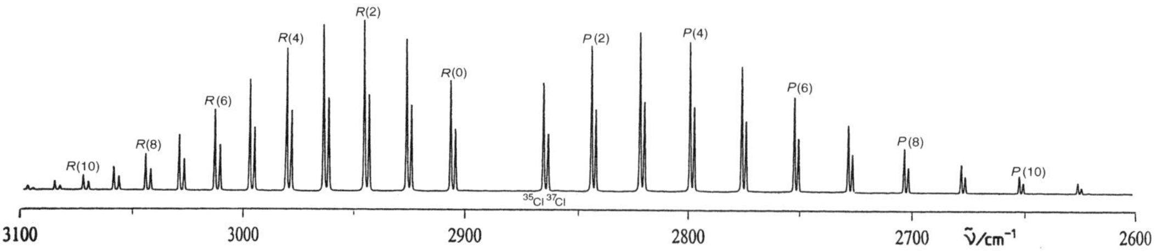

Figure 5.9 The $v = 1 - 0$ infrared spectrum of the $^1H^{35}Cl$ and $^1H^{37}Cl$ molecules showing the *P*- and *R*-branch rotational structure

numbers of the *P*-branch transitions, $\tilde{\nu}[P(J)]$, are given by

$$\tilde{\nu}[P(J)] = \omega_0 + B(J-1)J - BJ(J+1)$$
$$= \omega_0 - 2BJ \qquad (5.30)$$

From equations (5.29) and (5.30), it follows that the zero gap, $\tilde{\nu}[R(0)] - \tilde{\nu}[P(1)]$, is $4B$ and that the spacing is $2B$ between adjacent *R*-branch lines and also between adjacent *P*-branch lines; hence the approximate symmetry of the band.

A close look at Figure 5.9 reveals that the bands are not exactly symmetrical but show a convergence in the *R*-branch and a divergence in the *P*-branch. This behaviour is due principally to the inequality of B_0 and B_1 and there is enough information in the band to be able to determine these two quantities separately. The method used is called the method of combination differences, which employs a principle common in spectroscopy. The principle is that, if we wish to derive information about a series of lower states and a series of upper states, between which transitions are occurring, then differences in wavenumber between transitions with a common upper state are dependent on properties of the lower states only. Similarly, differences in wavenumber between transitions with a common lower state are dependent on properties of the upper states only.

In the case of a vibration–rotation band it is clear from Figure 5.8(a) that, since *R*(0) and *P*(2) have a common upper state with $J' = 1$, then $\tilde{\nu}[R(0)] - \tilde{\nu}[P(2)]$ must be a function of B'' only. The transitions *R*(1) and *P*(3) have $J' = 2$ in common and, in general, $\tilde{\nu}[R(J-1)] - \tilde{\nu}[P(J+1)]$, usually written as $\Delta_2'' F(J)$,† is given by

$$\Delta_2'' F(J) = \tilde{\nu}[R(J-1) - \tilde{\nu}[P(J+1)]$$
$$= \omega_0 + B'J(J+1) - B''(J-1)J$$
$$- [\omega_0 + B'J(J+1) - B''(J+1)(J+2)]$$
$$= 4B''(J+\tfrac{1}{2}) \qquad (5.31)$$

After assignment and measurement of the wavenumbers of the rotational lines, a graph of $\Delta_2'' F(J)$ versus $(J+\frac{1}{2})$ is a straight line of slope $4B''$.

Similarly, since all pairs of transitions *R*(*J*) and *P*(*J*) have common lower states, $\tilde{\nu}[R(J) - \tilde{\nu}[P(J)]$ is a function of B' only and we have

$$\Delta_2' F(J) = \tilde{\nu}[R(J)] - \tilde{\nu}[P(J)]$$
$$= \omega_0 + B'(J+1)(J+2) - B''J(J+1)$$
$$- [\omega_0 + B'(J-1)J - B''J(J+1)]$$
$$= 4B'(J+\tfrac{1}{2}) \qquad (5.32)$$

and a graph of $\Delta_2' F(J)$ versus $(J+\frac{1}{2})$ is a straight line of slope $4B'$.

The band centre is not exactly midway between *R*(0) and *P*(1) but its wavenumber ω_0 can be obtained from

$$\omega_o = \tilde{\nu}[R(0)] - 2B'$$
$$= \tilde{\nu}[P(1)] + 2B'' \qquad (5.33)$$

Any effects of centrifugal distortion will show up as slight curvature of the $\Delta_2 F(J)$ versus

†The reason for the subscript 2 in the $\Delta_2'' F(J)$ symbol is that these are the differences between rotational term values, in a particular vibrational state, with *J* differing by 2.

$(J+\frac{1}{2})$ graphs. If the term $-DJ^2(J+1)^2$ is included in the rotational term value expression we obtain

$$\Delta_2''F(J) = (4B'' - 6D'')(J+\tfrac{1}{2}) - 8D''(J+\tfrac{1}{2})^3 \tag{5.34}$$

and

$$\Delta_2'F(J) = (4B' - 6D')(J+\tfrac{1}{2}) - 8D'(J+\tfrac{1}{2})^3 \tag{5.35}$$

A graph of $\Delta_2''F(J)/(J+\frac{1}{2})$ versus $(J+\frac{1}{2})^2$ is a straight line of slope $8D''$ and intercept $4B''$ [or, strictly, $(4B''-6D'')$, but $6D$ may be too small to affect the intercept]. Similarly, a graph of $\Delta_2'F(J)/(J+\frac{1}{2})$ versus $(J+\frac{1}{2})^2$ is a straight line of slope $8D'$ and intercept $4B'$.

If B_υ can be obtained for at least two vibrational levels, say B_0 and B_1, then B_e and the vibration–rotation interaction constant α can be obtained from equation (4.54) if the third and subsequent terms are neglected. Values for B_e and α, together with other constants, are given for $^1H^{35}Cl$ in Table 5.2.

It is important to remember that, perhaps contrary to expectation, the constant α is *not* zero in the harmonic oscillator. This point is illuminated by the expression

$$\alpha = -(6B_e^2/\omega_e) - [2B_e^2 k_3/hc\omega_e(2a^3 B_e\omega_e)^{\frac{1}{2}}] \tag{5.36}$$

where a is the same quantity as in equation (5.27). The first term is the harmonic oscillator term and arises from the fact that it is not the mean value of the displacement x (zero in a harmonic oscillator) which determines the variation of the rotational constant B_υ with vibrational state but the mean value of $(r^{-2} - r_e^{-2})$, which is non-zero. The second term in equation (5.36), involving k_3, clearly represents the anharmonic contribution to α. Both terms are of comparable magnitudes but the second is almost always larger than the first. The fact that k_3 is always negative means that α is almost invariably positive.

Table 5.2 Rotational and vibrational constants for $^1H^{35}Cl$[a]

$\upsilon = 0$		$\upsilon = 1$	
B_0	$10.440\,254\ \text{cm}^{-1}$	B_1	$10.136\,228\ \text{cm}^{-1}$
D_0	$5.2828 \times 10^{-4}\ \text{cm}^{-1}$	D_1	$5.2157 \times 10^{-4}\ \text{cm}^{-1}$
ω_0 (for $\upsilon = 1$–0 transition) $2885.9775\ \text{cm}^{-1}$			
B_e $10.593\,42\ \text{cm}^{-1}$			
α_e $0.307\,18\ \text{cm}^{-1}$			

[a]Data taken from Rank, D. H., Rao, B. S. and Wiggins, T. A. (1965). *J. Mol. Spectrosc.*, **17**, 122.

The intensity distribution in the rotational transitions in a vibration–rotation band is governed principally by the Boltzmann population distribution among the initial states of the transitions, namely

$$N_{J''}/N = (hcB''/kT)(2J''+1)\exp[-hcB''J''(J''+1)/kT] \tag{5.37}$$

where $N_{J''}$ is the number of molecules with quantum number J'', N is the total number of molecules, kT/hcB'' is the rotational partition function (see equation 4.27) and B'' has the dimensions of wavenumber. For bands showing only a P- and R-branch the intensities I_a of vibration–rotation transitions in absorption are given by

$$I_a = C_a\tilde{\nu}(hcB''/kT)(J'+J''+1)\exp[-hcB''J''(J''+1)/kT] \tag{5.38}$$

where C_a is a constant, if we neglect induced emission, and depends on the absolute intensity of the vibrational transition and the number of molecules absorbing. The factor $(J'+J''+1)$ replaces the factor $(2J''+1)$ in equation (5.37) to take account of a slight variation of the transition moment with J. Unlike pure rotation spectra (Section 4.2.2), not only is induced emission relatively unimportant but also the wavenumber which appears in equation (5.38) is essentially constant since the spread of the band is typically small compared with the wavenumber at which it occurs.

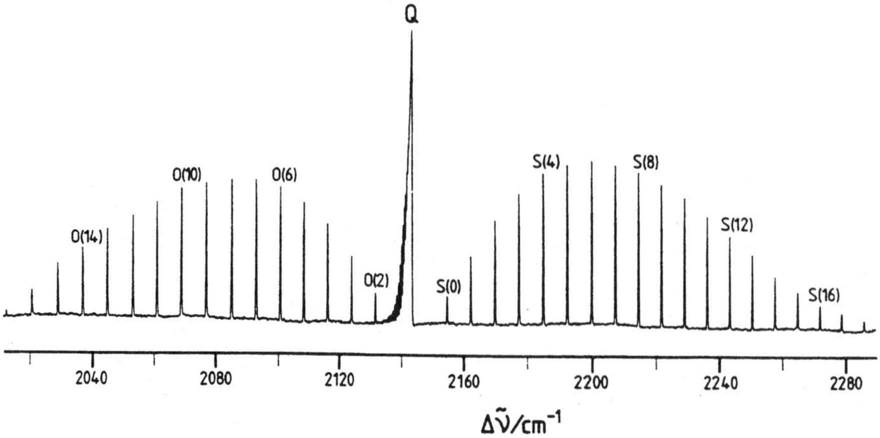

Figure 5.10 The $\upsilon = 1\text{–}0$ Stokes Raman spectrum of the CO molecule showing the *O*-, *Q*- and *S*-branch rotational structure

Note that the intensity distribution I_e in vibration–rotation bands observed in emission is given by

$$I_e = C_e\tilde{\nu}^4(hcB'/kT)(J' + J'' + 1)\exp[-hcB'J'(J' + 1)/kT] \quad (5.39)$$

where the initial state is now the upper state. In this equation it has been assumed that, after the molecules have been excited to an excited vibrational state, they have had time to reach rotational equilibrium before emission occurs.

5.1.3.2 *Raman vibration–rotation spectra*

The rotational selection rule for vibration–rotation Raman transitions in diatomic molecules is

$$\Delta J = 0, \ \pm 2$$

giving a $Q(\Delta J = 0)$, an $S(\Delta J = +2)$ and an $O(\Delta J = -2)$ branch, as shown in Figure 5.8(b).

Figure 5.10 shows the resulting rotational structure of the $\upsilon = 1\text{–}0$ Stokes Raman transition of CO. The approximate symmetry of the band is due, as in the infrared vibration–rotation spectrum, to the fact that $B_1 \simeq B_0$. If we assume that $B_1 = B_0 = B$ then the wavenumbers $\tilde{\nu}[S(J)]$, $\tilde{\nu}[O(J)]$ and $\tilde{\nu}[Q(J)]$ of the *S*-, *O*- and *Q*-branch lines, respectively, are given by

$$\tilde{\nu}[S(J)] = \omega_0 + B(J + 2)(J + 3) - BJ(J + 1) = \omega_0 + 4BJ + 6B \quad (5.40)$$

$$\tilde{\nu}[O(J)] = \omega_0 + B(J - 2)(J - 1) - BJ(J + 1) = \omega_0 - 4BJ + 2B \quad (5.41)$$

$$\tilde{\nu}[Q(J)] = \omega_0 \quad (5.42)$$

The *Q*-branch lines are coincident in this approximation. The separation of the first *S*-branch line $S(0)$ and the first *O*-branch line $O(2)$ is $12B$ and the separation of adjacent *S*-branch lines and of adjacent *O*-branch lines is $4B$.

More accurately, we can use the method of combination differences, while still neglecting centrifugal distortion, to obtain B'' and B'. Transitions having wavenumbers $\tilde{\nu}[S(J - 2)]$ and $\tilde{\nu}[O(J + 2)]$ have a common upper state so that the corresponding combination difference $\Delta_4''F(J)$ is a function of B'' only:

$$\Delta_4''F(J) = \tilde{\nu}[S(J - 2)] - \tilde{\nu}[O(J + 2)] = 8B''(J + \tfrac{1}{2}) \quad (5.43)$$

Similarly, transitions having wavenumbers $\tilde{\nu}[S(J)]$ and $\tilde{\nu}[O(J)]$ have a common lower state and

$$\Delta_4' F(J) = \tilde{\nu}[S(J)] - \tilde{\nu}[O(J)] = 8B'(J + \tfrac{1}{2}) \qquad (5.44)$$

Plotting graphs of $\Delta_4'' F(J)$ versus $(J + \frac{1}{2})$ and $\Delta_4' F(J)$ versus $(J + \frac{1}{2})$ gives straight lines with slopes $8B''$ and $8B'$, respectively.

Nuclear spin statistical weights have been discussed in Section 4.4 and the effects on the populations of the rotational levels in the $\upsilon = 0$ states of $^1H_2$, $^{19}F_2$, $^2H_2$, $^{14}N_2$ and $^{16}O_2$ illustrated as examples in Figure 4.22. The effect of these statistical weights in the vibration–rotation Raman spectra is to cause a J'' even:odd intensity alternation of 1:3 for $^1H_2$ and $^{19}F_2$, 6:3 for $^2H_2$ and $^{14}N_2$ and, for $^{16}O_2$, all transitions with J'' even are absent. It is for the purposes of Raman spectra of such homonuclear diatomics that the 's' or 'a' labels, indicating symmetry or antisymmetry of the rotational wave function ψ_r with respect to interchange of nuclei, are attached to Figure 4.22.

5.2 POLYATOMIC MOLECULES

5.2.1 Group vibrations

An N-atomic molecule has $3N - 5$ normal modes of vibration if it is linear and $3N - 6$ if it is non-linear; these expressions were derived in Section 1.3.1.

Classically we can think of the vibrational motions of a molecule as being those of a set of balls, representing the nuclei, of various masses and connected by Hooke's law springs, representing the various forces acting between the nuclei. Such a model for the H_2O molecule is illustrated in Figure 5.11. The stronger forces between the bonded O and H nuclei are represented by strong springs which provide resistance to stretching the bonds. The weaker force between the non-bonded hydrogen nuclei is represented by a weaker spring which provides resistance to an increase or decrease of the HOH angle.

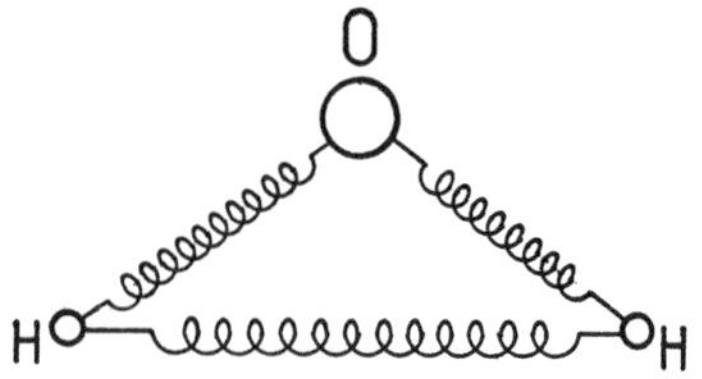

Figure 5.11 Ball and spring model of the H_2O molecule, the springs representing forces between the nuclei

Even with this simple model it is clear that if one of the nuclei is given a sudden displacement it is very likely that the whole molecule will undergo a very complicated motion, a so-called Lissajous motion, consisting of a mixture of angle bending and bond stretching. The Lissajous motion can always be broken down into a combination of the normal vibrations of the system which, in the Lissajous motion, are superimposed in varying proportions.

A normal mode of vibration is one in which all the nuclei undergo harmonic motion, have the same frequency of oscillation and move in-phase but generally with different amplitudes.

The H_2O molecule has three normal vibrations which are labelled conventionally† ν_1, ν_2 and ν_3 and are illustrated in Figure 5.12. The arrows attached to the nuclei are vectors representing their relative amplitudes and directions of motion. The way in which the form of the normal vibrations can be obtained from the nuclear masses and the force constants, which are a measure of the strengths of the springs in Figure 5.11, will be considered in detail in Section 5.2.3.

In an approximation which is analogous to that which we have used in a diatomic molecule, each of the vibrations of a polyatomic molecule can be regarded as being an approximately harmonic motion. Quantum mechanical treatment in the harmonic oscillator approximation shows that the vibrational term values $G(\upsilon_i)$ associated with each normal vibration i are

†See footnote on p. 213 for convention.

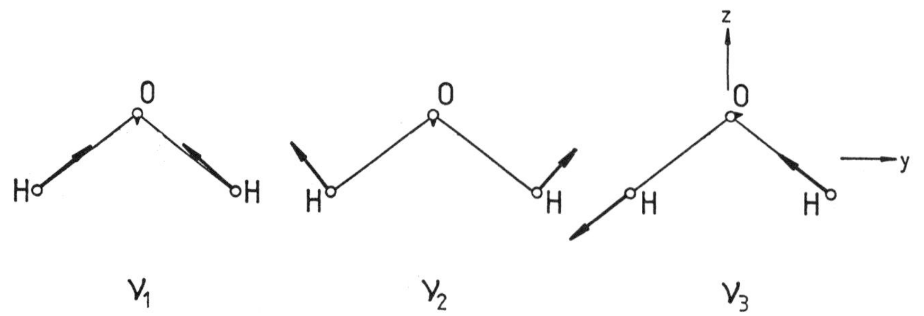

Figure 5.12 The three normal vibrations of the H_2O molecule

given, when all the vibrations are non-degenerate, by

$$G(v_i) = \omega_i(v_i + \tfrac{1}{2}) \qquad (5.45)$$

where ω_i is the classical vibration wavenumber and v_i the vibrational quantum number which can take the values 0, 1, 2, 3, ... In general, for vibrations with a degree of degeneracy d_i, equation (5.45) becomes

$$G(v_i) = \omega_i(v_i + d_i/2) \qquad (5.46)$$

As for a diatomic molecule the harmonic oscillator selection rule is

$$\Delta v_i = \pm 1$$

with $\Delta v_i = \pm 2, \ \pm 3 \ldots$ vibrational overtone transitions allowed, but generally weak, when account is taken of anharmonicity.

In addition, there is the possibility of combination tones which involve transitions to vibrationally excited states in which more than one fundamental is excited. Fundamental, overtone and combination tone transitions involving two normal vibrations ν_i and ν_j are illustrated in Figure 5.13.

For vibrational transitions to be allowed in the infrared spectrum there is the additional requirement that there must be an accompanying change of dipole moment and, in the Raman spectrum, an accompanying change of the amplitude of the induced dipole moment. These requirements represent further selection rules which depend on the symmetry properties of the molecule concerned and will be discussed

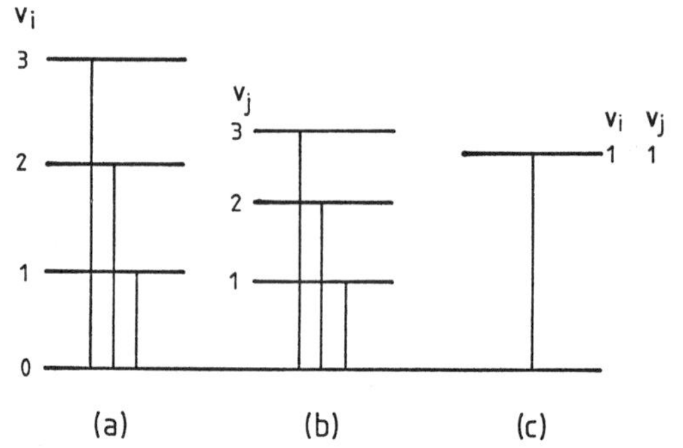

Figure 5.13 (a, b) Fundamental and overtone transitions involving two vibrations ν_i and ν_j. (c) A combination tone transition involving one quantum each of ν_i and ν_j

in detail in Section 5.2.4. However, in this discussion of group vibrations we shall be concerned primarily with molecules of low symmetry for which the requirements of a change of dipole moment or of amplitude of induced dipole moment are not very restrictive. For example, in a molecule with only a plane of symmetry such as 1-chloro-2-fluorobenzene (Figures 5.14) in which the plane of the nuclei is the plane of symmetry, *all* the 30 normal modes involve a change of dipole moment and of amplitude of induced dipole moment. Therefore, all $\Delta v_i = \pm 1$ transitions are allowed in the infrared and Raman spectra. However, their intensities depend on the magnitudes of the changes of the dipole moment and of the induced dipole moment which may be small for some vibrations.

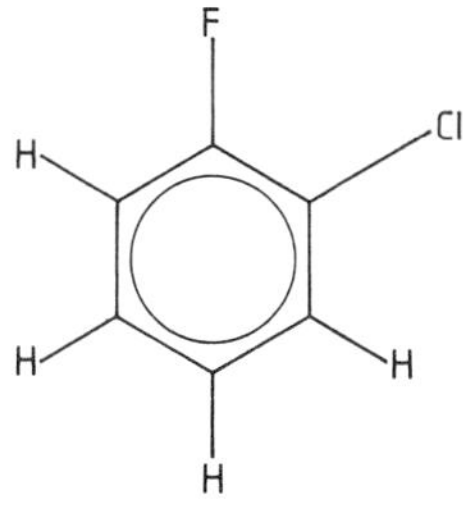

Figure 5.14 The 1-chloro-2-fluorobenzene molecule

From the observed fundamental vibration wavenumbers[†] it is possible (see Section 5.2.3) to derive force constants which are associated with the molecular vibrations. There are several possible ways of making this association, each corresponding to a particular force field. It is the valence force field which is most suitable for our present purposes; this associates a force constant with the stretching of each bond and the bending of each angle.

It has been found, not surprisingly perhaps, that the force constant associated with a particular bond, such as stretching of the O—H bond, is not much affected by the molecule in which the bond is present. This is equivalent to saying that, in the ball and spring picture of internuclear forces, the spring representing, say, the O—H bond in H_2O (Figure 5.11) is of a similar strength to that representing the O—H bond in, say, CH_3OH methanol or C_6H_5OH (phenol). Of course, there are variations due to the immediate structural environment, the effect of which may be appreciable and characteristic. A good example of this is the C—H stretching force constant. Table 5.3 shows that a typical value is 5.85 aJ Å^{-2} when the carbon atom is attached by a triple bond to the other part of the molecule, 5.1 aJ Å^{-2} when attached by a double and a single bond and 4.79 aJ Å^{-2} when attached by three single bonds. The increase of the C—H stretching force constant with the

[†]The fundamental vibration wavenumber is that of the $v = 1$–0 transition.

Table 5.3 Typical bond-stretching and angle-bending force constants, k

Bond stretching		Angle bending	
Group	k/aJ Å^{-2}	Group	(k/r_1r_2)/aJ Å^{-2}
≡C—H	5.85	≡C—H	0.21
=C⟨H	5.1	=C⟨H,H	0.30
—C—H	4.79	=C⟨H,H	0.51
—C≡C—	15.59	—C⟨H,H,H	0.46
C=C	9.6	—C⟨O,H	1.5
—C—C—	4.50	≡C—C	0.16
C=O	12.1	H—C—F	0.76
—C≡N	17.73		
—C—F	5.96	H—C—Cl	0.58
—C—Cl	3.64		
—C—Br	3.13	H—C—Br	0.52
—C—I	2.65	H—C—I	0.45
—O—H	7.66		
N—H	6.35		

multiplicity of the other bonds attached to the carbon atom can be rationalized using the hybridization picture of bonding. When carbon is at one end of a triple bond, as in H—C≡C—H (acetylene), the hybridization on carbon is *sp*. When it is at one end of a double bond, as in H_2C=CH_2 (ethylene), the hybridization is *sp^2* and, when involved in only single

bonds, as in CH_4 (methane), the hybridization is sp^3. Since greater s-character in a bond leads to a stronger bond, the C—H stretching force constant increases with the multiplicity of the other bonds in which carbon is involved.

Other typical stretching force constants are also given in Table 5.3.

Angle-bending force constants are more sensitive than stretching force constants to the structural environment, but typical values are still useful and some are listed in Table 5.3.

It should be noted that it is usual to quote bending force constants in the same units as stretching force constants, here in aJ Å^{-2}. It may be thought that a more natural unit would be aJ rad^{-2} but, since the radian is dimensionless (it is distance divided by distance), the bending force constants would then have different dimensions from the stretching force constants, which is inconvenient in calculations. For this reason, bending force constants k are replaced by k/r_1r_2, where r_1 and r_2 are the equilibrium bond lengths of the two bonds which contain the angle.

The fact that force constants for stretching and bending within particular groups of atoms are largely invariant to the neighbouring atoms of the molecule makes it useful to use the approximation of transferring force constants from one molecule to another.

It is less immediately obvious than for force constants that fundamental vibration wavenumbers may also be characteristic of a particular group of atoms and may be transferred, to a degree of approximation, to other molecules. Since it is the wavenumbers, rather than the force constants, which are observed experimentally, the approximate transferability of wavenumbers results in vibrational spectroscopy being an important analytical tool.

Although, in general, a normal mode of vibration involves movement of all the atoms in the molecule there are circumstances in which normal modes are localized in a part of the molecule. For example, if the vibration involves the stretching or bending of a terminal —X—Y group, where X is heavy compared with Y, as in the O—H group of CH_3CH_2OH (ethanol), then the corresponding fundamental wavenumbers are almost independent of the rest of the molecule to which —X—Y is attached. In ethanol the motions of the hydrogen atom of the O—H group are approximately those which it would have if it were attached to an infinite mass by a bond whose force constants are those of a typical O—H bond. For this reason, we can speak of a typical wavenumber of an O—H stretching vibration, for which the symbol ν(O—H) is used,[†] of 3590–3650 cm^{-1} in the absence of any hydrogen bonding. The wavenumber range is small and reflects the relatively slight dependence on the nature of the molecule in the immediate neighbourhood of the group. Such a typical wavenumber is called a group wavenumber.[‡] Another group wavenumber of the O—H group is the bending, or deformation, vibration, which is typically in the 1050–1200 cm^{-1} range.

Other general circumstances in which normal vibrations tend to be localized in a particular group of atoms obtain when there is a chain of atoms in which the force constant between two atoms is very different from those between other atoms in the chain. For example, in the molecule HC≡C—CH=CH_2 the force constants in the C—C, C=C and C≡C bonds are dissimilar. It follows that the stretching of the bonds are not strongly coupled and that each stretching vibration wavenumber is typical of the C—C, C=C, or C≡C group.

Table 5.4 lists a number of group vibration wavenumbers for both bond stretching and angle bending vibrations.

Not all parts of a molecule are characterized by group vibrations. Many normal modes involve strong coupling between stretching or bending motions of atoms in a straight chain, a

[†]It is common to use the symbols ν(X–Y), δ(X–Y), and γ(X–Y) for stretching, in-plane bending and out-of-plane bending vibrations, respectively, in the X–Y group. In addition, the word 'deformation' is often used to imply a bending motion.

[‡]It is also called commonly, but strictly erroneously, a group frequency.

Table 5.4 Typical bond stretching and angle bending group vibration wavenumbers, $\tilde{\nu}$

Bond stretching		Angle bending	
Group	$\tilde{\nu}$/cm^{-1}	Group	$\tilde{\nu}$/cm^{-1}
≡C—H	3300	≡C⌒—H	700
=C(—H)<	3020	=C⌒(H)(H)	1100
except O=C(—H)<	2800	—C⌒(H)(—H)(H)	1000
>C—H (—C—H)	2960	>C(H)(H)	1450
—C≡C—	2050	C≡C⌒—C	300
>C=C<	1650		
—C—C— (>C—C<)	900		
—Si—Si— (>Si—Si<)	430		
>C=O	1700		
—C≡N	2100		
—C—F (>C—F)	1100		
—C—Cl (>C—Cl)	650		
—C—Br (>C—Br)	560		
—C—I (>C—I)	500		
—O—H	3600[a]		
>N—H	3350		
—P=O (>P=O)	1295		
>S=O	1310		

[a]May be reduced by hydrogen bonding.

branched chain, or a ring. Such vibrations are skeletal vibrations and tend to be specific to a particular molecule. For this reason, the region where skeletal vibrations mostly occur, from about 1300 cm^{-1} to low wavenumber, is sometimes called the fingerprint region. On the other hand, the region 1500–3700 cm^{-1}, where many transferable group vibrations occur, is the functional group region.

In addition to the descriptions of group vibrations as stretch and bend (or deformation), the terms rock, twist, scissors, wag, torsion, ring breathing and inversion (or umbrella) are frequently used; these motions are illustrated in Figure 5.15.

The use of group vibrations as an important tool in qualitative analysis of a molecular sample was, until the advent of laser Raman spectroscopy, confined largely to infrared spectroscopy. In the infrared spectrum, the intensity of absorption due to a particular vibration depends on the change of dipole moment during the vibration, much as for a diatomic molecule (see Figure 5.1). For example, the stretching vibration of the strongly polar C=O bond gives a strong absorption band whereas that of the C=C bond gives a weak band. Indeed, if the C=C bond is in a symmetrical molecule such as $H_2C{=}CH_2$, there is no change of dipole moment at all and the vibration is infrared inactive. If the C=C bond is in, say, $HFC{=}CH_2$ there will be a small change of dipole moment due to stretching of the bond, but clearly not as large a change as that due to stretching of the C—F bond.

Just as group vibration wavenumbers are fairly constant from one molecule to another, so are their intensities. For example, if a molecule were being tested for the presence of a C—F bond there must not only be an infrared absorption band due to bond stretching at about 1100 cm^{-1}, but it must also be intense. A weak band in this region might be attributable to another normal mode.

Infrared spectra for purposes of qualitative analysis may be obtained from a pure liquid, the solid in a KBr disk or a solution using a

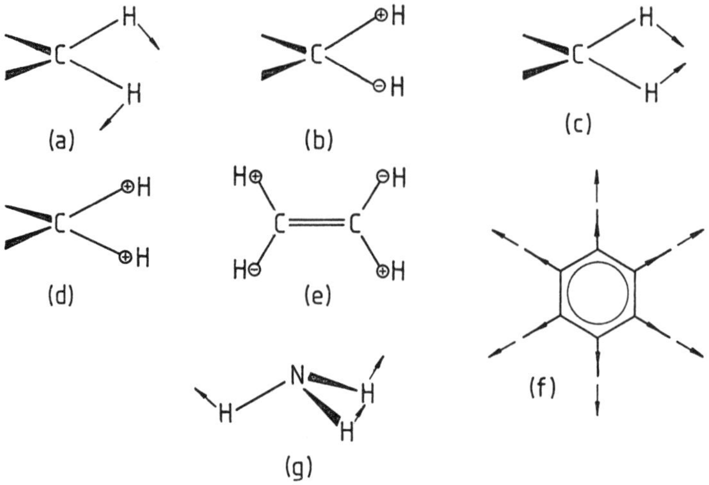

Figure 5.15 Illustration of the (a) rocking, (b) twisting, (c) scissoring and (d) wagging vibrations in a CH_2 group. Also shown are (e) the torsional vibration in ethylene, (f) the ring breathing vibration in benzene and (g) the inversion, or umbrella, vibration in ammonia

non-polar solvent; polar solvents may affect both group vibration wavenumbers and band intensities and may be involved in hydrogen-bonding with the solute. For example, the O—H stretching vibration of phenol in solution in hexane has a wavenumber of 3622 cm^{-1}, but this is reduced to 3344 cm^{-1} in diethyl ether because of the hydrogen bonding illustrated in Figure 5.16.

The use of vibrational Raman spectroscopy in qualitative analysis has increased greatly since the introduction of lasers as monochromatic sources. Although a laser Raman spectrometer is more expensive than a typical near infrared spectrometer used for qualitative analysis, it does have the advantage that low and high wavenumber vibrations can be observed with equal ease, whereas in the infrared a different, far infrared, spectrometer may be required for observations below about 400 cm^{-1}.

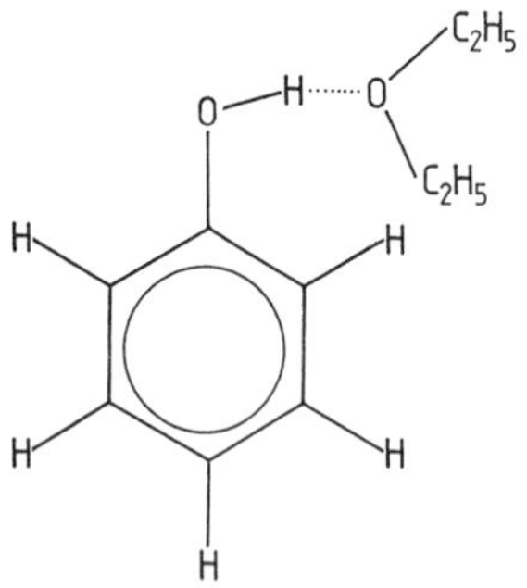

Figure 5.16 Hydrogen bonding between phenol and diethyl ether

The observation of a vibrational band in the Raman spectrum depends on there being an accompanying change of the amplitude of the induced dipole moment and the intensity, as in the case of a diatomic molecule (see Figure 5.2), depends on the magnitude of this change. It seems that, in general, this change is less sensitive than the change of dipole moment to the environment of the vibrating group. As a result, group vibration intensities are more accurately transferable from one molecule to another and from one phase or solvent to another in the Raman than in the infrared spectrum.

Figure 5.17(a) shows the near infrared spectrum of *s-trans*-crotonaldehyde, illustrated in Figure 5.18, and Figure 5.17(b) shows the laser Raman spectrum. The infrared spectrum is mostly of a solution in carbon tetrachloride but partly of a thin film of the pure liquid in the

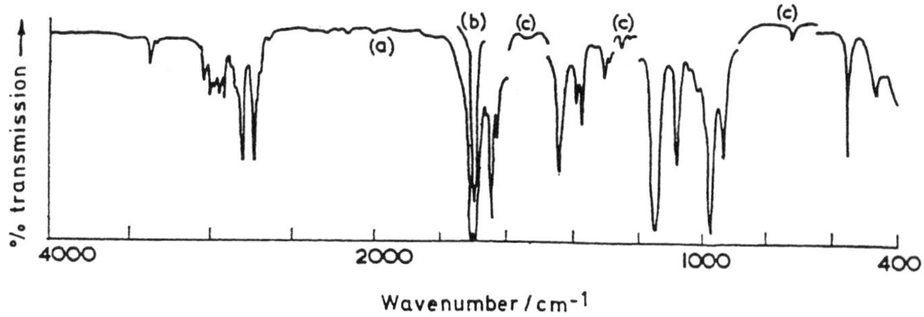

Figure 5.17(a) The near infrared vibrational spectrum of crotonaldehyde. The parts marked (a), (b) and (c) refer to a 10% (by volume) solution in CCl_4, a 1% solution in CCl_4 and a thin liquid film, respectively [Reproduced, with permission, from Bowles, A. J., George, W. O. and Maddams, W. F. (1969). *J. Chem. Soc.*, 810]

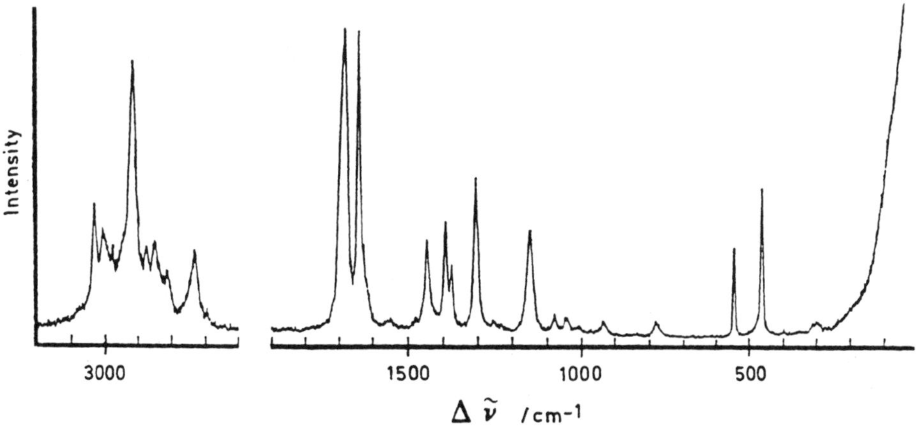

Figure 5.17(b) Laser Raman vibrational spectrum of liquid crotonaldehyde [Reproduced, with permission, from Durig, J. R., Brown, S. C., Kalasinsky, V. F. and George, W. O. (1976). *Spectrochim. Acta*, **32A**, 807]

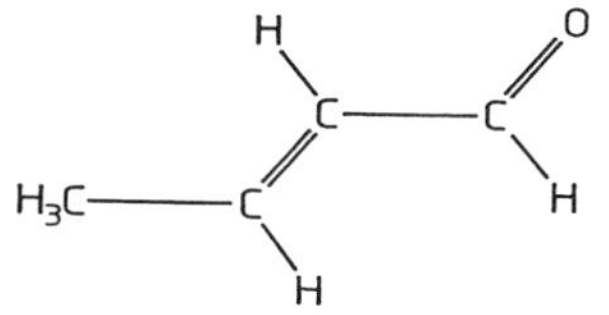

Figure 5.18 The *s-trans*-crotonaldehyde molecule

region where carbon tetrachloride itself absorbs. The Raman spectrum is of the pure liquid. Table 5.5 records the vibration wavenumbers of all 27 normal modes, together with an approximate description of the vibrational motions. A comparison of this table with Table 5.4 shows that ν_1, ν_2, ν_5, ν_6, ν_7 and ν_{15} are all well behaved group vibrations. A comparison of the infrared and Raman spectra shows many similarities of intensities but also some large differences. For example ν_{15}, the C—CH_3 stretching vibration, is strong in the infrared but very weak in the Raman, whereas ν_3, the CH_3 antisymmetric stretching vibration, is very strong in the Raman but weak in the infrared.

The Raman spectrum can be used to give additional information regarding the symmetry properties of vibrations. This information derives from the measurement of the depolarization ratio ρ for each Raman band. The quantity ρ is a measure of the degree to which the polarization properties of the incident

Table 5.5 Fundamental vibration wavenumbers of crotonaldehyde obtained from the infrared and Raman spectra

		$\tilde{\nu}/cm^{-1}$	
Vibration[a]	Approximate description	Infrared	Raman
In-plane			
ν_1	CH antisymmetric stretch on C=C	3042	3032
ν_2	CH symmetric stretch on C=C	3002	3006
ν_3	CH_3 antisymmetric stretch	2944	2949
ν_4	CH_3 symmetric stretch	2916	2918
ν_5	CH stretch on CHO	2727	2732
ν_6	C=O stretch	1693	1682
ν_7	C=C stretch	1641	1641
ν_8	CH_3 antisymmetric deformation	1444	1445
ν_9	CH rock (in-plane bend) on CHO	1389	1393
ν_{10}	CH_3 symmetric deformation	1375	1380
ν_{11}	CH symmetric deformation on C=C	1305	1306
ν_{12}	CH antisymmetric deformation on C=C	1253	1252
ν_{13}	CH_3 in-plane rock	1075	1080
ν_{14}	C—CHO stretch	1042	1046
ν_{15}	C—CH_3 stretch	931	931
ν_{16}	CH_3—C=C bend	542	545
ν_{17}	C=C—C bend	459	464
ν_{18}	C—C=O bend	216	230
Out-of-plane			
ν_{19}	CH_3 antisymmetric stretch	2982	2976
ν_{20}	CH_3 antisymmetric deformation	1444	1445
ν_{21}	CH_3 rock	1146	1149
ν_{22}	CH antisymmetric[b] deformation on C=C	966	—
ν_{23}	CH symmetric[b] deformation on C=C	—	780
ν_{24}	CH wag (out-of-plane bend) on CHO	727	—
ν_{25}	CH_3 bend	297	300
ν_{26}	CH_3 torsion	173	—
ν_{27}	CHO torsion	121	—

[a]See footnote on page 213 for vibrational numbering convention.
[b]To inversion of the two hydrogens through the centre of the C—C bond.

radiation may be changed after scattering has occurred.

The incident radiation is usually unpolarized or plane or circularly polarized. Plane polarized radiation was defined in Section 2.1, the plane being that which contains the direction of propagation and that of the oscillating electric vector. Circularly polarized radiation is composed of radiation polarized in the *xz*-plane, where *z* is the direction of propagation, and radiation polarized in the perpendicular *yz*-plane with the two associated waves having identical amplitudes and being out-of-phase by an odd multiple of $\pi/2$. In general, when these conditions of phase difference and amplitude are not met, elliptical polarization results. Both circularly and elliptically polarized radiation may be right handed or left handed. It is said to be right handed when an observer looking towards the radiation sees the tip of the electric vector tracing out a circle, or an ellipse, in a clockwise direction. Similarly, it is left handed when the circle or ellipse is traced out in an anticlockwise direction.

Raman and Rayleigh scattering occur in all directions but it is useful to define the scattering plane as that containing the direction of propagation of the incident radiation and the direction of observation of the scattered radiation. We define also the angle θ as that between the direction of propagation and observation.

When the incident radiation is plane polarized and the electric vector is parallel to the scattering plane the corresponding depolarization ratio $\rho_{\parallel}(\theta)$ is defined as

$$\rho_{\parallel}(\theta) = \frac{{}^{\parallel}I_{\perp}(\theta)}{{}^{\parallel}I_{\parallel}(\theta)} \tag{5.47}$$

where ${}^{\parallel}I_{\perp}(\theta)$ and ${}^{\parallel}I_{\parallel}(\theta)$ are the intensities of scattered radiation, at an observation angle θ, with the electric vector perpendicular and parallel to the scattering plane, respectively. The superscript indicates the polarization of the incident radiation. Figure 5.19(a) illustrates the case where z is the direction of propagation and the radiation is polarized in the xz scattering plane.

For incident radiation which is plane polarized with the electric vector perpendicular to the scattering plane, the depolarization ratio $\rho_{\perp}(\theta)$ is defined as

$$\rho_{\perp}(\theta) = \frac{{}^{\perp}I_{\parallel}(\theta)}{{}^{\perp}I_{\perp}(\theta)} \tag{5.48}$$

Figure 5.19(b) illustrates this when the direction of propagation and the scattering plane are the same as in Figure 5.19(a).

For natural (unpolarized) incident radiation the depolarization ratio $\rho_n(\theta)$ is defined as

$$\rho_n(\theta) = \frac{{}^{n}I_{\parallel}(\theta)}{{}^{n}I_{\perp}(\theta)} \tag{5.49}$$

and is illustrated in Figure 5.19(c).

For circularly polarized incident radiation the reversal factor $P(\theta)$ may be used to define the change of polarization in the scattered radiation, where

$$P(\theta) = \frac{{}^{\mathrm{R}}I_{\mathrm{L}}(\theta)}{{}^{\mathrm{R}}I_{\mathrm{R}}(\theta)} = \frac{{}^{\mathrm{L}}I_{\mathrm{R}}(\theta)}{{}^{\mathrm{L}}I_{\mathrm{L}}(\theta)} \tag{5.50}$$

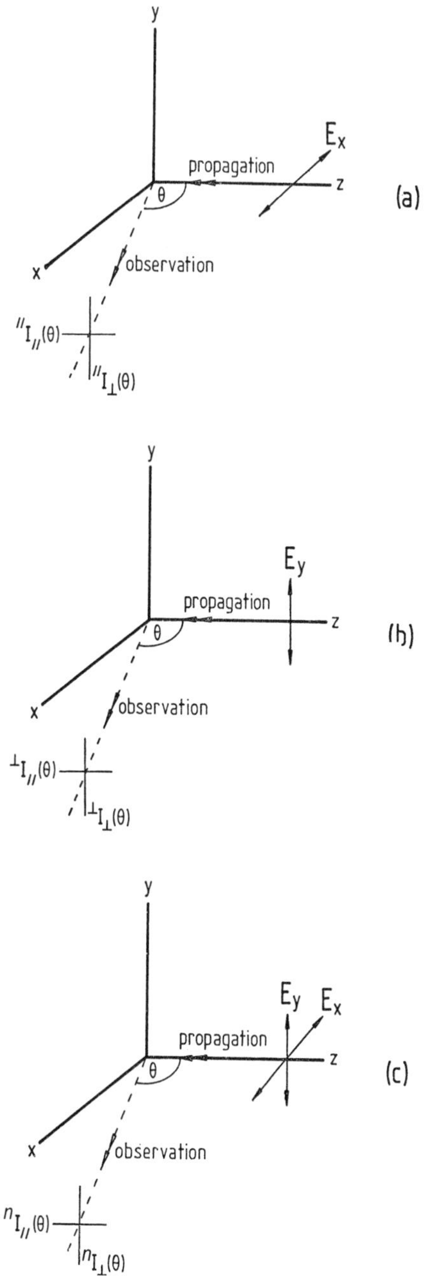

Figure 5.19 Three experimental configurations for the observation of scattering in the xz-plane at an angle θ to the incident radiation which is propagated along the z-axis. The incident radiation is (a) plane polarized with the electric vector parallel to the xz-plane, (b) plane polarized with the electric vector perpendicular to the xz-plane and (c) unpolarized

in which ${}^{\mathrm{R}}I_{\mathrm{L}}(\theta)$, for example, indicates incident and scattered radiation which are right and left circularly polarized, respectively. Alternatively, the degree of circularity $C(\theta)$ may be used, which is defined as

$$C(\theta) = \frac{I_{\mathrm{R}}(\theta) - I_{\mathrm{L}}(\theta)}{I_{\mathrm{T}}(\theta)} \tag{5.51}$$

where $I_{\mathrm{T}}(\theta)$ is the total intensity of scattered radiation. For right or left circularly polarized incident radiation

$${}^{\mathrm{R}}C(\theta) = -{}^{\mathrm{L}}C(\theta) \tag{5.52}$$

When plane polarized incident radiation is used, it is usual for the scattering to be observed in a direction normal to the incident radiation, i.e. $\theta = \pi/2$. For circularly polarized incident radiation the forward $(\theta = \pi)$ or backward $(\theta = 0)$ scattering is observed.

For Rayleigh scattering, the depolarization ratios, reversal factor and degree of circularity are restricted as follows:

$$\rho_{\parallel}\left(\frac{\pi}{2}\right) = 1,\ 0 \leqslant \rho_{\perp}\left(\frac{\pi}{2}\right) < \frac{3}{4},\ 0 \leqslant \rho_{n}\left(\frac{\pi}{2}\right) < \frac{6}{7},$$
$$0 \leqslant P(0) < 6,\ -\frac{5}{7} < {}^{\mathrm{R}}C(0) \leqslant 1$$

and, for Raman scattering.

$$\rho_{\parallel}\left(\frac{\pi}{2}\right) = 1,\ 0 \leqslant \rho_{\perp}\left(\frac{\pi}{2}\right) \leqslant \frac{3}{4},\ 0 \leqslant \rho_{n}\left(\frac{\pi}{2}\right) \leqslant \frac{6}{7}$$
$$0 \leqslant P(0) \leqslant 6,\ -\frac{5}{7} \leqslant {}^{\mathrm{R}}C(0) \leqslant 1$$

Because $\rho_{\parallel}(\pi/2)$ cannot be different from 1, this depolarization ratio is not used.

For Raman scattering the radiation is said to be depolarized when

$$\rho_{\perp}\left(\frac{\pi}{2}\right) = \frac{3}{4} \quad \text{or} \quad \rho_{n}\left(\frac{\pi}{2}\right) = \frac{6}{7}$$

polarized when

$$0 < \rho_{\perp}\left(\frac{\pi}{2}\right) < \frac{3}{4} \quad \text{or} \quad 0 < \rho_{n}\left(\frac{\pi}{2}\right) < \frac{6}{7}$$

and completely polarized when

$$\rho_{\perp}\left(\frac{\pi}{2}\right) = 0 \quad \text{or} \quad \rho_{n}\left(\frac{\pi}{2}\right) = 0$$

When circularly polarized incident radiation is used, the scattered radiation is said to be completely unreversed when

$$P(0) = 0,\ {}^{\mathrm{R}}C(0) = 1 \quad \text{or} \quad {}^{\mathrm{L}}C(0) = -1$$

and partly reversed when

$$0 < P(0) \leqslant 6,\ -\frac{5}{7} \leqslant {}^{\mathrm{R}}C(0) < 1$$
$$\text{or} \quad -1 < {}^{\mathrm{L}}C(0) \leqslant \frac{5}{7}$$

In vibrational Raman spectroscopy, the depolarization ratio, reversal factor and degree of circularity depend on the symmetry of the vibrational motion. If the vibration preserves all the symmetry of the molecule it is said to be totally symmetric but, if it does not, it is said to be non-totally symmetric. The polarization characteristics are capable of distinguishing these two types when the sample is in the gas or liquid phase. (In a single crystal different orientations of the crystal can be used to distinguish various types of non-totally symmetric vibration.)

For a totally symmetric vibration

$$\rho_{\perp}\left(\frac{\pi}{2}\right) < \frac{3}{4},\ \rho_{n}\left(\frac{\pi}{2}\right) < \frac{6}{7} \quad \text{or} \quad P(0) < 6$$

and, for a non-totally symmetric vibration,

$$\rho_{\perp}\left(\frac{\pi}{2}\right) = \frac{3}{4},\ \rho_{n}\left(\frac{\pi}{2}\right) = \frac{6}{7} \quad \text{or} \quad P(0) = 6$$

Measurement of $\rho_{\perp}$ or ρ_{n}, rather than $P(0)$, is more usual for the purpose of distinguishing totally symmetric and non-totally symmetric vibrations.

In crotonaldehyde, totally symmetric vibrations, $\nu_1 - \nu_{18}$ in Table 5.5, are in-plane†

†Apart from the hydrogen atoms of the CH_3 group we can regard crotonaldehyde as a planar molecule.

vibrations and non-totally symmetric vibrations, $\nu_{19} - \nu_{27}$ in Table 5.5, are out-of-plane vibrations. Clearly, measurements of depolarization ratios are a useful aid to such assignments but caution is sometimes necessary. Although a value of $\rho_\perp$ or ρ_n which, allowing for the uncertainty of measurement, is less than $\frac{3}{4}$ or $\frac{6}{7}$, respectively, shows that the vibration is totally symmetric, a value of $\rho_\perp$ or ρ_n close to $\frac{3}{4}$ or $\frac{6}{7}$ has to be interpreted with caution since it suggests that the vibration is nontotally symmetric but does not exclude the possibility that it is totally symmetric.

The use of group vibrations is very often concerned with molecules which do not have a high degree of symmetry but, for a molecule, such as ethylene ($H_2C{=}CH_2$), with a centre of symmetry i (see p. 95), the mutual exclusion rule holds. This rule states that vibrational fundamentals which are allowed in the infrared spectrum are forbidden in the Raman spectrum and vice versa; that is, the infrared and Raman spectra are mutually exclusive. However, in many molecules with a centre of symmetry there are vibrations which are forbidden in *both* the infrared *and* Raman spectra.

A detailed discussion of selection rules for vibrational spectra of polyatomic molecules will be given in Section 5.2.4.

In addition to bands in the infrared and Raman spectra due to $\Delta v = 1$ transitions of fundamental vibrations, combination and overtone bands may also occur with appreciable intensity particularly in the infrared. Care must be taken not to confuse such bands as being due to a weakly active fundamental. Occasionally combinations and, more often, overtones may be used to aid identification of group vibrations.

There is an important vibrational interaction, called Fermi resonance, which may interfere with a group vibrational analysis. Such a resonance occurs if a fundamental and an overtone, or a combination, vibrational level would, in the absence of any interaction, be at a very similar energy; an additional requirement is that the two vibrational wave functions have the same symmetry (see Section 5.2.4). The effects of the resonance are (i) to push apart the two vibrational levels and (ii) to mix the intensities of the two transitions. The result is that a fundamental band may be far removed in the spectrum from where it would be expected and that a combination band may also be shifted and be more intense than usual.

If the infrared spectrum of the sample in the gas phase can be obtained, determination of the rotational selection rules from the rotational contour of each band can be an important aid in identifying fundamentals. This will be discussed further in Section 5.2.5.

Most of the discussion of group vibrations up to this point has been confined to organic molecules because it is to such molecules that their use is most commonly applied. In inorganic molecules and ions the approximations involved in the concepts of transferable force constants and group vibration wavenumbers are not so often valid. Nevertheless, vibrations involving stretching of M—H, M—C, M—X and M=O bonds, where M is a metal atom and X a halogen atom, are examples of useful group vibrations.

Inorganic complexes containing organic ligands exhibit at least some group vibrations which are characteristic of the ligands.

Because of the heavy atoms which may be present in inorganic molecules, there is often a need to obtain the vibrational spectrum in the low wavenumber region. For this reason it is more important than for organic molecules to obtain the far infrared spectrum.

5.2.2 Molecular symmetry

5.2.2.1 Elements of symmetry

In Chapter 4, and up to this point in the present chapter, we have drawn increasingly on arguments involving the symmetry of the molecule concerned. In Section 4.1 we used the concept of an S_n rotation–reflection axis and also of a C_n axis of symmetry, in the definition of a symmetric rotor. In Section 2.2 we used C_n axes

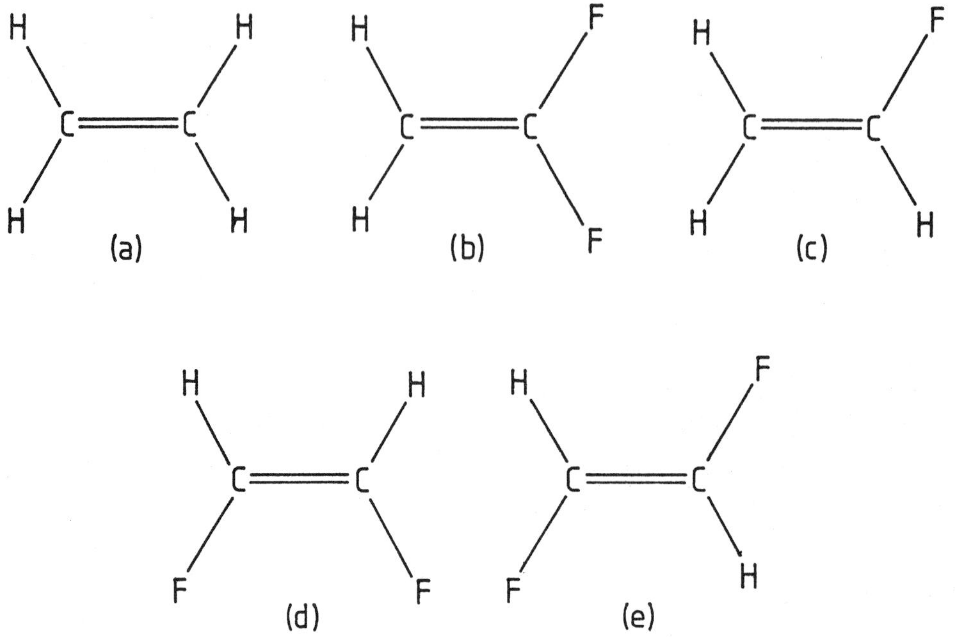

Figure 5.20 The molecules of (a) ethylene, (b) 1,1-difluoroethylene, (c) fluoroethylene, (d) *cis*-1,2-difluoroethylene and (e) *trans*-1,2-difluoroethylene

in discussing polarization of a transition moment and in the same section we encountered also a plane of symmetry, σ. In Section 5.2.1 the importance of a centre of symmetry i was noted in considering the possible mutual exclusion of Raman- and infrared-active vibrations.

Before going on to consider how the normal modes of vibration, for example those of H_2O in Figure 5.12, can be determined quantitatively and what the vibrational selection rules are in polyatomic molecules, it is necessary to treat the subject of molecular symmetry in greater depth than has been required so far.

Whatever intuitive feeling we may have for molecular symmetry, it is clear that there is a lowering of symmetry along the series of planar molecules ethylene, 1,1-difluoroethylene and fluoroethylene, illustrated in Figure 5.20 (a–c). However, the degree of symmetry of 1,1-difluoroethylene, *cis*-1,2-difluoroethylene and *trans*-1,2-difluoroethylene, shown in Figure 5.20(b), (d) and (e), respectively, cannot be distinguished so easily. These last examples make it clear that intuitive judgement of symmetry will not take us very far in the application of symmetry to molecular structure and properties.

In classifying the symmetry of molecules we start by considering the nuclei in their equilibrium configuration. Of course, it is essential, before we can do this, to know the equilibrium structure of the molecule concerned. In addition to many spectroscopic methods, such techniques as X-ray, electron and neutron diffraction may be used to determine this structure.

We shall be concerned only with the symmetry of the free molecule, in which any interactions with neighbouring molecules can be neglected; this condition is realized in the gas phase at low pressure and in a supersonic jet.

In order to classify systematically the symmetry of molecules we use elements of symmetry of which there are just five types—*n*-fold

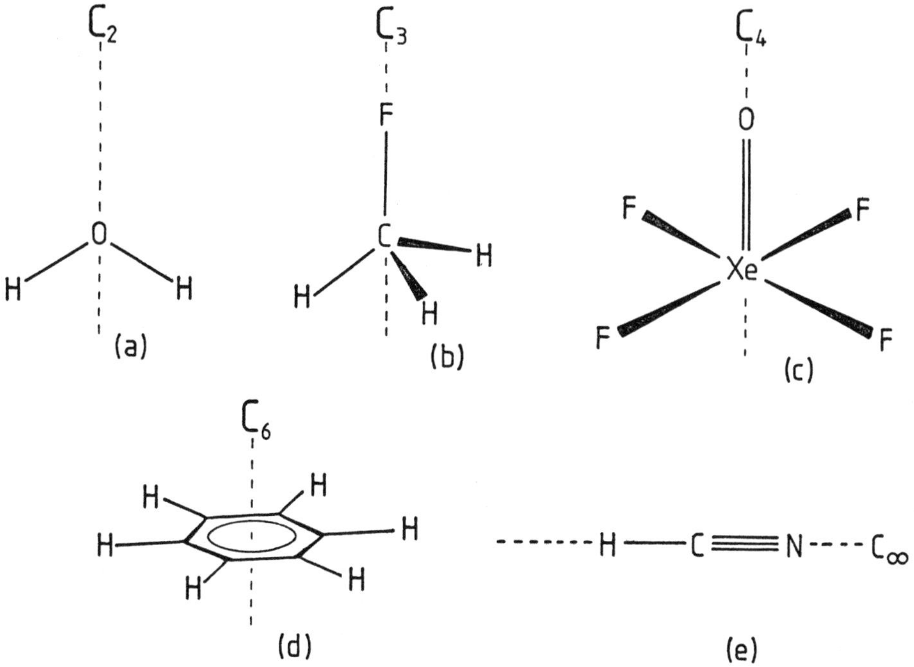

Figure 5.21 (a) C_2 axis in H_2O, (b) C_3 axis in CH_3F, (c) C_4 axis in $XeOF_4$, (d) C_6 axis in C_6H_6 and (e) C_∞ axis in HCN

axes of symmetry, a plane of symmetry, a centre of symmetry, n-fold rotation–reflection axes of symmetry and the identity element of symmetry—and one of these, the identity element, is in many ways trivial. We will now consider each of these types in detail.

5.2.2.1(i) *n-Fold axis of symmetry, C_n*

If a molecule has an n-fold axis of symmetry, for which the symbol is C_n, rotation of the molecule by $2\pi/n$ rad about the axis produces a configuration which, to a stationary observer, is indistinguishable from the initial one. Figure 5.21(a–e) illustrate[†] a C_2 axis in H_2O, a C_3 axis in CH_3F, a C_4 axis in $XeOF_4$, a C_6 axis in C_6H_6, and a C_∞ axis in HCN; a C_∞ axis is an axis about which a rotation by *any* angle produces an indistinguishable configuration and is possessed by all linear molecules.

Corresponding to every symmetry element there is a symmetry operation for which the symbol is always the same as that for the symmetry element. For example, C_n is a label not only for indicating an n-fold rotation axis, but also for the operation of carrying out a clockwise[‡] rotation of the molecule by $2\pi/n$ rad about the axis.

5.2.2.1(ii) *Plane of symmetry, σ*

If a molecule has a plane of symmetry, for which the symbol is σ, reflection of all the nuclei through the plane to an equal distance on the opposite side produces a configuration indistinguishable from the initial one.

[†]A fairly accurately constructed molecular model is a very useful aid in identifying elements of symmetry.

[‡]Some authors take this rotation to be anti-clockwise.

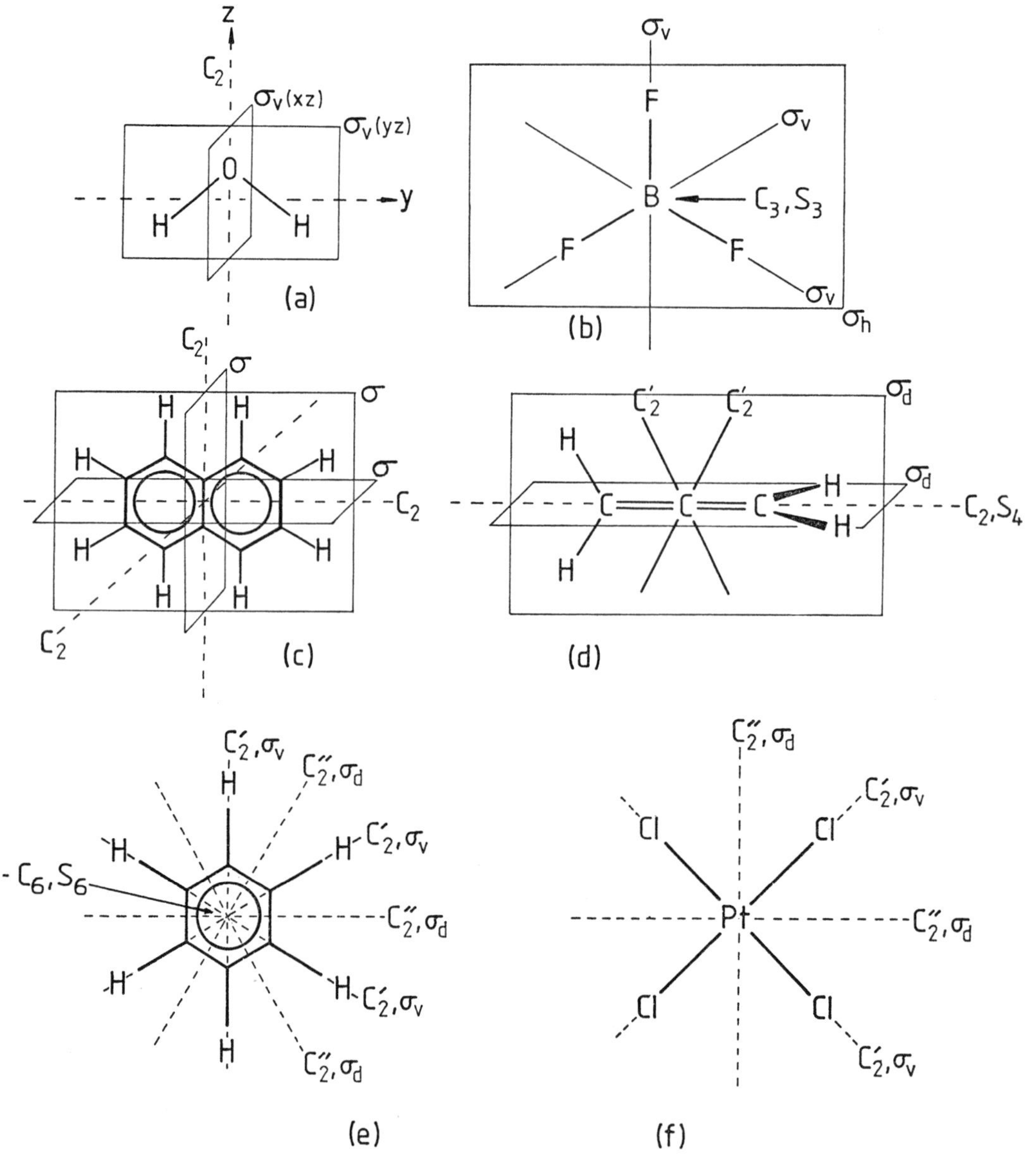

Figure 5.22 Planes and axes of symmetry in (a) H_2O, (b) BF_3, (c) naphthalene, (d) allene, (e) benzene and (f) $[PtCl_4]^{2-}$

Figure 5.22(a) shows the two planes of symmetry of H_2O, $\sigma_v(xz)$ and $\sigma_v(yz)$. With the axes labelled conventionally, as shown, $\sigma_v(yz)$ is the plane of the molecule. Since all the nuclei are in this plane, reflection through the plane has no effect and the resulting configuration is indistinguishable from the initial one; similarly, for *all* planar molecules the plane of the molecule is a plane of symmetry. The subscript v stands for 'vertical' and is a useful indication that the plane is vertical, where the vertical direction is defined as that of the highest-fold C_n axis, here C_2.

In the planar molecule BF_3 shown in Figure 5.22(b), the highest-fold C_n axis is the C_3 axis through B and perpendicular to the figure. Therefore, the three planes of symmetry, perpendicular to the figure and each containing a B—F bond, are σ_v planes. On the other hand, the plane of the molecule is labelled σ_h where the subscript h implies that the plane is 'horizontal' with respect to C_3 being vertical.

In a molecule such as naphthalene, shown in Figure 5.22(c), the highest-fold axis is C_2, of which there are three, at right-angles to each other. In these circumstances a vertical direction is not defined and no subscript is used.

A third subscript d, which stands for 'dihedral', is sometimes useful. Its use is illustrated by the allene molecule shown in Figure 5.22(d). The axes labelled C_2' are twofold axes at 90° to each other and at 45° to the plane of the figure. These are called dihedral axes which, in general, are C_2 axes at equal angles to each other. The σ_d planes bisect the angles between the dihedral axes. In molecules such as benzene and the square planar $[PtCl_4]^{2-}$ ion there is a choice of σ_v or σ_d labels for the planes perpendicular to the plane of the molecule. Figure 5.22(e) shows that for benzene there are two sets of planes perpendicular to the plane of the molecule, one set of planes bisecting each set of axes. All these planes could be labelled σ_v or σ_d, according to the previous definitions of such planes. The normal convention in benzene is to label σ_v those planes which pass through atoms and σ_d those which bisect carbon–carbon bonds. Similarly, in $[PtCl_4]^{2-}$ shown in Figure 5.22(f), the convention is to label σ_v those planes which pass through atoms and σ_d those which bisect bond angles.

The symmetry operation σ is the operation of reflecting the nuclei across the plane.

5.2.2.1(iii) Centre of inversion, or centre of symmetry, i

If a molecule has a centre of inversion, for which the symbol is i, then reflection of each nucleus through the centre of the molecule to a position which is an equal distance on the opposite side of the centre produces a configuration indistinguishable from the initial one. Figure 5.23 shows two examples, *s-trans*-buta-1,3-diene and sulphur hexafluoride, of molecules having a centre of inversion. Note also that *trans*-1,2-difluoroethylene [Figure 5.20(e)] has a centre of inversion but the *cis* isomer [Figure 5.20(d)] does not.

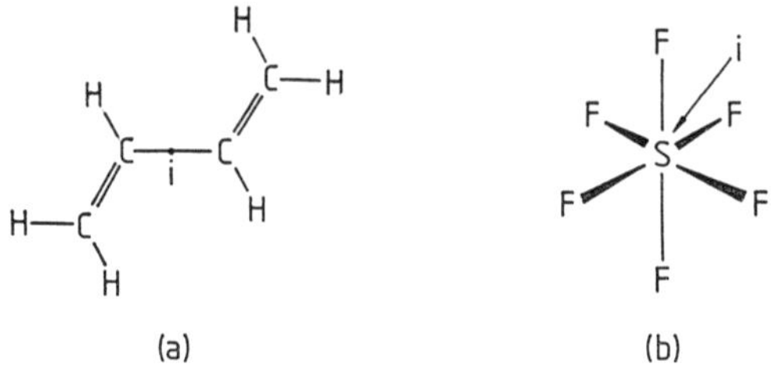

Figure 5.23 Centre of inversion in (a) *s-trans*-buta-1,3-diene and (b) SF_6

5.2.2.1(iv) n-Fold rotation–reflection axis of symmetry, S_n

If a molecule has an n-fold rotation–reflection axis of symmetry, for which the symbol is S_n, rotation by $2\pi/n$ rad about the axis followed by reflection of all the nuclei through a plane perpendicular to the axis and through the centre of the molecule produces a configuration indistinguishable from the initial one. Figure 5.22(d) illustrates the allene molecule, which has an S_4 axis; rotation by $2\pi/4$ rad about this axis followed by reflection through a plane perpendicular to the axis and through the central carbon atom produces an indistinguishable configuration. The plane of reflection need not necessarily be a plane of symmetry. In allene it is not a plane of symmetry but in BF_3 [Figure 5.22(b)], which has an S_3 axis, the plane of reflection is also a σ_h plane.

In BF_3 it is clear that the presence of the C_3 axis and the σ_h plane necessitate the C_3 axis being also an S_3 axis. This is always true of any combination of a C_n axis and a σ_h plane and we say that C_n and σ_h *generate* S_n or, simply,

$$\sigma_h \times C_n = S_n \tag{5.53}$$

This equation can be interpreted also in terms of the corresponding symmetry operations as meaning that, if we carry out a C_n operation followed by a σ_h operation, the effect is the same as carrying out an S_n operation. (The convention is that carrying out the operation B followed by A is written as A × B: in the case of C_n and σ_h the order does not matter but for other operations it may, as we shall see later.)

It is easily seen that $S_1 = \sigma$ and $S_2 = i$, so that the S_1 and S_2 symbols are not used.

The S_n element is often referred to as an improper rotation axis, whereas C_n is a proper rotation axis. Similarly, the S_n operation is an improper operation, whereas C_n is a proper operation.

5.2.2.1(v) *The identity element of symmetry, I (or E)*[†]

All molecules possess the identity element of symmetry, for which the symbol is *I*. The symmetry operation *I* is the operation of doing nothing to the molecule. At this stage this symmetry element may appear too trivial to be of any importance, but we shall see that the properties of groups in general, discussed in Section 5.2.2.3, require such an element.

Since the C_1 operation, a rotation by 2π rad, is equivalent to doing nothing to the molecule, $C_1 = I$ and the C_1 symbol is not used.

5.2.2.1(vi) *Generation of elements*

We have seen already [equation (5.53)] how C_n and σ_h elements generate S_n.

Figure 5.24 illustrates, using the CH_2F_2 (difluoromethane) molecule as an example, that

$$\sigma_v \times C_2 = \sigma_v' \tag{5.54}$$

so that C_2 and one of the σ_v generate the other σ_v.

[†]Some authors use *E* for this element, but this may cause confusion with its use as a symmetry species label (p. 196).

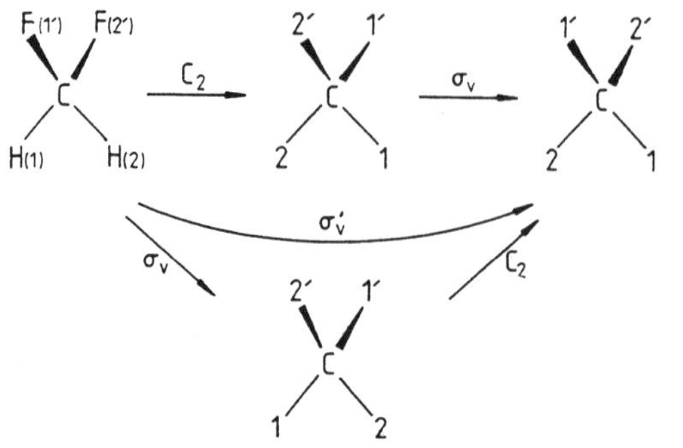

Figure 5.24 Illustration, using the CH_2F_2 molecule, that C_2 and σ_v generate σ_v' and that the C_2 and σ_v operations commute. The CF_2 plane is σ_v' and the CH_2 plane σ_v

From a C_n element we can generate other elements by raising it to the powers 1, 2, 3, . . ., $(n-1)$. For example, if there is a C_3 symmetry element there must also be C_3^2, where

$$C_3^2 = C_3 \times C_3 \tag{5.55}$$

The C_3^2 operation is a clockwise rotation by $2 \times 2\pi/3$. Also, if there is a C_6 element there must necessarily be $C_6^2(= C_3)$, $C_6^3(= C_2)$, $C_6^4(= C_3^2)$ and C_6^5; the C_6^5 operation is equivalent to an anti-clockwise rotation by $2\pi/6$, an operation which is given the symbol C_6^{-1}. Similarly, C_3^2 is equivalent to C_3^{-1} and, in general,

$$C_n^{n-1} = C_n^{-1} \tag{5.56}$$

where C_n^{-1} is called the inverse of C_n.

Similarly, from an S_n element we can generate S_n to the powers 1, 2, 3, . . . $(n-1)$. For example, Figure 5.25 illustrates, using allene as an example, the S_4^2 and S_4^3 operations and shows that

$$S_4^2 = C_2 \tag{5.57}$$

and

$$S_4^3 = S_4^{-1} \tag{5.58}$$

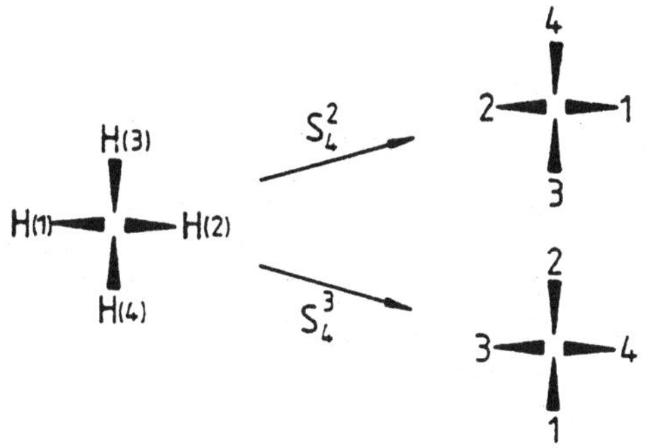

Figure 5.25 Illustration of the S_4^2 and S_4^3 operations on the allene molecule

where S_4^{-1} implies an anti-clockwise rotation by $2\pi/4$ followed by a reflection (note that a reflection is the same as its inverse).

5.2.2.1(vii) *Commutation and non-commutation of operations*

Figure 5.24 shows that, for the elements C_2 and σ_v, it does not matter in which order we carry out the corresponding operations, i.e.

$$\sigma_v \times C_2 = C_2 \times \sigma_v \tag{5.59}$$

If this is so for any pair of operations, those operations (or elements) are said to commute or to be commutative. On the other hand, C_3 and σ_v do not commute, or are non-commutative, i.e.

$$C_3 \times \sigma_v \neq \sigma_v \times C_3 \tag{5.60}$$

as Figure 5.26 shows, for BF_3.

5.2.2.2 **Point groups**

All the elements of symmetry which any conceivable molecule may have constitute a point group. Examples are the I, C_2, $\sigma_v(xz)$, $\sigma_v(yz)$ elements of H_2O and the I, S_4, S_4^{-1}, C_2, C_2', C_2', σ_d, σ_d elements of allene.

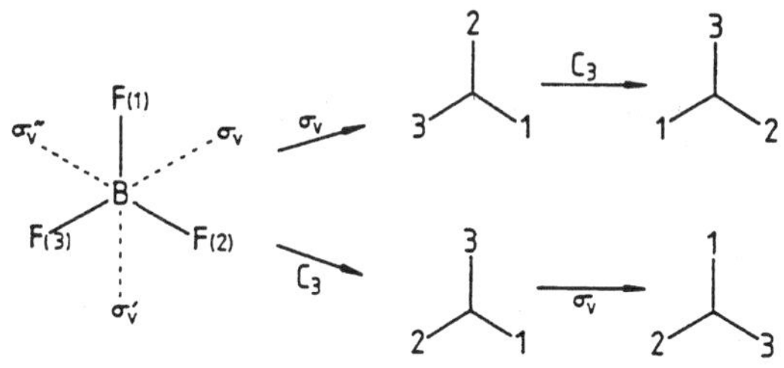

Figure 5.26 Illustration of the fact that, in BF_3, the σ_v and C_3 operations do not commute

Point groups should be distinguished from space groups. Point groups are so called because, when all the operations of the group are carried out, at least one point is unaffected; in the case of allene it is the point at the centre of the molecule, whereas in H_2O any point on the C_2 axis is unaffected. Space groups are appropriate to the symmetry properties of regular arrangements of molecules in space such as are found in crystals, and we shall not be concerned with them here.

Many point groups are necessary to cover all possible molecules and it is convenient to collect together point groups which have certain types of elements in common. It is also convenient in defining point groups not to list all the elements of the group. In theory it is necessary to list only the generating elements from which all the other elements may be generated. In practice, it is more helpful to give rather more than the generating elements and this is what we shall do here. Of course, all point groups contain the identity element and this will not be included in the point group definitions.

5.2.2.2(i) *C_n point groups*

A C_n point group contains a C_n axis of symmetry. By implication it does not contain σ, i or S_n elements. However, it must contain $C_n^2, C_n^3, \ldots, C_n^{n-1}$.

Examples of molecules belonging to a C_n point group, for which $n = 1, 2, 3, \ldots$, are not

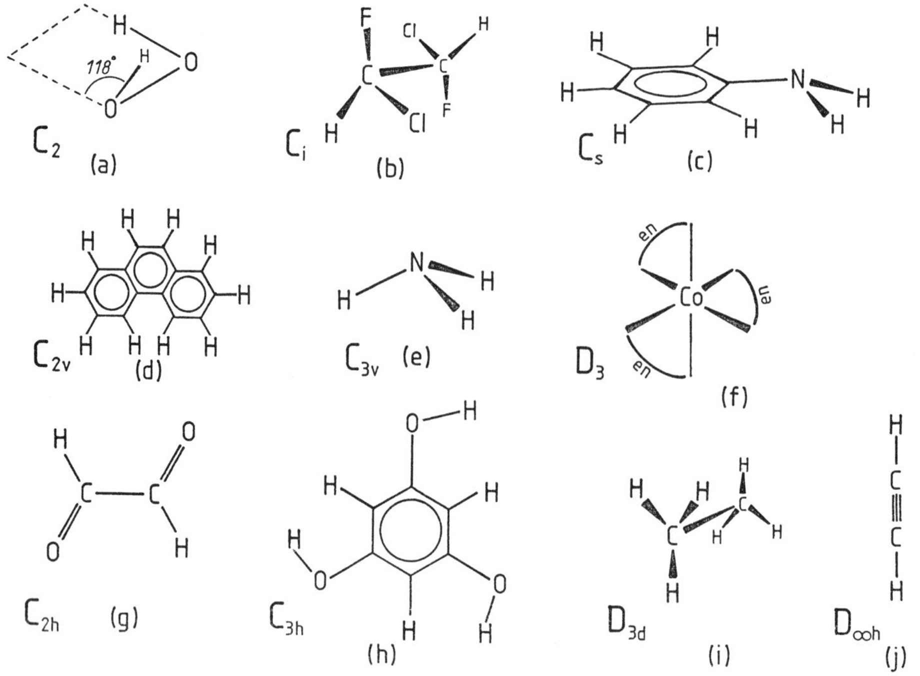

Figure 5.27 Examples of molecules belonging to various point groups

very common, but hydrogen peroxide (H_2O_2), shown in Figure 5.27(a), belongs to the C_2 point group; the only symmetry element, apart from I, is the C_2 axis bisecting the 118° angle between the two O—O—H planes. CHBrClF, shown in Figure 4.6(a), has no symmetry element, apart from I, and belongs to the C_1 point group.

5.2.2.2(ii) S_n point groups

An S_n point group contains an S_n axis of symmetry. The group must contain also S_n^2, $S_n^3, \ldots, S_n^{n-1}$.

Examples are rare except for the S_2 point group. This point group has only an S_2 axis but, since $S_2 = i$, it has only a centre of inversion and the symbol usually used for this point group is C_i. The isomer of the molecule ClFHC—CHFCl in which all pairs of identical H, F or Cl atoms are *trans* to each other, shown in Figure 5.27(b), belongs to the C_i point group.

5.2.2.2(iii) C_{nv} point groups

A C_{nv} point group contains a C_n axis of symmetry and n σ_v planes of symmetry, all of which contain the C_n axis. It also contains other elements which may be generated from these.

Many molecules belong to the C_{1v} point group. Apart from $I(= C_1)$ they have only a plane of symmetry. Fluoroethylene [Figure 5.20(c)] and aniline [Figure 5.27(c)], which has a pyramidal configuration about the nitrogen atom and a plane of symmetry bisecting the HNH angle, are examples. Instead of C_{1v} the symbol C_s is usually used.

H_2O [Figure 5.21(a)], difluoromethane (Figure 5.24), 1,1-difluoroethylene [Figure 5.20(b)], *cis*-1,2-difluorethylene [Figure 5.20(d)] and phenanthrene [Figure 5.27(d)] have a C_2 axis and two σ_v planes and therefore belong to the C_{2v} point group.

NH_3 [Figure 5.27(e)] and methyl fluoride [Figure 5.21(b)] have a C_3 axis and three σ_v planes and belong to the C_{3v} point group. $XeOF_4$ [Figure 5.21(c)] belongs to the C_{4v} point group.

$C_{\infty v}$ is an important point group since all unsymmetrical linear molecules, without a σ_h plane, belong to it. HCN [Figure 5.21(e)] has a C_∞ axis and an infinite number of σ_v planes, all containing the C_∞ axis and at any angle to the page, and therefore belongs to the $C_{\infty v}$ point group.

5.2.2.2(iv) D_n point groups

A D_n point group contains a C_n axis and nC_2 axes. The C_2 axes are perpendicular to C_n and at equal angles to each other. It also contains other elements which may be generated from these.

Molecules belonging to D_n point groups are not common. They can be visualized as being formed by two identical C_{nv} fragments being connected back-to-back in such a way that one fragment is staggered with respect to the other by any angle other than $m\pi/n$, where m is an integer.

The complex ion $[Co(ethylenediamine)_3]^{3+}$, shown in Figure 5.27(f), in which en is the $H_2NCH_2CH_2NH_2$ group,[†] belongs to the D_3 point group and can be thought of as consisting of two staggered C_{3v} fragments.

5.2.2.2(v) C_{nh} point groups

A C_{nh} point group contains a C_n axis and a σ_h plane, perpendicular to C_n. For n even the point group contains a centre of inversion i. It also contains other elements which may be generated from these.

The point group C_{1h} contains only a plane of symmetry, in addition to I. It is therefore the same as C_{1v} and is usually given the symbol C_s.

Examples of molecules having a C_2 axis, a σ_h plane and a centre of inversion i and therefore belonging to the C_{2h} point group are *trans*-1,2-difluoroethylene [Figure 5.20(e)], *s-trans*-buta-1,3-diene [Figure 5.23(a)] and *s-trans*-glyoxal [Figure 5.27(g)].

1,3,5-Trihydroxybenzene [Figure 5.27(h)] belongs to the C_{3h} point group.

5.2.2.2(vi) D_{nd} point groups

A D_{nd} point group contains a C_n axis, an S_{2n} axis, n C_2 axes perpendicular to C_n and at equal angles to each other and n σ_d planes bisecting the angles between the C_2 axes. For n odd the point group contains a centre of inversion i. It also contains other elements which may be generated from these.

For $n = 1$, D_{1d} is the same as C_{2v}.

Molecules belonging to other D_{nd} point groups can be visualized as consisting of two identical fragments of C_{nv} symmetry back-to-back with one staggered at an angle of π/n to the other.

Allene [Figure 5.22(d)] belongs to the D_{2d} point group and ethane [Figure 5.27(i)], which has a staggered configuration in which each of the C—H bonds at one end of the molecule bisects an HCH angle at the other, belongs to the D_{3d} point group.

Apart from these two important examples not many molecules belong to D_{nd} point groups.

5.2.2.2(vii) D_{nh} point groups

A D_{nh} point group contains a C_n axis, n C_2 axes perpendicular to C_n and at equal angles to each other, a σ_h plane and n other σ planes. For n even the point group contains a centre of inversion i. It also contains other elements which may be generated from these.

A D_{nh} point group is related to the corresponding C_{nv} point group by the inclusion of a σ_h plane.

For $n = 1$, D_{1h} is the same as C_{2v}.

Ethylene [Figure 5.20(a)] and naphthalene [Figure 5.22(c)] belong to the D_{2h} point group in which, because of the equivalence of the three

[†] The puckered nature of this substituent has been neglected.

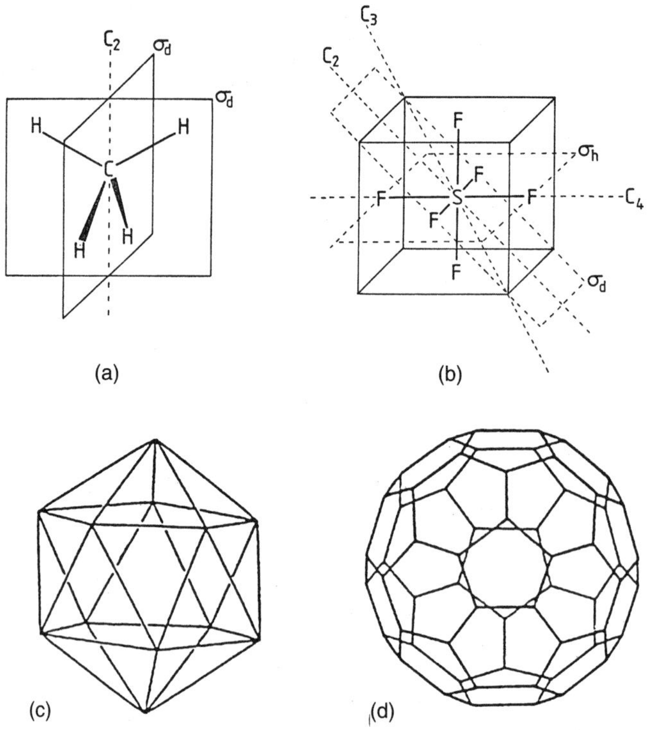

Figure 5.28 (a) Some of the C_2 and σ_d elements in the methane molecule. (b) Some of the symmetry elements of the SF_6 molecule. (c) A regular icosahedron. (d) Buckminsterfullerene (C_{60})

mutually perpendicular C_2 axes, no subscripts are used for the planes of symmetry.

BF_3 [Figure 5.22(b)] belongs to D_{3h}, $[PtCl_4]^{2-}$ [Figure 5.22(f)] belongs to D_{4h} and benzene [Figure 5.22(e)] belongs to the D_{6h} point group.

The $D_{\infty h}$ point group is derived from $C_{\infty v}$ by the inclusion of σ_h and therefore all linear molecules with a plane of symmetry perpendicular to the C_∞ axis belong to $D_{\infty h}$. Acetylene [Figure 5.27(j)], for example, and all homonuclear diatomic molecules belong to this point group.

5.2.2.2(viii) T_d point group

The T_d point group contains four C_3 axes, three C_2 axes and six σ_d planes. It also contains elements generated from these.

This is the point group to which all regular tetrahedral molecules, such as methane [Figure 5.28(a)], silane (SiH_4) and nickel tetracarbonyl [$Ni(CO)_4$], belong.

The C_3 axes, in the case of methane, are the directions of each of the C—H bonds. The σ_d planes are the six planes of all the possible CH_2 fragments. The C_2 axes are not quite so easy to see but Figure 5.28(a) shows that the line of intersection of any two mutually perpendicular σ_d planes is a C_2 axis.

5.2.2.2(ix) O_h point group

The O_h point group contains three C_4 axes, four C_3 axes, six C_2 axes, three σ_h planes, six σ_d planes and a centre of inversion i. It also contains elements generated from these.

This is the point group to which all regular octahedral molecules, such as SF_6 [Figure 5.28(b)] and $[Fe(CN)_6]^{3-}$, belong.

The main symmetry elements in SF_6 can be shown, as in Figure 5.28(b), by considering the sulphur atom at the centre of a cube and a fluorine atom at the centre of each face. The three C_4 axes are the three F—S—F directions, the four C_3 axes are the body diagonals of the cube, the six C_2 axes join the midpoints of diagonally opposite edges, the three σ_h planes are each half way between opposite faces and the six σ_d planes join diagonally opposite edges of the cube.

5.2.2.2(x) K_h point group

The K_h point group contains an infinite number of C_∞ axes and a centre of inversion i. It also contains elements generated from these.

This is the point group to which a sphere, and therefore all atoms, belong.†

5.2.2.2(xi) I_h point group

The I_h point group is that to which a regular icosahedron, illustrated in Figure 5.28(c), belongs. It contains 20 equilateral triangles arranged in a three-dimensional structure. This is the conformation of the $B_{12}H_{12}^{-}$ anion in which there is a boron atom, with a hydrogen atom attached, at each of the 12 vertices.

A regular dodecahedron, consisting of 12 regular pentagons in a three-dimensional structure, also belongs to this point group. An example is the dodecahedrane molecule, $C_{20}H_{20}$, in which there is a carbon atom, with a hydrogen atom attached, at each of the 20 vertices of adjoining pentagons. This molecule also belongs to the I_h point group.

A more recently discovered, and outstandingly important, example of a molecule belonging to the I_h point group is C_{60}, known as buckminsterfullerene, shown in Figure 5.28(d). There is a carbon atom at each of the 60 identical vertices, each vertex connecting two hexagons and one pentagon. There are 12 regular pentagons and 20 regular hexagons in this structure, the 12 (and only 12) pentagons being essential for closure of the three-dimensional structure. There are only two types of carbon–carbon bonds, one joining two hexagons and the other joining a pentagon and a hexagon.

Molecules belonging to the I_h point group are very highly symmetrical, having 15 C_2 axes, 10 C_3 axes, six C_5 axes, 15 σ planes, 10 S_6 axes, six S_{10} axes and a centre of inversion i. In addition to these symmetry elements are other elements which can be generated from them.

The point groups discussed here are all those which one is likely to use, but there are a few very uncommon ones which have not been included; they can be found in several of the books on molecular symmetry mentioned in the Bibliography at the end of this chapter.

5.2.2.3 Point groups as examples of groups in general

Group theory was developed by mathematicians in the early nineteenth century and it was much later, in the 1920s and 1930s, that the theory was applied to problems associated with molecular symmetry.

Point groups are special examples of groups in general, the elements of which obey a simple set of rules which, with illustrative examples from point groups, are as follows:

(1) The product R of any two elements P and Q of a group must also be an element of the group. As Figure 5.24 shows, in the C_{2v} point group

$$\sigma_v \times C_2 = \sigma_v' \tag{5.61}$$

and σ_v' is indeed an element of the group.

†Although an atom with partially filled orbitals, other than s orbitals, is not spherically symmetrical, the electronic wave function ψ_e is classified according to the K_h point group.

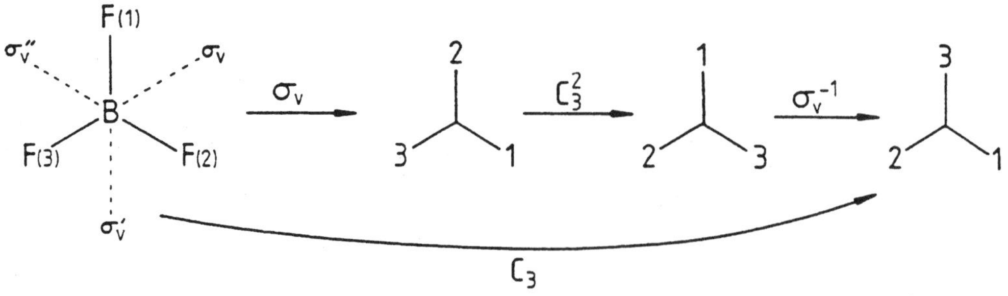

Figure 5.29 Illustration of the fact that, in the D_{3h} point group, C_3 and C_3^2 belong to the same class

(2) The product of any three elements P, Q and R is associative, i.e.

$$P \times (Q \times R) = (P \times Q) \times R \tag{5.62}$$

where the multiplication inside the parentheses is carried out first. It is clearly true that, in the C_{2v} point group,

$$C_2 \times (\sigma_v' \times \sigma_v) = (C_2 \times \sigma_v') \times \sigma_v \tag{5.63}$$

(3) A group must contain an element which, when it is multiplied by any other element, leaves it unchanged and which commutes with all other elements of the group. In general group theory this element is called the unit element but in point groups it is the identity element I. For example, in the C_{2v} point group, I commutes with C_2 and, when multiplying C_2, leaves it unchanged:

$$I \times C_2 = C_2 \times I = C_2 \tag{5.64}$$

(4) Every element of a group has an inverse which is also an element of the group. For example, in the C_{3v} point group, C_3 and C_3^{-1} $(= C_3^2)$ are both elements of the group. Note that I, σ, C_2 and i are equal to their own inverses.

(5) If all pairs of elements P and Q commute, i.e.

$$P \times Q = Q \times P \tag{5.65}$$

the group is said to be Abelian. If any of the pairs do not commute the group is non-Abelian. The example in Figure 5.26 shows that in D_{3h}, the point group to which BF_3 belongs, C_3 and σ_v do not commute and therefore the point group is non-Abelian. In fact, only the point groups C_n, S_n, C_{nh}, C_{2v}, D_2 and D_{2h} are Abelian; all others are non-Abelian.

(6) The order of a group is the total number of elements in the group. Therefore, the C_{2v} point group, with elements I, C_2, σ_v, σ_v', is of order four, and the C_{3v} point group with elements I, C_3, C_3^2, σ_v, σ_v', σ_v'', is of order six.

(7) Two elements P and Q are said to belong to the same class if there exists a third element R such that

$$P = R^{-1} \times Q \times R \tag{5.66}$$

Figure 5.29 shows that, in the D_{3h} point group,

$$C_3 = \sigma_v^{-1} \times C_3^2 \times \sigma_v \tag{5.67}$$

and therefore C_3 and C_3^2 belong to the same class.

If a point group contains either a C_n axis with $n > 2$ or an S_4 axis, the molecule may have degenerate properties, such as electronic or vibrational wave functions, and the point group is said to be a degenerate point group.[†] For example, HCN belongs to the degenerate point group $C_{\infty v}$ and one degenerate property is

[†]The reason why a molecule with an S_4 axis belongs to a degenerate point group is the same as that given in Section 4.1 for its being a symmetric rotor.

the bending vibration ν_2 shown in Figure 4.16. Point groups without an S_4 axis or a C_n axis with $n > 2$ are non-degenerate point groups. In a non-degenerate point group all the elements belong to separate classes and, in *any* point group, the elements I, i and σ_h are each in classes by themselves.

An example of a group which is not a point group is the set of elements 1, −1, i, −i, where $i = \sqrt{-1}$. This is an Abelian group of order four and the elements have properties which are analogous to those of the I, C_2 $(= C_4^2)$, C_4 C_4^3 elements of the C_4 point group. Table 5.6 gives the multiplication tables of these two groups consisting of the results of multiplying all possible pairs of elements. These tables demonstrate the one-to-one correlation of elements in the two groups; because of this correlation, the groups are said to be isomorphous.

5.2.2.4 *Point group character tables*

In Sections 5.2.2.1–5.2.2.2 we have seen how the symmetry properties of the equilibrium nuclear configuration of any molecule may be classified and the molecule assigned to a point group. However, a molecule may have properties, such as electronic and vibrational wave functions, in which not all of the elements of symmetry are preserved. For example, the normal vibration ν_3 of H_2O, shown in Figure 5.12, clearly does not preserve all the symmetry elements which the molecule has in its equilibrium configuration.

In considering the symmetry classification of such molecular properties we shall consider only three point groups in detail: (i) C_{2v}, a simple example of a non-degenerate point group, (ii) C_{3v}, a simple example of a degenerate point group, and (iii) $C_{\infty v}$, an example of an infinite group, which is a group having an infinite number of elements.

5.2.2.4(i) C_{2v} character table

A property such as a vibrational wave function of, say, H_2O may or may not preserve an element of symmetry. If it preserves the element, carrying out the corresponding symmetry operation, for example σ_v, has no effect on the wave function, which we write as

$$\psi_v \xrightarrow{\sigma_v} (+1)\psi_v \tag{5.68}$$

and we say that ψ_v is symmetric to σ_v. The only other possibility, in a non-degenerate point group, is that the wave function may be changed in sign by carrying out the operation

$$\psi_v \xrightarrow{\sigma_v} (-1)\psi_v \tag{5.69}$$

in which case we say that ψ_v is antisymmetric to σ_v. The +1 in equation (5.68) or the −1 in equation (5.69) is known as the character of, in this case, ψ_v with respect to σ_v.

We have seen that any two of the C_2, $\sigma_v(xz)$, $\sigma_v'(yz)$ elements may be regarded as generating elements. There are four possible combinations of +1 or −1 characters with respect to these generating elements: they are +1 and +1, +1 and −1, −1 and +1, −1 and −1 with respect to C_2 and $\sigma_v(xz)$. These combinations are entered in columns 3 and 4 of the C_{2v} character table in Table 5.7. The character with respect to I must

Table 5.6 Multiplication tables for the C_4 and (1, −1, i, −i) groups

	C_4					(1, −1, i, −i)			
	I	C_2	C_4	C_4^3		1	−1	i	−i
I	I	C_2	C_4	C_4^3	1	1	−1	i	−i
C_2	C_2	I	C_4^3	C_4	−1	−1	1	−i	i
C_4	C_4	C_4^3	C_2	I	i	i	−i	−1	1
C_4^3	C_4^3	C_4	I	C_2	−i	−i	i	1	−1

Table 5.7 C_{2v} character table

C_{2v}	I	C_2	$\sigma_v(xz)$	$\sigma'_v(yz)$		
A_1	1	1	1	1	T_z	$\alpha_{xx}, \alpha_{yy}, \alpha_{zz}$
A_2	1	1	−1	−1	R_z	α_{xy}
B_1	1	−1	1	−1	T_x, R_y	α_{xz}
B_2	1	−1	−1	1	T_y, R_x	α_{yz}

always be +1 and, just as $\sigma'_v(yz)$ is generated from C_2 and $\sigma_v(xz)$, the character with respect to $\sigma'_v(yz)$ is the product of characters with respect to C_2 and $\sigma_v(xz)$. Each of the four rows of characters is called an irreducible representation of the group and, for convenience, each is represented by a symmetry species A_1, A_2, B_1 or B_2. The A_1 species is said to be totally symmetric since all the characters are +1; the other three species are non-totally symmetric.

The symmetry species labels are conventional: A and B indicate symmetry or antisymmetry, respectively, to C_2 and the subscripts 1 and 2 indicate symmetry or antisymmetry, respectively, to $\sigma_v(xz)$.

In the sixth column of the character table is indicated the symmetry species of translations (T) of the molecule along, and rotations (R) about, the cartesian axes. In Figure 5.30 vectors attached to the nuclei of H_2O represent these motions which are assigned to symmetry species by their behaviour under the operations C_2 and $\sigma_v(xz)$. Figure 5.30(a) shows that

$$\Gamma(T_x) = B_1; \quad \Gamma(T_y) = B_2; \quad \Gamma(T_z) = A_1 \tag{5.70}$$

and Figure 5.30(b) shows that

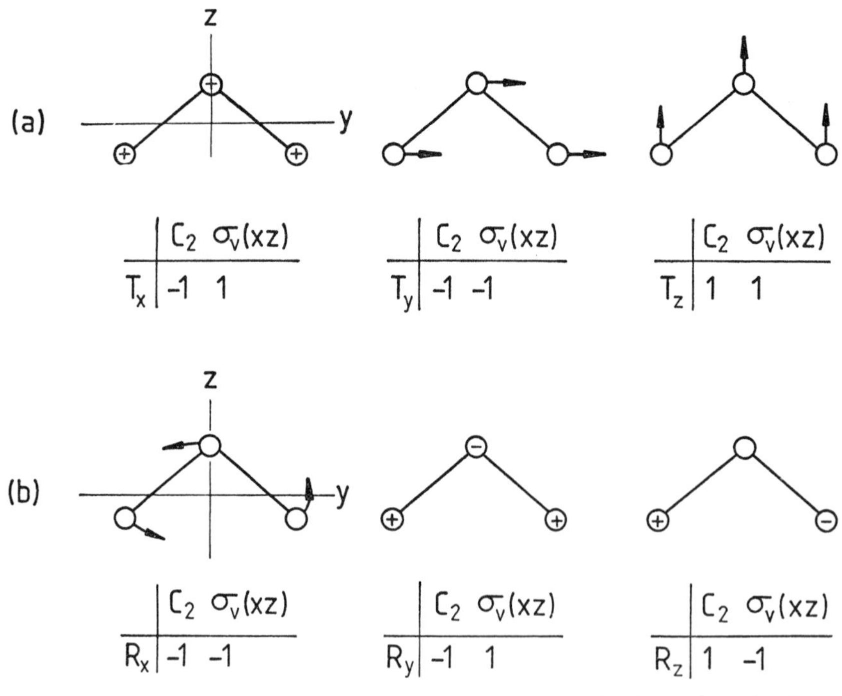

Figure 5.30 Symmetry species of (a) translations and (b) rotations in the H_2O molecule

Table 5.8 C_{3v} character table

C_{3v}	I	$2C_3$	$3\sigma_v$		
A_1	1	1	1	T_z	$\alpha_{xx}+\alpha_{yy}, \alpha_{zz}$
A_2	1	1	−1	R_z	
E	2	−1	0	$(T_x, T_y), (R_x, R_y)$	$(\alpha_{xx}-\alpha_{yy}, \alpha_{xy}), (\alpha_{xz}, \alpha_{yz})$

$$\Gamma(R_x) = B_2;\quad \Gamma(R_y) = B_1;\quad \Gamma(R_z) = A_2 \tag{5.71}$$

The symbol Γ, in general, stands for 'representation of . . .'. Here it is an irreducible representation, or symmetry species. In Section 5.2.4.1, where we derive the infrared vibrational selection rules for polyatomic molecules, we shall require the symmetry species of the translations.

In the final column of the character table are given the assignments to symmetry species of α_{xx}, α_{yy}, α_{zz}, α_{xy}, α_{yz} and α_{xz}, which are the components of the symmetric polarizability tensor $\boldsymbol{\alpha}$ [equation (4.117)]. The symmetry species assignments of the components of $\boldsymbol{\alpha}$ are important in deriving vibrational Raman selection rules, discussed in Section 5.2.4.2.

We can use the C_{2v} character table to classify the vibrational wave functions ψ_v for each of the three normal vibrations of H_2O shown in Figure 5.12. The characters of the three vibrations under the operations C_2 and $\sigma_v(xz)$ are, respectively, +1 and +1 for ν_1, +1 and +1 for ν_2 and −1 and −1 for ν_3. Therefore,

$$\Gamma[\psi_{v(1)}] = A_1;\quad \Gamma[\psi_{v(2)}] = A_1;\quad \Gamma[\psi_{v(3)}] = B_2 \tag{5.72}$$

It is important to realize that the assignment of symmetry species to molecular properties depends on the labelling of axes. The axis labels used for the Ts, Rs and αs in the last two columns of a character table are always the same, but the choice of axis labels for the particular molecule under consideration is not easily standardized. Mulliken[48] has suggested conventions for axis labelling in many cases and these conventions are widely used. For a planar C_{2v} molecule the convention is that the C_2 axis is taken to be the z-axis and the x-axis is perpendicular to the plane. This is the convention used here for H_2O. The importance of using a convention can be illustrated by the fact that if, say, we were to exchange the x- and y-axis labels then $\Gamma(\psi_{v(3)}]$ would be B_1 rather than B_2. However, for a non-planar C_{2v} molecule, such as CH_2F_2 (Figure 5.24), although it is natural to take the C_2 axis to be the z-axis, the choice of x- and y-axes is arbitrary. This example serves to demonstrate how important it is to indicate the axis notation being used.

There will be many occasions when we shall need to multiply symmetry species or, in the language of group theory, to obtain their direct product. For example, if H_2O is vibrationally excited simultaneously with one quantum each of ν_1 and ν_3, the symmetry species of the wave function for this vibrational combination state is

$$\Gamma(\psi_v) = A_1 \times B_2 = B_2 \tag{5.73}$$

In order to obtain the direct product of two species we multiply the characters under each symmetry element using the rules

$$(+1)\times(+1) = 1;\quad (+1)\times(-1) = -1;\quad (-1)\times(-1) = 1 \tag{5.74}$$

In this way, the result in equation (5.73) is obtained. If H_2O is excited with, say, two quanta of ν_3 then

$$\Gamma(\psi_v) = B_2 \times B_2 = A_1 \tag{5.75}$$

The results of multiplication in equations (5.73) and (5.75) that (a) the direct product of any species with the totally symmetric species leaves it unchanged and (b) the direct product of any species with itself gives the totally symmetric

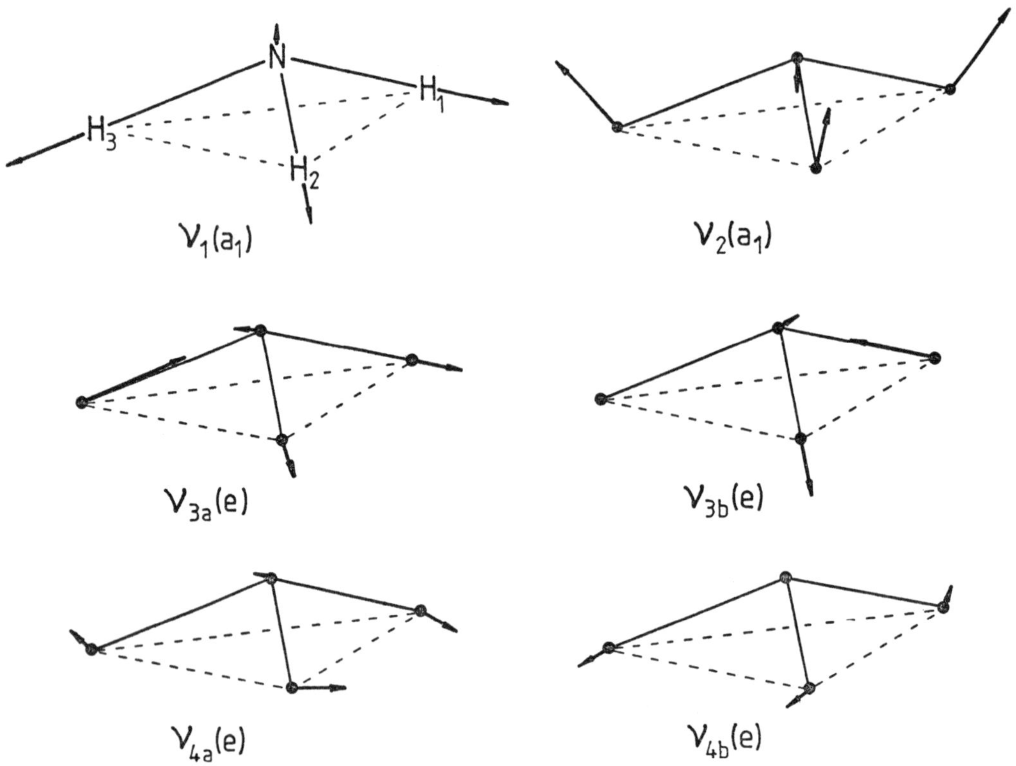

Figure 5.31 Normal vibrations of the NH_3 molecule

species are general for all non-degenerate point groups.

Using the rules for forming direct products it can be shown also that, in the C_{2v} point group,

$$A_2 \times B_1 = B_2; \quad A_2 \times B_2 = B_1 \qquad (5.76)$$

The character tables for all important point groups, degenerate and non-degenerate, are given in the Appendix.

5.2.2.4(ii) C_{3v} character table

Inspection of this character table, given in Table 5.8, shows two obvious differences from a character table for any non-degenerate point group. The first is the grouping together of all elements of the same class, namely C_3 and C_3^2 as $2C_3$, and σ_v, σ_v' and σ_v'' as $3\sigma_v$. Elements belonging to the same class have the same characters and the number of symmetry species is equal to the number of classes, as is the case in non-degenerate point groups.

The second difference is the appearance of a doubly degenerate E symmetry species whose characters are not always either the $+1$ or -1 that we have encountered in non-degenerate point groups.

The $+1$ and -1 characters of the A_1 and A_2 species have the same significance as in a non-degenerate point group. The characters of the E species may be understood by using the normal vibrations of NH_3, shown in Figure 5.31, as examples.† Vibrations ν_1 and ν_2 are clearly a_1.‡ The two vibrations ν_{3a} and ν_{3b} are

†See footnote on p. 213 for vibrational numbering convention.

‡Lower case letters are recommended for the symmetry species of a vibration (and for an electronic orbital) whereas upper case letters are recommended for the symmetry species of wave functions.

degenerate; although it is not obvious, it requires the same energy to excite one quantum of either of them and, clearly, they have different wave functions. Similarly, ν_{4a} and ν_{4b} are degenerate.

The symmetry properties of a fundamental vibrational wave function ψ_v are the same as those of the corresponding normal coordinate Q. For example, when the C_3 operation is carried out on Q_1, the normal coordinate for ν_1, it is transformed into Q_1':

$$Q_1 \xrightarrow{C_3} Q_1' = (+1)Q_1 \tag{5.77}$$

On the other hand, when a symmetry operation is carried out on a degenerate normal coordinate, it does not simply remain the same or change sign; in general, it is transformed into a linear combination of the two degenerate normal coordinates. Thus, on carrying out a symmetry operation S:

$$\begin{aligned} Q_{3a} &\xrightarrow{S} Q_{3a}' = d_{aa}Q_{3a} + d_{ab}Q_{3b} \\ Q_{3b} &\xrightarrow{S} Q_{3b}' = d_{ba}Q_{3a} + d_{bb}Q_{3b} \end{aligned} \tag{5.78}$$

In matrix notation this can be written as

$$\begin{pmatrix} Q_{3a}' \\ Q_{3b}' \end{pmatrix} = \begin{pmatrix} d_{aa} & d_{ab} \\ d_{ba} & d_{bb} \end{pmatrix} \begin{pmatrix} Q_{3a} \\ Q_{3b} \end{pmatrix} \tag{5.79}$$

The quantity $d_{aa} + d_{bb}$ is the trace of the matrix and this is the character of the property, here a normal coordinate, under the symmetry operation S.

The character of the E species under the I operation is obtained from

$$\begin{aligned} Q_{3a} &\xrightarrow{I} Q_{3a}' = 1 \times Q_{3a} + 0 \times Q_{3b} \\ Q_{3b} &\xrightarrow{I} Q_{3b}' = 0 \times Q_{3a} + 1 \times Q_{3b} \end{aligned} \tag{5.80}$$

or

$$\begin{pmatrix} Q_{3a}' \\ Q_{3b}' \end{pmatrix} = \begin{pmatrix} 1 & 0 \\ 0 & 1 \end{pmatrix} \begin{pmatrix} Q_{3a} \\ Q_{3b} \end{pmatrix} \tag{5.81}$$

The trace of the matrix is 2 and this is the character of E under I.

Of the two components of ν_3 one, ν_{3a}, is symmetric to reflection across the σ_v plane bisecting the angle between H_1 and H_2, and the other is antisymmetric giving

$$\begin{pmatrix} Q_{3a}' \\ Q_{3b}' \end{pmatrix} = \begin{pmatrix} 1 & 0 \\ 0 & -1 \end{pmatrix} \begin{pmatrix} Q_{3a} \\ Q_{3b} \end{pmatrix} \tag{5.82}$$

Therefore, the character of E under σ_v is 0. Because of the equivalence of all the σ_vs the character must be the same under σ_v' and σ_v''.

When the operation is a rotation by an angle ϕ about a C_n axis, in this case by an angle of $2\pi/3$ rad about the C_3 axis, the resulting coordinate transformation, a result which will not be derived here, is given by

$$\begin{aligned} \begin{pmatrix} Q_{3a}' \\ Q_{3b}' \end{pmatrix} &= \begin{pmatrix} \cos\phi & \sin\phi \\ -\sin\phi & \cos\phi \end{pmatrix} \begin{pmatrix} Q_{3a} \\ Q_{3b} \end{pmatrix} \\ &= \begin{pmatrix} -\frac{1}{2} & \sqrt{\frac{3}{2}} \\ -\sqrt{\frac{3}{2}} & -\frac{1}{2} \end{pmatrix} \begin{pmatrix} Q_{3a} \\ Q_{3b} \end{pmatrix} \end{aligned} \tag{5.83}$$

The trace of the matrix is -1 and this is the character of E under C_3.

Except for the multiplication of E by E we follow the rules for forming direct products used in non-degenerate point groups; the characters under the various symmetry operations are obtained by multiplying the characters of the species being multiplied, giving

$$\begin{aligned} A_1 \times A_2 = A_2; \quad A_2 \times A_2 = A_1; \\ A_1 \times E = E; \quad A_2 \times E = E \end{aligned} \tag{5.84}$$

In multiplying E by E we use, again, examples of the vibrations of NH_3. The result depends on whether we require $\Gamma(\psi_v)$ when (a) one quantum of each of two *different* e vibrations is excited, i.e. a combination level, or (b) two quanta of the *same* e vibration are excited, i.e. an overtone level. In case (a), such as for the combination $\nu_3 + \nu_4$, the product is written $E \times E$ and the result is obtained by first squaring the characters under each operation, giving

	I	$2C_3$	$3\sigma_v$
$E \times E$	4	1	0

(5.85)

The characters 4, 1, 0 form a reducible representation in the C_{3v} point group and we require to reduce it to a set of irreducible representations, the sum of whose characters under each operation is equal to that of the reducible representation. We can express this algebraically as

$$\chi_C(k) \times \chi_D(k) = \chi_F(k) + \chi_G(k) + \ldots \qquad (5.86)$$

where χ is a character of any operation k, and the result of multiplying the degenerate species C and D is

$$C \times D = F + G + \ldots \qquad (5.87)$$

The reduction of the $E \times E$ representation, like all such reductions, gives a unique set of irreducible representations, which is

$$E \times E = A_1 + A_2 + E \qquad (5.88)$$

We can see from Table 5.8 that the sum of the characters of A_1, A_2 and E under I, C_3 and σ_v gives the reducible representation of equation (5.85).

In case (b), where two quanta of the same e vibration are excited, such as $2\nu_3$, the product is written $(E)^2$ where

$$(E)^2 = A_1 + E \qquad (5.89)$$

which is the symmetric part of $E \times E$, being the part which is symmetric to particle exchange. We arrive at the result of equation (5.89) by first obtaining the product $E \times E$. Then application of the Pauli principle (Section 4.4) forbids one of the species of the $E \times E$ product. In degenerate point groups in general this is an A species and, where possible, one which is non-totally symmetric: in this case it is the A_2 species which is forbidden and this is the antisymmetric part of $E \times E$.

Table 5.9 gives the symmetry species of vibrational combination states resulting from the excitation of one quantum of each of two *different* degenerate vibrations in various point groups. Table 5.10 gives the species of vibrational overtone states resulting from the excitation of two quanta of the *same* degenerate vibration.

5.2.2.4(iii) $C_{\infty v}$ *character table*

In this character table, given in Table 5.11, there are an infinite number of classes since rotation about the C_∞ axis may be by *any* angle ϕ and each C_∞^ϕ element belongs to a different class. However, $C_\infty^{-\phi}$, an anti-clockwise rotation by ϕ, belongs to the same class as C_∞^ϕ. Since there is an infinite number of classes there is an infinite number of symmetry species. Labels for these are A_1, A_2, E_1, E_2, . . ., E_∞ if we follow conventions used in other character tables. Unfortunately, before symmetry species labels had been adopted generally, another convention had grown up and was being used, particularly in electronic spectroscopy of diatomic molecules. Electronic states were given labels Σ, Π, Δ, Φ, . . . corresponding to the value of an electronic orbital angular momentum quantum number Λ (see Section 6.2.3.1) which can take the values 0, 1, 2, 3, ... It is mainly this system of labelling which is used in the $C_{\infty v}$, as well as the $D_{\infty h}$, point group although both systems are given in Table 5.11.

Multiplication of symmetry species is carried out using the usual rules so that, for example,

$$\Sigma^+ \times \Sigma^- = \Sigma^-; \quad \Sigma^- \times \Pi = \Pi; \quad \Sigma^+ \times \Delta = \Delta \qquad (5.90)$$

For the product $\Pi \times \Pi$, the reducible representation is

	I	$2C_\infty^\phi$	$\infty\sigma_v$
$\Pi \times \Pi$	4	$4\cos^2\phi$ $(= 2 + 2\cos 2\phi)$	0

(5.91)

which reduces to give

$$\Pi \times \Pi = \Sigma^+ + \Sigma^- + \Delta \qquad (5.92)$$

Table 5.9 Symmetry species of vibrational states resulting from one quantum of each of two different degenerate vibrations

Point group	Symmetry species of two quanta vibrational states
C_3	$e \times e = 2A + E$
C_4	$e \times e = 2A + 2B$
C_5	$e_1 \times e_1 = 2A + E_2$, $e_2 \times e_2 = 2A + E_1$, $e_1 \times e_2 = E_1 + E_2$
C_6	$e_1 \times e_1 = 2A + E_2$, $e_2 \times e_2 = 2A + E_2$, $e_1 \times e_2 = 2B + E_1$
C_7	$e_1 \times e_1 = 2A + E_2$, $e_2 \times e_2 = 2A + E_3$, $e_3 \times e_3 = 2A + E_1$, $e_1 \times e_2 = E_1 + E_3$, $e_1 \times e_3 = E_2 + E_3$, $e_2 \times e_3 = E_1 + E_2$
C_8	$e_1 \times e_1 = 2A + E_2$, $e_2 \times e_2 = 2A + 2B$, $e_3 \times e_3 = 2A + E_2$, $e_1 \times e_2 = E_1 + E_3$, $e_1 \times e_3 = 2B + E_2$, $e_2 \times e_3 = E_1 + E_3$
C_{3v}	$e \times e = A_1 + A_2 + E$
C_{4v}	$e \times e = A_1 + A_2 + B_1 + B_2$
C_{5v}	$e_1 \times e_1 = A_1 + A_2 + E_2$, $e_2 \times e_2 = A_1 + A_2 + E_1$, $e_1 \times e_2 = E_1 + E_2$
C_{6v}	$e_1 \times e_1 = A_1 + A_2 + E_2$, $e_2 \times e_2 = A_1 + A_2 + E_2$, $e_1 \times e_2 = B_1 + B_2 + E_1$
$C_{\infty v}$	$\pi \times \pi = \Sigma^+ + \Sigma^- + \Delta$, $\delta \times \delta = \Sigma^+ + \Sigma^- + \Gamma$, $\pi \times \delta = \Pi + \Phi$
D_3	$e \times e = A_1 + A_2 + E$
D_4	$e \times e = A_1 + A_2 + B_1 + B_2$
D_5	$e_1 \times e_1 = A_1 + A_2 + E_2$, $e_2 \times e_2 = A_1 + A_2 + E_1$, $e_1 \times e_2 = E_1 + E_2$
D_6	$e_1 \times e_1 = A_1 + A_2 + E_2$, $e_2 \times e_2 = A_1 + A_2 + E_2$, $e_1 \times e_2 = B_1 + B_2 + E_1$
C_{3h}	$e' \times e' = 2A' + E'$, $e'' \times e'' = 2A' + E'$, $e' \times e'' = 2A'' + E''$
C_{4h}	$e_g \times e_g = 2A_g + 2B_g$, $e_u \times e_u = 2A_g + 2B_g$, $e_g \times e_u = 2A_u + 2B_u$
C_{5h}	$e_1' \times e_1' = 2A' + E_2'$, $e_2' \times e_2' = 2A' + E_1'$, $e_1'' \times e_1'' = 2A' + E_2'$, $e_2'' \times e_2'' = 2A' + E_1'$, $e_1' \times e_1'' = 2A'' + E_2''$, $e_1' \times e_2'' = E_1'' + E_2''$, $e_1' \times e_2' = E_1' + E_2'$, $e_2' \times e_2'' = 2A'' + E_1''$, $e_1'' \times e_2' = E_1'' + E_2''$, $e_1'' \times e_2'' = E_1' + E_2'$
C_{6h}	$e_{1g} \times e_{1g} = 2A_g + E_{2g}$, $e_{2g} \times e_{2g} = 2A_g + E_{2g}$, $e_{1u} \times e_{1u} = 2A_g + E_{2g}$, $e_{2u} \times e_{2u} = 2A_g + 2E_{2g}$, $e_{1g} \times e_{1u} = 2A_u + E_{2u}$, $e_{1g} \times e_{2g} = 2B_g + E_{1g}$, $e_{1g} \times 2_{2u} = 2B_u + E_{1u}$, $e_{1u} \times e_{2g} = 2B_u + E_{1u}$, $e_{1u} \times e_{2u} = 2B_g + 2E_{1g}$, $e_{2g} \times e_{2u} = 2A_u + E_{2u}$
D_{2d}	$e \times e = A_1 + A_2 + B_1 + B_2$
D_{3d}	$e_g \times e_g = A_{1g} + A_{2g} + E_g$, $e_u \times e_u = A_{1g} + A_{2g} + E_g$, $e_g \times e_u = A_{1u} + A_{2u} + E_u$
D_{4d}	$e_1 \times e_1 = A_1 + A_2 + E_2$, $e_2 \times e_2 = A_1 + A_2 + B_1 + B_2$, $e_3 \times e_3 = A_1 + A_2 + E_2$, $e_1 \times e_2 = E_1 + E_3$, $e_1 \times e_3 = B_1 + B_2 + E_2$, $e_2 \times e_3 = E_1 + E_3$
D_{5d}	$e_{1g} \times e_{1g} = A_{1g} + A_{2g} + E_{2g}$, $e_{2g} \times e_{2g} = A_{1g} + A_{2g} + E_{1g}$, $e_{1u} \times e_{1u} = A_{1g} + A_{2g} + E_{2g}$, $e_{2u} \times e_{2u} = A_{1g} + A_{2g} + E_{1g}$, $e_{1g} \times e_{2g} = E_{1g} + E_{2g}$, $e_{1g} \times e_{1u} = A_{1u} + A_{2u} + E_{2u}$, $e_{1u} \times e_{2u} = E_{1g} + E_{2g}$, $e_{1g} \times e_{2u} = E_{1u} + E_{2u}$, $e_{2g} \times e_{1u} = E_{1u} + E_{2u}$, $e_{2g} \times e_{2u} = A_{1u} + A_{2u} + E_{1u}$
D_{6d}	$e_1 \times e_1 = A_1 + A_2 + E_2$, $e_2 \times e_2 = A_1 + A_2 + E_4$, $e_3 \times e_3 = A_1 + A_2 + B_1 + B_2$, $e_4 \times e_4 = A_1 + A_2 + E_4$, $e_5 \times e_5 = A_1 + A_2 + E_2$, $e_1 \times e_2 = E_1 + E_3$, $e_1 \times e_3 = E_2 + E_4$, $e_1 \times e_4 = E_3 + E_5$, $e_1 \times e_5 = B_1 + B_2 + E_4$, $e_2 \times e_3 = E_1 + E_5$, $e_2 \times e_4 = B_1 + B_2 + E_2$, $e_2 \times e_5 = E_3 + E_5$, $e_3 \times e_4 = E_1 + E_5$, $e_3 \times e_5 = E_2 + E_4$, $e_4 \times e_5 = E_1 + E_3$
D_{3h}	$e' \times e' = A_1' + A_2' + E'$, $e'' \times e'' = A_1' + A_2' + E'$, $e' \times e'' = A_1'' + A_2'' + E''$
D_{4h}	$e_g \times e_g = A_{1g} + A_{2g} + B_{1g} + B_{2g}$, $e_u \times e_u = A_{1g} + A_{2g} + B_{1g} + B_{2g}$, $e_g \times e_u = A_{1u} + A_{2u} + B_{1u} + B_{2u}$
D_{5h}	$e_1' \times e_1' = A_1' + A_2' + E_2'$, $e_1'' \times e_1'' = A_1' + A_2' + E_2'$, $e_2' \times e_2' = A_1' + A_2' + E_1'$, $e_2'' \times e_2'' = A_1' + A_2' + E_1'$, $e_1' \times e_1'' = A_1'' + A_2'' + E_2''$, $e_1' \times e_2' = E_1' + E_2'$, $e_1' \times e_2'' = E_1'' + E_2''$, $e_1'' \times e_2' = E_1'' + E_2''$, $e_1'' \times e_2'' = E_1' + E_2'$, $e_2' \times e_2'' = A_1'' + A_2'' + E_1''$

(continued)

Table 5.9 (*continued*)

Point group	Symmetry species of two quanta vibrational states
D_{6h}	$e_{1g}\times e_{1g}=A_{1g}+A_{2g}+E_{2g}$, $e_{2g}\times e_{2g}=A_{1g}+A_{2g}+E_{2g}$, $e_{1u}\times e_{1u}=A_{1g}+A_{2g}+E_{2g}$, $e_{2u}\times e_{2u}=A_{1g}+A_{2g}+E_{2g}$, $e_{1g}\times e_{1u}=A_{1u}+A_{2u}+E_{2u}$, $e_{2g}\times e_{2u}=A_{1u}+A_{2u}+E_{2u}$, $e_{1g}\times e_{2g}=B_{1g}+B_{2g}+E_{1g}$, $e_{1g}\times e_{2u}=B_{1u}+B_{2u}+E_{1u}$, $e_{1u}\times e_{2g}=B_{1u}+B_{2u}+E_{1u}$, $e_{1u}\times e_{2u}=B_{1g}+B_{2g}+E_{1g}$
$D_{\infty h}$	$\pi_g\times\pi_g=\Sigma_g^++\Sigma_g^-+\Delta_g$, $\delta_g\times\delta_g=\Sigma_g^++\Sigma_g^-+\Gamma_g$, $\pi_u\times\pi_u=\Sigma_g^++\Sigma_g^-+\Delta_g$, $\delta_u\times\delta_u=\Sigma_g^++\Sigma_g^-+\Gamma_g$, $\pi_g\times\pi_u=\Sigma_u^++\Sigma_u^-+\Delta_u$, $\delta_g\times\delta_u=\Sigma_u^++\Sigma_u^-+\Gamma_u$, $\pi_g\times\delta_g=\Pi_g+\Phi_g$, $\pi_u\times\delta_u=\Pi_g+\Phi_g$, $\pi_g\times\delta_u=\Pi_u+\Phi_u$, $\pi_u\times\delta_g=\Pi_u+\Phi_u$
S_4	$e\times e=2A+2B$
S_6	$e_g\times e_g=2A_g+E_g$, $e_u\times e_u=2A_g+E_g$, $e_g\times e_u=2A_u+E_u$
S_8	$e_1\times e_1=2A+E_2$, $e_2\times e_2=2A+2B$, $e_3\times e_3=2A+E_2$, $e_1\times e_2=E_1+E_3$, $e_1\times e_3=2B+E_2$, $e_2\times e_3=E_1+E_3$
T_d	$e\times e=A_1+A_2+E$, $t_1\times t_1=A_1+E+T_1+T_2$, $t_2\times t_2=A_1+E+T_1+T_2$, $e\times t_1=T_1+T_2$, $e\times t_2=T_1+T_2$, $t_1\times t_2=A_2+E+T_1+T_2$
T	$e\times e=2A+E$, $t\times t=A+E+2T$, $e\times t=2T$
O_h	$e_g\times e_g=A_{1g}+A_{2g}+E_g$, $e_u\times e_u=A_{1g}+A_{2g}+E_g$, $e_g\times e_u=A_{1u}+A_{2u}+E_u$, $t_{1g}\times t_{1g}=A_{1g}+E_g+T_{1g}+T_{2g}$, $t_{1u}\times t_{1u}=A_{1g}+E_g+T_{1g}+T_{2g}$, $t_{1g}\times t_{1u}=A_{1u}+E_u+T_{1u}+T_{2u}$, $t_{2g}\times t_{2g}=A_{1g}+E_g+T_{1g}+T_{2g}$, $t_{2u}\times t_{2u}=A_{1g}+E_g+T_{1g}+T_{2g}$, $t_{2g}\times t_{2u}=A_{1u}+E_u+T_{1u}+T_{2u}$, $e_g\times t_{1g}=T_{1g}+T_{2g}$, $e_u\times t_{1u}=T_{1g}+T_{2g}$, $e_g\times t_{1u}=T_{1u}+T_{2u}$, $e_u\times t_{1g}=T_{1u}+T_{2u}$, $e_g\times t_{2g}=T_{1g}+T_{2g}$, $e_u\times t_{2u}=T_{1g}+T_{2g}$, $e_g\times t_{2u}=T_{1u}+T_{2u}$, $e_u\times t_{2g}=T_{1u}+T_{2u}$, $t_{1g}\times t_{2g}=A_{2g}+E_g+T_{1g}+T_{2g}$, $t_{1u}\times t_{2u}=A_{2g}+E_g+T_{1g}+T_{2g}$, $t_{1u}\times t_{2g}=A_{2u}+E_u+T_{1u}+T_{2u}$, $t_{1g}\times t_{2u}=A_{2u}+E_u+T_{1u}+T_{2u}$
O	$e\times e=A_1+A_2+E$, $t_1\times t_1=A_1+E+T_1+T_2$, $t_2\times t_2=A_1+E+T_1+T_2$, $e\times t_1=T_1+T_2$, $e\times t_2=T_1+T_2$, $t_1\times t_2=A_2+E+T_1+T_2$

Table 5.10 Symmetry species of vibrational states resulting from two quanta of one degenerate vibration

Point group	Symmetry species of two quanta vibrational states
C_3	$(e)^2=A+E$
C_4	$(e)^2=A+2B$
C_5	$(e_1)^2=A+E_2$, $(e_2)^2=A+E_1$
C_6	$(e_1)^2=A+E_2$, $(e_2)^2=A+E_2$
C_7	$(e_1)^2=A+E_2$, $(e_2)^2=A+E_3$, $(e_3)^2=A+E_1$
C_8	$(e_1)^2=A+E_2$, $(e_2)^2=A+2B$, $(e_3)^2=A+E_2$
C_{3v}	$(e)^2=A_1+E$
C_{4v}	$(e)^2=A_1+B_1+B_2$
C_{5v}	$(e_1)^2=A_1+E_2$, $(e_2)^2=A_1+E_1$
C_{6v}	$(e_1)^2=A_1+E_2$, $(e_2)^2=A_1+E_2$
$C_{\infty v}$	$(\pi)^2=\Sigma^++\Delta$, $(\delta)^2=\Sigma^++\Gamma$
D_3	$(e)^2=A_1+E$
D_4	$(e)^2=A_1+B_1+B_2$
D_5	$(e_1)^2=A_1+E_2$, $(e_2)^2=A_1+E_1$

Table 5.10 (*continued*)

Point group	Symmetry species of two quanta vibrational states
D_6	$(e_1)^2 = A_1 + E_2$, $(e_2)^2 = A_1 + E_2$
C_{3h}	$(e')^2 = A' + E'$, $(e'')^2 = A' + E'$
C_{4h}	$(e_g)^2 = A_g + 2B_g$, $(e_u)^2 = A_g + 2B_g$
C_{5h}	$(e'_1)^2 = A' + E'_2$, $(e'_2)^2 = A' + E'_1$, $(e''_1)^2 = A' + E'_2$, $(e''_2)^2 = A' + E'_1$
C_{6h}	$(e_{1g})^2 = A_g + E_{2g}$, $(e_{2g})^2 = A_g + E_{2g}$, $(e_{1u})^2 = A_g + E_{2g}$, $(e_{2u})^2 = A_g + E_{2g}$
D_{2d}	$(e)^2 = A_1 + B_1 + B_2$
D_{3d}	$(e_g)^2 = A_{1g} + E_g$, $(e_u)^2 = A_{1g} + E_g$
D_{4d}	$(e_1)^2 = A_1 + E_2$, $(e_2)^2 = A_1 + B_1 + B_2$, $(e_3)^2 = A_1 + E_2$
D_{5d}	$(e_{1g})^2 = A_{1g} + E_{2g}$, $(e_{2g})^2 = A_{1g} + E_{1g}$, $(e_{1u})^2 = A_{1g} + E_{2g}$, $(e_{2u})^2 = A_{1g} + E_{1g}$
D_{6d}	$(e_1)^2 = A_1 + E_2$, $(e_2)^2 = A_1 + E_4$, $(e_3)^2 = A_1 + B_1 + B_2$, $(e_4)^2 = A_1 + E_4$, $(e_5)^2 = A_1 + E_2$
D_{3h}	$(e')^2 = A'_1 + E'$, $(e'')^2 = A'_1 + E'$
D_{4h}	$(e_g)^2 = A_{1g} + B_{1g} + B_{2g}$, $(e_u)^2 = A_{1g} + B_{1g} + B_{2g}$
D_{5h}	$(e'_1)^2 = A'_1 + E'_2$, $(e'_2)^2 = A'_1 + E'_1$, $(e''_1)^2 = A'_1 + E'_2$, $(e''_2)^2 = A'_1 + E'_1$
D_{6h}	$(e_{1g})^2 = A_{1g} + E_{2g}$, $(e_{2g})^2 = A_{1g} + E_{2g}$, $(e_{1u})^2 = A_{1g} + E_{2g}$, $(e_{2u})^2 = A_{1g} + E_{2g}$
$D_{\infty h}$	$(\pi_g)^2 = \Sigma_g^+ + \Delta_g$, $(\pi_u)^2 = \Sigma_g^+ + \Delta_g$, $(\delta_g)^2 = \Sigma_g^+ + \Gamma_g$, $(\delta_u)^2 = \Sigma_g^+ + \Gamma_g$
S_4	$(e)^2 = A + 2B$
S_6	$(e_g)^2 = A_g + E_g$, $(e_u)^2 = A_g + E_g$
S_8	$(e_1)^2 = A + E_2$, $(e_2)^2 = A + 2B$, $(e_3)^2 = A + E_2$
T_d	$(e)^2 = A_1 + E$, $(t_1)^2 = A_1 + E + T_2$, $(t_2)^2 = A_1 + E + T_2$
T	$(e)^2 = A + E$, $(t)^2 = A + E + T$
O_h	$(e_g)^2 = A_{1g} + E_g$, $(e_u)^2 = A_{1g} + E_g$, $(t_{1g})^2 = A_{1g} + E_g + T_{2g}$, $(t_{2g})^2 = A_{1g} + E_g + T_{2g}$, $(t_{1u})^2 = A_{1g} + E_g + T_{2g}$, $(t_{2u})^2 = A_{1g} + E_g + T_{2g}$
O	$(e)^2 = A_1 + E$, $(t_1)^2 = A_1 + E + T_2$, $(t_2)^2 = A_1 + E + T_2$

Use of the same rule as that for deriving $(E)^2$ from $E \times E$ gives

$$(\Pi)^2 = \Sigma^+ + \Delta \qquad (5.93)$$

5.2.3 Determination of normal modes of vibration

5.2.3.1 Introduction

The quantitative determination of normal modes of vibration, such as those for H_2O and NH_3 shown in Figures 5.12 and 5.31, is a problem of classical mechanics provided that we are concerned only with the classical vibration wavenumber [as in equation (5.45)] of each of the normal modes; it follows that each normal mode is treated in the harmonic oscillator approximation. The problem becomes a quantum mechanical one only if we require vibrational wave functions or need to allow for the effects of anharmonicity, but we shall not be concerned here with quantum mechanical methods.

In general there are two useful approaches to normal mode calculations. Either (i) we start from known fundamental vibration wavenumbers and wish to obtain the force constants and the form of the normal modes, or (ii) we assume values of force constants which, as we have seen in Section 5.2.1, can be transferred with reasonable accuracy to a structurally similar molecule, and need to obtain estimates of the vibration wavenumbers and the form of the normal modes. In both these approaches we shall require to solve Lagrange's equation of classical mechanics:

$$(\mathrm{d}/\mathrm{d}t)(\partial T/\partial \dot{q}_i) + \partial V/\partial q_i = 0 \qquad (5.94)$$

Table 5.11 $C_{\infty v}$ character table

$C_{\infty v}$	I	$2C_{\infty}^{\phi}$	...	$\infty\sigma_v$		
$A_1 \equiv \Sigma^+$	1	1	...	1	T_z	$\alpha_{xx}+\alpha_{yy}, \alpha_{zz}$
$A_2 \equiv \Sigma^-$	1	1	...	−1	R_z	
$E_1 \equiv \Pi$	2	$2\cos\phi$	...	0	$(T_x, T_y), (R_x, R_y)$	$(\alpha_{xz}, \alpha_{yz})$
$E_2 \equiv \Delta$	2	$2\cos 2\phi$	...	0		$(\alpha_{xx}-\alpha_{yy}, \alpha_{xy})$
$E_3 \equiv \Phi$	2	$2\cos 3\phi$	...	0		
⋮	⋮	⋮		⋮		

which involves the kinetic energy T and the potential energy V due to motion along a coordinate q_i; the quantity $\dot{q}_i$ is $\partial q_i/\partial t$.

The kinetic energy T. If cartesian axes are attached to each nucleus with origin at the equilibrium position of the nucleus the kinetic energy is given by

$$T = \tfrac{1}{2}\sum_{j=1}^{N} m_j(\dot{x}_j^2 + \dot{y}_j^2 + \dot{z}_j^2) \tag{5.95}$$

where x_j, y_j, z_j are the coordinates of the jth nucleus of mass m_j and N is the total number of nuclei. The orthogonality of the cartesian displacement coordinates in equation (5.95) ensures that it contains only squared terms in the coordinates and no cross-products.

Translational and rotational, in addition to vibrational, motions of the molecule contribute to the kinetic energy of equation (5.95). In order to exclude the translational motion we impose the requirements

$$\sum_{j=1}^{N} m_j\dot{x}_j = 0; \quad \sum_{j=1}^{N} m_j\dot{y}_j = 0;$$
$$\sum_{j=1}^{N} m_j\dot{z}_j = 0 \tag{5.96}$$

To exclude the rotational motion we impose the requirements that there must be no angular momentum, namely

$$\sum_{j=1}^{N} m_j(y_j\dot{z}_j - z_j\dot{y}_j) = 0$$
$$\sum_{j=1}^{N} m_j(z_j\dot{x}_j - x_j\dot{z}_j) = 0 \tag{5.97}$$
$$\sum_{j=1}^{N} m_j(x_j\dot{y}_j - y_j\dot{x}_j) = 0$$

The potential energy V. The potential energy of a vibrating molecule is due to distortion from the equilibrium configuration which can be described in terms of changes of bond lengths and angles. It is natural, therefore, to express the potential energy in terms of internal displacement coordinates rather than the cartesian coordinates used for the kinetic energy. One set of internal coordinates consist of the displacements δr_{ij} and δr_{kl} of all pairs of nuclei i, j and k, l. In the harmonic oscillator approximation the potential energy is then given by

$$V = \tfrac{1}{2}\sum_{ij}\sum_{kl} k_{ij,kl}\,\delta r_{ij}\,\delta r_{kl} \tag{5.98}$$

where the $k_{ij,\,kl}$ are force constants. Application of this equation to the water molecule:

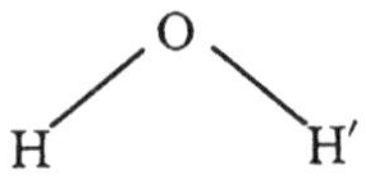

gives

$$V = \tfrac{1}{2}k_{OH}(\delta r_{OH})^2 + \tfrac{1}{2}k_{OH'}(\delta r_{OH'})^2 + \tfrac{1}{2}k_{HH'}(\delta r_{HH'})^2 + k_{OH,OH'}\delta r_{OH}\delta r_{OH'} + k_{OH,HH'}\delta r_{OH}\delta r_{HH'} + k_{OH',HH'}\delta r_{OH'}\delta r_{HH'} \quad (5.99)$$

where $k_{OH,OH}$, for example, has been abbreviated to k_{OH}. This expression for V assumes a general force field which takes into account all possible interactive forces.

Although it is assumed that $k_{OH} = k_{OH'}$ there are, unfortunately, four independent force constants in equation (5.99) whereas there are only three normal modes whose fundamental vibration wavenumbers may be found experimentally. This problem of there being more force constants in a molecule than there are normal modes is a general one, and one which gets much worse for larger molecules. Even for small molecules the problem is worse when they are less symmetrical. For example, in the bent molecule FOH there are six force constants compared with four in H_2O.

In calculating force constants from vibration wavenumbers, this means that the general force field is usually impractical and some approximations have to be made.

One such approximation is the central force field, for which the potential energy includes only terms involving the squares of changes of internuclear distances. For H_2O this force field gives

$$V = \tfrac{1}{2}k_{OH}(\delta r_{OH})^2 + \tfrac{1}{2}k_{OH}(\delta r_{OH'})^2 + \tfrac{1}{2}k_{HH'}(\delta r_{HH'})^2 \quad (5.100)$$

Comparing this with equation (5.99) we can see that the central force field assumes $k_{OH,OH'} = k_{OH,HH'} = 0$ but the force constants are now fewer than the number of normal modes. One difficulty with such a force field is that it cannot be used for bending vibrations of linear molecules because in, for example, HCN small displacements in the bending vibration do not alter the H . . . N distance.

A more generally satisfactory approximation is the valence force field, for which the potential energy includes only terms involving the squares of the changes of distances between valence-bonded nuclei and of angles between pairs of valence bonds. For H_2O this force field gives

$$V = \tfrac{1}{2}k_{OH}(\delta r_{OH})^2 + \tfrac{1}{2}k_{OH}(\delta r_{OH'})^2 + \tfrac{1}{2}k_{\alpha}(\delta\alpha)^2 \quad (5.101)$$

where $\alpha = \angle$HOH and $\delta\alpha$ is the change from the equilibrium value. Angle bending force constants such as k_α are often redefined so that they have the same dimensions as bond stretching force constants; if this is done in equation (5.101) the last term becomes $\tfrac{1}{2}k'_\alpha(r_{OH}\delta\alpha)^2$, where r_{OH} is the equilibrium O—H bond length. From any two of the three fundamental vibration wavenumbers k_{OH} and k_α (or k'_α) may be found. The success of the approximate force field may then be judged from the third vibration wavenumber predicted from the two force constants. On the other hand, the third wavenumber might be used to calculate a third force constant. In the case of H_2O, the quantity $k_{OH,\alpha}$, a force constant representing interaction between bond stretching and angle bending, might be introduced, modifying the potential energy to

$$V = \tfrac{1}{2}k_{OH}(\delta r_{OH})^2 + \tfrac{1}{2}k_{OH}(\delta r_{OH'})^2 + \tfrac{1}{2}k_{\alpha}(\delta\alpha)^2 + \tfrac{1}{2}k_{OH,\alpha}(\delta\alpha\delta r_{OH} + \delta\alpha\delta r_{OH'}) \quad (5.102)$$

In general, if we define $3N-6$ internal displacement coordinates R_i (or $3N-5$ for a linear molecule), which is the minimum number necessary to describe the changes of nuclear configuration, the potential energy is given by

$$V = \tfrac{1}{2}\sum_{i=1}^{3N-6}\sum_{j=1}^{3N-6} k_{ij}R_iR_j \quad (5.103)$$

A comparison of the expressions for the kinetic energy [equation (5.95)] and for the potential energy [equation (5.103)] shows that the former involves cartesian and the latter internal coordinates; therefore, we cannot use them as they stand in the Lagrange equation [equation (5.94)]. It is necessary first to express one set of coordinates in terms of the other:

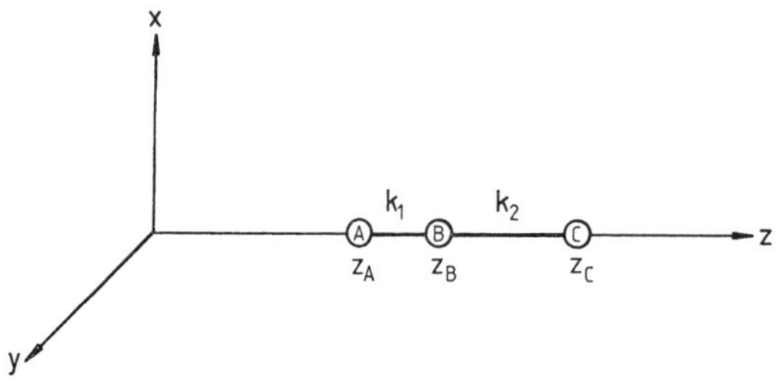

Figure 5.32 Force constants and nuclear coordinates for the stretching vibrations of a linear molecule ABC

either the Rs can be expressed as functions of the cartesian coordinates,

$$R_i = f_i(x_1, y_1, z_1, \ldots, x_{3N}, y_{3N}, z_{3N}) \quad (5.104)$$

or the cartesian coordinates as functions of the internal coordinates,

$$x_j = f_i(R_1, R_2, \ldots, R_{3N-6}) \quad (5.105)$$

Although it is generally easier to express the Rs as functions of cartesian coordinates, it leads to $3N$ Lagrange equations, whereas expressing cartesian coordinates as functions of the Rs leads to only $3N-6$ equations. This advantage greatly favours the use of the coordinate transformation of equation (5.105).

Before going on to introduce the simplifying effects of the symmetry of the molecule on the force field, we will apply what has been discussed so far to the vibrational motions of a linear molecule along the line of the C_∞ axis.

5.2.3.2 The stretching vibrations of a linear molecule ABC

Motion of the three nuclei of a linear unsymmetrical triatomic molecule ABC, such as HCN or OCS, along the C_∞ axis of the molecule, which is conventionally labelled the z-axis, represents three degrees of freedom; two of these are vibrational, involving stretching and compression of the bonds, and the other is translational, the molecule as a whole being translated along the z-axis with no change of bond lengths. The molecule is illustrated in Figure 5.32.

The potential energy, for which the general expression is given by equation (5.103), can be expressed as

$$V = \tfrac{1}{2}k_1R_1^2 + \tfrac{1}{2}k_2R_2^2 \quad (5.106)$$

where $R_1 = \delta r_{AB}$, the displacement of the bond A—B from its equilibrium value r_{AB}, and $R_2 = \delta r_{BC}$; k_1 and k_2 are the corresponding stretching force constants.

The kinetic energy, for which the general expression is given by equation (5.95), due to motion of the three nuclei along the space-fixed z-axis is given by

$$T = \tfrac{1}{2}m_A\dot{z}_A^2 + \tfrac{1}{2}m_B\dot{z}_B^2 + \tfrac{1}{2}m_C\dot{z}_C^2 \quad (5.107)$$

The internal coordinates R_1 and R_2 can be expressed in terms of the cartesian coordinates

$$R_1 = \delta r_{AB} = (z_B - z_A) - r_{AB} \quad (5.108)$$

$$R_2 = \delta r_{BC} = (z_C - z_B) - r_{BC} \quad (5.109)$$

Differentiating with respect to time, we obtain

$$\dot{R}_1 = \dot{z}_B - \dot{z}_A \quad (5.110)$$

$$\dot{R}_2 = \dot{z}_C - \dot{z}_B \quad (5.111)$$

since r_{AB} and r_{BC} are independent of time. These equations can be written in matrix notation:

$$\begin{pmatrix} \dot{R}_1 \\ \dot{R}_2 \end{pmatrix} = \begin{pmatrix} -1 & 1 & 0 \\ 0 & -1 & 1 \end{pmatrix} \begin{pmatrix} \dot{z}_A \\ \dot{z}_B \\ \dot{z}_C \end{pmatrix} \quad (5.112)$$

However, we wish to change not from Rs to zs, but from zs to Rs, and clearly this cannot be done with the two equations (5.110) and (5.111). We require a third equation and this can be derived from the requirement that, if we are concerned with vibrational motion only, there must be no translation along the z-axis and therefore no net linear momentum. This is expressed by

$$m_A\dot{z}_A + m_B\dot{z}_B + m_C\dot{z}_C = 0 \quad (5.113)$$

where m_A, m_B and m_C are the masses of the nuclei A, B and C, respectively. Equations

(5.110), (5.111) and (5.113) can all be expressed by the matrix equation

$$\begin{pmatrix} \dot{R}_1 \\ \dot{R}_2 \\ 0 \end{pmatrix} = \begin{pmatrix} -1 & 1 & 0 \\ 0 & -1 & 1 \\ m_A & m_B & m_C \end{pmatrix} \begin{pmatrix} \dot{z}_A \\ \dot{z}_B \\ \dot{z}_C \end{pmatrix} \quad (5.114)$$

Algebraic solution of equations (5.110), (5.111) and (5.113) will give the zs in terms of the Rs but matrix algebra more quickly and elegantly gives the result

$$\begin{pmatrix} \dot{z}_A \\ \dot{z}_B \\ \dot{z}_C \end{pmatrix} = \begin{pmatrix} -(m_B + m_C)/M & -m_C/M & 1/M \\ m_A/M & -m_C/M & 1/M \\ m_A/M & (m_B + m_B)/M & 1/M \end{pmatrix} \begin{pmatrix} \dot{R}_1 \\ \dot{R}_2 \\ 0 \end{pmatrix} \quad (5.115)$$

where $M = m_A + m_B + m_C$ and the 3×3 matrix in equation (5.115) is the *inverse*† of that in equation (5.114). This process of matrix inversion is an important feature of all transformations from cartesian to internal coordinates which are involved in determinations of normal modes.

Substituting the expressions for the zs, given by equation (5.115), into the kinetic energy equation (5.107) gives

$$\begin{aligned} T = {} & \frac{m_A}{2M^2}(m_B^2 + 2m_B m_C + m_C^2 + m_A m_B + m_A m_C)\dot{R}_1^2 \\ & + \frac{m_A m_C}{2M^2}(2m_C + 2m_A + 2m_B)\dot{R}_1 \dot{R}_2 \\ & + \frac{m_C}{2M^2}(m_B^2 + 2m_A m_B + m_A^2 + m_B m_C + m_A m_C)\dot{R}_2^2 \end{aligned} \quad (5.116)$$

Putting this expression for T and that of equation (5.106) for V into the Lagrange equation (5.94), for which $q_i = R_1$ or R_2, gives

$$\begin{aligned} & \frac{m_A}{M^2}(m_B^2 + 2m_B m_C + m_C^2 + m_A m_B + m_A m_C)\ddot{R}_1 \\ & \quad + \frac{m_A}{2M^2}(2m_C^2 + 2m_A m_C + 2m_B m_C)\ddot{R}_2 + k_1 R_1 = 0 \end{aligned} \quad (5.117)$$

and

$$\begin{aligned} & \frac{m_C}{M^2}(m_B^2 + 2m_A m_B + m_A^2 + m_B m_C + m_A m_C)\ddot{R}_2 \\ & + \frac{m_A}{2M^2}(2m_C^2 + 2m_A m_C + 2m_B m_C)\ddot{R}_1 + k_2 R_2 = 0 \end{aligned} \quad (5.118)$$

Solutions of equations (5.117) and (5.118) are of the form

$$R_1 = A_1 \cos 2\pi c\tilde{\nu}t \quad (5.119)$$

$$R_2 = A_2 \cos 2\pi c\tilde{\nu}t \quad (5.120)$$

From equations (5.119) and (5.120), it follows that

$$\ddot{R}_1 = -4\pi^2 c^2 \tilde{\nu}^2 R_1 \quad (5.121)$$

$$\ddot{R}_2 = -4\pi^2 c^2 \tilde{\nu}^2 R_2 \quad (5.122)$$

Substitution of equations (5.119)–(5.122) into equations (5.117) and (5.118) gives the simultaneous secular equations

†A matrix A has an inverse A^{-1} only if it is a square matrix and only if the determinant of A, symbolized by $|A|$, is non-zero. The jith element of A^{-1}, where j and i are row and column numbers, respectively, is given by

$$(A^{-1})_{ji} = (-1)^{i+j} \frac{|M_{ij}|}{|A|}$$

where M_{ij} is called the minor of the matrix and is obtained from the matrix A by striking out the ith row and jth column.

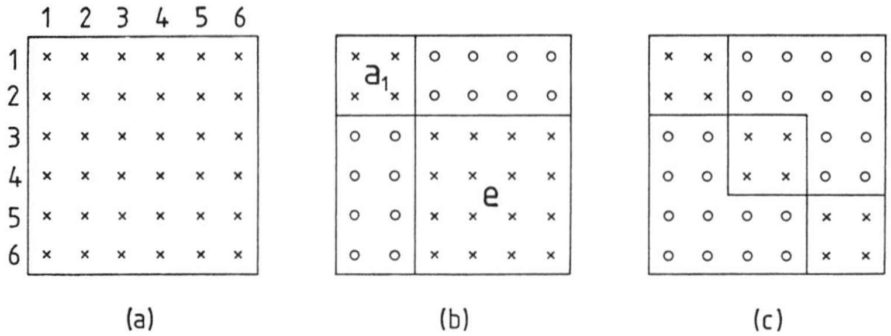

Figure 5.33 (a) A 6 × 6 secular determinant for the vibrations of the NH_3 molecule. (b) Factorization into a_1 and e blocks. (c) Further factorization of the 4 × 4 e block into two identical 2 × 2 blocks

$$-\left[4\pi^2c^2\tilde{\nu}^2\frac{m_A}{M^2}(m_B^2+2m_Bm_C+m_C^2+m_Am_B+m_Am_C)+k_1\right]A_1 - \left[4\pi^2c^2\tilde{\nu}^2\frac{m_A}{2M^2}(2m_C^2+2m_Am_C+2m_Bm_C)\right]A_2=0 \quad (5.123)$$

$$-\left[4\pi^2c^2\tilde{\nu}^2\frac{m_A}{2M^2}(2m_C^2+2m_Am_C+2m_Bm_C)\right]A_1 - \left[4\pi^2c^2\tilde{\nu}^2\frac{m_C}{M^2}(m_B^2+2m_Am_B+m_A^2+m_Bm_C+m_Am_C)+k_2\right]A_2=0 \quad (5.124)$$

If we rewrite these last two equations as

$$aA_1+bA_2=0 \quad (5.125)$$
$$cA_1+dA_2=0 \quad (5.126)$$

then the condition that they have non-trivial solutions is that

$$\begin{vmatrix} a & b \\ c & d \end{vmatrix}=0 \quad (5.127)$$

The expansion of this determinant, known as the secular determinant, gives a quadratic in $\tilde{\nu}^2$ from which two values of $\tilde{\nu}^2$ and, taking only the positive values of $\tilde{\nu}$, two values of $\tilde{\nu}$ can be obtained. On the other hand, if we know two fundamental vibration wavenumbers, we can find values of k_1 and k_2 from the secular determinant.

If we take the two stretching vibration wavenumbers of HCN to be $\tilde{\nu}_1 = 2089\ \mathrm{cm}^{-1}$ and $\tilde{\nu}_3 = 3312\ \mathrm{cm}^{-1}$, then we obtain

$$k_{CH}=5.8\ \mathrm{aJ\ \AA^{-2}};\quad k_{CN}=17.9\ \mathrm{aJ\ \AA^{-2}} \quad (5.128)$$

These values agree well with the 'typical' values quoted in Table 5.3.

From equations (5.123) and (5.124), the ratios of the As, the amplitudes corresponding to the internal coordinates for the vibrations ν_1 and ν_3, can be obtained:

$$\left.\begin{array}{ll}\text{for } \nu_1, & A_1/A_2=0.405/1.00\\ \text{for } \nu_2 & A_1/A_2=1.00/-0.137\end{array}\right\} \quad (5.129)$$

From these ratios, it is clear that ν_1 involves predominantly stretching of the C≡N bond and ν_3 stretching of the C—H bond. These results are consistent with the approximate transferability of group vibration wavenumbers from one molecule to another, as discussed in Section 5.2.1.

5.2.3.3 Use of symmetry coordinates

Symmetry coordinates are certain linear combinations of the internal coordinates R_i. Their use may greatly simplify calculation of force constants or vibration wavenumbers: the more

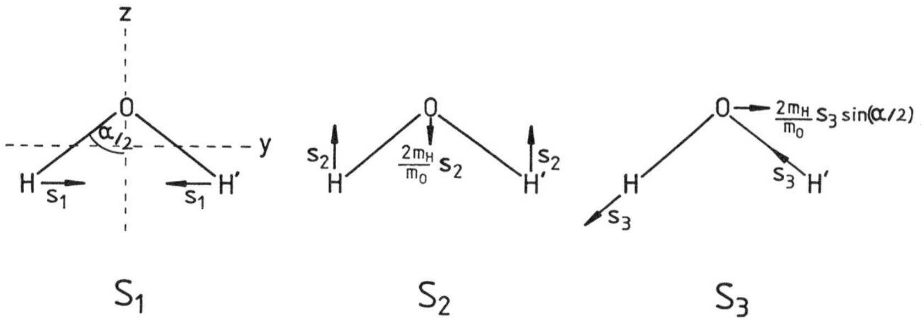

Figure 5.34 One set of symmetry coordinates for vibrations of the H_2O molecule

symmetrical the molecule the greater is the possible simplification.

The effect of using symmetry is to permit factorization of the secular determinant of the type illustrated by the simple 2×2 determinant of equation (5.127). In NH_3, for example, there are six vibrations and the secular determinant to be solved, without using symmetry coordinates, would be 6×6, as shown in Figure 5.33(a). If symmetry coordinates are used the secular determinant can be factorized into blocks whose elements are concerned only with vibrations of the same symmetry. In the case of NH_3 this means that use of symmetry coordinates results in a determinant of the type shown in Figure 5.33(b) consisting of two blocks, a 2×2 and a 4×4, corresponding to the two a_1 and the four e vibrations, of non-zero elements into which the determinant is factorized. If a determinant of this kind is equal to zero then each block produced by factorization is also equal to zero. In addition, a block involving degenerate vibrations with a degree of degeneracy d can be factorized further into d identical sub-blocks. In the NH_3 example, the 4×4 block can be factorized into two identical 2×2 blocks, as in Figure 5.33(c), so that this treatment of the six vibrations requires the solution of only two 2×2 determinants.

As the number of zeros in the secular determinant of Figure 5.33(c) illustrates, the use of symmetry coordinates reduces the number of unknowns in the problem. For example, if two symmetry coordinates S_a and S_b are, respectively, symmetric and antisymmetric to any symmetry operation K, i.e. $S_a \overset{K}{\rightarrow} S_a$ and $S_b \overset{K}{\rightarrow} -S_b$, then the coefficients of cross-terms $S_a S_b$ in the potential energy and $\dot{S}_a \dot{S}_b$ in the kinetic energy must be zero; otherwise, because $S_a S_b \overset{K}{\rightarrow} -S_a S_b$, the potential (and the kinetic) energy would be changed by carrying out a symmetry operation, and this cannot happen.

The simplification due to symmetry is even greater in a large symmetrical molecule such as benzene. The symmetry species of the 30 normal vibrations of benzene are

$$2a_{1g} + 1a_{2g} + 1a_{2u} + 2b_{1u} + 2b_{2g} + 2b_{2u} + 1e_{1g} + 3e_{1u} + 4e_{2g} + 2e_{2u}$$

counting the two components of each of the e vibrations.[†] Therefore, what would be a 30×30 secular determinant without using symmetry coordinates becomes one 4×4, one 3×3, five 2×2 and three 1×1 determinants.

Benzene, having a high degree of symmetry, is an ideal example of the effect of using symmetry coordinates but for a molecule such as 1-chloro-2-fluorobenzene (Figure 5.14) factorization of the secular determinant is possible

[†]The number of normal vibrations belonging to each symmetry species can be derived, for any molecule, using tables given in *Infrared and Raman Spectra* (Herzberg), and *Symmetry in Molecules* (Hollas) and some of the other books listed in the Bibliography at the end of this chapter.

only to a much smaller extent. The 30 normal vibrations of this molecule consist of

$$21a' + 9a''$$

and the determinant factorizes only into one 21×21 and one 9×9.

A relatively simple illustration of the use of symmetry coordinates is the vibrational problem in H_2O. Of the three normal vibrations of H_2O, shown in Figure 5.12, two are a_1 and one is b_2 [equation (5.72)]. Using this information symmetry coordinates can be chosen. One such choice is shown in Figure 5.34. The symmetry coordinate S_1 represents motion of the two hydrogen nuclei a distance S_1 towards each other parallel to the y-axis.† S_2 and S_3 involve motions of all three nuclei as shown. The coordinates S_1 and S_2 are A_1 and S_3 is B_2 and they have all been chosen so that the motions result in no overall translation or rotation of the molecule.

Derivation of symmetry coordinates in more complex molecules is clearly more difficult.

In cartesian coordinates, the expression for the kinetic energy in H_2O is

$$T = \tfrac{1}{2}m_{\mathrm{H}}(\dot{y}_{\mathrm{H}}^2 + \dot{z}_{\mathrm{H}}^2) + \tfrac{1}{2}m_{\mathrm{O}}(\dot{y}_{\mathrm{O}}^2 + \dot{z}_{\mathrm{O}}^2) + \tfrac{1}{2}m_{\mathrm{H}}(\dot{y}_{\mathrm{H'}}^2 + \dot{z}_{\mathrm{H'}}^2) \tag{5.130}$$

and, using a valence force field, that for the potential energy is

$$V = \tfrac{1}{2}k_{\mathrm{OH}}(\delta r_{\mathrm{OH}})^2 + \tfrac{1}{2}k_{\mathrm{OH}}(\delta r_{\mathrm{OH'}})^2 + \tfrac{1}{2}k_\alpha(\delta\alpha)^2 \tag{5.131}$$

The velocities $\dot{y}$ and $\dot{z}$ of the nuclei can be expressed in terms of $\dot{S}_1$, $\dot{S}_2$ and $\dot{S}_3$ using Figure 5.34:

$$\dot{y}_{\mathrm{H}} = \dot{S}_1 - \dot{S}_3 \sin\frac{\alpha}{2};\quad \dot{z}_{\mathrm{H}} = \dot{S}_2 - \dot{S}_3 \cos\frac{\alpha}{2} \tag{5.132}$$

$$\dot{y}_{\mathrm{H'}} = -\dot{S}_1 - \dot{S}_3 \sin\frac{\alpha}{2};\quad \dot{z}_{\mathrm{H'}} = \dot{S}_2 + \dot{S}_3 \cos\frac{\alpha}{2} \tag{5.133}$$

$$\dot{y}_{\mathrm{O}} = \frac{2m_{\mathrm{H}}}{m_{\mathrm{O}}}\dot{S}_3 \sin\frac{\alpha}{2};\quad \dot{z}_{\mathrm{O}} = -\frac{2m_{\mathrm{H}}}{m_{\mathrm{O}}}\dot{S}_2 \tag{5.134}$$

Substitution of equations (5.132)–(5.134) into equation (5.130) gives

$$T = m_{\mathrm{H}}\dot{S}_1^2 + m_{\mathrm{H}}\left(1 + \frac{2m_{\mathrm{H}}}{m_{\mathrm{O}}}\right)\dot{S}_2^2 + m_{\mathrm{H}}\left(1 + \frac{2m_{\mathrm{H}}}{m_{\mathrm{O}}}\sin^2\frac{\alpha}{2}\right)\dot{S}_3^2 \tag{5.135}$$

To obtain V in terms of S_1, S_2 and S_3, we express first the internal coordinates in terms of cartesian displacements:

$$\delta r_{\mathrm{OH}} = (\delta y_{\mathrm{O}} - \delta y_{\mathrm{H}})\sin\frac{\alpha}{2} + (\delta z_{\mathrm{O}} - \delta z_{\mathrm{H}})\cos\frac{\alpha}{2} \tag{5.136}$$

$$\delta r_{\mathrm{OH'}} = -(\delta y_{\mathrm{O}} - \delta y_{\mathrm{H'}})\sin\frac{\alpha}{2} + (\delta z_{\mathrm{O}} - \delta z_{\mathrm{H'}})\cos\frac{\alpha}{2} \tag{5.137}$$

$$\delta\alpha = \frac{1}{r_{\mathrm{OH}}}\left[(\delta y_{\mathrm{H'}} - \delta y_{\mathrm{H}})\cos\frac{\alpha}{2} + (\delta z_{\mathrm{H'}} + \delta z_{\mathrm{H}})\sin\frac{\alpha}{2} - 2\delta z_{\mathrm{O}}\sin\frac{\alpha}{2}\right] \tag{5.138}$$

These equations can then be written as

†We use S_1, S_2, S_3 to represent the hydrogen atom displacements in addition to the complete symmetry coordinates.

$$\delta r_{\mathrm{OH}} = -\left(\sin\frac{\alpha}{2}\right)S_1 - \left[\left(1+\frac{2m_{\mathrm{H}}}{m_{\mathrm{O}}}\right)\cos\frac{\alpha}{2}\right]S_2 + \left(1+\frac{2m_{\mathrm{H}}}{m_{\mathrm{O}}}\sin^2\frac{\alpha}{2}\right)S_3 \tag{5.139}$$

$$\delta r_{\mathrm{OH'}} = -\left(\sin\frac{\alpha}{2}\right)S_1 - \left[\left(1+\frac{2m_{\mathrm{H}}}{m_{\mathrm{O}}}\right)\cos\frac{\alpha}{2}\right]S_2 - \left(1+\frac{2m_{\mathrm{H}}}{m_{\mathrm{O}}}\sin^2\frac{\alpha}{2}\right)S_3 \tag{5.140}$$

$$\delta\alpha = \left(\frac{2}{r_{\mathrm{OH}}}\cos\frac{\alpha}{2}\right)S_1 + \left[\frac{2}{r_{\mathrm{OH}}}\left(1+\frac{2m_{\mathrm{H}}}{m_{\mathrm{O}}}\sin\frac{\alpha}{2}\right)\sin\frac{\alpha}{2}\right]S_2 \tag{5.141}$$

The expression for V given in equation (5.131) now becomes

$$\begin{aligned} V = {} & \left(k_{\mathrm{OH}}\sin^2\frac{\alpha}{2} + \frac{2k_\alpha}{r_{\mathrm{OH}}^2}\cos^2\frac{\alpha}{2}\right)S_1^2 \\ & + \left\{k_{\mathrm{OH}}\left(1+\frac{2m_{\mathrm{H}}}{m_{\mathrm{O}}}\right)^2\cos^2\frac{\alpha}{2} + \frac{2k_\alpha}{r_{\mathrm{OH}}^2}\left[\left(1+\frac{2m_{\mathrm{H}}}{m_{\mathrm{O}}}\sin\frac{\alpha}{2}\right)^2\sin^2\frac{\alpha}{2}\right]\right\}S_2^2 \\ & + \left\{2k_{\mathrm{OH}}\left(1+\frac{2m_{\mathrm{H}}}{m_{\mathrm{O}}}\right)\sin\frac{\alpha}{2}\cos\frac{\alpha}{2} + \frac{4k_\alpha}{r_{\mathrm{OH}}^2}\left[\left(1+\frac{2m_{\mathrm{H}}}{m_{\mathrm{O}}}\sin\frac{\alpha}{2}\right)\sin\frac{\alpha}{2}\cos\frac{\alpha}{2}\right]\right\}S_1S_2 \\ & + k_{\mathrm{OH}}\left(1+\frac{2m_{\mathrm{H}}}{m_{\mathrm{O}}}\sin^2\frac{\alpha}{2}\right)^2 S_3^2 \end{aligned} \tag{5.142}$$

For simplicity, we rewrite T and V in equations (5.135) and (5.142) as

$$T = a_{11}\dot{S}_1^2 + a_{22}\dot{S}_2^2 + a_{33}\dot{S}_3^2 \tag{5.143}$$

$$V = b_{11}S_1^2 + 2b_{12}S_1S_2 + b_{22}S_2^2 + b_{33}S_3^2 \tag{5.144}$$

Application of the Lagrange equation [equation (5.94)] gives three equations, one in each of S_1, S_2 and S_3. These equations, analogous to equations (5.117) and (5.118), have solutions of the form

$$S_1 = A_1\cos 2\pi c\tilde{\nu}t \tag{5.145}$$

etc. Substitution of these solutions into the three Lagrange equations gives three secular equations in A_1, A_2 and A_3. The condition that the secular equations are all simultaneously valid is that the secular determinant is zero:

$$\begin{vmatrix} b_{11}-\lambda a_{11} & b_{12} & 0 \\ b_{12} & b_{22}-\lambda a_{22} & 0 \\ 0 & 0 & b_{33}-\lambda a_{33} \end{vmatrix} = 0 \tag{5.146}$$

where

$$\lambda = 4\pi^2c^2\tilde{\nu}^2 \tag{5.147}$$

As we intended, the determinant can be factorized:

$$\begin{vmatrix} b_{11}-\lambda a_{11} & b_{12} \\ b_{12} & b_{22}-\lambda a_{22} \end{vmatrix} = 0 \tag{5.148}$$

giving two values of λ and hence wavenumbers of two a_1 fundamental vibrations, and

$$b_{33} - \lambda a_{33} = 0 \tag{5.149}$$

giving the wavenumber of the b_2 vibration.

5.2.3.4 *Use of Wilson's FG matrix method*

Although in principle the method described in Section 5.2.3.3 is mathematically simple, in practice it is long and tedious even for such a small molecule as H_2O.

Wilson has produced a much more generally useful and elegant method known as the 'FG matrix method', which, as the name implies, involves considerable use of matrix algebra. We shall not describe the method here but the reader is referred to books mentioned in the Bibliography at the end of the chapter and, in particular, to *Introduction to the Theory of Molecular Vibrations and Vibrational Spectroscopy* by Woodward.

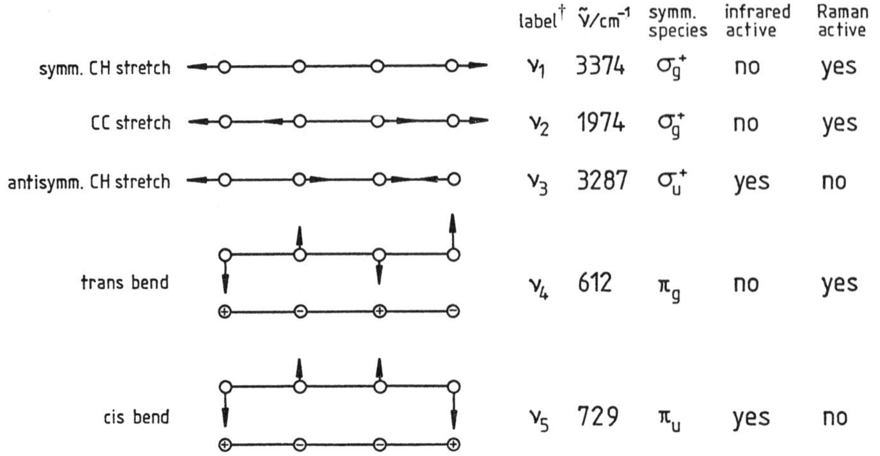

†See footnote on p. 213

Figure 5.35 The normal modes of vibration of the acetylene molecule

5.2.4 Vibrational selection rules

5.2.4.1 Electric dipole selection rules

In the process of absorption or emission of radiation involving transitions between a pair of vibrational states of a molecule, the interaction is usually between the molecule and the electric, rather than the magnetic, component of the electromagnetic radiation. For this reason, the selection rules which operate are known as electric dipole selection rules or, simply, as dipole selection rules.

Transitions between vibrational states involving absorption or emission (as opposed to scattering—see Section 5.2.4.2) occur mostly, but not always, in the infrared region of the spectrum and the selection rules are therefore often referred to as infrared selection rules.

In diatomic molecules (Section 5.1.2) we have seen that the anharmonic oscillator selection rule is $\Delta v = \pm 1, \pm 2, \pm 3, \ldots$ with the $\Delta v = \pm 1$ transitions, obeying the harmonic oscillator selection rule, being very much the strongest. In addition, there must be a change of dipole moment accompanying the vibration, which means that transitions are allowed in heteronuclear but not homonuclear diatomic molecules.

In polyatomic molecules (Section 5.2.1) we saw that the selection rule for *each* of the normal vibrations is the same as in a diatomic:

$$\Delta v_i = \pm 1, \pm 2, \pm 3, \ldots$$

and that there must be a change of dipole moment during the vibration.

In the case of the water molecule, it is easy to see from the form of the normal modes shown in Figure 5.12 that all the vibrations ν_1, ν_2 and ν_3 involve a change of dipole moment and are infrared active and, therefore, that $v = 1\text{–}0$ transitions in each are allowed by dipole selection rules.

Acetylene (H—C≡C—H), being a linear four-atomic molecule, has seven normal modes of vibration as illustrated in Figure 5.35. The *trans* and *cis* bending vibrations, ν_4 and ν_5, are each doubly degenerate. This is rather like ν_3 and ν_4 of ammonia (Figure 5.31) except that in acetylene the degeneracy of the two components is more obvious. For example, in the ν_4 mode the molecule can bend in any two mutually perpendicular planes but the vectors representing the motion in one plane have zero components in the other. The two motions clearly have different wave functions but equal energies.

Inspection of the normal modes shows that ν_3 and ν_5 involve a change of dipole moment and are therefore infrared active but ν_1, ν_2 and ν_4 are infrared inactive.

Symmetry considerations have an important part to play in formulating dipole selection rules not only for 1–0 transitions in fundamentals, where simple deductions like those above for H_2O and C_2H_2 may be made, but also for vibrational overtones and combinations.

In Section 2.2 we have discussed in general terms intensities of transitions between a lower state m and an upper state n. In cases where the lower and upper state wave functions are ψ_v'' and ψ_v', respectively, the expression for the electric dipole transition moment of equation (2.21) becomes

$$\boldsymbol{R}^{v'v''} = \int \psi_v'^* \boldsymbol{\mu} \psi_v'' \mathrm{d}\tau_v \qquad (5.150)$$

where the ψ_vs are vibrational wave functions, $\boldsymbol{\mu}$ is the electric dipole moment operator given by equation (2.22) and the integration is over nuclear coordinates. Since the intensity of a transition is proportional to the square of the magnitude of the transition moment, $|\boldsymbol{R}^{v'v''}|^2$, this means that

$$\begin{aligned} \boldsymbol{R}^{v'v''} &= 0 \text{ for a forbidden transition} \\ \boldsymbol{R}^{v'v''} &\neq 0 \text{ for an allowed transition} \end{aligned} \qquad (5.151)$$

There are simple symmetry requirements for the integral of equation (5.150) to be non-zero and therefore for the transition to be allowed. The requirement for transitions between states, neither of which is degenerate, is that the symmetry of the quantity to be integrated is totally symmetric; this can be written as

$$\Gamma(\psi_v') \times \Gamma(\mu) \times \Gamma(\psi_v'') = A \qquad (5.152)$$

where A is used here to denote the totally symmetric species of *any* point group.

The symmetry requirement for an allowed transition between two states, of which at least one is degenerate, has to be modified since if, say, the symmetry species $\Gamma a(\psi_v')$ and $\Gamma(\mu)$ were

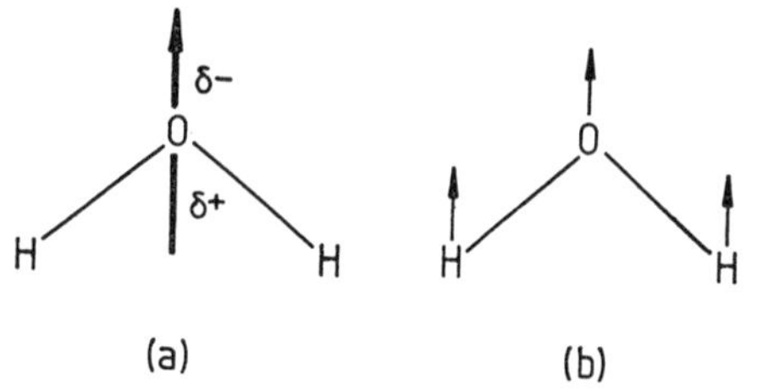

Figure 5.36 (a) The dipole moment vector and (b) a translation in the same direction, have the same symmetry species

each degenerate the product of equation (5.152) would give more than one symmetry species, as in equation (5.88) and (5.92). The modified requirement is that the symmetry species of the quantity to be integrated contains the totally symmetric symmetry species; this can be written as

$$\Gamma(\psi_v') \times \Gamma(\mu) \times \Gamma(\psi_v'') \supset A \qquad (5.153)$$

where the boolean symbol $\supset$ means 'contains'. For example, in the C_{3v} point group we can see from equation (5.88) that

$$E \times E \supset A_1 \qquad (5.154)$$

As we saw in equation (2.28), the electric dipole moment vector has components μ_x, μ_y, μ_z along the cartesian axes. Correspondingly, the transition moment can be resolved into three components:

$$\begin{aligned} R_x^{v'v''} &= \int \psi_v'^* \mu_x \psi_v' \,\mathrm{d}\tau_v \\ R_y^{v'v''} &= \int \psi_v'^* \mu_y \psi_v' \,\mathrm{d}\tau_v \\ R_z^{v'v''} &= \int \psi_v'^* \mu_z \psi_v' \,\mathrm{d}\tau_v \end{aligned} \qquad (5.155)$$

and since, as in equation (2.30),

$$|\boldsymbol{R}^{v'v''}|^2 = (R_x^{v'v''})^2 + (R_y^{v'v''})^2 + (R_z^{v'v''})^2 \qquad (5.156)$$

the transition is allowed if any of $R_x^{v'v''}$, $R_y^{v'v''}$ or $R_z^{v'v''}$ is non-zero.

Since the dipole moment is a vector in a particular direction it has the same symmetry

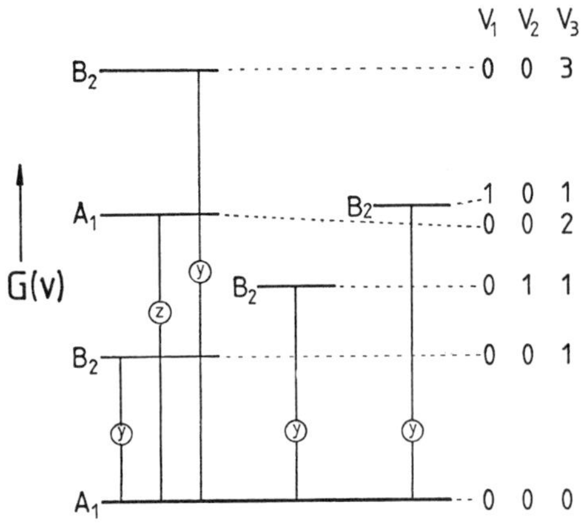

Figure 5.37 Symmetry species of various overtone and combination tone levels of the H_2O molecule together with directions of polarization of transition moments. The vibration wavenumbers are $\tilde{\nu}_1 = 3657.1\ \text{cm}^{-1}$, $\tilde{\nu}_2 = 1594.8\ \text{cm}^{-1}$, $\tilde{\nu}_3 = 3755.8\ \text{cm}^{-1}$

species as a translation of the molecule in the same direction. Figure 5.36 illustrates this in the case of H_2O, in which both the dipole moment and the translation in the same direction are totally symmetric, A_1.

In general, we have

$$\begin{aligned} \Gamma(\mu_x) &= \Gamma(T_x) \\ \Gamma(\mu_y) &= \Gamma(T_y) \\ \Gamma(\mu_z) &= \Gamma(T_z) \end{aligned} \tag{5.157}$$

For transitions between non-degenerate vibrational states equation (5.152) becomes

$$\begin{aligned} &\Gamma(\psi'_v) \times \Gamma(T_x) \times \Gamma(\psi''_v) = A \\ \text{and/or}\quad &\Gamma(\psi'_v) \times \Gamma(T_y) \times \Gamma(\psi''_v) = A \\ \text{and/or}\quad &\Gamma(\psi'_v) \times \Gamma(T_z) \times \Gamma(\psi''_v) = A \end{aligned} \tag{5.158}$$

where the 'and/or' implies that, for an allowed transition, one or more of the components of the transition moment $\boldsymbol{R}^{v'v''}$ may be non-zero.

If we consider initially only transitions for which the lower state is the zero-point level, in which there is no vibrational excitation, then $\Gamma(\psi''_v) = A$ and equation (5.158) becomes

$$\begin{aligned} &\Gamma(\psi'_v) \times \Gamma(T_x) = A \\ \text{and/or}\quad &\Gamma(\psi'_v) \times \Gamma(T_y) = A \\ \text{and/or}\quad &\Gamma(\psi'_v) \times \Gamma(T_z) = A \end{aligned} \tag{5.159}$$

If multiplication of two symmetry species gives the totally symmetric species then the two species must be the same. Therefore, it follows from equation (5.159) that

$$\Gamma(\psi'_v) = \Gamma(T_x) \text{ and/or } \Gamma(T_y) \text{ and/or } \Gamma(T_z) \tag{5.160}$$

This equation enables us to write down, on inspection of the relevant character table, the vibrational selection rules for transitions to and from the zero-point level for any molecule. The procedure is: (a) assign the molecule to a point group; (b) look up the translational symmetry species in the relevant character table; (c) the allowed transitions are

$$\Gamma(T_x) - A;\quad \Gamma(T_y) - A;\quad \Gamma(T_z) - A \tag{5.161}$$

where we conventionally indicate a transition between an upper state Y and a lower state X as Y–X.

The selection rules apply irrespective of whether the transition is observed as an absorption or an emission process.

Using the C_{2v} character table in Table 5.7, we can immediately write down the allowed transitions involving the zero-point level as

$$A_1\text{–}A_1;\quad B_1\text{–}A_1;\quad B_2\text{–}A_1 \tag{5.162}$$

These are polarized along the z-, x- and y-axes, respectively. It follows that 1–0 transitions in the vibrations $\nu_1(a_1)$, $\nu_2(a_1)$ and $\nu_3(b_2)$ of H_2O, shown in Figure 5.12, are allowed.

We had derived this result previously simply by observing that all the vibrations produce a change of dipole moment, but the rules of equation (5.161) enable us also to derive selection rules for overtone and combination transitions. Equation (5.75) indicates that if H_2O is vibrating with two quanta of ν_3 then $\Gamma(\psi_v)$ is A_1 and, in general, if it is vibrating with

n quanta of a vibration with symmetry species S then

$$\Gamma(\psi_v) = S^n \tag{5.163}$$

Figure 5.37 shows, for example, that the symmetry species of vibrational fundamental and overtone levels for ν_3 alternate, being A_1 for v even and B_2 for v odd. It follows that the transitions 3_0^1, 3_0^2, 3_0^3, ... are allowed and polarized along the y-, z-, y-, ... axes.[†]

Equation (5.73) gives the result that, if H_2O is vibrating with one quantum each of ν_1 and ν_3, then $\Gamma(\psi_v) = B_2$. The result is the same if it is vibrating with one quantum each of ν_2 and ν_3, as indicated in Figure 5.37. We deduce, therefore, that the $1_0^1 3_0^1$ and $2_0^1 3_0^1$ transitions are both allowed and polarized along the y-axis.

The H_2O molecule has no a_2 or b_1 vibrations but selection rules for another C_{2v} molecule such as CH_2F_2, which has vibrations of *all* symmetry species, could be treated in an analogous way.

If we use as an example a molecule with much lower symmetry, such as phenol (C_6H_5OH), belonging to the C_s point group, the symmetry species may correspond to translations along more than one cartesian axis. Table A.1 in the Appendix gives the C_s character table and Figure 5.38 illustrates two of the 33 normal modes; Figure 5.38(a) and (b) show the ring-OH stretching vibration, ν_{13}, and the out-of-plane (torsional) vibration of the hydrogen of the OH group, ν_{32}.[‡] Since ν_{13} is symmetric to the plane of symmetry σ (the plane of the molecule) it is an a' vibration whereas ν_{32} is an a'' vibration. Since, according to Table A.1, $\Gamma(T_x)$ and $\Gamma(T_y)$ are both a' the 13_0^1 transition is allowed and the transition moment lies in the xy-plane, the plane of the molecule; the particular direction in this plane is dependent on the form of the normal mode.

[†] Vibrational states in small polyatomic molecules are often labelled $(v_1 v_2 v_3 \ldots)$ and the transitions $(v_1' v_2' v_3' \ldots)$–$(v_1'' v_2'' v_3'' \ldots)$ but the system can be very cumbersome. More generally useful is the system in which, for example, the (001)–(000) transition is indicated by 3_0^1 and, in general, the symbol $n_{v''}^{v'}$ is used for a transition involving vibration n in which v'' quanta are excited in the lower and v' in the upper state. A transition to a combination level, such as (110)–(000), is indicated by $1_0^1 2_0^1$.

In cases of degenerate vibrations with l-doubling, the value of $|l|$ is indicated by a superscript as in $(02^0 0)$–(000), in the first notation, but separated from the corresponding quantum number by a comma, as in $2_0^{2,0}$, in the second notation. We shall be using the more economical second notation throughout.

In electronic transitions in polyatomic molecules a similar convention (see footnote on p. 385) is used whereby 3_0^1, for example, implies a transition between electronic states in which $v_3 = 0$ in the lower and $v_3 = 1$ in the upper electronic state. It will always be clear from the context whether, for example, 3_0^1 implies a pure vibrational or an electronic–vibrational (vibronic) transition.

[‡] A systematic scheme for numbering normal modes of vibration in polyatomic molecules has been recommended by Mulliken ['Report on notation for the spectra of polyatomic molecules', *J. Chem. Phys.*, **23**, 1997 (1955)] and is that adopted by Herzberg [*Infrared and Raman Spectra*, pp. 271–272. Van Nostrand, New York (1945)]

The vibrations are grouped according to their symmetry species which are taken to be in the order used for the character table in Herzberg's book (this is important since not all authors use the same order). For each symmetry species the vibrations are sorted in order of decreasing wavenumber. The totally symmetric species is always the first in the character table and the totally symmetric vibration with the highest wavenumber is labelled ν_1, the next one ν_2, and so on.

For example, when the scheme is applied to H_2O, the three vibrations with wavenumbers of 3657.1 cm^{-1} (a_1), 1594.8 cm^{-1}(a_1) and 3755.8 cm^{-1}(b_2) are labelled ν_1, ν_2 and ν_3, respectively. Figure 5.35 illustrates the use of the scheme in labelling the vibrations of acetylene. However, in linear XY_2 and XYZ molecules (such as CO_2 and HCN) there is an important exception: the doubly degenerate bending vibration (π_u or π) is *always* labelled ν_2 because of common usage prior to any recommended scheme.

In unsymmetrical linear triatomics, such as HCN, there is a further accepted violation of the Mulliken recommendations. In these molecules there are two σ^+ stretching vibrations and the *lower* wavenumber one is labelled ν_1, such as the CN-stretching vibration in HCN, and the *higher* is ν_3, such as the CH-stretching vibration in HCN.

There are also many cases where a certain flexibility is desirable in the numbering scheme adopted. The case of phenol illustrates this well because there are two possible reasons for departing from a scheme based on the Mulliken recommendation for the C_s point group: these are, first, that most of the vibrations except those of the OH group strongly resemble those of a C_{2v} monosubstituted benzene, such as fluorobenzene, and, second, that many of the vibrations resemble those of benzene itself. If direct comparisons are being made between the vibrations of phenol and fluorobenzene or benzene then it is sensible to use a numbering scheme based on that for one or other of these molecules.

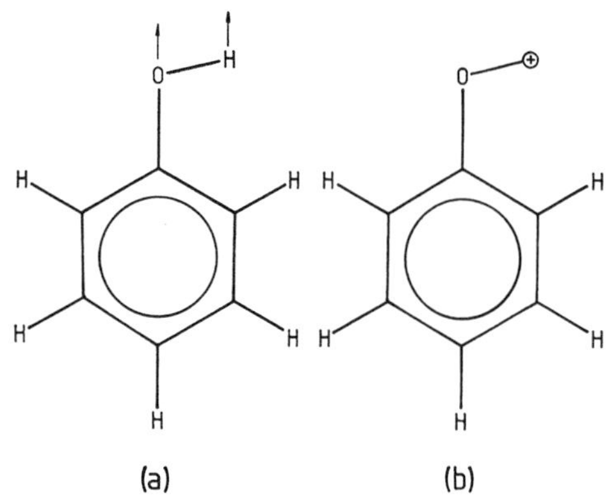

Figure 5.38 (a) The ring-OH stretching vibration, ν_{13}, and (b) the OH torsional vibration, ν_{32}, in the phenol molecule

On the other hand, ν_{32} is an a'' vibration and the transition moment for the allowed 32^1_0 transition is confined by symmetry to the z-axis, perpendicular to the plane of the molecule.

Not all vibrational transitions have the zero-point level as the lower state. The population N_1 of the $v = 1$ level of a particular vibrational mode relative to the population N_0 of the zero-point level is obtained from the Boltzmann distribution law [equation (2.8)] giving, for a non-degenerate vibration,

$$N_1/N_0 = \exp(-\Delta E/kT) \qquad (5.164)$$

where ΔE is the energy difference, $hc[G(1) - G(0)]$, between the $v = 1$ and the zero-point level and $G(v)$ is the vibrational term value. If, for example, $G(1) - G(0)$ is $400\,\text{cm}^{-1}$, which is not an unusually low vibration wavenumber, then N_1/N_0 is 0.14 at $T = 293$ K; if $G(1) - G(0)$ is as low as $200\,\text{cm}^{-1}$ then N_1/N_0 is 0.37.

From these examples at room temperature we can see that, in molecules with one or more low wavenumber fundamental vibrations, there is appreciable population of some excited vibrational levels.

Transitions from such levels give rise, in an absorption spectrum, to hot bands, so called because the hotter the sample, the more intense are such bands.

The selection rules for hot bands follow directly from equation (5.158), which may be rewritten as

$$\Gamma(\psi'_v) \times \Gamma(\psi''_v) = \Gamma(T_x) \text{ and/or } \Gamma(T_y) \text{ and/or } \Gamma(T_z) \qquad (5.165)$$

If, for example, in a molecule such as CH_2F_2 in the C_{2v} point group $\Gamma(\psi''_v) = B_2$ and $\Gamma(\psi'_v) = B_1$, their product is A_2 and the transition is forbidden. On the other hand, if $\Gamma(\psi''_v) = A_2$ and $\Gamma(\psi'_v) = B_1$, their product is B_2 and the transition is allowed, polarized along the y-axis.

If a molecule belongs to a degenerate point group and the vibrational transition is between the zero-point level and the $v = 1$ level of a degenerate vibration, then $\Gamma(\psi''_v) = A$ and equations (5.153) and (5.157) give

$$\begin{aligned} &\Gamma(\psi'_v) \times \Gamma(T_x) \supset A \\ \text{and/or}\quad &\Gamma(\psi'_v) \times \Gamma(T_y) \supset A \\ \text{and/or}\quad &\Gamma(\psi'_v) \times \Gamma(T_z) \supset A \end{aligned} \qquad (5.166)$$

These reduce to

$$\Gamma(\psi'_v) \supset \Gamma(T_x) \text{ and/or } \Gamma(T_y) \text{ and/or } \Gamma(T_z) \qquad (5.167)$$

which is analogous to equation (5.160) for non-degenerate states.

In NH_3 (Figure 5.31), for example, the transitions 1^1_0 and 2^1_0 are allowed; it can be confirmed from Table 5.8 that $\Gamma(T_z) = A_1$ and therefore the transition moment is along the z-axis. The 3^1_0 and 4^1_0 transitions are allowed also and, since $\Gamma(T_x, T_y) = E$, the transition moment is in the xy-plane in both cases.

If the upper state is a combination or overtone level then $\Gamma(\psi'_v)$ may not give a single symmetry species and we have to use equation (5.166). For example, for the combination band $3^1_0 4^1_0$ in NH_3 we have, according to equation (5.88),

$$\Gamma(\psi'_v) = E \times E = A_1 + A_2 + E \qquad (5.168)$$

Since $\Gamma(\psi'_v) \supset \Gamma(T_x, T_y)$ and $\Gamma(T_z)$, the transition is allowed, or, to be more specific, transitions involving two of the three components (A_1 and E) of the combination state are allowed but that involving the A_2 component is forbidden.

For the overtone transition 3_0^2 we have, according to equation (5.89),

$$\Gamma(\psi'_v) = (E)^2 = A_1 + E \tag{5.169}$$

and both components of the transition are allowed.

Acetylene (HC≡CH) belongs to the $D_{\infty h}$ point group, whose character table is given in Table A.37, and its vibrations are illustrated in Figure 5.35. Since ν_3 is a σ_u^+ vibration and $\Gamma(T_z) = \Sigma_u^+$, the 3_0^1 transition is allowed with the transition moment along the z-axis. Similarly, since ν_5 is a π_u vibration, the 5_0^1 transition is allowed with the transition moment in the xy-plane.

The symmetry species of the $4^1 5^1$ combination state is given by

$$\Gamma(\psi'_v) = \Pi_g \times \Pi_u = \Sigma_u^+ + \Sigma_u^- + \Delta_u \tag{5.170}$$

a result which can be obtained in a similar way to that of equation (5.92) for the $C_{\infty v}$ point group. Therefore, one component of the $4_0^1 5_0^1$ transition, that involving the Σ_u^+ level, is allowed.

The symmetry species of the 5^2 state is given by

$$\Gamma(\psi'_v) = (\Pi_u)^2 = \Sigma_g^+ + \Delta_g \tag{5.171}$$

a result similar to that in equation (5.93). Therefore, the 5_0^2 transition is forbidden.

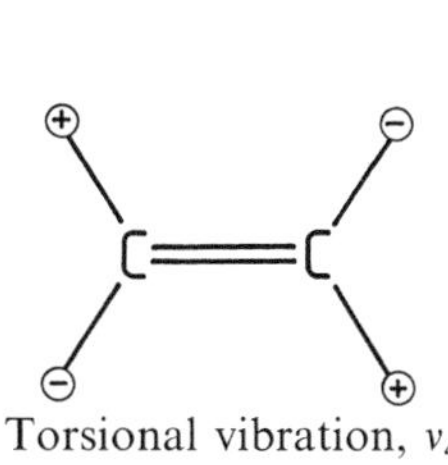

Figure 5.39 Torsional vibration, ν_4, of the ethylene molecule

As a final example, we can show that the hot band transition $3_0^1 4_1^0$ is allowed since

$$\Gamma(\psi'_v) \times \Gamma(\psi''_v) = \Sigma_u^+ \times \Pi_g = \Pi_u = \Gamma(T_x, T_y) \tag{5.172}$$

It is important to remember that selection rules in general tell us nothing about transition intensities other than their being zero or non-zero. In the case of a vibrational electric dipole transition the intensity is determined by the transition moment of equation (5.150) and it is possible that, even though this may be non-zero, it may be so small as to give an immeasurably low intensity.

It should also be remembered that the selection rules derived here are relevant to the free molecule and may break down in the liquid or solid phase to the extent that transitions which are forbidden in the free molecule may become weakly allowed in a high pressure gas or in a condensed phase. This is the case, for example, with the electric dipole forbidden 4_0^1 transition in ethylene, where ν_4 is the a_u torsional vibration shown in Figure 5.39. It is not observed in the infrared spectrum of the gas but is observed weakly in the liquid and solid phases.

5.2.4.2 *Raman selection rules*

In rotational Raman spectroscopy we have seen [equation (4.118)] that intense monochromatic radiation falling on a sample induces in it an electric dipole $\boldsymbol{\mu}$ which is related to the oscillating electric field $\boldsymbol{E}$ of the radiation by

$$\boldsymbol{\mu} = \boldsymbol{\alpha}\boldsymbol{E} \tag{5.173}$$

In a discussion of vibrational Raman scattering $\boldsymbol{\alpha}$ is replaced by $\boldsymbol{\alpha}' Q_k$, where $\boldsymbol{\alpha}'$ is the derivative polarizability tensor. The elements of $\boldsymbol{\alpha}'$ are derivatives, $\mathrm{d}\alpha/\mathrm{d}Q_k$, with respect to the normal coordinate Q_k for vibration k rather than those of the polarizability tensor [equation (4.117)] which are appropriate to Rayleigh and rotational Raman scattering. Equation (5.173) can be written in matrix form as

$$\begin{pmatrix} \mu_x \\ \mu_y \\ \mu_z \end{pmatrix} = Q_k \begin{pmatrix} \alpha'_{xx} & \alpha'_{xy} & \alpha'_{xz} \\ \alpha'_{yx} & \alpha'_{yy} & \alpha'_{yz} \\ \alpha'_{zx} & \alpha'_{zy} & \alpha'_{zz} \end{pmatrix} \begin{pmatrix} E_x \\ E_y \\ E_z \end{pmatrix} \tag{5.174}$$

where the x-, y- and z-axes for the polarizability components are molecule-fixed and chosen according to the usual conventions. The tensor is symmetric about the leading diagonal so that $\alpha'_{xy} = \alpha'_{yx}$, etc.

The electric field $\boldsymbol{E}$ is oscillating rapidly with time and, therefore, so is the induced dipole $\boldsymbol{\mu}$. Since, here, we require only to derive the selection rules it will be sufficient to treat the problem semi-classically. Then we need consider only the amplitude $\boldsymbol{E}^0$ of the electric field. The amplitude $\boldsymbol{\mu}^0$ of the induced dipole moment is related to $\boldsymbol{E}^0$ by

$$\boldsymbol{\mu}^0 = \alpha' Q_k \boldsymbol{E}^0 \tag{5.175}$$

The intensity of the Raman-scattered radiation is proportional to the square of the transition moment or, as it is sometimes called, transition polarizability $\boldsymbol{R}^{v'v''}$, which is given by

$$\boldsymbol{R}^{v'v''} = \int \psi_v'^* \boldsymbol{\mu}^0 \psi_v'' \mathrm{d}\tau_v \tag{5.176}$$

However, equation (5.175) gives

$$\mu_x^0 = (\alpha'_{xx} E_x^0 + \alpha'_{xy} E_y^0 + \alpha'_{xz} E_z^0) Q_k \tag{5.177}$$

$$\mu_y^0 = (\alpha'_{yx} E_x^0 + \alpha'_{yy} E_y^0 + \alpha'_{yz} E_z^0) Q_k \tag{5.178}$$

$$\mu_z^0 = (\alpha'_{zx} E_x^0 + \alpha'_{zy} E_y^0 + \alpha'_{zz} E_z^0) Q_k \tag{5.179}$$

which, with equations (5.175) and (5.176), give

$$\begin{aligned} R_x^{v'v''} = \Big[& E_x^0 \int \psi_v'^* \alpha'_{xx} \psi_v'' \mathrm{d}\tau_v \\ & + E_y^0 \int \psi_v'^* \alpha'_{xy} \psi_v'' \mathrm{d}\tau_v \\ & + E_z^0 \int \psi_v'^* \alpha'_{xz} \psi_v'' \mathrm{d}\tau_v \Big] Q_k \end{aligned} \tag{5.180}$$

$$\begin{aligned} R_y^{v'v''} = \Big[& E_x^0 \int \psi_v'^* \alpha'_{yx} \psi_v'' \mathrm{d}\tau_v \\ & + E_y^0 \int \psi_v'^* \alpha'_{yy} \psi_v'' \mathrm{d}\tau_v \\ & + E_z^0 \int \psi_v'^* \alpha'_{yz} \psi_v'' \mathrm{d}\tau_v \Big] Q_k \end{aligned} \tag{5.181}$$

$$\begin{aligned} R_z^{v'v''} = \Big[& E_x^0 \int \psi_v'^* \alpha'_{zx} \psi_v'' \mathrm{d}\tau_v \\ & + E_y^0 \int \psi_v'^* \alpha'_{zy} \psi_v'' \mathrm{d}\tau_v \\ & + E_z^0 \int \psi_v'^* \alpha'_{zz} \psi_v'' \mathrm{d}\tau_v \Big] Q_k \end{aligned} \tag{5.182}$$

For the vibrational transition to be allowed the transition moment $\boldsymbol{R}^{v'v''}$ of equation (5.176) must be non-zero. This is so if *any* of the nine integrals of equations (5.180)–(5.182) is non-zero and, of these, only six are different, since $\alpha'_{yx} = \alpha'_{xy}$, etc.

For any of the integrals to be non-zero the symmetry requirement, for transitions between non-degenerate vibrational states, is that the product of the symmetry species of the quantities inside the integral must be totally symmetric. Since $\Gamma[(\mathrm{d}\alpha_{ij}/\mathrm{d}Q_k)Q_k] = \Gamma(\alpha_{ij})$, this can be written as

$$\Gamma(\psi_v') \times \Gamma(\alpha_{ij}) \times \Gamma(\psi_v'') = A \tag{5.183}$$

where i and j can be x, y or z and A is a totally symmetric species.

For transitions between two vibrational states, at least one of which is degenerate, the requirement becomes

$$\Gamma(\psi_v') \times \Gamma(\alpha_{ij}) \times \Gamma(\psi_v'') \supset A \tag{5.184}$$

In the most usual cases where the lower vibrational level is the zero-point level $\Gamma(\psi_v'') = A$ and therefore

$$\Gamma(\psi_v') = \Gamma(\alpha_{ij}) \tag{5.185}$$

for a vibration, degenerate or non-degenerate, to be Raman active.

We shall not derive here the symmetry species of the components of the polarizability tensor: derivations can be found in some of the books listed in the Bibliography at the end of the chapter. Whenever they are required they are to be found in the relevant character table, in Tables 5.7, 5.8 and 5.11 or in the Appendix.†

In degenerate point groups, such as C_{3v}, symmetry species may not be assigned to all individual components of α but only to certain linear combinations of some of them.

Inspection of the C_{2v} and C_{3v} character tables (Tables 5.7 and 5.8) shows that 1–0 transitions of all the vibrations of H_2O (Figure 5.12) and of NH_3 (Figure 5.31) are allowed in the Raman spectrum.

The $D_{\infty h}$ character table (Table A.37) shows that, of the 1–0 transitions of acetylene (Figure 5.35), only the 1_0^1, 2_0^1 and 4_0^1 are allowed in the Raman spectrum since

$$\Sigma_g^+ = \Gamma(\alpha_{xx} + \alpha_{yy}) \text{ and } \Gamma(\alpha_{zz}) \tag{5.186}$$

and

$$\Pi_g = \Gamma(\alpha_{xz}) \text{ and } \Gamma(\alpha_{yz}) \tag{5.187}$$

The 3_0^1 and 5_0^1 transitions are forbidden.

Figure 5.35 provides an example of the rule of mutual exclusion. The rule states that, for a molecule with a centre of inversion, the fundamental vibrations which are active in the Raman spectrum are inactive in the infrared spectrum and vice versa, i.e. the two spectra are mutually exclusive. However, the converse is not necessarily true: there are some vibrations which are inactive in *both* the Raman and infrared spectra, such as the a_u torsional fundamental vibration ν_4 of ethylene, in Figure 5.39. The reason for this is demonstrated in the D_{2h} character table (Table A.32). The A_u species is that of neither a translation nor a component of the polarizability.

†In character tables in some other texts the symmetry species of the functions x^2, y^2, z^2, xy, yz and xz are given but these transform like α_{xx}, α_{yy}, α_{zz}, α_{xy}, α_{yz} and α_{xz}, respectively.

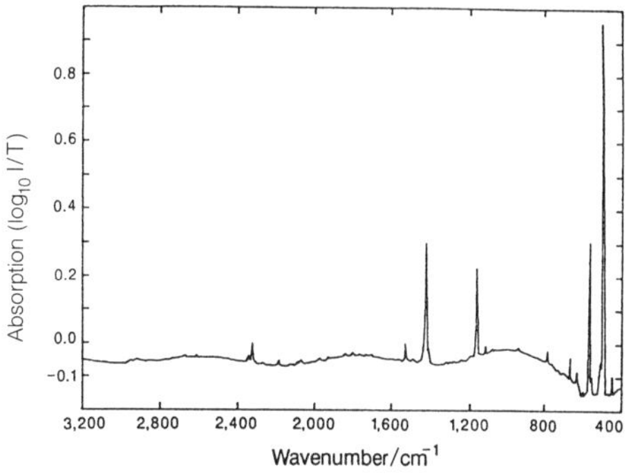

Figure 5.40 The vibrational infrared spectrum of a 2 μm thick coating of solid C_{60} on a silicon substrate. [Reproduced, with permission, from Kratschmer, W., Lamb, L. D., Fostiropoulos, K. and Huffman, D. R. (1990). *Nature (London)*, **347**, 354]

An example of a molecule with a centre of inversion and which, therefore, should obey the rule of mutual exclusion is C_{60}, buckminsterfullerene. By a brilliant process of deduction, based largely on the evidence of its extreme stability, it was proposed[49] that this molecule has the truncated icosahedral cage structure shown in Figure 5.28(d) and discussed in Section 5.2.2.2(xi). Various experimental techniques were used subsequently to prove this structure correct. One of these was ^{13}C NMR spectroscopy. The spectrum shows[50] a single line consistent with all the carbon atoms being in structurally indistinguishable positions at each of the 60 identical vertices each of which connects two hexagons and one pentagon.

Further important experimental evidence for this structure came from observations of the infrared[51] and the Raman[52] vibrational spectra. The truncated icosahedral structure has a centre of inversion which should result in the infrared and Raman spectra being mutually exclusive.

For this structure, belonging to the I_h point group, the 174 vibrations belong to the various symmetry species as follows:

$$2a_g + 3t_{1g} + 4t_{2g} + 6g_g + 8h_g + a_u + 4t_{1u} + 5t_{2u} + 6g_u + 7h_u$$

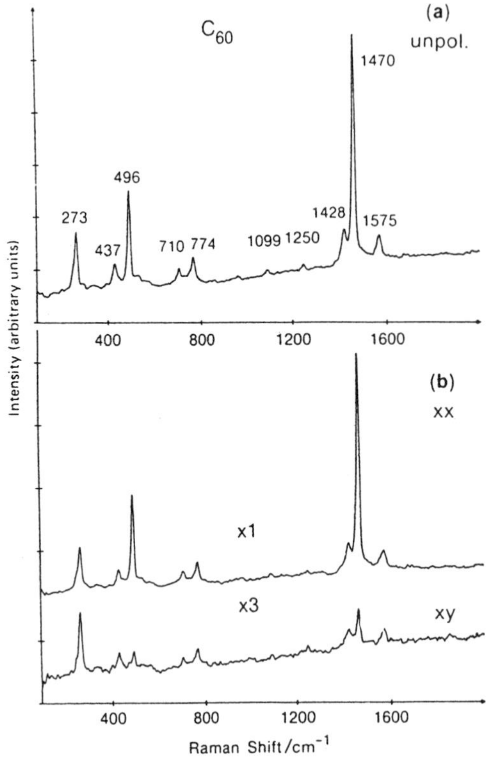

Figure 5.41 Vibrational Raman spectra of a solid film of C_{60} on a Suprasil substrate with (a) unpolarized and (b) polarized radiation. [Reproduced, with permission, from Bethune, D. S., Meijer, G., Tang, W. C., Rosen, H. J., Golden, W. G., Seki, H., Brown, C. A. and de Vries, M. S. (1991). *Chem. Phys. Lett.*, **179**, 181]

(Note that t, g, and h vibrations are three-, four- and fivefold degenerate, respectively.) The I_h character table, given in Table A.46 in the Appendix, indicates that only the t_{1u} vibrations are infrared allowed and only the a_g and h_g vibrations are Raman allowed. Therefore, if the molecule belongs to this point group, we should see a maximum of four vibrations in the infrared spectrum and 10 vibrations in the Raman spectrum.

Figure 5.40 shows the infrared spectrum and Figure 5.41 the Raman spectrum. The infrared spectrum shows the expected four bands at 528, 577, 1183 and 1429 cm^{-1} due to the t_{1u} vibrations. The weaker bands in the spectrum may be due to other species, such as C_{70}. The Raman spectrum, obtained using an argon ion laser as a monochromatic source, shows the expected 10 bands at 273, 437, 496, 710, 774, 1099, 1250, 1428, 1470 and 1575 cm^{-1}. Typical of a vibrational Raman spectrum the totally symmetric (a_g) vibrations, at 496 and 1470 cm^{-1}, are stronger than the other eight non-totally symmetric (h_g) vibrations. The spectra in Figure 5.41(b) show that the depolarization ratio (see Section 5.2.1) confirms the a_g or h_g assignments.

The infrared and Raman spectra of C_{60} are mutually exclusive with the apparent exception of bands at 1429 cm^{-1} (infrared) and 1428 cm^{-1} (Raman). In fact, these t_{1u} and h_g vibrations correspond to different normal modes which are accidentally almost degenerate.

5.2.5 Vibration–rotation spectroscopy

As in diatomic molecules, there are stacks of rotational energy levels associated with all vibrational levels of polyatomic molecules. The resulting term values S are given by the sum of the rotational and vibrational term values:

$$S = F_{v_i} + G(v_i) \qquad (5.188)$$

where i refers to a particular vibration. When each vibration is treated in the harmonic oscillator approximation (anharmonicity will not concern us until Section 5.2.6) the vibrational term values are given by

$$G(v_i) = \omega_i(v_i + d_i/2) \qquad (5.189)$$

We encountered this previously as equation (5.46): ω_i is the classical vibration wavenumber, v_i the quantum number and d_i the degree of degeneracy of the vibration.

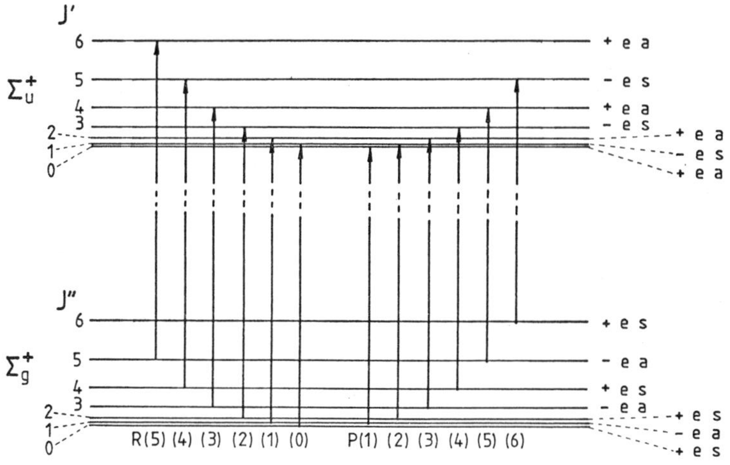

Figure 5.42 Rotational transitions accompanying a Σ_u^+–Σ_g^+ vibrational transition in a linear molecule. The $\Delta J = \pm 1$ selection rule results in a P and an R branch. If the molecule has no centre of inversion the g and u subscripts and the s and a labels should be dropped

5.2.5.1 *Linear molecules*

Linear molecules belong to either the $D_{\infty h}$ or $C_{\infty v}$ point groups, depending on whether they do or do not have a centre of inversion.

It follows from the discussion of electric dipole selection rules in Section 5.2.4.1, and in particular from equation (5.165), that vibrational transitions in the infrared spectrum are allowed if

$$\Gamma(\psi_v') \times \Gamma(\psi_v'') \supset \Sigma_u^+ \text{ or } \Pi_u \tag{5.190}$$

in the $D_{\infty h}$ point group, or

$$\Gamma(\psi_v') \times \Gamma(\psi_v'') \supset \Sigma^+ \text{ or } \Pi \tag{5.191}$$

in the $C_{\infty v}$ point group. If the lower level is the zero-point level then $\Gamma(\psi_v'')$ is Σ_g^+ in $D_{\infty h}$ and Σ^+ in $C_{\infty v}$ and the allowed transitions are

$$\Sigma_u^+–\Sigma_g^+, \quad \Pi_u–\Sigma_g^+$$

in $D_{\infty h}$ and

$$\Sigma^+–\Sigma^+, \ \Pi–\Sigma^+$$

in $C_{\infty v}$.

It follows from equation (5.184) that a vibrational transition in the Raman spectrum is allowed if

$$\Gamma(\psi_v') \times \Gamma(\psi_v'') \supset \Gamma(\alpha_{ij}) \tag{5.192}$$

If the lower state is the zero-point level the allowed transitions are

$$\Sigma_g^+–\Sigma_g^+, \ \Pi_g–\Sigma_g^+, \ \Delta_g–\Sigma_g^+$$

in $D_{\infty h}$ and

$$\Sigma^+–\Sigma^+, \ \Pi–\Sigma^+, \ \Delta–\Sigma^+$$

in $C_{\infty v}$.

In Section 4.2.6.2 rotation of linear molecules in excited vibrational states has been considered in some detail; this was in the context of pure rotational transitions in excited vibrational states. It may be necessary for the reader at this stage to refer back to that section.

The rotational term values for a particular vibrational state are given by

$$F_v(J) = B_v J(J+1) - D_v J^2(J+1)^2 \tag{5.193}$$

Here we are neglecting small additional terms in higher powers of $J(J+1)$ and assuming the vibration to be non-degenerate.

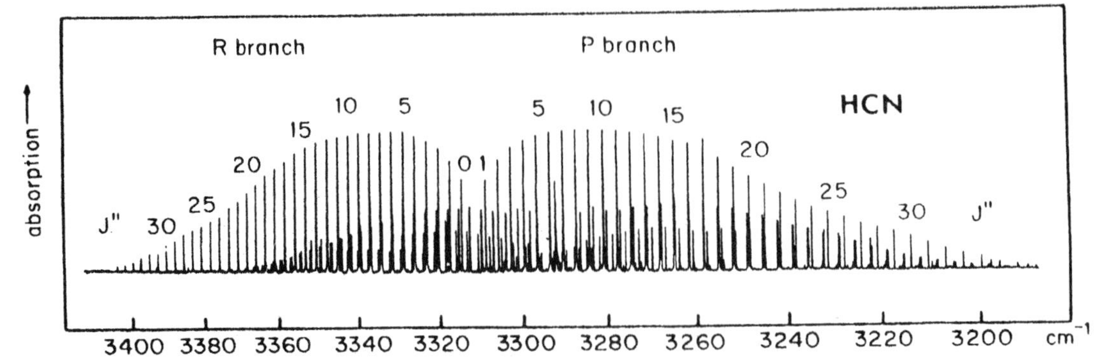

Figure 5.43 The 3_0^1, Σ^+–Σ^+ infrared band of the HCN molecule showing a P and an R branch. Overlapping this is the weaker $2_{1,1}^{1,1}3_0^1$, Π–Π hot band, the Q branch being at about 3292 cm^{-1}, and also the $2_2^2 3_0^1$ band with its origin at about 3290 cm^{-1}. [From Cole, A. R. H. (1977). *Tables of Wavenumbers for the Calibration of Infrared Spectrometers*, 2nd edn, p. 28. Pergamon Press, Oxford. Reproduced by permission]

5.2.5.1(i) Infrared spectra

The selection rules for transitions between the rotational levels in vibrational states, between which a transition is allowed by electric dipole (infrared) selection rules, are

$$\Delta J = \pm 1 \quad \text{for a } \Sigma\text{–}\Sigma \text{ transition}$$
$$\Delta J = 0, \pm 1 \quad \text{for a } \Pi\text{–}\Sigma \text{ transition}$$

A Σ–Σ transition gives rise to what is often called a parallel band because the transition moment is parallel to the internuclear axis and a Π–Σ transition gives rise to a perpendicular band because the transition moment is perpendicular to the same axis. However, great care must be exercised when trying to do the converse, namely associating the selection rule $\Delta J = \pm 1$ with a parallel band and $\Delta J = 0, \pm 1$ with a perpendicular band because the association is not always valid. We shall see, for example, that a Π–Π transition is a parallel band because of the direction of the transition moment but the rotational selection rule is $\Delta J = 0, \pm 1$.

The rovibrational transitions for a Σ–Σ transition are shown in Figure 5.42. The example chosen is Σ_u^+–Σ_g^+ in a $D_{\infty h}$ molecule for which the label s or a, indicating that ψ_r is symmetric or antisymmetric to nuclear exchange, apply. For a Σ^+–Σ^+ transition in a $C_{\infty v}$ molecule the figure would be the same except for the absence of the s and a labels. This figure is very similar to Figure 5.8(a) for a diatomic molecule in which the only vibration is necessarily Σ^+ (or Σ_g^+) and therefore *all* vibrational transitions, in a closed shell diatomic, are of the Σ–Σ type.

In Figure 5.42 the parity labels, + or −, and the alternative e or f labels (see p. 109) are attached and it is clear that the transitions obey the selection rules†

$$+ \longleftrightarrow -, \quad + \not\longleftrightarrow +, \quad - \not\longleftrightarrow -$$

or, alternatively,

$$e \longleftrightarrow f,\ e \not\longleftrightarrow e,\ f \not\longleftrightarrow f \quad \text{for } \Delta J = 0$$
$$e \not\longleftrightarrow f,\ e \longleftrightarrow e,\ f \longleftrightarrow f \quad \text{for } \Delta J = \pm 1$$

However, these selection rules are superfluous until we consider π vibrations, since the $\Delta J = \pm 1$ transitions automatically obey the parity selection rules.

As an example of a Σ–Σ transition we consider the 3_0^1 transition of HCN, where ν_3 is the C—H stretching vibration, illustrated in Figure 4.16. The spectrum is shown in Figure

† $\longleftrightarrow$ and $\not\longleftrightarrow$ indicate allowed and forbidden transitions, respectively, whichever is the upper state.

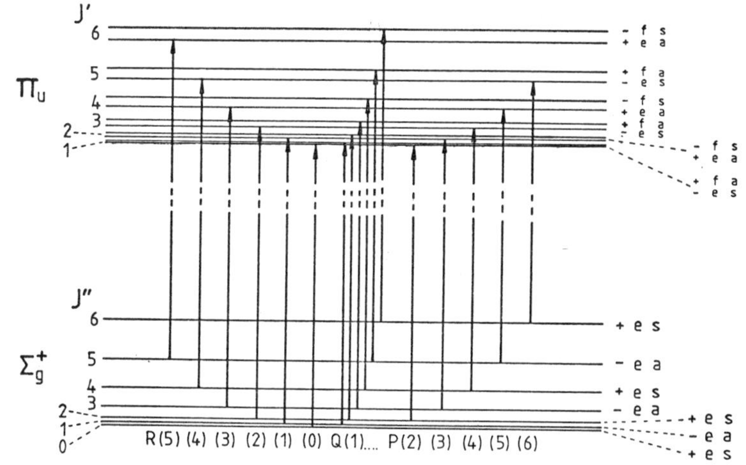

Figure 5.44 Rotational transitions accompanying a Π_u–Σ_g^+ vibrational transition in a linear molecule. The $\Delta J = 0, \pm 1$ selection rule results in a P, a Q and an R branch. If the molecule has no centre of inversion the g and u subscripts and the s and a labels should be dropped

5.43. Apart from some weak overlapping hot bands (see below), the band strongly resembles that of a diatomic such as HCl (Figure 5.9). The approximate spacings within the R and P branches are $2B$, as equations (5.29)–(5.30) predict for a diatomic, with a zero gap between $R(0)$ and $P(1)$ of about $4B$.

The slight convergence to high J in the R branch and divergence in the P branch are due to the fact that B' is slightly smaller than B''. The method of combination differences, plotting $\Delta_2''F(J)$ versus $(J+\frac{1}{2})$ or $\Delta_2'F(J)$ versus $(J+\frac{1}{2})$ as described in Section 5.1.3.1, can be used to obtain B'' or B', respectively [see equations (5.31)–(5.32)]. At a higher level of accuracy, including centrifugal distortion, $\Delta_2''F(J)/(J+\frac{1}{2})$ or $\Delta_2'F(J)/(J+\frac{1}{2})$ versus $(J+\frac{1}{2})^2$ can be plotted to give B'', D'', B' and D' [equations (5.34)–(5.35)].

The intensity distribution among the rovibrational transitions in absorption is given by equation (5.38).

An example of a Π–Σ transition is the Π_u–Σ_g^+, $1_0^1 5_0^{1,1}$ transition of acetylene. The upper level here is a vibrational combination level involving one quantum each of the symmetric CH stretching and the *cis* bending vibrations illustrated in Figure 5.35.

The stack of rotational energy levels in the Π_u state is like that in Figure 4.18 and shows a first-order Coriolis interaction giving rise to l-doubling. The reader is referred to Section 4.2.6.2 for a discussion of this.

There is a rule proposed by Jahn which says that Coriolis interaction between two (or more) vibrational states may occur if the product of the symmetry species of the vibrational states, with wave functions ψ_{v_i} and ψ_{v_j}, contains that of a rotation; that is,

$$\Gamma(\psi_{v_i}) \times \Gamma(\psi_{v_j}) \supset \Gamma(R_x) \text{ and/or } \Gamma(R_y) \text{ and/or } \Gamma(R_z) \tag{5.194}$$

If the two interacting states are the two components of a doubly degenerate state (or the three components of a triply degenerate state) then the interaction is large and first order. In the case of the two components of a Π vibrational state in a linear molecule

$$\Pi \times \Pi = \Sigma^+ + \Sigma^- + \Delta \supset \Sigma^- = \Gamma(R_z) \tag{5.195}$$

leading to a first-order Coriolis interaction. The term values are modified to

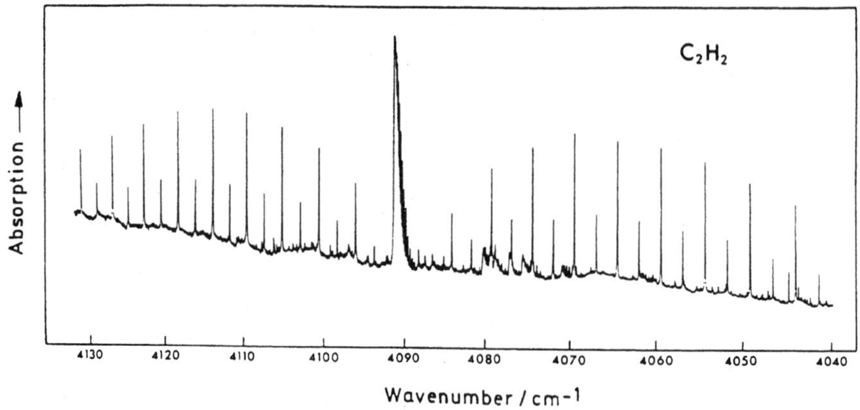

Figure 5.45 The $1_0^1 5_0^{1,1}$, Π_u–Σ_g^+ infrared band of the acetylene molecule showing a P, a Q and an R branch and an intensity alternation of 1:3 for J'' even:odd. [From Cole, A. R. H. (1977). *Tables of Wavenumbers for the Calibration of Infrared Spectrometers*, 2nd edn, p. 12. Pergamon Press, Oxford. Reproduced by permission]

$$
\begin{aligned}
F_v(J, l^{\pm}) = B_v[J(J+1) - l^2] \\
\pm (q_i/4)(v_i + 1)J(J+1) \quad (5.196) \\
- D_v[J(J+1) - l^2]^2
\end{aligned}
$$

Figure 5.44 shows the stacks of rotational levels associated with Π_u and Σ_g^+ states and the allowed transitions obeying the $\Delta J = 0, \pm 1$ and parity (or e–f) selection rules already given. Figure 5.45 shows the $1_0^1 5_0^{1,1}$ absorption band of acetylene, which obeys these selection rules. In this spectrum the Q-branch is only partially resolved.

The clear intensity alternation of 1:3 for J'' even:odd is due to the exchange of the two hydrogen atoms on rotation of the molecule by π rad in exactly the same way as occurs in $^1H_2$ and illustrated in Figure 4.22.

If either the l-doubling constant q_i is small enough to be neglected, or the resolution of the spectrum is not sufficiently high to merit its inclusion, the values of B_v and D_v can be obtained for both states in a Π–Σ^+ transition from the R and P branches alone; the method of combination differences is used exactly as in a Σ–Σ band. On the other hand, if the Q branch is resolved q_i can be obtained. Since all the R- and P-branch transitions go to the lower components of l-doubled levels, the combination difference method applied to these transitions

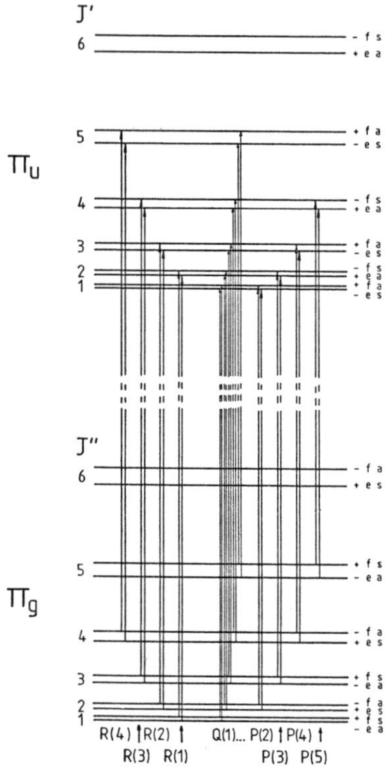

Figure 5.46 Rotational transitions accompanying a Π_u–Π_g vibrational transition in a linear molecule. The $\Delta J = 0, \pm 1$ selection rule results in P, Q and R branches. If the molecule has no centre of inversion the g and u subscripts and the s and a labels should be dropped

gives an effective B value for the upper components. The Q-branch transitions, going only to the upper components, give another effective B value and the two effective B values can be used to obtain q_i.

If the molecule whose spectrum is being observed has one or more vibrations of sufficiently low wavenumber that a $v = 1$ level is appreciably populated at the temperature of the experiment, then hot bands having this as their lower state may be observed. Hot bands are so-called because their intensities increase considerably with increasing temperature. On the other hand, cold bands have the zero-point level as their lower state and their intensities are more or less temperature independent, although their intensities decrease slightly with increasing temperature due to a small depopulation of the zero-point level.

Overlapping the 3_0^1 band of HCN in Figure 5.43 is the $2_{1,1}^{1,1}3_0^1$ hot band, which is fairly intense since $\tilde{\nu}_2$ is only $712\,\mathrm{cm}^{-1}$. The transition is of the Π–Π type. Since

$$\Gamma(\psi_v') \times \Gamma(\psi_v'') = \Pi \times \Pi = \Sigma^+ + \Sigma^- + \Delta \supset \Sigma^+ \tag{5.197}$$

and Σ^+ is $\Gamma(T_z)$, the transition is allowed with the transition moment polarized along the z-axis giving a parallel band. The rotational selection rule is $\Delta J = 0, \pm 1$, producing a P, a Q and an R branch but, as Figure 5.43 shows, the Q branch is weaker, relative to the P and R branches, than in a Π–Σ transition. Figure 5.46 shows the allowed rotational transitions for a Π_u–Π_g transition of a $D_{\infty h}$ molecule or, if the g and u subscripts are dropped, a Π–Π transition of a $C_{\infty v}$ molecule. The figure shows that, in comparison with a Π–Σ transition, the $R(0)$ line is missing and l-type doubling in both states can produce doubling of *all* P-, Q- and R-branch lines. The line doubling increases with J and is greater in the Q branch, where it is the *sum* of the l-doubling in both states, than in the P and R branches, where it is the *difference*. The doubling in P and R branches can be observed in the wings of the $2_{1,1}^{1,1}3_0^1$ band in Figure 5.43.

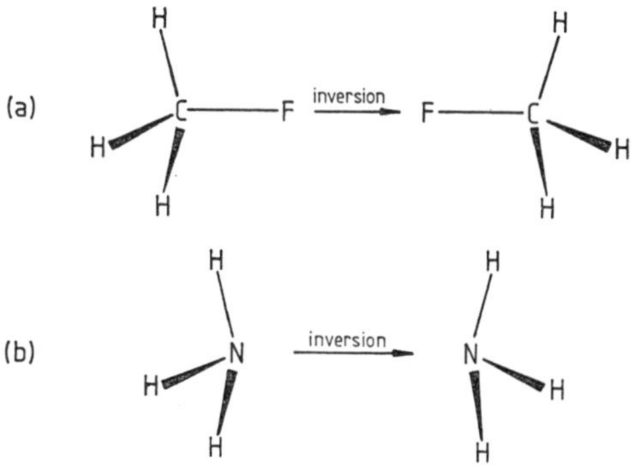

Figure 5.47 (a) A non-feasible inversion in the CH_3F molecule and (b) a feasible inversion in the NH_3 molecule

Figure 5.46 shows for a $D_{\infty h}$ molecule the s and a properties, indicating symmetry or antisymmetry with respect to nuclear exchange, of the rotational levels. The figure makes it clear that, if l-doubling is not resolved in the spectrum, pairs of s and a levels for each value of J are effectively combined and no intensity alternation is observed.

In Figure 5.43 there can be seen a third, weaker, band overlapping the other two; this is the hot band $2_2^2 3_0^1$ which has two components, $2_{2,0}^{2,0}3_0^1$ and $2_{2,2}^{2,2}3_0^1$. The separation of all three bands is due to anharmonicity which will be discussed in Section 5.2.6.

5.2.5.1(ii) Raman spectra

In vibration–rotation Raman bands the rotational selection rules[†] are

(a) $\Delta J = 0, \pm 2$ either for a Σ–Σ transition or, in general, when $l = 0$ in both states;

(b) $\Delta J = 0, \pm 1, \pm 2$ either for a Π–Σ transition or, in general, when $l \neq 0$ in either state.

5.2.5.2 Symmetric rotor molecules

5.2.5.2(i) Infrared spectra

If a vibrational transition from the zero-point level is to be electric dipole (infrared) allowed

[†]For general derivations of Raman selection rules, see Placzek, G. and Teller, E. (1933). *Z. Phys.*, **81**, 209.

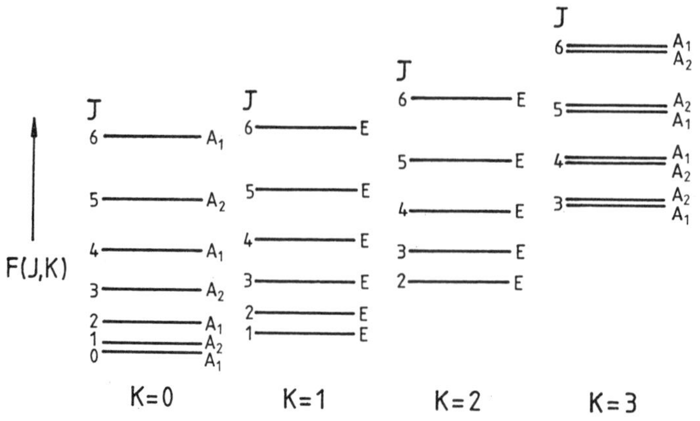

Figure 5.48 Rotational term values associated with an A_1 vibrational state of a prolate symmetric rotor

there must be a change of dipole moment during the transition which is either along the top axis, the z-axis, or perpendicular to it, in the xy-plane. For a change along the top axis

$$\Gamma(\psi'_v) = \Gamma(T_z) \tag{5.198}$$

which is a non-degenerate symmetry species. For a change perpendicular to the top axis

$$\Gamma(\psi'_v) = \Gamma(T_x, T_y) \tag{5.199}$$

which is a doubly degenerate E species. For example, $\Gamma(T_z)$ and $\Gamma(T_x, T_y)$ are A_1 and E, respectively, in the C_{3v} point group, A''_2 and E' in D_{3h} and B_2 and E in D_{2d}.

Hot bands are electric dipole allowed provided the general selection rule

$$\Gamma(\psi'_v)\times\Gamma(\psi''_v) \supset \Gamma(T_x) \text{ and/or } \Gamma(T_y) \text{ and/or } \Gamma(T_z) \tag{5.200}$$

is obeyed.

Rotational energy levels of symmetric rotors in both the zero-point and excited vibrational states have been discussed in Section 4.3.

For the purposes of the discussion here we shall consider in detail vibrational transitions only in symmetric rotors for which there is a very high energy barrier to inversion through the centre of mass. Examples of such molecules are the methyl halides, CH_3X, and the haloforms, CHX_3, all of which belong to the C_{3v} point group. Figure 5.47(a) illustrates the process of inversion in CH_3F in which it is a so-called *non-feasible* operation because of the extremely high energy required. Figure 5.47(b) shows the corresponding inversion in NH_3 which requires relatively little energy and is a *feasible* operation: we shall discuss the case of NH_3 inversion in more detail in Section 5.2.7.1.

For a prolate symmetric rotor which is either in the zero-point level or in any non-degenerate vibrational state, the rotational term values are given by

$$F_v(J, K) = B_vJ(J+1) + (A_v - B_v)K^2 - D_JJ^2(J+1)^2 - D_{JK}J(J+1)K^2 - D_KK^4 \tag{5.201}$$

and for an oblate symmetric rotor by

$$F_v(J, K) = B_vJ(J+1) + (C_v - B_v)K^2 - D_JJ^2(J+1)^2 - D_{JK}J(J+1)K^2 - D_KK^4 \tag{5.202}$$

as in equations (4.69) and (4.70). As in those equations, the v dependence of the centrifugal distortion constants has not been indicated.

From this point, detailed discussion will be confined mainly to prolate symmetric rotors and particularly those belonging to the C_{3v}

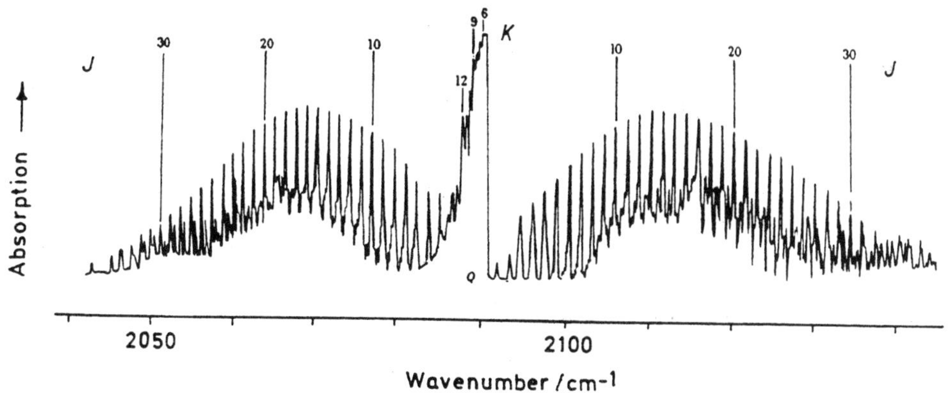

Figure 5.49 The 1_0^1, A_1–A_1 vibrational infrared parallel band of the C^2H_3F molecule [Reproduced, with permission, from Jones, E. W., Popplewell, R. J. and Thompson, H. W. (1966). *Proc. Roy. Soc.*, **A290**, 490]

point group. In the infrared spectrum, transitions of the type A_1–A_1 and E–A_1, involving A_1 or E fundamentals, will be considered. We shall follow common usage and refer to an A_1–A_1 band as a 'parallel' band and E–A_1 as a 'perpendicular' band, although we must be aware of the dangers, mentioned in Section 5.2.5.1(i), of associating a certain set of rotational selection rules with a particular polarization of the transition moment.

The rotational levels associated with an A_1 vibrational state are illustrated in Figure 5.48. This is similar to Figure 4.20(a) except for two additional features. The first is a slight splitting of all the levels with $K = 3n$, where $n = 1, 2, 3, \ldots$ This a very small effect due to high order Coriolis interaction which is not always large enough to be observed except in microwave spectroscopy. The second feature is the attachment to the levels of symmetry species of the rovibrational wave functions $\psi_v\psi_r$.

For an A_2–A_1 transition the rotational selection rules are

$$\text{for } K = 0,\ \Delta K = 0 \ \text{ and } \ \Delta J = \pm 1 \tag{5.203}$$

$$\text{for } K \neq 0,\ \Delta K = 0 \ \text{ and } \ \Delta J = 0, \pm 1 \tag{5.204}$$

for transitions between two sets of rotational levels such as those in Figure 5.48. If the small splittings due to Coriolis interaction are neglected these selection rules are sufficient.

The wavenumbers of the rovibrational transitions in a parallel band are given by

$$\tilde{\nu} = \omega_0 + F'(J, K) - F''(J, K) \tag{5.205}$$

where ω_0 is the wavenumber of the band origin (the pure vibrational transition) and the rotational term values are given by equation (5.201) and (5.202).

It is clear from the selection rules that there is a P, Q and R branch for every value of K except $K = 0$, for which the Q branch is missing. Because of the restriction that $J \geqslant K$ a number of lines are missing from each branch and the number increases with K. Each set of P, Q and R (or P and R) branches is said to comprise a sub-band. A sub-band origin is the position of the unobserved $J = 0$ to $J = 0$ transition of a particular sub-band.

The wavenumbers, $\tilde{\nu}_0^{\text{sub}}$, of the sub-band origins are given by

$$\tilde{\nu}_0^{\text{sub}} = \omega_0 + F'_v(0, K) - F''_v(0, K) = \omega_0 + [(A' - A'') - (B' - B'')]K^2 \tag{5.206}$$

which is obtained from equations (5.205) and 5.201), neglecting centrifugal distortion.

As in many vibrational transitions, there is commonly little change of rotational constants

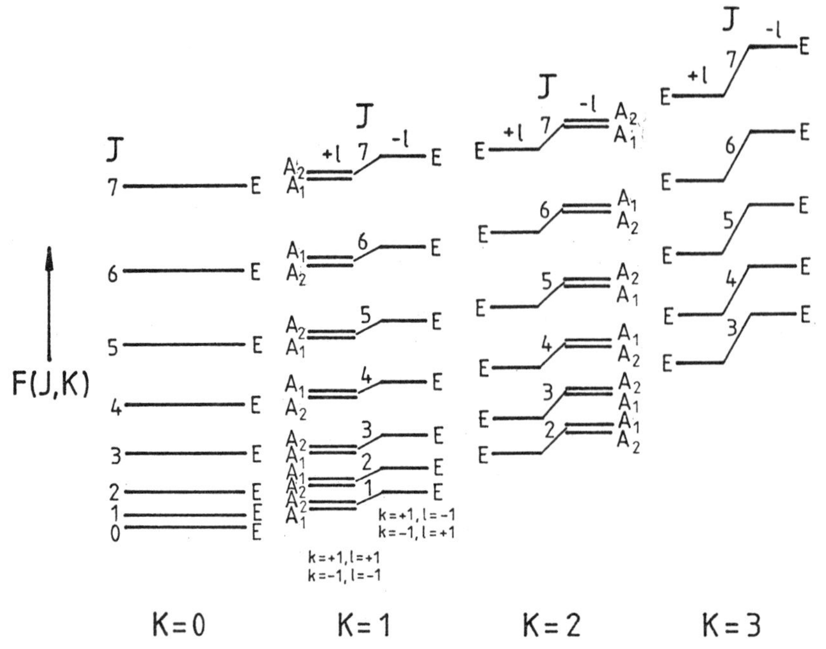

Figure 5.50 Rotational term values associated with an E vibrational state of a prolate symmetric rotor

on vibrational excitation so that, if $A' \approx A''$ and $B' \approx B''$, the sub-band origins, and therefore the P, Q and R branches of each sub-band, are almost coincident. The result is that the general appearance of the band is often similar to that of a Π–Σ band of a linear molecule. As an example Figure 5.49 shows the 1_0^1 band of C^2H_3F, where ν_1 is the a_1 C—^{2}H stretching vibration. The sub-band origins are not exactly coincident, as indicated by the slight separation of the Q branches with different values of K.

The intensities I_a of rovibrational transitions in a parallel, A_1–A_1, band in absorption are given by

$$I_a = C_a A_{KJ} \tilde{\nu} g_{KJ} \exp[-hcF(J,K)/kT] \quad (5.207)$$

where C_a is a constant, if we neglect induced emission, dependent on the strength of the pure vibrational transition of wavenumber $\tilde{\nu}$. The degeneracy factor g_{KJ} is given by

$$g_{KJ} = (2J+1) \text{ for } K=0$$
$$\text{and } 2(2J+1) \text{ for } K \neq 0 \quad (5.208)$$

The factor A_{KJ} is a function of both quantum numbers and has been determined by Hönl and London:

for $\Delta J = +1$,

$$A_{KJ} = [(J+1)^2 - K^2]/(J+1)(2J+1) \quad (5.209)$$

for $\Delta J = 0$,

$$A_{KJ} = K^2/J(J+1) \quad (5.210)$$

for $\Delta J = -1$,

$$A_{KJ} = [J^2 - K^2]/J(2J+1) \quad (5.211)$$

As discussed in Section 4.4, there is also a nuclear spin statistical weight factor which affects the intensities, these being proportional to the statistical weight factor for the initial state of a transition. The magnitude of this factor depends on the value of the nuclear spin quantum number I for the equivalent nuclei.

For $I = \frac{1}{2}$, as in C^1H_3F, the statistical weight factors for A_1 levels with $K = 0, 1, 2, 3, 4, 5, 6, \ldots$ are 4, 2, 2, 4, 2, 2, 4, ... In the case of C^2H_3F, where $I = 1$, the factors are 11, 8, 8, 11, 8, 8, 11, . . .

In a parallel band observed with only moderate resolution, as in Figure 5.49, the statistical weights may have no apparent effect because of the near superposition of P and R branches of all sub-bands, but the alternation may be observed in the Q branches if they are sufficiently resolved.

In a planar symmetric rotor there may also be an intensity alternation for even and odd J when $K = 0$; this was discussed in Section 4.4.

For a symmetric rotor in an E vibrational state there is a first-order Coriolis interaction which splits all rotational levels with $K > 0$. The effect is to modify the term values for a prolate symmetric rotor from those of equation (5.201) to

$$F_v(J, K) = B_vJ(J+1) + (A_v - B_v)K^2 - 2(A\zeta_i)_v kl - D_J J^2(J+1)^2 - D_{JK}J(J+1)K^2 - D_K K^4 \qquad (5.212)$$

The quantity ζ_i is the Coriolis coupling constant introduced in Section 4.2.6.2. The quantum number l, which takes the values $v_i, v_i - 2, v_i - 4, \ldots - v_i$, was also introduced in Section 4.2.6.2 and the signed quantum number k, equal to $\pm K$, in Section 4.3.

From this point we shall neglect the centrifugal distortion terms.

The rotational term values of equation (5.212) are illustrated in Figure 5.50. Since $v = 1$, l can take the values $+1$ and -1. Except for those with $K = 0$, all the levels are split into two components, the splitting being proportional to k (and, therefore, K) and also to ζ_i, which lies in the range $0 \leqslant |\zeta_i| \leqslant 1$. In the figure the labels $+l$ and $-l$ attached to the components indicate that k and l have the same or different signs, respectively. For example, if $K = 4$ and $l = \pm 1$, the levels with $k = 4, l = 1$ and $k = -4, l = -1$ are labelled $+1$, whereas those with $k = 4, l = -1$ and $k = -4, l = +1$ are labelled $-l$. From this example, it is clear that, even when this splitting has been taken into account, a double degeneracy of all levels with $K > 0$ still remains. This degeneracy appears in Figure 5.50, the wave function $\psi_v\psi_r$ for each rovibronic level having symmetry species E or $A_1 + A_2$.†

The degeneracy of the E levels cannot be removed but that of the A_1, A_2 pairs can. The splitting of these pairs is called l-type doubling and the effect is small except when $K = 1$. The effects of such doubling have been observed in high resolution vibration–rotation spectra and also in microwave spectra of symmetric rotors in degenerate vibrational states, but we shall subsequently disregard it.

For an E–A, perpendicular, band the rotational selection rules are

$$\Delta K = \pm 1, \qquad \Delta J = 0, \pm 1$$

and, in addition,

for $\Delta K = +1$, transitions involve only the $l = +1$ levels

for $\Delta K = -1$, transitions involve only the $l = -1$ levels

Taking account of the symmetry species of the rovibrational levels, the selection rules can be stated more succinctly by saying that electric dipole transitions are allowed between two rovibrational states with wave functions ψ''_{vr} and ψ'_{vr} provided that

$$\Gamma(\psi''_{vr}) \times \Gamma(\psi'_{vr}) \supset \Gamma(T_z) \times \Gamma(R_z) \qquad (5.213)$$

In the C_{3v} point group, $\Gamma(T_z) = A_1$ and $\Gamma(R_z) = A_2$ and the rovibrational selection rules are that only

$$A_1 \longleftrightarrow A_2 \text{ and } E \longleftrightarrow E$$

transitions are allowed. This not only is consistent with the selection rules expressed in

†For justification of the symmetry species, see Hougen, J. T. (1962). *J. Chem. Phys.*, **37**, 1433.

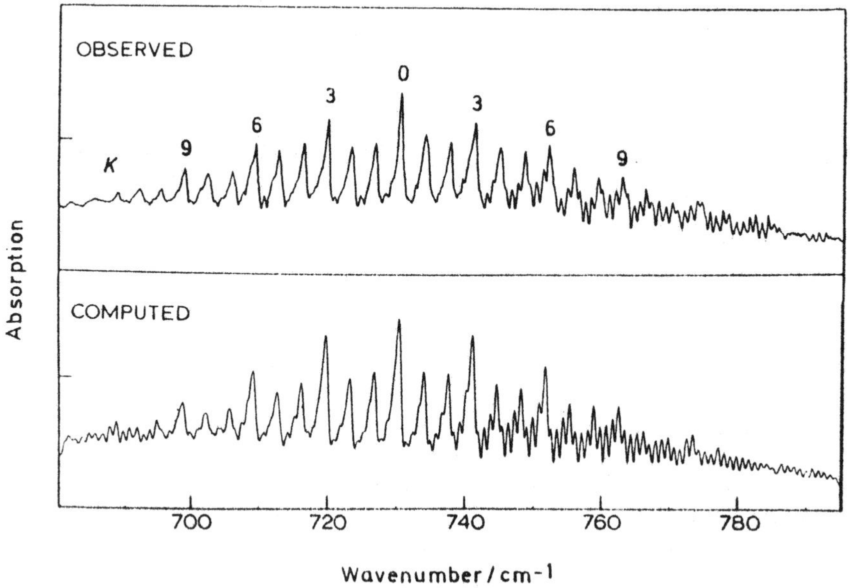

Figure 5.51 The 6_0^1 E–A_1 vibrational infrared perpendicular band of the SiH_3F molecule [Reproduced with permission, from Robiette, A. G., Cartwright, G. J., Hoy, A. R. and Mills, I. M. (1971). *Mol. Phys.*, **20**, 541]

terms of quantum numbers but also is more useful if l-doubling is resolved.

As a result of the selection rules there is, as in a parallel band, a set of sub-bands but which *all* have a P, a Q and an R branch. A more important difference from parallel bands is that, because of the $\Delta K = \pm 1$ selection rule, the sub-band origins are widely separated. Their wavenumbers are given by

$$\begin{aligned}\tilde{\nu}_0^{\text{sub}} &= \omega_0 + F'_v(0, K \pm 1) - F''_v(0, K) \\ &= \omega_0 + (A' - B') \pm 2(A' - B')K \\ &\quad + [(A' - A'') - (B' - B'')]K^2 \end{aligned} \tag{5.214}$$

if $\zeta_i = 0$, and by

$$\begin{aligned}\tilde{\nu}_0^{\text{sub}} &= \omega_0 + [A'(1 - 2\zeta_i) - B'] \\ &\quad \pm 2[A'(1 - \zeta_i) - B']K + [(A' - A'') \\ &\quad - (B' - B'')]K^2 \end{aligned} \tag{5.215}$$

if $\zeta_i \neq 0$, where the upper and lower sign refer to $\Delta K = +1$ and -1, respectively, in both equations.

If $B' \approx B''$ and $A' \approx A''$, the resulting band is dominated by Q branches with unresolved, or partially resolved, J structure. The separation of adjacent Q branches with K differing by 1 is due to the terms linear in K in equations (5.214) and (5.215) and is $2(A' - B')$ if $\zeta_i = 0$ and $2[(A'(1 - \zeta_i) - B']$ if $\zeta \neq 0$; this contrasts with a parallel band for which equation (5.206) shows that there is no term linear in K and the Q branches are almost coincident.

There is a useful addition to the symbolism used so far in which the value of ΔK which, in general, can be $\ldots +2, +1, 0, -1, -2 \ldots$ is indicated by $\ldots$ $o, p, q, r, s,$ $\ldots$ as a pre-superscript to the symbol implying the value of ΔJ. For example, a ^qR branch is one in which $\Delta K = 0$ and $\Delta J = +1$.

In a prolate symmetric rotor $A' > B'$ so, when $\zeta_i = 0$, the ^rQ branches are to high wavenumber of the band origin and ^pQ branches to low wavenumber. In an oblate symmetric rotor we have to replace A by C in all the equations and, since $C' < B'$, the ^rQ and ^pQ branches are to low and high wavenumber, respectively, of the origin. In a prolate symmetric rotor the separation can be very small if ζ_i is positive. In an extreme case, where $A'(1-\zeta_i) < B'$, the positions of the ^rQ and ^pQ branches relative to the band origin are reversed. Similarly, they are reversed in an oblate symmetric rotor if ζ_i is negative and $C'(1-\zeta_i) > B'$.

Figure 5.51 shows the 6^1_0 E–A_1 band of the prolate symmetric rotor silyl fluoride, SiH_3F, where $\tilde{\nu}_6$ is the degenerate rocking vibration of the SiH_3 group. Since $\zeta_6 = +0.2$, the ^rQ branches are to high wavenumber of ω_0, which is at 729 cm^{-1}. Most of the resolved structure consists of Q branches but there is some resolved J structure at the high wavenumber end of the band.

For the three equivalent protons with $I = \frac{1}{2}$ in SiH_3F, the ground-state statistical weight factors for levels with $K = 0$, 1, 2, 3, 4, 5, 6, . . . are 4, 2, 2, 4, 2, 2, 4, ... The effect on the Q-branch intensities is clear in Figure 5.51.

Also important, so far as the intensity distribution in a perpendicular band is concerned, is the A_{KJ} factor of equation (5.207). For a perpendicular band this factor is given by:

for $\Delta J = +1$,

$$A_{KJ} = (J+2 \pm K)(J+1 \pm K)/(J+1)(2J+1)$$

for $\Delta J = 0$,

$$A_{KJ} = (J+1 \pm K)(J \mp K)/J(J+1)$$

for $\Delta J = -1$,

$$A_{KJ} = (J-1 \mp K)(J \mp K)/J(2J+1) \qquad (5.216)$$

where the upper and lower signs refer to $\Delta J = +1$ and -1, respectively.

It is clear from equation (5.215) that not all the parameters that we require, in particular ζ_i, can be obtained from a Q-branch analysis of a single band. However, at a level of approximation appropriate to medium resolution spectra, such as that in Figure 5.51, there is a useful method of obtaining ζ_i using a ζ sum rule. For C_{3v} molecules of the type XYZ_3 the sum rule is

$$\zeta_4 + \zeta_5 + \zeta_6 = B/2A \qquad (5.217)$$

where ν_4, ν_5 and ν_6 are the e vibrations involving stretching, deformation and rocking, respectively, of the YZ_3 group. Strictly this equation is valid only for equilibrium values of B and A. If the vibrational dependence of A and B is neglected and the separation $\Delta\tilde{\nu}_i$ of Q branches in all of the 4^1_0, 5^1_0 and 6^1_0 bands can be measured, the sum of the average separations is given by

$$\begin{aligned}\Delta\tilde{\nu}_4 + \Delta\tilde{\nu}_5 + \Delta\tilde{\nu}_6 = {} & 2[A(1-\zeta_4) - B] \\ & + 2[A(1-\zeta_5) - B] \\ & + 2[A(1-\zeta_6) - B] \qquad (5.218)\end{aligned}$$

Together with equation (5.217), this gives

$$\Delta\tilde{\nu}_4 + \Delta\tilde{\nu}_5 + \Delta\tilde{\nu}_6 = 6A - 7B \qquad (5.219)$$

If B can be obtained from the P and R branches of a parallel band, A can be obtained from equation (5.219) and hence the various ζ_i from the Q-branch separations in perpendicular bands.

With the information available from present-day high resolution infrared, microwave and Raman spectra, the method outlined above for obtaining ζs is now used only for very approximate values.

In considering Coriolis interactions in linear molecules, we have seen in equation (4.61) that, in a linear triatomic molecule, there is a contribution to the parameter q_2 due to Coriolis interactions between the bending vibration ν_2 and the stretching vibrations ν_1 and ν_3. These interactions are reflected in the values of the Coriolis coupling constants ζ_{21} and ζ_{23}. The effect on the spectrum is manifest through the value of q_2 and equation (4.61) shows that this is usually small owing to ω_2 being different from ω_1 and ω_3.

In symmetric rotor molecules, also, this kind of Coriolis interaction can take place and is more likely to be greater in a molecule with a larger number of vibrations because of the higher probability of two interacting vibrations being of a similar wavenumber. However, not all vibrations can undergo Coriolis interaction;

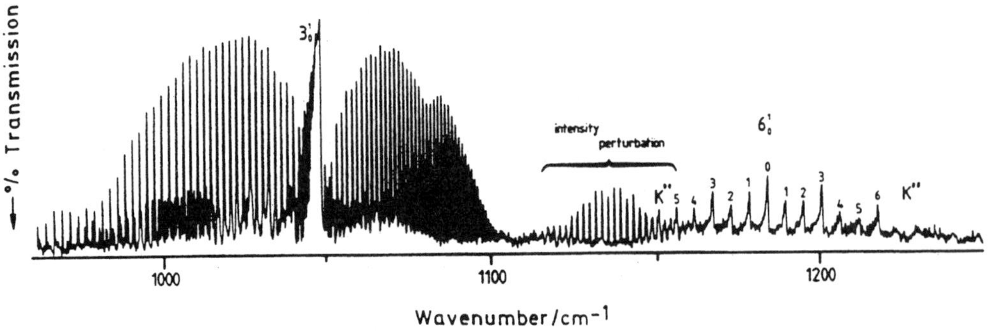

Figure 5.52 Effects of Coriolis interaction between the vibrations $\nu_3(a_1)$ and $\nu_6(e)$ of CH_3F show in the intense 3^1_0 and the weaker 6^1_0 bands in the infrared spectrum. The most obvious effect is an intensity perturbation in the low wavenumber wing of the 6^1_0 band

the general requirement is given by Jahn's rule, stated in equation (5.194).

An example of Coriolis interaction between two vibrations in a symmetric rotor is that in methyl fluoride (CH_3F) between ν_3, the a_1 C—F stretching vibration, and ν_6, the e CH_3 rocking vibration. Since

$$\Gamma(\psi_{\nu_3}) \times \Gamma(\psi_{\nu_6}) = A_1 \times E \supset \Gamma(R_x, R_y) \quad (5.220)$$

the Coriolis interaction is symmetry allowed but the wavenumbers of ν_3 (1048.60 cm^{-1}) and ν_6 (1182.35 cm^{-1}) are sufficiently different for the interaction to be fairly weak.

Figure 5.52 shows the effects of the interaction in the observed spectrum. One of the effects is to change the A' and B' rotational constants in both the $\upsilon_3 = 1$ and $\upsilon_6 = 1$ states from what they would be in the absence of the interaction. Because the interaction is small it can be treated adequately by second-order Coriolis interaction and it shows up as contributions to the vibration–rotation interaction constants α_3^B and α_6^B [see equation (4.73) for the definition of α]. The reason why only α^B is affected is that the rotation connecting ν_3 and ν_6 is about the x- and y-axes [equation (5.220)] and these are the axes to which the rotational constant B refers. The Coriolis contributions to the αs, which we will denote by α_{Cor}, are given by

$$\begin{aligned}\alpha^B_{\text{Cor}}(\nu_3) &= -2\alpha^B_{\text{Cor}}(\nu_6)\\ &= 4B^2\zeta^2_{3,6a}/(\tilde{\nu}_6 - \tilde{\nu}_3)\end{aligned} \quad (5.221)$$

These contributions can be estimated only if $\zeta_{3,6a}$ is known; this, or in fact its modulus, can be calculated from the force field and normal coordinates of the molecule.

The second, and more immediately obvious, effect of the interaction is the intensity perturbation in the low wavenumber wing of the 6^1_0 band. Figure 5.52 shows how the intensity of the ^PP sub-branches is enhanced to such an extent that the ^PQ sub-branches with $K = 5$ and 6 are almost swamped. di Lauro and Mills[53] have shown that this intensity enhancement occurs only if $\zeta_{3,6a}(\partial\mu_z/\partial Q_3)/(\partial\mu_x/\partial Q_{6a})$ is *positive*, where $\partial\mu/\partial Q$ is the change of dipole moment with normal coordinate. If, in this case, the quantity had been *negative*, the intensity enhancement would have been in the ^rR branches.

5.2.5.2(ii) *Raman spectra*

In vibration–rotation bands in the Raman spectra of symmetric rotors the selection rule in the quantum number J is always

$$\Delta J = 0, \pm 1, \pm 2 \text{ and } (J' + J'') \geqslant 2$$

similar to that for the pure rotational Raman spectrum except for the $(J' + J'')$ restriction.

The selection rule in K includes, in general,

$$\Delta K = 0, \pm 1, \pm 2$$

but, for a transition involving the totally symmetric zero-point level, it depends on the component of the polarizability to which the symmetry species of the excited state vibrational wave function corresponds. The rules have been determined, as for linear molecules, by Placzek and Teller (see reference in the footnote on p. 223) and are

$$\begin{aligned} &\Delta K = 0 && \text{if } \Gamma(\psi'_v) = \alpha_{zz} \text{ and/or } (\alpha_{xx} + \alpha_{yy}) \\ &\Delta K = \pm 1 && \text{if } \Gamma(\psi'_v) = \alpha_{xz} \text{ and/or } \alpha_{yz} \\ &\Delta K = \pm 2 && \text{if } \Gamma(\psi'_v) = (\alpha_{xx} - \alpha_{yy}) \text{ and/or } \alpha_{xy} \end{aligned}$$

The following examples illustrate the use of these rules.

(1) For C_{3v} molecules:

$$\Delta K = 0 \text{ for } A_1\text{–}A_1;$$
$$\Delta K = \pm 1, \pm 2 \text{ for } E\text{–}A_1$$

(2) For D_{3h} molecules:

$$\Delta K = 0 \text{ for } A'_1\text{–}A'_1;$$
$$\Delta K = \pm 1 \text{ for } E''\text{–}A'_1;$$
$$\Delta K = \pm 2 \text{ for } E'\text{–}A'_1$$

(3) For D_{2d} molecules:

$$\Delta K = 0 \text{ for } A_1\text{–}A_1;$$
$$\Delta K = \pm 1 \text{ for } E\text{–}A_1;$$
$$\Delta K = \pm 2 \text{ for } B_1\text{–}A_1 \text{ and } B_2\text{–}A_1$$

The selection rules for hot bands follow similarly from the symmetry species of $\Gamma(\psi'_v) \times \Gamma(\psi''_v)$.

For a doubly degenerate fundamental with a first-order Coriolis interaction, causing splitting into $+l$ and $-l$ components of all levels with $K > 1$, transitions with $\Delta K = +1$ and -2 involve only the $+l$ levels and those with $\Delta K = -1$ and $+2$ involve only the $-l$ levels.

An important result of these selection rules is that if $\Delta K = \pm 2$ for a transition involving a degenerate excited state, such as in an $E' - A'_1$ transition in the D_{3h} point group, it is possible to obtain ζ_i from a combination of infrared and Raman observations of the transition. Like the infrared band, the corresponding Raman band is dominated by line-like Q branches. In the Raman band their positions are those of the $\Delta K = \pm 2$ sub-band origins given by

$$\begin{aligned} \tilde{\nu}_0^{\text{sub}} &= \omega_0 + 4[A'(1 + \zeta_i) - B'] \\ &\pm 4\left[A'\left(1 + \frac{\zeta_i}{2}\right) - B'\right]K \\ &+ [(A' - A'') - (B' - B'')]K^2 \qquad (5.222) \end{aligned}$$

where the + or − sign refers to $\Delta K = +2$ or -2, respectively.

The spacing of the Q branches is now $4[A'(1 + \zeta_i/2) - B']$ compared with $2[A'(1 - \zeta_i) - B']$ in the corresponding infrared band [see equation (5.215)] and both A' and ζ_i can be obtained if the bands can be observed in the infrared and Raman spectra.

5.2.5.3 *Spherical rotor molecules*

Spherical rotor molecules commonly belong to either the T_d or O_h point groups as in the examples of methane (CH_4), and sulphur hexafluoride (SF_6).

In the infrared spectrum the only vibrational transitions from the zero-point level which are allowed are T_2–A_1, in the T_d point group, and T_{1u}–A_{1g}, in the O_h point group.[†]

In the Raman spectrum, A_1–A_1, E–A_1 and T_2–A_1 transitions are allowed in T_d and A_{1g}–A_{1g}, E_g–A_{1g} and T_{2g}–A_{1g} in O_h.

[†]Workers in the field of vibration–rotation spectroscopy of spherical rotors mostly use F rather than T to indicate the symmetry species of a triply degenerate state. However, I have taken the view that, in the whole field of spectroscopy which includes, particularly, spectroscopy of inorganic complexes, probably more people use T than F and have therefore used T throughout.

5.2.5.3(i) Infrared spectra

In a discussion of the rotational transitions accompanying vibrational transitions we shall consider examples from the T_d point group only.

Neglecting the effects of centrifugal distortion, the rotational term values of equation (4.115), for an A_1 vibrational state, become

$$F_\upsilon(J) = B_\upsilon J(J+1) \tag{5.223}$$

As discussed in Section 4.6, there is a $(2J+1)^2$-fold degeneracy associated with each level, one $(2J+1)$ factor being due to the fact that M_J can take $(2J+1)$ values—the space degeneracy factor—and the second $(2J+1)$ factor being due to the fact that, compared with a symmetric rotor, all the levels with different values of K are degenerate—the K degeneracy factor.

Each of the $(2J+1)$ levels caused by the K degeneracy can be assigned a symmetry species of the T_d point group.† The result, for an A_1 vibrational state, is the set of rovibrational levels shown in Figure 5.53. This set may appear complex but, in the absence of second-order Coriolis interaction with other vibrational states, the A_1, A_2, E, T_1 and T_2 levels for a given value of J remain degenerate and the simple term value expression of equation (5.223) applies.

Figure 5.53 shows how the $(2J+1)$-fold K degeneracy of each J level is made up: for example, the $J=9$ level contains $1A_1 + 1A_2 + 1E + 3T_1 + 2T_2$ levels, giving a total degeneracy of 19. Except in the presence of an external field, this degeneracy can never be completely removed. Coriolis interaction can never split the two- or threefold degeneracy within an E, T_1 or T_2 state.

Rovibrational levels in an E vibrational state of a molecule belonging to a T_d point group are simpler, at the same level of approximation as we used for an A_1 vibrational state, than those for a symmetric rotor because there is no first-order Coriolis interaction. The reason for this is that

$$E \times E = A_1 + A_2 + E \tag{5.224}$$

which does not contain a rotational symmetry species.

The rovibrational levels for an E vibrational state are shown in Figure 5.54. The degeneracy of each J level is now $2(2J+1)$, excluding the $(2J+1)$ space degeneracy factor, the additional factor of 2 being due to the double degeneracy associated with an E vibration. For example, the $J=9$ level consists of $1A_1 + 1A_2 + 3E + 5T_1 + 5T_2$, giving a total degeneracy of 38.

In a T_2 vibrational state Jahn's rule, stated in equation (5.194), applies, resulting in a splitting of the threefold degeneracy, by a first-order Coriolis interaction, since

$$T_2 \times T_2 = A_1 + E + T_1 + T_2 \supset \Gamma(R_x, R_y, R_z) \tag{5.225}$$

The three components of the T_2 state are labelled T^+, T^0 and T^- and the term values for these components are given by‡

$$F_\upsilon^+(J) = B_\upsilon J(J+1) + 2B_\upsilon \zeta_i J \tag{5.226}$$

$$F_\upsilon^0(J) = B_\upsilon J(J+1) - 2B_\upsilon \zeta_i \tag{5.227}$$

$$F_\upsilon^-(J) = B_\upsilon J(J+1) - 2B_\upsilon \zeta_i (J+1) \tag{5.228}$$

Such a set of rovibrational levels is illustrated in Figure 5.55. One feature of these is the absence of $J=0$ for the T^- and T^0 components and another is that there is a $3(2J+1)$-fold degeneracy for a particular value of J; for example, for $J=8$ there are $2A_1 + 2A_2 + 4E + 6T_1 + 7T_2$ sub-levels, corresponding to a degeneracy of 51.

If the four identical nuclei at the corners of the tetrahedron have zero nuclear spin ($I=0$), as in nickel tetracarbonyl, $Ni(CO)_4$, in which $I=0$ for both ^{12}C and ^{16}O, the E, T_1 and T_2 levels are absent, as shown in Table 5.12. This table also gives the statistical weight factors for $I=\frac{1}{2}$, as in C^1H_4, and for $I=1$, as in C^2H_4.

†For the derivations of these symmetry species, the reader is referred to Hougen, J. T. (1963) *J. Chem Phys.*, **39**, 358.

‡A term $2B_\upsilon\zeta_i$ may be either added to each of these term values or, as here, assumed to be included in the vibrational term value.

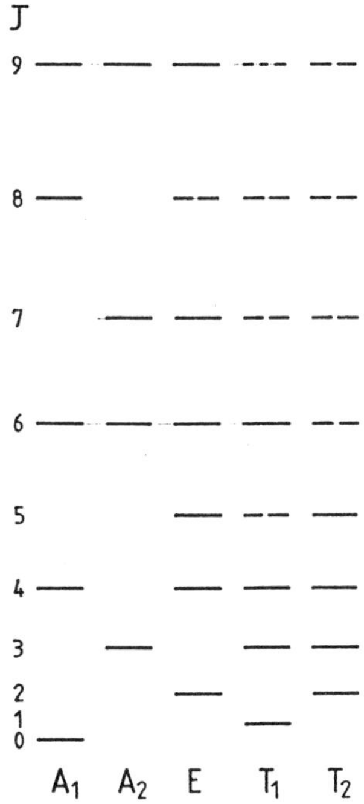

Figure 5.53 Rovibrational levels for a T_d spherical rotor in an A_1 vibrational state

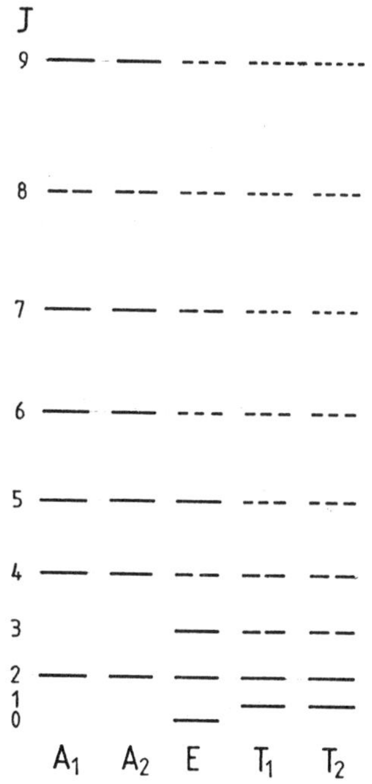

Figure 5.54 Rovibrational levels for a T_d spherical rotor in an E vibrational state

In the infrared spectrum the rotational selection rules for a T_2–A_1 transition, the only type of transition allowed for an A_1 lower state, are

$$\Delta J = +1 \quad \text{for } T^- - A_1;$$
$$\Delta J = 0 \quad \text{for } T^0 - A_1;$$
$$\Delta J = -1 \quad \text{for } T^+ - A_1$$

The result is that the band appears similar to a Π–Σ^+ infrared band of a linear molecule showing a central Q branch, and P and R branches with spacings of about $2B(1-\zeta_i)$, if $B' \approx B''$. In the case of the 3_0^1, T_2–A_1, band of C^1H_4, $\zeta_3 = 0.055$ and this small value leads to spacings of about $2B$. On the other hand, in the 4_0^1, T_2–A_1, band of C^1H_4, $\zeta_4 = 0.45$ and there is also considerable Coriolis interaction between ν_4 and $\nu_2(e)$. Both these factors lead to a much more complex appearance of the band.

Table 5.12 Nuclear spin statistical weight factors for tetrahedral molecules XY_4

$\Gamma(\psi_v\psi_r) =$	A_1	A_2	E	T_1	T_2
$I^Y = 0$	1	1	0	0	0
$I^Y = \frac{1}{2}$	5	5	2	3	3
$I^Y = 1$	15	15	12	18	18

5.2.5.3(ii) Raman spectra

In the Raman spectrum of a spherical rotor the general rotational selection rule is

$$\Delta J = 0, \ +1, \ \pm 2$$

with the restriction that

$$(J' + J'') \geqslant 2$$

For an A_1–A_1 transition the selection rule is $\Delta J = 0$, giving only a Q branch, and, for an E–

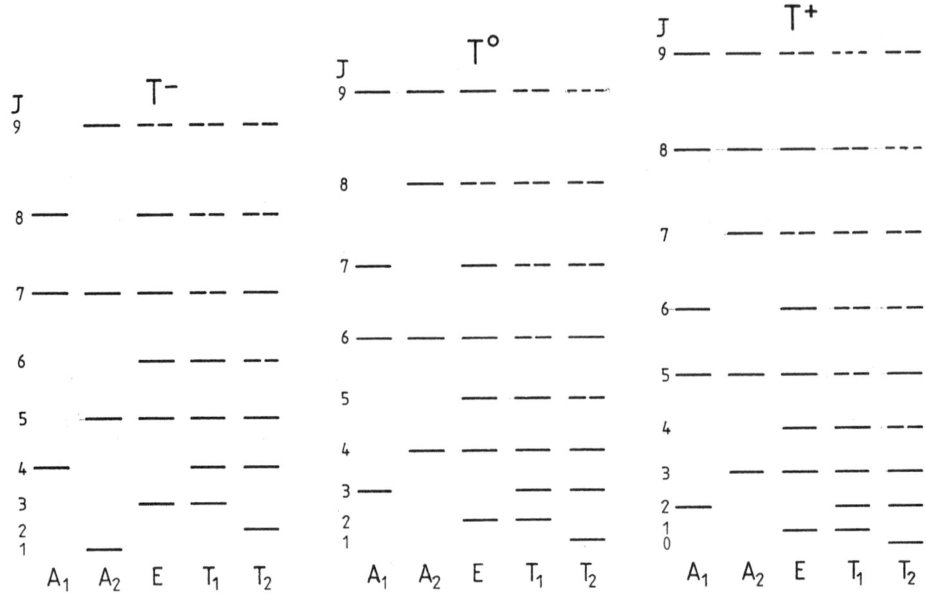

Figure 5.55 Rovibrational levels for a T_d spherical rotor in a T_2 vibrational state

A_1 transition, $\Delta J = 0, \pm 1, \pm 2$, giving O, P, Q, R and S branches.

For a T_2–A_1 Raman transition $\Delta J = 0, \pm 1, \pm 2$ and, since there is no restriction on which component of T_2 combines with A_1, there are five branches for each of the T_2^+–A_1, T_2^0–A_1 and T_2^-–A_1 transitions, giving 15 branches in all; these are labelled S^+, S^0, S^-, etc.

Figure 5.56 shows the 3_0^1, T_2–A_1, Raman band of C^2H_4 for which $\zeta_3 = 0.166$. The intensities of the various transitions will not be discussed in detail here[†] but are proportional to the Boltzmann factor, the degeneracy, the nuclear spin statistical weights and a quantity analogous to A_{KJ} of equations (5.209)–(5.211) for symmetric rotors. This latter factor results in the strongest branches being S^+ and O^-, which is clearly the case in Figure 5.56.

[†]For further discussion, see Herranz, J. and Stoicheff, B. P. (1963). *J. Mol. Spectrosc.*, **10**, *448*.

5.2.5.4 Asymmetric rotor molecules

For a vibrational transition from the zero-point level to be electric dipole (infrared) allowed, there must be a change of dipole moment during the transition. In a point group of sufficiently high symmetry, such as C_{2v} in which three mutually perpendicular axes are uniquely defined by symmetry, the dipole moment change must be along one of these axes and we have, for transitions from the zero-point level,

$$\Gamma(\psi_v') = \Gamma(T_x) \text{ or } \Gamma(T_y) \text{ or } \Gamma(T_z) \qquad (5.229)$$

if the axes are labelled x, y and z. For the purposes of formulating rotational selection rules for rovibrational transitions in asymmetric rotors it is not convenient to use the x-, y- and z-axis labels; instead, we use the a, b and c inertial axis labels (see Section 4.1) so that equation (5.229) becomes

$$\Gamma(\psi_v') = \Gamma(T_a) \text{ or } \Gamma(T_b) \text{ or } \Gamma(T_c) \qquad (5.230)$$

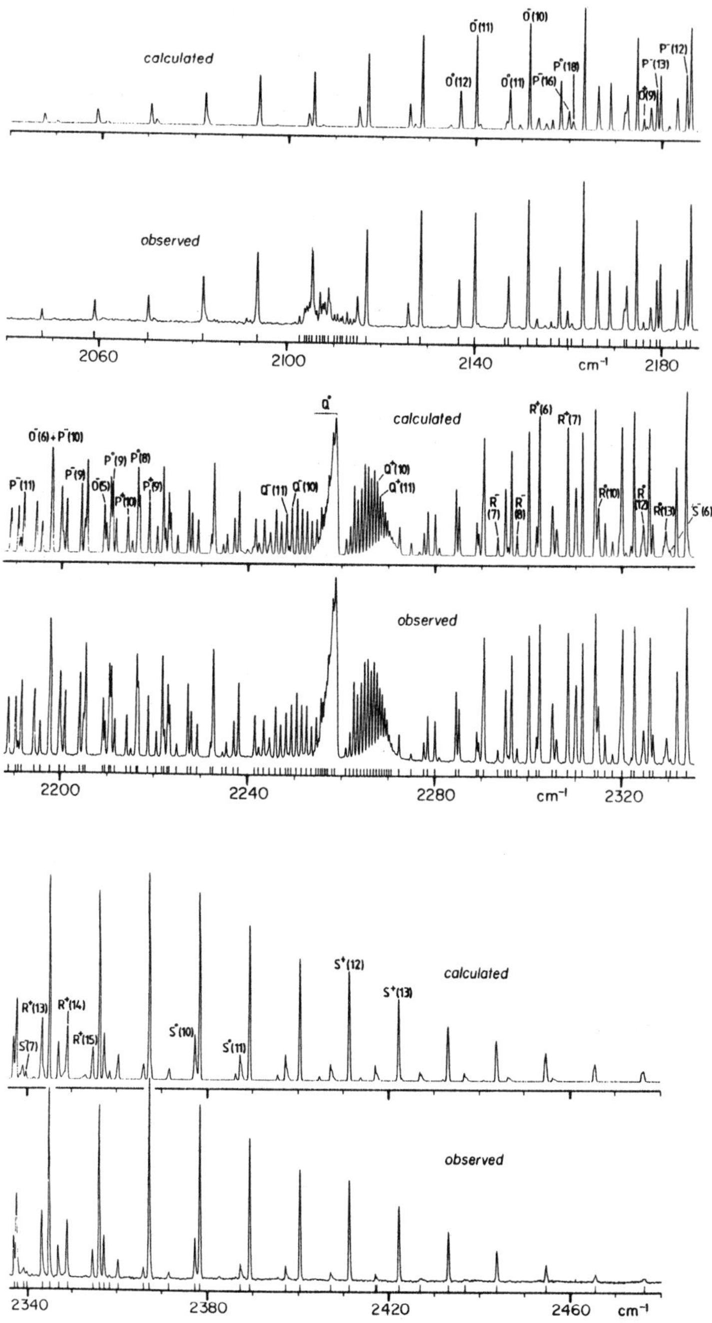

Figure 5.56 The 3_0^1, T_2–A_1, Raman band of the C^2H_4 molecule, a spherical rotor. Above is the computed and, below, the observed band. The weak lines near 2110 cm^{-1} in the observed band are due to the overlapping 1_0^1 band [Reproduced, with permission, from Brodersen, S., Gray, D. L. and Robiette, A. G. (1977). *Mol. Phys.*, **34**, 617]

where no particular correspondence between the x, y, z and a, b, c labels is intended. For vibrational hot bands equation (5.230) becomes

$$\Gamma(\psi'_v) \times \Gamma(\psi''_v) = \Gamma(T_a) \text{ or } \Gamma(T_b) \text{ or } \Gamma(T_c) \tag{5.231}$$

For molecules of lower symmetry a change of dipole moment accompanying a vibrational transition may have components along two, or even all three, axes. For example, in a planar C_s molecule, where the c-axis is perpendicular to the symmetry plane,

$$\Gamma(\psi'_v) = \Gamma(T_c), \text{ or } \Gamma(T_a) \textit{ and } \Gamma(T_b) \tag{5.232}$$

When the dipole moment change has components along two or three of the axes the resulting band is said to be a hybrid band.

For allowed vibrational transitions from the zero-point level in the Raman spectrum

$$\Gamma(\psi'_v) = \Gamma(\alpha_{ij}) \tag{5.233}$$

and, for hot bands,

$$\Gamma(\psi'_v) \times \Gamma(\psi''_v) = \Gamma(\alpha_{ij}) \tag{5.234}$$

In asymmetric rotors of fairly low symmetry the transition moment may involve more than one of the α_{ij} components giving the Raman analogue of an infrared hybrid band.

The rotational term values for asymmetric rotors have been discussed in Section 4.5. There are, in general, no closed formulae for these term values and Figure 4.25 shows schematically how they change from a prolate to an oblate symmetric rotor as the asymmetry parameter κ (or b), defined in equation (4.109)—or (4.98) and (4.105)—varies from -1 to $+1$ (or 0 to -1). For a particular molecule there is a set of rotational term values, obtainable accurately only by matrix diagonalization, for each vibrational state: the variations of the rotational constants A, B and C with vibration are expressed by equations (4.112)–(4.114).

The classification of the rotational wave functions in terms of the evenness of oddness

Table 5.13 Character table of the $V(a, b, c)$ or four group

	I	C_2^a	C_2^b	C_2^c	K_aK_c notation
A	1	1	1	1	ee
B_a	1	1	-1	-1	eo
B_b	1	-1	1	-1	oo
B_c	1	-1	-1	1	oe

Table 5.14 Rotational selection rules for electric dipole rovibrational transitions in asymmetric rotors

Symmetry species of $V(a, b, c)$ group and alternative K_aK_c label		A ee	B_a eo	B_b oo	B_c oe
A	ee	—	a	b	c
B_a	eo	a	—	c	b
B_b	oo	b	c	—	a
B_c	oe	c	b	a	—

of the corresponding prolate and oblate symmetric rotor quantum numbers K_a and K_c was discussed in Section 4.5. The wave function labelled eo, for example, corresponds to K_a even and K_c odd.

We shall introduce here an alternative set of labels for the sake of completeness. These labels are used frequently in discussion of asymmetric rotor energy levels but not very often in formulating selection rules.

The alternative labelling uses the point group which contains the identity element I and three C_2 elements corresponding to rotation by π rad of the momental ellipsoid (see Figure 4.1) about the a, b and c inertial axes. This group is equivalent to the D_2 point group with the axis labels changed from x, y, z to a, b, c, but not necessarily in that order. The old notation for the D_2 group was $V^\dagger$ and the notation often used now is $V(a, b, c)$. It is also sometimes called the four group, because it has four symmetry elements and species. The character table is given in Table 5.13, which makes it clear that a rotational wave function for a state with

† From the German 'vierer Gruppe' meaning 'group of four'.

K_a even is symmetric to the C_2^a operation and antisymmetric for K_a odd; and similarly for K_c and C_2^c. Alternatively, the species B_a can be used to indicate that the rotational wave function is symmetric to the C_2^a operation and antisymmetric to C_2^b and C_2^c, and similarly for B_b and B_c. The species A is symmetric to all operations.

5.2.5.4(i) *Infrared spectra*

The rotational selection rules for rovibrational transitions between stacks of rotational energy levels associated with two vibrational levels between which there is an allowed transition are the same as those given in Table 4.5 for pure rotational electric dipole transitions except that now we are concerned with a *change* of dipole moment along the *a*-, *b*- or *c*-axis. These selection rules are given in Table 5.14 in a form which will correspond to those (in Table 5.16) for the more complex Raman spectra. In Table 5.14 *a*, *b* and *c* indicate a transition moment along the corresponding axis and so-called type *A*, type *B* and type *C* selection rules result, respectively. For example, for a transition moment along the *b*-axis, the type *B* selection rules are

$$\text{ee} \longleftrightarrow \text{oo}$$
$$\text{oe} \longleftrightarrow \text{eo}$$

in addition to

$$\Delta J = 0, \ \pm 1$$

For a hybrid band, such as that resulting from a 1–0 transition of the C—F stretching vibration in $CH_2{=}CFBr$ [Figure 4.26(b)], the transition moment is in the *ab*-plane with components along both the *a*- and *b*-axes. The result is that type *A and* type *B* selection rules apply.

If a transition moment $\boldsymbol{R}$ is in the *ab*-plane and makes an angle θ with, say, the *a*-axis, the components R_a and R_b along the *a*- and *b*-axes are given by

$$R_a = |\boldsymbol{R}| \cos\theta$$
$$R_b = |\boldsymbol{R}| \sin\theta \qquad (5.235)$$

and, since the intensity I is proportional to the square of the transition moment,

$$I(B)/I(A) = R_b^2/R_a^2 = \tan^2\theta \qquad (5.236)$$

where $I(B)$ and $I(A)$ are the intensities of the type *B* and type *A* components of the band.

From the energy level diagram of Figure 4.25 and equations (4.96)–(4.97) and (4.103)–(4.104), it is apparent that the energy levels of an asymmetric rotor, and hence the vibration–rotation spectrum, can be extremely irregular in appearance. This is particularly true of a light molecule with a small value of the asymmetry parameter κ.

All asymmetric rotors tend to behave like symmetric rotors in certain ranges of J and K_a or K_c. We can see from Figure 4.25 that, in a prolate (or oblate) symmetric rotor, all levels except those with $K_a = 0$ (or $K_c = 0$) are doubly degenerate. In an asymmetric rotor the degeneracies may be split. In a prolate asymmetric rotor the splitting increases with J for constant K_a and decreases with K_a for constant J. In particular, at low values of K_a (prolate) or K_c (oblate) and high values of J the molecule is tending to rotate about the *c*-axis (prolate) or the *a*-axis (oblate). Under these circumstances a prolate asymmetric rotor is tending to behave like an oblate symmetric rotor and an oblate asymmetric rotor like a prolate symmetric rotor. The general usefulness of this 'high J, low K approximation' was first recognized by Gora.[54]

Table 5.15 Correlation between asymmetric and symmetric rotor band types

Band type of asymmetric rotor	Band type of symmetric rotor Prolate	Oblate
Type *A*	$\parallel$	$\perp$
Type *B*	$\perp$	$\perp$
Type *C*	$\perp$	$\parallel$

It is also a useful approximation that at high values of K_a (prolate) or K_c (oblate), and especially when $J \approx K_a$ or K_c, the rotational angular momentum is mainly about the a (prolate) or c (oblate) axis. Under these circumstances a prolate asymmetric rotor tends to behave like a prolate symmetric rotor and, similarly, an oblate asymmetric rotor like an oblate symmetric rotor; we shall call this the 'high K, low J approximation'. It is this approximation to which the term value expression of equation (4.100), for a prolate, and equation (4.106), for an oblate, asymmetric rotor apply. These equations become valid for a larger range of J and K_a or K_c for molecules with $\kappa \approx \pm 1$, i.e. for prolate or oblate near-symmetric rotors. Table 5.15 shows how the type A, B and C bands of asymmetric rotors go over to parallel or perpendicular bands as the moment of inertia I_b approaches either I_c or I_a. Since

$$I_a \leqslant I_b \leqslant I_c \tag{5.237}$$

this produces a prolate or oblate symmetric rotor, respectively.

The difficulty with light prolate asymmetric rotor molecules such as H_2O, so far as these approximations are concerned, is that only rotational levels with relatively low values of J and K_a are normally appreciably populated so that approximations which apply to high J or K are not particularly useful. However, the fact that the rotational levels are rather widely separated leads to vibration–rotation bands whose rotational structure can easily be resolved experimentally and a line-by-line analysis carried out. On the other hand, for an oblate asymmetric rotor the coefficient of K_c^2 in equation (4.106) is never very large and transitions with fairly high K_c are observed in *all* such molecules.

Type A bands. As Table 5.15 shows, a type A band correlates with a parallel band in the prolate symmetric rotor limit for which the selection rules are

$$\text{for } K_a = 0, \quad \Delta K_a = 0, \quad \Delta J = \pm 1$$
$$\text{for } K_a \neq 0, \quad \Delta K_a = 0, \quad \Delta J = 0, \ \pm 1$$

and with a perpendicular band in the oblate symmetric rotor limit, for which the selection rules are

$$\Delta K_c = \pm 1, \qquad \Delta J = 0, \ \pm 1$$

These two sets of selection rules combine to give

$$\text{for } K_a = 0, \quad \Delta K_a = 0, \quad \Delta K_c = \pm 1, \quad \Delta J = \pm 1$$
$$\text{for } K_a \neq 0, \quad \Delta K_a = 0, \quad \Delta K_c = \pm 1, \quad \Delta J = 0, \ \pm 1$$

Although, in an asymmetric rotor, the selection rules relax to include

$$\Delta K_a = 0, \ \pm 2, \ \pm 4, \ \pm 6, \ldots$$
$$\text{and } \Delta K_c = \pm 1, \ \pm 3, \ \pm 5, \ldots$$

those transitions obeying the symmetric rotor selection rules, $\Delta K_a = 0$ and $\Delta K_c = \pm 1$, are much the most important and account for the bulk of the intensity.

Of course, in whatever form the selection rules are expressed, they are all consistent with the general type A selection rules:

$$\text{ee} \longleftrightarrow \text{eo}$$
$$\text{eo} \longleftrightarrow \text{oo}$$

given in Table 5.14.

Just as with the rotational term values, there are no closed formulae for transition intensities in asymmetric rotors. Exact intensities can be obtained only through the matrix diagonalization procedures necessary in determining the term values; they will not be discussed here but can be found in several of the books referred to in the Bibliography at the end of both this chapter and Chapter 4. However, it is often a useful approximation to use the corresponding symmetric rotor intensities for transitions which obey symmetric rotor selection rules.

If the molecule has one or more C_2 axes, nuclear spin statistical weight factors must be

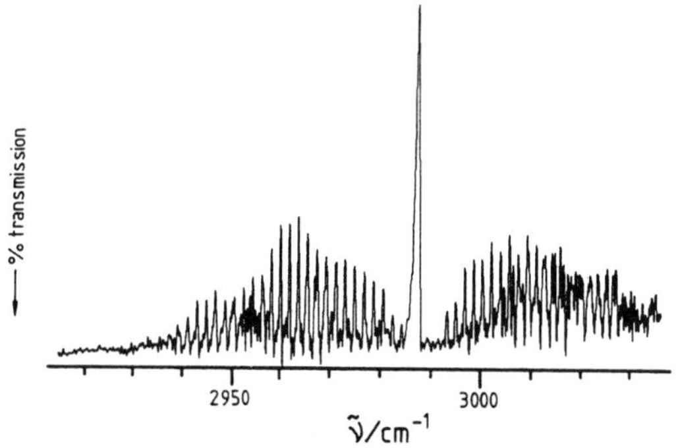

Figure 5.57 The 11^1_0 type A infrared band of the ethylene molecule, a prolate asymmetric rotor. This band overlaps the 9^1_0 band of Figure 5.60

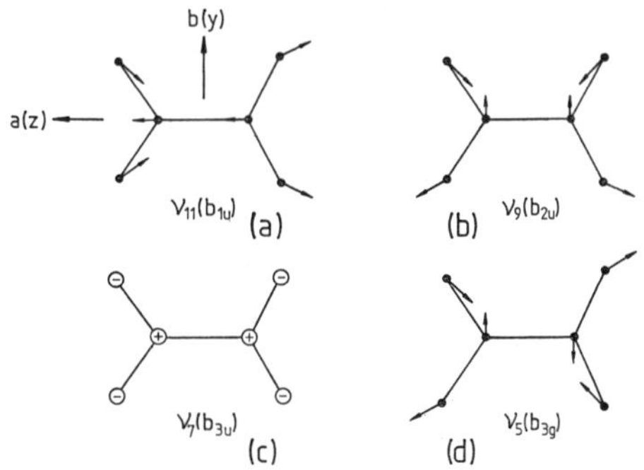

Figure 5.58 The normal modes of vibration (a) ν_{11}, (b) ν_9, (c) ν_7 and (d) ν_5 of the ethylene molecule

included. These have been discussed for asymmetric rotors in Section 4.5.

An example of a type A band of a prolate asymmetric rotor is the 11^1_0 band of ethylene shown in Figure 5.57, where ν_{11} is the b_{1u} C—H stretching vibration shown in Figure 5.58(a), using the conventional (Mulliken) axis notation. The dipole moment change is clearly along the a-axis giving a type A band.

Ethylene has a κ value of -0.913 so the type A band strongly resembles a parallel band of a prolate symmetric rotor with coincident Q branches in the centre and P and R lines, coincident for several values of K_a, on either side. It is only for low values of K_a, particularly 0, 1 and 2, and relatively high values of J that transitions in this band resemble those of an oblate symmetric rotor. Ethylene is not a light molecule and there is appreciable population of levels with $K_a = 0$–10. The sub-bands with $K_a = 0$–2 are not sufficiently exposed in this band for the effect of asymmetry to be obvious at this resolution.

The band in Figure 5.57 is only partially resolved and what is observed is called a band contour. The characteristic shape of contours can often be used to identify the band type without resorting to analysis of a highly resolved spectrum. This technique is particularly useful in spectra of molecules which are so heavy, and, therefore, the transitions so closely spaced, that a resolved spectrum is virtually unobtainable.

Figure 5.59 shows part of the far infrared spectrum of perdeuteronaphthalene, $C_{10}{}^2H_8$, for which κ is -0.68. There is a band with a contour characteristic of a type A band of a prolate asymmetric rotor at 594 cm^{-1}. The molecule belongs to the D_{2h} point group and this band is due to a b_{2u} fundamental.†

In oblate asymmetric rotors there is a strong tendency for a type A band to resemble a perpendicular band of an oblate symmetric rotor. Such a band shows prominent Q branches at the sub-band origins whose wavenumbers, $\tilde{\nu}_0^{\text{sub}}$, are given by

$$\tilde{\nu}_0^{\text{sub}} \approx \omega_0 + (C - \bar{B}) \pm 2(C - \bar{B})K_c \qquad (5.238)$$

This equation is obtained by applying the selection rules $\Delta J = 0$, $\Delta K_c = \pm 1$ to sets of rotational term values, given in equation (4.106), for the two combining vibrational states and taking C and $\bar{B}$ to be unchanged in the excited state. Adjacent Q branches are separated by $2(C - \bar{B})$ but those with low values of K_c are distorted by the effects of asymmetry. This distortion near the band centre is analogous to that in a type C band of a prolate asymmetric rotor which will be discussed in more detail below. The result is a general

†This symmetry species results from the choice of axes in which x is perpendicular to the molecular plane and z is along the C(9)–C(10) bond.

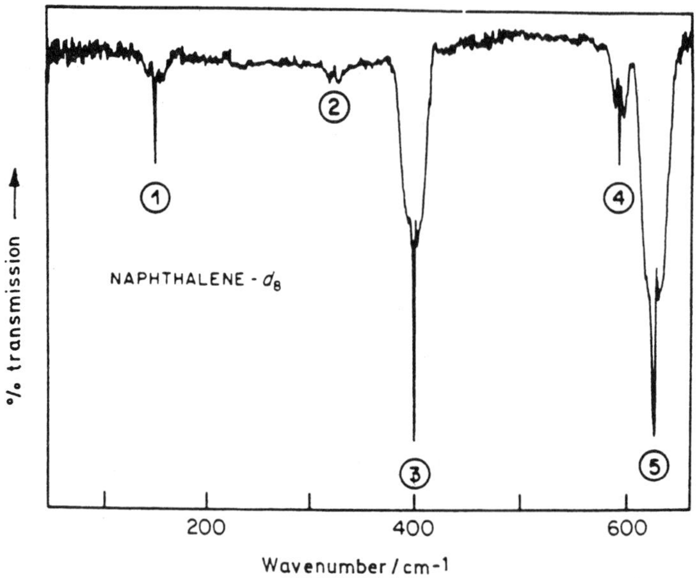

Figure 5.59 Part of the far infrared spectrum of perdeuteronaphthalene. Band 4 is a typical type *A* band, band 2 a typical type *B* band and bands 1, 3 and 5 typical type *C* bands of a large prolate asymmetric rotor [Reproduced, with permission, from Duckett, J. A., Smithson, T. L. and Wieser, H. (1978). *J. Mol. Struct.*, **44**, 97]

resemblance between a type *A* band of a prolate and a type *C* band of an oblate asymmetric rotor.

Type B bands. Table 5.15 shows that a type *B* band correlates with a perpendicular band in both the prolate and oblate symmetric rotor limits. The selection rules at these limits are

$$\Delta K_a = \pm 1, \qquad \Delta K_c = \pm 1, \qquad \Delta J = 0, \pm 1$$

In an asymmetric rotor the selection rules relax to include

$$\Delta K_a = \pm 1, \pm 3, \pm 5, \ldots$$
$$\text{and } \Delta K_c = \pm 1, \pm 3, \pm 5, \ldots$$

but those transitions with $\Delta K_a = \pm 1$ and $\Delta K_c = \pm 1$ are the most intense.

These selection rules are consistent with the general type *B* selection rules:

$$\text{ee} \longleftrightarrow \text{oo}$$
$$\text{oe} \longleftrightarrow \text{eo}$$

in Table 5.14.

For a prolate asymmetric rotor a type *B* band is characterized by prominent *Q* branches corresponding to values of K_a for which the near-symmetric rotor term values of equation (4.100) are a reasonable approximation. These *Q* branches are at the sub-band origins whose wavenumbers, $\tilde{\nu}_0^{\text{sub}}$, are given by

$$\tilde{\nu}_0^{\text{sub}} \simeq \omega_0 + (A - \bar{B}) \pm 2(A - \bar{B})K_a \qquad (5.239)$$

which is analogous to equation (5.238) for an oblate rotor. The *Q* branches are separated by $2(A - \bar{B})$.

Figure 5.60 shows the 9_0^1 type *B* band of ethylene, a prolate asymmetric rotor, in which ν_9 is the b_{2u} C—H stretching vibration shown in Figure 5.58(b). The *Q* branches are clearly observed in the wings of the band but, close to the band centre, there is some distortion from the kind of symmetric rotor perpendicular band shown in Figure 5.49. This is due to asymmetry of the molecule, one effect of which is to produce a depletion of intensity near the band centre. Because of the nuclear spin statistical weight factors, given in Table 4.6 for $^{12}C_2{}^1H_4$, there is a 10 : 6 intensity alternation for K_a even : odd which can be observed in the Q branches in which K_a is sufficiently high that the effects of asymmetry are small.

Figure 5.59 shows a weak type B band at 326 cm^{-1} due to a b_{1u} fundamental in the far infrared spectrum of perdeuteronaphthalene, $C_{10}{}^2H_8$, which shows a characteristic type B contour of a large prolate asymmetric rotor.

A type B band of an oblate asymmetric rotor shows similar characteristics except that the wavenumbers of the Q branches in the wings are given by equation (5.238) when K_c is sufficiently high that the molecule behaves as an oblate near-symmetric rotor.

Type C bands. A type C band correlates, as shown in Table 5.15, with a perpendicular band in the prolate and a parallel band in the oblate symmetric rotor limits. As the selection rules at these limits are

$$\text{for } K_c = 0, \quad \Delta K_c = 0, \quad \Delta J = \pm 1$$
$$\text{for } K_c \neq 0, \quad \Delta K_c = 0, \quad \Delta J = 0, \pm 1$$

for an oblate symmetric rotor and

$$\Delta K_a = \pm 1, \quad \Delta J = 0, \pm 1$$

for a prolate symmetric rotor, these two sets of selection rules combine to give

$$\text{for } K_c = 0, \quad \Delta K_c = 0, \quad \Delta K_a = \pm 1, \quad \Delta J = \pm 1$$
$$\text{for } K_c \neq 0, \quad \Delta K_c = 0, \quad \Delta K_a = \pm 1, \quad \Delta J = 0, \pm 1$$

In an asymmetric rotor, the selection rules relax to include

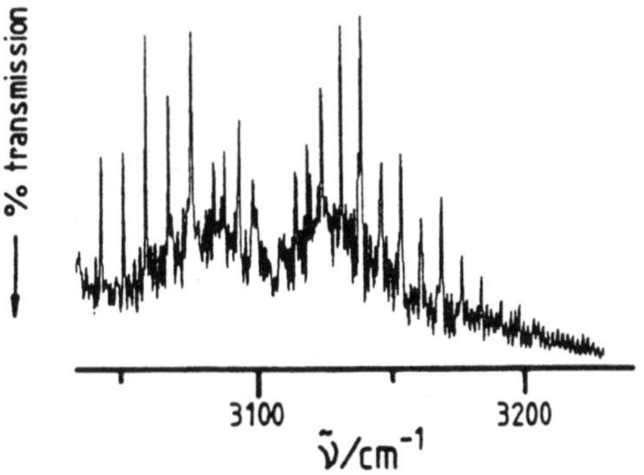

Figure 5.60 The 9_0^1 type B infrared band of the ethylene molecule, a prolate asymmetric rotor

$$\Delta K_c = 0, \pm 2, \pm 4, \pm 6, \ldots$$
$$\text{and } \Delta K_a = \pm 1, \pm 3, \pm 5, \ldots$$

but those transitions obeying the symmetric rotor selection rules, $\Delta K_a = \pm 1$ and $\Delta K_c = 0$, are the most important and account for the bulk of the intensity.

In whatever form the selection rules are expressed, they are all consistent with the general type C selection rules:

$$\text{ee} \longleftrightarrow \text{oe}$$
$$\text{eo} \longleftrightarrow \text{oo}$$

given in Table 5.14.

A type C band of a prolate asymmetric rotor resembles a type B band in that it shows sharp Q branches for K_a values which are sufficiently large that the near-symmetric rotor term values of equation (4.100) are a reasonable approximation. The wavenumbers of these Q branches are given by equation (5.239).

Figure 5.61 shows the 7_0^1 type C band of the prolate asymmetric rotor ethylene. The vibration ν_7 is the b_{3u} out-of-plane bending vibration shown in Figure 5.58(c). The wings of the band are dominated by Q branches which show a 10 : 6 intensity alternation for K_a even : odd, as in the type B band. They are separated approximately by $2(A - \bar{B})$. The centre of the band shows considerable distortion from what

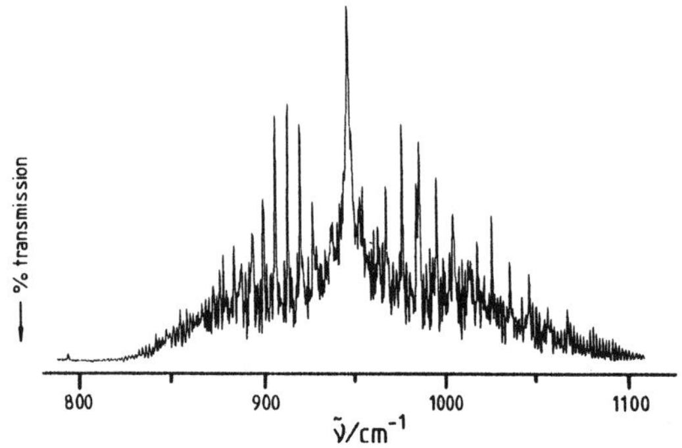

Figure 5.61 The 7_0^1 type C infrared band of the ethylene molecule, a prolate asymmetric rotor

is found in a perpendicular band of a symmetric rotor. The effect of asymmetry is to produce a piling-up of transitions near the band centre. The transitions responsible are Q branches with low values of K_a in which the molecule is tending to behave like an oblate near-symmetric rotor for which the Q branches are nearly coincident at the band centre.

Figure 5.59 shows three bands in the far infrared spectrum of perdeuteronaphthalene, $C_{10}{}^2H_8$, at 154, 401 and 629 cm^{-1} whose contours are typical of type C bands of a large prolate asymmetric rotor. These bands are due to b_{3u} fundamentals.

It is generally true, and this is illustrated by the bands in Figure 5.59, that contours of type A and type C bands of large asymmetric rotors may sometimes be difficult to distinguish. In $C_{10}{}^2H_8$ the central intense peak of the type A band is weaker, relative to the wings, than in a type C band, but this distinguishing feature varies with the asymmetry of the molecule and, in any case, may be masked if there is overlapping of bands. It is true of all asymmetric rotors that it is the type B bands which are most easily identified.

In oblate asymmetric rotors type C bands, as indicated in Table 5.15, resemble type A bands of a prolate asymmetric rotor since both kinds of bands correlate with parallel bands in the symmetric rotor limit.

5.2.5.4(ii) Raman spectra

The rotational selection rules for rovibrational Raman transitions in asymmetric rotors are summarized in Table 5.16.[†] In this table the six different components of the polarizability [see equation (4.117)] are referred, for ease of use, to the unambiguously defined a, b and c inertial axes rather than the x, y and z axes, We can see, for example, from this table that, if a vibration involves the polarizability component α_{bc}, the rotational selection rules are

$$\text{eo} \longleftrightarrow \text{ee}$$
$$\text{oo} \longleftrightarrow \text{oe}$$

or, using the alternative four group symmetry species (see Table 5.13),

$$B_a \longleftrightarrow A$$
$$B_b \longleftrightarrow B_c$$

and the band is called a type B_a band by analogy with a type A infrared band. Similarly, there are type B_b, type B_c and type A bands. Raman bands fall into these distinct categories only if the symmetry of the molecule is sufficiently high that the symmetry species of α_{ab}, α_{bc}, α_{ac} and (α_{aa}, α_{bb}, α_{cc}) are all different; this is so in only the D_{2h}, D_2 and C_{2v} point groups.

For molecules of lower symmetry, belonging to the C_{2h}, C_2 or C_s point groups, vibrational transition polarizabilities may involve two components of $\boldsymbol{\alpha}$. For example, if they are α_{ab} and α_{bc}, the selection rules from Table 5.16 are

$$\text{oe} \longleftrightarrow \text{ee}$$
$$\text{eo} \longleftrightarrow \text{oo}$$

and

$$\text{ee} \longleftrightarrow \text{eo}$$
$$\text{oe} \longleftrightarrow \text{oo}$$

[†]For a discussion of Raman vibration–rotation selection rules, see Hills, G. W., Foster, R. B. and Jones, W. J. (1977). *Mol. Phys.*, **33**, 1571

Table 5.16 Rotational selection rules for rovibrational transitions in Raman spectra of asymmetric rotors

Symmetry species of $V(a, b, c)$ group and alternative K_aK_c label		A ee	B_a eo	B_b oo	B_c oe
A	ee	$\alpha_{aa}, \alpha_{bb}, \alpha_{cc}$	α_{bc}	α_{ac}	α_{ab}
B_a	eo	α_{bc}	$\alpha_{aa}, \alpha_{bb}, \alpha_{cc}$	α_{ab}	α_{ac}
B_b	oo	α_{ac}	α_{ab}	$\alpha_{aa}, \alpha_{bb}, \alpha_{cc}$	α_{bc}
B_c	oe	α_{ab}	α_{ac}	α_{bc}	$\alpha_{aa}, \alpha_{bb}, \alpha_{cc}$

or, alternatively,

$$B_c \longleftrightarrow A$$
$$B_a \longleftrightarrow B_b$$

and

$$B_a \longleftrightarrow A$$
$$B_c \longleftrightarrow B_b$$

and a type B_c–type B_a hybrid band results in which the relative intensities of the two components depend on the relative magnitudes of the transition polarizabilities involving α_{ab} and α_{bc}.

Considerable care must be exercised in using Table 5.16 to determine the rotational selection rules appropriate to a particular vibrational transition. This can be demonstrated by considering the rotational selection rules for the 5^1_0 Raman band of ethylene. The C—H stretching vibration ν_5 is illustrated in Figure 5.58(d). Using the Mulliken axis notation in Figure 5.58(a), this is a b_{3g} vibration. According to the D_{2h} character table, given in Table A.32 in the Appendix, $b_{3g} = \Gamma(\alpha_{yz})$ which, with the axis notation we are using, is the same as α_{ab}. Therefore, 5^1_0 is a type B_c band and the selection rules follow from Table 5.16. Particular pitfalls to be wary of in reading existing literature can be avoided by (a) taking care to note which x, y, z axis notation the author is using in attributing symmetry species to vibrations (many authors do not use the Mulliken conventions) and (b) noting which x, y, z axis notation the author is using in discussing rotational energy levels and selection rules.

Conventions regarding axis notation for rotational levels are varied. In principle there are six ways of correlating the a, b, c with the x, y, z axes and they are all given in Table 5.17. Each combination is called a representation and the symbol I, II or III indicates that the a, b or c axis, respectively, is the z axis while the superscript R or L indicates a right- or left-handed set of axes. Often, but by no means always, the $\mathrm{I^R}$ representation is used when dealing with a prolate and $\mathrm{III^R}$ for an oblate asymmetric rotor. Whatever representation is used for a rotational problem, it may not be the same as that for the vibrational problem.

The general selection rules in Table 5.16 involving the parity of K_a and K_c correlate with the general selection rules

$$\Delta K_a = 0, \ \pm 1, \ \pm 2$$
$$\Delta K_c = 0, \ \pm 1, \ \pm 2$$

in symmetric rotors, so that some of the most intense transitions in asymmetric rotors are those which obey these rules.

In addition to the selection rules involving K_a and K_c there is the selection rule

$$\Delta J = 0, \ \pm 1, \ \pm 2 \text{ for } (J' + J'') \geqslant 2$$

Type A Raman bands constitute a special case since they have no analogues in infrared spectra. As Table 5.16 shows, they involve a change of α_{aa}, α_{bb}, α_{cc} and the selection rules are

Table 5.17 Possible associations of x, y, z axis labels with a, b, c inertial axes

Inertial axis	Representation of x, y, z axes					
	I^R	I^L	II^R	II^L	III^R	III^L
a	z	z	y	x	x	y
b	x	y	z	z	y	x
c	y	x	x	y	z	z

$$\text{ee} \longleftrightarrow \text{ee}$$
$$\text{eo} \longleftrightarrow \text{eo}$$
$$\text{oo} \longleftrightarrow \text{oo}$$
$$\text{oe} \longleftrightarrow \text{oe}$$

showing no change in the parity of either K_a or K_c. In the two symmetric rotor limits the selection rules become

$$\Delta K_a = 0, \ \pm 2$$
$$\Delta K_c = 0, \ \pm 2$$

In symmetric rotors we have seen, in Section 5.2.5.2(ii), that $\Delta K = 0$ and $\Delta K = \pm 2$ transitions result from the involvement of different polarizability components α_{xx}, α_{yy} and α_{zz}. As a consequence, the intensities of ΔK_a (or ΔK_c) = 0 and ΔK_a (or ΔK_c) = ±2 transitions in asymmetric rotors are governed by different factors and the ratio R of the amplitudes of the ΔK_a (or ΔK_c) = ±2 to the ΔK_a (or ΔK_c) = 0 transitions is given by

$$R = -2^{-\frac{1}{2}}(\alpha'_{bb} - \alpha'_{cc})/6^{-\frac{1}{2}}(\alpha'_{bb} + \alpha'_{cc} - 2\alpha'_{aa}) \tag{5.240}$$

for a prolate asymmetric rotor, where α'_{ii} represents the derivative of the polarizability with respect to the normal coordinate [equation (5.174)]. For an oblate asymmetric rotor the corresponding expression is

$$R = -2^{-\frac{1}{2}}(\alpha'_{bb} - \alpha'_{aa})/6^{-\frac{1}{2}}(\alpha'_{bb} + \alpha'_{aa} - 2\alpha'_{cc}) \tag{5.241}$$

5.2.6 Anharmonicity

5.2.6.1 Potential energy surfaces

In general terms, the result of anharmonicity in a single oscillator is that the period of the oscillation varies with amplitude, just as in the case of a pendulum swinging with different amplitudes. The effect of this on the vibrational energy levels of a molecule undergoing vibrational oscillation is to cause them to be unequally spaced.

In a diatomic molecule the potential energy V plotted against the internuclear distance r is shown in Figure 5.4. The effect of anharmonicity is to cause the energy levels to close up smoothly with increasing v and the separations of adjacent levels becomes zero at the limit of dissociation.

It requires a two-dimensional plot in order to illustrate the variation of V with the vibrational coordinate in the case of a diatomic molecule: one dimension for the energy and a second for the vibrational coordinate. In the case of polyatomic molecules, with $3N - 6$ (non-linear) or $3N - 5$ (linear) normal vibrations, it requires a $[(3N - 6) + 1]$- or $[(3N - 5) + 1]$-dimensional surface to illustrate completely the variation of V with all the normal coordinates. For any molecule with more than two atoms the surface is in more than three dimensions and is called a hypersurface. Of course, such a surface cannot be shown in diagrammatic form but we can take sections in two dimensions, corresponding to V and each of the normal coordinates in turn, thereby producing a potential energy curve of each normal coordinate.

Using a contour map to represent a three-dimensional surface, with each contour line representing constant potential energy, *two* vibrational coordinates can be considered. Figure 5.62 shows such a contour map for the linear molecule CO_2. The coordinates plotted here are not normal coordinates but the two CO bond lengths r_1 and r_2 shown in Figure 5.63(a) and it is assumed that the molecule remains linear.

In Figure 5.62 the regions labelled A_1 and A_2 are valleys corresponding to energy minima when one or other oxygen atom is removed to leave a CO molecule. Region B is a closed-in valley being the region in which CO_2 is stable. This valley is much deeper than A_1 and A_2 reflecting the thermodynamic stability of CO_2 relative to CO + O. In an atom–molecule reaction

$$O + CO \longrightarrow OCO \longrightarrow OC + O \qquad (5.242)$$

the reaction coordinate represents the pathway of minimum energy for the reaction in going from A_1 to B to A_2. The variation of energy along this coordinate is illustrated in Figure 5.64(a).

In a reaction such as

$$H + H_2 \longrightarrow H \cdots H \cdots H \longrightarrow H_2 + H \qquad (5.243)$$

in which the linear triatomic molecule is unstable there is an energy *maximum* corresponding to the region B in Figure 5.62 and the variation of potential energy with reaction coordinate is like that in Figure 5.64(b).

In order to see how V varies with the two normal coordinates Q_1 and Q_3, corresponding to the symmetric and antisymmetric stretching vibrations ν_1 and ν_3 of CO_2 in Figure 5.63(b) and (c), we need to proceed along the dashed line labelled Q_1 in Figure 5.62, which corresponds to changing r_1 and r_2 identically, or along that labelled Q_3, which corresponds to increasing r_1 and decreasing r_2 (or vice versa) by equal amounts. The resulting potential energy curves are shown in Figure 5.65(a) and (b). That for ν_1 appears similar to that for a diatomic molecule. The horizontal part of the curve corresponds to region C in Figure 5.62 in which dissociation to O + C + O has occurred, a very high energy process involving the breaking of two double bonds.

On the other hand, the curve for ν_3 in Figure 5.65(b) is symmetrical about the centre, showing steep sides corresponding to the reluctance of the oxygen nuclei to approach the carbon

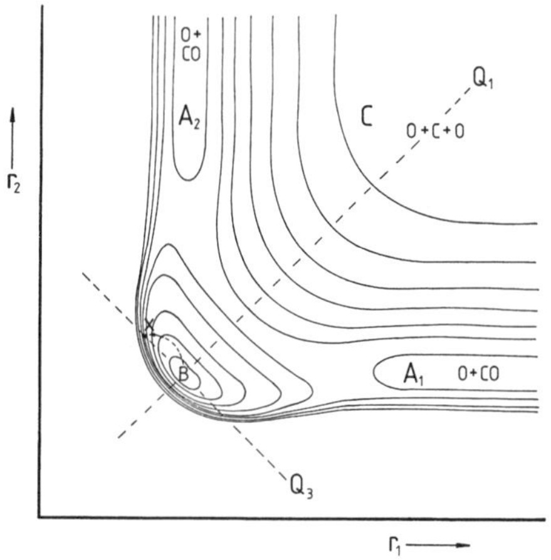

Figure 5.62 Contour map of potential energy as a function of the two C—O bond lengths, r_1 and r_2, in the CO_2 molecule

closely at either extreme of the vibrational motion.

The vibrations ν_1 and ν_3 of CO_2 illustrate an important general point in polyatomic molecules, namely that some vibrations are dissociative, such as ν_1, and others are non-dissociative such as ν_3. The bending vibration ν_2 is also non-dissociative.

Figure 5.62 illustrates, in a classical sense, anharmonicity connecting the vibrations ν_1 and ν_3 of CO_2. If the molecule starts from point X on the potential energy surface it will tend to follow the line of maximum slope to the equilibrium point B. In doing so it deviates considerably from the dashed line representing Q_3 and so involves an admixture of Q_1 and Q_3.

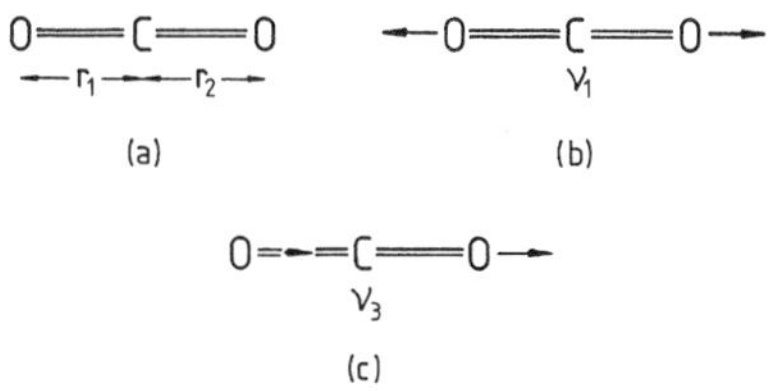

Figure 5.63 (a) The coordinates r_1 and r_2, (b) the symmetric stretching vibration ν_1, and (c) the antisymmetric stretching vibration ν_3 of the CO_2 molecule

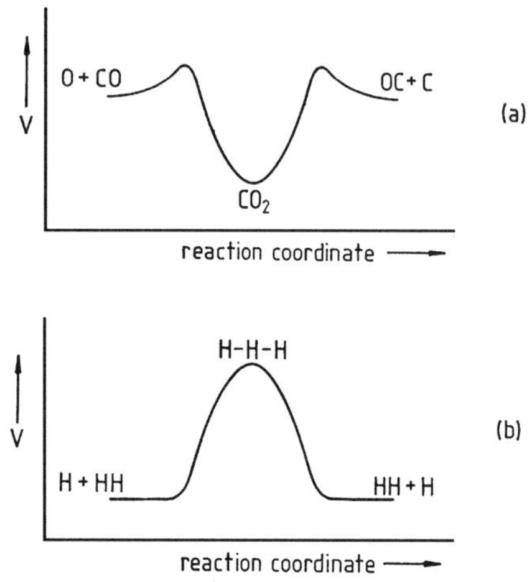

Figure 5.64 Potential energy variation along the reaction coordinate for the processes (a) $O+CO \longrightarrow OC+O$ and (b) $H+HH \longrightarrow HH+H$

5.2.6.2 Anharmonic vibrations

Whatever the form of the potential function for a particular vibration, the energy levels are always, to a greater or lesser extent, anharmonic. The effect is often, but not always, to cause the levels to close up smoothly with increasing quantum number v, rather as they do in a diatomic molecule.

Anharmonicity is taken account of in the Schrödinger equation by modifying the harmonic oscillator potential function of equation (5.103) to

$$V = \tfrac{1}{2}\sum_{i\leqslant j} k_{ij}R_iR_j + \sum_{i\leqslant j\leqslant k} k_{ijk}R_iR_jR_k + \sum_{i\leqslant j\leqslant k\leqslant l} k_{ijkl}R_iR_jR_kR_l + \ldots \tag{5.244}$$

where it is conventional to use the k symbols for force constants when the sums over $i, j,\ldots$ are restricted by $i\leqslant j, \ldots$. An alternative formulation is

$$V = \frac{1}{2!}\sum_{i,j} f_{ij}R_iR_j + \frac{1}{3!}\sum_{i,j,k} f_{ijk}R_iR_jR_k + \frac{1}{4!}\sum_{i,j,k,l} f_{ijkl}R_iR_jR_kR_l + \ldots \tag{5.245}$$

when the f force constants are used when the sum over $i, j, \ldots$ is unrestricted. The first term in the potential is the quadratic term of the harmonic force field. The second and third terms correspond to the cubic and quartic contributions to the force field. Higher terms are not normally taken into account.

Solution of the Schrödinger equation for a molecule in which all vibration are non-degenerate gives the vibrational term values

$$\sum_i G(v_i) = \sum_i \omega_i(v_i + \tfrac{1}{2}) + \sum_{i\leqslant j} x_{ij}(v_i+\tfrac{1}{2})(v_j+\tfrac{1}{2}) + \ldots \tag{5.246}$$

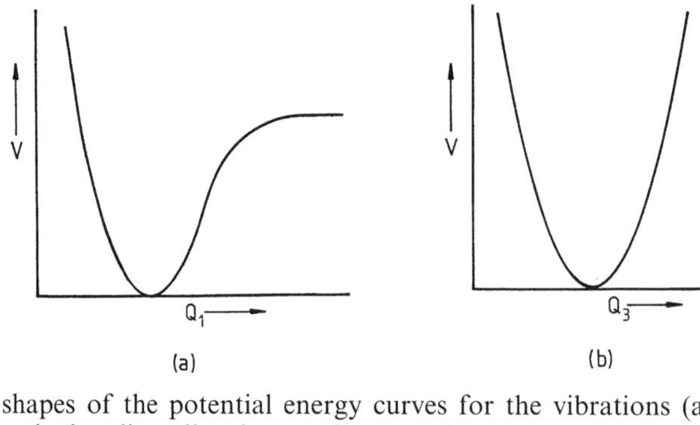

Figure 5.65 General shapes of the potential energy curves for the vibrations (a) ν_1 and (b) ν_3 of the CO_2 molecule. The curve for the bending vibration ν_2 also resembles that in (b) in that there is no dissociation

where the ω_i are the fundamental vibration wavenumbers for infinitely small displacement from the equilibrium configuration—analogous to ω_e for a diatomic molecule. The x_{ii} are anharmonic constants and are analogous to $-\omega_e x_e$ in a diatomic[†] but the x_{ij} have no such analogues.

For a molecule in which there are degenerate vibrations, equation (5.246) is replaced by

$$\sum_i G(v_i) = \sum_i \omega_i(v_i + d_i/2) + \sum_{i \leqslant j} x_{ij}(v_i + d_i/2)(v_j + d_j/2) + \sum_{i \leqslant j} g_{ij} l_i l_j + \ldots \quad (5.247)$$

where d_i is the degree of degeneracy of vibration i, l is the angular momentum quantum number introduced in Sections 4.2.6.2 and 4.3 and g_{ij} are additional anharmonic constants.

The inclusion of the x_{ij} (and g_{ij}) terms for cases when $i \neq j$ demonstrates that one effect of anharmonicity in polyatomic molecules is to cause a degree of mixing, which is usually small, of normal modes of vibration. The vibrations of CO_2 can be used to illustrate this effect.

In a linear triatomic molecule, in which ν_2 is the doubly degenerate bending mode, equation (5.247) reduces to

$$\begin{aligned} G(v_1, v_2, v_3) = {} & \omega_1(v_1 + \tfrac{1}{2}) + \omega_2(v_2 + 1) \\ & + \omega_3(v_3 + \tfrac{1}{2}) + x_{11}(v_1 + \tfrac{1}{2})^2 \\ & + x_{22}(v_2 + 1)^2 + x_{33}(v_3 + \tfrac{1}{2})^2 \\ & + x_{12}(v_1 + \tfrac{1}{2})(v_2 + 1) \\ & + x_{23}(v_2 + 1)(v_3 + \tfrac{1}{2}) \\ & + x_{13}(v_1 + \tfrac{1}{2})(v_3 + \tfrac{1}{2}) \\ & + g_{22} l_2^2 + \ldots \end{aligned} \quad (5.248)$$

In CO_2, ν_1 is the symmetric stretching vibration shown in Figure 5.63(b). If the normal mode is not mixed with any others then $\Delta r_1 = \Delta r_2$ and the potential energy curve is like that in Figure 5.65(a). The closing up of the vibrational energy levels as v_1 increases is taken account of by the $x_{11}(v_1 + \frac{1}{2})^2$ term in equation (5.248). If we assume, for the moment, that $x_{12} = x_{13} = 0$ then the term values for ν_1 are independent of ν_2 and ν_3 and can be written as

$$G(v_1) \simeq \omega_1(v_1 + \tfrac{1}{2}) + x_{11}(v_1 + \tfrac{1}{2})^2 \quad (5.249)$$

which is analogous to the first two terms of equation (5.14) for a diatomic molecule: x_{11} is a negative quantity. However, if we do not neglect x_{12} and x_{13}, not only are the term values for each normal vibration no longer independent but it is now clear that we have made an approximation in assuming that sections can be made in the potential energy hypersurface to give potential curves such as those in Figure 5.65. The approximation in drawing such curves is the neglect of the $x_{ij}(v_i + d_i/2)(v_j + d_j/2)$ terms.

Table 5.18 summarizes the fundamental vibration wavenumbers and anharmonic constants for $^{12}C^{16}O_2$.

For diatomic molecules we have seen, as for example in equation (5.36), that anharmonicity is responsible for an important contribution to the vibration–rotation interaction constant α. There is also a harmonic contribution which is normally somewhat smaller. All this is true for polyatomic molecules also but, in addition, there are contributions from Coriolis interaction, such as that illustrated by equation (5.221), so that, in general,

$$\alpha_i = \alpha_i^{\text{harm}} + \alpha_i^{\text{anh}} + \alpha_i^{\text{Cor}} \quad (5.250)$$

5.2.6.3 *Fermi and Darling–Dennison resonances*

We have seen how normal vibrations are mixed by anharmonicity as evidenced by the x_{ij} and other high order cross-terms in the term value expression of equation (5.247). In fact, *all* vibrational energy levels, whether they correspond to fundamentals, overtones or

[†]The illogicality of using $-\omega_e x_e$ instead of simply x_e for this anharmonic constant of a diatomic molecule is now evident, but this is a convention which stubbornly remains.

Table 5.18 Vibration wavenumbers and anharmonic constants for $^{12}C^{16}O_2$ (all in cm^{-1})[a]

ω_1	1354.07	x_{12}	−5.37
ω_2	672.95	x_{13}	−19.27
ω_3	2396.30	x_{23}	−12.51
x_{11}	−3.10	g_{22}	−0.62
x_{22}	1.59	W_e	−52.84
x_{33}	−12.50		

[a]Quoted by Suzuki, I. (1968). *J. Mol. Spectrosc.*, **25**, 479.

combinations, are perturbed from their harmonic positions by anharmonicity. All the perturbations are rigidly restricted by symmetry so that only levels of the same symmetry can perturb each other. The effect is that the lower in energy of any two mutually perturbing levels is pushed down and the upper one pushed up by the perturbation: the closer the levels in the harmonic approximation, the greater is the perturbation. It is the sum total of all these symmetry-restricted perturbations which is reflected in the x_{ij} and higher order cross-terms.

Occasionally, however, two (or more) levels of the same symmetry occur particularly close together in the harmonic approximation giving rise to an unusually large perturbation. Such an effect is called Fermi resonance[55] if it involves a fundamental and an overtone or combination of the same symmetry. The usual procedure is to treat such a large effect separately in 'deperturbing' the levels involved in the Fermi resonance and quoting the x_{ij}, etc., for the deperturbed levels.

The energy levels can be deperturbed using the standard quantum mechanical techniques of perturbation theory. The result of applying first-order perturbation theory to a case of two very close-lying levels with unperturbed energies E_i^0 and E_j^0 is that the energy shift ΔE of the levels is given by the determinant

$$\begin{vmatrix} E_i^0 - \Delta E & W_e \\ W_e & E_j^0 - \Delta E \end{vmatrix} = 0 \tag{5.251}$$

where W_e is the Fermi coupling constant. It follows, from solving the determinant, that

$$\Delta E = [(E_i^0 + E_j^0)/2] \pm (4W_e^2 + \delta^2)^{\frac{1}{2}}/2 \tag{5.252}$$

where δ is the energy separation, $E_i^0 - E_j^0$, of the unperturbed levels. The first term is the mean energy of the unperturbed levels and the second term shows that the two levels are perturbed equally in magnitude but that one is pushed up in energy and the other down. If the separation δ of the unperturbed levels is small compared with $2W_e$ then the shift of the levels due to the perturbation is W_e up or down in energy. If, on the other hand, δ is large compared with $2W_e$, equation (5.252) becomes

$$\Delta E \approx [(E_i^0 + E_j^0)/2] \pm [(\delta/2) + W_e^2/\delta] \tag{5.253}$$

This equation represents the usual state of affairs for any pair of fairly well separated vibrational levels of the same symmetry. Being a second-order effect the interaction is not called Fermi resonance. Each pair of interacting levels contributes a term W_e^2/δ and the net effect of all such terms results in the x_{ij} terms in equation (5.247).

The pair of levels resulting from a Fermi resonance is called a Fermi diad and, in general, in a multiple level resonance it is a Fermi polyad.

As a result of Fermi resonance, the wave functions of the unperturbed states are mixed. In a Fermi diad the wave functions ψ_i and ψ_j of the perturbed states are linear combinations of the wave function ψ_i^0 and ψ_j^0 of the unperturbed states, as in the equations

$$\begin{aligned} \psi_i &= a\psi_i^0 - b\psi_j^0 \\ \psi_j &= b\psi_i^0 + a\psi_j^0 \end{aligned} \tag{5.254}$$

where a and b are numerical coefficients.

In an extreme case where $\delta = 0$, the wave functions ψ_i and ψ_j are both equal mixtures of ψ_i^0 and ψ_j^0. In such a case we cannot make separate vibrational assignments of each component of the observed diad.

There is a classic case in CO_2 of Fermi resonance in which δ is very small. Table 5.18

shows that the (unperturbed) levels 1^1 and 2^2 are only 8.17 cm^{-1} apart in the harmonic approximation. The symmetry species of the 1^1 level is Σ_g^+ and that of the 2^2 is $\Sigma_g^+ + \Delta_g [= (\pi_g)^2]$, where Σ_g^+ applies to the $2^{2,0}$ (i.e. $l_2 = 0$) and Δ_g to the $2^{2,2}$ (i.e. $l_2 = 2$) components. Because of the symmetry requirement for Fermi resonance, only the 1^1 and $2^{2,0}$ levels interact and the corresponding value of W_e is given in Table 5.18; it is clearly large compared with δ. In fact, δ is so small that it has proved difficult to show[56] that it is the lower of the deperturbed levels which involves the greatest contribution from 1^1 and the upper one the greatest contribution from $2^{2,0}$.

A direct result of the strong mixing is that the vibrational Raman spectrum shows transitions involving *both* Σ_g^+ components of the diad with *similar* intensities, whereas normally an overtone would be very weak. No Raman or infrared transitions are allowed between the zero-point and the Δ_g component of 2^2 so this unperturbed level cannot be located directly.

Because of the Fermi resonance between the 1^1 and $2^{2,0}$ levels there are further resonances high up the various stacks of vibrational levels. For example, the $1^1 2^{1,1}$ and $2^{3,1}$ levels are very close in the harmonic approximation and both have Π_u symmetry. More complex is the three-level resonance between the Σ_g^+ levels 1^2, $1^1 2^{1,0}$ and $2^{4,0}$.

A rather less common resonance occurring between close-lying vibrational energy levels of the same symmetry is Darling–Dennison resonance.[57] This is most likely to be encountered when the first overtones of two different vibrations, having different symmetries, are of similar energies.

The example which brought this type of resonance to the attention of Darling and Dennison occurs in H_2O. The symmetric OH-stretching vibration ν_1, shown in Figure 5.12, has a fundamental wavenumber 3657 cm^{-1} and the antisymmetric stretching vibration ν_3, also shown in the figure, has a fundamental wavenumber of 3756 cm^{-1}. Because they have different symmetries, a_1 for ν_1 and b_2 for ν_3, there can be no vibrational resonance between them. However, the symmetries of their overtone levels ($\upsilon_1 = 2$ and $\upsilon_3 = 2$) are the same since $a_1 \times a_1 = a_1$ and $b_2 \times b_2 = a_1$. These levels are expected to be fairly close together, giving rise to a Darling–Dennison resonance. In a similar way to a Fermi resonance, the effect is to push up in energy the higher of the two levels and to push down the lower.

The first overtone levels of ν_1 and ν_3 are observed at 7201 and 7445 cm^{-1}, respectively. Two effects cause these to be different from those predicted at 7314 and 7512 cm^{-1}, assuming both vibrations to be harmonic: (a) anharmonicity, causing the levels to close up with increasing υ, and (b) Darling–Dennison resonance pushing the resulting two levels further apart.

5.2.6.4 *Local mode treatment of vibrations*

Normal modes of vibration, with their corresponding normal coordinates, are satisfactory in describing the low-lying vibrational levels, usually those with $\upsilon = 1$ or 2, which can be investigated by traditional infrared absorption or Raman spectroscopy. For certain types of vibration, particularly stretching vibrations involving more than one symmetrically equivalent terminal atom, this description becomes much less satisfactory as υ increases.

Consider the CH-stretching vibrations of benzene, for example. Since there are six identical C—H bonds there are six CH-stretching vibrations. These belong to various symmetry species but only one, ν_2[†] illustrated in Figure 5.66, is totally symmetric a_{1g}.

It might be supposed that, since the potential energy curve for ν_2 is of a similar shape to that in Figure 5.65(a), if we excite the molecule with sufficiently high energy it will eventually dissociate losing six hydrogen atoms in the process:

[†]Using the Wilson numbering system which is frequently used for benzene (see the Bibliography at the end of the chapter).

$$C_6H_6 \longrightarrow C_6 + 6H \qquad (5.255)$$

This seems reasonable when we think only in terms of normal vibrations but intuition suggests that, since the dissociation in equation (5.255) would require something like six times the C—H bond dissociation energy (*ca* $6 \times 412\,\mathrm{kJ\,mol^{-1}}$), the process

$$C_6H_6 \longrightarrow C_6H_5 + H \qquad (5.256)$$

is surely more likely to occur. It is now known that this is what happens, but the difficulty is that no normal vibration of benzene leads to CH stretching being localized in only one bond. These considerations, together with a very early (1929) observation of up to eight quanta of CH-stretching in benzene, led to the concept of a local mode of vibration.[58]

There is a similar problem concerning the H_2O molecule. No normal vibration leads to the dissociation into H + OH. However, this case is simpler because there are only two OH-stretching vibrations, the symmetric stretch, ν_1, and the antisymmetric stretch, ν_3, shown in Figure 5.12.

The fact that, at high levels of excitation of bond stretching vibrations in H_2O, the motion becomes concentrated in only one bond requires, again, the concept of a local mode of vibration. The model which has been adopted to describe this, and other, local modes is that of harmonically coupled anharmonic oscillators.[59] In H_2O there are two anharmonic oscillators representing the stretching motion of either of the O—H bonds.

The anharmonic oscillator potential which has been adopted is the Morse potential given in equation (5.22) and which takes account of the molecule dissociating. The vibrational term values arising from this potential are given by equation (5.23), which can be rewritten as

$$G(\upsilon) = \omega_m(\upsilon + \tfrac{1}{2}) + x_m(\upsilon + \tfrac{1}{2})^2 \qquad (5.257)$$

where ω_m and x_m are the vibration wavenumber and anharmonic constant in the Morse approximation. At high vibrational energies the term

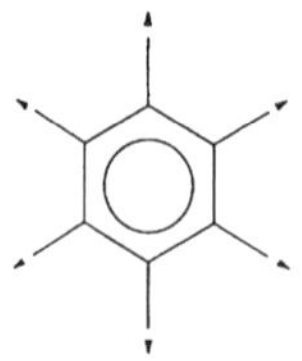

Figure 5.66 The totally symmetric CH-stretching vibration ν_2 of benzene

values for vibration in both of the bonds are given, to a good approximation, by this equation. At lower vibrational energies the motions of the bonds are no longer independent, with the energy levels doubly degenerate, but they are coupled. An interbond coupling parameter λ has been defined where, for H_2O,

$$\lambda = \frac{1}{2}(\upsilon_3 - \upsilon_1) = \frac{1}{2}(3756 - 3657) = 49.5\,\mathrm{cm^{-1}} \qquad (5.258)$$

When the effects of anharmonicity, represented by the parameter x_m in equation (5.257), exceed the strength of the interbond coupling a local mode description is more appropriate while, when the opposite is the case, a normal mode description is more appropriate.

There must be a smooth correlation between these two extremes and this is illustrated schematically in Figure 5.67 for a molecule such as H_2O with two bond-stretching vibrations ν_1 and ν_3.

On the left-hand side are shown the vibrational energy levels, in the normal mode extreme, when the Morse parameter x_m is zero but the coupling parameter λ is non-zero. The levels are labelled $(\upsilon_1\upsilon_3)$ with the vibrational quantum numbers υ_1 and υ_3. In addition to the overtone levels of ν_1 and ν_3 all the possible combination levels are also shown. For example, the (11) combination level is close to the (20) and (02) overtone levels.

On the right-hand side are the energy levels in the local mode extreme when λ is zero but x_m is non-zero. The labels m and n, used as in $[mn]$, refer to the number of quanta of (Morse) vibration in each of the two bonds. For

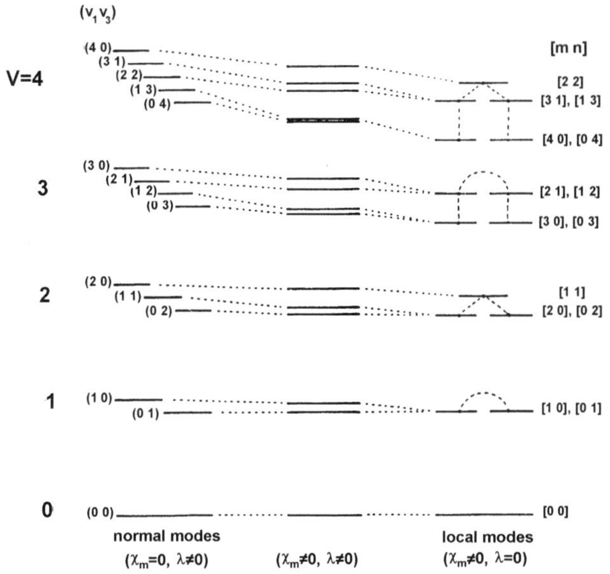

Figure 5.67 Correlation of the vibrational energy levels of two coupled Morse oscillators between the extremes of a normal mode and a local mode treatment [Reproduced, with permission, from Mills, I. M. (1992). *Making Light Work: Advances in Near Infrared Spectroscopy*, Eds Murray, I. and Cowe, I. VCH, Weinheim]

example, corresponding to the normal mode limit for which there are two separated levels, (10 and (01), there are two degenerate levels [10] and [01], in the local mode limit corresponding to excitation in either of the identical bonds. Further up the stack of local mode levels it can be seen that there are many degenerate pairs arising when there is identical excitation in both bonds.

In the centre of Figure 5.67 are shown the vibrational energy levels which result from an intermediate situation where both x_m and λ are non-zero. For each value of the total vibrational quantum number V where

$$V = v_1 + v_3 = m + n \qquad (5.259)$$

there is a manifold of levels the lower two of which become more nearly degenerate with increasing V.

The figure shows that, in general, the normal mode model is more successful for the lower energy levels and the local mode model for the high levels. Where in the stack of vibrational levels the region of switch-over from one model to the other occurs depends on the molecule. For example,[59] the local mode model gives a better fit to the experimentally observed lower vibrational levels of the bond-stretching vibrations of H_2O and $C_2{}^1H_2$ (acetylene), whereas the normal mode model gives a better fit to the lower levels of $C_2{}^2H_2$ and SO_2.

This local mode behaviour applies to vibrations of many other molecules with two or more equivalent terminal atoms, and CO_2 is such an example.

The probability of going, in Figure 5.62, from region B, where CO_2 is in the equilibrium configuration, to region C, in which it has lost both oxygen atoms, by exciting more and more quanta of ν_1 and proceeding along the dashed line Q_1, seems energetically unlikely. The molecule will prefer to go from B to A_1 (or A_2), losing one oxygen atom only, by local mode behaviour. Then it will go from A_1 (or A_2) to C by dissociation of CO. Therefore, the high overtone levels of CO stretching in CO_2 will be better described by a local mode model, involving anharmonic stretching of one C=O bond, rather than a normal mode model.

Although any stretching vibration involving symmetrically equivalent atoms should show local mode behaviour at high v, the CH-stretching vibrations are more amenable to investigation. The reason for this is that their vibrational quanta are large and, therefore, a relatively low value of v is nearer to the dissociation energy than, say, for CF-stretching vibrations, and the transition probability decreases rapidly as v increases.

5.2.7 Vibrational potential functions with more than one minimum

Figure 5.65(a) shows a potential energy curve for a stretching vibration of a polyatomic molecule; this qualitatively resembles that for a diatomic.

Figure 5.65(b) shows a potential energy curve for the non-dissociative bending vibration of CO_2. This kind of curve applies equally to any bending vibration of a linear molecule, such as HCN or acetylene (C_2H_2), where the minimum in the curve corresponds to the linear configuration.

Both types of potential curve are typical in that they show only a single minimum and the energy levels can be fitted to a term value expression such as that in equation (5.247).

There are, however, some vibrations whose potential functions do not resemble either of those of Figure 5.65 and whose term values are not given by equation (5.247). These vibrations have potential functions showing more than one minimum. Such vibrations can be separated into various types which will now be discussed individually.

5.2.7.1 *Inversion vibrations*

The inversion vibration, ν_2, of ammonia is illustrated in Figure 5.31. It has the symmetry species a_1 in the C_{3v} point group. The equilibrium configuration of NH_3 is pyramidal with $\angle HNH = 106.7^\circ$ but, for large amplitude motion in this vibrational mode, the molecule may go through the planar configuration to a pyramidal configuration which is identical with the initial one except that the pyramid has been inverted. The planar and the two equivalent pyramidal configurations are shown in Figure 5.68. The pyramidal configurations (i) and (iii) obviously correspond to two identical minima in the potential energy curve and the planar configuration (ii) to an energy maximum. The energy difference V_1 between the minima and maximum is the energy barrier and is the classical energy which is required to invert the pyramid.

In a quantum mechanical system it may not be necessary to surmount the barrier in order to go from one equilibrium configuration to an equivalent one. The phenomenon of quantum mechanical tunnelling allows a degree of penetration of the barrier. If the barrier is sufficiently low or narrow (or both), the penetration may be so great that interaction may occur between the identical sets of vibrational levels into two components. The splitting becomes greater towards the top of the barrier where tunnelling is more effective.

The splitting of levels is shown in Figure 5.69(b). Above the barrier they eventually become evenly spaced but those just above the barrier show a slight staggering.

The term non-rigid or, sometimes, floppy is applied to NH_3 and, generally, to any molecule with a low barrier between two or more equivalent minima in the potential energy surface and through which tunnelling is appreciable.

Figure 5.69(a) shows how the splittings are removed and the levels are evenly spaced in the case where the barrier is infinitely high (or infinitely wide). When the barrier is reduced to zero the levels again become evenly spaced, as shown in Figure 5.69(c).

The effect of tunnelling on the vibrational wave functions is shown in Figure 5.70. The wave functions in Figure 5.70(a) are those for a potential function like that of Figure 5.69(a) with an infinite barrier. In Figure 5.70(a) the

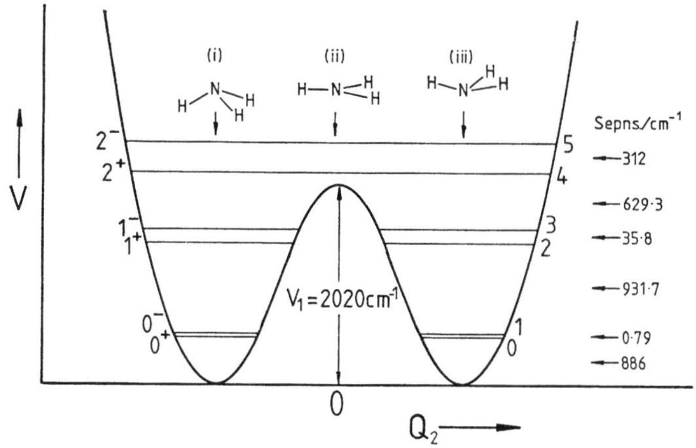

Figure 5.68 Potential energy curve for the inversion vibration ν_2 of the NH_3 molecule. The equilibrium configuration, (i) or (iii), is pyramidal and, classically, it requires a potential energy V_1, the height of the barrier, to go to the planar configuration (ii)

energy is the same, whether the wave functions for each pair of levels are in-phase or out-of-phase, because there is no overlap.

As the barrier decreases from Figure 5.69(a) to 5.69(b), the wave functions for each vibrational level start to overlap. When this happens the relative signs of the overlapping wave functions are important: they overlap either with the signs in-phase, as in the 0^+, 1^+, 2^+, . . . levels in Figure 5.70(b), or out-of-phase, as in the 0^-, 1^-, 2^-, . . . levels.

In Figure 5.70(b) we can see that all the wave functions arising from the out-of-phase combinations have a node, where $\psi_v = 0$, at the centre whereas the others have not. The fact that all the out-of-phase combinations result in an energy higher than that of the corresponding in-phase combinations follows from the general rule that additional nodes increase the energy.

The rate of tunnelling through the barrier, i.e. the rate at which the molecule inverts from one form to the other, is directly related to the frequency separation $\Delta\nu$ of the vibrational levels in which the molecule finds itself. The time τ for inversion is given by

$$\tau = (2\Delta\nu)^{-1} \qquad (5.260)$$

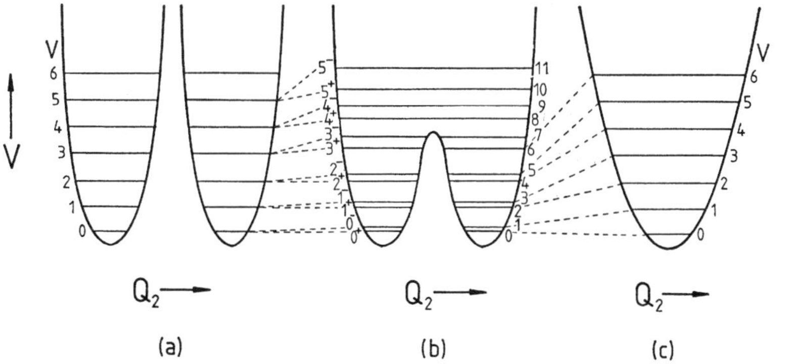

Figure 5.69 Potential energy curves and energy levels for the inversion vibration ν_2 of NH_3 when the barrier to planarity is (a) infinite, (b) moderately low and (c) zero

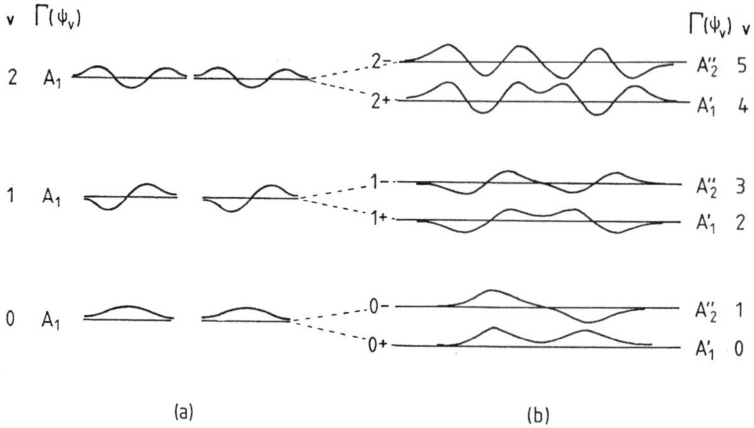

Figure 5.70 The forms of the vibrational wave functions for an inversion vibration when the barrier to planarity is (a) infinite and (b) moderately low

The labelling of the vibrational levels on the left-hand side of Figures 5.69(b) and 5.70(b) uses the superscript + or − to indicate an in-phase or out-of-phase combination, respectively. The wave functions ψ_v^+ and ψ_v^- corresponding to these levels are given by

$$\psi_v^+ = \psi_v(Q_2) + \psi_v(-Q_2)$$
$$\psi_v^- = \psi_v(Q_2) - \psi_v(-Q_2) \qquad (5.261)$$

where $Q_2 = 0$ corresponds to the planar configuration. The vibrational quantum number used in the label is that which is appropriate to the infinite barrier limit.

On the right-hand side of Figures 5.69(b) and 5.70(b) the labelling used is the vibrational quantum number appropriate to the zero barrier limit.

Which labelling system is used depends generally on whether the molecule under consideration approximates to the high or low barrier extreme. However there is a tendency for the zero barrier labelling system to be used in most cases where the splitting of levels is observed, even if the barrier is fairly high.

Figure 5.70(b) shows that tunnelling through the barrier in ammonia effectively introduces a plane of symmetry at $Q_2 = 0$ and the + and − wave functions are symmetric or antisymmetric, respectively, to reflection in this plane. Therefore, in attaching symmetry species to the wave function, we should use a point group of higher order than C_{3v} to which rigid ammonia belongs. That group of higher order is obtained by adding the σ_h element to the C_3 and $3\sigma_v$ of C_{3v} and is therefore D_{3h}. Comparison of the C_{3v} (Table 5.8) and D_{3h} (Table A.33 in the Appendix) character tables shows that, in progressing from D_{3h} to C_{3v} the symmetry elements which are retained are C_3 and $3\sigma_v$ so that the symmetry species of the two groups correlate in the way shown in Table 5.19. This correlation shows that the symmetry species of + and − levels in the D_{3h} point group are A_1' and A_2'', respectively, as indicated in Figure 5.70(b). They both correlate with A_1 in C_{3v}.

In ammonia the height of the barrier V_1 is 2020 cm^{-1} and is, like all potential function parameters, isotope independent.† On the other hand, the vibrational term values and the effectiveness of tunnelling are reduced for heavier isotopes. Thus the $v_2 = 1$ to $v_2 = 0$ or (0^- to 0^+) separation is 23.786, 22.705 and 1.600 GHz for $^{14}N^1H_3$, $^{15}N^1H_3$ and $^{14}N^2H_3$,

†This is true so long as the form of the normal coordinate is unaffected by isotopic substitution. If, for example, Q_2 were pure inversion in $^{14}N^1H_3$ but, in $^{14}N^2H_3$, was slightly mixed with Q_1, the N–H stretching coordinate of the same symmetry, the effective value of V_1 obtained from the observed levels would be different in the two species.

Table 5.19 Correlation of symmetry species in the C_{3v} and D_{3h} point groups

C_{3v}	D_{3h}
A_1	A_1', A_2''
A_2	A_2', A_1''
E	E', E''

respectively. These separations are all so small that the 2_0^1 vibrational transition occurs in the microwave region of the spectrum. It is an A_2''–A_1' transition and, since $\Gamma(A_2'') = \Gamma(T_z)$, is electric dipole allowed. Figure 5.68 shows separations for some of the lower vibrational levels of $^{14}N^1H_3$.

From equation (5.260) it follows that the time τ for inversion is 2.1×10^{-11}, 2.2×10^{-11} and 3.1×10^{-10} s for $^{14}N^1H_3$, $^{15}N^1H_3$ and $^{14}N^2H_3$, respectively.

The rotational transitions accompanying the 2_0^1 transition show that separation of the vibrational levels varies with the quantum numbers J and K. When the molecule is rotating about the top axis, centrifugal effects tend to flatten the pyramid, decrease the barrier and increase the splitting of vibrational levels. Rotation about an axis perpendicular to the top axis steepens the pyramid and therefore decreases the vibrational level separation.

The nuclear spin statistical weights for the rotational levels associated with $v_2 = 0$ and 1 and all other vibrational levels are those of a D_{3h} molecule. For $^{14}N^1H_3$ those levels with K a multiple of three have a statistical weight of 2 while the rest have a weight of 1. In addition, for $K = 0$, only levels with odd J occur for the $v_2 = 0$ (A_1') state and, for the $v_2 = 1$ (A_2'') state, only those with even J. If we can imagine the splitting of the $v_2 = 0$ and 1 levels smoothly approaching zero, as it is effectively in AsH_3, the $K = 0$ levels for $v_2 = 0$ and 1 merge and then the rotational levels with both even and odd J have the same statistical weights. This is consistent with using the C_{3v} point group when tunnelling through the barrier is negligible compared with the resolution of a particular experiment.

There have been several suggested potential functions which reproduce the general W-shape, as in Figure 5.68, but the usefulness of such a function is reflected in its ability to reproduce the observed vibrational term values in various molecules of this type, particularly NH_3 for which more experimental data are available. Perhaps the most successful potential function is

$$V(Q) = \tfrac{1}{2}aQ^2 + b\exp(-cQ^2) \qquad (5.262)$$

proposed by Swalen and Ibers.[60] this is a parabolic potential (first term) perturbed by a gaussian barrier (second term).

The dipole moment of NH_3 has been determined experimentally as 1.47 D (4.90×10^{-30} C m). It might be thought that a non-zero dipole moment is inconsistent with the D_{3h} classification of the vibrational levels, but this is not so. In the infinitely high barrier limit of Figure 5.69(a) a measured dipole moment is the permanent dipole moment in the $v_2 = 0$ level. When the $v_2 = 0$ level is slightly split by tunnelling, as in Figure 5.70(b), the measured dipole moment is really the transition moment [equation (5.4)] for the electric dipole transition between the two components of the split level. When the barrier becomes zero, as in Figure 5.69(c), the splitting is so large that the transition moment becomes very small and the measured dipole moment essentially zero, as it must be in a planar molecule.

Some other molecules which have a vibration with a W-shaped potential function are formamide (NH_2CHO) and aniline ($C_6H_5NH_2$) with barriers of 370 and 547 cm^{-1}, respectively, for inversion of the NH_2 group. Formaldehyde (H_2CO), although planar in its ground electronic state, is pyramidal in its first excited triplet and singlet states (see Section 6.3.1.5) with barrier heights of 783 and 350 cm^{-1}, respectively, for inversion of the CH_2 group.

In a case where the barrier is non-zero, but so low that it is below the zero-point level, the molecule is referred to as quasi-planar.

5.2.7.2 *Ring-puckering vibrations*

Cyclic molecules which are at least partly saturated and contain such groups as —CH_2—, —O— or —S— have low wavenumber vibrations involving a bending motion of the group out of the plane of the ring (or what would be the plane if it were a planar molecule). Two such examples are cyclobutane and cyclopentene, shown in Figure 5.71(a) and (b), respectively. In cyclobutane the puckering vibration is an out-of-plane bending of the molecule about a line joining two opposite carbon atoms. Bell[61] proposed that the potential energy for such a vibration, in a molecule whose equilibrium configuration is planar, is that of a quartic oscillator:

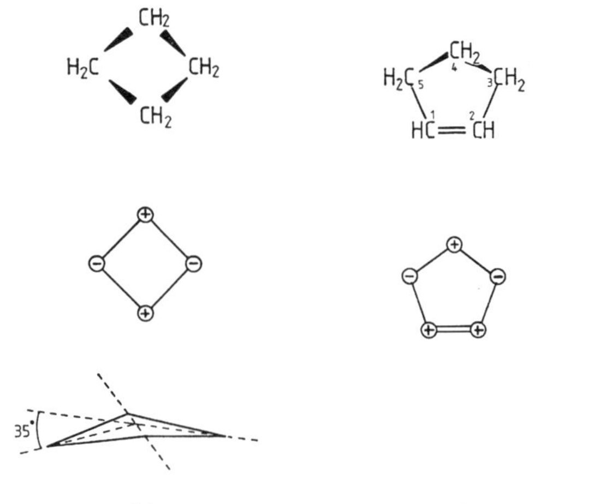

(a) (b)

Figure 5.71 (a) The cyclobutane molecule, the ring-puckering vibration and the equilibrium out-of-plane angle of the ring. (b) The cyclopentene molecule, having a pseudo-four-membered ring, and the ring-puckering vibration

$$V(Q) = aQ^4 \tag{5.263}$$

where Q is the puckering coordinate. The reason for this is that, for small displacements of the nuclei in this normal coordinate, the changes in bond angles and lengths are proportional to the *square* of the displacement; this results in the potential energy being proportional to its fourth power. Such a potential curve is flat-bottomed and steep-sided compared with a parabolic potential. The vibrational energy levels for a quartic potential function diverge smoothly with increasing quantum number v.

In the case of cyclobutane the equilibrium configuration of the carbon atoms is non-planar. The dihedral angle, shown in Figure 5.71(a), is 35°. The potential curve for the puckering vibration shows, therefore, two identical minima corresponding to the ring being puckered 'downwards', as in Figure 5.71(a), or 'upwards'.

The introduction of a barrier is accomplished by including a quadratic term in the potential, giving

$$V = aQ^4 + bQ^2 \tag{5.264}$$

Since the parameter b is negative the barrier is produced by adding an inverted parabola to a quartic curve.

The W-shaped potential functions for an inversion vibration (Section 5.2.7.1) and for a ring-puckering vibration are more than qualitatively similar. The potential function in equation (5.264) gives fairly good inversion vibration energy levels in NH_3, but agreement with the observed levels is not as good as with the potential in equation (5.262).

This may seem surprising because, for large vibrational displacements, the potential function in equation (5.262) is dominated by the quadratic term whereas that in equation (5.264) is dominated by the quartic term. The result is that, above the barrier, the vibrational energy levels for the function of equation (5.262) settle down to being equally separated whereas, for the function of equation (5.264), they diverge smoothly with increasing V. However, since observations are usually confined to relatively low-lying energy levels, this is not an important

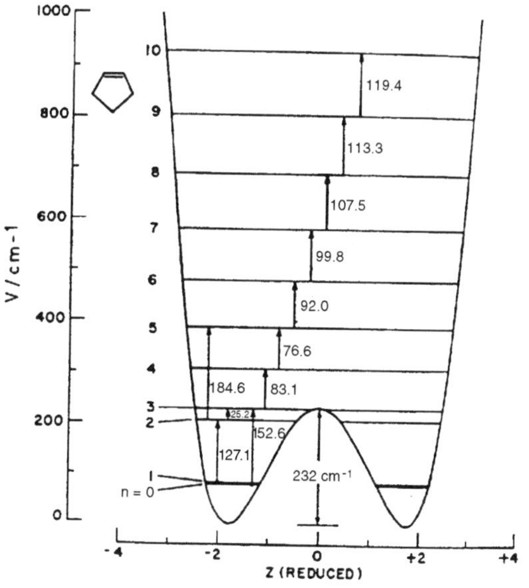

Figure 5.72 Puckering potential function for cyclopentene. Potential energy V is plotted against a reduced coordinate z which is proportional to the normal coordinate but takes into account the reduced mass for the puckering motion [Reproduced, with permission, from Laane, J. and Lord, R. C. (1967). *J. Chem. Phys.*, **47**, 4941]

difference in practice. Behaviour close to equilibrium and near to the top of the barrier is fairly well reproduced by both functions.

Some larger ring molecules containing unsaturated groups may behave, in respect of ring puckering, as pseudo-four-membered rings. Such an example is cyclopentene, shown in Figure 5.71(b), in which the —HC=CH— part of the ring is so resistant to twisting that the low wavenumber puckering vibration is essentially a motion only of the CH_2 group in position 4. In the equilibrium configuration the C(4) atom is not coplanar with the other carbon atoms: the out-of-plane angle corresponding to that in cyclobutane is 23° and the molecule belongs to the C_s point group.

The puckering potential function for cyclopentene is shown in Figure 5.72. This has been derived from the observation of many $\Delta v = 1$ and a few $\Delta v = 3$ transitions in the far infrared spectrum which is reproduced in Figure 5.73.

If splitting of the levels by tunnelling through the barrier is resolved C_{2v} classification is used, the $v = 0, 1, 2, 3, \ldots$ levels are A_1, B_1, A_1, B_1, . . ., and the bands are B_1–A_1 or A_1–B_1 type C bands of an asymmetric rotor dominated by an intense maximum in the rotational contour at the band centre. Further confirmation of the interpretation of this spectrum has been provided by the observation[62] of many $\Delta v = 2$ and one $\Delta v = 4$ A_1–A_1 or B_1–B_1 transitions in the vibrational Raman spectrum. Although the $\Delta v = 1$ or 3 transitions are allowed by both Raman and infrared selection rules, they are expected to be weak in the Raman spectrum and have not been observed.

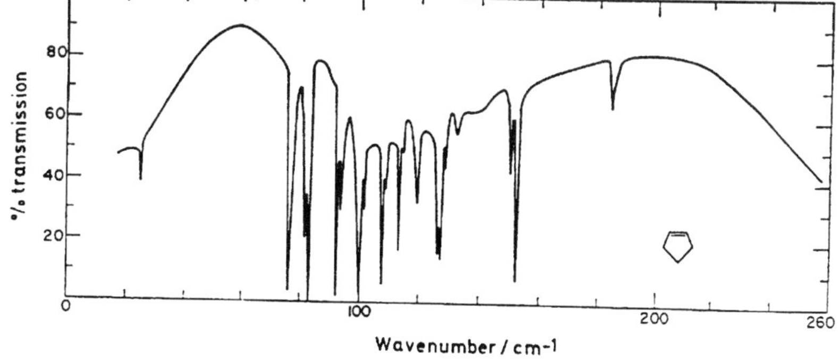

Figure 5.73 Far infrared spectrum of cyclopentene in which percentage transmission is plotted against wavenumber [Reproduced, with permission, from Laane, J. and Lord, R. C. (1967). *J. Chem. Phys.*, **47**, 4941]

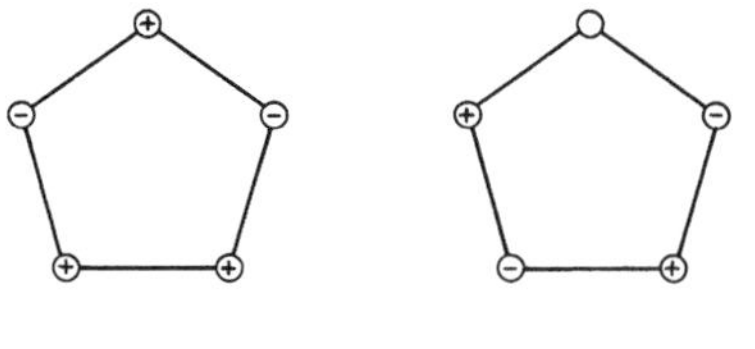

Figure 5.74 The two components of the doubly degenerate puckering vibration in the cyclopentane molecule

In cases where the barrier is non-zero but below the zero-point level the molecule is said to be quasi-planar, as for inversion. This is the case in the four-membered ring molecule oxetane, $\overline{CH_2-CH_2-CH_2}$ (O bridging), in which the barrier is only 15 cm^{-1}.

In a five-membered ring in which there is no part which is particularly resistant to twisting there are *two* low wavenumber ring-puckering vibrations. We take cyclopentane as our example and make an initial approximation that the carbon atoms are coplanar. The two ring-puckering modes, shown in Figure 5.74, are degenerate. That in Figure 5.74(a) is analogous to the ring-puckering vibration and that in Figure 5.74(b) to the C=C twisting vibration of cyclopentene in which they can be treated independently. In cyclopentane they are coupled through a vibration in which each CH_2 group undergoes a ring-puckering motion the phase of which rotates around the ring. This vibration is known as pseudorotation but there is little or no rotational angular momentum associated with it. The pseudorotational term values, with dimensions of wavenumber, are given by

$$F(n) = n^2 B = n^2 h^2 / 8\pi^2 cI \qquad (5.265)$$

where $n = 0, \pm 1, \pm 2, \ldots$ and I is the moment of inertia association with the pseudorotation.

The potential function for the degenerate puckering vibration is represented by a three-dimensional surface in which potential energy is plotted against the normal coordinates Q_a and Q_b corresponding to the two components of the vibration. This surface is bowl-shaped with

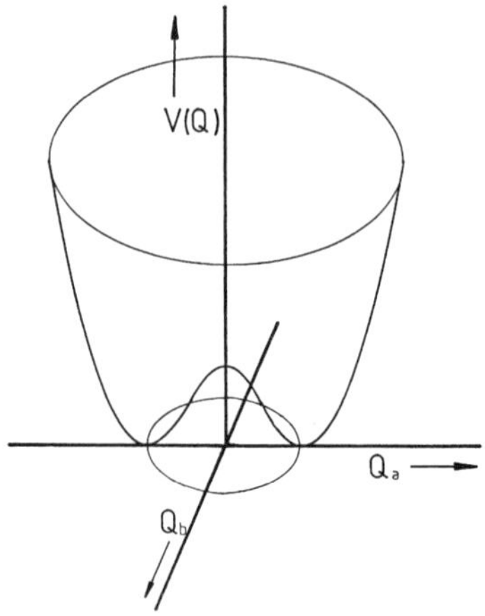

Figure 5.75 The potential function for the degenerate puckering vibration in cyclopentane. The barrier to pseudorotation around the circular valley is assumed to be zero

steep (quartic) sides. In fact, however, the equilibrium configuration of cyclopentane is that of a puckered ring and this results in there being a central circular hump in the potential surface as shown in Figure 5.75. The height of the hump is the barrier to planarity.

Although in cross-section the resulting W-shaped potential resembles that for the non-degenerate puckering vibration of cyclobutane, there is one major difference. In cyclobutane the molecule can move from one minimum to the other only by tunnelling through the barrier. In cyclopentane it can reach any point in the circular valley by pseudorotation without tunnelling through the circular hump. A consequence of this is that there is no splitting of levels due to tunnelling, even close to the top of the barrier.

If the bottom of the circular valley is smooth with no humps encountered in travelling around it there is free pseudorotation, a motion which is different from that of actual rotation. Alternatively, there may be maxima and minima in the valley and, provided that the maxima are small compared with the barrier to planarity, the motion around the circular valley can be treated independently of the puckering vibrational motion.

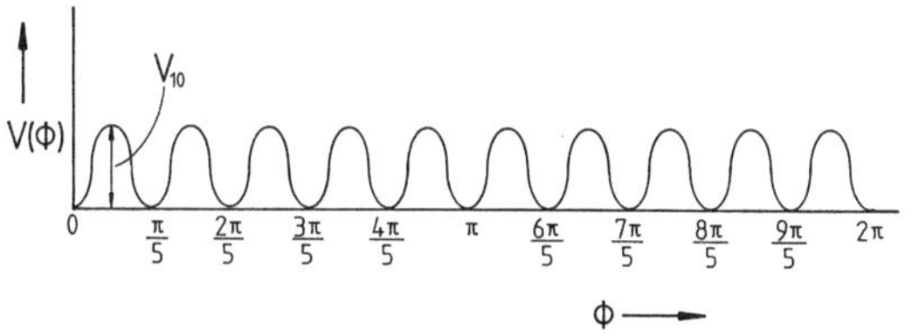

Figure 5.76 The 10 maxima and minima in the potential function for cyclopentane representing motion around the circular valley in Figure 5.75 and assuming a small barrier to pseudorotation

In cyclopentane there are 10 maxima and 10 minima. A potential function which will produce this behaviour is

$$V(\phi) = \tfrac{1}{2}V_{10}(1 - \cos 10\phi) \qquad (5.266)$$

where ϕ is the angle of pseudorotation and V_{10} is the height of the 10-fold barrier. This function is shown in Figure 5.76, which shows the maxima and minima encountered during pseudorotation through 2π rad. The height of each maximum is V_{10} and it is only when the molecule has sufficient energy to surmount the barrier that true pseudorotation occurs.

In general, the potential function for hindered pseudorotation, in cases where the barrier to pseudorotation is small compared with the barrier to planarity, is given by

$$V(\phi) = \tfrac{1}{2}\sum_{n} V_n(1 - \cos n\phi) \qquad (5.267)$$

In cyclopentane, strictly, $V(\phi)$ should include all terms of this summation with n a multiple of 10, but the V_{10} term is likely to be the most important.

The nature of the function in equation (5.267) shows, by comparison with that for torsional vibrations discussed in Section 5.2.7.4, that hindered pseudorotation could also be described as pseudotorsional vibration.

Cyclopentane has no permanent dipole moment so that pure pseudorotational transitions cannot be studied by either far infrared or microwave spectroscopy. To overcome this difficulty, molecules such as silacyclopentane ($\overline{SiH_2CH_2CH_2CH_2CH_2}$) have been investigated. In this case, where the planar molecule would have C_{2v} symmetry, all terms in equation (5.267) with n even are non-zero, although terms with low n are dominant.

5.2.7.3 *Bending vibrations in non-linear triatomic molecules*

The bending vibration, ν_2, of a nonlinear triatomic molecule such as H_2O shows a double minimum in the potential function $V(Q_2)$ of the type shown in Figure 5.77. The maximum in the curve corresponds to the unstable linear configuration and the minima to the two equivalent bent configurations. The barrier to linearity is very high, about $10\,400\ \mathrm{cm}^{-1}$.

This potential function bears some resemblance to that for inversion in that it has two identical minima, and a function of the form

$$V(Q_2) = \tfrac{1}{2}aQ_2^2 + b\exp(-cQ_2^2) \qquad (5.268)$$

identical with that of equation (5.262) for inversion, has been found to be reasonably satisfactory [but a function like that for ring-puckering, in equation (5.264), may also be used]. There is one major difference from inversion in that the bending vibration is, in the linear molecule, a doubly degenerate π vibration with components ν_{2a} and ν_{2b}. Therefore, figure 5.77 is only a section of a three-dimensional potential energy surface with a circular valley. Any two sections of the surface taken perpendicular to each other and including the $V(Q_2)$ axis show two identical curves corresponding to ν_{2a} and ν_{2b}.

It follows that the Q_2 surface resembles that of Figure 5.75 for cyclopentane but, for the bending vibration, there can be no humps in the circular valley. Because we can proceed from a minimum in Q_{2a} to a minimum in Q_{2b} without surmounting the barrier, there is no splitting of levels due to tunnelling.

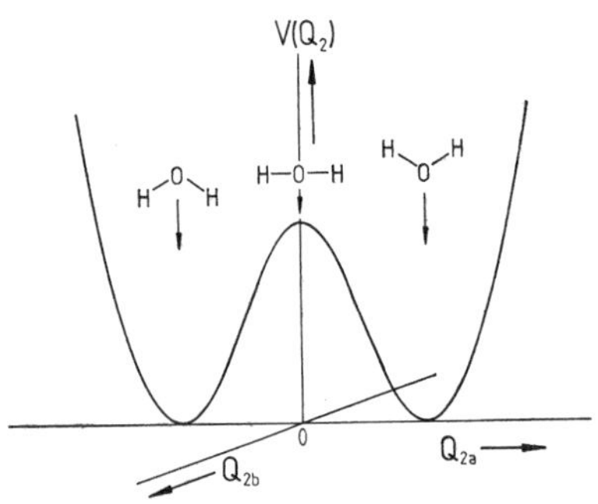

Figure 5.77 Potential function for the bending vibration, ν_2, of the H_2O molecule

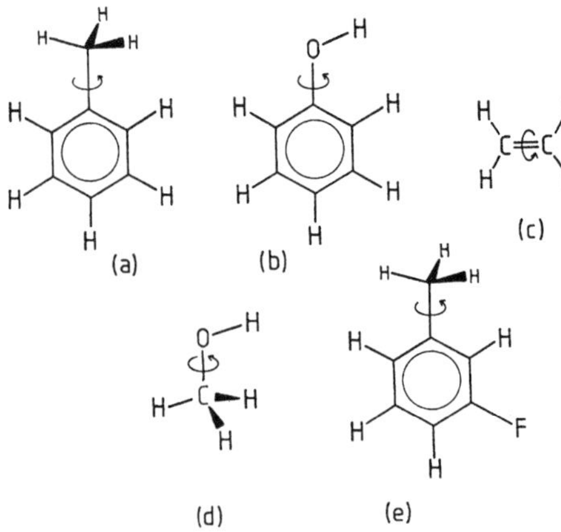

Figure 5.78 Torsional vibrations in the (a) toluene, (b) phenol, (c) ethylene, (d) methanol and (e) 3-fluorotoluene molecules

What happens to the vibrational levels as $V(Q_2)$ increases is complex in detail but fairly simple in principle.

In the rigidly bent molecule it is rotation about the a inertial axis (in-plane and perpendicular to C_2) which takes it from a minimum in Q_{2a} to one in Q_{2b} and this gives rise to an angular momentum. The quantum number associated with this angular momentum is K_a but, since H_2O is a prolate asymmetric rotor, this is not a good quantum number.

In the linear molecule, the corresponding angular momentum is about the C_∞ axis; it is due to vibration in the ν_2 bending mode and is described by the quantum number l.

As the top of the barrier is approached there is a complex set of rovibrational levels resulting from the correlation of levels with values of K_a and l appropriate to the bent and linear forms. Dixon[63] has shown that, from this set of levels, those with Σ overall symmetry, corresponding to K_a or $l = 0$, show fairly simple behaviour: the levels converge slowly as the top of the barrier is reached and then diverge slowly above the barrier. This means that, if a sufficient number of Σ levels can be identified, the barrier height can be obtained directly from the separations of these levels.

5.2.7.4 *Torsional vibrations*

A torsional vibration involves a twisting motion between two rigid parts of a molecule. We consider, somewhat arbitrarily, one part, called the top, twisting relative to the other part which is fixed and called the frame. For example, in toluene ($C_6H_5CH_3$) shown in Figure 5.78(a), we might take the CH_3 group to be the top and the phenyl group to be the frame. In some molecules, such as ethane (CH_3CH_3) and ethylene ($CH_2{=}CH_2$), where the torsional motion is about the C—C or C=C bond, respectively, the frame and the top are identical.

The general form of the potential function for torsional motion is expressed as

$$V(\phi) = \tfrac{1}{2}\sum_n V_n(1 - \cos n\phi) \qquad (5.269)$$

This equation is the same as equation (5.267) for hindered pseudorotation but here ϕ is the angle of twist of the top relative to the frame.

Which terms are important in the expression for $V(\phi)$ depends on the molecule concerned. For example, in phenol, shown in Figure 5.78(b), the potential function for torsion about the C—O bond must repeat for $\phi = n\pi$, where $n = 0, 1, 2, \ldots$. The minima in $V(\phi)$ at $\phi = 0, \pi, 2\pi \ldots$ correspond to the planar configuration which is stabilized by some overlap (conjugation) between the π-orbitals of the ring and the $2p$ orbital on the oxygen atom and

Table 5.20 Barrier heights (V) in some non-rigid molecules

Molecule	V/cm^{-1}	
NH_3	2020	(inversion)
$C_6N_5NH_2$	547	(NH_2 inversion)
C_6H_5OH	1207	(C—O torsion)
$CH_2{=}CH_2$	14000 }	(C=C torsion)
	22750 }	independent estimates
H_2O	10400	(bending)
CH_3OH	375	(C—O torsion)
CH_3CH_3	960	(C—C torsion)
$C_6H_5CH_3$	4.9	(C—CH_3 torsion)
CH_3NO_2	2.1	(C—N torsion)
$(CH_3)_3CNO_2$	71	(C—N torsion)
$(CHO)_2$	1771	(C—C torsion) *s-trans*→*s-cis*
$CH_2{=}CH{-}CH{=}CH_2$	2660	(C—C torsion) *s-trans*→*s-cis*

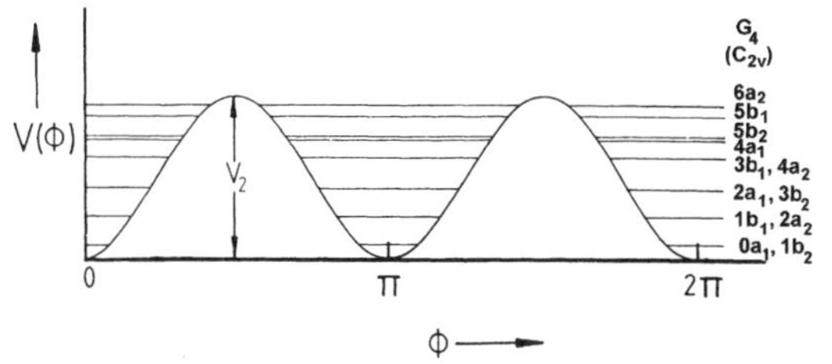

Figure 5.79 Schematic potential function of the form $V(\phi) = \frac{1}{2}V_2(1 - \cos 2\phi)$

perpendicular to the ring. This results in the term with $n = 2$ being dominant and $V(\phi)$ being given approximately by

$$V(\phi) \approx \tfrac{1}{2}V_2(1 - \cos 2\phi) \qquad (5.270)$$

which is shown in Figure 5.79.

Although the V_2 term is dominant n can, in principle, take any even value. For instance, it might be argued that if there were any steric hindrance in the planar configuration between the hydrogen atom on the OH group and those in the 2- (or *ortho*-) positions in the ring, this would be relieved in the $\phi = \pi/2$ configuration. Any secondary, shallower minima or even broadened maxima at $\phi = \pi/2, 3\pi/2, \ldots$ could be taken account of by a small V_4 term in the potential.

For the potential function, shown in Figure 5.79 for the torsional vibration of a molecule such as phenol, V_2 is the barrier height which is given for phenol, together with that for several other molecules, in Table 5.20.

Whereas the motion involved below the barrier is described as torsional vibration, that above the barrier is internal rotation.

Below the barrier, the vibrational energy levels within such a potential function converge with increasing v and, when tunnelling becomes appreciable near the top of the barrier, are split. In this case of a twofold barrier they are split into two components. If the symmetry classification of these components is required we have to use a group of order four (the group G_4) rather than the rigid molecule group C_s, of order two. This is analogous to the case of NH_3 (Section 5.2.7.1) in which levels split by tunnelling must be classified using a group of order six (D_{3h}) rather than that of the rigid molecule (C_{3v}) of order three.

Longuet-Higgins[64] described the additional symmetry operations of inversion and nuclear permutation which are necessary in deriving the groups to which non-rigid molecules in general belong.

We will consider the relatively simple case of a molecule such as phenol [Figure 5.78(b)] with a twofold torsional barrier and a torsional potential of the type shown in Figure 5.79.

Phenol, considered as a rigid molecule in which torsional motion occurs but with no tunnelling through the torsional barrier, belongs to the C_s point group and all symmetry classifications are of the type a' or a'' (see the character Table A.1 in the Appendix). However, when the molecule is considered as non-rigid with the possibility of tunnelling through the barrier, the symmetry operations of permutation, P, and inversion, I^* (or E^* if E is being used for the identity element) must be introduced. They are illustrated in Figure 5.80.

The permutation P is of the nuclei labelled 2 and 6 and of those labelled 3 and 5. These are the nuclei which are effectively exchanged when the hydrogen atom of the O—H group moves, during a torsional motion, to the opposite side of the z-axis. It is important to realize that the only permutations which are included in non-rigid molecule symmetry groups are those which are feasible through a relatively low energy process. The permutation in Figure 5.80 is feasible through a torsional motion but no other atom permutations are feasible.

The inversion I^* is an inversion of all the nuclei through the centre of gravity of the molecule. This is analogous to an inversion operation in a rigid molecule but phenol, considered as a rigid molecule, does not have a centre of inversion.

The four elements of the G_4 group are then I, P, I^* and P^*, where P^* is a permutation-inversion given by

$$P^* = P \times I^* \tag{5.271}$$

The character table for the G_4 group is given in Table 5.21. Comparison of this table with the character table for the C_{2v} point group (Table A.11 in the Appendix) shows that the G_4 and C_{2v} groups are isomorphous: there is a one-to-one relation between the elements I, P, I^* and P^* of G_4 with I, C_2, σ'_v and σ_v of C_{2v}. For this reason, the symmetry species labels A_1, A_2, B_1 and B_2, used for C_{2v}, are used for G_4 also.

The torsional energy levels in Figure 5.79 are assigned symmetry species relevant to the G_4 group. When the $v = 0$ level is not split by tunnelling it consists of a degenerate pair of levels with symmetry species a_1 and b_2. This

Table 5.21 Character table for the G_4 group

G_4	I	P	I^*	P^*		
A_1	1	1	1	1	T_z	$\alpha_{xx}, \alpha_{yy}, \alpha_{zz}$
A_2	1	1	−1	−1	R_z	α_{xy}
B_1	1	−1	−1	1	T_x, R_y	α_{xz}
B_2	1	−1	1	−1	T_y, R_x	α_{yz}

applies to all levels with v even but, if any of these levels is split by tunnelling, the b_2 component is always above the a_1 component. All levels with v odd comprise a b_1, a_2 degenerate pair. If such a pair is split by tunnelling the a_2 component is always above the b_1 component. (The integer preceding the symmetry species of the torsional levels is explained in the context of Figure 5.83.)

It is a characteristic of the vibrational energy levels of a non-rigid molecule that reduction of the symmetry produces degeneracies. In the case in Figure 5.79, reduction of the symmetry from G_4 (C_{2v}) to C_s, in which there is no tunnelling splitting, results in the a_1, b_2 and b_1, a_2 degeneracies. This effect is the opposite of what we encounter in rigid molecule groups. For example, all properties are non-degenerate in molecules belonging to the C_{2v} point group but there are doubly degenerate (E) properties in molecules belonging to the higher symmetry C_{3v} point group.

There is a torsional vibration about the C=C bond in ethylene [Figure 5.74(c)] and the potential function is again given by equation (5.270). Because of the high resistance of the double bond to twisting, the barrier V_2 is very much higher than that in phenol (Table 5.20).

In methanol, shown in Figure 5.78(d), the dominant term is that with $n = 3$ giving

$$V(\phi) \approx \tfrac{1}{2}V_3(1 - \cos 3\phi) \tag{5.272}$$

The top of the threefold barrier corresponds to the O—H bond eclipsing one of the C—H bonds. In ethane (CH_3CH_3) the dominant term is also that with $n = 3$. Barrier heights for both molecules are given in Table 5.20.

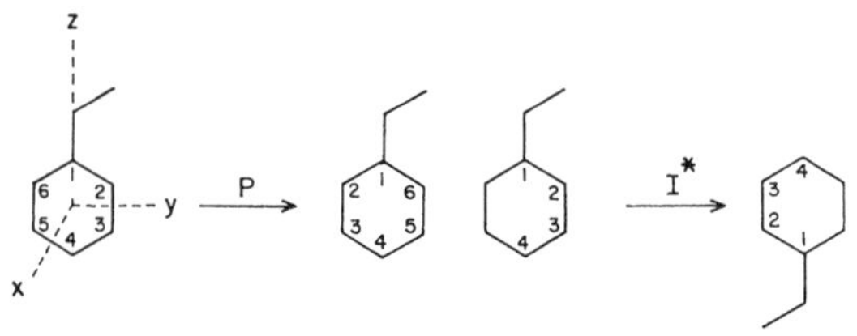

Figure 5.80

Toluene [Figure 5.78(a)] and nitromethane (CH_3NO_2) both have a predominantly sixfold barrier. In toluene the top of the barrier corresponds to a C—H bond eclipsing an adjacent C—C bond of the ring. Whatever the torsional angle the C—H bonds can never get very far away from eclipsing the C—C bonds and so the barrier is very low (Table 5.20). It is generally true that the higher the order of the barrier to rotation about a formally single bond, the lower is its height, although the result for $(CH_3)_3CNO_2$ in Table 5.20 shows that there are considerable variations.

In molecules with torsional vibrations whose potential functions can be represented fairly well by just one term in the summation in equation (5.269), there is an approximate method which can be used to obtain the barrier height when only the $\upsilon = 1$–0 torsional transition has been observed. In many cases experimental measurements have been restricted to this single observation. This transition often has a low wavenumber and has been observed for many molecules in the far infrared region.

We consider here the example of a molecule, such as ethylene, with a torsional potential of the form given in equation (5.270). We can expand cos 2ϕ as a power series:

$$\cos 2\phi = 1 - \frac{1}{2!}(2\phi)^2 + \frac{1}{4!}(2\phi)^4 - \ldots \quad (5.273)$$

giving

$$V(\phi) = V_2(\phi^2 - \tfrac{1}{3}\phi^4 + \ldots) \quad (5.274)$$

If the wavenumber of the $\upsilon = 1$–0 transition is small compared with the barrier height V_2, the torsional amplitude ϕ is small and the second and subsequent terms in equation (5.274) can be neglected, giving

$$V(\phi) \approx V_2\phi^2 \quad (5.275)$$

This is a harmonic oscillator potential function for which

$$V_2 = 2\pi^2 c^2 \tilde{\nu}^2 I_r \quad (5.276)$$

where $\tilde{\nu}$ is the wavenumber of the $\upsilon = 1$–0 transition and I_r is the reduced moment of inertia for the torsional motion. I_r can be obtained either approximately by the method of Pitzer[65] or, more accurately, from the appropriate element of the *G*-matrix in a normal coordinate treatment.

There are many cases of molecules for which there are two terms in the summation of equation (5.269) which are comparable in magnitude. The result is that there are *two* sets of equivalent minima in $V(\phi)$. The set with the lower energy corresponds to the stable configuration and that with the higher to a metastable configuration. Some examples, shown in Figure 5.81, are buta-1,3-diene, glyoxal and acrolein. In all three molecules the central C—C bond is formally single but has some double bond character due to conjugation. In all cases the so-called *s-trans* rotamer,† referring to the *trans* configuration of the double bonds about the single bond, is the stable configuration and the *s-cis*, shown for buta-1,3-diene in Figure 5.81(d), is metastable. In all three molecules the dominant terms in $V(\phi)$ are given by

$$V(\phi) \approx \tfrac{1}{2}V_1(1 - \cos\phi) + \tfrac{1}{2}V_2(1 - \cos 2\phi) \quad (5.277)$$

and the shape of the potential function is like that shown in Figure 5.82. This particular $V(\phi)$ has been plotted for $V_1 = -100$, $V_2 = 200$ and $F = 1.0$ units (say cm^{-1}), where F is the internal rotation constant for free rotation about the C—C and is given by

† A rotamer is an isomer formed by rotational isomerization.

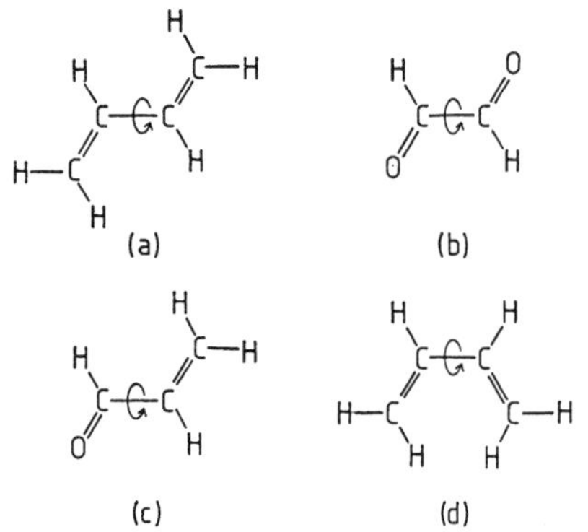

Figure 5.81 Torsional vibration about the carbon–carbon single bond in (a) *s-trans*-buta-1,3-diene, (b) *s-trans*-glyoxal, (c) acrolein, and (d) *s-cis*-buta-1,3-diene

$$F = h/8\pi^2 c I_r \tag{5.278}$$

Figure 5.82 also shows the calculated vibrational wave functions for the lowest 18 levels. It can be seen that up to the fourteenth level the wave functions are concentrated in either the *s-cis* or *s-trans* potential well and a molecule in any of these levels can be said to be an *s-cis* or *s-trans* molecule undergoing torsional vibration. From the fourteenth level upwards tunnelling through the barrier is appreciable so that in the sixteenth level, for example, there is a high probability of finding the molecule in the *s-cis* well but a small probability of finding it in the *s-trans* well. The eighteenth level, although classically above the barrier, still shows a higher probability in the *s-cis* well. Further above the barrier the levels settle down to become those of free internal rotation.

Experimental data are not sufficiently easy to obtain for potential functions to be established in all but a few cases. In glyoxal the data have been fitted to give values of V_1 and V_2 of 1719.4 and 1063.5 cm^{-1}, respectively[66]. The *s-trans*→*s-cis* barrier height is 2077 cm^{-1}. The wavenumber difference between the *s-trans* and *s-cis* minima is 1688 cm^{-1}, reduced to 1060 cm^{-1} in buta-1,3-diene and 660 cm^{-1} in acrolein.

In all three molecules the possibility has been considered that the metastable rotamer is not *s-cis* but *gauche*, with a configuration which is non-planar, i.e. $\phi \neq 0$ or 2π in Figure 5.82. In this case *two* equivalent *gauche* minima would be encountered during one complete revolution and the energy level scheme becomes even more complex. However, this is not the case in acrolein and glyoxal and is probably not in buta-1,3-diene either.

In molecules, such as 1,2-difluoroethylene, having isomers which are *cis* or *trans* about a double bond, the potential function is also of the form given in equation (5.277) but the barrier to interconversion is so high, presumably similar to that in ethylene itself, that the two forms can be prepared separately and, under normal circumstances, interconversion never occurs.

In a molecule such as 3- (or *meta*-) fluorotoluene, shown in Figure 5.78(e), there are dominant V_6 and V_3 terms in the potential function for torsional motion of the CH_3 group. The V_6 term is due to the resistance to torsion provided by the ring, as in toluene, and the V_3 term is due to the differing resistance of the adjacent C–C bonds. From analysis of the experimental data it appears that there are two possible pairs of values of V_6 and V_3, 7.97 and 15.8 cm^{-1} or −5.28 and 16.9 cm^{-1}, respectively, all of them very small[67] (see, however, Ref. 392).

No matter how complex the torsional vibrational potential function may be in a particular molecule, once all energy barriers are surmounted the molecule undergoes free internal rotation. Under these circumstances a vibrational degree of freedom has been lost and a rotational one gained. There is an additional rotational angular momentum and an associated quantum number m. The effect on the term values of, say, a prolate symmetric rotor is that they are modified from those of equation (5.201) to

$$F_v(J, k, m) = B_v J(J+1) + (A_v^F - B_v)k^2 + (A_v^F + A_v^T)m^2 - 2A_v^F mk \tag{5.279}$$

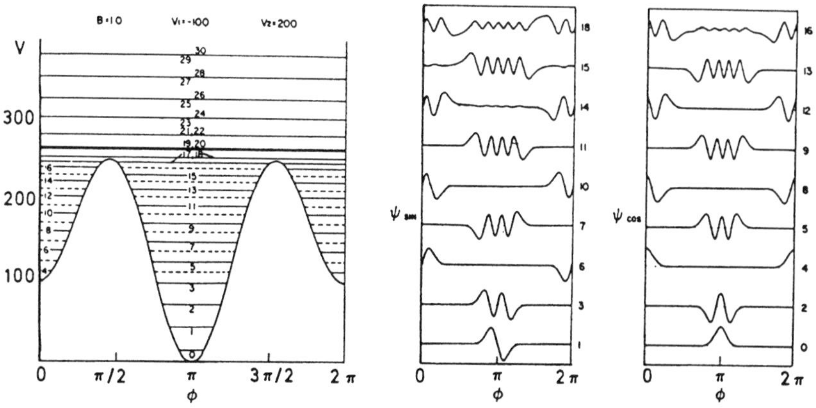

Figure 5.82 Potential function, $V(\phi) = \frac{1}{2}V_1(1 - \cos\phi) + \frac{1}{2}V_2(1 - \cos 2\phi)$, energy levels and wave functions for *s-trans* and *s-cis* rotamers. $\phi = \pi$ corresponds to *s-trans* for those molecules in Figure 5.81. [Reproduced, with permission, from Lewis, J. D., Malloy, T. B., Jr, Chao, T. H. and Laane, J. (1972). *J. Mol. Struct.*, **12**, 427]

where k is the signed quantum number ($K = |k|$), m is the internal rotation quantum number which can take values 0, ± 1, $\pm 2, \ldots \pm \infty$, A_v^{F} and A_v^{T} are the A_v rotational constants of the frame and top, respectively, and centrifugal distortion is neglected. It is assumed that the internal rotation axis is the top axis of the molecule. For example, in CH_3CCl_3 the CCl_3 group is regarded as the frame and

$$A^{\mathrm{F}} = h/8\pi^2 c I_a^{\mathrm{F}} \qquad (5.280)$$

where I_a^{F} is the moment of inertia of the frame about the top axis of the molecule. A similar equation holds for the top, which is the CH_3 group.

Selection rules are the same as in rigid and symmetric rotors with, in addition, $\Delta m = 0$.

We have seen in Figure 5.79 how the torsional energy levels in phenol must be classified according to the non-rigid molecular symmetry group G_4, of order four, if there is observable splitting due to tunnelling through the twofold torsional barrier. In the case of molecules with a three- or sixfold barrier the relevant molecular symmetry group to be used for classifying torsional energy levels is of a higher order.

In the case of a threefold barrier, in a molecule such as 2- (or *ortho*-) fluorotoluene, the higher order group (see the book by Bunker in the Bibliography) is G_6, of order six. It is isomorphous with the C_{3v} point group and the symmetry species labels used for the G_6 group are the same as those for C_{3v}.

Figure 5.83 shows a torsional potential given by equation (5.272) with a threefold barrier repeating every 120°. Some of the torsional

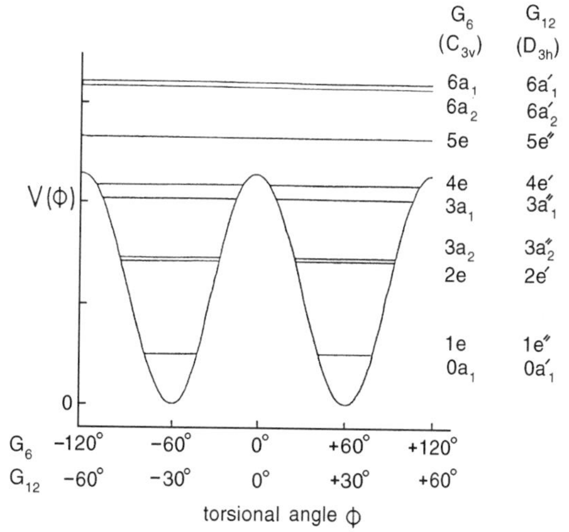

Figure 5.83 Torsional potential with a three- or sixfold barrier

energy levels are shown together with their symmetry classification according to the G_6 group. In this case the two lowest energy levels, $0a_1$ and $1e$, are degenerate as there is no effective tunnelling through the barrier. Above the sixth level the situation approaches that of free rotation. The number 0, 1, 2, 3, 4, 5 or 6 which precedes the symmetry species is the modulus of the internal rotation quantum number m which is applicable in the limit of free internal rotation. It can be seen, for example, that the $6a_1$ and $6a_2$ levels, which are above the barrier, are close together; in the free internal rotation limit they would be degenerate corresponding to $m = \pm 6$.

For a molecule, such as toluene shown in Figure 5.78(a), with a sixfold barrier repeating every 60° the torsional potential is given by

$$V(\phi) = \tfrac{1}{2}V_6(1 - \cos 6\phi) \tag{5.281}$$

The appropriate molecular symmetry group is G_{12} of order 12 and isomorphous with the D_{3h} point group whose symmetry species are used. Figure 5.83 gives the symmetry species assignments, together with the corresponding value of the modulus of the internal rotation quantum number m.

BIBLIOGRAPHY

ALLEN, H. C., JR and CROSS, P. C. (1963). *Molecular Vib-Rotors*. Wiley, New York

BLANKE, J. F. and OVEREND, J. (1978). 'Normal coordinates and the vibrations of polyatomic molecules', Chapter 3 in *Vibrational Spectra and Structure*, Vol. 7 (Ed. J. R. Durig). Elsevier, Amsterdam

BUNKER, P. R. (1979). *Molecular Symmetry and Spectroscopy*. Academic Press, New York

CALIFANO, S. (1976). *Vibrational States*. Wiley, London

COTTON, F. A. (1971). *Chemical Applications of Group Theory*. Wiley, New York

FERRARO, J. R. and NAKAMOTO, K. (1994). *Introductory Raman Spectroscopy*. Academic Press, New York.

GANS, P. (1971). *Vibrating Molecules*. Chapman and Hall, London.

HERZBERG, G. (1945). *Infrared and Raman Spectra*. Van Nostrand, New York

HERZBERG, G. (1950). *Spectra of Diatomic Molecules*. Van Nostrand, New York

HOLLAS, J. M. (1972). *Symmetry in Molecules*. Chapman and Hall, London

LONG, D. A. (1977). *Raman Spectroscopy*. McGraw-Hill, London

MILLS, I. M. (1963). 'Force constant calculations for small molecules', Chapter 5 in *Infrared Spectroscopy and Molecular Structure* (Ed. M. Davies). Elsevier, Amsterdam

MILLS, I. M. (1974). 'Harmonic and anharmonic force fie[illegible] calculations', Chapter 4 in *Theore[illegible] Chemistry*, Vol. 1 (*Quantum Chemistry*) *Specialist Periodical Report*. Chemical Society, London

SCHONLAND, D. (1965). *Molecular Symmetry*. Van Nostrand, London

STOICHEFF, B. P. (1959). 'High resolution Raman spectroscopy', Chapter 5 in *Advances in Spectroscopy*, Vol. 1 (Ed. H. W. Thompson). Interscience, London

WEBER, A. (1973). 'High resolution Raman studies of gases', Chapter 9 in *The Raman Effect*, Vol. 2 (Ed. A. Anderson). Dekker, New York

WILSON, E. B. (1934). *Phys. Rev.*, **45**, 706

WILSON, E. B., DECIUS, J. C. and CROSS, P. C. (1955). *Molecular Vibrations*. McGraw-Hill, New York

WOODWARD, L. A. (1972). *Introduction to the Theory of Molecular Vibrations and Vibrational Spectroscopy*. Oxford University Press, Oxford

6 ELECTRONIC SPECTROSCOPY

6.1 ELECTRONIC SPECTROSCOPY OF ATOMS

Electronic spectroscopy is the study of transitions, in absorption or emission, between different electronic states of an atom or molecule.

Apart from translation and nuclear spin, atoms have only electronic degrees of freedom. For this reason, their electronic spectra are very much simpler than those of diatomic or polyatomic molecules which have, in addition, vibrational and rotational degrees of freedom: these add greatly to the complexity of their electronic spectra.

6.1.1 The periodic table

The Schrödinger equation has been discussed in Section 1.2 and the time-independent equation given, in its most general form, in equation (1.45).

For the hydrogen atom, and for hydrogen-like ions such as He^+, Li^{2+}, . . . , with a single electron in the field of a nucleus with charge $+Ze$, the expression for the hamiltonian H is given by

$$H = -(\hbar^2\nabla^2/2\mu) - (Ze^2/4\pi\varepsilon_0 r) \qquad (6.1)$$

This is analogous to equation (1.48) for the hydrogen atom. The symbol ∇^2 stands for the laplacian operator of equation (1.36), μ is the reduced mass of the nucleus and electron and r is the distance of the electron from the nucleus. The first term of this hamiltonian is the quantum mechanical equivalent of the classical kinetic energy and the second term is the potential energy.

Solution of the Schrödinger equation gives the energy levels in equation (1.64) and which are illustrated for hydrogen in Figure 1.5. At this level of approximation the only quantum number involved in the energy is the principal quantum number n, which can take values of 1, 2, 3, . . . , ∞. The fact that the energy is approximately independent of the azimuthal quantum number l makes hydrogen and one-electron ions unique among all atoms and ions. The energy is independent of the magnetic quantum number m_l (equation 1.52) in the absence of a magnetic or electric field as, indeed, it is in all atoms.

In the helium atom there are two electrons in orbitals around a nucleus of charge $+2e$. We label these electrons 1 and 2 and take them to be at distances r_1 and r_2, respectively, from the nucleus and a distance r_{12} apart, as shown in Figure 6.1. The electronic hamiltonian for this system is

$$H = -(\hbar^2/2m_e)(\nabla_1^2 + \nabla_2^2) - (2e^2/4\pi\varepsilon_0 r_1) - (2e^2/4\pi\varepsilon_0 r_2) + (e^2/4\pi\varepsilon_0 r_{12}) \qquad (6.2)$$

in the approximation of a fixed nucleus. The first term is the quantum mechanical kinetic

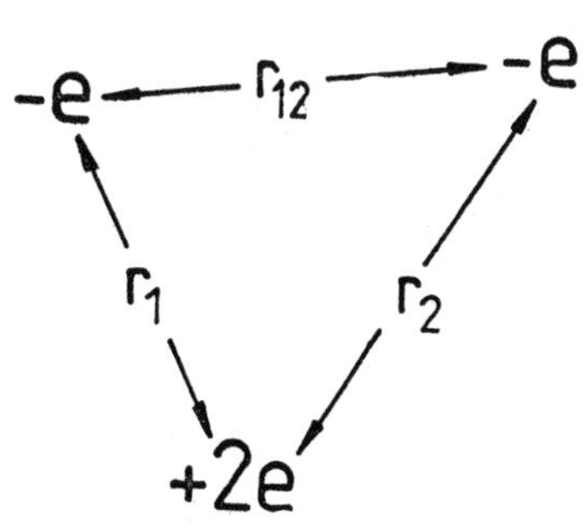

Figure 6.1 Electron coordinates in the helium atom

energy of the electrons and ∇_1^2 and ∇_2^2 refer to the coordinates of electrons 1 and 2, respectively. The second and third terms are the contributions to the potential energy by the coulombic attractions between the nucleus and each of the electrons.

All the first three terms are analogous to terms in equation (6.1) but the fourth term is of a type not encountered in the one-electron atom. This term is the contribution to the potential energy of the coulombic repulsion between the electrons.

For a polyelectronic atom in general the electronic hamiltonian is given by

$$H = -(\hbar^2/2m_e)\sum_i \nabla_i^2 - \sum_i (Ze^2/4\pi\varepsilon_0 r_i) + \sum_{i<j} (e^2/4\pi\varepsilon_0 r_{ij}) \tag{6.3}$$

where the summation is over all electrons and the restriction $i < j$ in the last term ensures that the repulsion between electrons i and j is not counted twice.

Whereas the Schrödinger equation is exactly soluble for one-electron atoms using the hamiltonian of equation (6.1), this is no longer true for a polyelectronic atom with a hamiltonian of the form given in equation (6.3). The reason for this is that the hamiltonian for such an atom cannot be broken down exactly into separate contributions from each electron, i.e.

$$H \neq \sum_i H_i \tag{6.4}$$

and it is the third term in the hamiltonian which causes this inequality.

It is obvious that the effect of mutual repulsions among all the electrons cannot be broken down into contributions from each, but Hartree managed to go some way towards this. He rewrote equation (6.3) as

$$H \approx -(\hbar^2/2m_e)\sum_i \nabla_i^2 - \sum_i (Ze^2/4\pi\varepsilon_0 r_i) + \sum_i V(r_i) \tag{6.5}$$

where $V(r_i)$ is an averaged contribution to the potential field due to repulsions experienced by electron i. The problem of obtaining the best form of this contribution to the potential is the basis of the Hartree self-consistent field (or SCF) method.

The total electronic wave function ψ_e is given, in this approximation, by the product of the atomic orbital wave functions χ_i of all occupied orbitals:

$$\psi_e = \chi_1 \times \chi_2 \times \chi_3 \ldots \times \chi_N = \prod_i \chi_i \tag{6.6}$$

where N is the total number of electrons. If we use a very approximate form, H^0, of the hamiltonian of equation (6.3), in which the electron repulsion term is neglected, the Schrödinger equation factorizes into

$$H_i^0 \chi_i = E_i \chi_i \tag{6.7}$$

for each electron. Then the average potential field in which the jth electron moves is estimated from all the χ_i except χ_j. This gives an improved wave function χ_j, and so on for all the χ_i. All the χ_i can then be improved further using the first-improved functions. The process is continued until the optimum $V(r_i)$ is obtained.

Although the total wave function is given, in this approximation, by the product of orbital wave functions [equation (6.6)], the total energy E is not given by $\Sigma_i E_i$ but

$$E = \sum_i E_i - \sum_{i<j} (J_{ij} - K_{ij}) \tag{6.8}$$

where J_{ij} is the Coulomb integral, representing the total repulsion between the charge clouds of electrons i and j. K_{ij} is the exchange integral; it has no classical analogue and appears only in the Hartree–Fock modified treatment in which the possibility of electron exchange is included.

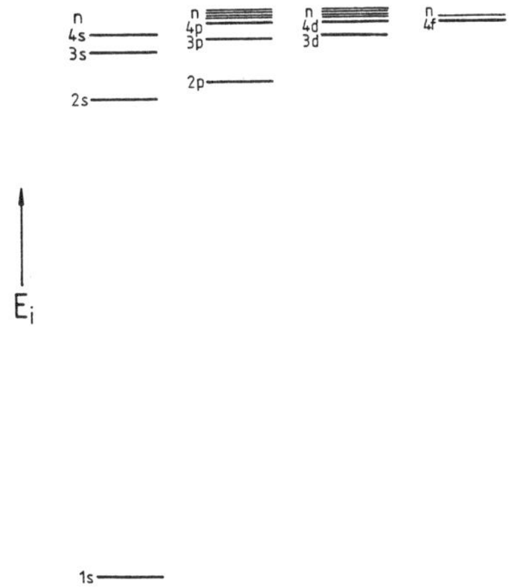

Figure 6.2 Typical order of orbital energies, E_i, in a polyelectronic atom

An important effect of the coulombic repulsion term on the energies of self-consistent field atomic orbitals is that they are dependent on the quantum number l, in addition to n. In the absence of an electric or magnetic field they are independent of m_l and also of m_s, the quantum number arising from space quantization of electron spin (equation 1.91).

The dependence of orbital energies on l is shown in Figure 6.2. This figure illustrates a typical order of orbital energies E_i in a polyelectronic atom. It should be emphasized that the energy scale is quantitatively different for each atom or ion. This is illustrated by, for example, the fact that the energy required to remove an electron from the $1s$ orbital, i.e. the ionization energy, is 13.6 eV in H and 870.4 eV in Ne.

The Pauli exclusion principle states that no two electrons can have the same set of quantum numbers n, l, m_l, m_s. Since m_l can take $(2l+1)$ values and $m_s = \pm\frac{1}{2}$, each orbital characterized by particular values of n and l has the same energy for $2(2l+1)$ electrons. It follows that any ns orbital, with $l=0$, can accommodate two electrons, an np orbital, with $l=1$, can take six, an nd orbital, with $l=2$, can take ten, and so on.

It is worth mentioning here that, although we often refer loosely to a $2p$, $3d$, . . . 'orbital', the word is more strictly used to imply a particular set of values of n, l and m_l, as in, for example, $2p_x$ and $3d_{z^2}$ orbitals. When a complete set of quantum numbers n, l, m_l and m_s is referred to, the term spin-orbital is used.

A shell refers to all the orbitals having the same value of n. For $n = 1, 2, 3, 4, \ldots$ the shells are labelled $K, L, M, N, \ldots$. The term sub-shell is used to refer to all the orbitals having the same n and l, as in a $3d$ sub-shell.

Following the *aufbau* or building-up principle, the lowest energy, or ground, configuration for a particular atom is obtained by feeding the available electrons into the orbitals, filling them if possible, in order of increasing energy. It follows, for example, that the carbon atom has the ground configuration $1s^2 2s^2 2p^2$.

The distinction between a configuration and a state is an important one. A configuration consists of a description of the way in which the electrons are distributed among various orbitals. As we shall see in Section 6.1.2, a single configuration can give rise to several states.

Table 6.1 illustrates features which are common to the electron configurations of elements which are known to be similar in their chemical properties. The alkali metals, Li, Na, K, Rb and Cs, all have an outer ns^1 configuration consistent with their being monovalent, and the alkaline earth metals, Be, Mg, Ca, Sr and Ba, all have an outer ns^2 configuration consistent with their being divalent. The noble gases Ne, Ar, Kr, Xe and Rn all have an outer np^6 configuration, the filled sub-shell conferring chemical inertness, and the filled K shell in He has a similar effect.

The stability associated with filled shells and sub-shells is illustrated in Figure 6.3, which shows the first ionization energy for each of the first 18 elements. The ionization process involved, for an atom A, is

$$A \rightarrow A^+ + e \qquad (6.9)$$

where the electron is being removed from the outermost occupied orbital. Pronounced peaks in the ionization energies occur for He, Ne and

Table 6.1 Ground configurations and ground states of atoms

Atom	Atomic number (Z)	Ground configuration	First ionization energy/eV[a]	Ground state
H	1	$1s^1$	13.598	$^2S_{\frac{1}{2}}$
He	2	$1s^2$	24.587	$^1S_0$
Li	3	$K2s^1$	5.392	$^2S_{\frac{1}{2}}$
Be	4	$K2s^2$	9.322	$^1S_0$
B	5	$K2s^22p^1$	8.298	$^2P^o_{\frac{1}{2}}$
C	6	$K2s^22p^2$	11.260	$^3P_0$
N	7	$K2s^22p^3$	14.534	$^4S^o_{\frac{3}{2}}$
O	8	$K2s^22p^4$	13.618	$^3P_2$
F	9	$K2s^22p^5$	17.422	$^2P^o_{\frac{3}{2}}$
Ne	10	$K2s^22p^6$	21.564	$^1S_0$
Na	11	$KL3s^1$	5.139	$^2S_{\frac{1}{2}}$
Mg	12	$KL3s^2$	7.646	$^1S_0$
Al	13	$KL3s^23p^1$	5.986	$^2P^o_{\frac{1}{2}}$
Si	14	$KL3s^23p^2$	8.151	$^3P_0$
P	15	$KL3s^23p^3$	10.486	$^4S^o_{\frac{3}{2}}$
S	16	$KL3s^23p^4$	10.360	$^3P_2$
Cl	17	$KL3s^23p^5$	12.967	$^2P^o_{\frac{3}{2}}$
Ar	18	$KL3s^23p^6$	15.759	$^1S_0$
K	19	$KL3s^23p^64s^1$	4.341	$^2S_{\frac{1}{2}}$
Ca	20	$KL3s^23p^64s^2$	6.113	$^1S_0$
Sc	21	$KL3s^23p^63d^14s^2$	6.54	$^2D_{\frac{3}{2}}$
Ti	22	$KL3s^23p^63d^24s^2$	6.82	$^3F_2$
V	23	$KL3s^23p^63d^34s^2$	6.74	$^4F_{\frac{3}{2}}$
Cr	24	$KL3s^23p^63d^54s^1$	6.766	$^7S_3$
Mn	25	$KL3s^23p^63d^54s^2$	7.435	$^6S_{\frac{5}{2}}$
Fe	26	$KL3s^23p^63d^64s^2$	7.870	$^5D_4$
Co	27	$KL3s^23p^63d^74s^2$	7.86	$^4F_{\frac{9}{2}}$
Ni	28	$KL3s^23p^63d^84s^2$	7.635	$^3F_4$
Cu	29	$KLM4s^1$	7.726	$^2S_{\frac{1}{2}}$
Zn	30	$KLM4s^2$	9.394	$^1S_0$
Ga	31	$KLM4s^24p^1$	5.999	$^2P^o_{\frac{1}{2}}$
Ge	32	$KLM4s^24p^2$	7.899	$^3P_0$
As	33	$KLM4s^24p^3$	9.81	$^4S^o_{\frac{3}{2}}$
Se	34	$KLM4s^24p^4$	9.752	$^3P_2$
Br	35	$KLM4s^24p^5$	11.814	$^2P^o_{\frac{3}{2}}$
Kr	36	$KLM4s^24p^6$	13.999	$^1S_0$
Rb	37	$KLM4s^24p^65s^1$	4.177	$^2S_{\frac{1}{2}}$
Sr	38	$KLM4s^24p^65s^2$	5.695	$^1S_0$

(continued)

Table 6.1 *(continued)*

Atom	Atomic number (Z)	Ground configuration	First ionization energy/eV[a]	Ground state
Y	39	$KLM4s^2 4p^6 4d^1 5s^2$	6.38	$^3D_{\frac{3}{2}}$
Zr	40	$KLM4s^2 4p^6 4d^2 5s^2$	6.84	$^3F_2$
Nb	41	$KLM4s^2 4p^6 4d^4 5s^1$	6.88	$^6D_{\frac{1}{2}}$
Mo	42	$KLM4s^2 4p^6 4d^5 5s^1$	7.099	$^7S_3$
Tc	43	$KLM4s^2 4p^6 4d^5 5s^2$	7.28	$^6S_{\frac{5}{2}}$
Ru	44	$KLM4s^2 4p^6 4d^7 5s^1$	7.37	$^5F_5$
Rh	45	$KLM4s^2 4p^6 4d^8 5s^1$	7.46	$^4F_{\frac{9}{2}}$
Pd	46	$KLM4s^2 4p^6 4d^{10}$	8.34	$^1S_0$
Ag	47	$KLM4s^2 4p^6 4d^{10} 5s^1$	7.576	$^2S_{\frac{1}{2}}$
Cd	48	$KLM4s^2 4p^6 4d^{10} 5s^2$	8.993	$^1S_0$
In	49	$KLM4s^2 4p^6 4d^{10} 5s^2 5p^1$	5.786	$^2P_{\frac{1}{2}}{}^o$
Sn	50	$KLM4s^2 4p^6 4d^{10} 5s^2 5p^2$	7.344	$^3P_0$
Sb	51	$KLM4s^2 4p^6 4d^{10} 5s^2 5p^3$	8.641	$^4S^o_{\frac{3}{2}}$
Te	52	$KLM4s^2 4p^6 4d^{10} 5s^2 5p^4$	9.009	$^3P_2$
I	53	$KLM4s^2 4p^6 4d^{10} 5s^2 5p^5$	10.451	$^2P_{\frac{3}{2}}{}^o$
Xe	54	$KLM4s^2 4p^6 4d^{10} 5s^2 5p^6$	12.130	$^1S_0$
Cs	55	$KLM4s^2 4p^6 4d^{10} 5s^2 5p^6 6s^1$	3.894	$^2S_{\frac{1}{2}}$
Ba	56	$KLM4s^2 4p^6 4d^{10} 5s^2 5p^6 6s^2$	5.212	$^1S_0$
La	57	$KLM4s^2 4p^6 4d^{10} 5s^2 5p^6 5d^1 6s^2$	5.577	$^2D_{\frac{3}{2}}$
Ce	58	$KLM4s^2 4p^6 4d^{10} 4f^1 5s^2 5p^6 5d^1 6s^2$	5.47	$^1G_4{}^o$
Pr	59	$KLM4s^2 4p^6 4d^{10} 4f^3 5s^2 5p^6 6s^2$	5.42	$^4I_{\frac{9}{2}}{}^o$
Nd	60	$KLM4s^2 4p^6 4d^{10} 4f^4 5s^2 5p^6 6s^2$	5.49	$^5I_4$
Pm	61	$KLM4s^2 4p^6 4d^{10} 4f^5 5s^2 5p^6 6s^2$	5.55	$^6H_{\frac{5}{2}}{}^o$
Sm	62	$KLM4s^2 4p^6 4d^{10} 4f^6 5s^2 5p^6 6s^2$	5.63	$^7F_0$
Eu	63	$KLM4s^2 4p^6 4d^{10} 4f^7 5s^2 5p^6 6s^2$	5.67	$^8S_{\frac{7}{2}}{}^o$
Gd	64	$KLM4s^2 4p^6 4d^{10} 4f^7 5s^2 5p^6 5d^1 6s^2$	6.14	$^9D_2{}^o$
Tb	65	$KLM4s^2 4p^6 4d^{10} 4f^9 5s^2 5p^6 6s^2$	5.85	$^6H_{\frac{15}{2}}{}^o$
Dy	66	$KLM4s^2 4p^6 4d^{10} 4f^{10} 5s^2 5p^6 6s^2$	5.93	$^5I_8$
Ho	67	$KLM4s^2 4p^6 4d^{10} 4f^{11} 5s^2 5p^6 6s^2$	6.02	$^4I_{\frac{15}{2}}{}^o$
Er	68	$KLM4s^2 4p^6 4d^{10} 4f^{12} 5s^2 5p^6 6s^2$	6.10	$^3H_6$
Tm	69	$KLM4s^2 4p^6 4d^{10} 4f^{13} 5s^2 5p^6 6s^2$	6.18	$^2F_{\frac{7}{2}}{}^o$
Yb	70	$KLMN5s^2 5p^6 6s^2$	6.254	$^1S_0$
Lu	71	$KLMN5s^2 5p^6 5d^1 6s^2$	5.426	$^2D_{\frac{3}{2}}$
Hf	72	$KLMN5s^2 5p^6 5d^2 6s^2$	7.0	$^3F_2$
Ta	73	$KLMN5s^2 5p^6 5d^3 6s^2$	7.89	$^4F_{\frac{3}{2}}$
W	74	$KLMN5s^2 5p^6 5d^4 6s^2$	7.98	$^5D_0$
Re	75	$KLMN5s^2 5p^6 5d^5 6s^2$	7.88	$^6S_{\frac{5}{2}}$
Os	76	$KLMN5s^2 5p^6 5d^6 6s^2$	8.7	$^5D_4$

(continued)

Table 6.1 *(continued)*

Atom	Atomic number (Z)	Ground configuration	First ionization energy/eV[a]	Ground state
Ir	77	$KLMN5s^25p^65d^76s^2$	9.1	$^4F_{\frac{9}{2}}$
Pt	78	$KLMN5s^25p^65d^96s^1$	9.0	$^3D_3$
Au	79	$KLMN5s^25p^65d^{10}6s^1$	9.225	$^2S_{\frac{1}{2}}$
Hg	80	$KLMN5s^25p^65d^{10}6s^2$	10.437	$^1S_0$
Tl	81	$KLMN5s^25p^65d^{10}6s^26p^1$	6.108	$^2P_{\frac{1}{2}}^{\rm o}$
Pb	82	$KLMN5s^25p^65d^{10}6s^26p^2$	7.416	$^3P_0$
Bi	83	$KLMN5s^25p^65d^{10}6s^26p^3$	7.289	$^4S_{\frac{3}{2}}^{\rm o}$
Po	84	$KLMN5s^25p^65d^{10}6s^26p^4$	8.42	$^3P_2$
At	85	$KLMN5s^25p^65d^{10}6s^26p^5$	–	$^2P_{\frac{3}{2}}^{\rm o}$
Rn	86	$KLMN5s^25p^65d^{10}6s^26p^6$	10.748	$^1S_0$
Fr	87	$KLMN5s^25p^65d^{10}6s^26p^67s^1$	–	$^2S_{\frac{1}{2}}$
Ra	88	$KLMN5s^25p^65d^{10}6s^26p^67s^2$	5.279	$^1S_0$
Ac	89	$KLMN5s^25p^65d^{10}6s^26p^66d^17s^2$	6.9	$^2D_{\frac{3}{2}}$
Th	90	$KLMN5s^25p^65d^{10}6s^26p^66d^27s^2$	–	$^3F_2$
Pa	91	$KLMN5s^25p^65d^{10}5f^26s^26p^66d^17s^2$	–	$^4K_{\frac{11}{2}}$
U	92	$KLMN5s^25p^65d^{10}5f^36s^26p^66d^17s^2$	–	$^5L_6^{\rm o}$
Np	93	$KLMN5s^25p^65d^{10}5f^46s^26p^66d^17s^2$	–	$^6L_{\frac{11}{2}}$
Pu	94	$KLMN5s^25p^65d^{10}5f^66s^26p^67s^2$	5.8	$^7F_0$
Am	95	$KLMN5s^25p^65d^{10}5f^76s^26p^67s^2$	6.0	$^8S_{\frac{7}{2}}^{\rm o}$
Cm	96	$KLMN5s^25p^65d^{10}5f^76s^26p^66d^17s^2$	–	$^9D_2^{\rm o}$
Bk	97	$KLMN5s^25p^65d^{10}5f^96s^26p^67s^2$	–	$^6H_{\frac{15}{2}}^{\rm o}$
Cf	98	$KLMN5s^25p^65d^{10}5f^{10}6s^26p^67s^2$	–	$^5I_8$
Es	99	$KLMN5s^25p^65d^{10}5f^{11}6s^26p^67s^2$	–	$^4I_{\frac{15}{2}}^{\rm o}$
Fm	100	$KLMN5s^25p^65d^{10}5f^{12}6s^26p^67s^2$	–	$^3H_6$
Md	101	$KLMN5s^25p^65d^{10}5f^{13}6s^26p^67s^2$	–	$^2F_{\frac{3}{2}}^{\rm o}$
No	102	$KLMNO6s^26p^67s^2$	–	$^1S_0$
Lr	103	$KLMNO6s^26p^66d^17s^2$	–	$^2D_{\frac{3}{2}}$
–	104	$KLMNO6s^26p^66d^27s^2$	–	$^3F_2$

[a]For the process $A \rightarrow A^+ + e$, where A^+ is in its ground state.

Ar corresponding to removal of an electron from a filled 1*s* shell or *np* sub-shell. Lesser peaks occur for Be and Mg, where the electron is removed from a filled *ns* sub-shell. Small peaks occur also for N and P. These both have an np^3 configuration and there is an innate stability associated with a half-filled sub-shell. There is no simple reason for this but, when the sub-shell is half-filled or less than half-filled, all the electrons have parallel spins, i.e. all have either $m_s = +\frac{1}{2}$ or $-\frac{1}{2}$. Under these circumstances the electron–nucleus interactions are modified so as to lower the energy.

The first transition series of elements, Sc, Ti, V, Cr, Mn, Fe, Co, Ni, Cu and Zn, is characterized by the filling up of the 3*d* sub-shell. Figure 6.2 indicates that the 3*d* and 4*s* orbitals are very similar in energy. The situation is complicated further by the fact that the 3*d*–4*s* separation changes along the series. The result is

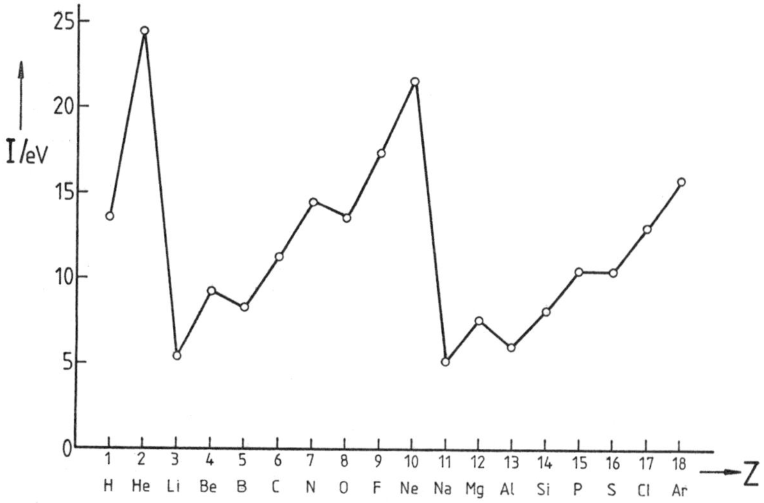

Figure 6.3 Ionization energies, I, for each of the first 18 elements in the periodic table

that, although there is a preference by most of these elements for having two electrons in the $4s$ sub-shell, Cr has a ... $3d^5 4s^1$ and Cu a ... $3d^{10}4s^1$ ground configuration because of the stability associated with a half-filled or filled $3d$ sub-shell.

In the second transition series, consisting of Y to Cd, in which $4d$ is being filled and $5s$ is very similar in energy, the elements Nb, Mo, Ru, Rh and Ag prefer the $5s^1$ configuration, while Pd has the ground configuration ... $4d^{10}5s^0$.

In the third transition series, consisting of Lu to Hg, in which the $5d$ sub-shell is being filled, the $5d$ and $6s$ orbitals are similar in energy with two electrons in $6s$ being preferred by all except Pt, which has the ground configuration ... $5d^9 6s^1$.

Filling up of the $4f$ sub-shell is a feature of the lanthanides. The $4f$ and $5d$ orbitals are of similar energy so that occasionally, as in La, Ce and Gd, one electron goes into $5d$ instead of $4f$.

Similarly in the actinides, Ac to No, the $5f$ sub-shell is being filled in competition with $6d$. In their ground configurations Ac($6d^1$), Th($6d^2$), Pa($6d^1$), U($6d^1$), Np($6d^1$) and Cm($6d^1$), all prefer at least one electron in the $6d$ sub-shell but, for the rest, successive electrons go into $5f$.

6.1.2 Vector representation of momenta and vector coupling approximations

6.1.2.1 Angular momenta and magnetic moments

Figure 1.7 illustrates how the orbital angular momentum of an electron can be represented by a vector. The direction of the vector is determined by the right-hand screw rule.

Every electron in an atom has two possible kinds of angular momenta, one due to its orbital and the other to its spin motion. The magnitude of the orbital angular momentum vector for a single electron is given, as in equation (1.70), by

$$[l(l+1)]^{\frac{1}{2}}\hbar = l^*\hbar \tag{6.10}$$

where $l = 0, 1, 2, \ldots, (n-1)$. (We shall come across the quantity $[Q(Q+1)]^{\frac{1}{2}}$, where Q is a quantum number, so often that it is convenient to abbreviate it to Q^*.) Similarly, the magnitude of the spin angular momentum vector for a single electron is, as in equation (1.90),

$$[s(s+1)]^{\frac{1}{2}}\hbar = s^*\hbar \tag{6.11}$$

where $s = \frac{1}{2}$ only.

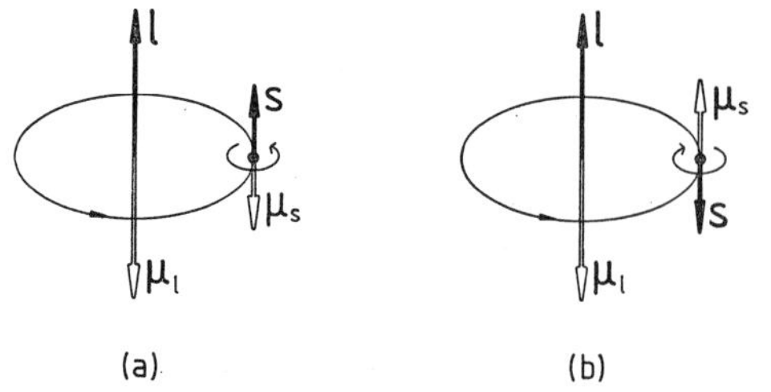

Figure 6.4 Vectors $\boldsymbol{l}$ and $\boldsymbol{s}$, and magnetic moments $\boldsymbol{\mu}_l$ and $\boldsymbol{\mu}_s$, associated with orbital and spin angular momenta, respectively, of an electron when the orbital and spin motions are (a) in the same direction, and (b) in opposite directions

For an electron which has both orbital and spin angular momentum there is a quantum number j associated with the total (orbital+ spin) angular momentum. This is also a vector quantity whose magnitude is given by

$$[j(j+1)]^{\frac{1}{2}}\hbar = j^*\hbar \tag{6.12}$$

where j can take the values

$$j = l+s,\ l+s-1,\ \ldots,\ |l-s| \tag{6.13}$$

This is called a Clebsch–Gordan series, of which we shall encounter several more in Section 6.1.2.2.

A charge of $-e$ circulating in an orbit is equivalent to a current flowing in a wire and therefore causes a magnetic moment. The symbol for that due to orbital motion is $\boldsymbol{\mu}_l$ and it is related to the orbital angular momentum $\boldsymbol{l}$ by

$$\begin{aligned}\boldsymbol{\mu}_l &= (-e/2m_e)\boldsymbol{l}\\ &= \gamma \boldsymbol{l}\end{aligned} \tag{6.14}$$

where γ is called the magnetogyric ratio. The relationship between $\boldsymbol{\mu}_l$ and $\boldsymbol{l}$ can be derived from classical mechanics and shows that they are parallel but opposed vectors, as illustrated in Figure 6.4(a).

The classical picture of an electron spinning on its own axis indicates that there is a magnetic moment associated with this angular momentum also.

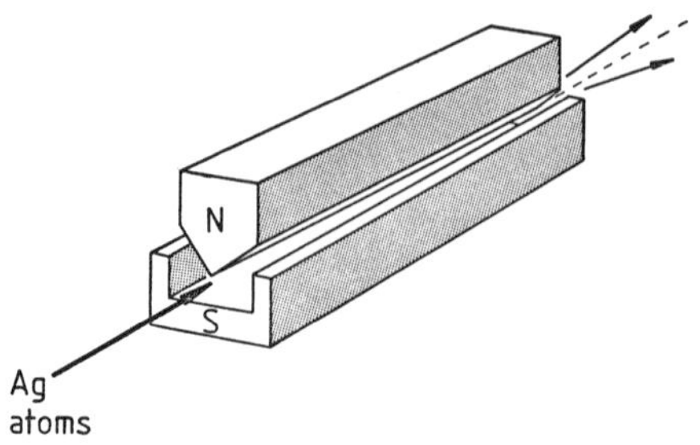

Figure 6.5 The Stern–Gerlach experiment

It was an experiment performed by Stern and Gerlach in 1921 which was instrumental in determining the magnitude of the magnetic moment $\boldsymbol{\mu}_s$ due to electron spin. Figure 6.5 illustrates the Stern–Gerlach experiment in which a low intensity beam of silver atoms, passed between the poles of an inhomogeneous magnet, was split into two components. Table 6.1 shows that all the electrons except one in a silver atom are in filled sub-shells which (see Section 6.1.2.2) have zero net orbital angular momentum. The only electron which is not in a filled sub-shell is in a 5s orbital for which $l=0$ and also has zero orbital angular momentum. Hence the angular momentum of a silver atom is due only to electron spin† and, since all the momenta of electrons with paired spins cancel, this must be due to that of the 5s electron only. For one electron the spin quantum number $s=\frac{1}{2}$ and $m_s = \pm\frac{1}{2}$ [see equations (1.90) and (1.91)] and the splitting of the beam in the magnetic field is into those atoms with $m_s = +\frac{1}{2}$ and those with $m_s = -\frac{1}{2}$. The magnitude of the splitting was interpreted in terms of the relationship

$$\begin{aligned}\boldsymbol{\mu}_s &= g_e\gamma \boldsymbol{s}\\ &= -g_e(\mu_B/\hbar)\boldsymbol{s}\end{aligned} \tag{6.15}$$

†Both the less abundant naturally occurring isotopes ^{107}Ag and ^{109}Ag also have a nuclear spin angular momentum ($I=\frac{1}{2}$) but, in a strong magnetic field, no additional splitting of the beam due to this could be observed.

where $\boldsymbol{\mu}_s$ is the magnetic moment due to electron spin, γ is the magnetogyric ratio, as in equation (6.14), and g_e is simply referred to as the g-value of the electron. The value of g_e was thought, from early experiments, to be exactly 2, but its value is 2.0023. Dirac quantum mechanics is inadequate here as it predicts a value of exactly 2 and it requires the more complete theory of quantum electrodynamics to account for the discrepancy. There will be a further occasion, when discussing the hydrogen atom in Section 6.1.4, when we shall refer to this theory.

The quantity μ_B is called the Bohr magneton and is a measure of magnetic moment. It is a fundamental constant given by

$$\mu_B = e\hbar/2m_e \tag{6.16}$$

Equation (6.15) shows that, as with orbital angular momentum, the magnetic dipole moment vector due to electron spin angular momentum is parallel, but opposed, to the spin momentum vector.

Figure 6.4(a) and (b) show that the two magnetic moment vectors $\boldsymbol{\mu}_l$ and $\boldsymbol{\mu}_s$ may be parallel or opposed. We shall return to the consequence of these alternatives in discussing spin–orbit coupling in Section 6.1.2.2.

Nuclear spin angular momentum has been discussed briefly in Section 1.3.3 and the similarities with electron spin angular momentum are exemplified by equations (1.90)–(1.93). The magnitude of the nuclear spin angular momentum is

$$[I(I+1)]^{\frac{1}{2}}\hbar = I^*\hbar \tag{6.17}$$

where the quantum number I can be zero, integral or half-integral depending on which particular nucleus we are considering—Table 1.3 gives a few examples.

The magnetic moment $\boldsymbol{\mu}_I$ due to the nuclear spin angular momentum $\boldsymbol{I}$ is given by

$$\begin{aligned}\boldsymbol{\mu}_I &= \gamma_I \boldsymbol{I}\\ &= g_I(\mu_N/\hbar)\boldsymbol{I}\end{aligned} \tag{6.18}$$

where γ_I is the nuclear magnetogyric ratio. Whereas electrons have only negative magnetogyric ratios, resulting in the angular momentum and magnetic moment vectors being opposed, this is not so for nuclei. In fact, most have a positive γ_I, only ^{29}Si, ^{117}Sn and ^{119}Sn of the more common nuclei having negative values.

The quantity μ_N is the nuclear magneton, defined by

$$\mu_N = e\hbar/2m_p \tag{6.19}$$

where m_p is the mass of the proton. From equations (6.16) and (6.19), it follows that

$$\mu_B/\mu_N = m_p/m_e = 1837 \tag{6.20}$$

which demonstrates how much smaller are typical magnetic moments due to nuclear spin than those due to electron spin.

Analogous to g_e for an electron, g_I is called the nuclear g-factor. Its value varies for different nuclei and is 5.585 for ^{1}H.

6.1.2.2 *Coupling of angular momenta*

Each type of angular momentum, whether it is due to electron spin, electron orbital motion or nuclear spin, produces a magnetic moment. Each magnetic moment behaves like a tiny bar magnet. Just as a number of bar magnets interact with each other, so do the magnetic moments. The interaction is of the dipole–dipole type and the effect is referred to as coupling of the angular momenta. The strength of the couplings between various momenta covers a wide range and some are so weak that they can often be neglected.

Coupling between the spin motions of the protons and neutrons contained in the nucleus is very strong and results in the value of the nuclear spin quantum number I.

We have seen that the magnetic moment due to nuclear spin, for nuclei with $I\neq0$, is small compared with those due to spin and orbital momenta of the electron. It follows that the coupling of these to the nuclear spin angular

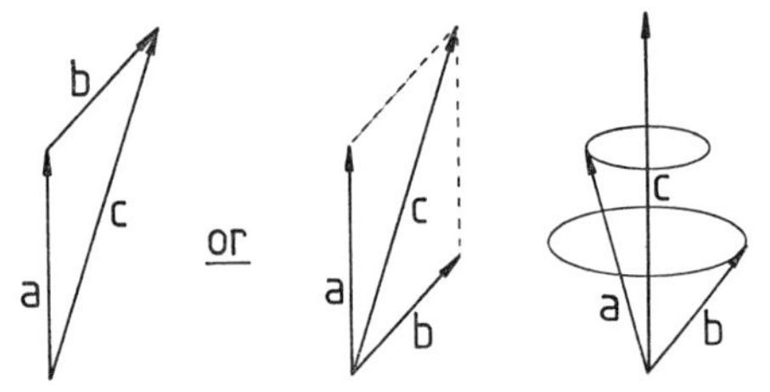

Figure 6.6 (a) Alternative ways of representing the addition of two vectors $\boldsymbol{a}$ and $\boldsymbol{b}$ to give a third, $\boldsymbol{c}$. (b) Precession of vectors $\boldsymbol{a}$ and $\boldsymbol{b}$ around the resultant $\boldsymbol{c}$

momentum is small and we shall not take this into account until Section 6.1.2.2(iii).

Coupling between any two vectors $\boldsymbol{a}$ and $\boldsymbol{b}$ produces a resultant vector $\boldsymbol{c}$ as shown in Figure 6.6(a). If the vectors represent angular momenta then $\boldsymbol{a}$ and $\boldsymbol{b}$ precess (see Section 1.3.3) around the resultant $\boldsymbol{c}$, as in Figure 6.6(b). Precession is the process by which the tip of the vector, say $\boldsymbol{a}$, traces out at a constant rate, the precession frequency, a circular path around the resultant vector $\boldsymbol{c}$. The rate of precession increases with the strength of coupling.

In practice, the resultant angular momentum vector $\boldsymbol{c}$ is precessing about an arbitrary direction in space. When an electric or magnetic field is introduced in this direction the effects of space quantization (see Section 1.3.3) can be observed.

The coupling between the spin and orbital motions of the electrons, the so-called spin–orbit coupling, can be variable in strength, according to the atom we are considering.

In a polyelectronic atom the spin of one electron can interact with (a) the spins of other electrons, (b) its own orbital motion and (c) the orbital motions of the other electrons. This last is called spin–other-orbit interaction and is normally too small to be taken into account. The interactions (a) and (b) are more important and the methods of treating them involve two types of approximation representing two extremes of coupling.

The first approximation neglects coupling between the spin of an electron and its own orbital momentum. Coupling between the orbital momenta is strong and that between spin momenta is relatively weak. This is the Russell–Saunders coupling approximation and is much the most important as it serves as a useful basis for the description of most states of most atoms.

The second approximation is the opposite extreme to Russell–Saunders coupling. Coupling between spin momenta is assumed to be sufficiently small to be neglected, as also is the coupling between orbital momenta. The coupling between the spin of an electron and its own orbital momentum is assumed to be strong. This is the *jj* coupling approximation and is useful in a limited number of cases, these being confined mainly to a few states of heavy atoms.

6.1.2.2(i) *Russell–Saunders coupling approximation*

We start by considering coupling of what are called non-equivalent electrons. These are electrons which have different values of either n or l so that the electrons in a $3p^13d^1$ or a $3p^14p^1$ configuration are non-equivalent but those in a $2p^2$ configuration are equivalent.

The strong coupling between the orbital momenta is referred to as *ll* coupling and can best be illustrated with an example.

Consider just two non-equivalent electrons in an atom, for example the helium atom in the highly excited configuration $2p^13d^1$. We shall label the $2p$ and $3d$ electrons '1' and '2', respectively, so that $l_1 = 1$ and $l_2 = 2$. The $\boldsymbol{l}_1$ and $\boldsymbol{l}_2$ vectors representing these orbital angular momenta are of magnitudes $2^{\frac{1}{2}}\hbar$ and $6^{\frac{1}{2}}\hbar$, respectively (see equation 6.10). These vectors couple, as in Figure 6.6(a), to give a resultant, $\boldsymbol{L}$, of magnitude

$$[L(L+1)]^{\frac{1}{2}}\hbar = L^*\hbar \qquad (6.21)$$

However, L, the total orbital angular momentum quantum number, can take only certain values or, in other words, l_1 and l_2 can take only certain orientations relative to each other. The orientations which they can take are governed by the values that the quantum number L can take. L is associated with the total orbital angular momentum for the two electrons and is restricted to the values

$$L = l_1 + l_2, l_1 + l_2 - 1, \ldots, |l_1 - l_2| \quad (6.22)$$

In the present case $L = 3$, 2 or 1 and the magnitude of $\boldsymbol{L}$ is $12^{\frac{1}{2}}\hbar$, $6^{\frac{1}{2}}\hbar$, or $2^{\frac{1}{2}}\hbar$. These are illustrated by the vector diagrams of Figure 6.7(a).

The terms of the atom are labelled S, P, D, F, G, . . . corresponding to $L = 0$, 1, 2, 3, 4, . . . , analogous to the labelling of one-electron orbitals s, p, d, f, g, . . . according to the value of l.

It follows that the $2p^1 3d^1$ configuration gives rise to P, D and F terms.

In a similar way, the coupling of a third vector to any of the $\boldsymbol{L}$ in Figure 6.7(a) will give the terms arising from three non-equivalent electrons, and so on.

It can be shown fairly easily that, for a filled sub-shell, such as $2p^6$ or $3d^{10}$, $L = 0$. Space quantization of the total orbital angular momentum produces $(2L + 1)$ components with $M_L = L, L-1, \ldots, -L$, analogous to space quantization of l. In a filled sub-shell $\Sigma_i (m_l)_i = 0$, where the sum is over all electrons in the sub-shell. Since $M_L = \Sigma_i (m_l)_i$, it follows that $L = 0$. Therefore, the excited configurations

$$\begin{array}{ll} \text{C} & 1s^2 2s^2 2p^1 3d^1 \\ \text{Si} & 1s^2 2s^2 2p^6 3s^2 3p^1 3d^1 \end{array} \quad (6.23)$$

of C and Si both give P, D and F terms.

The coupling between the spin momenta is referred to as ss coupling.

The result of coupling of the $\boldsymbol{s}$ vectors can be obtained in a similar way to ll coupling with the difference that, since s is always $\frac{1}{2}$, the vector for each electron is *always* of magnitude $3^{\frac{1}{2}}\hbar/2$ according to equation (6.11). The two $\boldsymbol{s}$ vectors can only take orientations relative to each other such that the resultant $\boldsymbol{S}$ is of magnitude

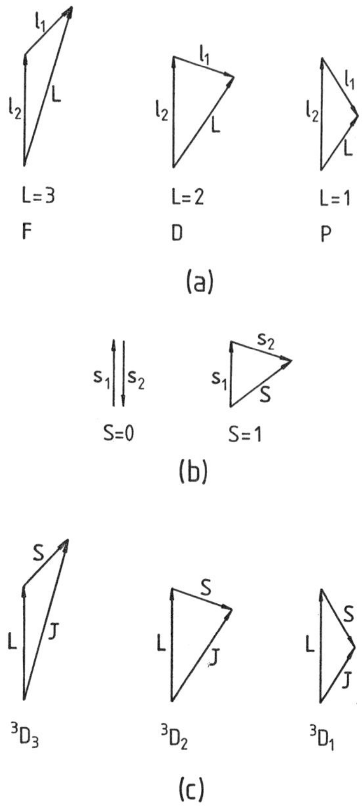

Figure 6.7 Vector representations of the coupling, in the Russell–Saunders approximation, of (a) the orbital angular momenta $\boldsymbol{l}_1$ and $\boldsymbol{l}_2$, (b) the spin angular momenta $\boldsymbol{s}_1$ and $\boldsymbol{s}_2$ and (c) the total orbital and total spin angular momenta, $\boldsymbol{L}$ and $\boldsymbol{S}$, respectively, of a p and a d electron

$$[S(S + 1)]^{\frac{1}{2}}\hbar = S^*\hbar \quad (6.24)$$

where S, the total spin quantum number,† is restricted to the values

$$S = s_1 + s_2, s_1 + s_2 - 1, \ldots, |s_1 - s_2| \quad (6.25)$$

In the case of two electrons this means that $S = 0$ or 1 only. The vector sums, giving resultant $\boldsymbol{S}$ vectors of magnitude 0 and $2^{\frac{1}{2}}\hbar$, are illustrated in Figure 6.7(b).

†This use of S should not be confused with its use as a term symbol to imply that $L = 0$.

Table 6.2 Terms arising from some configurations of non-equivalent and equivalent electrons

Non-equivalent electrons		Equivalent electrons	
Configuration	Terms	Configuration	Terms[a]
s^1s^1	$^{1,3}S$	p^2	^1S, ^3P, ^1D
s^1p^1	$^{1,3}P$	p^3	^4S, ^2P, ^2D
s^1d^1	$^{1,3}D$	d^2	^1S, ^3P, ^1D, ^3F, ^1G
s^1f^1	$^{1,3}F$	d^3	^2P, ^4P, $^2D(2)$, ^2F,
p^1p^1	$^{1,3}S$, $^{1,3}P$, $^{1,3}D$		^4F, ^2G, ^2H
p^1d^1	$^{1,3}P$, $^{1,3}D$, $^{1,3}F$	d^4	$^1S(2)$, $^3P(2)$, $^1D(2)$,
p^1f^1	$^{1,3}D$, $^{1,3}F$, $^{1,3}G$		^3D, ^5D, ^1F, $^3F(2)$,
d^1d^1	$^{1,3}S$, $^{1,3}P$, $^{1,3}D$, $^{1,3}F$, $^{1,3}G$		$^1G(2)$, ^3G, ^3H, ^1I
d^1f^1	$^{1,3}P$, $^{1,3}D$, $^{1,3}F$, $^{1,3}G$, $^{1,3}H$	d^5	^2S, ^6S, ^2P, ^4P, $^2D(3)$,
f^1f^1	$^{1,3}S$, $^{1,3}P$, $^{1,3}D$, $^{1,3}F$, $^{1,3}G$,		^4D, $^2F(2)$, ^4F, $^2G(2)$,
	$^{1,3}H$, $^{1,3}I$		^4G, ^2H, ^2I

[a]The numbers in parentheses indicate that a particular term occurs more than once.

The labels for the terms indicate the value of S by having $(2S+1)$ as a pre-superscript to the $S, P, D, \ldots$ label. The value of $(2S+1)$ is the multiplicity and is the number of values that M_S can take; these are

$$M_S = S, S-1, \ldots, -S \qquad (6.26)$$

Since, for two electrons, $S=0$ or 1, the value of $(2S+1)$ is 1 or 3 and the resulting terms are called singlet or triplet, respectively.

Just as $L=0$ for a filled sub-shell $S=0$ also, since $M_S = \Sigma_i (m_s)_i = 0$.

It follows from this that the excited configurations of C and Si in equation (6.23) give ^1P, ^3P, ^1D, ^3D, ^1F and ^3F terms. It follows also that the noble (inert) gases, in which all occupied sub-shells are filled, have only ^1S terms arising from their ground configurations.

Table 6.2 lists the terms which arise from various combinations of two non-equivalent electrons.

There is appreciable coupling between the resultant orbital and resultant spin momenta. This is referred to as LS coupling and is due to spin–orbit interaction.† This interaction is caused by the positive charge Z on the nucleus and is proportional to Z^4. The coupling between $\boldsymbol{L}$ and $\boldsymbol{S}$ gives the total angular momentum vector $\boldsymbol{J}$.

†We should really refer to this as *total* spin–*total* orbital interaction to distinguish it from spin–orbit interaction in a single electron, as occurs in *jj* coupling.

The phrase 'total angular momentum' is commonly used to refer to a number of different quantities. Here it implies 'orbital plus electron spin' but later we shall use it to imply 'orbital plus electron spin plus nuclear spin'.

The resultant vector $\boldsymbol{J}$ has the magnitude

$$[J(J+1)]^{\frac{1}{2}}\hbar = J^*\hbar \qquad (6.27)$$

where J is restricted to the values

$$J = L+S, L+S-1, \ldots, |L-S| \qquad (6.28)$$

from which it follows that, if $L \geqslant S$, then J can take $(2S+1)$ values but, if $L \leqslant S$, J can take $(2L+1)$ values.

As an example, we consider LS coupling in a ^3D term. Since $S=1$ and $L=2$, $J=3$, 2 or 1 and Figure 6.7(c) illustrates the three ways of coupling $\boldsymbol{L}$ and $\boldsymbol{S}$. The value of J is attached to the term symbol as a post-subscript so that the three components of ^3D are $^3D_3$, $^3D_2$ and $^3D_1$.

The total number of states arising from the C or Si configuration of equation (6.23) is now seen to comprise

$$^1P_1, {}^3P_0, {}^3P_1, {}^3P_2, {}^1D_2, {}^3D_1, {}^3D_2,$$
$${}^3D_3, {}^1F_3, {}^3F_2, {}^3F_3, {}^3F_4$$

At this stage it is appropriate to digress briefly on the subject of 'configurations', 'terms' and 'states'.

It is important that electron configurations, such as those in Table 6.1 and equation (6.23), should never be confused with terms or states. An electron configuration represents a gross but useful approximation in which the electrons have been fed into orbitals whose energies have been calculated neglecting the last term in equation (6.5). Nearly all configurations (all those with at least one unfilled sub-shell) give rise to more than one term or state and so it is wrong to speak of, for example, the $1s^12s^1$ state of helium; this is properly called a configuration and it gives rise to two states, ${}^1S_0$ and ${}^3S_1$.

The use of the words 'term' and 'state' is not so clearcut. The word 'term' was originally used in the early days of spectroscopy in the sense of equation (1.3), where the frequency of a line in an atomic spectrum is expressed as the difference between two terms which are simply terms in an equation.

Nowadays there is a tendency to use the word 'term' to describe what arises from an approximate treatment of an electron configuration, whereas the word 'state' is used to describe something which is observable experimentally. For example, we can say that the $1s^22s^22p^13d^1$ configuration of C gives rise to a ${}^3P$ term which, when spin–orbit coupling is taken into account, splits into ${}^3P_1$, ${}^3P_2$ and ${}^3P_3$ states. Since spin–orbit coupling can be excluded only in theory, but never in practice, there can be no experimental observation associated with the ${}^3P$ term.†

We shall see later in this section that, if the nucleus possesses a spin angular momentum, these states are further split and therefore, perhaps, should not have been called states in the first place! However, the splitting due to nuclear spin is small and it is normal to refer to nuclear spin *components* of states.

The Russell–Saunders coupling scheme for two, or more, equivalent electrons, i.e. with the same n and l, is rather more tedious to apply. We shall use the example of two equivalent p electrons, as in the ground configuration of carbon:

$$\text{C} \quad 1s^22s^22p^2 \tag{6.29}$$

Again, for the closed sub-shells $L=0$ and $S=0$, so we have to consider only the $2p$ electrons. Since $n=2$ and $l=1$ for both electrons the Pauli exclusion principle is in danger of being violated unless the two electrons have different values of m_l and/or m_s. For non-equivalent electrons we do not have to consider the values of these two quantum numbers because, as either n or l is not the same for both electrons, there is no danger of violation.

For the $2p$ electron which we shall label 1, $l_1=1$ and $(m_l)_1=+1$, 0 or -1 and, in addition, $s_1=\frac{1}{2}$ and $(m_s)_1=+\frac{1}{2}$ or $-\frac{1}{2}$; and similarly for electron 2. The Pauli exclusion principle requires that the pair of quantum numbers $(m_l)_1$ and $(m_s)_1$ cannot simultaneously have the same values as $(m_l)_2$ and $(m_s)_2$. The result is that there are 15 allowed combinations of values and they are all given in Table 6.3.

It should be noted that the indistinguishability of the electrons has been taken into account in the table so that, for example, the combination $(m_l)_1=(m_l)_2=1$, $(m_s)_1=-\frac{1}{2}$, $(m_s)_2=\frac{1}{2}$ cannot be included in addition to $(m_l)_1=(m_l)_2=1$, $(m_s)_1=\frac{1}{2}$, $(m_s)_2=-\frac{1}{2}$ which is obtained from the first by electron exchange.

The values of M_L $[=\Sigma_i(m_l)_i]$ and M_S $[=\Sigma_i(m_s)_i]$ are given in Table 6.3. The highest value of M_L is 2 and this indicates that this is also the highest value of L and that there is a D term. Since $M_L=2$ is associated only with $M_S=0$ it must be a ${}^1D$ term. This term accounts for five of the combinations, as shown at the bottom of the table. In the remaining combinations, the highest value of L is 1 and, since this

†It is unfortunately true, however, that it is not uncommon to speak of, for example, a ${}^3P$ state.

Table 6.3 Derivation of terms arising from two equivalent p electrons

Quantum number	Values														
$(m_l)_1$	1	1	1	1	1	1	1	1	1	0	0	0	0	0	-1
$(m_l)_2$	1	0	0	0	0	-1	-1	-1	-1	0	-1	-1	-1	-1	-1
$(m_s)_1$	$\frac{1}{2}$	$\frac{1}{2}$	$\frac{1}{2}$	$-\frac{1}{2}$	$-\frac{1}{2}$	$\frac{1}{2}$	$\frac{1}{2}$	$-\frac{1}{2}$	$-\frac{1}{2}$	$\frac{1}{2}$	$\frac{1}{2}$	$\frac{1}{2}$	$-\frac{1}{2}$	$-\frac{1}{2}$	$\frac{1}{2}$
$(m_s)_2$	$-\frac{1}{2}$	$\frac{1}{2}$	$-\frac{1}{2}$	$\frac{1}{2}$	$-\frac{1}{2}$	$\frac{1}{2}$	$-\frac{1}{2}$	$\frac{1}{2}$	$-\frac{1}{2}$	$-\frac{1}{2}$	$\frac{1}{2}$	$-\frac{1}{2}$	$\frac{1}{2}$	$-\frac{1}{2}$	$-\frac{1}{2}$
$M_L = \sum_i (m_l)_i$	2	1	1	1	1	0	0	0	0	0	-1	-1	-1	-1	-2
$M_S = \sum_i (m_s)_i$	0	1	0	0	-1	1	0	0	-1	0	1	0	0	-1	0
	Pairs of values of M_L and M_S can be arranged as follows:														
M_L	2	1	0	-1	-2	1	0	-1	1	0	-1	1	0	-1	0
M_S	0	0	0	0	0	1	1	1	0	0	0	-1	-1	-1	0
	^1D					^3P									^1S

is associated with $M_S = 1$, 0 and -1, there is a ^3P term. This accounts for a further nine combinations leaving only $M_L = 0$, $M_S = 0$ which implies a ^1S term.

It is interesting to note that of the ^1S, ^3S, ^1P, ^3P, ^1D and ^3D terms which arise from two non-equivalent p electrons, as in the $1s^22s^22p^13p^1$ configuration of the carbon atom, only ^1S, ^3P and ^1D are allowed for two equivalent p electrons: the Pauli exclusion principle forbids the other three.

Terms arising from three equivalent p electrons and also from various equivalent d electrons can be derived using the same methods but this can be a very lengthy operation. The results are given in Table 6.2.

In deriving the terms arising from non-equivalent or equivalent electrons there is a very useful rule that, in this respect, a vacancy in a sub-shell behaves like an electron. For example, the ground configurations of C and O:

$$\left.\begin{array}{l} \mathrm{C}\ \ 1s^22s^22p^2 \\ \mathrm{O}\ \ 1s^22s^22p^4 \end{array}\right\} {}^1S,\ {}^3P,\ {}^1D \tag{6.30}$$

give rise to the same terms, as do the excited configurations of C and Ne:

$$\left.\begin{array}{l} \mathrm{C}\ \ 1s^22s^22p^13d^1 \\ \mathrm{Ne}\ \ 1s^22s^22p^53d^1 \end{array}\right\} {}^{1,3}P,\ {}^{1,3}D,\ {}^{1,3}F \tag{6.31}$$

In 1927, Hund formulated two empirical rules which enable us to determine which of the terms arising from equivalent electrons lies lowest in energy. This means that for a ground configuration with only equivalent electrons in partly filled sub-shells we can determine the lowest energy, or ground, term. These rules are as follows:

1. Of the terms arising from equivalent electrons, those with the highest multiplicity lie lowest in energy.
2. Of these, the lowest is that with the highest value of L.

Using these rules, it follows that, for the ground configurations of both C and O in equation (6.30), the ^3P term is the lowest in energy.

The ground configuration of Ti is

$$\mathrm{Ti}\ \ KL3s^23p^63d^24s^2 \tag{6.32}$$

From the terms arising from the d^2 configuration given in Table 6.2 the rules indicate that ^3F is the lowest in energy.

The splitting of a term by spin–orbit interaction is proportional to J:

$$E_J - E_{J-1} = AJ \tag{6.33}$$

and produces a multiplet. If A is positive, the component with the smallest value of J lies lowest in energy and the multiplet is said to be normal whereas, if A is negative, the multiplet is said to be inverted.

There are two further rules for ground terms which tell us whether a multiplet arising from equivalent electrons is normal or inverted:

3. Normal multiplets arise from equivalent electrons when an incomplete sub-shell is *less* than half full.
4. Inverted multiplets arise from equivalent electrons when an incomplete sub-shell is *more* than half full.

It follows that the lowest energy term of Ti, ^3F, is split by spin–orbit coupling into a normal multiplet and therefore the ground state is $^3F_2$. Similarly, the lowest energy term of C, ^3P, splits into a normal multiplet, resulting in a $^3P_0$ ground state, whereas that of O, with an inverted multiplet, is $^3P_2$.

Atoms with a ground configuration in which a sub-shell is exactly half-filled, e.g. in N($2p^3$), Mn($3d^5$) and Eu($4f^7$), always have an S ground state. Since such states have only one component, the problem of a normal or inverted multiplet does not arise.

Table 6.1 gives the ground states of all atoms in the periodic table.

There is one further addition to the state symbolism which we have not mentioned so far. This is the superscript o as in the ground state of boron, $^2P^{\text{o}}_{\frac{1}{2}}$. The symbol implies that the arithmetic sum $\Sigma_i l_i$ for *all* the electrons in the atom is an odd number, 1 in this case. When there is no such superscript, this implies that the sum is an even number, e.g. 4 in the case of oxygen.

The ordering of terms and states arising from non-equivalent electrons does not have such simplifying rules as the four mentioned above for equivalent electrons. The reader is referred, for details, to the book by Condon and Shortley, listed in the Bibliography at the end of the chapter, but it is worth looking briefly at one example.

For a p^1p^1 configuration of non-equivalent electrons, e.g. C $1s^22s^22p^13p^1$, the terms which arise are $^{1,3}S$, $^{1,3}P$ and $^{1,3}D$. Their energies are given by

$$E(^{1,3}S) = F_0 + 10F_2 \pm (G_0 + 10G_2) \quad (6.34)$$

$$E(^{1,3}P) = F_0 - 5F_2 \mp (G_0 - 5G_2) \quad (6.35)$$

$$E(^{1,3}D) = F_0 + F_2 \pm (G_0 + G_2) \quad (6.36)$$

where the upper sign refers to the singlet term and the lower to the triplet. The quantity G_2 is usually so small that in most cases G_0, a positive quantity, determines the splitting of the singlet and triplet terms. Equations (6.34)–(6.36) show that for the S and D terms the singlet state is above the triplet but, for the P term, it is below. This kind of alternation with L occurs generally for configurations having two non-equivalent electrons.

For excited terms split by spin–orbit interaction there are no general rules regarding normal or inverted multiplets. For example, in He, excited states form mostly inverted multiplets while in the alkaline earth metals Be, Mg, Ca, . . . , they are mostly normal.

6.1.2.2(ii) jj Coupling approximation

In this approximation, coupling between the orbital and spin momenta for each electron, referred to as *ls* coupling, is strong. Coupling between the resulting total (orbital plus spin) momenta $\boldsymbol{j}$ for each electron, referred to as *jj* coupling, is appreciable but less strong; *ll* and *ss* coupling are neglected.

As an example, we shall consider the case of an s and a p electron with $l_1 = 0$, $s_1 = \frac{1}{2}$ and $l_2 = 1$, $s_2 = \frac{1}{2}$, respectively.

The orbital and spin momentum vectors, $\boldsymbol{l}$ and $\boldsymbol{s}$, for each electron can only take orientations relative to each other such that the magnitude of the resultant $\boldsymbol{j}$ is $[j(j+1)]^{\frac{1}{2}}\hbar$, as in equation (6.12). The total (orbital plus spin) angular momentum quantum number, j for a single electron, can take the values given by the Clebsch–Gordan series of equation (6.13).

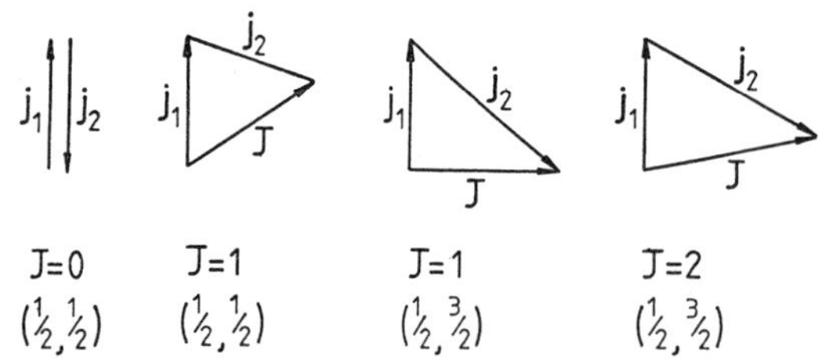

Figure 6.8 Vector representations of the four states which result from coupling, in the *jj* approximation, between spin and orbital angular momenta of an *s* and a *p* electron

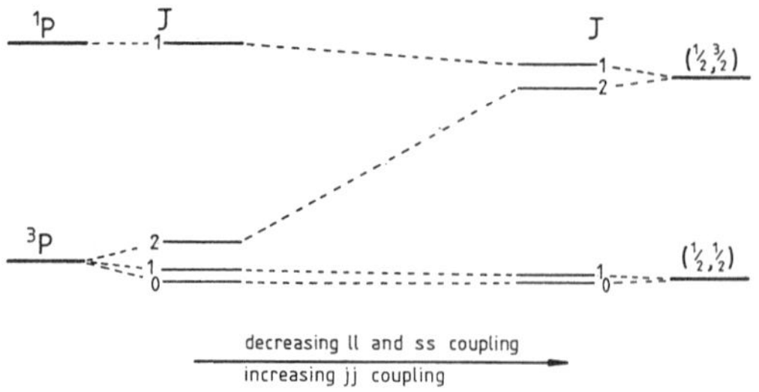

Figure 6.9 Correlation of the four states which arise from coupling between spin and orbital angular momenta in the Russell–Saunders (left-hand side) and *jj* (right-hand side) approximations for an *s* and a *p* electron

Therefore, for electron 1, $j_1 = \frac{1}{2}$ and, for electron 2, $j_2 = \frac{3}{2}, \frac{1}{2}$. The $\boldsymbol{j}_1$ and $\boldsymbol{j}_2$ vectors can only take relative orientations such that the magnitude of the resultant $\boldsymbol{J}$ is given by

$$[J(J+1)]^{\frac{1}{2}}\hbar = J^*\hbar \qquad (6.37)$$

where J can take the values

$$J = j_1 + j_2, j_1 + j_2 - 1, \ldots, |j_1 - j_2| \qquad (6.38)$$

When $j_1 = \frac{1}{2}$ and $j_2 = \frac{1}{2}$ we have $J = 1$, 0 and when $j_1 = \frac{1}{2}$ and $j_2 = \frac{3}{2}$ we $J = 2$, 1. The four corresponding vector sums are illustrated in Figure 6.8. The resulting states are labelled by the value of J and, in parentheses, j_1 and j_2. L and S are not good quantum numbers in this coupling approximation.

The states resulting from an *s* and a *p* electron in the Russell-Saunders approximation are $^1P_1$, $^3P_2$, $^3P_1$ and $^3P_0$. Figure 6.9 shows how these four states and the four resulting from *jj* coupling correlate as, in moving from left to right of the figure, *ll* and *ss* coupling decreases and *jj* coupling increases. J is a good quantum number in both approximations so it must be the same for correlated levels. It is assumed that the ^3P multiplet is normal.

On the extreme left of the figure there is zero spin–orbit coupling. When this increases the ^3P term splits into three components, the splitting being given by equation (6.33).

The first sign of states not conforming well to the Russell–Saunders approximation is that multiplet splitting is no longer given by equation (6.33).

Extreme *jj* coupling is rare except in a few excited states of a few atoms with high nuclear charge, such as the lowest ^3P term of Pb. Most coupling cases where the Russell–Saunders approximation is inadequate are intermediate cases falling in the region about half-way between the two extremes.

Ground states of *all* atoms are best described by Russell–Saunders coupling.

6.1.2.2(iii) *Coupling of nuclear spin to other angular momenta*

In Section 6.1.2.1 we saw that, for nuclei with the nuclear spin quantum number $I \neq 0$, there is an angular momentum of magnitude $[I(I+1)]^{\frac{1}{2}}\hbar$ due to this motion [equation (6.17)]. The associated magnetic moment μ_I, given by equation (6.18), is small compared with those due to spin and orbital motions of the electrons, so that coupling between these angular momenta and nuclear spin is quite weak.

The coupling is pictured in the usual kind of vector diagram in which the vector $\boldsymbol{J}$, of magnitude $[J(J+1)]^{\frac{1}{2}}\hbar$ and representing the total (orbital plus spin) angular momentum of the electrons, and the vector $\boldsymbol{I}$, representing the nuclear spin angular momentum, can take only certain orientations relative to each other.

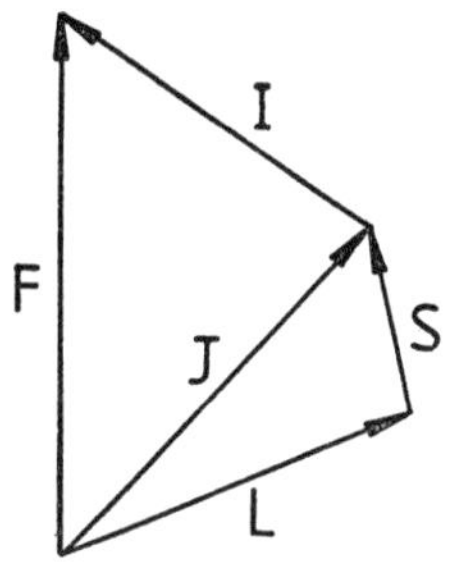

Figure 6.10 Vector representation of coupling of the nuclear spin angular momentum ***I*** to the orbital-plus-spin angular momentum ***J***

These orientations are the ones which produce a resultant vector ***F*** of magnitude

$$[F(F+1)]^{\frac{1}{2}}\hbar = F^{*}\hbar \qquad (6.39)$$

F is the quantum number associated with the total angular momentum, which now includes those due to orbital and spin motion of the electrons and spin motion of the nucleus. The fact that the orientations of ***J*** and ***I*** are restricted is due to the limitations on the value of F given by

$$F = J+I, J+I-1, \ldots, |J-I| \qquad (6.40)$$

Figure 6.10 illustrates the complete vector diagram for coupling of all the angular momenta. Russell–Saunders coupling has been assumed here for the electrons but the coupling of ***J*** and ***I*** is unaffected by the type of coupling by which ***J*** is obtained.

The magnitudes of the magnetic moments involved in the various coupling mechanisms are reflected in the precession rates. Since the ratio of the Bohr magneton to the nuclear magneton is 1837 [equation (6.20)], the precession of ***J*** and ***I*** about ***F*** is very much slower than that of ***L*** and ***S*** about ***J***.

A state with a particular value of J is split by nuclear spin into $(2J+1)$ components if $J \leqslant I$ and $(2I+1)$ if $J \geqslant I$. The splitting is given by

$$E_F - E_{F-1} = A'F \qquad (6.41)$$

A' is a constant for a particular value of J but it decreases rapidly with J. This equation is analogous to equation (6.33) for splitting due to spin–orbit interaction but $A' \ll A$. As a result, the splitting due to spin–orbit interaction gives rise to what is called fine structure in an atomic spectrum, whereas the much more closely spaced structure caused by nuclear spin is called nuclear hyperfine structure.

The fact that many elements contain more than one isotope in natural abundance causes lines in the spectra of such atoms to appear as multiplets in which each component corresponds to a single isotope. The existence of many isotopes, notable ^{2}H, was first demonstrated by their atomic spectra. Separations of these components are typically small and comparable to those of nuclear hyperfine components. For this reason, the isotope components have also been referred to as hyperfine structure, but the phrase is probably better not used in this context since the cause is completely unrelated to nuclear spin.

6.1.3 Spectra of the alkali metal atoms

The hydrogen atom and one-electron ions are the simplest systems in the sense that, having only one electron, there are no inter-electron repulsions. However, this unique property leads to degeneracies, or near degeneracies, which are absent in all other atoms and ions. The result is that the spectrum of the hydrogen atom, although very simple in its coarse structure (Figure 1.5), is more complex in its fine structure than polyelectronic atoms. For this reason, we shall defer a discussion of its spectrum to the next section.

The alkali metal atoms all have one valence electron in an outer ns orbital, where $n=2$, 3, 4, 5, 6 for Li, Na, K, Rb and Cs, respectively. If we consider only promotions of the electron to and from this orbital, the behaviour is expected to resemble that of the hydrogen atom. The reason for this is that the core, consisting of the nucleus of charge $+Ze$ and filled sub-shells

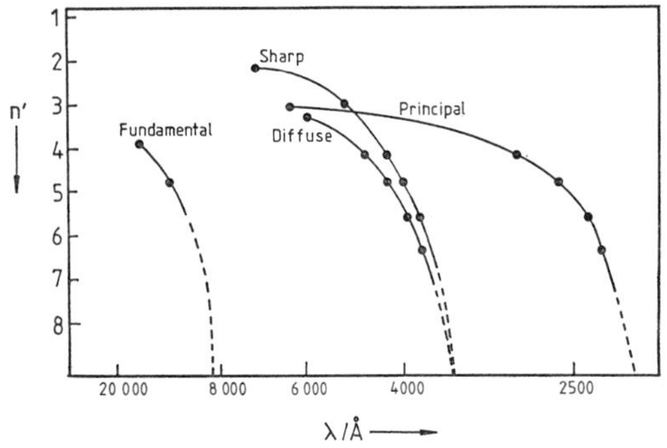

Figure 6.11 Principal, sharp, diffuse and fundamental series in the emission spectrum of the lithium atom

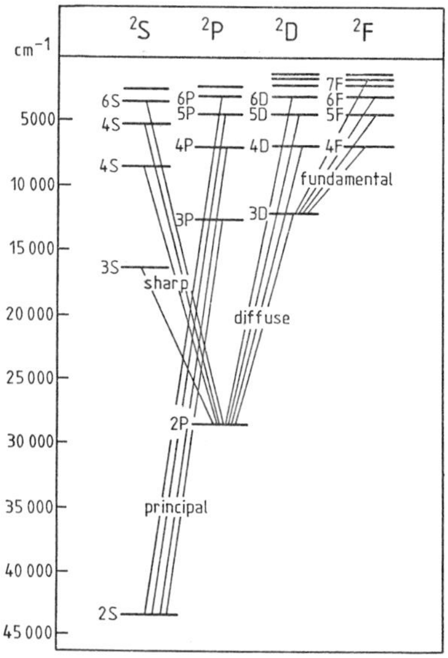

Figure 6.12 Grotrian diagram for the lithium atom

containing $Z-1$ electrons, has a net charge of $+e$ and therefore has an effect on the valence electron which is similar to that of the nucleus of the hydrogen atom on its electron.

Whereas the emission spectrum of the hydrogen atom shows only one series, the Balmer series (see Figure 1.5), in the visible region the alkali metals show at least three. The spectra can be excited in a discharge lamp containing a sample of the appropriate metal. One series was called the principal series because it could also be observed in absorption through a column of the vapour. The other two were called sharp and diffuse because of their general appearance. A part of a fourth series, the fundamental series, can sometimes be observed.

Figure 6.11 shows these series schematically for Li. All such series converge smoothly towards high energy (low wavelength) in a way that resembles the series in the hydrogen spectrum.

Figure 6.12 shows an energy level diagram, a so-called Grotrian diagram, for Li for which the ground configuration is $1s^2 2s^1$. The lowest energy level corresponds to this configuration. The higher energy levels are labelled according to the orbital to which the valence electron has been promoted; for example, the level labelled $4p$ corresponds to the configuration $1s^2 4p^1$.

The relatively large energy separation between configurations in which the valence electron is in orbitals differing only in the value of l, e.g. $1s^2 3s^1$, $1s^2 3p^1$, $1s^2 3d^1$, is characteristic of all atoms except hydrogen (and one-electron ions).

The selection rules governing the promotion of the electron to an excited orbital, and also its falling back from an excited orbital, are (a) Δn is unrestricted, as in the hydrogen atom, and (b) $\Delta l = \pm 1$.

These selection rules lead to the sharp, principal, diffuse and fundamental series, shown in Figures 6.11 and 6.12, in which the promoted electron is in an s, p, d and f orbital, respectively. These curious orbital symbols originate from the first letters of the names of the corresponding series observed in the spectrum.

The sharp, principal and diffuse series are coincident in the hydrogen atom spectrum and are analogous to the Balmer series.

Because of the qualitative similarities between the converging series in the alkali metal spectra and those of the hydrogen atom, attempts were made to fit the wavenumbers of the members of each series to an expression of the type

Table 6.4 Quantum defects, η, for some terms of the sodium atom

Excited orbital	n = 3	4	5	6	7	8
s	1.373	1.357	1.353	1.351	1.350	1.350
p	0.883	0.867	0.862	0.860	0.858	0.857
d	0.010	0.012	0.013	0.014	0.014	0.014
f	–	0.001	0.000	0.000	0.000	0.000

$$\tilde{\nu} = \tilde{R}_\infty Z^2\{[1/(n_e'')^2] - [1/(n_e')^2]\} \quad (6.42)$$

This is analogous to equations (1.17)–(1.19) for the hydrogen atom. Z is the atomic number and $\tilde{R}_\infty$ is the Rydberg constant for an atom with an infinitely heavy nucleus given by

$$\tilde{R}_\infty = m_e e^4/8h^3\varepsilon_0^2 c \quad (6.43)$$

to be compared with $\tilde{R}_\mathrm{H}$ in equation (1.19).

In equation (6.42), n_e'' and n_e' are the effective quantum numbers for the lower and upper levels, respectively. In general they do not have integral values but can be related to the principal quantum number n by

$$n_e = n - \eta \quad (6.44)$$

where η is the quantum defect.

Table 6.4 gives some values of η for the sodium atom. These values show that η is zero when the promoted electron is in an f orbital. The reason for this is that the electron density in an f orbital is far from the nucleus and an electron in such an orbital tends to see the core, consisting of the nucleus of charge $+11e$ and 10 electrons in filled K and L shells, as a net charge of $+e$ concentrated in a volume which is small relative to the orbital size; hence the hydrogen-like behaviour.

Table 6.4 shows also that η increases rapidly as l decreases. Figure 6.13 shows how the radial charge density function $4\pi r^2 R_{nl}^2$, discussed in Section 1.3.2, varies with the distance r from the nucleus for a $3s$ and a $3p$ orbital and also for all the core electrons. The $3s$ electron density overlaps the core density more than $3p$. The

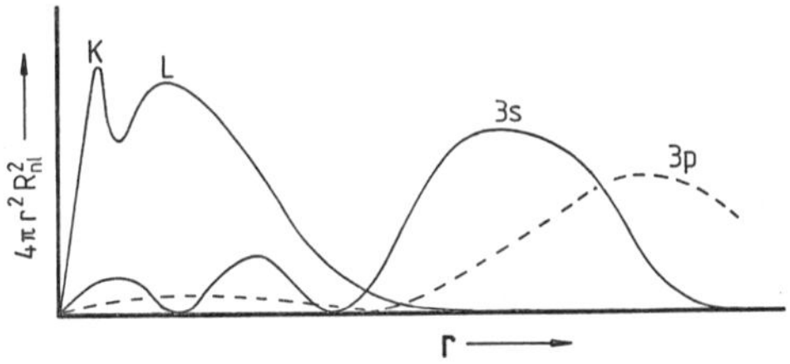

Figure 6.13 Variation of the radial charge density, $4\pi r^2 R_{nl}^2$, with distance r from the nucleus for K and L core electrons and a $3s$ and $3p$ valence electron in the sodium atom

effect is called penetration of the core by the valence electron and increases as l decreases. As l decreases there is also a decrease in the shielding of the nucleus by the core electrons; complete shielding results in a net charge of $+e$ for the core-plus-nucleus, whereas incomplete shielding increases this effective net charge. It is the shielding effects which result in the $s, p, d, \ldots$ orbitals for a given n having different energies and the degree of penetration which determines their magnitudes. It is therefore the greater penetration of the s orbitals which leads to their having the largest values of η. Similar results are found for all the alkali metals.

Some excited configurations of the lithium atom, involving promotion of only the valence electron, are given in Table 6.5, which also lists the states arising from these configurations. Similar states can easily be derived for other alkali metals.

Spin–orbit coupling splits apart the two components of the ^2P, ^2D, ^2F, . . . terms. The splitting decreases with L and n and increases with atomic number but is not large enough, for any terms of the lithium atom, to show on the Grotrian diagram of Figure 6.12. The resulting fine structure is difficult to resolve in the lithium spectrum but is more easily observed in other alkali metal spectra.

In the sodium atom pairs of $^2P_{\frac{1}{2}}$, $^2P_{\frac{3}{2}}$ states result from the promotion of the $3s$ valence electron to any np orbital with $n>2$. It is convenient to label the states with this value of

Table 6.5 Configurations and states of the lithium atom

Configuration	States
$1s^2 2s^1$	$^2S_{\frac{1}{2}}$
$1s^2 ns^1$ ($n = 3, 4, \ldots$)	$^2S_{\frac{1}{2}}$
$1s^2 np^1$ ($n = 2, 3, \ldots$)	$^2P_{\frac{1}{2}}$, $^2P_{\frac{3}{2}}$
$1s^2 nd^1$ ($n = 3, 4, \ldots$)	$^2D_{\frac{3}{2}}$, $^2D_{\frac{5}{2}}$
$1s^2 nf^1$ ($n = 4, 5, \ldots$)	$^2F_{\frac{5}{2}}$, $^2F_{\frac{7}{2}}$

n, as $n^2P_{\frac{1}{2}}$ and $n^2P_{\frac{3}{2}}$, the n label being helpful for states which arise when only one electron is promoted and the unpromoted electrons are either in filled sub-shells or in an s orbital. The n label can be used, therefore, for hydrogen, the alkali metals, helium and the alkaline earths. In other atoms it is usual to precede the state symbols by the configuration of the electrons in unfilled sub-shells, as in the $2p3p\,^1S_0$ state of carbon.

The splitting of the $3^2P_{\frac{1}{2}}$, $3^2P_{\frac{3}{2}}$ states of sodium is 17.2 cm^{-1} and this reduces to 5.6, 2.5 and 1.3 cm^{-1} for $n = 4$, 5 and 6, respectively. The splitting decreases rapidly with L, as exemplified by the splitting of only 0.1 cm^{-1} for the $3^2D_{\frac{3}{2}}$, $3^2D_{\frac{5}{2}}$ states. All these ^2P and ^2D multiplets are normal, the state with lowest J lying lowest in energy.

The fine structure selection rule is

$$\Delta J = 0, \ \pm 1 \text{ except } J = 0 \nleftrightarrow J = 0$$

The result is that the principal series consists of pairs of $^2P_{\frac{1}{2}}$–$^2S_{\frac{1}{2}}$, $^2P_{\frac{3}{2}}$–$^2S_{\frac{1}{2}}$ transitions,† as illustrated in Figure 6.14(a). The pairs are known as simple doublets. The first member of this series in sodium, called the sodium D-line, appears in the yellow region of the spectrum with components at 589.592 and 588.995 nm.

The $3^2P_{\frac{1}{2}}$, $3^2P_{\frac{3}{2}}$ excited states involved in the sodium D-line are the lowest energy excited states of the atom. Consequently, in a discharge

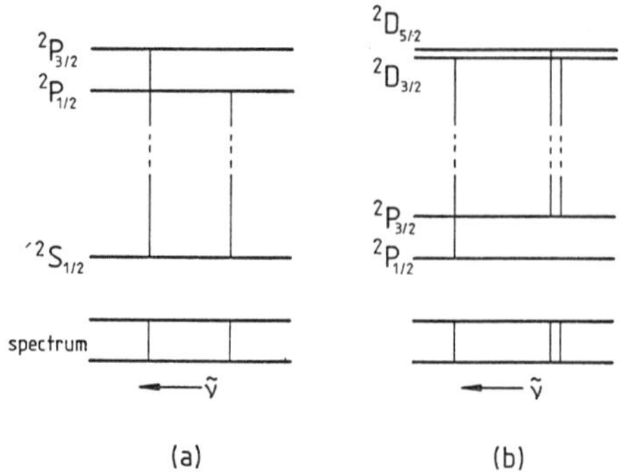

Figure 6.14 (a) A simple doublet and (b) a compound doublet in the spectrum of, for example, the sodium atom

in the vapour at a pressure which is sufficiently high for collisional deactivation of excited states to occur readily, a majority of atoms find themselves in these states before emission of radiation has taken place. Therefore, the D-line is prominent in emission and this explains the predominantly yellow colour of sodium discharge lamps.

The sharp series members are all simple doublets which all show the same splitting, namely, that of the $3^2P_{\frac{1}{2}}$ and $3^2P_{\frac{3}{2}}$ states.

All members of a diffuse series consist of compound doublets, as illustrated in Figure 6.14(b), but the splitting of the $^2D_{\frac{3}{2}}$, $^2D_{\frac{5}{2}}$ states may be too small for the close pair of transitions to be resolved. It is for this reason that the set of three transitions has become known as a compound doublet rather than a triplet.

Many of the alkali metal atoms have a nuclear spin angular momentum and consequently show nuclear hyperfine structure in their spectra. Sodium is a good example of this because it is isotopically pure as ^{23}Na in its natural form. For this nucleus $I = \frac{3}{2}$ so, for a $^2P_{\frac{1}{2}}$ or $^2S_{\frac{1}{2}}$ state, the quantum number F, given by equation (6.40), can take the values 2 and 1 whereas, for a $^2P_{\frac{3}{2}}$ state, it can be 3, 2, 1 and 0. The result of applying the selection rule

$$\Delta F = 0, \ \pm 1 \text{ except } F = 0 \nleftrightarrow F = 0$$

†The convention of indicating a transition involving an *upper* electronic state N and a *lower* electronic state M by N–M is analogous to that used in rotational and vibrational spectroscopy.

is that, as shown in Figure 6.15, the $3^2P_{\frac{1}{2}}$–$3^2S_{\frac{1}{2}}$ and $3^2P_{\frac{3}{2}}$–$3^2S_{\frac{1}{2}}$ components of the sodium D-line should have four and six hyperfine components, respectively. In practice, however, the nuclear hyperfine splitting of the $3^2P_{\frac{1}{2}}$ and $3^2P_{\frac{3}{2}}$ states is so small compared with that in the $3^2S_{\frac{1}{2}}$ state that each component can only be observed, with conventional spectroscopic techniques, as having two hyperfine components. These reflect the splitting in the $3^2S_{\frac{1}{2}}$ state but even this is only 0.06 cm^{-1}.

By using a laser and saturation techniques, all the nuclear hyperfine structure of the 2^3P–2^3S transitions in Li^+ has been resolved.† Similar resolution of nuclear hyperfine structure of many atoms and ions, including that of the sodium D-line, has been obtained.

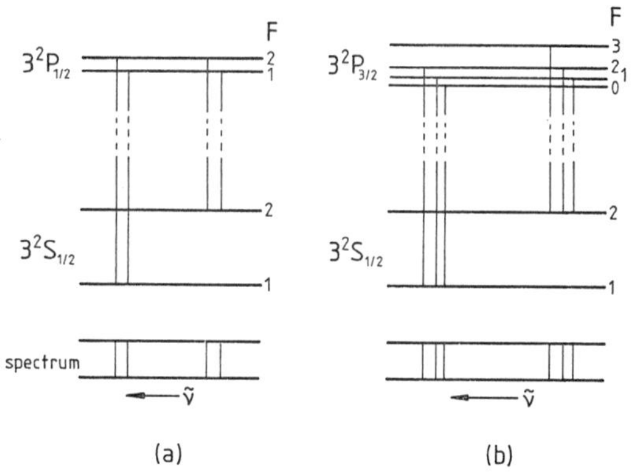

Figure 6.15 Nuclear hyperfine components of (a) the $3^2P_{\frac{1}{2}}$–$3^2S_{\frac{1}{2}}$, and (b) the $3^2P_{\frac{3}{2}}$–$3^2S_{\frac{1}{2}}$ components of the sodium D-line

6.1.4 Spectrum of the hydrogen atom

The hydrogen atom presented a unique opportunity in the development of quantum mechanics. The single electron moves in a coulombic field, free from the effects of inter-electron repulsions. This has two important consequences which do not apply to any atom with two or more electrons and which are

1. The Schrödinger equation (1.45) is exactly soluble with the hamiltonian of equation (1.48).
2. The orbital energies are, at this level of approximation, independent of the quantum number l, as shown by Figure 1.5, which is to be contrasted with the energy diagram for lithium in Figure 6.12.

Figure 1.5 shows the first few of an infinite number of series of transitions in the hydrogen atom, all smoothly converging to the ionization limit at $n = \infty$. Under various conditions many of the series may be observed in emission but, in the absorption spectrum under normal experimental conditions, only the Lyman series is observed because only the $n = 1$ level is populated.

All stars, including the sun, contain enormous quantities of hydrogen atoms. It is easily the most abundant element, the next being helium, which is formed from hydrogen by thermonuclear reaction and is present to the extent of about one fifth of the quantity of hydrogen. The temperature of the interior of a star is of the order of 10^6 K but that of the exterior, the photosphere, is much lower, of the order of 10^3 K. The absorption spectrum observed from a star involves the interior acting as the source of continuous radiation and the photosphere as the absorber. Such a spectrum shows the Balmer series of hydrogen in absorption. The ratio of the populations of the $n = 2$ and $n = 1$ levels at a temperature of 10^3 K can be obtained from equation (2.8) and is only 2.9×10^{-5}, but the concentration of hydrogen atoms, together with the long path-length of absorption in the photosphere, combine to make the observation possible.

The interstellar medium (see Section 4.7) also contains large quantities of atomic hydrogen. Scanning of these regions with radiotelescopes has resulted in the observation in emission of a number of transitions in the hydrogen atom involving very high values of n. The first member of each of the series with $n'' = 90$,

†zu Putlitz, G. (1979). *Laser Spectroscopy IV* (Eds. H. Walther and K. W. Rothe), p. 539, Springer-Verlag, Berlin).

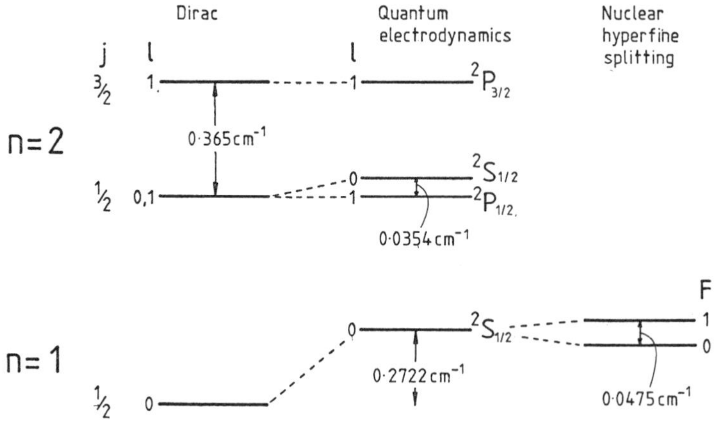

Figure 6.16 The $n=1$ and $n=2$ levels of the hydrogen atom according to Dirac quantum mechanics and quantum electrodynamics. The latter results in Lamb shifts of the $1^2S_{\frac{1}{2}}$ and $2^2S_{\frac{1}{2}}$ states. Nuclear hyperfine splitting of the $1^2S_{\frac{1}{2}}$ state is also shown

104, 109, 126, 156, 158, 159 and 166 has been observed.

When observed at high resolution the members of the Balmer and Paschen series show closely spaced fine structure. With the Bohr theory such fine structure could not be explained and an attempt by Sommerfeld was quantitatively successful but for the wrong reasons (see Section 1.1).

Figure 6.16 shows how Dirac relativistic quantum mechanics predicts the splitting of the $n=2$ level into two components $0.365\,\text{cm}^{-1}$ apart. For $n=2$, l can be 0 or 1. Equation (6.13) shows that, since $s=\frac{1}{2}$, j can be $\frac{3}{2}$ or $\frac{1}{2}$ for $l=1$ and $\frac{1}{2}$ for $l=0$. One of the components of the $n=2$ level has $j=\frac{3}{2}$, $l=1$ and the other is doubly degenerate with $j=\frac{1}{2}$, $l=0$, 1.

In 1947, Lamb and Retherford[68] observed the $2^2P_{\frac{3}{2}}$–$2^2S_{\frac{1}{2}}$ transition using microwave techniques. They found it to have a wavenumber $0.0354\,\text{cm}^{-1}$ less than predicted. The corresponding shift of the energy level is known as the Lamb shift and is shown in Figure 6.16; the $2^2P_{\frac{1}{2}}$ level is not shifted. Later, they observed the $2^2S_{\frac{1}{2}}$–$2^2P_{\frac{1}{2}}$ transition directly with a wavenumber of $0.0354\,\text{cm}^{-1}$.

In Figure 6.16 it is made clear, from the example of $n=2$, how strikingly different from the alkali metals is the pattern of levels in the hydrogen atom. In the alkali metals the separation of $^2P_{\frac{1}{2}}$ and $^2P_{\frac{3}{2}}$ states, due to spin–orbit coupling, is very small compared with the ^2P–^2S splitting but, in hydrogen, they are comparable.

Dirac theory had to be modified to account for the Lamb shift and this theory is called quantum electrodynamics. This takes into account the interaction, neglected in Dirac theory, between the electron and the radiation field. It is the currently accepted theory and is applicable to all atomic spectra, but it is only in accounting for extremely accurate data on a few states of a few atoms and ions that it is necessary.

The Lamb shift is proportional to n^{-3} and therefore decreases rapidly with n. Figure 6.16 shows that it is $0.2722\,\text{cm}^{-1}$ for the $n=1$ level but, since there is only one component, $^2S_{\frac{1}{2}}$, the shift cannot be observed as directly as it can for $n=2$. It shows up as a small shift of the members of the Lyman series compared with predictions neglecting the effect (see Section 8.2.18 for its measurement).

The hydrogen nucleus has spin angular momentum with $I=\frac{1}{2}$ and splitting of the $n=1$ level due to nuclear spin is appreciable.

In this level $j = s = \frac{1}{2}$ and therefore F can take the values 0 and 1 [equation (6.40)]. These two states can be described as having the electron and nuclear spins anti-parallel and parallel, respectively.

The $F = 1$–0 transition can be observed directly in the radiofrequency region with a wavelength of 21 cm. The frequency of the transition depends on the value of g_e [equation (6.15)] and, again, it is only quantum electrodynamics which correctly predicts this value.

The large quantities of hydrogen atoms in stars and the interstellar medium leads to the 21 cm transition being observed with considerable intensity by radiotelescopes scanning the appropriate regions of the sky. The transition is observed either as an emission process or by absorption of the background continuum. Results show the highest concentrations of hydrogen atoms in regions, such as the Milky Way, where there are large concentrations of stars.

For the $n = 2$ levels the hyperfine splitting is considerably smaller than for $n = 1$. The $F = 0$ and 1 components of the $2^2S_{\frac{1}{2}}$ state are split by 0.005 921 cm^{-1} and the splitting of the components of $2^2P_{\frac{1}{2}}$ and $2^2P_{\frac{3}{2}}$ is even smaller.

6.1.5 Spectra of the helium atom and the alkaline earth metal atoms

The emission spectrum of a discharge in helium gas in the visible and near ultraviolet regions appears rather like the spectrum of two alkali metals. It shows series of lines converging smoothly to high energy (low wavelength) which can be divided into two groups. One group consists of single lines and the other, at low resolution, double lines. Derivation of an energy level diagram shows that it consists of two sets of energy levels, one corresponding to the single lines and the other to the double lines, and that no transitions between the two sets of levels are observed. For this reason it was suggested that helium exists in two separate forms. That giving rise to the single line spectrum was called parahelium and that giving the double line spectrum was called orthohelium. In 1925 it became clear that, when account is taken of electron spin, what had been called parahelium and orthohelium are really singlet helium and triplet helium, respectively.

For hydrogen and the alkali metal atoms in their ground configurations, or excited configurations involving promotion of the valence electron, there is only one electron with an unpaired spin. For this electron $m_s = +\frac{1}{2}$ or $-\frac{1}{2}$ and the corresponding electron spin part of the total wave function is conventionally given the symbol α or β, respectively. All states arising are doublet states since $2S + 1 = 2$.

In helium we need to look more closely at the consequences of electron spin, since this is the prototype of all atoms and molecules having easily accessible states with two different multiplicities.

If the spin–orbit coupling is small, as it is in helium, the total electronic wave function ψ_e can be factorized into an orbital part ψ_e^o and a spin part ψ_e^s.

$$\psi_e = \psi_e^o \psi_e^s \tag{6.45}$$

The spin part ψ_e^s can be derived by labelling the electrons 1 and 2 and remembering that, in general, each can have an α or β spin wave function, giving four possible combinations: $\alpha(1)\beta(2)$, $\beta(1)\alpha(2)$, $\alpha(1)\alpha(2)$ and $\beta(1)\beta(2)$. Because the first two are neither symmetric nor antisymmetric to exchange of electrons, which is equivalent to exchange of the labels 1 and 2, they must be replaced by linear combinations giving

$$\psi_e^s = 2^{-\frac{1}{2}}[\alpha(1)\beta(2) - \beta(1)\alpha(2)] \tag{6.46}$$

and

$$\begin{aligned}\psi_e^s &= \alpha(1)\alpha(2) \\ &\text{or } \beta(1)\beta(2) \\ &\text{or } 2^{-\frac{1}{2}}[\alpha(1)\beta(2) + \beta(1)\alpha(2)]\end{aligned} \tag{6.47}$$

where the factors $2^{-\frac{1}{2}}$ are normalization constants. The wave function ψ_e^s in equation (6.46) is antisymmetric to electron exchange and is that of a singlet state, whereas the three in equation (6.47) are symmetric and are those of a triplet state.

Turning to the orbital part of ψ_e, we consider the electrons in two different atomic orbitals χ_a and χ_b as, for example, in the $1s^1 2p^1$ configuration of helium. There are two ways of placing electrons 1 and 2 in these orbitals giving wave functions $\chi_a(1)\chi_b(2)$ and $\chi_a(2)\chi_b(1)$ but, once again, we have to use, instead, the linear combinations

$$\psi_e^{\,o} = 2^{-\frac{1}{2}}[\chi_a(1)\chi_b(2) + \chi_a(2)\chi_b(1)] \quad (6.48)$$

$$\psi_e^{\,o} = 2^{-\frac{1}{2}}[\chi_a(1)\chi_b(2) - \chi_a(2)\chi_b(1)] \quad (6.49)$$

The wave functions in equations (6.48) and (6.49) are symmetric and antisymmetric, respectively, to electron exchange.

The most general statement of the Pauli principle for electrons and other fermions is that the total wave function must be antisymmetric to electron (or fermion) exchange. For bosons it must be symmetric to exchange.

For helium, therefore, the singlet spin wave function of equation (6.46) can combine only with the orbital wave function of equation (6.48) giving, for singlet states,

$$\psi_e = 2^{-1}[\chi_a(1)\chi_b(2) + \chi_a(2)\chi_b(1)] \times [\alpha(1)\beta(2) - \beta(1)\alpha(2)] \quad (6.50)$$

Similarly, for triplet states,

$$\begin{aligned}\psi_e = {} & 2^{-\frac{1}{2}}[\chi_a(1)\chi_b(2) - \chi_a(2)\chi_b(1)]\alpha(1)\alpha(2) \quad (6.51)\\ & \text{or } 2^{-\frac{1}{2}}[\chi_a(1)\chi_b(2) - \chi_a(2)\chi_b(1)]\beta(1)\beta(2)\\ & \text{or } 2^{-1}[\chi_a(1)\chi_b(2) - \chi_a(2)\chi_b(1)]\\ & \quad \times [\alpha(1)\beta(2) + \beta(1)\alpha(2)]\end{aligned}$$

For the ground configuration, $1s^2$, the orbital wave function is given by

$$\psi_e^{\,o} = \chi_a(1)\chi_a(2) \quad (6.52)$$

which is symmetric to electron exchange. This configuration leads, therefore, only to a singlet term, whereas each excited configuration arising from the promotion of one electron gives rise to a singlet and a triplet term. Each triplet term lies lower in energy than the corresponding singlet.

The Grotrian diagram in Figure 6.17 gives the energy levels for all the terms arising from the promotion of one electron in helium to an excited orbital.

The selection rule for the promoted electron is, as for all atoms, $\Delta l = \pm 1$. In addition, the selection rule $\Delta S = 0$ is rigidly obeyed. This rule tends to break down for atoms of higher nuclear charge but, in the helium spectrum, no singlet–triplet transitions, which would violate the $\Delta S = 0$ selection rule, have been observed.

Because of this selection rule, atoms which get into the lowest triplet state, 2^3S_1, do not easily revert to the ground 1^1S_0 state; the transition is forbidden by both the orbital and spin selection rules. The lowest triplet state is therefore metastable. In a typical discharge it has a lifetime of the order 1 ms before a collisional process takes it to the ground state.

The first excited singlet state, 2^1S_0, is also metastable in the sense that a transition to the ground state is forbidden by the $\Delta l \neq 0$ selection rule but, because the transition is not spin forbidden, this state is not as long-lived as the 2^3S_1 metastable state.

Owing to the effects of spin–orbit coupling, all the triplet terms, except ^3S, are split into three components. For example, in the case of a ^3P term, with $L = 1$ and $S = 1$, J can take the values 2, 1, 0 [equation (6.28)].

The splitting of triplet terms of helium is unusual in two respects. First, multiplets may be inverted and, second, the splitting of the multiplet components does not obey equation (6.33). Figure 6.18 shows how this applies to the 2^3P and 3^3P terms. Heisenberg has shown that equation (6.33) does not apply when the nuclear charge is small.

For this reason we shall discuss fine structure due to spin–orbit coupling in the context of the

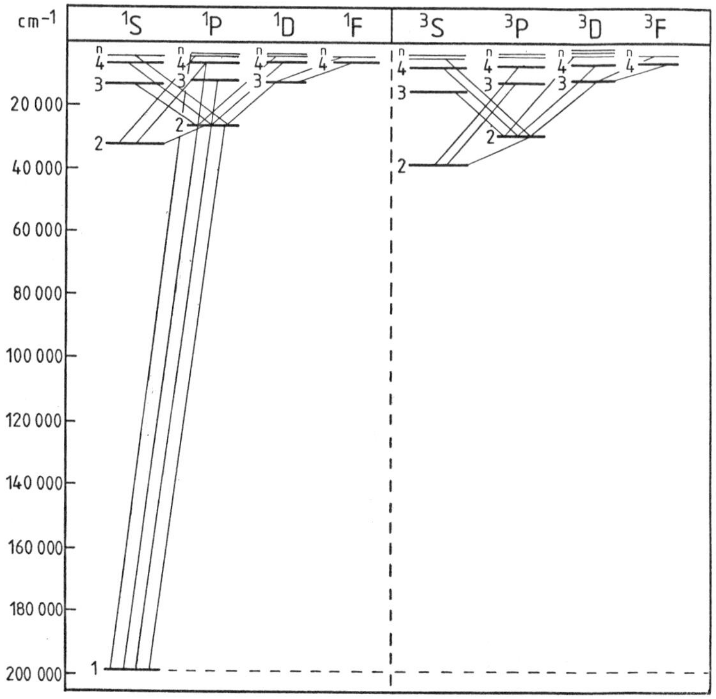

Figure 6.17 Grotrian diagram for the helium atom showing separately the manifolds of singlet and triplet states. The scale is too small to show the effects of spin–orbit coupling

alkaline earth atomic spectra where multiplets are usually normal and also obey equation (6.33).

The alkaline earth metal spectra resemble that of helium fairly closely. All the atoms have an ns^2 outer subshell structure and, when one of these electrons is promoted, give a series of singlet and triplet states.

The fine structure of a ${}^3P$–${}^3S$ transition of an alkaline earth metal is illustrated in Figure 6.19(a). The ΔJ selection rule (p. 286) results in a simple triplet. The very small separation of 2^3P_1 and 2^3P_2 in helium (Figure 6.18) accounts for the early description of the low resolution spectrum of triplet helium as consisting of 'doublets'.

A ${}^3D$–${}^3P$ transition, shown in Figure 6.19(b), has six components. As with doublet states, the multiplet splitting decreases rapidly with L so the resulting six lines in the spectrum appear, at medium resolution, as a triplet. For this reason, the fine structure is often called a compound triplet.

6.16 Spectra of other polyelectronic atoms

So far we have considered only hydrogen, helium, the alkali metals and the alkaline earth metals but the selection rules and general principles encountered can be extended straightforwardly to any other atom.

An obvious difference between the emission spectra of most atoms and those we have considered so far is their complexity, the spectra showing very many lines and no obvious series. An extreme example is the spectrum of an open iron arc, which is so rich in lines that it is commonly used as a calibration spectrum throughout the visible and ultraviolet regions.

However complex the atom, we can usually use the Russell–Saunders coupling (or the *jj*

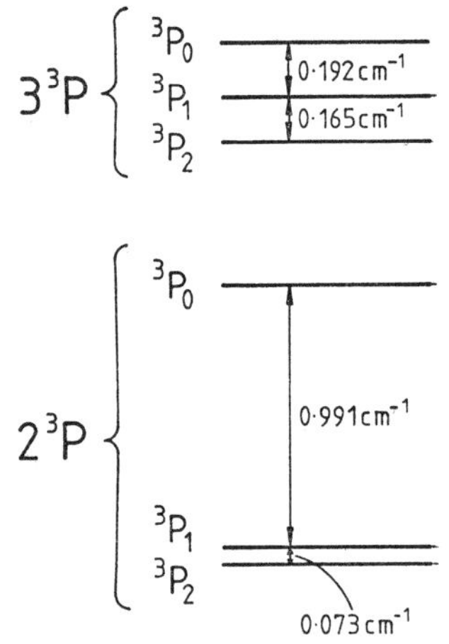

Figure 6.18 The unusual 2^3P and 3^3P multiplet splittings in the helium atom

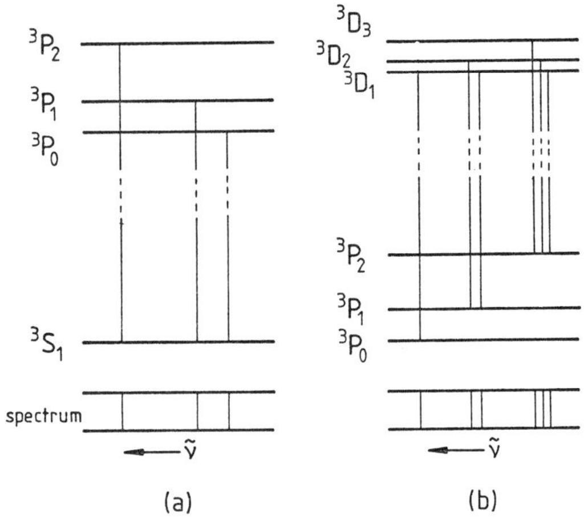

Figure 6.19 (a) A simple triplet and (b) a compound triplet in the spectrum of, for example, an alkaline earth metal atom

coupling) approximation to derive the states which arise from any configuration. The general selection rules which apply to transitions between these states are

1. $\Delta L = 0, \pm 1$ except $L = 0 \not\leftrightarrow L = 0$

Previously we have considered the promotion of only one electron, for which $\Delta l = \pm 1$ applies, but the general rule given here involves the total orbital angular momentum quantum number L and applies to the promotion of any number of electrons.

2. even $\not\leftrightarrow$ even, odd $\not\leftrightarrow$ odd, even $\longleftrightarrow$ odd

Here 'even' and 'odd' refer to the arithmetic sum $\Sigma_i l_i$ over all the electrons and this selection rule is called the Laporte rule. An important result of this is that transitions are forbidden between states arising from the same configuration. For example, of the states given in equation (6.31) arising from the $1s^2 3s^2 2p^1 3d^1$ configuration of the carbon atom, a ^1P–^1D transition would be allowed if we considered only the ΔS (see rule 4 below) and ΔL selection rules but the Laporte rule forbids it. Similarly, any transitions between states arising from the $1s^2 2s^2 2p^1 3d^1$ configuration, with $\Sigma_i l_i = 3$, and those arising from the $1s^2 2s^2 3d^1 4f^1$ configuration, with $\Sigma_i l_i = 5$, are also forbidden.† The Laporte rule is consistent with $\Delta l = \pm 1$ when only one electron is promoted from the ground configuration.

3. $\Delta J = 0, \pm 1$ except $J = 0 \not\leftrightarrow J = 0$

This rule is the same for all atoms.

4. $\Delta S = 0$

This rule applies only to atoms with a small nuclear charge. In atoms with a larger nuclear charge it breaks down so that the factorization of ψ_e, in equation (6.45), no longer applies due to spin–orbit interaction and states can no longer be described accurately as singlet, doublet, etc.

The mercury atom provides a good example of the breakdown of the $\Delta S = 0$ selection rule. Having the ground configuration $KLMN5s^2 5p^6 5d^{10} 6s^2$, it is rather like an alkaline earth metal. Promotion of an electron from the $6s$ to the $6p$

†The $2^2P_{\frac{3}{2}}$–$2^2S_{\frac{1}{2}}$ and $2^2S_{\frac{1}{2}}$–$2^2P_{\frac{1}{2}}$ transitions observed in the hydrogen atom violate the Laporte rule because they are magnetic dipole transitions and the rule applies only to electric dipole transitions.

orbital produces 6^1P_1, 6^3P_0, 6^3P_1 and 6^3P_2 states, the three components of 6^3P being widely split by spin–orbit interaction. This interaction also breaks down the spin selection rule to such an extent that the 6^3P_1–6^1S_0 transition at 253.652 nm is one of the strongest in the mercury emission spectrum and has been used in Raman spectroscopy as a source of intense, monochromatic radiation.

6.1.7 Symmetry selection rules

For vibrational transitions of the electric dipole type we have seen in Section 5.2.4.1 that the selection rules can be expressed in terms of the symmetry properties of the two combining states, illustrated by equation (5.165). For electric dipole transitions between electronic states of atoms we have seen that the selection rules can be expressed entirely in terms of quantum numbers. When we consider diatomic and polyatomic molecules we shall see that there are fewer quantum numbers available to describe electronic states and it is necessary to consider the symmetry properties of the electronic wave functions, ψ_e, of the combining states. In the extreme case of a non-linear polyatomic molecule there is, at best, only one good quantum number associated with electronic motion and that is S, the electron spin quantum number.

The division which is apparent in discussing selection rules, in atoms, in terms of quantum numbers and, in non-linear polyatomic molecules, in terms of symmetry properties may seem somewhat artificial and, indeed, it is. Once the theory of point groups, discussed in Section 5.2.2, has been understood the selection rules (a) $\Delta l = \pm 1$ for a one-electron promotion and (b) $\Delta L = 0$ except $L = 0 \nleftrightarrow L = 0$, can be derived using the full rotation–inversion point group K_h to which all atoms belong.

For an electronic transition between lower and upper states with wave functions ψ''_e and ψ'_e, respectively, to be electric dipole allowed it is required that, analogous to equations (5.153) and (5.157) for vibrational transitions,

$$\begin{aligned} &\Gamma(\psi'_e) \times \Gamma(T_x) \times \Gamma(\psi''_e) \supset A \\ \text{and/or } &\Gamma(\psi'_e) \times \Gamma(T_y) \times \Gamma(\psi''_e) \supset A \\ \text{and/or } &\Gamma(\psi'_e) \times \Gamma(T_z) \times \Gamma(\psi''_e) \supset A \end{aligned} \tag{6.53}$$

where A is a totally symmetric species.

Atomic orbitals are labelled s, p, d, f, . . . corresponding to $l = 0$, 1, 2, 3, . . . , respectively, but if they are given the full symmetry species labels according to the K_h point group, whose character table is given in Table A.45 in the Appendix, they should be labelled s_g, p_u, d_g, f_u, . . . , the g or u character being obvious from the form of the orbitals.

In the K_h point group all the three translations are degenerate and

$$\Gamma(T_x, T_y, T_z) = P_u \tag{6.54}$$

From this, and equation (6.53), it follows that, for an allowed transition,

$$\Gamma(\psi'_e) \times \Gamma(\psi''_e) \supset P_u \tag{6.55}$$

To apply this selection rule we need to multiply symmetry species of the K_h point group. This can be achieved either by reducing the resulting reducible representation, as we have done in other degenerate point groups, or simply by remembering that s, p, d, f, . . . correspond to values of l and using the Clebsch–Gordan series of equation (6.22).

We will consider, as an example, the orbital promotion

$$KL4s_g^1 \leftarrow KL3s_g^1 \tag{6.56}$$

in the sodium atom. Since $L = 0$ for all filled sub-shells or, in other words, $\Gamma(\psi_e) = S_g$, the symmetry species for the configuration as a whole is simply that of the orbital occupied by the valence electron. Therefore, for the promotion of equation (6.56),

$$\Gamma(\psi'_e) \times \Gamma(\psi''_e) = S_g \times S_g = S_g \tag{6.57}$$

Since this does not contain P_u the promotion is forbidden. Similarly, for the promotion

$$KL4p_u^1 \leftarrow KL3p_u^1 \qquad (6.58)$$

it follows, by using the Clebsch–Gordan series and the usual rules for multiplying g and u, that

$$\Gamma(\psi_e') \times \Gamma(\psi_e'') = P_u \times P_u = D_g + P_g + S_g \qquad (6.59)$$

Since this does not contain P_u the orbital promotion of equation (6.58) is forbidden and, in the same way, it follows that all $\Delta l = 0$ promotions are forbidden.

On the other hand, the promotion

$$KL3d_g^1 \leftarrow KL3p_u^1 \qquad (6.60)$$

gives

$$\Gamma(\psi_e') \times \Gamma(\psi_e'') = F_u + D_u + P_u \supset P_u \qquad (6.61)$$

and the promotion is allowed. Similarly, it can be shown that all $\Delta l = \pm 1$ promotions are allowed.

We will now consider Laporte-allowed transitions between states arising from the excited configurations $1s_g^2 2p_u^1 3d_g^1$ and $1s_g^2 3d_g^1 4d_g^1$ of the beryllium atom. For the first, lower, configuration

$$\Gamma(\psi_e'') = P_u \times D_g = F_u + D_u + P_u \qquad (6.62)$$

and for the second, upper, configuration

$$\Gamma(\psi_e') = D_g \times D_g = G_g + F_g + D_g + P_g + S_g \qquad (6.63)$$

The transition P_g–P_u is allowed since

$$P_g \times P_u = D_u + P_u + S_u \qquad (6.64)$$

which contains P_u. The D_g–D_u and F_g–F_u transitions are also allowed as their products also contain P_u. This example shows how $\Delta L = 0$ transitions may be allowed when more than one electron is promoted. However, S–S transitions must *always* be forbidden since $S \times S = S$ and can never contain P_u.

6.1.8 Atoms in a magnetic field

6.1.8.1 The Zeeman effect

In equation (6.14), the relationship is given between the orbital angular momentum for one electron and the magnetic dipole moment which results. The corresponding expression for a many-electron atom in a singlet state and, therefore, with no other electronic angular momentum is

$$\boldsymbol{\mu}_L = \gamma \boldsymbol{L} \qquad (6.65)$$

and the magnitudes of these vector quantities are related by

$$\mu_L = \gamma L^* \hbar = -\mu_B L^* \qquad (6.66)$$

where $L^* = [L(L+1)]^{\frac{1}{2}}$ and μ_B is the Bohr magneton introduced in equation (6.15).

When a magnetic field is introduced along an axis which we define as the z-axis, the vector $\boldsymbol{L}$ precesses about that axis and can take up only certain orientations relative to it so that the component along the axis is $M_L\hbar$, where $M_L = L, L-1, \ldots, -L$ (see Section 1.3.3 for a discussion of space quantization). The component of μ_L along the z-axis is then

$$(\mu_L)_z = -\mu_B M_L \qquad (6.67)$$

In a magnetic field $\boldsymbol{B}$ the energy E' of a magnetic dipole $\boldsymbol{\mu}$ is given by $-\boldsymbol{\mu}.\boldsymbol{B}$ so we have

$$E' = B\mu_B M_L \qquad (6.68)$$

The splitting of adjacent M_L levels is $B\mu_B$ and therefore independent of the atom and of the state, provided it is a singlet state.

Figure 6.20 shows how a $^1D_2$ state is split into five components and a $^1P_1$ state into three components in a magnetic field. The selection rules for transitions between these components are $\Delta M_L = 0$, for which the transition moment is *parallel* to the field direction, and $\Delta M_L = \pm 1$, for which it is *perpendicular* to the field. Because of the identical magnitude of the splitting in all singlet states, all transitions between singlet

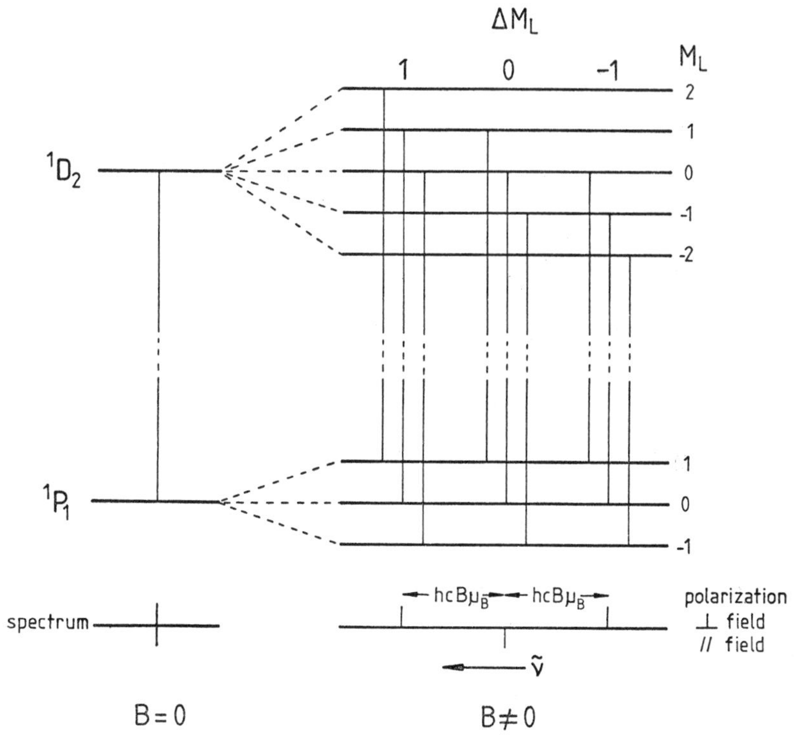

Figure 6.20 Zeeman effect on a $^1D_2$–$^1P_1$ transition

states are split into three components, with $\Delta M_L = 1, 0$ or -1, as demonstrated by the $^1D_2$–$^1P_1$ example in Figure 6.20. The splitting of singlet transitions in a magnetic field is called the normal Zeeman effect.

The normal Zeeman effect is not particularly useful, other than for distinguishing transitions between singlet states from any others or for determining the magnitude of B between the poles of a magnet.

Much more interesting and informative is the Zeeman effect on transitions involving states with the spin quantum number $S \neq 0$.

When an atom has orbital *and* electron spin angular momentum, the magnitude of the magnetic moment μ_J is given by

$$\mu_J = g_J \gamma J^* \hbar = -g_J \mu_B J^* \qquad (6.69)$$

where $J^* = [J(J+1)]^{\frac{1}{2}}$. This is analogous to equation (6.66) except for the inclusion of g_J, the Landé g-factor. The component of μ_J along the z-direction is

$$(\mu_J)_z = -g_J \mu_B M_J \qquad (6.70)$$

where the quantum number M_J can take the values $J, J-1, \ldots, -J$. The energy E' associated with this magnetic moment is given by an expression analogous to equation (6.68), namely

$$E' = g_J B \mu_B M_J \qquad (6.71)$$

and the splitting of adjacent M_J levels is $g_J B \mu_B$. The Landé g-factor is given by

$$g_J = 1 + [(J^{*2} + S^{*2} - L^{*2})/2J^{*2}] \qquad (6.72)$$

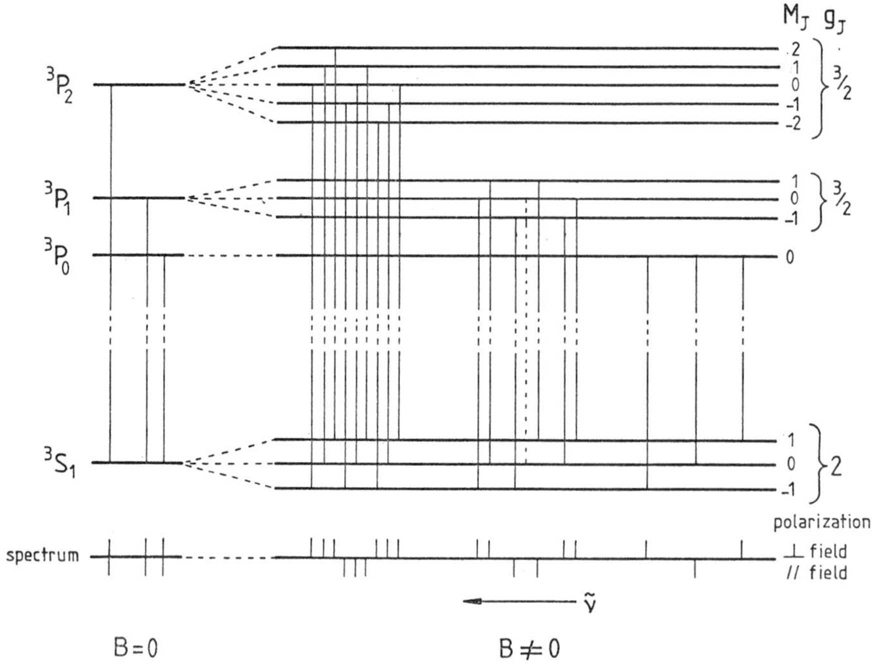

Figure 6.21 Zeeman splitting of the three components of a ^3P–^3S transition

when g_e [equation (6.15)] is assumed to be exactly 2. Since g_J depends on S and L, as well as J, it varies from one state to another, as does the splitting of adjacent M_J levels.

The selection rules for transitions between M_J components of different electronic states are

$$\Delta M_J = 0$$

except that $M_J = 0 \not\leftrightarrow M_J = 0$ for $\Delta J = 0$, for which the transition moment is parallel to the field direction, and

$$\Delta M_J = \pm 1$$

for which the transition moment is perpendicular to the field direction.

Figure 6.21 illustrates the splitting of the three components of a ^3P–^3S transition into 18 components in a magnetic field. This Zeeman pattern is characteristic only of a ^3P–^3S transition and is therefore a much more useful aid in assigning transitions between states with $S \neq 0$ than for those between singlet states.

The $^2P_{\frac{3}{2}}$–$^2S_{\frac{1}{2}}$ and $^2P_{\frac{1}{2}}$–$^2S_{\frac{1}{2}}$ components of the sodium D-line are further split in a magnetic field into six and four components, respectively, thereby confirming their assignments. In fact Zeeman, in 1896, first observed the effect, subsequently named after him, when the two components of the sodium D-line were found to be broadened in the spectrum of a sodium flame placed between the poles of a magnet.

The Zeeman effect for transitions between singlet states has been called the normal Zeeman effect and that for transitions between states with $S \neq 0$ the anomalous Zeeman effect. These distinctions date back to early experimental observations and are no longer particularly useful. This point is brought out by putting $S = 0$ in equation (6.72), resulting in $J = L$ and $g_J = 1$, and the so-called anomalous effect goes over to the normal Zeeman effect.

If an atom has nuclear spin, in addition to electron orbital and spin, momentum the vector $\boldsymbol{F}$, in the presence of a *very weak* magnetic field, precesses about the z-axis, the direction of the field, and can take up only certain orientations relative to it. These orientations are such that the component along the z-axis is $M_F\hbar$, where $M_F = F, F-1, \ldots, -F$. Each hyperfine level is split into $(2F+1)$ components and selection rules governing transitions between them are

$$\Delta M_F = 0$$

except that $M_F = 0 \nleftrightarrow M_F = 0$ for $\Delta F = 0$, and

$$\Delta M_F = \pm 1$$

However, the nuclear spin angular momentum is easily uncoupled from the angular momenta of the electron by even a weak magnetic field, resulting in a nuclear Paschen–Back effect discussed in the following section.

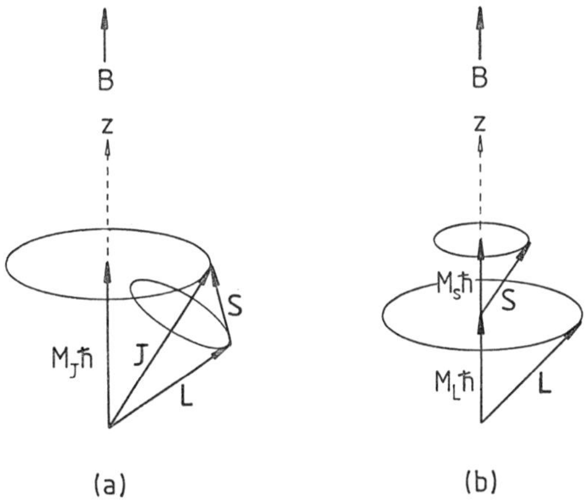

Figure 6.22 (a) Precession of $\boldsymbol{J}$ around the direction of a magnetic field along the z-axis. (b) The Paschen–Back effect in which $\boldsymbol{L}$ and $\boldsymbol{S}$ are uncoupled and precess around the direction of the magnetic field

6.1.8.2 The Paschen–Back effect

In the anomalous Zeeman effect the magnetic field along the z-axis causes the $\boldsymbol{J}$ vector to precess about that axis, the greater the field the greater the precession frequency. This is illustrated in Figure 6.22(a).

As the magnetic field B increases, $\boldsymbol{L}$ and $\boldsymbol{S}$ become more strongly coupled to the external field than they are to each other. This results, eventually, in $\boldsymbol{L}$ and $\boldsymbol{S}$ being uncoupled and precessing independently about the field direction, as shown in Figure 6.22(b), and is called the Paschen–Back effect.

The energy E' associated with the magnetic moment due to the total spin and that due to the total orbital momenta of the electrons is the sum of the two contributions. The orbital contribution is given by equation (6.68) and the spin contribution, $g_e B\mu_B M_S$, follows from the expression in equation (6.15) for the spin magnetic moment giving the sum as

$$E' = B\mu_B M_L + g_e B\mu_B M_S \qquad (6.73)$$

Selection rules for transitions between the components are

$$\Delta M_L = 0$$

for which the transition moment is parallel to the field direction,

$$\Delta M_L = \pm 1$$

for which the transition moment is perpendicular to the field direction, and

$$\Delta M_S = 0$$

If $\boldsymbol{L}$ and $\boldsymbol{S}$ are completely uncoupled, the Paschen–Back effect results in each atomic line being split into three components, with $\Delta M_L = 0, \pm 1$, just as in the normal Zeeman effect. Any residual coupling between $\boldsymbol{L}$ and $\boldsymbol{S}$ can be taken account of by modifying equation (6.73) to

$$E' = B\mu_B(M_L + g_e M_S) + AM_L M_S \qquad (6.74)$$

where A is constant for a particular multiplet.

The anomalous Zeeman effect and the Paschen–Back effect represent low and high field extremes, respectively. With an intermediate field a more complex multiplet structure results.

In an atom which has a nuclear spin angular momentum, in addition to orbital and electron spin angular momenta, a magnetic field which is large enough to produce a resolvable nuclear Zeeman effect is usually sufficient to uncouple $\boldsymbol{J}$ and $\boldsymbol{I}$ so that they precess independently about the field direction. This is the nuclear Paschen–Back effect and is illustrated in Figure 6.23.

The energy E' associated with the magnetic moments is

$$E' = g_J B \mu_B M_J - g_I B \mu_N M_I + A M_J M_I \quad (6.75)$$

where $M_I = I, I-1, \ldots, -I$. The second term is always small because of the magnitude of μ_N and derives from equation (6.18). The third term is usually large compared with the second and represents the residual coupling between electron and nuclear momenta. The selection rules for transitions are

$\Delta M_J = 0$ (parallel polarization) and ± 1 (perpendicular polarization)

and

$$\Delta M_I = 0$$

6.1.9 Atoms in an electric field

The effect of an electric field on atomic spectra was first observed by Stark in 1913 using the Balmer series of hydrogen and the splitting of lines is known as the Stark effect. Later, the technique was extended to molecular spectroscopy and particularly microwave and millimetre wave rotational spectroscopy, as discussed in Section 4.2.4.

It turns out that, as in so many respects, the hydrogen atom behaves uniquely by showing a first-order Stark effect. A first-order Stark effect is one in which the splitting of levels is proportional to the field $\mathcal{E}$. The levels which are split in this way are those with the same n but differing in l by one.

More usual is the second-order Stark effect observed in other atoms in which the splitting

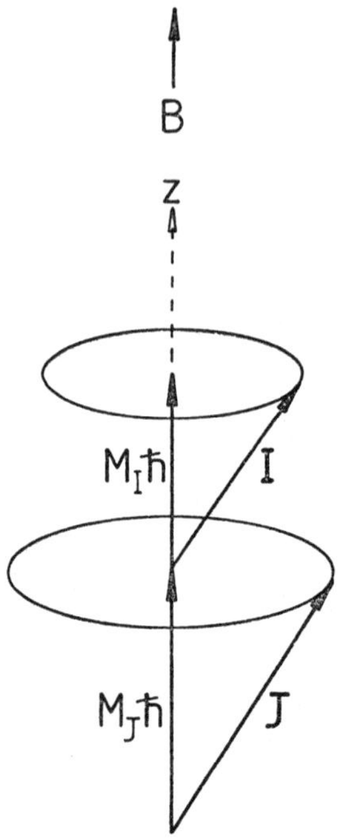

Figure 6.23 The nuclear Paschen–Back effect in which $\boldsymbol{J}$ and $\boldsymbol{I}$ precess around the direction of the magnetic field along the z-axis

of levels is proportional to $\mathcal{E}^2$. The vector diagram of Figure 6.22(a) is relevant here except that, because of the $\mathcal{E}^2$ dependence of the splitting, levels with the same magnitude, but different sign, of M_J remain degenerate. For example, a level with $J = \frac{5}{2}$ is split into three components with $|M_J| = \frac{5}{2}, \frac{3}{2}, \frac{1}{2}$.

In general, the Stark effect is a much less useful tool than the Zeeman effect in assigning transitions in atomic spectra.

6.2 ELECTRONIC SPECTROSCOPY OF DIATOMIC MOLECULES

6.2.1 Electronic structure

6.2.1.1 Homonuclear diatomic molecules

It is not the intention for this book to be a primary source on the general subject of valence theory; such sources are referred to in the

Bibliography at the end of the chapter. Nevertheless, it is necessary here to remind the reader briefly of the results of a molecular orbital treatment of the electronic structure of diatomic molecules which is very approximate but which still is conceptually useful.

With regard to their electronic structure, atoms occupy a unique position, rather like hydrogen is unique among atoms themselves. In helium, and all other atoms except hydrogen, the presence of more than one electron introduces repulsions between electrons which have a profound effect on the electronic structure. In the hydrogen molecule and all other molecules, diatomic or polyatomic, the electronic structure is substantially different from that of atoms because of the additional attractions and repulsions when more than one nucleus is present.

Conceptually there are two methods of approach which can be used in making the step from atomic to molecular systems. The first is the united atom approach, in which the molecule is considered first as the atom formed in the limit of zero internuclear distance. For example, starting with the hydrogen molecule and allowing r to become zero the nuclei coalesce to give a single nucleus with a mass which is twice that of the proton and the two electrons move in its field. The united atom in this case is ^{2}He, an unstable isotope but one for which atomic orbitals can be calculated. Similarly, the united atom limit for the nitrogen molecule is ^{28}Si. The electronic structure of the molecule is obtained by studying the changes in the motions of the electrons as the nucleus of the united atom is split into two equal portions and these pulled apart until they are separated by the equilibrium internuclear distance in the diatomic molecule.

The second method is the separated atom approach. Here the molecule is considered first in the limit of its constituent atoms being infinitely far apart and then allowing them to approach, eventually settling down so that the nuclei are separated by the equilibrium internuclear distance.

Complete correlation diagrams can be constructed in which the atomic orbitals of the united atom are correlated through the orbitals of the molecule with those of the separated atoms. Figure 6.24 illustrates how the $1s$ atomic orbital (AO) of ^{2}He correlates with the $\sigma_g 1s$ molecular orbital (MO) of $^{1}H_2$—this MO symbolism will be discussed later—and with the $1s$ AOs of ^{1}H. Similarly, one of the $2p$ AOs of ^{2}He correlates with a $\sigma_u 1s$ MO of $^{1}H_2$ and the $1s$ AOs of ^{1}H.

Correlation diagrams are useful in correlating orbitals, as here, and also states, but must always take account of the non-crossing rule which states that, in such diagrams, lines representing orbitals or states of the same symmetry must not cross. In constructing such diagrams for correlations of states there are additional difficulties if there is appreciable spin–orbit coupling.

These diagrams are conceptually helpful and quantitatively useful in considering dissociation.

In obtaining molecular electronic wave functions of polyatomic molecules in general, and diatomic molecules in particular, there are two general techniques, the valence bond (VB) and molecular orbital (MO) methods. Both start from extremes and various degrees of approximation give wave functions of varying accuracy but it should be stressed that both methods converge, in the limits where no approximations are made, on the *same* wave function, namely the exact wave function of the molecule.

The approach adopted in the valence bond method is to imagine bringing, for example, two hydrogen atoms, complete with their electrons, from infinity to a distance apart known from the equilibrium molecular geometry. Then the electrons and nuclei are allowed to interact, resulting in modification of the AO wave functions for the hydrogen atoms. In the hydrogen molecule there are only two electrons which can interact but, in molecules with more electrons, choices can be made. This is the so-called perfect pairing approximation and calls on chemical intuition to decide which electron

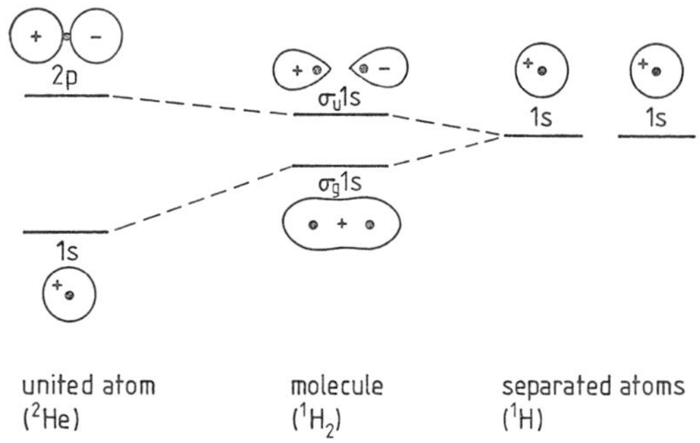

Figure 6.24 Correlation of some AOs between the united atom ^{2}He and two separated ^{1}H atoms via the MOs of $^1H_2$

pairs are important in bond formation. Then the wave function is expressed as a linear combination of those representing likely pairing schemes.

The valence bond approach is rather cumbersome in a discussion of molecular spectra and from this point we shall be using the MO treatment almost exclusively.

The approach adopted in the MO method is to place the nuclei, without their electrons, a distance apart equal to the equilibrium internuclear distance and construct MOs around them. Electrons can be fed into these orbitals in pairs (with $m_s = \pm\frac{1}{2}$) in order of increasing energy until the electrons are used up. This will give the ground configuration of the molecule, using the *aufbau* principle, just as for atoms (Section 6.1.1).

The basis of constructing the MOs is the linear combination of atomic orbitals (LCAO) method. This takes account of the fact that, in the region close to a nucleus, the MO wave function resembles an AO wave function for the atom of which the nucleus is a part. It is reasonable, then, to express an MO wave function ψ as a linear combination of AO wave functions χ_i:

$$\psi = \sum_i c_i \chi_i \tag{6.76}$$

where c_i is the coefficient of the wave function χ_i. However, not all linear combinations are effective in the sense of producing an MO which is appreciably different from the AOs from which it is formed. For effective linear combinations, (a) the energies of the AOs must be comparable, (b) the AOs should overlap as much as possible and (c) the AOs must have the same symmetry properties with respect to certain symmetry elements of the molecule.

For a homonuclear diatomic molecule with nuclei labelled 1 and 2 the LCAO method gives the MO wave function

$$\psi = c_1\chi_1 + c_2\chi_2 \tag{6.77}$$

Using the N_2 molecule as an example we can see that, for instance, the nitrogen 1s AOs satisfy condition (a), since their energies are identical, but not (b), because the high nuclear charge causes the 1s AOs to be close to the nuclei resulting in little overlap. On the other hand, the 2s AOs satisfy both conditions and, since they are spherically symmetrical, also condition (c). Examples of AOs which satisfy (a) and (b) but not (c) are the 2s and 2p_x orbitals in Figure 6.25. The 2s AO is symmetric to reflection across a σ_v plane containing the internuclear z-axis and perpendicular to the figure, whereas the 2p_x AO is antisymmetric to this operation. In fact, we can easily see that any overlap between 2s and the positive lobe of 2p_x is exactly cancelled by that involving the negative lobe.

Important properties of the MO are the energy, E, associated with it and the values of c_1 and c_2 in equation (6.77). They are obtained from the Schrödinger equation:

$$H\psi = E\psi \tag{6.78}$$

Multiplying both sides by ψ^*, which is the complex conjugate of ψ and is obtained from it by replacing all i ($= \sqrt{-1}$) by $-$i, and integrating over all space gives

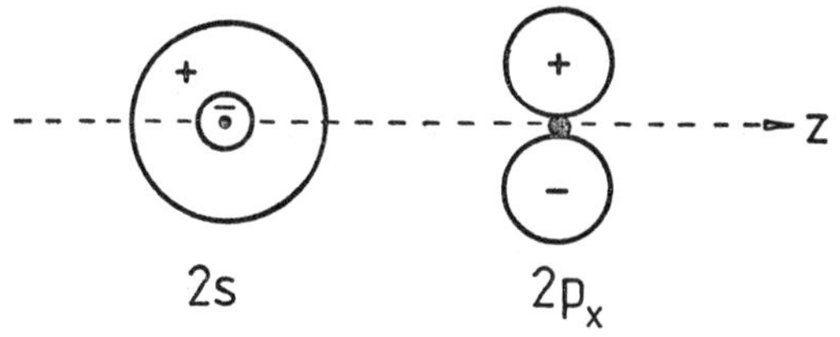

Figure 6.25 Illustration of why there is no overlap between a 2s and a 2p_x (or 2p_y) AO

$$E = \int \psi^* H\psi \mathrm{d}\tau \bigg/ \int \psi^*\psi \mathrm{d}\tau \qquad (6.79)$$

E can be calculated only if ψ is known, so what is done is to make an intelligent guess at the MO wave function, say ψ_n, and calculate the corresponding expectation value of the energy $\bar{E}_n$ from equation (6.79). A second guess at ψ, say ψ_m, gives a corresponding energy $\bar{E}_m$. The variation principle states that, if $\bar{E}_m < \bar{E}_n$, then ψ_m is closer than ψ_n to the true MO wave function. This applies to the ground state only. In this way the true ground state wave function can be approached as closely as we choose, but this is usually done by varying parameters in the chosen wave function until they have their optimum values.

Combining equations (6.77) and (6.79), and assuming that χ_1 and χ_2 are not complex, gives

$$\bar{E} = \frac{\int(c_1^2\chi_1 H\chi_1 + c_1c_2\chi_1 H\chi_2 + c_1c_2\chi_2 H\chi_1 + c_2^2\chi_2 H\chi_2)\mathrm{d}\tau}{\int(c_1^2\chi_1^2 + 2c_1c_2\chi_1\chi_2 + c_2^2\chi_2^2)\mathrm{d}\tau} \qquad (6.80)$$

If the AO wave functions are normalized

$$\int \chi_1^2 \mathrm{d}\tau = \int \chi_2^2 \mathrm{d}\tau = 1 \qquad (6.81)$$

and, as H is a hermitian operator,

$$\int \chi_1 H\chi_2 \mathrm{d}\tau = \int \chi_2 H\chi_1 \mathrm{d}\tau = H_{12} \text{ (say)} \qquad (6.82)$$

The quantity

$$\int \chi_1\chi_2 \mathrm{d}\tau = S \qquad (6.83)$$

and is called the overlap integral as it is a measure of the degree to which χ_1 and χ_2 overlap each other. In addition, integrals such as $\int \chi_1 H\chi_1 \mathrm{d}\tau$ are abbreviated to H_{11}. All these simplifications and abbreviations reduce equation (6.80) to

$$\bar{E} = \frac{c_1^2 H_{11} + 2c_1c_2H_{12} + c_2^2H_{22}}{c_1^2 + 2c_1c_2S + c_2^2} \qquad (6.84)$$

Using the variation principle to optimize c_1 and c_2 we obtain $\partial\bar{E}/\partial c_1$ and $\partial\bar{E}/\partial c_2$ from equation (6.84) and put them equal to zero, giving

$$\begin{aligned} c_1(H_{11} - E) + c_2(H_{12} - ES) &= 0 \\ c_1(H_{12} - ES) + c_2(H_{22} - E) &= 0 \end{aligned} \qquad (6.85)$$

where we have replaced $\bar{E}$ by E since, although it is probably not the true energy, it is the nearest approach to it with the wave function of equation (6.77). The equations (6.85) are the secular equations and the two values of E which satisfy them are obtained from the secular determinant

$$\begin{vmatrix} H_{11} - E & H_{12} - ES \\ H_{12} - ES & H_{22} - E \end{vmatrix} = 0 \qquad (6.86)$$

H_{12} is the resonance integral, usually symbolized by β. In a homonuclear diatomic molecule $H_{11} = H_{22} = \alpha$, which is known as the Coulomb integral, and the secular determinant becomes

$$\begin{vmatrix} \alpha - E & \beta - ES \\ \beta - ES & \alpha - E \end{vmatrix} = 0 \qquad (6.87)$$

giving

$$E_{\pm} = (\alpha \pm \beta)/(1 \pm S) \qquad (6.88)$$

If we are interested only in very approximate MO wave functions and energies, we can assume that $S = 0$ (a typical value is about 0.2) and that the hamiltonian H is the same as

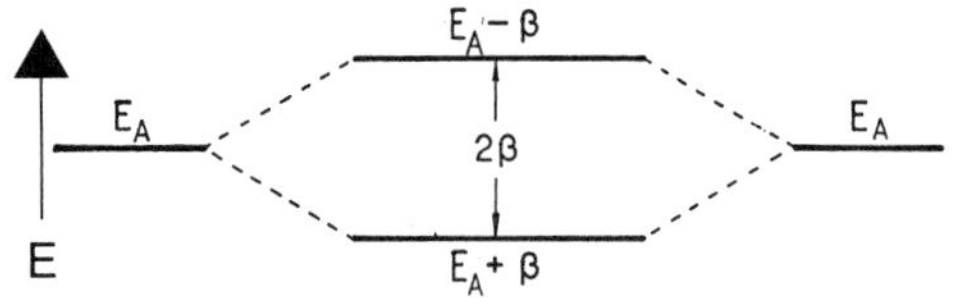

Figure 6.26 Formation of two MOs from two AOs of energy E_A. The separation of the MOs is twice the resonance energy β

in the atom giving $\alpha = E_A$, the AO energy. These assumptions result in

$$E_{\pm} = E_A \pm \beta \tag{6.89}$$

At this level of approximation the two MOs are symmetrically displaced from E_A with a separation of 2β. Since β is a negative quantity the orbital with energy $(E_A + \beta)$ lies lowest, as shown in Figure 6.26.

With the approximation that $S = 0$, the secular equations of equation (6.85) become

$$\begin{aligned} c_1(\alpha - E) + c_2\beta &= 0 \\ c_1\beta + c_2(\alpha - E) &= 0 \end{aligned} \tag{6.90}$$

Putting $E = E_+$ or E_-, we obtain $c_1/c_2 = 1$ or -1, respectively, and therefore the corresponding wave functions ψ_+ and ψ_- are given by

$$\begin{aligned} \psi_+ &= N_+(\chi_1 + \chi_2) \\ \psi_- &= N_-(\chi_1 - \chi_2) \end{aligned} \tag{6.91}$$

N_+ and N_- are normalization constants obtained from the conditions

$$\int \psi_+^2 \mathrm{d}\tau = \int \psi_-^2 \mathrm{d}\tau = 1 \tag{6.92}$$

Neglecting $\int \chi_1\chi_2 \mathrm{d}\tau$, the overlap integral, gives $N_+ = N_- = 2^{-\frac{1}{2}}$ and

$$\psi_{\pm} = 2^{-\frac{1}{2}}(\chi_1 \pm \chi_2) \tag{6.93}$$

Every linear combination of AOs gives two MOs, one higher and the other lower in energy than the AOs. Figure 6.27 illustrates the MOs from $1s$, $2s$ and $2p$ AOs showing the approximate forms of the MO wave functions.

In two infinitely separated atoms the electrons and nucleus of one experience no interactions with those of the other. As the separation is decreased, repulsions between the electrons, repulsions between the nuclei and attractions between the electrons and nuclei of the two atoms begin to be effective. When the nuclei are separated by the equilibrium distance, all the electrons experience the electrostatic field due to the two nuclei and the result is that their behaviour resembles that of electrons in an atom placed in an electric field (the Stark effect mentioned in Section 6.1.9). Here we consider just one electron as in, for example, H_2^+. Unless the spin–orbit coupling is large, as would be the case if at least one of the nuclei had a high charge, the effect of the electrostatic field is to uncouple the $\boldsymbol{l}$ and $\boldsymbol{s}$ vectors. These vectors then precess independently around the internuclear z-axis, as illustrated in Figure 6.28. The velocity of precession of $\boldsymbol{l}$ is so great that its magnitude is no longer defined and l is not a good quantum number. However, the component of $\boldsymbol{l}$ along the z-axis is defined and is given by $\lambda\hbar$, where

$$\lambda = |m_l| = 0, 1, 2, \ldots, l \tag{6.94}$$

The quantum number λ is analogous to $|m_l|$ since, as we have seen in Section 6.1.9, an electric field does not separate levels with the same value but different sign of m (or M in polyelectronic atoms). Therefore, all MOs with $\lambda > 0$ are doubly degenerate.

The MO symbolism in Figure 6.27 includes the value of λ. This is indicated by the symbols σ, π, δ, ϕ, ... corresponding to $\lambda = 0$, 1, 2, 3, ... The Greek letters used here correspond to s, p, d, f, ... used in atoms to indicate the value of the quantum number l.

As in atoms, the coupling between $\boldsymbol{l}$ and $\boldsymbol{s}$ is likely to be very small in most molecules but, even if they are not completely uncoupled by the electrostatic field, the orbitals are classified

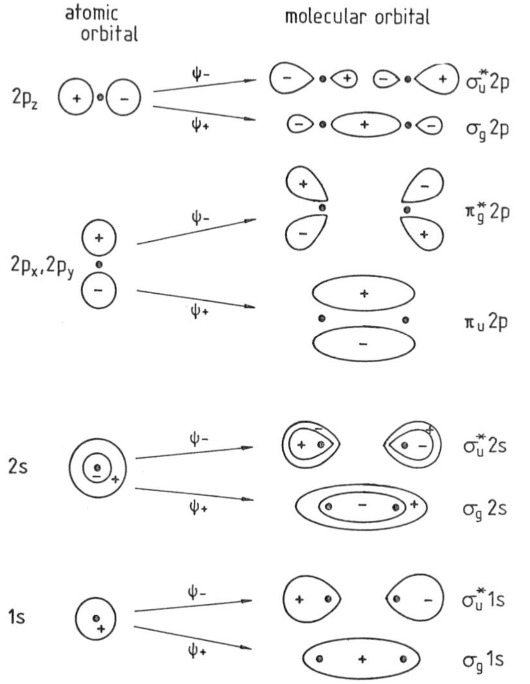

Figure 6.27 The formation of MOs from $1s$, $2s$ and $2p$ AOs

as σ, π, δ, ϕ, . . . according to the value of λ in the uncoupled approximation.

The σ, π, δ, ϕ, . . . symbols are also symmetry species of the $D_{\infty h}$ point group, just as in atoms s, p, d, f, . . . are symmetry species of the K_h point group. Therefore, MOs can be assigned an appropriate symbol either by determining the value of λ or by using the $D_{\infty h}$ character table. In assigning symmetry species to MOs there is a simple short cut which can often be used because most MOs we encounter are either σ or π and the former are cylindrically symmetrical about the z-axis whereas the latter are not.

Strictly, the σ orbitals should all be labelled σ^+ as they are symmetric to all σ_v but, as there are no σ^- orbitals, the superscript + is usually omitted.

Added to the σ, π, δ, . . . part of the MO symbolism is the g or u property, corresponding to symmetry or asymmetry to inversion, of the orbital which is easily justified from, for

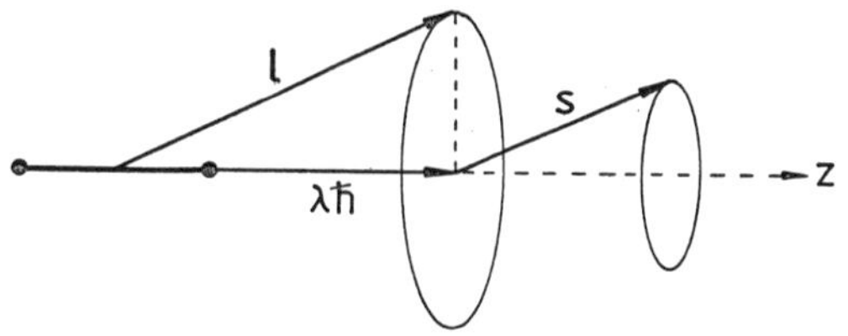

Figure 6.28 Precession of the uncoupled $\boldsymbol{l}$ and $\boldsymbol{s}$ for a single electron about the direction of the internuclear axis of a diatomic molecule

example, Figure 6.27. An alternative to the g or u subscript which is sometimes convenient is the asterisk to indicate antibonding character of the orbital caused by a nodal plane between the nuclei as in, for example, σ^*1s and π^*2p. The asterisk is particularly useful in heteronuclear diatomic molecules which do not have a centre of inversion and for which g or u cannot be used.

In Figure 6.27 both types of symbol are attached to the σ or π labels.

The MO symbolism is completed by the AO symbol, such as $1s$ and $2p$, from which the MO has been formed.

The MOs from $1s$, $2s$ and $2p$ AOs are arranged in order of increasing energy in Figure 6.29. However, the order of energies of these MOs is not the same for all molecules, just as the order of AOs is not the same for all atoms. As Figure 6.27 shows, the degenerate $2p_x$ and $2p_y$ AOs form both π and π^* MOs each of which is doubly degenerate and can accommodate four electrons. However, the z-axis is unique in the molecule and the $2p_z$ AOs form σ and σ^* MOs for which β, the resonance integral, is larger than for the MOs formed from $2p_x$ and $2p_y$ AOs. Consequently, the anticipated order of MO energies is

$$\begin{aligned}&\sigma_g 1s < \sigma_u^* 1s < \sigma_g 2s < \sigma_u^* 2s\\&< \sigma_g 2p < \pi_u 2p < \pi_g^* 2p < \sigma_u^* 2p\end{aligned} \tag{6.95}$$

However, this order is maintained for only O_2 and F_2.

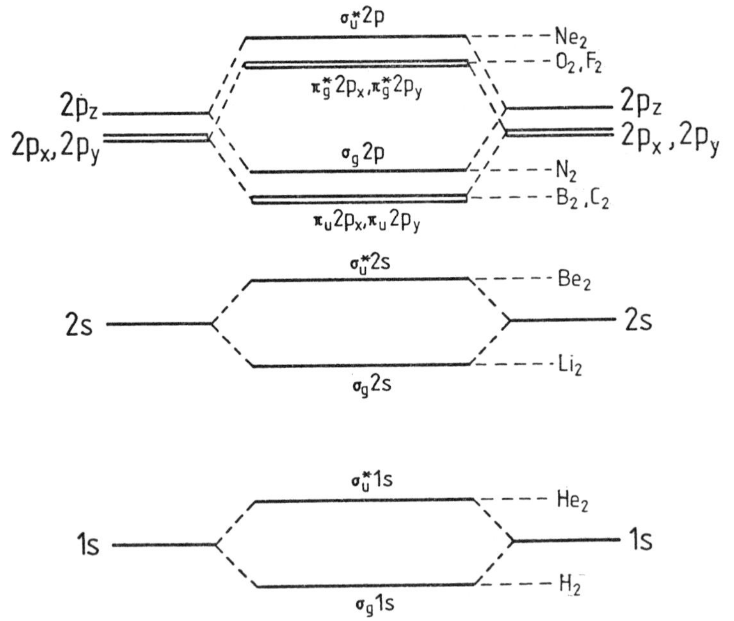

Figure 6.29 MO energy level diagram for first-row homonuclear diatomic molecules. (In O_2 and F_2 the order of $\sigma_g 2p$ and $\pi_u 2p$ is reversed)

In general, orbitals of the same symmetry derived at this level of approximation can interact. This results in a mixing and consequent pushing apart of the orbitals, especially if they are not widely separated before interaction. The most important interaction is between $\sigma_g 2p$ and $\sigma_g 2s$ and, except for oxygen and fluorine, this results in $\sigma_g 2p$ being pushed *above* $\pi_u 2p$ giving the order of MOs in Figure 6.29. There is a similar interaction between $\sigma_u^* 2p$ and $\sigma_u^* 2s$ but this does not affect the order of MO energies.

The electronic structure of any first-row diatomic molecule can be obtained by feeding the available electrons into the MOs in order of increasing energy. For example, the ground configuration of the 14-electron nitrogen molecule is

$$(\sigma_g 1s)^2(\sigma_u^* 1s)^2(\sigma_g 2s)^2(\sigma_u^* 2s)^2(\pi_u 2p)^4(\sigma_g 2p)^2 \tag{6.96}$$

Some of these electrons are in bonding and some in antibonding orbitals. There is a general rule that the bonding character of an electron in a bonding orbital is approximately cancelled by the antibonding character of an electron in the antibonding orbital constructed from the same AOs as the bonding orbital. In nitrogen, therefore, the bonding by the two electrons in the $\sigma_g 1s$ MO is cancelled by the antibonding of the two in the $\sigma_u^* 1s$ MO and similarly with the $\sigma_g 2s$ and $\sigma_u^* 2s$ electrons. The remaining six electrons are in bonding orbitals. This gives the bond order, which is the net number of bonding electrons divided by two, as three in nitrogen. This molecular orbital picture is consistent with nitrogen having a triple bond and attributes two electrons to a σ bond and four to two π bonds.

We can derive the ground state from the ground configuration of nitrogen in an analogous way to that used for atoms in Section 6.1.7. In general, the symmetry species $\Gamma(\psi_e^o)$ of the total orbital electronic wave function corresponding to a particular configuration in which molecular orbitals ψ_i are occupied is given by

$$\Gamma(\psi_e^{\circ}) = \prod_i \Gamma(\psi_i) \tag{6.97}$$

where $\prod_i \Gamma(\psi_i)$ is the product of the symmetry species of all occupied MOs. As for atoms, the product for filled orbitals gives the totally symmetric species. Since all electron spins are paired, $S = 0$ for filled orbitals. In nitrogen all occupied orbitals are filled, therefore the state arising from the ground configuration is $^1\Sigma_g^+$ since Σ_g^+ is the totally symmetric species for the $D_{\infty h}$ point group. Upper case letters $\Sigma, \Pi, \Delta, \Phi, \Gamma, \ldots$ are used for states and lower case for orbitals, the same convention as that used for atoms.

For the excited configuration of nitrogen:

$$\ldots (\pi_u 2p)^4(\sigma_g 2p)^1(\pi_g^* 2p)^1 \tag{6.98}$$

the orbital symmetry species, derived using equation (6.97), is

$$\Gamma(\psi_e{}^{\circ}) = \sigma_g^+ \times \pi_g = \Pi_g \tag{6.99}$$

using the rules for multiplication discussed in Section 5.2.2.4. The two electrons in partly filled orbitals may have their spins parallel or antiparallel so there are two states arising from this configuration, $^1\Pi_g$ and $^3\Pi_g$.

If an electron is promoted from the $\pi_u 2p$ orbital of nitrogen to give the excited configuration

$$\ldots (\pi_u 2p)^3(\sigma_g 2p)^2(\pi_g^* 2p)^1 \tag{6.100}$$

and we recall that the vacancy in the $\pi_u 2p$ orbital can be treated like a single electron, equation (6.97) gives

$$\Gamma(\psi_e^{\circ}) = \pi_u \times \pi_g = \Sigma_u{}^+ + \Sigma_u{}^- + \Delta_u \tag{6.101}$$

Direct products of degenerate symmetry species, such as that in equation (6.101), are the same irrespective of whether we are concerned with the excitation of one quantum each of a π_u and π_g vibration in a vibrational combination, as in Section 5.2.2.4, or, as here, a single electron in each of a π_u and π_g orbital. Therefore, we can use the direct products given in Table 5.9 in determining $\Gamma(\psi_e)$ as well as $\Gamma(\psi_v)$.

It follows that the states arising from the configuration of equation (6.100) are $^{1,3}\Sigma_u{}^+$, $^{1,3}\Sigma_u{}^-$, $^{1,3}\Delta_u$ since the two electrons (or one electron and one vacancy) may have parallel or antiparallel spins.

The ground configuration of oxygen, a 16-electron molecule, is

$$(\sigma_g 1s)^2(\sigma_u^* 1s)^2(\sigma_g 2s)^2(\sigma_u^* 2s)^2$$
$$(\sigma_g 2p)^2(\pi_u 2p)^4(\pi_g^* 2p)^2 \tag{6.102}$$

All the filled orbitals give the totally symmetric species and the spins of electrons in these orbitals are all paired. As in the case of nitrogen, we have to consider only the unfilled orbitals, in this case $(\pi_g^* 2p)^2$.

In order to obtain $\Gamma(\psi_e^{\circ})$ when two, or more, electrons are in a degenerate orbital, we use the direct products of the type in Table 5.9 for vibrational combination levels. In the case of the $(\pi_g^* 2p)^2$ configuration the result, using Table 5.9, is

$$\Gamma(\psi_e^{\circ}) = \pi_g \times \pi_g = \Sigma_g^+ + \Sigma_g^- + \Delta_g \tag{6.103}$$

This is the same result as we would have obtained for the two electrons in *different* π_g orbitals. The difference lies in the treatment of the two electron spins. When the electrons are in the same degenerate orbital the Pauli principle forbids some orbital and spin combinations. This is similar to the problem encountered in Section 5.2.2.4 in determining $\Gamma(\psi_v)$ when a molecule is vibrating with two quanta of the same degenerate vibration: the Pauli principle forbids the antisymmetric part of the direct product.

The analogy is even closer when the situation in oxygen is compared to that in excited configurations of the helium atom discussed in Section 6.1.5 and summarized in equations

(6.50) and (6.51). According to the Pauli principle for electrons the total wave function must be antisymmetric to electron exchange.

It has been explained in Section 5.2.2.4 that the direct product of two identical symmetry species contains a symmetric part and an antisymmetric part. The antisymmetric part is an A species and, where possible, *not* the totally symmetric species so that, in the direct product of equation (6.103), Σ_g^- (or, using the alternative symmetry species convention, A_2) is the antisymmetric and $\Sigma_g^+ + \Delta_g$ the symmetric part. In the present context symmetric and antisymmetric refer to behaviour of ψ_e^{o} to electron exchange and the reader is referred to Schonland's *Molecular Symmetry* referred to in the Bibliography of Chapter 5 for justification of these properties.

Equation (6.45) expresses the total electronic wave function ψ_e as the product of the orbital and spin parts. Since ψ_e must be antisymmetric to electron exchange the Σ_g^+ and Δ_g orbital wave functions of oxygen combine only with the antisymmetric spin wave function, which is the same as that of equation (6.46) for helium. Similarly, the Σ_g^- orbital wave function combines only with the three symmetric spin wave functions which are the same as those of equation (6.47) for helium.

Hence the states that arise from the ground configuration of oxygen are $^3\Sigma_g^-$, $^1\Sigma_g^+$ and $^1\Delta_g$ and the Pauli principle forbids $^1\Sigma_g^-$, $^3\Sigma_g^+$ and $^3\Delta_g$.

Table 6.6 lists the states arising from configurations in which a degenerate MO, in a molecule belonging to *any* point group, has more than one electron (or more than one vacancy).

The rule proposed by Hund that, of the states arising from the ground configuration, those with the highest multiplicity lie lowest in energy, applies to molecules as well as atoms. It follows that the ground state of the oxygen molecule is $^3\Sigma_g^-$ and, therefore, paramagnetic. This property is demonstrated in the familiar experiment in which a stream of liquid oxygen is deviated when flowing between the poles of a magnet.

The electronic structure of homonuclear diatomic molecules containing elements in the third and subsequent rows of the periodic table can be discussed, at this level of approximation, using arguments analogous to those used for first-row elements.

6.2.1.2 *Heteronuclear diatomic molecules*

Some heteronuclear diatomic molecules, such as nitric oxide (NO), carbon monoxide (CO) and the short-lived CN molecule, contain atoms which are sufficiently similar that the MOs resemble fairly closely those of homonuclear diatomics. In nitric oxide the 15 electrons can be fed into MOs, in the order relevant to O_2 and F_2, to give the ground configuration

$$(\sigma 1s)^2(\sigma^* 1s)^2(\sigma 2s)^2(\sigma^* 2s)^2(\sigma 2p)^2(\pi 2p)^4(\pi^* 2p)^1 \tag{6.104}$$

where the g and u subscripts have been dropped from the orbital symbols because of the lack of a centre of inversion. The net number of bonding electrons is five, giving a bond order of two and a half. Two electrons are attributed to a σ bond and three to π bonds.

The single unpaired electron in the π^*2p MO leads to the ground state being a doublet state, $^2\Pi$, and to the observed paramagnetism of nitric oxide.

Promotion of an electron from, for example, the $\sigma 2p$ into the π^*2p orbital produces the excited configuration

$$\ldots (\sigma 2p)^1(\pi 2p)^4(\pi^* 2p)^2 \tag{6.105}$$

There are several states which arise from this configuration but they can be derived in an analogous way to that used in Section 6.2.1.1. The $(\pi^*2p)^2$ part of the configuration gives, similar to the ground configuration of oxygen, $^3\Sigma^-$, $^1\Sigma^+$ and $^1\Delta$. The $(\sigma 2p)^1$ electron is taken account of in two stages. First, the orbital symmetry species already obtained are multiplied by σ, which leaves them unchanged. Second, the spin of the electron in the $\sigma 2p$

Table 6.6 Electronic states arising from ground configurations in which there is more than one electron or vacancy in a degenerate orbital

Point group	Configuration	States
C_3	$(e)^2$	$^3A+{}^1A+{}^1E$
C_4	$(e)^2$	$^3A+{}^1A+{}^1B+{}^1B$
C_5	$(e_1)^2$	$^3A+{}^1A+{}^1E_2$
	$(e_2)^2$	$^3A+{}^1A+{}^1E_1$
C_6	$(e_1)^2$ and $(e_2)^2$	$^3A+{}^1A+{}^1E_2$
C_7	$(e_1)^2$	$^3A+{}^1A+{}^1E_2$
	$(e_2)^2$	$^3A+{}^1A+{}^1E_3$
	$(e_3)^2$	$^3A+{}^1A+{}^1E_1$
C_8	$(e_1)^2$ and $(e_3)^2$	$^3A+{}^1A+{}^1E_2$
	$(e_2)^2$	$^3A+{}^1A+{}^1B+{}^1B$
C_{3v}	$(e)^2$	$^3A_2+{}^1A_1+{}^1E$
C_{4v}	$(e)^2$	$^3A_2+{}^1A_1+{}^1B_1+{}^1B_2$
C_{5v}	$(e_1)^2$	$^3A_2+{}^1A_1+{}^1E_2$
	$(e_2)^2$	$^3A_2+{}^1A_1+{}^1E_1$
C_{6v}	$(e_1)^2$ and $(e_2)^2$	$^3A_2+{}^1A_1+{}^1E_2$
$C_{\infty v}$	$(\pi)^2$	$^1\Sigma^+ + {}^3\Sigma^- + {}^1\Delta$
	$(\delta)^2$	$^1\Sigma^+ + {}^3\Sigma^- + {}^1\Gamma$
D_3	$(e)^2$	$^3A_2+{}^1A_1+{}^1E$
D_4	$(e)^2$	$^3A_2+{}^1A_1+{}^1B_1+{}^1B_2$
D_5	$(e_1)^2$	$^3A_2+{}^1A_1+{}^1E_2$
	$(e_2)^2$	$^3A_2+{}^1A_1+{}^1E_1$
D_6	$(e_1)^2$ and $(e_2)^2$	$^3A_2+{}^1A_1+{}^1E_2$
C_{3h}	$(e')^2$ and $(e'')^2$	$^3A'+{}^1A'+{}^1E'$
C_{4h}	$(e_g)^2$ and $(e_u)^2$	$^3A_g+{}^1A_g+{}^1B_g+{}^1B_g$
C_{5h}	$(e'_1)^2$ and $(e''_1)^2$	$^3A'+{}^1A'+{}^1E'_2$
	$(e'_2)^2$ and $(e''_1)^2$	$^3A'+{}^1A'+{}^1E'_1$
C_{6h}	$(e_{1g})^2$, $(e_{1u})^2$, $(e_{2g})^2$ and $(e_{2u})^2$	$^3A_g+{}^1A_g+{}^1E_{2g}$
D_{2d}	$(e)^2$	$^3A_2+{}^1A_1+{}^1B_1+{}^1B_2$
D_{3d}	$(e_g)^2$ and $(e_u)^2$	$^3A_{2g}+{}^1A_{1g}+{}^1E_g$
D_{4d}	$(e_1)^2$ and $(e_3)^2$	$^3A_2+{}^1A_1+{}^1E_2$
	$(e_2)^2$	$^3A_2+{}^1A_1+{}^1B_1+{}^1B_2$
D_{5d}	$(e_{1g})^2$ and $(e_{1u})^2$	$^3A_{2g}+{}^1A_{1g}+{}^1E_{2g}$
	$(e_{2g})^2$ and $(e_{2u})^2$	$^3A_{2g}+{}^1A_{1g}+{}^1E_{1g}$
D_{6d}	$(e_1)^2$ and $(e_5)^2$	$^3A_2+{}^1A_1+{}^1E_2$
	$(e_2)^2$ and $(e_4)^2$	$^3A_2+{}^1A_1+E_4$
	$(e_3)^2$	$^3A_2+{}^1A_1+{}^1B_1+{}^1B_2$
D_{3h}	$(e')^2$ and $(e'')^2$	$^3A'_2+{}^1A'_1+{}^1E'$
D_{4h}	$(e_g)^2$ and $(e_u)^2$	$^3A_{2g}+{}^1A_{1g}+{}^1B_{1g}+{}^1B_{2g}$
D_{5h}	$(e'_1)^2$ and $(e''_1)^2$	$^3A'_2+{}^1A'_1+{}^1E'_2$
	$(e'_2)^2$ and $(e''_2)^2$	$^3A'_2+{}^1A'_1+{}^1E'_1$
D_{6h}	$(e_{1g})^2$, $(e_{1u})^2$, $(e_{2g})^2$ and $(e_{2u})^2$	$^3A_{2g}+{}^1A_{1g}+{}^1E_{2g}$
$D_{\infty h}$	$(\pi_g)^2$ and $(\pi_u)^2$	$^3\Sigma_g^- + {}^1\Sigma_g^+ + {}^1\Delta_g$
	$(\delta_g)^2$ and $(\delta_u)^2$	$^3\Sigma_g^- + {}^1\Sigma_g^+ + {}^1\Gamma_g$
S_4	$(e)^2$	$^3A+{}^1A+{}^1B+{}^1B$
S_6	$(e_g)^2$ and $(e_u)^2$	$^3A_g+{}^1A_g+{}^1E_g$
S_8	$(e_1)^2$ and $(e_3)^2$	$^3A+{}^1A+{}^1E_2$
	$(e_2)^2$	$^3A+{}^1A+{}^1B+{}^1B$
T_d	$(t_1)^2$, $(t_1)^4$, $(t_2)^2$ and $(t_2)^4$	$^3T_1+{}^1A_1+{}^1E+{}^1T_2$
	$(t_1)^3$	$^4A_1+{}^2E+{}^2T_1+{}^2T_2$
	$(t_2)^3$	$^4A_2+{}^2E+{}^2T_1+{}^2T_2$
	$(e)^2$	$^3A_2+{}^1A_1+{}^1E$
T	$(t)^2$ and $(t)^4$	$^3T+{}^1A+{}^1E+{}^1T$
	$(t)^3$	$^4A+{}^2E+{}^2T+{}^2T$
	$(e)^2$	$^3A+{}^1A+{}^1E$
O_h	$(t_{1g})^2$, $(t_{1u})^2$, $(t_{1g})^4$, $(t_{1u})^4$, $(t_{2g})^2$, $(t_{2u})^2$, $(t_{2g})^4$, $(t_{2u})^4$	$^3T_{1g}+{}^1A_{1g}+{}^1E_g+{}^1T_{2g}$
	$(t_{1g})^3$	$^4A_{1g}+{}^2E_g+{}^2T_{1g}+{}^2T_{2g}$
	$(t_{1u})^3$	$^4A_{1u}+{}^2E_u+{}^2T_{1u}+{}^2T_{2u}$
	$(t_{2g})^3$	$^4A_{2g}+{}^2E_g+{}^2T_{1g}+{}^2T_{2g}$
	$(t_{2u})^3$	$^4A_{2u}+{}^2E_u+{}^2T_{1u}+{}^2T_{2u}$
	$(e_g)^2$ and $(e_u)^2$	$^3A_{2g}+{}^1A_{1g}+{}^1E_g$
O	$(t_1)^2$, $(t_1)^4$, $(t_2)^2$ and $(t_2)^4$	$^3T_1+{}^1A_1+{}^1E+{}^1T_2$
	$(t_1)^3$	$^4A_1+{}^2E+{}^2T_1+{}^2T_2$
	$(t_2)^3$	$^4A_2+{}^2E+{}^2T_1+{}^2T_2$
	$(e)^2$	$^3A_2+{}^1A_1+{}^1E$

MO can be parallel or antiparallel to those of the electrons in the π^*2p MO, giving the states $^{2,4}\Sigma^-$, $^2\Sigma^+$ and $^2\Delta$.

Even molecules such as the short-lived SO and PO molecules can be treated, at the present level of approximation, rather like homonuclear diatomics even though the two atoms are very different. The reason is that outer shell LCAO MOs can be constructed from $2s$ and $2p$ AOs on the oxygen atom and $3s$ and $3p$ AOs on the sulphur or phosphorus atom and they will appear qualitatively similar to those in Figure 6.27. These linear combinations, such as $2p$ on oxygen with $3p$ on sulphur, obey all the rules given on p. 300 and notably the first one, which states that the energies of the AOs should be comparable. This comparability is clear from their first ionization energies, given in Table 6.1.

There is, in principle, no reason why linear combinations should not be made between AOs which have the correct symmetry but have very different energies; an example of such orbitals is $1s$ on the oxygen atom and $1s$ on the phosphorus atom. The result would be that the resonance integral β would be extremely small so that the MOs would be virtually unchanged from the AOs from which they were constructed and the linear combination would be ineffective. It follows from this that, in the SO and PO molecules, the $1s$ electrons on the sulphur and phosphorus atoms remain as core electrons taking virtually no part in bonding and remaining in orbitals almost unchanged from AOs.

Similarly, in the short-lived molecule AsO, outer shell MOs can be constructed from $2s$ and $2p$ on the oxygen atom and $4s$ and $4p$ on the arsenic atom.

In a molecule such as hydrogen chloride (HCl), and most other heteronuclear diatomics, the MOs are different from those in Figure 6.27, but the rules for making effective linear combinations of atomic orbitals still hold good.

The electron configuration of the chlorine atom is $KL3s^23p^5$ and it is only the $3p$ electrons (ionization energy 12.967 eV) which are comparable in energy to the hydrogen $1s$ electron (ionization energy 13.598 eV). Of the $3p$ orbitals it is only $3p_z$ which has the correct symmetry to form a linear combination with the hydrogen $1s$ orbital, as shown in Figure 6.30(a). The MO wave function is of the form given in equation (6.77), where χ_1 and χ_2 are the H $1s$ and Cl $3p_z$ AO wave functions but c_1/c_2 is no longer ±1. Because of the higher electronegativity of chlorine compared with hydrogen there is considerable concentration in the region of the chlorine atom of the two electrons which go into the resulting σ bonding orbital. The two electrons in this orbital form the single bond in hydrogen chloride.

The $3p_x$ and $3p_y$ orbitals of chlorine cannot overlap with the $1s$ orbital of hydrogen for reasons of symmetry and the two electrons remaining in each of the $3p_x$ and $3p_y$ orbitals are referred to as lone pairs since they do not take part in bonding, at least in this degree of approximation. These are illustrated in Figure 6.30(b).

The ground state of hydrogen chloride is $^1\Sigma^+$ since all occupied orbitals are filled.

Promotion of an electron from one of the lone pair orbitals, of π symmetry, into the σ^* orbital formed by the linear combination of $1s$ and $3p_z$ produces $^1\Pi$ and $^3\Pi$ excited states.

6.2.2 Rydberg orbitals

Experimentally it is found that the higher energy molecular orbitals converge to a limit which represents, just as in an atom, the removal of an electron in an ionization process. The analogy with atoms extends to the molecular orbital energies E_n being given by

$$E_n/hc = -\tilde{R}/(n-\eta)^2 \qquad (6.106)$$

for high values of n. In this equation, $\tilde{R}$ is the Rydberg constant for the molecule concerned, n is an integer and η is the quantum defect. These term values are analogous to those for the alkali metal atoms discussed in Section 6.1.3.

Any MO resembles an AO when the MO is large compared with the size of the molecular core. For example, an electron in a high energy

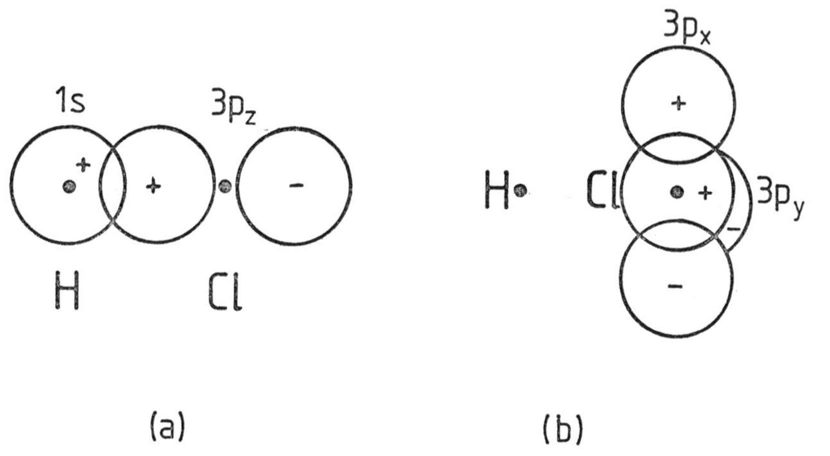

Figure 6.30 In the HCl molecule, (a) the single bond MO is formed by a linear combination of $1s$ on H and $3p_z$ on Cl and (b) electrons in the $3p_x$ and $3p_y$ AOs on Cl remain as lone pairs

orbital in the oxygen molecule tends to see the O_2^+ core as being concentrated at a point. The total nuclear charge is $+16e$ and the net charge of nuclei and electrons is $+1e$, so the O_2^+ core resembles, in the united atom approximation (Section 6.2.1.1), S^+ and the Rydberg MOs of the oxygen molecule resemble the high energy AOs of the sulphur atom. Therefore, we should be able to correlate the MOs of the oxygen molecule with the AOs of the sulphur atom and similarly for any homonuclear (or heteronuclear) diatomic molecule.

It should be obvious from Figure 6.27 that, as the nuclei coalesce in the united atom, the $\sigma_g 1s$ MO becomes a spherically symmetrical 1s AO, the $\sigma_u^* 1s$ MO becomes a $2p_z$ AO and the $\sigma_g 2s$ MO becomes a $2s$ AO. However, the correlation of some of the MOs with united atom AOs is not so obvious. For example, $\sigma_g 2p$ correlates with $3s$ and the doubly degenerate $\pi_g^* 2p$ MO correlates with the $3d_{xz}$ and $3d_{yz}$ AOs shown in Figure 1.15.

Some correlations between MOs and united atom AOs are given in Table 6.7.

The distinction between a Rydberg and a valence orbital is not always a clear one and this is particularly true of low-lying 'Rydberg' orbitals, which can be regarded as being of mixed Rydberg and valence character.

An alternative and more generally useful way of correlating MOs and united atom AOs, whatever the point group of the molecule concerned, is to consider the united atom AOs as being slightly perturbed by the symmetry of the core. In reclassifying the AOs according to any molecular point group the following rules for s, p and d orbitals are useful:

1. An s AO always has the totally symmetric species of the point group.
2. The p_x, p_y and p_z AOs have the species of the translations T_x, T_y and T_z, respectively.
3. In degenerate point groups, such as $D_{\infty h}$ and $C_{\infty v}$, the d_{z^2}, $d_{x^2-y^2}$, d_{xy}, d_{yz} and d_{xz} AOs have the species of the components of the polarizability α_{zz} (or $\alpha_{xx}+\alpha_{yy}-2\alpha_{zz}$), $\alpha_{xx}-\alpha_{yy}$, α_{xy}, α_{yz} and α_{xz}, respectively. These species are given in the relevant character table. In non-degenerate point groups the species are easily obtained from the form of the d orbitals and the appropriate character table.

With these rules applied to the $D_{\infty h}$ point group we can make the correlations between MOs and united atom AOs for homonuclear diatomic molecules given in Table 6.8.

Table 6.7 Correlation of some homonuclear diatomic molecule MOs with united atom AOs

Diatomic molecule MO	United atom AO	Diatomic molecule MO	United atom AO
$\sigma_g 1s$	$1s$	$\sigma_u 2p$	$4p_z$
$\sigma_u 1s$	$2p_z$	$\sigma_g 3s$	$3d_{z^2}$
$\sigma_g 2s$	$2s$	$\sigma_u 3s$	$5p_z$
$\sigma_u 2s$	$3p_z$	$\sigma_g 3p$	$4s$
$\sigma_g 2p$	$3s$	$\pi_u 3p$	$3p_x$, $3p_y$
$\pi_u 2p$	$2p_x$, $2p_y$	$\pi_g 3p$	$4d_{xz}$, $4d_{yz}$
$\pi_g 2p$	$3d_{xz}$, $3d_{yz}$		

Table 6.8 Symmetry species in the $D_{\infty h}$ point group of MOs derived from s, p and d united atom AOs

United atom AO	MO
s	σ_g^+
p_z	σ_u^+
p_x, p_y	π_u
d_{z^2}	σ_g^+
d_{xz}, d_{yz}	π_g
$d_{x^2-y^2}$, d_{xy}	δ_g

In alkali metal atoms the quantum defect η of equation (6.44) is a measure of the degree of penetration of the core by a valence orbital and decreases in the order $s > p > d > f$. This is also the case for the Rydberg MO quantum defect η of equation (6.106), which decreases in the order $s > p > d > f$ corresponding to the united atom AO.

6.2.3 Classification of electronic states: selection rules

6.2.3.1 Classification of states

When a molecule is undergoing end-over-end rotation there is, as we have seen in Chapters 1 and 4, an angular momentum associated with that motion. The coupling of this angular momentum to those associated with electronic motion will not concern us until Section 6.2.5. Here we consider the vector picture of momenta due to the orbital and spin motions of the electrons when the molecule is not rotating.

We have seen in Section 6.2.1 and, in particular, in Figure 6.28 how the orbital and spin angular momenta for a single electron may be uncoupled by the electrostatic field of the nuclei.

For all many-electron diatomic molecules the coupling approximation which best describes electronic states is that which is analogous to the Russell–Saunders approximation in atoms. All the orbital angular momenta are coupled to give a resultant $\boldsymbol{L}$ and all the electron spin momenta to give a resultant $\boldsymbol{S}$. However, if there is no highly charged nucleus in the molecule, the spin–orbit coupling between $\boldsymbol{L}$ and $\boldsymbol{S}$ is weak and the electrostatic field of the nuclei uncouples them, as shown in Figure 6.31(a). (Since the electrostatic field produces an effect analogous to the Stark effect in atoms, this uncoupling of $\boldsymbol{L}$ and $\boldsymbol{S}$ is the Stark equivalent of the Paschen–Back effect illustrated in Figure 6.22(b).)

The vector $\boldsymbol{L}$ is so strongly coupled to the electrostatic field and, consequently, the precession frequency is so high that the magnitude of $\boldsymbol{L}$ is not defined; in other words, L is not a good quantum number. Only the component $\Lambda\hbar$ along the z-axis is defined where Λ can take the values

$$\Lambda = 0, 1, 2, \ldots \tag{6.107}$$

In an atom, states with the same magnitudes but different sign of M_L remain degenerate in an electric field. Since Λ is equivalent to $|M_L|$, all states in a diatomic molecule with $\Lambda > 0$ are also doubly degenerate.

Classically this degeneracy can be thought of as being due to the electrons, with non-zero orbital angular momenta, orbiting clockwise or anticlockwise around the internuclear axis, the energy being the same in both cases.

Analogous to the labelling of one-electron orbitals, electronic states are designated Σ, Π, Δ, Φ, Γ, . . . corresponding to $\Lambda = 0$, 1, 2, 3, 4, . . . The dual interpretation of Σ, Π, etc., as symmetry species of a point group, as discussed in Section 6.2.1, as well as implying orbital angular momentum quantum numbers, is analogous to the dual interpretation of S, P, etc., in atoms discussed in Sections 6.1.2 and 6.1.7.

Figure 6.31(a) shows also that the component of $\boldsymbol{S}$ along the internuclear axis is $\Sigma\hbar$. The coupling of $\boldsymbol{S}$ to the internuclear axis is not caused by the electrostatic field, which has no effect on it, but is due to the internal magnetic field along the axis due to the orbital motion of the electrons. For this reason the quantum number Σ is analogous to M_S in an atom and can take the values

$$\Sigma = S, S - 1, \ldots, -S \tag{6.108}$$

S remains a good quantum number and, for states with $\Lambda > 0$, there are $(2S+1)$ components corresponding to the number of values that Σ can take. The multiplicity of the state is the value of $(2S+1)$ and is indicated, as in atoms, by a pre-superscript as, for example, in $^3\Pi$.

The component of the total angular momentum of the electrons along the internuclear axis is $\Omega\hbar$, indicated in Figure 6.31(a), where the quantum number Ω is given by

$$\Omega = |\Lambda + \Sigma| \tag{6.109}$$

The value of $\Lambda + \Sigma$, *not* $|\Lambda + \Sigma|$, is attached to the term symbol as a post-subscript. For example, for a $^4\Pi$ term, $\Lambda = 1$, $S = \frac{3}{2}$, $\Sigma = \frac{3}{2}, \frac{1}{2}, -\frac{1}{2}, -\frac{3}{2}$ and $\Omega = \frac{5}{2}, \frac{3}{2}, \frac{1}{2}, \frac{1}{2}$, but the components are labelled $^4\Pi_{\frac{5}{2}}$, $^4\Pi_{\frac{3}{2}}$, $^4\Pi_{\frac{1}{2}}$, $^4\Pi_{-\frac{1}{2}}$.

Spin–orbit interaction splits the components of a multiplet so that the energy level before interaction is shifted by the interaction energy E', where

$$E' = A\Lambda\Sigma \tag{6.110}$$

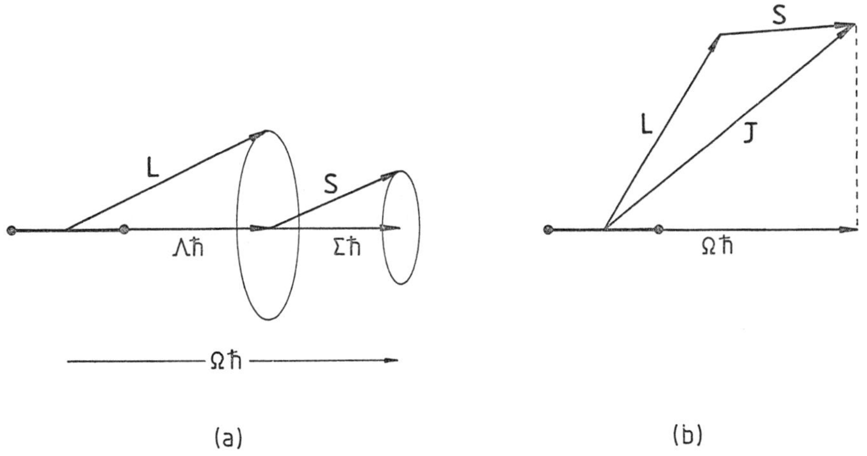

Figure 6.31 (a) Uncoupling of $\boldsymbol{L}$ and $\boldsymbol{S}$ in a diatomic molecule in which spin–orbit coupling is weak, as in Hund's case (a). (b) When spin–orbit coupling is strong, as in Hund's case (c), $\boldsymbol{L}$ and $\boldsymbol{S}$ are sufficiently strongly coupled that they are not uncoupled by the field of the nuclei

in which A is the spin–orbit coupling constant. For a given value of Λ, within a given multiplet, the splitting between adjacent components is constant, unlike the case with atoms. The sign of A determines whether the multiplet is normal or inverted, i.e. whether the energy increases or decreases, respectively, with Ω.

For Σ states there is no orbital angular momentum and therefore no internal magnetic field to couple $\boldsymbol{S}$ to the axis. For this reason, when spin–spin coupling is neglected, the quantum number Σ is not defined and a Σ state, whatever its multiplicity, has only one component. This is no longer true when spin–spin coupling is included.

The situation pictured in Figure 6.31(a), in which $\boldsymbol{L}$ and $\boldsymbol{S}$ are completely uncoupled, is the most commonly encountered and is an example of what is called Hund's case (a) coupling. This case will be discussed more generally in Section 6.2.5.3 when the rotational angular momentum will be included.

Like all assumed coupling of angular momenta, Hund's case (a) represents an approximation, albeit a commonly useful one. Another extreme coupling approximation is Hund's case (c), illustrated, again assuming no rotational angular momentum, in Figure 6.31(b). Here the spin–orbit coupling is sufficiently large that $\boldsymbol{L}$ and $\boldsymbol{S}$ are not uncoupled and Λ is no longer a good quantum number. Nevertheless, the Σ, Π, Δ, . . . labels for states are still used. Examples of this coupling case are to be found in molecules with at least one highly charged nucleus and it will be discussed in more detail in Section 6.2.5.5.

6.2.3.2 *Electric dipole selection rules for transitions with $\Delta S = 0$*

In atoms there are sufficient good quantum numbers to make a discussion of selection rules in terms of symmetry properties merely a more complex alternative.

For Hund's case (a) coupling in diatomic molecules, the quantum numbers Λ, S and, in states other than those with $\Lambda = 0$, Σ are insufficient for specifying selection rules. In addition, two symmetry properties of ψ_e must be used for homonuclear and one for heteronuclear diatomic molecules.

For homonuclear diatomics ψ_e may be symmetric or antisymmetric to inversion and this is indicated by a post-subscript g or u, respectively.

For all diatomics in Σ states ψ_e may be symmetric or antisymmetric to reflection across any σ_v plane. This is indicated, as shown in the character tables for $C_{\infty v}$ (Table 5.11) and $D_{\infty h}$ (Table A.37 in the Appendix), by a post-superscript plus or minus sign, respectively, which is referred to as the parity of the state. Although Π, Δ, Φ, . . . states do not normally have such parity symbols attached, it should be understood that, of the two components of these doubly degenerate states, one is plus and the other is minus. This is indicated occasionally by $\Pi^{\pm}$, $\Delta^{\pm}$, etc.

The electronic selection rules for the case (a) coupling approximation, assuming electric dipole transitions, are:

1. $\Delta\Lambda = 0, \pm 1$: for example Σ–Σ, Π–Σ, Δ–Π transitions are allowed but not Δ–Σ or Φ–Π.
2. $\Delta S = 0$ in molecules with no highly charged nuclei, analogous to the rule in atoms. For example, triplet–singlet transitions are highly forbidden in H_2 but, in CO, the $a^3\Pi$–$X^1\Sigma^+$ transition† is observed weakly.

†There is a convention, which is commonly but not always used, regarding the labelling of electronic states in diatomic molecules. The ground state is labelled X and higher states of the same multiplicity are labelled A, B, C, . . . in order of increasing energy. States of multiplicity different from that of the ground state are labelled a, b, c, . . . in order of increasing energy. Therefore the 'a' state of carbon monoxide is the first excited triplet state.

In some cases which we shall encounter, for example N_2 and I_2, some commonly used labels are adhered to which do not follow this convention.

In polyatomic molecules the same convention is used except that a tilde is placed over the label, as in $\tilde{A}$ and $\tilde{a}$ states, to remove any confusion with symmetry species. This convention applies even to linear polyatomics since they may be non-linear in excited electronic states.

3. $\Delta\Sigma = 0$: this selection rule, and the following one, refer to transitions between multiplet components.
4. $\Delta\Omega = 0, \pm 1$.
5. $+ \not\leftrightarrow -$, $+ \longleftrightarrow +$, $- \longleftrightarrow -$: this is relevant only for Σ–Σ transitions. It follows that only Σ^+–Σ^+ and Σ^-–Σ^- transitions are allowed.
6. $g \longleftrightarrow u$, $g \not\leftrightarrow g$, $u \not\leftrightarrow u$: for example, a Σ_g^+–Σ_g^+ transition is forbidden.

The selection rules for the orbital part of the electronic wave function, namely the rules involving $\Delta\Lambda$, + and −, g and u, can all be expressed in terms of the symmetry properties of ψ_e for each of the combining electronic states. The general symmetry requirements are the same as those given for atoms in equation (6.53) and can be summarized as

$$\Gamma(\psi_e') \times \Gamma(\psi_e'') \supset \Gamma(T_x) \text{ and/or } \Gamma(T_y) \text{ and/or } \Gamma(T_z) \qquad (6.111)$$

A homonuclear diatomic molecule belongs to the $D_{\infty h}$ point group, for which $\Gamma(T_z) = \Sigma_u^+$ and $\Gamma(T_x, T_y) = \Pi_u$. Therefore, equation (6.111) becomes

$$\Gamma(\psi_e') \times \Gamma(\psi_e'') \supset \Sigma_u^+ \text{ or } \Pi_u \qquad (6.112)$$

for an allowed transition and the transition moment is polarized (see Section 2.2) along the z-axis or in the xy-plane, respectively.

We have shown already that, for example, a Δ_g–Π_u transition is allowed since $\Delta\Lambda = 1$ and it is g–u. Use of Table 5.9 to obtain a direct product gives the result that

$$\Gamma(\psi_e') \times \Gamma(\psi_e'') = \Delta_g \times \Pi_u = \Pi_u + \Phi_u \qquad (6.113)$$

Since the product contains Π_u, the transition is allowed with the transition moment polarized in the xy-plane.

Similar arguments can be used to determine selection rules for heteronuclear diatomic molecules. These belong to the $C_{\infty v}$ point group for which $\Gamma(T_z) = \Sigma^+$ and $\Gamma(T_x, T_y) = \Pi$.

For case (c) coupling, all the selection rules given for case (a) on p. 312–313 apply, except for (1), (3) and (5). Since Λ and Σ are not good quantum numbers, rules (1) and (3) are not relevant. In the case of rule (5) the + or − label now refers to the symmetry or antisymmetry of ψ_e with respect to σ_v when $\Omega = 0$ [not $\Lambda = 0$, as in case (a)] and the rule is

$$0^+ \longleftrightarrow 0^+, \; 0^- \longleftrightarrow 0^-, \; 0^+ \not\leftrightarrow 0^-$$

where 0 refers to the value of Ω.

The lowest $^3\Pi_u$ term of iodine approximates more closely to case (c) than case (a). Since, in a case (a) description, Λ and S would be 1 and Σ would be 1, 0 or −1, the value of Ω can be 2, 1 or 0. The $\Omega = 0$ and 1 components are split by 3881 cm^{-1}, indicating large spin–orbit interaction and a case (c) approximation. The transition† $B^3\Pi_{0^+u}$–$X^1\Sigma_g^+$ involves the $\Omega = 0$ component for which ψ_e is symmetric to σ_v and the transition is allowed by the $0^+ \longleftrightarrow 0^+$ selection rule. Note also that the selection rule $\Delta S = 0$ breaks down completely in a molecule with such a high nuclear charge. Indeed, the transition is fairly intense; it occurs in the visible region of the spectrum and is responsible for the violet colour of iodine vapour.

6.2.3.3 *Electric dipole selection rules for transitions with* $\Delta S \neq 0$

If one or both of the nuclei have a high charge, the selection rule $\Delta S = 0$ may break down. Many transitions with $\Delta S = \pm 1$ have been observed and these are mainly between singlet and triplet states, as in the $a^3\Pi$–$X^1\Sigma^+$ transition of carbon monoxide. A few other types of $\Delta S = \pm 1$ transitions such as the $a^4\Sigma^-$–$X^2\Pi$ transition of the short-lived molecule SiF have been observed but we shall not consider these further. Nor shall we consider transitions with

†The labelling of the $B^3\Pi_{0^+u}$ state of iodine follows general usage rather than the usual convention which would label it $b^3\Pi_{0^+u}$.

$|\Delta S| > 1$, for example between a singlet and a quintet state.

The $\Delta S = 0$ selection rule is broken down by spin–orbit interaction.† In a molecule which has, in the approximation of zero spin–orbit interaction, a manifold of singlet states and a manifold of triplet states, shown in Figure 6.32, the presence of such an interaction causes a mixing of the singlet and triplet states—the greater the interaction the greater the mixing.

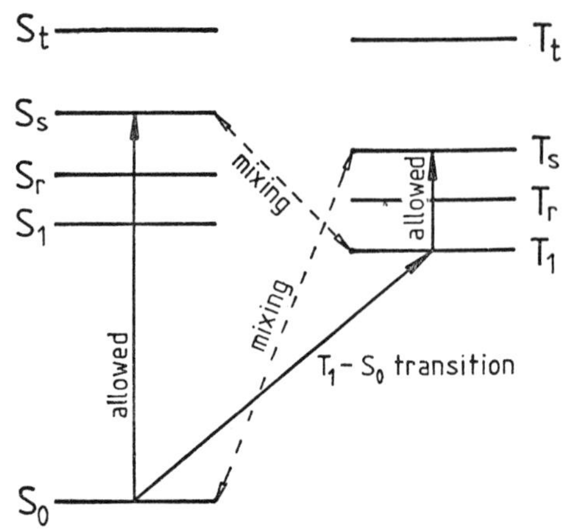

Figure 6.32 Processes by which a triplet–singlet transition T_1–S_0 may gain intensity by spin–orbit interaction, which has the effect of mixing singlet and triplet states

We shall consider here the transition from the ground singlet state, S_0, to the first excited triplet state, T_1. This T_1–S_0 transition may gain intensity either by S_0 being mixed by spin–orbit interaction with triplet states or by T_1 being mixed with singlet states, or both.

In principle, S_0 may mix with several triplet states but, because of restrictions of symmetry and energy separation, it usually mixes preferentially with only one, say T_s. Similarly, T_1 may mix with several singlet states but usually it mixes preferentially with one, say S_s. If, as in Figure 6.32, transitions S_s–S_0 and T_s–T_1 are allowed, the T_1–S_0 transition may gain intensity by intensity stealing from either S_s–S_0 or T_s–T_1. It might steal intensity from both but one mechanism usually dominates and this is often the one resulting from the S_s and T_1 type of mixing.

It should be realized that this concept of intensity stealing by one transition from another is a completely artificial one arising only because of our initial neglect of spin–orbit coupling. It is only because of the general validity of this neglect that the concept of intensity stealing arises.

Whether spin–orbit interaction can mix states may be expressed entirely in terms of symmetry arguments.

The symmetry species of ψ_e for a triplet state is the product of the orbital and spin wave functions [as in equation (6.45) for atoms]. Each of the spin components has the symmetry species of a rotation, so that a triplet–singlet transition may be observed if either the triplet state is connected by a rotational symmetry species to a nearby singlet state or the singlet state is similarly connected to a nearby triplet state. We can apply this rule to the T_1–S_0 transition in Figure 6.32 and deduce that the transition may have non-zero intensity if one or more of the following symmetry requirements is met:

$$\begin{aligned} \Gamma(S_0) \times \Gamma(R_x) &\supset \Gamma(T_s) \\ \Gamma(S_0) \times \Gamma(R_y) &\supset \Gamma(T_s) \\ \Gamma(S_0) \times \Gamma(R_z) &\supset \Gamma(T_s) \end{aligned} \quad (6.114)$$

$$\begin{aligned} \Gamma(S_s) \times \Gamma(R_x) &\supset \Gamma(T_1) \\ \Gamma(S_s) \times \Gamma(R_y) &\supset \Gamma(T_1) \\ \Gamma(S_s) \times \Gamma(R_z) &\supset \Gamma(T_1) \end{aligned} \quad (6.115)$$

†For a discussion of spin-forbidden transitions, applied particularly to asymmetric rotors, see Hougen, J. T. (1964). *Can. J. Phys.*, **42**, 433.

where S_0, etc., implies the orbital part of the electronic wave function for that state.

We shall consider the application to the $A^3\Sigma_u^+$–$X^1\Sigma_g^+$ Vegard–Kaplan system of nitrogen† which has been observed in the 506–210 nm region.

In the $D_{\infty h}$ point group $\Gamma(R_x, R_y) = \Pi_g$ and $\Gamma(R_z) = \Sigma_g^-$. Therefore, the $X^1\Sigma_g^+$ state may mix with triplet states, given by equation (6.114), with orbital symmetry

$$\begin{aligned} \Sigma_g^+ \times \Pi_g &= \Pi_g \\ \Sigma_g^+ \times \Sigma_g^- &= \Sigma_g^- \end{aligned} \tag{6.116}$$

and the A–X transition may steal intensity from $^3\Pi_g$–$A^3\Sigma_u^+$ or $^3\Sigma_g^-$–$A^3\Sigma_u^+$ transitions. Since the latter are parity forbidden, intensity may be stolen only from the former.

Alternatively, or in addition, the $A^3\Sigma_u^+$ state may mix with singlet states. Equation (6.115) gives

$$\begin{aligned} \Pi_u \times \Pi_g &= \Sigma_u^+ + \Sigma_u^- + \Delta_u \\ \Sigma_u^- \times \Sigma_g^- &= \Sigma_u^+ \end{aligned} \tag{6.117}$$

and the A–X transition may steal intensity from $^1\Pi_u$–$X^1\Sigma_g^+$ or $^1\Sigma_u^-$–$X^1\Sigma_g^+$ transitions. Again, the latter are parity forbidden so intensity may be stolen only from the former type of transition.

We conclude that the $A^3\Sigma_u^+$–$X^1\Sigma_g^+$ transition may steal its intensity from a $^3\Pi_g$–$A^3\Sigma_u^+$ transition by spin–orbit interaction between the X state and a $^3\Pi_g$ state or from a $^1\Pi_u$–$X^1\Sigma_g^+$ transition by spin–orbit interaction between the A state and a $^1\Pi_u$ state, or both.

It is an important rule that the polarization of the transition moment in a spin-forbidden transition is that of the transition from which the intensity is stolen.

In the case of the A–X transition of nitrogen, both types of transition from which intensity may be stolen are polarized in the xy-plane and this is the polarization of the A–X transition moment. In polyatomic molecules of lower symmetry a triplet–singlet transition moment may be of mixed polarization (see, for example, the discussion of formaldehyde transitions in Section 6.3.1.5).

†The labelling of the $A^3\Sigma_u^+$ state of nitrogen follows general usage rather than the usual convention which would label it a $^3\Sigma_u^+$.

6.2.3.4 Magnetic dipole transitions

It was mentioned in Section 2.1 that we are concerned in this book more with interaction of an atom or molecule with the electric component of the electromagnetic radiation than with the magnetic component. The reason for this is that electric dipole interaction is typically 10^5 times stronger than magnetic dipole interaction. Nevertheless, a few magnetic dipole transitions have been observed in diatomic molecules and we consider them briefly here. Also, in Section 6.3.1.5, we shall encounter magnetic dipole transitions in formaldehyde and thioformaldehyde.

The components m_x, m_y and m_z of a magnetic dipole moment have the symmetry species of the rotations R_x, R_y and R_z, respectively. Therefore, analogous to equation (6.111) for an electric dipole transition, an electronic transition between states of the same multiplicity is allowed by magnetic dipole selection rules if

$$\Gamma(\psi_e') \times \Gamma(\psi_e'') \supset \Gamma(R_x) \text{ and/or } \Gamma(R_y) \text{ and/or } \Gamma(R_z) \tag{6.118}$$

The first excited singlet state of nitrogen, $a^1\Pi_g$, results from the excited configuration $\ldots(\pi_u 2p)^4(\sigma_g 2p)^1(\pi_g{}^*2p)^1$, as in equation (6.98). The transition $a^1\Pi_g$–$X^1\Sigma_g^+$ has been observed in both absorption and emission and gives rise to the so-called Lyman–Birge–Hopfield bands in the ultraviolet and far ultraviolet regions. Table A.37 in the Appendix shows that, in the $D_{\infty h}$ point group,

$$\Gamma(R_x, R_y) = \Pi_g \tag{6.119}$$

and therefore the a–X transition is magnetic dipole allowed.

The rotational transitions accompanying a magnetic dipole transition are the same as those for the analogous electric dipole transition (see Section 6.2.5) so the rotational fine structure of a $^1\Pi_g$–$^1\Sigma_g^+$ is the same as for a $^1\Pi_u$–$^1\Sigma_g^+$ transition.

It has been shown in Section 6.2.1.1. that the states arising from the ground configuration of the oxygen molecule [equation (6.102)] are $^3\Sigma_g^-$, $^1\Sigma_g^+$ and $^1\Delta_g$. The $X^3\Sigma_g^-$ state is the ground state and the singlet states are low-lying excited states. Of these, $a^1\Delta_g$ is the lowest and the $a^1\Delta_g$–$X^3\Sigma_g^-$ transition has been observed in the absorption spectrum of the earth's atmosphere with the sun's radiation as continuum—these bands are called the infrared atmospheric absorption bands. The band system is extremely weak because not only is it a magnetic dipole transition but it also defies the $\Delta S = 0$ selection rule. It is observed only because of the enormous quantity of oxygen in the absorbing path.

The $b^1\Sigma_g^+$–$X^3\Sigma_g^-$ transition is also observed, giving rise to the visible atmospheric absorption bands. This electronic transition also defies the $\Delta S = 0$ selection rule and is magnetic dipole allowed.

6.2.4 Vibrational coarse structure

6.2.4.1 Excited state potential energy curves

In Section 5.1 we have discussed the form of the potential energy curve in the ground electronic state of a diatomic molecule. Figure 5.4 shows a typical curve with a minimum potential energy when the internuclear distance is r_e. Dissociation occurs at high energy, the dissociation energy being D_e, relative to the minimum in the curve, or D_0, relative to the zero-point level.

The vibrational term values, $G(v)$, are given in equation (5.3), for the harmonic oscillator approximation, and in equation (5.14) for the anharmonic oscillator.

The total term value S for a molecule in an electronic state with electronic term value T, corresponding to the equilibrium configuration, and with vibrational and rotational term values $G(v)$ and $F(J)$, is given by

$$S = T + G(v) + F(J) \qquad (6.120)$$

In each excited electronic state of a diatomic molecule there is a corresponding potential energy curve. Most of these appear quantitatively similar to that for the ground electronic state.

Figure 6.33 shows potential energy curves for the ground state, $X^1\Sigma_g^+$, and several excited states of the short-lived molecule C_2.

The ground configuration, labelled 1 in Table 6.9, is obtained by feeding the 12 electrons into the MOs in Figure 6.29. All occupied orbitals are filled and, therefore, the ground state is $X^1\Sigma_g^+$. The excited configuration 2 in Table 6.9 gives rise to the $a^3\Pi_u$ and $A^1\Pi_u$ excited states. In the case of atoms [Section 6.1.2.2(i)] one of Hund's rules states that 'of the states which arise from equivalent electrons those with the highest multiplicity lie lowest in energy'. This holds for molecules also but, as for atoms, it does not hold for non-equivalent electrons. Nevertheless, it is usual in molecules for the state of highest multiplicity to lie lowest in energy whether or not the electrons are equivalent. Figure 6.33 shows how far the $a^3\Pi_u$ state lies below the $A^1\Pi_u$ state.

The three states arising from configuration 3 in Table 6.9 are obtained in the same way as for the ground configuration of oxygen [equation (6.103) *et seq*.]. The $b^3\Sigma_g^-$ state is fairly low-lying but the $E^1\Sigma_g^+$ and $^1\Delta_g$ states are much higher.

The information contained in Figure 6.33 was obtained using various experimental techniques to observe absorption or emission spectra. Table 6.10 lists the transitions which have been observed, the names of the workers associated with their discovery, and after whom the transitions are named, the region of the spectrum where they are observed, and the nature of the source. The mixture of techniques for observing the spectra, including a high temperature furnace, flames, arcs, discharges and, for a short-lived species, flash photolysis, is

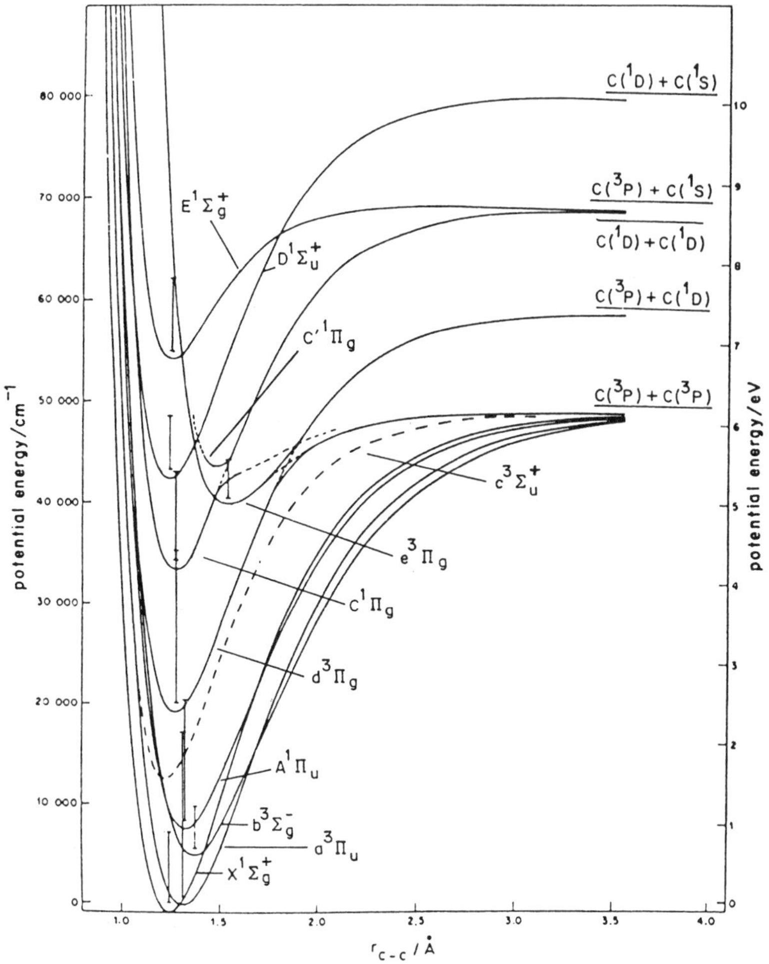

Figure 6.33 Potential energy curves for the ground and several excited states of the C_2 molecule [Reproduced, with permission, from Ballik, E. A. and Ramsay, D. A. (1963). *Astrophys. J.*, **137**, 84]

typical of molecular electronic spectroscopy, as is the observation of both emission and absorption processes.

Table 6.10 shows that several systems of C_2 are observed in absorption with $a^3\Pi_u$ as the lower state. It is unusual for an absorption system to have any other than the ground electronic state as its lower state. However, in C_2, the $a^3\Pi_u$ state is only $716\,cm^{-1}$ above the ground state so that the population of the $a^3\Pi_u$ state, given for equilibrium conditions by the Boltzmann equation (2.8), is appreciable at only moderate temperatures.

In addition to these laboratory-based experiments, it is interesting that the Swan bands of C_2 are also important in astrophysical problems. They have been observed in the emission spectra of comets and also in the absorption spectra of stellar atmospheres, including that of our own sun, in which the interior of the star acts as the continuum source.

All the transitions in Table 6.10 are allowed by the electric dipole selection rules of Hund's case (a) given on p. 312–313.

The products of dissociation are indicated on the right of Figure 6.33.

Table 6.9 Configurations and states of the C_2 molecule

Configuration	States
1 $(\sigma_g 1s)^2(\sigma_u^* 1s)^2(\sigma_g 2s)^2(\sigma_u^* 2s)^2(\pi_u 2p)^4$	$X^1\Sigma_g^+$
2 $\ldots\ldots\ldots(\sigma_u^* 2s)^2(\pi_u 2p)^3(\sigma_g 2p)^1$	$a^3\Pi_u$, $A^1\Pi_u$
3 $\ldots\ldots\ldots(\sigma_u^* 2s)^2(\pi_u 2p)^2(\sigma_g 2p)^2$	$b^3\Sigma_g^-$, $E^1\Sigma_g^+$, $^1\Delta_g$
4 $\ldots\ldots\ldots(\sigma_u^* 2s)^1(\pi_u 2p)^4(\sigma_g 2p)^1$	$c^3\Sigma_u^+$, $D^1\Sigma_u^+$
5 $\ldots\ldots\ldots(\sigma_u^* 2s)^1(\pi_u 2p)^3(\sigma_g 2p)^2$	$d^3\Pi_g$, $C^1\Pi_g$

Table 6.10 Electronic transitions observed in C_2

Transition	Names associated	Spectral region/nm	Source of spectrum
$b^3\Sigma_g^- \to a^3\Pi_u$	Ballik–Ramsay	2700–1100	King furnace
$A^1\Pi_u \rightleftarrows X^1\Sigma_g^+$	Phillips	1549–672	Discharges
$d^3\Pi_g \rightleftarrows a^3\Pi_u$	Swan	785–340	Numerous, including carbon arc
$C^1\Pi_g \to A^1\Pi_u$	Deslandres–d'Azambuja	411–339	Discharges, flames
$e^3\Pi_g \to a^3\Pi_u$	Fox–Herzberg	329–237	Discharges
$D^1\Sigma_u^+ \rightleftarrows X^1\Sigma_g^+$	Mulliken	242–231	Discharges, flames
$E^1\Sigma_g^+ \to A^1\Pi_u$	Freymark	222–207	Discharge in acetylene
$f^3\Sigma_g^- \leftarrow a^3\Pi_u$	—	143–137	$\left.\begin{array}{c} \\ \\ \\ \end{array}\right\}$
$g^3\Delta_g \leftarrow a^3\Pi_u$	—	140–137	Flash photolysis of mixture of a
$F^1\Pi_u \leftarrow X^1\Sigma_g^+$	—	135–131	hydrocarbon and an inert gas

It will be remembered from Section 6.1.2.2(i) that the ground configuration $1s^2 2s^2 2p^2$ of the carbon atom gives three terms, ^3P, ^1D and ^1S. The ground term is ^3P, and ^1D and ^1S are successively higher in energy. Figure 6.33 shows that six states of C_2 all dissociate to give two ^3P carbon atoms. Other states give dissociation products involving at least one excited carbon atom.

Most of the potential energy curves of C_2, and of other diatomic molecules, have vibrational energy levels associated with them which converge smoothly towards the dissociation limit. The vibrational term values, $G(v)$, relevant to any electronic state can be expressed, exactly as in equation (5.14), by

$$G(v) = \omega_e(v + \tfrac{1}{2}) - \omega_e x_e(v + \tfrac{1}{2})^2$$
$$+ \omega_e y_e(v + \tfrac{1}{2})^3 + \ldots \qquad (6.121)$$

The vibration wavenumber ω_e, the anharmonic constants $\omega_e x_e$, $\omega_e y_e$, ..., and the equilibrium internuclear distance r_e are all dependent on the electronic structure and therefore are constant only for a particular electronic state.

6.2.4.2 Progressions and sequences

Figure 6.34 shows sets of vibrational energy levels associated with two electronic states between which we shall assume an electronic transition is allowed. The vibrational levels of the upper and lower states are labelled by the quantum numbers v' and v'', respectively. We shall be discussing both absorption and emission processes and it will be assumed, unless stated otherwise, that the lower electronic state is the ground state.

In electronic spectra there is no restriction on the values that Δv can take but, as we shall see in Section 6.2.4.3, the Franck–Condon principle imposes restrictions on the intensities of the transitions.

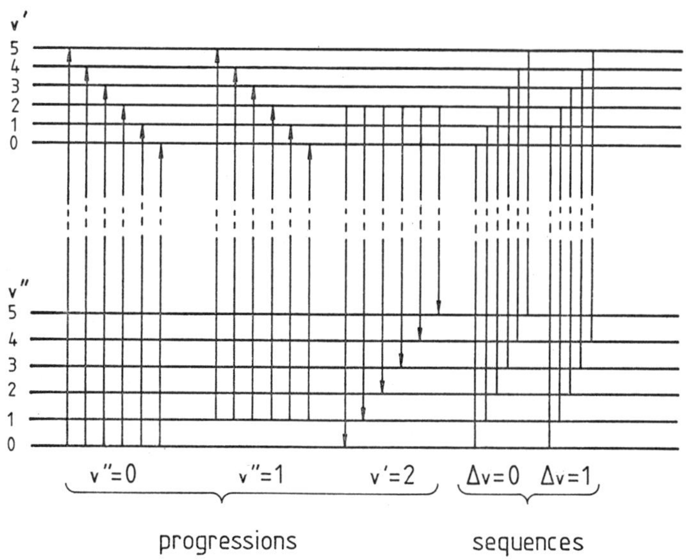

Figure 6.34 Vibrational progressions and sequences in the electronic spectrum of a diatomic molecule

Vibrational transitions accompanying an electronic transition are called vibronic transitions. These vibronic transitions, with their accompanying rotational or, strictly, rovibronic transitions, give rise to bands in the spectrum and the set of bands associated with a single electronic transition is an electronic band system. This terminology is usually adhered to in high resolution electronic spectroscopy but, in low resolution work, particularly in the liquid phase, vibrational structure may not be resolved and the whole band system is often referred to as an electronic band.

Vibronic transitions may be divided conveniently into progressions and sequences.

A progression, as Figure 6.34 shows, involves a series of vibronic transitions with a common lower or upper level. For example, the $\upsilon''=0$ progression members all have the $\upsilon''=0$ level in common.

Apart from the necessity for Franck–Condon intensities of vibronic transitions to be appreciable, it is essential for the initial state of a transition to be sufficiently highly populated for a transition to be observed. Under equilibrium conditions the population N of the υ''th level is related to that of the $\upsilon''=0$ level by

$$N_{\upsilon''}/N_0 = \exp -\{[G(\upsilon'') - G(0)]hc/kT\} \quad (6.122)$$

which follows from the Boltzmann equation [equation (2.6)].

Because of the relatively high population of the $\upsilon''=0$ level, the $\upsilon''=0$ progression is likely to be prominent in the absorption spectrum. In emission the relative populations of the υ' levels depend on the method of excitation. In a low pressure discharge, in which there are not many collisions to provide a channel for vibrational deactivation, the populations may be somewhat random. On the other hand, higher pressure may result in most of the molecules being in the $\upsilon'=0$ state and the $\upsilon'=0$ progression being prominent.

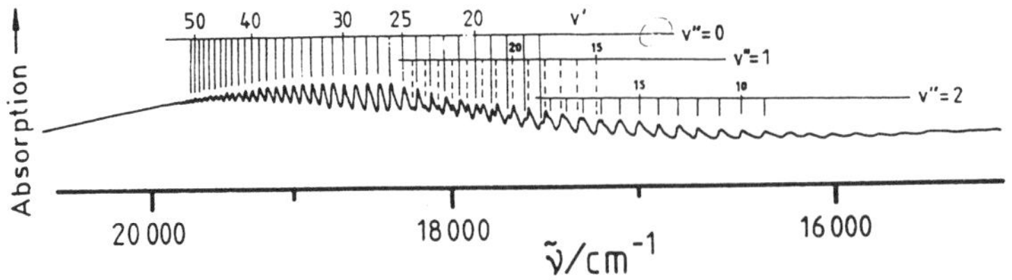

Figure 6.35 Progressions with $\upsilon'' = 0$, 1 and 2 in the $B^3\Pi_{0^+u}$–$X^1\Sigma_g^+$ system of the I_2 molecule

The progression with $\upsilon'' = 1$ may also be observed in absorption but only in a molecule with a vibration wavenumber low enough for the $\upsilon'' = 1$ level to be sufficiently populated. This is the case in, for example, iodine for which $\omega_e'' = 214.50\,\text{cm}^{-1}$. As a result, the $B^3\Pi_{0^+u}$–$X^1\Sigma_g^+$ visible system shows, in absorption at room temperature, not only a $\upsilon'' = 0$ but also a $\upsilon'' = 1$ and a $\upsilon'' = 2$ progression, as shown in Figure 6.35.

A progression with $\upsilon' = 2$, illustrated in Figure 6.34, could be observed only in emission. Its observation could result from a random population of υ' levels or it could be observed on its own under rather special conditions involving monochromatic excitation from $\upsilon'' = 0$ to $\upsilon' = 2$ with no collisions occurring before emission. This kind of excitation may be achieved with radiation involving an intense line in an atomic spectrum, whose wavenumber matches the transition exactly, or with a tunable laser.

If the emission is between states of the same multiplicity it is called fluorescence and, if it is from only one vibrational level of the upper electronic state, it is single vibronic level fluorescence.

A group of transitions with the same value of $\Delta\upsilon$ is referred to as a sequence. Because of the population requirements long sequences are observed mostly in emission. For example, sequences of five or six members are observed in the $C^3\Pi_u$–$B^3\Pi_g$ band system of N_2 in the emission spectrum in the visible and near ultraviolet from a low pressure discharge in nitrogen gas. The vibration wavenumber ω_e is high ($2035.1\,\text{cm}^{-1}$) in the C state and equilibrium population of the vibrational levels is not achieved before emission.

It is clear from Figure 6.34 that progressions and sequences are not mutually exclusive. Each member of a sequence is also a member of two progressions. However, the distinction is useful because of the nature of typical patterns of bands found in a band system. Progression members are generally widely spaced with approximate separations of ω_e' in absorption and ω_e'' in emission. On the other hand, sequence members are closely spaced with approximate separations of $\omega_e' - \omega_e''$.

The general symbolism for indicating a vibronic transition between an upper and lower level with vibrational quantum numbers υ' and υ'', respectively, is υ'–υ'', consistent with the general spectroscopic convention. Hence the electronic transition is labelled 0–0.

6.2.4.3 *Franck–Condon principle*

In 1925, the year before the development of the Schrödinger equation, Franck put forward qualitative arguments to explain the various types of intensity distributions found in vibronic transitions. His conclusions were based on an appreciation of the fact that an electronic transition in a molecule takes place so much more rapidly than a vibrational transition that, in a vibronic transition, the nuclei have very nearly the same position and velocity before and after the transition.

Possible consequences of this are illustrated in Figure 6.36(a), which shows potential curves for the lower state, which is the ground state if we are considering an absorption process, and

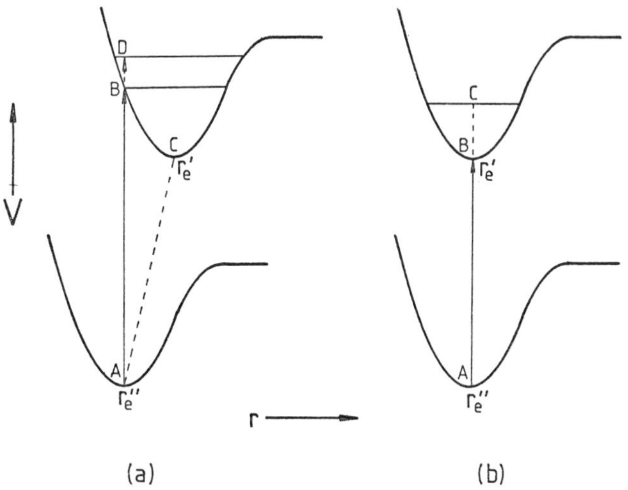

Figure 6.36 Illustration of the semi-classical Franck principle. The vibronic transition B–A is the most probable and is shown for (a) $r_e' > r_e''$ and (b) $r_e' \approx r_e''$

the upper state. The curves have been drawn so that $r_e' > r_e''$. When the lower state is the ground state this is very often the case since the electron promotion involved is often from a bonding orbital to an orbital which is less bonding, or even antibonding, leading to an increase in bond length. For example, in nitrogen, promotion of an electron from the $\sigma_g 2p$ to the $\pi_g^* 2p$ orbital (Figure 6.29) gives two states, $a^1\Pi_g$ and $B^3\Pi_g$, in which r_e is 1.2203 and 1.2126 Å, respectively, considerably increased from 1.0977 Å in the $X^1\Sigma_g^+$ ground state.

In absorption from point A of the ground state in Figure 6.36(a) (zero-point energy can be neglected in considering Franck's semi-classical arguments) the transition will be to point B of the upper state. The requirement that the nuclei have the same position before and after the transition means that the transition is between points which lie on a vertical line in the figure; this means that r remains constant and such a transition is often referred to as a vertical transition. The second requirement, that the nuclei have the same velocity before and after the transition, means that a transition from A, where the nuclei are stationary, must go to B as this is the classical turning-point of a vibration, where the nuclei are also stationary.

A transition from A to C is highly improbable because, although the nuclei are stationary at A and C, there is a large change of r. An A to D transition is also unlikely because, although r is unchanged, the nuclei are in motion at the point D.

Figure 6.36(b) illustrates the case where $r_e' \approx r_e''$. An example of such a transition is the $D^1\Sigma_u^+ - X^1\Sigma_g^+$ Mulliken band system of C_2 (see Table 6.10 and Figure 6.33). The value of r_e is 1.2380 Å in the D state and 1.2425 Å in the X state. Here the most probable transition is from A to B with no vibrational energy in the upper state. The transition from A to C maintains the value of r but the nuclear velocities are increased owing to their having kinetic energy equivalent to the distance BC.

In 1928, Condon treated the intensities of vibronic transitions quantum mechanically.

The intensity of a vibronic transition is proportional to the square of the transition moment $\boldsymbol{R}_{ev}$ which is given by [see equation (2.21)]

$$\boldsymbol{R}_{ev} = \int \psi_{ev}'^{*} \boldsymbol{\mu} \psi_{ev}'' \mathrm{d}\tau_{ev} \tag{6.123}$$

where $\boldsymbol{\mu}$ is the electric dipole moment operator and ψ'_{ev} and ψ''_{ev} are the vibronic wave functions of the upper and lower states, respectively. The integration is over electronic and vibrational coordinates. Assuming that the Born–Oppenheimer approximation (Section 1.3.3) holds, ψ_{ev} can be factorized into $\psi_e \psi_v$. Then equation (6.123) becomes

$$\boldsymbol{R}_{ev} = \iint \psi_e'^* \psi_v' \boldsymbol{\mu} \psi_e'' \psi_v'' \, \mathrm{d}\tau_e \mathrm{d}r \qquad (6.124)$$

First we integrate over electron coordinates giving

$$\boldsymbol{R}_{ev} = \int \psi_v' \boldsymbol{R}_e \psi_v'' \, \mathrm{d}r \qquad (6.125)$$

where

$$\boldsymbol{R}_e = \int \psi_e' \boldsymbol{\mu} \psi_e'' \, \mathrm{d}\tau_e \qquad (6.126)$$

and is the electronic transition moment. Our ability to do this integration is a consequence of the Born–Oppenheimer approximation, which assumes that the nuclei can be regarded as stationary in relation to the much more fast-moving electrons. This also allows us to take $\boldsymbol{R}_e$ out of the integral in equation (6.125) and regard it as a constant to an approximation which is good enough for our purposes here. Hence we have

$$\boldsymbol{R}_{ev} = \boldsymbol{R}_e \int \psi_v' \psi_v'' \, \mathrm{d}r \qquad (6.127)$$

The quantity $\int \psi_v' \psi_v'' \, \mathrm{d}r$ is the vibrational overlap integral, and is a measure of the degree to which the two vibrational wave functions overlap; its square is known as the Franck–Condon factor.

In carrying out the integration, the requirement that r remains constant during the transition is necessarily taken into account.

The classical turning-point of a vibration, where nuclear velocities are zero, is replaced in quantum mechanics by a maximum, or minimum, in ψ_v near to this turning-point. As is illustrated in Figure 1.22, the larger is v the closer is the maximum, or minimum, in ψ_v to the classical turning-point.

Figure 6.37 illustrates a particular case where the maximum of the $v' = 4$ wave function near to the classical turning-point is vertically above that of the $v'' = 0$ wave function. The maximum contribution to the vibrational overlap integral is indicated by the solid line but appreciable contributions extend to values of r within the dashed lines. Clearly overlap integrals for v' close to four are also appreciable and give an intensity distribution in the $v'' = 0$ progression like that in Figure 6.38(b).

If $r_e' \gg r_e''$ there may be appreciable intensity involving the continuum of vibrational levels above the dissociation limit. This results in a $v'' = 0$ progression like that in Figure 6.38(c) where the intensity maximum is at a high value of v; or it may be in the continuum. An example of this is the $B^3\Pi_{0^+u}$–$X^1\Sigma_g^+$ transition of iodine. In the B and X states r_e is 3.025 and 2.666 Å, respectively, leading to the broad intensity maximum close to the continuum, as observed in Figure 6.35.

Figure 6.38(a) shows the intensity maximum at $v' = 0$ for the case when $r_e' \approx r_e''$. The intensity usually falls off rapidly in such a case.

Occasionally we encounter a case where $r_e' < r_e''$. When the lower state is the ground state this is unusual, but it can happen when the electron promotion is from an antibonding to a non-bonding or bonding orbital. The situation is more likely to arise in a transition between two excited states. Qualitatively the situation is similar to that in Figure 6.37 except that the upper potential curve is displaced to low r so that the right-hand maximum of, for example, $v' = 4$ is above the $v'' = 0$ maximum. The result is, again, an intensity distribution like that in Figure 6.38(b) so that an observation of a long $v'' = 0$ progression with an intensity maximum at $v' > 0$ indicates qualitatively an appreciable change in r_e from the lower to the upper state but does not indicate the sign of the change. This would be true, even quantitatively, if the

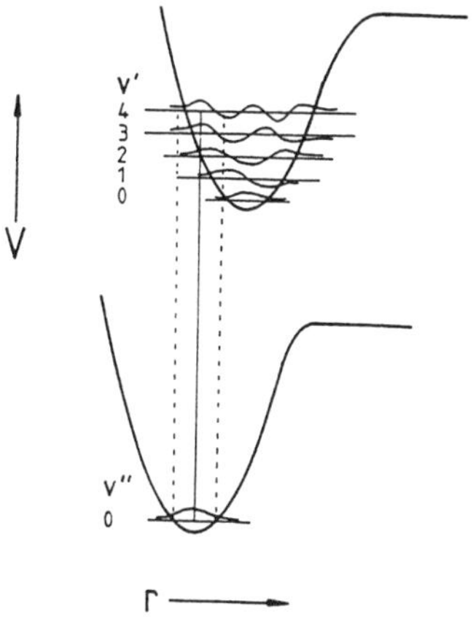

Figure 6.37 Illustration of the quantum mechanical Franck–Condon principle for a case in which $r_e' > r_e''$ and the 4–0 vibronic transition is the most probable

molecule behaved as a harmonic oscillator but, owing to anharmonicity, the intensity distribution along the progression is slightly different for $r_e' > r_e''$ than for $r_e' < r_e''$.

In the case where $r_e' > r_e''$ there is, when anharmonicity is taken into account, a relatively steep part of the excited state potential curve above $\upsilon'' = 0$ giving a relatively broad maximum in the progression intensity. On the other hand, for $r_e' < r_e''$, there is a shallower part of the excited state potential curve above $\upsilon'' = 0$ and a sharper intensity maximum results.

Accurate intensity measurements have been made in many cases and calculations of r_e'–r_e'' made, including the effects of anharmonicity and even allowing for breakdown of the Born–Oppenheimer approximation.

In emission spectra, the possible intensity distributions along a $\upsilon' = 0$ progression can also be divided into the three categories illustrated in Figure 6.38. The $\upsilon'' = 0$ and $\upsilon' = 0$ progressions have an approximate mirror image relationship about the 0–0 band, as shown in Figure 6.39 for the case where $r_e' > r_e''$. The relationship is only approximate because of the differences in ω_e and in the anharmonicity in the two electronic states.

The $\upsilon' = 0$ progression in emission, in the case where $r_e' \gg r_e''$, is the approximate mirror image of the $\upsilon'' = 0$ progression in absorption shown in Figure 6.38(c), but there may be a large gap between them, in regions of low values of υ' or υ'', due to very small vibrational overlap integrals. In cases like this, where the intensity maximum is at a high value of υ and early progression members are not observed, it may be difficult to determine the vibrational numbering exactly. This was the case in the $B^3\Pi_{0+u}$–$X^1\Sigma_g^+$ system of iodine and it was not until 1965 that the numbering was settled unambiguously.[69]

For a progression observed in emission with $\upsilon' > 0$ and $r_e' > r_e''$ (or $r_e' < r_e''$), there may be *two* intensity maxima. This is demonstrated in Figure 6.40 for a $\upsilon' = 6$ progression and $r_e' > r_e''$. The two maxima in the progression are, in this example, at $\upsilon'' = 4$ and 10. If $r_e' \gg r_e''$ the maximum towards high υ'' may be in the continuum. In the case where $r_e'' \approx r_e'$ the two maxima coincide.

In detail, the intensity distribution along a progression, particularly when the initial state is

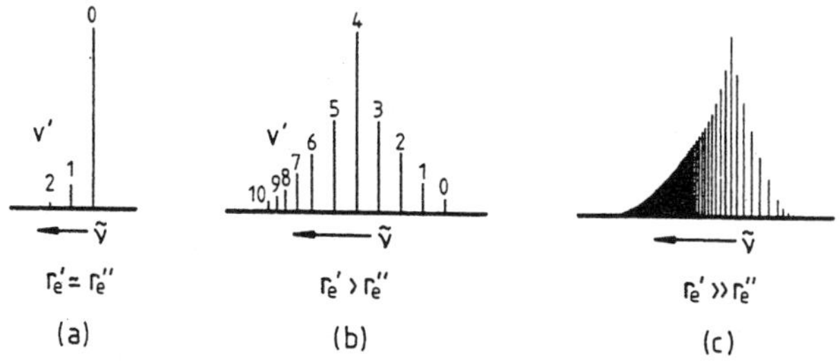

Figure 6.38 Typical vibrational progression intensity distributions

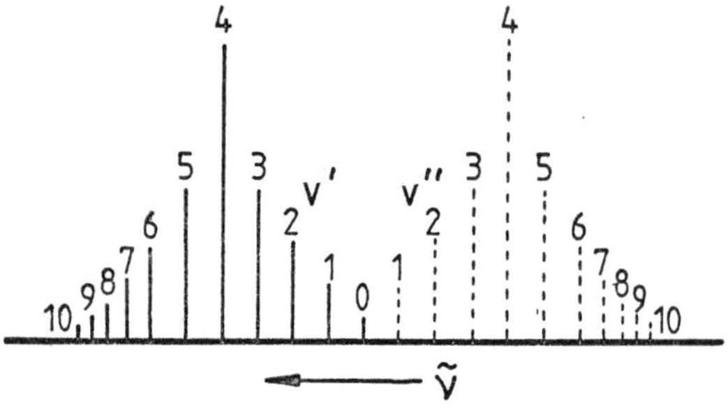

Figure 6.39 Approximate mirror image relationship between the absorption (solid lines) and emission (dashed lines) spectra of a diatomic molecule

a highly excited vibrational state, may not be so straightforward. Generally, there may be two reasons for this. The first is that subsidiary maxima and minima, other than those near the classical turning-points (see Figure 1.22), have been neglected whereas vibrational overlap involving some of them may be appreciable. The second is that anharmonicity has also been neglected and this causes the undulations in the vibrational wave function to be closer together at small values of r and further apart for large values.

An example of the irregular intensity distribution which may result is in the B–X emission of iodine excited by 546.08 nm mercury radiation and in which the initial state is the $\upsilon' = 25$ level of the B state. There are many intensity undulations along the $\upsilon' = 25$ progression.[69]

6.2.4.4 *Deslandres tables*

The illustration of various types of vibronic transitions in Figure 6.34 suggests that we can use the method of combination differences to obtain the separations of vibrational levels from observed transition wavenumbers. This method was introduced in Section 5.1.3.1 and was applied to obtaining rotational constants for two combining vibrational states. The method works on the simple principle that, if two transitions have an upper level in common,

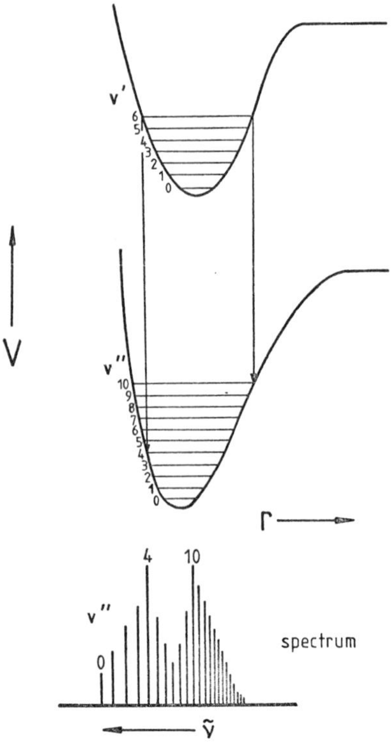

Figure 6.40 Two maxima may be observed in a progression in emission from a single vibronic level with $\upsilon' > 0$

their wavenumber difference is a function of lower state parameters only, and vice versa if they have a lower level in common.

In using the combination difference method to obtain vibrational parameters ω_e, $\omega_e x_e$, etc., for two electronic states between which vibronic transitions are observed, the first step is to organize all the vibronic transition wavenumbers into a Deslandres table. An example is shown in Table 6.11 for the $A^1\Pi$–$X^1\Sigma^+$ system of carbon monoxide. The electronic transition results from an electron promotion which, because of the similarity of the two nuclei, can be described approximately in terms of the MO diagram for homonuclear diatomic molecules in figure 6.29. Carbon monoxide is isoelectronic with the nitrogen molecule so the lowest energy electron promotion is from $\sigma 2p$ to π^*2p (the g and u subscripts do not apply to

Table 6.11 Deslandres table for the $A^1\Pi$–$X^1\Sigma^+$ system of carbon monoxide[a]

	v''												
v'	0		1		2		3		4		5		6
0	64 758	(2145)	62 613	(2117)	60 496	(2092)	58 404	(2063)	56 341	(2037)	54 304		–
	(1476)		(1485)				(1487)		(1486)		(1487)		
1	66 234	(2136)	64 098		–		59 891	(2064)	57 827	(2036)	55 791	(2010)	53 781
	(1448)		(1441)				(1444)				(1443)		(1443)
2	67 682	(2143)	65 539	(2115)	63 424	(2089)	61 335		–		57 234	(2010)	55 224
	(1407)		(1413)		(1414)						(1410)		
3	69 089	(2137)	66 952	(2114)	64 838		–		60 683	(2039)	58 644		–
	(1378)		(1382)		(1370)				(1379)				
4	70 467	(2133)	68 334	(2126)	66 208	(2085)	64 123	(2061)	62 062		–		58 011
	(1341)		(1338)		(1350)		(1343)						(1340)
5	71 808	(2136)	69 672	(2114)	67 558	(2092)	65 466		–		61 365	(2014)	59 351
	(1307)		(1305)		(1303)		(1299)				(1307)		
6	73 115	(2138)	70 977	(2116)	68 861	(2096)	66 765	(2053)	64 712	(2040)	62 672		–

[a]The table, truncated at $v' = v'' = 6$, has been constructed from data in Rosen (1970). *Spectroscopic Data Relative to Diatomic Molecules*, referred to in the Bibliography at the end of the chapter. Units are cm^{-1} throughout. Measurements are of band heads, not band origins.

heteronuclear diatomic molecules). The promotion gives two states, $A^1\Pi$ and $a^3\Pi$. The $A^3\Pi$–$X^1\Sigma^+$ band system lies in the far ultraviolet region of the spectrum, with the 0–0 band at 154.4 nm.

In the Deslandres table in Table 6.11 all the vibronic transition wavenumbers correspond to band heads (regions of maximum intensity in the rotational fine structure—see Section 6.2.5) rather than to vibronic band origins. A band origin corresponds to the position of the $J' = 0$–$J'' = 0$ transition (see Figure 6.48) between two states in which there is no rotation. If either a rotational analysis has not been carried out or the resolution is too low for rotational fine structure to be resolved, band head measurements have to be used and a constant head–origin separation is assumed for all bands.

In Table 6.11, all the transition wavenumbers have been arranged in rows and columns so that the differences between wavenumbers in adjacent columns correspond to vibrational level separations in the lower (ground) electronic state and the differences between adjacent rows to separations in the upper electronic state. These differences are given in parentheses. The variations of the differences, for example between the first two columns, are a result of uncertainties in the experimental measurements or differences in head–origin separations.

From the table a series of averaged values of vibrational term value differences, $G(v+1)-G(v)$, can be obtained for both electronic states and, from equation (5.15), values of ω_e, $\omega_e x_e$, etc., for both states. For example, in the $A^1\Pi$ and $X^1\Sigma^+$ states of CO, ω_e is 1518.2 and 2169.8 cm^{-1}, respectively. The large decrease in ω_e in the A state is a consequence of promoting an electron from a bonding to an antibonding orbital, greatly reducing the force constant.

6.2.4.5 *Dissociation energies*

If a sufficient number of vibrational term values are known in any electronic state, the dissociation energy, D_0, can be obtained from a Birge–Sponer extrapolation, as discussed in Section 5.1.2 and illustrated in Figure 5.5. The possible inaccuracies of the method were made clear and it was stressed that these are reduced by obtaining term values near to the dissociation

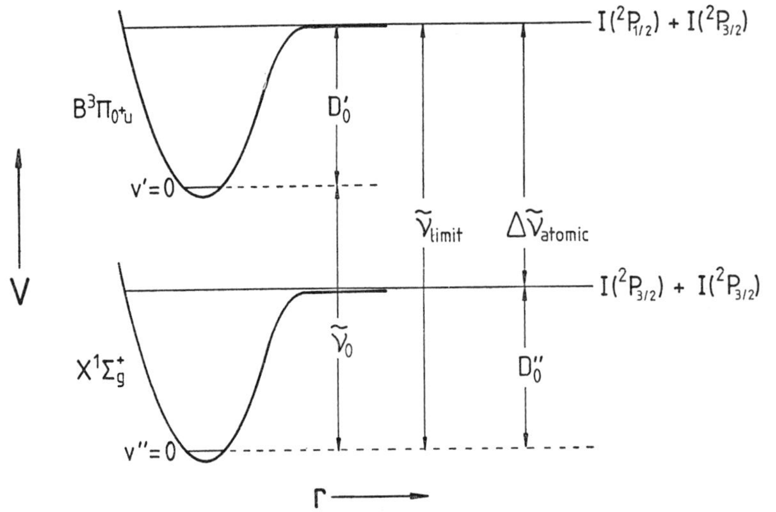

Figure 6.41 Illustration of how the dissociation energies D_0'' and D_0' can be obtained from the value of $\tilde{\nu}_{\text{limit}}$, the wavenumber of the onset of a continuum in a progression

limit. Whether this can be done depends very much on the relative dispositions of the various potential curves in a particular molecule and whether electronic transitions between them are allowed. How many ground state vibrational term values can be obtained from an emission spectrum is determined by the Franck–Condon principle. If $r_e' \approx r_e''$ then progressions in emission are very short and few term values result, but if r_e' is very different from r_e'', as in the $\text{A}^1\Pi$–$\text{X}^1\Sigma^+$ system of carbon monoxide, then long progressions are observed in emission and a more accurate value of D_0'' can be obtained.

To obtain an accurate value of D_0'' for the ground electronic state is virtually impossible by vibrational spectroscopy because of the problems of rapidly decreasing population with increasing v. In fact, most determinations are made from electronic emission spectra from one, or more, excited electronic states to the ground state.

Obtaining D_0' for an excited electronic state depends on observing progressions either in absorption, usually from the ground state, or in emission from higher excited states. Again, the length of a progression limits the accuracy of the dissociation energy.

If the values of r_e in the combining states are very different the dissociation limit of a progression may be observed directly as an onset of diffuseness. However, the onset is not always particularly sharp; this is the case in the $\text{B}^3\Pi_{0^+u}$–$\text{X}^1\Sigma_g^+$ absorption system of iodine shown in Figure 6.35, where the wavenumber $\tilde{\nu}_{\text{limit}}$ illustrated in Figure 6.41 is obtained more accurately by extrapolation than by direct observation.

Figure 6.41 shows that

$$\tilde{\nu}_{\text{limit}} = D_0' + \tilde{\nu}_0 = D_0'' + \Delta\tilde{\nu}_{\text{atomic}} \tag{6.128}$$

Hence D_0' can be obtained from $\tilde{\nu}_{\text{limit}}$ if $\tilde{\nu}_0$, the wavenumber of the 0–0 band, is known. Figure 6.35 shows that extrapolation may be required to obtain $\tilde{\nu}_0$, limiting the accuracy of D_0'.

Equation (6.128) also shows that D_0'' may be obtained from $\tilde{\nu}_{\text{limit}}$ since $\Delta\tilde{\nu}_{\text{atomic}}$ is the wavenumber difference between two atomic states, the ground state, $^2P_{\frac{3}{2}}$, and first excited state, $^2P_{\frac{1}{2}}$, of the iodine atom, known accurately from the atomic spectrum. Hence the accuracy of D_0'' is limited only by that of $\tilde{\nu}_{\text{limit}}$.

D'_e and D''_e, the dissociation energies relative to the minima in the potential curves, are obtained from D'_0 and D''_0 by

$$D_e = D_0 + G(0) \qquad (6.129)$$

where $G(0)$ is the zero-point term value given by equation (5.19).

The absorption spectrum of the hydrogen molecule in the 80–90 nm region in the far ultraviolet shows[70] a much sharper onset of diffuseness due to dissociation than that of iodine. The dissociation products are $H(n=1)+H(n=2)$ and, although three singlet states of the molecule, labelled $B^1\Sigma_u^+$, $C^1\Pi_u$ and $B'^1\Sigma_u^+$, all give these products, the observed onset of the absorption continuum involves only the B′ state. For $^1H_2$ the onset is observed in the range $118\,376.6 \pm 1.0\,\text{cm}^{-1}$ for $J''=0$ in the ground electronic state (the rotational constant B is large in H_2 and well separated continuum edges are observed for different values of J''). The uncertainty in the wavenumber is necessarily large compared with the accuracy of measurement because of a discrete absorption line overlapping the continuum edge. Measurements of continuum edges for other values of J'', followed by extrapolation, give

$$\tilde{\nu}_{\text{limit}} = 118\,376.4 \pm 1\,\text{cm}^{-1} \qquad (6.130)$$

The $n=2$ term of the hydrogen atom gives $^2P_{\frac{3}{2}}$, $^2S_{\frac{1}{2}}$, and $^2P_{\frac{1}{2}}$ states (Section 6.1.4) but these all lie within $0.365\,\text{cm}^{-1}$ (Figure 6.16), which is small compared with the uncertainty of $\tilde{\nu}_{\text{limit}}$. The separation of the $n=2$ and $n=1$ terms is

$$\Delta\tilde{\nu}_{\text{atomic}} = 82\,259.10 \pm 0.15\,\text{cm}^{-1} \qquad (6.131)$$

the uncertainty being due to the spread of $n=2$ states.

Combining equation (6.128) with the values for $\tilde{\nu}_{\text{limit}}$ and $\Delta\tilde{\nu}_{\text{atomic}}$ in equations (6.130)–(6.131) gives

$$D''_0 = 36\,117.3 \pm 1.0\,\text{cm}^{-1} \qquad (6.132)$$

This result is especially important in its relation to theory. Quantum mechanical calculations in molecules, as in atoms, become extremely difficult when the molecule has more than one electron. Calculations by Kolos and Wolniewicz† on the hydrogen molecule are a milestone in this field and the agreement between their calculated value for D''_0 of $36\,117.4\,\text{cm}^{-1}$ and the experimental value in equation (6.132) is an important vindication of both techniques. Indeed, this result represents a particular theoretical triumph since it is so accurate that it prompted a reappraisal of previous experimental measurements.

6.2.4.6 *Repulsive states, continuous spectra and predissociation*

The ground configuration of the He_2 molecule is, according to Figure 6.29, $(\sigma_g 1s)^2(\sigma_u^* 1s)^2$ and is expected to be unstable because of the cancelling of the bonding character of a $\sigma_g 1s$ by the antibonding character of a $\sigma_u^* 1s$ orbital. The potential energy curve for the resulting $X^1\Sigma_g^+$ state shows no minimum but the potential energy decreases smoothly as r increases, as shown in Figure 6.42(a). Such a state is known as a repulsive state since the atoms repel each other. In this type of state, either there are no discrete vibrational levels or there may be a few in a very shallow minimum—see Section 8.1.8.9. All, or most, of the vibrational states form a continuum of levels.

Promotion of an electron in He_2 from $\sigma_u^* 1s$ to a bonding orbital produces bound states of the molecule of which several have been characterized in emission spectroscopy. For example, the configuration $(\sigma_g 1s)^2(\sigma_u^* 1s)^1(\sigma_g 2s)^1$ gives rise to the $A^1\Sigma_u^+$ and $a^3\Sigma_u^+$ bound states. Figure 6.42(a) shows the form of the potential curve for the $A^1\Sigma_u^+$ state. The A–X transition is allowed and gives rise to an intense continuum

†See Kolos, W. and Wolniewicz, L. (1968). J. *Chem. Phys.*, **49**, 404, and references cited therein.

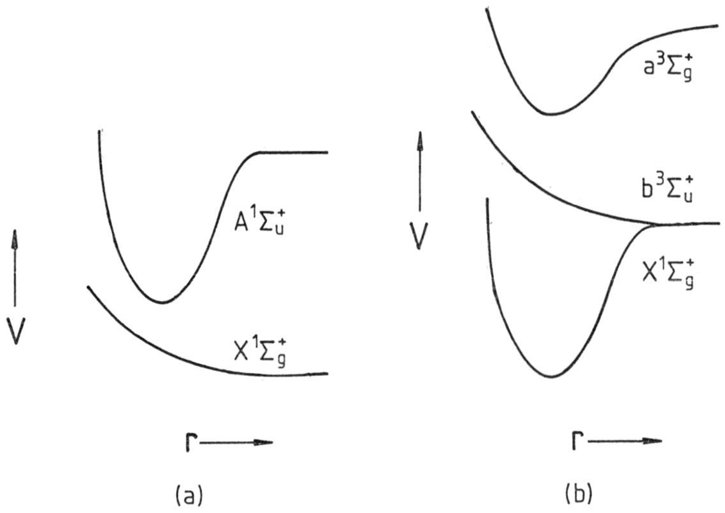

Figure 6.42 (a) The repulsive ground state, $X^1\Sigma_g^+$, and an excited state, $A^1\Sigma_u^+$, of the He_2 molecule. (b) Two bound states, $X^1\Sigma_g^+$ and $a^3\Sigma_g^+$, and one repulsive excited state, $b^3\Sigma_u^+$, of the H_2 molecule

in emission between 60 and 100 nm. This is used as a far ultraviolet continuum source (see Section 3.9), as are the corresponding continua from other inert gas diatomic molecules.

Another example of a continuous emission spectrum is that from a discharge in molecular hydrogen. It covers the range 160–500 nm and is used as a near ultraviolet continuum source (see Section 3.8.1). The transition involved is from the bound $a^3\Sigma_g^+$ to the repulsive $b^3\Sigma_u^+$ state, shown in figure 6.42(b).

The $b^3\Sigma_u^+$ state arises from the excited configuration $(\sigma_g 1s)^1(\sigma_u^* 1s)^1$ and is a repulsive state. The dissociation products are two ground state (1^2S) hydrogen atoms, the same as for the $X^1\Sigma_g^+$ ground state.

The $a^3\Sigma_g^+$ state arises from the configuration $(\sigma_g 1s)^1(\sigma_g 2s)^1$.

There are cases where two potential energy curves, in the absence of any perturbation between them, would cross; this is shown for the dashed curves AD and CB in Figure 6.43(a), which cross at the point X. If the two electronic states are of the same symmetry there is a homogeneous perturbation which results in an avoided crossing and the potential curves observed are AB and CD. This is an example of the non-crossing rule. There is such an avoided crossing between the $C^1\Pi_g$ and $C'^1\Pi_g$ states of C_2 (Figure 6.33). The result is that vibrational levels of both states are perturbed in the region of the avoided crossing.

If the two states are not of the same symmetry but differ by one in Λ, the curves cross and remain as AD and CB but there is a heterogeneous perturbation in the region of crossing where perturbations occur in the rotational, but not the vibrational, levels of each electronic state.

Figure 6.43(b) illustrates the case where the second potential curve CD is a repulsive curve. Once again the perturbation may be homogeneous or heterogeneous and the effect is that there is a non-zero probability of the molecule in the electronic state defined by the curve AB transferring to CD when it is in the region of crossing. The effect of this is to broaden the rotational levels to such an extent that no rovibronic transitions are observed and to broaden the vibrational levels so that vibronic transitions appear diffuse. The effect is called predissociation since, if a molecule is in the electronic state defined by the curve AB, it may

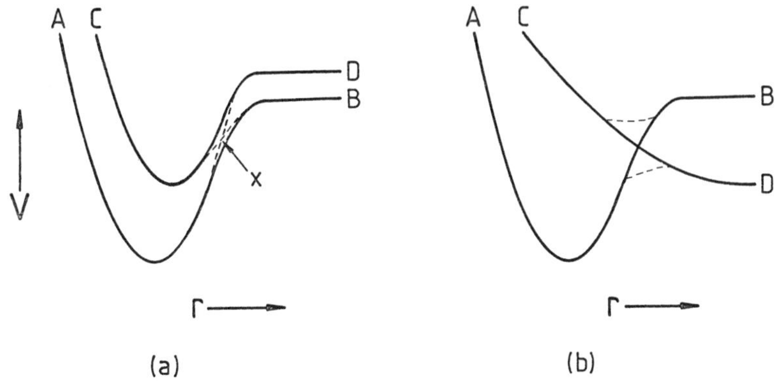

Figure 6.43 (a) A homogeneous perturbation between curves AD and CB results in an avoided crossing (solid lines); a heterogeneous perturbation leaves the curves unaltered (dashed lines). (b) Crossing of two potential curves of which one, CD, is repulsive leads to predissociation

dissociate at a lower energy when the repulsive curve CD crosses it than it would otherwise.

6.2.4.7 Ion-pair states

Figure 6.41 shows that I_2, in its ground electronic state, dissociates into neutral iodine atoms. In fact, all ground state diatomic molecules dissociate into neutral atoms—even those heteronuclear diatomics, such as HCl, which have a strongly polar bond.

If we consider a molecule AB the difference $I(\mathrm{A})-E(\mathrm{B})$, where $I(\mathrm{A})$ is the ionization energy of atom A and $E(\mathrm{B})$ is the electron affinity of atom B, must be small to give the greatest possibility of the ground state dissociating into ions. For example,

for H_2 $I(\mathrm{H})-E(\mathrm{H}) = 12.85\,\mathrm{eV}$
for HCl $I(\mathrm{H})-E(\mathrm{Cl}) = 9.88\,\mathrm{eV}$

both values being sufficiently large that both H_2 and HCl dissociate into atoms.

However, there are many molecules which dissociate into ions when they are in some excited electronic states. Such states are called ion-pair states.

The potential energy curve for an ion-pair state is qualitatively like that for any bound state but quantitatively it is completely different.

For two ions of charge e a distance r apart, the potential $V(r)$ is coulombic:

$$V(r) = -\frac{e^2}{4\pi e_0 r} \tag{6.132a}$$

but, when r is small, the electron clouds of the ions interpenetrate. This can be taken account of by adding a negative exponential term, giving

$$V(r) = -\frac{e^2}{4\pi\varepsilon_0 r} + B\exp(-\frac{r}{\rho}) \tag{6.132b}$$

where B and ρ are constants.

Figure 6.44(a) shows a typical potential for a molecule AB which dissociates, in its ground state, into neutral atoms and also for an excited ion-pair state which dissociates into ions. The ion-pair potential shows behaviour at large values of r typical of a coulombic potential. The curve becomes asymptotic to the energy of $A^+ + B^-$ only at very large inter-ion distances, beyond the extent of the diagram and long after the ground state has become asymptotic to the energy of A + B. The figure also shows the energy difference $I(\mathrm{A})-E(\mathrm{B})$.

There is another important difference between the two types of potential energy curves. That of an ion-pair state supports an infinite number of vibrational levels whereas

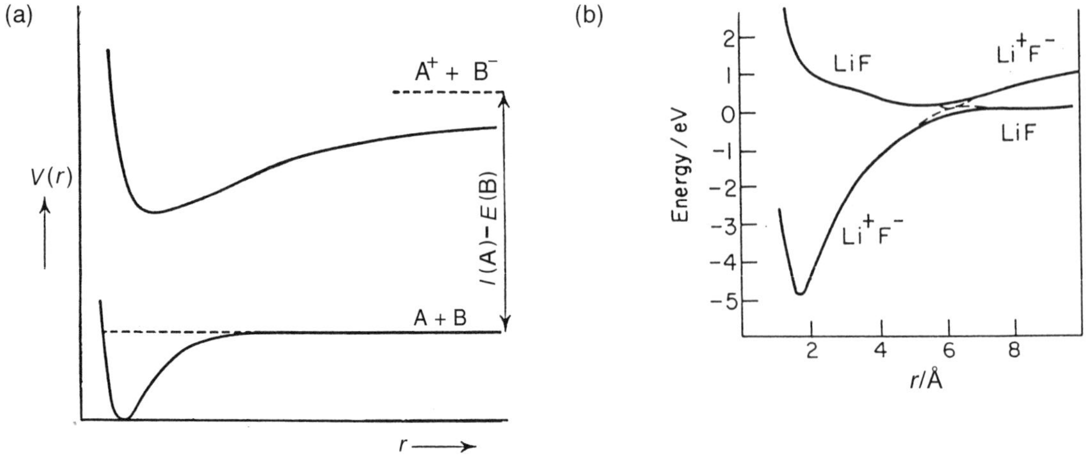

Figure 6.44 (a) Ground state potential and an ion-pair excited state potential of a diatomic molecule AB. (b) Ground state ion-pair potential of LiF undergoing intersystem crossing with that of a low-lying excited repulsive state potential

states dissociating into neutral atoms support only a finite number, as illustrated by the intercept on the horizontal axis in the Birge–Sponer extrapolation in Figure 5.5.

The reason why an ion-pair potential supports an infinite number of vibrational levels is the same as the reason why there are an infinite number of electronic energy levels below the ionization limit in atomic hydrogen, as shown in Figure 1.5, or, indeed, in any atom or molecule. In both cases the potential is dominated, at large inter-particle distances, by a coulombic force of attraction whereas this is not the case for dissociation of a molecule to give neutral atoms.

For those molecules for which $I(\mathrm{A})-E(\mathrm{B})$ is small, the ground state potential is, at least in the region of the minimum, that of an ion-pair state. Such examples are CsCl, NaCl and LiF for which $I(\mathrm{A})-E(\mathrm{B})$ is 0.28, 1.42 and 1.94 eV, respectively. Even in these cases, however, the ground state still dissociates into neutral atoms.

The reason for this is illustrated, for the case of LiF, in Figure 6.44(b). The molecule is highly polar and, for the ground state, there is a typical ion-pair potential, showing a very slow rise from the minimum to the production of the ions Li^+ and F^- at a large inter-ion distance. At $r \approx 6$ Å, however, there is an intersection with a repulsive potential resulting from the promotion of an electron from the bonding σ orbital to the lowest antibonding σ^* orbital. Because both the ground state and this low-lying excited state are $^1\Sigma^+$ states, the non-crossing rule applies and, as shown in the figure, LiF in its ground state dissociates into Li and F.

The molecules CsCl and NaCl, and other molecules with small values of $I(\mathrm{A})-E(\mathrm{B})$, dissociate in their ground states into neutral atoms for the same reason as LiF.

6.2.5 Rotational fine structure

6.2.5.1 $^1\Sigma$–$^1\Sigma$ *electronic and vibronic transitions*

The rotational fine structure accompanying electronic and vibronic transitions of the type $^1\Sigma$–$^1\Sigma$ is simpler than that found with all other types of transitions. The reason for this is that there is no angular momentum in a $^1\Sigma$ state due to the orbital or spin motions of the electrons and the only angular momentum is that due to overall rotation about an axis perpendicular to the internuclear axis and through the centre of gravity [Figure 1.12(a)]. The rotational angular

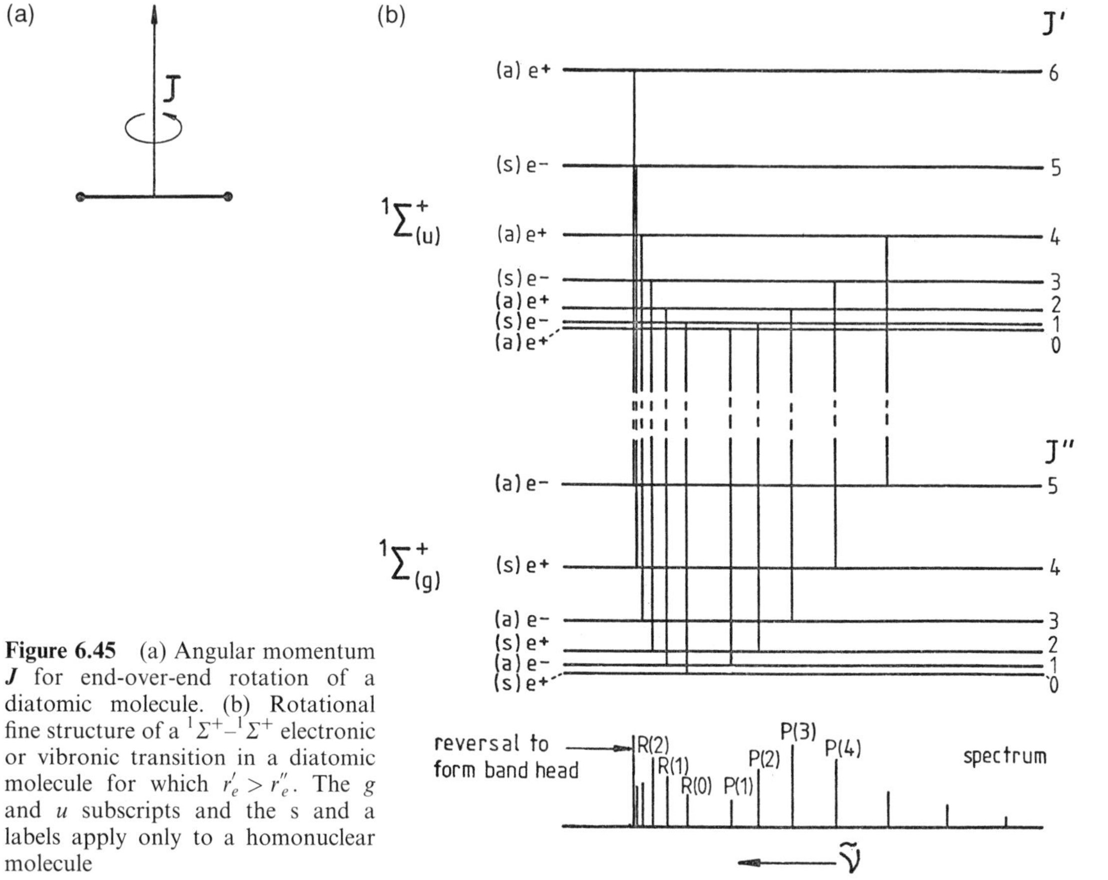

Figure 6.45 (a) Angular momentum $\boldsymbol{J}$ for end-over-end rotation of a diatomic molecule. (b) Rotational fine structure of a $^1\Sigma^+$–$^1\Sigma^+$ electronic or vibronic transition in a diatomic molecule for which $r_e' > r_e''$. The g and u subscripts and the s and a labels apply only to a homonuclear molecule

momentum vector $\boldsymbol{J}$ points along the axis of rotation, as shown in Figure 6.45(a).

It should be noted here that $\boldsymbol{J}$ is used consistently throughout atomic and molecular spectroscopy to represent the total angular momentum, excluding nuclear spin, of the atom or molecule. Its use in $^1\Sigma$ states of diatomic molecules is consistent with this convention since the *only* angular momentum is due to overall rotation.

The rotational term values for any $^1\Sigma$ electronic state are given by

$$F_\upsilon(J) = B_\upsilon J(J+1) - D_\upsilon J^2(J+1)^2 + H_\upsilon J^3(J+1)^3 + \ldots \quad (6.133)$$

where B_υ is the rotational constant [equations (4.18) and (4.19)], D_υ and H_υ are centrifugal distortion constants and the υ subscripts indicate the vibrational dependence of these parameters. This equation applies to any $^1\Sigma$ excited electronic state just as it applies to a $^1\Sigma$ ground electronic state in equation (4.42).

The variations of B_υ and D_υ with υ are given by equations (4.54) and (4.55) and the quantities B_e, α, γ, . . ., D_e, β, . . . are all characteristic of a particular electronic state.

Figure 6.45(b) shows the two stacks of rotational levels of the lower and upper electronic (or vibronic) states of a $^1\Sigma^+$–$^1\Sigma^+$ transition.

The rotational levels are labelled + or − and are said to be of even or odd parity, respectively. The label indicates the symmetry or antisymmetry of the total wave function $\psi_e \psi_v \psi_r$ to inversion. The parity of ψ_v is always even so that, for a $^1\Sigma^+$ vibronic state, of even parity, the parity of the rotational levels is simply that of ψ_r. The parity of ψ_r is $(-1)^J$ and this is indicated in Figure 6.45(b).

It follows that the parity of the rotational levels is $(-1)^{J+1}$ for a $^1\Sigma^-$ vibronic state.

The selection rules governing the rotational transitions for a $^1\Sigma^+$–$^1\Sigma^+$ or $^1\Sigma^-$–$^1\Sigma^-$ electronic or vibronic transition in a heteronuclear diatomic molecule are $\Delta J = \pm 1$ and[†]

$$+ \longleftrightarrow -, \quad + \not\longleftrightarrow +, \quad - \not\longleftrightarrow -$$

or, using the alternative 'e' and 'f' labels discussed in Sections 4.2.6.2 and 5.2.5.1(i),

$$\begin{array}{llll} e \longleftrightarrow f, & e \not\longleftrightarrow e, & f \not\longleftrightarrow f & \text{for } \Delta J = 0 \\ e \not\longleftrightarrow f, & e \longleftrightarrow e, & f \longleftrightarrow f & \text{for } \Delta J = \pm 1 \end{array}$$

These apply also to transitions in homonuclear diatomics for which a g or u subscript must be added to the term symbol.

The parity or e–f selection rules apply generally to electronic and vibronic transitions and we have seen in Sections 4.2.6.2 and 5.2.5.1(i) that they also apply to rotational and rovibrational transitions.

In fact, however, these selection rules are superfluous in all $^1\Sigma$–$^1\Sigma$ transitions. Figure 6.45(b) shows that transitions with $\Delta J = \pm 1$ necessarily obey the parity rules and that, for example, transitions with $\Delta J = 0$ would not.

If the lower and upper states in Figure 6.45(b) are the ground and an excited electronic state respectively then, for reasons discussed in Section 6.2.4.3, it is likely that $r' > r''$ and, therefore, that $B' < B''$. This means that the rotational levels in the excited state diverge with increasing J more slowly than in the ground state. The figure has been drawn for such a case and shows that the resulting spectrum is very asymmetric about the band centre. There is a strong convergence to form a band head, due to reversal of the R branch ($\Delta J = \pm 1$), and a divergence in the P branch ($\Delta J = -1$). The band is said to be degraded, or shaded. In this case it is degraded to high wavelength, or to the red; if the P branch forms a head it is degraded to the blue.

The wavenumbers of R-branch transitions, $\tilde{\nu}(R)$, are given by differences of term values of the type in equation (6.120):

$$\tilde{\nu}(R) = [T' + G'(v)] - [T'' + G''(v)] + B'(J+1)(J+2) - B''J(J+1) \quad (6.134)$$

Differentiating $\tilde{\nu}(R)$ with respect to J and putting the result equal to zero gives J_{rev}, the value of J at the point of reversal, as

$$J_{\text{rev}} = (B'' - 3B')/2(B' - B'') \quad (6.135)$$

Similarly, for P-branch transitions,

$$\tilde{\nu}(P) = (T' + G'(v)] - [T'' + G''(v)] + B'(J-1)J - B''J(J+1) \quad (6.136)$$

and, in cases where $B' > B''$, reversal of the P branch occurs at

$$J_{\text{rev}} = (B'' + B')/2(B' - B'') \quad (6.137)$$

Figure 6.45(b) should be contrasted with Figures 5.8(a) and 5.9 for a vibration–rotation band of a diatomic molecule which involves a Σ^+–Σ^+ vibrational transition. Whereas the change in the rotational constant B from one vibrational state to another is typically small, depending as it does only on α, the change from one electronic state to another is typically much larger.

Figure 6.46 shows the rotational fine structure of the $A^1\Sigma^+$–$X^1\Sigma^+$ electronic transition of the short-lived molecule CuH. This can be observed in absorption at 428 nm by heating metallic copper in hydrogen gas in a high temperature furnace.

There is strong degradation to the red due to the increase of r_e from 1.463 Å in the $X^1\Sigma^+$ state to 1.572 Å in the $A^1\Sigma^+$ state. Since B_0'' and B_0' are 7.813 and 6.743 cm^{-1}, respectively, it follows,

[†]Note that the rotational parity selection rules are the opposite of the selection rules concerned with the + or − property of electronic states.

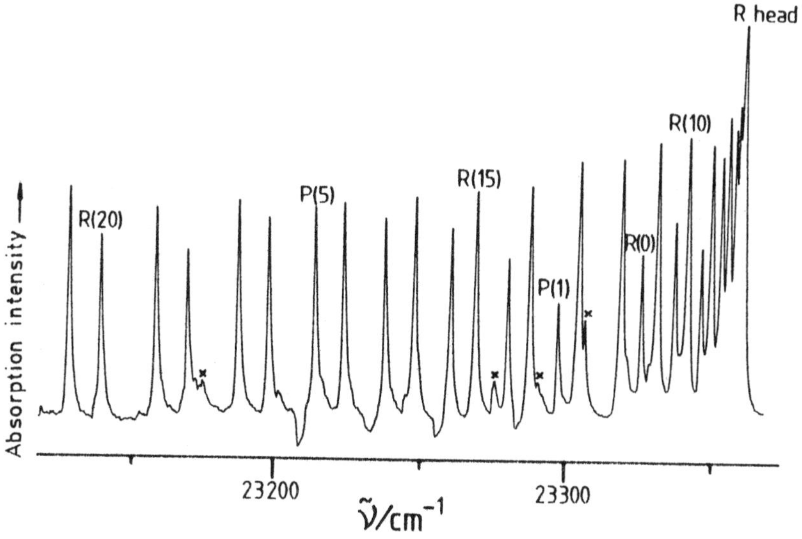

Figure 6.46 The P- and R-branch structure of the $A^1\Sigma^+$–$X^1\Sigma^+$ electronic transition of CuH in absorption. Lines marked with a cross are not due to CuH

from equation (6.135), that the reversal of the R branch is at $J_{rev} \approx 6$.

It is now apparent that, when we use band head instead of band origin measurements in rotationally unresolved spectra to obtain vibrational parameters from an electronic band system, we are assuming that the head–origin separation remains constant for all bands in the system.

The intensity distribution among the rotational transitions is governed by the population distribution among the rotational levels in the *initial* electronic or vibronic state of the transition. In absorption from the ground state the intensities depend on rotational populations in that state. If there is population equilibrium at a temperature T, the relative populations are given by the Boltzmann distribution law [equation (4.22)].

If the spectrum is observed in emission it is the rotational populations of the upper state which determine relative intensities. They may or may not be equilibrium populations within that state, depending on the condition of formation of the state. Even if the molecules have attained rotational equilibrium in the upper state, the so-called rotational temperature to which the equilibrium populations correspond may not be the temperature of the molecules as recorded by conventional means. The molecules may be produced in the excited state in a rotationally 'hot' or 'cold' form.

The spectrum of CuH in Figure 6.46 shows that either the temperature is much higher or the rotational constant B much lower than in the example in Figure 6.45(b). As a result, the P and R branches are much longer in CuH and there is considerable overlap between the P branch and the reversed R branch.

Obtaining the rotational constants B and D for each of the two combining states in a ${}^1\Sigma$–${}^1\Sigma$ transition proceeds exactly as for an infrared vibration–rotation band described in Section 5.1.3.1. The method of combination differences is used and, if only B'' and B' are required, $\Delta_2'' F(J)$ or $\Delta_2' F(J)$ is plotted against $(J+\frac{1}{2})$ and the slope of the straight line is $4B''$ or $4B'$, respectively [see equations (5.31) and (5.32)]. To obtain B and D for either state the corresponding $\Delta_2 F(J)/(J+\frac{1}{2})$ is plotted against $(J+\frac{1}{2})^2$ to give a straight line of slope $8D$ and intercept $(4B-6D)$, as shown by equations (5.34) and (5.35).

It is interesting that electronic spectroscopy provides the fifth method, for heteronuclear diatomic molecules, of obtaining the rotational constants and, from these, the internuclear distance in the ground electronic state. The other four arise through the techniques of pure rotation spectroscopy (microwave, millimetre wave or far infrared, and Raman) and vibration–rotation spectroscopy (infrared and Raman). In homonuclear diatomics, only the Raman techniques may be used. However, if the molecule is short-lived, as with, for example, CuH and C_2, electronic spectroscopy may be the only means of determining the ground state geometry.

In a homonuclear molecule there may be an intensity alternation with J for the same reasons that were discussed in Section 4.4 and illustrated in Figure 4.22.

The letters a and s, indicating that the rotational wave function ψ_r is antisymmetric or symmetric, respectively, to nuclear exchange, are attached to the rotational levels in Figure 4.22 with the assumption that the electronic state of H_2, F_2 or N_2 is ${}^1\Sigma_g^+$. In this case the levels with J even are s and those with J odd are a. The opposite is true for either a ${}^1\Sigma_g^-$ or a ${}^1\Sigma_u^+$ state, since ψ_e is antisymmetric to nuclear exchange in both cases. For a ${}^1\Sigma_u^-$ state ψ_e is symmetric to nuclear exchange and, again, J even levels are s and J odd levels are a.

The selection rules are

$$\mathrm{s} \longleftrightarrow \mathrm{s}, \quad \mathrm{a} \longleftrightarrow \mathrm{a}, \quad \mathrm{a} \not\longleftrightarrow \mathrm{s}$$

and are general for all electronic and vibronic transitions.

To illustrate the effects of nuclear spin, we can use Figure 4.22 to show that in a ${}^1\Sigma_u^+ - {}^1\Sigma_g^+$ transition in ${}^1H_2$ there would be an intensity alternation of 1 : 3 for J'' even : odd in absorption but of 3 : 1 for J' even : odd in emission.

6.2.5.2 ${}^1\Pi - {}^1\Sigma$ *electronic and vibronic transitions*

In a ${}^1\Pi$ state there are two angular momenta, orbital and overall rotational. Figure 6.47 shows how the resultant total angular momentum $\boldsymbol{J}$ is obtained from $\Lambda\hbar$ and the angular momentum $\boldsymbol{R}$ due to overall rotation.† From Figure 6.47, it is clear that

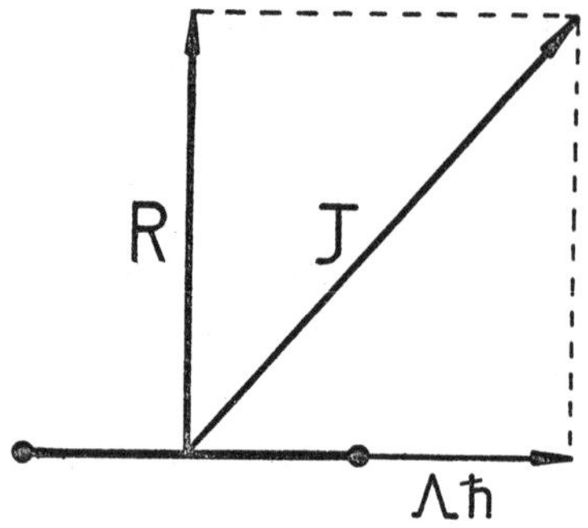

Figure 6.47 Resultant, $\boldsymbol{J}$, of the overall rotational angular momentum, $\boldsymbol{R}$, and the component, $\Lambda\hbar$, of the orbital angular momentum

$$\boldsymbol{J} = \boldsymbol{R} + \boldsymbol{\Lambda} \qquad (6.138)$$

In a ${}^1\Pi$ state, therefore, the quantum number J cannot be less than Λ and can take the values 1, 2, 3, . . . Such a stack of rotational levels is shown in the upper part of Figure 6.48.

It has been mentioned in Sections 6.2.3.1 and 6.2.3.2 that all states with $\Lambda > 0$ are doubly degenerate. The components can be labelled + and − depending on whether or not they are symmetric to σ_v and the degeneracy can be thought of, classically, as being caused by the same energy being associated with clockwise or anticlockwise orbital motion of the electrons about the internuclear axis. Alternatively, the e and f labels (Section 4.2.6.2) can be used. At this level of approximation there are two

†This vector has commonly been labelled $\boldsymbol{O}$. Since the quantum number associated with it, which is a good quantum number for Hund's case (d) only, is R it is logical to use $\boldsymbol{R}$ for the corresponding vector. This was done in Herzberg's *Spectra of Diatomic Molecules* but only for case (d).

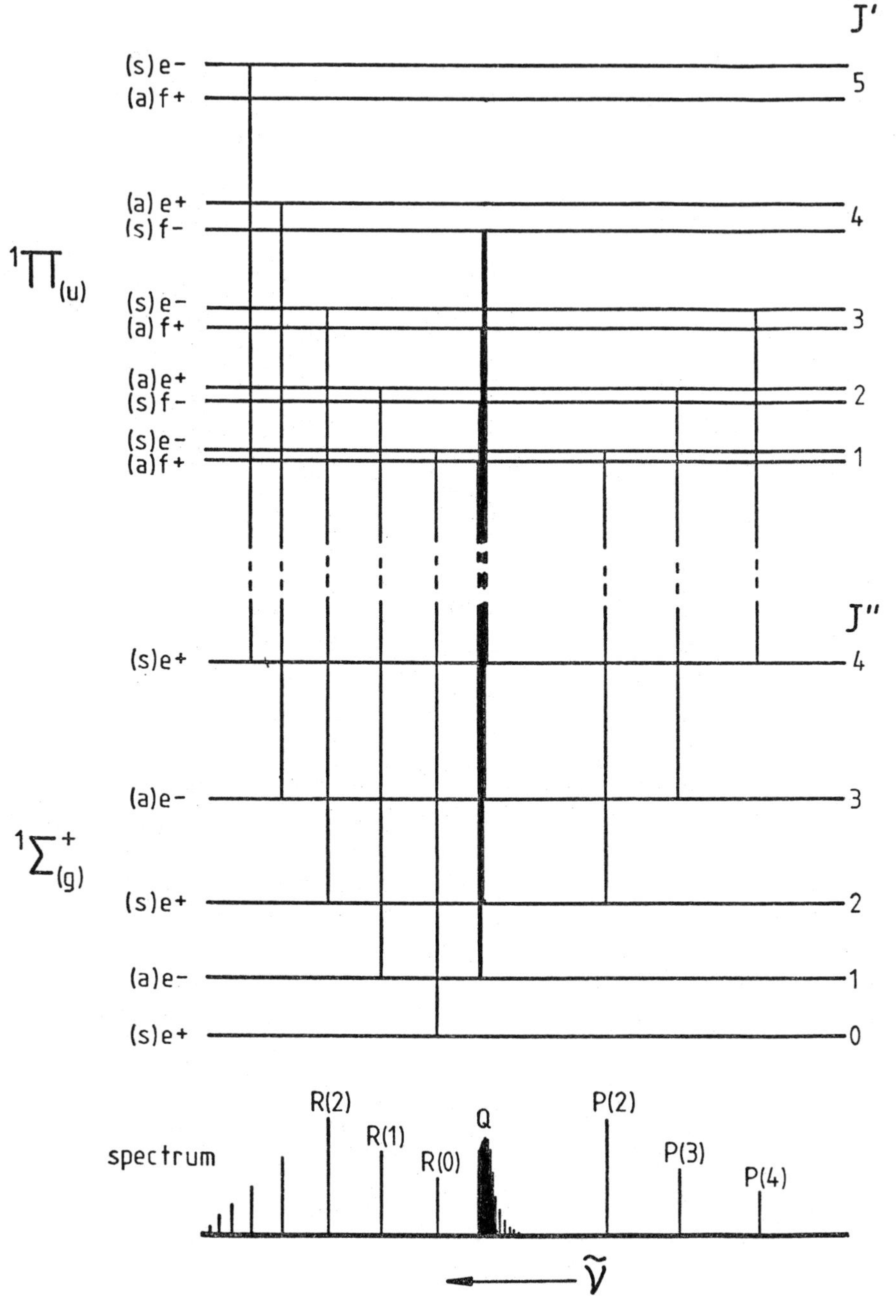

Figure 6.48 Rotational fine structure of a $^1\Pi$–$^1\Sigma^+$ electronic or vibronic transition in a diatomic molecule for which $r'_e > r''_e$

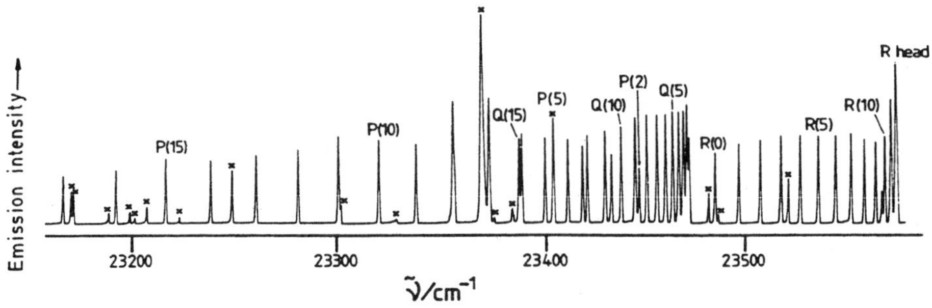

Figure 6.49 The *P*-, *Q*- and *R*-branch structure of the $A^1\Pi$–$X^1\Sigma^+$ electronic transition of AlH in emission. Lines marked with a cross are not due to AlH

coincident stacks of rotational levels for a $^1\Pi$ state.

However, this is true only if the interaction between the orbital motion and the overall rotation of the nuclei is neglected. When this is taken into account, the degeneracy is split. The interaction increases with the speed of overall rotation and the splitting, $\Delta F(J)$, of term values due to this is given by

$$\Delta F(J) = qJ(J+1) \qquad (6.139)$$

where q is a constant for a particular electronic state. The effect is known as Λ-type doubling and is shown, exaggeratedly, in Figure 6.48, where it is assumed that the Π_e stack of levels is higher in energy than the Π_f stack. Which way round these are depends on the sign of q.

The splitting of the two stacks of levels is typically very small and is even smaller in states with $\Lambda > 1$.

Whatever the order of the + and − components for $J=1$, it alternates with J, whereas, as Figure 6.48 shows, there is no alternation when the e and f labels are used.

For homonuclear diatomics Figure 6.48 refers to a $^1\Pi_u$ upper state and it shows that the s and a character of the rotational levels is the same for the Π_u^+ and Π_u^- components as for Σ_u^+ and Σ_u^- states, respectively.

The rotational selection rules for the $^1\Pi$–$^1\Sigma^+$ (or $^1\Pi_u$–$^1\Sigma_g^+$) transition in Figure 6.48 are the same as for a $^1\Sigma$–$^1\Sigma$ transition except for $\Delta J = 0, \pm 1$.

The $\Delta J = 0$ transitions form the Q branch and clearly are allowed by the parity or e–f selection rules given in Section 6.2.5.1.

The stacks of levels have been drawn for B' slightly less than B'', giving a converging R branch and a diverging P branch. Under these conditions the Q branch diverges to low wavenumber.

Figure 6.49 shows, as an example, the $A^1\Pi$–$X^1\Sigma^+$ electronic transition at 424 nm of the short-lived molecule AlH. The band is degraded more strongly to the red than the example in Figure 6.48, leading to considerable overlap of the P and Q branches. The degradation is due not to an appreciable geometry change, since r_e is 1.6478 Å in the $X^1\Sigma^+$ state and 1.648 Å in the $A^1\Pi$ state, but to the relatively large value of q of 0.0080 cm^{-1} in the $A^1\Pi$ state.

The method of combination differences applied to the P and R branches gives the lower state rotational parameters B'' and D'', just as in a $^1\Sigma$–$^1\Sigma$ transition. These branches also give rotational parameters relating to the *upper* components of the $A^1\Pi$ rotational levels. The parameters for the lower components can be obtained from the Q branch. The effective B' and D' for a $^1\Pi$ state are usually taken as the average of those for the upper and lower components.

All that has been said here about $^1\Pi$–$^1\Sigma$ electronic or vibronic transitions is very reminiscent of what was said in Section 5.2.5.1(i) regarding infrared vibration–rotation spectra of linear polyatomic molecules. In a Π

vibrational state of a linear molecule there is a vibrational angular momentum in addition to that due to overall rotation, whereas in a Π electronic or vibronic state of a diatomic molecule there is also an additional angular momentum, but it is electronic in nature.

The l-type doubling in linear molecules in Π vibrational states is analogous to Λ-type doubling in a diatomic molecule. Indeed, the complete rotational term value expression for a Π state, including the first centrifugal distortion term, is

$$F_v(J, \Lambda) = B_v[J(J+1) - \Lambda^2] \pm (q/2)[J(J+1)] - D_v[J(J+1) - \Lambda^2]^2 \quad (6.140)$$

This is analogous to equation (5.196) if we replace l by Λ and put $v_i = 1$ in that equation. In Λ-type doubling, the term $-B_v\Lambda^2$ (which is simply $-B_v$ in a Π state) is incorporated into the electronic or vibronic term value so that the first term of $F_v(J, \Lambda)$ becomes the usual rotational term, with J given by equation (6.138), and the second term includes the effects of Λ-type doubling previously expressed separately in equation (6.139).

We have considered in detail only a $^1\Pi$–$^1\Sigma^+$ transition but, in a $^1\Pi$–$^1\Sigma^-$ transition, the same selection rules apply.

These selection rules apply also to all other singlet–singlet transitions (except $^1\Sigma$–$^1\Sigma$). They all show P, Q and R branches but when $\Delta\Lambda = 0$ as in $^1\Pi$–$^1\Pi$ and $^1\Delta$–$^1\Delta$ transitions, the Q branch is weak compared with transitions with $\Delta\Lambda = \pm 1$, such as $^1\Pi$–$^1\Sigma$ and $^1\Delta$–$^1\Pi$. If Λ-type doubling is resolved then $^1\Pi$–$^1\Pi$, $^1\Delta$–$^1\Pi$ and $^1\Delta$–$^1\Delta$ transitions all show doubled P, Q and R branches, although doubling is very small for $^1\Delta$–$^1\Delta$.

6.2.5.3 *Hund's coupling case (a)*

When there is an electron spin angular momentum, in addition to orbital and overall rotational angular momenta, there is a wide range of extreme coupling approximations which may apply to a particular state of a particular molecule. Hund proposed five coupling approximations and these are commonly referred to as Hund cases (a), (b), (c), (d) and (e).

There are many examples of states which do not correspond sufficiently closely to any of the coupling cases but may be discussed in terms of their being intermediate between two of them. Such intermediate coupling cases will not be discussed.

The reader is referred to the book *Spectra of Diatomic Molecules* by G. Herzberg, listed in the Bibliography at the end of this chapter, for a discussion of intermediate cases.

Hund's case (a) is an extension of the orbital–rotational coupling in Figure 6.47. In case (a) coupling the electrostatic field of the nuclei is sufficient to uncouple $\boldsymbol{L}$ and $\boldsymbol{S}$, the orbital and spin vectors of the electrons. They both precess around the internuclear axis, as shown in Figure 6.50. $\boldsymbol{L}$ precesses so rapidly that the quantum number L is not defined. The resultant of the angular momentum $\boldsymbol{R}$ due to nuclear rotation and the component of the total electronic angular momentum along the internuclear axis, $\Omega\hbar$, is the total angular momentum $\boldsymbol{J}$. The $\boldsymbol{R}$ vector precesses around $\boldsymbol{J}$ and therefore the molecule as a whole rotates about $\boldsymbol{J}$, which is regarded as fixed in space.

The good quantum numbers are S, Λ, Σ, Ω and J.

The rotational term values, neglecting Λ-type doubling and centrifugal distortion, are given by

$$F_v(J, \Omega) = B_v[J(J+1) - \Omega^2] \quad (6.141)$$

which is analogous to the first term of equation (6.140) for a singlet state. Corresponding to equation (6.138)

$$\boldsymbol{J} = \boldsymbol{R} + \Omega \quad (6.142)$$

and J clearly cannot be less than Ω.

For a $^3\Pi$ state the three components, with $\Omega = 0$, 1 and 2, are split by the spin–orbit interaction term of equation (6.110) and there is

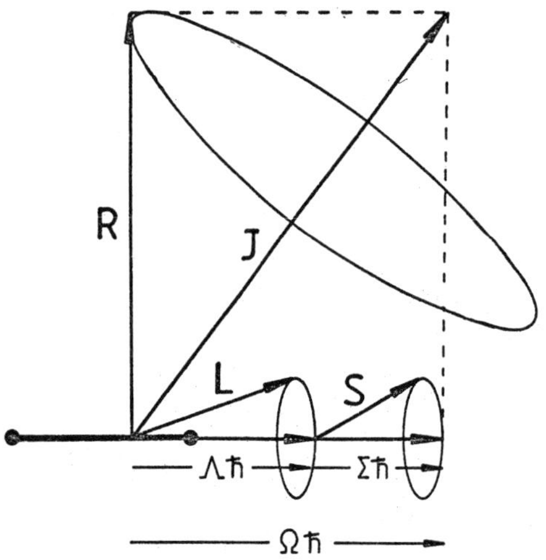

Figure 6.50 Vector sum of the overall rotational angular momentum, ***R***, and the components of the electron orbital and spin angular momenta along the internuclear axis to give the resultant ***J*** in Hund's case (a)

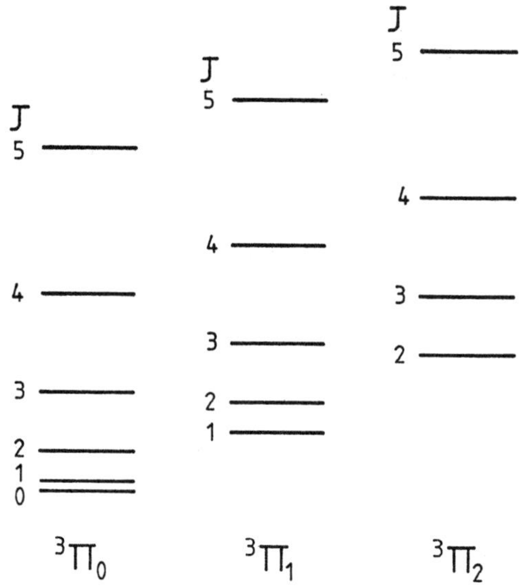

Figure 6.51 Rotational levels for a $^3\Pi$, Hund's case (a), state

a stack of rotational levels associated with each component, as illustrated in Figure 6.51.

If Λ-type doubling is taken into account each level in the three stacks is split into two. The splitting is small and increases with J, except for the $^3\Pi_0$ component, for which the splitting is relatively large and almost independent of J.

The electronic selection rules for case (a) states have been discussed in Section 6.2.3.2.

6.2.5.4 Hund's coupling case (b)

This coupling case always applies to Σ states ($\Lambda = 0$) with $S \neq 0$, provided that spin–spin coupling is not large.

As discussed in Section 6.2.3.1, the ***S*** vector by itself cannot be coupled to the internuclear axis. It can only be coupled to the axis by the magnetic field due to the orbiting electrons but, if $\Lambda = 0$, there is no field and therefore no coupling.

In some states with $\Lambda \neq 0$, ***S*** may be coupled only very weakly to the axis. This tends to happen in light molecules with few electrons and Hund's case (b) is an appropriate approximation.

Figure 6.52 shows the resulting vector diagram for $\Lambda \neq 0$. The resultant of ***R*** and ***Λ*** is ***N*** and coupling of ***N*** and ***S*** produces the total angular momentum ***J***. The only good quantum numbers are Λ, S, N and J. The values that J can take are given by

$$J = N + S, N + S - 1, \ldots, |N - S| \quad (6.143)$$

Therefore, J can be integer or half-integer and each rotational level has $(2S+1)$ components when $N \geqslant S$.

In, for example, a $^2\Sigma$ state, the rotational term values $F_1(N)$ and $F_2(N)$, corresponding to J having values $(N+\frac{1}{2})$ and $(N-\frac{1}{2})$, respectively, are given by

$$F_1(N) = B_v N(N+1) + \tfrac{1}{2}\gamma N$$
$$F_2(N) = B_v N(N+1) - \tfrac{1}{2}\gamma(N+1) \quad (6.144)$$

where γ is the spin–rotation coupling constant, which is usually very small. The coupling of spin with overall rotation may contain small contributions due to direct magnetic coupling and also to coupling through the orbital motion of the electrons.

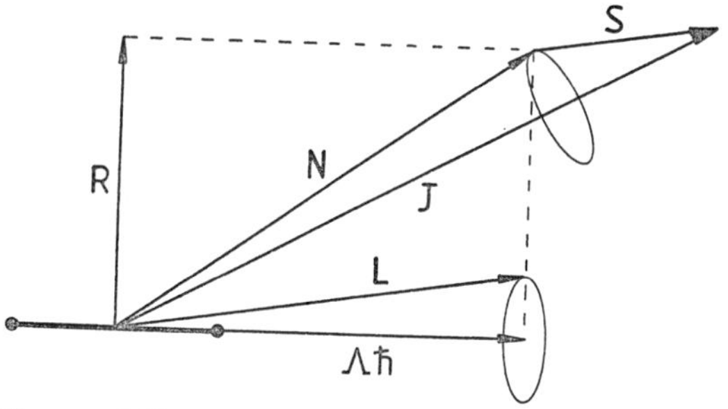

Figure 6.52 Coupling of angular momenta in Hund's case (b)

In a $^3\Sigma$ state there is an additional interaction to be taken into account, that of the two unpaired electron spins. The rotational term values are given by

$$\begin{aligned} F_1(N) &= B_v N(N+1) + (2N+3)B_v + \gamma(N+1) \\ &\quad - \lambda - [(2N+3)^2 B_v{}^2 + \lambda^2 - 2\lambda B_v]^{\frac{1}{2}} \\ F_2(N) &= B_v N(N+1) \\ F_3(N) &= B_v N(N+1) - (2N+1)B_v - \gamma N \\ &\quad - \lambda + [(2N-1)^2 B_v{}^2 + \lambda^2 - 2\lambda B_v]^{\frac{1}{2}} \end{aligned} \tag{6.145}$$

where $F_1(N)$, $F_2(N)$ and $F_3(N)$ correspond to $J=(N+1)$, N and $(N-1)$, respectively, and λ is the spin–spin coupling constant.

The electronic selection rules are the same as for case (a), given in Section 6.2.3.2, except that Σ, and therefore Ω, is not a good quantum number and the selection rules relating to them do not apply.

Although states with $\Lambda = 0$ are mostly case (b) states, the exceptions being those for which λ is large, states with $\Lambda \neq 0$, such as $^2\Pi$, are often best described as a transition case between (a) and (b), in which case (a) is a good approximation for low N and case (b) for high N. This happens in the $A^2\Pi$ state of the short-lived CN molecule which has been observed in the $A^2\Pi$–$X^2\Sigma^+$ 'red' system in emission in a discharge in cyanogen, C_2N_2.

Any singlet state with $\Lambda \neq 0$ can be described as either case (a), as was done in Section 6.2.5.2, or case (b) since they become identical when $S=0$.

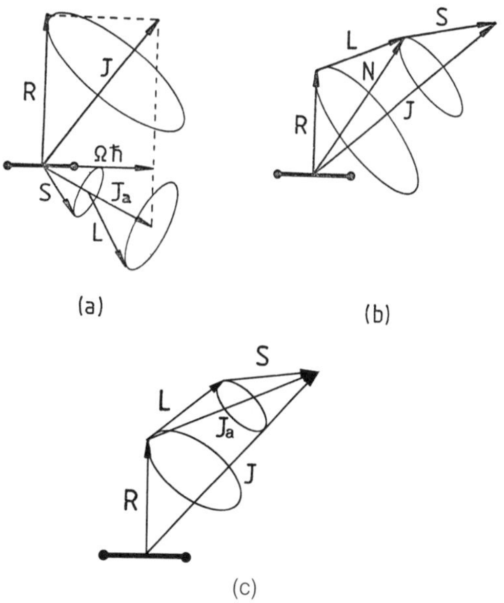

Figure 6.53 (a) Hund's case (c), (b) Hund's case (d) and (c) Hund's case (e) coupling

6.2.5.5 *Hund's coupling case (c)*

In molecules with either or both nuclei having high nuclear charges, the spin–orbit coupling may be so large that $\boldsymbol{L}$ and $\boldsymbol{S}$ are not uncoupled by the field of the nuclei. This is the basis of Hund's case (c) coupling and is found, for example, in some states of the heavier halogens such as the lowest $^3\Pi_u$ term of iodine.

Case (c) coupling of electron spin and orbital motions has been discussed in Section 6.2.3.1 and illustrated in Figure 6.31(b).

Coupling of the overall rotational to the electronic angular momentum is illustrated, in the case (c) approximation, in Figure 6.53(a). The vectors $\boldsymbol{L}$ and $\boldsymbol{S}$ couple to give the resultant $\boldsymbol{J}_a$ with component $\Omega\hbar$ along the internuclear axis. The component couples with $\boldsymbol{R}$ to give $\boldsymbol{J}$ about which the molecule rotates. The good quantum numbers are S, Ω and J but not Λ and Σ. Even though Λ is not a good quantum number it is only for states which approximate very closely to case (c) that the Σ, Π, Δ, . . . symbols are dropped and the value of Ω used instead as, for example, in $^30_u^+$.

The electronic selection rules for case (c) states have been discussed in Section 6.2.3.2. The rotational term values are given by equations (6.140) and (6.141), just as for case (a).

The $^3\Pi_{0^+u}$, $^3\Pi_{1u}$ and $^3\Pi_{2u}$ components† of the lowest $^3\Pi_u$ term of iodine are so widely separated by spin–orbit interaction that they can best be regarded as separate electronic states, with their characteristic rotational constants and stacks of rotational levels, rather than an example of case (c) coupling.

6.2.5.6 *Hund's coupling case (d)*

Hund's case (d) approximation is commonly a useful one for Rydberg states (see Section 6.2.2) in which the promoted electron is in an orbital which is so far from the nuclei that it tends to see them and the other electrons as a point positive charge.

Under these circumstances, the orbital angular momentum $\boldsymbol{L}$ is not coupled to the internuclear axis but to the overall rotation angular momentum. The resultant is $\boldsymbol{N}$, as shown in Figure 6.53(b). $\boldsymbol{N}$ couples to the spin angular momentum $\boldsymbol{S}$ to give the resultant $\boldsymbol{J}$.

The rotational term values are given by the same expression as for a $^1\Sigma$ state [equation (6.133)] except that J is replaced by the quantum number R which is associated with overall rotation *only*. Since, when $R \geqslant L$, the quantum number N can take $(2L+1)$ values, each rotational level is split into $(2L+1)$ components. Any additional splitting due to coupling of $\boldsymbol{S}$ with $\boldsymbol{N}$ is very small owing to the weakness of the coupling.

6.2.5.7 *Hund's coupling case (e)*

Hund's case (e) approximation is illustrated in Figure 6.53(c) but is extremely rare.

†The labelling of states which are only approximately case (c) includes the value of Ω as a subscript.

The orbital angular momentum $\boldsymbol{L}$ and the spin angular momentum $\boldsymbol{S}$ are strongly coupled to form the resultant $\boldsymbol{J}_a$. This is then coupled to the overall rotation angular momentum $\boldsymbol{R}$ to give the resultant $\boldsymbol{J}$.

The only good quantum number in the case (e) coupling approximation is J. However, there may be weak coupling of $\boldsymbol{L}$ and $\boldsymbol{S}$ to the internuclear axis in which case Ω can be used as an approximate quantum number. In going from the approximation of case (e) to that of case (c) coupling, Ω becomes a good quantum number.

An example of this unusual coupling case (e) has been found[71] in the weakly bound van der Waals molecule $HeAr^+$.

6.2.5.8 *Rotational fine structure of electronic transitions involving states with non-zero multiplicity*

Transitions involving only Σ electronic states are generally the most straightforward because most Σ states approximate to Hund's case (b).

For a $^2\Sigma$–$^2\Sigma$ transition the selection rule in the quantum number N [see equation (6.143)] is $\Delta N = \pm 1$ and J can take values of $N \pm \frac{1}{2}$. As a consequence there is a P and an R branch (P and R referring now to $\Delta N = -1$ and $+1$, respectively) which, unless the small splitting due to spin–orbit coupling is resolved, appear similar to those of a $^1\Sigma$–$^1\Sigma$ transition. In fact, each N-level is split into two components and the selection rules result in each P- and R-branch line being split into a closely spaced triplet. Since one component is very weak the lines appear usually as doublets. A good example of such behaviour is in the $B^2\Sigma^+$–$X^2\Sigma^+$ system of CN in the violet region.

In a $^3\Sigma$–$^3\Sigma$ transition the selection rule $\Delta N = \pm 1$ again applies and results in each P- and R-branch line being split into three components, each of which is of sufficient intensity to be observed.

A $^2\Pi$ state may belong to Hund's case (a) or (b) or an intermediate case. Case (b) gives the simplest result so that, for example, a $^2\Pi$(b)–$^2\Sigma$

transition appears similar to a $^1\Pi$–$^1\Sigma$ transition except that each *P*-, *Q*- and *R*-branch line is split into two components. This is the case in the $A^2\Pi$–$X^2\Sigma^+$ transition of MgH.

A $^2\Pi$ case (a) state can be regarded as consisting of two components, $^2\Pi_{\frac{1}{2}}$ and $^2\Pi_{\frac{3}{2}}$. The selection rule $\Delta N = 0, \pm 1$ which would apply to a $^1\Pi$–$^1\Sigma$ transition does not apply to a $^2\Pi$–$^2\Sigma$ transition but all the transitions obey the $\Delta J = 0, \pm 1$ selection rule. The result is that twelve branches are possible and all have been observed in the $A^2\Pi$–$X^2\Sigma^+$ transition of CdH.

More discussion of similar transitions in linear polyatomic molecules is given in Section 6.3.5.1(ii).

Transitions which disobey the $\Delta S = 0$ selection rule have been observed but are most commonly of the triplet–singlet type such as the $A^3\Sigma_u^+$–$X^1\Sigma_g^+$ transition of N_2 and the $a^3\Pi$–$X^1\Sigma^+$ transition of CO. A $^3\Sigma$–$^1\Sigma$ transition shows four rotational branches, a $^3\Pi(b)$–$^1\Sigma$ transition shows nine branches and a $^3\Pi(a)$–$^1\Sigma$ transition shows five branches.

The mechanism by which transitions with $\Delta S \neq 0$ obtain their intensity has been discussed in Section 6.2.3.3.

6.2.5.9 *Nuclear hyperfine structure*

If one or both of the nuclei have spin angular momentum consequent nuclear hyperfine structure may be observed in the electronic spectrum.

Many of the early observations of this effect were on metal hydrides containing a metal atom with a relatively high value of the nuclear spin quantum number I and therefore [equation (6.18)] a large magnetic moment $\boldsymbol{\mu}_I$. Although the hydrogen nucleus in these molecules also has a magnetic moment ($I = \frac{1}{2}$), that of the metal atom is very much larger so that, with the experimental resolution which can normally be obtained, the nuclear spin of the hydrogen atom can be ignored.

As well as having the advantage of large nuclear magnetic moments, the metal hydrides also have small moments of inertia. This leads to well separated rotational fine structure and easier observation of hyperfine splitting.

Later observations were not confined to hydrides but included oxides and nitrides.

The nuclear spin angular momentum due to a single nucleus can be represented by a vector of magnitude $[I(I+1)]^{\frac{1}{2}}\hbar$, as was discussed for atoms in Section 6.1.2.1. In molecules also the coupling of the nuclear spin to the other angular momenta in the system can be represented by a vector diagram in which $\boldsymbol{I}$ is added vectorially to the other angular momenta to give a resultant total angular momentum $\boldsymbol{F}$. However, in molecules, the situation is complicated by the fact that not only are there several different ways of coupling the other angular momenta, giving rise to Hund's various coupling cases, but there are also different ways of coupling the nuclear spin to these angular momenta. This is discussed in terms of subdivisions of Hund's coupling cases which we limit to cases (a), (b) and (c).

In Hund's case (a), illustrated in Figure 6.50, the components of the orbital and spin angular momenta of the electrons along the internuclear axis couple to the overall rotational angular momentum $\boldsymbol{R}$ to give the resultant $\boldsymbol{J}$.

In case (a_α) the nuclear spin is not coupled to the orbital or electron spin momenta but is coupled to the internuclear axis with component $I_z\hbar$, as shown in Figure 6.54. This is the simplest possible coupling case but is unlikely to be important because the nuclear spin magnetic moment is so small that coupling to the axis is very weak. Even if it occurs it is likely to be easily uncoupled from the axis by rotation.

Case (a_β) is more important and is illustrated in Figure 6.55. Here $\boldsymbol{I}$ is not coupled to the internuclear axis but to $\boldsymbol{J}$ to form the resultant $\boldsymbol{F}$. The $A^2\Pi_{\frac{1}{2}}$ state of ^{201}HgH and ^{199}HgH, in which $I(^{201}\text{Hg}) = \frac{3}{2}$ and $I(^{199}\text{Hg}) = \frac{1}{2}$, has been shown from observation of the $A^2\Pi_{\frac{1}{2}}$–$X^2\Sigma^+$ system in emission to conform to case (a_β).

A $^3\Phi$–$^3\Delta$ transition, attributed to NbN or, possibly, NbO^+, has been observed in emission in the red region of the spectrum and has three components, $^3\Phi_4$–$^3\Delta_3$, $^3\Phi_3$–$^3\Delta_2$ and $^3\Phi_2$–$^3\Delta_1$.

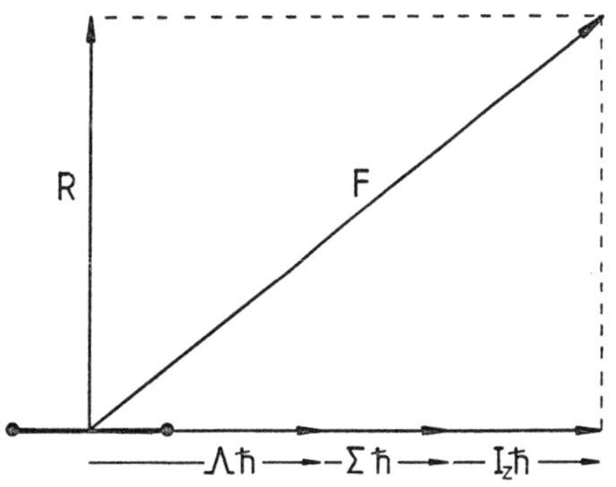

Figure 6.54 Hund's case (a_α) coupling of the overall rotation angular momentum ***R*** to the components of the orbital and spin angular momenta of the electrons and the nuclear spin angular momentum along the internuclear axis

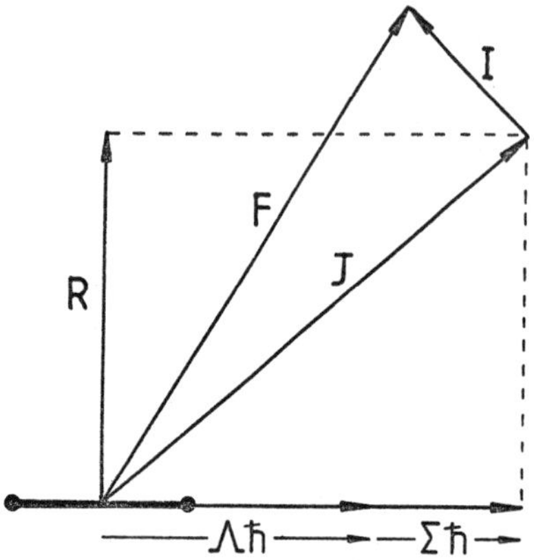

Figure 6.55 Hund's case (a_β) coupling of nuclear spin ***I*** to other angular momenta

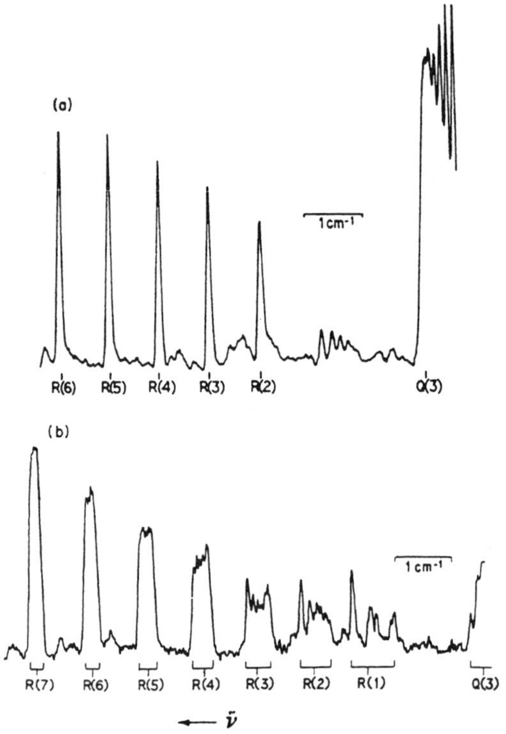

Figure 6.56 Part of an *R* branch in (a) the ${}^3\Phi_3$–${}^3\Delta_2$ and (b) the ${}^3\Phi_2$–${}^3\Delta_1$ transitions of NbN (or, possibly, NbO^+) showing nuclear hyperfine splitting [Reproduced, with permission, from *Molecular Spectroscopy: Modern Research* (Eds K. N. Rao and C. W. Mathews), p. 249. Academic Press, New York, 1972]

Only the ${}^3\Delta_3$ and ${}^3\Delta_1$ states show nuclear hyperfine splitting and they are best described as case (a_β). Figure 6.56 shows part of an *R* branch in the ${}^3\Phi_3$–${}^3\Delta_2$ transition and in the ${}^3\Phi_2$–${}^3\Delta_1$ transition. The former shows no nuclear hyperfine splitting whereas the latter shows considerable splitting which decreases as *J* increases.

In discussing the subdivisions of Hund's case (b) we consider only the most common example, that of a Σ state in which $\Lambda = 0$, for which the vector diagram is shown in Figure 6.52.

When neither ***S*** nor ***I*** is coupled to the internuclear axis there are three possible coupling extremes:

1. ***I*** is coupled to ***R*** and the resultant $\boldsymbol{F}_1$ is coupled to ***S*** to give the resultant ***F***. This is called case ($b_{\beta R}$) and is illustrated in Figure 6.57. The symbol β implies that ***I*** is not coupled to the axis, as in case (a_β), and the symbol *R* implies that ***I*** is coupled directly to ***R***. Cases of ($b_{\beta R}$) coupling are rare.
2. ***I*** is coupled to ***S*** and the resultant ***G*** is coupled to ***R*** to give the resultant ***F***. This is known as case ($b_{\beta S}$) for the reasons given in (1) and is illustrated in Figure 6.58. An example of such coupling is found in the

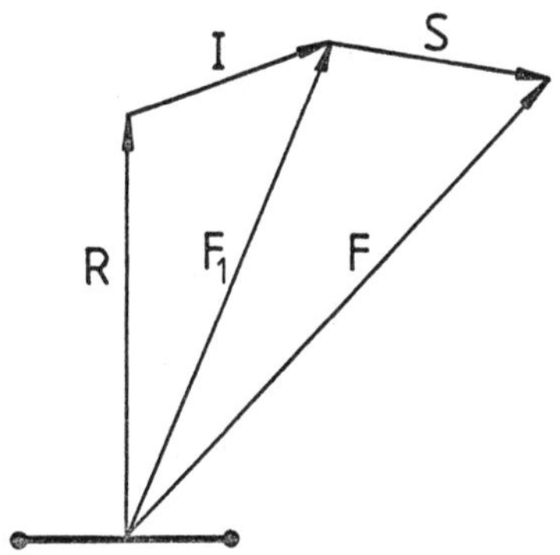

Figure 6.57 Hund's case ($b_{\beta R}$) coupling of nuclear spin ***I*** to other angular momenta in a Σ electronic state

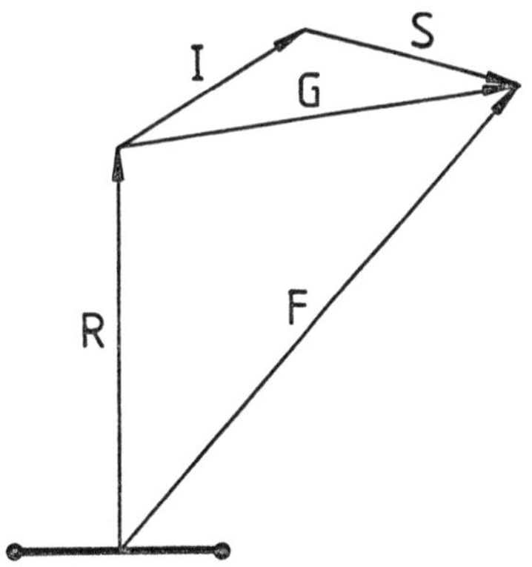

Figure 6.58 Hund's case ($b_{\beta S}$) coupling of nuclear spin ***I*** to other angular momenta in a Σ electronic state

ground state $X^2\Sigma^+$ of ScO, in which $I(^{45}Sc)=\frac{7}{2}$. The fact that identical nuclear hyperfine splitting is observed in the $B^2\Sigma^+$–$X^2\Sigma^+$ and $A^2\Pi$–$X^2\Sigma^+$ systems confirms that the nuclear hyperfine splitting of levels is confined to the X state.

3. ***S*** is coupled to ***R*** to give the resultant ***J***, as in case (b). ***I*** is coupled to ***J*** to give the resultant ***F***. This is case ($b_{\beta J}$) and is illustrated in Figure 6.59. There are several known examples of states which are best described as case ($b_{\beta J}$). The $X^2\Sigma^+$ state of HgH, mentioned earlier as the lower state of a transition for which the upper state is case (a_β), is such an example. Other examples are the $X^4\Sigma^-$ state of VO, observed in the $B^4\Pi$–$X^4\Sigma^-$ and $C^4\Sigma^-$–$X^4\Sigma^-$ transitions, and the $X^4\Sigma^-$ state of NbO, observed in the $A^4\Sigma^-$–$X^4\Sigma^-$ transition and in a second transition which probably involves a $^4\Pi$ excited state.

As with Hund's cases (a) and (b), the only important coupling of nuclear spin angular momentum appropriate to case (c) is that in which ***I*** is not coupled to the internuclear axis. This is labelled case (c_β), in which ***I*** is coupled to ***J***, and is illustrated in Figure 6.60. This case has been shown to apply to the $X_1{}^2\Pi_{\frac{1}{2}}$ ground state† of BiO through observation of the $A^2\Pi_{\frac{1}{2}}$–$X_1{}^2\Pi_{\frac{1}{2}}$, $B^4\Sigma^-$–$X_1{}^2\Pi_{\frac{1}{2}}$ and $C^2\Delta_{\frac{3}{2}}$–$X_1{}^2\Pi_{\frac{1}{2}}$ transitions.

All the coupling cases mentioned here are extreme cases to which real molecules often only approximate. Dunn, in the review referred to in the Bibliography at the end of this chapter, has discussed and given references to the various cases in detail, including term values and the effects of Λ-type doubling and nuclear quadrupoles.

6.3 ELECTRONIC SPECTROSCOPY OF POLYATOMIC MOLECULES

6.3.1 Orbitals, states and electronic transitions

Polyatomic molecules cover such a wide range of electronic structures that it is not practical to discuss here more than a few general examples. This discussion of molecular orbitals and electronic states will be at the same fairly elementary level as that for diatomic molecules in Section 6.2.1. The aim here is to give the reader some feeling for the orbitals involved in electronic transitions in some important types

†The other component of the $^2\Pi$ term gives the low-lying excited state $X_2{}^2\Pi_{\frac{3}{2}}$.

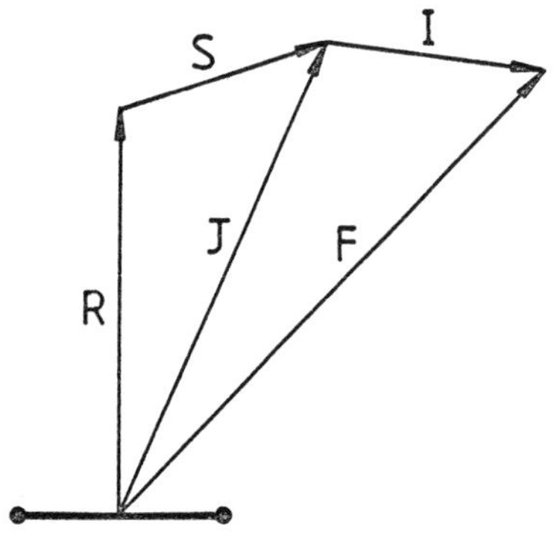

Figure 6.59 Hund's case ($b_{\beta J}$) coupling of nuclear spin ***I*** to other angular momenta in a Σ electronic state

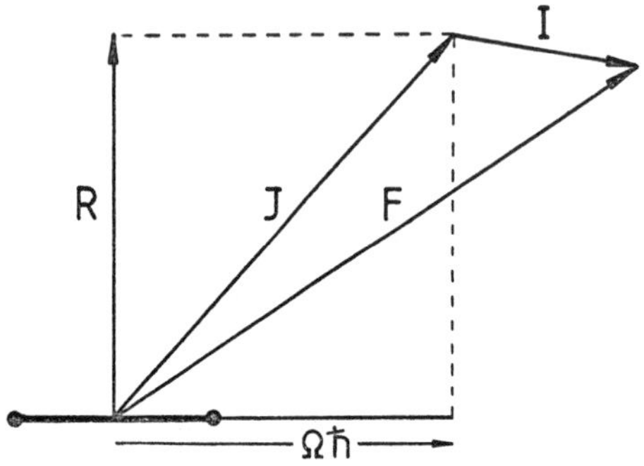

Figure 6.60 Hund's case (c_{β}) coupling of nuclear spin ***I*** to other angular momenta

of molecules without any of the rigour involved in *ab initio* calculations. At the expense of quantitative accuracy we shall be able to draw conclusions regarding orbitals involved in a particular grouping of atoms and to treat the orbitals as transferable from one molecule to another.

As in the case of diatomic molecules we shall neglect (a) *electron correlation*, which takes account of the fact that the motion of an electron is not independent but is correlated with the motions of the other electrons; (b) *electron reorganization*, which is the effect of all the molecular orbital energies and wave functions being slightly changed when an electron is promoted to another orbital, or removed in an ionization process; and (c) *configuration interaction*, which takes account of the interaction between electron configurations of the same symmetry. Any good, quantitative molecular orbital (MO) calculation should take all of these into account.

6.3.1.1 AH_2 *molecules*

In this section we consider the molecules represented by the general formula AH_2. In most cases A will be a first-row element (Li to Ne) but some examples in which it is a second- or third-row element will be mentioned. The molecular orbital treatment of these molecules was first suggested by Walsh,† who also treated the molecules with general formula AB_2, BAC, HAB, HAAH, AH_3, AB_3 and H_2AB, where A, B and C are usually first- or second-row elements.

In this treatment of AH_2 molecules the atomic orbitals of A and H which are important are the valence AOs since these form the valence MOs of AH_2. These valence AOs are 1*s* on H and 2*s* and 2*p* on A [or, in general, $(n+1)s$ and $(n+1)p$ for an *n*th-row atom]. In MO language these orbitals are referred to as the basis set from which the MOs are constructed.

The total electron density contributed by all the electrons in an AH_2, or indeed any, molecule is a property which is easy to visualize and it is possible to imagine an experiment in which this could be observed. Total electron density is, at least in principle, an experimentally observable quantity. It is when we try to break this down into a contribution from each of the electrons that problems arise. The methods employing hybridization or equivalent orbitals are useful in certain circumstances such as rationalizing properties of a localized part of the molecule. But the promotion of an electron

†Walsh, A. D. (1953). *J. Chem. Soc.*, 2260; see also pp. 2266–2317 for other types of molecules.

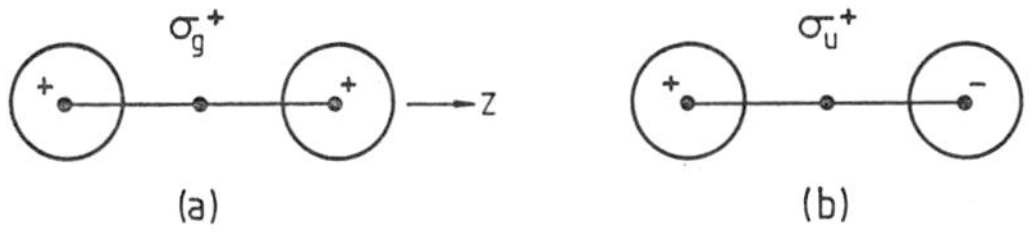

Figure 6.61 Symmetry MOs (a) σ_g^+ and (b) σ_u^+ formed from $1s$ AOs in a linear AH_2 molecule

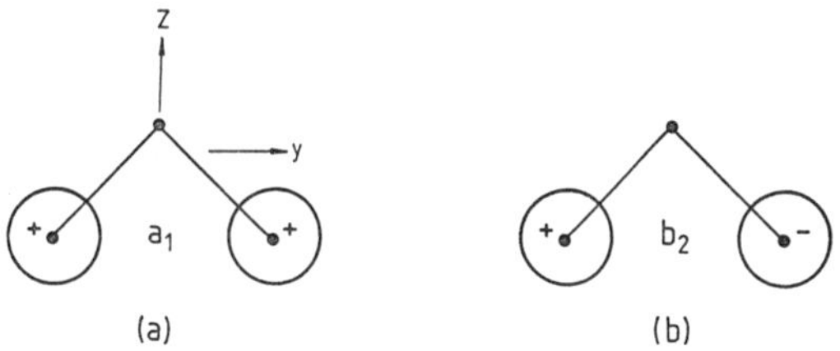

Figure 6.62 Symmetry MOs (a) a_1 and (b) b_2 formed from $1s$ AOs in a bent AH_2 molecule

from one orbital to another, in an electronic transition, or the complete removal of it, in an ionization process, both obey symmetry selection rules. For this reason, the orbitals used to describe the difference between either two electronic states of the molecule or an electronic state of the molecule and an electronic state of the ion must be molecular orbitals which belong to symmetry species of the point group to which the molecule belongs. Such orbitals are called symmetry orbitals, and it is these with which we shall be concerned here.

The $2s$ and $2p$ orbitals on A must be assigned to symmetry species of the $D_{\infty h}$ point group in the case of linear AH_2. The $2s$ AO belongs to the species σ_g^+, the $2p_z$ AO, where z is the internuclear axis, belongs to σ_u^+, and the $2p_x$ and $2p_y$ AOs belong to π_u. On the other hand, a $1s$ AO on H does not, by itself, belong to a symmetry species. We get over this problem by taking in-phase and out-of-phase combinations of them, as shown in Figure 6.61(a) and (b). The resulting orbitals belong to the σ_g^+ and σ_u^+ symmetry species, respectively.

Similarly, in the case of bent AH_2 belonging to the C_{2v} point group, the $2s$, $2p_x$, $2p_y$ and $2p_z$ AOs belong to the a_1, b_1, b_2 and a_1 species, respectively, where the axes are labelled as in Figure 6.62(a). The $1s$ AOs on the H atoms must be delocalized to give the in-phase and out-of phase combinations, which belong to a_1 and b_2, respectively, shown in Figure 6.62.

The MOs, which must belong to symmetry species of the $D_{\infty h}$ and C_{2v} point groups, can be constructed only from the AOs which belong to the same symmetry species and are comparable in energy. For linear AH_2, $\angle$ HAH = 180° and, for bent AH_2, the minimum angle is 90°.

∠ HAH = 180°

The formation of MOs for linear AH_2 is indicated on the right-hand side of Figure 6.63.

The symmetry and energy requirements for the mixing of AOs allows the $1s + 1s$ orbital of Figure 6.61(a) to mix only with $2s$ on A. A σ_g^+ MO results, but the superscript + is usually omitted as no MOs belong to a σ^- species. The MO is labelled $2\sigma_g$ according to the convention of numbering MOs of the same symmetry in order of increasing energy. The corresponding antibonding orbital is $3\sigma_g$. The $1\sigma_g$ MO is not shown in the figure; it is virtually a $1s$ AO on A which, because of its low energy, cannot mix appreciably with the $1s$ AOs on H.

Symmetry and energy requirements allow the $1s$–$1s$ orbital of Figure 6.61(b) to mix only with $2p_z$ on A to form the $1\sigma_u$ and $2\sigma_u$ MOs which are bonding and antibonding, respectively.

The other two $2p$ AOs on A cannot mix, for reasons of symmetry, with either $1s + 1s$ or $1s - 1s$ and remain as unchanged, doubly degenerate, AOs on A and are classified as $1\pi_u$ in $D_{\infty h}$.

Arranging the MOs in order of increasing energy is straightforward using the general principle that a decrease of s character or of the number of nodes in an MO increases the energy. The $2\sigma_g$ and $1\sigma_u$ MOs are both bonding between A and H but the nodal plane through A results in $1\sigma_u$ being higher in energy. Similar arguments place $2\sigma_u$ above $3\sigma_g$.

∠ HAH = 90°

In bent AH_2, belonging to the C_{2v} point group, the $1s + 1s$ orbital is a_1 [Figure 6.62(a)] and can

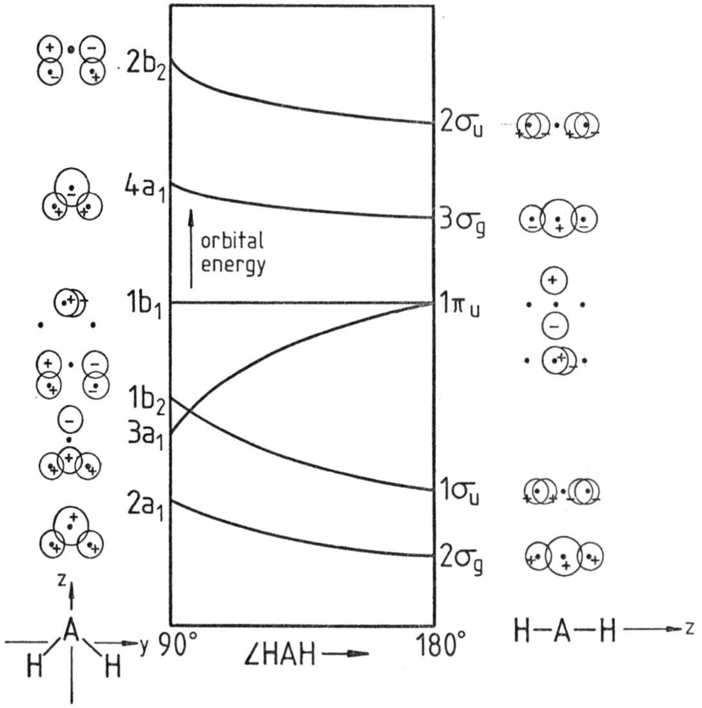

Figure 6.63 Walsh MO diagram for AH_2 molecules

mix with $2s$ and $2p_z$ on A to give the $2a_1$, $3a_1$ and $4a_1$ MOs shown on the left-hand side of Figure 6.63. The $1a_1$ MO is the unchanged $1s$ AO on A and is not shown in the figure. The $2p_y$ AO on A can mix only with the $1s$–$1s$ orbital and the $2p_x$ AO on A cannot mix with any other AOs but becomes a b_1 lone pair orbital in the bent molecule.

The ordering of the MOs in terms of energy follows the general rules used for linear AH_2 but the details rely on experimental data.

Correlation of the MOs as the HAH angle changes from 90° to 180° is shown in Figure 6.63. It should be noted that the Mulliken axis notation has been used for both C_{2v} and $D_{\infty h}$ point groups and this results in an exchange of the y- and z-axes in making the correlations.

The correlations of all MOs are obvious from their shapes, which are indicated in Figure 6.63. One particularly important correlation is $3a_1$–$1\pi_u$. Because of relaxation of symmetry restrictions in the C_{2v} compared with the $D_{\infty h}$ point group, bending of the molecule results in some mixing between the $3a_1$ and $2a_1$ MOs which are now of the same symmetry. Since $2a_1$ is strongly bonding one effect of the mixing is to impart some bonding character to $3a_1$. This causes the very steep energy increase as the angle changes from 90° to 180° and results in the observation in some AH_2 molecules of not only a change of geometry but a change of point group from the ground to an excited electronic state or to a state of AH_2^+. A change of point group can occur only in polyatomic molecules and this is the first of many examples that we shall encounter.

Table 6.12 shows the results of an MO calculation for the ground configuration of H_2O which has an HOH angle of 104.5° and two electrons in each of the $2a_1$, $3a_1$, $1b_2$ and $1b_1$ MOs of Figure 6.63. The results show that the $2a_1$ and $3a_1$ MOs are not so simple as indicated

Table 6.12 Results of an MO calculation for H_2O[a]

AO[b]	Coefficients of AOs in MO[c]				
	$1a_1$	$2a_1$	$1b_2$	$3a_1$	$1b_1$
$1s$	+0.997	−0.222	0	+0.093	0
$2s$	+0.015	+0.843	0	−0.516	0
$2p_z$	+0.003	+0.132	0	+0.787	0
$2p_y$	0	0	+0.624	0	0
$2p_x$	0	0	0	0	+1.000
$1s+1s$	−0.004	+0.152	0	+0.264	0
$1s-1s$	0	0	+0.423	0	0
MO energy/eV	−559.1	−35.0	−17.0	−12.7	−11.0

[a]Results taken from Aung, S., Pitzer, R. M. and Chan, S. I. (1968). *J. Chem. Phys.*, **49**, 2071.
[b]The $1s$, $2s$ and $2p$ AOs are on the oxygen atom: $1s+1s$ and $1s-1s$ are on the hydrogen atoms.
[c]The signs of the coefficients indicate the relative phases of the AO wave functions in the MO.

in the figure. The fact that they belong to the same symmetry species allows some mixing. The coefficients in Table 6.12 show that, in the case of H_2O the $3a_1$ MO involves a considerable contribution from the $2s$ AO. It is this contribution from $2s$ which is assumed to be generally responsible, in all AH_2 molecules, for the steep fall in energy of the $1\pi_u$–$3a_1$ orbital as the HAH angle decreases. The $2a_1$ MO is also seen to involve an appreciable contribution from $2p_z$.

Some examples of AH_2 molecules will now be considered.

BeH_2 and MgH_2

Neither of these molecules is known in the gas phase but, for the ground configuration of BeH_2, there are four electrons (two from Be and two from the hydrogen atoms) to be fed into the MOs of Figure 6.63, giving a configuration

$$\ldots (2\sigma_g)^2(1\sigma_u)^2 \tag{6.146}$$

and an $\tilde{X}^1\Sigma_g^+$ ground state.[†] Figure 6.63 shows that both occupied orbitals have minimum energy when the HAH angle is 180°, and the molecule is expected to be linear.

Promotion of an electron from $1\sigma_u$ to the next highest orbital gives the first excited configuration

$$\ldots (2a_1)^2(1b_2)^1(3a_1)^1 \tag{6.147}$$

The effect of one electron in an orbital which strongly favours the bent molecule is expected to more than counterbalance the effect of three electrons in orbitals which slightly favour the linear molecule. Therefore the molecule is expected to be bent in both the $\tilde{a}^3B_2$ and $\tilde{A}^1B_2$ states which arise[‡] from the configuration in equation (6.147).

Similar conclusions apply to MgH_2, which in an argon matrix, has been shown to be linear in its ground state.

BH_2 and AlH_2

The ground configuration of BH_2 is

$$\ldots (2a_1)^2(1b_2)^2(3a_1)^1 \tag{6.148}$$

giving only one state, the $\tilde{X}^2A_1$ ground state, in which the molecule is expected to be bent, the

[†]A tilde is placed above the label for the state for *all* polyatomic molecules, whether linear or not, in order to distinguish, for example, an A or B label from an *A* or *B* symmetry species.

[‡]See equation (6.97) for obtaining state from orbital symmetry species.

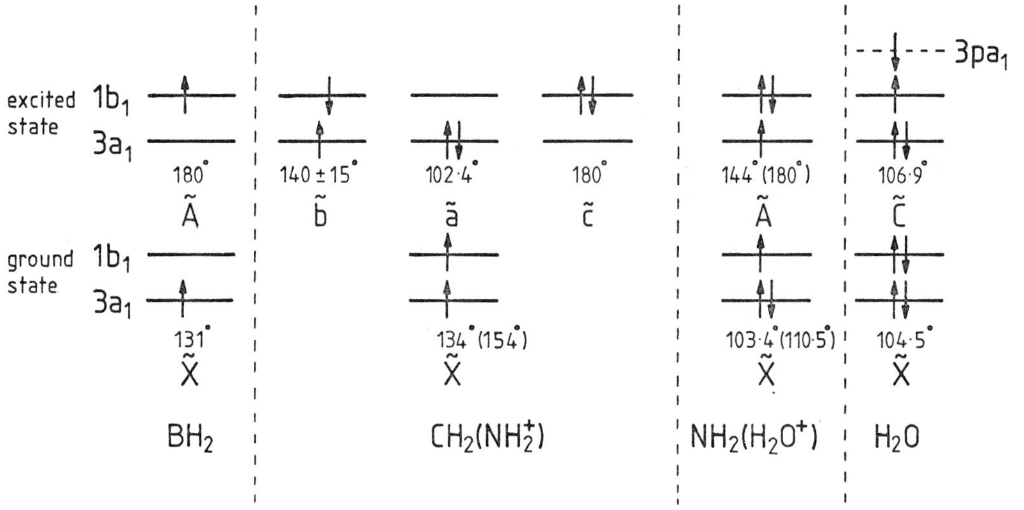

Figure 6.64 Valence electron MO configurations for some AH_2 molecules in ground and low-lying excited electronic states

effect of one electron in $3a_1$ more than counterbalancing the effect of the other four.

The first excited configuration is

$$\ldots (2\sigma_g)^2(1\sigma_u)^2(1\pi_u)^1 \quad (6.149)$$

Here the species for the linear molecule are used because the promotion of the electron from the $3a_1$ to the $1b_1$ orbital, an orbital which favours neither the linear nor the bent molecule, results in the molecule becoming linear.

There is only one excited state, $\tilde{A}^2\Pi_u$, arising from the configuration of equation (6.149).

The $\tilde{A}^2\Pi_u$–$\tilde{X}^2A_1$ system was observed† in absorption following the flash photolysis (see Section 3.8.1) of borine carbonyl, BH_3CO. The BH_2 molecule is a short-lived species which exists for about 30 μs under typical conditions of production.

In a transition like this, where there is a change of point group from one state to the other, the selection rules are those appropriate to the point group of lower symmetry. In the $\tilde{A}$–$\tilde{X}$ transition of BH_2 this point group is C_{2v} for which the first excited configuration of equation (6.149) becomes

$$\ldots (2a_1)^2(1b_2)^2(1b_1)^1 \quad (6.150)$$

This configuration results in the $\tilde{A}^2B_1$ excited state.

The transition $\tilde{A}^2B_1$–$\tilde{X}^2A_1$ is allowed by the spin selection rule $\Delta S = 0$. It is also allowed by the orbital selection rule of equation (6.111) since $B_2 \times A_1 = B_2 = \Gamma(T_y)$.

The band system falls in the 640–865 nm (red) region of the spectrum. The HBH angle is 131° and 180° in the $\tilde{X}$ and $\tilde{A}$ states, respectively, as shown in Figure 6.64.

The corresponding $\tilde{A}^2B_1$–$\tilde{X}^2A_1$ system of AlH_2 is allowed and has been observed in absorption following the flash photolysis of aluminium borohydride, $Al(BH_4)_3$. The HAlH angle is 119° and 180° in the $\tilde{X}$ and $\tilde{A}$ states, respectively. Since Al is a second-row element the s and p AOs involved in an MO diagram analogous to that of Figure 6.63 are $3s$ and $3p$.

†See Herzberg's *Electronic Spectra of Polyatomic Molecules*, referred to in the Bibliography at the end of the chapter, for references to this and many of the other AH_2 spectra discussed here.

There is a marked tendency, shown by the $\tilde{X}$ states of BH_2 and AlH_2, for the bond angle of a bent molecule to be considerably reduced when A is a second-row rather than a first-row element.

CH_2, NH_2^+, SiH_2 and GeH_2

In CH_2 there are six electrons to be fed into the MOs of Figure 6.63. Four of them fill the two lowest MOs and one low energy configuration results from placing the other two in the $3a_1$ orbital, with antiparallel spins

$$\ldots(2a_1)^2(1b_2)^2(3a_1)^2 \qquad (6.151)$$

giving only one state, $\tilde{a}^1A_1$, in which the molecule is bent.

There is a second low energy configuration

$$\ldots(2a_1)^2(1b_2)^2(3a_1)^1(1b_1)^1 \qquad (6.152)$$

which gives rise to two states $\tilde{X}^3B_1$ and $\tilde{b}^1B_1$ with parallel and antiparallel spins, respectively. The $\tilde{X}^3B_1$ state lies lower in energy than $\tilde{b}^1B_1$; this follows from one of Hund's rules for two electrons in a degenerate orbital. In CH_2 the $3a_1$ and $1b_1$ orbitals are strictly degenerate only in the linear molecule but the rule is still expected to apply.

It is not possible to predict from the Walsh diagram of Figure 6.63 whether the $\tilde{X}^3B_1$ or the $\tilde{a}^1A_1$ state is lower in energy, and therefore the true ground state, but there is incontrovertible evidence that it is $\tilde{X}^3B_1$ and that $\tilde{a}^1A_1$ is a very low-lying excited state (see Section 8.2.20) only $3147\,cm^{-1}$ ($37.65\,kJ\,mol^{-1}$), for zero-point level, or $3223\,cm^{-1}$ ($38.56\,kJ\,mol^{-1}$), for equilibrium separations, above the ground state.[72]

Because the $\tilde{a}$–$\tilde{X}$ transition is spin forbidden, CH_2 in the $\tilde{a}$ state is metastable and, under favourable conditions in the gas phase, has a lifetime which is sufficiently long that the kinetics of its reactions with other species can be followed before it reverts to the ground state.

In the absorption spectrum of CH_2, produced by the flash photolysis of, for example, diazomethane, $CH_2{=}N{=}N$, a band system has been observed in the red region of the spectrum. This is due to the $\tilde{b}^1B_1$–$\tilde{a}^1A_1$ allowed transition. The HCH angle is 102.4° in the $\tilde{a}$ state and 140±15° in the $\tilde{b}$ state.

A second weak absorption system involving $\tilde{a}$ as the lower state has been observed in the 362–330 nm region and is assigned to the $\tilde{c}^1\Sigma_g^+$–$\tilde{a}^1A_1$ allowed transition. The molecule is linear, or nearly linear, in the $\tilde{c}$ state which arises from the configuration

$$\ldots(2\sigma_g)^2(1\sigma_u)^2(1\pi_u)^2 \qquad (6.153)$$

In the far ultraviolet, at about 142 nm, there is a band system which has been assigned to the $\tilde{B}^3A_2$–$\tilde{X}^3B_1$ allowed transition. The $\tilde{B}$ state results from the promotion of an electron from the $3a_1$ to a Rydberg orbital (Section 6.2.2) which correlates with a $3d$ AO in oxygen, the corresponding united atom. The geometry in this state is not known but the HCH angle is about 134° in the $\tilde{X}$ state.†

The electron configurations and bond angles in the $\tilde{X}$, $\tilde{a}$, $\tilde{b}$ and $\tilde{c}$ states are summarized in Figure 6.64. This shows how similar is the angle in the $\tilde{X}$ state of BH_2 and the $\tilde{X}$ and $\tilde{b}$ states of CH_2, in all of which there is just one electron in the $3a_1$ orbital. In the $\tilde{a}$ state of CH_2 the angle is greatly decreased because there are now two electrons in $3a_1$.

The amidogen cation, NH_2^+, has been produced in a glow discharge in a gaseous mixture of nitrogen, hydrogen and helium.[73] Like CH_2, with which it is isoelectronic, it has a triplet ground state $\tilde{X}^3B_1$ but with a much larger angle calculated[74] to be 153.8°. The barrier to linearity is so low, $155\,cm^{-1}$, that the molecule can be regarded as quasi-linear. The first excited state, $\tilde{a}^1A_1$, is estimated[75] to be about $10\,500\,cm^{-1}$ above the ground state, much higher in energy than that of CH_2, and with an angle of about 110°.

†For a discussion of the geometry in the $\tilde{X}$ state, see Herzberg, G. and Johns, J. W. C. (1971). *J. Chem. Phys.*, **54**, 2276.

The short-lived molecule SiH_2 is formed by flash photolysis of phenylsilane, $C_6H_5SiH_3$, and by flash discharge in silane, SiH_4. The latter technique is similar to flash photolysis except that a short-lived discharge between electrodes inside the sample cell precedes the flash of the continuum source.

An absorption system has been observed in the red region. The allowed electronic transition has been assigned as $\tilde{A}^1B_1$–$\tilde{X}^1A_1$, similar to that of the $\tilde{b}^1B_1$–$\tilde{a}^1A_1$ transition of CH_2 in the red region, except that there is reason to believe, from theoretical calculations, that the ground state of SiH_2 is singlet rather than triplet. The HSiH angle is 92.1° in the $\tilde{X}$ state and 123° in the $\tilde{A}$ state.[76] The increase in angle is analogous to that from the $\tilde{a}$ to the $\tilde{b}$ state of CH_2 and the smaller angles in both states of SiH_2, compared with those of CH_2, are typical of a second-row element.

The GeH_2 molecule has been made by laser photolysis of phenylgermane, $C_6H_5GeH_3$, and by an electrical discharge in GeH_4.

Calculations have shown[77] that the trend of an increasing separation of the lowest $^1A_1$ and $^3B_1$ states continues from CH_2, in which it is negative since the ground state is $^3B_1$, to SiH_2 and GeH_2 in which it is increasingly positive. It is estimated to be 8000–10 000 cm^{-1} in GeH_2 compared with 6000–8000 cm^{-1} in SiH_2 with $^1A_1$ being the ground state in both molecules.

The $\tilde{A}^1B_1$–$\tilde{X}^1A_1$ transition has been observed[78,79] and the bond angle found to be 91.2° in the $\tilde{X}$ state and 123.4° in the $\tilde{A}$ state, very similar values to those for the corresponding states of SiH_2.

NH_2, H_2O^+, PH_2, H_2S^+ and AsH_2

The short-lived molecule NH_2 is produced by flash photolysis of ammonia, NH_3.

The ground configuration of NH_2 is

$$\ldots (2a_1)^2(1b_2)^2(3a_1)^2(1b_1)^1 \qquad (6.154)$$

giving the $\tilde{X}^2B_1$ ground state. The first excited configuration is

$$\ldots (2a_1)^2(1b_2)^2(3a_1)^1(1b_1)^2 \qquad (6.155)$$

giving the first excited state $\tilde{A}^2A_1$. The allowed $\tilde{A}$–$\tilde{X}$ transition occurs, like those of BH_2 and CH_2 resulting from a $1b_1$–$3a_1$ promotion, in the red region and has been observed in absorption. The HNH angle is 103.4° in the $\tilde{X}$ state, a value very similar to that of 102.4° for the $\tilde{a}$ state of CH_2 in which there are also two electrons in the $3a_1$ orbital, as shown in Figure 6.64. In the $\tilde{A}$ state of NH_2 the angle is 144° which Figure 6.64 shows to be a fairly characteristic value when only one electron is in the $3a_1$ orbital.

The ion H_2O^+ is isoelectronic with NH_2. The $\tilde{A}^2A_1$–$\tilde{X}^2B_1$ emission spectrum of H_2O^+, analogous to the $\tilde{A}$–$\tilde{X}$ absorption spectrum of NH_2, has been observed in the visible region using an arc discharge in low pressure water vapour. The HOH angle in the $\tilde{X}^2B_1$ state is 110.5°, similar to that in the $\tilde{X}$ state of NH_2, but the angle in the $\tilde{A}^2A_1$ state[80] is 180°, in disagreement with that in the $\tilde{b}$ state of CH_2 and the $\tilde{A}$ state of NH_2.

The emission spectrum of H_2O^+, at a rotational temperature of less than 100 K, was identified in the tail of the comet Kohoutek in 1973.[81] A similar spectrum had been observed in the comet Ikeya in 1963 but was not identified at that time.

The $\tilde{A}^2A_1$–$\tilde{X}^2B_1$ allowed system of PH_2 has been observed in absorption following the flash photolysis of phosphine, PH_3. This is analogous to the $\tilde{A}$–$\tilde{X}$ transition of NH_2 and the HPH angle is 91.5° in the $\tilde{X}$ and 123.1° in the $\tilde{A}$ state. The large increase in the $\tilde{A}$ state parallels that in NH_2 but the angles in both states are considerably less than in NH_2. This parallels the behaviour of SiH_2 compared with CH_2.

The ion H_2S^+ is isoelectronic with PH_2 and an HSH angle of 92.9° has been found for the $\tilde{X}$ state and *ca* 127° for the $\tilde{A}$ state from analysis of the $\tilde{A}^2A_1$–$\tilde{X}^2B_1$ band system.[82]

The species AsH_2 contains a third-row element and it has been shown,[83] from analysis of the $\tilde{A}^2A_1$–$\tilde{X}^2B_1$ allowed absorption system, that the HAsH angle is 90.7° in the $\tilde{X}$ and 123.0° in the $\tilde{A}$ state, very similar values to those for PH_2.

H_2O, H_2S, H_2Se and H_2Te

The ground configuration of H_2O is

$$\ldots (2a_1)^2(1b_2)^2(3a_1)^2(1b_1)^2 \quad (6.156)$$

Compared with that of NH_2 there is one more electron in the $1b_1$ orbital, which favours neither the linear nor the bent molecule. Therefore, the angle in the $\tilde{X}^1A_1$ ground state of H_2O is expected to be similar to that of the $\tilde{X}^2B_1$ state of NH_2. Figure 6.64 shows that this is what is observed.

Some of the lower-lying excited states of H_2O involve promotion of an electron from the $1b_1$ orbital to orbitals which are predominantly Rydberg in character (see Section 6.2.2). Since a Rydberg orbital is virtually non-bonding, very little geometry change is expected. This is borne out by the rotational analysis of the $\tilde{C}^1B_1$–$\tilde{X}^1A_1$ allowed transition, which showed that the HOH angle is 106.9° in the $\tilde{C}$ state, in which an electron has been promoted to a $3p$ Rydberg orbital, compared with 104.5° in the ground state.

In H_2S, H_2Se and H_2Te, the angle in the $\tilde{X}^1A_1$ ground state is 92.1°, 90.6° and 90.3°, respectively, showing, again, the tendency for the angle to decrease when the atom A of an AH_2 molecule is lower in the periodic table.

Rydberg transitions of H_2S and H_2Se have been observed in which there is very little change of angle from the ground state, as in the corresponding transition of H_2O.

The results summarized in Figure 6.64 for various states of BH_2, CH_2, NH_2^+, NH_2, H_2O^+ and H_2O demonstrate the remarkable degree to which the predictions from the Walsh diagram of Figure 6.63 are fulfilled. Values for the HAH angle for the $\tilde{a}$ state of CH_2, the $\tilde{X}$ state of NH_2 and the $\tilde{X}$ and $\tilde{C}$ states of H_2O, in all of which there are two electrons in the $3a_1$ orbital, show an impressive constancy. This applies also to the $\tilde{X}$ state of BH_2, the $\tilde{X}$ and $\tilde{b}$ states of CH_2, and the $\tilde{A}$ state of NH_2, in all of which there is only one electron in the $3a_1$ orbital.

Similarly, for second-row elements, there is close agreement between the angle for the $\tilde{X}$ states of SiH_2 (92.1°), PH_2 (91.5°) and H_2S (92.1°), in all of which there are two electrons in the analogue of the $3a_1$ orbital. There is agreement also for the $\tilde{X}$ state of AlH_2 (119°), the $\tilde{A}$ state of SiH_2 (123°) and the $\tilde{A}$ state of PH_2 (123°), in all of which there is only one electron in that orbital.

The quantitative agreement is all the more remarkable when we consider all the approximations involved in building up Figure 6.63. In particular, the MOs portrayed are one-electron orbitals taking no account of the change in inter-electron repulsions and of nuclear charge on atom A as A changes. All this is in addition to the neglect of the more subtle effects of electron correlation, electron reorganization and configuration interaction. Nevertheless, the results for AH_2 molecules show that Walsh diagrams can be used very successfully to rationalize or predict experimental observations and they command an important position in the theory of electronic spectra of small, polyatomic molecules.

H_3 and H_3^+

H_3 and H_3^+ are special cases of AH_2 molecules in that neither of them has the linear or bent shape already discussed but they both exist as cyclic D_{3h} molecules.

Although the H_3 molecule, in its ground state $\tilde{X}^2A_1'$, is unstable with respect to the dissociation products $H+H_2$, it is stable, and has been observed, in several excited states.† This stability in excited states is due to the promoted electron being in an orbital which has considerable

†See, for example, Herzberg, G. and Watson, J. K. G. (1980). *Can. J. Phys.*, **58**, 1250.

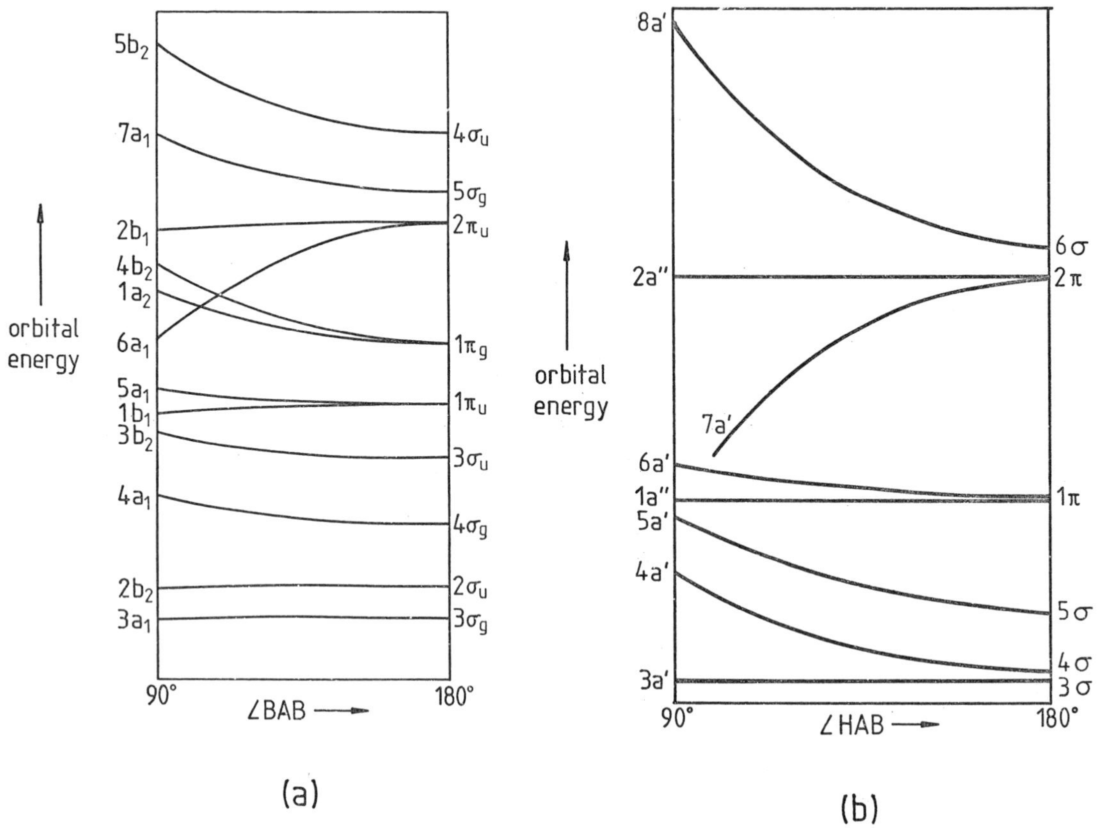

Figure 6.65 (a) Walsh MO diagram for AB_2 molecules. The z-axis is the C_2 axis in bent AB_2 and the C_∞ axis in linear AB_2. In bent AB_2 the x-axis is perpendicular to the plane of the molecule. (b) Walsh MO diagram for HAB molecules

Rydberg character (Section 6.2.2) while the structure of the core resembles that of the ground state of H_3^+. Owing to the high proton affinity of H_2, the H_3^+ species is stable in its $\tilde{X}^1A_1'$ ground state.[84] It has been found in the atmospheres of the planets Jupiter,[85] Uranus[86] and Saturn[87] and is likely to be an important component of the interstellar medium.

Because H_3 is stable only in Rydberg states, united atom orbitals (Section 6.2.1.1) are more useful in considering the electronic structure than the MOs we have used so far in this section.

In the D_{3h} point group, to which cyclic H_3 belongs, the ns united atom atomic orbitals become nsa_1' Rydberg orbitals, the np orbitals become npa_2'' and npe' and the nd orbitals become nda_1', nde' and nde''. Some of the lower energy Rydberg states are, in order of increasing energy, $\tilde{A}^2A_1'(2s)$, $\tilde{B}^2A_2''(2p)$, $\tilde{C}^2E'(3p)$, $\tilde{D}^2A_1'(3s)$, $\tilde{E}^2A_2''(3p)$, $\tilde{F}^2E'(3d)$, $\tilde{G}^2E''(3d)$ and $\tilde{H}^2A_1'(3d)$, where the united atom Rydberg orbital occupied by the promoted electron is indicated in parentheses and the other two electrons are in the 1s united atom orbital. The transitions $\tilde{C}$–$\tilde{A}$ (710 nm), $\tilde{D}$–$\tilde{B}$ (602.5 nm), $\tilde{G}$–$\tilde{B}$ (590 nm) and $\tilde{E}$–$\tilde{A}$ (560 nm) have been observed in the visible region and $\tilde{D}$–$\tilde{C}$ (3550 cm^{-1}) and $\tilde{G}$–$\tilde{C}$ (3950 cm^{-1}) in the near infrared.[88] The transitions were all observed in emission in the negative glow of a hollow cathode discharge in hydrogen gas.

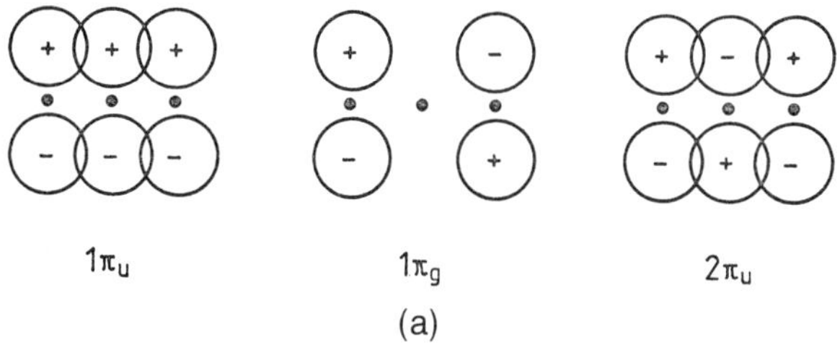

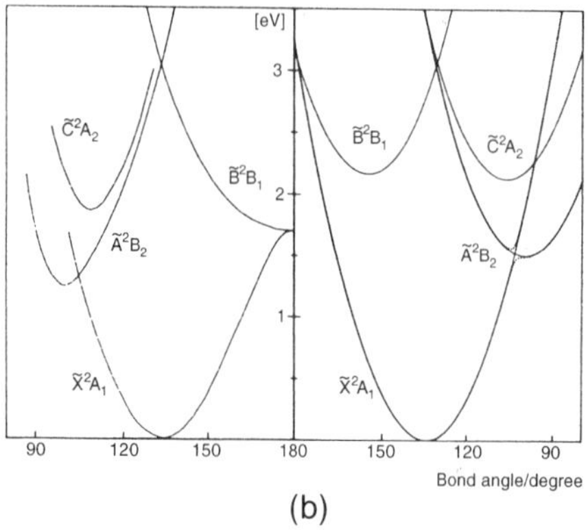

Figure 6.66 (a) Some π MOs in a linear AB_2 molecule (only one component is shown). (b) On the right are shown calculated potential energy curves for the lower electronic states of NO_2. These are compared with, on the left, an earlier calculation [Reproduced, with permission, from Haller, E., Koppel, H. and Cederbaum, L. S. (1985). *J. Mol. Spectrosc.*, **111**, 377]

6.3.1.2 AB_2 *and* HAB *molecules*

Examples of AB_2 molecules, in which A and B are first-row elements, are the stable species carbon dioxide, CO_2, nitrogen dioxide, NO_2 and ozone, O_3.

The AOs on A and B which are important in bond formation are $2s$ and $2p$. In linear AB_2 six AOs, consisting of the $2s$ and $2p_z$ AOs on A and B, mix to give six MOs of σ symmetry. These are labelled $3\sigma_g$, $4\sigma_g$, $5\sigma_g$ and $2\sigma_u$, $3\sigma_u$, $4\sigma_u$ on the right-hand side of Figure 6.65(a). The $1s$ AOs on A and B are unchanged in the molecule and become $1\sigma_u$, $1\sigma_g$, $2\sigma_g$. As in AH_2 molecules, all the σ orbital energies increase slightly or are unchanged from the linear to the bent configuration.

Figure 6.66(a) shows the three different ways in which the $2p_x$ or $2p_y$ AOs on A and B overlap to form three MOs in the linear molecule. Since the x- and y-axes remain indistinguishable in the molecule, each MO is doubly degenerate. The $1\pi_g$ MO cannot involve any contribution from the A atom as the $2p$ AOs on the B atoms are out-of-phase.

The orbital energies increase with the number of nodal planes and they are arranged accordingly in Figure 6.65(a).

In the bent molecule the degeneracies of all the π orbitals are split. This is clear from Figure 6.66(a) if we imagine a second component of each MO similar to the one shown except that the lobes of the wave function are above and below the plane of the figure. If the molecule bends in the plane of the figure, say, the energies of the two components of each MO are no longer equal. As Figure 6.65(a) shows, the splitting is not large for $1\pi_u$ and $1\pi_g$ although, for both components of the latter, the linear molecule is favoured. For the $2\pi_u$ orbital, bending out of the plane of Figure 6.66(a) does not affect the energy of the component shown: it becomes the $2b_1$ orbital in the bent molecule with energy almost unchanged.† However, the energy of the other component of $2\pi_u$ falls sharply in the bent molecule, becoming the $6a_1$ orbital. Mixing of the $6a_1$ MO with other bonding MOs of the same symmetry leads to the introduction of s character into $6a_1$ and to an appreciable lowering of the energy, rather like the $1\pi_u$–$3a_1$ orbital in AH_2 molecules. Occupation of the $2\pi_u$–$6a_1$ MO is particularly important in determining the bond angle since this is the only valence orbital for which the bent molecule is strongly favoured.

†As in AH_2, the C_2 axis is the z-axis and the other in-plane axis is the y-axis.

Table 6.13 Ground and low energy excited configurations of some AB_2 molecules[a]

Molecule	No. of valence electrons	Electron configuration											State	$\angle BAB/°$
C_3	12	...	$(3\sigma_g)^2$	$(2\sigma_u)^2$	$(4\sigma_g)^2$	$(1\pi_u)^4$	$(3\sigma_u)^2$						$\tilde{X}^1\Sigma_g^+$	180
		...	...	...	...	...	$(3\sigma_u)^1$	$(1\pi_g)^1$					$\tilde{A}^1\Pi_u$	180
CNC	13	...	...	...	...	...	$(3\sigma_u)^2$	$(1\pi_g)^1$					$\tilde{X}^2\Pi_g$	180
		...	...	...	...	...	$(3\sigma_u)^1$	$(1\pi_g)^2$					{ $\tilde{A}^2\Delta_u$ { $\tilde{B}^2\Sigma_u$	180 180
NCN	14	...	...	...	...	...	$(3\sigma_u)^2$	$(1\pi_g)^2$					$\tilde{X}^3\Sigma_g^-$	180
		...	...	...	...	...	$(3\sigma_u)^1$	$(1\pi_g)^3$					$\tilde{A}^3\Pi_u$	180
BO_2, CO_2^+, N_3	15	...	...	...	...	$(3\sigma_u)^2$	$(1\pi_u)^4$	$(1\pi_g)^3$					$\tilde{X}^2\Pi_g$	180
		...	...	...	...	...	$(1\pi_u)^3$	$(1\pi_g)^4$					$\tilde{A}^2\Pi_u$	180 (not known for N_3)
		...	...	...	...	$(3\sigma_u)^1$	$(1\pi_u)^4$	$(1\pi_g)^4$					$\tilde{B}^2\Sigma_u^+$	180
CO_2	16	...	...	...	...	...	$(1\pi_u)^4$	$(1\pi_g)^4$					$\tilde{X}^1\Sigma_g^+$	180
		...	...	...	...	...	...	$(1a_2)^2$	$(4b_2)^1$	$(6a_1)^1$			$\tilde{A}^1B_2$	122
NO_2, CO_2^-	17	...	...	...	...	...	...	...	$(4b_2)^2$	$(6a_1)^1$			$\tilde{X}^2A_1$	134.1 (NO_2), 134 (CO_2^-)
		...	...	...	...	...	...	...	$(4b_2)^1$	$(6a_1)^2$			$\tilde{A}^2B_2$	102–118 (NO_2)
		...	...	...	...	...	...	...	$(4b_2)^2$	$(2b_1)^1$			$\tilde{B}^2B_1$	*ca* 150 (NO_2)
CF_2, O_3	18	...	...	...	...	...	...	...	$(4b_2)^2$	$(1a_2)^2$	$(6a_1)^2$		$\tilde{X}^1A_1$	116.8 (O_3) 105.0 (CF_2)
		...	...	...	...	...	...	...	$(4b_2)^2$	$(1a_2)^2$	$(6a_1)^1$	$(2b_1)^1$	$\tilde{A}^1B_1$	122.3 (CF_2)

[a]Note that the order of $1\pi_u$ and $3\sigma_u$ is reversed in some molecules compared with that in Figure 6.65(a).

Table 6.13 lists examples of known first-row AB_2 molecules together with some of their electron configurations, states and bond angles. All the molecules except CO_2, NO_2 and O_3 are short-lived. The molecules C_3 and NCN have both been observed by flash photolysis of diazomethane, CH_2N_2, but C_3 is formed also by flash photolysis of diazopropyne, $HC{\equiv}CCH{=}N_2$, and NCN by flash photolysis of cyanogen azide, NCN_3. The CNC molecule is observed by flash photolysis of diazoacetonitrile, $HC(CN)N_2$.

These examples serve to show how difficult it is to predict the products of flash photolysis from the structure of the parent molecule. The case of the discovery of C_3 is a particularly interesting example.†

In 1882 strong bands were observed in the 405 nm region of the emission spectra of comets. They were thought originally to be due to CH_2, a molecule which was unknown at that time. Attempts were made to reproduce the spectrum in the laboratory by passing a high voltage discharge through methane, CH_4. The spectrum was not found when the discharge was continuous but only when it was repeatedly turned on and off. The fact that the spectrum was obtained with methane as the parent molecule appeared to confirm the assignment to CH_2. In 1949 the experiment was repeated with C^2H_4 and the 405 nm spectrum was identical with that obtained with C^1H_4, making the CH_2 assignment untenable. It was in 1951 that Douglas obtained the 405 nm spectrum in absorption from the flash photolysis of diazomethane and showed that it was due to C_3. Ironically, the CH_2 molecule was also observed by the flash photolysis of diazomethane.

Flash photolysis of a mixture of BCl_3 and O_2 was used to obtain the spectrum of BO_2. The spectrum also appears in emission giving the characteristic green colour of boron-containing flames. Flash photolysis of CF_2Br_2 and also several fluorocarbons including tetrafluoroethylene, C_2F_4, has been used to obtain the spectrum of CF_2.

N_3 was produced by the flash photolysis of hydrazoic acid, HN_3.

Discharges are a useful source of spectra of short-lived molecules and, particularly, positive ions. The emission spectrum of CF_2 is observed in a discharge in CF_4 and that of CO_2^+ in a discharge in gaseous mixtures containing either CO_2 or CO.

The CO_2^- ion is a special case in that it has been observed only in the solid state and the ground state bond angle obtained in an electron spin resonance experiment.

Excluding the $1s$ core electrons of each atom, 16 valence electrons are the maximum that can be accommodated in the MOs of Figure 6.65(a) if the molecule is to be linear. The experimentally determined angles in Table 6.13 illustrate this. The table shows also that the order of the $3\sigma_u$ and $1\pi_u$ MOs is switched from that in Figure 6.65(a) for molecules with fewer than 15 electrons.

In the excited $\tilde{A}$ state of CO_2 the molecule is bent with an angle of 122°. This is a result of promoting an electron from the $4b_2$ orbital, which favours a linear molecule, to the $6a_1$ orbital, which strongly favours a bent molecule.

The ground configurations of NO_2 and CO_2^- result in a larger bond angle of 134° because of the extra electron in $4b_2$ compared with the $\tilde{A}$ state of CO_2. A second electron in $6a_1$, as in the ground state of ozone, gives a smaller angle of 116.8°.

The situation with regard to the lower energy excited states of NO_2 is complicated by the strong angle dependence of the $4b_2$ and $6a_1$ orbital energies. As shown in Table 6.13, the lowest excited state, $\tilde{A}^2B_2$, arises from the configuration in which there are two electrons in $6a_1$ and one in $4b_2$. At the decreased equilibrium bond angle of 102–118° the $6a_1$ orbital is lower in energy than the $4b_2$ orbital. In the $\tilde{B}^2B_1$ excited state the electron in the $6a_1$

†See Herzberg's *Electronic Spectra of Polyatomic Molecules*, referred to in the Bibliography at the end of the chapter, for references to this and most of the other AB_2 spectra discussed here.

orbital has been promoted to the $2b_1$ orbital, resulting in an increase in bond angle to *ca* 150°.

The potential energy curves which have been calculated[89] for some of the lower electronic states are shown on the right-hand side of Figure 6.66(b). This calculation shows curves for the $\tilde{X}$, $\tilde{A}$ and $\tilde{B}$ states and also for a $^2A_2$ excited state which arises from the . . . $(1a_2)^1$ $(4b_2)^2$ $(6a_1)^2$ configuration.

The visible absorption spectrum of NO_2 at room temperature involves, principally, the $\tilde{A}$–$\tilde{X}$ band system and gives rise to the orange–brown colour of the gas. The spectrum is so extremely dense that very little of it could be assigned until it was observed under the conditions of drastic cooling in a supersonic jet.[90] The unexpectedly high density of bands is attributable to an interaction in the region where, as Figure 6.66(b) shows, the $\tilde{X}^2A_1$ and $\tilde{A}^2B_2$ potential curves would cross. This will be discussed further in Section 8.2.16.

By applying the electric dipole selection rules of equations (6.111) it can be seen that all the $\tilde{A}$–$\tilde{X}$ transitions of the molecules in Table 6.13 (and the $\tilde{B}$–$\tilde{X}$ transition of NO_2) are allowed.

In diatomic molecules we are able to use, for heteronuclear molecules, the MOs constructed for homonuclear molecules, provided that the atoms are sufficiently similar. In the same way, we can use the MOs of Figure 6.65(a) for such molecules as CCN, CCO, NCO, N_2O^+ and N_2O. The CCN, CCO and NCO molecules are formed by flash photolysis of diazoacetonitrile [HC(CN)N_2], carbon suboxide (C_3O_2) and isocyanic acid (HNCO), respectively. There are 13 valence electrons in CCN and 14 in CCO; both have linear ground states. So also do the 15 valence electron molecules NCO, N_2O^+, whose emission spectrum is observed in a discharge in N_2O, and the N_2O molecule, which has 16 valence electrons.

Some AB_2 molecules involving second-row elements, such as CS_2, CS_2^+ and SiF_2, are known and their structures can be explained similarly in terms of a simple MO picture.

In molecules of the type HAB the AOs involved in the valence MOs are 1*s* on H and 2*s*, 2*p* on both A and B, if they are first-row atoms. This type of molecule may be linear or bent, with an extreme bond angle of 90°.

The Walsh diagram in Figure 6.65(b) is constructed in a similar way to that for AB_2 molecules in Figure 6.65(a) and shows how the energies of the resulting MOs change, and how the MOs are correlated, as the HAB angle changes from 180° to 90°.

In the linear HAB molecule the 1*s* AOs on A and B are assumed to be little changed and form the 1σ and 2σ MOs, not shown in Figure 6.65(b). The 3σ MO consists mainly of the 2*s* AO on B. The 4σ MO is formed from the 1*s* AO on H and, principally, the 2*s* AOs on A and B, while the 5σ MO is formed from the 1*s* AO on H and, principally, the $2p_z$ AOs on A and B. The 1π MO is the doubly degenerate π orbital which is bonding between A and B and is formed by the in-phase overlap of the $2p_x$ and $2p_y$ AOs on A and B. The 2π MO is also doubly degenerate but is antibonding, resulting from the out-of-phase overlap of the $2p_x$ and $2p_y$ AOs. The 6σ MO is an antibonding σ orbital.

When the molecule bends, the orbital energies of the σ MOs mainly increase, as in AH_2 and AB_2 molecules. The degeneracy of the π MOs is split but, similar to the case of the bonding $1\pi_u$ MO in AB_2 molecules, shown in Figure 6..66(a), the 1π MO is not split appreciably. Much more important is the splitting of the antibonding 2π MO. The component which is out of the plane of bending is unaffected and becomes the $2a''$ MO, classified according to the C_s point group to which the bent HAB molecule belongs. The other component, in the plane of bending, attains much more 2*s* character on A leading to a steep fall in the orbital energy. As a result, molecules with electron configurations in which this $7a'$–2π MO is singly or doubly occupied are bent whereas other molecules are linear.

Table 6.14 illustrates this point. All the known HAB molecules which have more than 10 valence electrons have a non-linear ground state. The table shows that one electron in the $7a'$ MO results in an HAB angle in the range 114–125°, although the $\tilde{A}^2A'$ state of HO_2 is an

Table 6.14 Ground and low energy excited configurations of some HAB molecules

Molecule	No. of valence electrons	Electron configuration							State	∠ HAB/°
C_2H	9	... $(3\sigma)^2$	$(4\sigma)^2$	$(5\sigma)^1$	$(1\pi)^4$				$\tilde{X}^2\Sigma^+$	180
		... ...	...	$(5\sigma)^2$	$(1\pi)^3$				$\tilde{A}^2\Pi$	180° (?)
HCN, HNC, N_2H^+, HCO^+	10	... ...	...	$(5\sigma)^2$	$(1\pi)^4$				$\tilde{X}^1\Sigma^+$	180
		... $(3a')^2$	$(4a')^2$	$(5a')^2$	$(1a'')^1$	$(6a')^2$	$(7a')^1$		$\tilde{A}^1A''$	125.0 (HCN)
		... ...	...	...	$(1a'')^2$	$(6a')^1$	$(7a')^1$		$\tilde{B}^1A''$	114.5 (HCN)
HCO	11	... ...	...	...	$(1a'')^2$	$(6a')^2$	$(7a')^1$		$\tilde{X}^2A'$	119.5
		... $(3\sigma)^2$	$(4\sigma)^2$	$(5\sigma)^2$	$(1\pi)^4$	$(2\pi)^1$			$\tilde{A}^2\Pi$	180
		... $(3a')^2$	$(4a')^2$	$(5a')^2$	$(1a'')^2$	$(6a')^1$	$(7a')^2$		$\tilde{B}^2A'$	111±4
HNO, HCF	12	... $(3a')^2$	$(4a')^2$	$(5a')^2$	$(1a'')^2$	$(6a')^2$	$(7a')^2$		$\tilde{X}^1A'$	108.6 (HNO), 101.6 (HCF)
		... ...	...	...	$(6a')^2$	$(7a')^1$	$(2a'')^1$		$\tilde{A}^1A''$	116.3 (HNO), 127.6 (HCF)
HO_2, HNF	13	... ...	...	...	$(6a')^2$	$(7a')^2$	$(2a'')^1$		$\tilde{X}^2A''$	104.0 (HO_2), 105 (HNF)
		... ...	...	...	$(6a')^2$	$(7a')^1$	$(2a'')^2$		$\tilde{A}^2A'$	102.7 (HO_2), 125 (HNF)
HO_2^-	14	... ...	...	...	$(6a')^2$	$(7a')^2$	$(2a'')^2$		$\tilde{X}^1A'$	104

exception. When there are two electrons in the $7a'$ MO the angle is reduced and, for the few examples known, it lies in the range 101–111°. The ranges of angle for the $(7a')^1$ and $(7a')^2$ configurations are much wider than we have encountered for the corresponding configurations of AH_2 and AB_2 molecules. It is clear that the Walsh diagram for HAB molecules is less useful in a quantitative sense. Indeed, the general theory used in justifying the diagram seems to be violated in the $\tilde{A}$ state of HO_2 but the insertion of a second electron in the $2a''$ MO, as in the ground state $\tilde{X}^1A'$ of HO_2^-, produces,[92] as expected, no change in bond angle from that of 104° for HO_2.

The C_2H molecule, having nine valence electrons, is linear in the ground state, $\tilde{X}^2\Pi$. This molecule has been observed by its pure rotation spectrum in the interstellar medium[91] (Section 4.7).

The HCN molecule is of particular interest as it is linear in its ground state and bent in those excited states which are known and in which there is one electron in the $7a'$ MO. The HCP molecule, containing a second-row atom, shows similar behaviour, the angle being 128° in the $\tilde{A}^1A''$ state.

The HNC, N_2H^+ and HCO^+ molecules are all isoelectronic with HCN and were all first observed, by their pure rotation spectra, in the interstellar medium.[93–95] These three molecules are known only in their ground electronic states in which all are linear. All the molecules were later observed in discharges in the laboratory.[96–98]

The HCO molecule is formed by the flash photolysis of several aldehydes, including acetaldehyde, CH_3CHO. The $\tilde{A}^2\Pi$–$\tilde{X}^2A'$ absorption spectrum shows that the ground state is bent and the $\tilde{A}$ state is linear due to promotion of an electron from the $7a'$ to the $2a''$ MO. The $\tilde{X}$ and $\tilde{A}$ states both derive from a $^2\Pi$ state of the linear molecule which is split by the Renner–Teller effect (Section 6.3.4.4) giving a bent lower state and a linear upper state, as in Figure 6.99(c).

It had long been thought that the bands observed in the near ultraviolet emission spectra of several hydrocarbon flames in air, the so-called hydrocarbon flame bands, were due to the HCO molecule. The bands were later identified[99] as belonging to the $\tilde{B}$–$\tilde{X}$ and $\tilde{C}$–$\tilde{X}$ systems of this molecule and the HCO angle in the $\tilde{B}$ state was found to be 111±4°, the reduction from the $\tilde{X}$ state being consistent with an electron promotion to the $7a'$ MO.

The HNO and HCF molecules are produced by the flash photolysis of a mixture of NH_3 and NO, to give HNO, and of $CHFBr_2$, to give HCF. Both molecules are strongly bent in their ground states, in which there are two electrons in the $7a'$ MO and less so in their $\tilde{A}$ states, in which an electron has been promoted from the $7a'$ to the $2a''$ MO.

The search for an electronic spectrum of HO_2 proved, for a long time, unrewarding. The species was known to exist from mass spectrometry but no spectrum of any kind was observed until 1974 when the pure rotation spectrum was detected by the very sensitive technique of laser magnetic resonance (Section 8.2.6). The production of HO_2 for this purpose by, for example, passing O_2 through a microwave discharge and then into a cell containing acetylene led to a renewed search for an electronic spectrum, expected to be in the near infrared. This was first observed in 1974 and the emission spectrum recorded at high resolution and analysed in 1979.[100] The bond angle in the ground state is consistent with that of other HAB molecules having two electrons in the $7a'$ MO, but the fact that there is not an appreciable increase in the $\tilde{A}^2A'$ state is completely at variance with the geometry of other molecules which have only one electron in the $7a'$ MO.

Table 6.14 shows that the geometry in the $\tilde{X}$ and $\tilde{A}$ state of the isoelectronic HNF molecule,[101] produced by flash photolysis of HNF_2, is in good agreement with the prediction from the Walsh diagram.

6.3.1.3 AH_3 *molecules*

Molecules of this type may be either planar or non-planar belonging to either the D_{3h} or C_{3v} point group, respectively. In both point groups the C_3 axis is the z-axis.

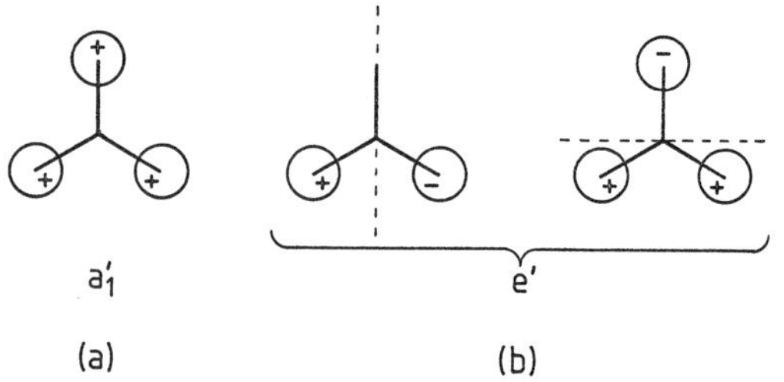

Figure 6.67 (a) An a_1' and (b) an e' MO formed from hydrogen $1s$ AOs in a planar AH_3 molecule

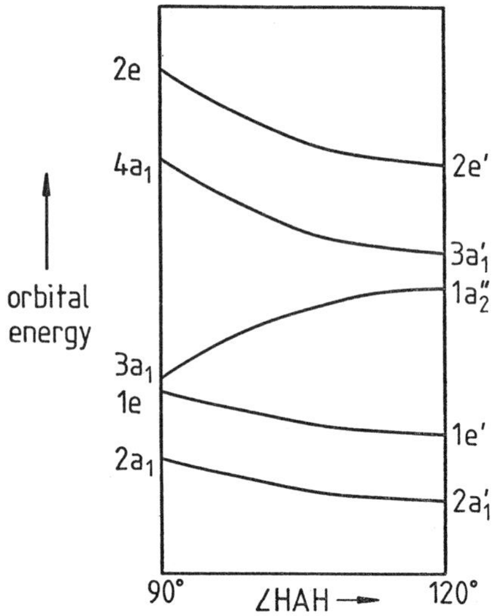

Figure 6.68 Walsh MO diagram for AH_3 molecules

The AOs to be considered in the formation of MOs are the outer s and p valence orbitals on A, $2s$ and $2p$ if it is a first-row element, and $1s$ on H.

In the planar molecule the $2p_z$ orbital on A belongs to the a_2'' symmetry species (see Table A.33 in the Appendix for the D_{3h} character table) and cannot mix with any other AOs since none of them belongs to this species.

The $2s$ orbital on A belongs to the a_1' species and $2p_x$ and $2p_y$ are doubly degenerate and belong to the e' species.

The three hydrogen $1s$ orbitals may be either all in-phase to form the a_1' combination in Figure 6.67(a) or there may be two orthogonal nodal planes, one choice being shown in Figure 6.67(b), resulting in an e' pair of degenerate orbitals.

Bonding occurs by overlap of the a_1' orbital of Figure 6.67(a) with the $2s$ a_1' orbital on A, and of the e' orbitals of Figure 6.67(b) with $2p_x$ and $2p_y$ e' orbitals on A. In-phase overlap produces a $2a_1'$ and a $1e'$ bonding MO and out-of-phase overlap a $3a_1'$ and a $2e'$ anti-bonding MO. These, together with the $2p_z$ $1a_2''$ lone pair orbital, are arranged in order of increasing energy on the right-hand side of Figure 6.68. The $1a_1'$ MO is the unchanged $1s$ AO on A.

In non-planar AH_3 the e' MOs become e and the a_1' MOs become a_1 (see Table 5.8 for the C_{3v} character table). However, the $1a_2''$ MO, which is non-bonding in the planar configuration, also becomes a_1 and can mix with the other a_1 MOs. The mixing leads to an increase in s character in $3a_1$ and produces a steep fall in the energy of the $1a_2''$–$3a_1$ orbital as the HAH angle decreases from 120° to 90°.

The only AH_3 molecules known, in which A is a first-row element, are CH_3, NH_3 and H_3O^+.

The CH_3 molecule is a short-lived species whose absorption spectrum was first observed following the flash photolysis of dimethylmercury, $Hg(CH_3)_2$. It is also produced by flash photolysis of acetaldehyde, CH_3CHO, acetone, $(CH_3)_2CO$, and some methyl halides, but greater concentrations result from flash photolysis of a mixture of diazomethane, CH_2N_2, and molecular hydrogen; the former produces CH_2, which reacts with hydrogen atoms produced from the latter.

The seven valence electrons of CH_3, fed into the D_{3h} MOs of Figure 6.68, give the ground configuration.

$$\ldots (2a_1')^2(1e')^4(1a_2'')^1 \tag{6.157}$$

The only state arising from this configuration is the ground state $\tilde{X}^2A_2''$.

The only electronic transitions observed in CH_3 are from the ground state to various excited states resulting from the promotion of the $1a_2''$ electron to Rydberg orbitals. Evidence from these band systems indicates very strongly that CH_3 is planar in the $\tilde{X}$ state [see Section 6.3.5.2(i)] and also in the various Rydberg states. Therefore, the effect of one electron in the $1a_2''$ orbital, for which a non-planar molecule is favoured, is not sufficient to counteract the effect of those in the $2a_1'$ and $1e'$ orbitals, for which a planar molecule is favoured.

Two low-lying excited configurations of CH_3 are

$$\ldots (2a_1')^2(1e')^4(3a_1')^1 \tag{6.158}$$

giving a $^2A_1'$ excited state, and

$$\ldots (2a_1')^2(1e')^3(1a_2'')^2 \tag{6.159}$$

giving a $^2E'$ excited state. The molecule should be planar in the $^2A_1'$ state. The $^2A_1'$–$\tilde{X}^2A_2''$ transition is allowed since, from equation (6.111),

$$\Gamma(\psi_e') \times \Gamma(\psi_e'') = A_1' \times A_2'' = A_2'' = \Gamma(T_z) \tag{6.160}$$

but it has not been observed.

The $^2E'$–$\tilde{X}^2A_2''$ transition is forbidden in the planar molecule but the configuration of equation (6.159), in which there are two electrons in the $1a_2''$ MO, should result in a non-planar molecule. In the C_{3v} point group the transition becomes ^2E–$\tilde{X}^2A_1$ (see Table 5.19 for C_{3v}–D_{3h} correlations), which is allowed but expected to be weak; it has not been observed.

The ground configuration of NH_3 is

$$\ldots (2a_1)^2(1e)^4(3a_1)^2 \tag{6.161}$$

and the resulting state is $\tilde{X}^1A_1$.† The non-planarity in this state is expected because of the

†$\tilde{X}^1A_1'$, when tunnelling through the inversion barrier is taken into account and D_{3h} classification is appropriate.

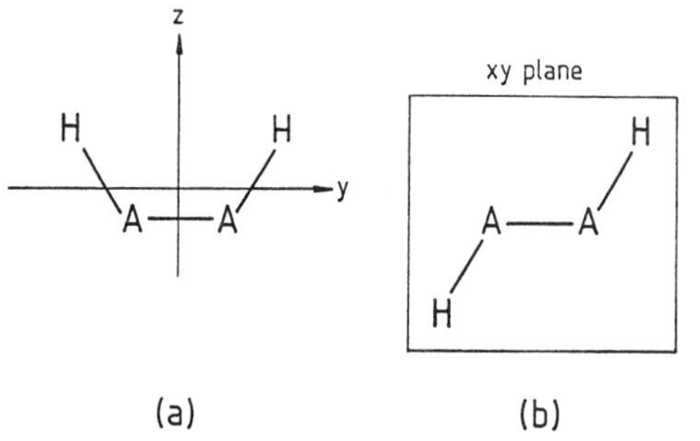

Figure 6.69 An HAAH molecule in the (a) *cis* and (b) *trans* planar configurations

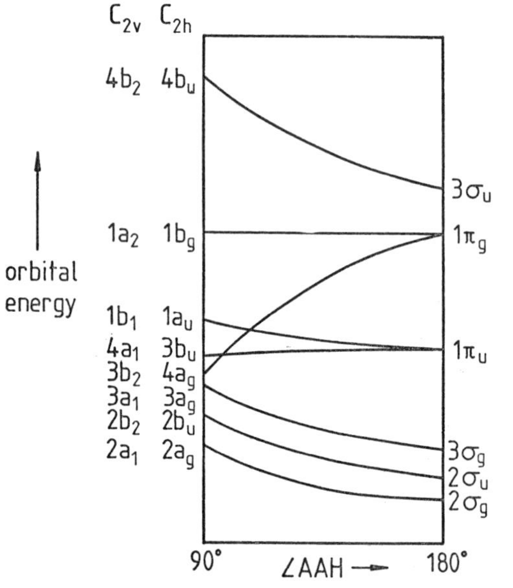

Figure 6.70 Walsh MO diagram for HAAH molecules

effect of two electrons in the $3a_1$–$1a_2''$ orbital. The HNH angle is 107.8°, slightly less than the tetrahedral angle of 109.5°. The angle is similar to the isoelectronic H_3O^+.

As in the case of CH_3, no transitions have been observed other than Rydberg transitions involving the promotion of an electron from $3a_1$ to a Rydberg orbital. The result of such a promotion is to produce a planar molecule, consistent with the ground state of CH_3.

No electronic spectrum of BH_3 has yet been observed. An attempt to observe the absorption spectrum following the flash photolysis of borine carbonyl, BH_3CO, resulted in the observation of spectra of B, BH and BH_2, but not BH_3. The expected ground configuration of BH_3 is

$$\ldots (2a_1')^2(1e')^4 \tag{6.162}$$

and it should be planar in the resulting $\tilde{X}^1A_1'$ ground state, analogous to the Rydberg states of CH_3.

The only other AH_3 molecules for which the HAH angle is known are phosphine, PH_3, arsine, AsH_3, and stibine, SbH_3. They have ground state angles of 93.3°, 92.1° and 91.6°, respectively. This decrease of the angle going down the periodic table parallels the observations in AH_2 molecules.

6.3.1.4 HAAH *molecules*

Molecules of this type, such as acetylene, HC≡CH, may have an AAH angle of 180° or less. When the molecule is nonlinear it may exist in either the *cis* or *trans* form, belonging to the C_{2v} or C_{2h} point group, respectively, as shown in Figure 6.69. In assigning the bent molecule to either the C_{2v} or C_{2h} point group we are assuming that the molecule is planar, a point to which we shall return.

In linear HAAH the $1\sigma_g$ and $1\sigma_u$ MOs are formed from almost unchanged 1*s* orbitals of the A atoms. These are not included in the MO diagram in Figure 6.70.

There are six σ MOs resulting from the overlap of the 2*s* and 2*p* AOs on A and the 1*s* AOs on H. Three of them are strongly bonding and are labelled $2\sigma_g$, $2\sigma_u$ and $3\sigma_g$.

In-phase overlap of the $2p_x$ and $2p_y$ AOs on the A atoms gives the doubly degenerate $1\pi_u$ bonding MO, while out-of-phase overlap gives the $1\pi_g$ antibonding MO, with a nodal plane perpendicular to the A–A bond.

In the bent C_{2h} molecule, with the axis notation in Figure 6.69(b), the σ_g orbitals become a_g and the σ_u become b_u and, for reasons similar to those given for other molecules, the linear molecule is favoured by them. The degeneracy of the $1\pi_u$ orbital is split, but only slightly, in the bent molecule. It is the behaviour of the $1\pi_g$ orbital which is important if a bent molecule is to be favoured. The component of $1\pi_g$ perpendicular to the xy plane of the C_{2h} molecule is unaffected by bending but the component in the plane can mix with the $3\sigma_g$ bonding orbital, thereby taking on some $2s$ character and resulting in the steep energy decrease with AAH angle shown in Figure 6.70.

In the C_{2v} bent molecule the MOs are different from those in the C_{2h} molecule because of the different symmetry requirements allowing orbitals in the linear molecule to mix on bending. For example, the component of the $1\pi_g$ orbital, which has a nodal plane perpendicular to the plane of the bent molecule, mixes with $2\sigma_u$ and $3\sigma_u$, as they all become b_2 in C_{2v}. This mixing imparts s character to the $1\pi_g$ orbital but the mechanism differs from that in C_{2h}.

In acetylene there are 10 valence electrons to be fed into the MOs of Figure 6.70, giving the ground configuration

$$\ldots (2\sigma_g)^2(2\sigma_u)^2(3\sigma_g)^2(1\pi_u)^4 \tag{6.163}$$

The ground state is $\tilde{X}^1\Sigma_g^+$ and the molecule is linear. However, promotion of an electron from the $1\pi_u$ to the $1\pi_g$–$4a_g$ orbital causes the molecule to become bent with the electron configuration

$$\ldots (2a_g)^2(2b_u)^2(3a_g)^2(3b_u)^2(1a_u)^1(4a_g)^1 \quad \text{in the } trans \text{ form}$$

$$\ldots (2a_1)^2(2b_2)^2(3a_1)^2(4a_1)^2(1b_1)^1(3b_2)^1 \quad \text{in the } cis \text{ form} \tag{6.164}$$

The states arising from these configurations include $\tilde{a}^3A_2$ (*cis*) and $\tilde{A}^1A_u$ (*trans*). The $\tilde{a}^3A_2$–$\tilde{X}^1\Sigma_g^+$ transition is spin-forbidden and has been observed only in the presence of oxygen at high pressure which breaks down the spin selection rule. The geometry in the $\tilde{a}$ state is thought, from observation of the $\tilde{b}^3B_2$–$\tilde{a}^3A_2$ transition to be *cis* bent.[102]

The $\tilde{A}^1A_u$–$\tilde{X}^1\Sigma_g^+$ transition has been observed in absorption, in the 210–237 nm region, and the molecule shown to have a C_{2h} configuration in the $\tilde{A}$ state with a CCH angle of 120°. Coupled with this angle change there is a large increase in the CC bond length, from 1.208 Å in the $\tilde{X}$ to 1.39 Å in the $\tilde{A}$ state, consistent with promotion of an electron from an orbital which is bonding to one which is antibonding in the CC bond.

In hybrid orbital language we could describe the carbon atoms as sp^2 hybridized in the $\tilde{A}$ state, consistent with the 120° angle.

Diimide, N_2H_2, is a reactive molecule but has a lifetime of about 30 min in favourable conditions. It has 12 valence electrons and the ground configuration

$$\ldots (2a_g)^2(2b_u)^2(3a_g)^2(3b_u)^2(1a_u)^2(4a_g)^2 \tag{6.165}$$

As a consequence of having two electrons in the $4a_g$ orbital, it is more strongly bent in the $\tilde{X}^1A_g$ ground state than is C_2H_2 in the $\tilde{A}^1A_u$ excited state; it has an NNH angle of 107°. As is the case for the $\tilde{A}$ state of acetylene, only the *trans* configuration is known.

The first excited configuration of diimide is

$$\ldots (2a_g)^2(2b_u)^2(3a_g)^2(3b_u)^2(1a_u)^2(4a_g)^1(1b_g)^1 \tag{6.166}$$

giving an $\tilde{a}^3B_g$ and an $\tilde{A}^1B_g$ state. The $\tilde{A}^1B_g$–$\tilde{X}^1A_g$ electronic transition is forbidden [see equation (6.111)], but a band system observed in the 300–430 nm region has been attributed to this transition.[103] In a polyatomic molecule a band system may be observed, even though the pure electronic transition is forbidden. The transition is said to be vibrationally induced, an effect which will be discussed in Section 6.3.4.1.

In the $\tilde{A}$ state of diimide the NNH angle is increased to about 132°, consistent with electron promotion from the $4a_g$ to the $1b_g$ orbital.

Hydrogen peroxide, H_2O_2, has 14 valence electrons, giving a ground configuration

$$\ldots (2a_g)^2(2b_u)^2(3a_g)^2(3b_u)^2(1a_u)^2(4a_g)^2(1b_g)^2 \tag{6.167}$$

The implication from the MOs of Figure 6.70 is that the shape should be the same as that of the ground state of N_2H_2 since the $1b_g$–$1\pi_g$ orbital favours neither the linear nor the bent molecule. However, we must remember that this orbital is strongly antibonding between the two A atoms and, if there are electrons in this orbital, the energy could be lowered by twisting the molecule about the A–A bond. This happens in the ground state of H_2O_2. As figures 5.27(a) shows, the dihedral angle between the two OOH planes is 118° and the molecule belongs to the C_2 point group. The OOH angle is 94.8°, rather lower than in the $\tilde{X}$ state of N_2H_2.

The H_2O_2 molecule does not show a discrete electronic spectrum so there is no information on the geometry in excited states.

Hydrogen persulphide, H_2S_2, shows a similar structure to H_2O_2 in its ground electronic state. The dihedral angle is 90.6° and the SSH angle is 91.3°.

6.3.1.5 H_2AB *molecules*

The only stable molecule of this type is formaldehyde, H_2CO, which is planar in its ground electronic state $\tilde{X}$ and belongs to the C_{2v} point group, as shown in Figure 6.71(a). In the first excited triplet and singlet states, $\tilde{a}$ and $\tilde{A}$, the molecule is pyramidal with an out-of-plane angle, defined by ϕ as shown in Figure 6.71(b), of 43° in the $\tilde{a}$ state and 38° in the $\tilde{A}$ state.† In these states the molecule belongs to the C_s point group with the plane of symmetry bisecting the HCH angle.

These observations suggest that, in a Walsh diagram for H_2AB molecules, orbital energies should be plotted as a function of the HAB angle, which is taken to be equal to the HAH angle. In the planar configuration the angle is 120° and, in the extreme non-planar configuration, it is 90°.

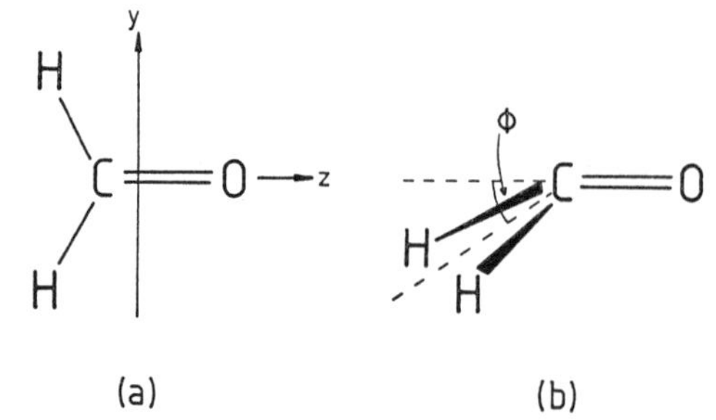

Figure 6.71 The formaldehyde molecule is (a) planar in its ground electronic state and (b) non-planar in the $\tilde{a}$ and $\tilde{A}$ excited electronic states

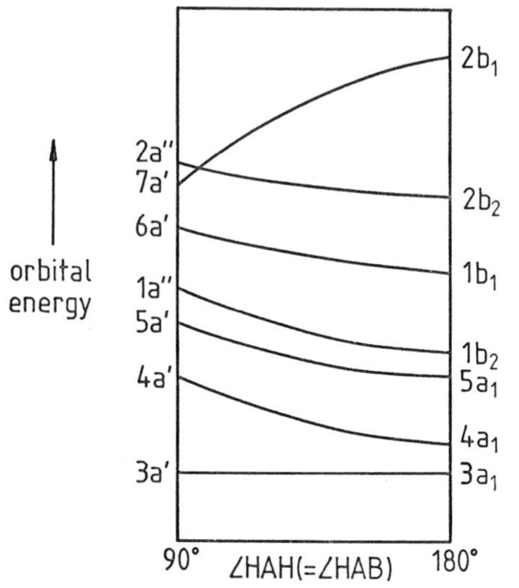

Figure 6.72 Walsh MO diagram for H_2AB molecules

†These values were obtained by Coon, J. B., Naugle, N. W. and McKenzie, R. D. (1966). *J. Mol. Spectrosc.*, **20**, 107 from the vibrational term values in these states and are larger than those obtained from rotational constants for the zero-point levels.

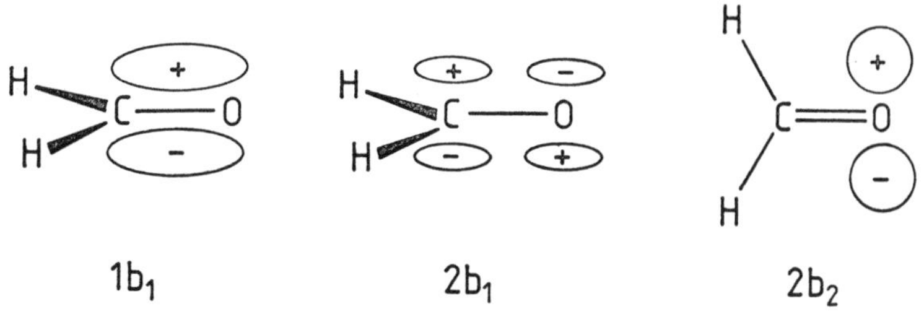

Figure 6.73 The $1b_1(\pi)$, $2b_1(\pi^*)$ and $2b_2(n)$ MOs in formaldehyde

The Walsh diagram is shown in Figure 6.72.

In the planar molecule the $3a_1$, $4a_1$, $5a_1$ and $1b_2$ MOs are all bonding orbitals of the σ bond framework of the molecule. The orbital energies are either little changed or slightly increased in the non-planar molecule.

In formaldehyde the $1b_1$ orbital is a π-type orbital, bonding between C and O. The $2b_1$ orbital is also a π-type orbital but antibonding between C and O, and the $2b_2$ orbital is the non-bonding $2p_y$ orbital on O [see Figure 6.71(a) for axis notation]. All three MOs are illustrated in Figure 6.73. The $1b_1$ orbital increases in energy slightly in the non-planar molecule, as does the $2b_2$ non-bonding orbital, but the energy of $2b_1$ falls steeply, as Figure 6.72 shows. The reason for this is that the $2b_1$ orbital between C and O is able to overlap with the hydrogen $1s+1s$ orbital, thereby gaining some C–H bonding character, and is also able to mix with the $2s$ AO on C. The result is an increase in s character and a substantial lowering of the energy in the non-planar molecule.

The H_2CO molecule has 12 electrons to be fed into the MOs of Figure 6.72, giving the ground configuration

$$\ldots (1b_2)^2(1b_1)^2(2b_2)^2 \tag{6.168}$$

leading to an $\tilde{X}^1A_1$ ground state and a planar molecule.

The lowest energy electron promotion is from the $2b_2$ non-bonding, or n, orbital into the $2b_1$ anti-bonding, or π^*, orbital, giving the configuration

$$\ldots (1a'')^2(6a')^2(2a'')^1(7a')^1 \tag{6.169}$$

resulting in $\tilde{a}^3A''$ and $\tilde{A}^1A''$ states. C_s point group symmetry species are used because of the non-planarity in these states.

Orbital promotions of this type give rise to states, such as the ã and Ã states of formaldehyde, which are commonly referred to as $n\pi^*$ states. In addition, transitions to such states, for example the ã–X̃ and Ã–X̃ transitions of formaldehyde, are referred to colloquially as π^*–n, or n-to-π^*, transitions.

There is a useful way of distinguishing a transition of the π^*–n type from one of, say, the π^*–π type. The former is blue shifted, i.e. shifted to lower wavelength, in a hydrogen-bonding solvent. The reason is that such a solvent, e.g. ethanol, forms a hydrogen bond by weak MO formation between the n orbital and the $1s$ orbital on the hydrogen atom of the OH group of the solvent. This lowers the energy of the n orbital and therefore increases the energy of the π^*–n transition.

As we have seen, formaldehyde is non-planar in both the ã and Ã $n\pi^*$ states and this is rationalized as being the effect of the steeply falling energy of the $2b_1$–$7a'$ orbital as the out-of-plane angle increases.

We shall see in Section 6.3.4.2(ii) that, although the molecule is non-planar in the ã and Ã states, there is tunnelling through the barrier which splits the zero-point level and requires the use of C_{2v} classification in both excited states (see Section 5.2.7.1 for a discussion of inversion). Therefore, the lowest energy electronic transitions are more usefully designated $\tilde{a}^3A_2$–$\tilde{X}^1A_1$ and $\tilde{A}^1A_2$–$\tilde{X}^1A_1$.

Since A_2 is not a translational symmetry species, equation (6.111) shows that the $\tilde{A}$–$\tilde{X}$ transition is electric dipole forbidden. The fact that the pure electronic transition is observed weakly, at about 355 nm, is because it is allowed by magnetic dipole selection rules,[104] given by equation (6.118). The $\tilde{A}^1A_2$–$\tilde{X}^1A_1$ transition is magnetic dipole allowed because $A_2 = \Gamma(R_z)$ and is one of few observed examples of magnetic dipole transitions in polyatomic molecules. In addition, stronger vibronic transitions are observed as electric dipole transitions which are allowed by vibronic interaction (see Section 6.3.4.1).

The $\tilde{a}$–$\tilde{X}$ system is observed in absorption in the 360–397 nm region. It is spin forbidden but appreciable spin–orbit coupling due to the relatively high charge on the oxygen nucleus breaks down the $\Delta S = 0$ selection rule sufficiently for it to be observed. This spin–orbit coupling may mix either the $\tilde{X}$ state (S_0) with an excited triplet state (T_s) or the $\tilde{a}$ (T_1) with an excited singlet state (S_s), or both. These possibilities are summarized in equations (6.114) and (6.115) and Figure 6.32.

It is more usual for the mixing of T_1 with S_s to be the dominant mechanism. In the case of the $\tilde{a}$ state of formaldehyde this gives, from equation (6.115),

$$\Gamma(S_s) \times B_2 = A_2, \therefore \Gamma(S_s) = B_1 \qquad (6.170)$$

$$\Gamma(S_s) \times B_1 = A_2, \therefore \Gamma(S_s) = B_2 \qquad (6.171)$$

$$\Gamma(S_s) \times A_2 = A_2, \therefore \Gamma(S_s) = A_1 \qquad (6.172)$$

Raynes[105] has shown that the mechanism of equation (6.172) is dominant, which means that the $\tilde{a}^3A_2$–$\tilde{X}^1A_1$ transition obtains most of its intensity by mixing of the $\tilde{a}^3A_2$ with the $^1A_1$ state. The intensity is stolen from a $^1A_1$–$\tilde{X}^1A_1$ transition but the energy location of the $^1A_1$ excited state is not certain. Since $A_1 = \Gamma(T_z)$, the transition moment of the $^1A_1$–$\tilde{X}^1A_1$ transition is polarized along the z-axis and therefore the

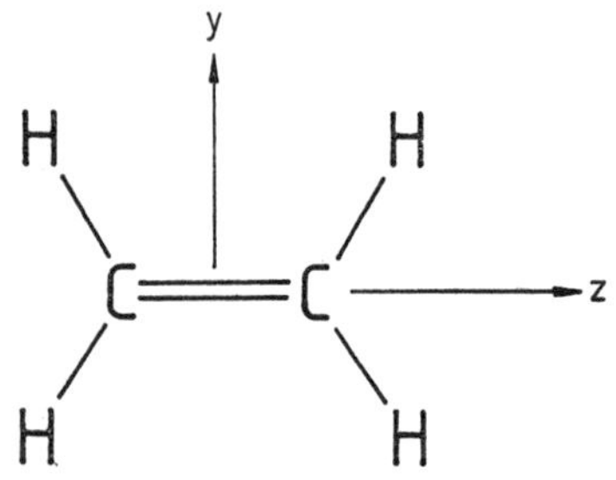

Figure 6.74 Axis notation in the ethylene molecule

transition moment of the $\tilde{a}^3A_2$–$\tilde{X}^1A_1$ transition is polarized along the same axis.

The absorption spectrum of the short-lived molecule thioformaldehyde, H_2CS, in the 700–400 nm region shows $\tilde{a}^3A_2$–$\tilde{X}^1A_1$ and $\tilde{A}^1A_2$–$\tilde{X}^1A_1$ systems analogous to those of formaldehyde but H_2CS is more nearly planar in both the $\tilde{a}$ and $\tilde{A}$ states.† The pure electronic transition of the $\tilde{A}$–$\tilde{X}$ system is a magnetic dipole transition.

6.3.1.6 H_2AAH_2 *molecules*

The only molecule of this type for which information on the geometry in the ground and some excited states is available is ethylene, $H_2C{=}CH_2$. This is planar in its ground state and belongs to the D_{2h} point group; it is shown, with the recommended axis notation, in Figure 6.74.

One way of picturing the MOs in a planar H_2AAH_2 molecule is to regard them as in-phase and out-of-phase combinations of the orbitals for bent AH_2 shown in Figure 6.63. The out-of-phase combination always leads to a higher energy orbital than the in-phase combination because of the additional node. The resulting ordering of MO energies is indicated in Figure

†The inversion barrier is only about 80 cm^{-1} in the $\tilde{A}$ state—see Judge, R. H. and King, G. W. (1979). *J. Mol. Spectrosc.*, **74**, 175.

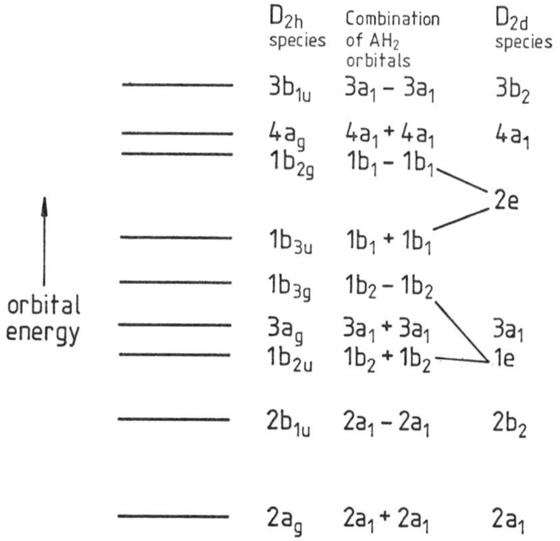

Figure 6.75 MOs in H_2AAH_2 molecules in the D_{2h} and D_{2d} configurations

6.75 and, for the planar molecule, they are classified according to the D_{2h} point group.

Ethylene has 12 electrons to be fed into these orbitals, giving the ground configuration

$$\ldots (1b_{2u})^2(3a_g)^2(1b_{3g})^2(1b_{3u})^2 \qquad (6.173)$$

and an $\tilde{X}^1A_g$ ground state.

The first excited configuration is

$$\ldots (1b_{2u})^2(3a_g)^2(1b_{3g})^2(1b_{3u})^1(1b_{2g})^1 \qquad (6.174)$$

giving the $\tilde{a}^3B_{1u}$ and $\tilde{A}^1B_{1u}$ states.

The $\tilde{a}^3B_{1u}$–$\tilde{X}^1A_g$ transition is spin-forbidden and extremely weak. It has been observed in the liquid phase, and also in the gas phase but only in the presence of oxygen at high pressure.

The $\tilde{A}^1B_{1u}$–$\tilde{X}^1A_g$ band system occurs in absorption in the 160–210 nm region and it has been shown† that the molecule is twisted in the $\tilde{A}$ state into a D_{2d} configuration in which the two CH_2 groups are perpendicular to each other.

†See Merer, A. J. and Mulliken, R. S. (1969). *Chem. Rev.*, **69**, 639 and references cited therein.

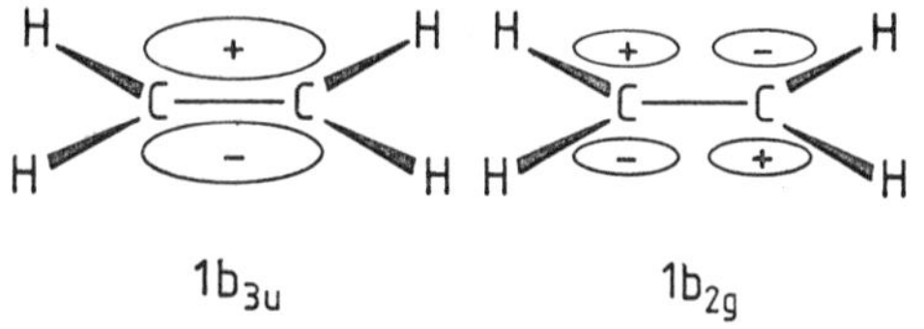

Figure 6.76 The $1b_{3u}$ (π) and $1b_{2g}(\pi^*)$ MOs in the ethylene molecule

The explanation for this is that the $1b_{3u}$ and $1b_{2g}$ orbitals involved in the promotion are the bonding and antibonding π-type orbitals shown in Figure 6.76. When an electron is in the $1b_{2g}$ antibonding orbital the energy can be lowered by the molecule twisting about the C–C bond, thereby minimizing the repulsion across the xy nodal plane in the orbital. The reason for twisting of ethylene in the $\tilde{A}$ state is, therefore, very similar to that for the twisting of hydrogen peroxide in the $\tilde{X}$ state.

The MOs of H_2AAH_2 are classified in Figure 6.75 according to the D_{2d} point group. The excited configuration of ethylene in equation (6.174) becomes

$$\ldots (1e)^4(3a_1)^2(2e)^2 \qquad (6.175)$$

In the D_{2d} point group

$$e \times e = A_1 + A_2 + B_1 + B_2 \qquad (6.176)$$

so the configuration of equation (6.175) gives rise to four states, $^1A_1$, $^3A_2$, $^1B_1$ and $^1B_2$ (see Table 6.6). The $\tilde{A}^1B_{1u}$ state in D_{2h} correlates with $\tilde{A}^1B_2$ in D_{2d}.

Hydrazine, N_2H_4, has two more electrons than ethylene and is twisted, with the two NH_2 groups perpendicular to each other, in the ground state, $\tilde{X}^1A_1$. The electron configuration for this state is

$$\ldots (1e)^4(3a_1)^2(2e)^4 \qquad (6.177)$$

In addition, there is a pyramidal configuration about each nitrogen atom so that, strictly, the point group is C_2.

The geometry of none of the excited states is known.

We have seen, in ethylene, that the molecular orbitals involved in the lowest energy electron promotion are the π orbitals shown in Figure 6.76. If we are concerned with low energy promotions, then only the π-type MOs, constructed from the $2p$ orbitals on the carbon atoms and perpendicular to the plane of the molecule, have to be considered and the σ MOs, constituting the single bond framework of the molecule, can be neglected. This separate treatment of the π MOs is one of the approximations used originally by Hückel and the resulting orbitals are referred to as Hückel molecular orbitals.

The Hückel method is based on the LCAO method discussed for diatomic molecules in Section 6.2.1.

In a polyatomic molecule the secular determinant, given by equation (6.86) for a diatomic, becomes

$$\begin{vmatrix} H_{11}-E & H_{12}-ES_{12} & \dots & H_{1n}-ES_{1n} \\ H_{12}-ES_{12} & H_{22}-E & \dots & H_{2n}-ES_{2n} \\ \vdots & \vdots & & \vdots \\ H_{1n}-ES_{1n} & H_{2n}-ES_{2n} & \dots & H_{nn}-E \end{vmatrix} = 0 \tag{6.178}$$

where the symbols have the same significance as in equation (6.86). This secular determinant can be written in an abbreviated form as

$$|H_{mn} - ES_{mn}| = 0 \tag{6.179}$$

where $S_{mn} = 1$ when $m = n$.

The Hückel method has been modified and extended but originally there were five approximations involved:

1. Only electrons in π orbitals are considered, those in σ orbitals being neglected. In a molecule, such as ethylene, with sufficiently high symmetry the σ–π separation is not an approximation since the σ and π MOs have different symmetry species and therefore cannot mix. In a molecule of low symmetry, such as but-1-ene, $CH_3CH_2CH{=}CH_2$, the σ and π MOs have the *same* symmetry species but, in the Hückel treatment, the σ MOs are still neglected since they can be assumed to be much lower in energy than the π MOs.
2. It is assumed that, for $m \neq n$, the overlap integral

$$S_{mn} = 0 \tag{6.180}$$

implying zero overlap of AOs even for nearest-neighbour atoms.

3. When $m = n$, the quantity H_{nn}, the Coulomb integral, is assumed to be the same for each atom and, as in Section 6.2.1, is given the symbol α:

$$H_{nn} = \alpha \tag{6.181}$$

For example, in formaldehyde, α is assumed to be the same for oxygen as for carbon.

4. When $m \neq n$, the quantity H_{mn}, the resonance integral, is assumed to have the same value for any pair of atoms directly bonded. The symbolism

$$H_{mn} = \beta \tag{6.182}$$

is used, as for diatomic molecules

5. When m and n are not directly bonded,

$$H_{mn} = 0 \tag{6.183}$$

The π-electron wave functions in the Hückel method are given by

$$\psi = \sum_i c_i \chi_i \tag{6.184}$$

as in equation (6.76) for LCAO MOs of a diatomic molecule, but now the χ_i are only those AO wave functions, usually $2p$, which are involved in the π MOs.

In the case of ethylene the χ_i involved are the $2p$ AOs on the carbon atoms and perpendicular to the plane of the molecule. Therefore, from equation (6.184),

$$\psi = c_1\chi_1 + c_2\chi_2 \tag{6.185}$$

and the determinant of equation (6.178) becomes

$$\begin{vmatrix} \alpha - E & \beta \\ \beta & \alpha - E \end{vmatrix} = 0 \tag{6.186}$$

using the Hückel assumptions (1)–(5). Dividing through by β and putting

$$(\alpha - E)/\beta = x \tag{6.187}$$

we obtain

$$\begin{vmatrix} x & 1 \\ 1 & x \end{vmatrix} = 0 \tag{6.188}$$

Solving this determinant gives

$$E = \alpha \pm \beta \tag{6.189}$$

These two values of E correspond to two MOs with energies symmetrically disposed about α, as shown in Figure 6.77. The resonance integral β is a negative quantity.

The MO wave functions corresponding to these two values of E can be shown, from the secular equations corresponding to the determinant of equation (6.188) and the normalization condition

$$c_1^2 + c_2^2 = 1 \tag{6.190}$$

to be

$$\psi_1 = 2^{-\frac{1}{2}}(\chi_1 + \chi_2) \tag{6.191}$$

and

$$\psi_2 = 2^{-\frac{1}{2}}(\chi_1 - \chi_2) \tag{6.192}$$

The ψ_1 function has no node between the carbon atoms and is the $1b_{3u}$ MO in Figure 6.76; ψ_2 has such a node and is the $1b_{2g}$ MO shown also in the figure.

The Hückel treatment of ethylene has told us little that we had not deduced already. It confirms that the energy associated with ψ_1 is lower than with ψ_2, but the indication from Figure 6.77 that the promotion energy from $1b_{3u}$ to $1b_{2g}$ is 2β is of little quantitative value. Nor do estimates of β provide any reasonably consistent value which can be transferred to other molecules. However, the treatment is generally useful in conjugated π-electron systems.

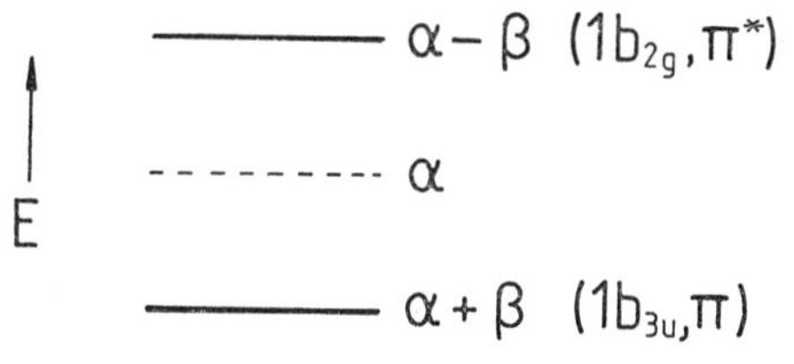

Figure 6.77 Energies of Hückel MOs in the ethylene molecule

6.3.1.7 XHC–CHY *molecules*

In this type of molecule the groups X and Y may be $=CH_2$ or $=O$. When X and Y are both $=CH_2$, the molecule is buta-1,3-diene, when they are both $=O$, it is glyoxal and, when one is $=CH_2$ and the other $=O$, it is acrolein.

These molecules are illustrated in Figure 5.81 and are all examples of conjugated π-electron systems, having alternate double and single bonds. It is for molecules like these, and also cyclic conjugated molecules such as benzene, that Hückel MOs are much more useful than for ethylene.

The *s-trans*, *s-cis* or, possibly *gauche* conformations of buta-1,3-diene, glyoxal and acrolein have been discussed in Section 5.2.7.4. Although the *s-trans* conformation is lowest in energy in their ground electronic states, Hückel treatment of the π MOs of these molecules is not sufficiently rigorous to be affected by the conformation. Indeed, for convenience, we can regard the chains of carbon and oxygen atoms as linear in all three molecules.

The MOs in buta-1,3-diene are formed from linear combinations of four $2p$ AOs, one on each carbon atom and perpendicular to the molecular plane, giving the MO wave functions

$$\psi = c_1\chi_1 + c_2\chi_2 + c_3\chi_3 + c_4\chi_4 \qquad (6.193)$$

The resulting secular determinant is

$$\begin{vmatrix} x & 1 & 0 & 0 \\ 1 & x & 1 & 0 \\ 0 & 1 & x & 1 \\ 0 & 0 & 1 & x \end{vmatrix} = 0 \qquad (6.194)$$

analogous to that of equation (6.188) for ethylene. The determinant can be factorized using symmetry arguments, but this is scarcely worthwhile since the solutions

$$x = \pm 1.62 \text{ or } \pm 0.62 \qquad (6.195)$$

are obtained easily by solving the determinant as it stands.

Figure 6.78 illustrates the energies of the resulting four π MOs.

The coefficients c_i in equation (6.193) can be obtained from the four secular equations and the normalization condition for ψ. The resulting MO wave functions are

$$\begin{aligned} \psi_1 &= 0.37\chi_1 + 0.60\chi_2 + 0.60\chi_3 + 0.37\chi_4 \\ \psi_2 &= 0.60\chi_1 + 0.37\chi_2 - 0.37\chi_3 - 0.60\chi_4 \\ \psi_3 &= 0.60\chi_1 - 0.37\chi_2 - 0.37\chi_3 + 0.60\chi_4 \\ \psi_4 &= 0.37\chi_1 - 0.60\chi_2 + 0.60\chi_3 - 0.37\chi_4 \end{aligned} \qquad (6.196)$$

The numbering of these MOs is in order of increasing energy. Their approximate forms are illustrated in Figure 6.79, in which symmetry species are assigned to them according to the C_{2h} (buta-1,3-diene and glyoxal in the *s-trans* form), C_s (acrolein in the *s-trans* or *s-cis* form) or C_{2v} (buta-1,3-diene and glyoxal in the *s-cis* form) point group.

In *s-trans* buta-1,3-diene the ground configuration, excluding the σ MOs, is

$$\ldots (1a_u)^2(1b_g)^2 \qquad (6.197)$$

giving the $\tilde{X}^1A_g$ state. The first excited configuration is

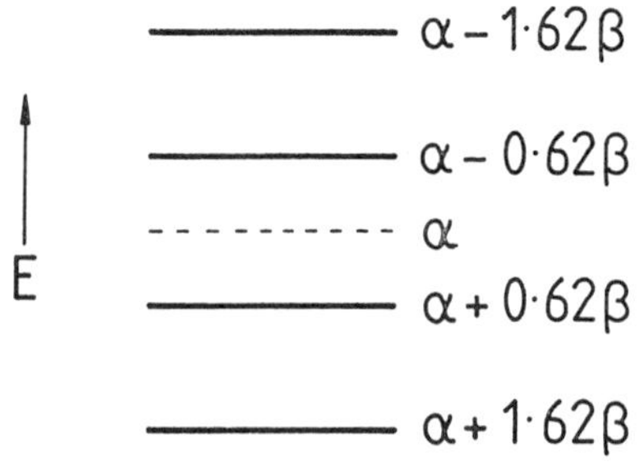

Figure 6.78 Energies of Hückel MOs in the buta-1,3-diene molecule

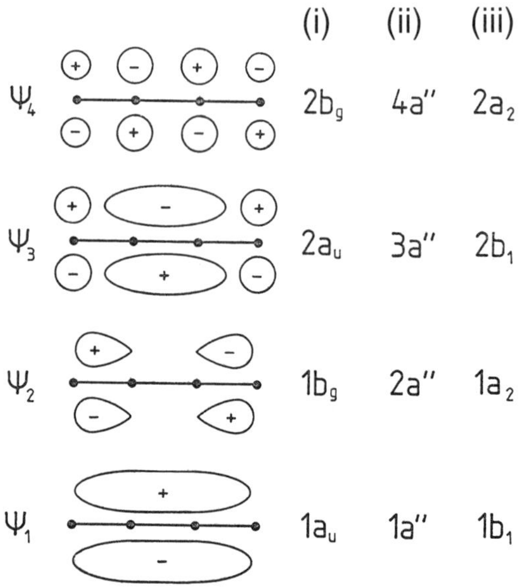

Figure 6.79 Symmetry species of Hückel MOs in (i) *s-trans*-buta-1,3-diene (C_{2h}) and *s-trans*-glyoxal (C_{2h}), (ii) *s-cis*- and *s-trans*-acrolein (C_s) and (iii) *s-cis*-buta-1,3-diene and *s-cis*-glyoxal (C_{2v})

$$\ldots (1a_u)^2(1b_g)^1(2a_u)^1 \qquad (6.198)$$

giving the $\tilde{A}^1B_u$ and $\tilde{a}^3B_u$ states. The $\tilde{A}^1B_u$–$\tilde{X}^1A_g$ transition falls in the 197–217 nm region.

In glyoxal there are two more MOs with energy comparable to the π MOs of buta-1,3-diene. These are the n (non-bonding) MOs formed by the in-phase and out-of-phase combination of the lone pair (l.p.) $2p$ orbitals on each oxygen atom. The in-phase combination gives the l.p.b_u. MO and the out-of-phase combination the higher energy l.p.a_g MO. They both have energies which lie between those of the $1b_g$ and $2a_u$ MOs giving the ground configuration

$$\ldots (1a_u)^2(1b_g)^2(\text{l.p.}b_u)^2(\text{l.p.}a_g)^2 \qquad (6.199)$$

and the $\tilde{X}^1A_g$ ground state.

The first excited configuration is

$$\ldots (1a_u)^2(1b_g)^2(\text{l.p.}b_u)^2(\text{l.p.}a_g)^1(2a_u)^1 \qquad (6.200)$$

giving $\tilde{A}^1A_u$ and $\tilde{a}^3A_u$ excited states. Both the $\tilde{a}^3A_u$–$\tilde{X}^1A_g$ and $\tilde{A}^1A_u$–$\tilde{X}^1A_g$ transitions have been observed. The former occurs in the 515–575 nm region and is weak, obtaining intensity by spin–orbit interaction due largely to the presence of oxygen atoms. The latter transition is more intense occurring in the 390–540 nm region and causing glyoxal vapour to appear green.

Both the $\tilde{A}$–$\tilde{X}$ and $\tilde{a}$–$\tilde{X}$ transitions are referred to as π^*–n transitions, analogous to those in formaldehyde (Section 6.3.1.5), because of the types of orbitals involved in the electron promotion.

The electronic structure of acrolein is similar to that of glyoxal except that there is only one n orbital, namely the one due to the l.p. $2p$ orbital on the oxygen atom; this gives the l.p.a' MO. The ground configuration is

$$\ldots (1a'')^2(2a'')^2(\text{l.p.}a')^2 \qquad (6.201)$$

giving the $\tilde{X}^1A'$ ground state. The first excited configuration

$$\ldots (1a'')^2(2a'')^2(\text{l.p.}a')^1(3a'')^1 \qquad (6.202)$$

gives the $\tilde{A}^1A''$ and $\tilde{a}^3A''$ excited states. The $\tilde{a}^3A''$–$\tilde{X}^1A'$ transition is very weak and lies in the 403–412 nm region while the $\tilde{A}^1A''$–$\tilde{X}^1A'$ transition is stronger and lies in the 300–387 nm region.

In the $\tilde{A}$ states of glyoxal[106] and acrolein[107] it has been shown experimentally that the *s-cis* is lower in energy than the *s-trans* configuration, whereas the opposite is the case in the $\tilde{X}$ states; but it requires a far more accurate MO calculation than we have considered here to reproduce the *s-cis–s-trans* energy differences in any electronic state.

6.3.1.8 *Benzene and substituted benzenes*

Benzene is a planar molecule with the carbon atoms in the form of a regular hexagon; it is a cyclic, conjugated π-electron system.

Hückel treatment of the six $2p$ orbitals on the carbon atoms and perpendicular to the plane of the molecule leads to the secular determinant

$$\begin{vmatrix} x & 1 & 0 & 0 & 0 & 1 \\ 1 & x & 1 & 0 & 0 & 0 \\ 0 & 1 & x & 1 & 0 & 0 \\ 0 & 0 & 1 & x & 1 & 0 \\ 0 & 0 & 0 & 1 & x & 1 \\ 1 & 0 & 0 & 0 & 1 & x \end{vmatrix} \qquad (6.203)$$

Solution of this determinant gives

$$x = \pm 1, \ \pm 1 \text{ or } \pm 2 \qquad (6.204)$$

The fact that the $x = \pm 1$ solution appears twice implies that the MOs with $x = +1$ are doubly degenerate, as are those with $x = -1$, as Figure 6.80 shows.

The six MO wave functions are

$$\begin{aligned}
\psi_1 &= 6^{-\frac{1}{2}}\chi_1 + 6^{-\frac{1}{2}}\chi_2 + 6^{-\frac{1}{2}}\chi_3 + 6^{-\frac{1}{2}}\chi_4 + 6^{-\frac{1}{2}}\chi_5 + 6^{-\frac{1}{2}}\chi_6 \\
\psi_2 &= \tfrac{1}{2}\chi_2 + \tfrac{1}{2}\chi_3 - \tfrac{1}{2}\chi_5 - \tfrac{1}{2}\chi_6 \\
\psi_3 &= 3^{-\frac{1}{2}}\chi_1 + 12^{-\frac{1}{2}}\chi_2 - 12^{-\frac{1}{2}}\chi_3 - 3^{-\frac{1}{2}}\chi_4 - 12^{-\frac{1}{2}}\chi_5 + 12^{-\frac{1}{2}}\chi_6 \\
\psi_4 &= -\tfrac{1}{2}\chi_2 + \tfrac{1}{2}\chi_3 - \tfrac{1}{2}\chi_5 + \tfrac{1}{2}\chi_6 \\
\psi_5 &= 3^{-\frac{1}{2}}\chi_1 - 12^{-\frac{1}{2}}\chi_2 - 12^{-\frac{1}{2}}\chi_3 + 3^{-\frac{1}{2}}\chi_4 - 12^{-\frac{1}{2}}\chi_5 - 12^{-\frac{1}{2}}\chi_6 \\
\psi_6 &= 6^{-\frac{1}{2}}\chi_1 - 6^{-\frac{1}{2}}\chi_2 + 6^{-\frac{1}{2}}\chi_3 - 6^{-\frac{1}{2}}\chi_4 + 6^{-\frac{1}{2}}\chi_5 - 6^{-\frac{1}{2}}\chi_6
\end{aligned} \qquad (6.205)$$

For the doubly degenerate pairs of MOs ψ_2, ψ_3 and ψ_4, ψ_5 other forms of the wave functions

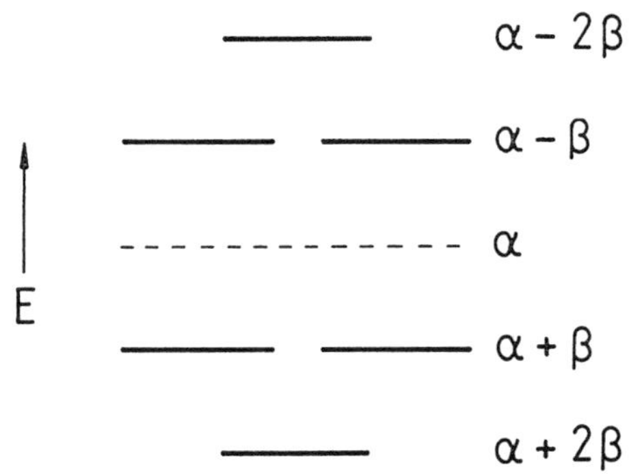

Figure 6.80 Energies of Hückel MOs in the benzene molecule

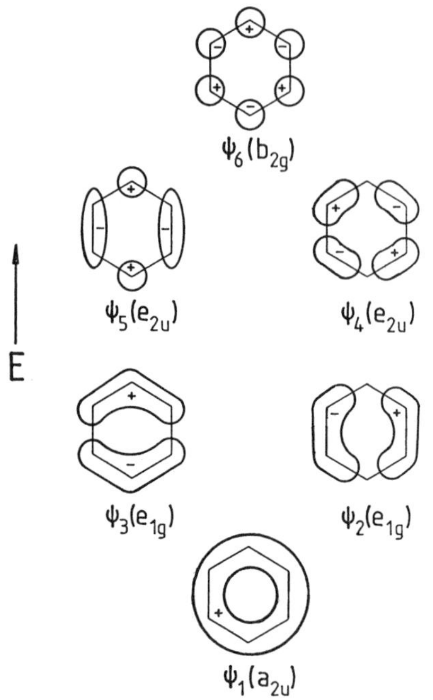

Figure 6.81 Hückel MOs in the benzene molecule

may be chosen which have complex rather than real coefficients.

Figure 6.81 illustrates the MO wave functions and gives the symmetry species, according to the D_{6h} point group, which may be confirmed using the character table in the Appendix (Table A.36). In the figure only the parts of the wave functions above the plane of the molecule are shown. The parts below are identical in form but opposite in sign, which means that, like all π-type MOs in planar molecules, they are antisymmetric to reflection in the plane of the molecule. As usual, the energy increases with the number of nodal planes.

The ground electron configuration of benzene is obtained by feeding the six electrons into the lower energy MOs, giving

$$\ldots (1a_{2u})^2(1e_{1g})^4 \tag{6.206}$$

and the ground state is $\tilde{X}^1A_{1g}$.

The first excited configuration is

$$\ldots (1a_{2u})^2(1e_{1g})^3(1e_{2u})^1 \tag{6.207}$$

The states arising from this configuration are the same as those from $\ldots (1e_{1g})^1(1e_{2u})^1$ since the single vacancy can be treated like a single electron. Following the method used for the derivation of states arising from the excited configuration of the nitrogen molecule in equation (6.100), we obtain, for benzene, using Table 5.9,

$$\Gamma(\psi_e^0) = e_{1g} \times e_{2u} = B_{1u} + B_{2u} + E_{1u} \tag{6.208}$$

and the resulting states are ${}^{1,3}B_{1u}$, ${}^{1,3}B_{2u}$ and ${}^{1,3}E_{1u}$. The singlet states are, in order of increasing energy, $\tilde{A}^1B_{2u}$, $\tilde{B}^1B_{1u}$ and $\tilde{C}^1E_{1u}$ although there is some doubt about the identification of the $\tilde{B}$ state. Some calculations suggest it may be the ${}^1E_{2g}$ state arising from the configuration

$$\ldots (1a_{2u})^2(1e_{1g})^3(1b_{2g})^1 \tag{6.209}$$

The $\tilde{A}$–$\tilde{X}$ and $\tilde{B}$–$\tilde{X}$ systems fall in the 227–267 nm and 185–205 nm regions, respectively. Both are electronically forbidden since neither B_{2u} nor B_{1u} (or E_{2g}) is a translational symmetry species (see Table A.36 in the Appendix), but vibronic transitions occur through vibronic interaction (see Section 6.3.4.1).

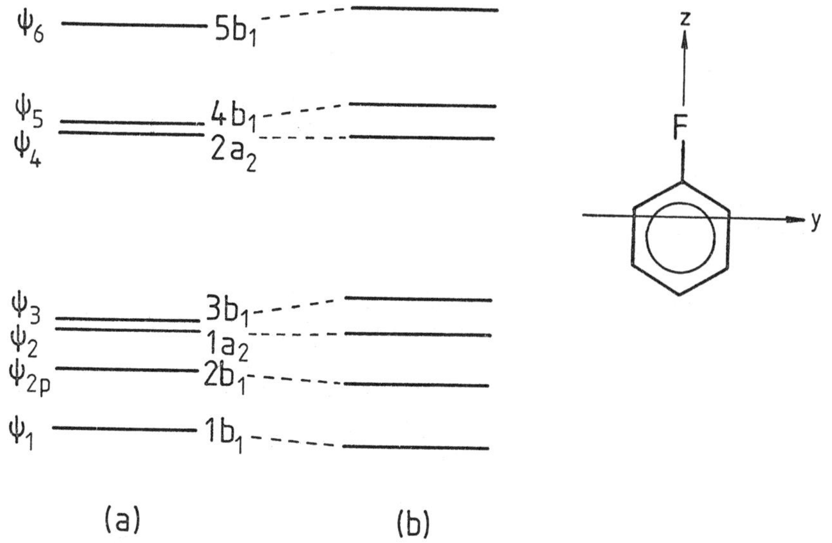

Figure 6.82 Perturbation of Hückel MOs of the benzene molecule by a fluorine substituent

The triplet states are,[108] in order of increasing energy, $\tilde{a}^3B_{1u}$, $\tilde{b}^3E_{1u}$ and $\tilde{c}^3B_{2u}$. The $\tilde{a}$–$\tilde{X}$ system has been observed in absorption in the 300–340 nm region but only in the presence of oxygen.

There are many types of substituted benzenes but we consider here only the monosubstituted benzenes and use fluorobenzene as an example.

Fluorobenzene belongs to the C_{2v} point group. In Figure 6.82(a) the π orbitals of benzene are reclassified according to this point group, with the axis notation shown. The only AO on the fluorine atom which is of energy comparable to the π MOs of the benzene ring and of the same symmetry is the $2p_x$ lone pair AO which has b_1 symmetry in the molecule. This orbital has been inserted in Figure 6.82(a) as ψ_{2p} and placed between ψ_1 and the degenerate pair ψ_2, ψ_3.

In Figure 6.82(b) the perturbing effect of the $2b_1$ ($2p_x$) orbital on the other b_1 orbitals is shown. The effect is to push b_1 orbitals of higher energy still higher and to push the $1b_1$ orbital lower.

The ground configuration of fluorobenzene is

$$\ldots (1b_1)^2(2b_1)^2(1a_2)^2(3b_1)^2 \qquad (6.210)$$

giving the $\tilde{X}^1A_1$ ground state, and the lowest excited configuration is

$$\ldots (1b_1)^2(2b_1)^2(1a_2)^2(3b_1)^1(2a_2)^1 \qquad (6.211)$$

giving the $\tilde{A}^1B_2$ and $\tilde{a}^3B_2$ excited states. The $\tilde{A}^1B_2$–$\tilde{X}^1A_1$ system is electronically allowed and falls in the 238–270 nm region. The pure electronic transition has a wavenumber of 37 814 cm^{-1} compared with 38 086 cm^{-1} in benzene, showing the shift to low wavenumber on substitution due to the perturbation by ψ_{2p}. This is an effect typical of similar substituents (e.g. NH_2, OH, Cl) but a much more accurate MO method is needed to obtain quantitative agreement.

6.3.1.9 Naphthalene

Naphthalene is the bicyclic conjugated π-electron molecule shown in Figure 6.83.

The ten $2p_x$ AOs one on each carbon atom, give ten Hückel MOs. These are illustrated in

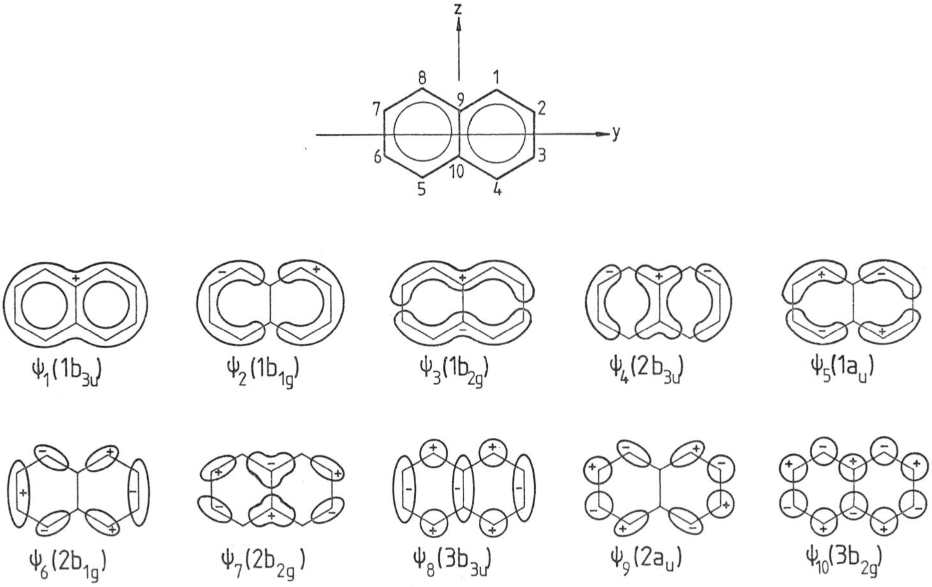

Figure 6.83 Hückel MOs in the naphthalene molecule ($C_{10}H_8$)

Figure 6.83 and assigned to symmetry species of the D_{2h} character table (Table A.32 in the Appendix) using the Mulliken axis labelling shown.

There are only three types of carbon atom in naphthalene, of which examples are those in positions 1, 2, and 9. Instead of giving the wave functions ψ_1–ψ_{10} in full, Table 6.15 gives the coefficients c_1, c_2 and c_9 for each wave function and also the MO energies. The coefficients for the other positions can be obtained from these by using the sign characteristics of the wave functions shown in Figure 6.83 and the fact that the coefficients for two carbon atoms of the same type must either be identical or differ only in sign.

The ground configuration of naphthalene is

$$\ldots(1b_{3u})^2(1b_{1g})^2(1b_{2g})^2(2b_{3u})^2(1a_u)^2 \quad (6.212)$$

giving the $\tilde{X}^1A_g$ ground state.

The first excited configuration is

$$\ldots(1b_{3u})^2(1b_{1g})^2(1b_{2g})^2(2b_{3u})^2(1a_u)^1(2b_{1g})^1 \quad (6.213)$$

giving a ${}^1B_{1u}$ and a ${}^3B_{1u}$ state. However, these are not the lowest excited states. The excited configurations

$$\ldots(1b_{3u})^2(1b_{1g})^2(1b_{2g})^2(2b_{3u})^1(1a_u)^2(2b_{1g})^1 \quad (6.214)$$

$$\ldots(1b_{3u})^2(1b_{1g})^2(1b_{2g})^2(2b_{3u})^2(1a_u)^1(2b_{2g})^1 \quad (6.215)$$

both give ${}^1B_{2u}$ and ${}^3B_{2u}$ states.

There is an important phenomenon, which we have neglected elsewhere but which is particularly important here, called configuration interaction, involving the two configurations of equations (6.214) and (6.215), which have identical energies in the Hückel approximation. As a result of the interaction one pair

Table 6.15 Hückel MO wave function coefficients and energies for naphthalene

	c_1	c_2	c_9	$x\ [= (\alpha - E)/\beta]$
ψ_1	0.301	0.231	0.461	−2.30
ψ_2	0.263	0.425	0	−1.62
ψ_3	0.400	0.174	0.347	−1.30
ψ_4	0	−0.408	0.408	−1.00
ψ_5	−0.425	−0.263	0	−0.62
ψ_6	0.425	−0.263	0	+0.62
ψ_7	0	0.408	−0.408	+1.00
ψ_8	0.400	−0.174	−0.347	+1.30
ψ_9	0.263	−0.425	0	+1.62
ψ_{10}	−0.301	0.231	0.461	+2.30

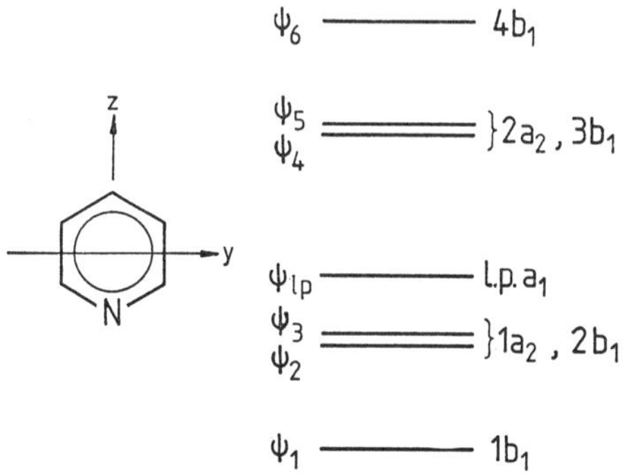

Figure 6.84 Perturbation of Hückel MOs of the benzene molecule by a nitrogen substituent in the ring

of $^{1,3}B_{2u}$ states is pushed down and the other pushed up in energy. The pair that is pushed down is then below the $^{1,3}B_{1u}$ states from the configuration of equation (6.213) so that the lowest energy singlet state is $\tilde{A}^1B_{2u}$ and second highest is $\tilde{B}^1B_{1u}$. The next excited singlet states are $\tilde{C}^1B_{3g}$ and $\tilde{D}^1A_g$, which arise from other configurations. The second $^1B_{2u}$ state, pushed higher in energy by configuration interaction, is the $\tilde{E}^1B_{2u}$ state.

The $\tilde{A}$–$\tilde{X}$, $\tilde{B}$–$\tilde{X}$ and $\tilde{C}$–$\tilde{X}$ systems are observed in the 290–320, 240–290 and 200–230 nm regions, respectively.

6.3.1.10 Aza-aromatic molecules

The cyclic, conjugated series of molecules, such as benzene and naphthalene, are known as 'aromatic'. The molecules in which one or more of the C—H groups is replaced by a nitrogen atom are known as the aza-aromatics, of which the simplest example is pyridine, shown in Figure 6.84.

The π orbitals of pyridine are similar to those of benzene except that the presence of the nitrogen requires use of the C_{2v} classification and removes the ψ_2,ψ_3 and ψ_4,ψ_5 degeneracies in a similar way to that encountered in fluorobenzene. The main difference from fluorobenzene is that, in pyridine, instead of an extra $2p_x$ AO being added to the scheme, there is a nitrogen non-bonding lone pair orbital. This orbital, labelled 'l.p.', is of a_1 symmetry and of higher energy than the ψ_2,ψ_3 pair of π orbitals. The resulting orbital scheme is shown in figure 6.84.

The ground configuration of pyridine is

$$\ldots (1b_1)^2(1a_2)^2(2b_1)^2(\text{l.p.}a_1)^2 \qquad (6.216)$$

giving the $\tilde{X}^1A_1$ ground state.

The excited configuration

$$\ldots (1b_1)^2(1a_2)^2(2b_1)^2(\text{l.p.}a_1)^1(2a_2)^1 \qquad (6.217)$$

gives $^{1,3}A_2$ states. The $^1A_2$–$\tilde{X}^1A_1$ transition is electric dipole forbidden and no associated band system has been observed. The alternative first excited configuration

$$\ldots (1b_1)^2(1a_2)^2(2b_1)^2(\text{l.p.}a_1)^1(3b_1)^1 \qquad (6.218)$$

gives $^{1,3}B_1$ states. The $\tilde{A}^1B_1$–$\tilde{X}^1A_1$ system is observed in the 265–290 nm region and is a π^*–n type of transition.

In the diazabenzenes, pyridazine (1,2-diaza-), pyrimidine (1,3-diaza-) and pyrazine (1,4-diaza-) there are *two* lone-pair orbitals. From these, MOs can be constructed by making in-phase and out-of-phase combinations of them and assigning them to the appropriate symmetry species.

Similar procedures can be carried out for the triazabenzenes and the tetraazabenzenes, of which *sym*-tetrazine, 1,2,4,5-tetraazabenezene, is the most commonly encountered.

There are three isomers of azanaphthalene and many isomers of di- and polyazanaphthalenes which can be treated similarly.

6.3.1.11 *Crystal field and ligand field molecular orbitals*

Transition metal atoms are distinguished from other atoms by the fact that their valence orbitals include *d* orbitals. The 3*d*, 4*d* and 5*d* sub-shells can each accommodate 10 electrons, resulting in there being 10 atoms in each of the first, second and third transition series.

Here we shall consider in detail only the first transition series, Sc, Ti, V, Cr, Mn, Fe, Co, Ni, Cu and Zn, in which the 3*d* valence orbital is involved.

Transition metals readily form complexes, such as $[Fe(CN_6]^{4-}$, the hexacyanoferrate(II) ion, $Ni(CO)_4$, nickel tetracarbonyl, and $[CuCl_4]^{2-}$, the copper(II) tetrachloride ion. There has been a tendency for the MO treatment of such molecules and ions to be developed separately from that of other types of molecules and it is for this reason that the terms 'crystal field theory' and 'ligand field theory' have arisen. Unfortunately, the use of these terms tends to disguise the fact that they are both simply particular facets of MO theory.

The use of the word 'ligand' to describe an atom, or group of atoms, attached to the central metal atom can also lead to some confusion. The word has arisen because of the fact that the type of bonding found in complexes is often different from that in a molecule such as, say, H_2O. However, the difference is quantitative rather than qualitative so that there is no reason why we should not refer to the hydrogen atoms of H_2O as ligands attached to the oxygen atom—it just happens that we rarely do.

Ligands in a transition metal complex are usually arranged in a highly symmetrical way. For example, six ligands often take up an octahedral configuration, as in $[Fe(CN)_6]^{4-}$, and four ligands a tetrahedral configuration, as in $Ni(CO)_4$; both are shown in Figure 6.85. It is the octahedral complexes which we shall consider in the greatest detail.

In a transition metal complex the higher energy occupied MOs are perturbed *d* orbitals of the metal atom. In an octahedral complex, if the perturbation is weak, the ligands can be treated approximately as point charges at the corners of a regular octahedron. This is reminiscent of the perturbation of the orbitals of an Na^+ ion by six octahedrally arranged nearest-neighbour Cl^- ions in the NaCl lattice and it is for this reason that this type of MO theory is referred to as crystal field theory.

In fact, a crystal field treatment even in molecules where the ligands perturb the metal atom orbitals weakly is a rather gross approximation, akin to a Hückel treatment of conjugated π-electron molecules, but is nevertheless a conceptually useful starting point.

In molecules in which the ligands interact more strongly with the metal atom, they cannot be treated as point charges in any useful approximation. The interaction between the MOs of the ligand and the metal atom *d* orbitals must be taken into account. This type of MO theory is known as ligand field theory.

6.3.1.11(i) *Crystal field theory*

In the presence of six point charges arranged octahedrally on the cartesian axes, the five *d* orbitals of Figure 1.15(a) are perturbed and they must be classified according to the O_h point group.

In degenerate point groups the *d* orbitals have the same symmetry species as those of the corresponding components of the polarizability $\boldsymbol{\alpha}$. In the O_h point group the symmetry species of the d_{z^2}, $d_{x^2-y^2}$, d_{xy}, d_{yz} and d_{xz} orbitals are the same as those of $\alpha_{xx}-\alpha_{yy}-2\alpha_{zz}$, $\alpha_{xx}-\alpha_{yy}$, α_{xy}, α_{yz} and α_{xz}, respectively. Table A.43 in the Appendix gives the O_h character table and shows that these species are e_g, e_g, t_{2g}, t_{2g} and t_{2g}, respectively. Therefore, the fivefold degenerate *d* orbitals are split into doubly and triply degenerate sets. Since d_{z^2} and $d_{x^2-y^2}$ orbitals

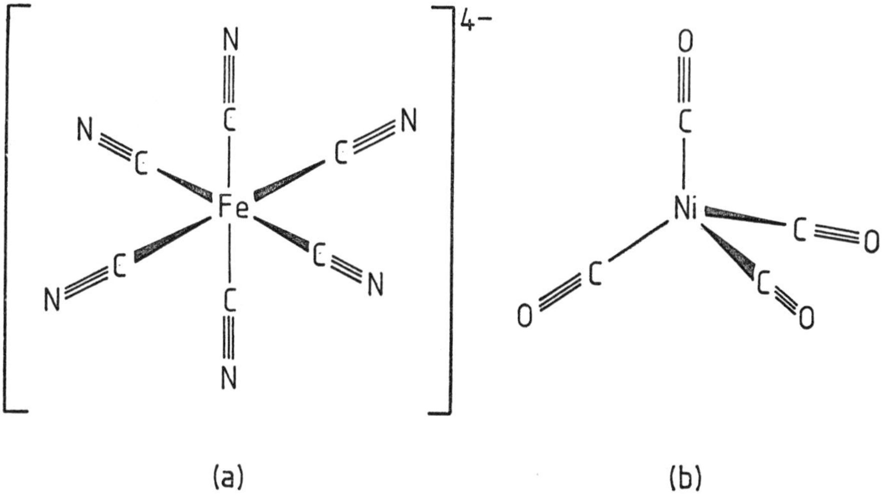

Figure 6.85 (a) Octahedral $[Fe(CN)_6]^{4-}$ and (b) tetrahedral $Ni(CO)_4$ molecules

have much of their electron density along the line of the metal–ligand bonds, the electrons in these orbitals experience more repulsion by the ligand electrons than electrons in the d_{xy}, d_{yz} or d_{xz} orbitals. The result is that the e_g orbitals are higher in energy than t_{2g}, as shown on the left in Figure 6.86. The e_g orbitals are pushed up in energy by $\frac{3}{5}\Delta_o$ and the t_{2g} orbitals pushed down by $\frac{2}{5}\Delta_o$, where Δ_o is the e_g–t_{2g} splitting in an octahedral field.

In a tetrahedral field the metal d orbitals are split into e and t_2 orbitals but, as the right of Figure 6.86 shows, the e orbital is pushed *down* by $\frac{3}{5}\Delta_t$ and the t_2 orbital pushed *up* by $\frac{2}{5}\Delta_t$, where Δ_t is the e–t_2 splitting in a tetrahedral field.

The values of Δ_t and Δ_o, which is usually larger than Δ_t, are typically such that promotion of an electron from a lower to an upper orbital leads to an absorption spectrum in the visible region and the characteristic property of transition metal complexes being coloured.

Table 6.16 indicates the way in which the d orbitals split in fields of various symmetries.

We shall now consider, in an octahedral complex, how the electrons of a transition metal

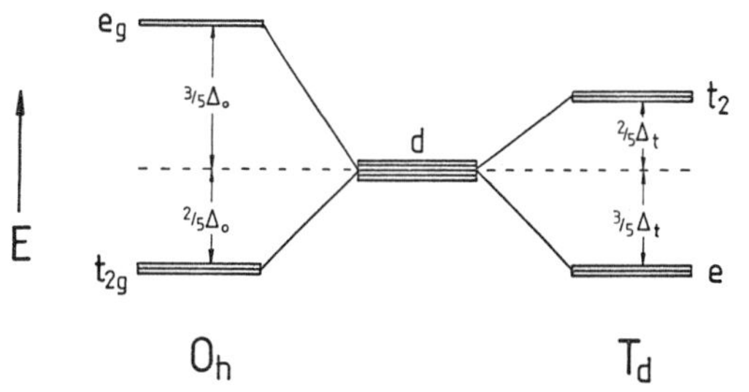

Figure 6.86 Splitting of d orbitals on the central atom in an O_h (left) and a T_d (right) crystal field

atom can be fed into the MOs on the left of Figure 6.86 to give the ground configuration. This is illustrated for d^1, d^2, d^3, d^8, d^9 and d^{10} configurations in Figure 6.87, in which the components of the degenerate MOs have been separated for convenience.

For a metal ion with a d^1 configuration and in an octahedral field, such as $Ti^{3+\dagger}$ in the

†As in all these examples, the two 4s electrons are always removed in the formation of the ion.

Table 6.16 Symmetry species of orbitals resulting from the splitting of *d* orbitals by various ligand arrangements

Point group	*d* orbitals				
	d_{z^2}	$d_{x^2-y^2}$	d_{xy}	d_{yz}	d_{xz}
O_h	$\longleftarrow e_g \longrightarrow$		$\longleftarrow t_{2g} \longrightarrow$		
T_d	$\longleftarrow e \longrightarrow$		$\longleftarrow t \longrightarrow$		
D_{3h}	a_1'	$\longleftarrow e' \longrightarrow$		$\longleftarrow e'' \longrightarrow$	
D_{4h}	a_{1g}	b_{1g}	b_{2g}	$\longleftarrow e_g \longrightarrow$	
$D_{\infty h}$	σ_g^+	$\longleftarrow \delta_g \longrightarrow$		$\longleftarrow \pi_g \longrightarrow$	
C_{2v}	a_1	a_1	a_2	b_2	b_1
C_{3v}	a_1	a_1	a_2	$\longleftarrow e \longrightarrow$	
C_{4v}	a_1	b_1	b_2	$\longleftarrow e \longrightarrow$	
D_{2d}	a_1	b_1	b_2	$\longleftarrow e \longrightarrow$	
D_{4d}	a_1	$\longleftarrow e_2 \longrightarrow$		$\longleftarrow e_3 \longrightarrow$	

complex ion $[Ti(H_2O)_6]^{3+}$, the single electron goes into a t_{2g} orbital, as shown in Figure 6.87, and the $(t_{2g})^1$ configuration gives the $\tilde{X}^2T_{2g}$ ground state of the ion. The first excited state is $\tilde{A}^2E_g$ and arises from the $(e_g)^1$ excited configuration. The transition is forbidden since (see Table 5.9).

$$T_{2g} \times E_g = T_{1g} + T_{2g} \tag{6.219}$$

and this does not contain the translational species T_{1u} (see Table A.43 in the Appendix). The observation of the $\tilde{A}$–$\tilde{X}$ system weakly in absorption at about 500 nm is due to vibronic interaction (see Section 6.3.4.1) as, for example, in the 260 nm system of benzene. As low energy electronic transitions in all transition metal complexes involve electron promotions between orbitals which have *g* symmetry, i.e. symmetric to inversion through the centre, they are *all* forbidden by the $g \leftrightarrow\!\!\!\!/\; g$ electric dipole selection rule.

For a metal ion such as Cu^{2+} with a d^9 configuration, the ground MO configuration in an octahedral complex such as $[Cu(H_2O)_6]^{2+}$ is $(t_{2g})^6(e_g)^3$, as Figure 6.87 shows. Remembering that a vacancy in an orbital can be treated like an electron, we can see that the ground state is $\tilde{X}^2E_g$. The first excited configuration is $(t_{2g})^5(e_g)^4$ giving the $\tilde{A}^2T_{2g}$ excited state. The weak $\tilde{A}$–$\tilde{X}$ system is observed in absorption at about 790 nm and is responsible for the blue colour of aqueous Cu^{2+} solutions. This system also is electronically forbidden and observed through vibronic interaction, but there is an additional complication of Jahn–Teller distortion (see Section 6.3.4.5) in the $\tilde{X}$ state.

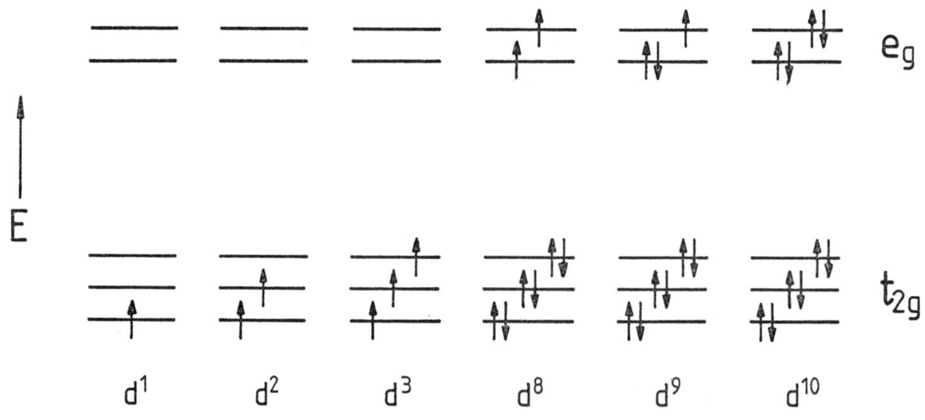

Figure 6.87 Electron configurations in d^1, d^2, d^3, d^8, d^9 and d^{10} octahedral complexes

Since, in d^1 and d^9 complexes, only one state arises from each of the ground and first excited configurations, the $\tilde{A}$–$\tilde{X}$ transition energy is a direct measure of Δ_o. In d^2, d^3, . . . , d^8 complexes more than one state arises from each configuration, the spectrum is much more complex and Δ_o cannot be found directly but only estimated following identification of the various transitions.

In a d^2 configuration one of Hund's rules indicates that the lowest energy state arising, which is therefore the ground state, is that in which the two electrons are in different t_{2g} orbitals with parallel spins, as for the ground state of O_2. This situation is illustrated in Figure 6.87.

The states arising from a $(t_{2g})^2$ configuration are (see Table 6.6) $^3T_{1g} + ^1A_{1g} + ^1E_g + ^1T_{2g}$ and the ground state is therefore $\tilde{X}^3T_{1g}$; the other three states are low-lying excited states.

The first excited configuration is $(t_{2g})^1(e_g)^1$ which gives, using the result of $t_{2g} \times e_g$ in Table 5.9, the states $^{1,3}T_{1g} + ^{1,3}T_{2g}$.

In the visible absorption spectrum of $[V(H_2O)_6]^{3+}$, in which the central metal atom has a d^2 configuration, there is a system at about 590 nm attributed to $\tilde{A}^3T_{2g}$–$\tilde{X}^3T_{1g}$ and a second at about 400 nm attributed to $\tilde{B}^3T_{1g}$–$\tilde{X}^3T_{1g}$. These are spin allowed but orbitally forbidden transitions and involve vibronic interaction.

In the ion $[Ni(H_2O)_6]^{2+}$ the central metal atom has a d^8 configuration. The lowest energy MO configuration in an octahedral field is that shown in Figure 6.87, in which the two e_g electrons are in different orbitals with parallel spins.

The states arising from the $(t_{2g})^6(e_g)^2$ configuration are, according to Table 6.6, $^3A_{2g} + ^1A_{1g} + ^1E_g$, of which the ground state is $\tilde{X}^3A_{2g}$.

The first excited configuration, $(t_{2g})^5(e_g)^3$, gives (Table 5.9) $^{1,3}T_{1g} + ^{1,3}T_{2g}$ excited states. The lowest energy systems in the absorption spectrum of $[Ni(H_2O)_6]^{2+}$, at about 1150 and 690 nm, are attributed to $\tilde{A}^3T_{2g}$–$\tilde{X}^3A_{2g}$ and $\tilde{B}^3T_{1g}$–$\tilde{X}^3A_{2g}$, respectively. Both systems are electronically forbidden and involve vibronic interaction.

The $[Cr(H_2O)_6]^{3+}$ ion has a central metal atom with a d^3 configuration. The lowest energy MO configuration, shown in Figure 6.87, is $(t_{2g})^3$ with all spins parallel. The states arising from this are (Table 6.6) $^4A_{2g} + ^2E_g + ^2T_{1g} + ^2T_{2g}$ of which $\tilde{X}^4A_{2g}$ is the ground state. The first excited configuration is $(t_{2g})^2(e_g)^1$ and the states which arise from this† are $^{2,4}T_{1g} + ^{2,4}T_{2g} + ^2E_g + ^2A_{1g} + ^2A_{2g} + ^2E_g + ^2T_{1g} + ^2T_{2g}$. The only excited states which concern us here are $^4T_{1g}$ and $^4T_{2g}$ since the $\tilde{A}^4T_{2g}$–$\tilde{X}^4A_{2g}$ and $\tilde{B}^4T_{1g}$–$\tilde{X}^4A_{2g}$ transitions are spin-allowed and observed, through vibronic interaction, at about 570 and 400 nm, respectively.

For molecules in which the metal atom has four, five, six or seven d electrons there is a further complication. When the field due to the ligands is weak, Δ_o is relatively small and it may be energetically favourable for electrons to be promoted from t_{2g} to e_g MOs with the maximum number of parallel spins rather than suffer a positive contribution to the energy due to the electrons being in MOs in pairs with antiparallel spins. The resulting configuration is called a high spin configuration. Examples are illustrated in Figure 6.88(a).

When the field due to the ligands is stronger, Δ_o is larger and as many electrons as possible are in the t_{2g} orbitals with antiparallel spins giving a low spin configuration. Examples are shown in Figure 6.88(b).

The ion $[Cr(H_2O)_6]^{2+}$ has a metal atom with a d^4 configuration leading to a ground MO configuration $(t_{2g})^3(e_g)^1$ of the high spin type shown in Figure 6.88(a).

The states arising from this are numerous but we shall, for this example, go through the necessary steps for deriving them.

The $(t_{2g})^3$ part of the configuration gives (Table 6.6) $^4A_{2g} + ^2E_g + ^2T_{1g} + ^2T_{2g}$ states. These symmetry species have then all to be multiplied

†See below for the method of deriving these states, using the example of $(t_{2g})^3(e_g)^1$ in $[Cr(H_2O)_6]^{2+}$.

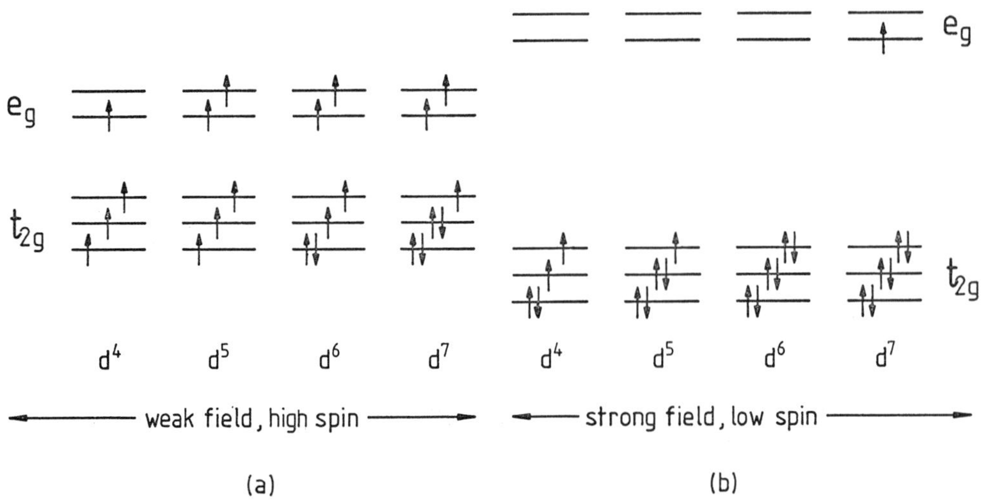

Figure 6.88 (a) Weak field, high spin and (b) strong field, low spin electron configurations in d^4, d^5, d^6 and d^7 octahedral complexes

by e_g (Table 5.9) and the multiplicity is obtained by bearing in mind that the spin of the e_g electron may be either parallel to the unpaired spins of the state to which it is coupling or antiparallel to one of them. Hence

$$^4A_{2g} \times e_g \rightarrow {}^{3,5}E_g$$
$$^2E_g \times e_g \rightarrow {}^{1,3}A_{1g} + {}^{1,3}A_{2g} + {}^{1,3}E_g$$
$$^2T_{1g} \times e_g \rightarrow {}^{1,3}T_{1g} + {}^{1,3}T_{2g}$$
$$^2T_{2g} \times e_g \rightarrow {}^{1,3}T_{1g} + {}^{1,3}T_{2g}$$

Hund's rules show that $\tilde{X}^5E_g$ is the ground state.

The first excited configuration, $(t_{2g})^2(e_g)^2$, also gives several states which can be derived similarly but there is only one quintet state, $\tilde{A}^5T_{2g}$. The $\tilde{A}^5T_{2g}$–$\tilde{X}^5E_g$ transition occurs at about 700 nm and is allowed only through vibronic interaction.

The low spin ground MO configuration for a complex with a d^4 metal atom is, as shown in Figure 6.88(b), $(t_{2g})^4$, giving rise to $^3T_{1g} + {}^1A_{1g} + {}^1E_{1g} + {}^1T_{2g}$ states (see Table 6.6), of which $\tilde{X}^3T_{1g}$ is the ground state. Examples of such complexes are rare.

For a d^5 metal atom the high spin ground MO configuration is $(t_{2g})^3(e_g)^2$, as in Figure 6.88(a). By arguments similar to those for the $(t_{2g})^3(e_g)^1$ configuration it can be shown that the ground state is $\tilde{X}^6A_{1g}$. Since all the t_{2g} and e_g MOs are singly occupied in the $(t_{2g})^3(e_g)^2$ configuration there is no sextet state arising from the $(t_{2g})^2(e_g)^3$ excited configuration and therefore no spin-allowed transition in the visible region. This explains why the high spin complex ion $[Mn(H_2O)_6]^{2+}$ is colourless.

Low spin d^5 complexes are rare but Figure 6.88(b) illustrates their $(t_{2g})^5$ configuration.

The $[Fe(H_2O)_6]^{2+}$ and $[CoF_6]^{3-}$ ions are examples of high spin complexes whose metal atoms have a d^6 configuration. The ground MO configuration is $(t_{2g})^4(e_g)^2$, shown in Figure 6.88(a), giving an $\tilde{X}^5T_{2g}$ ground state. The only quintet state arising from the $(t_{2g})^3(e_g)^3$ excited configuration is $\tilde{A}^5E_g$. In $[Fe(H_2O)_6]^{2+}$ the $\tilde{A}^5E_g$–$\tilde{X}^5T_{2g}$ transition is observed in absorption at about 1100 nm.

Examples of d^6 low spin complexes are $[Fe(CN)_6]^{4-}$ and $[Co(NH_3)_6]^{3+}$. The ground configuration $(t_{2g})^6$, shown in Figure 6.88(b), gives only one state, the $\tilde{X}^1A_{1g}$ ground state. The excited configuration $(t_{2g})^5(e_g)^1$ gives the states $^{1,3}T_{1g} + ^{1,3}T_{2g}$. The $\tilde{A}^1T_{1g}$–$\tilde{X}^1A_{1g}$ absorption system, and, in some ions, the $\tilde{B}^1T_{2g}$–$\tilde{X}^1A_{1g}$ system also, fall in the visible regions.

In a d^7 high spin complex, such as $[Co(H_2O)_6]^{2+}$, the ground configuration is $(t_{2g})^5(e_g)^2$, as in Figure 6.88(a). The ground state is $\tilde{X}^4T_{1g}$, and the only quartet states arising from the $(t_{2g})^4(e_g)^3$ excited configuration are $\tilde{A}^4T_{2g}$ and $\tilde{C}^4T_{1g}$. The $\tilde{A}^4T_{2g}$–$\tilde{X}^4T_{1g}$ and $\tilde{C}^4T_{1g}$–$\tilde{X}^4T_{1g}$ systems occur at about 1300 and 520 nm, respectively, in absorption. [The extremely weak $\tilde{B}^4A_{2g}$–$\tilde{X}^4T_{1g}$ transition occurs at about 640 nm and involves the lowest of several excited states arising from the *second* excited configuration $(t_{2g})^3(e_g)^4$.]

The ground MO configuration of a d^7 low spin complex is $(t_{2g})^6(e_g)^1$ as shown in Figure 6.88(b). The only state arising from this is the $\tilde{X}^2E_g$ ground state. Examples of such complexes are rare.

6.3.1.11(ii) Ligand field theory

When ligands interact so strongly with the central metal atom that they can no longer be treated as negative point charges the crystal field approximation breaks down and the MOs of the ligand, L, must be taken into account in deriving the MOs for the whole molecule.

The ligand MOs can be divided into those that are cylindrically symmetrical about the metal–ligand bond and classed as σ-type, and those that are not cylindrically symmetrical and are π-type. The σ-type of metal–ligand bonding is invariably the stronger and π-type bonding will be neglected here. We shall consider in detail only octahedral complexes.

Figure 6.89 shows six ligand σ orbitals arranged octahedrally about the metal atom M. These orbitals might be, for example, the *sp* hybrid orbitals on the carbon atoms of CN ligands or the sp^3 hybrid orbitals on the nitrogen atoms of NH_3 ligands. Table 6.17 shows that, in the O_h point group, the six ligand orbitals of ML_6 form an a_{1g}, an e_g and a t_{1u} MO. The table also gives the classifications of ligand σ orbitals in complexes belonging to other point groups.

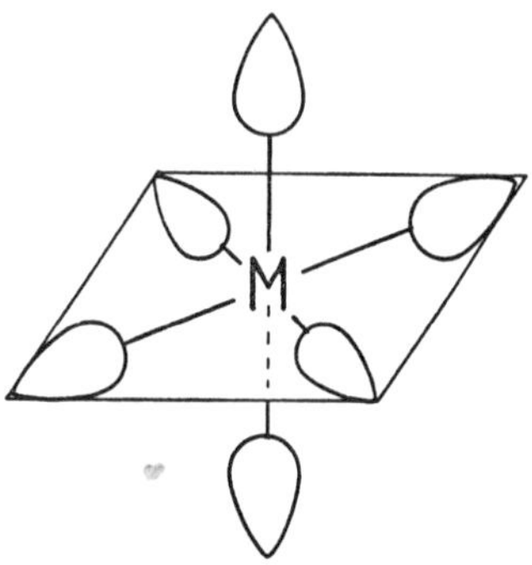

Figure 6.89 Central atom M surrounded octahedrally by six σ-type ligand orbitals

Table 6.17 Classification of σ ligand orbitals in various point groups

Point group	σ orbital symmetry species
O_h	$a_{1g} + e_g + t_{1u}$ (in octahedral ML_6)
T_d	$a_1 + t_2$ (in tetrahedral ML_4)
D_{3h}	$2a_1' + a_2'' + e'$ (in trigonal bipyramidal ML_5)
D_{4h}	$a_{1g} + b_{1g} + e_u$ (in square planar ML_4)
	$2a_{1g} + a_{2u} + b_{1g} + e_u$ (in *trans*-octahedral ML_4L_2')
$D_{\infty h}$	$\sigma_g + \sigma_u$ (in linear ML_2)
C_{2v}	$a_1 + b_2$ (in nonlinear ML_2)
	$2a_1 + b_1 + b_2$ (in tetrahedral ML_2L_2')
	$3a_1 + a_2 + b_1 + b_2$ (in *cis*-octahedral ML_4L_2')
C_{3v}	$2a_1 + e$ (in tetrahedral ML_3L')
	$2a_1 + 2e$ (in all-*cis*-octahedral ML_3L_3')
C_{4v}	$2a_1 + b_1 + e$ (in square pyramidal ML_4L')
	$3a_1 + b_1 + e$ (in octahedral ML_5L')
D_{2d}	$2a_1 + 2b_2 + 2e$ (in dodecahedral ML_8)
D_{4d}	$a_1 + b_2 + e_1 + e_2 + e_3$ (in square antiprism ML_8)

The O_h ligand σ MOs shown on the right-hand side of Figure 6.90 are lower in energy than the metal atom *d* orbitals shown on the left-hand side. For weak (crystal field) interaction the orbitals are split as shown towards the left in the figure but a strong interaction between orbitals of the same symmetry affects specifically the e_g orbitals of the metal atom and

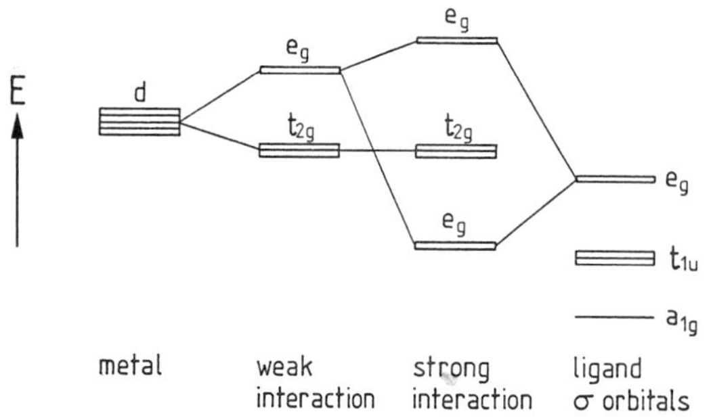

Figure 6.90 Ligand field MOs resulting from interaction between the σ-type ligand orbitals and the d orbitals of the central atom in an octahedral complex

the ligands, the former being pushed higher in energy and the latter lower. The t_{2g} orbitals are unaffected as the ligands do not have any σ orbitals with t_{2g} symmetry. The result, as shown in the figure, is that Δ_o is increased relative to the case of weak interaction. This increase in Δ_o leads, in turn, to a tendency towards low spin rather than high spin complexes, as in $[Fe(CN)_6]^{4-}$ and $[Co(NH_3)_6]^{3+}$.

6.3.2 Selection rules and intensities for electronic transitions

The symmetry selection rules for electronic dipole–dipole transitions in polyatomic molecules are precisely those given for atoms in equation (6.53). These selection rules are summarized in equation (6.111) as

$$\Gamma(\psi_e') \times \Gamma(\psi_e'') \supset \Gamma(T_x) \text{ and/or } \Gamma(T_y) \text{ and/or } \Gamma(T_z) \qquad (6.220)$$

i.e. the product of the symmetry species of the wave functions in the upper (ψ_e') and lower (ψ_e'') electronic states must contain the symmetry species of a translation. If the molecule belongs to a point group of sufficiently high symmetry, such as C_{2v}, the product in equation (6.220) is uniquely that of $\Gamma(T_x)$, $\Gamma(T_y)$ or $\Gamma(T_z)$ as shown in Table A.11 in the Appendix. If it belongs to a point group of low symmetry, such as C_s, the product may give either A', which is, as shown in Table A.1 in the Appendix, $\Gamma(T_x)$ and $\Gamma(T_y)$, or A'', which is $\Gamma(T_z)$.

In linear polyatomic molecules, as in diatomic molecules, there are a number of good quantum numbers and the selection rules may be expressed, at least partly, in terms of these. These rules have been described in Sections 6.2.3.2 and 6.2.3.3.

In non-linear polyatomic molecules the only possible good electronic quantum number is S, the electron spin quantum number. However, as in atoms, diatomic molecules and all linear molecules, this quantum number may become 'less good' and the $\Delta S = 0$ selection rule breaks down if one or more of the atoms has a high nuclear charge. All other electronic selection rules must be expressed in terms of symmetry and derived from equation (6.220).

In a molecular electronic spectrum the intensity of the electronic transition is obtained by integrating the intensities of all associated vibronic and rovibronic transitions. This intensity is then analogous to the integrated intensity of a single line, or multiplet, in an atomic spectrum.

Intensities of electronic transitions are discussed conveniently in terms of the oscillator strength f which was introduced in Section 2.2 and related to the molar absorption coefficient ε by equation (2.26).

The highest value f can have is approximately 1 and, in atomic spectra, all f-values are of this order of magnitude. There is a useful sum rule that, for a one-electron atom,

$$\sum_n f_{nm} = 1 \qquad (6.221)$$

where the sum is over transitions involving all upper states n and a single lower state m.

Rydberg transitions in molecules involve excited orbitals resembling those of an atom and therefore have high f-values. The first member of a Rydberg series has an f-value close to 1 and the continuum beyond the series limit has an f-value of the order of 10^{-2}.

Transitions for which the excited state correlates, not with an excited state of the corresponding united atom, but with the ground state are sometimes called sub-Rydberg transitions and are generally less intense. Exceptions to this general rule are charge-transfer transitions. These are transitions in which an electron is promoted from a bonding to the corresponding antibonding orbital and are so called because of the considerable movement of electronic charge involved. Mulliken labelled such charge-transfer states as V states; he also labelled the ground state N and therefore charge-transfer transitions are sometimes called V–N transitions.

A simple example of a charge-transfer transition is the $B^1\Sigma_u^+$–$X^1\Sigma_g^+$ transition of H_2 involving promotion of an electron from the $\sigma_g 1s$ to the $\sigma^*_u 1s$ MO (Figure 6.27 and 6.29).

The $B^3\Sigma_u^-$–$X^3\Sigma_g^-$ Schumann–Runge system of O_2 involves an electron promotion from the $\pi_u 2p$ to the $\pi^*_g 2p$ MO and is therefore a charge-transfer transition and has a typically high f-value of 0.19.

In ethylene the $\tilde{A}^1B_{1u}$–$\tilde{X}^1A_g$ transition involves a $1b_{2g}$–$1b_{3u}$ (π^*–π) promotion (see Section 6.3.1.6 and Figure 6.76). It is therefore charge-transfer in nature with a high f-value; indeed, this system is analogous to the Schumann-Runge system of O_2, with which ethylene is isoelectronic.

Intense charge-transfer spectra occur also in complexes formed between an electron donor and an electron acceptor, the charge-transfer transition involving the transfer of an electron from the donor to the acceptor.

In general the classification of spectra as being of a charge-transfer type is not clearcut since *all* electron promotions involve the transfer of some charge. However, the term 'charge-transfer spectra' is usually restricted to sub-Rydberg transitions characterized by an f-value of the order of 0.1–1.0.

The $\tilde{A}$–$\tilde{X}$, $\tilde{B}$–$\tilde{X}$ and $\tilde{C}$–$\tilde{X}$ systems of naphthalene (see Section 6.3.1.9) all involve π^*–π electron promotions. The f-values are 0.002, 0.2 and 1.7, respectively, and it is typical of aromatic molecules that they show a weak, medium intense and very intense system in the near ultraviolet/visible region. The anomalous f-value of 1.7 for the $\tilde{C}$–$\tilde{X}$ system of naphthalene is a consequence of the configuration interaction mentioned in Section 6.3.1.9.

Promotions of the π^*–n type lead to weak transitions. For example, the f-values of the $\tilde{A}$–$\tilde{X}$ transitions of acrolein (see Section 6.3.1.7) and pyridine (see Section 6.3.1.10) are 4×10^{-4} and 3×10^{-3}, respectively.

The f-value for the complete $\tilde{A}$–$\tilde{X}$ system of formaldehyde, resulting from a π^*–n promotion is 2.4×10^{-4} but for the 0_0^0 band alone, which is allowed only by magnetic dipole selection rules (see Section 6.3.1.5), the f-value is only 1×10^{-5}.

Spin-forbidden transitions are typically weak, unless the molecule contains a heavy atom, with the intensity depending on the degree of spin–orbit interaction. In the $\tilde{a}$–$\tilde{X}$ system of formaldehyde, for example, the f-value is 1.2×10^{-6}.

The transition metal complexes, which we have considered in Section 6.3.1.11, have low energy transitions which are electronically forbidden but which gain intensity through vibronic interaction, as does the $\tilde{A}$–$\tilde{X}$ system of benzene (see Section 6.3.4.1). Such transitions are weak with f-values comparable in magnitude to those resulting from π^*–n promotions.

6.3.3 Chromophores

We have seen in Section 5.2.1 how the concept of a group vibration leads to the important use of vibrational spectroscopy as a qualitative analytical tool.

In a similar way, a group of atoms in a molecule may show an electronic spectrum at a characteristic wavelength and with a typical intensity irrespective of the nature of the rest of the molecule in which the group of atoms occurs. Such a group is called a chromophore since it results in a characteristic colour of the compound due to absorption of visible or, using the word ‘colour’ in its general sense, ultraviolet radiation.

The ethylenic group, $>C=C<$, is an example of a chromophore. A molecule containing this group, such as $XHC=CH_2$, $XHC=CHX$, $X_2C=CH_2$ and cyclohexene, shows an intense absorption system with a maximum intensity at about 180 nm, very similar in intensity and wavelength to the $\tilde{A}$–$\tilde{X}$ system of ethylene itself (Section 6.3.1.6). However, the group can act as a chromophore only if it is not conjugated to any other π-electron system. We have seen in buta-1,3-diene (Section 6.3.1.7) and benzene (Section 6.3.1.8) that conjugated ethylenic groups show absorption at much longer wavelengths.

On the other hand, the benzene ring itself can be treated as a chromophore showing a characteristic, fairly weak absorption at about 260 nm as in, for example, phenylcyclohexane.

Similarly, the acetylenic group, $-C\equiv C-$, shows an intense absorption system at about 190 nm, corresponding to the $\tilde{A}$–$\tilde{X}$ system of acetylene (Section 6.3.1.4). This group can be regarded as a chromophore as can the allylic group, $>C=C=C<$, which shows characteristic absorption at about 225 nm.

Transitions involving π^*–n promotions are useful in identifying chromophores as they give characteristically weak absorption systems which are usually to high wavelenth of those due to π^*–π promotions and are not interfered with by them. The aldehyde group, –CHO, is a useful chromophore showing a weak absorption system at about 280 nm, like formaldehyde itself (Section 6.3.1.5). However, in molecules such as glyoxal and acrolein (Section 6.3.1.7) and benzaldehyde (C_6H_5CHO), the aldehyde group is part of a conjugated π-electron system and can no longer be treated as a chromophore.

6.3.4 Vibrational coarse structure

6.3.4.1 Herzberg–Teller intensity stealing

The transition moment $\boldsymbol{R}_{ev}$ for an electric dipole transition between an upper and a lower vibronic state, with vibronic wave functions ψ'_{ev} and ψ''_{ev}, respectively, is given by

$$\boldsymbol{R}_{ev} = \int \psi'^{*}_{ev} \boldsymbol{\mu} \psi''_{ev} \mathrm{d}\tau_{ev} \tag{6.222}$$

where $\boldsymbol{\mu}$ is the electric dipole moment operator and the integration is over electronic and vibrational coordinates. As in the case of a pure electronic transition the requirement for $\boldsymbol{R}_{ev}$ to be non-zero, and therefore for the transition to be allowed, is that the product of the symmetry species of the quantities inside the integral is totally symmetric or, in a degenerate point group, contains the totally symmetric symmetry species.

In the case of a diatomic molecule the integral in equation (6.222) can be expressed as (see equation (6.127)]

$$\boldsymbol{R}_{ev} = \boldsymbol{R}_{e} \int \psi''_{v} \psi'_{v} \mathrm{d}\tau_{v} \tag{6.223}$$

namely the product of the electronic transition moment, assumed here to be independent of nuclear motion, and the vibrational overlap integral. The normal mode of vibration is necessarily totally symmetric (σ_g^+ in a homonuclear and σ^+ in a heteronuclear diatomic molecule) so that, if the electronic transition moment integral is totally symmetric, then so are all vibronic transition moment integrals. It follows that *all* vibronic transitions associated with allowed electronic transitions are allowed by symmetry but, as discussed in Section 6.2.4.3, their intensities are governed by the Franck–Condon principle.

In a polyatomic molecule the same arguments can be applied to progressions involving vibrations in the upper or lower electronic state but only if the vibrations are totally symmetric and the transition is electronically allowed. The intensities of bands in such progressions are proportional to the square of the vibronic transition moment of equation (6.223) and will be discussed further in Section 6.3.4.2. However, there may be another factor contributing to the progression intensities due to the intensity stealing mechanism to be discussed in this section.

What is new in polyatomic, compared with diatomic, molecules is that there are many vibronic transitions which are allowed by symmetry involving non-totally symmetric vibrations. The general symmetry requirement for a vibronic transition to be allowed by electric dipole selection rules is that the integral of equation (6.222) contains the totally symmetric species. Since the symmetry species of $\boldsymbol{\mu}$ is that of a translation it follows that

$$\Gamma(\psi'_{ev}) \times \Gamma(\psi''_{ev}) \supset \Gamma(T_x) \text{ and/or } \Gamma(T_y) \text{ and/or } \Gamma(T_z) \qquad (6.224)$$

This requirement is analogous to that of equation (6.220) for an electronic transition.

The symmetry species for a vibronic state is always given by

$$\Gamma(\psi_{ev}) = \Gamma(\psi_e) \times \Gamma(\psi_v) \qquad (6.225)$$

irrespective of whether the Born–Oppenheimer approximation [equation (1.80)] holds so that equation (6.224) becomes

$$\Gamma(\psi'_e) \times \Gamma(\psi'_v) \times \Gamma(\psi''_e) \times \Gamma(\psi''_v) \supset \Gamma(T_x) \text{ and/or } \Gamma(T_y) \text{ and/or } \Gamma(T_z) \qquad (6.226)$$

Figure 6.91 illustrates two examples of the application of this equation to a molecule belonging to the C_{2v} point group. In Figure 6.91(a) the transition 1 from the ground electronic state $\tilde{X}^1A_1{}^e$ (the post-superscipt e or ev may be used, when necessary, to distinguish electronic and vibronic states)[†] to the excited electronic state ${}^1A_2{}^e$ is forbidden. The vibronic transition 2, involving the excitation of one quantum of a b_1 vibration in the ${}^1A_2{}^e$ state, is allowed, however, since $\Gamma(\psi'_e) = A_2$, $\Gamma(\psi'_v) = B_1$ and $\Gamma(\psi''_e) = A_1$ giving the product on the left-hand side of equation (6.226) as B_2; this is the species of T_y (Table A.11 in the Appendix) so the vibronic transition moment is polarized along the y-axis. Similarly, it can be shown that transition 3, involving vibrational excitation in the $\tilde{X}^1A_1{}^e$ state is symmetry allowed with the transition moment polarized along the y-axis. In an absorption process transition 3 gives rise to a hot band, so called because it involves vibrational excitation in the lower electronic state and therefore its intensity increases with temperature.

The example in Figure 6.91(b) shows an allowed electronic transition 1 polarized along the y-axis with vibronic transitions 2 and 3 also allowed but with their transition moments polarized along the z-axis.

These two examples serve to illustrate how a band system which is electronically either forbidden or allowed may show vibrationally induced bands involving non-totally symmetric vibrations. However, we must be careful to distinguish the fact that a vibronic transition is allowed by symmetry from whether it has any observable intensity. For all electronic transitions we can deduce many vibronic transitions, such as those in Figure 6.91, which are symmetry allowed but it turns out that most of them have insufficient intensity to be observed. For those which do have sufficient intensity the explanation of how it is derived was due initially to Herzberg and Teller[109] but has been amplified to include a second mechanism.[110,111]

The phenomenon of obtaining intensity by such vibronic transitions is known as Herzberg–Teller intensity stealing.[‡] It was first explained by them as being due to the fact that, when the vibrational overlap integral of equation (6.223) is non-totally symmetric, the electronic transition moment $\boldsymbol{R}_e$ [equation (6.127)] is no longer independent of vibrational motion. This implies a breakdown of the Franck–Condon

[†]When it is not necessary to include the multiplicity it is usual to use e or ev as a *pre*-superscript.

[‡]Also known as intensity borrowing—but it is never returned!

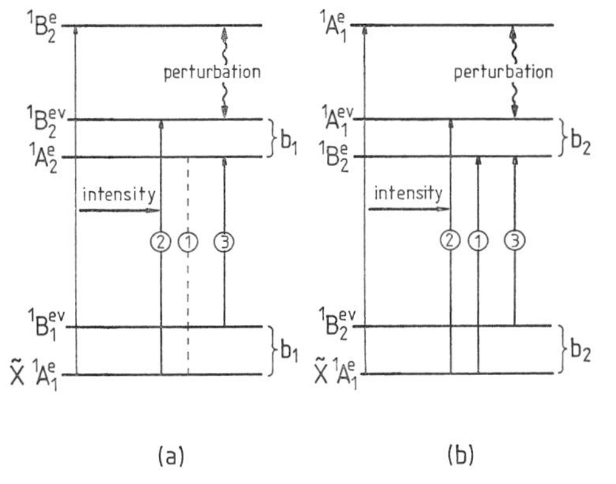

Figure 6.91 Herzberg-Teller intensity stealing by (a) a $^1B_2{}^{ev}$–$^1A_1{}^e$ vibronic transition from a $^1B_2{}^e$–$^1A_1{}^e$ electronic transition, and (b) a $^1A_1{}^{ev}$–$^1A_1{}^e$ vibronic transition from a $^1A_1{}^e$–$^1A_1{}^e$ electronic transition

approximation (Section 6.2.4.3) in which nuclei are considered to be stationary during an electronic transition. The effect of vibrational motion is taken account of by expanding $\boldsymbol{R}_e$ as a Taylor series, in the normal coordinates Q_k, Q_l, . . . of the vibrations concerned, as follows:

$$\begin{aligned}\boldsymbol{R}_e = {} & (\boldsymbol{R}_e)_{\text{eq}} && \text{zero order}\\ & + \sum_k \left(\frac{\partial \boldsymbol{R}_e}{\partial Q_k}\right)_{\text{eq}} Q_k && \text{first order}\\ & + \frac{1}{2!}\sum_{k,l}\left(\frac{\partial^2 \boldsymbol{R}_e}{\partial Q_k \partial Q_l}\right)_{\text{eq}} Q_k Q_l && \text{second order}\\ & + \frac{1}{3!}\sum_{k,l,m}\left(\frac{\partial^3 \boldsymbol{R}_e}{\partial Q_k \partial Q_l \partial Q_m}\right)_{\text{eq}} Q_k Q_l Q_m && \text{third order}\\ & + \ldots && (6.227)\end{aligned}$$

where the subscript eq refers to the equilibrium configuration of the molecule and the order of each term with respect to normal coordinates is indicated.

What this expansion implies in practice is that, if we calculate $\boldsymbol{R}_e$ for the equilibrium configuration and then for a configuration in which nuclear displacement has occurred, a different value will result.

The first-order term in equation (6.227) is much more important than higher order terms. For the example in Figure 6.91(a), if the b_1 vibrational symmetry were a result of the combination of one quantum each of an a_2 and a b_2 vibration, the vibronic transition moment would involve a second-order term and would normally be very small. Similarly, a combination of three different nontotally symmetric vibrations involves a third-order term. In the $\tilde{A}\,^1B_{2u}$–$\tilde{X}\,^1A_{1g}$ system of benzene, second- and third-order terms have been found to result in very weak, but comparably intense, vibronic transitions.[112]

Truncating the series at the second term and incorporating into equation (6.222) gives

$$\boldsymbol{R}_{ev} = \int \psi'_v \left[(\boldsymbol{R}_e)_{\text{eq}} + \sum_k \left(\frac{\partial \boldsymbol{R}_e}{\partial Q_k}\right)_{\text{eq}} Q_k\right] \psi''_v \, \mathrm{d}\tau_v \qquad (6.228)$$

At this stage it is convenient to separate the vibrations into those which are totally symmetric (s) and those which are nontotally symmetric (a). Then equation (6.228) becomes

$$\begin{aligned}\boldsymbol{R}_{ev} = {} & (\boldsymbol{R}_e)_{\text{eq}} \int \psi'_v \psi''_v \, \mathrm{d}\tau_v + \sum_s \left(\frac{\partial \boldsymbol{R}_e}{\partial Q_s}\right)_{\text{eq}} \int \psi'_v Q_s \psi''_v \, \mathrm{d}\tau_v \\ & + \sum_a \left(\frac{\partial \boldsymbol{R}_e}{\partial Q_a}\right)_{\text{eq}} \int \psi'_v Q_a \psi''_v \, \mathrm{d}\tau_v\end{aligned} \qquad (6.229)$$

For an electronically forbidden system, like that illustrated in Figure 6.91(a), $(\boldsymbol{R}_e)_{\text{eq}} = 0$ and $(\partial \boldsymbol{R}_e / \partial Q_s)_{\text{eq}} = 0$ and the first two terms on the right-hand side of equation (6.229) are zero; only the third term is non-zero and the system is said to be electronically forbidden but vibronically allowed, or vibrationally induced.

For an electronically allowed system, as in Figure 6.91(b), the first two terms are non-zero but the second term is usually small. The third term may also be non-zero and accounts for any part of an electronically allowed system which is induced by non-totally symmetric vibrations.

A useful way of visualizing the result of the non-zero values of the first-, second-, third- and higher order terms in equation (6.227) is in terms of vibronic coupling between electronic and vibronic states of the same symmetry. This coupling is implied by the dependence of $\boldsymbol{R}_e$ on vibrational coordinates. Figure 6.91(a) illustrates how the $^1B_2^{ev}$–$\tilde{X}^1A_1^e$ transition obtains its intensity, if any, through the coupling between the $^1B_2^{ev}$ vibronic state and a nearby $^1B_2^e$ electronic state of the same symmetry; the closer are the states, the greater is the coupling. Some of the intensity which, in the absence of any interaction, would all have been in the $^1B_2^e$–$\tilde{X}^1A_1^e$ allowed electronic transition, is stolen by the $^1B_2^{ev}$–$\tilde{X}^1A_1^e$ vibronic transition. In the example shown in Figure 6.91(b) intensity is stolen by the $^1A_1^{ev}$–$\tilde{X}^1A_1^e$ vibronic transition from the $^1A_1^e$–$\tilde{X}^1A_1^e$ allowed electronic transition.

In the examples in Figure 6.91 the *upper* electronic state is coupled to a higher electronic state through a non-totally symmetric vibration. It is also possible that the *ground* electronic state couples to a higher electronic state through a similar vibration. In this case the intensity of the vibronic transition is stolen from the transition between two excited electronic states.

A second mechanism for intensity stealing depends on the breakdown of the Born–Oppenheimer approximation, rather than of the Franck–Condon approximation. In the Born–Oppenheimer separation of the hamiltonian of equation (1.75) into the electronic [equation (1.77)] and nuclear [equation (1.78)] components the motion of the electrons was assumed to be independent of the kinetic energy of the nuclei, T_n. This is not quite true and if account is taken of T_n the result again may be that intensity stealing occurs. The selection rules are exactly the same whichever mechanism causes intensity stealing but there are important quantitative effects which differ according to which mechanism dominates.

One of the most studied examples of a band system which is electronically forbidden is the $\tilde{A}^1B_{2u}$–$\tilde{X}^1A_{1g}$ 260 nm system of benzene (see Section 6.3.1.8). Figure 6.92 shows this system in absorption at low resolution. The position of the missing 0_0^0 band† is indicated.

To high wavenumber of the 0_0^0 position is an intense band 6_0^1,‡ where ν_6 is an e_{2g} vibration. Since (Table 5.9)

$$B_{2u} \times e_{2g} = E_{1u} \tag{6.230}$$

and $E_{1u} = \Gamma(T_x, T_y)$ in the D_{6h} point group (Table A.36 in the Appendix) the 6_0^1 transition is symmetry allowed. The intensity of the transition derives from Herzberg–Teller intensity stealing from the $\tilde{C}^1E_{1u}$–$\tilde{X}^1A_{1g}$ electronic transition caused by coupling of the 6^1, $^1E_{1u}^{ev}$, vibronic state with the $\tilde{C}^1E_{1u}^e$ electronic state. The hot band 6_1^0 also derives intensity from the $\tilde{C}$–$\tilde{X}$ transition but is weakened, in absorption, by the Boltzmann factor ($\tilde{\nu}_6'' = 608\,\text{cm}^{-1}$). In the $\tilde{A}$–$\tilde{X}$ system in emission (fluorescence) the 6_1^0 band is intense and the 6_0^1 band weakened, if there is equilibrium population among vibrational levels, by the Boltzmann factor ($\tilde{\nu}_6' = 521\,\text{cm}^{-1}$).

Bands such as 6_0^1 and 6_1^0 are sometimes called false origins since the vibrational structure built on to them, which will be discussed in subsequent sections of this chapter, resembles that which might be built on to a 0_0^0 band of an electronically allowed system.

Benzene has four e_{2g} vibrations all of which could, like ν_6, be involved in Herzberg–Teller intensity stealing. In fact the other three, ν_7, ν_8

†In electronic spectroscopy of polyatomic molecules a vibronic transition between states in which vibration 12, say, is excited to the $\upsilon'' = 3$ level in the lower and the $\upsilon' = 1$ level in the upper electronic state is indicated by 12_3^1. Hence 0_0^0 indicates the pure electronic transition. There is an alternative but more cumbersome system which is commonly used for small molecules in which, for example, (010)–(110) indicates a vibronic transition in a triatomic molecule in which $\upsilon_1' = 0$, $\upsilon_2' = 1$, $\upsilon_3' = 0$ and $\upsilon_1'' = 1$, $\upsilon_2'' = 1$, $\upsilon_3'' = 0$.

‡The vibrational numbering scheme for benzene which is most commonly used is that due to Wilson, E. B. (1934). *Phys. Rev.*, **45**, 706. Unfortunately, this does not comply with the Mulliken recommendations but there is so much published material on benzene using the Wilson scheme that its use is likely to continue.

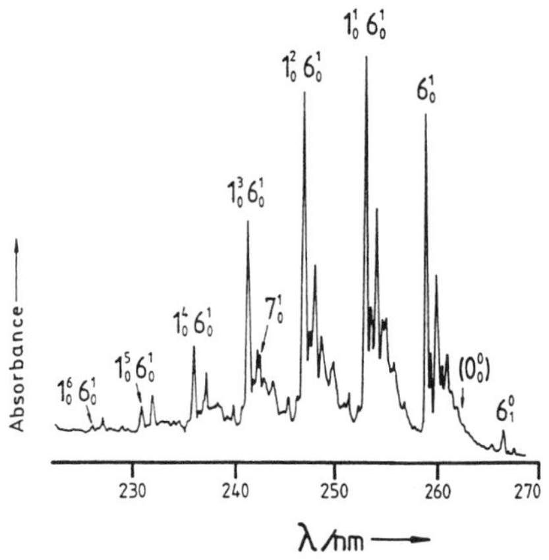

Figure 6.92 Low resolution $\tilde{A}^1B_{2u}$–$\tilde{X}^1A_{1g}$ absorption spectrum of a 1 cm path of benzene vapour above the liquid at room temperature and atmospheric pressure

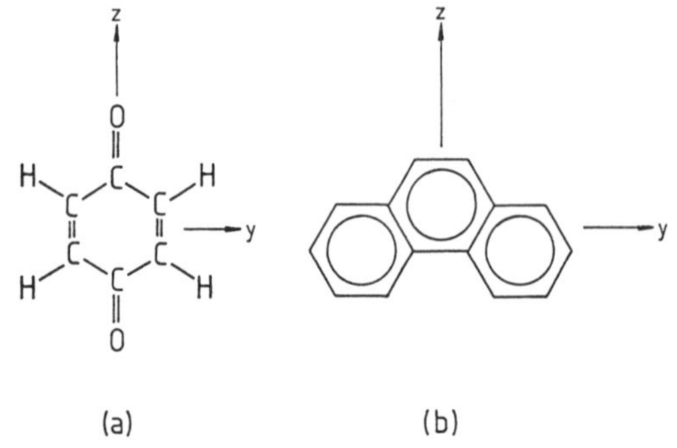

Figure 6.93 The (a) p-benzoquinone and (b) phenanthrene molecules

and ν_9, lead to false origins at least an order of magnitude weaker than those due to ν_6. This is because $(\partial \boldsymbol{R}_e/\partial Q_k)_{eq}$ is much smaller for the other three vibrational modes. The weak 7_0^1 band can be seen in Figure 6.92.

The $\tilde{A}^1B_{1g}$–$\tilde{X}^1A_g$ system of p-benzoquinone, which belongs to the D_{2h} point group and whose structure is shown in Figure 6.93(a), occurs in the 400–500 nm region and is electronically forbidden by electric dipole selection rules.† However, the 0_0^0 band has been observed weakly[113] as a magnetic dipole transition in a supersonic jet. This molecule has two a_u vibrations, ν_{22} and ν_{23}, which induce the vibronic transitions 22_0^1, 23_0^1, 22_1^0 and 23_1^0. Since

$$B_{1g} \times a_u = B_{1u} = \Gamma(T_z) \qquad (6.231)$$

the vibronic transition moments are polarized along the z-axis. Intensity is stolen by these transitions from a ${}^1B_{1u}$–$\tilde{X}^1A_g$ electronic transition.

In addition one, or possibly two, of the three b_{3u} vibrations induce vibronic transitions. Their transition moments are polarized along the y-axis since

$$B_{1g} \times b_{3u} = B_{2u} = \Gamma(T_y) \qquad (6.232)$$

Intensity is stolen from a ${}^1B_{2u}$–$\tilde{X}^1A_g$ electronic transition.

The $\tilde{A}^1A_2$–$\tilde{X}^1A_1$ electronic transition of formaldehyde is forbidden by electric dipole selection rules (see Section 6.3.1.5) but vibronic transitions are induced by, principally, the b_1 out-of-plane deformation (inversion) vibration ν_4.‡ The pattern of the false origin bands 4_0^1 and 4_1^0 in the absorption spectrum is similar to that illustrated in Figure 6.91(a) and the intensity of these bands is stolen from a ${}^1B_2$–$\tilde{X}^1A_1$ electronic transition. The two b_2 vibrations ν_5, the antisymmetric CH stretching vibration, and ν_6, the CH_2 rocking vibration, also induce vibronic transitions but the total intensity due to these vibrations is only about 25% of the total and is stolen from a ${}^1B_1$–$\tilde{X}^1A_1$ electronic transition. Whereas the transition moment of a band induced by ν_4 is polarized along the y-axis, that of a band induced by ν_5 or ν_6 is polarized along the x-axis.

Although the 0_0^0 transition of the $\tilde{A}$–$\tilde{X}$ system of formaldehyde is forbidden by electric dipole selection rules the transition is observed very weakly. The reason for this is that it is allowed

†See Hollas, J. M. (1964). *Spectrochim. Acta*, **20**, 1563.

‡This numbering of vibrations follows from the Mulliken axis notation shown in Figure 6.71(a). Much of the older literature on formaldehyde interchanges the x- and y-axes.

by magnetic dipole selection rules[114] (see Section 6.2.3.4) since $A_2 = \Gamma(R_z)$. It is unusual for such a transition to be sufficiently intense to be observed and this was the first example in a polatomic molecule.

The $\tilde{A}^1B_{2u}$–$\tilde{X}^1A_g$ system of naphthalene (see Section 6.3.1.9 for discussion and Figure 6.83 for axis notation) is unusual in that, although it is electronically allowed, with the transition moment polarized along the y-axis, there are some vibrationally induced bands which are about 10 times more intense than 0^0_0. Naphthalene has eight b_{3g} vibrations and the two of lowest wavenumber are involved in intense vibronic transitions whose intensity is stolen from the $\tilde{B}^1B_{1u}$–$\tilde{X}^1A_g$ electronic transition with a transition moment polarized along the z-axis.†

The fact that the vibrationally induced bands are more intense than the electronically allowed ones is due not to unusually large intensity stealing but to an unusually small electronic transition moment.

For an electronically allowed transition the first two terms of equation (6.229) are non-zero. So far we have neglected the second term in these circumstances but it can happen, in cases where the electronic transition moment is small, that the second term is comparable to, or even larger than, the first. This situation results in intensity stealing induced by totally symmetric vibrations. The vibronic state involved is coupled to a nearby electronic state of the *same* symmetry resulting in intensity being stolen by the vibronic from the electronic transition.

The first clear example of this effect was found in the $\tilde{A}^1A_1$–$\tilde{X}^1A_1$ system of the C_{2v} molecule phenanthrene,[115] illustrated in Figure 6.93(b). An a_1 vibration, of wavenumber 671 cm^{-1} in the $\tilde{A}$ state, is involved in an intense A^1_0 band (we label the vibration A for convenience). The A^2_0, A^3_0, . . . bands are very weak which distinguishes a case of intensity stealing by a totally symmetric vibration from one in which there is a long progression in that vibration due to a large geometry change (see Section 6.3.4.2(i)).

For appreciable intensity stealing to occur the vibronic and electronic states which interact should not be too far apart. If the vibration is totally symmetric it follows that two electronic states of the same symmetry should not be too widely separated. This is clearly more likely in a molecule of low symmetry.

†See, for example, Craig, D. P., Hollas, J. M., Redies, M. F. and Wait, S. C. (1961). *Philos. Trans. R. Soc.*, **A253**, 543.

6.3.4.2 *Progressions: Franck–Condon principle*

6.3.4.2(i) *Totally symmetric vibrations*

In a polyatomic molecule the application of the Franck–Condon principle in the case of totally symmetric vibrations is similar to that for the case of a diatomic molecule (Section 6.2.4.3) except that it applies to each vibration separately. The Franck–Condon principle takes account of the fact that an electronic transition is rapid compared with vibrational motion. It follows that, if there is an appreciable change of geometry from the lower to the upper electronic state in the direction of one (or more) of the normal coordinates, then there is a progression in one (or more) of the corresponding vibrations.

In the $\tilde{A}^1B_{2u}$ state of benzene there is an increase in r(CC) of 0.0345 Å‡ compared with the ground state, for which r(CC) = 1.3974 Å, and r(CH) is the same in both states, 1.084 Å. The result of the geometry change is that there is a long progression in the ring-breathing vibration ν_1; this is an a_{1g} vibration in which the carbon ring uniformly expands and contracts and involves very little stretching of the CH bonds. Figure 6.92 shows that the progression in ν_1 in the $\tilde{A}$–$\tilde{X}$ absorption spectrum has about six members built on the false origin 6^1_0 with the maximum intensity in the $1^1_0\,6^1_0$ band.

‡See McClean, W. M. and Harris, R. A. (1977). Ch. 1 in *Excited States*, vol. 3. Academic Press, New York, and references cited therein.

Just as for diatomic molecules it is possible to interpret quantitatively the intensity distribution along a progression in terms of the change in the equilibrium value of the normal coordinate from one electronic state to another. The problems in attempting to do this are, however, greater in polyatomic molecules because

1. Band intensities are likely to be less reliable because of the greater possibility of bands overlapping.
2. The form of the vibrational normal coordinate may not be known accurately.
3. Even if the normal coordinate is known it is likely to involve changes in more than one geometrical parameter so that interpretation of a change of a normal coordinate in terms of a change of a single geometrical parameter may not be a valid approximation.

The $\tilde{A}$–$\tilde{X}$ system of benzene is one of few in which these problems have been overcome.[†] The intensities of the progression members involving ν_1 are fairly reliable and, more important, the molecule is of such high symmetry that, assuming a regular hexagonal structure in the $\tilde{A}$ state, only two parameters, r(CC) and r(CH), are required for the complete structure. Because of the high symmetry there are only two a_{1g} vibrations. These are ν_1 and ν_2, the symmetric C–H stretching vibration.

Quantitative Franck–Condon calculations in the $\tilde{A}$–$\tilde{X}$ system of benzene have given values for the increase of r(CC) from the $\tilde{X}$ to the $\tilde{A}$ state of 0.036–0.037 Å, in good agreement with the more accurate value of 0.0345 Å obtained by rotational analysis.

An important technique, which can result in progressions which are relatively free from overlapping bands and therefore whose intensities can be measured more reliably, is that of single vibronic level fluorescence (SVLF), sometimes called dispersed fluorescence (DF), a technique first used in the gas phase[‡] and, later, in a supersonic jet (see Section 8.2.17).

Emission of radiation due to electronic transitions is divided into two types—fluorescence and phosphorescence. Fluorescence involves emission due to a transition between states of the same multiplicity ('spin allowed') and phosphorescence between states of different multiplicity ('spin forbidden'). In general, lifetimes of excited states involved in fluorescence are shorter than those involved in phosphorescence, because of the spin allowedness or forbiddenness, but this distinction tends to break down in molecules with atoms having large nuclear charges and, consequently, large spin–orbit coupling.

SVLF involves the use of monochromatic radiation which is absorbed by the molecule in a specific vibronic transition, thereby populating a single vibronic level in the excited electronic state. The monochromatic visible or near ultraviolet radiation may be obtained either from a high pressure xenon arc, using a monochromator to tune the wavelength, or, more commonly since the advent of lasers, a tunable dye laser (Section 8.2.17). The minimum bandwidth of the radiation is about 5 cm^{-1} with a xenon arc but very much less with a laser.

Monochromatic excitation is distinct from either broad band excitation, using a large wavelength range, or excitation in a high voltage discharge, both of which result in more or less random population of vibronic levels in an excited electronic state.

If, in the gas phase, the pressure p of the emitting sample is sufficiently low a molecule will not collide with any others in the time it takes for the emission to take place. This is the free molecule condition. The number of collisions z which a single molecule makes per unit time follows from the kinetic theory of gases and is given by

† See, for example, Craig, D. P. (1950). *J. Chem. Soc.*, 2146.

‡ See the review by Atkinson and Parmenter, referred to in the Bibliography at the end of the chapter.

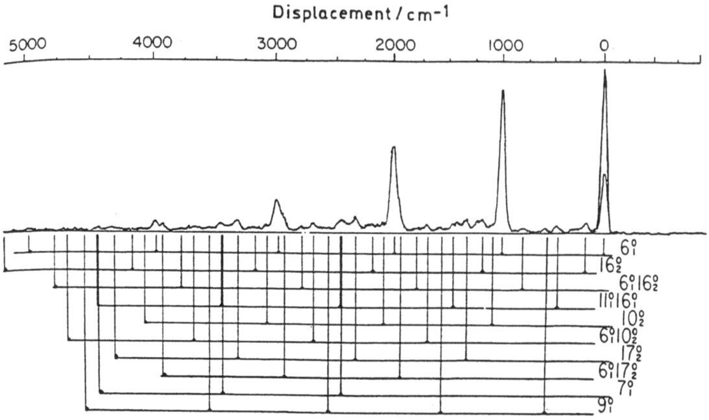

Figure 6.94 Single vibronic level fluorescence spectrum of benzene following excitation in the $6_1{}^0$ band of the $\tilde{A}\,{}^1B_{2u}$–$\tilde{X}\,{}^1A_{1g}$ system. The intensity of the 6_1^0 transition is enhanced by scattered exciting radiation which was from a xenon arc and monochromator. [Reproduced, with permission, from Parmenter, C. S. and Schuyler, M. W. (1970). *J. Chem. Phys.*, **52**, 5366]

$$z = \pi d^2 \left(\frac{8kT}{\pi m}\right)^{\frac{1}{2}} \frac{p}{kT} \tag{6.233}$$

where d is the collision diameter and m the molecular mass. If the pressure is sufficiently high for a number of collisions to take place before emission, vibrational relaxation occurs and a Boltzmann population distribution results. Under these conditions emission from the zero-point level of the excited electronic state is the most intense. In SVLF the pressure must be sufficiently low that collisions do not occur within the lifetime of the emitting vibronic level. For example, in benzene, taking the collision diameter to be 5 Å, we obtain $z = 7.29 \times 10^5\,\mathrm{s}^{-1}$ for $T = 293\,\mathrm{K}$ and $p = 1$ Torr. The fluorescence lifetime of the first excited electronic state is about 80 ns and, in this time, each molecule undergoes 0.06 collisions so that a pressure of 1 Torr would apparently be sufficiently low for collision-free conditions to prevail. However, it is found, in practice, that a pressure of only 0.1 Torr can be used in order to avoid vibrational relaxation. This corresponds to a typically large effective collision diameter for vibrational relaxation of 16 Å, a value which illustrates the point that the application of the kinetic theory of gases to molecular relaxation processes is somewhat artificial and does not constitute a very useful physical picture.

Figure 6.94 shows SVLF from the zero-point level of the $\tilde{A}\,{}^1B_{2u}$ state of benzene in the gas phase at low pressure, obtained by exciting monochromatically in the 6_1^0 band. Comparison of this spectrum with the absorption spectrum in Figure 6.92 shows how free of other bands is the $6_1^0 1_n^0$ progression in emission compared with the $6_0^1 1_0^n$ progression in absorption.

In the $\tilde{A}\,{}^1A_2$–$\tilde{X}\,{}^1A_1$ system of formaldehyde (Section 6.3.1.5) there is a long progression in the a_1 C$=$O stretching vibration v_2 as a result of the π^*–n electron promotion. This results in an increase in r(CO) from 1.210 Å in the $\tilde{X}$ to 1.312 Å in the $\tilde{A}$ state.

Long progressions in the CO stretching vibration are characteristic of π^*–n transitions in molecules containing the C$=$O chromophore and are found in the $\tilde{A}$–$\tilde{X}$ systems of, for example, acetaldehyde (CH_3CHO), glyoxal (OHCCHO), acrolein ($H_2C{=}CHCHO$) and propynal (HC≡CCHO).

In the cyclic molecule diazirine ($H_2\overline{C{-}N{=}N}$) and substituted diazirines the π^*–n, $\tilde{A}$–$\tilde{X}$ transition in the N$=$N chromophore shows long progressions in the N$=$N stretching vibration because of the decreased bonding character in the N=N group in the $\tilde{A}$ state.

Long progressions in the a_1 bending vibrations, ν_2, of NH_2, PH_2 and AsH_2 are found in the $\tilde{A}^2A_1$–$\tilde{X}^2B_1$ systems because of the large increase of bond angle in the $\tilde{A}$ state, as discussed in Section 6.3.1.1. There is a similar progression in the $\tilde{b}^1B_1$–$\tilde{a}^1A_1$ system of CH_2. These will be discussed further in Section 6.3.4.4.

Long progressions are likely to arise when there is a change of point group of the equilibrium configuration from one electronic state to another. An example of this is found in the $\tilde{A}^1A_u$–$\tilde{X}^1\Sigma_g^+$ system of acetylene (Section 6.3.1.4) in which the molecule is linear in the $\tilde{X}$ state and *trans*-bent (C_{2h}) in the $\tilde{A}$ state. There is a long progression in the *trans*-bending vibration, ν_3. (It is ν_4 in the linear ground state.)

The equilibrium configuration of the NH_3 molecule is pyramidal in the ground state, $\tilde{X}^1A_1'$, but planar in several excited Rydberg states (see Section 6.3.1.3). The result is that, in the corresponding band systems in absorption, there are long progressions in the inversion vibration ν_2. Although this vibration is totally symmetric in the C_{3v} classification of the $\tilde{X}$ state, we have seen in Section 5.2.7.1 that tunnelling through the inversion barrier results in a splitting of the $v_2 = 0$ and 1 levels of about $0.8\,\mathrm{cm}^{-1}$ and consequent D_{3h} classification of all levels. In this point group ν_2 is an a_2'' vibration and, although it is strictly a non-totally symmetric vibration, progression members involve alternately the $v_2 = 0$ and 1 levels in the $\tilde{X}^1A_1'$ state, as shown in Figure 6.95. In the example shown, involving the $\tilde{A}^1A_2''$ Rydberg state,

$$\Gamma(\psi'_{ev}) \times \Gamma(\psi''_{ev}) = A_2'' \qquad (6.234)$$

for all the transitions indicated. The separation of the $v_2'' = 0$ and 1 levels is so small that there is almost identical population of both at room temperature.

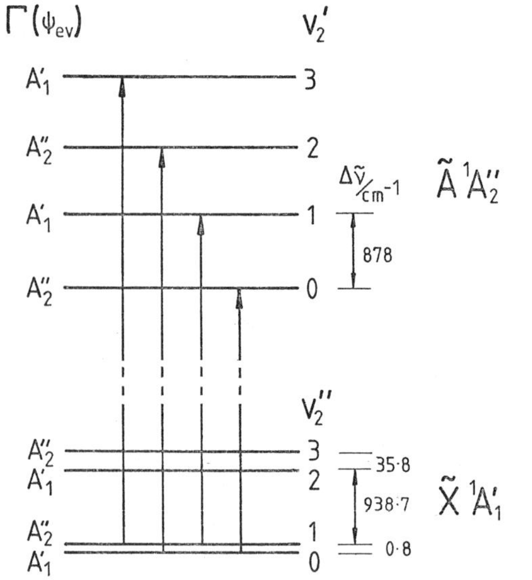

Figure 6.95 Vibrational progression in the inversion vibration ν_2 in the $\tilde{A}^1A_2''$–$\tilde{X}^1A_1'$ system of the NH_3 molecule

6.3.4.2(ii) Non-totally symmetric vibrations

In the $\tilde{A}^1A_2$–$\tilde{X}^1A_1$ system of formaldehyde there is not only a long progression in the C=O stretching vibration, ν_2, due to an increase in the C=O bond length in the $\tilde{A}$ state but, more interestingly, fairly long progressions in ν_4, the b_1 inversion vibration, due to a non-planar equilibrium configuration in the $\tilde{A}$ state (Section 6.3.1.5).

In the $\tilde{A}$ state the potential function for ν_4 shows a double minimum similar to that for NH_3 in the $\tilde{X}$ state (see Figures 5.68–5.70). However, tunnelling through the barrier is greater in formaldehyde resulting in $v_4' = 1$–0, 2–1 and 3–2 separations of 124, 359 and $407\,\mathrm{cm}^{-1}$, respectively.

Figure 6.96 illustrates some of the most important vibronic transitions in the $\tilde{A}$–$\tilde{X}$ system observed in absorption and emission.

The 0_0^0 transition, and others with Δv_4 even, are allowed by magnetic dipole selection rules, as discussed in Section 6.3.1.5, and are observed weakly. The bulk of the intensity is induced by

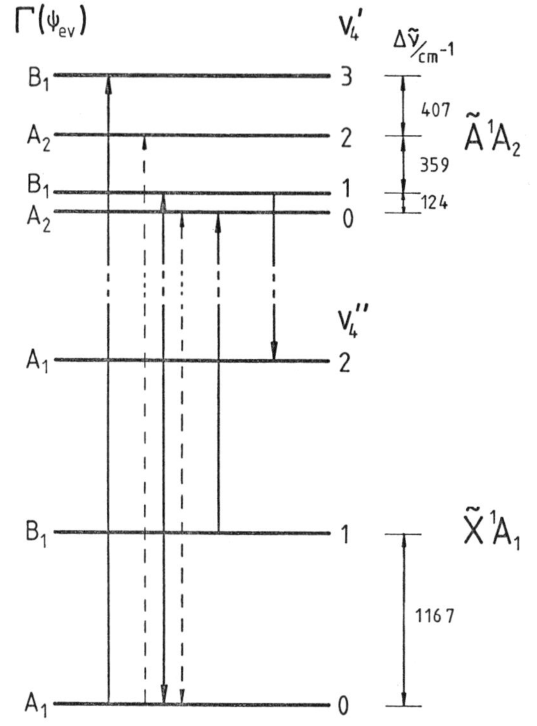

Figure 6.96 Some vibronic transitions in the $\tilde{A}^1A_2$–$\tilde{X}^1A_1$ system of the formaldehyde molecule involving the inversion vibration ν_4. The dashed transitions 0_0^0 and 4_0^2 are weak, being allowed only by magnetic dipole selection rules

ν_4 which is involved in Herzberg–Teller intensity stealing. Built on to the 4_0^1 and 4_1^0 bands are others with Δv_4 odd such as the 4_0^3 and 4_2^1 bands shown in the figure.

It is very unusual for there to be a progression of appreciable length in a non-totally symmetric vibration when there is *no* change of point group between electronic states but this does occur in the $\tilde{A}^1B_{3u}$–$\tilde{X}^1A_g$ system of pyrazine (1,4-diazabenzene).† There is a progression in the b_{1g} out-of-plane vibration ν_{10a} which, in the emission spectrum, shows about four members. The vibrational intervals in the $\tilde{X}$ state can be fitted to those of a potential function of the type in equation (5.259) which applies to puckering vibrations but, for ν_{10a}, there is only one minimum.

Smith[116] has derived expressions for intensity distributions along progressions in non-totally symmetric vibrations assuming that the breakdown of the Franck–Condon approximation accounts for intensity stealing.

6.3.4.3 *Sequences and cross-sequences*

The larger a molecule is, the more normal vibrations it has and the greater is the likelihood of there being several of these with sufficiently low wavenumbers to give rise to prominent sequences (see Section 6.2.4.2 for discussion of sequences in diatomic molecules).

In the $\tilde{A}^1B_{2u}$–$\tilde{X}^1A_g$ system of naphthalene (Section 6.3.1.9), for example, there are more than six sequences observable in absorption at room temperature. Two are very intense, each having five members, and are due to vibrations with wavenumbers of 166 and 195 cm^{-1}. All the sequences are negative running which means that they run to low wavenumber. This is often the case because of the tendency for the electron promotion to be a less strongly bonding upper orbital leading to a tendency for force constants to decrease from the ground to an excited electronic state. However, positive running sequences are observed in cases where a vibration involves a part of the molecule in which there is an increase of force constants. Such examples are found in the $\tilde{A}^1B_{1g}$–$\tilde{X}^1A_g$ system of *p*-benzoquinone [Figure 6.93(a)], in which there is, in the $\tilde{A}$ state, an increase in double bond character of those CC bonds which are nominally single in the $\tilde{X}$ state.

Low wavenumber vibrations are mostly non-totally symmetric, except in non-planar molecules of low symmetry, and they are observed mainly in $\Delta v = 0$ sequences. If the vibrational overlap integral of equation (6.223) is approximately unity, which is very often the case for sequence-forming vibrations, sequence band intensities are simply the product of the

†See, for example, Udugawa, Y., Ito, M. and Suzuka, I. (1978). *Chem. Phys. Lett.*, **60**, 25.

electronic transition intensity and the vibrational Boltzmann factor (equation (6.122)].

In cases where there is an appreciable change of geometry in the direction of a normal coordinate involved in a sequence the vibrational overlap integral is not unity. Such an example occurs in the $\tilde{A}^1B_2$–$\tilde{X}^1A_1$ system of aniline, $C_6H_5NH_2$ [see Figure 5.27(c)]. In this molecule the hydrogen atoms of the NH_2 group are not coplanar with the rest of the molecule. The angle between the bisector of ∠ HNH and the plane of the rest of the molecule is 39.4°, determined from the microwave spectrum in the $\tilde{X}$ state, but the molecule is planar in the $\tilde{A}$ state.[117] The resulting pattern of vibrational levels for ν_I, the NH_2-inversion vibration, is shown in Figure 6.97. There is a W-shaped potential function in the $\tilde{X}$ state with a barrier of 547 cm^{-1} and tunnelling through it to give the observed, irregularly spaced levels.

The observed I_1^1 and I_2^2 sequence bands, as well as bands with $\Delta v_I = 2$, are indicated in the figure. The intensity of the I_1^1 band is equal to that of the 0_0^0 band whereas if it depended only on the Boltzmann factor it would be 20% less. The effect is even more pronounced in $C_6H_5N^2H_2$.[118]

More dramatic deviations from the Boltzmann intensity distribution are possible in two further cases. The first case arises if a vibration A is involved in Herzberg–Teller intensity stealing through bands A_1^0 and A_0^1 and also in sequence bands A_2^1, A_3^2, . . . and A_1^2, A_2^3, . . . The sequence band intensities, I_v^{v+1}, are given by

$$I_v^{v+1} \propto (v+1)\exp-\{[G(v'')-G(0)]hc/kT\} \quad (6.235)$$

The factor $(v+1)$ in addition to the Boltzmann factor, results in an enhancement of the sequence band intensities.

The second case arises if a vibration A is involved in an A_0^2 band in addition to being active in sequences. Then the sequence bands A_1^3, A_2^4,... show intensities, I_v^{v+2}, given by

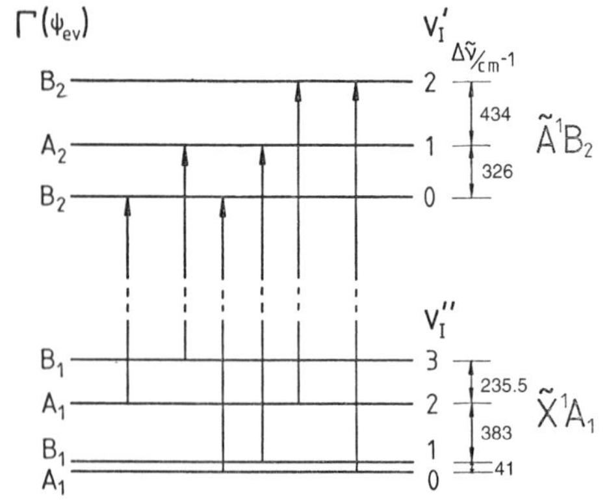

Figure 6.97 Vibronic transitions involving the NH_2-inversion vibration, ν_I, in the $\tilde{A}^1B_2$–$\tilde{X}^1A_1$ system of the aniline molecule

$$I_v^{v+2} \propto (v+1)(v+2)\exp-\{[G(v'')-G(0)]hc/kT\} \quad (6.236)$$

The factor $(v+1)(v+2)$ gives an intensity enhancement larger than in the previous case.

Examples of both these cases have been identified in the $\tilde{A}^1B_{1g}$–$\tilde{X}^1A_g$ and $\tilde{a}^3A_u$–$\tilde{X}^1A_g$ systems of *p*-benzoquinone.[119] In this molecule the vibration involved in *both* cases has such a low wavenumber (87 cm^{-1} in the $\tilde{X}$ state) that intensities along the sequences built on A_1^0 and A_0^1 rise to a maximum, due to the $(v+1)$ factor in equation (6.235), and then fall off as v increases further. The sequence built on A_0^2 shows a more pronounced effect due to the $(v+1)(v+2)$ factor in equation (6.236).

In an excited electronic state the molecular geometry and force constants are, in general, different from those in the ground state. Since the difference is usually small it is useful to treat the normal modes in the excited state as resulting from a slight mixing of those in the ground state. Just as symmetry considerations allow factorization of the vibrational secular determinant (Section 5.2.3.3), they also restrict the mixing of normal modes from the ground to an excited electronic state, and, in general, between *any* two electronic states, to those belonging to the same symmetry species. This was first pointed out by Duschinsky[120] and is known as the Duschinsky effect.

Sequences of the type $A_1^0B_0^1$, $A_2^0B_0^2$, . . . in vibrations A and B which belong to the same symmetry species are called cross sequences, with $\Delta v = 0$ in this example. Any intensity which they may have is due to the Duschinsky effect mixing the A and B modes. Intensities are usually low, although the $41_0^1 42_1^0$ band in the $\tilde{A}^1A'$–$\tilde{X}^1A'$ system of styrene is an exception.[121] A good example is the family of cross sequences $4_0^1 16b_1^0$, $17b_0^1 16b_1^0$, $10b_0^1 16b_1^0$, $\tau_0^1 16b_1^0$ and $11_0^1 16b_1^0$, in the $\tilde{A}^1B_2$–$\tilde{X}^1A_1$ system of phenol,[122] C_6H_5OH (Figure 5.38). This family involves ν_{16b} and all other b_1 vibrations† and the intensities are 3, 0.2, 0.1, 0.05 and 0.05, respectively, relative to the intensity of the $16b_1^1$ sequence band of 12. These intensities show that there is appreciable mixing of only ν_{16b} with ν_4 in the $\tilde{A}$ state.

Puckering vibrations of saturated, or partially saturated, ring molecules are often of low wavenumber (Section 5.2.7.2) and may be active in sequences. The puckering vibration of the CH_2 group in position 2 in indane, shown in Figure 6.98, shows sequence members with $\Delta v = 0$ and $v = 0$–6 in the $\tilde{A}^1A_1$–$\tilde{X}^1A_1$ system. Splitting of the sequence bands, increasing with v, indicates an increasing tunnelling through the barrier in a W-shaped potential and a lower barrier in the $\tilde{A}$ than in the $\tilde{X}$ state.[123]

Sequences in degenerate vibrations are complicated by the fact that each band may have more than one component due to the multiplicity of states which arise when a degenerate vibration is excited to levels with $v > 1$. These states may be split by the anharmonic term $\Sigma_{i \leqslant j}(g_{ij} l_i l_j)$ of equation (5.247), where l is the quantum number associated with the angular momentum of a degenerate vibration and g_{ij} is an anharmonic constant. Splitting of the states may then result in sequence bands with more than one component. Examples have been found in the $\tilde{A}^1B_{2u}$–$\tilde{X}^1A_{1g}$ system of benzene[124] in which the sequence-forming vibration ν_{16} is the lowest wavenumber (399 cm^{-1}) vibration in the $\tilde{X}$ state and has e_{2u} symmetry.

†C_{2v} symmetry species are useful for most vibrations although, strictly, the molecule belongs to the C_s point group.

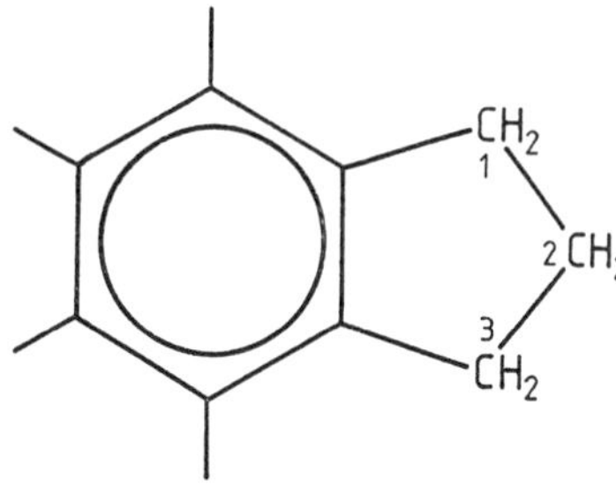

Figure 6.98 The indane molecule

Sequences in degenerate π vibrations of a linear molecule may be important in providing evidence for a Renner–Teller effect, discussed in the next section.

6.3.4.4 *The Renner–Teller effect*

We have seen, in considering the Walsh diagrams for AH_2, AB_2 and HAB molecules in Section 6.3.1.1–6.3.1.2, that the double degeneracy of a π orbital in the linear configuration of the molecule is split when the molecule bends. In an analogous way the degeneracy of a Π (and also a Δ, Φ, . . .) electronic state of a linear molecule is split when the bending vibration is excited.

In *all* linear triatomic molecules a Π electronic state may be split in any of three general ways when the bending vibration ν_2 is excited. The resulting potential functions $V(Q_2)$ are shown in Figure 6.99 in which they are compared with that in Figure 6.99(a) for a non-degenerate electronic state. In Figure 6.99(b) the two potential curves, with potential functions V^+ for the inner and V^- for the outer curve, both have an identical minimum in the linear configuration ($Q_2 = 0$). Figure 6.99(c) and

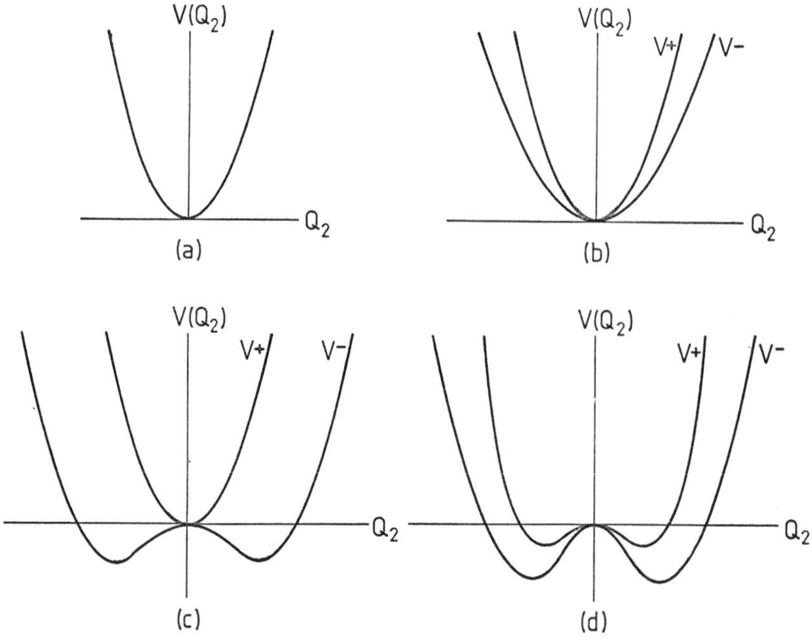

Figure 6.99 Potential functions for the bending vibration, ν_2, of a triatomic molecule (a) in a non-degenerate electronic state, and in a degenerate electronic state split by the Renner–Teller effect in which the molecule is linear in the equilibrium configuration (b) for both components of the state, (c) for only the upper component and (d) for neither component

(d) illustrate examples where V^-, or both V^+ and V^-, show W-shaped potential curves with minima corresponding to a bent molecule.

An example corresponding to Figure 6.99(b) is the $\tilde{A}^1\Pi_u$ excited state of the C_3 molecule (see Table 6.13) in which it is linear. A further example is the $\tilde{G}^1\Pi_u$ state of acetylene, C_2H_2, but there is a complication here in that either or both of the bending vibrations ν_4 and ν_5 (see Figure 5.35) may split the degeneracy of the $^1\Pi_u$ state.

In the case shown in Figure 6.99(d), the two potential curves represent two separate electronic states of a bent molecule which become degenerate in the non-equilibrium linear configuration. The $\tilde{A}^2A_1$–$\tilde{X}^2B_1$ transition of NH_2 (also PH_2 and AsH_2) is a transition between two such states which both correlate with a $^2\Pi_u$ state of the linear molecule. This state would result from the configuration

$$\ldots (1\sigma_u)^2(1\pi_u)^3 \qquad (6.237)$$

obtained using the MO diagram in Figure 6.63.

The $\tilde{b}^1B_1$–$\tilde{a}^1A_1$ transition of CH_2, discussed in Section 6.3.1.1, also provides an example of the case in Figure 6.99(d) while the $\tilde{A}^2\Pi_u$–$\tilde{X}^2A_1$ transition of BH_2 is an example of that in Figure 6.99(c).

For a linear molecule in a Π, Δ, Φ, . . . electronic state there is an electronic orbital angular momentum with a component $\Lambda\hbar$ along the internuclear axis, where $\Lambda = 1, 2, 3, \ldots$ (Section 6.2.3.1). If the bending vibration is excited there is also a vibrational angular momentum $l\hbar$ about the internuclear axis where $l = \pm 1, \pm 2, \pm 3, \ldots$ corresponding to Π, Δ, Φ, . . . vibrational states (Section 4.2.6.2). The resulting vibronic angular momentum about the axis is $K\hbar$, where[†]

$$K = |\pm \Lambda + l| \qquad (6.238)$$

This quantum number K is the same as the rotational quantum number K_a in a bent triatomic molecule which is a prolate near-symmetric rotor since both refer to the angular momentum about the same axis.

For a Π electronic state in which one quantum of v_2 is excited, $\Lambda = 1$, $v_2 = 1$ and, since $l = v_i, v_i - 2, \ldots -v_i$, we have $l = \pm 1$, giving

$$\begin{aligned} K = &|+1+1| = 2 \\ &|-1-1| = 2 \\ &|+1-1| = 0 \\ &|-1+1| = 0 \end{aligned} \qquad (6.239)$$

The vibronic states with $K = 0, 1, 2, \ldots$ are labelled Σ, Π, Δ, . . .

The result of equation (6.239) is in accord with that obtained by multiplying the electronic and vibrational symmetry species to give the vibronic symmetry species. For example, for the $v_2 = 1$ state of a linear AH_2 molecule $\Gamma(\psi_v) = \Pi_u$ and, if it is in a Π_u electronic state,

$$\Gamma(\psi_{ev}) = \Pi_u \times \Pi_u = \Sigma_g^+ + \Sigma_g^- + \Delta_g \qquad (6.240)$$

The $\Sigma_g{}^+$ and $\Sigma_g{}^-$ states correspond to the two $K = 0$ states of equation (6.239) and the two components of the doubly degenerate Δ_g state to the two $K = 2$ states.

[†]The problem of whether to use signed or unsigned quantum numbers K, Λ and l is particularly apparent here. I have chosen to use l as a signed quantum number, as is usually done, for the treatment of l-type doubling in Section 4.2.6.2 and elsewhere in this book. On the other hand, Λ and K (except when specified as k) are taken as unsigned, as is commonly the case. If K, Λ and l are *all* unsigned, equation (6.238) becomes $K = |\pm \Lambda \pm l|$ whereas, if they are *all* signed, it becomes $k = \Lambda + l$.

For an electronic state with non-zero electron spin which is strongly coupled to the orbital angular momentum, the component of the orbital-plus-spin angular momentum along the internuclear axis is $\Omega\hbar$ [equation (6.109)] and addition of a vibrational angular momentum $l\hbar$ gives the vibronic angular momentum $P\hbar$ about the internuclear axis where

$$P = |\pm \Omega + l| \qquad (6.241)$$

When the bending vibration is excited in a degenerate electronic state of a linear molecule, or in either of the electronic states of a bent molecule which correlate with a degenerate state of a linear molecule, the Born–Oppenheimer approximation of equation (1.80) breaks down. The reason for this is that wave functions of the electronic states which become degenerate in the linear molecule are strongly dependent on Q_2. One result is that Λ and l (or Ω and l) are no longer good quantum numbers but K (or P) is. A second result is that vibronic states within the two electronic states are mixed to such an extent that it may no longer be possible to say that a vibronic state belongs to one or other of the electronic states. These results of the Born–Oppenheimer breakdown are known as the Renner–Teller effect.[125,126]

A useful potential function for the bending coordinate Q_2 is of the form

$$V = aQ_2^2 + bQ_2^4 + \ldots \qquad (6.242)$$

where the quartic and higher terms make the otherwise parabolic potential curve more steep-sided for large displacements. The splitting of the potential curve into V^+ and V^- components is also given by a similar expression

$$V^+ - V^- = \alpha Q_2^2 + \beta Q_2^4 + \ldots \qquad (6.243)$$

The magnitude of α relative to a determines whether the V^+ and V^- potential curves resemble those in Figure 6.99(b) or (c).

Renner treated only the case illustrated in Figure 6.99(b) for a $^1\Pi$ electronic state and neglected all but the quadratic terms in equations (6.242) and (6.243). The resulting vibrational term values $G^{\pm}(v_2, K)$ for the V^+ and V^- potential functions for the Σ vibronic states, with $K = 0$, are

$$G^{\pm}(v_2, 0) = \omega_2(1 \pm \varepsilon)^{\frac{1}{2}}(v_2 + 1) \quad (6.244)$$

where $v_2 = 1, 3, 5, \ldots$, since only these states have Σ components. For $K \neq 0$ and $v_2 = K - 1$,

$$G(v_2, K) = \omega_2[(v_2 + 1) - \tfrac{1}{8}\varepsilon^2 K(K + 1)] \quad (6.245)$$

whereas for $K \neq 0$ and $v_2 > K - 1$,

$$G^{\pm}(v_2, K) = \omega_2(1 - \tfrac{1}{8}\varepsilon^2)(v_2 + 1) \pm \tfrac{1}{2}\omega_2\varepsilon[(v_2 + 1)^2 - K^2]^{\frac{1}{2}} \quad (6.245a)$$

The quantity ε is the Renner parameter and is given by

$$\varepsilon = \frac{\alpha}{2a} \quad (6.246)$$

where α and a are the coefficients in equations (6.242) and (6.243).

Only the Σ vibronic states can be said to belong to either the V^+ or V^- potential curve. This is not so for Π, Δ, ... states, which may be said to belong to both.

The $\tilde{A}^1\Pi_u$–$\tilde{X}^1\Sigma_g^+$ band system of C_3 shows evidence for a Renner-Teller effect in the $\tilde{A}^1\Pi_u$ state.[127] Since the molecule is linear in the $\tilde{A}$ and $\tilde{X}$ states there is no progression in ν_2 and, in addition, transitions with Δv_2 odd are forbidden by the $g \leftrightarrow\!\!\!\!\!/\; g$ selection rule. However, since ν_2'' has the unusually low value of $63.1\ \mathrm{cm^{-1}}$, sequences in this vibration are intense and all information on the Renner–Teller effect in the $\tilde{A}$ state came from such bands.

Figure 6.100 illustrates the 0_0^0 and 2_1^1 transitions, the latter being split into three widely spaced components because of the large value of ε of 0.537. It should be noticed that all the transitions in the figure obey the vibronic selection rule of equation (6.224).

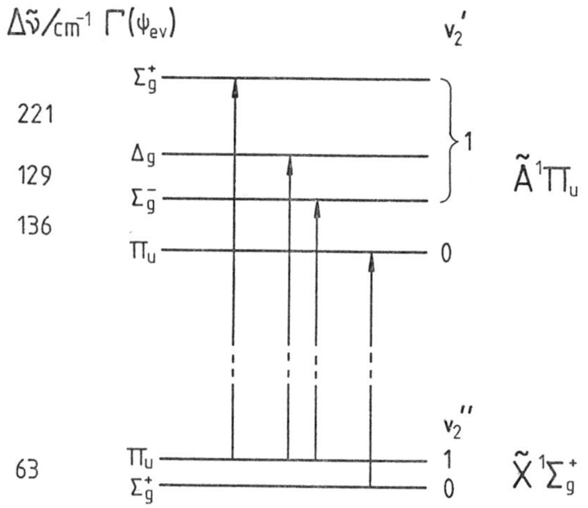

Figure 6.100 The Renner–Teller effect in the upper state of the $\tilde{A}^1\Pi_u$–$\tilde{X}^1\Sigma_g^+$ system of the C_3 molecule splits the $v_2' = 1$ level into three components

The $\tilde{A}^2A_1$–$\tilde{X}^2B_1$ transition of NH_2 is an example of the case in Figure 6.99(d) and, as all such examples do, provides experimental information on more vibronic levels involving ν_2 than a transition, such as the $\tilde{A}$–$\tilde{X}$ transition in C_3, between two electronic states in which the molecule is linear. The reason for this is the large change of equilibrium angle from the $\tilde{X}$ to the $\tilde{A}$ state. It is 103.4° in the $\tilde{X}$ and 144° in the $\tilde{A}$ state of NH_2 (Figure 6.64) and the result is a long progression in ν_2 with observation† in the absorption spectrum of bands up to 2_0^{18}.

Pople and Longuet–Higgins[128] have derived vibrational term value expressions for the V^+ state for the case in Figure 6.99(c). It was necessary to include quartic terms in the potential energy expressions of equations (6.242) and (6.243) and, although the $\tilde{A}^2A_1$ state of NH_2 is non-linear, the observed vibrational term values can be fitted reasonably well to the theory.

The lower-lying vibrational levels of ν_2 in the V^- state, which, in absorption, are the only ones with sufficient population to be involved in

†See Johns, J. W. C., Ramsay, D. A. and Ross, S. C. (1976). *Can. J. Phys.*, **54**, 1804, and references cited therein.

transitions, behave normally for a bent molecule (Section 5.2.7.3).

More general, and more accurate, theoretical treatments, such as that by Jungen and Merer,[129] are applicable to all the cases in Figure 6.99. These treatments recognize that the case when the molecule is linear in both components [Figure 6.99(b)] or the case when it is bent in either the lower or both components [Figure 6.99(c) or (d)] represent two limiting cases. Between these limits there must be a smooth correlation of the associated vibronic and rovibronic energy levels.

When the equilibrium configuration of the molecule is linear, the orbital angular momentum in the Π electronic state splits each level of the π bending vibration into several components, as illustrated by equation (6.239). When the equilibrium configuration is bent, in either the lower state or both, there can be no orbital degeneracy near equilibrium. Instead, there is a strong Coriolis interaction (see Section 4.2.6.2) caused by a vibrational angular momentum. These two types of angular momenta are then two limiting descriptions of the same thing.

Calculated rovibronic energy levels and transitions in the $\tilde{b}^1B_1$–$\tilde{a}^1A_1$ visible band system of CH_2, for example, show[130] that high quality *ab initio* calculations have advanced to such an extent that they are becoming a vital aid to experimental assignments.

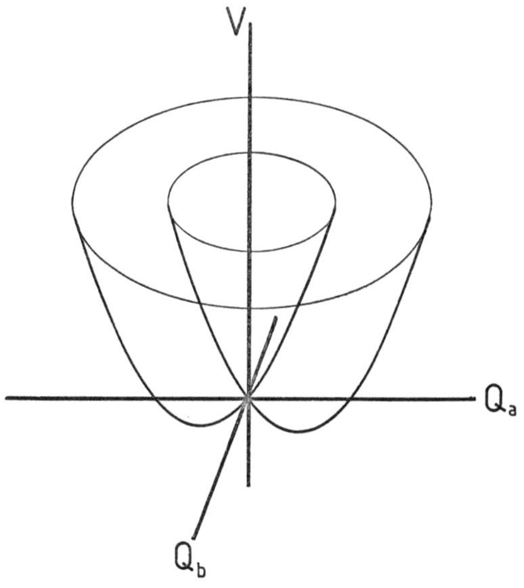

Figure 6.101 The static Jahn–Teller effect results in the splitting of electronic degeneracy in a non-linear molecule by a degenerate vibration with normal coordinates Q_a and Q_b for the two components. The resulting potential energy surface is shown

6.3.4.5 The Jahn–Teller effect

We have seen in the previous section that distortion in the direction of a degenerate normal coordinate in a degenerate electronic state of a linear molecule causes a splitting of the electronic state—the Renner–Teller effect. This splitting may or may not result in a bent equilibrium nuclear configuration (Figure 6.99).

The Jahn–Teller effect[131] applies to those non-linear molecules which may have degenerate E or T electronic states, i.e. symmetric rotors and spherical rotors. Like the Renner–Teller effect, it is due to electronic–vibrational interaction following a breakdown of the Born–Oppenheimer approximation when the molecule is distorted in the direction of a non-totally symmetric normal coordinate when it is in a degenerate electronic state. The main difference is that, whereas a non-zero Renner–Teller effect *may* result in a molecule being distorted in the direction of the non-totally symmetric normal coordinate, a non-zero Jahn–Teller effect *must* result in there being at least one such coordinate which removes the electronic degeneracy and results in an equilibrium configuration distorted in the direction of that coordinate. In most of the examples we shall consider, taken from the C_{3v} or D_{3h} point groups, the non-totally symmetric vibration will be a degenerate one but in, for example, the D_{4h} point group a non-degenerate b_{1g} or b_{2g} vibration may have a similar effect.

Figure 6.101 shows the potential energy surface resulting from the splitting of the electronic degeneracy due to the distortion of the molecule in the direction of the doubly degenerate normal coordinate Q with

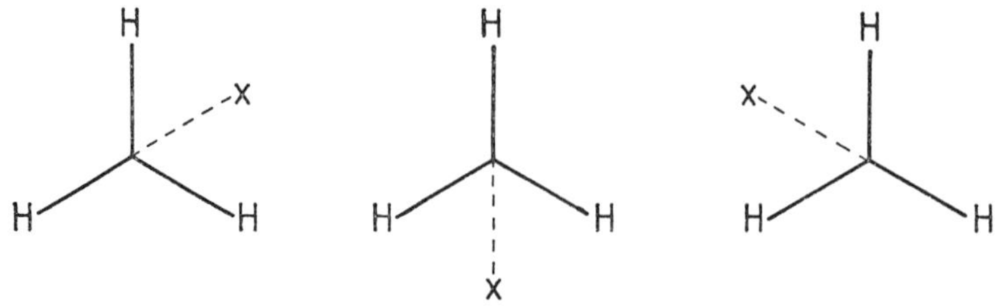

Figure 6.102 The ν_6, CH_3 rocking vibration of the CH_3X molecule can involve a bending of the molecule about the carbon atom in the three equivalent ways shown

components Q_a and Q_b. The effect of the distortion on the potential energy surface is the static Jahn–Teller effect. In the VQ_a plane there are two parabolas intersecting at $Q_a = 0$ and the complete surface is obtained by rotation of these by 180° about the V axis.

The intersection at $Q = 0$ at an angle other than zero distinguishes this from the Renner–Teller case but there may be, among the e vibrations not giving a Jahn-Teller effect, one or more for which the split potential energy curves are tangential to each other, as in Figure 6.99(b), and a Renner–Teller effect results.

If we consider as an example a methyl halide CH_3X in a ^1E electronic state the degeneracy may be removed by the distortion of the molecule in the direction of the e normal coordinate corresponding to the CH_3 rocking vibration, ν_6. As shown in Figure 6.102, this results in a bending of the molecule about the carbon atom in three equivalent ways. The result is that the circular valley in Figure 6.101 is not smooth but has three identical minima corresponding to X in each of the staggered configurations in Figure 6.102. The positions of these minima in Figure 6.101 are at 120° to each other on rotation about the V axis. (This is similar to the case of pseudorotation discussed in Section 5.2.7.2.) However, minima of this kind are usually neglected in Jahn–Teller theory.

When ν_6 of CH_3X is excited with one quantum in a ^1E electronic state the resulting vibronic states have symmetry species $A_1 + A_2 + E$ (see Table 5.9) and, when excited with two quanta, $A_1 + A_2 + E + E$ (see Tables 5.9 and 5.10). Similar derivations can be made for any number of quanta. Breakdown of the Born–Oppenheimer approximation and consequent vibrational–electronic interaction split the components of the vibrational levels; this is known as the dynamic Jahn–Teller effect.

The Jahn–Teller coupling parameter D is a measure of the strength of coupling of vibronic states. For small coupling, with $D<0.05$, Child[132] has derived vibrational term values, excluding anharmonicity, for cases which include the following.

(a) For a doubly degenerate E electronic state, in any molecule not containing a C_4 axis, in which the degeneracy is removed by a doubly degenerate e vibration, ν_2, the vibrational term values are given by

$$G(v_2, j_2) = [(v_2 + 1) - (1 \pm 2j_2)D_2]\omega_2 \quad (6.247)$$

where ω_2 is the wavenumber of the vibration and j_2 is a vibronic quantum number introduced† because l, associated with the angular momentum of degenerate vibrations, is no longer a good quantum number. In general,

†See, for example, Longuet-Higgins, H. C. (1961). 'Some recent developments in the theory of molecular energy levels', ch. 9 in *Advances in Spectroscopy*, vol. 2 (Ed. H. W. Thompson). Interscience, New York.

$$j = \pm\tfrac{1}{2}, \pm\tfrac{3}{2}, \ldots, \pm(v + \tfrac{1}{2}) \qquad (6.248)$$

where, consistent with the treatment of l, we take j to be a signed quantum number. (As in the case of l, some authors take it to be unsigned.) In general,

$$j = l \pm \tfrac{1}{2} \qquad (6.249)$$

and the upper signs of equations (6.247) and (6.249) go together, as do the lower signs. Since l can take the values $v, v-2, v-4, \ldots -v$, we have, for $v_2 = 1$, $l_2 = \pm 1$. Taking the upper sign of equation (6.249) gives $j_2 = \frac{3}{2}, -\frac{1}{2}$ and, with the upper sign of equation (6.247), these values give

$$G(1, j_2) = (2 - 4D_2)\omega_2, \text{ or } 2\omega_2 \qquad (6.250)$$

Taking the lower signs gives identical term values. The term value $2\omega_2$ is the unperturbed value for a doubly degenerate vibration [equation (5.46)] and, in general, vibrational levels are split into $(v+1)$ components. The number of components is the number of values that $|j|$ can take. For example, for $v_6 = 1$ in CH_3X, $|j| = \frac{1}{2}, \frac{3}{2}$ and the level is split into two, an E level and the unsplit A_1, A_2 pair. Similarly, for $v_6 = 2$, $|j| = \frac{1}{2}, \frac{3}{2}, \frac{5}{2}$ and the level is split into two separate E components and an A_1, A_2 pair.

(b) For an E electronic state in a molecule containing a C_4 axis, belonging to the C_4, C_{4v}, C_{4h}, D_4 or D_{4h} point group, the electronic degeneracy may be removed by two non-degenerate vibrations ν_1 and ν_2, with b_1- and b_2-type symmetry species, respectively. In the case where $\omega_1 \approx \omega_2 = \omega$ the term values are

$$G(v, j) = [v + 1 - D_1 + D_2 + (1 \pm 2j)(D_1 D_2)^{\frac{1}{2}}] \qquad (6.251)$$

where v and j are the same for both vibrations and D_1 and $D_2 < 0.05$.

For the case where ω_1 and ω_2 are widely different and $(\omega_1 - \omega_2)/(\omega_1 + \omega_2) \gg (D_1 D_2)^{\frac{1}{2}}$,

$$G(v_1, v_2) = (v_1 + \tfrac{1}{2} - D_1)\omega_1 + (v_2 + \tfrac{1}{2} - D_2)\omega_2 + [v_1(v_2+1) - v_2(v_1+1)]D_1 D_2(\omega_1 + \omega_2)^2/(\omega_1 - \omega_2) \qquad (6.252)$$

For greater vibronic coupling (larger D)

$$G(u, j) = \omega(u + \tfrac{1}{2}) + A^* j^2 \qquad (6.253)$$

The first term on the right-hand side of this equation is due to radial vibration, to which the quantum number u refers, along the degenerate normal coordinate, and the second to rotation around the circular potential well. The rotational constant A^* for this rotation is given by

$$A^* = h/8\pi^2 c\mu r^2 \qquad (6.254)$$

where μ is the reduced mass for the motion and r is the distance of the circular minimum from the V axis.

In the $\tilde{C}^1E$–$\tilde{X}^1A_1$ system of CF_3I there is some evidence from the 6_1^1 sequence band of Jahn–Teller splitting in the $v_6' = 1$ level of the doubly degenerate CF_3 rocking vibration ν_6.

The $\tilde{B}^1E''$–$\tilde{X}^1A_1'$ system of NH_3 presents a clearer example of the Jahn–Teller effect. This system involves a long progression in the inversion vibration ν_2 due to the change of geometry from a pyramidal ground state to a planar $\tilde{B}$ state. Rotational analysis[133] of these bands showed an unexpectedly large l-type doubling and the possibility of this being due to a slight Jahn–Teller distortion of the molecule in the $\tilde{B}$ state was suggested.

It was shown later[134] that there is a dynamic Jahn–Teller effect in the $\tilde{B}$ state. What had been previously assigned as the $\tilde{C}$ electronic state (the $\tilde{C}$–$\tilde{X}$ band system consisting of a progression of parallel bands involving the ν_2 inversion vibration) was reassigned as a vibronically induced component of the $\tilde{B}$ state. The progression is built on one quantum of the degenerate e' vibration ν_3 (see Figure 5.31) in the $\tilde{B}$ state. Table 5.9 gives the result that

$$E'' \times e' = A_1'' + A_2'' + E''$$

but the only allowed transitions from the A_1' ground electronic state are to the A_2'' component of each vibronic level giving (see Table A.33 in the Appendix) a progression of parallel bands. The distortion parameter D was estimated to have a very small value of 0.05 for the $\tilde{B}$ state.

Examples of a stronger Jahn–Teller effect have been found[135] in the $\tilde{B}^2A_2''$–$\tilde{X}^2E''$ allowed electronic band systems of the symmetrical (D_{3h}) 1,3,5-trifluoro- and 1,3,5-trichlorobenzene positive ions, $C_6H_3F_3^+$ and $C_6H_3Cl_3^+$. Both show a static and a dynamic Jahn–Teller effect in the ground electronic state.

The ground states of these ions arise from the configuration in which two π-electrons are in the benzene-like ψ_1 MO (a_2 in D_{3h}) of Figure 6.81 and three in the ψ_2 and ψ_3 MOs (e'' in D_{3h}). In the excited state there is one in ψ_1 and four in ψ_2 and ψ_3.

In both of these ions there are seven e' vibrations, all of which, in principle, may be Jahn–Teller active modes. The most strongly active in progressions is ν_{13}, a C–C–C bending vibration, but ν_9, a C–C stretching vibration, is also fairly strongly active. A C–F bending vibration, ν_{14}, is only weakly active but was included in a three-mode calculation[135] which took account of quadratic vibronic coupling and mode mixing; the former turned out not to be very important but the effects of mode mixing, which means that ν_{13}, ν_9 and ν_{14} cannot be treated independently, are appreciable. The resulting values of the distortion parameter D for ν_{13}, ν_9 and ν_{14} are 0.62, 0.18 and 0.025, respectively, for $C_6H_3F_3{}^+$, and 0.80, 0.38 and 0.015 for $C_6H_3Cl_3{}^+$. The geometrical Jahn–Teller distortions of these ions in their ground states are 2.8° in the C–C–C bond angles and 0.026 Å in the C–C bond lengths for $C_6H_3F_3^+$, and 2.6° and 0.018 Å for $C_6H_3Cl_3^+$. The energy barrier to be surmounted on going to the undistorted molecule is 1001 cm^{-1} in $C_6H_3F_3^+$ and 547 cm^{-1} in $C_6H_3Cl_3^+$.

In a similar way, investigation of the $\tilde{B}^2A_{2u}$–$\tilde{X}^2E_{1g}$ system of the $C_6F_6^+$ ion has shown[136] that it also suffers Jahn–Teller distortion in the degenerate ground state.

The higher (D_{6h}) symmetry of $C_6F_3^+$ limits the number of vibrations capable of exhibiting Jahn–Teller activity to just the four e_{2g} vibrations. These vibrations are ν_{15}, ν_{16}, ν_{17} and ν_{18}, described approximately as C–C stretch, C–F stretch, C–C–C bend and C–F bend, respectively. Their distortion parameters, D, in the $\tilde{X}$ state were found to be 0.23, 0.05, 0.68 and 0.38, respectively, and the corresponding energy barriers to the undistorted state were 370, 61, 289 and 101 cm^{-1}.

Using symmetry coordinates to estimate the ring distortions gave a C–C bond length distortion of 0.011 Å and a C–C–C angle distortion of 2.0°. These compare with 0.020 Å and 2.7° with the approximation of using normal coordinates, an approximation which was used[135] to estimate the corresponding distortions in $C_6H_3F_3^+$ and $C_6H_3Cl_3^+$.

From their infrared spectra there is evidence of a small Jahn–Teller effect in the degenerate ground states of TcF_6, RuF_6, ReF_6 and OsF_6.[137]

6.3.5 Rotational fine structure

6.3.5.1 *Linear molecules*

6.3.5.1(i) *Transitions between singlet states*

For a $^1\Sigma$ electronic state of a linear molecule the rotational term value expression is exactly the same as that for a diatomic molecule, given in equation (6.133). The main quantitative difference is that the rotational constant B_v is likely to be smaller in a polyatomic molecule as, for example, in dicyanoacetylene, N≡C–C≡C–C≡N, with a long chain of nuclei and a resulting closely spaced stack of rotational levels.

The rotational selection rules for a $^1\Sigma$–$^1\Sigma$ electronic transition are the same as those given in Section 6.2.5.1 for a diatomic molecule and result in simple P- and R-branch rotational fine structure. Degradation of bands is usual and results in an R-branch head if $B' < B''$ and a P-branch head if $B' > B''$.

Excitation of totally symmetric, σ^+ or σ_g^+, vibrations in either or both of the combining electronic states gives rise to vibronic transitions which are also of the $^1\Sigma$–$^1\Sigma$ type and show similar P- and R-branch structure.

Excitation of a non-totally symmetric π, π_g or π_u vibration in a $^1\Sigma$ type of electronic state results in a first-order Coriolis interaction and the term values are modified to those of equation (5.196). The rotational selection rules for a $^1\Pi$–$^1\Sigma$ type of vibronic transition are just the same as those for a Π–Σ infrared vibrational transition. The resulting rotational fine structure shows a P, Q and R branch. The band structure is like that shown for a vibrational transition in Figure 5.44 except that more severe degradation is typical of a vibronic transition.

If the molecule belongs to the $D_{\infty h}$ point group there may be a nuclear spin statistical weight alternation associated with the rotational levels and a consequent alternation of intensities along each rotational branch. For example, in the $\tilde{A}^1\Pi_u$–$\tilde{X}^1\Sigma_g^+$ system of $^{12}C_3$ rotational transitions with alternate values of J are missing since the nuclear spin quantum number I is zero for ^{12}C.

Similar alternations are found in electronic transitions between states of higher multiplicity also. One such example is the 2:1 intensity alternation with J in the $\tilde{B}^2\Sigma_u^+$–$\tilde{X}^2\Pi_g$ system of $^{14}N_3$ resulting from the fact that $I=1$ for ^{14}N (see Figure 6.104).

Electronic transitions of the type $^1\Pi$–$^1\Sigma$ have been observed and analysed. Two such examples are the $\tilde{A}^1\Pi_u$–$\tilde{X}^1\Sigma_g^+$ transition in C_3 and the $\tilde{G}^1\Pi_u$–$\tilde{X}^1\Sigma_g^+$ transition of acetylene, H–C≡C–H.

In a $^1\Pi$ electronic state of a linear molecule the phenomenon of Λ-type doubling occurs, just as in diatomic molecules (see Section 6.2.5.2). This is caused by interaction of rotational and electronic motion. The value of the Λ-doubling constant q, defined in equation (6.139), is a measure of this interaction. It is very small in the $\tilde{A}^1\Pi_u$ state of C_3 ($q=0.0004\,\text{cm}^{-1}$) compared with the $A^1\Pi$ state of AlH ($q=0.0080\,\text{cm}^{-1}$) in which it is unusually large.

Excitation of a π vibration, such as the bending vibration in C_3, in a $^1\Pi$ type of electronic state gives rise to the Renner–Teller effect (Section 6.3.4.4). Vibrational–electronic interaction results in Λ and l no longer being good quantum numbers and they are replaced by the vibronic quantum number K. The rotational term value expression of equation (6.133), neglecting the small term involving H_v, becomes

$$F_v(J)=B_v[J(J+1)-K^2]-D_v[J(J+1)-K^2]^2 \tag{6.255}$$

for the resulting vibronic states, these being doubly degenerate for $K>0$. The degeneracy may be split by interaction between the rotational and electronic motions. The effect is called K-type doubling; this is similar to Λ-type doubling and the splitting of the term values is given by

$$\Delta F(J)=q_{ev}J(J+1) \tag{6.256}$$

for $K=1$, namely Π, vibronic states. If, in addition, there is Λ-type doubling in a $^1\Pi$ electronic state, q_{ev} contains an electronic and vibrational contribution, and

$$q_{ev}=\tfrac{1}{2}q_e \pm q_v \tag{6.257}$$

6.3.5.1(ii) *Transitions involving states with non-zero multiplicity*

For Σ electronic states with $S\neq 0$, Hund's coupling case (b) usually applies (see Section 6.2.5.4). The quantum number J can take the values given by equation (6.143). For $^2\Sigma$ and $^3\Sigma$ states the rotational term values are given by equations (6.144) and (6.145), respectively, as for a diatomic molecule. For $^2\Sigma$–$^2\Sigma$ and $^3\Sigma$–$^3\Sigma$ transitions the general rotational selection rules given in Section 6.2.5.7 apply.

For $^2\Pi$ electronic states Hund's case (a) (see Section 6.2.5.3) coupling applies for low values of J. The orbital and spin angular momenta are uncoupled from each other and coupled to the

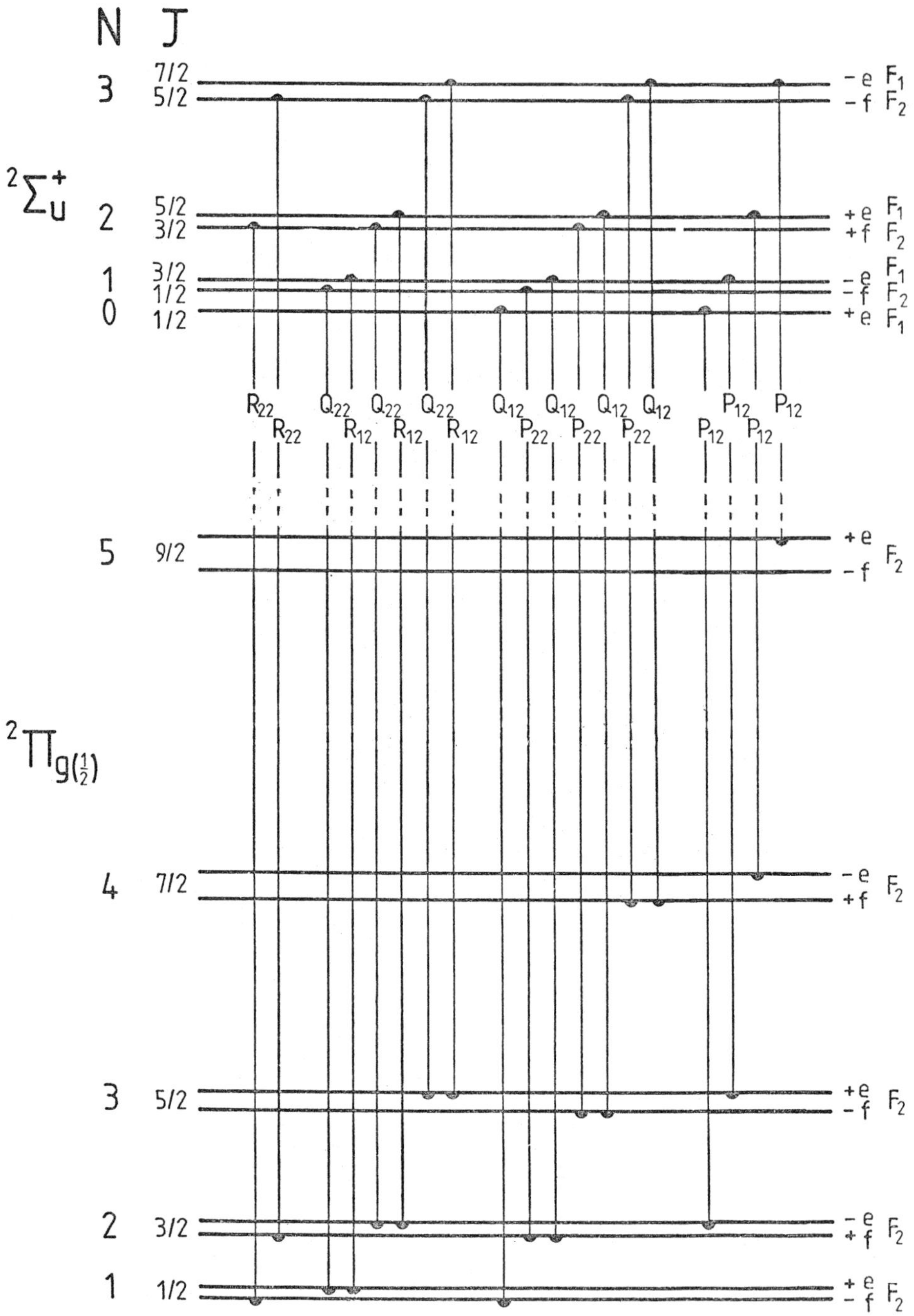

Figure 6.103 Rotational transitions accompanying a $^2\Sigma_u^+ - {}^2\Pi_{g(1/2)}$ transition of a linear molecule, such as the $\tilde{B}-\tilde{X}$ transition of N_3. Spin doubling in the $^2\Sigma_u^+$ state and Λ-type doubling in the $^2\Pi_{g(1/2)}$ component of the $^2\Pi_g$ state are shown

internuclear axis. The rotational term values are given by equation (6.141).

At higher values of J, the spin momentum is uncoupled from the internuclear axis and there is a changeover to Hund's case (b) coupling as in $^2\Pi$ states of diatomic molecules. Whatever the extent of the uncoupling, the rotational term values for the two components of a $^2\Pi$ state are given by

$$F_{1,2}(J) = B_v\{(J+\tfrac{1}{2})^2 - \Lambda^2 \mp \tfrac{1}{2}[4(J+\tfrac{1}{2})^2 + Y(Y-4)\Lambda^2]^{\frac{1}{2}}\} \quad (6.258)$$

where the $-$ and $+$ signs refer to the F_1 and F_2 components, respectively. The constant $Y = A/B_v$, where A is the spin–orbit coupling constant. Centrifugal distortion and Λ-type doubling have been neglected.

Figure 6.103 illustrates the allowed rotational transitions accompanying a $^2\Sigma_u^+$–$^2\Pi_g$ electronic transition, such as the $\tilde{B}$–$\tilde{X}$ transition of N_3 (see Table 6.13). Case (a) coupling is assumed and the $^2\Pi_{\frac{3}{2}}$ component of the $^2\Pi_g$ state, which is independent of and well separated from the $^2\Pi_{\frac{1}{2}}$ component, is not shown.† The spin doubling in the $^2\Sigma_u^+$ state [Hund's case (b)] is shown, in addition to Λ-type doubling in the $^1\Pi_{\frac{1}{2}}$ component. Application of the rotational selection rules shows that there are six branches allowed, R_{22}, R_{12}, Q_{22}, Q_{12}, P_{22} and P_{12}, where the subscripts refer to the involvement of the F_1 or F_2 spin component in the upper state (first) and in the lower state (second). The spectrum involving the $^2\Pi_{\frac{3}{2}}$ component shows another six branches.

Figure 6.104 shows the two P branches, P_{22} and P_{12}, of the 0_0^0 band of the $\tilde{B}^2\Sigma_u^+$–$\tilde{X}^2\Pi_{g(\frac{1}{2})}$ transition of N_3. In the $\tilde{X}$ state the Λ doubling and, in the $\tilde{B}$ state, the spin splitting could not be detected from the spectrum. As a result of this small spin splitting the R_{12}:Q_{22} and P_{22}:Q_{12} pairs of branches could not be resolved so that only four branches were observed involving the $^2\Pi_{\frac{3}{2}}$ component and another four involving the $^2\Pi_{\frac{1}{2}}$ component.

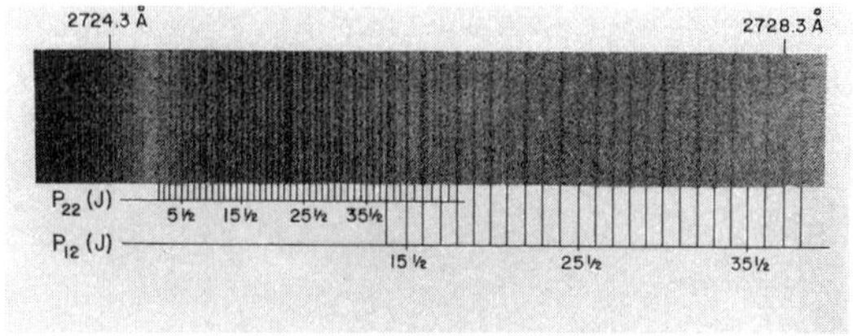

Figure 6.104 Part of the photographically recorded 0_0^0 band of the $\tilde{B}^2\Sigma_u^+$–$\tilde{X}^2\Pi_{g(1/2)}$ transition of N_3 [Reproduced, with permission, from Douglas, A. E. and Jones, W. J. (1965). *Can. J. Phys.*, **43**, 2216]

The $\tilde{A}^3\Pi_u$ state of NCN (see Table 6.13) has been shown, from an analysis of the $\tilde{A}^3\Pi_u$–$\tilde{X}^3\Sigma_g^-$ absorption system, to approximate fairly closely to Hund's case (a) at low values of J but, at high values, there is a smooth change to case (b) in which the electron spin $\boldsymbol{S}$ is uncoupled from the internuclear axis. For any degree of uncoupling the rotational term values for the three components of a $^3\Pi$ state are given approximately by

$$\begin{aligned} F_1(J) &= B_v[J(J+1) - Z_1^{\frac{1}{2}} - 2Z_2] \\ F_2(J) &= B_v[J(J+1) + 4Z_2] \\ F_3(J) &= B_v[J(J+1) + Z_1^{\frac{1}{2}} - 2Z_2] \end{aligned} \quad (6.259)$$

neglecting centrifugal distortion and Λ-type doubling. The quantities Z_1 and Z_2 are given by

$$Z_1 = \Lambda^2 Y(Y-4) + \tfrac{4}{3} + 4J(J+1) \quad (6.260)$$

$$Z_2 = (3Z_1)^{-1}[\Lambda^2 Y(Y-1) - \tfrac{4}{9} - 2J(J+1)] \quad (6.261)$$

where $Y = A/B_v$, as in equation (6.258).

In the $\tilde{A}$ state of NCN the spin–orbit coupling constant A is large ($-37.56\,\text{cm}^{-1}$) and the rotational constant B is small ($B_0 = 0.3962\,\text{cm}^{-1}$). the resulting value of Y (94.80) is

†The $\tilde{X}^2\Pi_g$ state of N_3 is inverted, i.e. $^2\Pi_{\frac{3}{2}}$ is below $^2\Pi_{\frac{1}{2}}$. A consequence of this is that the rotational levels of $^2\Pi_{\frac{1}{2}}$ are the F_2 and of $^2\Pi_{\frac{3}{2}}$ the F_1 spin components [equation (6.258)].

unusually large and the case (a) approximation an unusually good one for moderate values of J.

As for diatomic molecules (see Section 6.2.5.7), examples of transitions with $\Delta S \neq 0$ are mainly confined to the triplet–singlet type. Not many examples of these are known but the $\tilde{a}^3\Sigma_u^+-\tilde{X}^1\Sigma_g^+$ transition of cyanogen, C_2N_2, shows the four rotational branches expected.

For $^3\Pi(b)-^1\Sigma$ and $^3\Pi(a)-^1\Sigma$ types of transitions, where (b) and (a) refer to the coupling case, nine and five branches, respectively, should be observed.

6.3.5.2 *Symmetric rotor molecules*

6.3.5.2(i) Transitions between singlet states

The rotational transitions accompanying electronic or vibronic electric dipole transitions in symmetric rotors have very much in common with those accompanying infrared vibrational transitions. These have been discussed at length in Section 5.2.5.2(i) and the reader is referred there for details.

The rotational term values for prolate and oblate symmetric rotors are given by equations (5.201) and (5.202).

For the rotational transitions accompanying an electronic transition, the selection rules depend on the direction of polarization of the electronic transition moment (whereas for a vibrational transition they depend on the polarization of the vibrational transition moment). If

$$\Gamma(\psi_e') \times \Gamma(\psi_e'') \supset \Gamma(T_z) \tag{6.262}$$

the electronic transition moment is polarized along the z-axis, which is the top axis. The rotational selection rules are

for $K = 0$, $\Delta K = 0$ and $\Delta J = \pm 1$
for $K \neq 0$, $\Delta K = 0$ and $\Delta J = 0, \pm 1$

The resulting band is a parallel band, just as for an infrared vibrational transition.

For a symmetric rotor with a C_3 axis, such as CH_3 and NH_3, there may be a splitting of the rotational levels with $K = 3n$, where $n = 1, 2, 3, \ldots$, due to high order Coriolis interaction. This is illustrated in the rotational energy level diagram in Figure 5.48 but is often sufficiently small to be neglected in electronic spectra.

An example of a parallel band is to be found in the $\tilde{A}^1A_2''-\tilde{X}^1A_1'$ system of N^2H_3 (the rotational fine structure of the corresponding transition in N^1H_3 is too diffuse to be analysed). As discussed in Section 5.2.7.1, the zero-point level, and all other vibrational levels, in the $\tilde{X}$ state of NH_3 must be classified according to the D_{3h} point group because of tunnelling through the inversion barrier. The planarity or near planarity of the molecule in the $\tilde{A}$ state (see Section 6.3.1.3) requires the use of the D_{3h} point group in that state also.

Although the selection rules for a parallel band are the same whether the band involves an infrared vibrational or an electronic transition, the general appearance of the band is typically very different. The symmetrical appearance of the parallel band in Figure 5.49 is typical of an infrared vibrational transition in which the changes of rotational constants A and B (prolate) or C and B (oblate) are very small, resulting in almost coincident sub-band origins [equation (5.206)].

Considerably different rotational constants in the combining states are typical of an electronic transition producing, as in diatomic and linear polyatomic molecules, degradation of the electronic and vibronic bands. For example, in N^2H_3, an oblate symmetric rotor, $B'' = 5.1426\,\mathrm{cm}^{-1}$ and $C'' = 3.117\,\mathrm{cm}^{-1}$ in the zero-point level of the $\tilde{X}$ state, whereas $B' = 4.78\,\mathrm{cm}^{-1}$ and $C' = 2.32\,\mathrm{cm}^{-1}$ in the $v_2 = 1$ level of the $\tilde{A}$ state. The result is that the parallel band 2_0^1 of the $\tilde{A}-\tilde{X}$ system has a rather irregular appearance[138] and is strongly degraded to low wavenumber. In addition, sub-band origins, given by equation (5.206) but replacing $(A'-A'')$ by $(C'-C'')$ for an oblate symmetric rotor, are no longer nearly coincident.

In the CH_3 molecule (see Section 6.3.1.3) the transition $\tilde{B}^2A_1'-\tilde{X}^2A_2''$ has been observed in

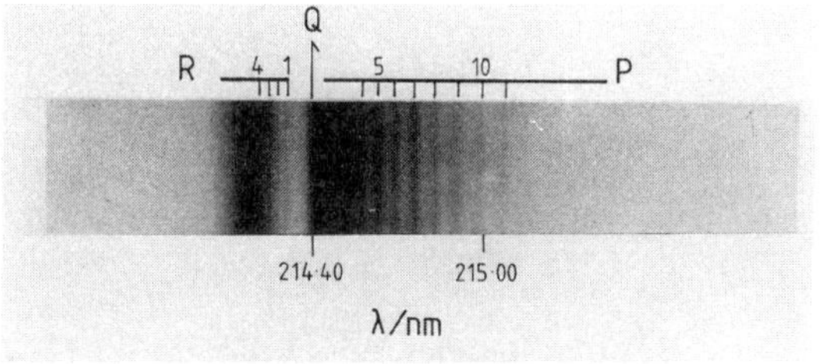

Figure 6.105 The 0_0^0 parallel band of the $\tilde{B}^2A_1'$–$\tilde{X}^2A_2''$ system of C^2H_3, recorded photographically

absorption following flash photolysis of various parent molecules. The system is dominated by one intense band assigned as the 0_0^0 band. The lack of progressions is attributed to the fact that the $\tilde{B}$ state is a Rydberg state resulting from the promotion of an electron from the $1a_2''$ non-bonding orbital [see equation (6.157)] into a $3sa_1'$ Rydberg orbital; little change of geometry is expected in such a transition.

The 0_0^0 band of C^1H_3 is rotationally diffuse but the corresponding band of C^2H_3 is sufficiently sharp for some important results to be obtained.

The fact that the transition is between doublet, and not singlet, states need not concern us here because no effects of electron spin are resolved in the spectrum which appears just like that of a $^1A_1'$–$^1A_2''$ transition. Since $\Gamma(T_z) = A_2''$ in the D_{3h} point group the 0_0^0 band is a parallel band and is shown in Figure 6.105.

The relatively simple appearance of the band compared with that of N^2H_3 is due to the near coincidence of sub-band origins caused by the coefficient of K^2 in equation (5.206) being almost zero. The result is that the transitions within each sub-band are nearly coincident for all values of K.

Important features of the rotational fine structure are (a) there is no $R(0)$ transition observed, and (b) there is a slight intensity alternation which can just be observed in the P-branch lines: those with odd J are more intense than those with even J. The reason for both features is the nuclear spin statistical weight alternation with J for levels with $K=0$, as discussed in Section 4.4 and summarized in equations (4.86) and (4.87).

For a molecule whose electronic and vibronic states are classified according to the D_{3h} point group, the statistical weight factor, for levels with $K=0$, is given by

$$\tfrac{1}{3}(2I+1)(2I+3)(I+1) \tag{6.263}$$

for J even and

$$\tfrac{1}{3}(2I+1)(2I-1)I \tag{6.264}$$

for J odd, provided that the three identical nuclei exchanged by the C_3 operation are bosons ($I=0$, 1, 2, . . .), for example ^2H with $I=1$. For fermions ($I=\frac{1}{2}$, $\frac{3}{2}$, $\frac{5}{2}$, . . .), for example ^1H with $I=\frac{1}{2}$, equations (6.263) and (6.264) apply to J odd and J even, respectively. However, all this is true only when the rotational levels are associated with A_1' or A_1'' electronic or vibronic states. For A_2' or A_2'' states the alternation with J is reversed.

Figure 6.106 gives the statistical weight factors for the $K=0$ levels of C^1H_3 and C^2H_3 in the ground electronic state $\tilde{X}^2A_2''$ for the vibrational states $v_2=0$ and 1, where v_2 is the a_2'' inversion vibration. On the left-hand side of the figure the $v_2=0$ and 1 levels are separated, the separation being either large or small corresponding to the cases of a planar molecule or a non-planar molecule with a lower inversion barrier, illustrated in Figure 5.69(c) or (b), respectively. In both cases D_{3h} symmetry species and selection rules apply and there is a statistical weight alternation with J, for $K=0$, the alternation being reversed for $v_2=1$ compared with $v_2=0$.

If we imagine the inversion barrier being raised smoothly in going from the situation in Figure 5.69(c) to that in Figure 5.69(a), the $v_2=0$ and 1 levels come together smoothly as shown in the correlation between rotational levels in Figure 6.106(a) and (b). In other words, the inversion of the molecule is changing

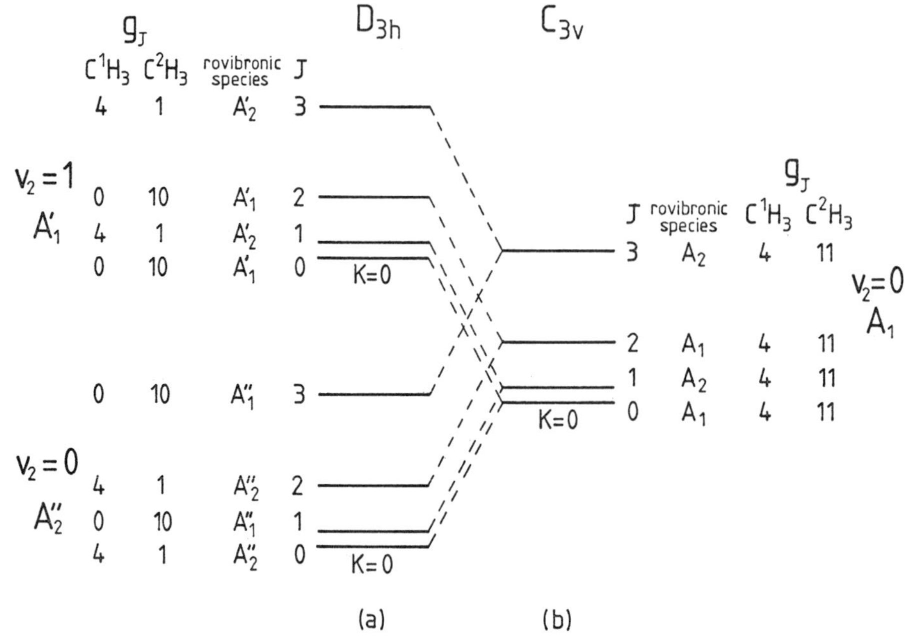

Figure 6.106 Correlation of energy levels and nuclear spin statistical weight factors g_J for (a) planar and (b) non-planar (with a high barrier to planarity) C^1H_3 and C^2H_3

from a *feasible* [Figure 5.69(c,b)] to a *non-feasible* [Figure 5.69(a)] process.

In figure 6.106(b) the molecule is rigidly C_{3v} and there is no longer any statistical weight alternation. This occurs in practice when the $v_2 = 0$ and 1 levels become coincident within the resolution obtained in the spectrum concerned, which may be limited by either the instrumentation or the natural line width.

In an electronic transition in an AH_3 molecule, such as CH_3 or NH_3, there are four possible cases of feasibility of inversion in the combining electronic states:

1. Inversion non-feasible in both states. In this case the symmetry species used for the rovibronic levels in both states are those of the C_{3v} point group. There is no intensity alternation with J for $K=0$.
2. Inversion feasible in both states. The rovibronic levels are classified, in both electronic states, according to the D_{3h} point group. There is an intensity alternation with J for $K=0$.
3. Inversion feasible in the lower electronic state only. In this case the electronic transition is (in absorption) from a lower state with a stack of rotational levels like those in Figure 6.106(a) to an upper state with a stack of levels like those in Figure 6.106(b). Under these circumstances it is useful to retain the D_{3h} classification of the rovibronic levels in the upper electronic state even though inversion is non-feasible. Then the rotational selection rules can be applied in the usual way and they result in an intensity alternation with J for $K=0$.
4. Inversion feasible in the upper electronic state only. This case is very similar to that of the previous one and an intensity alternation with J for $K=0$ again results.

We can see now that the observation of an intensity alternation with J in the 0_0^0 band of

C^2H_3 in Figure 6.105 implies that inversion is a feasible operation in at least one of the electronic states $\tilde{B}$ and $\tilde{X}$. The intensity alternation results from the 1:10 alternation in the $K=0$ sub-band; this sub-band is almost coincident with all the other sub-bands which do not show any alternation. The fact that the $R(0)$ line is not observed is because only the $K=0$ sub-band contributes to it and then the statistical weight factor g_J is only 1 (Figure 6.106)

To go further and draw conclusions regarding the planarity, or otherwise, of the molecule in these states requires additional information. Ideally the rotational constants B and C could be obtained in both electronic states, from the rotational fine structure of a perpendicular band† of the system, and then, if $B=\frac{1}{2}C$ for an electronic state, the molecule is planar in that state.

In the $\tilde{B}$–$\tilde{X}$ system of C^2H_3 no perpendicular bands have been observed and, in any case, the rotational fine structure is not sufficiently well resolved for B and C to be obtained in both states. Hence, in this as in many similar cases, the arguments regarding planarity require vibrational information to augment the observation of a rotational intensity alternation.

The system shows only very weak progressions from which it follows that the geometry is similar in the $\tilde{B}$ and $\tilde{X}$ states. If, in both states, the molecule were non-planar but with tunnelling through the barrier sufficient to separate the $v_2=0$ and 1 levels by only a small amount, the $v_2=1$ level in the $\tilde{X}$ state would be highly populated at room temperature and the 2_1^1 sequence band would be expected to be intense. The fact that such a band is not observed indicates that the $v_2=0$–1 separation in the $\tilde{X}$ state is large enough to indicate that the molecule is either planar, or non-planar with a low barrier in the $\tilde{X}$ state. From what has been said previously this applies to the $\tilde{A}$ state also.

†From a parallel band B', B'' and ΔC ($=C'-C''$) can be obtained, but not C' and C'' independently.

To make more specific deductions regarding the planarity required the observation of several inversion vibration levels. This has been achieved for the $\tilde{X}$ state from a diode laser study,[139] in the infrared region, involving rotational analyses of the $v_2=1$–0, 2–1 and 3–2 bands of the inversion vibration ν_2. The C^1H_3 was generated by a 60 Hz discharge in di-*tert*-butyl peroxide, $(CH_3)_3COOC(CH_3)_3$.

For vibrational levels with even values of v_2, rotational levels with odd values of the rotational quantum number N are missing for $K=0$ whereas, for levels with odd v_2, levels with even values of N are missing for $K=0$. This demonstrates that each vibrational level is non-degenerate and that there cannot be such a high barrier to inversion, as in Figure 5.69(a), that there is essentially no tunnelling through it; there must be either a low barrier, as in figure 5.69(b), or no barrier at all, as in Figure 5.69(c).

Fitting the lowest three vibrational levels to a potential showed that there is no barrier and, therefore, that the molecule is planar. In these circumstances the 'inversion' vibration becomes the out-of-plane bending vibration.

In general, the potential function constructed from a set of observed vibrational term values is one of the most sensitive indicators of molecular geometry.

In the $\tilde{A}$–$\tilde{X}$ system of N^2H_3, also, there is an intensity alternation with J in the $K=0$ sub-band. In this case the molecule is planar in the $\tilde{A}$ state and non-planar in the $\tilde{X}$ state but with appreciable tunnelling through the inversion barrier splitting the vibrational levels.

In CH_3 and NH_3 there is also a nuclear spin statistical weight factor which varies with K (see also Section 4.4). For $K=3n$, where $n=0,1,2,\ldots$ this factor is given by

$$\tfrac{1}{3}(2I+1)(4I^2+4I+3) \tag{6.265}$$

and, for $K\neq 3n$, by

$$\tfrac{1}{3}(2I+1)(4I^2+4I) \tag{6.266}$$

It follows that, for C^1H_3 and N^1H_3, the statistical weights for $K=0, 1, 2, 3, 4, 5, 6, \ldots$ are 4, 2, 2, 4, 2, 2, 4, ... and, for C^2H_3 and N^2H_3, they are 11, 8, 8, 11, 8, 8, 11, ... In the band of C^2H_3 in Figure 6.105 the effect of these statistical weights is not apparent because of the coincidence of sub-bands.

Intensities of the rotational transitions accompanying an electronic or vibronic transition in a symmetric rotor are given, for an absorption process, by equation (5.207), just as for rovibrational transitions. For a parallel band the A_{KJ} intensity factors, derived by Hönl and London, are given by equations (5.209)–(5.211).

For a prolate symmetric rotor in an E electronic state the rotational term values are given by the expression

$$F_v(J, K)=B_vJ(J+1)+(A_v-B_v)K^2-2(A_v\zeta_t)k - D_JJ^2(J+1)^2-D_{JK}J(J+1)K^2-D_KK^4 \tag{6.267}$$

For an oblate symmetric rotor A_v is replaced by C_v. This equation is similar to equation (5.212) for an E vibrational state of a prolate symmetric rotor except that, here, the Coriolis coupling constant ζ_t is due to vibronic rather than vibrational angular momentum and the quantum number l is not included in the $2(A_v\zeta_t)k$ term since there is no degenerate vibration excited. If a degenerate vibration is excited in the E electronic state there is an electronic *and* a vibrational contribution to ζ_t, which is then given by

$$|\zeta_t| = |\zeta_e \pm \zeta_v| \tag{6.268}$$

Whereas ζ_v lies in the range 0–1 (see Section 4.3), ζ_e is unrestricted. $|\zeta_e|$ is analogous to Λ in a linear molecule (Section 6.2.3.1), which is the quantum number associated with the component of orbital angular momentum along the internuclear axis.

Analogous to l-type doubling in a degenerate vibrational state [Section 5.2.5.2(i)] there is j-type doubling in a degenerate vibronic state. This doubling refers to the splitting of the A_1 and A_2 levels for $K=1$ (see Figure 5.50) and j is the quantum number encountered in the Jahn–Teller effect (Section 6.3.4.5).

When there is no excitation of an e vibration in an E electronic state $l=0$ and $j=\pm\frac{1}{2}$ [see equation (6.249)]. If an e vibration is excited l is no longer a good quantum number, due to the Jahn–Teller effect (Section 6.3.4.5). Indeed, j-type doubling in a state with $l=0$ is an indication of a Jahn–Teller distortion of the molecule.

Such an effect has been observed[133] in the $\tilde{B}^1E''$–$\tilde{X}^1A_1'$ system of NH_3, an oblate symmetric rotor, indicating some Jahn–Teller distortion. There is, in this system, large j-type doubling which is analogous to what has been called giant l-type doubling in vibrational bands. In Section 4.3 l-type doubling was assumed to be small and independent of J. For giant l-type (or j-type) doubling this is not so and an extra term has to be added to the term value expression in equation (6.267) giving, for an oblate symmetric rotor and neglecting centrifugal distortion,

$$F_v(J, K)=B_v(J+1)+(C_v-B_v)K^2-2(C_v\zeta_t)k \pm \frac{q_v{}^2[J(J+1)-K(K\mp 1)][J(J+1)-(K\mp 1)(K\mp 2)]}{16(K\mp 1)[(1-\zeta_t)C_v-B_v]} \tag{6.269}$$

where the upper and lower signs refer to the $+j$ (or $+l$) and $-j$ (or $-l$) levels, respectively, and q_v is the j-type doubling constant.

The rotational selection rules for an E–A_1 electronic perpendicular band are the same as those for a vibrational perpendicular band discussed in Section 5.2.5.2(i). The intensity factors A_{KJ} are given by the Hönl–London formulae of equation (5.216).

The major difference in appearance between an electronic (or vibronic) and a vibrational perpendicular band is due to the typically larger changes of rotational constants for an electronic transition. The consequent degradation of the electronic band removes the approximate symmetry about the band centre typical of an infrared band and, as is the case for a parallel band, can cause it to appear very irregular.

A later rotational analysis[140] of bands in the $\tilde{B}^1E''$–$\tilde{X}^1A_1'$ system of NH_3 resulted in a value of 0.826 for ζ_t (which is equal to ζ_e, since no *e* vibration is excited). The *j*-type doubling constant q_v decreases rapidly from 0.887 for $v_2' = 1$ to 0.193 for $v_2' = 8$ in the long progression in the inversion vibration ν_2 which results from the change from a pyramidal ground state to a planar excited state.

The $\tilde{A}^1B_{2u}$–$\tilde{X}^1A_{1g}$ system of benzene is electronically forbidden and gains most of its intensity by the Herzberg–Teller intensity stealing mechanism involving principally the e_{2g} vibration ν_6 (see Section 6.3.4.1). The 6^1_0 transition is therefore of the type ${}^1E^{ev}_{1u}$–${}^1A^e_{1g}$. Since $E_{1u} = \Gamma(T_x, T_y)$, this vibronic transition is electric dipole allowed with the transition moment polarized in the plane of the molecule. As a result, the band is a perpendicular band of an oblate symmetric rotor. The rotational term values and selection rules are the same as for the $\tilde{B}^1E''$–$\tilde{X}^1A_1'$ system of NH_3 except that ζ_t refers now to the vibrational angular momentum of ν_6 in the $\tilde{A}^1B_{2u}$ state and *l* replaces the quantum number *j*. Additionally, the *l*-type (or *j*-type) doubling constant q_v is very small, which is typical for vibrational motion, as has been mentioned in Section 5.2.5.2(i), compared with the value in the $\tilde{B}$–$\tilde{X}$ system of NH_3.

The 6^1_0 band has been rotationally analysed and the parameters *C*, *B* and ζ_6 obtained for the $\tilde{A}^1B_{2u}$ state. The constant q_v was neglected.

The analysis of the band was achieved by the method of band contour analysis. This method must be used when there are insufficient rotational lines resolved for a line-by-line analysis to be made using combination difference techniques. Figure 6.107 shows a computed rotational contour consisting of an intensity envelope of all the underlying, unresolved rotational transitions associated with the 6^1_0 band.

In the computed contour it was necessary to include transitions with *J* and $K = 0$–100 in the $\tilde{X}^1A_{1g}$ state. On average, there are about 50 transitions in each branch and therefore 150 in each sub-band, which consists of a *P*, *Q* and *R* branch for a particular value of *K*. In a perpendicular band $\Delta K = \pm 1$ so there are 200 sub-bands making a total of about 30 000 transitions. These are crowded into the 25 cm^{-1} spread of the 6^1_0 band giving an average separation of about 0.0008 cm^{-1} between transitions. If we assume the Doppler line width to be 0.02 cm^{-1} there are, on average, about 25 transitions within the Doppler width. Therefore, individual rotational lines can never be resolved unless the Doppler width can be reduced by, for example, forming a molecular beam, or unless the rotational temperature can be substantially reduced, as in a supersonic jet (Section 3.4.3).

This result is general in large molecules and really defines what a spectroscopist means by a large molecule, namely one which is so large that the rotational constants are sufficiently small that resolution of individual rotational lines is limited, under normal temperature conditions, by the Doppler effect.

In general there are two other reasons why band contour analysis may be the only method possible. The first is that the experimental resolution is insufficiently high, and the second is that the line width is greater than the Doppler line width due to the short lifetime for decay of the excited state (see Section 2.3).

For the computed contour in Figure 6.107 it was assumed that the benzene molecule is a planar, regular hexagon in both the $\tilde{A}$ and $\tilde{X}$ states. More specifically, the molecule was taken to be a symmetric rotor with zero inertial defect Δ $(= I_c - I_a - I_b)$ so that there are only three rotational parameters required for the contour calculation, B' and ζ_6' for the $\tilde{A}$ state and B'' for the $\tilde{X}$ state. The value of 0.1896 cm^{-1} for B'' was known from rotational Raman spectroscopy, so the only parameters to be varied in fitting a computed to the observed contour were B' and ζ_6'. The optimum values of 0.1810 ± 0.0005 cm^{-1} and 0.60 ± 0.05, respectively, differ slightly from those used in the computed contour shown in the figure.

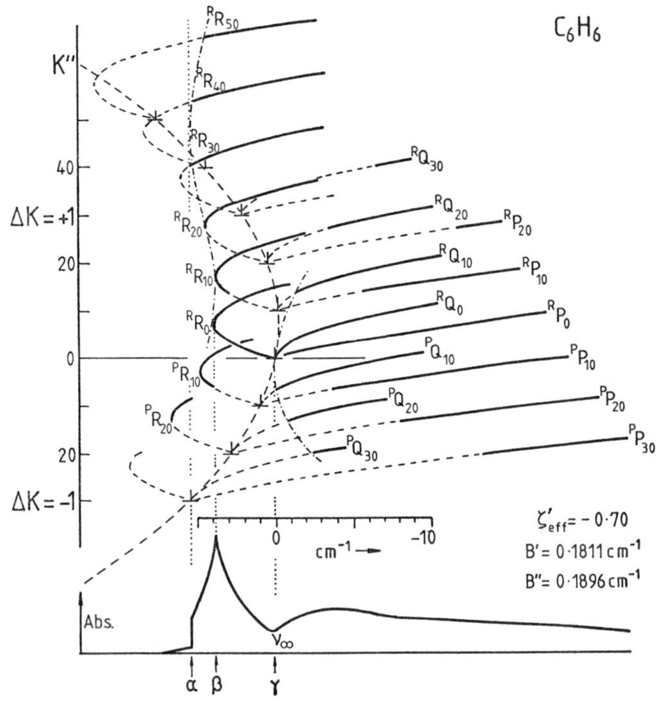

Figure 6.107 A computed rotational contour to match the observed contour of the 6^1_0 band of the $\tilde{A}^1B_{2u}$–$\tilde{X}^1A_{1g}$ system of benzene. The breakdown of the contour into contributions from various P, Q and R branches is shown in the upper part of the figure [Reproduced, with permission from Callomon, J. H., Dunn, T. M. and Mills, I. M. (1966). *Philos. Trans. Roy. Soc.*, **259A**, 499]

Figure 6.107 illustrates, by means of Fortrat curves showing the P, Q and R branches for a few values of K'', the way in which three important features of the contour, labelled α, β and γ, are formed.

The position of the band origin, where the hypothetical $J'' = K'' = 0$ to $J' = K' = 0$ transition would be, corresponds fairly closely to the intensity minimum γ; this is a very useful, but by no means general, feature of rotational contours.

Because ΔB is negative the R branches reverse at a J value of J_{rev} which, since it is independent of K, is given by equation (6.135); J_{rev} is 21 in this case. Figure 6.107 shows that the heads of the ${}^rR$ branches,† which according to the Hönl–London A_{KJ} intensity factors of equation (5.216) are more intense than ${}^pR$ branches, themselves pile up around $K = 10$. This is the cause of the most intense peak, labelled β, in the contour.

The formation of the sharp edge α is due to a turning-point of the first lines of ${}^rR$ branches with K in the region 25–35. This turning-point is caused by the combination of ${}^rR$-branch degradation to low wavenumber together with the loss of one line in the branch every time K increases by one because of the restriction that $J \geqslant K$.

†See p. 228 for explanation of symbols.

6.3.5.2(ii) Transitions involving states with non-zero multiplicity

For a symmetric rotor in an electronic state which is not orbitally degenerate, the spin–orbit coupling is small since there is no orbital angular momentum and the state approximates to Hund's case (b), discussed in Section 6.2.5.4. In a doublet state the rotational term values $F(N, K)$, analogous to those of equations (5.201) and (5.202), are split into two components given by

$$F_1(N, K)=F(N, K)+\frac{N}{2}\left[\frac{\kappa K^2}{N(N+1)}+\mu\right]$$
$$F_2(N, K)=F(N, K)-\frac{(N+1)}{2}\left[\frac{\kappa K^2}{N(N+1)}+\mu\right] \tag{6.270}$$

where κ and μ are spin–rotation coupling constants.

For a triplet state the three components into which the term values are split are given by

$$F_1(N, K)=F(N, K)+(N+1)\left[\frac{\kappa K^2}{(N+1)^2}+\mu\right]-\frac{2\lambda[(N+1)^2-K^2]}{(N+1)(2N+3)}$$
$$F_2(N, K)=F(N, K)-\frac{\kappa K^2}{N(N+1)}-\frac{2\lambda K^2}{N(N+1)}$$
$$F_3(N, K)=F(N, K)-N\left[\frac{\kappa K^2}{N^2}+\mu\right]-\frac{2\lambda(N^2-K^2)}{N(2N-1)} \tag{6.271}$$

where λ is the spin–spin coupling constant analogous to that for a diatomic molecule in a $^3\Sigma$ state encountered in equation (6.145).

For doublet–doublet and triplet–triplet transitions the selection rules $\Delta N=0, \pm 1$ and $\Delta J=0, \pm 1$ hold as for linear molecules.

Hougen[141] has shown that the rotational selection rules for triplet–singlet transitions are

$$\Delta N=0, \pm 1, \pm 2; \quad \Delta K=0, \pm 1, \pm 2$$

These selection rules are also important in near-symmetric rotors [see Section 6.3.5.3(ii)].

Larger spin–orbit coupling is expected in degenerate electronic states.

6.3.5.3 Asymmetric rotor molecules

6.3.5.3(i) Transitions between singlet states

The rotational term values of an asymmetric rotor cannot be expressed as simple closed formulae.

The term values have been discussed in Section 4.5 in which the asymmetry parameters b [equations (4.98) and (4.105)] and κ [equation (4.109)] were introduced. We shall use here only κ, which lies between -1 and $+1$, the prolate and oblate symmetric rotor values, respectively. Approximate term value expressions applicable to near-symmetric rotors were also discussed.

Exactly the same rotational selection rules apply to electronic and vibronic electric dipole transitions between singlet states of asymmetric rotors as to infrared vibrational transitions [see Section 5.2.5.4(i)]. As a consequence, type A, type B, type C or hybrid bands may result depending on whether the oscillating electric dipole is set up along the a, b, or c inertial axis or between them. The main difference between electronic and infrared vibrational transitions is, as always, that there is likely to be a much larger change of rotational constants due to an electronic than a vibrational transition. For this reason we cannot speak, in electronic spectroscopy, of a typical type A, B or C band as we could in infrared vibrational spectroscopy. Bands are likely to be strongly degraded and the degradation varies widely from one transition to another and from one molecule to another.

As has been discussed in Section 5.2.5.4(i), all asymmetric rotors tend to behave like symmetric rotors for certain ranges of J and K_a or K_c. A prolate asymmetric rotor (κ negative) tends to behave like an oblate symmetric rotor for low K_a and high J whereas an oblate asymmetric rotor (κ positive) tends to behave like a prolate symmetric rotor for low K_c and high J—the 'high J, low K' approximation recognized by Gora. In addition, and perhaps more generally important, a prolate asymmetric rotor tends to behave like a prolate symmetric rotor for high K_a and, in particular, when $J \approx K_a$. Similarly, an oblate asymmetric rotor tends to behave like an oblate symmetric rotor for high K_c and, in particular, when $J \approx K_c$.

For these reasons it is usual, rather than unusual, for the rotational fine structure of an asymmetric rotor to show regions of regularity. We have to go to a light prolate asymmetric rotor, rather than one with a particularly small value of $|\kappa|$, to find extremely irregular rotational transitions. The reason for this is that, in a light prolate rotor molecule, there is appreciable population, under normal conditions of Boltzmann population, of rotational levels with only low values of K_a. Under these conditions rotational transitions with high K_a, resembling those of a symmetric rotor, are not observed. For an oblate rotor the coefficient of K_c^2 in the term value expression [see equation (4.106)] is never very large so that transitions with high K_c are observed in all such molecules.

The H_2O molecule is a prolate asymmetric rotor with $\kappa = -0.438$ in the $\tilde{X}^1A_1$ ground state and -0.533 in the $\tilde{C}^1B_1$ excited state. In this case it is both the rather small value of $|\kappa|$ and the lightness of the molecule which cause the rotational structure of the type C 0_0^0 band of the $\tilde{C}$–$\tilde{X}$ absorption system to appear very irregular.[142] At room temperature, levels with $K_a \leqslant 5$ only are appreciably populated.

The bent molecule HNO, with $\angle$ HNO = 108.6° and $\kappa = -0.988$ in the ground electronic state, is a prolate near-symmetric rotor. The 0_0^0 band of the $\tilde{A}^1A''$–$\tilde{X}^1A'$ system is a type C band with the transition moment polarized perpendicular to the molecular plane. In the symmetric rotor limit this becomes a perpendicular band with $\Delta K_a = \pm 1$ and $\Delta J = 0, \pm 1$ so that, except for transitions involving low values of K_a, the 0_0^0 band of HNO strongly resembles such a band. This point is illustrated in Figure 6.108, which shows the ^rR and ^rQ branches, where r implies $\Delta K_a = +1$, for $K'_a = 3 \leftarrow K''_a = 2$ and for $K'_a = 4 \leftarrow K''_a = 3$. The former sub-band shows a clear doubling of rotational lines for $J'' > 8$ due to the asymmetry splitting of the rotational levels (see, for example, Figure 4.25), whereas the latter shows no such doubling.

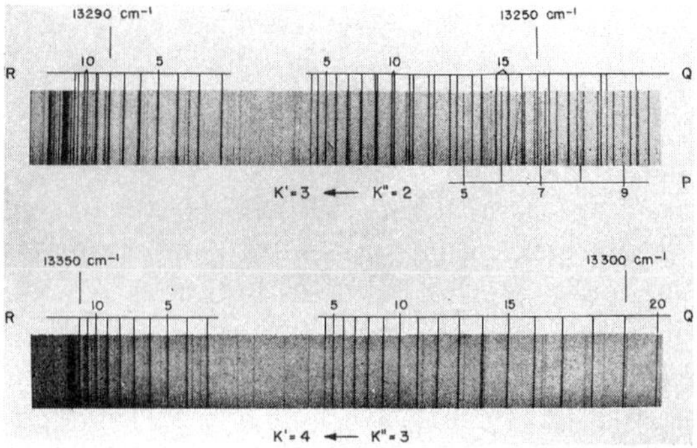

Figure 6.108 The ^rR and ^rQ branches for $K'_a = 3 \leftarrow K''_a = 2$ and $K'_a = 4 \leftarrow K''_a = 3$ transitions of the type C 0_0^0 band of the $\tilde{A}^1A''$–$\tilde{X}^1A'$ system, recorded photographically in absorption, of the prolate, near-symmetric rotor HNO. Asymmetry doubling shows in the former sub-band for $J'' > 8$ but not in the latter [Reproduced, with permission, from Dalby, F. W. (1958). *Can. J. Phys.*, **36**, 1336]

In solving the Schrödinger equation to determine rotational term values for any molecule, the set of axes used is fixed to the molecule; these are the principal inertial axes a, b and c. In a transition from one electronic state to another the geometry change may be such that either two or all of the axes a', b' and c' in the upper state may be rotated relative to that in the lower state. This phenomenon has been called axis switching† but is more appropriately referred to as axis tilting and this term will be used subsequently.

It is clear that such an effect cannot occur in a molecule which is a symmetric rotor in both states nor in an asymmetric rotor of sufficiently

†See Hougen, J. T. and Watson, J. K. G. (1965). *Can. J. Phys.*, **43**, 298, and the footnote on p. 380 of *J. Mol. Spectrosc.*, **66** (1977).

high symmetry, e.g. C_{2v}. However, in a bent triatomic molecule such as HSiCl, belonging to the C_s point group, a rotation of the in-plane *a*- and *b*-axes accompanies any geometry change. In the $\tilde{A}^1A''$–$\tilde{X}^1A'$ system of HSiCl, the rotation of the *a*- and *b*-axes from the $\tilde{X}$ to the $\tilde{A}$ state is by only 0.70°, but the result is to allow $\Delta K_a = 0$, ± 2 sub-bands with observable intensity in the type C 0_0^0 band in which, in the absence of axis tilting, the selection rule would be $\Delta K_a = \pm 1$, predominantly, with other ΔK_a odd transitions being weak.

In general, the main effect of axis tilting is to allow transitions with $\Delta K_a \neq \pm 1$ for a type B or C band and $\Delta K_a \neq 0$ for a type A band of a prolate asymmetric rotor and with $\Delta K_c \neq \pm 1$ for a type A or B band and $\Delta K_c \neq 0$ for a type C band of an oblate asymmetric rotor, provided the molecule is of sufficiently low symmetry in at least one of the combining states.

In the $\tilde{A}^1A_u$–$\tilde{X}^1\Sigma_g^+$, bent-linear transition of acetylene (see Section 6.3.1.4) the rotation, in the C_{2h} *trans*-bent $\tilde{A}$ state, of the in-plane *a*- and *b*-axes is by 2.95°. The result is that the 0_0^0 band shows weak transitions with $\Delta K_a = 0$, ± 2 in addition to the strong ones with $\Delta K_a = \pm 1$.

The $\tilde{A}^1A''$–$\tilde{X}^1\Sigma^+$, bent-linear transition of HCN shows a similar effect due to an *a*- and *b*-axis rotation by 1.47°.

Although axis tilting modifies the ΔK selection rule, a similar modification may also result from Coriolis perturbation or the transition being spin-forbidden [see Sections 6.3.5.2(ii) and 6.3.5.3(ii)]. The former alternative is difficult to distinguish but the latter may be distinguished either by the fact that intensities of electronic transitions with $\Delta S \neq 0$ are expected to be low or by the different J dependence of intensity in the rotational structure.

Difluorodiazirine, $F_2\overline{C{-}N{=}N}$, has a three-membered ring and belongs to the C_{2v} point group. The $\tilde{A}^1B_1$–$\tilde{X}^1A_1$ transition involves an electron promotion from a non-bonding n orbital on the nitrogen atoms to a π^* orbital in the N=N bond. The 0_0^0 band is a type B band with the transition moment polarized along the b (also labelled x) axis which is perpendicular to the plane of the ring. Even though the molecule has a very high degree of asymmetry ($\kappa'' = -0.0865$), it is sufficiently heavy for population of rotational levels with $K_a = 0$–30 to be appreciable. Because the rotational transitions in the 0_0^0 band are congested, only a band contour can be observed as is the case in the $\tilde{A}$–$\tilde{X}$ system of benzene [discussed in Section 6.3.5.2(i)].

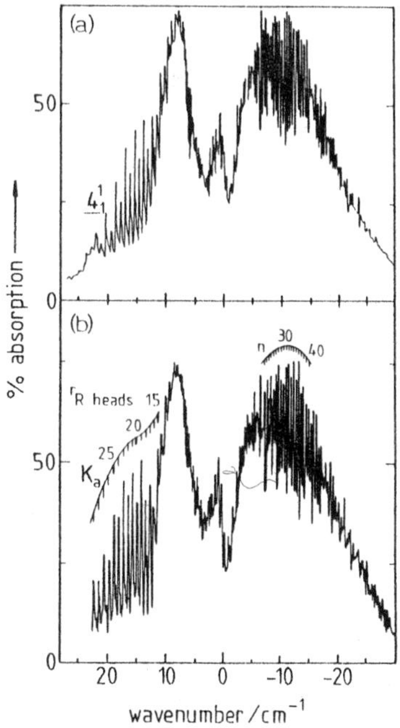

Figure 6.109 (a) The observed and (b) the best computed rotational contour of the type B 0_0^0 band of the $\tilde{A}^1B_1$–$\tilde{X}^1A_1$ system of the prolate asymmetric rotor difluorodiazirine, $F_2\overline{C{-}N{=}N}$ [Reproduced, with permission, from Hepburn, P. H. and Hollas, J. M. (1974). *J. Mol. Spectrosc.*, **50**, 126]

The 0_0^0 contour, together with that of a very weak overlapping sequence band 4_1^1, is shown in Figure 6.109(a), and the computed contour which best matches it in Figure 6.109(b). The latter shows that a regular series of ${}^rR$-branch heads (rather like those shown in Figure 6.108 for HNO) with $K_a > 15$ is formed. There is an intensity alternation of 5:7 for K_a even:odd because of the two nitrogen nuclei ($I=1$) symmetrically disposed about the C_2 axis, which is also the a-axis. Although the ${}^rR$ heads are shifted from where they would be in a symmetric rotor they do not show any resolvable asymmetry splitting in spite of the very small value of $|\kappa|$.

When the method of computer simulation of a rotational band contour is applied to asymmetric rotor spectra, as illustrated by Figure 6.109, there are five important pieces of information to be derived and therefore which must be input into a computer program which is to be used.† These are the rotational selection rules (type A, B or C or hybrid), the rotational constants A'_0, B'_0 and C'_0 in the excited electronic state (assuming that they are known in the ground state) and the rotational line width (usually quoted at half intensity).

The easiest of these to determine is the rotational line width. The degree of sharpness of features in the computed contour is very sensitive to this parameter, which can be determined typically to an accuracy of $\pm 0.02\,\mathrm{cm}^{-1}$.

In general, band contours, for a particular set of rotational constants, are different for different rotational selection rules, although cases are known where there is some possible ambiguity.

Determination of the rotational constants presents more of a problem. Ideally the ground state constants (assuming that the lower state is the ground state) are known from microwave, electron diffraction or X-ray diffraction but, even if they are not, a reasonable guess at the molecular geometry may suffice. The reason for this is that computed rotational contours are much more sensitive to the *changes* of rotational constants ΔA_0, ΔB_0 and ΔC_0, where $\Delta A_0 = A'_0 - A''_0$, etc., than to their absolute values.

In a molecule which is planar in its excited state the problem of determining A'_0, B'_0 and C'_0 is reduced to one of finding only two constants because of the relationship

$$I_c^{\,0} - I_a^{\,0} - I_b^{\,0} = \Delta_0 = 0 \qquad (6.272)$$

where the inertial defect Δ_0 for the zero-point level is assumed to be zero within the accuracy of the method.

In a band contour of a symmetric or near-symmetric rotor three kinds of regular series of features, with one member for each value of K_a (or K_c), may stand out. These are:

1. Line-like Q branches in which many Q-branch lines are nearly coincident. This occurs when $\bar{B}' \approx \bar{B}''$, where $\bar{B} = (B+C)/2$.
2. R-branch heads for which J_{rev} is given by equation (6.135), replacing B by $\bar{B}$. For head formation $\bar{B}' < \bar{B}''$. For the observation of a long series of heads J_{rev} must not be too high, otherwise the intensity at the turning-point is insufficient, and it must not be too low otherwise, since $J \geqslant K$, not many heads would be observed.
3. P-branch heads for which J_{rev} is given by equation (6.137), replacing B by $\bar{B}$. For head formation $\bar{B}' > \bar{B}''$ and the same restrictions on J_{rev} apply as for R-branch heads if a long series is to be observed.

For a symmetric or near-symmetric rotor the separation of each R- or P-branch head from its sub-band origin is independent of K_a (or K_c), and line-like Q branches are at sub-band origins. Therefore, whatever the type of feature forming an observed series, their wavenumbers fit an expression similar to that for sub-band origins, namely, for a parallel-type band,

$$\tilde{\nu} = \tilde{\nu}_0 + \Delta(A - \bar{B})K_a^{\,2} \qquad (6.273)$$

for a prolate rotor, or

†The method of band contour analysis has been summarized by Hollas in the review referred to in the Bibliography at the end of the chapter.

$$\tilde{\nu} = \tilde{\nu}_0 + \Delta(C - \bar{B})K_c^2 \tag{6.274}$$

for an oblate rotor, where $\tilde{\nu}_0$ is the wavenumber of the feature with $K = 0$. For a perpendicular-type band, their wavenumbers are given by

$$\tilde{\nu} = \tilde{\nu}_0 + (A' - \bar{B}') \pm 2(A' - \bar{B}')K_a + \Delta(A - \bar{B})K_a^2 \tag{6.275}$$

for a prolate rotor, or

$$\tilde{\nu} = \tilde{\nu}_0 + (C' - \bar{B}') \pm 2(C' - \bar{B}')K_c + \Delta(C - \bar{B})K_c^2 \tag{6.276}$$

for an oblate rotor, where the upper sign refers to $\Delta K = +1$ and the lower to $\Delta K = -1$. Therefore, for a symmetric or near-symmetric rotor observation of such a series of K-dependent features in a parallel- or perpendicular-type band would yield a value for $\Delta(A - \bar{B})$, or $\Delta(C - \bar{B})$, and hence $(A' - \bar{B}')$, or $(C' - \bar{B}')$.

In *any* asymmetric rotor one parameter can be obtained from such a series. However, there are important differences from a symmetric or near-symmetric rotor. One is that a Q-branch head may be observed at the turning-point of a Q branch whereas such heads cannot be formed for a symmetric rotor. The second difference is that the separation of a P-, Q- or R-branch head from the sub-band origin is not independent of K and, although the symmetric rotor approach is useful semi-quantitatively, a much better quantitative approach is through a second-order perturbation calculation of head positions using, for example, the method of Polo.[143] Then the parameter $(A' - \bar{B}')$, or $(C' - \bar{B}')$, can be varied until there is quantitative agreement between observed and calculated head positions. The second-order perturbation method is, like the symmetric rotor approximation, more valid quantitatively when J is not very much larger than K_a or K_c.

In general, the effect of asymmetry is to split each P, Q and R branch of a symmetric rotor into two sub-branches, as Figure 6.108 illustrates. However, the extreme effect is to cause pairs of sub-branches to come together and this occurs in the high J, low K domain. In this domain a prolate asymmetric rotor, for example, behaves like an oblate symmetric rotor and the clustering of sub-branches is due to the clustering of rotational energy levels, as shown on the right-hand side of Figure 4.25. For a planar, or nearly planar, prolate asymmetric rotor the wavenumbers of the members of these coincident sub-branches are given by:[144]

$$\text{(a)} \quad \tilde{\nu} = \tilde{\nu}_0 - \frac{m_2}{4}(\Delta A + \Delta B) - (2n+1)C'' + n^2 \Delta C \tag{6.277}$$

for coincident ${}^{qp}P^e$ and ${}^{qp}P^o$ sub-branches of a type A band where the pre-superscripts refer to ΔK_a and ΔK_c, respectively, and the post-superscript e or o indicates that $J + K_a + K_c$ in the lower state is even or odd, respectively. The quantity $\tilde{\nu}_0$ is the wavenumber of the band origin, $m_2 = (m+1)^2 + m^2$, $m = J - K_c$ and $n = 2J - K_c$. Equation (6.277) applies also to coincident ${}^{rp}P$ and ${}^{pp}P$ sub-branches in a type B band.

$$\text{(b)} \quad \tilde{\nu} = \tilde{\nu}_0 - \frac{m_2}{2}(\Delta A + \Delta B) + (2n+3)C' + (n+1)^2 \Delta C \tag{6.278}$$

for coincident ${}^{qr}R^e$ and ${}^{qr}R^o$ sub-branches in a type A band and ${}^{pr}R$ and ${}^{rr}R$ sub-branches in a type B band.

$$\text{(c)} \quad \tilde{\nu} = \tilde{\nu}_0 - \frac{m_2}{4}(\Delta A + \Delta B) + (n+1)^2 \Delta C \tag{6.279}$$

for coincident ${}^{rq}Q$ and ${}^{pq}Q$ sub-branches in a type C band.

Equations (6.277)–(6.279) apply also to oblate asymmetric rotors but then they apply to oblate symmetric rotor features.

Transitions given by equations (6.277)–(6.279) are intense because they involve only low values of K_a. In addition, it is common for several transitions with different values of m_2 to be almost coincident since the separation within such a group having the same value of n is approximately $(\Delta A + \Delta B)$ which is usually small compared with the experimental resolution. The result is that a series of intense, n-dependent features may stand out in a rotational contour.

It is the last term, $n^2\Delta C$ or $(n+1)^2\Delta C$, in equations (6.277)–(6.279) which is important in determining rotational constants. If such a series is observed in a contour then it is the coefficient of n^2, i.e. ΔC, which determines the degree of convergence or divergence and which can be obtained from the wavenumbers of the members of the series.

There is a pronounced series of coincident ^{rp}P and ^{pp}P sub-branches, with $n \approx 20$–40, in the type B 0_0^0 band of difluorodiazirine shown in Figure 6.109(b). In this case equation (6.277) does not apply exactly because the molecule is non-planar.

In type A and B bands of a prolate asymmetric rotor there is another, and more generally useful, way of obtaining ΔC from a rotational contour. If ΔC is negative the sub-branches given by equation (6.277) form a turning-point whose wavenumber, $\tilde{\nu}_{\text{rev}}$, is given by

$$\tilde{\nu}_{\text{rev}} = \tilde{\nu}_0 + C'(1 - C'/\Delta C) \qquad (6.280)$$

an expression which can be obtained from equation (6.277). Since equation (6.280) applies to type A and B bands, there must be an exactly similar turning-point in each band type for the same rotational constants. The turning-point is likely to be a very intense feature in a contour because it consists of many individually intense transitions.

In a type B band of a prolate asymmetric rotor there is a possible turning-point of the first lines of ^{r}R branches given by

$$\tilde{\nu}_{\text{rev}} = \tilde{\nu}_0 + A'(1 - A'/\Delta A) \qquad (6.281)$$

which is approximate but becomes exact in a symmetric rotor.

It is no accident that equations (6.280) and (6.281) have the same form. They both apply

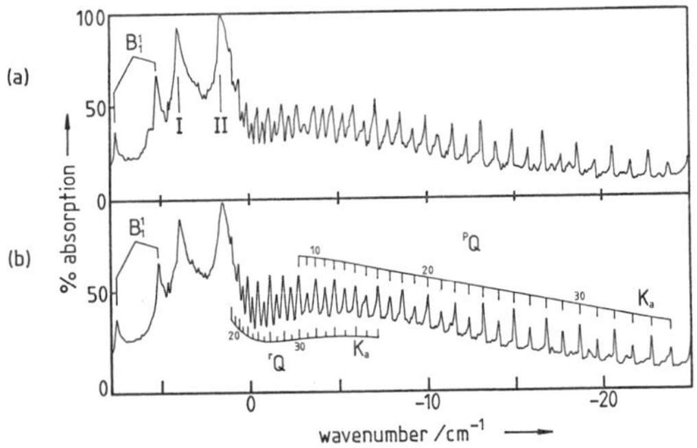

Figure 6.110 (a) The observed and (b) the best computed rotational contour of the type B 0_0^0 band of the $\tilde{\text{A}}^1B_{2u}$–$\tilde{\text{X}}^1A_g$ system of 1,4-difluorobenzene. There is a weak overlapping sequence band labelled B_1^1 [Reproduced, with permission, from Cvitaš, T. and Hollas, J. M. (1970). *Mol. Phys.*, **18**, 793]

exactly to oblate and prolate symmetric rotors, respectively, and approximately when a prolate asymmetric rotor is tending to behave like an oblate symmetric rotor, when equation (6.280) applies, or like a prolate symmetric rotor, when equation (6.281) applies.

Figure 6.110(a) shows the type B 0_0^0 band of the $\tilde{\text{A}}^1B_{2u}$–$\tilde{\text{X}}^1A_g$ system of 1,4-difluorobenzene with a weak overlapping sequence band B_1^1. The molecule is a prolate asymmetric rotor with $\kappa'' = -0.872$. Figure 6.110(b) shows a matching computed contour in which the long series of ^{p}Q- and ^{r}Q-branch heads are assigned values of K_a. The 10:6 intensity alternation for K_a even:odd, due to two pairs of hydrogen nuclei which are exchanged by rotation about the a-axis, is clear. The feature labelled II in Figure 6.110(a) is the turning-point of the ^{r}Q branches and that labelled I is the turning-point of coincident ^{rp}P and ^{pp}P sub-branches.

Figure 6.111(a) and (b) shows type A and C contours, respectively, computed with the same parameters as were used for the type B contour in Figure 6.110(b). These demonstrate the pronounced effect on the contour of type A, B

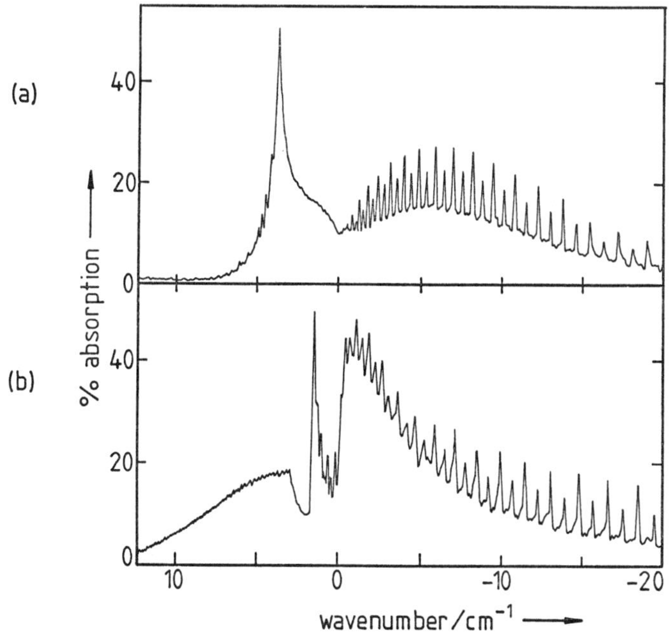

Figure 6.111 Computed (a) type *A* and (b) type *C* rotational contours for 1,4-difluorobenzene using the same rotational constants as for Figure 6.110(b) [Reproduced, with permission, from Cvitaš, T. and Hollas, J. M. (1970). *Mol. Phys.*, **18**, 793]

or *C* selection rules. Figures 6.110(b) and 6.111(a) show also how the intense peak at $\tilde{\nu}_0+3.90\,\mathrm{cm}^{-1}$, given by equation (6.280), appears in *both* a type *B* and a type *A* band.

6.3.5.3(ii) *Transitions involving states with non-zero multiplicity*

Asymmetric rotors cannot have electronically degenerate states so, as in the case of a non-degenerate state of a symmetric rotor, there cannot be any orbital angular momentum. There can be, however, either a small interaction between the spin magnetic moment and the small magnetic moment due to molecular rotation or a small instantaneous interaction between spin and orbital angular momentum in spite of the fact that the average value of the latter is zero.

For a near-symmetric rotor in a doublet state the term values $F(J_\tau)$ of equation (4.108) are split into two components $F_1(N_\tau)$ and $F_2(N_\tau)$, where

$$\begin{aligned} F_1(N_\tau) &= F(N_\tau) + \tfrac{1}{2}\gamma N \\ F_2(N_\tau) &= F(N_\tau) - \tfrac{1}{2}\gamma(N+1) \end{aligned} \tag{6.282}$$

The spin splitting parameter γ is given approximately by

$$\gamma = \frac{\kappa K^2}{N(N+1)} + \mu \pm \frac{\eta_K}{2} \tag{6.283}$$

where $K = K_a$ or K_c for a prolate or oblate asymmetric rotor, respectively, and κ, not to be confused with the asymmetry parameter, and μ are the spin–rotation coupling constants introduced for a symmetric rotor in equation (6.270). η_K can be neglected for $K > 1$ but, for $K = 1$, it is the same as a spin–rotation coupling constant representing the difference in coupling about the b and c inertial axes. Alternatively, some authors use the constants $a = -\kappa/3$, $a_0 = -(\kappa/3) - \mu$ and $b = \eta_1$.

Treating NH_2 as a near-symmetric rotor, the values $\kappa = 0.335\,\text{cm}^{-1}$, $\mu = 0.02\,\text{cm}^{-1}$ and $\eta_1 = 0.016\,\text{cm}^{-1}$ have been obtained for the $\tilde{X}^2B_1$ state. However, since the asymmetry parameter $\kappa'' = -0.386$ in NH_2, these values can be only approximate.

Dixon, Duxbury and Ramsay[145] have shown that the approximate expressions in equations (6.282) and (6.283) are inadequate for both the upper ($\kappa' = -0.839$) and lower ($\kappa'' = 0.576$) electronic states of the $\tilde{A}^2A_1$–$\tilde{X}^2B_1$ transition of PH_2. In the lower state the inadequacy is due to the asymmetry of the molecule and the asymmetric rotor spin–rotation constants ε_{aa}, ε_{bb} and ε_{cc}, where a, b and c refer to the inertial axes, had to be used. They obtained values of $\varepsilon_{aa} = -0.30\,\text{cm}^{-1}$ and $\frac{1}{2}(\varepsilon_{bb} + \varepsilon_{cc}) = -0.07\,\text{cm}^{-1}$; ε_{bb} and ε_{cc} could not be determined independently. In the upper state the molecule is less asymmetric but a large value of the symmetric rotor centrifugal distortion constant D_K introduces an appreciable dependence of ε_{aa} on K_a.

For the most commonly observed type of multiplicity forbidden transitions, the triplet–singlet transition, Hougen[141] has shown that, as for symmetric rotors [see Section 6.3.5.2(ii)], the rotational selection rules for near-symmetric rotors are

$$\Delta N = 0,\ \pm 1,\ \pm 2;\ \Delta K = 0,\ \pm 1,\ \pm 2$$

In addition, the intensities depend on the magnitudes of three transition moments R_x, R_y and R_z. These are independent parameters and derive from the relative amount of intensity stealing from a singlet–singlet or triplet–triplet transition, polarized along the x-, y- or z-axis, by spin–orbit coupling (see Section 6.2.3.3).

It may appear that determination of the relative values of R_x, R_y and R_z, in addition to the rotation and spin–rotation constants, may present a formidable problem but, fortunately, in triplet–singlet transitions which have been rotationally analysed, one of the three transition moments is dominant. For example, in the $\tilde{a}^3A_2$–$\tilde{X}^1A_1$ system of formaldehyde, R_z [where $\Gamma(T_z) = A_1$] is the dominant transition moment. This means that the transition steals its intensity from a $^1A_1$–$\tilde{X}^1A_1$ or a $^3A_2$–$\tilde{a}^3A_2$ transition, or both.

For an asymmetric rotor belonging to the C_{2v} or D_{2h} point group there are three spin–rotation and two spin–spin constants.[146] In the $\tilde{a}^3A_2$ state of formaldehyde the spin–rotation constants a, a_0 and b and the two spin–spin constants α and β have been determined from rotational analysis of the 0^0_0 band of the $\tilde{a}$–$\tilde{X}$ system.†

6.3.5.4 *Spherical rotor molecules*

It is unusual for the electronic spectrum of a spherical rotor molecule to exhibit sharp rotational fine structure. Methane, for example, and the tetrahalomethanes have their first absorption systems in the far ultraviolet. The excited states involved are Rydberg states and the absorption spectra are almost continuous, not even showing much vibrational structure.

Regular octahedral molecules such as SF_6 and MoF_6, which have their first electronic band systems at lower energy, are similarly diffuse.

More promising are regular octahedral transition metal compounds which absorb at lower energies. For many of them there is a problem of low vapour pressure. Exceptions are ReF_6 and IrF_6, but systems observed in the near infrared and visible regions are not rotationally sharp. There is also the possibility of a Jahn–Teller effect and consequent distortion in these molecules.

Bands observed in emission from an electrical discharge in N^1H_3 (and N^2H_3) at about 565 and 663 nm (579 and 675 nm in N^2H_3), the so-called Schuster bands, have been attributed[147] to the spherical rotor N^1H_4 (or N^2H_4). The 565 nm (579 nm) band was later assigned[148] to the 2^1_1

†See, for example, Birss, F. W., Dong, R. Y. and Ramsay, D. A. (1973). *Chem. Phys. Lett.*, **18**, 11.

sequence band in the $\tilde{C}'$–$\tilde{A}$ emission between two excited electronic states of N^1H_3 (N^2H_3). The 663 nm (675 nm) band is also part of the same emission band system.

6.3.5.5 *Effects of magnetic and electric fields*

In the presence of a magnetic or electric field, energy levels may be split leading to the splitting of transitions. In general, as in atoms (Sections 6.1.8 and 6.1.9) and diatomic molecules, this is called the Zeeman effect for a magnetic field and the Stark effect for an electric field. In either type of field the split energy levels are labelled by a quantum number M arising from space quantization. The electric dipole selection rules are $\Delta M = 0$, except that $M = 0 \nleftrightarrow M = 0$, for $\Delta J = 0$ for which the transition moment is parallel to the field direction, and $\Delta M = \pm 1$ for which it is perpendicular to the field. For $S = 0$ (singlet states) the quantum number M is M_J, whereas for $S \neq 0$ it is M_N. In addition, when the spin is not strongly coupled to the rotational angular momentum, $\Delta M_S = 0$ for transitions with $\Delta S = 0$, and $\Delta M_N = 0, \pm 1$ for transitions with $\Delta S \neq 0$.

In electronic spectra of polyatomic molecules the splitting of rotational transitions by a magnetic field B is very small. The Zeeman effect often appears only as a broadening of rotational lines. The width of the line, δ, represents the spread of energies from $+M_J$ to $-M_J$ and is given by

$$\delta = 2\mu_J B \tag{6.284}$$

where μ_J is the effective magnetic moment for a particular value of J.

Douglas and Milton[149] have pointed out that the magnetic moment may arise in three ways:

1. Due to an electronic orbital angular momentum which occurs in any degenerate electronic state such as $^1\Pi$ or ^1E. In these cases μ_J decreases rapidly with J.
2. Due to an electron spin angular momentum, i.e. for states with $S \neq 0$. If the multiplet splitting due to spin–orbit coupling is small, μ_J is independent of J.
3. Due to rotation of a molecule in a non-degenerate singlet state which couples it to some other state which *does* have a magnetic moment. The resulting magnetic moment is usually small but increases linearly with J.

In the $\tilde{a}^3A_2$–$\tilde{X}^1\Sigma_g^+$, bent–linear transition of CS_2, a large Zeeman effect has been observed with fields of up to 0.53 T. The field was applied in pulses by discharging a condenser through a solenoid around the absorption cell, each pulse coinciding with that of a Lyman discharge continuum source (Section 3.9). The line broadening is linear in J and has been interpreted in terms of the only spin–orbit component of the triplet state observed being B_2, the transition obtaining intensity by spin–orbit coupling between the $\tilde{a}^3A_2$ and a $^1B_2$ state (see Section 6.2.3.3). The fact that the magnetic moment increases linearly with J is due apparently to coupling, by rotation, of the B_2 component with the other components, A_1 and B_1, of the $\tilde{a}^3A_2$ state. The result is that case 3, above, rather than case 2 is dominant.

In the upper state of the $\tilde{B}^1E''$–$\tilde{X}^1A'_1$ transition of NH_3 there is a large orbital angular momentum ($\zeta_e = 0.826$). As a result of this, some of the rotational lines in this band system show broadening in a magnetic field which is largest when $J \approx K_c$ and, as in case 1, decreases with J.

High resolution spectra of the $\tilde{B}^1E''$–$\tilde{X}^1A'_1$ system of N^2H_3, with the Doppler broadening removed, have been obtained[140] in the presence of a magnetic field. Rather than just a broadening of the lines these spectra show, for low values of J, the individual Zeeman components. For example, Figure 6.112 shows the three components of a $J' = 1$–$J'' = 3$ transition in the 2^3_0 vibronic band. The fact that ΔJ can be 2 is due to the fact that the transition is observed by a two-photon process for which the selection rules differ from those for a one-photon process (see Section 8.2.18). It is only the electronically

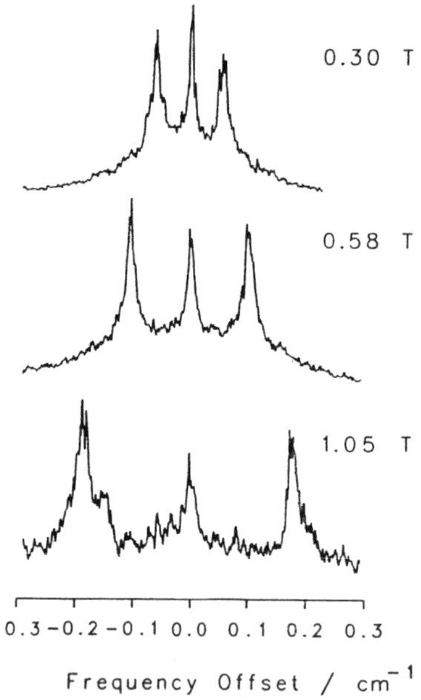

Figure 6.112 A $J' = 1 - J'' = 3$ transition in the 2_0^1 vibronic band of the $\tilde{B}^1E''$–$\tilde{X}^1A_1'$ system of N^2H_3 split into three components in a magnetic field of 0.30, 0.58 and 1.05 T [Reproduced, with permission, from Ashfold, M. N. R., Dixon, R. N., Little, N., Stickland, R. J. and Western, C. M. (1988). *J. Chem. Phys.*, **89**, 1754]

degenerate $\tilde{B}^1E''$ state which shows a Zeeman effect and the three components are due to the splitting of the $J' = 1$ level into the $M_J = +1$, 0, and −1 sub-levels.

The splitting, ΔE, of a rotational level of a doubly degenerate electronic state in a magnetic field B is given, to first order, by

$$\Delta E = \mu_B g_L B \frac{l M_J K}{J(J+1)} \tag{6.285}$$

where $l = \pm 1$ for the two components of the E'' electronic state and g_L is the Landé factor.

It was found that the value of g_L for each vibronic level is slightly less, by about 0.05–0.10, than the value of the Coriolis constant ζ. These small differences are attributed to the small Jahn–Teller effect in the $\tilde{B}^1E''$ state, discussed in Section 6.3.4.5.

Another kind of experiment employing the Zeeman effect results in a magnetic rotation spectrum. In this experiment two crossed linear polarizers are placed one before and one after an absorption cell. In the absence of a magnetic field no radiation is transmitted through the system. When a magnetic field of the order of 0.5 T is applied by winding wire coils around the absorption cell, radiation whose plane of polarization has been rotated is transmitted. Such a rotation occurs close to transitions in which there is a Zeeman effect. The resulting transmission spectrum appears like an emission rather than an absorption spectrum.

At first the technique was used as a diagnostic tool[150] to demonstrate the triplet–singlet character of, for example, the $\tilde{a}^3A_2$–$\tilde{X}^1A_1$ system of formaldehyde, the $\tilde{a}^3A_u$–$\tilde{X}^1A_g$ system of glyoxal, and the $\tilde{a}^3A_u$–$\tilde{X}^1A_g$ system of *p*-benzoquinone. Later the theory was developed further and applied to the $\tilde{a}$–$\tilde{X}$ system of formaldehyde.[151]

If a singlet state is perturbed by, and therefore slightly mixed with, a nearby triplet state, transitions to perturbed rotational levels of the singlet state should appear in a magnetic rotation spectrum. About 40 such levels have been identified in the $\tilde{A}^1A_2$ state of formaldehyde.[152]

We have seen in Sections 4.3 and 4.5 how an electric field splits the rotational energy levels and transitions in symmetric and asymmetric rotors—the Stark effect. Observation of this effect permits determination of the dipole moment.

The Stark effect on rotational transitions accompanying electronic transitions has been observed for a number of molecules from which the dipole moment in an excited electronic state can be found.

To observe the effect an electrostatic field of up to *ca* 20 kV cm^{-1} is applied across two parallel metal plates not more than a few centimetres apart. With radiation polarized parallel to the field there is zero intensity in the middle of a Q-branch, Stark-broadened line whereas, for perpendicular polarization, the same is true for *P*- and *R*-branch, Stark-broadened lines. In this way the dipole moments in the $\tilde{A}^1A_2$ and $\tilde{a}^3A_2$ states of formaldehyde have been found to be[153,154] 1.56±0.07 and 1.29±0.03 D, respectively, compared with 2.33±0.02 D in the $\tilde{X}^1A_1$ state. The decrease from the ground to each of these excited states is qualitatively reasonable for the promotion of an electron from an *n* orbital on the oxygen atom to a π^* orbital in the C=O bond.

Other fairly small molecules such as propynal (HC≡CCHO) and fluoroformaldehyde (HFCO) have been similarly investigated, in addition to several substituted benzene molecules[155] in which splitting of prominent features in the unresolved rotational contours were observed.

In all the molecules mentioned the Stark effect was observed in rotational lines where no asymmetry doubling was resolved so the symmetric rotor approximation [equation (4.72)] could be used.

The experiment with an electric field, analogous to that with a magnetic field and two crossed polarizers giving a magnetic rotation spectrum, produces a Kerr effect spectrum which has been observed in the $\tilde{A}^1A_2$–$\tilde{X}^1A_1$ system of formaldehyde.[156]

6.3.5.6 *Diffuse spectra*

A common cause of rotational diffuseness in absorption spectra of small polyatomic molecules is predissociation in an excited electronic state, as in diatomic molecules. There are many examples of such diffuseness varying in degree from cases in which the half intensity line width is slightly greater than that due to the Doppler effect to those in which the lines are so broad that the rotational structure cannot be used to obtain rotational constants. An example of the latter is in the $\tilde{B}$–$\tilde{X}$ system of C^1H_3, although the lines are much sharper in C^2H_3 [see Section 6.3.5.2(i)].

The rotational line width $\Delta\tilde{\nu}$ is related to the lifetime of the excited state τ and the first-order rate constant *k* for decay by

$$\Delta\tilde{\nu} = \frac{1}{2\pi c\tau} = \frac{k}{2\pi c} \tag{6.286}$$

derived from equations (2.39) and (2.35). If both radiative (r) and non-radiative (nr) decay of the excited state take place, then

$$\Delta\tilde{\nu} = \frac{1}{2\pi c}\left(\frac{1}{\tau_r} + \frac{1}{\tau_{nr}}\right) = \frac{k_r + k_{nr}}{2\pi c} \tag{6.287}$$

since the overall rate constant is the sum of the rate constants for all decay processes [equation (2.38), where k_r is equal to the Einstein coefficient A_{nm}]. If non-radiative processes are efficient relative to emission then $\tau_{nr} < \tau_r$ and the line width is increased compared with the natural line width, this being the width it would have if $\tau_{nr} \gg \tau_r$.

In practice, as discussed in Section 2.3.2, the Doppler line width is usually much greater than the natural line width. For this reason, τ_{nr} has to be *much less* than τ_r (*ca* 10–100 times less) for line broadening due to radiationless transitions to be observed. Therefore, broadening of rotational lines in the absorption spectrum is not a very sensitive indicator of predissociation.

The emission spectrum is much more sensitive. If, say, the two lifetimes τ_r and τ_{nr} are equal then, for energies above the predissociation limit, half the molecules in the excited electronic state predissociate before they have time to emit and the transition intensity is decreased by 50%. Therefore, for a rotational line involving an excited state above the predissociation energy there will be a *substantial* drop in intensity in the emission spectrum.

One such example in a polyatomic molecule is found in the $\tilde{A}^1A''$–$\tilde{X}^1A'$ emission spectrum of HNO. Figure 6.113 shows a dramatic fall in intensity beyond the ^rR-branch head with $K''_a = 12$ and $K'_a = 13$ in the 0^0_0 band and beyond that with $K''_a = 9$ and $K'_a = 10$ in the 2^1_0 band. These observations resulted in a value of

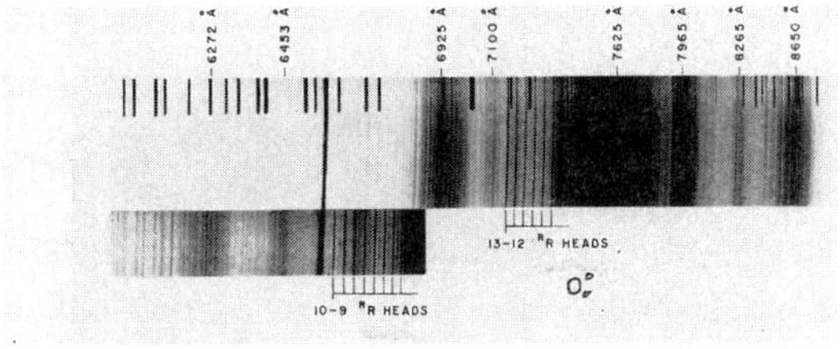

Figure 6.113 Predissociation in the $\tilde{A}^1A''$–$\tilde{X}^1A'$ photographically recorded emission spectrum of HNO results in a drop in intensity beyond the ^rR-branch head with $K''_a = 12$ and $K'_a = 13$ in the 0_0^0 band and with $K''_a = 9$ and $K'_a = 10$ in the 2_0^1 band [Reproduced with permission, from Clement, M. J. Y. and Ramsay, D. A. (1961). *Can. J. Phys.*, **39**, 205]

$D''_0 \leqslant 203\,\mathrm{kJ\,mol^{-1}}$ ($17\,000\,\mathrm{cm^{-1}}$), which is the predissociation energy relative to the zero-point level of the $\tilde{X}$ state. Corresponding to this predissociation energy a gradual increase in the rotational line width has been observed in the $1_0^1 3_0^1$ band of the absorption spectrum[157] but, for the reason explained above, the onset of this diffuseness is by no means as dramatic as the fall in intensity in the emission spectrum.

A high resolution, laser induced fluorescence excitation spectrum (see Section 8.2.16) of HNO shows, in greater detail, the breaking-off in the rotational structure due to predissociation.[158] In particular, it shows that the energy at which the breaking-off occurs is dependent only on the rotational quantum number J and not K_a or any vibrational quantum number. This evidence supports the attribution of the predissociation to intersystem crossing from the $\tilde{A}$ state to the continuum of vibrational levels above the $H(1^2S) + NO\,(X^2\Pi)$ dissociation limit of the $\tilde{X}$ state. Figure 6.114 shows part of the $1_0^1 3_0^1$ band and how it is dominated by the $^pP_1$ (1) line (with $\Delta K_a = -1, \Delta J = -1$ and $J'' = 1$). For this transition $J' = 0$ and the very low intensities of all other lines compared with those expected for a prolate near-symmetric rotor show that all rotational levels, other than that with $J' = 0$, suffer considerable predissociation. Such observations resulted in an improved estimate of the dissociation energy in the ground electronic state of $16\,450 \pm 10\,\mathrm{cm^{-1}}$.

There is a tendency, on going to larger molecules, towards increased diffuseness in their electronic spectra. There is also a tendency for emission, if observed at all, to be only from the lowest singlet or triplet excited state, commonly labelled S_1 and T_1, respectively. The S_1–S_0 fluorescence emission to the ground state S_0 is spin-allowed and relatively fast, the S_1 lifetime usually being in the range 10^{-8}–10^{-6} s. The T_1–S_0 phosphorescence emission is spin-forbidden and, when spin–orbit coupling is small, relatively slow with a T_1 lifetime which varies widely but which may be as long as a few seconds or more. The tendency to diffuseness means that it is generally true that, at best, only the S_1–S_0 and, if sufficiently intense to be observed, the T_1–S_0 systems are sufficiently rotationally sharp for a rotational analysis to be performed. At the worst, rotational diffuseness is total as in the S_1–S_0 systems of ethylene and buta-1,3-diene.

One of the simplest possible reasons for diffuseness is that of spectral congestion. In large molecules at room temperature there is extreme congestion of rotational transitions within each band contour so that, in anthracene for example, it has been estimated[159] that there are, on average, about 100 within the Doppler width.

In addition there is congestion of vibronic transitions, particularly of sequence bands, due to the large number of normal vibrations—anthracene, for example, has 66. Even though Franck–Condon factors for many vibronic transitions are very low, the sheer number of transitions may result in an intense quasi-continuum upon which the stronger bands are superimposed. The diffuseness is then apparent rather than real. This happens in the $\tilde{A}^1B_{1u}$–$\tilde{X}^1A_g$ system of anthracene.

However, in many cases simple congestion cannot be the reason for partial or total diffuseness. In particular, in molecules which are smaller than, say, anthracene the main cause of any appearance of diffuseness in electronic

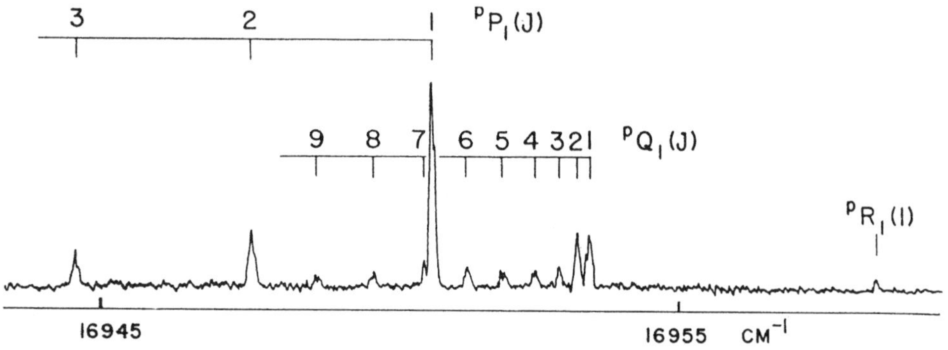

Figure 6.114 Part of the $1_0^1 3_0^1$ band in the laser induced fluorescence spectrum of HNO, showing a rapid drop in intensity of all lines with $J' > 0$ [Reproduced, with permission, from Dixon, R. N., Jones, K. B., Noble, M. and Carter, S. (1981). *Mol. Phys.*, **42**, 455]

and vibronic bands is a large line width of the accompanying rotational transitions and this does not bear any simple relation to the size of the molecule. For example, in the 0_0^0 band of the $\tilde{A}^1A''$–$\tilde{X}^1A'$ system of acrolein, $H_2C{=}CHCHO$, the half intensity rotational line width is 2 cm^{-1}, whereas in the 0_0^0 band of the $\tilde{A}^1B_{2u}$–$\tilde{X}^1A_g$ system of naphthalene it is only 0.1 cm^{-1}.

Although the phenomenon of predissociation may be important in a few cases of rotational diffuseness, even in low-lying electronic states, it cannot be the cause in the majority of cases. The most likely cause is due to channels for decay of the S_1, or, in some cases, the T_1, state being available in larger but not in smaller molecules. These channels are a result of nonradiative transitions other than predissociation.

Figure 6.115 shows the three lowest electronic states S_0, T_1 and S_1 of a polyatomic molecule. The molecule is assumed to be sufficiently large that the density of rovibronic states of both S_0 and T_1 is so high that, even at the zero-point level of S_1, there are pseudo-continua of both S_0 and T_1 degenerate with it. By pseudo-continuum we mean, in this context, continuous compared to the Doppler energy spread of a rovibronic level in S_1. The result is that the radiationless transitions S_1–T_1 and/or S_1–S_0 may be rapid compared to the S_1–S_0 fluorescence and the fluorescence quantum yield Φ_F, given by

$$\Phi_F = \frac{\text{number of molecules emitting radiation}}{\text{number of quanta absorbed}} \tag{6.288}$$

may be very small. For some molecules it is so small that no S_1–S_0 fluorescence has been detected. This is the case, for example, for the $\tilde{A}^1A''$–$\tilde{X}^1A'$ transition of acrolein.

The $\tilde{A}^1B_{2u}$–$\tilde{X}^1A_{1g}$, S_1–S_0 system of benzene has been closely studied with a view to estimating radiative and non-radiative decay rates for the S_1 state. The system is ideal in that it shows very little background absorption continuum and is readily observed in emission.

The fluorescence lifetime τ_F, which can be measured, is the lifetime of the S_1 state irrespective of how it decays. It is therefore related to k_r and k_{nr} by

$$\tau_F = \frac{1}{k_r + k_{nr}} \tag{6.289}$$

In addition, the quantum yield of fluorescence is given by

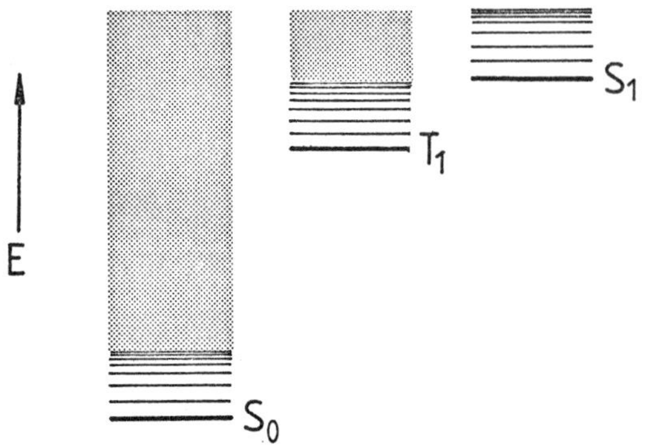

Figure 6.115 The three lowest electronic states, S_0, S_1 and T_1, of a polyatomic molecule showing regions of relatively low density of vibrational states and also pseudo-continua at higher energy

$$\Phi_F = \frac{k_r}{k_r + k_{nr}} \tag{6.290}$$

Therefore measurements of τ_F and, when possible, Φ_F can be translated into values of k_r and k_{nr}.

Table 6.18 shows some values of τ_F and k_{nr} obtained by the technique of single vibronic level fluorescence (see Section 6.3.4.2) from various vibronic levels of the S_1 state of benzene under the collision-free conditions which obtain in a supersonic jet (see Section 3.4.3). Values of Φ_F, where known previously, are also given. Where there were no available values, the radiative rate constant k_r was estimated from the normal mode character of the fluorescing state.

Table 6.18 shows that there is a tendency for τ_F to decrease and for k_{nr} to increase with increasing vibrational energy, but there are appreciable variations depending on the type of vibration(s) excited.

In the S_1–S_0 system of benzene there is a dramatic fall to almost zero in the fluorescence quantum yield when the energy of the vibronic levels populated prior to emission are somewhere in the range 33.5–39.5 kJ mol^{-1} (2800–3300 cm^{-1}) above the zero-point level of the S_1 state. This is illustrated by the data given in Table 6.18 for the fast-decaying $1^3 6^1$ state. The corresponding data for the nearby 7^1 state show, however, that, of the two e_{2g} vibrations ν_6 and ν_7, it is ν_6 which is much more effective in inducing the fast decay. As in the case of the dramatic reduction in the fluorescence of HNO, discussed above, the S_1–S_0 absorption spectrum of benzene shows[160] an onset of broadening of the rotational structure of vibronic bands in the region corresponding to the loss of fluorescence, with a similar distinction between the effects of ν_6 and ν_7.

These observations gave rise to the attribution of a so-called 'channel three' mechanism for non-radiative decay which opens up about 3000 cm^{-1} above the zero-point level of S_1. The other two channels would be intersystem crossing to the triplet state T_1 and internal conversion to the ground state S_0, both of which had been ruled out.

Observation[161] of the extremely weak fluorescence from vibronic levels up to 7000 cm^{-1} above the zero-point level showed that non-radiative lifetimes do show a marked increase in the channel three region but the effect is much less pronounced than had been deduced from the non-radiative lifetimes estimated from the line broadening in the absorption spectrum.[160] In a more detailed investigation,[162] it was shown that there is a considerable variation of non-radiative lifetimes for rotational states within the $1^2 14^1$ vibronic state, 3412 cm^{-1} above the zero-point level and observed in a two-photon process (see Section 8.2.18). These variations are due to Coriolis coupling within S_1. The coupling involved is an important mechanism which facilitates intramolecular vibrational redistribution (IVR) (see Section 8.2.9). The IVR involves coupling of an S_1 vibronic state, in this case $1^2 14^1$, to which a transition is allowed (an optically 'bright' state) to other nearby densely packed vibronic states within S_1 to which transitions are forbidden (optically 'dark' states).

These findings suggest that the channel three problem in benzene may not require a unique

Table 6.18 Wavenumbers above the zero-point level, $\Delta\tilde{\nu}$, fluorescence quantum yield, Φ_F, fluorescence lifetime, τ_F, and non-radiative rate constant, k_{nr}, for various vibronic levels of the $\tilde{A}^1B_{2u}$ state of benzene[a]

Vibronic level	$\Delta\tilde{\nu}/cm^{-1}$	Φ_F	τ_F/ns	k_{nr}/s^{-1}
0^0	0	0.22	106(1)	$7.4(4)\times10^{-6}$
6^1	521	0.27	87(1)	$8.4(6)\times10^{-6}$
4^16^1	898	–	63(1)	$12.8(13)\times10^{-6}$
1^1	923	0.21	95(2)	$8.3(5)\times10^{-6}$
9^1	1147	–	73(1)	$11.5(7)\times10^{-6}$
10^2	1164	–	59(2)	$14.8(9)\times10^{-6}$
5^110^1	1334	–	60(2)	$14.5(9)\times10^{-6}$
1^16^1	1444	0.24	82(2)	$9.3(6)\times10^{-6}$
1^2	1847	0.18	84(3)	$9.8(6)\times10^{-6}$
5^26^1	2017	–	47(3)	$18.2(19)\times10^{-6}$
1^18^1	2436	–	46(2)	$19.6(12)\times10^{-6}$
7^1	3073	0.01	68–76	14.1×10^{-6}
1^36^1	3284	$< 5 \times 10^{-4}$	< 3	$>300\times10^{-6}$

[a]Data taken from Stephenson, T. A. and Rice, S. A. (1984). *J. Chem. Phys.*, **81**, 1073.

solution but that the unusual behaviour may be a manifestation of IVR within S_1, a process which is important in all molecules, particularly large ones with high densities of vibronic states not too far above the zero-point level.

BIBLIOGRAPHY

ATKINSON, G. H. and PARMENTER, C. S. (1974). 'Single vibronic level fluorescence', chap. 2 in *Creation and Detection of the Excited State*, vol. 3 (Ed. W. R. Ware). Dekker, New York

CANDLER, C. (1964). *Atomic Spectra*. Hilger and Watts, London

CONDON, E. V. and SHORTLEY, G. H. (1953). *The Theory of Atomic Spectra*. Cambridge University Press, Cambridge

COULSON, C. A. (1961). *Valence*. Oxford University Press, Oxford

COULSON, C. A. and McWEENEY, R. (1979). *Coulson's Valence*. Oxford University Press, Oxford

DUNN, T. M. (1972). 'Nuclear hyperfine structure in the electronic spectra of diatomic molecules', chap. 4.4 in *Molecular Spectroscopy: Modern Research* (Eds K. N. Rao and C. W. Mathews). Academic Press, New York

HERZBERG, G. (1944). *Atomic Spectra and Atomic Structure*. Dover, New York

HERZBERG, G. (1950). *Spectra of Diatomic Molecules*. Van Nostrand, New York

HERZBERG, G. (1966). *Electronic Spectra of Polyatomic Molecules*. Van Nostrand, New York

HOLLAS, J. M. (1973). 'Electronic spectra of large molecules', chap. 2 in *Molecular Spectroscopy*, vol. 1, Specialist Periodical Report. Chemical Society, London

HUBER, K. P. and HERZBERG, G. (1979). *Constants of Diatomic Molecules*. Van Nostrand Reinhold, New York

KING, G. W. (1964). *Spectroscopy and Molecular Structure*. Holt, Rinehart and Winston, New York

KUHN, H. G. (1969). *Atomic Spectra*. Longmans, London

MURRELL, J. N. and HARGET, A. J. (1972). *Semi-empirical Self-consistent-field Molecular Orbital Theory of Molecules*. Wiley, London

MURRELL, J. N., KETTLE, S. F. A. and TEDDER, J. M. (1965). *Valence Theory*. Wiley, London

MURRELL, J. N., KETTLE, S. F. A. and TEDDER, J. M. (1978). *The Chemical Bond*. Wiley, London

ROSEN, B. (Ed.) (1970). *Spectroscopic Data Relative to Diatomic Molecules*. Pergamon, Oxford

STEINFELD, J. I. (1974). *Molecules and Radiation*. Harper and Row, New York

7 PHOTOELECTRON SPECTROSCOPY

7.1 INTRODUCTION

Photoelectron spectroscopy involves the ejection of electrons from atoms or molecules following bombardment by monochromatic photons. The ejected electrons are called photoelectrons and were discussed in Section 1.1 in the context of the photoelectric effect which was so important in the development of quantum theory.

The photoelectric effect was observed originally on surfaces of easily ionizable metals, such as the alkali metals. Bombardment of the surface with photons of tunable frequency does not produce any photoelectrons until the threshold frequency is reached (see Figure 1.2). At this frequency, ν_t, the photon energy is just enough to overcome the work function Φ of the metal, so that

$$h\nu_t = \Phi \tag{7.1}$$

At higher frequencies the excess energy of the photons becomes the kinetic energy of the photoelectrons:

$$h\nu = \tfrac{1}{2}m_e v^2 + \Phi \tag{7.2}$$

where m_e and v are their mass and velocity, respectively.

Work functions of alkali metal surfaces are only a few electronvolts so that the energy of near ultraviolet radiation is sufficient to produce ionization.

Photoelectron spectroscopy is a simple extention of the photoelectric effect involving the use of higher energy incident photons and, frequently, a gas-phase sample. Equations (7.1) and (7.2) apply also to photoelectron ejection from a gas-phase atom or molecule but the work function Φ is replaced by the ionization energy I.[†]

Even though Einstein developed the theory of the photoelectric effect as long ago as 1905, the experimental difficulties were such that it was not until the late 1950s and early 1960s that photoelectron spectroscopy, as we now know it, developed into an important branch of spectroscopy.

In this chapter we shall be concerned almost totally with gas-phase atoms and molecules and it will be useful to regard the electrons as being in atomic orbitals (AOs) or molecular orbitals (MOs), as discussed in Chapter 6.

Figure 7.1 shows a set of fully occupied, non-degenerate, orbitals of a gas-phase atom or molecule. Each orbital energy is relative to a zero of energy corresponding to removal of an electron from that orbital. The valence, or outer shell, electrons have higher orbital energies than the core, or inner shell, electrons. As indicated on the right-hand side of the figure, it is

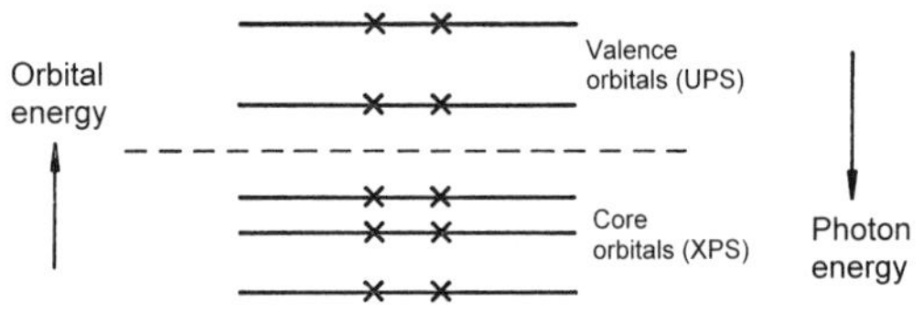

Figure 7.1 Filled, non-degenerate valence and core orbitals of an atom or molecule

[†] From equation (7.1), I is seen to have dimensions of energy and is properly called the ionization energy. What is measured experimentally is an electrical potential but this is easily converted to energy, the unit of energy often being the electronvolt (eV) which is the energy acquired by an electron when accelerated by a potential of 1 V. For this reason, I is often referred to, incorrectly, as the ionization potential.

necessary for an incident photon to have greater energy to remove an electron from a core orbital than from a valence orbital.

In general, a monochromatic source of soft (low energy) X-rays is used to remove core electrons and the technique is known as X-ray photoelectron spectroscopy, sometimes, as here, abbreviated to XPS. On the other hand, far ultraviolet radiation has sufficient energy to remove valence electrons from many atoms and molecules and a source of such monochromatic radiation is used in the technique known as ultraviolet photoelectron spectroscopy or UPS.

Although the division into XPS and UPS is conceptually artificial it is, nevertheless, a practically useful one because of the different experimental techniques used.

Photoelectron spectroscopy is a subject in which acronyms abound and it would be confusing to mention all of them but there is one, ESCA, which is encountered frequently. This stands for 'electron spectroscopy for chemical analysis'. Here, electron spectroscopy, as opposed to electronic spectroscopy, refers to the various branches of spectroscopy which involve the ejection of an electron from an atom or molecule. However, because ESCA was an acronym introduced by workers in the field of XPS, it is most often used to refer to XPS rather than to electron spectroscopy in general and we shall not use it subsequently.

Figure 7.2 illustrates the processes involved in five types of electron spectroscopy, of which we shall be concerned in detail with only two, UPS and XPS.

UPS and XPS [Figure 7.2 (a) and (b)] involve the ejection of an electron following interaction of an atom or molecule M with a photon of energy $h\nu$. In both cases a singly charged ion, M^+, is produced:

$$M + h\nu \rightarrow M^+ + e \tag{7.3}$$

Auger electron spectroscopy (AES) is illustrated in Figure 7.2(c) and involves more than one step. The initial step is, as in XPS, the ejection by a photon of a core electron. Following this, a valence electron falls down to fill the core vacancy, releasing sufficient energy to eject a valence electron—a so-called Auger electron. A doubly charged ion in an excited state results from this process:

$$\begin{aligned} M + h\nu \rightarrow & (M^{2+})^* + e \text{ (photoelectron)} \\ & + e \text{ (Auger)} \end{aligned} \tag{7.4}$$

Electron impact and Penning ionization spectroscopy are illustrated in Figure 7.2(d) and (e), respectively. In an electron impact process an electron imparts translational energy

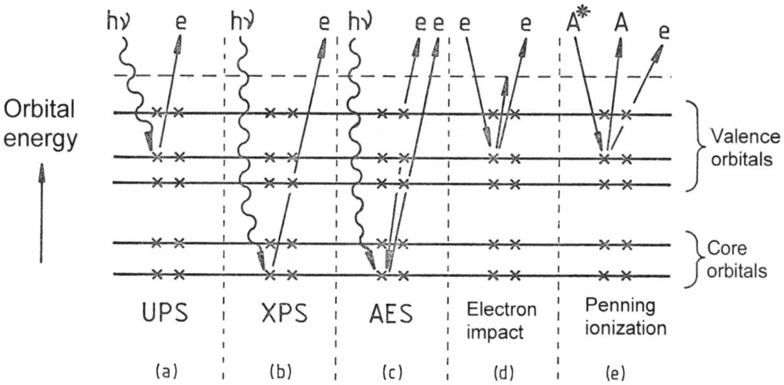

Figure 7.2 Processes occurring in (a) ultraviolet photoelectron spectroscopy (UPS), (b) X-ray photoelectron spectroscopy (XPS), (c) Auger electron spectroscopy (AES), (d) electron impact spectroscopy and (e) Penning ionization spectroscopy

to M putting it into an excited state and the energy of excitation is measured by the loss of energy of the electron:

$$M + e \rightarrow M^* + e \quad (7.5)$$

In Penning ionization an excited atom A* ionizes M on collision:

$$A^* + M \rightarrow A + M^+ + e \quad (7.6)$$

The development of gas-phase UPS followed the early work of Al-Joboury and Turner[163] and Vilesov, Kurbatov and Terenin[164] in 1961–62. XPS was, at first, applied mainly to solids and the early development is described in the book by Siegbahn *et al.* referred to in the Bibliography at the end of the chapter.

7.2 EXPERIMENTAL METHODS

Figure 7.3 illustrates in a symbolic way the important components of an ultraviolet or X-ray photoelectron spectrometer. The sample in the target chamber is bombarded with photons and the photoelectrons emerge from the exit slit of the chamber into the electron energy analyser. This separates the electrons according to their kinetic energy, rather in the way that ions are separated in a mass spectrometer.

The electrons pass through the exit slit of the analyser on to an electron detector and the spectrum recorded is the number of electrons per unit time (often counts s^{-1}) as a function of electron energy.

7.2.1 UPS spectrometers

Since the lowest ionization energy of an atom or molecule is typically of the order of 10 eV† the source of intense, monochromatic radiation for UPS must have at least this energy and, preferably, up to about 50 eV. Such sources are not very numerous but the most commonly used type is produced by a discharge in He, Ne, Ar or H_2 gas, resulting in emission of far ultraviolet radiation from an atom or ion. The most often used is the helium discharge which produces, predominantly, 21.22 eV radiation due to the $2^1P_1(1s^12p^1)$–$1^1S_0(1s^2)$ transition at 58.4 nm (see Figure 6.17); this is referred to as He$\mathrm{I}\alpha$, or simply HeI, radiation.

Helium discharge lamps are of three general types: (a) a direct current, high voltage capillary discharge, (b) a microwave discharge and (c) a high current arc discharge using a heated metal filament, such as nickel, as a source of electrons. Figure 7.4 shows a simple direct current, high voltage discharge lamp made out of glass with a quartz capillary and a cylindrical earthed

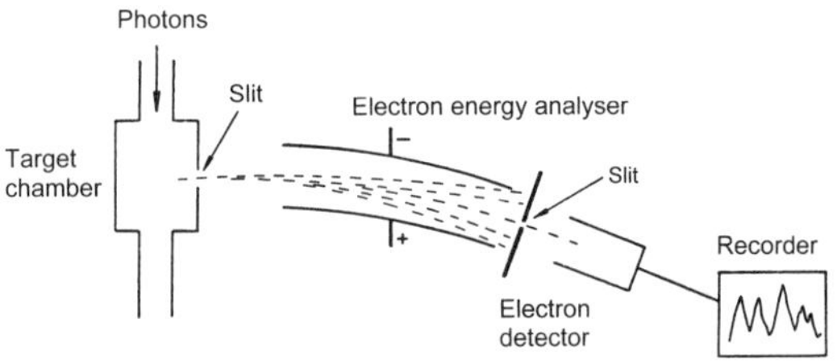

Figure 7.3 The principal components of a photoelectron spectrometer

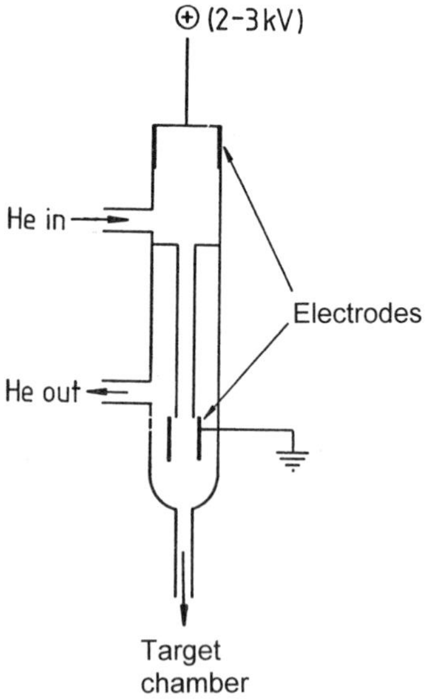

Figure 7.4 A helium discharge lamp photon source

† 1 eV≙96.4853 kJ mol^{-1}≙8065.54 cm^{-1}.

electrode through which the radiation passes to the target chamber and a second electrode sealed on to the top with a metal–glass seal. The helium pressure is about 1 Torr. Very thin (*ca* 100 nm) metal windows can be used at this wavelength but, more often, differential pumping is used to exclude helium from the rest of the spectrometer.

The width of the helium 58.4 nm line is sufficiently small that the resolution of the spectrometer is limited by other factors.

It is possible to change the conditions in the helium discharge lamp so that the helium is predominantly ionized to He^+ (HeII). The radiation is due mainly to the $n=2$–$n=1$ transition of HeII (analogous to the first member of the Lyman series of the hydrogen atom in Figure 1.5) at 30.4 nm with an energy of 40.81 eV. The conditions for producing this radiation include a higher current density and a lower gas pressure than for HeI. Thin aluminium foil can be used to filter any HeI from the HeII radiation.

When neon is used in a discharge lamp the radiation produced is predominantly in *two* lines, at 74.4 and 73.6 nm with energies of 16.67 and 16.85 eV, respectively. This makes the neon source less convenient than the more truly monochromatic HeI source. The radiation is known as NeI and the two components involve excited states of Ne which are better described by *jj* than by Russell–Saunders coupling [Section 6.1.2.2(ii)]. Both excited states arise from the configuration $1s^2 2s^2 2p^5 3s^1$. For the $2p$ vacancy $l_1 = 1$, $s_1 = \frac{1}{2}$ and $j_1 = \frac{3}{2}, \frac{1}{2}$ and, for the $3s$ electron, $j_2 = \frac{1}{2}$. There are four states arising, two with $j_1 = \frac{3}{2}$, $j_2 = \frac{1}{2}$, normally labelled $(\frac{3}{2}, \frac{1}{2})$ and having $J = 2$ or 1 [see equation (6.38)]. The other two are $(\frac{1}{2}, \frac{1}{2})$ states with $J = 1$ or 0. The 74.4 nm line has the lower intensity and is due to the transition $J = 1(\frac{1}{2}, \frac{1}{2})$–$^1S_0$ while the 73.5 nm line is due to the transition $J = 1(\frac{3}{2}, \frac{1}{2})$–$^1S_0$. The intensity ratio of the lines is about 1:10 but is pressure dependent.

Other wavelengths produced from discharge lamps containing neon, argon or hydrogen suffer from either being of low energy (H gives $n=2$–$n=1$ radiation predominantly with an energy of 10.20 eV) or consisting of closely spaced multiplets (Ne^+, Ar and Ar^+).

Another important source of ionizing radiation is the storage ring. This provides electromagnetic radiation which extends continuously from the millimetre wave to the X-ray region.

Figure 7.5 shows, as an example, the design of the storage ring at Daresbury (UK).

Electrons are generated in a small linear accelerator. Pulses of these electrons are injected tangentially into a booster synchroton. Here they are accelerated by 500 MHz radiofrequency radiation, at the same time forming bunches with a time interval between bunches of 2.0 ns, corresponding to the frequency of the accelerating radiation. The electrons are restricted to a circular orbit by dipole bending magnets.

When the electrons have achieved an energy of 600 MeV they are taken off tangentially from the synchrotron and injected tangentially into the storage ring where they are accelerated further to an energy of 2 GeV. Subsequently only a small amount of power is necessary to

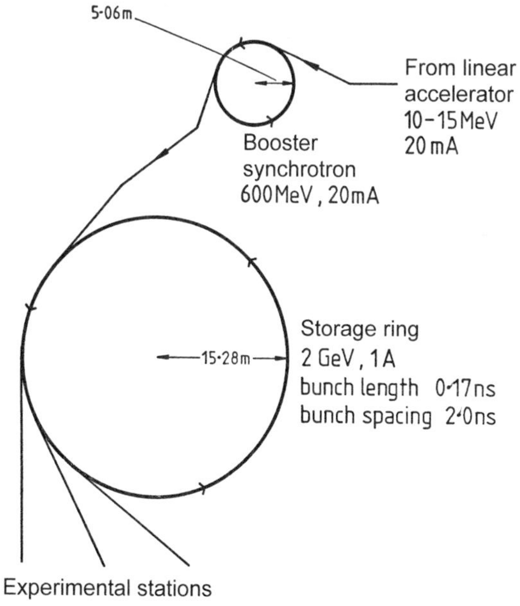

Figure 7.5 The Daresbury (UK) storage ring

keep the electrons circulating. Again, bending magnets restrict the electrons to a circular orbit.

The orbiting electrons lose energy continuously in the form of electromagnetic radiation. The energy, ΔE, radiated per revolution is related to the electron energy E by

$$\Delta E = 88.5\, E^4/R \qquad (7.7)$$

where R is the radius of the orbit and the constant has the value 88.5 provided that the units of ΔE, E and R are keV, GeV and m, respectively.

In fact, the distribution of power radiated also varies with E. As E increases the maximum power moves to higher energy; for example, it is at 10^3 eV for $E = 2$ GeV and at 10^4 eV for $E = 4$ GeV.

Since there is a continuous net loss of energy from the storage ring it can be used only for a limited time before the electrons have to be replenished.

The radiation is predominantly plane polarized and emerges tangentially in the plane of the orbit in pulses of length 0.17 ns and spacing 2.0 ns. Several radiation ports are arranged round the ring to serve experimental stations.

Synchrotrons themselves can also be used as sources of radiation but are much less stable and have been superseded by storage rings.

Whether the radiation is from a synchrotron or a storage ring the problem of designing a monochromator to produce a beam of small energy range has proved a difficult one. At low energy a grating is used with near normal incidence angles. At energies greater than about 40 eV gratings are also used but with grazing (very low angle) incidence. Above about 620 eV a crystal monochromator is used.

In UPS we shall be considering only gaseous samples and, if a sample is a stable gas, it can simply be leaked into the target chamber. Differential pumping must be employed to keep the pressure in the part of the spectrometer containing the analyser and detector down to about 10^{-5} Torr while allowing the sample pressure to be about 10^{-2} Torr.

The types of analysers used fall into two main categories, retarding grid and deflection analysers. In the former type a retarding grid is placed between the source of photoelectrons and an electron detecting plate. A retarding potential is applied to the grid and, when it corresponds to the energy of some of the emitted photoelectrons, there is a decrease in the electron current detected. This current shows a series of steps when plotted against retarding potential, each step corresponding to an ionization energy of the sample. A simple differentiating circuit converts the steps into peaks in the photoelectron spectrum.

The first UPS spectrometer, constructed by Al-Joboury and Turner,[163] used a cylindrical retarding grid and electron collector similar to that in Figure 7.6(a) but without the slotted grid. Price[165] made the modification of installing the slotted grid in order to restrict the photoelectrons falling on the collector to those emitted perpendicular to the photon beam. In the absence of a slotted grid the peaks in a spectrum are distorted on the side corresponding to lower photoelectron energy.

Frost, McDowell and Vroom[166] developed the spherical grid analyser shown in Figure 7.6(b). The advantage of spherical grids is that photoelectrons ejected in *all* directions are detected.

Figure 7.6(c) illustrates an analyser employing plane retarding grids. The use of plane grids is made possible by making the photoelectron beam parallel using an electron lens.

Deflection analysers make use of the property of electrons that those with different energies follow different paths when they are subjected to a magnetic or electric field. In photoelectron spectroscopy electric field analysers have been used predominantly and it is those which will be described here.

The simplest design is that of the parallel plate analyser in Figure 7.7(a). The photoelectrons enter the space between the two plates at an angle of less than 90° through an entrance slit. Varying the electrostatic field between the plates allows electrons of different energies to

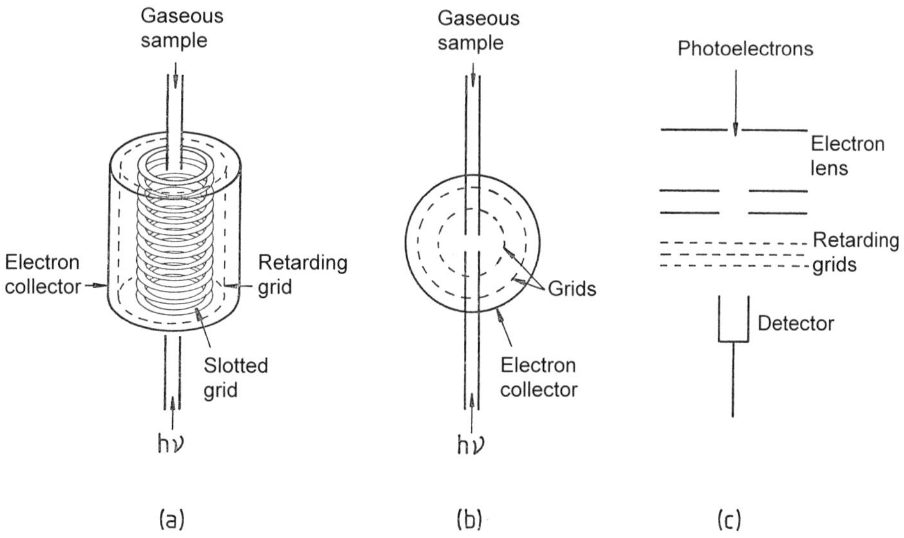

Figure 7.6 (a) Slotted grid, (b) spherical grid and (c) plane grid analysers

reach the exit slit and detector. Entry angles of 30° or 45° are used. The main disadvantage is that only a small proportion of the electrons enter the analyser.

More commonly used is the 127° cylindrical analyser developed by Turner and shown in Figure 7.7(b). The plates across which the field is applied are sections of concentric cylinders. The electrons enter through a slit and will emerge from the exit slit only if the field across the plates is such that they follow the path shown.

Hemispherical, like spherical, analysers have the advantage of collecting electrons over a greater angle. The hemispherical analyser is illustrated in Figure 7.7(c). As in the case of the 127° analyser, care has to be taken with the entrance and exit slits which are not in an equipotential plane and so may disturb the field. With a hemispherical analyser a lens system may be used to produce an image of the slit on the target chamber at the entrance to the analyser and a second lens system to focus the emerging beam on to the detector.

Analysers consisting of two concentric plates which are parts of hemispheres, so-called spherical sector plates, are also used.

For two plates of radii R_1 and $R_2 (R_2 > R_1)$ in any type of spherical, or spherical sector, analyser electrons of kinetic energy E will reach the detector when the voltage V between the plates is given by

$$V = \frac{E}{e}\left(\frac{R_2}{R_1} - \frac{R_1}{R_2}\right) \qquad (7.8)$$

The detector may be a simple electrometer when used with a cylindrical or spherical retarding grid analyser where the electrons are collected over a large surface. For use with other types of analysers an electron multiplier detector, which has much higher sensitivity, is necessary.

An electron multiplier consists of a number of dynodes each of which produces more electrons than it receives. For a measurable current to be produced, about 10–20 dynodes are required.

A multichannel electron multiplier in the focal plane of the analyser can be used to collect simultaneously electrons with a range of energies.

The resolving power R of the spectrometer is given by

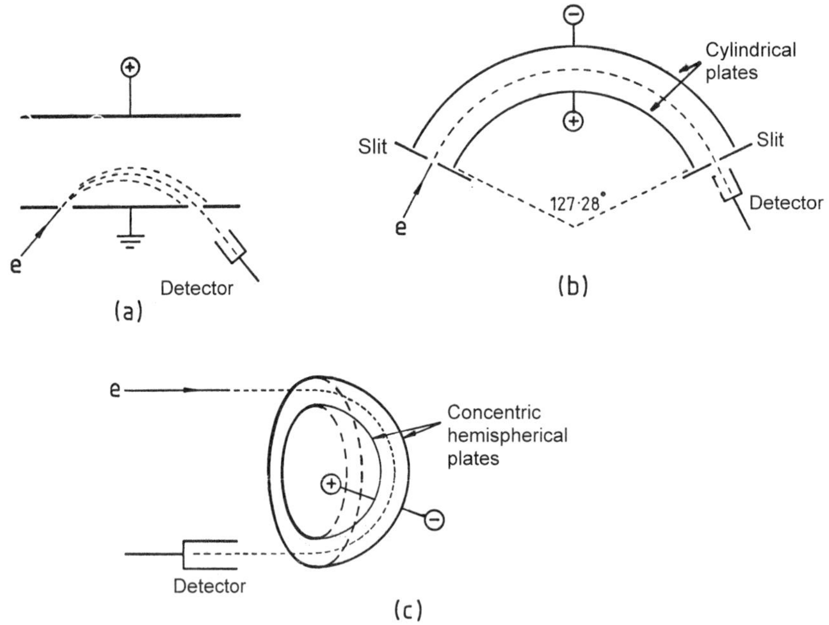

Figure 7.7 (a) Parallel plate, (b) 127° cylindrical and (c) hemispherical analysers

$$R = E/\mathrm{d}E = V/\mathrm{d}V \qquad (7.9)$$

analogous to equation (3.4). E is the kinetic energy of the photoelectrons and $\mathrm{d}E$, the resolution, is the energy difference which can just be detected. Alternatively, V is the voltage across the plates of the analyser and $\mathrm{d}V$ is the change in the voltage for which a change in electron current can just be detected.

In a photoelectron spectrometer the resolution obtainable depends very much on factors such as the efficiency of shielding from stray (including the earth's) magnetic fields by Mumetal or Helmholtz coils, the cleanliness of the analyser surfaces and the parameters used for construction. Resolution decreases also when the electron energy is below about 5 eV. The highest resolution which has been obtained is 4 meV, or 32 cm^{-1}, which means that the technique is a low resolution one compared with electronic spectroscopy.

The factor limiting the resolution in UPS spectra is the inability to measure the kinetic energy of the photoelectrons with sufficient accuracy. The problem can be solved, and much higher resolution spectra obtained, by producing the photoelectrons with zero (or near zero) kinetic energy, which is the basis of zero kinetic energy photoelectron (ZEKE-PE) spectroscopy. This requires the use of lasers as the source of ultraviolet radiation and will be discussed in Section 8.2.19.

7.2.2 XPS spectrometers

The most commonly used sources of X-ray radiation are Mg $K\alpha$ and Al $K\alpha$, where $K\alpha$ indicates that an electron has been ejected from the $K(n = 1)$ shell and the radiation is due to an electron falling back from the next highest energy shell, here the L $(n = 2)$ shell, to fill the vacancy. The Mg $K\alpha$ radiation consists primarily of a closely spaced doublet, $K\alpha_1$ and $K\alpha_2$, with energies of 1253.7 and 1253.4 eV and relative intensities of 67 and 33, respectively. The Al $K\alpha$ radiation consists of a doublet also;

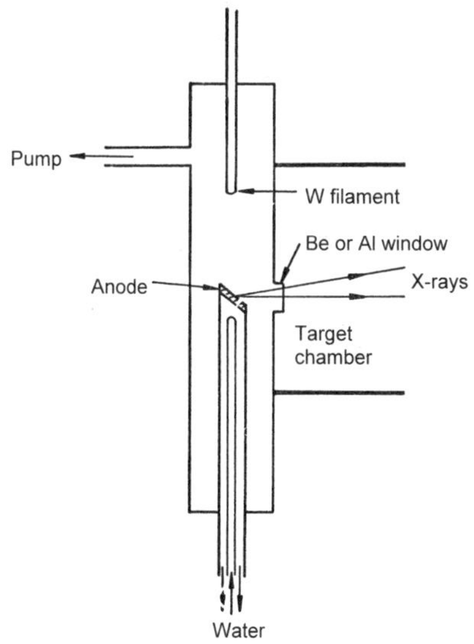

Figure 7.8 X-ray source

$K\alpha_1$ and $K\alpha_2$ have energies of 1486.7 and 1486.3 eV and relative intensities of 67 and 33, respectively. In both sources the reason for the doublet is that, on going from a $1s^1 2s^2 2p^6$ . . . to a $1s^2 2s^2 2p^5$ . . . configuration,† there are two transitions involved, ${}^2S_{\frac{1}{2}}-{}^2P_{\frac{1}{2}}$ and ${}^2S_{\frac{1}{2}}-{}^2P_{\frac{3}{2}}$. Since M_J can take $(2J+1)$ values (see Section 6.1.8.1), the statistical weight of the ${}^2P_{\frac{3}{2}}$ state is four while that of ${}^2P_{\frac{1}{2}}$ is two. For this reason the intensity of the ${}^2S_{\frac{1}{2}}-{}^2P_{\frac{3}{2}}$ transition is twice that of the ${}^2S_{\frac{1}{2}}-{}^2P_{\frac{1}{2}}$ transition.

Figure 7.8 shows how an X-ray source with an aluminium or magnesium water-cooled anode may be constructed and attached to the target chamber. The electrons, falling on the anode, are provided by a heated tungsten filament. A thin aluminium or beryllium window is used to isolate the source from the chamber. Instead of water cooling, rotation of the anode may be used to dissipate the heat evolved.

The background X-ray continuum, or *bremsstrahlung*, due to the rapid deceleration of the electrons on penetrating the anode surface, is not sufficiently intense to be a serious problem in XPS. More serious are five $K\alpha$ satellite lines in both Mg $K\alpha$ and Al $K\alpha$ radiation. The $K\alpha'$, $K\alpha_3$ and $K\alpha_4$ satellites are X-ray emission following the simultaneous creation of a vacancy in both the K and L shells to form a so-called double hole state. The $K\alpha_5$ and $K\alpha_6$ satellites involve initial triple hole states with one K- and two L-shell vacancies. There is a further $K\alpha$ satellite and, also, a $K\beta$ line involving an initial vacancy in the K shell which is filled by an electron from the M shell.

† In deriving states from configurations of Al, coupling between the $3p$ electron and the core electrons is ignored, i.e. the states are core states.

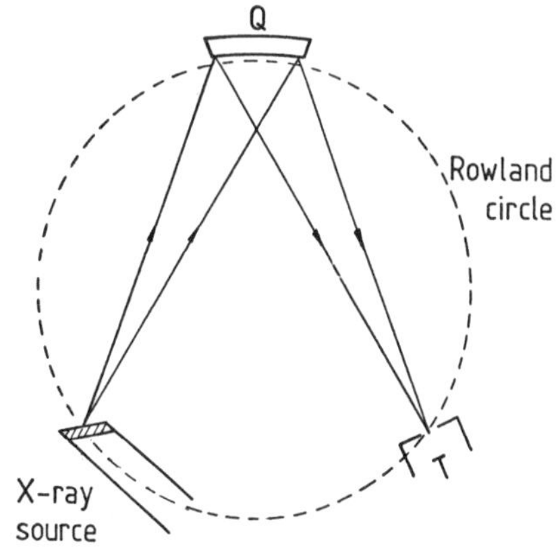

Figure 7.9 An X-ray monochromator using a bent quartz crystal Q

All of these satellites, and also the *bremsstrahlung*, can be removed with a monochromator.

An X-ray monochromator is illustrated in Figure 7.9. A quartz crystal Q is bent to form a concave X-ray diffraction grating. The X-ray source and the target chamber T are then placed on the Rowland circle (see Figure 3.34) whose diameter is equal to the radius of curvature of the grating. The target chamber receives X-rays of ~~only a~~ limited wavelength range.

A monochromator is useful not only for removing unwanted lines from the X-ray source but also for narrowing the otherwise broad lines. For example, each of the Al $K\alpha$ and Mg $K\alpha$ doublets is unresolved and about 1 eV wide at half intensity. A monochromator can reduce the width to about 0.2 eV. This is very important since the resolution in an XPS spectrum is limited by the width of the ionizing radiation.

X-rays of lower energy tend to have a smaller line width. Two such sources that have been used are Y $M\zeta$ and Zr $M\zeta$ produced with yttrium and zirconium anodes. The Y $M\zeta$ radiation has an energy of 132.3 eV and a line width of 0.47 eV whereas the Zr $M\zeta$ radiation has an energy of 151.4 eV and a line width of 0.77 eV. However, the energies are generally too low except for specific investigations and there is an additional problem with yttrium in that it oxidizes easily.

Synchrotron radiation is also a useful source in the X-ray region (see Section 7.2.1).

The analyser of an XPS spectrometer is, in principle, similar to that for a UPS spectrometer. In practice, the spherical sector type of analyser (see Section 7.2.1) has been used commonly and, indeed, spectrometers with such an analyser can be adapted to take both a helium discharge lamp and an X-ray source so that UPS and XPS spectra can be run on the same instrument.

The electron detector is an electron multiplier, single- or multichannel, as in UPS.

7.3 IONIZATION PROCESSES IN PHOTOELECTRON SPECTRA

The photoionization process with which we shall be concerned in both UPS and XPS is that in equation (7.3). The species produced is almost always the singly charged M^+ and no additional process involving excitation of electrons in M^+ has taken place.

The selection rules for such a photoionization process are trivial—*all* ionizations are allowed.

When M is an atom the total change in angular momentum for the process $M + h\nu \rightarrow M^+ + e$ must obey the electric dipole selection rule $\Delta l = \pm 1$ for the angular momentum change, but the electron can take away any amount of momentum. If, for example, the electron removed is from a d orbital of M it carries away one or three quanta of angular momentum depending on whether $\Delta l = -1$ or $+1$, respectively. The wave function of a free electron can be described, in general, by a mixture of $s, p, d, f, \ldots$ wave functions but, in this case, the ejected electron has just p and f character.

In molecules, also, the electric dipole selection rules apply to the photoionization process but, since the symmetry is lower, the MOs themselves are mixtures of $s, p, d, f, \ldots$ AOs and therefore the ejected electron is described by a more complex mixture of $s, p, d, f, \ldots$ character.

There are two processes other than that in equation (7.3) which may take place when a photon encounters an atom or molecule. These are

$$M + h\nu \rightarrow M^{2+} + 2e \qquad (7.10)$$

$$M + h\nu \rightarrow (M^+)^* + e \qquad (7.11)$$

That in equation (7.10) is called photo-double ionization since two electrons are ejected simultaneously by one photon. This process has been observed in inert gases and produces a broad peak in the spectrum but the photon energy required is so high that usually it can be observed only in XPS.

The process in equation (7.11) involves the promotion of an electron in M^+ simultaneously with the ejection of the photoelectron and has been observed, for example, in molecular hydrogen

$$H_2(\sigma_g 1s)^2 + h\nu \rightarrow H_2^+(\sigma_u 2s)^1 + e \qquad (7.12)$$

where the MOs are those in Figure 6.29.

The probabilities for the processes in equations (7.10) and (7.11) are very low. Indeed, in the one-electron AO or MO picture they are forbidden. They gain their intensity only through the effects of electron correlation (see Section 7.4).

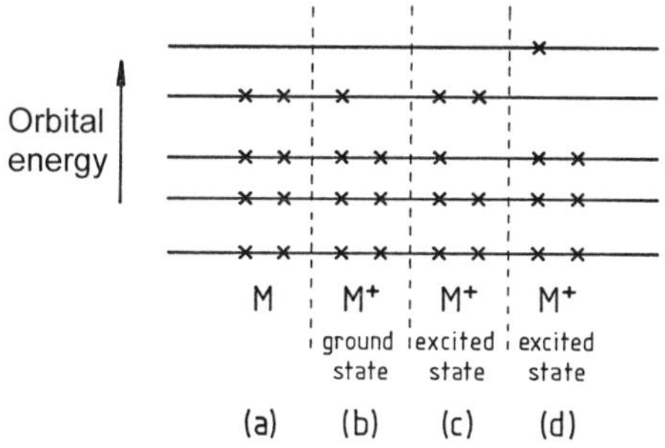

Figure 7.10 Electron orbital configurations in (a) the ground state of M, (b) the ground state of M^+ and (c) and (d) excited states of M^+

Figure 7.10(b) and (c) illustrates two examples of a vacancy created by ejection of a photoelectron from M. In Figure 7.10(b) M^+ is in its ground electronic state whereas in Figure 7.10(c) it is an excited electronic state. This excited state may or may not be the lowest excited state of M^+. It is possible that the excited state in Figure 7.10(d), inaccessible by a simple photoelectron ejection, may be the lowest. In any event, Figure 7.10 serves to illustrate the point that the first, second, third, . . . peaks (or band systems) in the photoelectron spectrum do not necessarily correspond to the ground, first excited, second excited, . . . states of M^+.

This discussion leads directly to the problem of what we call the various ionization energies recorded in a photoelectron spectrum. It would seem natural to call these the first, second, third, . . . ionization energies in order of their increasing values. Unfortunately, in atomic spectroscopy, the second ionization energy refers to the energy required for the process

$$M^+ \rightarrow M^{2+} + e \qquad (7.13)$$

and, in general, the nth ionization energy refers to

$$M^{(n-1)+} \rightarrow M^{n+} + e \qquad (7.14)$$

Using this definition, *all* the ionization energies in photoelectron spectroscopy are first ionization energies. To avoid any possible confusion we shall refer to the ionization energies observed in UPS as the lowest, second lowest, etc. In XPS we shall identify an ionization energy by the core atomic orbital from which the electron has been removed, e.g. C 1s on the carbon atom.

In addition to this problem of definition there is the problem of which energy to plot in a photoelectron spectrum. What is measured is the kinetic energy, $\frac{1}{2}m_e v^2$, of the photoelectrons and this is related to the ionization energy I by

$$h\nu = \tfrac{1}{2}m_e v^2 + I \qquad (7.15)$$

Since it is the ionization energy which is of importance in the interpretation of spectra it is this rather than the kinetic energy of the electrons which is usually plotted. In some photoelectron spectra I is also called the binding energy since it is this energy which can be regarded as binding the electron to the rest of the atom or molecule.

The type of experiment discussed so far is not concerned with the decay processes of the ions produced by photoionization. One such process which may occur when the ion is produced in an excited electronic state is fluorescence. This can be monitored at 90° to the incident photon beam in the target area. However, for an experiment of this kind, it is not necessary to use a monochromatic photon beam. Instead, an electron gun is used which produces a much higher ion concentration and, consequently, higher fluorescence intensity. In this way the $\tilde{A}$–$\tilde{X}$ and, in some cases, the $\tilde{B}$–$\tilde{X}$ emission spectra of many positive ions, such as $ClCN^+$, $BrCN^+$, ICN^+, mono- and dihaloacetylene positive ions, and the positive ions of many fluorobenzenes, have been observed.†

† See, for example, Maier, J. P. (1979). 'Decay processes of the lowest excited states of polyatomic radical cations', in *Kinetics of Ion–Molecule Reactions* (Ed. P. Ausloss). Plenum, New York.

7.4 KOOPMANS' THEOREM

From a diagrammatic representation of the photoionization process, like that in Figure 7.2(a) and (b), it would appear that the energy required to eject an electron from an orbital (atomic or molecular) is a direct measure of the orbital energy. This is approximately true and was proposed originally by Koopmans.[167] The theorem can be stated as follows: 'For a closed shell molecule the ionization energy for an electron in a particular orbital is approximately equal to the negative of the orbital energy calculated by an SCF *ab initio* method', or

$$I_i \approx -\varepsilon_i^{\mathrm{SCF}} \tag{7.16}$$

for the *i*th orbital. The negative sign is due to the convention that orbital energies, ε_i, are negative.

At the level of simple valence theory Koopmans' theorem seems to be so self-evident as to be scarcely worth stating. However, with a more accurate orbital treatment, this is no longer so and great interest attaches to why equation (7.16) is only approximately true.

The measured ionization energy, I_i, is the difference in energy between M and M^+ but the approximation is in equating this with an orbital energy. Orbitals are an entirely theoretical concept and their energies can be obtained exactly only by calculation and then, for a many-electron system, only with difficulty. Experimental ionization energies, such as those obtained by photoelectron spectroscopy, are the measurable quantities which most closely correspond to the orbital energies.

There are three important factors which may contribute to the fact that equation (7.16) is only approximately valid:†

1. *Electron reorganization.* When an electron is removed from an orbital, in the process $M \rightarrow M^+$, the orbitals in M^+ are different from those in M. This effect is called reorganization or, alternatively, relaxation or reorientation, and the measured ionization energy is the sum of the orbital energy and the reorganization energy.

   Reorganization is a consequence of there being one fewer electron and a net positive charge in the ion. The effect is often small but it can be large as in the extreme example of the ionization of H_2 to H_2^+.
2. *Electron correlation.* The effect of electron correlation, and the associated correlation energy, is due to the fact that electron motions are not independent but are correlated. This is caused by local distortion of an orbital by an adjacent electron. Neglect of the correlation energy is, like the neglect of reorganization energy, a fundamental defect of the Hartree–Fock SCF method (Section 6.1.1) and it has to be calculated separately by modifying orbitals in an average way to take account of the distortion.

   Koopmans' theorem assumes that the correlation energy is the same in M and in M^+.
3. *Relativistic effects.* The neglect of the effects of relativity on orbital energies is another defect of the Hartree–Fock method. The effect on orbital energies is particularly important for core electrons with large kinetic energies.

   Koopmans' theorem assumes that the relativistic energy is the same in M and in M^+.

In open shell systems, such as O_2 and NO, there are further difficulties in using the Hartree–Fock method to estimate orbital energies and Koopmans' theorem does not apply.

In most cases of closed shell molecules, Koopmans' theorem is a reasonable approximation but there are notable exceptions, such as N_2 (see Section 7.5.1.2).

† For a detailed discussion, see Richards, W. G. (1969). *Int. J. Mass Spectrom. Ion Phys.*, **2**, 419.

7.5 PHOTOELECTRON SPECTRA AND THEIR INTERPRETATION

The simplest, and perhaps the most important, information given by photoelectron spectra consists of values for ionization energies involving valence electrons (UPS) or core electrons (XPS).

Data on valence electron ionization energies of polyatomic molecules were, prior to the advent of UPS, very scant. At best, the lowest ionization energy might have been known to two- or three-figure accuracy from, for example, the convergence limit of transitions to Rydberg states observed in the far ultraviolet absorption spectrum. For diatomic molecules the situation was more satisfactory because their far ultraviolet absorption spectra are generally much sharper than those of polyatomic molecules, allowing ionization energies to be obtained, in favourable cases, with higher accuracy than in UPS.

Because of the general validity of Koopmans' theorem for closed shell species UPS ionization energies and, for molecules, the associated vibrational structure represent a vivid illustration of the validity of simple-minded MO theory of valence electrons.

Ionization energies for core electrons have been determined for the first time by XPS and lead to a closer knowledge of the extent to which the core orbitals differ from the pure atomic orbitals pictured by simple valence theory.

7.5.1 Ultraviolet Photoelectron Spectroscopy (UPS)

7.5.1.1 UPS of atoms

UPS of atoms is unlikely to produce any information which is not available with higher precision from other sources and, in any case, most atomic materials have such low vapour pressures that their UPS spectra can be recorded only at very high temperatures.

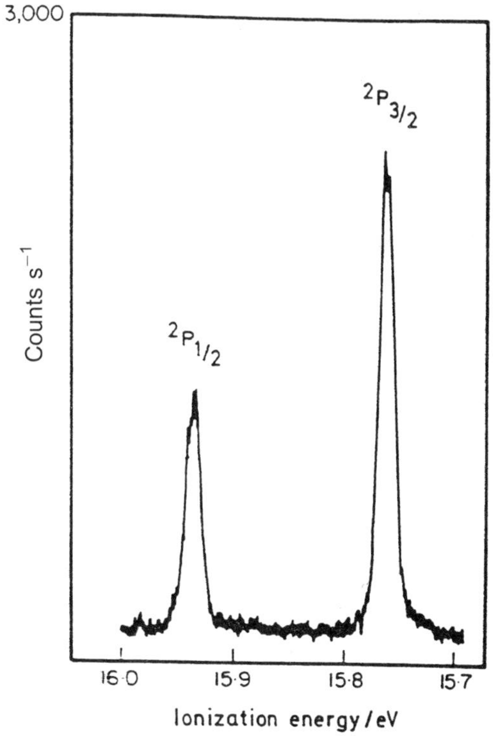

Figure 7.11 The HeI UPS spectrum of the argon atom
[Reproduced, with permission, from Turner, D. W., Baker, C., Baker, A. D. and Brundle, C. R. (1970). *Molecular Photoelectron Spectroscopy*, p. 41. Wiley, London]

Mercury and the inert gases are exceptions and their spectra illustrate aspects of their electronic structure which have been considered in Section 6.1.

The lowest ionization energies of He and Ne are too low (see Table 6.1) to be recorded with HeI (21.22 eV) radiation.

Figure 7.11 shows the HeI spectrum of argon. There are two peaks, both of them resulting from the removal of an electron from the $3p$ orbital:

$$\mathrm{Ar}\,(KL3s^2 3p^6) \rightarrow \mathrm{Ar}^+(KL3s^2 3p^5) \qquad (7.17)$$

The state arising from the ground configuration of Ar is $^1S_0$ but the ground configuration of Ar^+ gives two states, $^2P_{\frac{1}{2}}$ and $^2P_{\frac{3}{2}}$, since $L = 1$ and $S = \frac{1}{2}$, giving $J = \frac{1}{2}, \frac{3}{2}$ for a ^2P term. The two states are split by spin–orbit coupling. Since the unfilled $3p$ subshell is more than half filled the multiplet is inverted, the $^2P_{\frac{3}{2}}$ being below the $^2P_{\frac{1}{2}}$ state [see Section 6.1.2.2(i)]. This is clear from the spectrum in Figure 7.11, in which the peaks are identified by their intensities. As M_J can take $(2J+1)$ values [see equation (1.74)] the $^2P_{\frac{3}{2}}$ state is fourfold degenerate and the $^2P_{\frac{1}{2}}$ state twofold degenerate. Because of this the relative peak areas should be 2:1.

The fact that the ratio is slightly less is due to a difference of photoionization cross-section, σ. This is the effective cross-section of an atom, or molecule, within which collision with the photon produces ionization and is smaller for Ar^+ produced in the $^2P_{\frac{3}{2}}$ state than for the $^2P_{\frac{1}{2}}$ state when a HeI source is used.

Since removal of a $3p$ electron from Ar results in two ionization energies corresponding to the two possible states of Ar^+, this causes some difficulty with regard to the application of Koopmans' theorem [equation (7.16)]. Hartree-Fock SCF calculations do not take into account spin–orbit interaction so the effect of this must be estimated independently and the resulting single ionization energy, obtained after allowing for the interaction, related to the SCF orbital energy. Kr and Xe show similar HeI spectra to that of Ar but with the ionization energy decreasing with increasing atomic number. In addition, the splitting of the $^2P_{\frac{3}{2}}$ and $^2P_{\frac{1}{2}}$ states due to spin–orbit coupling increases with increasing atomic number. Both these effects are illustrated in Table 7.1.

Hg has the ground configuration $KLMN5s^2 5p^6 5d^{10} 6s^2$. Removal of a $6s$ electron gives a $^2S_{\frac{1}{2}}$ state whereas removal of a $5d$ electron gives $^2D_{\frac{5}{2}}$ and $^2D_{\frac{3}{2}}$ (inverted) states. Both ionizations are shown in the HeI spectrum in Figure 7.12.

Table 7.1 Lowest ionization energies of some noble gases

	I/eV	
	$^2P_{\frac{3}{2}}$	$^2P_{\frac{1}{2}}$
Ar	15.759	15.937
Kr	14.000	14.665
Xe	12.130	13.436

The splitting of the $^2D_{\frac{5}{2}}$ and $^2D_{\frac{3}{2}}$ states of Hg^+ by about 2 eV is large owing to the high nuclear charge resulting in large spin–orbit coupling in the ^2D term. The degeneracies of the $^2D_{\frac{5}{2}}$ and $^2D_{\frac{3}{2}}$ states are 6 and 4, respectively, owing to the values that M_J can take. In addition, there is a fivefold degeneracy of a d-orbital compared with the non-degeneracy of an s-orbital. All these degeneracy factors are reflected in the intensities in Figure 7.12.

The problem of obtaining UPS spectra of atoms (and molecules) with very low vapour pressures at room temperature may be overcome by using a high temperature furnace. Figure 7.13 shows, for example, the HeI spectrum of atomic uranium recorded with the uranium sample heated to about 2200 K.

Uranium has the ground configuration (see Table 6.1)

$$KLMN5s^2 5p^6 5d^{10} 5f^3 6s^2 6p^6 6d^1 7s^2$$

resulting in a $^5L_6$ ground state but with a $^5K_5$ excited state, arising from the same configuration, which is only 0.077 eV above the ground state.

The lines labelled B, C and D in Figure 7.12 have all been assigned as involving ionization from the ground state of uranium. Line D is due to U^+ in which a $5f$ electron has been removed resulting in a $^4I_{\frac{9}{2}}$ state, using the Russell–Saunders coupling description [see Section 6.1.2.2(i)]. Lines C and B are due to removal of a $7s$ electron creating U^+ in a $^6L_{\frac{13}{2}}$ and a $^6L_{\frac{11}{2}}$ state, respectively, although line B also contains a component due to U^+ in a $^4I_{\frac{9}{2}}$ state resulting from removal of a $6d$ electron.

Line A is assigned as arising from ionization of uranium in the low-lying $^5K_5$ excited state which is appreciably populated at the temperature of 2200 K. The state of U^+ involved in line A is the $^4I_{\frac{9}{2}}$ state resulting from the removal of a $6d$ electron.

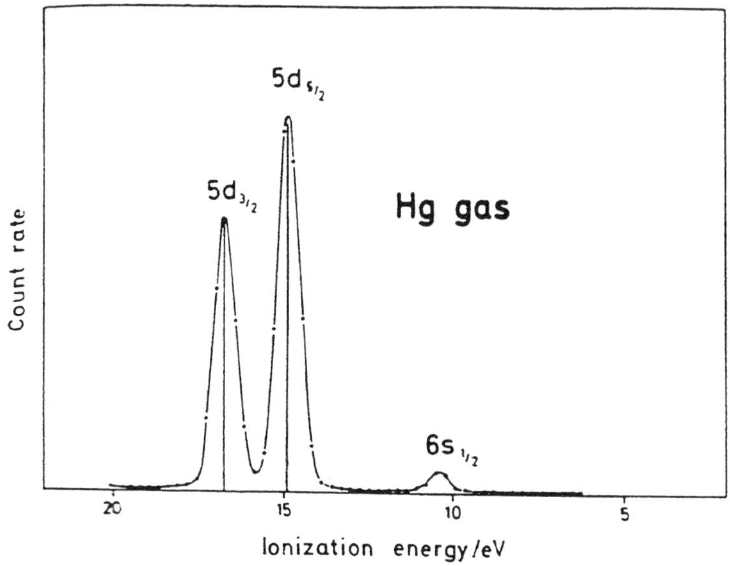

Figure 7.12 The HeI UPS spectrum of the mercury atom
[Reproduced, with permission, from Svensson, S., Mårtensson, N., Basilier, E., Malmqvist, P. Å., Gelius, U. and Siegbahn, K. (1976). *J. Electron Spectrosc.*, **9**, 51]

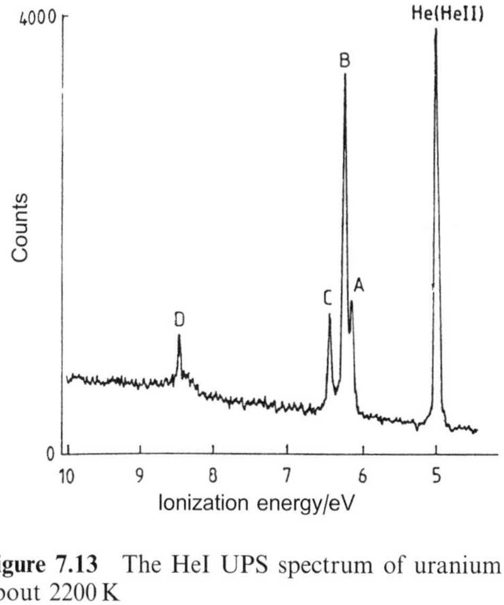

Figure 7.13 The HeI UPS spectrum of uranium at about 2200 K
[Reproduced, with permission, from Allen, G. C., Baerends, E. J., Vernooijs, P., Dyke, J. M., Ellis, A. M., Feher, M. and Morris, A. (1988). *J. Chem. Phys.*, **89**, 5363]

7.5.1.2 UPS of diatomic molecules

In considering a few typical UPS spectra of diatomic molecules we shall be using the MO theory developed in Sections 6.2.1.1 and 6.2.1.2.

7.5.1.2(i) Hydrogen

This is the simplest closed shell diatomic molecule and its HeI UPS spectrum is shown in Figure 7.14.

The MO configuration of H_2 is $(\sigma_g 1s)^2$, as shown in Figure 6.29. The spectrum in Figure 7.14 is caused by removing an electron from this MO to give the $X^2\Sigma_g^+$ ground state of H_2^+.

An obvious feature of the spectrum is that it does not consist of a single peak corresponding to the ionization but a vibronic band system, very similar to the system we might observe accompanying an electronic transition in a diatomic molecule (Section 6.2.4). The Franck–Condon principle (Section 6.2.4.3)

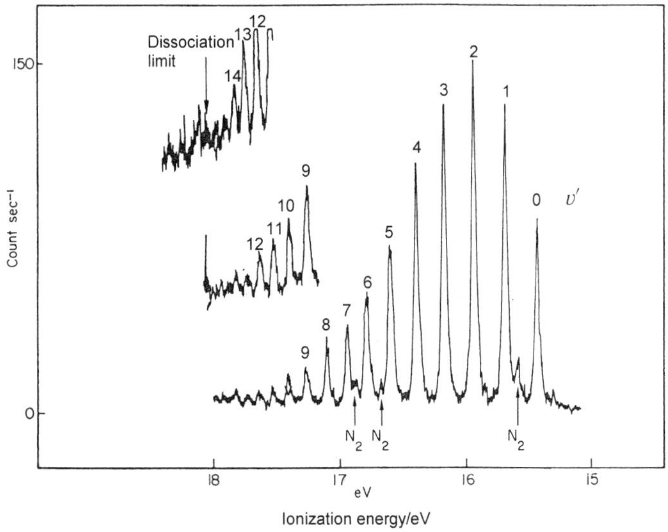

Figure 7.14 The HeI UPS spectrum of the hydrogen molecule
[Reproduced, with permission, from Turner, D. W., Baker, C., Baker, A. D. and Brundle, C. R. (1970). *Molecular Photoelectron Spectroscopy*, p. 44, Wiley, London]

applies to an ionization process just as it does to an electronic transition so that the most probable ionization is to a vibronic state of the ion in which the nuclear positions and velocities are the same as in the molecule. This is illustrated for $H_2^+(X^2\Sigma_g^+) \leftarrow H_2(X^1\Sigma_g^+)$ in Figure 7.15, which is similar to Figure 6.37 for an electronic transition. Removal of the electron from the $\sigma_g 1s$ bonding orbital results in an increase in the equilibrium internuclear distance, r_e, in the ion and the most probable transition is to $v' = 2$. Figure 7.14 shows that there is a long $v'' = 0$ vibrational progression which just reaches the dissociation limit for H_2^+ at $v' \approx 18$.

From the separations of the progression members the vibrational constants ω_e, $\omega_e x_e$, $\omega_e y_e$, . . . (equation 6.121) for H_2^+ can be obtained.

Comparison of values of ω_e of 2322 cm^{-1} in the $X^2\Sigma_g^+$ state of H_2^+ with 3115 cm^{-1} in the $X^1\Sigma_g^+$ state of H_2 shows that the vibrational force constant [see equation (5.2)] is much smaller in the ground state of H_2^+ than in H_2. This, in turn, implies a weaker bond in $H_2{}^+$ consistent with the removal of an electron from a bonding MO.

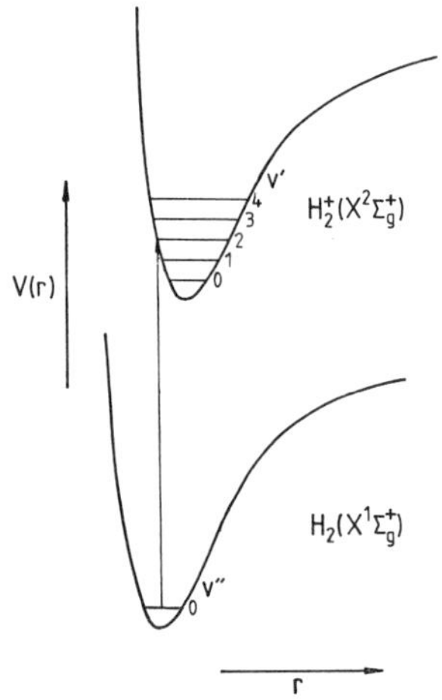

Figure 7.15 The Franck–Condon principle applied to the $H_2^+(X^2\Sigma_g^+) \leftarrow H_2(X^1\Sigma_g^+)$ ionization process

In addition the quantitative intensity distribution along the progression can be used to obtain the corresponding vibrational overlap integrals of equation (6.127) and the change in internuclear distance, $\Delta r_e = r_e(H_2^+) - r_e(H_2)$.

In many cases of diatomic molecular positive ions the vibrational constants and bond lengths can be obtained with much greater accuracy, because of higher resolution, from the electronic emission spectra of the ions produced, for example, in an electrical discharge. H_2^+ is an exception. No such discrete spectrum has been observed because all the low-lying excited states from which emission might occur are only very weakly bound.

Figures 7.14 and 7.15 illustrate the point that there are two ways in which we can define the ionization energy. One is the adiabatic ionization energy, which is defined as the energy of the $\upsilon' = 0 - \upsilon'' = 0$ ionization. This quantity can be subject to appreciable uncertainty if the progression is so long that its first members are observed only weakly, or not at all. The second is the vertical ionization energy, which is defined as the ionization energy corresponding to the intensity maximum in the intensity envelope of the $\upsilon'' = 0$ progression. This may lie *between* vibrational bands. Alternatively, the vertical ionization energy has been defined as that corresponding to the transition at the centre of gravity of the band system. This definition is useful only when the band system is not overlapped by another.

It is the vertical ionization energy which is applicable to Koopmans' theorem [equation (7.16)].

H_2^+ is one of the few ions which has a rotational constant B [equations (4.18) and (4.19)] which is sufficiently large for rotational fine structure to be observed in the UPS spectrum. To observe the structure the highest resolution is necessary. This has been achieved by Åsbrink with a NeI source consisting of a doublet with components of energy 16.67 and 16.85 eV and an intensity ratio of about 1:10.

The reason for using NeI rather than HeI radiation is that thermal motion of the target molecules broadens the peaks in the UPS spectrum. The width of the peak is proportional to $(\frac{1}{2}m_e\upsilon^2)^{\frac{1}{2}}$ where $\frac{1}{2}m_e\upsilon^2$ is the kinetic energy of the photoelectron. Whereas the half-intensity line width using HeI radiation is about 20 meV, because of thermal motion, it is only about 4–7 meV using NeI radiation.

Figure 7.16 shows the rotational fine structure of the $\upsilon' = 5$–$\upsilon'' = 0$ and $\upsilon' = 4$–$\upsilon'' = 0$ transitions excited with the 16.85 eV (73.6 nm) and 16.67 eV (74.4 nm) radiation, respectively.

The ground state of H_2^+, $X^2\Sigma_g^+$, is best described by Hund's case (b) (Section 6.2.5.4). The effect of any splitting of the two components of each term value due to the fact that $J = N + \frac{1}{2}$ and $N - \frac{1}{2}$ [equation (6.143)] is far too small to be observed in the UPS spectrum so we can treat the rotational quantum numbers J and N interchangeably.

For the ejection of a photoelectron there is, in general, no restriction on the change in J from the molecule to the ion. In this case of a Σ–Σ ionization the change is restricted by symmetry to ΔJ even. Rotational transition intensities are proportional to photoionization cross-sections for rotational transitions and have been shown[168] to decrease rapidly in the order $\Delta J = 0, \pm 2, \pm 4, \ldots$. The spectrum in Figure 7.16 agrees with this. Analysis of the rotational transitions for $\upsilon' = 5$ gives $B_5 = 2.75 \pm 0.03$ meV (22.2 ± 0.2 cm^{-1}). This agrees with values obtained by other techniques and, in particular, by theoretical calculation which, in the case of this simple one-electron molecule, is capable of higher precision than experiment.

7.5.1.2(ii) Nitrogen

The ground MO configuration of N_2 is given in equation (6.96) and the UPS spectrum, obtained with a HeI source, is shown in Figure 7.17.

According to the order of occupied MOs in equation (6.96), the lowest ionization energy (adiabatic value 15.58 eV) corresponds to the

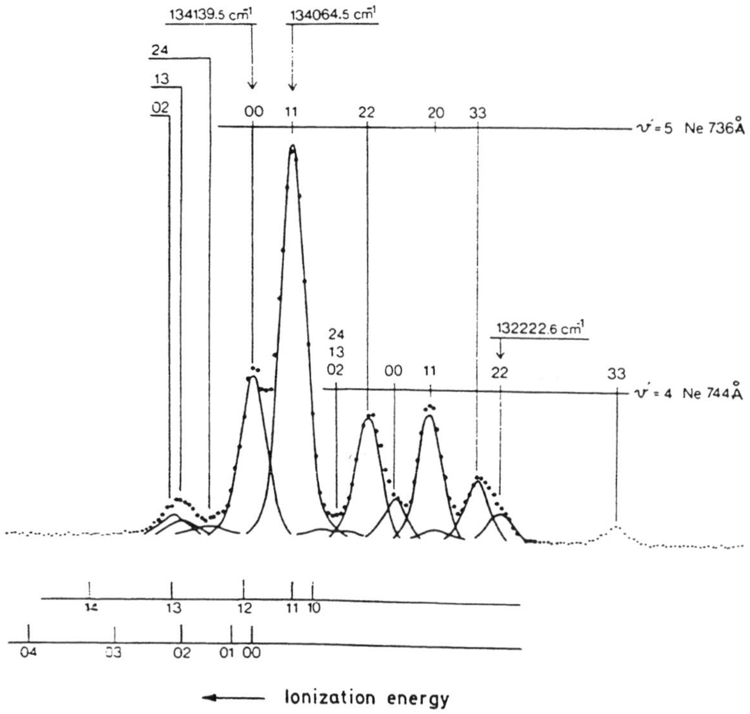

Figure 7.16 Rotational fine structure of the $v' = 5$–$v'' = 0$ and $v' = 4$–$v'' = 0$ transitions of the $H_2^+(X^2\Sigma_g^+) \leftarrow H_2(X^1\Sigma_g^+)$ band system of the UPS spectrum obtained with NeI 73.6 and 74.4 nm radiation, respectively. The lines are labelled with the quantum numbers $J''J'$
[Reproduced, with permission, from Åsbrink, L. (1970). *Chem. Phys. Lett.*, **7**, 549]

removal of an electron from the outermost, $\sigma_g 2p$, MO. The second (16.69 eV) and third (18.76 eV) lowest correspond to removal from $\pi_u 2p$ and $\sigma_u^* 2s$, respectively.

In the UPS of H_2 we saw how the extent of the vibrational progression accompanying the ionization is consistent, through the Franck–Condon principle, with the removal of an electron from a strongly bonding MO. Similarly, removal from a strongly antibonding MO would lead to a shortening of the bond in the ion and a long vibrational progression while removal from a non-bonding MO would result in very little change of bond length in the ion and a very short progression [see Figure 6.38(a)].

Using the Franck–Condon principle in this way we can see that the band system associated with the second lowest ionization energy, showing a long progression, is consistent with the removal of an electron from a bonding $\pi_u 2p$ MO. The progressions associated with both the lowest and third lowest ionization energies are short, indicating that the bonding and antibonding characteristics of the $\sigma_g 2p$ and $\sigma_u^* 2s$ MOs, respectively, are not as strong as simple MO theory would lead us to believe.

These conclusions are consistent with the bond length in N_2^+ in various excited electronic states found from the high resolution emission spectrum. In the $X^2\Sigma_g^+$ state [configuration $\ldots (\sigma_u^* 2s)^2(\pi_u 2p)^4(\sigma_g 2p)^1$], the $A^2\Pi_u$ state

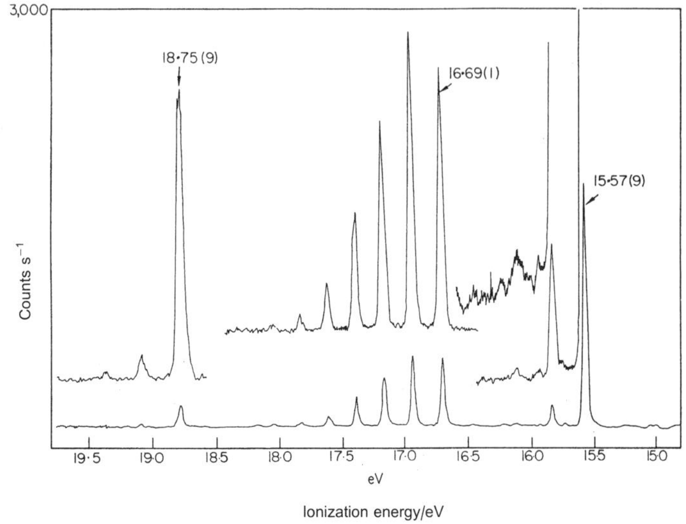

Figure 7.17 The HeI UPS spectrum of the nitrogen molecule
[Reproduced, with permission, from Turner, D. W., Baker, C., Baker, A. D. and Brundle, C. R. (1970). *Molecular Photoelectron Spectroscopy*, p. 46. Wiley, London]

[configuration ... $(\sigma_u^* 2s)^2(\pi_u 2p)^3(\sigma_g 2p)^2$] and the $B^2\Sigma_u^+$ state [configuration ... $(\sigma_u^* 2s)^1(\pi_u 2p)^4(\sigma_g 2p)^2$] r_e is 1.11642, 1.1749 and 1.074 Å, respectively, compared with 1.09769 Å in the ground state of N_2. In the X, A and B states of N_2^+ there is a very small increase, a large increase and a very small decrease, respectively.

In addition the values of ω_e in N_2^+ are consistent with this simple picture of the bonding or antibonding nature of the orbital from which the electron has been removed.

There is a degeneracy factor of two associated with a π orbital compared to the non-degeneracy of a σ orbital, so that it might be expected that the integrated intensity of the second band system would be twice that of each of the other two. The second system clearly is the most intense but the intensities depend also on the nature of the orbitals, particularly their size, nodal characteristics and degree of localization, as well as the experimental sensitivity to electrons of different kinetic energies. The result is that intensity ratios of band systems are only an approximate guide to orbital degeneracies. This is illustrated by the much greater intensity of the first compared with the third band system.

Hartree–Fock SCF calculations[169] for N_2^+ place the $A^2\Pi_u$ state *below* the $X^2\Sigma_g^+$ state. This discrepancy is due to deficiencies in the Hartree–Fock method and represents a breakdown of Koopmans' theorem.

7.5.1.2(iii) Oxygen, S_2, SO and nitric oxide

The ground configuration of O_2, given in Table 7.2, has been discussed in Section 6.2.1.1 and gives rise to the ground state, $X^3\Sigma_g^-$, and two excited states, $a^1\Delta_g$ and $b^1\Sigma_g^+$. Of the excited states only the low-lying $a^1\Delta_g$ state will concern us here.

In the triplet ground state the two electrons in the $\pi_g^* 2p$ orbital have parallel spins so that ejection of an electron from another orbital may result in a quartet or a doublet excited state of O_2^+ depending on whether the ejected electron had its spin parallel or antiparallel, respectively, to the two in $\pi_g^* 2p$. However, ejection of an

Table 7.2 States of O_2 and O_2^+ involved in the UPs spectrum

Species	Electron configuration	States	Adiabatic ionization energy/eV	r_e/Å	ω_e/cm^{-1}
O_2	. . . $(\sigma_u^*2s)^2(\sigma_g2p)^2(\pi_u2p)^4(\pi_g^*2p)^2$	$X^3\Sigma_g^-$	–	1.207 52[a]	1580[a]
		$a^1\Delta_g$	(0.9817)[a]	1.215 6[a]	1484[a]
O_2^+	. . . $(\sigma_u^*2s)^2(\sigma_g2p)^2(\pi_u2p)^4(\pi_g^*2p)^1$	$X^2\Pi_g$	12.071	1.116 4[a]	1905[a]
O_2^+	. . . $(\sigma_u^*2s)^2(\sigma_g2p)^2(\pi_u2p)^3(\pi_g^*2p)^2$	$a^4\Pi_u$	16.101	1.381 4[a]	1036[a]
		$A^2\Pi_u$ (+others)	17.045	1.409 1[a]	898[a]
O_2^+	. . . $(\sigma_u^*2s)^2(\sigma_g2p)^1(\pi_u2p)^4(\pi_g^*2p)^2$	$b^4\Sigma_g^-$	18.171	1.279 6[a]	1197[a]
		$B^2\Sigma_g^-$ (+others)	20.296	1.30	1156
O_2^+	. . . $(\sigma_u^*2s)^1(\sigma_g2p)^2(\pi_u2p)^4(\pi_g^*2p)^2$	$c^4\Sigma_u^-$	24.577	1.162	1545
		$(^2\Sigma_u^-)$ (+others)	–	–	–

[a] Data not obtained from UPS spectrum.

electron from the $\pi_g^* 2p$ orbital results in only one state, the $X^2\Pi_g$ ground state of O_2^+.

Table 7.2 lists various states of O_2^+ but only those arising from ionization of O_2 in its ground state. For example, the . . . $(\pi_u2p)^3(\pi_g^* 2p)^2$ configuration gives rise to *five* states, three $^2\Pi_u$, one $^4\Pi_u$ and one $^2\Pi_u$, but only two, $a^4\Pi_u$ and $A^2\Pi_u$, can result from ionization of O_2 in the $X^3\Sigma_g^-$ state.

The fact that, in an ion of an open shell molecule, many states may arise from a single electron configuration causes difficulties in applying Koopmans' theorem but the ionization energies in Table 7.2 justify the order of the σ_g2p and π_u2p orbital energies, discussed in Section 6.2.1.1.

Figure 7.18(a) shows the band system in the UPS spectrum of O_2 due to O_2^+ formed in its ground state, $X^2\Pi_g$. The source used was HeI (21.22 eV). The fairly long vibrational progression is consistent with a shortening of the bond by about 0.09 Å (Table 7.2) in O_2^+ due to the removal of an antibonding electron from $\pi_g^* 2p$. The peaks in the progression are split by about 25 meV due to spin–orbit coupling which splits the $^2\Pi_{\frac{1}{2}}$ and $^2\Pi_{\frac{3}{2}}$ components. The multiplet is normal so the more intense member of each doublet in the progression corresponds to the $^2\Pi_{\frac{1}{2}}$ component. The reason for this is that the photoionization cross-section, σ, is slightly

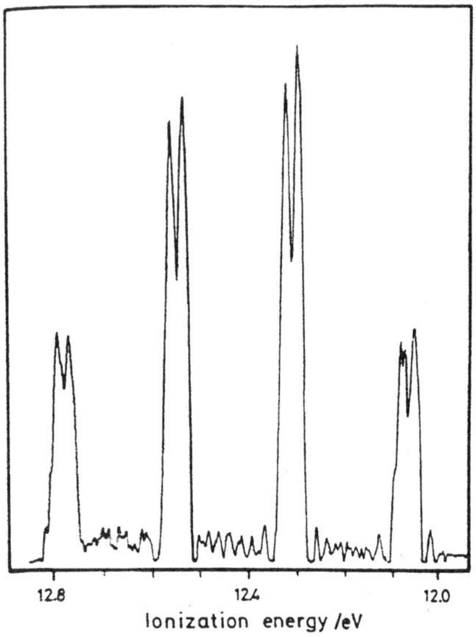

Figure 7.18(a) The $O_2^+(X^2\Pi_g) \leftarrow O_2(X^3\Sigma_g^-)$ UPs band system using a HeI source [Reproduced, with permission, from Dromey, R. G., Morrison, J. D. and Peel, J. B. (1973). *Chem. Phys. Lett.*, **23**, 30]

greater for $^2\Pi_{\frac{1}{2}}$ than for $^2\Pi_{\frac{3}{2}}$; the degeneracies of the two components are the same.

Figure 7.18(b) shows the remainder of the UPS spectrum, using a HeII (40.81 eV) source. Ionization to both the $a^4\Pi_u$ and $A^2\Pi_u$ states results in long, overlapping progressions with

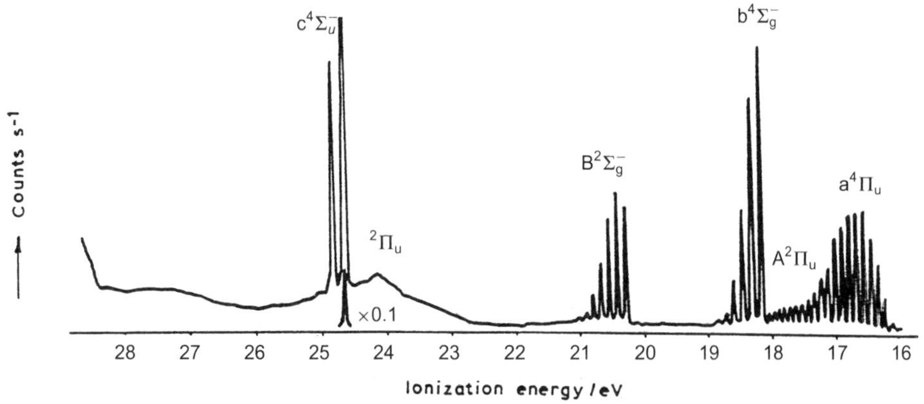

Figure 7.18(b) The UPS spectrum of the oxygen molecule in the 16–28 eV ionization energy region using a HeII source
[Reproduced, with permission, from Edqvist, O., Lindholm, E., Selin, L. E. and Åsbrink, L. (1970). *Phys. Scr.*, **1**, 25]

intensity maxima at $v' = 6$ and $v' = 7$, respectively. These are consistent with a considerable lengthening of the bond by about 0.18 Å in these two states due to removal of an electron from the bonding $\pi_u 2p$ orbital.

It is known from the electronic spectrum of O_2^+ that spin–orbit coupling is small in the $a^4\Pi_u$ and $A^2\Pi_u$ states resulting in splittings of 6 and 1 meV, respectively, too small to be detected in the UPS spectrum. Progressions involving the $b^4\Sigma_g^-$ and $B^2\Sigma_g^-$ states are shorter, consistent with a bond lengthening by about 0.08 Å due to removal of a bonding electron from the $\sigma_g 2p$ orbital. A slight shortening of the bond in the $c^4\Sigma_u^-$ state is due to removal of an electron from the antibonding $\sigma_u^* 2s$ orbital. The abrupt breaking off of the corresponding progression is due to predissociation.

The electronic emission spectrum of O_2^+ is extensive and has provided all except two of the values of r_e for the various states in Table 7.2. Those for the $B^2\Sigma_g^-$ and $c^4\Sigma_u^-$ states have been obtained only from Franck–Condon factors in the UPS spectrum.

Values of ω_e for O_2 and O_2^+ given in Table 7.2 are all consistent with the bonding or antibonding nature of the MO from which the electron has been removed.

Study of convergence limits of Rydberg transitions in the far ultraviolet absorption spectrum of O_2 has given adiabatic ionization energies[170] for several states of O_2^+. It is only for a few diatomic molecules that conventional spectroscopy has competed successfully with UPS in giving several ionization energies.

It is a noticeable feature of the band systems in Figure 7.18(b) that the ionization cross-section is considerably larger for the production of a quartet state than for the corresponding doublet state. This has been confirmed for the $a^4\Pi_u$ and $A^2\Pi_u$ states by calculation.[171] The ratio should be 2:1 if there were no spin–orbit coupling, the effect of which is to cause the three $^2\Pi_u$ states arising from the . . . $(\pi_u 2p)^3$ $(\pi_g^* 2p)^2$ configuration to interact and to change the ratio to 2:0.34.

The diffuse band system with a vertical ionization energy of about 24.0 eV is due to one of the other $^2\Pi_u$ states from the . . . $(\pi_u 2p)^3$ $(\pi_g^* 2p)^2$ configuration. The calculated intensity ratio, relative to $a^4\Pi_u$, is 0.64:2

Figure 7.19 shows the band system corresponding to O_2^+ in its $X^2\Pi_g$ state produced with NeI radiation which has a weak and an intense component with energies of 16.67 and 16.85 eV, respectively. Comparison with the same band

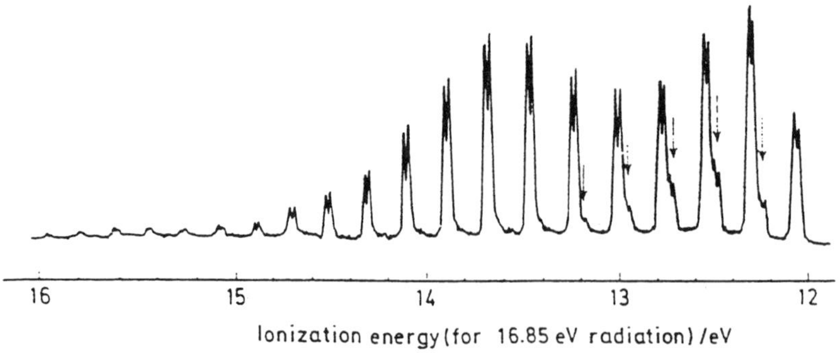

Figure 7.19 Autoionization in the $O_2^+(X^2\Pi_g) \leftarrow O_2$ $(X^3\Sigma_g^-)$ band system using a NeI source. The bands marked with arrows are due to the 16.67 eV radiation and do not show autoionization
[Reproduced, with permission, from Eland, J. H. D. (1974). *Photoelectron Spectroscopy*, p. 63, Butterworths, London]

system obtained with HeI radiation shows a pronounced difference in the intensity distribution in the vibrational progression. That with NeI radiation shows *two* maxima, one at $v' = 1$, as with HeI, and a second at $v' = 7$. The effect on the intensity distribution is due to autoionization.

Figure 7.18(b) shows that the energy of the NeI photons falls just below the $A^2\Pi_u$ ionization energy of O_2^+ of 17.045 eV. In the absorption spectrum of O_2 there are, in the region corresponding to the NeI photons, transitions to a series of Rydberg states converging to the 17.045 eV ionization energy. If the NeI photon corresponds to one of these transitions absorption of the radiation occurs, as shown in Figure 7.20, to a Rydberg state which is higher in energy than that required to ionize O_2 to O_2^+ in the $X^2\Pi_g$ state. By an autoionization process the molecule can revert from the Rydberg state to the $X^2\Pi_g$ state of O_2^+, a process involving simultaneous transfer of the electron from the Rydberg orbital to the $\pi_u 2p$ orbital and ejection of an electron from the $\pi_g^* 2p$ orbital.

In the UPS spectrum of O_2 only the intense NeI 16.85 eV component is involved in resonance absorption and autoionization. There is no Rydberg transition resonant with the weaker component, which produces a weak band system with normal intensity distribution, as Figure 7.19 shows.

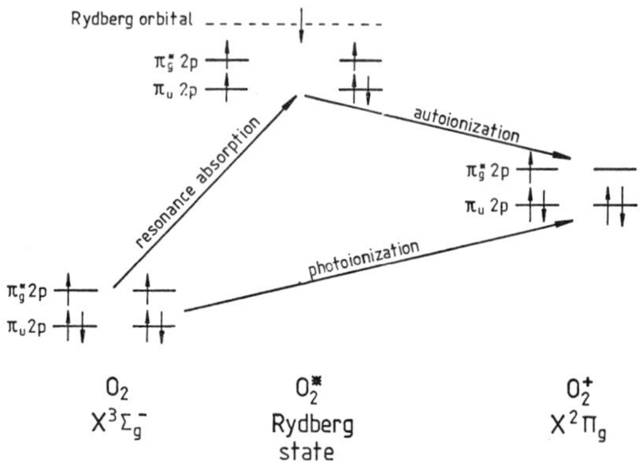

Figure 7.20 Two possible routes for the process $O_2^+(X^2\Pi_g) \leftarrow O_2(X^3\Sigma_g^-)$ by absorption and autoionization or by photoionization

Figure 7.20 illustrates how the process $O_2^+(X^2\Pi_g) \leftarrow O_2(X^3\Sigma_g^-)$ can be achieved either by direct photionization or by absorption and autoionization. The vibrational intensity distribution for direct photionization depends, in the usual way, on the difference in r_e between the X state of O_2^+ and the X state of O_2. On the other hand, the distribution in the absorption–autoionization process depends on the difference in r_e between the Rydberg state of O_2 and the X state of O_2^+; in this case the result is an intensity maximum at $\upsilon' = 7$.

The relative efficiencies of the two ways of producing O_2^+ in its ground state are reflected in the relative intensities of the progressions formed. Here the autoionization pathway is clearly efficient but in any band system of any molecule, in which its efficiency is low but not negligible, there may be a distortion of progression intensities and a corresponding error in a value of r_e obtained from them.

There has been a considerable amount of work done on the UPS of short-lived, or transient, species. These may be ground state atoms or molecules which are very reactive, or excited state species of stable atoms or molecules. The first UPS spectrum of a short-lived species to be observed[172] was that of O_2 in the low-lying $a^1\Delta_g$ excited state (see Table 7.2). This state is long-lived relative to excited states in general because the transition to the ground state, $X^3\Sigma_g^-$, is spin, orbitally and parity forbidden, violating the $\Delta S = 0$, $\Delta\Lambda = 0, \pm 1$, and $g \leftrightarrow u$ selection rules. As a result a concentration of about 1–20% of $O_2(a^1\Delta_g)$ can be built up in a microwave-induced discharge in O_2. The UPS spectrum of the resulting mixture shows the lowest ionization energy band system of $O_2(a^1\Delta_g)$, with an adiabatic ionization energy of 11.09 eV, corresponding to the formation of O_2^+ in its ground state, $X^2\Pi_g$. The vibrational progression shows an intensity maximum at $\upsilon' = 2$ and an intensity distribution consistent with a decrease in r_e of about 0.10 Å (Table 7.2).

With phase-sensitive detection, used to distinguish the $O_2(a^1\Delta_g)$ from the $O_2(X^3\Sigma_g^-)$ spectrum, ionization band systems were observed corresponding to O_2^+ in the $A^2\Pi_u$ state, the $C^2\Phi_u$ state arising from the ... $(\pi_u 2p)^3(\pi_g^* 2p)^2$ configuration and the $D^2\Delta_g$ state arising from the ... $(\sigma_g 2p)^1(\pi_u 2p)^4(\pi_g^* 2p)^2$ configuration.

The S_2 molecule is produced by heating flowers of sulphur. Depending on the temperature and pressure conditions, the vapour may also contain such species as S_6, S_7 and S_8 but, at temperatures in the range 250–600 K and pressures less than 1 Torr, S_2 is the predominant species.

The MO configurations of homonuclear diatomic molecules were discussed in Section 6.2.1.1 with concentration on those containing atoms in the first row of the periodic table. However, at the semi-quantitative level at which the discussion was given, molecules consisting of second-row atoms have analogous valence MOs made up from 3*s* and 3*p*, rather than 2*s* and 2*p*, AOs; it is assumed that 3*d* AOs do not play a significant part.

These expectations are borne out by the UPS spectrum of S_2.[173] This clearly shows band systems involving the $X^2\Pi_g$, $a^4\Pi_u$, $A^2\Pi_u$, $b^4\Sigma_g^-$ and $B^2\Sigma_g^-$ states and a further $^2\Pi_u$ state of S_2^+, with adiabatic ionization energies of 9.38 ($^2\Pi_{\frac{1}{2}}$) and 9.43 ($^2\Pi_{\frac{3}{2}}$), 11.53, 12.14, 13.20, 14.54 and (vertical) 15.58 eV, respectively. The systems involving each of these states resemble those involving the corresponding states of O_2^+ to a remarkable degree, including lengths of vibrational progressions and relative intensities of systems.

The bond lengths of S_2^+ in the X, a, A, b and B states have been obtained from Franck–Condon factors in the progressions and this remains the only source of such values; no electronic spectra of S_2^+ have been observed.

The main differences between the UPS spectra of O_2 and S_2 are that the ionization energies in S_2 are considerably lower than the corresponding ones in O_2, owing to the valence electrons in S_2 being further from the nuclei, and that the spin–orbit coupling in the $X^2\Pi_g$ state is much larger in S_2 because of the higher nuclear charge. No splitting of bands due to spin–orbit coupling has been observed in any other band system but some apparent broadening in the $a^4\Pi_u$ system of S_2 may be caused by this.

It was mentioned in Section 6.2.1.2 that the MOs of the short-lived molecule SO resemble those of O_2. This is because the 3*s* AO on sulphur has the same symmetry as, and comparable energy to, the 2*s* AO on oxygen and therefore can form MOs by linear combination; similarly with 3*p* on sulphur and

Table 7.3 States of NO and NO^+ involved in the UPS spectrum

Species	Electron configuration	States	Adiabatic ionization energy/eV	r_e/Å	ω_e/cm^{-1}
NO	. . . $(\sigma 2p)^2(\pi 2p)^4(\pi^* 2p)^1$	$X\left\{\begin{matrix}^2\Pi_{\frac{1}{2}} \\ ^2\Pi_{\frac{3}{2}}\end{matrix}\right.$	– (0.01486)†	1.150 77†	1904†
NO^+	. . . $(\sigma 2p)^2(\pi 2p)^4$	$X^1\Sigma^+$	9.262	1.063 22†	2376†
NO^+	. . . $(\sigma 2p)^2(\pi 2p)^3(\pi^* 2p)^1$	$a^3\Sigma^+$	15.667	1.284	1293
		$A'^1\Sigma^+$	17.811	1.287	1283
		$b'^3\Sigma^-$	17.59	1.290†	1284
		$W^1\Delta$	~18.08	1.288	1278
		$w^3\Delta$	16.863	1.280	1313
		$B'^1\Sigma^+$	~23(?)	–	730(?)
NO^+	. . . $(\sigma 2p)^1(\pi 2p)^4(\pi^* 2p)^1$	$A^1\Pi$	18.319	1.1931†	1602†
		$b^3\Pi$	16.562	1.175	1710†

† Data not obtained from UPS spectrum.

$2p$ on oxygen. Because SO has no centre of inversion the g and u property is not applicable.

The SO molecule can be formed in various ways, including a microwave-induced discharge in SO_2 and the reaction of atomic oxygen with H_2S or CS_2.

The UPS spectrum of SO shows[174] a strong resemblance to those of O_2 and S_2 showing band systems involving the states $X^2\Pi$, $a^4\Pi$, $A^2\Pi$, $b^4\Sigma^-$, $B^2\Sigma^-$ and another $^2\Pi$ state of SO^+ and with similar progression intensities. Spin–orbit coupling splits the two components of the $X^2\Pi$ state by 0.04 eV, intermediate between the splitting in O_2 and S_2. The ionization energy of each band system also falls between those for O_2 and S_2.

As for S_2^+, all experimental information regarding the electronic states of SO^+ has been obtained from the UPS spectrum.

The ground configuration . . . $(\sigma 2p)^2(\pi 2p)^4(\pi^* 2p)^1$ of NO [see also equation (6.104)] gives rise to a $^2\Pi$ term. The two components of this term, $X^2\Pi_{\frac{1}{2}}$ and $X^2\Pi_{\frac{3}{2}}$, are split by 121.1 cm^{-1} with $X^2\Pi_{\frac{1}{2}}$ being the lower in energy and therefore the true ground state.

The UPS spectrum has been obtained[175] with a resolution of about 25 meV using a HeII source and part of it with a resolution of 10 meV using a HeI source. Since 10 meV≡ 80 cm^{-1} the two components of the $X^2\Pi$ term can be regarded as coincident so far as the interpretation of the HeII spectrum is concerned, the only effect being a slight broadening of the bands. With the HeI source doubling of the bands due to the two $X^2\Pi$ states of NO is just resolved.

Loss of an electron from the outer orbital gives the configuration . . . $(\sigma 2p)^2(\pi 2p)^4$. This gives rise to only one state, the $X^1\Sigma^+$ ground state of NO^+, as given in Table 7.3. The adiabatic ionization energy is 9.262 eV and the vibrational progression shows about five members, consistent with a decrease in r_e of about 0.09 Å from NO (Table 7.3) due to the loss of an electron from an antibonding orbital. Table 7.3 shows also that six states arise from the configuration of NO^+ in which an electron has been ejected from the $\pi 2p$ orbital. Three of the states are singlet and three triplet as the ejected electron may have its spin parallel or antiparallel to that in the $\pi^* 2p$ orbital. All the states have been detected in the UPS spectrum, five of them for the first time. There is considerable overlapping between the corresponding band systems but all show long progressions consistent with an increase of r_e by about 0.14 Å from NO.

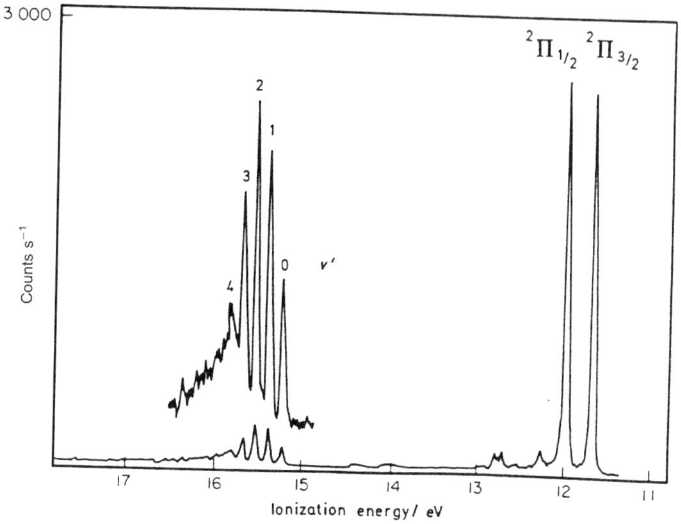

Figure 7.21 The HeI UPS spectrum of the HBr molecule
[Reproduced, with permission, from Turner, D. W., Baker, C., Baker, A. D. and Brundle, C. R. (1970). *Molecular Photoelectron Spectroscopy*, p. 57. Wiley, London]

Removal of an electron from the $\sigma 2p$ orbital results in two states, $A^1\Pi$ and $b^3\Pi$. The increases of bond length from NO are small, resulting in only short progressions.

Values of ω_e for NO and NO^+, given in Table 7.3, are consistent with the nature of the vacated orbital.

The difficulty of applying Koopmans' theorem to NO is evident from the range of ionization energies in Table 7.3 resulting from removal of an electron from the $\pi 2p$ or $\sigma 2p$ orbitals. The difficulty is not surprising when we recall (Section 6.2.1.1) that the order of these two orbitals is reversed on going from N_2 to O_2. They have very similar energies in NO.

No effects of splitting of triplet states in NO^+ due to spin–orbit coupling have been observed.

7.5.1.2(iv) *The hydrogen halides*

The MOs of the hydrogen halides are conceptually very simple. In the HCl molecule, for example, the occupied valence orbitals consist of $3p_x$ and $3p_y$ lone pair AOs on the chlorine atom and a σ-type bonding MO formed by linear combination of the $3p_z$ AO on chlorine and $1s$ AO on hydrogen (see Section 6.2.1.2 and Figure 6.30).

The UPS spectrum of HCl, and the other hydrogen halides, confirms this simple picture.

Figure 7.21 shows the HeI spectrum of HBr in which there are two band systems. The lower energy system shows a very short progression consistent with the removal of an electron from the doubly degenerate π_u lone pair ($4p_x$, $4p_y$) orbital on the bromine atom. The lone pair orbital picture and the expectation of a short progression are further confirmed by the fact that the bond length, r_e, is known, using other spectroscopic techniques, to be 1.4144 Å in the $X^1\Sigma^+$ ground state of HBr and has a fairly similar value of 1.4484 Å in the $X^2\Pi$ ground state of HBr^+.

The splitting by 0.33 eV of the $X^2\Pi$ state into an inverted multiplet, with the $^2\Pi_{\frac{3}{2}}$ below the $^2\Pi_{\frac{1}{2}}$ component, is relatively large due to the large spin–orbit coupling caused by the high nuclear charge on the bromine atom and also the fact that the orbital from which the electron has been removed is close to this atom. The corresponding splittings in the $X^2\Pi$ states of HCl and HI are 0.08 and 0.66 eV, respectively, consistent with the magnitude of the nuclear charges on Cl and I.

Figure 7.21 shows that the next highest energy band system of HBr consists of a fairly long progression with maximum intensity at $\upsilon' = 2$. A long progression is to be expected following the removal of an electron from the strongly bonding σ MO to form the $A^2\Sigma^+$ state of HBr^+, in which r_e has the value 1.6842 Å, an increase of about 0.27 Å from that of HBr.

In the second system the vibrational structure shows an onset of broadening between $\upsilon' = 3$ and 4. This is due to predissociation (see Section 6.2.4.6) caused by a repulsive potential curve due to a $^4\Pi$ state crossing that of the $A^2\Sigma^+$ state.

The second system appears qualitatively similar in HCl and HI.

Since $^2\Sigma^+$ states have only one component (Section 6.2.3.1), because there is no orbital angular momentum, there is no doubling of bands in the second system due to spin–orbit interaction.

7.5.1.3 UPS of triatomic molecules

7.5.1.3(i) Linear molecules

The $X–C \equiv N$ molecules, where X = H, F, Cl, Br and I, are all linear triatomics which are also linear in the low-lying states of the positive ion.

Considering only the MOs constructed from valence shell AOs, the ground configuration of the cyanogen halides can be written as

$$(1\sigma)^2(2\sigma)^2(3\sigma)^2(1\pi)^4(4\sigma)^2(2\pi)^4 \qquad (7.18)$$

The 1π and 2π MOs are formed from valence p_x and p_y AOs on all three atoms which are either all in-phase (1π) or in-phase between C and N but out-of-phase between X and C (2π). The 3σ and 4σ MOs can be regarded as axial lone-pair orbitals on the halogen and nitrogen atoms.

These simple MO pictures fit the observed UPS spectra well.† The lowest energy band system corresponds to the $\tilde{X}^2\Pi$ state of XCN^+. It is dominated by a progression in ν_1, the $C \equiv N$ stretching vibration, but ν_3, the X–C stretching vibration, is weakly excited. Spin–orbit coupling splits the $^2\Pi_{\frac{1}{2}}$ and $^2\Pi_{\frac{3}{2}}$ components observably when X = Cl, Br or I.

The second band system is typical of the removal of a non-bonding electron, there being only a very short progression in ν_3, and the third shows a long progression in ν_3, resolvable only in the case of ClCN.

In polyatomic molecules it becomes increasingly desirable, as the molecule becomes larger, to obtain the geometry of the positive ion from Franck–Condon factors in the photoelectron spectrum since the spectra of relatively few ions have been observed by other spectroscopic techniques. Unfortunately, an increasing number of atoms often leads to increasing vibrational complexity in the UPS spectrum, making it impossible to estimate molecular geometry from Franck–Condon factors.

Triatomics have only three vibrational modes (one of them doubly degenerate in a linear molecule) so that, in favourable cases, the ion geometry can be estimated. This has been achieved[176] for the $X^2\Pi$ states of $ClCN^+$, $BrCN^+$ and ICN^+, the $A^2\Sigma^+$ states of $ClCN^+$ and $BrCN^+$ and the $B^2\Pi$ state of $ClCN^+$. The geometry changes from the molecule to the ion broadly agree with expectations from simple MO theory.

The interpretation of the UPS spectrum of HCN is not so straightforward.‡ The ground MO configuration of HCN is

$$(1\sigma)^2(2\sigma)^2(3\sigma)^2(4\sigma)^2(5\sigma)^2(1\pi)^4 \qquad (7.19)$$

† See Bieri, G. (1977). *Chem. Phys. Lett.*, **46**, 107, and references cited therein.

‡ See Fridh, C. and Åsbrink, L. (1975). *J. Electron Spectrosc.*, **7**, 119, and references cited therein.

where 1π and 5σ are the bonding π and nitrogen lone pair MOs, respectively. Removal of an electron from these gives HCN^+ in the $\tilde{X}^2\Pi$ or $\tilde{A}^2\Sigma^+$ state, respectively. A Hartree–Fock MO calculation[177] predicts the $\tilde{X}$ state to be 0.85 eV below the $\tilde{A}$ state. Inclusion of the effect of electron correlation reduces the separation to 0.49 eV but does not reverse the order of the states, as happens for the isoelectronic N_2^+ [Section 7.5.1.2(ii)].

Experimentally, though, the identification of the second band system in the UPS spectrum has caused some difficulty. It seriously overlaps the first system and is not dominated by one intense band, as would normally be characteristic of the ejection of a non-bonding electron. There appears to be strong interaction between the $\tilde{X}$ and $\tilde{A}$ states involving the bending vibration ν_2. This results in the observation of $\Delta\upsilon_2 = 2$ transitions, although it has been suggested that $\Delta\upsilon_2 = 1$ transitions may be involved.

In UPS spectra, bands involving odd quanta of non-totally symmetric vibrations gain their intensity by Herzberg–Teller intensity stealing (Section 6.3.4.1). This entails mixing of the vibronic state of the ion with a nearby electronic state of the ion and of the same symmetry. The effect is more difficult to observe in UPS spectra than in electronic spectra because of the lower resolution and the lack of rotational fine structure to confirm the assignment. In electronic spectroscopy, intensity stealing is more prevalent in larger molecules but the UPS spectra of such molecules tend to be crowded with broad bands so as to mask any weaker bands involving the effect. Intensity stealing has been shown to occur in the UPS spectra of only a few molecules. An example is the first band system in the UPS spectrum of butatriene ($H_2C{=}C{=}C{=}CH_2$).[178]

7.5.1.3(ii) *Bent molecules*

The UPS spectra of H_2O and H_2S are good examples of the effect of a large change of angle from the molecule to the ion.

The ground MO configuration of H_2O is (see Section 6.3.1.1 and Figure 6.63)

$$\ldots (2a_1)^2(1b_2)^2(3a_1)^2(1b_1)^2 \tag{7.20}$$

and the HOH angle in the $\tilde{X}^1A_1$ state is 104.5°. The HeI UPS spectrum is shown in Figure 7.22.

Removal of an electron from the $1b_1$ orbital to give the $\tilde{X}^2B_1$ state of H_2O^+ should not affect the angle very much. Both ν_1, the symmetric stretching, and ν_2, the angle bending, vibrations show only short progressions. Franck–Condon factor calculations,[179] including the effects of vibrational anharmonicity, indicate that r(OH) increases from 0.9572 Å in the $\tilde{X}^1A_1$ state of H_2O to 0.995 ± 0.005 Å in the $\tilde{X}^2B_1$ state of H_2O^+. It was estimated that the HOH angle increases from 104.5° to a maximum possible value of 109°, although it was shown subsequently, from the high resolution emission spectrum of H_2O^+, that the angle in its ground state is rather larger, 110.5° (see Section 6.3.1.1).

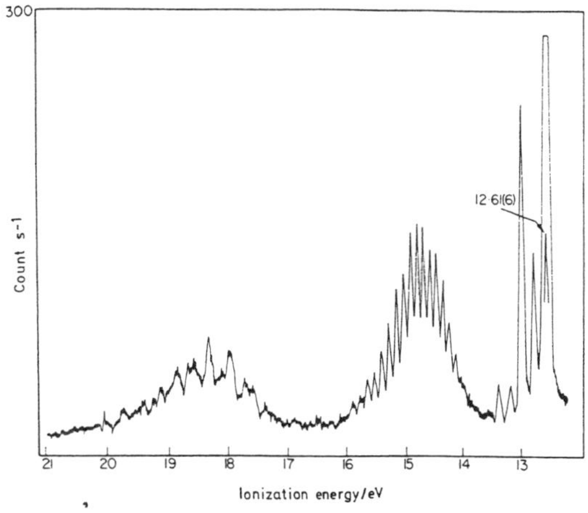

Figure 7.22 The HeI UPS spectrum of the H_2O molecule
[Reproduced, with permission, from Turner, D. W., Baker, C., Baker, A. D. and Brundle, C. R. (1970). *Molecular Photoelectron Spectroscopy*, p. 113. Wiley, London]

Removal of an electron from the $3a_1$ MO in H_2O should lead to a large increase in angle in the $\tilde{A}^2A_1$ state of H_2O^+. This was the conclusion from the emission spectrum of H_2O^+ from which an angle of 180° has been determined.[80] The observation of a long progression in the bending vibration, ν_2, in the second band system of the UPS spectrum is consistent with such a large increase of the angle.

Some irregularity in the first few members of the progression is consistent with the opinion that the ion is not exactly linear in the $\tilde{A}^2A_1$ state. The $\tilde{X}^2B_1$ and $\tilde{A}^2A_1$ states of bent H_2O^+ correlate with a $^2\Pi_u$ state in the linear ion and vibrational–electronic interaction between the $\tilde{X}$ and $\tilde{A}$ states results in a Renner–Teller effect (Section 6.3.4.4) and splitting of vibronic levels. The potential functions for these two states resemble those in Figure 6.99(d).

If $\angle$ HOH in the $\tilde{A}$ state of H_2O^+ is less than 180° the barrier to linearity is low and only the first few vibrational levels of ν_2 will be below it. These few levels will all be close to the barrier and therefore irregularly spaced (see Section 5.2.7.3), consistent with what is observed in the second band system. Above the barrier there is much greater regularity of levels, particularly of those which have Σ vibronic symmetry in the linear molecule.

The third band system, involving the removal of an electron from the $1b_2$ orbital, is vibrationally complex, consistent with the orbital being strongly bonding and the orbital energy having a minimum value when $\angle$HOH = 180° (Figure 6.63). Presumably both ν_1 and ν_2 are excited but the bands in this system are considerably broadened, making analysis unreliable.

High resolution investigation[180] of the 0_0^0 band of the first band system of H_2O, using NeI radiation, reveals a broad rotational contour which resembles that of the asymmetric rotor type $C\,0_0^0$ band of the $\tilde{C}$–$\tilde{X}$ Rydberg transition of H_2O [Section 6.3.5.3(i)]. The selection rules for a type C band due to an electric dipole transition include $\Delta K_a = \pm 1$ and $\Delta N = 0, \pm 1$, whereas those for photoionization include[181] $\Delta K_a = 0, \pm 1$ and $\Delta N = 0, \pm 1$, for an ejected electron described by a mixture of s and p wave functions. It appears that transitions with $\Delta K_a = \pm 1$ dominate the contour in the UPS spectrum.

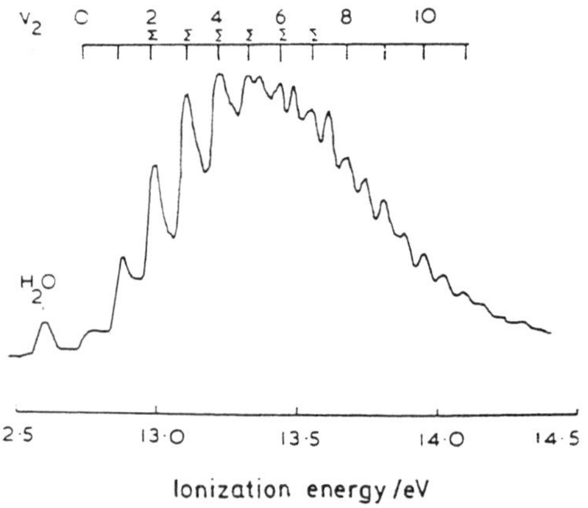

Figure 7.23 The HeI UPS spectrum of the H_2S molecule showing the second band system only [Reproduced, with permission, from Dixon, R. N., Duxbury, G., Horani, M. and Rostas, J. (1971). *Molc. Phys.*, **22**, 977]

The HeI UPS spectrum of H_2S strongly resembles that of H_2O, showing three band systems. The first is typical of the removal of an electron from a nonbonding orbital, analogous to $1b_1$ of H_2O, and involves the $\tilde{X}^2B_1$ state of H_2S^+. The third involves removal from a b_2 bonding orbital to give the $\tilde{B}^2B_2$ state and shows progressions in ν_1 and ν_2.

The second system of H_2S is shown in Figure 7.23. Like that of H_2O, it is dominated by a long progression in the bending vibration ν_2 because an electron has been removed from an orbital whose energy falls steeply as the bond angle decreases. The system of H_2S shows vibrational intervals of ν_2 decreasing smoothly as far as $v_2' = 4$, beyond which there is a small region of some confusion followed by a change to approximately half the vibrational spacing which then smoothly increases. This behaviour is expected for the Σ vibronic levels, with $K = 0$ (Section 6.3.4.4). The discontinuity between

$v_2' = 4$ and $v_2' = 6$ corresponds to the region at the top of the barrier to linearity. Both an energy barrier and Renner–Teller interaction are necessary to explain the observed band system.

The second system of H_2S is a good example of the contrast between the information which can be obtained from a high resolution ($< 1\,cm^{-1}$) electronic spectrum and from a UPS spectrum (*ca* $80\,cm^{-1}$). To anyone used to high resolution spectroscopy it is a daunting prospect to try, for example, to obtain accurate vibrational term values from a UPS spectrum such as that in Figure 7.23 or accurate rotational information from a broad contour such as that observed in the UPS spectrum of H_2O. Inevitably we have to set our sights at a much lower level of accuracy in trying to derive such information from UPS spectra.

7.5.1.4 UPS of larger polyatomic molecules

7.5.1.4(i) AH_3 molecules

In the UPS spectra of larger molecules there is a tendency towards greater, and in many cases unresolvable, vibrational complexity.

The first band system of NH_3, with an adiabatic ionization energy of 10.16 eV, is an exception being dominated by a long progression, of about 15 members, in the inversion vibration ν_2.† The reason for this is that the electron is removed from the outer $3a_1$ orbital [equation (6.161)] whose energy falls steeply as the HNH angle decreases (Figure 6.68). The result is that the ground state, $\tilde{X}^2A_2''$, of NH_3^+ is planar, as are Rydberg states of NH_3, and ν_2 is strongly excited on ionization.

The first band system of PH_3 shows a similar progression but much less well resolved.

The second band systems of NH_3 and PH_3, due to removal of an electron from the $1e$ orbital (or its equivalent in PH_3), do not show such well resolved vibrational structure as the first systems but that of NH_3, with an adiabatic ionization energy of 14.7 eV, shows some structure and evidence of a Jahn–Teller effect (Section 6.3.4.5). The static Jahn–Teller effect may split the degenerate $\tilde{A}^2E'$ state of NH_3^+ by distortion in the direction of either component of either of the e' vibrations ν_3 and ν_4, illustrated in Figure 5.31. Distortion in the direction of the normal coordinate Q_{3a} or Q_{4a} causes one NH bond to become different from the other two and the molecule assumes C_s symmetry. This results in the $^2E'$ state splitting into $^2A'$ and $^2A''$ states. Alternatively, distortion in the direction of Q_{3b} or Q_{4b} causes all NH bonds to be different. C_1 symmetry results and the $^2E'$ state splits into two ^2A states. Distortion to give C_s symmetry is thought to be more likely.[182]

The second band system of NH_3 shows two broad maxima separated by about 1 eV and it is assumed that these correspond to the two components of the split $\tilde{A}^2E'$ state. The vibrational structure is not sufficiently resolved for any incontravertible interpretation involving the dynamic Jahn–Teller effect to be made.

The UPS spectrum of CH_3 has been recorded,[183] the CH_3 being generated by the pyrolysis of azomethane ($CH_3N{=}NCH_3$) at about 1700 K. Removal of an electron from the outer $1a_2''$ orbital [equation (6.157)] to give the $\tilde{X}^1A_1'$ ground state of CH_3^+ produces a band system typical of a non-bonding electron. Evidence from the very short vibrational progressions favours a planar ground state of both CH_3^+ and CH_3.

7.5.1.4(ii) Ethylene

The HeI UPS spectrum of ethylene shows five band systems with adiabatic ionization energies of 10.51, 12.38, 14.47, 15.68 and 18.9 eV.

The first band system is shown in Figure 7.24. The most prominent features are (a) a progression in ν_2, the a_g C=C stretching vibration, (b) $\Delta v_4 = 2$ transitions in ν_4, the a_u torsional vibration (Figure 5.39) and (c) a short progression in ν_3, the a_g CH_2-scissoring vibration. The high intensity of bands involving $2\nu_4$ parallels

† See Harshbarger, W. R. (1972). *J. Chem. Phys.*, **56**, 177, and references cited therein.

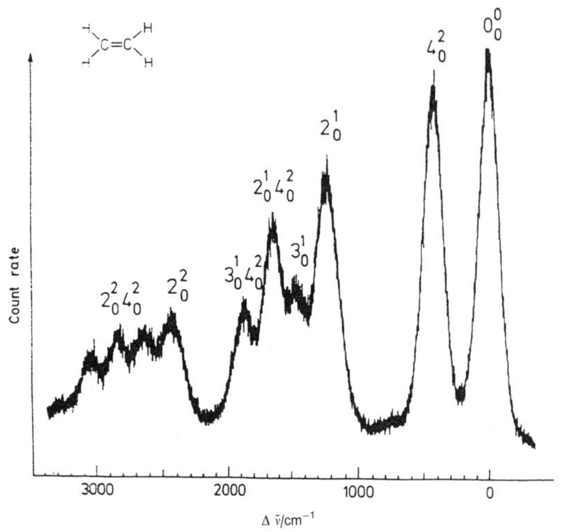

Figure 7.24 The HeI UPS spectrum of the ethylene molecule showing the first band system only. Along the abscissa is plotted the separation, in cm^{-1}, from the band of lowest ionization energy [Reproduced, with permission, from Cvitaš, T., Güsten, H. and Klasinc, L. (1979). *J. Chem. Phys.*, **70**, 57]

similar observations in the first Rydberg transition in ethylene.[184] In the first Rydberg state of C_2H_4 and the ground state of $C_2H_4^+$ the equilibrium configuration is non-planar, the CH_2 groups being twisted relative to each other by about 30°, and it is this geometry change which results, from the Franck–Condon principle, in the intense transitions with $\Delta v_4 = 2$.

In Section 6.3.1.6 we have seen that the ground configuration of ethylene is

$$\ldots (1b_{2u})^2(3a_g)^2(1b_{3g})^2(1b_{3u})^2(1b_{2g})^0 \quad (7.21)$$

where $1b_{3u}$ is a π bonding and $1b_{2g}$ a π^* antibonding orbital, shown in Figure 6.76. The ground state of C_2H_4 is $\tilde{X}^1A_g$ and the first excited singlet state, $\tilde{A}^1B_{1u}$, results from the promotion of an electron from $1b_{3u}$ to $1b_{2g}$. The molecule is twisted in this state, with an angle of 90° between the two CH_2 planes. In this configuration it belongs to the D_{2d} point group so that the state is really $\tilde{A}^1B_2$. The twisting is rationalized by the supposition that the repulsion in the $1b_{2g}$ π^* orbital across the xy nodal plane is minimized by this change of shape.

With this explanation in mind, it may seem surprising that removal of an electron from the $1b_{3u}$ orbital should also result in a twisted configuration, albeit by considerably less than 90°. The explanation of the slight twisting in the $\tilde{X}^2B_{3u}$ state of $C_2H_4^+$, and also in the first Rydberg state of C_2H_4, lies in the fact that, with a 30° angle of twist, the molecule belongs to the D_2 point group and the ground configuration of equation (7.21) becomes

$$\ldots (1b_2)^2(3a)^2(1b_3)^2(2b_3)^2(2b_2)^0 \quad (7.22)$$

Since the orbitals which were $1b_{2u}$ and $1b_{2g}$ in the D_{2h} point group now have the same symmetry, b_2, they can mix. This results in the $1b_2$ orbital gaining some π^* character. Similarly, the $1b_{3g}$ and $1b_{3u}$ orbitals now have b_3 symmetry and can also mix. Removal of an electron from $2b_3$, which is still largely π bonding in character, reduces the bonding considerably and the small amount of π^* character in $1b_2$ leads to a slight twisting in the ground state of $C_2H_4^+$, which becomes $\tilde{X}^2B_3$ in the D_2 point group.

7.5.1.4(iii) Methane

The four CH-bonding σ orbitals in CH_4 are classified (see Table 6.17) as a non-degenerate $2a_1$ and a triply degenerate $1t_2$ orbital in the T_d point group. These orbitals derive from the hydrogen $1s$ and the carbon $2s$ and $2p$ AOs. The carbon $1s$ AO is assumed not to be involved in bonding and becomes the $1a_1$ MO in CH_4. The ground configuration of CH_4 is, therefore,

$$(1a_1)^2(2a_1)^2(1t_2)^6 \quad (7.23)$$

resulting in the $\tilde{X}^1A_1$ ground state.

The second band system in the UPS spectrum is straightforward in its interpretation but has a high adiabatic ionization energy of 22.39 eV, necessitating the use of a HeII source.[185] The

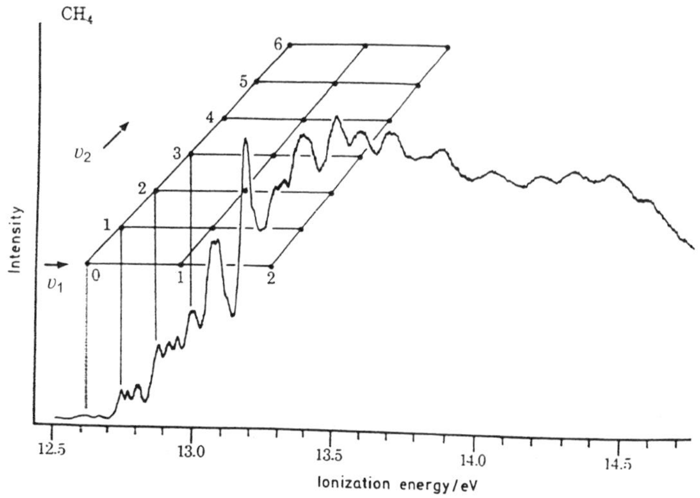

Figure 7.25 The NeI UPS spectrum of the methane molecule showing the first band system only [Reproduced, with permission, from Potts, A. W. and Price, W. C. (1972). *Proc. Roy. Soc.*, **A326**, 165]

system is dominated by a progression of about seven members in ν_1, the a_1 CH-stretching vibration, consistent with the removal of an electron from the $2a_1$ bonding orbital in the formation of the $\tilde{A}^2A_1$ excited state of CH_4^+.

The NeI UPS spectrum in Figure 7.25 shows the first band system resulting from removal of an electron from the $1t_2$ orbital to give CH_4^+ in its $\tilde{X}^2T_2$ ground state. The adiabatic ionization energy is *ca* 12.60 eV.

We have seen in Section 6.3.4.5 how a molecule in a degenerate electronic state is distorted in the direction of a normal coordinate in order to remove the degeneracy—the static Jahn–Teller effect. Dixon[186] has carried out *ab initio* calculations on the potential energy surface of CH_4^+ in the $\tilde{X}^2T_2$ state and has shown that distortion to a configuration which belongs to the D_{2d} point group leads to the minimum energy and therefore represents the true ground state of CH_4^+. The geometry of the molecule in this state is estimated to be such that, using the labelling in Figure 7.26, $\angle H_1CH_2 = \angle H_3CH_4 = 141.2^\circ$ and all others are 96.3°. This distortion is in the direction of the ν_{2a} component, shown in the figure, of the e vibration ν_2 and would, in the extremity, lead to a square planar configuration.

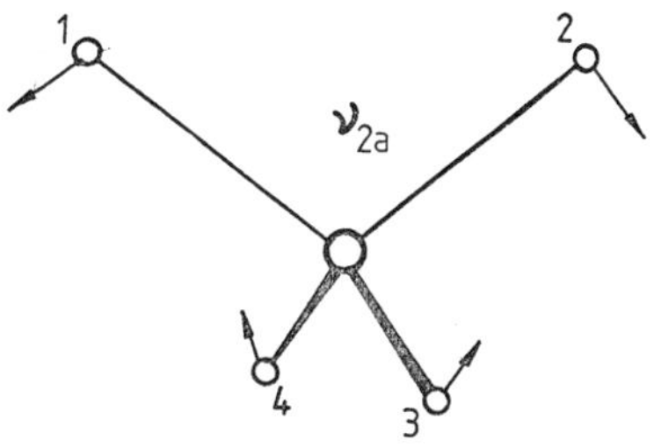

Figure 7.26 Distortion of CH_4^+, in its ground state, in the direction of the vibration ν_{2a}

In the D_{2d} configuration the degeneracy of the $1t_2$ orbital is lifted and it splits into a b_2 and an e orbital so that the ground configuration of CH_4^+ is

$$(1a_1)^2(2a_1)^2(1e)^4(1b_2)^1 \tag{7.24}$$

giving the $\tilde{X}^2B_2$ ground state. The first excited configuration is

$$(1a_1)^2(2a_1)^2(1e)^3(1b_2)^2 \tag{7.25}$$

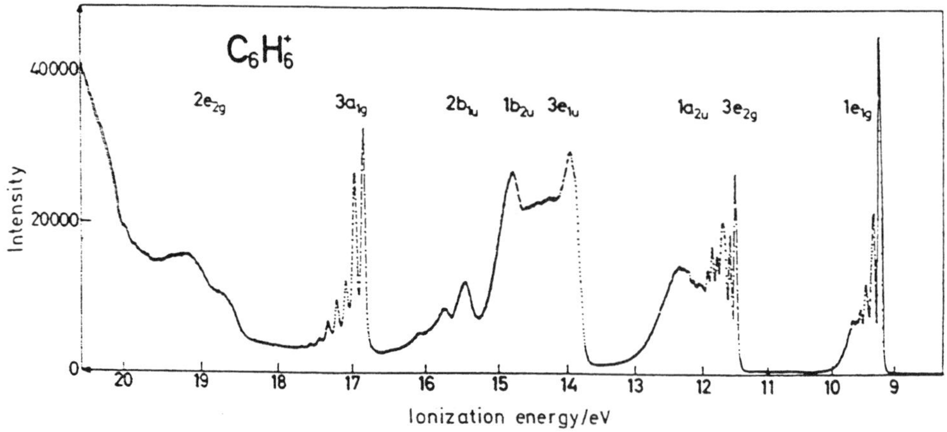

Figure 7.27 The HeI UPS spectrum of the benzene molecule
[Reproduced, with permission, from Karlsson, L., Mattsson, L., Jadrny, R., Bergmark, T. and Siegbahn, K. (1976). *Phys. Scr.*, **14**, 230]

giving the $\tilde{A}^2E$ state, which itself may be split by a further Jahn–Teller effect.

In the $\tilde{X}^2B_2$ state the vibration ν_{2a}, in T_d, becomes the a_1 vibration ν_2, in D_{2d}, and a progression in this vibration would be expected in the $\tilde{X}^2B_2(CH_4^+)$–$\tilde{X}^1A_1(CH_4)$ band system. As Figure 7.25 shows, this is what is observed in the region of the onset of the band system, starting at about 12.6 eV. The wavenumber of ν_2 in the progression is about 900 cm^{-1} compared with 1526 cm^{-1} in the $\tilde{X}$ state of CH_4.

The degeneracy of the $\tilde{X}^2T_2$ state of tetrahedral CH_4^+ may also be split by t_2 vibrations, of which there are two, ν_3 and ν_4. Dixon has shown that such activity of a t_2 vibration in a T_2–A_1 ionization process should lead to three broad, equally spaced maxima in the vibronic intensity contour. Experimentally the first band system does show such maxima at about 13.6, 14.4 and 15.0 eV, although they become much clearer in the corresponding band system of stannane, SnH_4.

The vibrational detail, even where it is as well resolved as in the region of the onset of the 12.6 eV system of CH_4, is extremely difficult to interpret with certainty. This is due, in part, to the Jahn–Teller effect and also to lack of sufficient resolution in the spectrum.

7.5.1.4(iv) *Benzene*

Figure 7.27 shows the HeI UPS spectrum of benzene. This spectrum poses problems of interpretation which are typical for large molecules. These problems concern the difficulty of identifying the onset of a band system when there is overlapping with neighbouring systems and, generally, a lack of resolved vibrational structure.

The recognition of the onset of a band system can be helped greatly by reliable MO calculations so that analysis of the spectrum of a large molecule is often a result of a combination of theory and experiment. In addition, study of the angular dependence of photoelectrons, which will be discussed in Section 7.5.1.5, can aid the identification of overlapping systems.

In Section 6.3.1.8 the π-electron MOs of benzene were obtained by the Hückel method using only the $2p_z$ AOs on the six carbon atoms. In the ground configuration [equation (6.206)]

Table 7.4 Calculated and experimental ionization energies of benzene

Orbital	Type	Calc. ionization energy/eV		Exp. ionization energy†/eV
		Excluding electron correlation	Including electron correlation	
$1e_{1g}$	$\pi 2p$	10.1	9.31	9.3
$3e_{2g}$	$\sigma 2p$	14.3	13.47	11.4
$1a_{2u}$	$\pi 2p$	14.6	13.79	12.1
$3e_{1u}$	$\sigma 2p$	16.9	16.14	13.8
$1b_{2u}$	$\sigma 2p$	17.8	17.02	14.7
$2b_{1u}$	$\sigma 2s$	18.0	17.42	15.4
$3a_{1g}$	$\sigma 2p$	20.1	19.47	16.9
$2e_{2g}$	$\sigma 2s$	23.0	22.43	19.2
$2e_{1u}$	$\sigma 2s$	28.2	27.66	22.5
$2a_{1g}$	$\sigma 2s$	31.8	31.43	25.9

† This is the vertical ionization energy.

four of the π electrons are in the outer $1e_{1g}$ and two in the $1a_{2u}$ MO.

An SCF calculation[187] of orbital energies for the ground configuration of C_6H_6, using as a basis set the $1s$, $2s$ and $2p$ AOs on carbon and the $1s$ AO on hydrogen, gives the results in Table 7.4, where they are compared with the ionization energies obtained from UPS spectra. Koopmans' theorem is assumed in such a comparison. The calculation does not take account of electron correlation and, partly owing to this, all the calculated ionization energies are considerably higher than those observed. Fortunately, however, the order of calculated orbital energies is much more reliable than their absolute values. The calculation clearly indicates that the UPS spectrum in the 11.5–12.5 eV region consists of two and, in the region 14–16 eV, three band systems. Inclusion of electron correlation[188] reduces the disagreement with experiment particularly for low ionization energies. These results are also given in Table 7.4.

An interesting feature of the calculations is that the $3e_{2g}$ orbital, which is a σ-type orbital, has a lower ionization energy than the $1a_{2u}$, π-type orbital. This clearly casts doubt on any kind of MO calculation which takes into account only π electrons.

Vibrational structure is, at best, only partially resolved in band systems of large molecules. The structure in the first, $(1e_{1g})^{-1}$, and second, $(3e_{2g})^{-1}$, systems† is complex which could be due, in part, to there being a Jahn–Teller effect in each of them. The only relatively simple structure is in the seventh $(3a_{1g})^{-1}$ system, which is dominated by a progression in v_1, the a_{1g} ring-breathing vibration.‡ This is consistent with the electron being ejected from a σ CC bonding orbital.

7.5.1.5 *Angular distribution of photoelectrons*

In Section 7.3 we have seen that, when an atom undergoes photoionization, it obeys the $\Delta l = \pm 1$ selection rule so that if, for example, an s electron is being removed the photoelectron is described by a p wave function. This has an angular distribution given by the angular wave function Y_{10} discussed in Section 1.3.2. The angular distribution $I(\theta)$ of photoelectrons is proportional to Y_{10}^2 and

†This nomenclature indicates removal of an electron from an orbital.

‡Here, as elsewhere, we use the Wilson vibrational numbering for benzene.

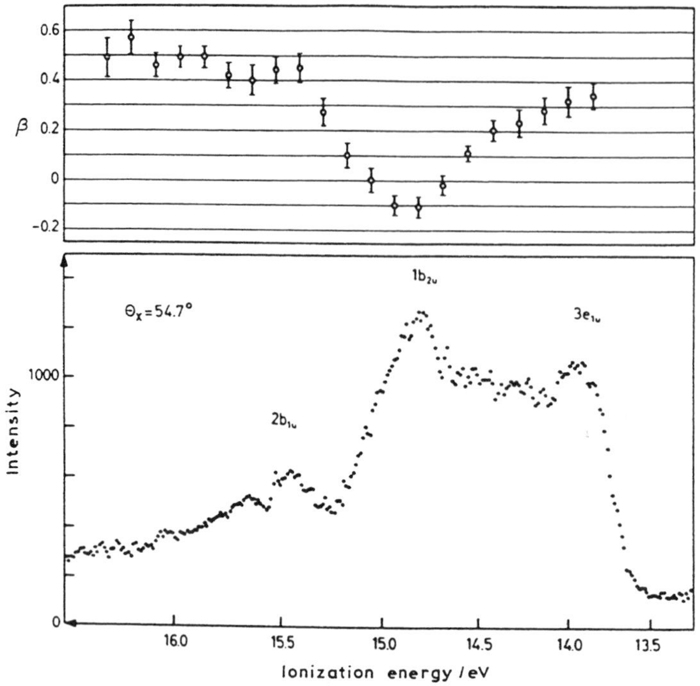

Figure 7.28 Variation of the photoelectron angular distribution parameter β over the range 14–16 eV of the HeI UPS spectrum of the benzene molecule
[Reproduced, with permission, from Mattsson, L., Karlsson, L., Jadrny, R. and Siegbahn, K. (1977). *Phys. Scr.*, **16**, 221]

$$I(\theta) \propto Y_{10}^2 = \frac{3}{4\pi}\cos^2\theta \tag{7.26}$$

where θ is the angle between the incident photons and the collected photoelectrons. It is clear from this equation that the design of the spectrometer is important with regard to the significance of measured intensities. For example, with a spherical grid analyser shown in Figure 7.6(b), photoelectrons are collected over all values of θ but, with the deflection analysers in Figure 7.7, they are collected for $\theta = 90°$ only.

In general the photoelectrons, whether ejected from atoms or molecules, are described by a mixture of $s, p, d, f, \ldots$ wave functions. Then, for an unpolarized photon source, the angular distribution of photoelectrons is given by†

$$I(\theta) = \frac{\sigma}{4\pi}\left[1 + \frac{\beta}{2}\left(\tfrac{3}{2}\sin^2\theta - 1\right)\right] \tag{7.27}$$

where σ is the photoionization cross-section, integrated over all θ, of the target atoms or molecules, and β is the anisotropy parameter. For a photoelectron described by a p wave function, $\beta = 2$ and, in general, $-1 \leqslant \beta \leqslant 2$.

In principle, β can be calculated for a particular orbital of an atom or molecule but, in practice, it is mostly the empirical fact that β is likely to have different values for different

†See, for example, Buckingham, A. D., Orr, B. J. and Sichel, J. M. (1970). *Philos. Trans. Roy. Soc.*, **A268**, 147.

orbitals which is used to distinguish one band system from another.

The experimental method must be modified in order to make angular distribution measurements. This entails a facility for rotating the source or analyser relative to the normal to the entrance slit of the analyser. If intensities are required which are independent of β, the so-called magic angle of $\theta = 54.7^\circ$ should be used.

Figure 7.28 shows how measurement of β over the complex 14–16 eV region of the benzene spectrum in Figure 7.27 confirms that there are three overlapping band systems. In order of increasing ionization energy they have values of β of about 0.35, -0.1 and 0.5, respectively.

7.5.2 X-ray photoelectron spectroscopy (XPS)

The technique of XPS has been applied to samples in the solid, liquid and vapour phases. Solid state and, especially, surface investigations represent a particularly important part of the subject but, in line with the general subject matter of this book, we shall be concerned only with the vapour phase.

7.5.2.1 The chemical shift

The ejection of an electron from the core of an atom, whether it is free or part of a molecule, requires a large amount of energy compared to ejection of a valence electron. This energy is supplied in XPS by an X-ray source of radiation, such as Al $K\alpha$ or Mg $K\alpha$ (Section 7.2.2), ideally with a monochromator to narrow the line width.

So far, in all aspects of valence theory of molecules that we have considered, the core electrons have been assumed to be in orbitals which are unchanged from the AOs of the corresponding atoms. XPS shows that this is almost true and the point is illustrated in Figure 7.29 for the 1s orbitals of the oxygen and carbon atoms in CO and CO_2. The figure shows the XPS spectrum of a 2:1 mixture of CO and CO_2 gases with Mg $K\alpha$ as the source. The C 1s ionization energy is 295.8 eV in CO and 297.8 eV in CO_2, while the O 1s ionization energy is 541.1 eV in CO and 539.8 eV in CO_2.

There are two features which stand out from these data. The first is that the ionization energy is much greater for O 1s than for C 1s owing to the effect of the larger nuclear charge of oxygen. The second effect is that the ionization for a particular kind of atom depends on its immediate environment in the molecule. This latter effect is known as the chemical shift. It is not to be confused with the chemical shift in NMR spectroscopy, which is the shift of the signal due to the nuclear spin of, for example, a proton due to neighbouring groups which may shield the proton from the applied magnetic field.

The chemical shift, ΔE_{nl}, for an atomic core orbital, with principal and orbital angular momentum quantum numbers n and l, respectively, in a molecule M is given by

$$\Delta E_{nl} = [E_{nl}(\mathrm{M}^+) - E_{nl}(\mathrm{M})] - [E_{nl}(\mathrm{A}^+) - E_{nl}(\mathrm{A})] \tag{7.28}$$

The shift is measured relative to the ionization energy, $[E_{nl}(\mathrm{A}^+) - E_{nl}(\mathrm{A})]$, of the corresponding orbital in the free atom A. In the approximation of Koopmans' theorem [equation (7.16)],

$$\Delta E_{nl} \approx -\varepsilon_{nl}(\mathrm{M}) + \varepsilon_{nl}(\mathrm{A}) \tag{7.29}$$

where ε_{nl} is the calculated orbital energy.

There has been much work done on the calculation of chemical shifts and comparison with experimental values.

In principle, a separate Hartree-Fock calculation of each of the four molecular or atomic energies on the right-hand side of equation (7.28) could be carried out to give the chemical shift but, more often, simpler methods are used.

A chemical shift is due to the core electron, which is to be ejected, experiencing a change in the electrostatic potential due to the valence electrons caused by a change in the immediate environment in the molecule. This may be expressed as

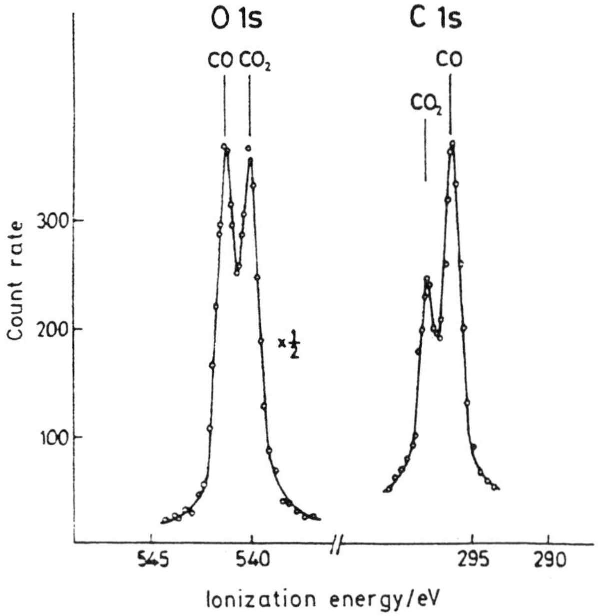

Figure 7.29 The Mg $K\alpha$ oxygen 1*s* and carbon 1*s* XPS spectrum of a 2:1 mixture of CO and CO_2 gases [Reproduced, with permission, from Allan, C. J. and Siegbahn, K. (November 1971). Publication No. UUIP-754, p. 48. Uppsala University Institute of Physics]

$$\Delta E_{nl} = \frac{1}{4\pi\varepsilon_0}\left[\frac{q_{\mathrm{v}}e^2}{r_{\mathrm{v}}}(\mathrm{M}) - \frac{q_{\mathrm{v}}e^2}{r_{\mathrm{v}}}(\mathrm{A})\right] = \frac{1}{4\pi\varepsilon_0}\Delta\left(\frac{q_{\mathrm{v}}e^2}{r_{\mathrm{v}}}\right) \tag{7.30}$$

where q_{v} is the charge due to the valence electrons of the atom concerned in its particular environment and r_{v} is the radius of the valence shell.

The recognition that it is only the valence electrons which are important in chemical shifts led to the conclusion that the MO and AO methods which were developed originally for outer shell electrons only would be suitable for chemical shift calculations.

We have seen in Section 7.4 that the neglect of electron reorganization in calculations of orbital energies can be important and this is particularly true for core orbitals. The reorganization energy, E_{R}, is given by

$$E_{\mathrm{R}} = -\varepsilon_{nl} - [E_{nl}(\mathrm{M}^+) - E_{nl}(\mathrm{M})] \tag{7.31}$$

In the series of molecules $NH_n(CH_3)_{3-n}$, for example, it has been shown[189] that, for $n = 0$–3, neglect of reorganization energy leads to a predicted chemical shift that *increases* with n, whereas the experimental value *decreases*. Inclusion of E_{R} in the calculations results in good agreement with experiment.

Since the chemical shift is related to the valence electron density on an atom, it is a natural extension to try to relate changes of chemical shift due to neighbouring atoms to the electronegativities of those atoms. Figure 7.30 shows the XPS carbon 1*s* spectrum of ethyl trifluoroacetate, $CF_3COOCH_2CH_3$, obtained with monochromatized Al $K\alpha$ ionizing radiation. This spectrum shows that, at least in this molecule, a very simple electronegativity argument leads to the correct C 1*s* assignments. The carbon atom of the CF_3 group is adjacent to three strongly electronegative fluorine atoms

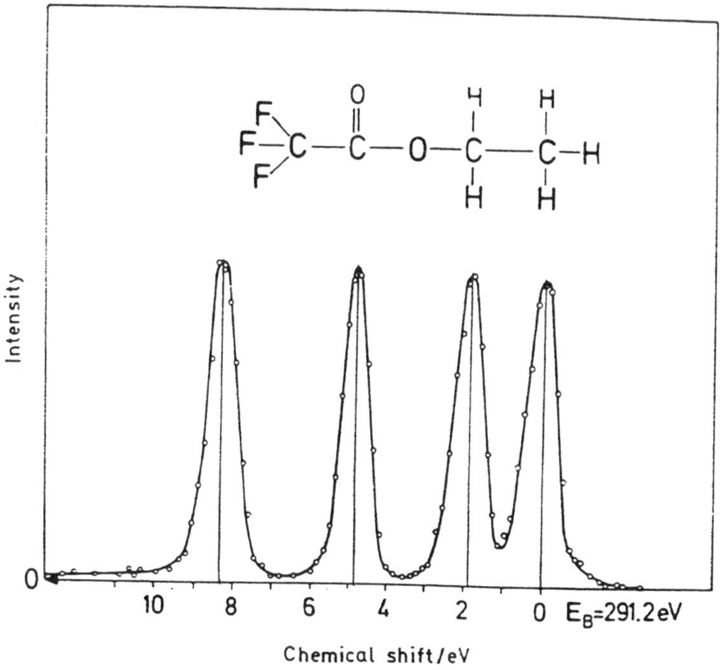

Figure 7.30 The monochromatized Al $K\alpha$ carbon 1*s* XPS spectrum of the ethyl trifluoroacetate molecule showing the chemical shifts relative to an ionization energy of 291.2 eV
[Reproduced, with permission, from Gelius, U., Basilier, E., Svensson, S., Bergmark, T. and Siegbahn, K. (1974). *J. Electron Spectrosc.*, **2**, 405]

which leads to a depletion of the electron density on the carbon atom and the fact that, consequently, it holds on to the 1*s* electron more strongly. The carbon atom of the C=O group has two neighbouring, fairly electronegative, oxygen atoms which have a similar but smaller effect. The carbon atom of the CH_2 group has two neighbouring, fairly electronegative, oxygen atoms which have a similar but smaller effect. The carbon atom of the CH_2 group has one neighbouring oxygen atom but that of the CH_3 group has none, leading to the lowest ionization energy.

However, for many molecules, this simple type of argument can lead to conclusions which are even qualitatively wrong.

Siegbahn *et al.* have attempted successfully to calculate changes of chemical shifts using quantitative electronegativities and the method is described in detail in their book referred to in the Bibliography at the end of the chapter.

The potential V experienced by a core electron in atom i includes contributions not only from its own valence shell but from all other atoms in the molecule. This is expressed by

$$V = \sum_j \left(\frac{q_j e^2}{4\pi\varepsilon_0 R_{ij}} \right) \tag{7.32}$$

where q_j is the net charge on atom j and can be treated as acting at the centre of the atom, and R_{ij} is the separation of atoms i and j. Including this additional contribution to the potential experienced by the core electron, the chemical shift of equation (7.30) becomes

$$\Delta E_{nl} = \frac{1}{4\pi\varepsilon_0} \Delta \left(\frac{q_v e^2}{r_v} \right) - \Delta V \tag{7.33}$$

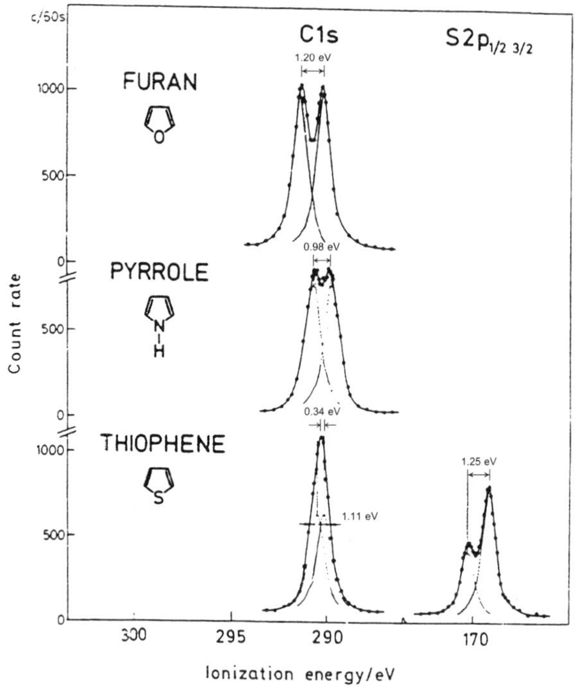

Figure 7.31 The carbon 1*s* XPS spectra of the furan, pyrrole and thiophene molecules. The sulphur 2*p* spectrum of thiophene is also shown
[Reproduced, with permission, from Gelius, U., Allan, C. J., Johansson, G., Siegbahn, H., Allison, D. A. and Siegbahn, K. (1971). *Phys. Scr.*, **3**, 237]

The change in the net charge, q_v, due to valence electrons of the atom from which the core electron is being removed, and also those charges, q_j, of the neighbouring atoms can be estimated using the expression

$$q = \sum_i I_i + Q \tag{7.34}$$

where Q is the formal charge, if any, on the atom and I_i is the ionic character of the ith bond. The method of Pauling[190] is used to calculate I_i from the expression

$$I_i = \frac{\chi_A - \chi_B}{|\chi_A - \chi_B|}\{1 - \exp[-0.25(\chi_A - \chi_B)^2]\} \tag{7.35}$$

for a bond between atoms A and B. Values for the electronegativity, χ, for all atoms are taken to be those on the Pauling scale and are given in his book.

For a molecule which can be thought of as having several resonance forms (in the valence bond theory sense), q is obtained from a weighted average of its values for the various forms.

The Siegbahn method has undergone several modifications. In particular, the net charge on an atom can be calculated by MO methods, and chemical shifts can be calculated directly by the *ab initio* Hartree–Fock method taking into account *all* electrons.

Figures 7.31 and 7.32 show further examples of the chemical shift effect in XPS spectra.

Figure 7.31 shows the C 1*s* spectra of furan, pyrrole and thiophene. Owing to the decreasing electronegativity of the heteroatom in the order $\chi_O > \chi_N > \chi_S$, the C 1*s* line is shifted to low

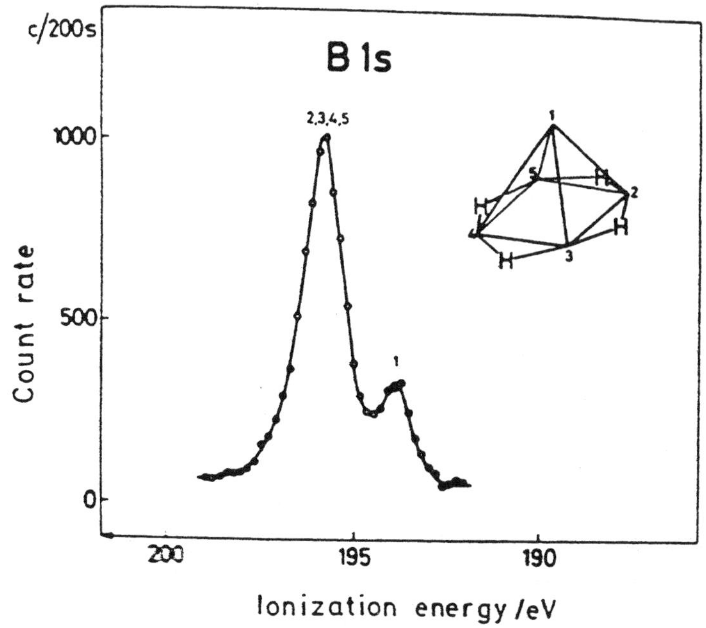

Figure 7.32 The Mg $K\alpha$ boron 1*s* XPS spectrum of the B_5H_9 molecule showing that only four of the boron atoms are in identical environments
[Reproduced, with permission, from Allison, D. A., Johansson, G., Allan, C. J., Gelius, U., Siegbahn, H., Allison, J. and Siegbahn, K. (1972–73). *J. Electron Spectrosc.*, **1**, 269]

ionization energy from furan to pyrrole to thiophene. In addition, the C 1*s* line is split into two components due to the two kinds of carbon atom environment. The carbon atom in position 2 is responsible for the C 1*s* component of higher ionization energy because it is nearer to the heteroatom than that in position 3. The splitting decreases with electronegativity and can only be observed for thiophene after deconvolution of the spectrum.

Also shown in Figure 7.31 is the line due to removal of a 2*p* electron from the sulphur atom in thiophene. Spin–orbit coupling is sufficient to split the resulting state into $^2P_{\frac{3}{2}}$ and $^2P_{\frac{1}{2}}$ components, the multiplet being inverted.

Figure 7.32 shows the B 1*s* spectrum of the B_5H_9 molecule. The boron atoms are situated at the corners of a square pyramid. There are four B—H—B bridging hydrogen atoms and there is also a terminal hydrogen attached to each boron. The four equivalent boron atoms at the base of the pyramid give rise to the intense component at higher ionization energy than the component due to the one at the apex.

The example of B_5H_9 serves to show how the chemical shift may be used as an aid to determining the structure of a molecule and, in particular, in deciding between alternative structures. There are many examples in the literature of this kind of application which is reminiscent of the way in which the chemical shift in NMR may be employed. However, there is one important difference in using the two kinds of chemical shift. In XPS there are no interactions affecting closely spaced lines in the spectrum, however close they may be. Figure 7.31 illustrates this for the C 1*s* lines of thiophene. In NMR, the spectrum becomes more complex, due to spin–spin interactions, when chemical shifts are similar.

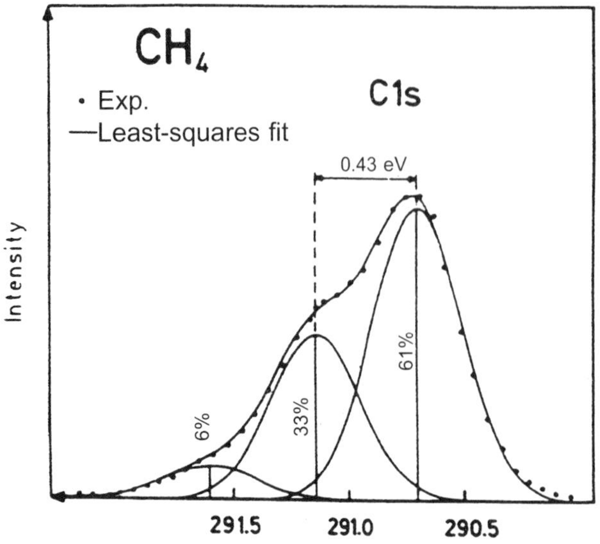

Figure 7.33 A short vibrational progression in the ν_1 vibration of methane in the carbon 1*s* XPS spectrum obtained with a monochromatized X-ray source [Reproduced, with permission, from Gelius, U., Svensson, S., Siegbahn, H., Basilier, E., Faxålv, Å. and Siegbahn, K. (1974). *Chem. Phys. Lett.*, **28**, 1]

7.5.2.2 Line widths in XPS

Without the use of an X-ray monochromator, the line width of, for example, Al $K\alpha$ radiation is about 1 eV due to doublet structure, satellite lines and *bremsstrahlung* (Section 7.2.2). Under these conditions the line width in the XPS spectrum is seriously limited and small effects on line shape cannot be observed.

With a monochromator, described in Section 7.2.2, a line width at half-intensity of 0.2 eV for Al $K\alpha$ radiation has been obtained. This makes it possible to investigate line width differences due to the lifetime of the state produced which, in turn, depends on such accompanying processes as Auger transitions (Section 7.1) and electron shakeup (Section 7.5.2.3).

The best resolution in XPS, limited to 0.2 eV (1600 cm^{-1}) by the source line width, is still very much worse than that of 4 meV (32 cm^{-1}) in UPS. Nevertheless, it is possible, in favourable circumstances, to observe some vibrational structure in XPS spectra. Figure 7.33 shows such structure in the C 1*s* spectrum of methane. The removal of a C 1*s* electron causes the other electrons to be more tightly bound, resulting in a reduction of C—H bond lengths and a general shrinkage of the molecule. The shrinkage is in the direction of ν_1, the symmetric C—H stretching vibration. The resulting progression in ν_1 can be understood in terms of the Franck–Condon principle [Section 6.3.4.2(i)]. The wavenumber of the vibration increases from 2917 cm^{-1} in CH_4 to about 3500 cm^{-1} in this state of CH_4^+, consistent with the shortening of the C—H bonds.

7.5.2.3 Electron shakeup and shakeoff satellites

Simultaneously with the removal of a core electron, another electron may be promoted to a higher energy orbital, in an electron shakeup process, or be removed altogether, in an

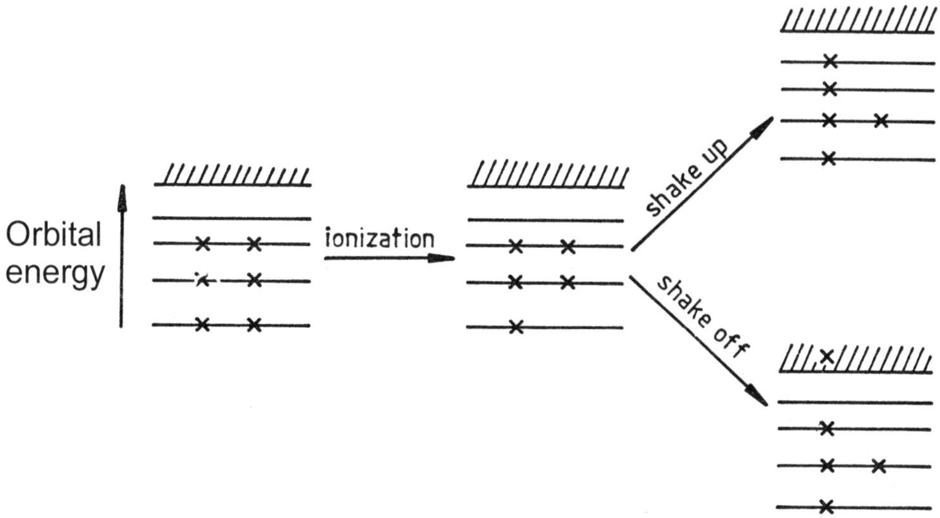

Figure 7.34 Electron shakeup and shakeoff processes following ionization

electron shakeoff process. Both processes are illustrated in Figure 7.34. The energy required for either process is provided from the kinetic energy of the photoelectron produced in the initial ionization, so the satellites appear on the low kinetic energy side of the main peak in the XPS spectrum.

Figure 7.35 shows shakeup and shakeoff satellites of the $1s$ XPS line of neon.

The promotion or removal of a second electron accompanying the removal of a core electron is a result of the second electron suddenly experiencing a change of potential. This is caused by a change in the shielding of the nucleus which, if it is a $1s$ electron that is removed initially, is considerable. The probability, P, of the second electron moving from an initial orbital i to a final orbital f is given by

$$P = \left[\int \chi_f^* \chi_i \mathrm{d}\tau \right]^2 \tag{7.36}$$

where χ_i and χ_f are the orbital wave functions. This expression for P is reminiscent of that for $|\boldsymbol{R}^{nm}|^2$ for an electric dipole transition, where the transition moment $\boldsymbol{R}^{nm}$ is given in equation (2.21), except that, here, the probability does not involve the electric dipole moment operator $\boldsymbol{\mu}$. The reason for this is that the shakeup, or shakeoff, process involves a monopole transition, the ion itself being a monopole.

The selection rules for monopole transitions are

$$\Delta l = 0,\ \Delta s = 0 \text{ and, therefore, } \Delta j = 0$$

for the electron involved. This means that a change of n only can occur in shakeup or shakeoff processes.

The most intense shakeup series in the Ne spectrum in Figure 7.35 consists of the lines labelled 3, 4, 5, 6, 7 and 9. It has been shown that the highest shakeup and shakeoff probabilities are for valence electrons. Taking this, and the selection rules, into account we expect the most intense shakeup series to involve promotion of a $2p$ electron to np orbitals with $n > 2$.

Table 7.5 indicates the assignments of lines in the spectrum to this series.

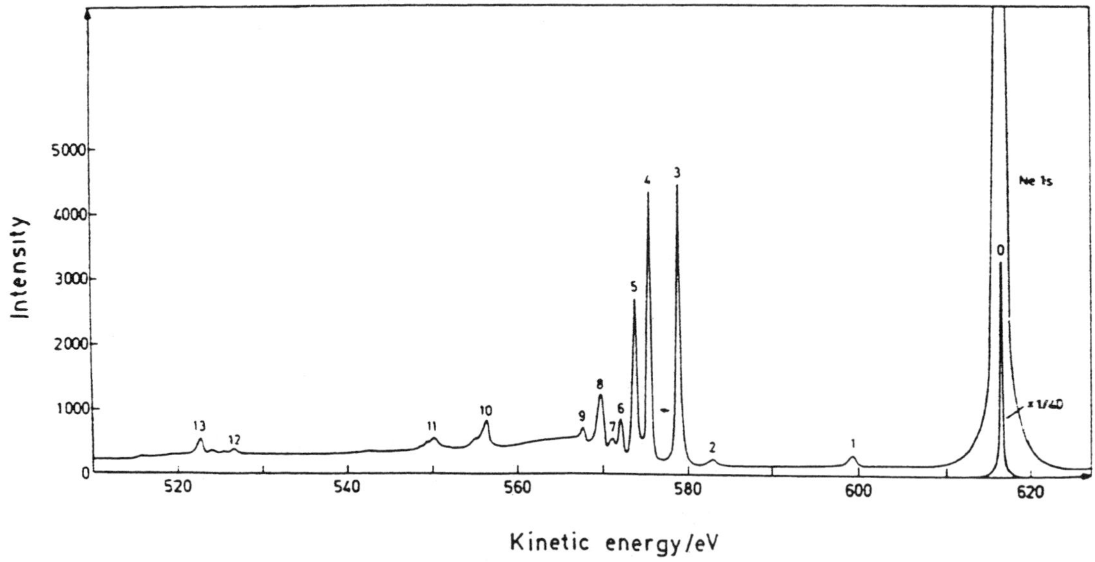

Figure 7.35 Shakeup and shakeoff satellites of the neon 1*s* XPS line using monochromatized Al *Kα* (1486 eV) radiation
[Reproduced, with permission, from Gelius, U., Basilier, E., Svensson, S., Bergmark, T. and Siegbahn, K. (1974). *J. Electron Spectrosc.*, **2**, 405]

Table 7.5 Shakeup and shakeoff spectrum of neon

Line in spectrum in Figure 7.35	Configuration of Ne^+(Ne^{2+} or Ne)
0	$1s^1 2s^2 2p^6$ (^2S)
1	(Ne) $1s^2 2s^2 2p^5 3s$
2	$1s^1 2s^2 2p^5 3s^1$(?)
3	$1s^1 2s^2 2p^5 3p^1$ (^2S, triplet)
4	$1s^1 2s^2 2p^5 3p^1$ (^2S, singlet)
5	$1s^1 2s^2 2p^5 4p^1$ (^2S, triplet)
6	$1s^1 2s^2 2p^5 5p^1$ (^2S, triplet)
7	{ $1s^1 2s^2 2p^5 6p^1$ (^2S, triplet); $1s^1 2s^2 2p^5 4p^1$ (^2S, singlet) }
8	(Ne^{2+}) $1s^1 2s^2 2p^5$ (^3P)
9	$1s^1 2s^2 2p^5 5p^1$ (^2S, singlet)
10	$1s^1 2s^1 2p^6 3s^1$ (^2S, triplet)
11	$1s^1 2s^1 2p^6 3s^1$ (^2S, singlet)
12, 13	Shakeup of two electrons (?)

For each $np \leftarrow 2p$ promotion, two states of Ne^+ are produced. We can regard these as a combination of a ^2S 'core state' with a triplet or a singlet 'valence state', as indicated in Table 7.5. The two states are appreciably split and the series converges towards a continuum. The continuum is due to the shakeoff of a 2*p* electron in Ne^+ to give Ne^{2+} in a ^3P state. Line 8 marks the onset of this continuum.

Lines 10 and 11 are due to the $3p \leftarrow 2s$ shakeup and are very much weaker than those involving shakeup of 2*p* electrons because they violate the $\Delta l = 0$ monopole selection rule.

Lines 12 and 13 may be due to a two-electron shakeup and line 2 to the monopole-forbidden $3s \leftarrow 2p$ promotion.

Line 1 is due to a process which itself forms another branch of spectroscopy—energy loss spectroscopy. The line is a result of impact between a photoelectron resulting from ejection from the

Ne 1*s* orbital and a neutral neon atom. The photoelectron suffers energy loss in promoting an electron from the 2*p* to the 3*s* orbital, in a process allowed by electric dipole selection rules.

Shakeup and shakeoff satellites have also been observed in the XPS spectra of molecules. Again, monopole selection rules are obeyed so that there must be no change of symmetry during an electron promotion. The main difficulty in interpreting satellites in molecular spectra is that there is no information from other sources, experimental or theoretical, on excited states of molecular unipositive ions in which the vacancy is in the core.

BIBLIOGRAPHY

BURHOP, E. H. S. (1952). *The Auger Effect and Other Radiationless Transitions*. Cambridge University Press, London

CARLSON, T. A. (1975). *Photoelectron and Auger Spectroscopy*. Plenum, New York

DEKOCK, R. L. and LLOYD, D. R. (1974). 'Vacuum ultraviolet photoelectron spectroscopy of inorganic molecules', in *Advances in Inorganic Chemistry and Radiochemistry*, vol. 16, pp. 65–107. Academic Press, New York

ELAND, J. H. D. (1974). *Photoelectron Spectroscopy*. Butterworths, London

RABALAIS, J. W. (1977). *Principles of Ultraviolet Photoelectron Spectroscopy*. Wiley, New York

SIEGBAHN, K., NORDLING, C., FAHLMAN, A., NORDBERG, R., HAMERIN, K., HEDMAN, J., JOHANSSON, G., BERGMARK, T., KARLSSON, S.-E., LINDGREN, I. and LINDBERG, B. (1967). *Electron Spectroscopy for Chemical Analysis—Atomic, Molecular, and Solid State Structure Studies by Means of Electron Spectroscopy*. Almqvist and Wiksells, Uppsala

TURNER, D. W., BAKER, C., BAKER, A. D. and BRUNDLE, C. R. (1970). *Molecular Photoelectron Spectroscopy*. Wiley, London

8 LASERS AND LASER SPECTROSCOPY

8.1 LASERS AND MASERS

8.1.1 General features of lasers and their design

The word 'laser' is an acronym derived from *l*ight *a*mplification by the *s*timulated *e*mission of *r*adiation. If the 'light' concerned happens to be in the microwave region then the word 'maser' is used, this also being an acronym with 'm' derived from *m*icrowave. Since we shall encounter only one maser, the ammonia maser, we shall use the word 'laser' when discussing these devices in general.

Lasers are sources of radiation with unique properties which will be discussed in Section 8.1.2. They operate by the process of induced emission which has been discussed in Chapter 2 where it is represented by equation (2.10) and illustrated in Figure 2.2. It is this process, rather than spontaneous emission [equation (2.12)], which is of prime importance in lasers. For induced emission, from the upper energy level n to the lower level m, shown in Figure 2.2, to dominate induced absorption [equation (2.11)] there must be a population inversion between the two levels, i.e. $N_n > N_m$, as discussed in Section 2.2. The equilibrium populations are given by the Boltzmann expression of equation (2.6) or, if there are degeneracies associated with either of the energy levels, equation (2.8). To disturb this Boltzmann distribution and create a population inversion requires an input of energy. The process by which such an inversion is brought about is known as pumping. A system in which a population inversion has been created is called an active medium and this medium may be gaseous, solid or liquid.

We consider an active medium of length dz under the influence of a photon flux F in the z direction, as in Figure 8.1. Provided the photons are of frequency ν, or wavenumber $\tilde{\nu}$, where, as in equation (2.4),

$$E_n - E_m = h\nu = hc\tilde{\nu} \tag{8.1}$$

and E_n and E_m are the energies of the upper and lower states, respectively, they will stimulate emission in the active medium. As a result, the outgoing flux is increased to $F + \mathrm{d}F$, and we have

$$\frac{\mathrm{d}F}{\mathrm{d}z} = \sigma_{nm} F (N_n - N_m) \tag{8.2}$$

where σ_{nm} is the stimulated emission cross-section of the atoms or molecules of the active medium. If there is a population inversion $N_n > N_m$ and $\mathrm{d}F/\mathrm{d}z > 0$. Under these circumstances the active medium acts as an amplifier having the property of photon amplification or gain. Otherwise, if $N_n < N_m$, the medium is not an active one but acts as an absorber.

To make an oscillator from an amplifier requires, in the language of electronics, positive

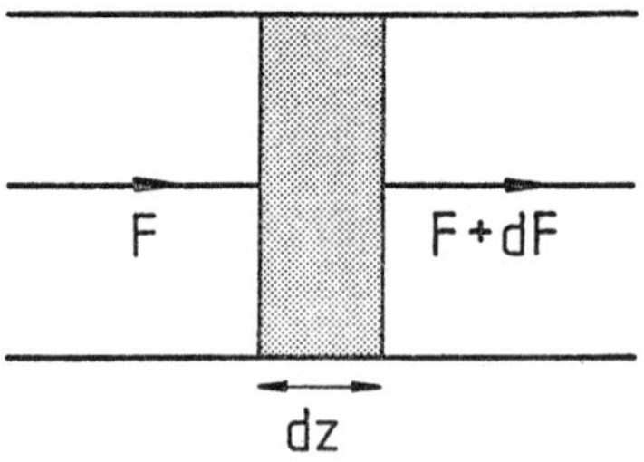

Figure 8.1 An active medium of length dz amplifies the ingoing photon flux F to $F + \mathrm{d}F$

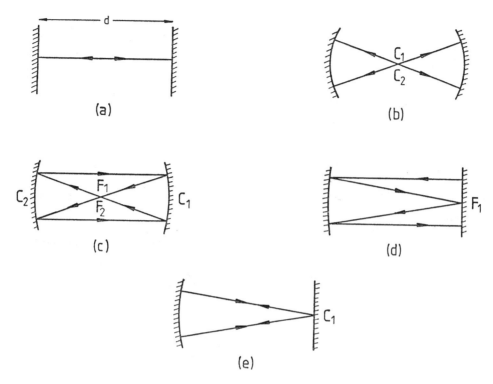

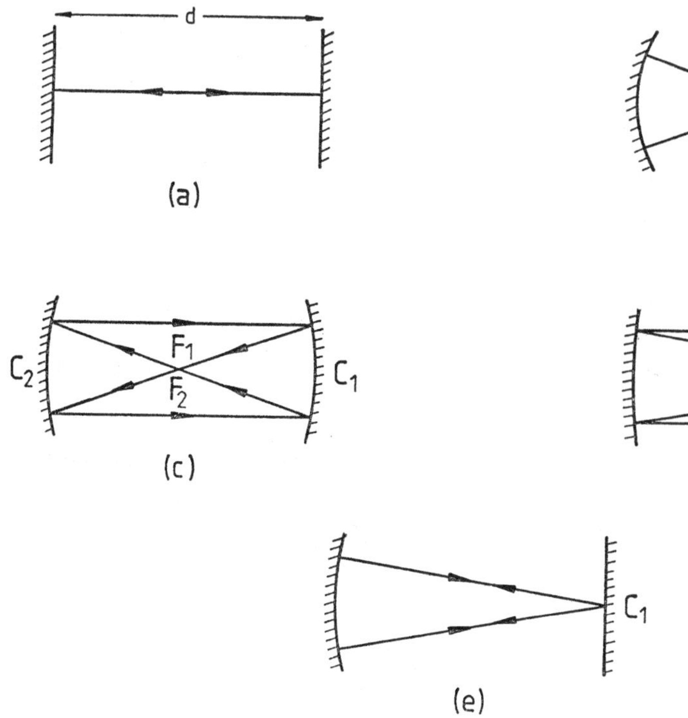

Figure 8.2 Laser cavity mirror systems: (a) two plane mirrors; (b) concentric resonator; (c) confocal resonator; (d) hemiconfocal resonator; (e) hemispherical resonator. The points C and F are the centre of curvature and focal point, respectively, of the concave mirrors

feedback. In a maser this is provided by the microwave cavity which is resonant at the maser frequency. For other lasers the feedback is produced, in most cases, by the active medium being between two mirrors, both highly reflecting but one rather less so in order to allow some of the laser radiation to leak out. The region bounded by the mirrors is the laser cavity.

Various mirror systems for the cavity are illustrated in Figure 8.2. The system shown in Figure 8.2(a) is the simplest employing two plane, parallel mirrors a distance d apart. This distance must be an integral number of half-wavelengths $n\lambda/2$, necessitating extremely accurate alignment. The resonant frequency, ν, of the cavity is thus given by

$$\nu = nc/2d \tag{8.3}$$

Figure 8.2(b) shows a system in which two spherical mirrors are placed with their centres of curvature, C_1 and C_2, coincident and half way between the mirrors. The resulting cavity is called a concentric resonator. Figure 8.2(c) shows a confocal resonator in which the focal points, F_1 and F_2, of the spherical mirrors are coincident and half way between them (the focal length is half the radius of curvature). Resonators with two spherical mirrors, with separation intermediate between that in Figure 8.2(b) and (c) are also used. Figure 8.2(d) and (e) show how a spherical and a plane mirror may be used in a hemiconfocal and hemispherical resonator, respectively.

In all these examples the separation, d, of the mirrors is $n\lambda/2$ and the resonant frequency of the cavity is given by equation (8.3).

The reflecting surfaces of the mirrors are specially coated for almost total reflection at the laser wavelength—the usual aluminium, silver or gold coatings are not sufficiently highly reflecting. These special coatings are made by depositing layers, each of thickness $\lambda/4$, of alternately high and low dielectric materials on to a highly polished substrate of an appropriate material such as quartz (visible and near ultraviolet) and zinc selenide or germanium (near infrared). Layers of zinc sulphide, ZnS, and cryolite, Na_3AlF_6, produce a soft coating which is relatively easily damaged, whereas titanium dioxide, TiO_2, and silicon monixide SiO, produce a hard coating. More than 20 layers are commonly used for a reflectivity of more than 99%. The output mirror, through which the laser radiation is allowed to leak, is made specifically to allow of the order of 1% of the radiation to leak out, for a low gain laser line, and of the order of 10% for a high gain line.

The stimulated emission in the active medium is produced initially by spontaneous emission which provides the photon flux. As Figure 8.2 shows for various types of cavity, the photon flux which falls on either of the mirrors is reflected backwards and forwards, thereby stimulating further emission. For each pass along the length of the cavity the gain, G, obtained by integrating equation (8.2), is given by

$$G = \exp[\sigma_{nm}(N_n - N_m)d] \tag{8.4}$$

For reflectivities R_1 and R_2 of the two mirrors, the threshold for oscillation occurs when

$$R_1 R_2 \exp[2\sigma_{nm}(N_n - N_m)d] = 1 \tag{8.5}$$

The value of $(N_n - N_m)$ which satisfies this requirement is called the critical inversion. This equation makes it clear that it is not only a population inversion which is necessary for a laser to operate.

8.1.2 Properties of laser radiation

The properties of laser radiation which make a laser such a unique source are that it is (a) highly directional, (b) monochromatic, (c) extremely bright and (d) both spatially and temporally coherent.

The directionality results in a highly parallel beam emerging from the output mirror and is a consequence of the strict requirements for the alignment of the two mirrors at the ends of the cavity. Deviation from being an exactly parallel beam is typically a few milliradians.

The monochromatic nature of the radiation may be due to either or both of two factors. If the energy levels n and m are sharp, as they are in a gaseous active medium, the Planck relation of equation (8.1) limits the wavelength range of the radiation. However, whatever the nature of the active medium, the fact that the laser cavity is resonant only for the frequencies given by equation (8.3) limits the wavelength range.

The brightness, defined as the power emitted from unit area of the output mirror per unit solid angle, is extremely high compared with a conventional source. The reason for this is that, although the power may be only modest, as in, for example, a low power 0.5 mW helium–neon gas laser, the solid angle over which it is distributed is very small.

Conventional sources of radiation are incoherent, which means that the electromagnetic waves associated with any two photons are, in general, out-of-phase. The spatial and temporal coherence of laser radiation, due to the waves associated with *all* photons being in-phase, is itself responsible for some of its remarkable properties such as its use as a source of intense local heating for metal cutting and welding, and in holography.

Spatial coherence implies that the electric field experienced at any two points on the wave front of a travelling electromagnetic wave remains constant for all positions of the wave. In cases where spatial coherence is not complete we may define a coherence area over which it is.

Temporal coherence implies that the wavelength, or frequency of a particular wave is constant with time. When temporal coherence is incomplete we may define a coherence time during which it is complete.

8.1.3 Methods of obtaining population inversion

A casual glance at equation (2.10) may give the impression that, in induced emission, we are getting something for nothing—putting in one quantum of energy and getting out two! Of course, this is not the case because we have, initially, to put in a quantum of energy to excite M to M*: this is the process of pumping. Not only are we not getting something for nothing, but it is also the case that most lasers operate at very low efficiency. For example, a nitrogen gas laser has a typical efficiency, which is the ratio of the pumping power to the output power of the laser radiation, of less than 0.1% and a semiconductor (diode) laser, one of the best in this respect, has an efficiency of about 30%.

Before we look at the various methods of pumping, we shall consider the types of energy level schemes encountered in lasing materials.

So far we have thought of the stimulated emission occurring in a lasing material as being in a simple two-level system like that in Figure 8.3(a), but it can easily be shown that a laser cannot be made to operate by pumping such a system. Since level 2 is higher in energy than level 1 it will, under equilibrium conditions, have a lower population. If it is a high-lying vibrational or an electronic energy level the population will be negligibly small. Pumping with energy $E_2 - E_1$ results, initially, in net absorption which continues until the populations N_1 and N_2 are equal, a condition known as saturation and encountered in Section 2.3.5. At this point further pumping results in absorption and induced emission occurring *at the same rate*, so that population inversion cannot be attained.

Commonly a three- or four-level system, illustrated in Figure 8.3(b) and (c), is necessary for population inversion between two of the levels to be obtained.

In the three-level system of Figure 8.3(b), population inversion between levels 2 and 1 is achieved by pumping the 3–1 transition. Level 2 is lower in energy than 3 and the 3–2 transition must be an efficient, fast process.

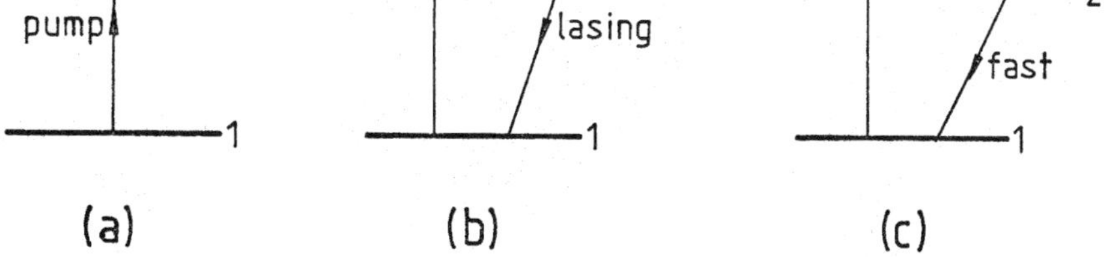

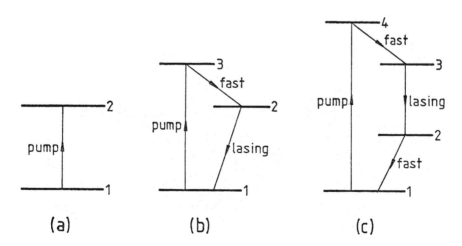

Figure 8.3 (a) Two-, (b) three- and (c) four-level systems. Only the three- and four-level systems can result in laser action by pumping the transitions indicated

The four-level system of Figure 8.3(c) is even more efficient in creating a population inversion, in this case between levels 3 and 2. The reason for the greater efficiency is that not only is level 3 populated through the fast 4–3 process, but also the population of level 2 is rapidly depleted by the fast 2–1 process.

In more complex systems there may be more levels between 4 and 3 and also 2 and 1, all involving fast processes to lower levels, but they are still referred to as four-level systems.

Population inversion is not only difficult to achieve but also to maintain. Indeed, for many laser systems there is no method of pumping which will maintain a continuous population inversion. For such systems inversion can only be brought about by means of a source which delivers energy in short, high energy pulses. The result is a pulsed laser as opposed to a continuous wave, or CW, laser.

Methods of pumping, irrespective of the system of levels involved and of whether lasing is to be pulsed or CW, fall into two general categories—optical and electrical pumping.

Optical pumping involves the transfer of energy to the system from a high intensity light source and is used particularly for solid and liquid lasers.

A very high photon density can be obtained over a short period of time from inert gas flashlamps of the type used in flash photolysis and described in Section 3.8.1. High capacity condensers are discharged between electrodes in the flashlamp containing an inert gas, usually Xe, at a pressure of about 100 Torr. The result is a pulsed laser, the repetition rate being that of the pumping source.

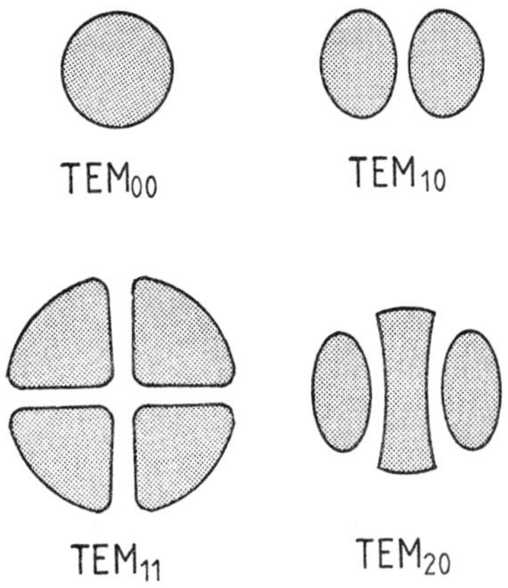

Figure 8.4 Four of the possible transverse modes, TEM_{ml}, of a laser cavity

CW optical pumping may be achieved by a continuously acting tungsten–iodine, krypton or high pressure mercury lamp.

Electrical pumping is used for gas and semiconductor lasers. In a gas laser this involves an electrical discharge in the gas which may be induced by microwave radiation outside the gas cell or by a high voltage across electrodes inside the cell.

In an electrical discharge electrons are produced by ionization in strong local electric fields. These electrons achieve high translational energy some of which may be transferred on collision to the atoms or molecules.

$$M + e \rightarrow M^* + e \tag{8.6}$$

which are thereby pumped, prior to laser action, into level 3 [Figure 8.3(b)] or level 4 [Figure 8.3(c)]. The process involved is the electron impact mechanism of excitation discussed in Section 7.1.

In some gas lasers it is preferable to use a mixture of gases M and N, where N serves only to transfer energy from the electrons to M. N is excited to N*, ideally a long-lived metastable state, by electron impact [equation (8.6)]. If the level of M to be pumped is of similar energy to that of N*, energy transfer between M and N* occurs on collision:

$$M + N^* \rightarrow M^* + N \tag{8.7}$$

8.1.4 Laser cavity modes

A few lasers, such as the nitrogen gas laser, are super-radiant, which means that the gain of the active medium is so high that stimulated

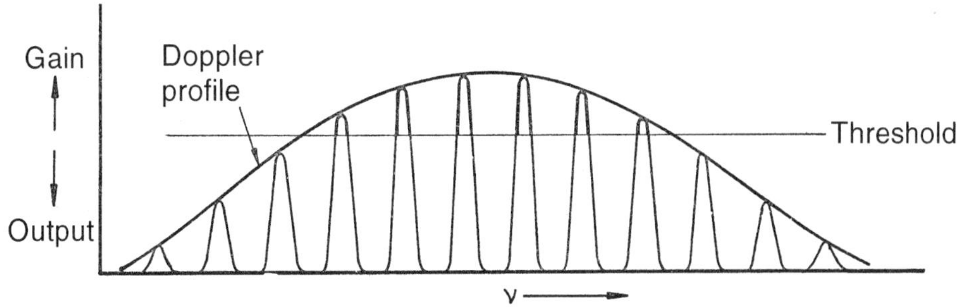

Figure 8.5 Gain profile of a Doppler-limited laser line showing 12 axial modes of the cavity within the Doppler profile

emission occurs without the necessity for any mirrors.† However, this is not normally the case and we have seen how the two mirrors, which are usually necessary, form a cavity which is resonant for a particular laser wavelength. For such a cavity there may be many modes of oscillation which have long been familiar to those working with resonant microwave cavities and waveguides. The cavity modes are of two types, the transverse and axial (or longitudinal) modes which are, respectively, normal to and along the direction of propagation of laser radiation in the cavity.

The transverse modes are labelled conventionally TEM_{ml} where TEM stands for *t*ransverse *e*lectric and *m*agnetic (field) and m and l are integers.

Transverse modes can be thought of as resulting from the various ways, apart from a straight line, in which a photon travelling from one mirror to the other can be amplified. By following a zig-zag path the gain will be higher than for a straight line. Figure 8.4 shows the cross-section of the laser beam when the cavity is oscillating in various TEM modes. As the figure is drawn, m and l correspond to the number of nodal planes in the vertical and horizontal directions, respectively. Usually it is preferable to use only the TEM_{00} mode which produces a laser line with a gaussian intensity distribution as a function of distance measured outwards from, and normal to, the optic axis of the laser. Other TEM modes may be suppressed in various ways, one of which involves placing an aperture stop inside the laser cavity. An aperture stop is simply a device, such as a piece of flat metal sheet with a hole cut in it, to restrict the passage of photons.

The fact that the cavity has axial modes leads to the laser radiation consisting of a number of frequencies given by equation (8.3), in which the integer n can take, in principle, any value. The frequency separation $\Delta\nu$ of the axial modes is given by

$$\Delta\nu = c/2d \qquad (8.8)$$

For a cavity length of, for example, 50 cm the mode separation is 300 MHz (0.01 cm^{-1}).

In practice the laser can operate only when n, in equation (8.3), takes values such that the corresponding resonant frequency ν lies within the width of the transition between the two energy levels involved. If the active medium is a gas the gain profile of the laser may well be limited by the Doppler line shape (Section 2.3.2). For the case illustrated in Figure 8.5 there are 12 axial modes within the Doppler profile. The number of modes in the laser radiation depends on the magnitude of the output which is allowed to leak out of the cavity through the output mirror. The so-called

†Strictly, super-radiance refers to amplified spontaneous emission but it has come to be used in referring to stimulated emission also.

threshold condition is that which obtains when gain in the active medium exceeds losses which occur inside the cavity as well as loss due to the output radiation. For the threshold indicated in Figure 8.5, six axial modes are observed. This number is increased if the output power is increased.

It is clear from Figure 8.5 that the laser line width for a single axial mode is much less than the Doppler line width. It may also be less than the natural line width (Section 2.3.1) and also less than the line width the laser would have if the cavity were simply acting as a Fabry–Perot interferomotric radiation filter (Section 3.5.4).

In most lasers the lower limit to the line width is imposed by the Q of the cavity (see Section 8.1.5). This line width is the $\Delta\nu$ in equation (8.9) and, according to equation (8.10), can be decreased by decreasing E_t/t, the energy leaked out from the cavity per unit time. By minimizing this, a line width of, for example, 1 Hz in a helium–neon laser can be obtained.

Unless special modifications are made to a laser it will normally operate in a multimode fashion, having several axial modes within the gain profile. On the other hand, there are ways of operating a laser in a single mode fashion. Since this results in a line width considerably less than the Doppler width it is desirable, for high resolution work, to employ such single mode operation.

One way of achieving this is clear from equation (8.8). If the cavity length d can be made sufficiently short that only one axial mode lies within the Doppler profile, single mode operation will result. In the infrared region the Doppler line width is relatively small, of the order of 100 MHz, and this is a standard method of obtaining single mode operation. However, the laser power is also decreased when a short cavity is used.

A visible or ultraviolet gas laser cannot be made to operate in a single mode by this method because of the larger Doppler line width. Similarly, the method cannot be used with solid or liquid lasing materials because of the large line width. In all such cases a Fabry–Perot etalon (Section 3.5.4) or a reflection interferometer inside the laser cavity is commonly used.

8.1.5 *Q*-switching

One important property of any laser cavity is its quality factor, Q. Since

$$Q = \nu/\Delta\nu \tag{8.9}$$

where $\Delta\nu$ is the line width, Q can be regarded as the 'resolving power' of the cavity, as defined in equation (3.4) for the dispersing element of a spectrometer. Single mode operation results in a smaller $\Delta\nu$ and, therefore, a higher Q than multimode operation. Q is related to the energy stored in the cavity, E_c, and the energy, E_t, allowed to leak out in time t by

$$Q = \frac{2\pi\nu E_c t}{E_t} \tag{8.10}$$

Q-switching is an operation which involves reducing the Q of a laser cavity for a short period of time by reducing the feedback of the mirrors. During this time the population of the upper of the two levels involved in laser action builds up to a much higher value than it would if the Q remained high. Then Q is allowed to increase rapidly and the result is a pulse which is shortened compared with the original one. Since the pulse power, P_p, is related to the pulse energy, E_p, by

$$P_p = E_p \Delta t \tag{8.11}$$

the power increases as the pulse duration, Δt, decreases. The resulting pulse is called a giant pulse. Although the pulse energy for a Q-switched laser is lower, owing to unavoidable losses in Q-switching, than if it were not Q-switched, the power is greatly increased owing to the reduction of Δt.

The brief suppression of mirror feedback is accomplished by some kind of shutter which can be switched on and off—a so-called Q-

switch. There are several types of Q-switch, of which three have been most commonly used.

(1) *Rotating mirror*. If the mirror at the end of the cavity opposite to the output mirror is rotated rapidly about a vertical axis, the Q of the cavity is large only during that short period of time when the mirror is in the normal laser cavity orientation. For the rest of the time the Q is low. This method of Q-switching may be used for a laser of any wavelength. Unfortunately, the result with this kind of Q-switch is often to produce a train of closely spaced pulses rather than a single pulse. For a visible or ultraviolet laser, introduction of a Lummer–Gehrke plate interferometer into the cavity converts to single pulse operation.

(2) *Pockels (or Kerr) cell.* An electro-optic material is one which, if an appropriate voltage is applied across it, becomes birefringent (doubly refracting). The result is that, if plane polarized radiation passes through the material, it emerges, in general, elliptically polarized. If the birefringence is proportional to the square of the applied electrical field the resulting effect is known as the Kerr effect, whereas if it is simply proportional to the electric field, it is the Pockels effect. A material which exhibits a large Kerr effect is liquid nitrobenzene. A Kerr cell containing nitrobenzene may be used as a Q-switch but nitrobenzene has the disadvantage of being unstable. Since it must also be very pure, it is not very suitable for Q-switching.

More often used is a Pockels cell, the Pockels effect being exhibited by such pure crystalline materials as ammonium dihydrogenphosphate [$(NH_4)H_2PO_4$] or ADP, potassium dihydrogenphosphate (KH_2PO_4) or KDP and potassium dideuteriumphosphate ($K^2H_2PO_4$) or KD*P. Only a non-centrosymmetric crystal, i.e. one whose unit cell does not have a centre of symmetry, exhibits a Pockels effect. The effect may be transverse or longitudinal, as shown in Figure 8.6(a) and (b), respectively. When used as a Q-switch, a Pockels cell is employed in the longitudinal mode and the voltage applied is such as to produce circular polarization.

Figure 8.7 shows how a Pockels cell (PC) can be used in a laser cavity. If the laser radiation is not naturally plane-polarized then it must be made so by the polarizer P. Radiation passing through this to the Pockels cell becomes circularly polarized when the voltage is applied. On reflection at the mirror M_1 the direction of circular polarization is reversed. On passing through the Pockels cell a second time it is plane-polarized, but at 90° to the original direction and, therefore, is not transmitted by the polarizer. It is only when the voltage is switched off that a Q-switched giant pulse emerges from the

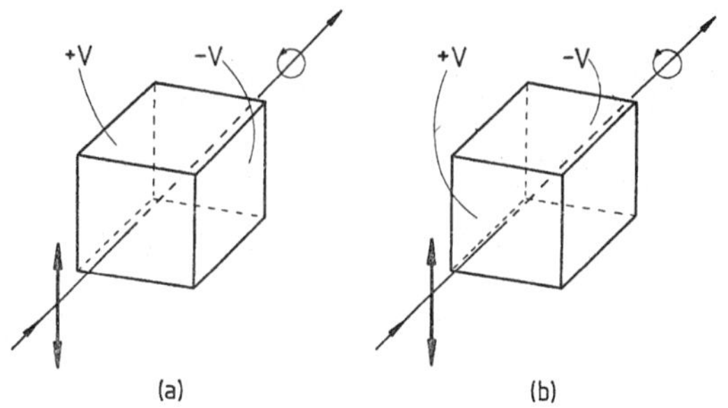

Figure 8.6 (a) Transverse, and (b) longitudinal Pockels effect on plane polarized light

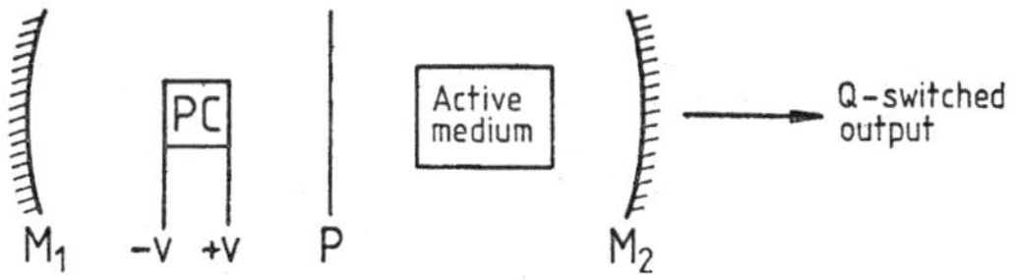

Figure 8.7 Use of a Pockels cell (PC) in a laser cavity to produce Q-switching

output mirror M_2. The timing of the voltage switching determines the power and length of the output pulses.

(3) *Saturable absorber*. A saturable dye is an example of a saturable absorber in the visible region and is one which rapidly bleaches when subjected to radiation of specific minimum intensity. Examples of such dyes are cryptocyanine and vanadium phthalocyanine. The dye is placed in a cell in the laser cavity. Before the dye is bleached it prevents radiation reaching one of the laser mirrors and reduces the Q of the cavity. When bleaching occurs the Q is rapidly increased and a giant pulse is output from the laser. For an infrared laser a saturable absorber such as formic acid (HCOOH) or methyl alcohol (CH_3OH) may be used.

8.1.6 Mode locking

When Q-switching is employed, as described in the previous section, the operation of the laser is normally multimode, in the axial mode sense, and the giant pulses are of the order of 10–200 ns long. We assume that the laser is operating in a single transverse mode, normally TEM_{00}. Even higher power, due to a much shorter pulse length, may be achieved by the technique of mode locking. The technique is applicable only to multimode operation and, since a larger number of modes can be excited only if the laser line is relatively wide, it is used principally with solid and liquid phase lasers.

Mode locking involves exciting many axial modes but with the correct amplitude and phase relationship. The amplitudes and phases of the various modes are normally random.

Each axial mode has its own characteristic pattern of nodal planes and the frequency separation $\Delta\nu$ between modes is given by equation (8.8). If the radiation in the cavity can be modulated at a frequency of $c/2d$, where the time t_r for the radiation to make one round-trip of the cavity (a distance of $2d$) is given by

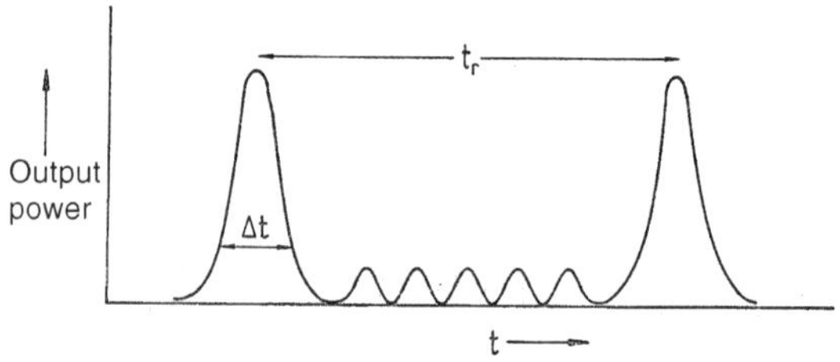

Figure 8.8 Suppression of five out of seven axial cavity modes by mode locking

$$t_r = 2d/c \tag{8.12}$$

then the modes of the cavity are locked in both amplitude and phase. The result is that, for the case of a cavity operated with, for example, seven modes, when the cavity is mode locked the time behaviour of the output is like that shown in Figure 8.8. Only the modes which have a node at one end of the cavity are output from the laser and all the others are suppressed.

The width Δt of the pulse at half power is given by

$$\Delta t = \frac{2\pi}{(2N+1)\Delta\nu} \tag{8.13}$$

where $(2N+1)$ is the number of axial modes excited and $\Delta\nu$ is the frequency separation of the modes. This equation shows that as more modes are excited in multimode operation the greater is the pulse shortening produced by mode locking.

A numerical example serves to show how dramatic the effect of mode locking may be.

For a cavity whose length d is 60 cm, t_r is 4 ns. Locking together of 100 modes gives a peak power which is 10^4 times that of each mode. The repetition rate of the mode locked pulses is 250 MHz and Δt is 40 ps.

Mode locking may be achieved by an acoustic modulator placed in the cavity and driven at a frequency of $c/2d$ or, alternatively, a saturable absorber may be used.

An important consequence of shortening a laser pulse is that the line width is increased.

This is a result of the uncertainty principle as it is stated in equation (1.29). When the width of the pulse is very small there arises a difficulty in measuring the energy (or frequency) precisely because of the rather small number of wavelengths in the pulse. For example, for a pulse width Δt of 40 ps there is a frequency spread of the laser radiation, given approximately by $(2\pi\Delta t)^{-1}$, of about 4.0 GHz (0.13 cm^{-1}).

8.1.7 Harmonic generation

In Section 4.8.1 the Raman effect was discussed. The Raman scattering results from the electric dipole $\boldsymbol{\mu}$ induced in the sample by the incident radiation which has an oscillating electric field $\boldsymbol{E}$. Equation (4.118) shows how $\boldsymbol{E}$ and $\boldsymbol{\mu}$ are related through the polarizability tensor α. In fact, however, this equation is only approximate. $\boldsymbol{\mu}$ should really be expressed as a power series in $\boldsymbol{E}$:

$$\begin{aligned}\boldsymbol{\mu} &= \boldsymbol{\mu}^{(1)} + \boldsymbol{\mu}^{(2)} + \boldsymbol{\mu}^{(3)} + \ldots \\ &= \alpha\boldsymbol{E} + \tfrac{1}{2}\boldsymbol{\beta}\boldsymbol{E}.\boldsymbol{E} + \tfrac{1}{6}\gamma\boldsymbol{E}.\boldsymbol{E}.\boldsymbol{E} + \ldots \end{aligned} \qquad (8.14)$$

where $\boldsymbol{\beta}$, the hyperpolarizability, is a third rank tensor and γ, the second hyperpolarizability, is a fourth rank tensor. Any effects due to the second (or higher) terms in the series are known as nonlinear effects because they arise from terms which are non-linear in $\boldsymbol{E}$.

The magnitude of the oscillating electric field is given by

$$E = A \sin 2\pi\nu t \qquad (8.15)$$

where A is the amplitude and ν the frequency. Since

$$E^2 = A^2(\sin 2\pi\nu t)^2 = \tfrac{1}{2}A^2(1 - \cos 2\pi 2\nu t) \quad (8.16)$$

the scattering contains, due to the $\boldsymbol{\mu}^{(2)}$ term, some radiation with *twice* the frequency (or *half* the wavelength) of the incident radiation. The phenomenon is called frequency doubling or second harmonic generation. In general, higher order terms in equation (8.14) can result in third, fourth and higher harmonic generation. Because the $\boldsymbol{\mu}^{(2)}$, $\boldsymbol{\mu}^{(3)}$, ... terms in equation (8.14) are very much smaller than the first, it is only with the extremely high photon densities which lasers produce that harmonic generation can be readily achieved.

There are several crystalline materials which can be used for harmonic generation, so-called non-linear crystals. Examples are ADP, KDP and KD*P, mentioned in Section 8.1.5 as materials used also for Pockels cells, KB_5O_8 (KPB), Li_3NbO_4 (LN), K_3NbO_4 (KN), $KTiOPO_4$ (KTP), $KTiOAsO_4$ (KTA), β-BaB_2O_4 (BBO), LiB_3O_5 (LBO) and $AgGaSe_2$. For second harmonic generation (frequency doubling) the crystal must not be centrosymmetric and each material is useful for only a limited frequency range.

In second harmonic generation, for example, the frequency doubled and undoubled radiation usually travel at different velocities in the crystal. Because the doubled radiation is generated as the undoubled radiation passes through the crystal, destructive interference occurs between the two types of radiation. If the crystal is birefringent, as are ADP and KDP, for example, both types of radiation travel with the same velocity. This is known as phase matching and, if it can be achieved, frequency doubling efficiency can be as high as 20–30% whereas, in a non-birefringent crystal, it may be as low as 1%.

8.1.8 Examples of lasers and masers

From the examples given in the following sections it will be clear that materials from which lasers or masers can be made are extremely varied, giving an impression of arbitrariness. The reason for this is that the detailed scheme of energy levels is unique for a particular atom or molecule in the solid, liquid or gas phase. In addition, the requirements for a level scheme to be suitable for the creation of a population inversion are fairly stringent. It is these two features which make the possibility of laser action accidental rather than something

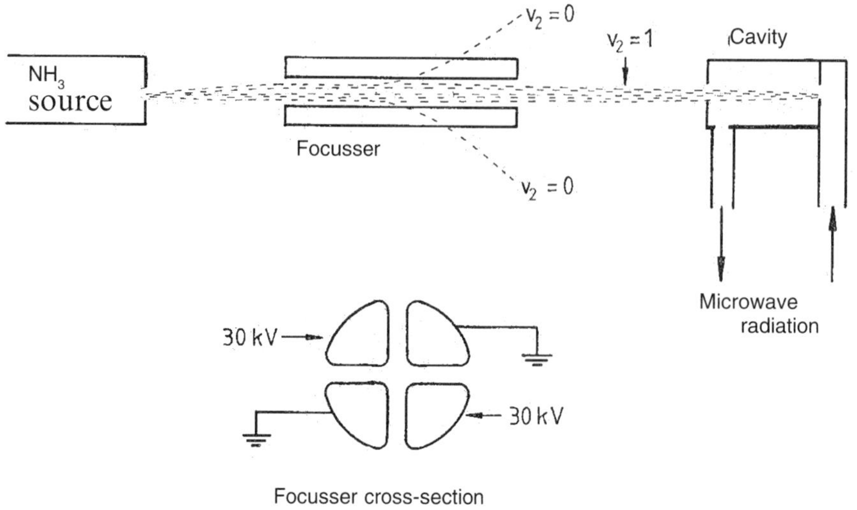

Figure 8.9 The ammonia maser, showing also the quadrupolar focuser in cross-section

which can be demanded of any atom or molecule.

8.1.8.1 The ammonia maser

In 1954 the first laser or maser device was constructed by Gordon, Zeiger and Townes.[191] The material used was ammonia gas and the population inversion created between the $\upsilon_2 = 1$ (or 0^-) and $\upsilon_2 = 0$ (or 0^+) levels of the inversion vibration ν_2 of $^{14}N^1H_3$. As described in Section 5.2.7.1 and illustrated in Figures 5.68 and 5.69, tunnelling through the inversion barrier causes a small splitting of the lowest vibrational level, which would not occur if the barrier were sufficiently high, into a lower, $\upsilon_2 = 0$, and an upper $\upsilon_2 = 1$, component. The separation of the two levels is only 23.786 GHz so the 2^1_0 (or $\upsilon_2 = 1$–0) transition occurs in the microwave region.

Population inversion between these two levels can be achieved by passing the molecules through a non-uniform, quadrupolar electrostatic field, as shown in Figure 8.9.

Ammonia molecules emerge from the source in the form of a unidirectional beam after passing through a number of parallel fine tubes. They then pass into an electrostatic field which is produced by applying a voltage of up to 30 kV to two opposite electrodes and earthing the other two. The cross-sections of the electrodes, shown in Figure 8.9, are such that each inner face is shaped like a segment of a hyperbola. The Stark effect (Section 4.2.4) produced by this field results in the molecules with $\upsilon_2 = 1$ having minimum energy on the centre line between the electrodes and therefore being brought to a focus. The molecules with $\upsilon_2 = 0$ experience the opposite effect. They are diverged by the field and lost from the beam.

The ammonia maser is unusual in that it involves only a two-level system, like that in Figure 8.3(a). However, population inversion has been achieved not by pumping—we have seen in Section 8.1.3 that this cannot be done—but by depletion of the population of the ground vibrational state. The ammonia maser is one of few such examples.

The beam is focused to a point in a resonant cavity where it is subjected to incoming microwave radiation which induces 2^1_0 transitions. Because of the population inversion the radiation emerging from the cavity has been amplified.

Associated with both the $v_2 = 0$ and 1 levels are stacks of rotational levels (Figure 5.48) involving the quantum numbers J and K. The 2_0^1 transition is allowed by electric dipole selection rules and the accompanying rotational transitions obey the selection rules

$$\Delta J = 0;\ \Delta K = 0,\ \text{except}\ K = 0 \not\leftrightarrow K = 0$$

Any transitions obeying these selection rules result in amplification. In the original experiments transitions with $J'' = K'' = 3$ and 2 were studied, the former being at 23.870 MHz and the more intense due to the nuclear spin statistical weight factor (Section 4.4).

The maser may act as an amplifier of microwave radiation of constant frequency (23.870 MHz, for example), or it may be used as a high resolution spectrometer by scanning the microwave input radiation. The resulting high resolution spectra of $^{14}N^{1}H_3$ show a very narrow line width, of the order of 5 Hz, and consequently hyperfine splitting due to the quadrupole moment of the ^{14}N nucleus (Section 4.2.5) and the spin of the ^{1}H nuclei can be resolved.

The power output of the ammonia maser is of the order of 10^{-10} W, very low compared with typical laser powers.

8.1.8.2 The ruby and alexandrite lasers

The ruby laser was first demonstrated in 1960 by Maiman[192]. It was the first such device after the ammonia maser and represented a major step forward. The fact that it is a solid state laser and that it operates in the visible region, where spontaneous emission competes more strongly with stimulated emission than in the microwave region [equation (2.20)], means that the technology involved bears little relation to that of the ammonia maser. Another major difference is that the ruby laser is a three-level laser [Figure 8.3(b)].

The ruby crystal used in lasers consists of aluminium oxide, Al_2O_3, with 0.5% by weight of Cr_2O_3. The crystal is pale pink rather than the deeper pink characteristic of the higher Cr_2O_3 concentration in most rubies.

The lasing constituent of ruby is the Cr^{3+} ions which are in such a low concentration as to act as free ions. The ground configuration of Cr^{3+} (see Table 6.1) is $KL3s^23p^63d^3$, which gives rise to eight terms (Table 6.2) of which ^4F is, according to Hund's rules [Section 6.1.2.2(i)], the lowest lying (ground) term. Of the others ^2G is the lowest excited term.

In the ruby crystal each Cr^{3+} ion is in a crystal field (Section 6.3.1.11) of approximately octahedral symmetry. In the octahedral point group, O_h, the ^4F ground term gives $^4A_2$, $^4T_1$ and $^4T_2$ states whereas the ^2G excited term gives $^2A_1$, ^2E, $^2T_1$ and $^2T_2$ states. Of these, the $^4A_2$ is the ground state and the $^4T_1$, $^4T_2$, ^2E and $^2T_2$ states are relatively low-lying excited states; all are shown in Figure 8.10.

The $^4T_1$ and $^4T_2$ states are broadened owing to slight variations of the crystal field. The $^2T_2$ and ^2E states are much sharper but the ^2E state is split into two components, 29 cm^{-1} apart, because the crystal field is distorted from regular octahedral.

Population inversion and consequent laser action occurs between the ^2E and $^4A_2$ states. This is achieved by optical pumping into the $^4T_2$ or $^4T_1$ states with 510–600 or 360–450 nm radiation, respectively. The broadness of the levels contributes to the efficiency of the pumping which is achieved with a flashlamp of the type described in Section 3.8.1. This may

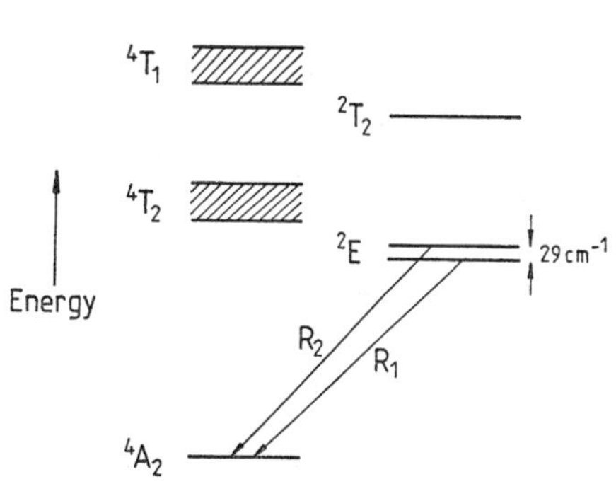

Figure 8.10 Energy levels of Cr^{3+} in the ruby crystal which are important in laser action

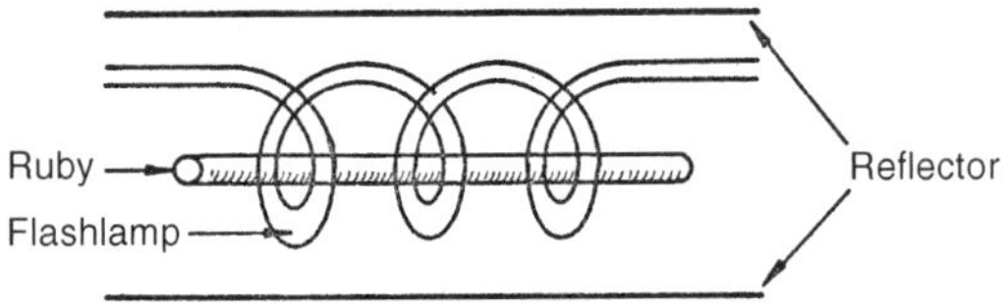

Figure 8.11 A design for a ruby laser consisting of a ruby crystal inside a helical flashlamp surrounded by a reflector

be in the form of a helix around the ruby crystal as shown in Figure 8.11. The crystal is cylindrical and may be as large as 2 cm in diameter and 20 cm in length. The crystal and lamp are placed inside a carefully designed reflector, often made from aluminium. The mirror surfaces are deposited directly on to the ends of the crystal.

Pumping to the $^4T_1$ and $^4T_2$ levels is followed by energy loss, in the form of heat, taking the Cr^{3+} ions to the ^2E levels, either directly or via $^2T_2$. Laser action involves principally the R_1 rather than the R_2 transition (Figure 8.10) This is because of the greater transition probability for R_1 which leads to more rapid depletion of population of the lower relative to the upper component of the ^2E level. The population ratio is quickly restored by transfer from the upper level. The transitions R_1 and R_2 are at 694.3 and 693.4 nm, respectively.

Generation of heat by the energy losses from the $^4T_1$ and $^4T_2$ states causes problems in the design and operation of the laser. The laser normally operates in the pulsed mode and the heat must be dissipated between each pulse. This is particularly important for constant wavelength output because the wavelength of the R_1 transition, for example, changes by about 0.7 nm over a range of 100 K. Cooling of the crystal by water or liquid nitrogen can be used to improve the rate of heat dissipation.

In the first ruby lasers, pulse rates were as low as one every few minutes, but a few pulses per minute is now typical. Pulse rates as high as 10–100 Hz have been obtained and, using narrow ruby rods, CW operation may be achieved.

In multimode operation the laser line width is about 300 GHz (*ca* 10 cm^{-1}) at room temperature and the pulse length is about 1 ms. *Q*-switching may reduce this to 10–20 ns and produce 100 MW peak power while mode locking can give a pulse length of 10 ps and peak power of the order of 1 GW. The efficiency is very low, less than 0.1%, and typical for a three-level laser.

Alexandrite, like ruby, contains Cr^{3+} ions but they are substituted in the lattice of chrysoberyl, $BeAl_2O_4$. The chromium ions occupy two symmetrically non-equivalent positions which would otherwise be occupied by aluminium ions. In this environment the $^4A_2$ ground state of Cr^{3+} is broadened, compared to that in ruby, by coupling to vibrations of the crystal lattice.

Lasing occurs at 680.4 nm in alexandrite, the transition involved being analogous to R_1 in ruby and not involving any vibrational excitation in the $^4A_2$ state. However, of much greater importance in alexandrite is laser action between the $^4T_2$ and $^4A_2$ states. Because both the $^4T_2$ and $^4A_2$ states are broadened by vibrational coupling the alexandrite laser is sometimes referred to as vibronic laser. As a result of this, broadening of both the electronic states involved in the $^4T_2$–$^4A_2$ transition the laser can be tuned over a wide wavelength range, 720–800 nm. This tunability gives the alexandrite laser a great advantage over the ruby laser which is limited to just two wavelengths.

A further advantage is the higher efficiency of the alexandrite laser due to its being a four-level laser. In the illustration in Figure 8.3(c), level 4 is a vibronic level and 3 the zero-point level of the $^4T_2$ state. Level 2 is a vibronic level of the $^4A_2$ state and 1 the zero-point level. Because of the excited nature of level 2 it is almost depopulated at room temperature, so that a population inversion between levels 3 and 2 is relatively easy to achieve. In fact, level 2 is a continuous band of vibronic levels covering a wide energy range resulting in the wide wavelength range over which the laser is tunable.

Pumping is with a flashlamp, as in the case of the ruby laser, and a pulse energy of the order 1 J may be achieved. Frequency doubling (second harmonic generation) can provide tunable radiation in the 360–400 nm region.

8.1.8.3 *The titanium–sapphire laser*

Despite the fact that the first laser to be produced (the ruby laser, Section 8.1.8.2) has the remarkable property of having all its power concentrated into one or two wavelengths, a property possessed by most lasers, it was soon realized that the inability to change these wavelengths appreciably, that is to tune the laser, is a serious drawback which limits the range of possible applications.

Historically, the first type of laser to be tunable over an appreciable wavelength range was the dye laser, to be described in Section 8.1.8.14. The alexandrite laser (Section 8.1.8.2), a tunable solid state laser, was first demonstrated in 1978 and then, in 1982, the titanium–sapphire laser. This is also a solid state laser but tunable over a larger wavelength range, 670–1100 nm than the alexandrite laser which has a range of 720–800 nm.

The lasing medium in the titanium–sapphire laser is crystalline sapphire (Al_2O_3) with about 0.1% by weight of Ti_2O_3. The titanium is present as Ti^{3+} and it is between energy levels of this ion that lasing occurs.

The ground configuration of Ti^{3+} (see Table 6.1) is $KL3s^23p^63d^1$. The crystal field experienced by the ion splits the $3d$ orbital into a triply degenerate lower energy t_2 orbital and a doubly degenerate higher energy e orbital (see Figure 6.86). If the electron is in the lower orbital a $^2T_2$ ground state results and, if it is in the upper orbital, a ^2E excited state results. These states are about 19 00 cm^{-1} apart but each is split into further components and is also coupled to the vibrations of the crystal lattice. In a similar way to that in alexandrite (Section 8.1.8.2), population inversion can be created between these two sets of levels resulting in a four-level vibronic laser with a tunable range of 675–1100 nm.

A further advantage, compared with the alexandrite laser, apart from a wider tuning range, is that it can operate in both the CW and the pulsed modes. In the CW mode the Ti^{3+}: sapphire laser may be pumped by a CW argon ion laser (see Section 8.1.8.7) and is capable of producing an output power of 5 W. In the pulsed mode pumping is usually achieved by a pulsed Nd^{3+}: YAG laser (see Section 8.1.8.4) and a pulse energy of 100 mJ may be achieved.

Frequency doubling of the radiation from a Ti^{3+}: sapphire laser, usually with a BBO (β-BaB_2O_4) or LBO (LiB_3O_5) crystal, produces tunable radiation in the 350–500 nm range. Frequency tripling, by adding the fundamental to the frequency doubled radiation in a BBO crystal, can produce 272–320 nm radiation. The missing range of 500–657 nm can be provided in a single laser by modifying a ring dye laser (see Section 8.1.8.14) to incorporate a Ti^{3+}: sapphire crystal. This is replaced by a dye cell for the 500–675 nm range.

Mode-locking, produced as a result of the Kerr effect in the crystal, can produce pulse lengths down to 50 fs. The use of 'chirped' quarter-wave dielectric mirrors, in which the multilayer period is modulated across the mirror, can reduce this to 8 fs.

8.1.8.4 *The neodymium-YAG laser*

Laser action can be induced in Nd^{3+} ions embedded in a suitable solid matrix. Several matrices, including crystalline YLF, which is yttrium aluminium fluoride ($YLiF_4$), and some amorphous glasses, are suitable but the most frequently used is yttrium aluminium garnet ($Y_3Al_5O_{12}$) which is commonly referred to as YAG. The advantages of YAG include high thermal conductivity, which is necessary for rapid heat dissipation, mechanical strength and resistance to optical damage.

The neutral neodymium atom has the electron configuration ... $4d^{10}4f^45s^25p^66s^2$ and a $^5I_4$ ground state (Table 6.1). The ground

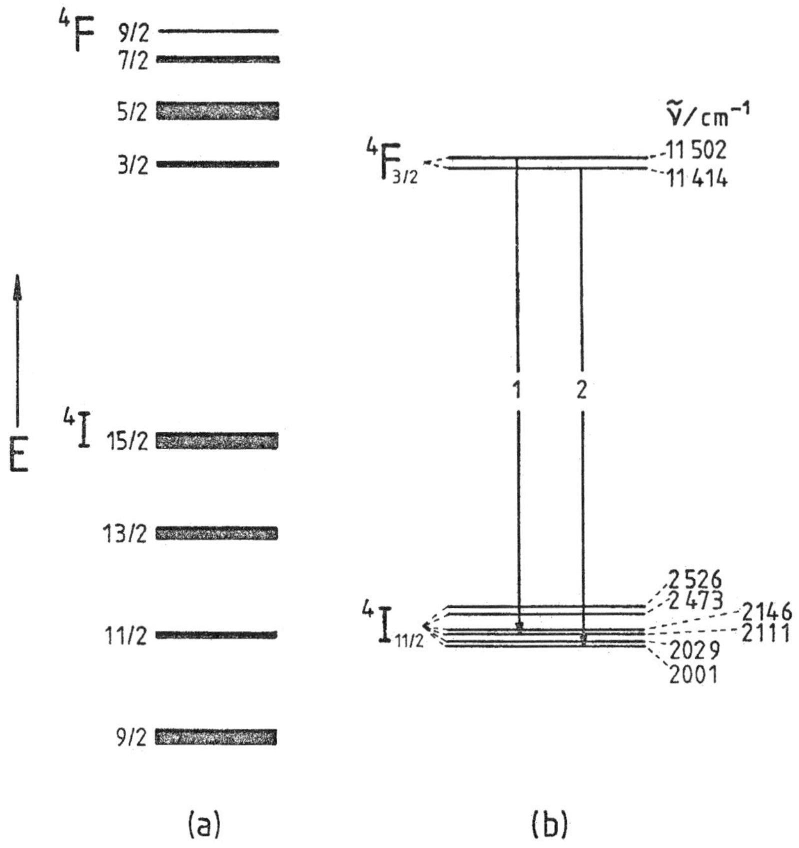

Figure 8.12 Energy levels in (a) free Nd^{3+}, and (b) Nd^{3+} split by crystal field interactions

configuration of Nd^{3+} is . . . $4d^{10}4f^3 5s^2 5p^6$. Of the terms arising from this configuration ^4I and ^4F are important in the laser. For the ^4I term $L = 6$ and $S = \frac{3}{2}$, giving $J = \frac{15}{2}, \frac{13}{2}, \frac{11}{2}, \frac{9}{2}$, in the Russell–Saunders approximation [Section 6.1.2.2(i)]. The multiplet is normal, i.e. the lowest value of J has the lowest energy, as shown in Figure 8.12(a). Also shown in the figure is the normal multiplet arising from the ^4F term.

Laser action involves the transitions $^4F_{\frac{3}{2}}$–$^4I_{\frac{13}{2}}$ (*ca* 1.35 μm), $^4F_{\frac{3}{2}}$–$^4I_{\frac{11}{2}}$ (*ca* 1.06 μm) and $^4F_{\frac{3}{2}}$–$^4I_{\frac{9}{2}}$ (*ca* 0.914 μm) but the $^4F_{\frac{3}{2}}$–$^4I_{\frac{11}{2}}$ transition is the most easily pumped and therefore a Nd^{3+}:YAG laser usually operates at *ca* 106 μm. It is only this transition that we shall consider here. Since $^4I_{\frac{11}{2}}$ is not the ground state, the laser employs a four-level system [see Figure 8.3(c)].

In the free Nd^{3+} ion the $^4F_{\frac{3}{2}}$–$^4I_{\frac{11}{2}}$ transition is doubly forbidden, violating the $\Delta L = 0, \pm 1$ and $\Delta J = 0, \pm 1$ selection rules (Section 6.1.6). In the YAG crystal the $^4I_{\frac{11}{2}}$ state of Nd^{3+} is split by crystal field interacitons into six and the $^4F_{\frac{3}{2}}$ state into two components, as shown in Figure 8.12(b), and the selection rules break down. There are eight transitions, grouped around 1.06 μm, between the components but only the two marked in the figure are important. At room temperature transition 1, at 1.0648 μm, is dominant but, at 77 K, transition 2, at 1.0612 μm is dominant. The reason for this is that, although 1 has a higher transition probability, the population of the lower $^4F_{\frac{3}{2}}$

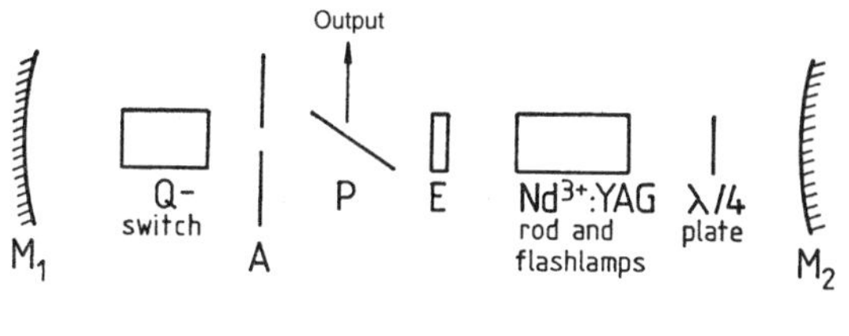

Figure 8.13 Fountain cavity design for an Nd^{3+}:YAG laser

sub-level is greatly increased at the lower temperature.

Absorption of the pumping radiation takes place to the $^4F_{\frac{7}{2}}$ and $^4F_{\frac{5}{2}}$, in addition to $^4F_{\frac{3}{2}}$, levels, thereby increasing the pumping efficiency. A tungsten lamp may be used for CW operation or a flashlamp for much higher power, pulsed operation. Pulsed or CW operation may be achieved by pumping with a diode laser (see Section 8.1.8.5).

The Nd^{3+} : YAG rod is a few centimetres long and contains about 0.5–2.0% by weight of Nd^{3+}. Several cavity designs have been used; that in Figure 8.13 is the so-called Fountain design, named after its developer, and uses flashlamp pumping. This design is unusual in that it employs an unstable, rather than a stable, resonant cavity. The cause of the instability is the convex mirror M_2, which increases the loss of the cavity. However, if this loss is employed as the laser output there can be advantages in using an unstable system particularly in the quality of the profile of the output beam which, ideally, should be gaussian.

To the left of the quarter-wave plate the radiation is plane polarized and the plate converts this to elliptical polarization. Reflection from M_2, and passing again through the plate, results in polarization which is 75% vertically (relative to the original horizontal plane of polarization) and 25% horizontally polarized. The polarizing plate P, inclined to the laser axis, couples out the vertical component, the horizontal component remaining to provide feedback. The etalon E ensures single mode operation and the *Q*-switch shortens the pulse length and increases the peak power. Symbol A indicates an aperture stop.

With this design it is possible to obtain a pulse energy of up to 1000 mJ at a repetition rate of 20 Hz, a pulse length of 15 ns and a line width of 0.01 cm^{-1}. The pulse energy is so high that harmonic generation (Section 8.1.7) with a suitable crystal is relatively efficient. For example, with a pulse energy of 1000 mJ for the fundamental at 1065 nm, second, third and fourth harmonics can be generated at 533, 355 and 266 nm with pulse energies of 350, 175 and 75 mJ, respectively, using non-linear crystals in temperature-controlled ovens.

8.1.8.5 Diode (or semiconductor), spin–flip Raman and colour centre lasers

A semiconductor belongs to a class of crystalline materials having electrons which, under certain conditions, are free to move throughout the crystal. In such a material the energy levels are not discrete, as in free atoms and molecules, but form continuous bands. Without going into detail, we can regard the wave functions of energy levels within these bands as resulting from the linear combination of the orbitals of the infinite number of particles in the crystal. In all the crystals we shall be considering the particles are atomic rather than molecular and the quantum mechanical treatment is an extension of the LCAO method (Section 6.2.1) to an infinite number of particles. According to the Pauli principle, each of the infinitely closely spaced energy levels within the band can take two electrons with anti-parallel spins. The electrons are fed into the energy levels in order of increasing energy.

In metallic copper, for example, the most important band of levels is formed from the 4*s* valence atomic orbitals on each copper atom which has the electron configuration $KLM4s^1$. As the valence orbital is not filled neither is the corresponding band of levels in the crystal. This is the situation illustrated in Figure 8.14(a) and the material is a conductor because the empty energy levels in the conduction band facilitate

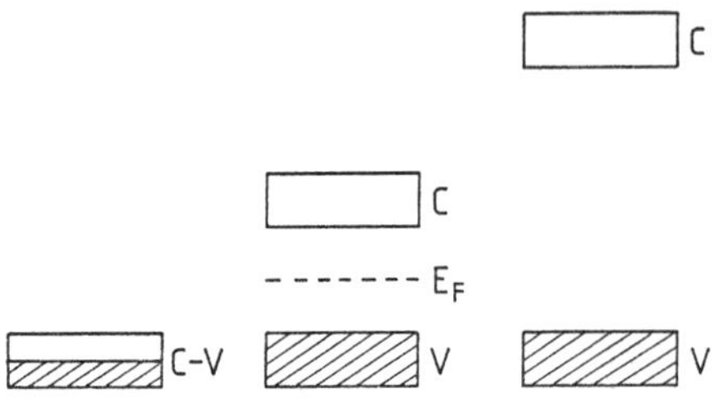

Figure 8.14 Conduction band, C, and valence band, V, in (a) a conductor, (b) a semiconductor and (c) an insulator

the transfer of electrons, resulting in high electrical conductivity.

Just as in molecules, when we carry out an LCAO treatment we obtain wave functions corresponding not only to occupied but also to unoccupied orbitals in the ground configuration. Figure 8.14(c) shows a case where the valence electrons completely fill a band of levels and the next highest energy band is empty. The lower band is called the valence band and the upper band the conduction band because, if the material is to conduct at all, some electrons must be promoted to it. (In a conductor, the band concerned in conduction is both the conduction and valence band.) However, if the band gap separating the bands is too large, greater than about 1 eV, this cannot be achieved and the material is an insulator. If the band gap is small, promotion of electrons by, say, heating the material is relatively easy and such a material is a semiconductor. This situation is illustrated in Figure 8.14(b). However, it is important to realize that the difference between an insulator and a semiconductor is one of degree and not type so that some materials may act as either insulators or semiconductors, depending on the temperature.

The populations, $f(E)$, of the energy levels are temperature dependent but, since electrons are fermions, the populations obey Fermi–Dirac statistics and

$$f(E) = \{1 + \exp[(E - E_F)/kT]\}^{-1} \quad (8.17)$$

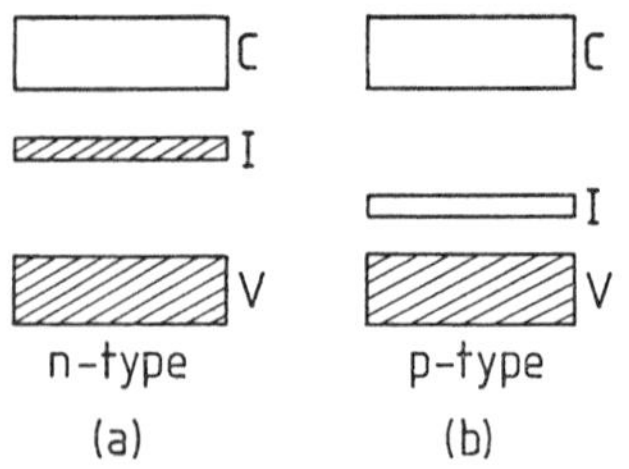

Figure 8.15 Impurity levels I in (a) an *n*-type, and (b) a *p*-type semiconductor

The quantity E_F is the Fermi energy and is characterized by the conditions that, at $T = 0$ K, all levels with $E < E_F$ are fully occupied, i.e. $f(E) = 1$, and those with $E > E_F$ are empty, i.e. $f(E) = 0$. For a semiconductor E_F is between the valence and conduction bands, as in Figure 8.14(b).

From the temperature dependence of energy level populations, it follows that a material may be an insulator at low temperatures and a semiconductor at higher temperatures. Such a material is called an intrinsic semiconductor and conduction is by electrons in the conduction band and also by vacancies (positive holes), in the valence band. Silicon, with a band gap of about 1 eV, is an example of an intrinsic semiconductor.

Semiconductors can also be produced from a material which is normally an insulator by introducing either lattice defects or impurities into the crystal. We shall consider the case only of an added impurity, a process known as doping.

Figure 8.15 illustrates two ways in which an impurity may increase semiconduction. In Figure 8.15(a) the dopant has one more valence electron per atom than the host and contributes a band of filled impurity levels I close to the conduction band C of the host. This characterizes an *n*-type semiconductor. An example is silicon ($KL3s^23p^2$) doped with phosphorus ($KL3s^23p^3$) which reduces the band gap to about 0.05 eV. Since kT at room temperature is about 0.025 eV, the phosphorus converts silicon from a high temperature semiconductor into a room temperature semiconductor.

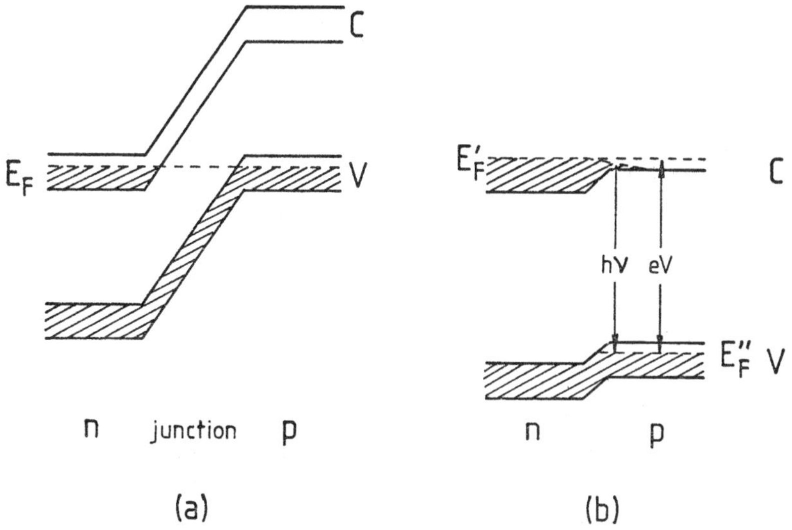

Figure 8.16 (a) At a junction between n- and p-type doped semiconductors energy levels are filled to the Fermi level, E_F. (b) The result of applying a voltage across a p–n junction is to change the Fermi energy on each side and produce population inversion between conduction and valence bands at the junction

Alternatively, as illustrated in Figure 8.15(b), a dopant with one valence electron per atom less than the host may be used to contribute an impurity b and I of empty energy levels accessible to electrons from the valence band V. An example of such a p-type semiconductor is silicon doped with aluminium ($KL3s^2 3p^1$) in which the band gap is about 0.08 eV.

A semiconductor laser takes advantage of the properties of a junction between a p-type and an n-type semiconductor made from the same host material doped with different impurities on either side of the junction. Such an n–p combination forms a semiconductor diode. A commonly used host is gallium arsenide (GaAs) doped with an acceptor element such as Zn, with a filled $4s^2$ valence orbital (Table 6.1) to form the p-type semiconductor and a donor element such as Te, with an unfilled $5p^4$ valence orbital, to form the n-type semiconductor.

Doping concentrations are fairly high and, as a result, the conduction and valence band energies of the host are shifted in the two semiconductors, as shown in Figure 8.16(a). Thermodynamic equilibrium across the junction ensures that energy levels are filled up to the Fermi level, although at $T > 0$ K this level is somewhat ill-defined.

If a voltage is applied to the p–n junction, with the negative and positive terminals attached to the n and p regions, respectively, electrons flow from the n to the p region and positive holes flow in the opposite direction. As shown in Figure 8.16(b), the levels are now displaced so that

$$eV = E'_F - E''_F \tag{8.18}$$

where V is the applied voltage and e is the electronic charge. The Fermi energies E'_F and E''_F of the n and p regions, respectively, are now different as there is no longer thermodynamic equilibrium. There is a large flow of current across the junction and, at the junction itself, electrons combine with holes to produce emission of photons. The population inversion between the lower conduction band levels and the upper valence band levels in the region of the junction leads to laser action. Like the ammonia maser, the diode laser is an example of a two-level system.

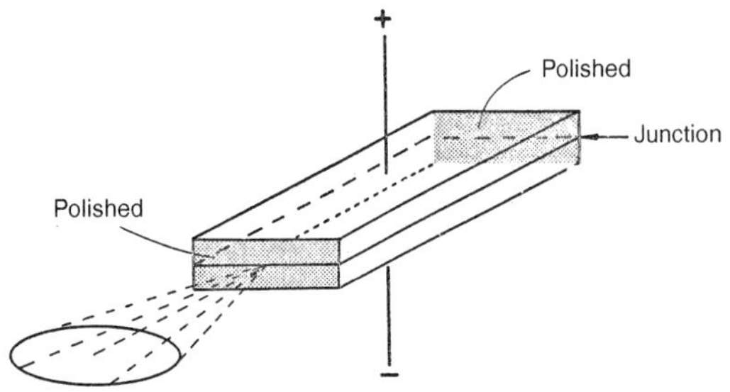

Figure 8.17 A semiconductor, or diode, laser

A typical diode laser, shown in Figure 8.17, is physically small, only a few millimetres long and with an effective thickness of *ca* 2 μm. Because the lifetime of an electron–hole pair before emission of the photon is only 1.0–0.1 ns, very high current densities are required to obtain population inversion. Cooling of the laser is beneficial in this respect, a typical current density required being 5×10^4 A cm^{-2} at 300 K and 200 A cm^{-2} at 20 K.

Diode lasers operate mostly in the near infrared but also in the visible region. For example, the GaAs laser wavelength is centred at about 0.84 μm and those of the type $GaAs_{1-x}P_x$ range in wavelength from 0.64 μm (for $x = 0.4$) to 0.84 μm (for $x = 0$). Shorter wavelength diode lasers include those of the types AlGaAs (650 nm), AlInGaP (620, 585 and 570 nm), GaP (555 nm), InGaN (514 nm), ZnTeSe (512 nm) and ZnCdSe (489 nm), the wavelengths given in parentheses referring to the peak power. The ZnTeSe laser is bright and the efficiency, defined as the ratio of the laser output power to the electrical input power, is relatively high (5.3%) for a laser operating in the green region. Changing to ZnCdSe produces a laser operating in the blue region but the efficiency is only 1.3%.

A variety of lead alloy semiconductors such as $Pb_{1-x}Sn_xSe$ and $PbS_{1-x}Se_x$ covers the range 2.8–30 μm, the region covered by a particular laser depending on the value of x.

The wavelength range of a diode laser is fairly small so that a whole series of them is required to cover a large wavelength range. The wavelength at which the laser intensity is a maximum varies considerably with temperature, which is typically less than 40 K. Gross tuning of the laser wavelength is achieved by surrounding the cavity by a refrigeration unit employing helium gas. At high resolution the laser radiation is found to show several discrete wavelengths due to cavity modes. Small, precise temperature changes, produced by varying the current density in the laser, enable the wavelength of each mode to be tuned but, to cover a reasonable wavelength range, it is necessary to transfer from one mode to another.

The two ends of the laser diode (Figure 8.17) are polished to increase internal reflection while the rest is left rough to avoid lasing in unwanted directions. As a consequence of the cavity geometry the laser beam is, unlike that from other lasers, highly divergent.

Since high current densities raise the temperature of the diode, necessitating even higher current densities for laser action, CW operation at room temperature poses problems of heat dissipation. However, high repetition rates of more than 10^{11} Hz, obtained by modulation of the applied current, can be used because of the very short lifetime of an electron–hole pair.

CW operation in single mode at low temperature gives a typical power of 0.1 mW.

CW operation at room temperature has been achieved by using a double heterojunction diode instead of a *p–n* diode. There are *two* lasing junctions in this kind of diode, for example those on either side of a thin layer of GaAs (*p*-type) sandwiched between layers of $Al_xGa_{1-x}As$ (*p*-type) and $Al_xGa_{1-x}As$ (*n*-type). A layer of tin plate acts as an efficient heat sink. Current densities of only 10^3 A cm^{-2} are required.

In single mode operation a laser line width of the order of 10^{-4} cm^{-1} is typical.

Some diode lasers are the most efficient of all lasers. Those operating in the near infrared may have an efficiency as high as 30%.

A different type of laser which also involves a semiconductor as the active medium is the spin–flip Raman laser. The semiconductor may be a crystal of indium antimonide (InSb) with polished ends which naturally reflect about 35% of the infrared laser radiation generated. A variable magnetic field of up to about 10 T (100 kG) is applied transversely across the cavity. The radiation is generated through stimulated Raman scattering (see Section 8.2.3) from the levels of the electrons in the semiconductor which are split due to the Zeeman effect caused by the magnetic field. The incident radiation which produces the stimulated Raman scattering is from a CO_2 or CO infrared laser directed into one end of the cavity. The stimulated Raman scattering emerges from the cavity almost collinear with the pump laser beam because, as explained in Section 8.2.3, stimulated Raman scattering is in the forward direction within about 10° of the incident radiation. Removal of the pump laser radiation from the emerging beam is achieved either with a small grating monochromator or by polarizers which make use of the fact that, if the pump radiation is polarized perpendicular to the magnetic field, the stimulated Raman scattering is polarized parallel to the field.

A tuning range of up to about 400 cm^{-1} is possible, using only one pumping frequency, by varying the magnetic field. With a CO or CO_2 pump laser a total range of about 5.3–6 or 9.3–10.8 μm, respectively, can be covered. As in the case of diode lasers, a spin–flip Raman laser must be operated at low temperature (4 K) to dissipate the heat generated by the pump laser.

Another type of solid state infrared laser, but not using a semiconductor as the active medium, is the F-centre, or colour centre, laser[193]. The active medium is an alkali metal halide crystal containing F centres. The simplest type of F centre is produced when an electron occupies a negative ion vacancy but the types used in lasers are usually those known as F_A and F_2^+.

An F_A type is a simple F centre in which one of the six nearest neighbour positive ions is replaced by a different alkali metal ion. For example, an F_A centre in KCl can be produced by a Cl^- ion being replaced by an electron and a nearest neighbour K^+ ion being replaced by Li^+.

An F_2^+ centre is produced from two adjacent simple F centres, along a (110) axis of the crystal, with one of the electrons removed.

F centres can be formed by various types of irradiation, e.g. with X-rays, ultraviolet radiation or an electron beam. The alternative name of 'colour centre' arises because the absorption spectrum of an F centre is mainly in the near infrared but has weak absorption in the red region giving a blue/violet colour to the crystal. Efficient relaxation to lower excited levels, with lifetimes of the order of 100 ns, following absorption, leads to population inversion and the possibility of laser action in the near infrared.

The crystal containing the F centres is placed in the laser cavity and pumping is usually achieved with a second laser directed on to the crystal at an angle to the optic axis of the cavity. An argon ion or krypton ion laser (Section 8.1.8.7) for CW operation, or a Nd^{3+}:YAG laser (Section 8.1.8.4) for pulsed operation is commonly used. Tuning outside the cavity may be achieved with a quartz prism.

Pumping of F_2^+ centres in KF with a Nd^{3+}:YAG laser produces tunable radiation in the 1.26–1.48 μm region and pumping of F_2^+ centres in LiF with a krypton ion laser produces radiation in the 0.82–1.07 μm region. With a suitable F centre radiation of higher wavelength, up to 3.5 μm, can be produced.

High resolution studies have shown that the line width for a single mode can be less than 25 kHz (10^{-6} cm^{-1}).

8.1.8.6 The helium–neon laser

The helium–neon laser is an example of a gas laser which is simple and reliable to operate and, for a laser of relatively low power, fairly inexpensive.

Laser action takes place between excited energy levels of the neon atom, the upper levels

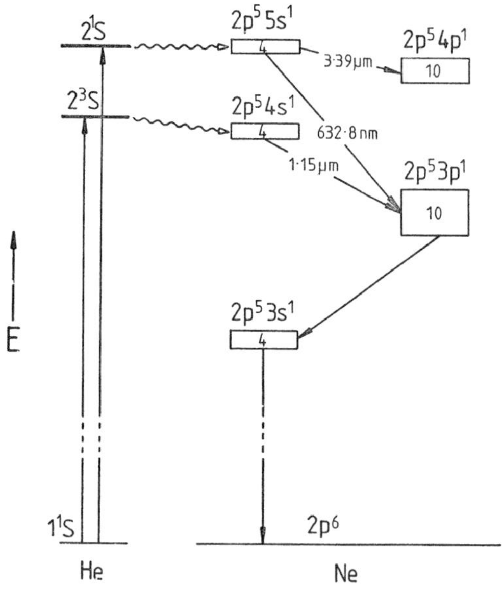

Figure 8.18 Energy levels of the He and Ne atoms relevant to the helium–neon laser. The number of states arising from each Ne configuration is given in the boxes

being populated partly through collisions with helium atoms in metastable excited states. The energy level scheme is shown in Figure 8.18.

An electrical discharge is created in a helium–neon gas mixture by applying either a high voltage or microwave radiation. In the discharge electrons are produced which collide with helium atoms promoting them to various excited states. Of these, the 2^3S and 2^1S states are metastable, and therefore long-lived, because transitions to the 1^1S ground state are forbidden (see Section 6.1.5). The 2^3S is the longer-lived state as the transition to 1^1S is both orbitally and spin forbidden.

The ground configuration of Ne is $1s^2 2s^2 2p^6$, giving a ^1S state. The excited configurations give rise to states to which the Russell–Saunders coupling approximation [Section 6.1.2.2(i)] does not apply. Nevertheless, any $\ldots 2p^5 ns^1$ or $\ldots 2p^5 np^1$ configuration, with $n > 2$, gives rise to four or ten states, respectively, as would be the case in the Russell–Saunders approximation. We shall not be concerned with the approximation which is appropriate for describing these states.

The states arising from the $\ldots 2p^5 5s^1$ configuration of Ne have very similar energy to the 2^1S state of He so that collisional energy transfer results in efficient population of these Ne states. Similarly, the states arising from the $\ldots 2p^5 4s^1$ configuration of Ne lie just below the 2^3S state of He and are also populated by collision. All the transitions in Ne involving an $s \rightarrow p$ orbital change are allowed by selection rules and all the $\ldots 2p^5 ns^1$ states have lifetimes of the order of 100 ns compared with 10 ns for the $\ldots 2p^5 np^1$ states. These conditions are ideal for four-level lasing in Ne with population inversion between $\ldots 2p^5 ns^1$ and $\ldots 2p^5 np^1$ states.

The first laser lines to be discovered in the He–Ne system were a group of five in the near infrared close to 1.15 μm and involving $\ldots 2p^5 4s^1 - \ldots 2p^5 3p^1$ transitions. Thirty transitions in this group are allowed but the strongest is at 1.1523 μm.

Similarly, the $\ldots 2p^5 5s^1 - \ldots 2p^5 3p^1$ transitions may give several laser lines in the red region of the spectrum but that at 632.8 nm shows the highest gain.

Infrared laser lines involving $\ldots 2p^5 5s^1 - \ldots 2p^5 4p^1$ transitions in the 3.39 μm region are not used very much but they cause some problems in a 632.8 nm laser. The gain of the infrared lines is so high that the laser is super-radiant at these wavelengths. In the 632.8 nm laser these infrared lines must be suppressed because they deplete the populations of the $\ldots 2p^5 5s^1$ states and decrease the 632.8 nm line intensity. This suppression may be achieved by using multilayer cavity mirrors designed specifically for the 632.8 nm wavelength or by placing a prism in the cavity orientated so as to reject the infrared radiation from the cavity.

Decay from the $\ldots 2p^5 3p^1$ states to the $\ldots 2p^5 3s^1$ states is rapid but the $\ldots 2p^5 3s^1$ states are relatively long-lived. Their population tends to build up and this increases the probability of the $\ldots 2p^5 3p^1 \rightarrow \ldots 2p^5 3s^1$ radiation being reabsorbed, a process called radiation trapping,

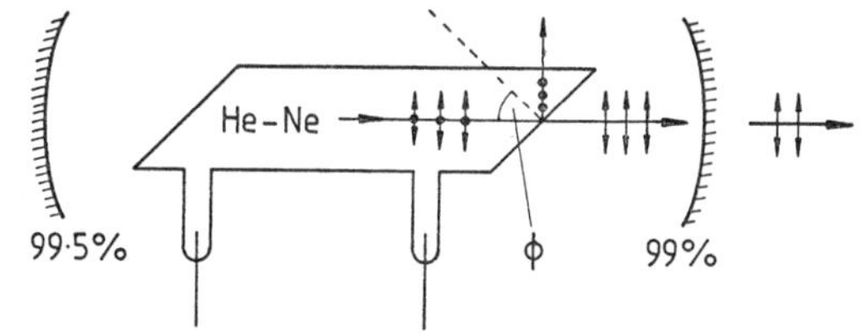

Figure 8.19 A helium–neon laser cavity showing the operation of Brewster angle windows

thereby increasing the . . . $2p^5 3p^1$ populations and decreasing the laser efficiency at 632.8 nm (and 1.15 μm). Depopulation of the . . . $2p^5 3s^1$ states is achieved by collisions with the walls of the discharge tube. For this reason narrow tubes, only a few millimetres in diameter, are used.

The fact that several of the laser lines can be obtained with pure neon shows that, although helium is effective in promoting population inversion, it does not provide the only mechanism.

Construction of the He–Ne laser, illustrated in Figure 8.19, is typical of a gas laser. The laser cavity is formed by two concave mirrors, one coated for as near 100% reflection at, say, 632.8 nm as possible. The other mirror allows about 1% transmission.

The discharge in the gas mixture may be excited by means of 30 MHz radiofrequency radiation emanating from metal sleeves around the discharge tube or, more commonly and as illustrated, directly by internal electrodes. The He–Ne gas mixture is in the ratio of about 10:1 and pressures of about 1 Torr of He and 0.1 Torr of Ne are typical.

The windows on the ends of the discharge tube are Brewster angle windows to prevent excessive light loss from multiple transmissions. If a window is at 90° to the optic axis of the laser a certain percentage of radiation is lost every time the radiation passes through. On the other hand, if the window is oriented so that the incident radiation is at Brewster's angle ϕ to the normal to the window, shown in Figure 8.19, some is lost by reflection on the first transmission but no more is lost on subsequent transmissions. As Figure 8.19 shows, for unpolarized radiation incident from inside the cavity the transmitted and reflected radiation are plane polarized, the planes being perpendicular to each other. The output beam is, therefore, plane polarized.

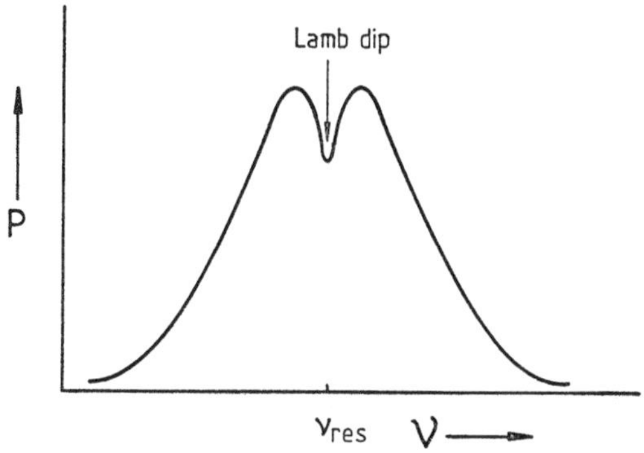

Figure 8.20 Power output curve for a single mode laser showing a Doppler line shape with a Lamb dip superimposed

Brewster's angle is given by

$$\tan \phi = n \qquad (8.19)$$

where n is the refractive index of the window material. Since n varies with wavelength, so does ϕ but, for glass in the visible region, ϕ is about 57° and varies little with wavelength.

The He–Ne laser is normally a CW laser with power up to about 100 mW but with very low efficiency. The laser may also be pulsed by pulsing the input power. Mode locking may be used to shorten the pulses to about 1 ns.

Figure 8.20 illustrates an effect on the power output P as a function of oscillating frequency ν which is commonly observed in a gas laser operating in single mode. The curve is obtained by changing the cavity length continuously over a half-wavelength. Since the line width in the He–Ne laser is Doppler limited, Figure 8.20 represents, apart from the central minimum, the Doppler line shape.

The central minimum is called a Lamb dip and is a consequence of circumstances very similar to those discussed in Section 3.4 in the context of Lamb dip absorption spectroscopy. Figures 3.8 and 3.9 illustrate how the introduction of a reflector results in hole burning in the velocity distribution of molecules in the *lower* state in an absorption experiment using monochromatic radiation. In a laser cavity it is emission, rather than absorption, which is occurring but the reflection backwards and forwards results in similar hole burning in the velocity distribution of the *upper* state. When the frequency of the laser cavity is tuned to ν_{res}, at the centre of the line, the lasing neon atoms are of the type represented by 3, 4 and 5 in Figure 3.8 with zero velocity component in the direction of propagation. As a result, the *same* atoms are stimulated to emit radiation in whichever direction the stimulating radiation is travelling and saturation (population equalization) of the two levels results more readily for this than for other frequencies. It follows that the gain is reduced at ν_{res} producing the Lamb dip. The width of the dip is the natural line width, which is very much smaller than the Doppler width (Section 2.3).

8.1.8.7 The argon ion and krypton ion lasers

Laser action occurs in the inert gas ions Ne^+, Ar^+, Kr^+ and Xe^+, but those most commonly used in lasers are Ar^+ and Kr^+.

For an ion laser much higher input energy is required than for an inert gas laser, such as the He–Ne laser, in which the lasing species is a neutral atom. The reason for this is that extra energy is required to ionize the atom. For example, the first ionization energies of argon and krypton are 15.76 and 14.00 eV, respectively. In addition, the ion has to be excited to the upper of the two levels involved in laser action. In Ar^+ it has been estimated that *two* collisions are required for the combined ionization and excitation processes.

Because so much power has to be put in, the problem of heat dissipation is serious. It has been solved by using a plasma tube made from beryllium oxide, BeO, which is an efficient heat conductor, and also by water cooling.

Early lasers were pulsed with a repetition rate of up to 2 kHz by periodically discharging condensers through the gas at a pressure of a few mTorr.

Most lasers are now CW and excited by a low voltage, high current discharge. A high current, I, is particularly important as the output power is proportional to I^6 near to threshold. A gas pressure of about 0.5 Torr is used in a plasma tube of 2–3 mm bore. Powers of up to 40 W can be obtained.

The spectroscopy of ion lasers is generally less well understood than that of neutral atom lasers because of the lack of detailed knowledge of ion energy level schemes. Indeed, ion lasers were first produced accidentally and attempts to interpret the transitions came later.

The ground configuration of Ar^+ is $KL3s^23p^5$, giving an inverted $^2P_{\frac{3}{2}}$, $^2P_{\frac{1}{2}}$ multiplet and the excited states involved in the laser involve promotion of an electron from the $3p$ into excited $4s$, $5s$, $4p$, $5p$, $3d$, $4d$,. . . orbitals. Similarly, excited states of Kr^+ arise from the promotion of an electron from the $4p$ orbital. In Ar^+ the $KL3s^23p^4$ core gives rise to ^1S, ^3P, ^1D terms [Section 6.1.2.2(i)]. Most laser transitions involve the core in one of the ^3P states and the promoted electron in the $4p$ orbital.[†]

The Ar^+ laser can produce about 10 lines in the region 454–529 nm, the most intense being at 488.0 and 514.5 nm. The Kr^+ laser produces about nine lines in the region 476–800 nm with the 647.1 nm line the most intense. A laser may contain a mixture of argon and krypton gases which produces a wide range of wavelengths involving each of the ions. Tuning is usually by rotation of a prism, inside the cavity, to select a particular wavelength.

[†]For tabulation and assignments of ion laser lines, see Bridges, W. B. and Chester, A. N. (1965). *IEEE J. Quantum Electron.*, **QE-1**, 66.

Table 8.1 Configurations and bond lengths for the X, A, B and C states of N_2

State	MO configuration	$r_e/\text{Å}$
$X^1\Sigma_g^+$	$\ldots(\sigma_u{*}2s)^2(\pi_u 2p)^4(\sigma_g 2p)^2$	1.0977
$A^3\Sigma_u^+$	$\ldots(\sigma_u{*}2s)^2(\pi_u 2p)^3(\sigma_g 2p)^2(\pi_g{*}2p)^1$	1.2866
$B^3\Pi_g$	$\ldots(\sigma_u{*}2s)^2(\pi_u 2p)^4(\sigma_g 2p)^1(\pi_g{*}2p)^1$	1.2126
$C^3\Pi_u$	$\ldots(\sigma_u{*}2s)^1(\pi_u 2p)^4(\sigma_g 2p)^2(\pi_g{*}2p)^1$	1.1487

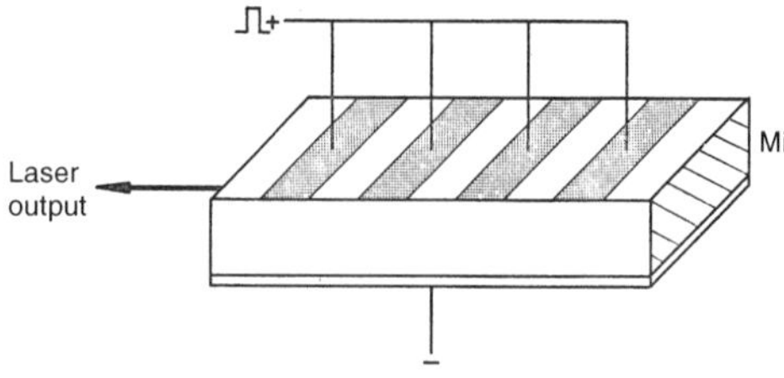

Figure 8.21 Nitrogen laser cavity

8.1.8.8 The nitrogen laser

Apart from the ammonia maser, this is the first molecular laser we have encountered.

The molecular orbital configuraiton of N_2 has been described in Section 6.2.1.1. The ground configuration is given in equation (6.96) and can be abbreviated to $\ldots(\sigma_u{*}2s)^2(\pi_u 2p)^4(\sigma_g 2p)^2$. It gives rise to the $X^1\Sigma_g^+$ ground state. When an electron is promoted to a higher energy orbital singlet and triplet states result. We shall be concerned here with only the triplet states and, in particular, the $A^3\Sigma_u^+$, $B^3\Pi_g$ and $C^3\Pi_u$ states.† The orbital configurations and values of r_e for these states are given in Table 8.1.

In a high voltage discharge through nitrogen gas there is a deep pink glow due mainly to two electronic band systems in emission. The B–A system, or so-called first positive system because it was thought initially to be due to N_2^+, stretches from the red to the green region whereas the C–B system, or so-called second positive system, stretches from the blue into the near ultraviolet.

Laser action has been obtained in a few transitions in both these systems but the C–B laser action has proved more important because it resulted in the first ultraviolet laser. It is only this system that we shall consider here.

The values of the equilibrium internuclear distance r_e for the various states in Table 8.1 indicate that the minimum of the potential for the C state lies almost vertically above that of the X state, as in Figure 6.36(b), whereas those of states B and A are shifted to high r. The result is that the electron–molecule collisional cross-section for the transition from $v''=0$ in the X state to $v'=0$ in the C state is greater than that for analogous transitions in the A–X and B–X systems. A population inversion is created between the $v=0$ level of the C state and the $v=0$ level of the B state. Lasing has been observed in the 0–0 transition, in addition to the 0–1 transition, of the C–B system. However, the laser action is self-terminating because the lifetime of the lower state B (10 μs) is *longer* than that of the upper state C (40 ns). This does not render laser action impossible but necessitates pulsing of the input energy with a pulse length shorter than the lifetime of the C state.

A design for a nitrogen laser is shown in Figure 8.21. A pulsed high voltage of about 20 kV, triggered by a spark gap or a thyratron, is applied transversely across the cavity. The laser is superradiant and a single mirror is used to double the output. Laser pulses of about 10 ns length are typical. Peak power can be as much as 1 MW with a pulse energy of about 10 mJ. The maximum repetition rate is about 100 Hz with longitudinally flowing gas. Much higher repetition rates are possible for transverse flow.

In the 0–0 band of the C–B transition several rotational lines show high gain. These all lie in the 337.044–337.144 nm region and involve *P*-

†The reader is reminded that the labels A, B, C, rather than a, b, c, for states of N_2 do not follow the usual convention.

branch transitions ($\Delta J = -1$) with, at room temperature, $J \approx 8$–9.

8.1.8.9 Excimer and exciplex lasers

An excimer is a dimer which is stable only in an excited electronic state but dissociates readily in the ground state. Examples are the inert gas dimers, of which Xe_2 is the most important from the point of view of laser action.

We do not normally think of Xe_2 as a stable molecule and might expect the potential energy curve for the $X^1\Sigma_g^+$ ground state to be repulsive (see Section 6.2.4.6) as shown for He_2 in Figure 6.42(a). In fact, there is a very shallow minimum with a depth of about 200 cm^{-1} (it is about 70 cm^{-1} in He_2) due to weak van der Waals attractive forces. The internuclear distance at the minimum is large, about 4.4 Å. The result is that the potential energy curves for the ground state and a strongly bound excited state are like those in Figure 8.22. If Xe_2 can be formed in the excited state there is always a population inversion between it and the ground state because of the extremely short ground state lifetime, typically a few picoseconds.

Emission at fairly high gas pressure is predominantly from the $\upsilon = 0$ level of the bound state and occurs at about 173 nm, in the far ultraviolet region. The Xe_2 laser was the first to operate at such a short wavelength. Owing to the spread of the $\upsilon = 0$ wave function the emission extends over about 10 nm and can be tuned with a prism or grating.

Of much wider applicability and covering a wide wavelength range are the inert gas halide exciplex lasers. An exciplex is a complex, consisting, in a diatomic molecule, of two *different* atoms, which is stable in an excited electronic state but dissociates readily in the ground state. In spite of this clear distinction between an excimer and an exciplex, it is now common for *all* such lasers to be called excimer lasers.

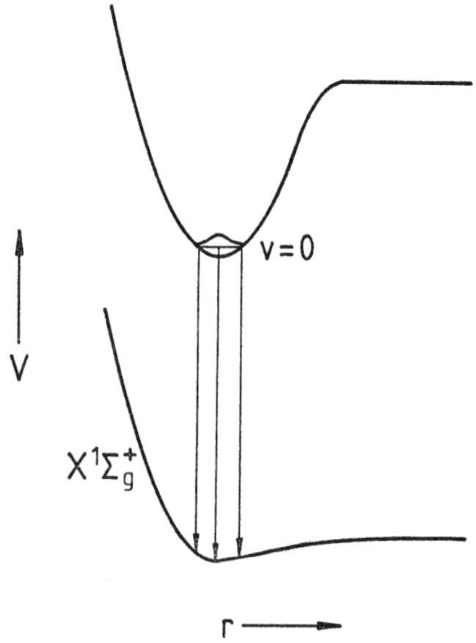

Figure 8.22 Potential curves for the weakly bound ground state and a strongly bound excited state of He_2 or Xe_2

Excimer lasers employing NeF, ArF, KrF, XeF, ArCl, KrCl, XeCl, ArBr, KrBr, XeBr, KrI and XeI as the active medium have been made.

The method of excitation was, in the early days, by an electron beam but now a transverse electrical discharge, like that for the nitrogen laser in Figure 8.21, is used. Indeed, such an excimer laser can be converted to a nitrogen laser by changing the gas.

In an excimer laser the mixture of inert gas, halogen gas and helium, used as a buffer, are pumped around a closed system comprising a reservoir and the cavity.

The examples of ArF (193 nm), KrF (248 nm), XeF (351 nm), KrCl (222 nm), XeCl (308 nm) and XeBr (282 nm) indicate the range of wavelengths from excimer lasers. Each of these wavelengths is, like that for Xe_2, the centre of a band which may be continuous or discrete depending on the nature of the ground state potential curve but these lasers are not tunable over as wide a wavelength range as Xe_2.

In the case of XeF the potential energy minimum in the $X^2\Sigma^+$ ground state is relatively deep, about 1150 cm^{-1}, and supports a number

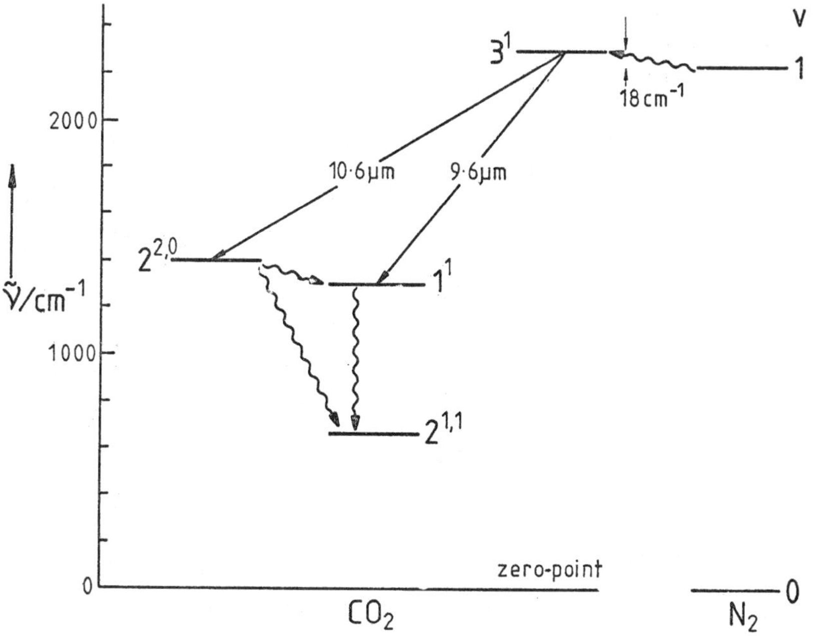

Figure 8.23 Vibrational levels of N_2 and CO_2 relevant to the CO_2 laser

of vibrational levels resulting in discrete emission with sharp rotational structure.†

The mechanism of excitation is not completely understood but, for the KrF laser, the following processes are probably the most important:

$$\begin{aligned} &\mathrm{Kr} + \mathrm{e}\ (\text{fast}) \rightarrow \mathrm{Kr}^+ + 2\mathrm{e} \\ &\mathrm{F}_2 + \mathrm{e}\ (\text{slow}) \rightarrow \mathrm{F}^- + \mathrm{F} \\ &\mathrm{F}^- + \mathrm{Kr}^+ + \mathrm{He} \rightarrow (\mathrm{KrF})^* + \mathrm{He} \\ &(\mathrm{KrF})^* + h\nu \rightarrow \mathrm{Kr} + \mathrm{F} + 2h\nu \end{aligned} \tag{8.20}$$

where (KrF)* represents the excimer and the final step is the stimulated emission process.

The excimer laser radiation is pulsed with a maximum rate of about 200 Hz. The pulse energy is high compared with a nitrogen laser and is of the order of 100 mJ with a peak power of about 5 MW.

†See, for example, Velazco, J. E., Kolts, J. H., Setser, D. W. and Coxon, J. A. (1977). *Chem. Phys. Lett.*, **46**, 99.

Excimer lasers are two-level lasers and have a high efficiency of about 20%.

If the normal broad band emission is not suitable for a particular experiment then some means of wavelength selection, employing a prism or grating, must be used.

The excited state lifetimes of the excimers are so short, for example about 6 ns for KrF, that excimer lasers are superradiant, and therefore only one mirror is used in the cavity, as in the nitrogen laser.

8.1.8.10 The carbon dioxide laser

The CO_2 laser is a near infrared gas laser capable of immense power and with an efficiency of about 20%, comparable to an excimer laser but less efficient than a diode laser.

CO_2 has three normal modes of vibration, ν_1, the symmetric stretch, ν_2, the bending vibration, and ν_3, the antisymmetric stretch with symmetry species σ_g^+, π_u and σ_u^+ and fundamental vibration wavenumbers of 1354, 673 and

2396 cm^{-1} (Figure 8.23), respectively. Figure 8.23 shows some of the vibrational levels, the nomenclature of which is explained in the footnote on p. 213, which are involved in the laser action. This occurs principally in the $3_0^1 2_{2,0}^{0,0}$ transition,† at about 10.6 μm, but may also be induced in the $3_0^1 1_1^0$ transition, at about 9.6 μm.

Population of the 3^1 level is partly by electron–molecule collisions and partly by energy transfer from nitrogen molecules in the $v = 1$ level, this being metastable due to the fact that the transition to $v = 0$ is electric dipole forbidden (Section 5.1.1). Energy transfer from nitrogen is particularly efficient because the $v = 1$ level is only 18 cm^{-1} below the 3^1 level of CO_2 (Figure 8.23). Because of near-degeneracies of higher vibrational levels of nitrogen and the ν_3 stack of CO_2, transfer to levels such as $3^2, 3^3, \ldots$ also occurs. Transitions down the ν_3 stack are fast until the 3^1 level is reached.

Decay of the 1^1 and $2^{2,0}$ lower levels of the laser transitions are rapid down to the $2^{1,1}$ level; this is depopulated mostly by collisions with helium atoms in the CO_2–N_2–He gas mixture which is used.

Lifetimes of upper and lower states are governed by collisions and that of the upper is always longer than that of the lower in the gas mixtures used.

The energy input into a CO_2 laser is in the form of an electrical discharge through the mixture of gases. The cavity may be sealed, in which case a little water vapour must be added in order to convert back to CO_2 any CO which is formed. More commonly longitudinal or, preferably, transverse gas flow through the cavity is used. The CO_2 laser can operate in a CW or pulsed mode.

One of the most successful designs for a high energy, pulsed CO_2 laser is the so-called TEA laser, TEA being an acronym for *t*ransverse

†See Section 5.2.6.3 for labelling of the components of the $2^{2,0}$, 1^1 Fermi diad, the order of which has been reversed[194] compared to the previous, long-standing assignment.

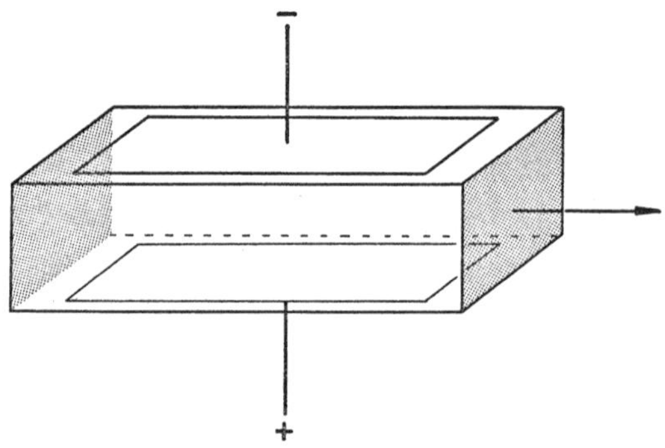

Figure 8.24 Transverse excitation across the cavity in a CO_2 TEA laser

*e*xcited *a*tmospheric. As this implies, the voltage for the discharge is applied transversely across the laser cavity, as shown in Figure 8.24. The laser is pulsed at rates of up to 1 kHz and the gas is flowed transversely although, at low repetition rates, flowing is not necessary. As the name TEA indicates, gas pressures of up to and above 1 atm can be used, resulting in very high pulse energy but the line width is large, about 3 GHz (0.1 cm^{-1}), due to pressure broadening.

Each of the lasing vibrational transitions has associated rotational fine structure, discussed for linear molecules in Section 5.2.5.1. The $3_0^1 1_1^0$ transition is Σ_u^+–Σ_g^+ with associated *P* and *R* branches, for which $\Delta J = -1$ and $+1$, respectively, similar to the 3_0^1 band of HCN in Figure 5.43. The $3_0^1 2_{2,0}^{0,0}$ band is, again, Σ_u^+–Σ_g^+ with a *P* and *R* branch.

Unless the cavity is tuned to a particular wavelength the vibration–rotation transition with the highest gain is the *P*-branch transition involving the rotational level which has the highest population in the 3^1 state. This is *P*(22), with $J'' = 22$ and $J' = 21$, at normal laser temperatures. The reason why this *P*-branch line is so dominant is that thermal redistribution of rotational level populations is faster than the population depletion due to emission.

P-branch transitions can be shown to have much higher gain than both *Q*- and *R*-branch transitions. *Q*- and *R*-branch transitions require population inversion in order to show gain but a *P*-branch transition may show gain even if the lower state population is slightly higher than that of the upper state. This is the case for an electronic transition also, such as that involved in the nitrogen laser (Section 8.1.8.8).

The cavity may be tuned by a prism or, preferably, by replacing one of the mirrors by a diffraction grating, as in the dye lasers (Section 8.1.8.14).

Mention should also be made of the gas dynamic CO_2 laser. In this laser heated CO_2 is injected into the cavity through supersonic nozzles, the gas molecules reaching velocities of around Mach 4. The rapid expansion causes the temperature of the molecules to be decreased in a period of time which is short compared with the lifetime of the upper laser level and long compared with that of the lower level. As a result there is very efficient population inversion and powers of 60 kW, multimode, and 30 kW, single mode, have been obtained, but with low efficiency. The laser can operate continuously but only over a period of a few seconds.

For use in high resolution spectroscopy a small line width and high stability are of paramount importance; the achievement of high power is a lower priority. For these purposes a low power, CW CO_2 laser operating in the TEM_{00} transverse mode and a single axial mode is used. The population inversion is produced by an electrical discharge in a mixture of gases consisting typically of CO_2 (3 Torr), N_2 (5 Torr) and He (14 Torr). The gas mixture may be flowed through the cell, which is equipped with Brewster angle windows, or the cell may be closed, particularly if isotopically substituted CO_2 is being used to achieve different laser wavelengths.

Replacement of one of the cavity mirrors by a grating, outside the gas cell, allows wavelength selection and the output mirror may be mounted on a piezoelectric translator so that the cavity length can be finely adjusted. Control of the translator by a lock-in stabilizer enables the maximum power output to be maintained. A laser line width of the order of 1 kHz can be obtained in this way.

8.1.8.11 The carbon monoxide laser

Unlike CO_2, the CO molecule has only one vibrational mode leading to a single stack of vibrational levels, each with a stack of rotational levels.

Population inversion in the vibrational levels may be achieved by several methods, including chemical reactions (see Section 8.1.8.13), but one of the most commonly used is an electrical discharge in a mixture of CO and N_2 gases. The laser cavity design is very similar to that described in Section 8.1.8.10 for the low power, CW CO_2 laser; indeed, the same apparatus may be used for either CO or CO_2 simply by changing the filling in the gas cell and rotating the grating.

The nitrogen serves the purpose, as in the CO_2 laser, of transferring vibrational energy on collision to the CO molecules. Population inversion between adjacent vibrational levels with $\upsilon''=3$–36 can be obtained and about 200 laser lines, all involving *P*-branch transitions, have been obtained† in the region 4.97–8.26 μm.

8.1.8.12 The water and hydrogen cyanide lasers

An electrical discharge in water vapour in a laser cavity produces more than a dozen laser lines when the discharge is continuous, and more than 100 lines when it is pulsed. Typical power produced is a few milliwatts (CW) and a few hundred milliwatts peak power in the pulsed mode. The laser lines occur in the infrared region and cover a wide range from 7 to 200 μm (1300–45 cm^{-1}), those which lie to longer wavelength of the CO_2 and CO lasers (Sections 8.1.8.10 and 8.1.8.11) being particularly useful sources of monochromatic radiation.

The transitions involved in laser action in the H_2O molecule[195] are rather random and it is not

†See, for example, Roh, W. B. and Rao, K. N. (1974). *J. Mol. Spectrosc.*, **49**, 317.

the case, as it is with the CO_2 and CO lasers, that a number of lines belong to the same vibrational band. Most of the observed laser transitions in $^1H_2{}^{16}O$ are between rovibrational states involving the 1^1, 2^2 and 3^1 vibrational levels where ν_1, ν_2 and ν_3 are the symmetric stretching, bending, and antisymmetric stretching vibrations, respectively. Sufficient gain for laser action results only when there is a strong perturbation between rovibronic levels; hence the random nature of the transitions.

Isotopically substituted water, such as $^1H_2{}^{18}O$ and $^2H_2{}^{16}O$, can be used to extend the number of lines obtained with the water vapour laser.

An electrical discharge in various gases containing carbon and nitrogen, such as a mixture of cyanogen (C_2N_2) and H_2O, in a laser cavity results in laser action in the wavelength region above 70 μm ($< 140\,\mathrm{cm}^{-1}$). The laser produces only a few lines when operated CW but many more lines in the pulsed mode.

At first, it was thought that the molecule responsible was the short-lived species CN, which is commonly observed in discharges, arcs, sparks, etc. In fact, the lasing molecule is ^{1}HCN,[196] and the most intense laser lines, which lie in the 337 μm region, involve transitions within and between the stacks of rotational levels associated with the $1^1 2^{1,1}$ and $2^{4,0}$ vibrational levels. The vibrational levels are only about $2.64\,\mathrm{cm}^{-1}$ apart and undergo Coriolis interaction (Section 4.2.6.2). As in the water vapour laser, it is a perturbation which is responsible for the population inversion and high gain obtained.

Other laser lines in the ^{1}HCN laser, in the 130 μm region, involve Coriolis interaction between the $1^1 2^{2,0}$, $1^1 2^{2,2}$ and $2^{5,1}$ vibrational levels, whereas the ^{2}HCN laser depends on Coriolis interaction between the $1^2 2^{2,0}$ and $2^{9,1}$ vibrational levels.[197]

In addition to population inversion being achieved in H_2O and HCN by a discharge, it has also been produced by optical pumping with a CO_2 laser[198] and this method produces far infrared laser radiation from many other molecules.[199]

8.1.8.13 Chemical lasers

A chemical laser is one in which the population inversion is created as a direct result of chemical reaction. This definition rather begs the question of what we mean by a chemical reaction but the hydrogen–fluorine and hydrogen–chlorine lasers clearly come into this category.

The overall reaction between hydrogen and fluorine is

$$^1H_2 + F_2 \rightarrow 2\,^1HF^* \qquad (8.21)$$

where $^1HF^*$ indicates that ^{1}HF is produced in vibrationally excited states. The reaction goes in two stages:

$$F + {}^1H_2 \rightarrow {}^1HF^*\,(\upsilon = 0\text{–}3) + {}^1H \qquad (8.22)$$

$$^1H + F_2 \rightarrow {}^1HF^*\,(\upsilon = 0\text{–}10) + F \qquad (8.23)$$

the fluorine atoms acting as a catalyst which can be produced in various ways, such as the electron-induced dissociation

$$SF_6 + e \rightarrow SF_5 + F + e \qquad (8.24)$$

Equations (8.22) and (8.23) show that each reaction produces ^{1}HF in a different range of vibrational states. For the reaction of equation (8.22) the distribution of population among the vibrational levels is typically in the ratios 1:2:10:5 for $\upsilon = 0, 1, 2, 3$, whereas for the reaction in equation (8.23), the ratios are 6:6:9:16:20:33:30:16:9:6:6 for $\upsilon = 0$–10.

Laser action occurs in vibration–rotation transitions of the 1–0, 2–1 and 3–2 bands in the range 2.7–3.2 μm. If stimulated emission occurs in, say, the 2–1 band the population of the $\upsilon = 2$ level is depleted, enhancing the laser action in the 3–2 band. In addition population of the $\upsilon = 1$ level is increased, enhancing the 1–0 laser action.

Laser gain may be obtained with $\Delta\upsilon = 2$ transitions such as the 3–1 transition. Such laser action covers the range 1.3–1.4 μm and considerable power can be generated at these wavelengths if the cavity optics are designed to be highly reflective for this range resulting in suppression of lasing on $\Delta v = 1$ transitions.

The design of a chemical laser is basically simple. In the case of the $^1H_2$–F_2 laser the reaction compartment is placed between the two mirrors forming the laser cavity. The fluorine atoms are generated from SF_6 in an electrical discharge tube outside the reaction compartment. Oxygen is added to the SF_6 to enhance the dissociation and to remove any free sulphur which is produced. The reaction mixture is pumped through the reaction compartment into which hydrogen is injected. Helium may be added to the mixture to act as a scavenger. The total pressure in the reaction compartment is only a few Torr. Commercial $^1H_2$–F_2 lasers may generate power up to 500 W, distributed over all wavelengths, reduced to 100 W for a single wavelength. For lasing on the $\Delta v = 2$ transitions the maximum multi-wavelength power is about 25 W, reduced to 10–15 W for a single wavelength. Pulsing of the electrical discharge can be employed to create a pulsed laser.

Replacement of $^1H_2$ by $^2H_2$ results in laser action in $^2HF^*$ in the 3.5–4.2 μm range with 60–80% of the power possible with $^1H_2$, although a 2 MW $^2H_2$–F_2 laser has been constructed for potential military purposes. No lasing occurs for the $\Delta v = 2$ transitions in $^2HF^*$.

It is interesting that in the $^1H_2$–F_2 reaction, population inversions are created between pairs of rotational energy levels within a particular vibrational state and laser action at longer wavelengths results. For example, lasing involving the $J = 12$–27 levels in the $v = 3$ state of ^1HF produces radiation in the range 11.54–21.79 μm.

The $^1H_2$–Cl_2 laser was the first chemical laser to be produced (in 1965). It operates in a similar way to the H_2–F_2 laser except that the initiation is by chlorine atoms which can be produced by the irradiation process:

$$Cl_2 + h\nu \rightarrow 2Cl \tag{8.25}$$

Reactions analogous to those in equations (8.22) and (8.23) then proceed but it is only the reaction analogous to equation (8.23) which produces vibrationally excited ^1HCl.

Laser action involves the 1–0 and 2–1 transitions in the range 3.5–4.1 μm.

In the ^2HF–CO_2 laser $^2HF^*$ is produced in vibrationally excited states by reaction between $^2H_2$ and F_2 in two steps analogous to those in equations (8.22) and (8.23). The initiating fluorine atoms may be produced by the reaction

$$NO + F_2 \rightarrow NOF + F \tag{8.26}$$

The CO_2 molecules are then vibrationally excited by collisional energy transfer:

$$^2HF^* + CO_2 \rightarrow {}^2HF + CO_2^* \tag{8.27}$$

Since the fundamental vibration wavenumber of ^2HF is 2998 cm^{-1} pumping of the 3^1 levels of CO_2 results ($\tilde{\nu}_3 = 2349$ cm^{-1}). Subsequently lasing occurs in CO_2 as described in Section 8.1.8.9.

The ^2HF–CO_2 laser is really a CO_2 laser initiated by chemical reaction and is sometimes called a hybrid laser.

The shortest wavelength, high power chemical laser is the O_2–I_2 laser. Reaction between Cl_2 and hydrogen peroxide (H_2O_2) generates O_2^* in a long-lived electronically excited state. The O_2^* molecules collide with, and dissociate, I_2 molecules creating electronically excited iodine atoms which lase at 1.315 μm. Power up to 25 kW has been achieved with CW operation.

8.1.8.14 Dye lasers

Dye lasers are unusual in that the active medium is a liquid and they have the important property of being continuously tunable over a large wavelength range.

Dyes cover a wide range of structural types of molecules but they all have in common the fact that the molecules have a planar skeleton with a conjugated π-electron system delocalized over a large part of the molecule. An important effect of a high degree of delocalization is to bring the

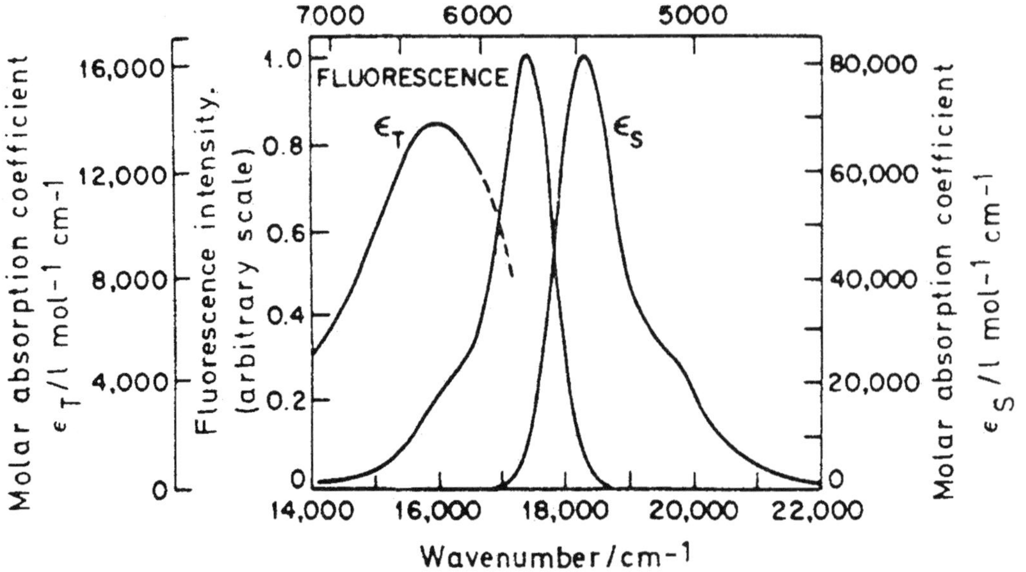

Figure 8.25 Absorption and fluorescence spectra of rhodamine B in methanol with a concentration of 5×10^{-5} mol L^{-1}. The curve marked ε_T is for the T_2–T_1 absorption (process 8 in Figure 8.26) and that marked ε_S for process 1
[Reproduced, with permission, from Dienes, A. and Shank, C. V. (1972). 'Dye lasers', chap. 4 in *Creation and Detection of the Excited State*, vol. 2 (Ed. W. R. Ware), p. 154. Marcel Dekker, New York]

highest occupied and the lowest unoccupied molecular orbitals close together. These MOs may be regarded as being of the Hückel type (see Sections 6.3.1.6–6.3.1.9) and the fact that they are close together results in the absorption from the ground state S_0 to the first singlet excited state S_1 being in the visible region of the spectrum. Also typical of a dye is a high absorbing power characterized by a value of the oscillator strength f close to 1 (see Sections 2.2 and 6.3.2), and also a value of the fluorescence quantum yield Φ_F (equation 6.288) close to 1.

Figure 8.25 illustrates these features in the case of the dye rhodamine B. The maximum of the typically broad S_1–S_0 absorption occurs at about 548 nm with a very high value of 80 000 L mol^{-1} cm^{-1} for ε_{max}, the maximum value of the molar absorption coefficient [equation (2.24)]. The fluorescence curve shows, as usual, an approximate mirror image relationship to the absorption curve. It has the additional property, important for all laser dyes, that the fluorescence and absorption maxima do not coincide: if they did, a large proportion of the fluorescence would be reabsorbed.

Figure 8.26 shows a typical energy level diagram of a dye molecule including the lowest electronic states S_0, S_1, and S_2 in the singlet manifold and T_1 and T_2 in the triplet manifold. Associated with each of these states are vibrational and rotational sub-levels broadened to such an extent by collisions in the liquid that they form a continuum. As a result the absorption spectrum, such as that in Figure 8.25, is typical of a liquid phase spectrum showing almost no structure within the band system.

Depending on the method of pumping the S_1–S_0 transition excitation may be by the S_1–S_0 or S_2–S_0 absorption process, labelled 1 and 2 in Figure 8.26, or both. Following either process collisional relaxation to the lower levels of S_1 is

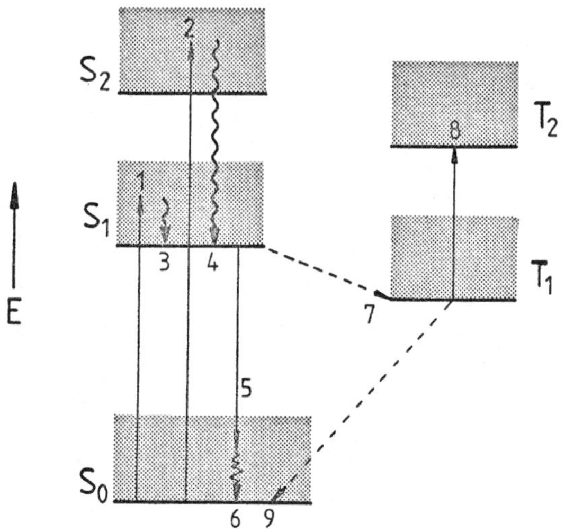

Figure 8.26 Typical energy level scheme for a dye molecule showing processes important in laser action

rapid by process 3 or 4; for example, the vibrational–rotational relaxation of process 3 takes of the order of 10 ps. Following relaxation the distribution among the levels of S_1 is that corresponding to thermal equilibrium, i.e. there is a Boltzmann population [equation (2.6)].

The state S_1 may decay by radiative (r) or non-radiative (nr) processes, labelled 5 and 7, respectively, in Figure 8.26. Process 5 is the fluorescence which forms the laser radiation and the figure shows it terminating in a vibrationally excited level of S_0. The fact that it does so is vital to the dye being usable as an active medium and is a consequence of the Frank–Condon principle (Section 6.3.4.3).

The shape of the broad absorption curve in Figure 8.25 is typical of that of any dye suitable for a laser. It shows an absorption maximum to low wavelength of the 0_0^0 band position, which is close to the absorption–fluorescence crossing point. The shape of the absorption curve results from a change of shape of the molecule, from S_0 to S_1, in the direction of one or more normal coordinates so that the most probable transition in absorption is to a vibrationally excited level of S_1. Similarly, in emission from the zero-point level of S_1 the most probable transition is to a vibrationally excited level of S_0. The fluorescence lifetime, τ_r, for spontaneous emission from S_1 is typically of the order of 1 ns while the relaxation process 6, like process 3, takes only about 10 ps. The result is that, following processes 1 and 3, there is a population inversion between the zero-point level of S_1 and vibrationally excited levels of S_0 to which emission may occur, provided that these levels are sufficiently highly excited to have negligible thermal population.

However, the population of S_1 may also be reduced by absorption of the fluorescence taking the molecule from S_1 into S_2, if the wavelengths of the two processes correspond, as well as by non-radiative transistions to either S_0 (internal conversion) or T_1 (intersystem crossing), as discussed in Section 6.3.5.6. In dye molecules it is the S_1–T_1 process, labelled 7 in Figure 8.26, which is the most important. This is a spin-forbidden process with a lifetime τ_{nr} of the order of 100 ns. The lifetime of the state S_1 is related to τ_{nr} and the radiative lifetime τ_r by

$$\frac{1}{\tau} = \frac{1}{\tau_r} + \frac{1}{\tau_{nr}} \tag{8.28}$$

Since τ_r is of the order of 1 ns, fluorescence is the dominant decay process for S_1.

The lifetime τ_T of the state T_1 is long because the T_1–S_0 transition, process 9 in Figure 8.26, is spin-forbidden. Depending on the molecule and on the conditions, particularly the amount of dissolved oxygen, it may be anywhere in the range 100 ns–1 ms. If $\tau_T > \tau_{nr}$ the result is that the concentration of molecules in T_1 can build up to a high level. It happens in many dye molecules that the intense, spin-allowed, T_2–T_1 absorption, process 8 in Figure 8.26, overlaps with, and therefore can be excited by, the S_1–S_0 emission thereby decreasing the efficiency of the laser considerably. Figure 8.25 shows how important this process is in rhodamine B.

In order to prevent this occurring a pulsed method of pumping is used with a repetition rate low enough to allow time for T_1–S_0 relaxation. For CW operation either τ_T must be sufficiently short or another dye must be used whose T_2–T_1 absorption does not overlap with the fluorescence.

There are many dyes available each of which can be used over a 20–30 nm range and which, together, cover the complete wavelength range from about 365 nm in the ultraviolet to about 930 nm in the near infrared. Dye concentrations are low, typically in the range 10^{-2}–10^{-4} mol L^{-1}.

Taking into account the possibility of frequency doubling (Section 8.1.7), dye lasers can provide tunable radiation throughout the range 220–930 nm but with varying levels of intensity and degrees of difficulty. The tunability and the extensive wavelength range make dye lasers probably the most generally useful of all visible or ultraviolet lasers.

A pulsed dye laser may be pumped with a flashlamp surrounding the cell through which the dye is flowing. With this method of excitation pulses from the dye laser about 1 μs long and with an energy of the order of 100 mJ can be obtained. Repetition rates are typically low—up to about 30 Hz.

More commonly a pulsed dye laser is pumped with a nitrogen, excimer or Nd^{3+}:YAG laser. The nitrogen laser, operating at 337 nm, and a xenon fluoride excimer laser, operating at 351 nm, both excite the dye initially into a singlet excited state higher in energy than S_1. The Nd^{3+}:YAG laser is either frequency doubled to operate at 532 nm or frequency tripled to operate at 355 nm depending on the dye which is being pumped.

In order to produce tunable ultraviolet radiation from a dye laser pumped with a Nd^{3+}:YAG laser there is, for some wavelength regions, a useful alternative to frequency doubling with appropriate crystals. These same crystals can be used for frequency mixing (see also Section 8.2.9) whereby if radiations of two different frequencies are combined in the crystal, some of the emerging radiation has a frequency equal to the sum or the difference of the incident frequencies. In order to produce ultraviolet radiation the dye laser radiation and part of the Nd^{3+}:YAG laser radiation, at either 1064, 532 or 355 nm, are combined in a nonlinear crystal so that the radiation with the sum of the two frequencies is in the ultraviolet.

When the excitation in a particular dye is at a lower wavelength than that covered by the S_1–S_0 system it may happen that the dye has no absorption at the wavelength of the pumping radiation. The problem may be solved by mixing it with a second dye which does absorb the pumping radiation. The energy is then transferred to the first dye by collisions. An example of this is the rhodamine 6G–cresyl violet perchlorate mixture used for the 645–680 nm region with a nitrogen pump laser. Rhodamine 6G by itself is used for the 570–600 nm region and, in the mixture, is merely used as an energy transfer agent.

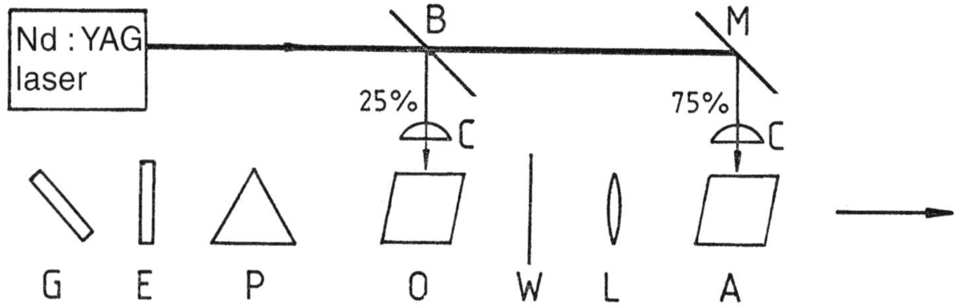

Figure 8.27 A Nd^{3+}:YAG laser pumped dye laser with an oscillator and amplifier dye cell, O and A, respectively

Figure 8.27 shows a typical experimental arrangement for a dye laser pumped with a Nd^{3+}:YAG laser. The pumping laser beam is divided by a beamsplitter B so that 25% is directed towards the oscillator dye cell O which is inside the laser cavity bounded by the reflection grating G and the output window W. The laser beam is focused into the cell by a cylindrical lens C to form a line in the dye solution.

P indicates a multiple prism system which serves as a beam expander in order to fill the grating with light, a condition which is necessary for maximum resolution by the grating (Section 3.5.3). Alternatively, a beam expanding telescope may be used but this takes up much more room in the laser cavity. Tuning is by rotation of the grating. An etalon, E, may be used for single mode operation.

Outside the cavity the laser radiation passes through the lens L into the amplifier dye cell A containing a similar dye solution. The oscillator–amplifier system of dye cells was devised by Hänsch[200] and results in improved output beam quality, particularly the divergence which may be as low as 0.5 mrad.

The dye solution must always be kept moving to prevent overheating and premature decomposition. This is usually achieved by pumping the solution continuously through the cells.

CW dye lasers are usually pumped with an argon ion laser. About 1 W of continuous power can be produced but the dye solution has to be pumped very efficiently and usually takes the form of a jet flowing rapidly across the laser cavity.

The ring dye laser† is a later modification of the CW dye laser which combines high power with extremely narrow line width. In this type of laser the cavity is not linear but is folded around on itself by a series of prisms or mirrors so that the stimulated radiation can travel continuously around the cavity which contains a device, based on the Faraday effect, which causes the wave to travel preferentially in one direction. The main difference, then, from a conventional laser is that there is a travelling rather than a standing wave in the cavity. The result is that burning of holes in the velocity distribution curve (Figure 3.9), which normally limits the power output, is avoided. This means that an efficiency as high as 20% can be achieved and more than 1 W of continuous power with a line width of about 1 MHz (3×10^{-5} cm^{-1}) obtained. Intracavity frequency doubling crystals may be used to produce CW ultraviolet radiation with power of up to 30 mW, a frequency doubling efficiency of 30%, whereas in a CW dye laser with a linear cavity, frequency doubling is much less efficient.

†See, for example, Schroder, H. W., Stein, L., Frolich, D., Fugger, B. and Welling, H. (1977). *Appl. Phys.*, **14**, 377.

8.1.8.15 The optical parametric oscillator

An optical parametric oscillator (OPO) resembles a laser in that the output radiation is from a resonant cavity and is coherent. However, the radiation is not stimulated as it does not result from a population inversion between an upper and a lower state. There is no upper state involved.

An OPO contains a non-linear crystal and is driven by a pump laser which directly drives the electrons in the non-linear medium. If the medium were linear the oscillating electric field generated by the electrons would be in the same direction and with the same phase and frequency as the pump laser radiation. In a non-linear medium other frequencies are generated. The OPO is, therefore, a source of coherent, tunable radiation which displaces, to some extent, the colour centre laser described in Section 8.1.8.5.

The non-linear crystal converts the frequency ν_p of the pump laser into two new frequencies, ν_s, the signal frequency and ν_i, the idler frequency, where

$$\nu_p = \nu_s + \nu_i \tag{8.29}$$

This equation suggests that a wide range of signal and idler frequencies is possible resulting

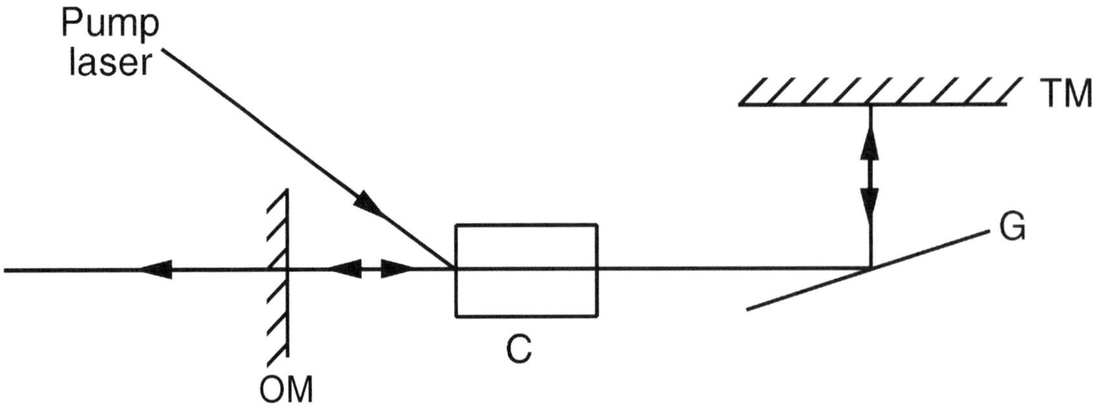

Figure 8.28 A typical optical parametric oscillator

in a wide tuning range. However, the tuning range is limited by the non-linear crystal employed, the more commonly used being BBO, LBO, KN and KTP (see Section 8.1.7 for their chemical formulae). Also used are LN, $AgGaS_2$, $AgGaSe_2$ and $ZnGeP_2$, but they have a low damage threshold which is a considerable limitation because of the necessarily high power of the pump laser.

The pump laser is usually either a Nd^{3+}:YAG or Ti^{3+}:sapphire laser. For example, with a BBO non-linear crystal tuning ranges of 300–2500, 410–2500 and 670–2500 nm may be achieved with 266, 355 and 532 nm pump radiation from frequency quadrupled, tripled and doubled radiation, respectively, from a Nd^{3+}:YAG laser.

In a typical arrangement, shown in Figure 8.28, the non-linear crystal, C, is placed inside a cavity bounded by an output mirror, OM, and a tuning mirror, TM, placed beyond a grazing incidence diffraction grating, G. The pump beam may be directed into the crystal either collinearly, through OM, or non-collinearly, as shown. Wavelength tuning is achieved by rotation of the tuning mirror. In contrast to the oscillator in a dye laser (see Section 8.1.8.14), a prism beam expander is not used to fill the grating with radiation because of the relatively low gain of the OPO. The low gain may also necessitate the use of an optical parametric amplifier (OPA) which operates on the same principle as an OPO but uses part of the pump radiation to increase the total conversion into signal and idler frequencies compared with using an OPO alone.

For efficient transfer of energy from the pump beam to the signal and idler beams these must be phase matched. This is achieved in the birefringent non-linear crystal by changing the crystal temperature or, more commonly, by changing the phase angle between the axis of the cavity and the crystal axis.

An OPO can be constructed so that it is resonant on either the signal or the idler frequency ranges, or both. The bandwidth of the coating for the cavity optics and phase matching range of the non-linear crystal must be correspondingly accommodating.

The pulse length of the output radiation from an OPO is always shorter than that of the pump radiation. This property makes it ideal for creating femtosecond pulses (<100 fs) covering a broad wavelength range. As described in Section 8.1.8.3, a mode-locked Ti^{3+}:sapphire laser may produce pulses as short as 50 fs with high power. The high power is necessary because the non-linear crystal in a femtosecond OPO must be short, around 1 mm. In addition, the Ti^{3}:sapphire laser has the advantage of being tunable over a large range (670–1100 nm). Pumping of an OPO with such a laser can produce femtosecond pulses over a wide wavelength range, 350–4000 nm, including second harmonic generation of the pump laser, with an average power of hundreds of milliwatts and a repetition rate of up to 100 MHz.

8.2 LASER SPECTROSCOPY

From 1960 onwards, the increasing availability of intense, monochromatic laser sources provided a tremendous impetus to a wide range of spectroscopic investigations.

The most immediately obvious application of early, essentially non-tunable, lasers was to all types of Raman spectroscopy in the gas, liquid or solid phase. The experimental techniques, employing laser radiation, were described in Section 4.8.2. Examples of the quality of spectra which can be obtained in the gas phase are to be found in Figure 4.35, which shows the pure rotational Raman spectrum of N_2, and Figures 5.10 and 5.56, which show the vibration–rotation Raman spectra of the $v = 1$–0 transition in CO and the 3_0^1 vibrational transition in CH_4, respectively. All of these spectra were obtained with an argon ion laser.

Laser radiation is very much more intense than that from, for example, a mercury arc (Figure 4.36) which was commonly used as a Raman source before 1960. As a result of the greater source intensity, much weaker Raman scattering can be observed. In addition, the narrower line width of laser radiation increases the resolution obtainable.

Resonance Raman spectroscopy, to be discussed in Section 8.2.1, requires intense, monochromatic sources covering a range of wavelengths. Prior to the development of lasers, studies were limited by the small number of sources available. Lasers provide a much wider choice of wavelengths and are ideally suited to observing the resonance Raman effect.

In addition to carrying out conventional Raman experiments with laser sources new kinds of Raman experiments became possible using *Q*-switched, giant pulse lasers to investigate effects which arise from the nonlinear relationship between the induced electric dipole and the oscillating electric field [equation (8.14)]. These are grouped under the general heading of nonlinear Raman effects.

For branches of spectroscopy other than Raman spectroscopy, most laser sources may appear to have a great disadvantage, that of non-tunability. In regions of the spectrum where a tunable laser is not available, ways have been devised for tuning the atomic or molecular energy levels. One way of achieving this is by applying an electric or magnetic field, as in laser Stark and laser magnetic resonance spectroscopy (Sections 8.2.6 and 8.2.7), of such a strength that the separation of a pair of energy levels matches the laser frequency.

In regions of the spectrum where a tunable laser is available, it may be practicable to use it in the way that a tunable klystron or backward wave oscillator is used in microwave or millimetre wave spectroscopy. Attenuation of the source radiation as a function of frequency produces the absorption spectrum. This technique is employed with a diode laser to produce an infrared absorption spectrum. When electronic transitions are being studied, greater sensitivity is usually achieved by monitoring secondary processes which follow, and are directly related to, the absorption which has occurred. Such processes include fluorescence, dissociation or predissociation and, following the absorption of one or more additional photons, ionization. The spectrum resulting from monitoring these processes as a function of laser frequency usually resembles the absorption spectrum very closely.

It may be apparent to the reader at this stage that, when lasers are being used as spectroscopic sources, we can no longer think in terms of a generally applicable experimental method, such as that illustrated in Figure 3.2 for a conventional absorption experiment. Many ingenious experiments and experimental techniques have been devised using laser sources and it will be possible here to describe only a representative selection of them.

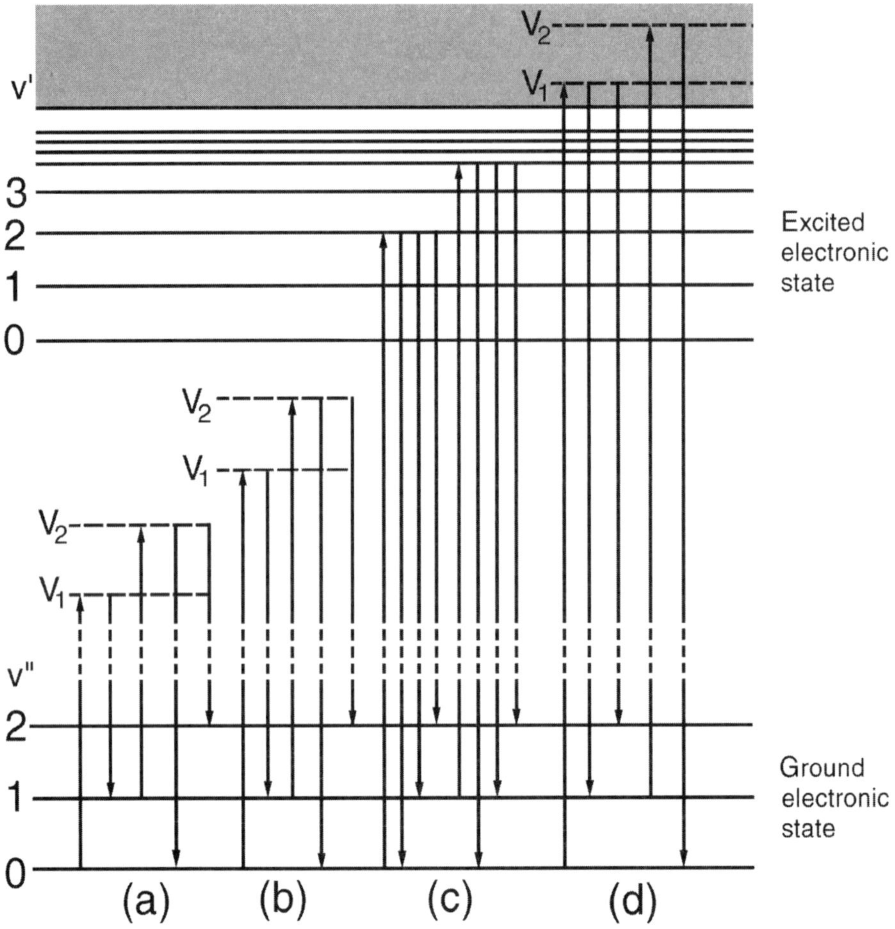

Figure 8.29 (a) Vibrational Raman effect, (b) pre-resonance Raman effect, (c) single vibronic level fluorescence and (d) resonance Raman effect

8.2.1 Resonance Raman and time-resolved Raman spectroscopy

Figure 8.29 indicates the various scattering and emission processes which may occur when a molecule is subjected to the intense, monochromatic, tunable radiation from, say, a dye laser.

Figure 8.29(a) illustrates the situation encountered in the vibrational Raman effect (see Figure 5.3), where the virtual states V_1 and V_2 are far away from any electronic state. The intensity I_{mn} of Raman scattering for an overall transition between states m and n is given by

$$I_{mn} = \frac{2^7\pi^5}{3^2} I_0(\tilde{\nu}_0 \pm \tilde{\nu}_R)^4 \sum_{ij} |(\alpha_{ij})_{mn}|^2 \qquad (8.30)$$

where I_0 and $\tilde{\nu}_0$ are the intensity and wavenumber of the laser radiation, $\tilde{\nu}_R$ is the vibrational Raman shift and the $(\alpha_{ij})_{mn}$, where i and j can be x, y or z, are the elements of the transition polarizability tensor (equation 4.117).

For the case shown in Figure 8.29(b), where the virtual states are fairly close to the vibrational and rotational eigenstates of an excited electronic state, equation (8.30) is no longer valid. The Raman intensities are distorted and the effect is called the pre-resonance Raman effect. When the virtual states become coincident with eigenstates, as in Figure 8.29(c), single vibronic level fluorescence, discussed in Section 6.3.4.2(i), occurs provided the substance is in the gas phase and the pressure is sufficiently low to avoid collisions before emission.

If the virtual states lie within a continuum of states, as in Figure 8.29(d), the true resonance Raman effect occurs. As a result of this requirement for a continuum of levels, resonance Raman scattering can occur when the exciting radiation lies within an electronic band system which has been broadened into a continuum by the molecules responsible being in solution as is the case, for example, in the absorption spectrum in Figure 8.25.

In obtaining a resonance Raman spectrum there is an experimental problem created by broad fluorescence, excited by the laser radiation, overlapping the resonance Raman spectrum. Fortunately there is usually a considerable difference in the lifetimes for the fluorescence and Raman scattering processes: these are typically of the order of 1 ns for fluorescence and 10 fs (1 fs $= 10^{-15}$ s) for scattering. A pulsed laser, mode-locked if necessary (Section 8.1.6) to give a very short pulse length, is used and a gated detection system which enables the detector to 'see' the scattering but not most of the fluorescence.

For both pre-resonance and resonance Raman scattering the tensor elements $(\alpha_{ij})_{mn}$ are given by the following equation:

$$(\alpha_{ij})_{mn} = \sum_r \left[\frac{R_j^{rm} R_i^{nr}}{E_r - E_m - E_0 + \mathrm{i}\Gamma_r} + \frac{R_i^{rm} R_j^{nr}}{E_r - E_n + E_0 + \mathrm{i}\Gamma_r} \right] \tag{8.31}$$

where i, j, m and n have the same significance as in equation (8.30), r is any intermediate state and

$$\mathrm{i} = \sqrt{-1}.$$

The quantities R_i, R_j are the magnitudes of transition moments [see equation (2.21)] along the direction i or j between the states indicated. E_m, E_n and E_r are the energies of the states and E_0 is the energy ($= hc\tilde{\nu}_0$) of the incident radiation. When $E_0 = E_r - E_m$ and the state r is in a continuum the incident radiation is absorbed and resonance Raman scattering occurs. The term Γ_r is a damping constant which prevents the denominator of the first term in equation (8.31) being zero under these circumstances and corresponds to the energy width of the intermediate state; the shorter the lifetime, the larger is Γ_r.

The intermediate states r in equation (8.31) may be above, below or between the initial and final states of the Raman transition so that their description as 'intermediate', although commonly used, has to be treated with care.

The summation in equation (8.31) over intermediate states is a consequence of the fact that any virtual state can always be described quantum mechanically in terms of a combination of eigenstates.

The summation is over vibronic, as well as electronic, states of the molecule but the equation gives no idea of which vibrations are likely to have their intensities enhanced in a resonance Raman spectrum. Albrecht[201] has treated this problem by expanding the electronic transition moments $\boldsymbol{R}_e$ as a Taylor series in the normal coordinates in exactly the same way as Herzberg and Teller did in the treatment of intensity stealing through vibrational activity in electronic band systems: this was described in Section 6.3.4.1 and the expansion given in equation (6.227). The conclusion is that vibrations which are enhanced in the pre-resonance Raman effect are those which are responsible for intensity stealing in the electronic band system involving the same excited electronic state as that concerned in the pre-resonance Raman effect. This has been shown to be the case for the $\tilde{\mathrm{A}}^1B_{3u}$–$\tilde{\mathrm{X}}^1A_g$ electronic system of pyrazine (1,4-diazabenzene), in which ν_{10a}, a b_{1g} vibration with a wavenumber of 919 cm^{-1} in the $\tilde{\mathrm{X}}$ state, is responsible for intensity stealing and

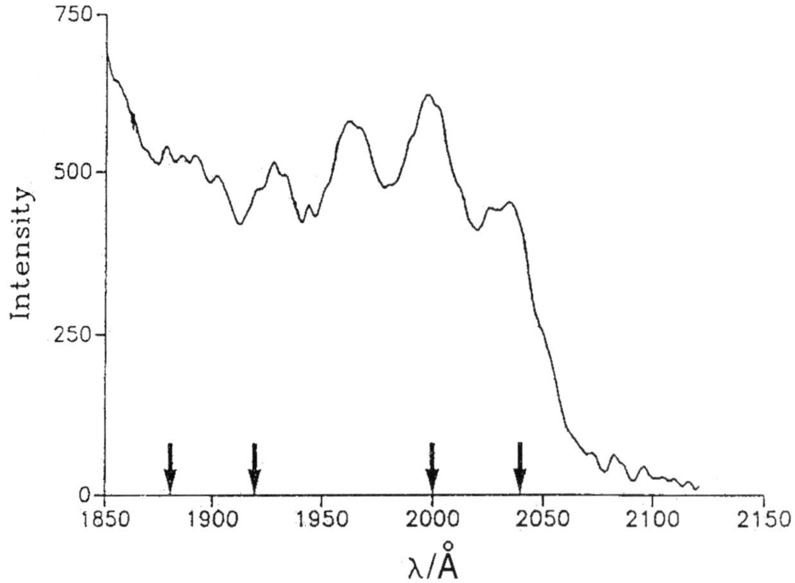

Figure 8.30 Absorption spectrum of benzene, in a supersonic jet, in the region of the $\tilde{B}^1B_{1u}$–$\tilde{X}^1A_{1g}$ system. [Reproduced, with permission, from Sension, R. J., Brudzynski, R. J., Li, S., Hudson, B. S., Zerbetto, F. and Zgierski, M. K. (1992). *J. Chem. Phys.*, **96**, 2617]

also shows intensity enhancement in the pre-resonance Raman effect;† this was observed with various exciting lines from an argon ion and krypton ion laser and two other lines from a mercury arc.

The $B^3\Pi_{0^+u} - X^1\Sigma_g^+$ absorption system of I_2, shown in Figure 6.35, is a good example of a system showing absorption right up to and beyond the dissociation limit of the upper state due to a large increase in bond length from the X to the B state (Section 6.2.4.3). The 488 nm radiation from an argon ion laser falls in the absorption continuum and the resonance Raman effect is observed.[202] This shows a long progression with $\upsilon'' = 1$–14 of fairly broad bands, each consisting of an S-branch ($\Delta J = +2$) head, and Q ($\Delta J = 0$) and O ($\Delta J = -2$) branches, according to the usual vibration–rotation Raman selection rules. The intensity decreases smoothly along the progression and the bands are polarized with the depolarization ratio $\rho_\perp < 0.45$ (see Section 5.2.1).

Contrasting with the resonance Raman spectrum is the single vibronic level fluorescence spectrum of I_2 using the 514.5 nm radiation from the same argon ion laser. The exciting radiation is absorbed in the $R(15)$ and $P(13)$ rotational lines of the $\upsilon = 43$–0 transition thereby selectively populating the $J = 16$ and 12 levels of the $\upsilon' = 43$ state. Single vibronic level fluorescence shows a long vibrational progression with a complex intensity distribution accounted for by the Franck–Condon principle (see, for example, Figure 6.40). The accompanying rotational transitions obey the $\Delta J = \pm 1$ electric dipole selection rules and the bands are depolarized, with $\rho_\perp = 0.75$.

In fairly large molecules the second, third, . . . lowest excited electronic states are mostly extremely broad and transitions to them, from the ground state, correspondingly diffuse, even in the gas phase or in a supersonic jet.

†See Thakur, S. N. and Innes K. K. (1974). *J. Mol. Spectrosc.*, **52**, 130, and references cited therein.

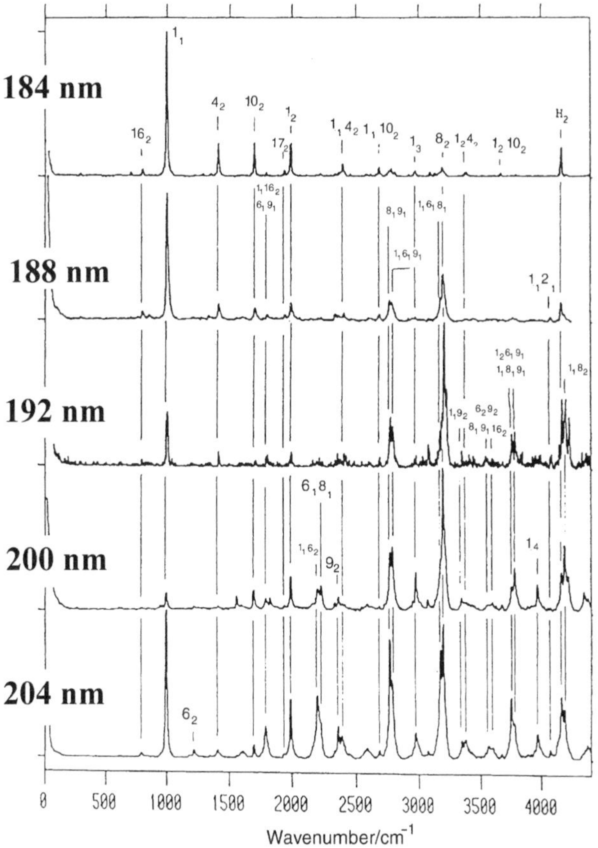

Figure 8.31 Resonance Raman spectra of benzene with laser excitation at 204, 200, 192, 188 and 184 nm [Reproduced, with permission, from Sension, R. J., Brudzynski, R. J., Li, S., Hudson, B. D., Zerbetto, F. and Zgierski, M. K. (1992). *J. Chem. Phys.*, **96**, 2617]

Figure 8.30 shows the $\tilde{B}^1B_{1u}$–$\tilde{X}^1A_{1g}$ absorption spectrum of benzene in the 185–205 nm region. Even though the sample was cooled in a supersonic jet, the $\tilde{B}$–$\tilde{X}$ system is characteristically diffuse, showing only some very broad vibronic structure. Figure 8.31 shows resonance Raman spectra[203] with laser excitation with wavelengths of 204, 200, 192, 188 and 184 nm. These wavelengths were obtained by anti-Stokes stimulated Raman shifting (see Section 8.2.3) by hydrogen gas of various harmonics of the output of a Nd^{3+} : YAG laser. These Raman shifted wavelengths are indicated on the absorption spectrum in Figure 8.30, except for 184 nm which lies just within the $\tilde{C}^1E_{1u}$–$\tilde{X}^1A_{1g}$ system.

The vibrational structure observed in a resonance Raman spectrum may contain two components. The first is the Franck–Condon component involving, usually, progressions in totally symmetric vibrations. The length of the progressions depends on the molecular geometry change from the ground electronic state to the superposition of states which describes the virtual state. The second component is the Herzberg–Teller component involving, usually, non-totally symmetric vibrations which vibronically couple an electronic state, to which a transition from the ground state is strongly electronically allowed, to the virtual state.

The spectra in Figure 8.31 show a progression in ν_1, the symmetrical ring-breathing

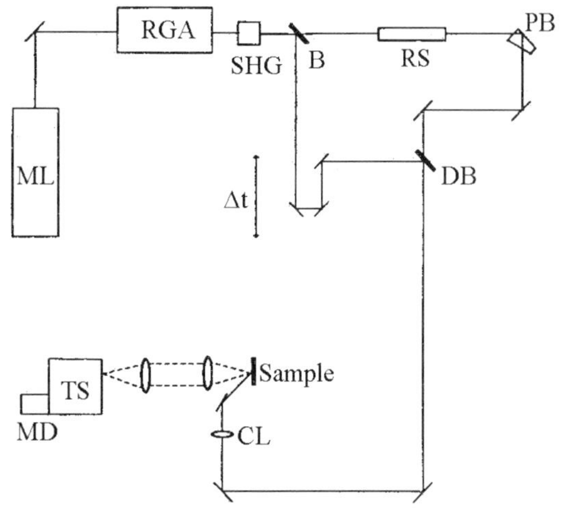

Figure 8.32 Experimental arrangement for time-resolved resonance Raman spectroscopy [Reproduced, with permission, from Kumble, R., Loppnow, G. R., Hu, S., Mukherjee, A., Thompson, M. A. and Spiro, T. G. (1995). *J. Phys. Chem.*, **99**, 5809]

vibration (using the Wilson vibrational numbering scheme referred to in the footnote on p. 385), which becomes particularly weak with 200 nm excitation. This progression is the Franck–Condon component and is a result of a ring expansion of the benzene molecule in excited electronic states.

Of particular interest is the Herzberg–Teller component comprising the overtone and combination bands involving the e_{2g} vibrations ν_6, ν_9 and, particularly, ν_8. The reason for the activity of these e_{2g} vibrations is that there is a Jahn–Teller effect (see Section 6.3.4.5) in the doubly degenerate $\tilde{C}^1E_{1u}$ state. Vibronic coupling between the $\tilde{C}$ and the $\tilde{B}$ states confers a pseudo-Jahn–Teller distortion on the $\tilde{B}$ state, 'pseudo' because a $^1B_{1u}$ state, being non-degenerate, could not normally experience such a distortion. The distortion, in the direction of ν_8, is to a D_{2h} geometry with four short and two long C—C bonds, the potential energy surface showing three equivalent minima. Calculations give an estimate of the barrier height to the D_{6h} conformation of $1000\,cm^{-1}$ and the pseudo-rotational barrier height between the equivalent D_{2h} conformations of $300\,cm^{-1}$.

This example is a good illustration of the power of resonance Raman spectroscopy in obtaining information about excited electronic states which could not be obtained from relatively featureless absorption from the ground state.

The high peak power of pulsed lasers presents the possibility of building up a sufficiently high population of an excited electronic state, say 1, that a resonance Raman spectrum may be obtained involving a broadened higher electronic state 2. This requires two lasers, or their equivalent. The first, the pump laser, takes the molecule from the ground state 0 to the excited state 1 and the second, the probe laser, produces a resonance Raman spectrum by exciting the molecule from state 1 to the broadened state 2, thereby giving vibration wavenumbers in state 1. If, as is often the case in this type of experiment, the excited state 1 is also very broad two types of resonance Raman spectra will be observed: the 1–0 and 2–1 resonance Raman spectra will be superimposed at the detector.

The problem of separation is overcome by time resolution and the technique is referred to as time-resolved resonance Raman, with the acronym TR^3, spectroscopy. Both the pump and probe lasers are pulsed and the probe laser must be delayed relative to the pump so that the population of state 1 has had time to build up to a maximum. A typical pulse beam delay is 100 ps.

Figure 8.32 shows an experimental arrangement used to obtain the time-resolved resonance Raman spectrum of the zinc complex of octaethylporphyrin, shown in Figure 8.33, in solution in tetrahydrofuran.[204] The pump and probe beams originate from the same laser, a mode-locked Nd^{3+} : YAG laser, ML, with an amplifier RGA. The 1056 nm radiation, produced in 100 ps pulses, is frequency doubled by a KTP crystal, SHG, and the 532 nm beam split into two parts by the beam splitter B. About 90% goes into an anti-Stokes Raman shifter, RS, (see Section 8.2.3) containing hydrogen (or

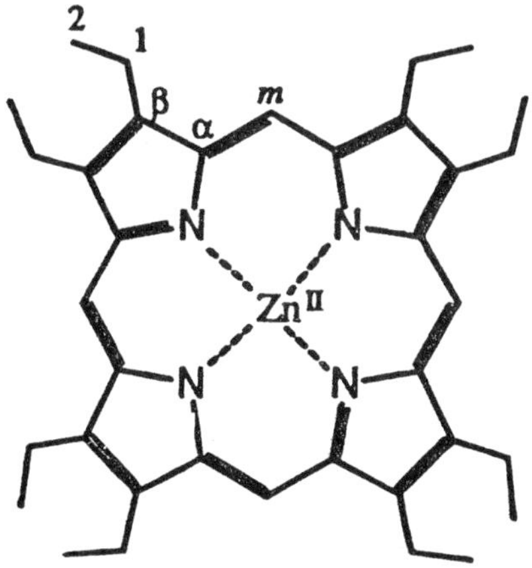

Figure 8.33 Zinc (II) octaethylporphyrin

deuterium) gas and forms the probe beam with a wavelength of 436 nm (or 459 nm). This radiation is separated from the Stokes Raman shifted radiation by the prism PB. The remaining 10% of the 532 nm radiation forms the pump beam and goes into a variable pathlength and, therefore, variable time, delay line. The pump and probe beams, with pulse energies of 0.1 and 0.02 mJ, respectively, are combined in a dichroic beam splitter DB. The combined beams are focused into the sample and the Raman scattered radiation focused into the spectrometer, TS, and detected by a multichannel detector, MD.

In this example the lower excited state is the lowest excited singlet state S_1. The 532 nm pump beam is absorbed in the S_1–S_0 system and there is also S_n–S_1 absorption of the 436 nm probe radiation.

The two spectra are accumulated and are resolved from each other by the time separation of the pump and probe pulses. The spectrum arising from the pump beam shows vibrations in only the ground state S_0. The spectrum arising from the probe beam shows vibrations in both S_0 and S_1 the relative intensities of these two components depending on the relative populations of S_0 and S_1 100 ps after excitation to S_1. The component showing only the S_1 vibrations is obtained by subtracting the pump beam from the probe beam spectrum until negative features are just avoided.

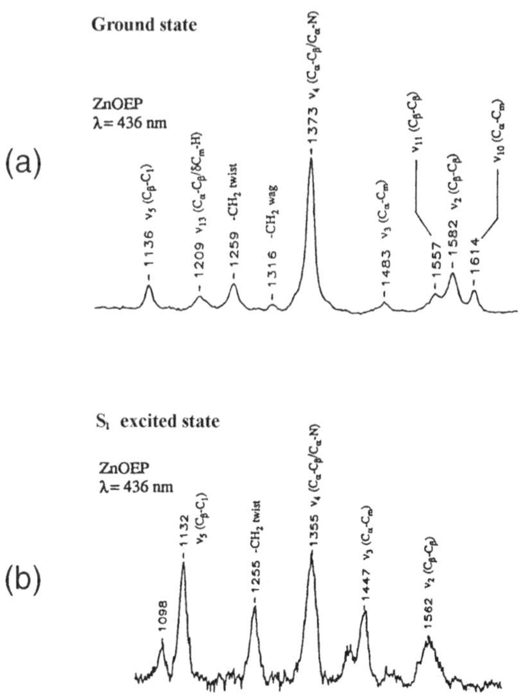

Figure 8.34 (a) S_0 and S_1 time-resolved resonance Raman spectra of zinc (II) octaethylporphyrin. The numbers attached to each band are vibration wavenumbers
[Reproduced, with permission, from Kumble, R., Loppnow, G. R., Hu, S., Mukherjee, A., Thompson, M. A. and Spiro, T. G. (1995). *J. Phys. Chem.*, **99**, 5809]

Figure 8.34(a) shows the resulting S_0 resonance Raman spectrum involving vibrations in only the ground state and Figure 8.34(b) shows the S_1 resonance Raman spectrum involving vibrations only in S_1. The vibrations which are active in the two spectra are seen to have appreciably different wavenumbers. Skeletal stretching wavenumbers are reduced in S_1 consistent with an overall expansion of the macrocycle in S_1, particularly in the C_α—C_m bonds indicated in Figure 8.33.

As with other Raman techniques, there may be a serious problem of fluorescence interfering with the resonance Raman scattered radiation. This may be overcome by using a probe beam, to produce the resonance Raman scattering, with a wavelength in the near infrared and also by using Fourier transform techniques, as in Fourier transform Raman spectroscopy (see Section 4.8.2).

In an example of the use of time-resolved Fourier transform resonance Raman spectroscopy[205] 1064 nm radiation from a Nd^{3+}:YAG laser was employed as the probe beam and 355 nm radiation, frequency tripled radiation from the same Nd^{3+}:YAG laser, formed the pump beam. The S_1–S_0 transition in anthracene, in liquid solution, was excited with 355 nm radiation and the resonance Raman scattering, giving vibration wavenumbers in the S_1 state, produced by the 1064 nm probe beam radiation exciting the S_3–S_1 transition.

8.2.2 Hyper Raman spectroscopy

We have seen in equation (8.14) how the dipole moment, induced in a material by incident radiation of the power density which a laser can produce, contains a small contribution which is proportional to the square of the oscillating electric field $\boldsymbol{E}$. This field can be sufficiently large when using a Q-switched laser focused on the sample that hyper Raman scattering, involving the hyperpolarizability β, introduced in equation (8.14), is sufficiently intense to be detected. This scattering is at a wavenumber $2\tilde{\nu}_0 \pm \tilde{\nu}_{HR}$, where $\tilde{\nu}_0$ is the wavenumber of the exciting radiation and $-\tilde{\nu}_{HR}$ and $+\tilde{\nu}_{HR}$ are the Stokes and anti-Stokes hyper Raman displacements, respectively. The hyper Raman is well separated from the Raman scattered radiation so there is no problem of overlapping but, even with a Q-switched laser, the hyper Raman scattering is extremely weak. One technique used for detection is photon counting with a single channel detector. The photons falling on a photomultiplier produce a signal which is amplified. The amplified signal passes to a discriminator which is set to reject signals which are at too low a level to be caused by photons. The remaining amplified pulses are then counted. Alternatively, a multichannel detector with an image intensifier and photon counting can be used to view the whole spectrum, thereby reducing the recording time.

The second term in equation (8.14) can be written in full as

$$\begin{pmatrix} \mu_x^{(2)} \\ \mu_y^{(2)} \\ \mu_z^{(2)} \end{pmatrix} = \frac{1}{2} \begin{pmatrix} \beta_{xxx} & \beta_{xyy} & \beta_{xzz} & \beta_{yxx} & \beta_{xyz} & \beta_{zxx} \\ \beta_{yxx} & \beta_{yyy} & \beta_{yzz} & \beta_{xyy} & \beta_{zyy} & \beta_{xyz} \\ \beta_{zxx} & \beta_{zyy} & \beta_{zzz} & \beta_{xyz} & \beta_{yzz} & \beta_{xzz} \end{pmatrix} \begin{pmatrix} E_x^2 \\ E_y^2 \\ E_z^2 \\ 2E_xE_y \\ 2E_yE_z \\ 2E_zE_x \end{pmatrix} \tag{8.32}$$

which is analogous to equation (4.119) involving the first term in equation (8.14).

Scattering of radiation of wavenumber $2\tilde{\nu}_0$ is known as hyper Rayleigh scattering by analogy with Rayleigh scattering of wavenumber $\tilde{\nu}_0$ (Section 4.8.1). However, whereas Rayleigh scattering *always* occurs, hyper Rayleigh scattering occurs only if the scattering material does not have a centre of inversion. Frequency doubled radiation, discussed in Section 8.1.7, consists of hyper Rayleigh scattering from a pure crystal and consequently a crystal used for frequency doubling must not have a centrosymmetric unit cell.

The selection rules for molecular vibrations giving rise to hyper Raman scattering are derived[206] in an analogous way to those for Raman scattering. The general requirement for a hyper Raman transition to be allowed between upper and lower vibrational states with wave functions ψ'_v and ψ''_v, respectively, is that

$$\Gamma(\psi'_v) \times \Gamma(\beta_{ijk}) \times \Gamma(\psi''_v) \supset A \tag{8.33}$$

where i, j and k can be x, y or z, and A is the totally symmetric symmetry species of the point group to which the molecule concerned belongs. This equation is analogous to equation (5.184) for Raman scattering. If, as is usually the case,

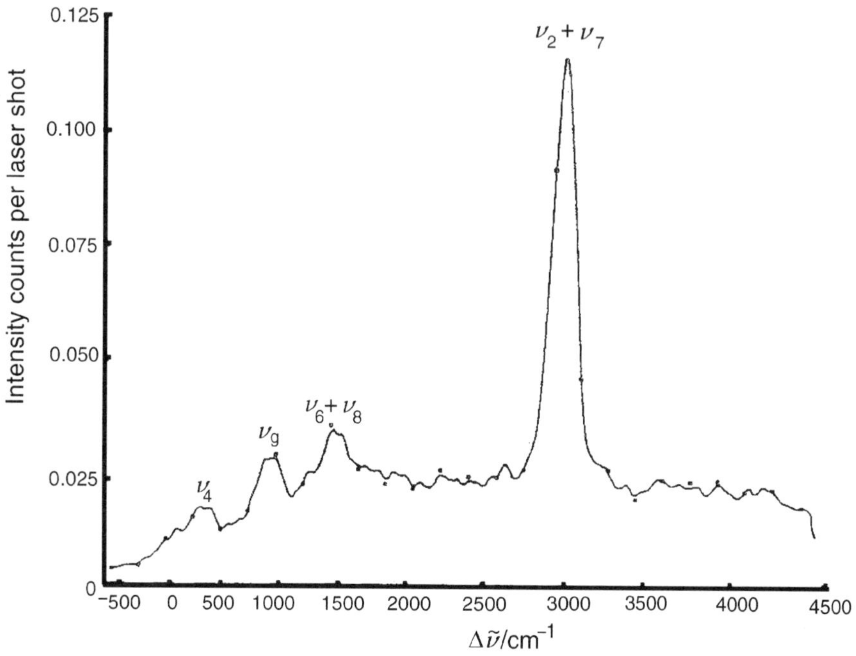

Figure 8.35 The hyper Raman spectrum of ethane
[Reproduced, with permission, from Verdick, J. F., Peterson, S. H., Savage, C. M. and Maker, P. D. (1970). *Chem. Phys. Lett.*, **7**, 219]

the lower vibrational state is the zero-point level $\Gamma(\psi''_v) = A$ and equation (8.33) becomes

$$\Gamma(\psi'_v) = \Gamma(\beta_{ijk}) \tag{8.34}$$

Cyvin, Rauch and Decius[206] have tabulated the symmetry species of β_{ijk}, or appropriate linear combinations, for all the common point groups. General conclusions to be drawn include

1. Vibrations allowed in the infrared, by electric dipole selection rules, are always allowed in the hyper Raman effect and are polarized, with $\rho_\perp \leqslant \frac{2}{3}$ for plane polarized incident radiation.
2. In a molecule with a centre of inversion all hyper Raman active vibrations are u vibrations.
3. Vibrations which are allowed in the hyper Raman effect but forbidden by electric dipole selection rules are depolarized, with $\rho_\perp = \frac{2}{3}$.
4. Some vibrations which are both Raman and infrared inactive may be allowed in the hyper Raman effect. Indeed, the appearance of some of these 'forbidden' vibrations in Raman spectra in a condensed phase has sometimes been attributed to an effect of the hyperpolarizability.

Figure 8.35 shows the hyper Raman spectrum of ethane, C_2H_6, which illustrates some of these points. The equilibrium configuration of the molecule is the staggered one shown in Figure 5.27(i) and it belongs to the D_{3d} point group. There is a centre of inversion and therefore hyper Rayleigh scattering is absent. In the D_{3d} point group vibrations of symmetry species a_{1u}, a_{2u} and e_u are allowed in the hyper Raman spectrum but only a_{2u} and e_u vibrations are electric dipole allowed (see Table A.28 in the Appendix). The intense band at about 3000 cm^{-1} from $2\tilde{\nu}_0$ contains contributions from the CH-stretching vibrations $\nu_2(a_{2u})$ and $\nu_7(e_u)$. That at about 1400 cm^{-1} involves the CH_3-deformation vibrations $\nu_6(a_{2u})$ and $\nu_8(e_u)$ and the one close to 900 cm^{-1} the bending vibration of the whole molecule $\nu_9(e_u)$. The band close to 300 cm^{-1} is the most interesting one as it involves ν_4, the a_{1u} torsional vibration which is forbidden in the Raman and infrared spectra.

Selection rules for the pure rotational hyper Raman effect[207] are, for a molecule of low symmetry,

$$\Delta J = 0, \pm 1, \pm 2, \pm 3$$

but are more restrictive for molecules of higher symmetry. For example, for a linear molecule, $\Delta J = \pm 1, \pm 3$ only.

Resonance hyper Raman spectroscopy is related to hyper Raman spectroscopy in the same way as resonance Raman spectroscopy is related to Raman spectroscopy. In hyper Raman spectroscopy the scattering occurs via a virtual state which is $2\tilde{\nu}_0$ above the ground state, where $2\tilde{\nu}_0$ is the wavenumber of the exciting radiation. If $2\tilde{\nu}_0$ takes the molecule into a virtual state which lies within a broadened electronic state resonance hyper Raman scattering results.

Since the initial absorption of the exciting radiation is a two photon process (see Section 8.2.18), for a molecule with a centre of inversion resonance hyper Raman spectroscopy probes the g eigenstates which contribute to the virtual state whereas resonance Raman spectroscopy probes the u states.

Figure 8.36 shows part of the resonance hyper Raman spectrum of naphthalene in solution in ethylene glycol.[208] The exciting radiation has a wavelength of 456 nm. Two quanta correspond to 228 nm which takes the naphthalene molecule into the overlapping $\tilde{D}^1A_g$ and $\tilde{E}^1B_{2u}$ states (see Section 6.3.1.9). The spectrum shows three prominent vibrations, ν_5, ν_8 and ν_{25} (ν_{36} using the Mulliken numbering). The ν_5 and ν_8 vibrations are totally symmetric, a_g, and constitute the Franck–

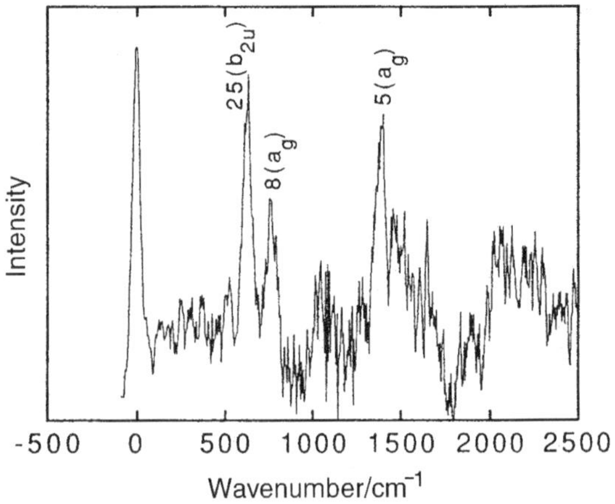

Figure 8.36 The resonance hyper Raman spectrum of naphthalene excited with 456 nm radiation [Reproduced, with permission, from Bonang, C. C. and Cameron, S. M. (1992). *J. Chem. Phys.*, **97**, 5377]

Condon contribution; they are thought to arise from the $\tilde{E}^1B_{2u}$ state. The activity of ν_{25} constitutes the Herzberg–Teller contribution and arises from the $\tilde{D}^1A_g$ state which is vibronically coupled to the overlapping $\tilde{E}^1B_{2u}$ state by ν_{25} which is a b_{2u} vibration.

8.2.3 Stimulated Raman and Raman gain spectroscopy

Stimulated Raman scattering is experimentally different from normal Raman scattering in that it is observed as forward scattering which emerges from the sample in the same direction as that of the emerging exciting radiation, or at a very small angle to this direction.

The probability P of absorption of the incident radiation of wavenumber $\tilde{\nu}_0$ and of emission of radiation of wavenumber $\tilde{\nu}_1 = \tilde{\nu}_0 \pm \tilde{\nu}_i$ in Stokes and anti-Stokes processes involving vibration i is given by

$$P = \hbar^{-4} \int |\boldsymbol{R}^{mn}|^2 \rho_0(\rho_1 + 8\pi h\tilde{\nu}_1^3)c\, d\tilde{\nu}_0 \quad (8.35)$$

where ρ_0 amd ρ_1 are the radiation density at $\tilde{\nu}_0$ and $\tilde{\nu}_1$ and $\boldsymbol{R}^{mn}$ is the transition moment for the two photon process involved in going from a

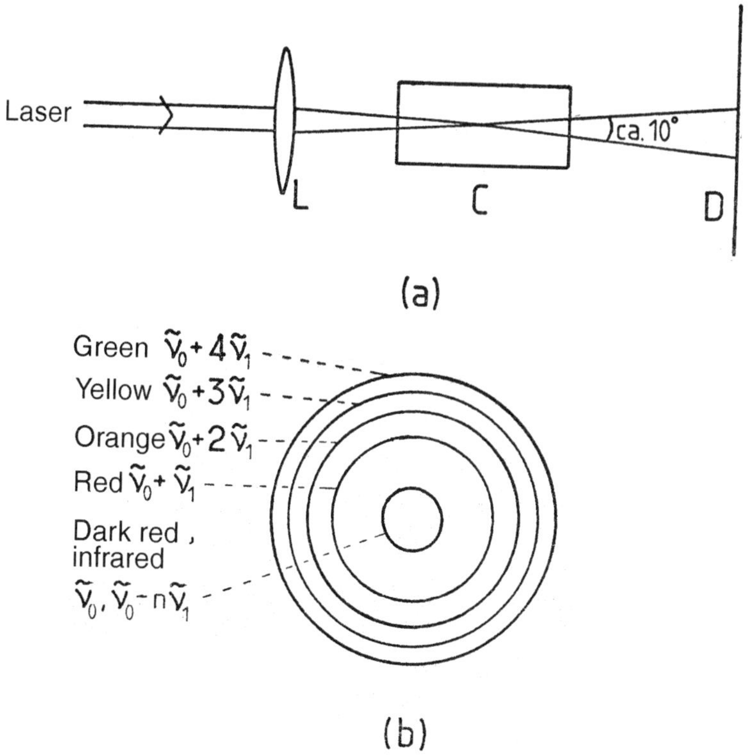

Figure 8.37 (a) Stimulated Raman scattering experiment using a Q-switched laser and a liquid sample in the cell C. (b) Concentric rings observed in the stimulated Raman scattering, in the forward direction, from liquid benzene

state m to a state n via a virtual state. The term $\rho_0 8\pi h\tilde{\nu}_1^3$ gives the intensity of the normal, or spontaneous, Raman scattering while $\rho_0\rho_1$ gives the intensity of the stimulated Raman scattering. Unless a high power laser source is used, ρ_1 is too small for the stimulated effect to be important. The effect was first observed, with liquid samples, in 1962.[209]

Figure 8.37(a) shows how stimulated Raman scattering can be observed by focusing radiation from a Q-switched ruby laser with a lens L into a cell C containing, for example, liquid benzene. The forward scattering occurs within an angle of about 10° and is collected by a detector D. If the detector is a photographic colour film broad concentric coloured rings, ranging from dark red in the centre to green on the outside, are observed as Figure 8.37(b) indicates. The wavenumbers corresponding to these rings range from $\tilde{\nu}_0$ (and $\tilde{\nu}_0 - n\tilde{\nu}_1$) in the centre to $\tilde{\nu}_0 + 4\tilde{\nu}_1$ on the outside. ν_1 is the ring-breathing vibration of benzene and the series of $\tilde{\nu}_0 + n\tilde{\nu}_1$ rings, with $n = 0$–4, shows a *constant* separation of ν_1, which is the $v_1 = 1$–0 interval of 992 cm^{-1}.

The reason why the spacings are equal, and are not the 1–0, 2–1, 3–2, . . . anharmonic vibrational intervals, is explained in Figure 8.38. The laser radiation of wavenumber $\tilde{\nu}_0$ takes benzene molecules into the virtual state V_1 from which they drop down to the $v_1 = 1$ level. The resulting Stokes scattering is, as mentioned above, extremely intense in the forward direction with about 50% of the incident laser radiation scattered at a wavenumber of $\tilde{\nu}_0 - \tilde{\nu}_1$. This radiation is sufficiently

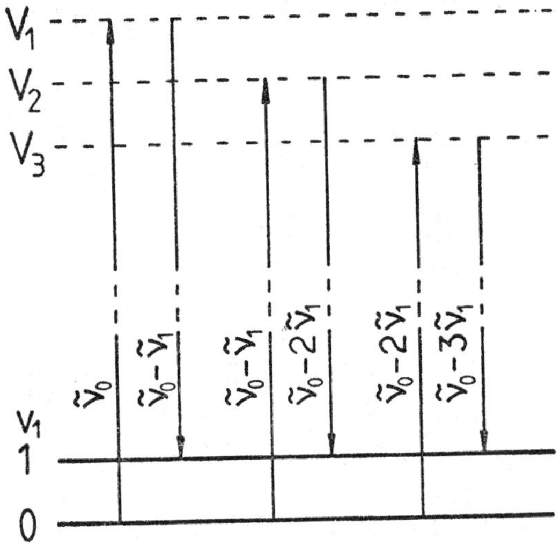

Figure 8.38 Transitions involved in the stimulated Stokes Raman effect

intense to take other molecules into the virtual state V_2 resulting in intense scattering at $\tilde{\nu}_0 - 2\tilde{\nu}_1$, and so on.

Stimulated anti-Stokes Raman scattering is produced in a similar way. The first step in Figure 8.38 results in a high population of the $\upsilon = 1$ level. A further laser photon then takes the molecule to a virtual state and the molecule reverting to the $\upsilon = 0$ level results in scattering at a wavenumber of $\tilde{\nu}_0 + \tilde{\nu}_1$. This anti-Stokes radiation may take the molecule to another virtual state; then reversion to the $\upsilon = 0$ state gives anti-Stokes scattering at a wavenumber of $\tilde{\nu}_0 + 2\tilde{\nu}_1$. Similarly, radiation of wavenumber $\tilde{\nu}_0 + n\tilde{\nu}_1$, for any value of n, may be produced. We can now see that the rings shown in Figure 8.37(b) are due to stimulated anti-Stokes Raman scattering.

In the stimulated Raman effect it is only the vibration which gives the most intense normal Raman scattering which is involved; this is the case with benzene. Only if vibrations give very nearly equally intense Raman scattering does the stimulated Raman scattering involve more than one vibration.

The high efficiency of conversion of the laser radiation $\tilde{\nu}_0$ into Stokes radiation of wavenumber $\tilde{\nu}_0 - n\tilde{\nu}_i$ and into anti-Stokes radiation of wavenumber $\tilde{\nu}_0 + n\tilde{\nu}_i$, where ν_i is the vibration involved, is the basis of the technique of Raman shifting of the wavelength of a powerful, Q-switched laser. The laser is directed into a gas cell containing a molecule with a suitable Raman active vibration, as, for example, in the experiment in Figure 8.32. The molecule is often $^1H_2$ or $^2H_2$ for which $\tilde{\nu}_i$ is 4158 or 2992 cm^{-1}, respectively. The high vibration wavenumbers of these molecules lead to large Stokes or anti-Stokes Raman shifting of the laser wavelength.

The stimulated Raman effect can also be used to measure vibrational lifetimes. The effect results in pumping of the $\upsilon_i = 1$ level, not to the extent of creating a population inversion, but of increasing the population far above the equilibrium Boltzmann distribution value. Because pumping is fast compared with vibrational relaxation to the $\upsilon_i = 0$ level the population can be monitored by measuring, as a function of time, the intensity of anti-Stokes Raman scattering at $\tilde{\nu}'_0 + \tilde{\nu}_i$ using a second laser of wavenumber $\tilde{\nu}'_0$. The second laser acts as a population probe and must be of sufficiently low power that it does not produce stimulated Raman scattering. In the gas phase vibrational lifetimes are relatively long, for example *ca* 10 μs in H_2. They are much shorter in liquids, of the order of 10 ps, necessitating the use of picosecond pulses from the pump and probe lasers.

In the stimulated Raman effect there is considerable line narrowing so that, with a gaseous sample, the line width in the resulting spectrum is Doppler limited.

The first gas phase stimulated Raman spectra were obtained in 1963 using $^1H_2$, $^2H_2$ and CH_4 sample gases.[210]

In 1977 a new type of Raman experiment, which is related to stimulated and, perhaps more closely, to inverse Raman spectroscopy (Section 8.2.4), was first reported using a liquid benzene sample.[211] Initially the technique was referred to, rather confusingly, as stimulated Raman spectroscopy but has subsequently been called Raman gain, Raman loss or Raman

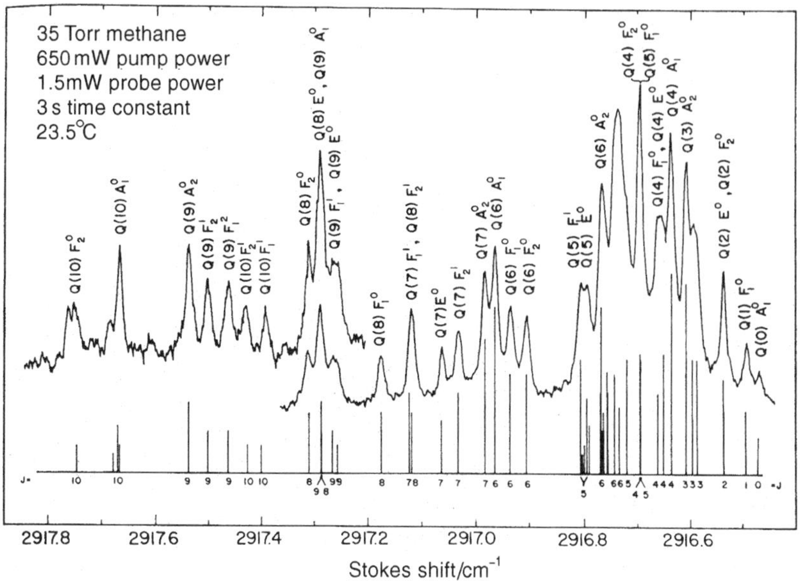

Figure 8.39 Raman gain spectrum showing part of the 1_0^1 vibrational band of methane with a resolution, limited by the Doppler width, of 300 MHz
[Reproduced, with permission, from Owyoung, A., Patterson, C. W. and McDowell, R. S. (1978). *Chem. Phys. Lett.*, **59**, 156]

absorption spectroscopy. An example of such a spectrum is that of methane gas shown in Figure 8.39 and it was obtained in the following way.

Two low power CW lasers were used. One of them, the pump laser, was a 650 mW single mode argon ion laser operating at 514.5 nm. This produced Raman scattering centred at $\tilde{\nu}_0 - \tilde{\nu}_i$, the first step in Figure 8.38, but, with the sample in the gas phase, there is associated rotational structure. The second laser, the probe laser, was a 1.5 mW single mode CW dye laser which was tunable around 605.4 nm, the region of the vibration–rotation band centred at $\tilde{\nu}_0 - \tilde{\nu}_i$. The pump laser was amplitude modulated and, when the probe laser wavelength corresponded to that of a vibration–rotation transition in the Raman scattering, the probe laser power, as recorded by a detector, showed an increase. In this mode, when the probe power is monitored, the description as Raman gain spectroscopy is appropriate. Alternatively, the pump power may be monitored, in which case there is a power decrease at the detector when the difference between the wavenumber of the pump and probe lasers corresponds to a vibration–rotation transition and the description as Raman loss spectroscopy is appropriate.

The effect of observing a change in the pump and probe laser powers when their wavenumber difference matches a transition in the sample molecule is due to three wave mixing. In this case the three waves are of wavenumber $\tilde{\nu}_0$, the pump radiation, $\tilde{\nu}_1$, the probe radiation, and $\tilde{\nu}_0 - \tilde{\nu}_1$. The mixing is a third order effect involving the third term in equation (8.14), the same term which is involved in third harmonic generation.

The spectrum in Figure 8.39 was recorded with a laser line width of 25 MHz (0.0008 cm^{-1}). The resolution in the spectrum shown is limited by the Doppler line width of *ca* 300 MHz (0.01 cm^{-1}), *appropriate to the wavenumber* $\tilde{\nu}_0 - \tilde{\nu}_1$. The region shown is that of the Q branches of the 1_0^1 vibrational transition in CH_4, where ν_1 is the a_1 symmetric CH-stretching vibration. For such an A_1–A_1 transition there are only Q branches since $\Delta J = 0$, as discussed in Section 5.2.5.3(ii). Prior to the observation of this region of the Raman spectrum at such high resolution there had been some confusion as to whether there is low-wavenumber or high-wavenumber degradation of the Q branches, corresponding to negative or positive ΔB ($= B_1 - B_0$), respectively. Analysis of the spectrum in Figure 8.39 gives $\Delta B = 0.010\,75 \pm 0.000\,06\ \text{cm}^{-1}$.

Whereas stimulated Raman spectroscopy is confined to just one vibration, that which is the most intense in the normal Raman effect, *all* the vibrations which are allowed by the Raman selection rules may be observed in a Raman gain or Raman loss spectrum.

The first step in the stimulated Raman process results in an appreciable decrease of the population of the ground state vibrational level and a corresponding increase of that of the excited vibrational level. As in other branches of spectroscopy, it is often desirable for the molecule under investigation to be seeded in a supersonic jet (see Section 3.4.3). Under the conditions of very low rotational and vibrational temperatures which obtain in the jet, the spectrum is greatly simplified and, in addition, new species, such as van der Waals or hydrogen bonded complexes and clusters, may be formed.

For a Raman experiment in a supersonic jet the basic weakness of the Raman effect must be overcome by increasing the sensitivity of detection. This may be done in the stimulated Raman effect by various techiques for probing the population changes of the vibrational levels concerned. One such method is illustrated in Figure 8.40 and has been called ionization loss, as in Figure 8.40(a), and ionization gain, as in Figure 8.40(b), stimulated Raman spectroscopy;[212] $\tilde{\nu}_i$ is the wavenumber of the vibrational level under investigation.

Both versions of the experiment employ lasers with three different wavenumbers, $\tilde{\nu}_0$, $\tilde{\nu}_0 - \tilde{\nu}_i$ and $\tilde{\nu}_\mathrm{I}$. In the ionization loss process in Figure 8.40(a) the laser of wavenumber $\tilde{\nu}_\mathrm{I}$ probes the population loss in the ground

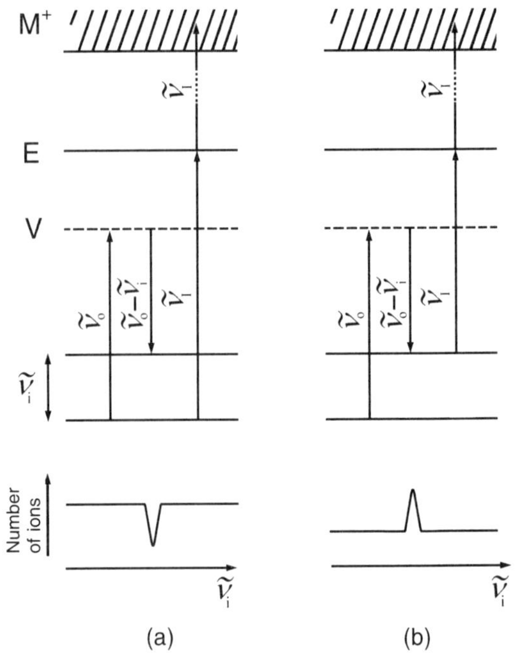

Figure 8.40 The processes involved in (a) ionization loss and (b) ionization gain stimulated Raman spectroscopy

vibrational state by resonant multiphoton ionization, a process which will be described in more detail in Section 8.2.18. In the example shown, two photons from the probe laser are absorbed, the first taking the molecule to an excited electronic state E and the second taking it above the ionization energy in a so-called 1 + 1 multiphoton ionization process. The detection process involves counting the positive ions produced as the $\tilde{\nu} - \tilde{\nu}_i$ radiation is tuned across the region of $\tilde{\nu}_0$. If there are a number of different species in the sample, such as a parent molecule and various cluster molecules formed in the jet with argon carrier gas, the ions may be mass analysed. In this way the ionization loss stimulated Raman spectrum of a selected species may be obtained.

Figure 8.40(b) shows the analogous processes involved in ionization gain stimulated Raman spectroscopy. In this case the $\tilde{\nu}_\mathrm{I}$ laser probes the gain in population of the excited vibrational level by a 1 + 1 multiphoton ionization process. When the $\tilde{\nu}_0 - \tilde{\nu}_i$ laser matches a vibrational interval there is an increase in the number of ions M^+ produced whereas, in the ionization loss process, in Figure 8.40(a), there is a decrease.

Figure 8.41 illustrates the importance of mass selection in seeing how the wavenumber of the ν_1 ring-breathing vibration of benzene changes when it forms van der Waals complexes with argon, methane and carbon dioxide. There is only an extremely small effect produced by argon and methane but carbon dioxide has a much larger effect, probably due to interaction between the π-electrons of carbon dioxide with those of benzene.

Using two lasers, of wavenumbers $\tilde{\nu}_0$ and $\tilde{\nu}_0 - \tilde{\nu}_i$, with band widths of the order of $0.002\,\mathrm{cm}^{-1}$, very high resolution ionization gain spectra of benzene have been obtained.[213] In this case the ionizing probe laser is tuned so that the first photon absorbed takes the molecule to the zero-point level of the $\tilde{A}^1B_{2u}$ electronic state. A larger band width for this laser is adequate because the Doppler broadening of an electronic transition is much larger than that of a vibrational Raman transition.

The vibrational levels of benzene investigated were the $\nu_8(e_{2g})$ and $\nu_1(a_g) + \nu_6(e_{2g})$ pair (using the Wilson numbering). These levels are close together, at 1609 and $1591\,\mathrm{cm}^{-1}$, respectively, and, because they have the same symmetry, form a Fermi diad with interlocking rotational structure. By tuning the ionizing laser so that the first photon populates different rotational levels of the $\tilde{A}^1B_{2u}$ state the rotational levels of the ν_8 and $\nu_1 + \nu_6$ vibrational states are separately depopulated. Figure 8.42(a) shows part of the ionization gain stimulated Raman spectrum of the 8^1_0 component of the diad, in particular ${}^oO$ branch transitions (with $\Delta K = -2$ and $\Delta J = -2$), with the first photon from the ionizing laser corresponding to a ${}^pP$ branch in the transition from the ν_8 vibrational level to the zero-point level of the $\tilde{A}^1B_{2u}$ electronic state. For the part of the spectrum shown in Figure 8.42(b) the first photon of the ionizing

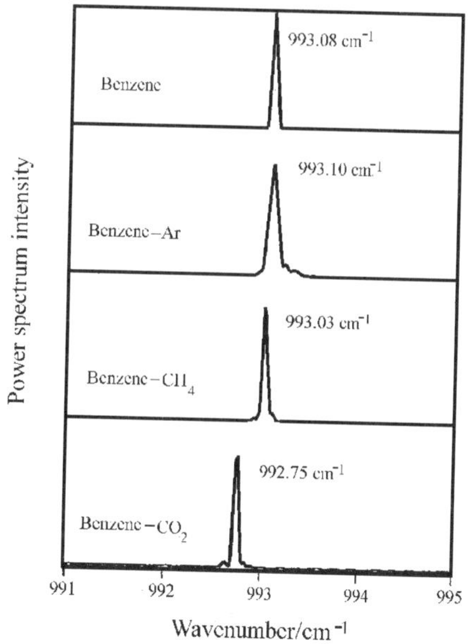

Figure 8.41 Ionization gain stimulated Raman spectra showing the ν_1 vibration of benzene and its van der Waals complexes with argon, methane and carbon dioxide.
[Reproduced, with permission, from Hartland, G. V., Henson, B. F., Venturo, V. A., Hertz, R. A. and Felker, P. M. (1990). *J. Opt. Soc. Am. B.*, **7**, 1950]

laser corresponds to an ${}^rR$ branch in the transition from the $\nu_1 + \nu_6$ level to the zero-point level of the $\tilde{A}$ electronic state with the result that ${}^sS$ transitions (with $\Delta K = 2$ and $\Delta J = 2$) in the $1^1_0 6^1_0$ component of the diad are observed.

In this way the two components of the Fermi diad were separated and rotationally analysed.

Other methods of detecting changes of vibrational level populations resulting from the stimulated Raman effect have been used, the most notable involving the probing of the vibrational population change by an increase or decrease of the fluorescence intensity following excitation from the ground or excited vibrational level to an excited electronic state.[214] In the benzene example just described an increase or decrease in the intensity of fluorescence from the $\tilde{A}^1B_{2u}$ state could be monitored instead of an increase in the number of ions produced by the second photon.

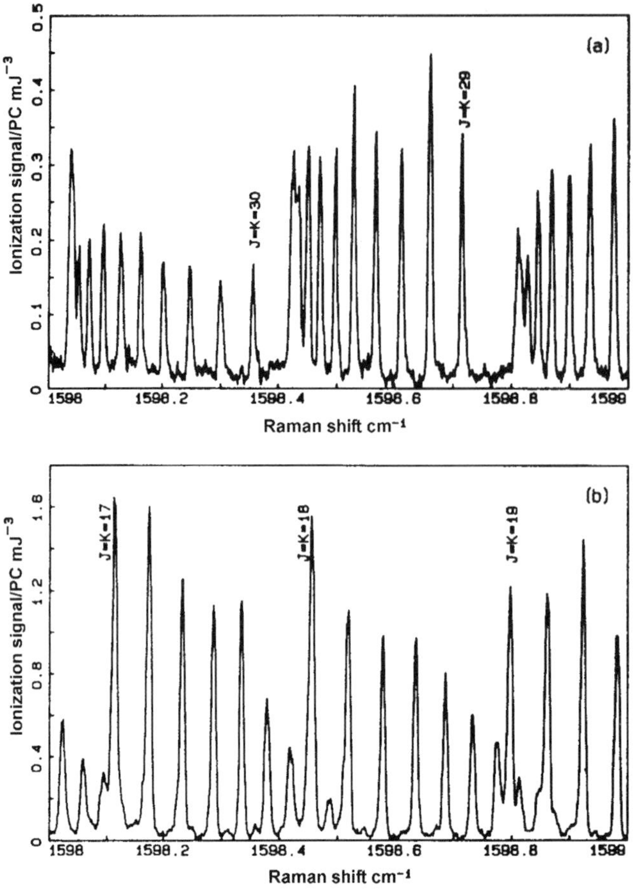

Figure 8.42 Ionization gain stimulated Raman spectra of benzene (a) with excitation in ${}^{s}S$-branch transitions from the ν_8 vibrational level to the zero-point level of the $\tilde{A}^1B_{2u}$ electronic state and showing ${}^{o}O$-branch transitions in the 8^1_0 vibrational transition, and (b) with excitation in ${}^{p}P$-branch transitions from the $\nu_1 + \nu_6$ vibrational level to the zero-point level of the $\tilde{A}^1B_{2u}$ electronic state and showing ${}^{s}S$-branch transitions in the $1^1_0 6^1_0$ vibrational transition
[Reproduced, with permission from Esherick, P., Owyoung, A. and Pliva, J. (1985). *J. Chem. Phys.*, **83**, 3311]

8.2.4 Inverse Raman spectroscopy

The inverse Raman effect is also a stimulated Raman effect, being due entirely to the term involving $\rho_0\rho_1$ in equation (8.35), but in which emission and absorption processes are reversed.

Figure 8.43(a) illustrates the stimulated Raman effect showing that the Raman and Rayleigh radiation involve emission processes from the virtual states V_1 and V_2. The Stokes and anti-Stokes processes are labelled S and AS, respectively. Figure 8.43(b) shows the inverse of these processes, in which the system is irradiated with radiation of wavenumber $\tilde{\nu}_0 \pm \tilde{\nu}_i$ together with $\tilde{\nu}_0$, and the Stokes and anti-Stokes processes appear as absorption of the incident $\tilde{\nu}_0 \pm \tilde{\nu}_i$ radiation. The probability of these inverse Raman processes is given by exchanging the subscripts 1 and 0 in equation (8.35).

In practice, instead of using radiation of wavenumber $\tilde{\nu}_0 \pm \tilde{\nu}_i$ it is more practicable to use a continuum which covers this range and then,

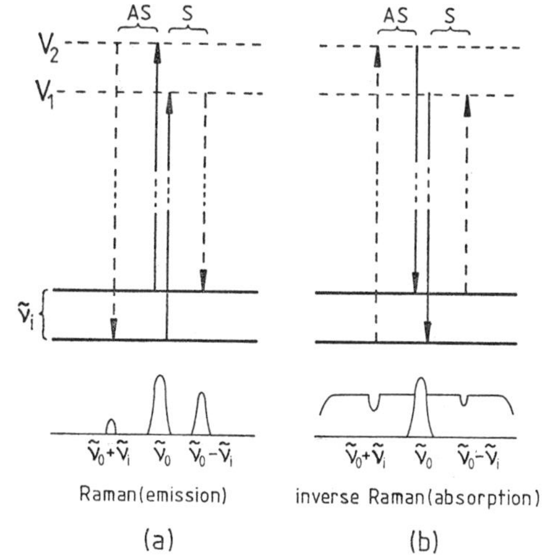

Figure 8.43 The stimulated Raman effect, illustrated in (a), is an emission process whereas the inverse Raman effect, illustrated in (b), is an absorption process

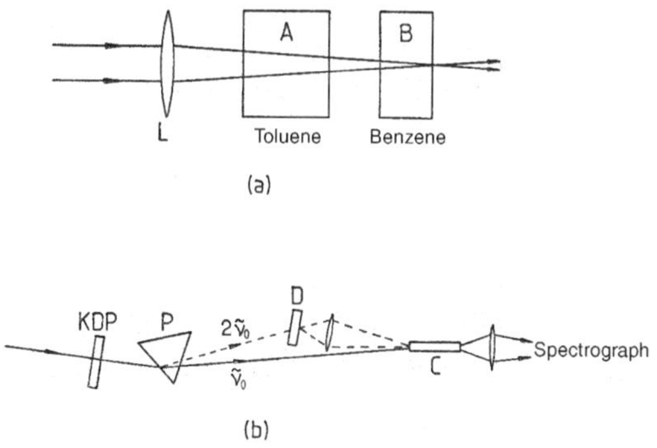

Figure 8.44 Methods for observing the inverse Raman effect using (a) broad Raman scattering from a molecule with a vibration of similar wavenumber and (b) broad fluorescence from a dye

as Figure 8.43(b) shows, the inverse Raman effect will show up as sharp absorption in the continuum.

In the first demonstration of the effect[215] the experimental arrangement in Figure 8.44(a) was used. Light from a *Q*-switched ruby laser was focused by a lens L into a cell B containing liquid benzene but first passing through another cell A containing liquid toluene. The vibration of toluene which corresponds to ν_1 in benzene has a wavenumber of $1003\,\text{cm}^{-1}$ and is the only one involved in the stimulated Raman effect. Therefore radiation emerging from cell A contains $\tilde{\nu}_0$ and $\tilde{\nu}_0 \pm 1003\,\text{cm}^{-1}$. It was found that, if the laser radiation was not sharp but contained more than one component, all Raman scattering, except the first Stokes transition, was broadened to several hundred cm^{-1}. As a result the anti-Stokes continuum extended sufficiently to show, with a photographic plate as detector, the $\tilde{\nu}_0 + 992\,\text{cm}^{-1}$ anti-Stokes transition of benzene in absorption.

An important requirement for an inverse Raman experiment is that the radiation $\tilde{\nu}_0$ and the continuum must be coherent, a condition which is clearly satisfied in this experiment.

The experiment illustrated in Figure 8.44(a) is very limited in that it is difficult to find suitable pairs of molecules. Figure 8.44(b) shows a more generally applicable method.[216] Radiation from a *Q*-switched ruby laser falls on a KDP frequency doubling crystal from which radiation of wavenumber $\tilde{\nu}_0$ and $2\tilde{\nu}_0$ emerges. These are separated by a prism P and radiation $\tilde{\nu}_0$ passes straight to the Raman cell C. The radiation $2\tilde{\nu}_0$ falls on a cell D containing a fluorescent dye. Rhodamine 6G or rhodamine B were used in the original experiments, both of them absorbing at 347.0 nm, the wavelength corresponding to $2\tilde{\nu}_0$. As Figure 8.25 shows, the fluorescence from a dye solution is a broad continuum; that from each of these dyes includes the region 560–680 nm which covers anti-Stokes Raman shifts in the range 300–$3500\,\text{cm}^{-1}$ from $\tilde{\nu}_0$. Temporal coherence of the two beams entering the cell C is assured because of the very short fluorescence lifetime of the dye, of the order of 10 ns, and spatial coherence is achieved by the cell being in the form of a capillary and acting as a light pipe with both beams being totally internally reflected. The spectra of diamond and of liquid benzene and toluene were obtained in this way.

Other methods of obtaining a continuum include passing high intensity laser pulses of only picosecond duration into a solid or liquid material when a continuum is created due to self-focusing which leads to self-phase modulation of the laser radiation. The result is that a continuum is created covering the whole of the visible region of the spectrum.[217]

With the sample, in this case methane, in a supersonic jet, with no carrier gas, an ultra-high resolution inverse Raman spectrum of the 1_0^1 band has been obtained.[218] The rotational line width was reduced in the jet to a sub-Doppler value of only 105 MHz.

An advantage of the inverse Raman effect over the stimulated Raman effect is that more vibrations are involved.

8.2.5 Coherent anti-Stokes and coherent Stokes Raman scattering spectroscopy

Both CARS and CSRS spectroscopy rely on the general phenomenon of wave mixing. We have encountered this previously in Raman gain spectroscopy and in ionization loss and ionization gain stimulated Raman spectroscopy, described in Section 8.2.3, in which there is three-wave mixing which results from the third-order term in equation (8.14).

In CARS and CSRS, radiation from two lasers of wavenumbers $\tilde{\nu}_1$ and $\tilde{\nu}_2$, where $\tilde{\nu}_1 < \tilde{\nu}_2$, fall on the sample. As a result of *four*-wave mixing, radiation of wavenumber $\tilde{\nu}_3$ is produced, where

$$\tilde{\nu}_3 = 2\tilde{\nu}_2 - \tilde{\nu}_1 = \tilde{\nu}_2 - (\tilde{\nu}_1 - \tilde{\nu}_2) \qquad (8.36)$$

or

$$\tilde{\nu}_3 = 2\tilde{\nu}_1 - \tilde{\nu}_2 = \tilde{\nu}_1 + (\tilde{\nu}_1 - \tilde{\nu}_2) \qquad (8.37)$$

The wave mixing is more efficient if $(\tilde{\nu}_1 - \tilde{\nu}_2) = \tilde{\nu}_i$, the wavenumber of a Raman-active vibrational or rotational transition of the sample.

When $\tilde{\nu}_3$ is given by equation (8.36) the Raman scattering is to low wavenumber of $\tilde{\nu}_2$, i.e. on the Stokes side, and, when it is given by equation (8.37) it is on the anti-Stokes side of $\tilde{\nu}_1$. Both types of scattering are coherent and, for this reason, the former is known as coherent Stokes Raman scattering, or CSRS, and the latter as coherent anti-Stokes Raman scattering, or CARS.

From the symmetry of equations (8.36) and (8.37) there seems to be no reason to favour the CSRS or CARS technique but, since $(2\tilde{\nu}_2 - \tilde{\nu}_1)$ is to low wavenumber of *both* $\tilde{\nu}_1$ and $\tilde{\nu}_2$, there is a strong tendency for CSRS to be overlapped by fluorescence. For this reason the CARS technique is more frequently used and subsequent discussion will be confined to this.

As a consequence of the coherence of CARS and the very high conversion efficiency to $\tilde{\nu}_3$ it forms a collimated and, therefore, laser-like beam.

The selection rules for CARS spectra are precisely the same as for spontaneous Raman scattering but CARS has the advantage of vastly increased intensity.

The experimental observation of a CARS spectrum involves the use of a laser of fixed wavenumber $\tilde{\nu}_1$ and a second laser of variable wavenumber $\tilde{\nu}_2$. In the set-up in Figure 8.45, $\tilde{\nu}_1$ is the wavenumber of a frequency doubled Nd^{3+}:YAG laser (Section 8.1.8.4) and $\tilde{\nu}_2$ is that of a tunable dye laser which is itself pumped by the Nd^{3+}:YAG laser. The two laser beams are focused with a lens L into the sample cell C so that they make a small angle 2α with each other. The collimated CARS emerges at an angle 3α to the optic axis and is spatially filtered from $\tilde{\nu}_1$ and $\tilde{\nu}_2$ by a filter F in the form of a pinhole. The CARS signal is detected with a semiconductor (PIN) diode D which may be preceded by a monochromator. The sample may be solid, liquid or gaseous.

Figure 8.46 shows part of the high resolution CARS spectrum of gaseous CH_4, in a particular part of the Q-branch region of the 1_0^1 vibrational band. The spectrum was obtained with a fixed frequency argon ion laser and a tunable dye laser. The part shown in the figure corresponds to the region in the Raman gain spectrum in Figure 8.39 from about 2916.52 to 2916.85 cm^{-1}. The resolution, like that in

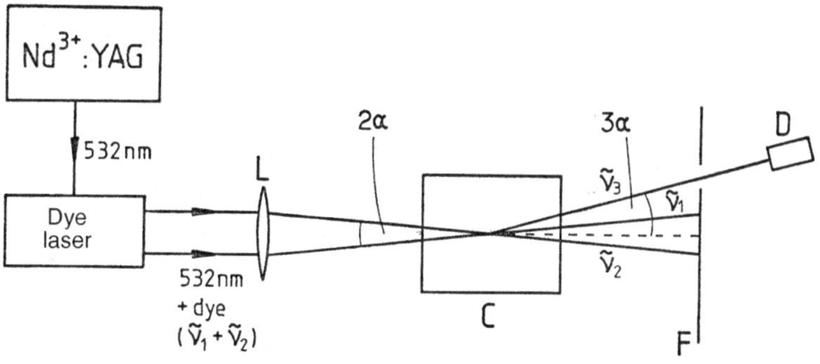

Figure 8.45 Experimental arrangement for obtaining a CARS spectrum

Figure 8.39, is limited by the Doppler width at wavenumber $\tilde{\nu}_i$ but, at the temperature of 80 K used for the spectrum in Figure 8.46, the Doppler line width is reduced to *ca* 120 MHz (0.004 cm^{-1}).

All Raman-active vibrations may be observed in the CARS spectrum.

CARS spectra, with the sample in a supersonic jet, have been obtained with medium[219] and high[220] resolution. Whereas the Raman intensity of a vibration–rotation transition from a lower state, $\upsilon''J''$, to an upper state, $\upsilon'J'$, is proportional to the population, $N_{\upsilon''J''}$, of the lower state, the CARS intensity is proportional to $(N_{\upsilon''J''} - N_{\upsilon'J'})^2$. Therefore, the rotational cooling in the jet, which increases $N_{\upsilon''J''}$, for low J, relative to $N_{\upsilon'J'}$, is ideal for increasing the CARS signal.

The CARS technique is used for monitoring fairly rapid processes such as those involved in combustion. It has also been used to study the products of photofragmentation.

The $\tilde{A}$–$\tilde{X}$ electronic absorption system of CF_3NO lies in the 500–720 nm region and photolysis, following absorption in this region, produces CF_3 and NO photofragments. CARS spectra of these photofragments have been obtained[221] using frequency doubled Nd^{3+}:YAG

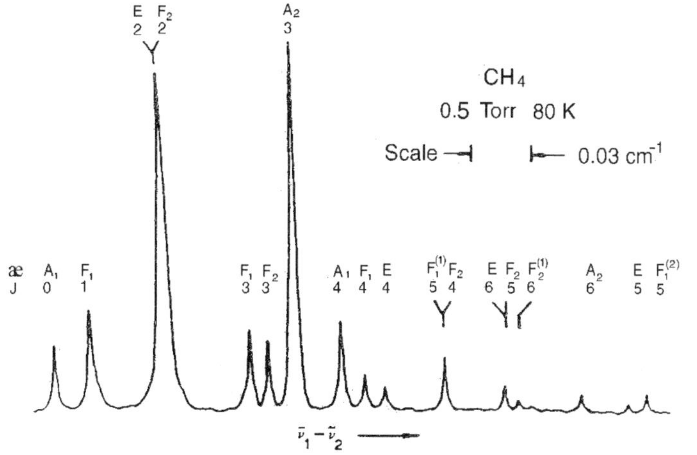

Figure 8.46 A high resolution CARS spectrum showing the 1_0^1 vibrational band of the methane molecule at a pressure of 0.5 Torr and at 80 K
[Reproduced, with permission, from Kozlov, D. N., Prokhorov, A. M. and Smirnov, V. V. (1979). *J. Mol. Spectrosc.*, **77**, 21]

radiation with a wavelength of 532 nm for the fixed wavenumber radiation, $\tilde{\nu}_1$. The variable wavenumber radiation, $\tilde{\nu}_2$, was provided by a dye laser which was used for simultaneously dissociating and detecting CF_3 and NO. Rotational analysis of the 1_0^1 band of CF_3 and the $v = 1$–0 band of NO indicated rotational temperatures in excess of 1100 K.

Most CARS experiments have been carried out with the $\tilde{\nu}_1$ and $\tilde{\nu}_2$ beams produced by pulsed lasers. However, it is possible to use CW lasers. One such system[222] employs an argon ion laser, in a ring configuration, for $\tilde{\nu}_1$, and a dye laser, pumped by the same argon ion laser, for $\tilde{\nu}_2$. The sample cell is in an intracavity position in the argon ion laser ring and the dye laser radiation is also directed into the cell. A CARS spectrum of ethane, with a resolution of less than 0.001 cm^{-1}, in the complex region of ν_1, a C—H stretching vibration, has been obtained.

8.2.6 Laser magnetic resonance (or laser Zeeman) spectroscopy

Most lasers cannot be tuned over an appreciable wavenumber range and this limitation seemed, in the early days of their development, to be a serious one. One way of overcoming this restriction to some extent is by 'tuning' the energy levels of the atom or molecule whose spectrum is being studied until a transition between the levels happens to match the laser wavenumber. At this point the laser beam suffers some attenuation.

The method of tuning of the levels to be discussed in this section involves the application of a magnetic field. The resulting splitting of energy levels and transitions is then due to the Zeeman effect, which was discussed in Sections 6.1.8.1 and 6.3.5.5. Consequently, this type of spectroscopy has been called laser Zeeman spectroscopy, a name which nicely parallels that of laser Stark spectroscopy, to be discussed in Section 8.2.7. Unfortunately, in my view, the technique is more often referred to as laser magnetic resonance, or LMR, spectroscopy, a name which implies a similarity to nuclear magnetic resonance (NMR) spectroscopy, which can be misleading.

The magnetic field is applied by placing the sample cell between the poles of a magnet. As in electron spin resonance (ESR) spectroscopy, with which LMR spectroscopy has much in common, there is a considerable gain in the signal-to-noise ratio by employing field modulation together with phase sensitive detection. Figure 8.47 illustrates the effect on a Doppler broadened transition observed by sweeping the magnetic field B while keeping the laser wavenumber fixed.

The field experienced by the sample is modulated by passing an alternating current through subsidiary coils on the pole faces. The modulation is sinusoidal and the amplitude must be small compared with the Doppler line width for a meaningful line shape but, for maximum sensitivity, it is only slightly less than the Doppler line width. On the 'up-slope' of the line shown in Figure 8.47(a) a small decrease in the modulated field produces a small decrease in the signal recorded by an ordinary detector, and a small increase in the field produces a small increase in the signal; in other words, the modulation and the signal are 'in phase'. Similarly, on the 'down-slope' of the line they are 'out of phase'. At the point of maximum intensity of the line there is no modulation-induced variation in the detected signal.

Figure 8.47(b) shows the effect of using a phase sensitive detector. A positive signal output from the phase sensitive detector results when the modulation and the ordinary signal are in phase, a negative output when they are out of phase and zero output corresponds to the intensity maximum of the line. The result is the first derivative of the true line shape and the centre of the line corresponds to the zero output position between the positive and negative lobes.

If the concentration of the absorbing species is large enough and the absorbance is sufficiently high, saturation of the transition may occur.

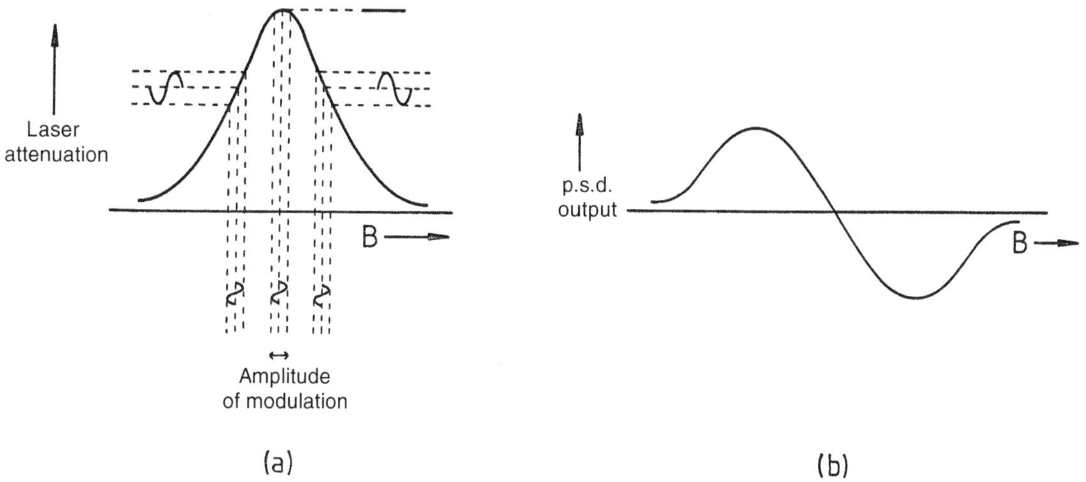

Figure 8.47 (a) A Doppler-limited line and (b) the effect on the line shape of modulation and phase sensitive detection

The phenomenon of saturation was discussed in Section 3.3 in the context of microwave and millimetre wave spectroscopy and referred, therefore, to the situation in which the two energy levels involved, m (lower) and n (upper), are relatively close together. Under these circumstances saturation occurs when the populations N_m and N_n are nearly equal. One important effect of saturation is that, if a reflecting mirror is placed at the end of an absorption cell, a Lamb dip may be observed in the absorption profile, as described in Section 3.4 and illustrated in Figures 3.8–3.10.

In general, saturation of a transition between two levels m and n refers to the situation in which the population N_m has been sufficiently decreased, and N_n sufficiently increased, relative to their equilibrium Boltzmann populations [see equation (2.6)] that the absorption intensity is non-linear; in other words, it is no longer given by the Beer–Lambert law [equation (2.24)]. Lamb[223] has shown that, when narrow-band radiation, such as that from a laser, is of wavenumber $\tilde{\nu}_L$ and is tuned across a Doppler-broadened absorption profile with a maximum at $\tilde{\nu}_{res}$ (see Figure 3.7), the absorbance $A'(\tilde{\nu}_L)$ is given by

$$A'(\tilde{\nu}_L) = A(\tilde{\nu}_L)\left[1 - \frac{2|\boldsymbol{\beta}|^2}{(\tilde{\nu}_L - \tilde{\nu}_{res})^2 + \gamma^2}\right] \quad (8.38)$$

in which $A(\tilde{\nu}_L)$ is the Beer–Lambert law (linear) absorbance and γ is the half-width at half maximum intensity of the Lamb dip. The quantity $\boldsymbol{\beta}$ is given by

$$\boldsymbol{\beta} = E\boldsymbol{R}^{nm}/2\hbar \quad (8.39)$$

where E is the amplitude of the electric field due to the incident laser radiation and $\boldsymbol{R}^{nm}$ is the electric dipole transition moment [see equation (2.21)].

In general, we say that the transition is saturated when the second (non-linear) term in brackets in equation (8.38) makes an appreciable contribution. It makes its greatest contribution, for any pair of levels, when $\tilde{\nu}_L = \tilde{\nu}_{res}$.

A very important feature from the point of view of obtaining saturation of transitions in the infrared, visible or ultraviolet regions is that the transition moment $\boldsymbol{R}^{nm}$ is much larger than for transitions in the microwave and millimetre wave regions. It follows that saturation can be obtained with a much larger population difference in the so-called 'optical' regions of

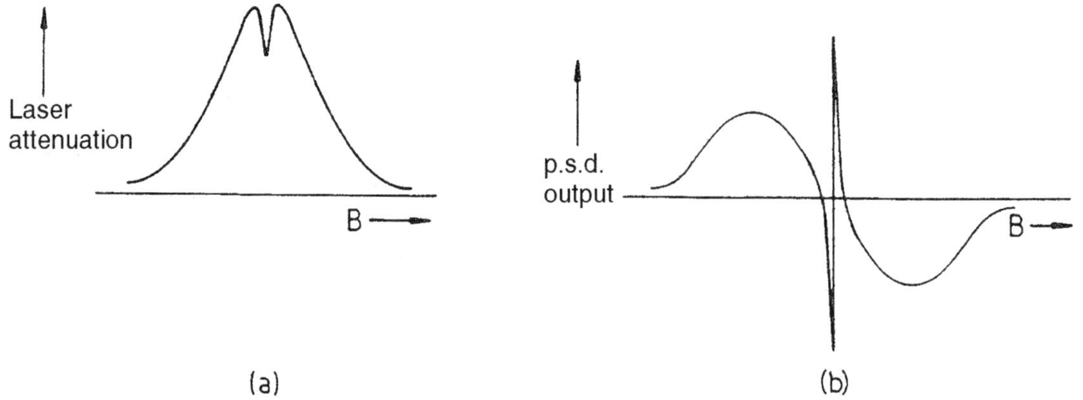

Figure 8.48 Absorption Lamb dip observed (a) without and (b) with modulation and phase sensitive detection

the spectrum and this makes Lamb dip spectroscopy a more sensitive technique than in the lower wavenumber regions.

Saturation may be achieved with the absorption cell outside the laser cavity, with a reflecting mirror at the end of the cell and normal to the laser beam, or inside the cavity. In either case measurement of laser attenuation (absorption) results in a line shape like that in Figure 8.48 which shows the result obtained without and with phase sensitive detection. It was explained in Section 3.4 that the width of the Lamb dip† is the natural line width. Doppler broadening is eliminated and the accuracy of measurement of transition wavenumbers greatly improved.

However, Lamb dips can be observed much more readily by placing the absorption cell inside the laser cavity.[224] Figure 8.49 illustrates the design of the cavity for use with a far infrared laser, such as an HCN or H_2O laser.

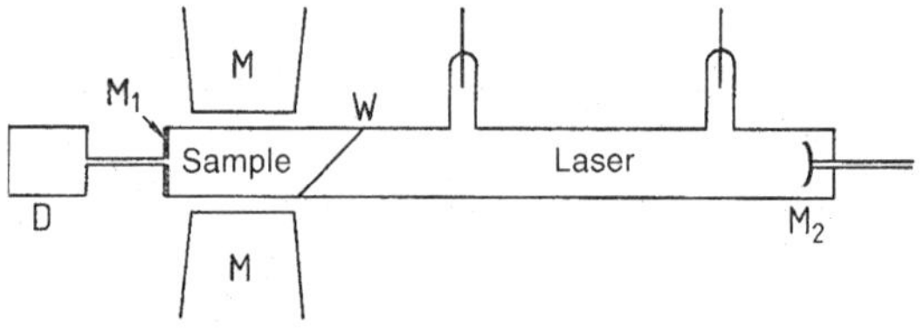

Figure 8.49 Experimental arrangement for obtaining an LMR spectrum in the far infrared with the absorption cell inside the laser cavity

The cavity is divided into two compartments by a Brewster angle window W made out of a material, such as polyethylene or Mylar, which is transparent in the far infrared. The laser compartment contains the active medium, such as HCN or H_2O, in which a discharge is created. The sample compartment is between the poles of a large electromagnet M capable of producing a field of about 2 T (20 kG). The cavity mirror, M_1, is planar with a small hole leading to the detector D. The other mirror, M_2, is concave and can be translated piezoelectrically to ensure that the laser is always operating at maximum gain.

For a similar intracavity LMR experiment in the near infrared[225] a CO_2 or CO laser is commonly used. The problem of finding a suitable material for the windows is not so

†There is some confusion in the literature concerning what have been called inverse Lamb dips. Because the absorption (attenuation) Lamb dip in Figure 8.48(a) is 'upside down' relative to the power output Lamb dip in Figure 8.20, the former has been called an inverse (or inverted) Lamb dip. However, since spectroscopic observations are more often of the laser power output type in Figure 8.20, we shall refer to the type of curve in Figure 8.48(a) as a Lamb dip and shall not use the expression 'inverse Lamb dip' subsequently. This is consistent with the use of the term 'Lamb dip' in Section 3.4.

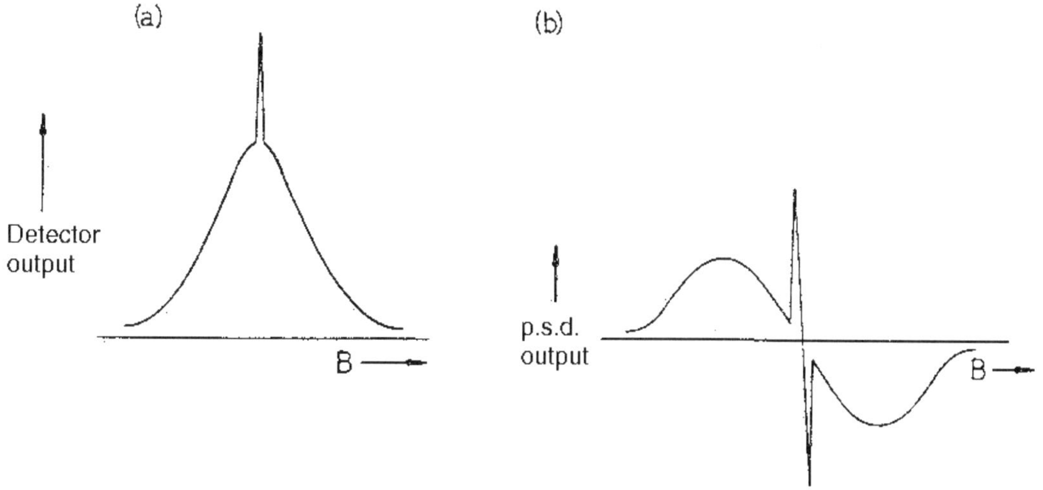

Figure 8.50 Sharp maximum, due to saturation, superimposed on the Doppler line shape of the laser power output (a) without and (b) with phase sensitive detection

serious. Zinc selenide, ZnSe, may be used and, since having more than one window in the cavity does not create transmission problems, the laser and sample compartments can be separate resulting in four Brewster windows inside the laser cavity. The two cells are placed between a grating, for laser line selection, and an output mirror.

Under conditions of saturation, measurement of laser power output, rather than attenuation, shows the type of curve in Figure 8.50(a). The sharp maximum in the output, corresponding to the minimum in the attenuation in Figure 8.48(a), is known as a Lamb peak. The first derivative curve, obtained by modulation and phase sensitive detection, is shown in Figure 8.50(b).

The first successful LMR experiment was carried out with an HCN laser focused into a sample cell containing O_2 gas.[226] The laser operates at about 27 cm^{-1}.

For a molecule to show a Zeeman effect it must have a magnetic moment. In most cases (see Section 6.3.5.5) this is due to the molecule having a non-zero electron spin angular momentum caused by one or more unpaired electrons. This is the case in the $X^3\Sigma_g^-$ ground state of O_2.

The LMR technique is particularly suited to studying the spectra of transient species since many of these have at least one unpaired electron. In addition, they are usually present in such low concentration that the extremely high sensitivity of the technique is necessary for detection.

The first transient species to be studied in this way was OH,[227] which was produced by reaction between hydrogen atoms and NO_2 in a flow system through the sample compartment. Although this was inside the laser cavity, saturation was not achieved and Lamb dips were not observed. The active medium in the laser was H_2O vapour lasing at a wavelength of about 79.1 μm (126 cm^{-1}). This lies fairly close to the $^2\Pi_{\frac{1}{2}}$–$^2\Pi_{\frac{3}{2}}$ transition between the two components of the inverted $X^2\Pi$ ground state of OH and, in particular, the $J' = \frac{1}{2}$–$J'' = \frac{3}{2}$ rotational transition. Both components of the $^2\Pi$ state show Λ-doubling (Section 6.2.5.2) and, in a magnetic field, each component of the $J = \frac{3}{2}$ level of the $^2\Pi_{\frac{3}{2}}$ state is split into four since $M_J = J, J-1, \ldots, -J = \frac{3}{2}, \frac{1}{2}, -\frac{1}{2}, \frac{3}{2}$. The Zeeman splitting in the $^2\Pi_{\frac{1}{2}}$ state is too small to be detected so that four transitions, obeying the general selection rule $\Delta M_J = 0, \pm 1$, were observed.

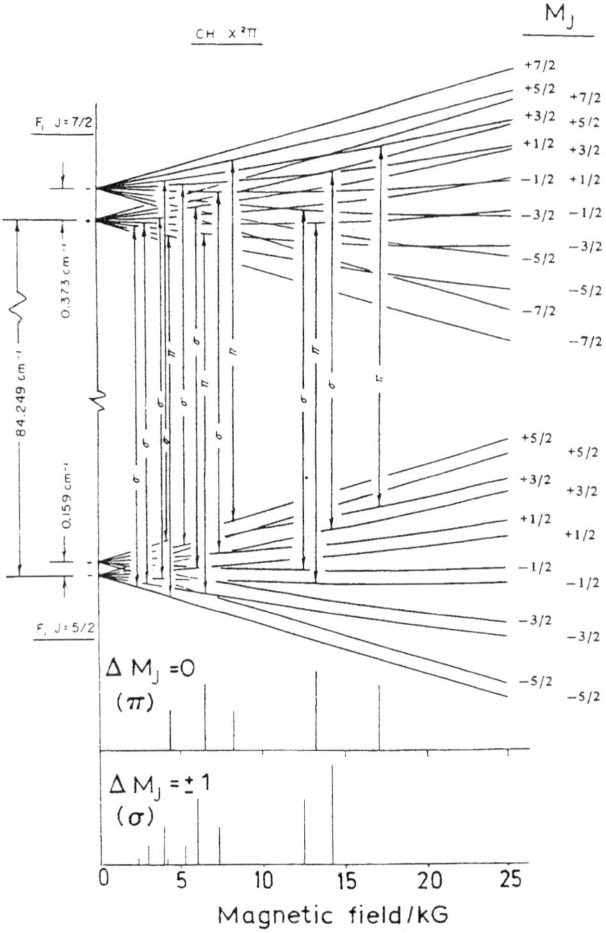

Figure 8.51 Zeeman splittings in the $J'=\frac{7}{2}$–$J''=\frac{5}{2}$ transition in the $X^2\Pi$ state of CH and the resulting transitions with parallel (π) and perpendicular (σ) polarization (1 G = 10^{-4} T)
[Reproduced, with permission, from Evenson, K. M., Radford, H. E. and Moran, M. M. Jr (1971). *Appl. Phys. Lett.*, **18**, 426]

The transitions with $\Delta M_J = 0$ and ± 1 are polarized parallel and perpendicular, respectively, to the field direction. The relative polarization can be distinguished by rotating Brewster angle windows inside the cavity. If the sample is outside the cavity these are the windows of the compartment containing the active medium. For the sample inside the cavity, as in Figure 8.49, it is W which is rotated. In the orientation shown, radiation reaching D is polarized parallel to the field and only $\Delta M_J = 0$ transitions are observed. Rotation of W by 90° about the laser axis results in D receiving perpendicularly polarized radiation and the observation of $\Delta M_J = \pm 1$ transitions.

The CH molecule is produced readily in hydrocarbon flames but the concentration is low. The ground state is $X^2\Pi$ with the $^2\Pi_{\frac{1}{2}}$ component, labelled F_1 [see equation (6.258)], slightly lower in energy than the $^2\Pi_{\frac{3}{2}}$ component, labelled F_2. Figure 8.51 shows that, at zero field, the $J'=\frac{7}{2}$–$J''=\frac{5}{2}$, F_1–F_1 transition lies at about 84 cm^{-1} and is almost coincident with a water vapour laser line at

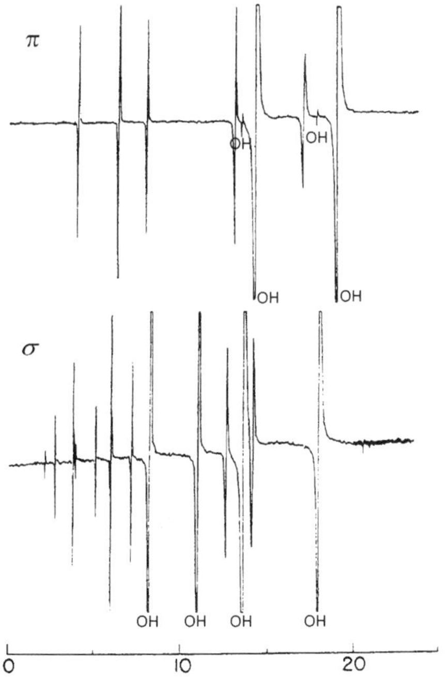

Figure 8.52 The LMR spectrum of CH showing the transitions indicated in Figure 8.51. There are some overlapping transitions due to OH. ($1 G = 10^{-4}$ T) [Reproduced, with permission, from Evenson, K. M., Radford, H. E. and Moran, M. M. Jr (1971). *Appl. Phys. Lett.*, **18**, 426]

118.6 μm ($84.3 cm^{-1}$). Splitting due to Λ-type doubling is 0.159 and $0.373 cm^{-1}$ in the $J = \frac{5}{2}$ and $\frac{7}{2}$ states, respectively.

The Zeeman splittings in Figure 8.51 were calculated as a function of field strength from parameters known from the electronic spectrum of CH. Hyperfine splitting due to the spin of the ^{1}H nucleus was neglected. Slight curvature of the split levels is due to magnetic interactions with nearby levels. As the figure shows, there are 14 transitions obeying the $\Delta M_J = 0, \pm 1$ selection rules which can be tuned into resonance with the laser radiation for field strengths in the range 0–2 T (0–20 kG). Five transitions, with $\Delta M_J = 0$, are polarized parallel to the field and nine, with $\Delta M_J = \pm 1$, perpendicular to the field.

The spectrum, shown in Figure 8.52, was obtained with an oxygen–acetylene flame in the sample compartment inside the laser cavity. It shows all 14 transitions with their polarization properties. No Lamb dips were observed because saturation was not achieved. Overlapping the CH spectrum are some very intense transitions due to OH ($^2\Pi_{\frac{3}{2}}$, $J' = \frac{5}{2}$–$J'' = \frac{3}{2}$) which is also produced in the flame.

A quantitative fitting of the observed to the calculated CH spectrum results in accurate values of (a) the rotational constant B_0, (b) the constant $Y(= A/B_0)$, where A is the spin–orbit coupling constant [see equation (6.258)] and (c) the difference between the Λ-doubling in the $J = \frac{7}{2}$ and $\frac{5}{2}$ states.

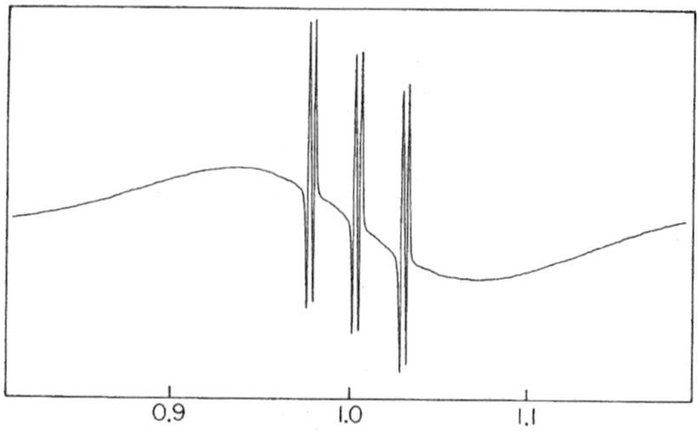

Figure 8.53 LMR spectrum of NO showing Lamb dips in the $M_J' = \frac{1}{2} - M_J'' = -\frac{1}{2}$ component of the $R(\frac{3}{2})$ transition of the $v = 1$–0 band. ($1 G = 10^{-4}$ T) [Reproduced, with permission, from Dale, R. M., Johns, J. W. C., McKellar, A. R. W. and Riggin, M. (1977). *J. Mol. Spectrosc.*, **67**, 440]

The LMR spectrum shown in Figure 8.53 is of NO using a CO laser. This laser produces a variety of vibration–rotation transitions and the one used here is the $P(13)$ line of the $v = 9$–8 transition at about $1884 cm^{-1}$. The sample was inside the laser cavity, saturation was obtained and Lamb dips observed.

The broad, S-shaped curve, upon which the dips are superimposed, is the Doppler broadened $M_J' = \frac{1}{2}$–$M_J'' = -\frac{1}{2}$ component of the $R(\frac{3}{2})$ transition of the $v = 1$–0 band involving only the $^2\Pi_{\frac{3}{2}}$ component of each vibrational state. The three pairs of Lamb dips constitute the hyperfine structure of this transition. The separation within each pair is the difference between the Λ-doubling in the $J = \frac{5}{2}$ and $J = \frac{3}{2}$ states involved in the $R(\frac{3}{2})$ transition and is only 2.7 MHz (0.000 09 cm^{-1}). The fact that there are three pairs is due to hyperfine splitting caused by the nuclear spin of nitrogen ($I = 1$). As in the case of atoms, explained in Section 6.1.8.2 and illustrated in Figure 6.23, the introduction of a magnetic field uncouples the nuclear spin angular momentum—the nuclear Paschen–Back effect. The spectrum in Figure 8.53 was obtained with polarization perpendicular to the magnetic field so that $\Delta M_J = \pm 1$ only. Since the selection rules for M_J and M_I must not be the same, $\Delta M_I = 0$ and the three doublets in the spectrum are due to $M_I = 1, 0, -1$.

One of the most spectacular early achievements of LMR spectroscopy was the detection in 1974 of the spectrum of HO_2.[228,229] This transient species had been identified previously by mass spectrometry and by its vibrational spectrum in a solid matrix, but no attempt to observe a high resolution gas phase spectrum of any kind had been successful. The LMR spectrum was obtained by passing O_2 through a microwave-induced discharge and then into the absorption cell containing acetylene, but several other oxygen atom reactions have proved productive.

The HO_2 molecule is nonlinear and a prolate near-symmetric rotor. Seven ^1H_2O or ^2H_2O far infrared laser lines were used which have near coincidences with various pure rotational transitions, in the region 50–150 cm^{-1}, in the ground electronic state, $\tilde{X}^2A''$, of HO_2 (Section 6.3.1.2). The ranges of quantum numbers involved in these transitions are $4 \leqslant N \leqslant 19$ and $1 \leqslant K_a \leqslant 4$.

The discovery of an efficient way of making HO_2 resulted in a renewed search for an electronic spectrum, expected to be in the near infrared. This was discovered in 1974 and the emission spectrum recorded at high resolution and analysed in 1979.[230]

Many transient species have been detected by LMR spectroscopy including atoms, such as O, Fe, S, Si, Se and Sn, diatomic molecules, such as SH, FO, NH, CF, NSe, SiH, SeH, NiH, CrH, FeH, AsH, CoH, TeH and GeH, triatomic molecules, such as NH_2, HCO, CH_2, HS_2, FeH_2, C_2H, N_3, CCN and NCN, and the polyatomic molecule CH_3O. Several of these, such as CH_3O[231] and FO,[232] were first discovered by this technique.

Many of these species have been produced in a flow system incorporating a microwave-induced discharge. For the production of CoH, for example,[233] hydrogen atoms, produced from H_2 in a microwave discharge, react with $CoNO(CO)_3$.

The ground electronic state of CoH is $X^3\Phi$. Since the quantum number $S = 1$, Σ can be 1, 0 or -1 and, since $\Lambda = 3$, Ω can be 4, 3 or 2 [see equation (6.109)]. The $^3\Phi_2$ component is the lowest in energy and the higher $^3\Phi_3$ and $^3\Phi_4$ components are separated by 728 cm^{-1}. Five rotational transitions with $\Delta J = +1$ have been observed[233] in the far infrared LMR spectrum, two of them in the $^3\Phi_3$ and three in the $^3\Phi_4$ component. For the far infrared laser it was necessary to use $^{12}C^2H_2F_2$, $^{13}C^1H_2F_2$, $^{12}C^1H_2F_2$, $^{12}C^1H_3O^2H$, $^{12}C^1H_3O^1H$ and $^{13}C^1H_3O^1H$ as lasing gases to cover the range of these transitions.

The LMR spectrum in Figure 8.54 shows the $J = 6$–5 rotational transition in the $^3\Phi_4$ spin component of CoH with parallel polarization. The selection rule $\Delta M_J = 0$ results in the five components shown. Each of these components shows further hyperfine splitting due to the nuclear spin of ^{59}Co, for which $I = \frac{7}{2}$. The eight components result from $M_I = \frac{7}{2}, \frac{5}{2}, \frac{3}{2}, \frac{1}{2}, -\frac{1}{2}, -\frac{3}{2}, -\frac{5}{2}, -\frac{7}{2}$ and the selection rule $\Delta M_I = 0$.

It is noticeable in Figure 8.54 that the high field octet with $M_J = 1$ shows constant spacings and equal intensities whereas those at lower field tend to show variable spacings and intensities. The reason for this is that, at low field, the coupling of the nuclear spin angular momentum approximates to Hund's case (a_β), as illustrated in Figure 6.55, whereas a higher field tends to uncouple the nuclear spin.

The highly accurate rotational constant, 7.313 713(67) cm^{-1}, obtained for CoH in this experiment, combined with a value of α_e from

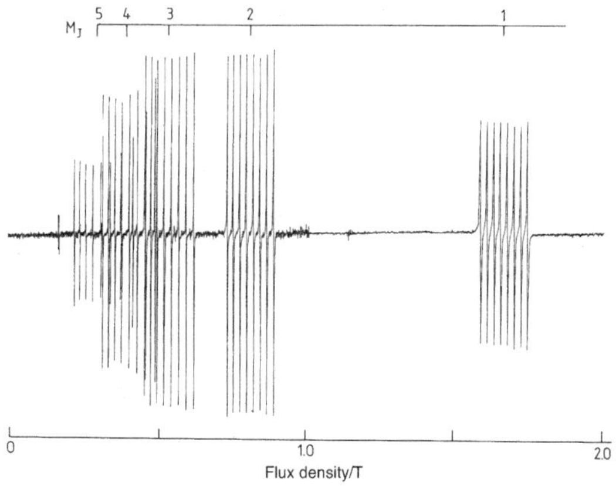

Figure 8.54 An LMR spectrum of CoH showing the five M_J components, with $\Delta M_J = 0$, of the $J = 6$–5 rotational transition in the $^3\Phi_4$ spin component. Each M_J component shows an octet hyperfine pattern due to the nuclear spin of ^{59}Co ($I = \frac{7}{2}$) [Reproduced, with permission, from Beaton, S. P., Evenson, K. M. and Brown, J. M. (1994). *J. Mol. Spectrosc.*, **164**, 395]

the infrared spectrum, gives a value of 1.513 843 5(80) Å for the equilibrium bond length r_e.

In addition to neutral transient species, positive ions are commonly produced in discharges. Using these sources LMR spectra of OH^+, HBr^+, N^+ and P^+, for example, have been obtained.

The species P^+ has been produced in a discharge containing PF_3. The ground electronic state of P^+ is a ^3P state split into a normal multiplet with the lowest component being $^3P_0$. The $^3P_1$ component is split by a magnetic field into three Zeeman components with $M_J = -1$, 0 and +1, in order of increasing energy. In the LMR spectrum the magnetic dipole transition from $^3P_0$ to the $M_J = -1$ component of $^3P_1$ has been observed,[234] a small nuclear hyperfine splitting into a doublet being due to the nuclear spin of ^{31}P for which $I = \frac{1}{2}$. The high accuracy of the frequency of 4 944 433.0 ± 1.7 MHz, or wavenumber of 164.928 53 cm^{-1}, obtained for the transition is typical of this kind of experiment.

Extension of far infrared LMR spectroscopy up to a wavenumber of 278 cm^{-1} has allowed the first observation[235] of a vibrational transition, that of the low wavenumber bending vibration of Fe^2H_2 at 226 cm^{-1}. The molecule has a linear $^5\Delta$ ground state and the spectrum reveals Renner–Teller coupling in the $v = 1$ vibrational state.

8.2.7 Laser Stark (or laser electric resonance) spectroscopy

We have seen in Section 4.2.4 how an electric field splits the rotational levels of a diatomic or linear polyatomic molecule—the Stark effect. Usually the additional energy term $E_{\mathscr{E}}$ due to the field $\mathscr{E}$ is proportional to $\mathscr{E}^2$ [equation (4.44)] so that it is a second-order, or quadratic, Stark effect. However, if the molecule is in an electronic state with $\Lambda \neq 0$ there is Λ-type doubling of the rotational levels (Section 6.2.5.2) and, if this is small compared with the Stark energy, there is a first-order, or linear, Stark effect,

$$E_{\mathscr{E}} = -\frac{\mu \mathscr{E} M_J \Omega}{J(J+1)} \tag{8.40}$$

where μ is the dipole moment and the total angular momentum quantum number Ω is given by equation (6.109).

For a symmetric rotor, equation (4.72) shows that there is a first- *and* second-order Stark effect for $K \neq 0$ but only a second-order effect for $K = 0$.

In Section 4.5 we saw that an asymmetric rotor shows only a second-order Stark effect if the asymmetry doubling is large compared with the Stark energy but, if it is not, the Stark effect is like that of a symmetric rotor.

The Stark effect, then, can be used to tune energy levels so that transitions between them, which are close to a laser line in the absence of an electric field, can be brought into coincidence with the laser line by introducing such a field.

This type of spectroscopy is usually called laser Stark spectroscopy but is also referred to as laser electric resonance spectroscopy, a name which parallels laser magnetic resonance discussed in the previous section. The laser Stark technique was first employed in 1969,[236] at about the same time as laser magnetic resonance.

Laser Stark closely resembles laser magnetic resonance spectroscopy but, as has often been the case where we have encountered the Zeeman and Stark effects, laser Stark spectroscopy has some important advantages. Whereas laser magnetic resonance spectroscopy requires the molecule to have a magnetic moment, usually due to unpaired electron spins, laser Stark spectroscopy requires only a permanent electric dipole moment, a much less stringent requirement. In addition, because the Stark effect depends quantitatively on the dipole moment, μ, this can be determined by laser Stark spectroscopy not only in the ground vibrational state, where microwave and millimetre wave spectroscopy can provide the same information, but also, with a near infrared laser, in vibrationally excited states.

One of the earliest experiments involved vibration–rotation transitions in NH_3,[237] using CO_2 and N_2O lasers in the 10 µm (1000 cm^{-1}) region and a grating at one end of the cavity for laser line selection. A second experiment allowed observation of the Stark components of the 6_{24}–6_{15} ($J = 6$, $K_a = 2$, $K_c = 4 - J = 6$, $K_a = 1$, $K_c = 5$) pure rotational Q-branch transition in ^2H_2O,[238] using a far infrared HCN laser at 337 µm (29.7 cm^{-1}).

In the 1969 experiment,[236] the sample cell was outside the laser cavity. A reflecting mirror was employed normal to the laser beam and Lamb dips observed. In 1977 Lamb dips were observed when experiments were carried out with the cell inside the cavity.[239] The experimental arrangement is shown in Figure 8.55. This is similar to that in Figure 8.49, with a window W dividing the cell into two compartments, one for the active medium of the laser and the other for the absorbing sample. The two Stark electrodes S, across which the electric field is applied, are typically of the order of

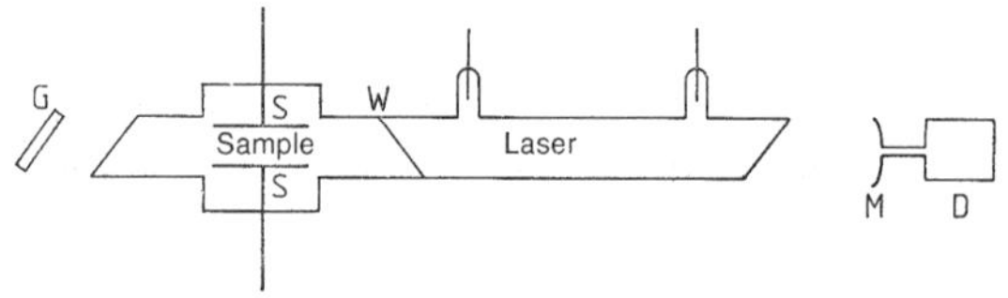

Figure 8.55 Laser cavity design for laser Stark spectroscopy with the sample cell inside the cavity

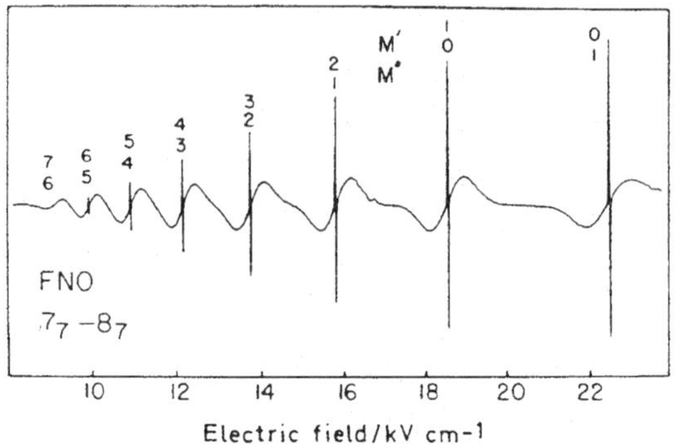

Figure 8.56 Laser Stark spectrum of FNO showing Lamb dips in the components of the $^qP_7(8)$ line of the 1_0^1 vibrational transition
[Reproduced, with permission, from Allegrini, M., Johns, J. W. C. and McKellar, A. R. W. (1978). *J. Mol. Spectrosc.*, **73**, 168]

20 cm long. They are only a few millimetres apart, in order to produce a large field between them, of the order of 50 kV cm^{-1}. The field is modulated and phase sensitive detection results in first derivative line shapes with Lamb dips. Alternatively, the so-called 2f detection method has been used[240] in which the phase sensitive detector operates at *twice* the frequency of Stark modulation whereas in the normal, so-called 1f, detection method it operates at the modulation frequency. 2f detection produces a normal Doppler line shape but with a Lamb dip, as in Figure 8.48(a).

Figure 8.56 shows part of the laser Stark spectrum of the bent triatomic molecule FNO obtained with the CO $v = 11$–10 $P(12)$ laser line at 1837.430 cm^{-1}. In this spectrum the signal from the phase sensitive detector is plotted against the Stark voltage per centimetre between the plates, the maximum voltage being limited by the possibility of electrical flashover. This also necessitates the use of low pressures, of the order of a few mTorr.

All the transitions in Figure 8.56 are Stark components of the ${}^qP_7(8)$ line of the 1_0^1 vibrational transition, where ν_1 is the N—F stretching vibration and the rotational symbolism implies that $\Delta K_a = 0$, $\Delta J = -1$, $K_a'' = 7$ and $J'' = 8$. FNO is a prolate near-symmetric rotor with $\kappa \approx -0.968$ but no effects of asymmetry doubling are observed. In an electric field each J level is split into $(J+1)$ components, each specified by its value of $|M_J|$, and the selection rule is $\Delta M_J = 0$, for parallel polarization, and $\Delta M_J = \pm 1$ for perpendicular polarization (as described in Section 4.2.4). The spectrum in Figure 8.56 was obtained with perpendicular polarization and eight of the Stark components are shown. All the transitions, except that for which $|M_J| = 7$–6, are saturated and show Lamb dips which are so sharp that the precision of measurement is about 3 MHz ($1 \times 10^{-4}\,\text{cm}^{-1}$). From these and other transitions it was shown that the dipole moment is 1.730 D in the $v_1 = 0$ state and is reduced by 5% in the $v_1 = 1$ state.

In general, Lamb dips are observed more readily in laser Stark than in laser magnetic resonance spectroscopy. The reason is that laser Stark spectroscopy requires low pressure at which saturation is more readily achieved, whereas laser magnetic resonance is frequently observed at higher pressure which causes fast collisional relaxation of the excited state molecule.

An example of the use of laser Stark spectroscopy in which dipole moment measurements in vibrationally excited states are of particular interest is that of NH_3.

In Section 5.2.7.1 it was pointed out that what we call the dipole moment, μ, of NH_3 is really the transition moment for the 0^-–0^+ (or 2_0^1) transition in the inversion vibration ν_2 [see Figure 5.69(b)]. The value of μ is 1.471 D. As the top of the inversion barrier, V_1 in Figure 5.68, is surmounted the vibrational levels become more evenly spaced and theory indicates that the transition moments should become very similar. A laser Stark investigation[241] of the 2^-–1^+ (or 2_2^5) hot band at about 10 μm using a CO_2 laser has shown, in conjunction with other transition moment measurements, that this is so: values for the 3^+–2^- (or 2_5^6) and 2^-–2^+ (or 2_4^5) transition moments are 1.05 and 1.02 D, respectively, consistent with the 2^+ (or 2^4) level being just above the barrier (Figure 5.68).

Figure 8.57 shows the laser Stark spectrum of an R-branch transition with $J'' = 9$ in the 4_0^1 perpendicular band of NH_3. The transition lies close to the CO $v = 13$–12, $P(15)$ laser line at $1775.259\,\text{cm}^{-1}$ and the spectrum shown was obtained with perpendicular polarization of the field relative to the laser radiation so that only $\Delta M_J = \pm 1$ transitions were observed. For low values of M_J the $\Delta M_J = +1$ and -1 transitions occur in pairs with $\Delta M_J = +1$ resulting in more intense transitions than $\Delta M_J = -1$. At higher values of M_J the intensity difference increases and several $\Delta M_J = -1$ transitions are too weak to be observed. Mid-way, in terms of wavenumber, between the members of each pair of Lamb dips is another dip, commonly known as a crossover resonance, or centre dip, and marked with an asterisk in Figure 8.57. Each crossover resonance has an intensity which is the geometric mean of the pair. In addition there is another set of crossover resonances consisting of those marked with a spot in the figure. This set is much weaker than the first one.

The first set of crossover resonances is caused by three-level double resonances of the type illustrated in Figure 8.58(a) and first discussed in 1966.[242] A requirement for a crossover resonance of this kind to occur is that the two transitions associated with it involve a common level. This is clearly the case for the asterisked crossover resonances in Figure 8.57 for which each pair of transitions involves a common *lower* level. It is also necessary that the two

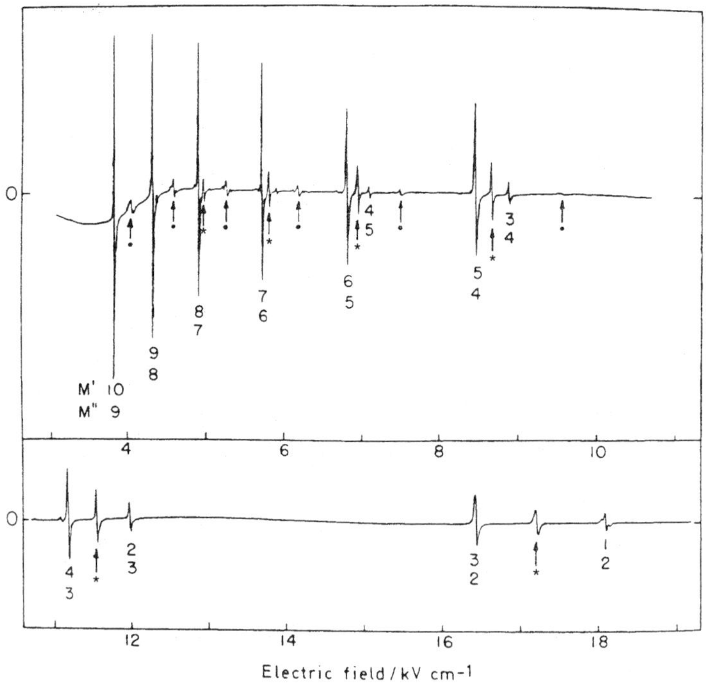

Figure 8.57 Laser Stark spectrum of NH_3 showing Lamb dips in components of an $R(9)$ line of the 4_0^1 vibrational transition
[Reproduced, with permission, from Johns, J. W. C., McKellar, A. R. W., Oka, T. and Römheld, M. (1975). *J. Chem. Phys.*, **62**, 1488]

transitions, with wavenumbers $\tilde{\nu}_1$ and $\tilde{\nu}_2$ in Figure 8.58(a), are not separated by much more than twice the Doppler line width; in other words, their Dopper profiles should overlap.

Three-level double resonance can be understood by referring to Figure 3.8. The source, now, is a laser of wavenumber $\tilde{\nu}_L$ and associated oscillating electric field E_+, from the source to the reflector R, and E_- after reflection from R. To an outside observer the wavenumber of both E_+ and E_- is $\tilde{\nu}_L$ but, to molecules with a velocity component v_a along the direction of the radiation, the wavenumber appears to be $\tilde{\nu}_L[1-(v_a/c)]$, for radiation travelling from the source to R, and $\tilde{\nu}_L[1+(v_a/c)]$ for radiation reflected from R. This is due to the Doppler effect.

In a simple two-level system, where the separation of the levels corresponds to a wavenumber $\tilde{\nu}_0$, we have seen in Section 3.4 that a Lamb dip is observed only when $\tilde{\nu}_L = \tilde{\nu}_0$ and $v_a = 0$. For a three-level system, as in Figure 8.58(a), Lamb dips are observed not only for $\tilde{\nu}_L = \tilde{\nu}_1$, $v_a = 0$ and $\tilde{\nu}_L = \tilde{\nu}_2$, $v_a = 0$ but

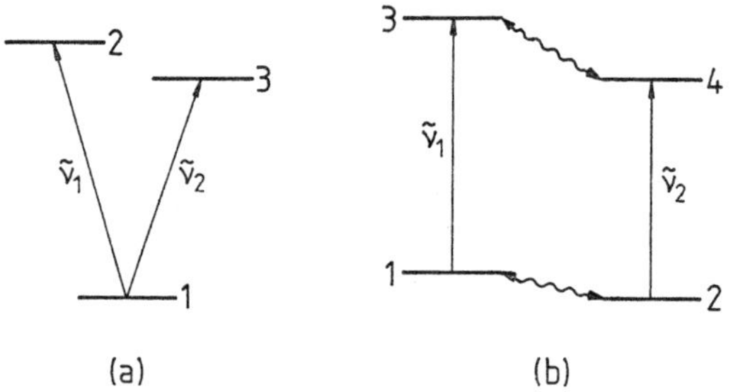

Figure 8.58 Crossover resonance can be caused by (a) a three-level double resonance or (b) a four-level system in which transfer from level 3 to 4, or 1 to 2, may occur by collision

also for another set of molecules with a velocity component v_a such that

$$\tilde{\nu}_L\left(1+\frac{v_a}{c}\right)=\tilde{\nu}_1 \text{ and } \tilde{\nu}_L\left(1-\frac{v_a}{c}\right)=\tilde{\nu}_2 \quad (8.41)$$

or

$$\tilde{\nu}_L\left(1+\frac{v_a}{c}\right)=\tilde{\nu}_2 \text{ and } \tilde{\nu}_L\left(1-\frac{v_a}{c}\right)=\tilde{\nu}_1 \quad (8.42)$$

From this, it is clear that radiation travelling in opposite directions induces different transitions in molecules with the same velocity component v_a. The resulting crossover resonances are observed at a wavenumber of $\frac{1}{2}(\tilde{\nu}_1+\tilde{\nu}_2)$ and the groups of molecules concerned have velocity components given by

$$v_a = \pm c(\tilde{\nu}_1-\tilde{\nu}_2)/2\tilde{\nu}_L \quad (8.43)$$

The set of Lamb dips indicated by spots in Figure 8.57 is due to a four-level system of the type in Figure 8.58(b) and in which molecules transfer from state 3 to 4 or 1 to 2 by collisional processes. The theory has been given in the paper from which Figure 8.57 is taken. In the case of the crossover resonance between the normal Lamb dips with $M_J''=9$ and 8 in Figure 8.57, it clearly cannot involve a three-level process, since there is no common level, but involves changes of M_J by ±1 in the upper and lower states. All the other collisional crossover resonances in the figure also involve $\Delta M_J=\pm1$. Many collisional crossover resonances have been observed with much higher values of $|\Delta M_J|$.

The sharpness of the collisional crossover resonances indicates that there is very little change of molecular velocity on collision.

There are cases where the collisional crossover resonances are midway between two normal dips in terms of wavenumber, or frequency, but not in terms of electric field. This happens when there is a second-order Stark effect, for example in a symmetric rotor such as NH_3.

One important use of crossover resonances, whether of the three- or four-level type, is in indicating accurately the position of one member of a pair of such resonances which is too weak to be observed. There are examples, involving three-level resonances, in Figure 8.57. Four-level crossover resonances find an important application in the study of collisional processes.

8.2.8 Spectroscopy with diode and colour centre lasers

Although there is a large variety of diode lasers, covering a wavelength range of about 0.5–30 μm (see Section 8.1.8.5), those that are useful for infrared spectroscopy cover a smaller range of about 3.3–30 μm (3000–330 cm^{-1}). Diode lasers are tunable, although the range covered by a particular diode laser is small, and have a typical line width of only about 15 MHz (0.0005 cm^{-1}). These properties make them ideal sources for obtaining very high resolution, mid-infrared absorption spectra.

Colour centre lasers (see Section 8.1.8.5) can cover the range 0.82–3.5 μm (12 200–2860 cm^{-1}) and are complementary to diode lasers in their ability to go above 3000 cm^{-1}. This region is particularly important for the observation of hydrogen stretching fundamental vibrations. Their use will be mentioned briefly here but at greater length in Section 8.2.11 in conjunction with the technique of optothermal spectroscopy.

Laser diode and colour centre laser spectroscopy can be used to investigate stable molecules but the technique has proved particularly powerful in the investigation of unstable species such as radicals, ions and van der Waals, or hydrogen bonded, clusters.

When molecules are entrained in a supersonic jet (see Section 3.4.3) their vibrational and rotational temperatures are reduced to typical values of 100 and 10 K, respectively. These temperatures are favourable for the formation of dimers, trimers, etc., of the molecule seeded into the jet or between that molecule and the atoms of the carrier gas. However, the concentrations of these cluster molecules, held together only by weak van der Waals or hydrogen bonding forces, are low and, in addition, transition probabilities in the infrared region are small compared with the visible and ultraviolet regions.

The low rotational temperature in the jet concentrates the population into rotational levels with low quantum numbers and consequently increases the sensitivity of detection. Nevertheless, in observing diode laser infrared absorption spectra of species present in low concentration, and even for stable monomer species, it is usually necessary to use as long an absorption path as possible.

The increased absorption path may be achieved by the nozzle of the jet being in the form of a slit rather than a pinhole. The width of the slit jet is about 50–120 μm, similar to the diameter of a typical pinhole nozzle. The length of the slit is typically about 5 cm, which results in a large gas throughput requiring a high capacity pumping system. The problem may be alleviated by pulsing the slit nozzle.

To increase the absorption pathlength still further, the diode laser beam is multiply passed, backwards and forwards, typically about 12 times, in a direction perpendicular to that of the beam, by a set of multiple reflection mirrors which may be in a White (see Section 3.7) or other configuration.

Even without any skimming away of the divergent edges of the jet, interrogation of the molecules by a laser beam perpendicular to the jet results in spectra whose rotational line width is sub-Doppler since the majority of molecules are travelling in a direction perpendicular to the laser beam. There are three contributions to the Doppler broadening which remain. The first is due to the molecules near to the edges of the jet travelling in a direction which is not exactly parallel to those nearer the centre, and the second is due to the multiple passes of the laser beam being not quite perpendicular to the jet. However, both of these contributions to Doppler broadening are less important than that due to the non-zero translational temperature in the jet.

The diode laser absorption spectrum of nickel tetracarbonyl, $Ni(CO)_4$, has been obtained[243] in a supersonic jet. The molecule is tetrahedral with T_d symmetry. The only vibrations which are infrared active are those of T_2 symmetry (see Table A.41 in the Appendix). The centre of the 5_0^1 band, where ν_5 is a T_2 C—O stretching mode, lies at $2061.314\,cm^{-1}$, a region well suited to diode laser investigation. For the monomer species a pinhole nozzle was adequate but the laser beam was multipassed 11 times across the jet.

Figure 5.55 shows how the three components T^-, T^0 and T^+ of a triply degenerate T_2 state are split by first order Coriolis interaction. The rotational selection rules for a T_2–A_1 transition are $\Delta J = +1$ for the T^-–A_1, $\Delta J = 0$ for the T^0–A_1 amd $\Delta J = -1$ for the T^+–A_1 component. In the case of ν_5 of $Ni(CO)_4$ the Coriolis interaction is negligibly small and the 5_0^1 band shows simple *P*-, *Q*- and *R*-branch structure similar to that of a Π–Σ^+ infrared band of a linear molecule.

Figure 8.59 shows the *P*-, *Q*- and *R*-structure near the 5_0^1 band centre. The *Q* branch is not resolved but it is noticeable that the *R*- and *P*-branch lines with $J'' = 1$, 2 and 5 are missing and that the intensities of the lines that are present show an unusual variation with *J*. These are a consequence of nuclear spin statistics. For $Ni(CO)_4$, with $I = 0$ for both the ^{12}C and ^{16}O nuclei, the rovibrational sub-levels, of each *J* level, with only A_1 or A_2 symmetry are allowed. For $J'' = 1$, 2 or 5 there is no such sub-level whereas, for $J'' = 7$, 8, 9, 10, 11 and 12, for example, there are 1, 1, 2, 2, 1 and 2 sub-levels, respectively. These are reflected in the intensities in Figure 8.59.

In a supersonic slit jet with HF seeded into argon, and 12 traversals of the jet, a laser diode absorption spectrum has been obtained[244] which shows rotationally resolved spectra, in the region of the H—F stretching vibration of the parent molecule, of Ar_nHF with $n = 1$–4. For free HF the band centre is at $3961.4229\,cm^{-1}$. On complexation the band centre is expected to shift to low wavenumber

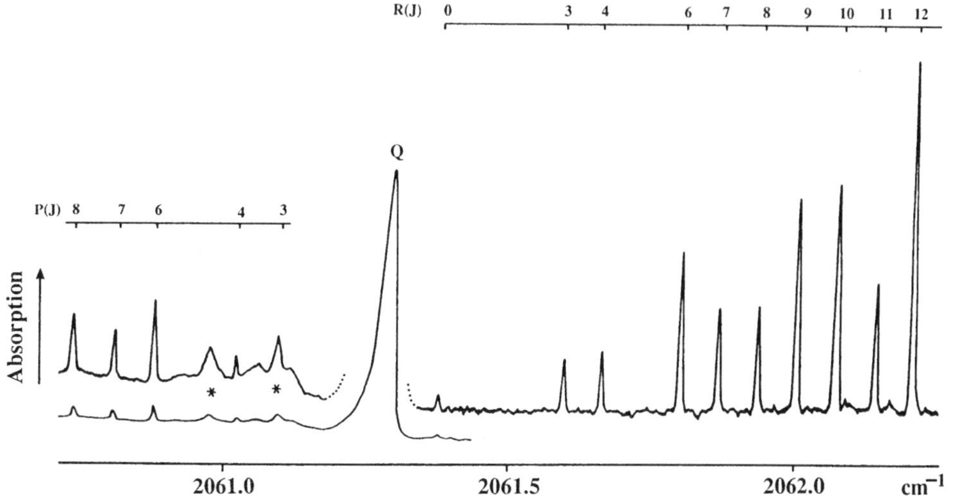

Figure 8.59 Region near the centre of the 5_0^1 band of $Ni(CO)_4$ in the diode laser absorption spectrum. Peaks marked with asterisks are Q branches of hot bands
[Reproduced, with permission, from Davies, P. B., Martin, N. A., Nunes, M. D., Pape, D. A. and Russell, D. K. (1990). *J. Chem. Phys.*, **93**, 1576]

because the weak van der Waals interaction weakens the H—F bond and, consequently, lowers the force constant. The shifts observed for Ar_nHF are 9.654, 14.827, 19.260, 19.697 cm^{-1} for $n = 1$, 2, 3, 4, respectively. There is an interesting comparison with the shift of 42.4 cm^{-1} for HF in a solid argon matrix in which as many argon atoms as possible surround the HF.

In general, as more argon atoms are added the H—F bond weakens further but these shifts show very little weakening from $n = 3$ to 4. This is consistent with the structures determined from both the laser diode infrared and the microwave[245] spectra. For $n = 1$ the structure is linear, for $n = 2$ it is T-shaped with the hydrogen atom nearest to the Ar—Ar line and, for $n = 3$, the molecule is an oblate symmetric rotor with the HF moiety perpendicular to a triangle of argon atoms towards which the hydrogen atom is pointing. For each step, as n goes from 1 to 3, there is one more argon atom bonding directly to HF but, for $n = 4$, the fourth argon atom is attached to the triangle of argon atoms in Ar_3HF on the opposite side from the HF. As it is not bonded directly to HF there is only a very small further shift of the band origin from $n = 3$ to 4.

Figure 8.60 shows a small part of the overlapping spectra of Ar_3HF and Ar_4HF. Both are symmetric rotors and these are parallel bands. Since the rotational constants change very little from $v = 0$ to $v = 1$ they show simple P-, Q- and R-branch structure. Figure 8.60 shows the Q branch and part of the P branch of Ar_3HF and the weaker Q branch of Ar_4HF.

Experimental evidence has proved that the structure of the carbon dioxide dimer, $(CO_2)_2$, is a 'slipped parallel' one in which the molecules are parallel to each other but one is displaced in a direction parallel to the other.[246] The infrared laser diode absorption spectrum of the trimer, $(CO_2)_3$, has been observed[247] in a supersonic slit jet and it has been shown that the structure is that of a planar oblate symmetric rotor with C_{3h} symmetry as shown in Figure 8.61.

Part of the spectrum of $(CO_2)_3$, in the region of the asymmetric stretching vibration, v_3, of CO_2, is shown in Figure 8.62, where it is overlapped by the spectrum of $(CO_2)_2$. Comparing the calculated spectrum of the dimer, a type A band of an asymmetric rotor, at the bottom

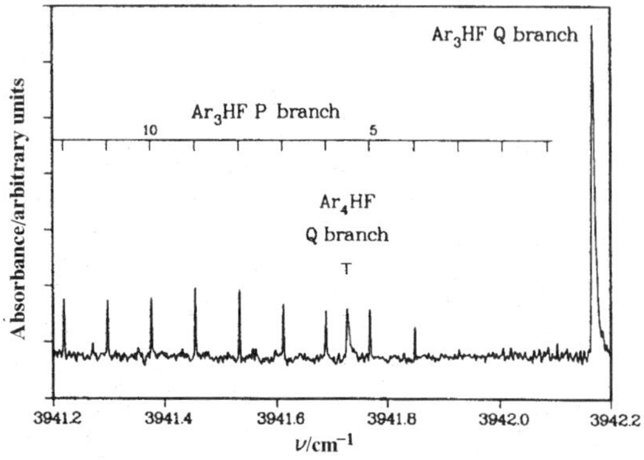

Figure 8.60 The Q branches of the HF $v = 1$–0 transition in Ar_nHF, with $n = 3$ and 4, and part of the P branch for $n = 3$ observed by laser diode infrared absorption. For this spectrum neon was used as the carrier gas which was seeded with an HF–Ar mixture
[Reproduced, with permission, from McIlroy, A., Lascola, R., Lovejoy, C. M. and Nesbitt, D. J. (1991). *J. Phys. Chem.*, **95**, 2636]

of the figure with the observed shows that there are many extra observed lines. The calculated spectrum for the trimer, based on the rotational constants determined from the analysis, shows that this species is responsible for the extra lines. The calculated intensities for rotational transitions in the band show that the symmetry of the trimer is C_{3h}. A value of $R_{CC} = 4.0376$ Å (see Figure 8.61) has been obtained from the rotational constants, and $\beta = 33.8°$ from the vibrational shift of 2.5755 cm^{-1} to high wavenumber from the value of $\tilde{\nu}_3$ for the monomer. The fact that no vibrational tunnelling splittings have been observed shows that the trimer is rigidly planar.

As with many other cases of the experimental observation of the spectrum of a cluster molecule, we can never be sure that the isomer which is observed is the only one. Others may exist but their spectra may not be observed because they are either outside the region of observation or are too broad. In such cases theory may be a reliable guide as to how many

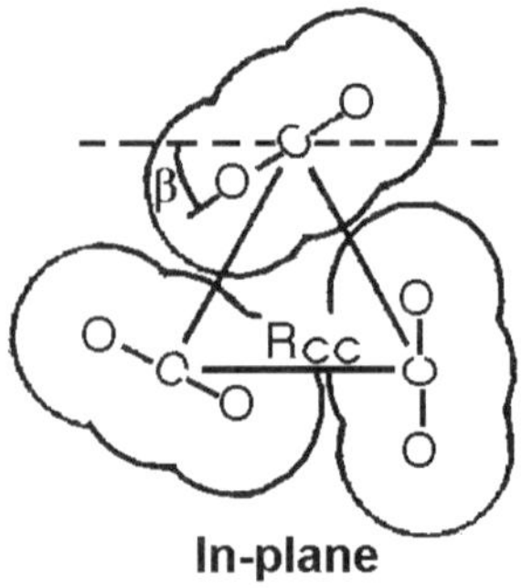

Figure 8.61 The C_{3h} structure of the carbon dioxide trimer
[Reproduced, with permission, from Weida, M. J., Sperhac, J. M. and Nesbitt, D. J. (1995), *J. Chem. Phys.*, **103**, 7685]

isomers there are and which of these is at the energy minimum in the complex potential energy surface connecting them.

In the case of $(CO_2)_3$ calculations suggest there is also a non-planar asymmetric rotor structure of comparable energy. Some extra lines observed in the spectrum may be due to this second form.

The weakly bound complex ArCO is T-shaped and a prolate near-symmetric rotor. The diode laser infrared absorption spectrum (and also spectra using other techniques) has been observed,[248] using a slit jet, in the C—O stretching region. This is a perpendicular band with $\Delta K_a = \pm 1$. Previous sub-band observations have been extended to include those with $K_a = 4$–3, 3–4, 5–4 and 6–5.

Molecules held together by such weak forces commonly show signs of non-rigidity. For example, there is a tendency for the CO moiety in ArCO to rotate in the plane of the complex in a so-called windmill motion. When the bending vibration, ν_2, is excited there is considerable interaction with this motion. However, the high K_a levels involve a 'propeller' motion about the CO axis and are not seriously perturbed as this motion cannot easily couple to other internal motions.

The radicals CH_3 and SiH_3 can be generated in an electrical discharge in gas mixtures containing CH_4 or SiH_4, respectively. Because

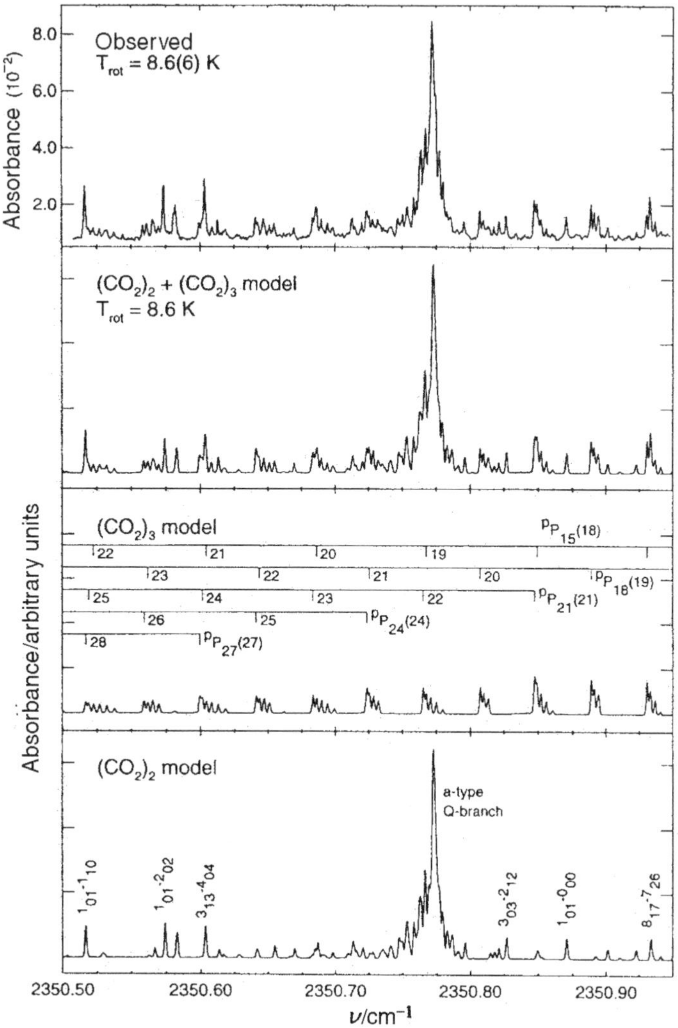

Figure 8.62 Part of the observed laser diode absorption spectrum of $(CO_2)_2$ and $(CO_2)_3$ in the region of the asymmetric stretching vibration of CO_2 compared with the calculated contributions from each species [Reproduced, with permission, from Weida, M. J., Sperhac, J. M. and Nesbitt, D. J. (1995). *J. Chem. Phys.*, **103**, 7685]

they each contain an unpaired electron Zeeman modulation[249] can be used for selective detection by diode laser absorption spectroscopy. Using a diode laser in the 14 μm (700 cm^{-1}) region, transitions involving the out-of-plane bending vibration, ν_2, of CH_3[250] and the inversion vibration, ν_2, of SiH_3[251] were obtained using a discharge tube 120 cm long with multiple reflection capability.

The difficulty of determining, from the electronic spectrum, whether CH_3 is planar in its ground electronic state has been discussed in Section 6.3.5.2(i), where it was stated that the observation of several out-of-plane bending (ν_2) levels was required to settle this question. The CH_3 produced in the discharge tube is at ambient temperature and rotationally resolved 2_1^2, 2_2^3 and 2_3^4 hot bands, in addition to 2_0^1, were analysed. The 2_0^1 band is at 606.4531 cm^{-1} and the vibrational levels are strongly divergent, as v_2 increases. The levels have been fitted to a potential which has a large quartic contribution but shows a single minimum—the molecule is planar.

The SiH_3 radical is pyramidal. The potential for ν_2 resembles that for NH_3 in Figure 5.68. The diode laser absorption spectrum[251] shows two components, 1^-–0^+ (or 2_0^3) and 1^+–0^- (or 2_1^2), of the ν_2 fundamental at 727.9438 and 721.0486 cm^{-1}, respectively. The splitting is due to tunnelling through the inversion barrier which has a height of 1868 cm^{-1}, slightly lower than that of 2020 cm^{-1} for NH_3.

Ions, in addition to radicals, are produced in electrical discharges, but both species are produced in very low concentrations. For example, ions may be present as only a few parts per million of neutral species. The absorption spectra of the neutrals are likely to mask that of the ion and some type of modulation technique must be used to detect the uncontaminated spectrum of the ion. Whereas Zeeman modulation is used for radical spectra, velocity modulation[252] has been used for ions.

In a discharge, positive ions drift towards the cathode, resulting in a Doppler shift of the absorption wavenumbers. Although this shift is extremely small, comparable to the Doppler line width, it indicates that the absorbing species is charged. When the polarity of the discharge is reversed, at a frequency of several kilohertz, the absorption is shifted alternately to high and low wavenumber. What the ion sees is an apparent modulation of the frequency of the laser. In this way, lock-in detection of the signal gives the spectrum of the ion, almost free from those of neutral species.

The technique of velocity modulation was first used[252] to observe *R*-branch lines in the ν_1 vibration–rotation spectrum of HCO^+, an important interstellar molecule (see Section 4.7), in the region 3091–3141 cm^{-1}. The ion was produced in a liquid nitrogen cooled discharge in H_2 and CO. The wavenumber region is too high for a diode laser and a colour centre laser was used. However, the lower limit of the range of the colour centre laser precluded the observation of the *P*-branch lines of HCO^+ in this experiment.

In 1985 an important chance discovery led to the observation of laser diode and colour centre laser spectra of anionic species which had previously proved elusive. It was found that a layer of sputtered metal on the wall of the discharge tube resulted in large anion signals. The velocity modulation technique was able not only to suppress the absorption of neutral species, but also to distinguish anionic and cationic species.[253] Because these species travel in opposite directions in the discharge tube their demodulated first derivative line shapes have opposite symmetry. This is shown in Figure 8.63, in which the lines due to OH^- and H_3O^+, both produced in a discharge in a gaseous mixture containing H_2 and O_2, have opposite symmetry.

Using this technique, infrared absorption spectra of anions such as SH^-, NH_2^-, N_3^-, NCO^- and C_2H^- have been recorded.[253]

A diode laser absorption spectrum of Si_2^-, produced in an air-cooled ac discharge in a mixture of hydrogen and silane (SiH_4), has been observed[254] using velocity modulation and a 1 m pathlength. Unusually, in the infrared region, this is an electronic spectrum, the transition being between the $X^2\Sigma_g^+$ ground state and the very low-lying $A^2\Pi_u$ state. The ground state corresponds to Hund's case (b) and the excited state to Hund's case (a) coupling. The $^2\Pi$ state is inverted with the $^2\Pi_{u(\frac{3}{2})}$ component below $^2\Pi_{u(\frac{1}{2})}$. Figure 8.64 shows the potential curves for the X and A states including the doubling, by spin–orbit coupling, of the vibrational levels of the A state.

The rotationally resolved vibronic transitions observed were the 1–0 band, in the 670–810 cm^{-1} region, and the 2–0 band, in the 1200–1340 cm^{-1} region. Typical of the small range covered by a diode laser, 12 lasers were required to cover the range of the 1–0 band and six to cover the 2–0 band.

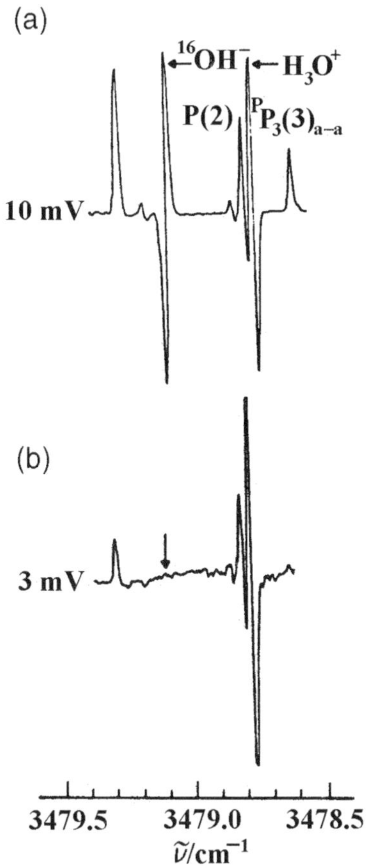

Figure 8.63 (a) Colour centre laser spectrum showing rovibrational transitions of OH^- and H_3O^+, in the region of an O—H stretching vibration, observed with copper metal sputtered onto the discharge tube wall. The opposite symmetry of the first derivative lines for the anion and cation is clear. (b) Spectrum in the same region but with a clean tube wall
[Reproduced, with permission, from Owrutsky, J., Rosenbaum, N., Tack, L., Gruebele, M., Polak, M. and Saykally, R. J. (1987). *Philos. Trans. Roy. Soc.*, **A234**, 97]

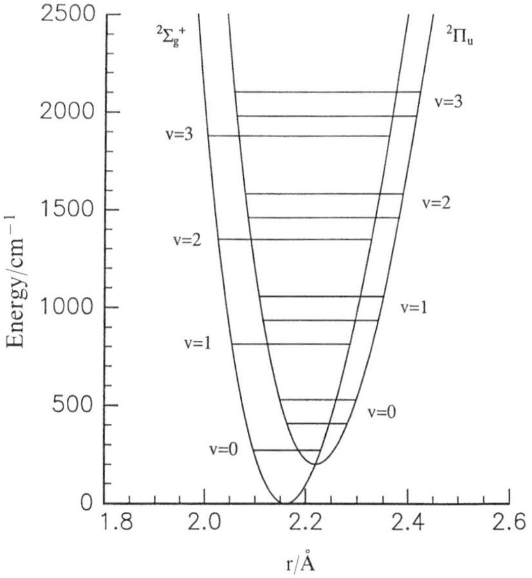

Figure 8.64 Potential energy curves and vibronic energy levels for the $X^2\Sigma_g^+$ and $A^2\Pi_u$ states of Si_2^- [Reproduced, with permission, from Liu, Z. and Davies, P. B. (1996). *J. Chem. Phys.*, **105**, 3443]

The method of obtaining laser diode spectra of ions and radicals in an electrical discharge has been combined with the added advantage of supersonic jet cooling by using a corona discharge slit expansion.[255] An electrical discharge is created in the carrier gas, seeded with the molecule of interest, immediately before the slit nozzle. The discharge is between a thin tungsten rod anode, just upstream of the slit nozzle, and a copper block cathode on to which the slit plates are secured. With a diode laser source rotational transitions in the 2_0^1 band of $^2H_3^+$ have been observed but spectra of other ions were more readily observed with a difference frequency laser (see Section 8.2.9).

A characteristic of the corona discharge slit expansion is a considerably higher temperature of the species produced in the discharge. The translational temperature of the $^2H_3^+$ was about 80 K resulting in a rotational line width of 140 MHz. The populations of the lower rotational levels gave a rotational temperature of about 77 K. However, as is commonly observed in jet expansions, higher energy rotational level populations deviate from a Boltzmann distribution and indicate a higher rotational temperature. The reason for this is the larger gap between levels which makes cooling by collisional processes less efficient.

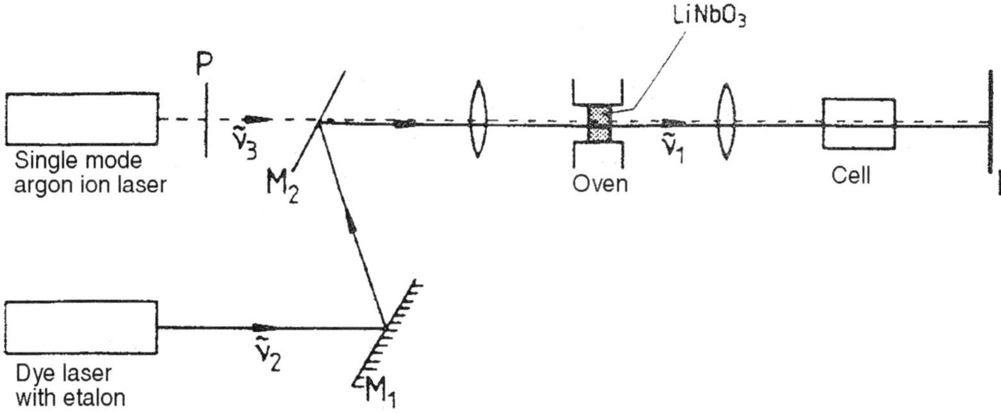

Figure 8.65 Design of a laser difference–frequency infrared spectrometer

8.2.9 Spectroscopy with a difference–frequency laser system

We have encountered previously in this chapter several examples of the change of frequency of laser radiation by nonlinear effects. In Section 8.1.7, for example, we saw how a crystalline material such as potassium dihydrogen phosphate or lithium niobate can produce radiation of wavenumber $2\tilde{\nu}$ by harmonic generation from the incident laser radiation $\tilde{\nu}$ by a three-wave mixing process in the crystal.

If two laser beams of wavenumber $\tilde{\nu}_2$ and $\tilde{\nu}_3$ ($\tilde{\nu}_3 > \tilde{\nu}_2$) are collinear and focused into a suitable non-linear crystal, three-wave mixing occurs and polarization conditions can be arranged so that only radiation of wavenumber $\tilde{\nu}_1$ emerges, where

$$\tilde{\nu}_1 = \tilde{\nu}_3 - \tilde{\nu}_2 \qquad (8.44)$$

If the $\tilde{\nu}_3$ radiation is from a fixed frequency laser and $\tilde{\nu}_2$ from a tunable laser, and $\tilde{\nu}_1$ lies in the infrared region, this system is capable of acting as a source of tunable infrared radiation.[256]

Figure 8.65 shows schematically the design of an infrared spectrometer using the difference–frequency technique to provide a source of tunable, narrow line width infrared radiation.

Radiation $\tilde{\nu}_2$ from a tunable single mode dye laser, with an intracavity etalon to reduce the line width, is combined collinearly, by a dichroic mirror M_2, with radiation $\tilde{\nu}_3$ from a single mode argon ion laser whose plane of polarization is rotated through 90° by a half-wave plate P so that $\tilde{\nu}_2$ and $\tilde{\nu}_3$ are polarized orthogonally. The beam is then focused into a suitably oriented lithium niobate crystal in a temperature-controlled oven. As discussed in Section 8.1.7, for efficient three-wave mixing between $\tilde{\nu}_1$, $\tilde{\nu}_2$ and $\tilde{\nu}_3$ the waves must be phase matched. This can be achieved only if the crystal is birefringent. The birefringence is temperature dependent so the crystsal has to be temperature tuned by the surrounding oven. The infrared radiation of wavenumber $\tilde{\nu}_1$ passes through the sample cell to the detector D.

The tuning range depends on the properties of the lithium niobate crystal and is 2.2–4.2 μm (4540–2380 cm^{-1}) with a line width of about 2 MHz. The power of the infrared radiation is of the order of 1 μW with powers of about 25–50 mW for $\tilde{\nu}_2$ and 50–100 mW for $\tilde{\nu}_3$.

The infrared line width, which determines the resolution, and the wavenumber range which can be covered by continuous tuning, are limited by the line width and tunable range attainable by the etalon in the dye laser.

The resolution of which this technique is capable is sub-Doppler, the Doppler line width at half maximum intensity being about 0.01 cm^{-1} in this spectral region, but the

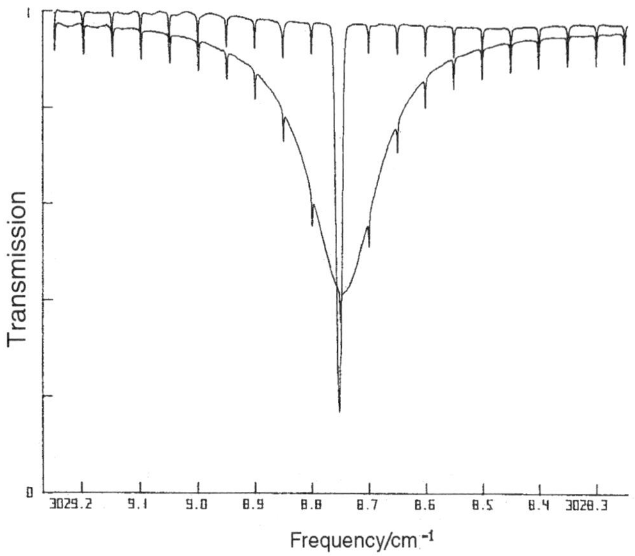

Figure 8.66 The $R(0)$ transition of the 3_0^1 vibrational band of the CH_4 molecule obtained with Doppler-limited resolution using a difference–frequency infrared spectrometer. The narrow line shows the true Doppler line shape at a pressure of 0.5 Torr, while the broad line shows the effects of pressure broadening with 1 Torr of CH_4 and 760 Torr of air. Calibration markers are superimposed on both spectra
[Reproduced, with permission, from Pine, A. S. (1976). *J. Opt. Soc. Am.*, **66**, 97]

power of about 1 μW is insufficient for saturation to be achieved and Lamb dips have not been observed.

The 3_0^1 band of CH_4, where ν_3 is an infrared-active t_2 vibration, has been observed in the 3020 cm^{-1} region with Doppler-limited resolution using this technique. Some 440 rovibrational transitions in the band have been fitted[257] to a theoretical model with a standard deviation of 0.0056 cm^{-1}. Figure 8.66 shows the $R(0)$ transition and demonstrates the true Doppler line shape and the effect of pressure broadening.

A homonuclear diatomic molecule has no dipole moment and, therefore, the vibration–rotation spectrum is forbidden by electric dipole selection rules. However, such a molecule has, in general, a non-zero quadrupole moment which changes when the molecule vibrates and shows a quadrupole vibration–rotation spectrum. Because the intensity of such a spectrum is of the order of 10^{-9} of that of a dipole allowed spectrum, it is very difficult to observe, requiring an extremely long absorbing path.

The rotational selection rule is

$$\Delta J = 0, \pm 2$$

resulting in an O, Q and S branch.

Such a spectrum of $^1H_2$ was first observed in 1949.[258] With absorption path lengths of up to 55 km, and the gas at a pressure of 1 atm, the 2–0 and 3–0 overtone bands were detected. These bands have an additional weakness characteristic of overtone transitions.

Using an argon ion–dye difference frequency laser system, 11 O-, Q- or S-branch lines were measured[259] in the 1–0 band of $^2H_2$ at 1 atm pressure. With a multiple reflection cell a path length of up to 80 m was obtained. The rotational constants agree well with those obtained, more readily, from the Raman spectrum.

The triangular cation H_3^+ has D_{3h} symmetry and the degenerate e' vibration ν_2, at 2521.564 cm^{-1}, is infrared active. Fifteen rovibrational transitions were observed[260] in this band with a difference frequency laser. This was

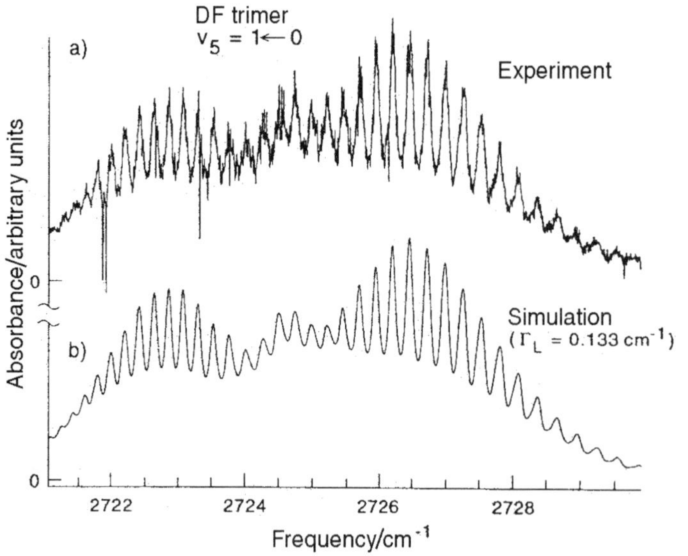

Figure 8.67 (a) The e' ^{2}H—F stretching fundamental of $(^2HF)_3$ and (b) a computer simulation of the rotational structure
[Reproduced, with permission, from Suhm, M. A., Farrell, J. T. Jr, Ashworth, S. A. and Nesbitt, D. J. (1993). *J. Chem. Phys.*, **98**, 5985]

the first spectroscopic observation of this species and rotational constants were obtained for both vibrational states.

Hydrogen fluoride forms a hydrogen bonded trimer which was predicted to be cyclic and planar. Investigation in the region of the e' H—F stretching vibration of $(^1HF)_3$ has revealed only a broad absorption with no rotational structure. However, the corresponding spectrum of the $(^2HF)_3$ isotopomer, investigated with an argon ion–dye difference frequency laser system[261] does show rotational structure.

Figure 8.67(a) shows the degenerate, e', ^{2}H—F stretching fundamental band of $(^2HF)_3$ formed in a slit jet. The band is centred at 2724.595 cm^{-1}, compared with 2906.67 cm^{-1} for the monomer. The computer simulation in Figure 8.64(b) shows that it is a perpendicular band of an oblate symmetric rotor, at a rotational temperature of 10.3 K, and shows a line width at half maximum intensity of 0.133 cm^{-1}. The molecule has a planar D_{3h} symmetry.

There is added interest in this spectrum concerning very much sharper structure, with a line width of about 0.0013 cm^{-1}, which is superimposed on the coarser structure and can be seen near the band centre in Figure 8.67(a). It has been suggested[261] that this is caused by an unusual case of intramolecular vibrational redistribution (IVR).

The process of IVR[262] is illustrated in Figure 8.68. On the left is what would be, in the absence of IVR, a single rovibrational level, v, J, in an anharmonic vibrational state v. A transition to this state is assumed to be allowed with a certain oscillator strength. It is known as a bright state. In a polyatomic molecule there will be many rotational levels belonging to other vibrational states v'. The larger is the molecule, the greater will be the density of these nearby states. These states are shown in the centre of the figure and, because transitions to them are assumed to be forbidden, they are known as dark states. Included in the descriptions of the bright and dark states are vibrational anharmonicity, centrifugal distortion and vibration–rotation interactions which result in the vibrational dependence of the rotational constants.

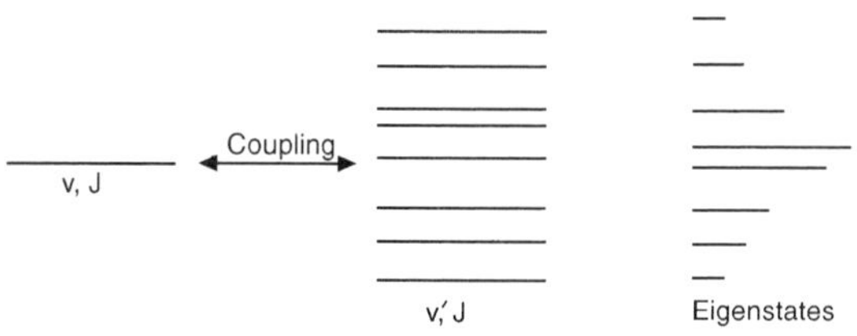

Figure 8.68 On the left is a single rotational level, υ, J, associated with an anharmonic vibrational state—a 'bright' state. In the centre is a stack of nearby rotational levels υ', J, associated with other vibrational states—the 'dark' states. On the right are the actual molecular eigenstates resulting from the coupling of the υ, J to the υ', J states. The oscillator strength of any transition to the υ, J state is now distributed among transitions to the eigenstates

IVR is caused by coupling between the bright and dark states, the latter also being known as bath states, the bright state being immersed in a bath of dark states. This coupling is due to additional rotational and vibrational interactions, such as Coriolis and Fermi interactions, which have not been taken into account in our descriptions of the bright and dark states. Not all the nearby bath states will undergo this coupling which is restricted, for example, by symmetry. The molecular eigenstates resulting from the coupling are shown on the right of the figure: the length of each line indicates the proportion of the oscillator strength which the eigenstate now carries.

One result of the IVR is that the oscillator strength of a transition to the bright state, υ, J, is now distributed among a number of molecular eigenstates. If the density of bath states is large the density of eigenstates may also be large in which case the resulting single line transition to the state υ, J which might have been expected may be broadened to an unresolvable continuum. However, if the density of bath states is not large, the transitions may be resolvable.

So far, we have considered effects of IVR when observations are in the frequency domain. However, another effect of IVR is that there is an energy flow, backwards and forwards, between the zero-order bright state and the molecular eigenstates. The width of the distribution of intensity among the eigenstates is a measure of the time scale of the energy flow.

The molecular eigenstates may be coherently excited by a pulse which is short compared with the time scale of energy flow and which has a wavenumber spread sufficient to cover all the eigenstates. Under these circumstances, the emission (usually fluorescence) from the eigenstates, measured as a function of time, shows a periodic increase and decrease of intensity known as quantum beats which are superimposed on the overall decay of the emission. The observation of quantum beats, in the time domain, complements the observation of the spectrum in the frequency domain. Quantum beats will be discussed further in Section 8.2.26.

In the case of the spectrum of $(^2HF)_3$ in Figure 8.67(a), the suggestion[261] that the very sharp structure superimposed on the broader structure which was simulated in Figure 8.67(b) is due to IVR involves a very unusual source of bath states. It is proposed that, since they are well above the energy required for opening of the $(^2HF)_3$ ring, they belong to floppy, chain-like conformations in which just one of the hydrogen bonds has been broken.

From the line width of $0.113\,\text{cm}^{-1}$ used for the computer simulation in Figure 8.67(b) it was estimated that the time scale of the IVR process is about 40 ps.

Using the corona discharge combined with a slit jet, as described in Section 8.2.8, an argon ion–dye difference frequency laser was used to extend the range of cationic species detected[254] to include H_3^+, N_2H^+, H_3O^+ and ArH^+.

Considerably higher densities of radicals or cations have been generated, also with a discharge source but pulsed and with a slit nozzle design which confines the discharge to a reigon upstream of the orifice.[263] This results in more efficient cooling to a rotational temperature of 30 K or less. With an absorption path, using multiple transversals, of 80 cm sub-Doppler spectra of OH, H_3O^+ and N_2H^+

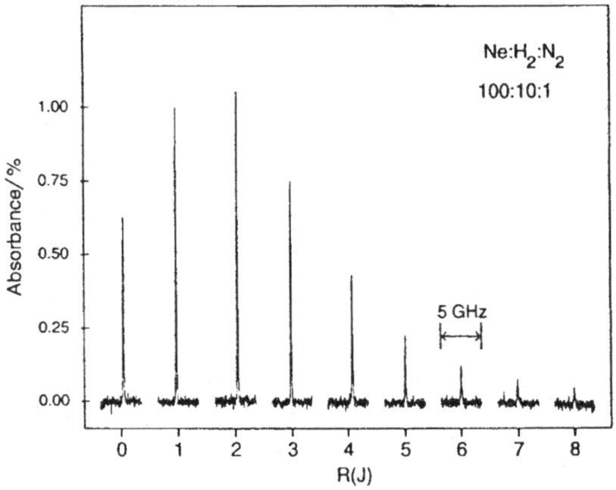

Figure 8.69 *R*-branch transitions in the $v_{NN} = 1$–0 band of N_2H^+ observed with a difference frequency laser
[Reproduced, with permission, from Anderson, D. T., Davis, S., Zwier, T. S. and Nesbitt, D. J. (1996). *Chem. Phys. Lett.*, **258**, 207]

were obtained with an argon ion–dye difference frequency laser.

Figure 8.69 shows nine *R*-branch transitions in the $v = 1$–0 band of the N—N stretching vibration of the linear N_2H^+ obtained[262] from a discharge in N_2 and H_2 with neon carrier gas. Each line was observed in a separate laser scan and the intensity distribution indicates a rotational temperature of 25 K.

The tuning range of 2.2–4.2 μm achieved with an argon ion and ring dye laser system has been extended[264] to cover the range 1.2–2.2 μm (8300–4540 cm^{-1}) by using a difference frequency laser system consisting of a CW Nd^{3+}:YAG laser, diode laser pumped and in a ring configuration, and a ring dye laser with a lithium niobate non-linear crystal. Significantly more power has been achieved than in the 2.2–4.2 μm range and this, together with the use of a 4 cm slit jet and up to 16 multipasses, has made it possible to observe weak first overtone transitions in van de Waals and hydrogen bonded complexes.

In ArHF, formed in an Ar–HF supersonic expansion, four bands have been observed[265] in the 7720–7810 cm^{-1} region with a Nd^{3+}:dye difference frequency laser system. In this region the first overtone of the HF moiety is expected.

The ArHF molecule is linear but, like ArHCl which was discussed in Section 4.10.1, it is very non-rigid with large amplitude Ar···HF stretching and degenerate bending vibrations. Because of this it has been suggested[266] that the label $(v_{HF}bKn)$ be used for vibrational levels to replace the conventional ones for a rigid linear triatomic molecule. Here, v_{HF} is the quantum number for H—F stretching, b for bending, n for van der Waals stretching and K is the quantum number associated with the projection of the total angular momentum along the internuclear axis.

The three bands observed[265] involve transitions from the zero-point level to the (2000), (2100) and (2110) levels. Although the transition to the (2000) level is much the strongest it is almost 100 times weaker than that to (1000).

Transitions to the (2000) and (2100) levels show simple *P*- and *R*-branch structure typical of a Σ–Σ transition of a linear, or nearly linear, molecule. However, the transition to the (2100) level, for which $K = 1$, shows a *P*-, *Q*- and *R*-branch structure typical of a Π–Σ transition, as shown in Figure 8.70. The perturbations observed in the *P* and *R* branches near to $J' = 7$ have been identified as due to the nearby (2002) state in which two quanta of the van der Waals stretching vibration are excited.

8.2.10 Spectroscopy with a tunable sideband laser

In Sections 8.2.6 and 8.2.7 we have seen how laser spectroscopy in the far infrared region may be carried out with various far infrared lasers. There is a wide variety of such lasers and a particular one may provide a number of different wavelengths but none of them is tunable. As has been described, this problem can be circumvented in some cases by tuning the energy levels of the molecule into resonance

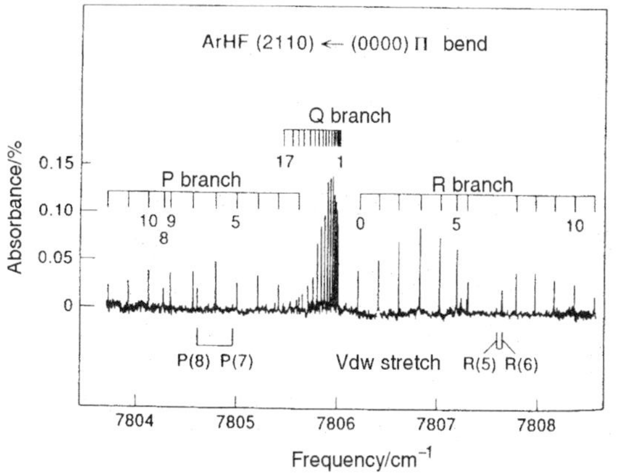

Figure 8.70 The (2110)–(0000) infrared band of ArHF, showing perturbations in the *P* and *R* branches near to $J' = 7$
[Reproduced, with permission, from Farrell, J. T., Jr, Sneh, O., McIlroy, A., Knight, A. E. W. and Nesbitt, D. J. (1992). *J. Chem. Phys.*, **97**, 7967]

with the laser by applying a magnetic or electric field.

However, it is more convenient, and more widely applicable, to use a tunable far infrared laser. Such a laser has been developed,[267] with more power than in earlier attempts, by non-linear mixing of tunable microwave or millimetre wave radiation with that from a fixed frequency, far infrared laser. In this way sidebands are attached to the laser radiation with frequencies corresponding to the sum and difference of the microwave (or millimetre wave) and far infrared frequencies.

A far infrared laser, pumped by a CO_2 laser, provides frequencies of 693 GHz (23 cm^{-1}) and 762 GHz (25 cm^{-1}), with formic acid (HCOOH) as the lasing medium, 1627 GHz (54 cm^{-1}), with difluoromethane (CF_2H_2), and 1839 GHz (61 cm^{-1}), with methanol (CH_3OH). Tunable millimetre wave radiation from a 93 GHz (3 cm^{-1}) klystron was mixed with the laser radiation in a gallium arsenide (GaAs) Shottky barrier diode. The laser radiation was coupled on to the diode by a phosphor bronze whisker antenna, 1.7 mm long. The millimetre wave radiation was also coupled onto the diode through the whisker.

Using this tunable source, rotational transitions of $^1H^2HO$ and formaldehyde (H_2CO) were observed[267] near to 600 GHz (20 cm^{-1}) and 800 GHz (27 cm^{-1}), respectively.

There is a very wide range of molecules producing more than 2000 far infrared laser frequencies in the range 10–200 cm^{-1} and sideband generation can provide tunable radiation associated with any of these frequencies. A typical range of 5 cm^{-1}, on both sides of the laser frequency, is covered by the sidebands.

The experimental set-up shown in Figure 8.71 was use to obtain a far infrared absorption spectrum of ArHCl in a planar supersonic jet with a 4 cm long slit and a single pass.[268]

The far infrared laser is pumped by a CO_2 laser and, in this experiment, the 23 and 25 cm^{-1} lines of HCOOH were used. Sidebands, generated in the GaAs mixer by combination with microwave radiation tunable in the range 2–40 GHz (0.07–1.3 cm^{-1}), provides tunable radiation in the region of the transition from the zero-point to the (0200) level,† in which two quanta of the bending vibration are excited, of the linear van der Waals molecule ArHCl. The sidebands are separated from the otherwise overwhelmingly intense radiation from the far infrared laser by the diplexer.

The bending vibration of ArHCl is of π symmetry but the first overtone level has Σ^+ and Δ components (see Table 5.10). The only allowed infrared transition from the Σ^+ zero-point level is to the Σ^+ component resulting in simple *P*- and *R*-branch structure. At a rotational temperature of 3 K more than 60 transitions for each of $ArH^{35}Cl$ and $ArH^{37}Cl$ were observed. Figure 8.72 shows the *P*(3) line of $ArH^{35}Cl$. This has a second derivative line shape, due to the lock-in detection being at twice the modulation frequency, and also shows nuclear hyperfine structure. Since $J'' = 3$ and, for the ^{35}Cl nucleus, $I = \frac{3}{2}$, the quantum number F'' [see equation (4.47)] can be $\frac{9}{2}, \frac{7}{2}, \frac{5}{2}$ and $\frac{3}{2}$. In the upper state $J' = 2$ and $F' = \frac{7}{2}, \frac{5}{2}, \frac{3}{2}$ and $\frac{1}{2}$. The

†See p. 544 for labelling of levels.

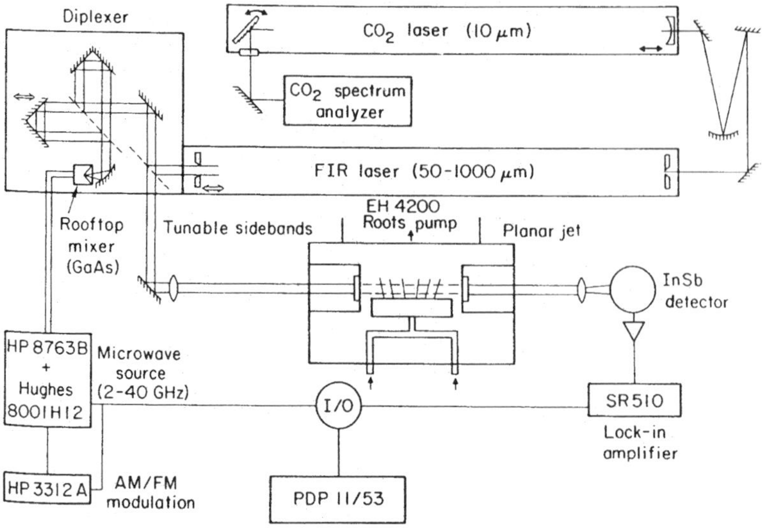

Figure 8.71 A tunable far infrared laser used to obtain absorption spectra in a planar supersonic jet [Reproduced, with permission, from Busarow, K. L., Blake, G. A., Laughlin, K. B., Cohen, R. C., Lee, Y. T. and Saykally, R. J. (1988). *J. Chem. Phys.*, **89**, 1268]

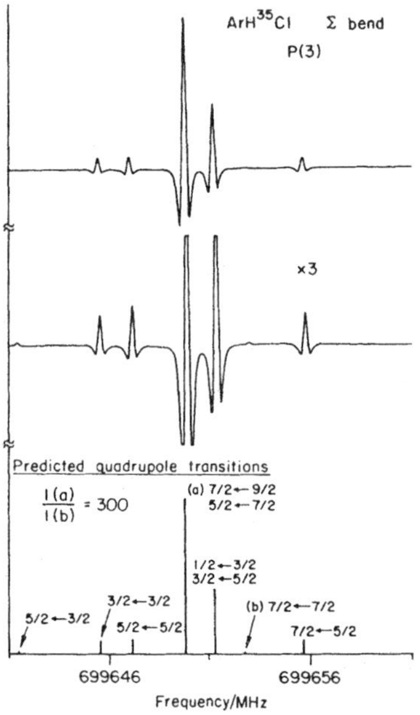

Figure 8.72 The *P*(3) line, showing the nuclear hyperfine structure, of $ArH^{35}Cl$ in the Σ^+–Σ^+ component of the first overtone of the bending vibration
[Reproduced, with permission, from Busarow, K. L., Blake, G. A., Laughlin, K. B., Cohen, R. C., Lee, Y. T. and Saykally, R. J. (1988). *J. Chem. Phys.*, **89**, 1268]

hyperfine components, some of them coincident, which result from the $\Delta F = 0, \pm 1$ selection rule are indicated at the bottom of the figure.

Rovibrational transitions were observed for much higher *J*-values (up to 16 or higher) than had been possible previously by millimetre wave spectroscopy and, in addition, the line widths were reduced to sub-Doppler values of 300–350 kHz (0.000 010–0.000 012 cm^{-1}) by observation normal to the planar jet. Consequently, more accurate rotational constants were obtained. For example, the band centre was determined to be at 709 223.654 MHz (23.657 154 6 cm^{-1}) for $ArH^{35}Cl$ and 707 838.288 9 MHz (23.610 943 84 cm^{-1}) for $ArH^{37}Cl$.

The extent of the detailed information which can be derived from such high resolution, sub-Doppler, jet-cooled spectra is exemplified by the spectrum of the water dimer, $(H_2O)_2$. The structure of this hydrogen bonded species is shown in Figure 4.44, but this is not a rigid structure. For example, the two equivalent hydrogen atoms in the H_2O moiety in a plane perpendicular to the figure, the acceptor moiety, may exchange by a feasible internal rotation of

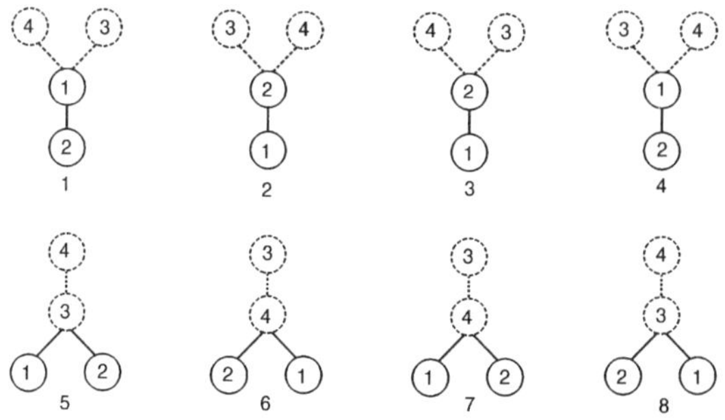

Figure 8.73 Eight energetically equivalent frameworks for $(H_2O)_2$ in which only the hydrogen atoms are shown and are labelled 1–4

that moiety. In addition, there may be tunnelling through the barrier to this internal rotation resulting in observable splitting of energy levels.

In total, there are eight energetically equivalent frameworks which represent the various tunnelling motions.[269] They are illustrated in Figure 8.73, in which only the hydrogen atoms are shown and are labelled 1–4. Fortunately, not all these tunnelling motions are feasible. The tunnelling between frameworks 1 and 4, so-called 1–4 tunnelling and also referred to as acceptor tunnelling since it occurs in the acceptor moeity, is judged to be the most important.[269] This is the internal rotational motion considered above; the tunnelling splitting is estimated to be 6 cm^{-1}. The 1–5 (or 6) motion is a geared internal rotation in which the hydrogen atom from the donor moiety is exchanged with one from the acceptor moiety. The resulting geared donor–acceptor tunnelling splitting[270] is only 0.63 cm^{-1}.

The dimer is a prolate near-symmetric rotor and the J levels for various K_a are split by a small amount by the 1–5 tunnelling. Four sub-levels result for each value of J but two are degenerate and the splitting is into three sub-levels. There is much larger splitting due to the 1–4 tunnelling. This splits the closely spaced sets of three levels into widely spaced pairs. The resulting sets of six sub-levels for each of the lowest two J values for $K_a = 0$, 1 and 2 are shown in Figure 8.74 in which asymmetry

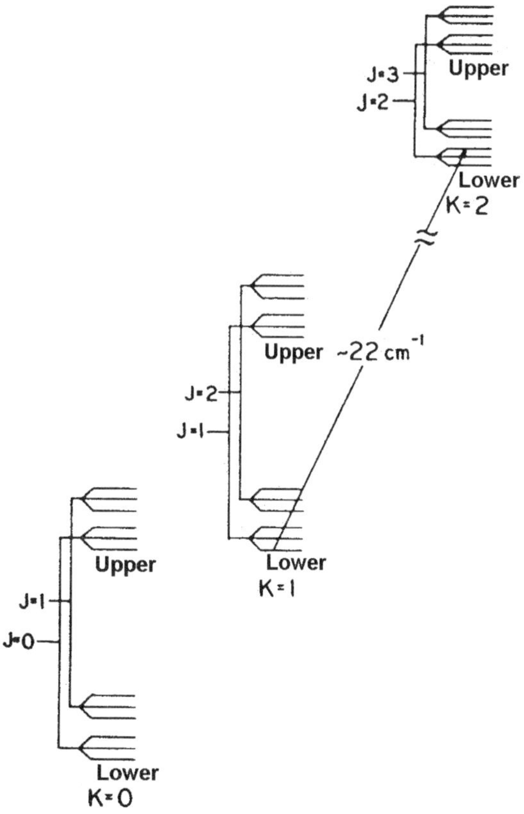

Figure 8.74 Energy level diagram for $(H_2O)_2$ showing the two lowest J-levels for $K_a = 0$, 1 and 2. Each J level is split into six sub-levels by tunnelling motions
[Reproduced, with permission, from Busarow, K, L., Cohen, R. C., Blake, G. A., Laughlin, K. B., Lee, Y. T. and Saykally, R. J. (1989). *J. Chem. Phys.*, **90**, 3937]

doubling of the levels with $K_a = 1$ and 2 is not shown.

The six sub-levels for each K_a and J value due to these two tunnelling motions is the maximum allowed by group theory. Any other feasible tunnelling motions, particularly 1–2 and 1–7, can produce additional shifts of the energy levels but no further splitting.

As indicated in Figure 8.74, transitions between levels with $K_a = 1$ and 2 lie in the region of 22 cm^{-1}. Using the tunable far infrared laser and planar supersonic jet shown in Figure 8.71, 56 tunnelling–rotation transitions associated with the $K_a = 2$–1 transition have been observed and assigned[271] and are in agreement with a proposed theoretical model[272] which includes the four different tunnelling motions discussed above.

8.2.11 Optothermal and vibrational predissociation spectroscopy

Optothermal spectroscopy is a technique which is distinguished from others by the method of detection.

When a molecule absorbs a quantum of infrared radiation it is vibrationally excited and becomes 'hot'. If the hot molecule falls on a detector in a time which is short compared with the time taken for the molecule to revert to the initial state by infrared fluorescence, the detector will experience heating. The temperature increase is very small but, if the detector is sufficiently sensitive, scanning the infrared source through a region in which a molecule absorbs, and monitoring the temperature changes in the detector, will produce an infrared absorption spectrum.

A bolometer, consisting of doped silicon and operating at liquid helium temperature, is a sufficiently sensitive temperature detector for this purpose. In fact, the limit of sensitivity is of the order of 1000 molecules per second.[273] A diode laser may be used as the infrared source but the increased power, typically tens of milliwatts, and the ease of tuning of a colour centre laser (see Section 8.1.8.5) have made this a preferred source. Most experiments have been done with the sample seeded into a supersonic jet from a circular nozzle. The jet is usually skimmed for sub-Doppler resolution and there may be a facility for multipassing the molecular beam.

Vibrational infrared fluorescence lifetimes are of the order of a few milliseconds. The distance from the point where the laser beam crosses the molecular beam must be sufficiently small, of the order of 20 cm, that the time the molecules take in travelling to the bolometer is short compared to the fluorescence lifetime. Of course, this distance depends on the pumping speed for evacuating the chamber containing the skimmer and bolometer.

Figure 8.75 shows an experimental arrangement[274] in which the colour centre laser is pumped by a CW, mode-locked Nd^{3+} : YAG laser and produces 150 mW of power in the 1.55 μm (6450 cm^{-1}) region where, for example, the first overtones of hydrogen stretching vibrations occur.

The fundamental and first overtone of the C—H stretching vibration, ν_1, of 1,1,1-trifluoropropyne (H—C≡C—CF_3) have been observed[274] in this way with sub-Doppler resolution. Both the 1_0^1 and 1_0^2 bands are parallel bands of a symmetric rotor but, although they both show the gross structure to be expected for such a band, they differ considerably in detail.

In such a heavy molecule as trifluoropropyne there are several low wavenumber vibrations and these contribute considerably to a high density of other vibrational states which are nearly degenerate with the 1^1 vibrational level. This density increases rapidly with increasing vibrational energy resulting in a much higher density of states which are nearly degenerate with the 1^2 level. These other vibrational states are, in the language of intramolecular vibrational energy redistribution, or IVR (see Section 8.2.9 and Figure 8.68), the bath states.

The 1_0^1 band lies in the region 3320–3360 cm^{-1}. The 1^1 state is of a_1 symmetry and the density of bath states which are also of a_1 symmetry, and therefore capable of interaction with the 1^1 state, has been estimated to be about 10 per cm^{-1}. However, interactions between the bright state and the bath states are restricted by other factors, such as similarity of rotational constants. In this case they are so restrictive that the 1_0^1 band has been analysed in terms of a perturbation by only a few bath states. Even so, these perturbations are regarded as evidence for the onset of IVR.

The density of a_1 bath states in the 6660 cm^{-1} region of the 1_0^2 band is greatly increased to about 1000 per cm^{-1}. Again, only a very small proportion of these are expected to perturb the

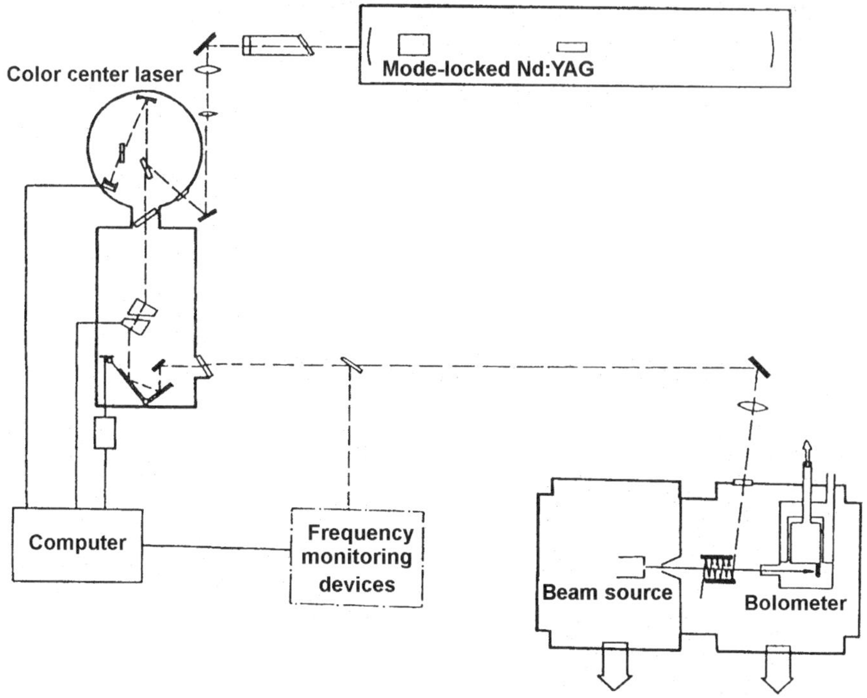

Figure 8.75 An optothermal spectrometer showing a colour centre laser, pumped by a Nd^{3+}:YAG laser, multipassing a skimmed molecular beam, and a bolometer detector
[Reproduced, with permission, from Pate, B. H., Lehmann, K. K. and Scoles, G. (1991). *J. Chem. Phys.*, **95**, 3891]

1^2 state but this is sufficient to create a dense and unassignable 1_0^2 band in which IVR is clearly very important.

The 1_0^2 band of propyne (H—C≡C—CH_3), where ν_1 is the acetylenic C—H stretching vibration, shows many fewer perturbations as a consequence of the much lower density, in the lighter molecule, of nearly degenerate bath states. Unlike the corresponding band of 1,1,1-trifluoropropyne, it can be rotationally analysed.[275] The $K_a = 0$ states show no perturbations but states with $K_a = 1$–3 are perturbed and the sub-band origins displaced. These results are interpreted in terms of a two-stage coupling, via a so-called doorway state, to the bath states.

The use of sub-Doppler resolution and rotational cooling in a skimmed supersonic jet to allow the analysis of infrared bands of fairly large molecules is also illustrated by the optothermal spectrum of the asymmetric rotor 2-fluoroethanol.[276] The lower energy conformation is the one in which the oxygen and fluorine atoms are in the *gauche* positions and the hydroxyl hydrogen atom is pointing towards the fluorine atom.

The band involving the asymmetric stretching of the CH_2 group containing the fluorine atom lies in the $2983\,cm^{-1}$ region and was observed using the optothermal technique. The colour centre laser was pumped by a dye laser which was itself pumped by an Ar^+ laser.

The rotational structure of the band was fitted using known lower state constants A'', B'' and C'' and an iterative procedure for the upper state constants. Only type C transitions were identified but, where a single line in the spectrum was expected, clusters of lines were observed and these were attributed to IVR.

Another example of the power of this experimental technique in making assignable those spectra which would be rotationally crowded and Doppler limited when observed in an absorption spectrum at room temperature is the 1_0^1 band of fluoroform, CHF_3, where ν_1 is the C—H stretching vibration. The band has been observed[277] with a colour centre laser and optothermal detection.

Although CHF_3 is a symmetric rotor the 1_0^1 band is so strongly perturbed that even cooling the sample to 87 K and using a difference frequency laser, in an earlier experiment, produced a spectrum which was only partly assignable. With sub-Doppler resolution and a rotational temperature of 4 K assignment was possible. The perturbations were assigned to two close-lying vibrational combination levels.

The use of a supersonic jet in these experiments with bolometric detection makes them ideal for obtaining spectra, with little rotational congestion, of small van der Waals or hydrogen bonded clusters.

A common problem in the infrared spectra of, say, a dimer molecule is that it is seriously overlapped by the corresponding spectrum of the more abundant monomer. Bolometric detection overcomes this because, whereas the bolometer detects an increase in temperature when the monomer absorbs the infrared radiation, it usually detects a decrease when the dimer, or any other cluster, absorbs radiation. The reason for this is that the absorption band being recorded is usually at a wavenumber greater than 3000 cm^{-1} which is above the dissociation energy of the weak van der Waals or hydrogen bond holding the cluster together. Therefore, when excitation of the cluster takes place, there is a decrease in the molecular beam flux resulting from the loss of the products of predissociation from the beam.

The structure of the hydrogen fluoride dimer, $(HF)_2$, is like that shown in Figure 4.44(a). In this molecule there are two H—F stretching vibrations in the region of 3900 cm^{-1} since one hydrogen atom is hydrogen bonded and the other is free. The vibration associated with the free hydrogen has been labelled ν_1 and the other vibration labelled ν_2. Using a difference frequency laser and a long path gas cell, these two vibrations were identified[278] but, whereas the rotational lines associated with ν_1 have a Doppler limited line width, those associated with ν_2 have about twice this width.

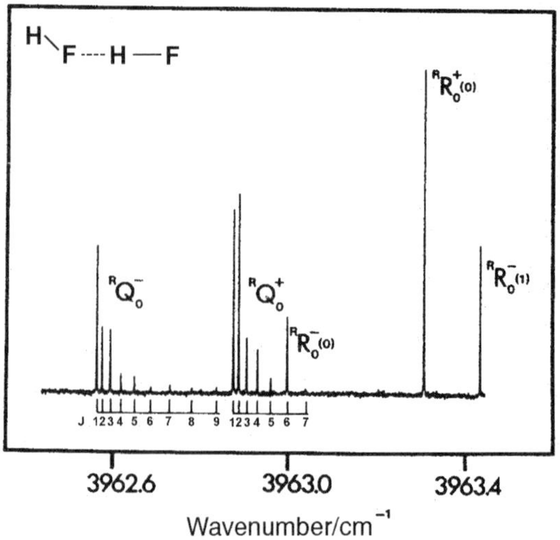

Figure 8.76 Part of the sub-Doppler 1_0^1 band of $(HF)_2$ showing parts of Q and R branches for $K_a = 1$–0. The two types of branches, labelled + and −, result from inversion splitting [Reproduced, with permission, from Huang, Z. S., Jucks, K. W. and Miller, R. E. (1986). *J. Chem. Phys.* **85**, 3338]

Figure 8.76 shows part of the 1_0^1 band of $(HF)_2$ obtained[279] by the optothermal method.

The $(HF)_2$ molecule is a prolate near-symmetric rotor and the 1_0^1 band is a type A/B hybrid band and Figure 8.76 shows part of the type B component.

The reason for the splitting of the branches into two components is that the molecule can invert, a process in which the free hydrogen atom becomes hydrogen bonded and the hydrogen bonded atom becomes free. Tunnelling through the inversion barrier produces this splitting.

In order to remove the Doppler broadening as much as possible spectra were obtained with only a single pass of the molecular beam. Figure 8.77 shows one line, obtained in this way, for the monomer and the dimer.

The lines in Figure 8.77 contain a component due to the remaining Doppler broadening,

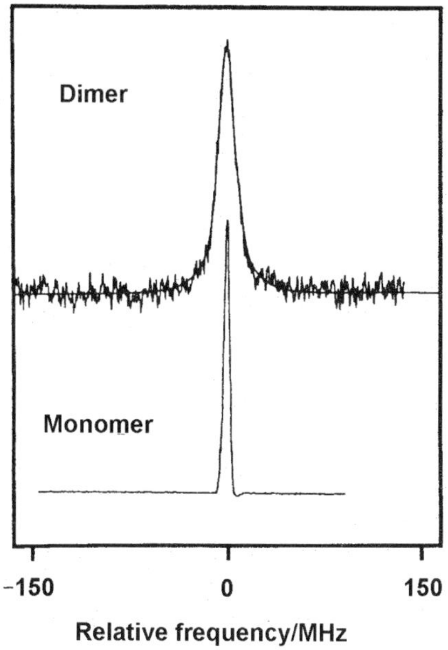

Figure 8.77 The upper and lower spectra show the ${}^{r}R_0^+(0)$ and $P(1)$ lines of the hydrogen fluoride dimer and monomer, respectively. The solid line drawn through the dimer spectrum is a calculated line with a Voigt profile
[Reproduced, with permission, from Huang, Z. S., Jucks, K. W. and Miller, R. E. (1986). *J. Chem. Phys.*, **85**, 3338]

which has a gaussian line shape (see Section 2.3.2), and a component due to the natural line broadening (see Section 2.3.1), which has a lorentzian line shape. A line which has both components is said to exhibit a Voigt profile. The full width, $\Delta\nu$, at half maximum intensity of the natural line broadening, lorentzian, component is related to the lifetime, τ, of the upper state by

$$\tau = \frac{1}{2\pi\Delta\nu} \tag{8.45}$$

which is derived from the uncertainty principle in the form given in equation (1.29).

The profile of the monomer line in Figure 8.77 was fitted to a gaussian line shape to determine the Doppler limited instrumental resolution. Then this line width was used as the gaussian contribution to the dimer line profile to determine the full width at half maximum intensity of the lorentzian component of the observed Voigt profile. This width is 13.4 ± 1.0 MHz. From this, equation (8.45) gives a lifetime of 12 ± 1 ns for the upper state of the dimer and compares with a value of 1.0 ± 0.1 ns for the upper state of the 2_0^1 band.[279] The fact that vibrational predissociation occurs more readily through vibration of the hydrogen bonded hydrogen than of the free hydrogen agrees nicely with expectation.

The 1_0^1 band of each of the isoelectronic molecules $N_2 \cdots HCN$ and $OC \cdots HCN$ have been observed[280] at sub-Doppler resolution using the optothermal technique. Both molecules are linear and ν_1 is the C—H stretching vibration of the HCN moiety. Both of the 1_0^1 bands are of the $\Sigma^+ - \Sigma^+$ type showing a simple

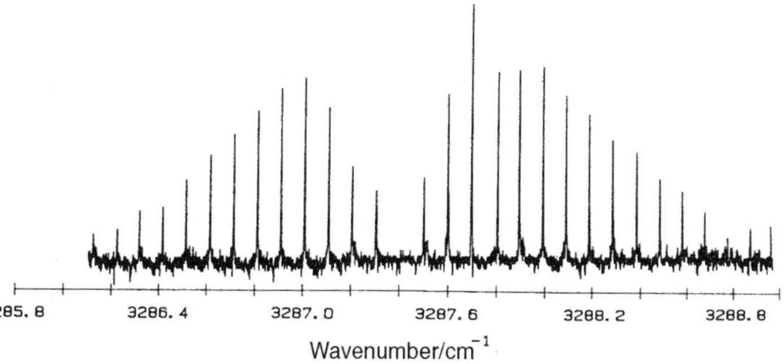

Figure 8.78 The 1_0^1 band of $OC \cdots HCN$ showing a simple *P*- and *R*-branch structure
[Reproduced, with permission, from Jucks, K. W. and Miller, R. E. (1988). *J. Chem. Phys.*, **89**, 1262]

P- and R-branch structure. This is shown for OC···HCN in Figure 8.78. The molecule is fairly heavy, having a value for B'' of only 0.049 213 cm^{-1}, and, as a result, the band extends over only about 3 cm^{-1}. The similarity of B'' and B' (= 0.049 194 cm^{-1}) results in an almost undegraded band.

As is the case in $(HF)_2$, the 1^1 level lies above the dissociation limit leading to vibrational predissociation and broadening of the lines of OC···HCN in comparison with HCN. The width of the lorentzian component in lines of the complex lead to a lifetime of the 1^1 state which is shortened by predissociation to 2.6 ns. For N_2···HCN the corresonding lifetime is 80.0 ns.

Cyanoacetylene, H—C≡C—C≡N, forms a dimer and a trimer in a supersonic jet. Both the 1^1_0 and 2^1_0 bands of the linear dimer, where v_1 is the free C—H stretching vibration and v_2 the hydrogen bonded C—H stretching vibration, have been observed[281] optothermally at sub-Doppler resolution. The 1^1_0 band of the linear trimer, involving the free C—H stretching vibration, has been observed[281,282] in a similar way.

Figure 8.79 shows the Σ^+–Σ^+, 1^1_0 band, with the band centre at 3324.438 cm^{-1}, of the linear dimer and the much weaker Σ^+–Σ^+, 1^1_0 band, with the band centre at 3323.682 45 cm^{-1}, of the linear trimer. These two bands are recorded upwards on the figure and result from a decrease in temperature of the bolometer due to vibrational predissociation following absorption. The lines recorded downwards are part of the 1^1_0 band of the monomer. This does not undergo predissociation and the temperature of the bolometer increases.

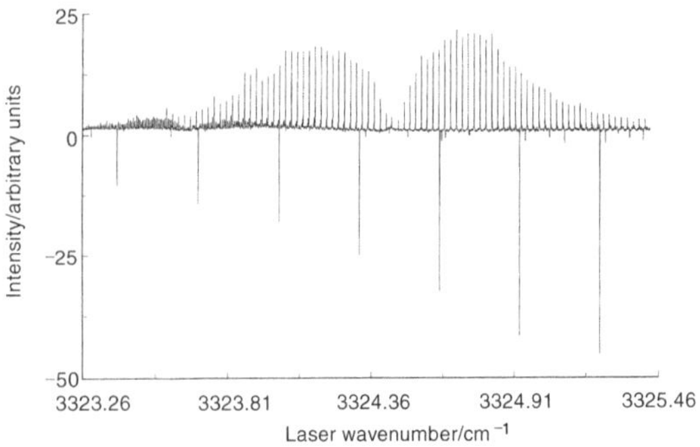

Figure 8.79 The 1^1_0 bands of $(HCCCN)_2$, centred at about 3324.4 cm^{-1}, and $(HCCCN)_3$, centred at about 3323.7 cm^{-1}. The downwards lines are part of the 1^1_0 band of HCCN
[Reproduced, with permission, from Kerstel, E. R. Th., Scoles, G. and Yang, X. (1993). *J. Chem. Phys.*, **99**, 876]

The linewidths in the 1^1_0 band of the dimer[281] were instrument limited and the predissociation lifetime could only be estimated to lie in the range 90 μs–16 ns. However, the lines of the 2^1_0 band are very much broader, at least 350 MHz, leading to an upper limit for the lifetime of only 450 ps.

The predissociation lifetime, estimated from the line widths in the 1^1_0 band of the linear trimer,[282] is 4.5 ns. A band at 3246.5 cm^{-1} was assigned as the in-phase C—H stretching vibration involving the two internal hydrogen bonded hydrogen atoms but was rotationally unresolved. The large effect of an electric field on this band showed that it belongs to a polar species. Other bands in the spectrum showed no such effect and may belong to a non-polar, cyclic isomer of the trimer.

The optothermal method has been extended to a higher wavenumber regime by using a different radiation source. A near infrared source was obtained[283] by Raman shifting the visible output of a dye laser pumped by a pulsed Nd^{3+}:YAG laser. With this system the optothermal spectrum of benzene, in a multi-passed supersonic jet, was obtained in the 8740–8870 cm^{-1} region. This is the region of the 20^3_0 second overtone band of v_{20} (using the Wilson numbering), an e_{1u} C—H stretching vibration. At least seven bands were observed in this region, all of them very broad and probably caused by Fermi resonances with combination levels near to the 20^3 level.

The region of the 20^4_0 band of benzene, estimated to be about 30 times weaker than 20^3_0, has also been observed.[284] This lies in the 11 400–11550 cm^{-1} region where there are several broad bands, three dominating the intensity. Again, it is possible that Fermi resonances, or more general interactions with the bath states, may account for the additional bands.

These spectra, observed in the regions of the 20^3_0 and 20^4_0 bands of benzene, represent a great improvement on previously obtained spectra at room temperature. Considerable interest attaches to these overtone bands as they show a tendency to local mode behaviour (see Section 5.2.6.4).

In Section 4.10.1 the technique of molecular beam electric resonance was discussed. As shown in Figure 4.42, molecules from a supersonic jet are subjected to a quadrupolar electric field. Only molecules in energy levels with a positive change of energy with increase of the field strength, that is with a positive Stark coefficient, reach the detector. Those with a negative Stark coefficient are rejected from the beam.

This technique has been modified by using the more sensitive optothermal detection and either microwave,[285] infrared, or a combination of microwave and infrared[286] radiation. The molecules in the jet are exposed to the radiation between the CW nozzle and a skimmer. The beam then passes through a 56 cm long state-selecting quadrupolar field with a central stopwire to prevent molecules which interact with the field weakly, or not at all, from reaching the detector.

Infrared absorption by a weakly bound complex is likely to lead to vibrational predissociation. In these cases molecules in the lower, rather than upper, state may be focused on the detector.

The microwave spectrum of $(H_2O)_2$ was measured[285] in this way in the range 14–110 GHz (0.47–3.7 cm^{-1}). Rotational transitions were observed in this prolate near-symmetric rotor in the $K_a = 0$–0, 1–1 and 1–0 sub-bands.

The rotational energy levels of $(H_2O)_2$ were discussed in Section 8.2.10 and are shown schematically in Figure 8.74. From the more extensive microwave data[285] it was shown, for example, that the overall splitting within the lower triplet of levels for $K_a = 0$ and $J = 0$ is 22.6 GHz (0.753 cm^{-1}), compared with 19.5 GHz (0.650 cm^{-1}) in the upper triplet, and that the splitting between the two sets of triplets is about 230 GHz (7.67 cm^{-1}).

A spectrometer with a microwave and infrared capability has been used to obtain spectra of the HF···HCl and HCl···HF complexes.[286] The latter is a higher energy (estimated to be about 70 cm^{-1}), metastable isomer and was observed for the first time. Unlike the case of HF···HF, the fact that the two structures are no longer energetically equivalent means that there is no doubling of vibrational levels caused by tunnelling.

Figure 8.80 shows equilibrium structures estimated for these two isomers from the microwave spectra of the ^{35}Cl and ^{37}Cl isotopomers of both molecules. They are both prolate near-symmetric rotors and Figure 8.81 shows the 3_{03}–2_{02} microwave transition ($J' = 3$, $K'_a = 0$, $K'_c = 3$ and $J'' = 2$, $K''_a = 0$, $K''_c = 2$) of the ^{35}Cl isotopomer of both molecules. These both show nuclear hyperfine components resulting from $I = \frac{3}{2}$, for both ^{35}Cl and ^{37}Cl,

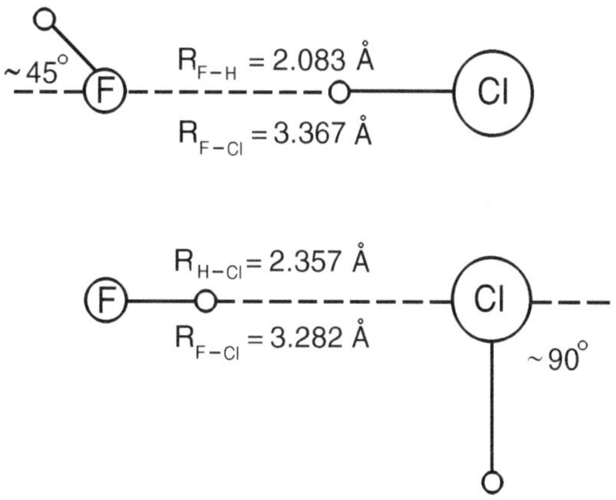

Figure 8.80 Equilibrium structures estimated for HF···HCl and HCl···HF

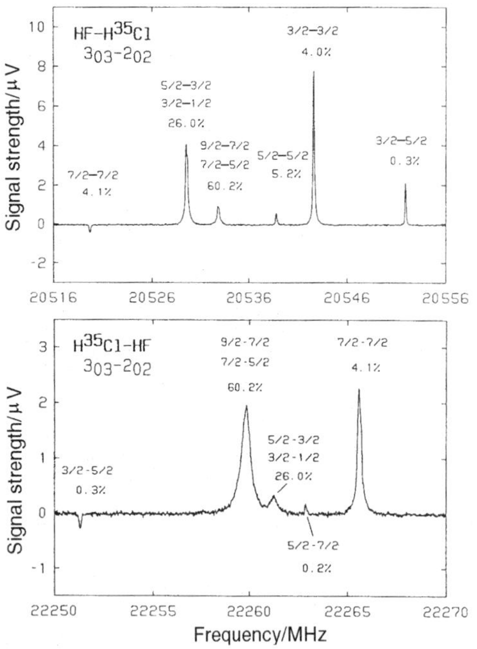

Figure 8.81 Electric resonance optothermal microwave spectra of the 3_{03}–2_{02} transition of $HF \cdots H^{35}Cl$ and $H^{35}Cl \cdots HF$ showing nuclear hyperfine splittings
[Reproduced, with permission, from Fraser, G. T. and Pine, A. S. (1989). *J. Chem. Phys.*, **91**, 637]

$F = J + I,\ J + I - 1, \ldots |J - I|$ and the selection rule $\Delta F = 0, \pm 1$.

Infrared absorption spectra were obtained using, instead of a microwave source, a colour centre laser directed normal to, and focused into, the jet before the skimmer. The $K_a = 0$–0 and 1–0 sub-bands of the type A band resulting from the H—F stretching vibration of $HF \cdots HCl$ and the $K_a = 0$–0 sub-band of the corresponding vibration of $HCl \cdots HF$ were analysed. There is a shift to low wavenumber of $21.0\,cm^{-1}$, in $HF \cdots HCl$, and $94.0\,cm^{-1}$, in $HCl \cdots HF$, relative to the vibration in free HF. The spectra of both complexes show line broadening due to vibrational predissociation, but with considerable, unexplained variations.

In order to obtain electric resonance optothermal infrared spectra at lower wavenumbers than are accessible with a colour centre laser, a tunable sideband laser has been used. In this way a Fermi triad near to $970\,cm^{-1}$ involving the fundamental ν_{10}, the e CH_3 rocking vibration of the C_{3v} molecule 1,1,1-trifluorethane (CF_3CH_3), has been observed and rotationally analysed.[287] The other two components of the triad arise from the $2\nu_6 + \nu_{11}$ and $\nu_5 + \nu_{12}$ combinations and, since ν_6 is the a_2 torsional vibration, splittings due to tunnelling through the torsional barrier have been resolved and explained.

We have seen how optothermal detection of the vibrational spectrum of a weakly bound complex depends on the fact that it falls apart when the wavenumber of the incident radiation lies above the limit of dissociation of the weak bond holding the complex together—detection depends on vibrational predissociation taking place. Another technique for obtaining vibrational spectra of weakly bound complexes, that of vibrational predissociation spectroscopy,[290] also depends on the complex falling apart. In this case, however, it is one of the products of the predissociation which is monitored as the source radiation is scanned through an absorption band.

The infrared spectrum of the ionic complex $H_2 \cdots HCO^+$ has been obtained in this way.[291] The HCO^+ ions are created by electron impact on a mixture of H_2 and CO and the products injected into a supersonic jet. The ionic complex is mass selected, by a quadrupole mass filter, and injected into an ion guide where they encounter pulsed infrared radiation from an optical parametric oscillator [OPO; see Section 8.1.8.15] combined with two optical parametric amplifiers (OPA) giving a tunable output from 2500 to $4200\,cm^{-1}$. The ionic product of vibrational predissociation, HCO^+, is selected by a second quadrupole mass filter and detected.

The $H_2 \cdots HCO^+$ complex is T-shaped with a C_{2v} structure and the H_2 moiety attached to the H of HCO^+. The dimensions of the H_2 and HCO^+ moieties are unchanged on complex formation and the intermolecular bond length is 1.75 Å. The H—H stretching vibration ν_1,

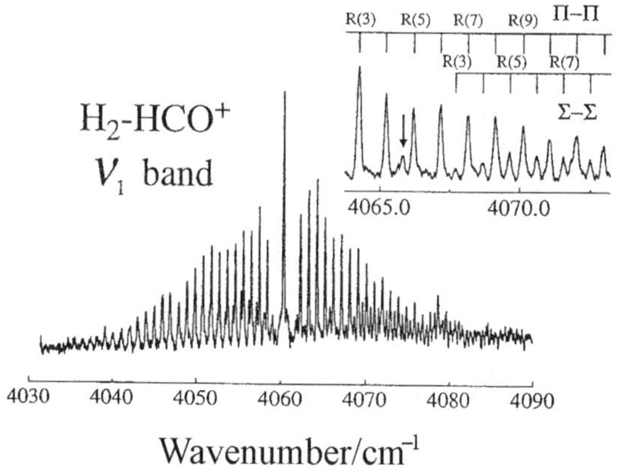

Figure 8.82 The ν_1, H—H stretching, vibration of the T-shaped $H_2 \cdots HCO^+$ complex and the vibrational predissociation spectrum near $4060\,cm^{-1}$, showing an expanded part of the *R*-branch region [Reproduced, with permission, from Bieske, E. J., Nizkorodov, S. A., Bennett, F. R. and Maier, J. P. (1995). *J. Chem. Phys.*, **102**, 5152]

which occurs at about $4060\,cm^{-1}$, is the only observed band which shows sharp rotational structure: this is shown in Figure 8.82. Other bands were observed at lower wavenumbers, particularly ν_2, the C—H stretching vibration, at $2840\,cm^{-1}$, but all are strongly predissociated. For example, an estimate of the rotational line width for ν_2 gives a predissociation lifetime of only 1 ps.

The complex is a prolate near-symmetric rotor with a very large A rotational constant due to the off-axis hydrogen atoms. The spectrum of the jet-cooled complex in Figure 8.82 shows the $\Delta K_a = 0$ transitions for $K_a'' = 0$ (a P and R branch) and $K_a'' = 1$ (a P, Q and R branch). They are labelled Σ–Σ and Π–Π, respectively, in the figure. The fact that the Σ–Σ component is considerably weaker than the Π–Π component is due to the 1 : 3 statistical weight alternation caused by the nuclear spin of the hydrogen atoms.

The estimated line width of $0.06\,cm^{-1}$ for the low J rotational lines in the ν_1 band indicate a predissociation lifetime of 90 ps. As expected, the H—H stretching vibration induces predissociation much less than the C—H stretching vibration.

8.2.12 Infrared laser induced fluorescence

We have seen, in Sections 8.2.9 and 8.2.11, how difference frequency laser and optothermal, colour centre laser spectroscopies can access high-lying vibrational states such as the first overtones of X—H stretching vibrations which tend to lie in the 6000–7000 cm^{-1} region.

Higher overtone levels are progressively more difficult to observe because of the rapidly decreasing transition probability. For example, in HF, that of the 3–0 transition is about 30 times less than that of the 2–0 transition. Because overtone transitions are so weak, excitation is well below saturation. As a result, the detector output increases linearly with laser power, and higher power is advantageous.

The second overtones of X—H stretching vibrations tend to lie in the 11 000–12 000 cm^{-1} region. This is accessible with a Ti^{3+}:sapphire laser (see Section 8.1.8.3), which has a tunable range of 675–1100 nm (14 800–9100 cm^{-1}).

A typical Ti^{3+}:sapphire laser, in a linear configuration, can produce a few watts of power when pumped by a 15 W Ar^+ laser but, in a ring configuration similar to that used in a ring dye laser (see Section 8.1.8.14), it can produce about 10 W inside the ring cavity when it is pumped by only a 5 W Ar^+ laser.

Such a Ti^{3+}:sapphire ring laser has been constructed and used to obtain infrared laser induced fluorescence spectra on van der Waals and hydrogen bonded complexes formed in a slit supersonic jet inserted into the laser cavity[290].

The experimental arrangement is shown in Figure 8.83(a). The Ar^+ laser (AIL) pumps the Ti^{3+}:sapphire crystal (TS) which is inside the 'figure of eight' ring bounded by two concave, dichroic mirrors (DC1, DC2) and two plane reflectors (PZTHR, OC). As in a ring dye laser,

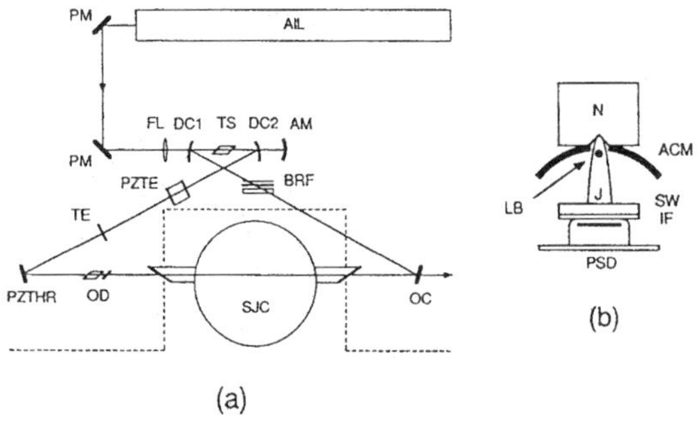

Figure 8.83 (a) A Ti^{3+}:sapphire ring laser, pumped by an Ar^+ laser, with an intracavity supersonic slit jet, SJC. (b) Infrared fluorescence collection, with a concave mirror ACM and lead sulphide detector PSD, from the supersonic jet, J, shown in a plane perpendicular to that of (a).
[Reproduced, with permission, from Chang, H.-C. and Klemperer, W. (1993). *J. Chem. Phys.*, **98**, 2497]

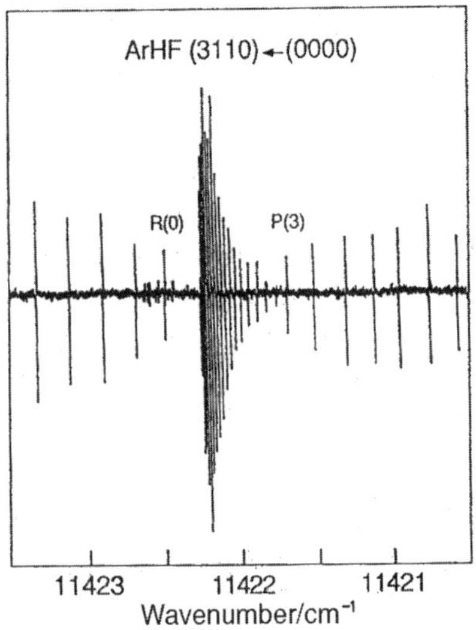

Figure 8.84 The *Q* branch and parts of the *P* and *R* branches of the (3110)–(0000) band of ArHF
[Reproduced, with permission, from Chang, H.-C. and Klemperer, W. (1993). *J. Chem. Phys.*, **98**, 2497]

there must be a device which causes the wave to travel in only one direction inside the cavity. Here it is the optical diode (OD), which consists of a Faraday rotator and a birefringent crystal. The laser linewidth is about 10 MHz.

The slit supersonic jet device (SJC) is arranged so that the jet is directed perpendicular to the figure. Figure 8.83(b) shows the jet (J) in cross section emerging from the nozzle (N) and traversed by the laser beam (LB). If the laser excites the molecule to, say, a $\upsilon = 3$ vibrational level, the energy will be lost by stepwise emission down to the zero-point level. This infrared fluorescence typically has a long decay time, e.g. 2 ms for the 3–2 emission in HF. As a result of this, and the high translational velocity of the molecules in the jet, only about 0.2% of the fluorescence is detected by the lead selenide detector (PSD) even with the enhancement by the aluminium concave mirror (ACM). There is an interference filter (IF), to cut out scattered laser radiation, and a sapphire window (SW) in front of the detector.

A spectrum obtained in this fashion is an infrared fluorescence excitation spectrum. A fluorescence excitation spectrum is usually associated with electronic excitation (see Section 8.2.17) but the principle is the same when the excitation is vibrational. The intensity of the total, undispersed fluorescence is monitored as the wavenumber of the source is scanned through the absorption.

Such a fluorescence excitation spectrum of the van der Waals complex ArHF has been obtained[290] in the region of absorption from the zero-point to the (3000) level, using the ($\upsilon_{\mathrm{HF}}bKn$) labelling explained in Section 8.2.9. The (3000) level involves three quanta of the H—F stretching vibration. In this region transitions to the (3100) and (3110) combination levels were also observed, analogous to the observation, by difference frequency laser spectroscopy, of transitions to the (2000), (2100) and (2110) levels[265] described in Section 8.2.9.

Figure 8.84 shows the central *Q* branch and parts of the *P* and *R* branches of the (3110)–(0000) transition. The first derivative line shapes are a result of frequency modulation. The transition is of the Π–Σ type, for which $\Delta J = 0, \pm 1$, and the band should be compared

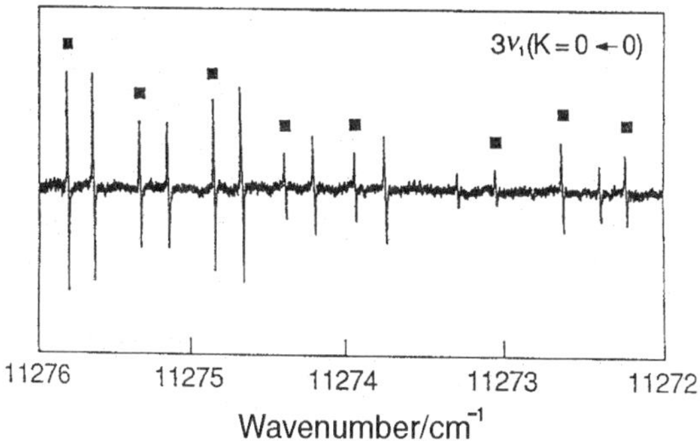

Figure 8.85 Part of the $K_a = 0$–0 sub-band, from $R(6)$ to $P(3)$, of the second overtone of the free H—F stretching vibration of $(HF)_2$. One component of the tunnelling doublets is indicated by a square [Reproduced, with permission, from Chang, H.-C. and Klemperer, W. (1993). *J. Chem. Phys.*, **98**, 9266]

with the (2110)–(0000) band shown in Figure 8.70. The band origin of the (3110)–(000) band is at 11 422.378 cm^{-1}. The rotational constants obtained from the three bands in this region agree well with extrapolations from results obtained for the H—F stretching fundamental and first overtone in ArHF. These constants, and the band origin positions, show evidence of a van der Vaals bond which strengthens with increasing vibrational energy.

The hydrogen bonded dimer $(HF)_2$, shown in Figure 8.76, has two different H—F stretching vibrations, ν_1, associated with the free hydrogen, and ν_2, associated with the hydrogen bonding hydrogen. These two fundamentals have been observed using optothermal detection, as described in Section 8.2.11. Transitions to the second overtone levels (30) and (03)[291] and to the combination levels (21) and (12)[292] have been observed with infrared laser induced fluorescence.

Figure 8.85 shows part of the $K_a = 0$–0 sub-band of the (30)–(00) band. The doubling of the P- and R-branch lines is due to splitting of vibrational levels due the tunnelling motion described in Section 8.2.11. The tunnelling splitting decreases from 19.747 03 (0.658 690 0 cm^{-1}) in the (10) state to 0.072 GHz (0.0024 cm^{-1}) in the (30) state showing that there is a surprising transformation from a non-rigid to a semi-rigid rotor as υ_1 increases. The lines are broadened by vibrational predissociation (see Section 8.2.11) and the lorentzian component of the Voigt profile indicates a full width at half maximum (FWHM) intensity of 100 MHz.

The $K_a = 1$–0 sub-band shows a very much larger FWHM of 1.9 GHz, giving a much smaller predissociation lifetime of 84 ps for $K'_a = 1$ than that of 1.6 ns for $K'_a = 0$. This K_a-dependent lifetime is unusual and is attributed to centrifugal interaction with the very close $K'_a = 1$ level of the vibrational combination level $3\nu_2 + \nu_5$, where ν_5 is an asymmetric bending vibration.

Considering only the $K'_a = 0$ levels as typical, the much larger FWHM of 10 GHz for the (03) state, compared with 100 Mz for the (30) state, agrees with the conclusion from the (01) and (10) states that, as expected, vibrational predissociation occurs more readily when the hydrogen bonding hydrogen atom is excited. This was the conclusion, also, for the (02) and (20) states with line widths of 1 GHz and 13.4 MHz, respectively.[279] The predissociation lifetimes measured for the (21) and (12) states[292] are intermediate between those for the (30) and (03) states.

8.2.13 Photoacoustic spectroscopy

The photoacoustic effect was first reported in 1880[293] but its application to spectroscopy began only in 1968 when part of the photoacoustic spectrum of water was obtained using a ruby laser light source.[294]

A molecule which has absorbed radiation may lose it by emission or by collision. In liquids, or in the gas phase at pressures of even a few Torr, loss of vibrational energy is mainly by collision because infrared fluorescence lifetimes are usually long, of the order of a few milliseconds. An effect of a collision is to

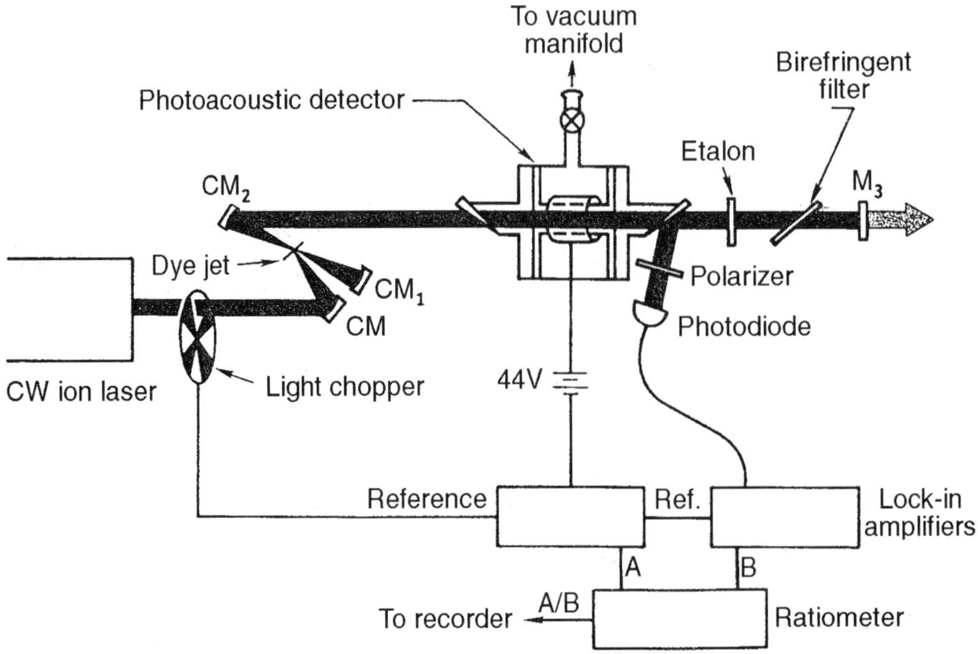

Figure 8.86 A photoacoustic spectrometer with an intracavity sample cell and an ion pumped CW dye laser [Reproduced, with permission, from Bray, R. G. and Berry, M. J. (1979). *J. Chem. Phys.*, **71**, 4909]

transfer the vibrational energy to the collision partner which receives it in the form of translational energy. The collision partner experiences an increase of temperature of less than 1 K. If the source of radiation exciting the molecule is modulated the temperature of the collision partner oscillates and a periodic pressure variation, in the form of a sound wave, travels through the medium. This may be detected by a microphone attached to the cell containing the sample. Scanning the wavenumber of the light source through a region of absorption and monitoring the intensity of the sound waves produce a photoacoustic spectrum.

We shall be concerned here only with photoacoustic vibrational spectra of gaseous, molecular samples in which the collision partner is usually another similar molecule. Mixing the sample with a buffer gas, particularly xenon, at pressure up to 300 Torr has been shown to increase the sensitivity[295] but high resolution spectra have been obtained mainly without a buffer gas.

As with optothermal spectroscopy, discussed in Section 8.2.11, one of the main advantages of photoacoustic spectroscopy is very high sensitivity. One of its most important uses is in the investigation of high vibrational overtone spectra in the near infrared and visible regions. Because the absorption is very weak, the signal increases linearly with the power of the source radiation, making it preferable for the sample to be inside the cavity of a high power laser.

An example of an experimental arrangement for obtaining photoacoustic spectra with an intracivity sample cell[296] is shown in Figure 8.86. The CW dye laser is pumped by an Ar^+ or Kr^+ laser. The gas cell has Brewster angle windows and, in this case, is surrounded by a coaxial microphone which detects the audio signal resulting from modulation of the radiation using a light chopper which can operate at a variable speed.

The photoacoustic method is not adaptable for use in conjunction with a supersonic jet and weakly bound complexes cannot be studied.

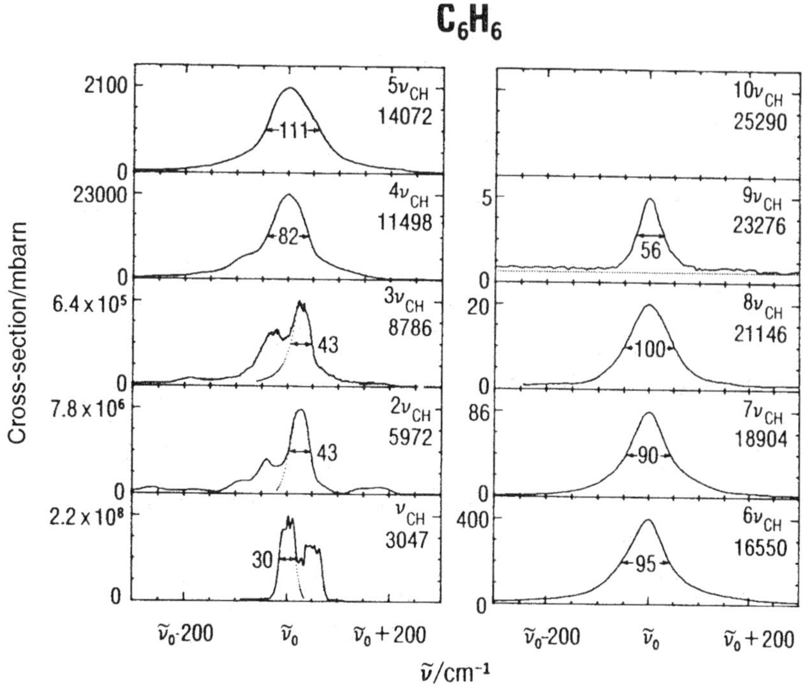

Figure 8.87 Photoacoustic spectra of benzene showing the e_{1u} C—H stretching fundamental and the first eight overtones (the ninth was not obtained)
[Reproduced, with permission, from Reddy, K. V., Heller, D. F. and Berry, M. J. (1982). *J. Chem. Phys.*, **76**, 2814]

Spectra are Doppler limited and, although a sample temperature as low as 77 K can be achieved,[297] there may still be appreciable rotational and vibrational congestion in spectra of larger molecules.

The high sensitivity of photoacoustic spectroscopy makes it ideal for the study of vibrational overtone transitions with high vibrational quantum numbers. Access to these vibrational levels opens up the possibility of studying two important effects which are expected to become more important as the vibrational energy increases. The first is a tendency towards local mode, rather than normal mode, vibrational behaviour (see Section 5.2.6.4) and the second is a tendency to show effects of intramolecular vibrational redistribution (IVR), (see Section 8.2.9).

Figure 8.87 shows the e_{1u} infrared active C—H stretching fundamental, υ_{20} (using the Wilson numbering), of benzene and the first eight overtones observed by photoacoustic spectroscopy.[298] The decline in intensity is indicated by the oscillator strength f [equation (2.26)] which decreases from 1.9×10^{-5} for the fundamental to only 5.0×10^{-13} for the eighth overtone. From the third overtone, the bands appear to be quite symmetrical, to have a lorentzian shape and a line width which varies in an oscillatory fashion.

The wavenumbers of the centres of the bands in Figure 8.87 were found to behave like those of a Morse potential, the energy levels for which are given in equation (5.257), with $\omega_m = 3157\,\mathrm{cm}^{-1}$ and $x_m = -57\,\mathrm{cm}^{-1}$. This supported a picture of the higher overtone states behaving according to a local mode model in which a CH group vibrates in an isolated fashion and like a diatomic Morse oscillator.

The width of the bands, and their approximately lorentzian shape, were interpreted as being a consequence of IVR. However, the spectra shown in Figure 8.87 were obtained with the sample at room temperature.

Optothermal spectra of the 20^3_0 and 20^4_0 bands of benzene have been recorded[283,284] in a supersonic jet, as we saw in Section 8.2.11. Even though the vibrational and rotational cooling was less than is usually obtained, the 20^3_0 band now shows at least seven components, compared with two in Figure 8.87, and the 20^4_0 band shows three components of similar intensity and several weaker ones. Although higher overtones have not been observed with this technique, the evidence from these two bands casts some doubt on the simple local mode interpretation in which the bands were assumed to be dominated by the C—H stretching overtones.

It has been shown from a study of the C—H stretching overtone spectra of a group of molecules CHX_3, where X is ^2H, F, Cl, Br or CF_3 and in which the C—H group is isolated, that there is a Fermi resonance between vibrational levels involving the C—H stretching and bending vibrations. For example, bands involving up to six quanta of C—H stretch of CHF_3 have been observed using both Fourier transform and photoacoustic methods[299] and the stronger bands have been interpreted in this way. The Fermi polyads observed result from the fact that the stretching vibration wavenumber, about 3000 cm^{-1}, is approximately twice that of the bending vibration, which has a wavenumber of about 1500 cm^{-1}.

There are many other bath states of these CHX_3 molecules which may be nearly degenerate with the C—H stretching levels, but those involving C—H bending are much more important. This is in agreement with a more sophisticated theory of IVR[300] in which the bath states are separated into tiers of decreasing importance, so far as the strength of coupling to the bright states is concerned, and of increasing density. In the CHX_3 molecules the C—H bending states constitute the first tier.

An attempt to interpret the benzene overtone spectra in Figure 8.87 uses a similar picture of a first tier of bath states involving overtone levels of the C—H in-plane bending vibration of the same CH group as is involved in the stretching levels.[301] In benzene, the bending vibration has a wavenumber (*ca* 1300 cm^{-1}) which is about half that of the stretching vibration (2530–2190 cm^{-1}) in the region of four to eight quanta of the stretching vibration. Anharmonicity of the C—H stretching vibration causes the Fermi resonance to vary and results in changes in width of the higher overtone bands, which can be seen in Figure 8.87.

In contrast with the benzene photoacoustic spectra those of HCN, obtained[302] with vibrational energies of between 15 000 and 18 500 cm^{-1}, are easily rotationally resolved. The 3^5_0, $1^3_0 3^3_0$, $1^5_0 3^2_0$, $1^2_0 3^4_0$, $1^1_0 3^5_0$ and 3^6_0 bands, as well as the $2^1_1 3^5_0$ hot sequence band, are all Σ^+–Σ^+ transitions. (ν_1, ν_2 and ν_3 are the C—N stretching, bending and C—H stretching vibrations, respectively.) These bands have all been observed for $^1H^{12}C^{14}N$. Rotational and vibrational analyses show that they are all unperturbed. Vibrational transition energies and rotational constants agree with predictions using first order anharmonic constants. Although there is a high density of states for HCN at these high vibrational energies, there are serious limitations on those which can perturb the observed states. In this small molecule there is no evidence for IVR occurring at vibrational energies below 18 500 cm^{-1}.

Observation of the higher vibrational levels of HCN provides a suitable test of the ability of a theoretically produced potential energy surface[303] to predict these levels, and there is good agreement.

The overtone photoacoustic spectrum of acetylene,[304] C_2H_2, contrasts with that of HCN[302] in two respects. First, the symmetrically equivalent C—H bonds give rise to the possibility of local mode behaviour[305] (see Section 5.2.6.4) in which bond stretching is localized in one bond. Second, the density of states is higher in the tetratomic molecule, increasing the possibility of IVR.

Vibrational levels observed in the 14 900–18 500 cm^{-1} region, involving up to six quanta

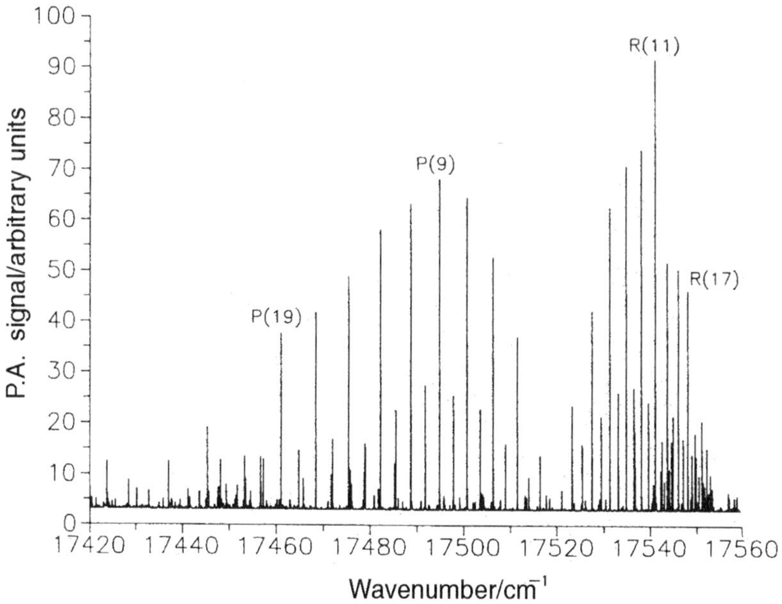

Figure 8.88 The overtone band of C_2H_2 involving five quanta of C—H stretching and one quantum of C—C stretching. Several weaker bands can also be seen
[Reproduced, with permission, from Zhan, X., Vaittinen, O., Kauppi, E. and Halonen, L. (1991). *Chem. Phys. Lett.*, **180**, 310]

of C—H stretching, are consistent with the theoretical prediction[59] that C_2H_2 is an intermediate case in which a local mode description is useful, but the true eigenstates result from mixed normal and local mode behaviour.

Unlike the unperturbed spectrum of HCN, that of C_2H_2 shows many rotational perturbations. Some of these could be caused by interaction with bath states with excited bending vibrational character. Several Fermi resonances with these bending bath states also occur but much less frequently than might be expected from the total density of such bath states. This is indicative of a strong selectivity for these resonances.

Photoacoustic spectra with increased sensitivity have been obtained using a ring dye laser,[306] or a ring laser in which the dye cell may be replaced by a Ti^{3+}:sapphire crystal,[307] thereby increasing the tuning range to 560–1100 nm (17 800–9090 cm^{-1}). The laser band width is about 20 MHz and the photoacoustic cell is placed inside the ring activity.

Figure 8.88 shows part of the photoacoustic spectrum of C_2H_2, in the region of the band involving five quanta of C—H stretching combined with one quantum of C—C stretching. The centre of this $\Sigma_u^+ - \Sigma_g^+$ band is at 17 518.8 cm^{-1}.[304] In the local mode limit, the symmetric (g) and antisymmetric (u) C—H stretching modes become degenerate. Here they are very nearly degenerate and the transition involves the u component. The band shows a simple P- and R-branch structure with convergence in the R and divergence in the P branch due to a decrease of the rotational constant B in the upper state. There is a 1 : 3 intensity alternation for J'' even : odd, just as in the band of C_2H_2 shown in Figure 5.45.

In the region of four quanta of the C—H stretching vibration of fluoroacetylene, HC_2F, seven $\Sigma^+ - \Sigma^+$ bands with the ground state as the lower state have been observed[308] using a Ti^{3+}:sapphire ring laser. Only one band, due to the uncoupled C—H stretching vibration, was expected, indicating strong coupling to

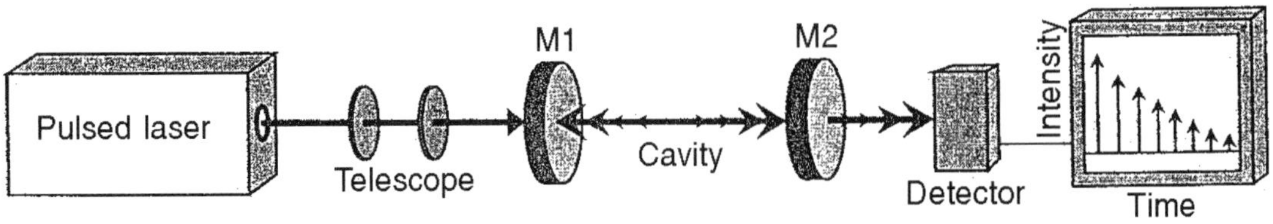

Figure 8.89 Schematic diagram illustrating the principles of cavity ring-down absorption spectroscopy using a pulsed laser and a ring-down cavity between mirrors M1 and M2
[Reproduced, with permission, from Scherer, J. J., Voelkel, D., Rakestraw, D. J., Paul, J. B., Collier, C. P., Saykally, R. J. and O'Keefe, A. (1995). *Chem. Phys. Lett.*, **245**, 273]

some of the bath states. In general, coupling to these states will be less restricted by symmetry than in C_2H_2.

8.2.14 Cavity ring-down absorption spectroscopy

The Beer–Lambert law, expressed previously in the form of equation (2.24), can be rewritten, in terms of napierian logarithms, as

$$\ln(I_0/I) = \kappa cl \tag{8.46}$$

where I_0 is the intensity of radiation entering an absorption cell of length l, I is the intensity of radiation emerging from the cell, c is the molar concentration of the absorbing species and κ is the molar (napierian) absorption coefficient. In cases where the concentration is not known, it is convenient to write equation (8.43) in the form

$$\ln(I_0/I) = \alpha l \tag{8.47}$$

where α is the linear (napierian) absorption coefficient. Alternatively, this can be expressed as

$$I/I_0 = \exp(-\alpha l) \tag{8.48}$$

This form of the Beer–Lambert law shows that the intensity of transmitted radiation declines exponentially as the length over which the absorption takes place increases. If the radiation, travelling with a speed c, takes a time t_l to travel along the absorbing path l then

$$I/I_0 = \exp(-\alpha c t_l) \tag{8.49}$$

The fact that the transmitted intensity decreases exponentially with time forms the basis of the technique of cavity ring-down absorption spectroscopy.[310]

Figure 8.89 illustrates the principle of the technique. A laser pulse enters a cavity, which constitutes the absorption cell, formed between two highly reflective mirrors M1 and M2. The mirrors are usually concave and in a confocal arrangement [see Figure 8.2(c)]. The position of the coincident focal points F_1 and F_2 is where a sample of limited extent, such as a molecular beam or an outlet from a flow tube, should be placed. Otherwise a gaseous sample fills the cavity. The mirrors are coated for maximum reflectance ($>99\%$) in the spectral region of interest so that there is minimal radiation entering and leaving the cavity.

Once the radiation enters the cavity it is reflected, or 'rings', backwards and forwards many times. With a typical cavity length of 1 m an absorption path length of the order of 10 000 m may be achieved.

The ring-down cavity imposes wavenumber restrictions on the output from M2. This is because the cavity can resonate only to wavenumbers corresponding to longitudinal modes of the cavity and is independent of whether the laser has a long or short coherence length.[310] The spacing, $\Delta\tilde{\nu}$, between these modes is given by

$$\Delta\tilde{\nu} = 1/2L \tag{8.50}$$

where L is the length of the cavity. For $L = 1$ m there is a spacing between modes of $0.005\,\text{cm}^{-1}$

and the resulting spectrum is effectively collected in a point-by-point manner, with the points a minimum of $0.005\,cm^{-1}$ apart. This would be the ultimate resolution capability but, in practice, Doppler and pressure broadening are likely to dominate.

As shown in Figure 8.89, what the detector sees initially, following a pulse of radiation entering the cavity, is radiation of relatively high intensity. If the cavity is 1 m long it will take 6.7 ns for more radiation to reach the detector, and so on for every backwards and forwards traversal of the cell. The periodic signal detected declines exponentially according to equation (8.49). If there is no absorbing sample in the cavity the absorption coefficient α is a consequence of the loss suffered on reflection. The time for the intensity I to decline to 1/e of the intensity I_0 of the initial output of M_2 is the ring-down time of the cavity. If there is a sample in the cavity which is absorbing the radiation the ring-down time is decreased: the weaker the absorption, the smaller is the decrease in the ring-down time.

A typical laser pulse length is 10 ns so that, with an interval of only about 6.7 ns between the detected signals, there is overlap between them, but this does not invalidate the method. The time taken for traversal of a 10 000 m path by the radiation is 33 μs so that, with a typical laser pulse repetition frequency of 10 Hz, there is no overlap between signals from successive laser pulses.

The sensitivity of the cavity ring-down method is very high, comparable to that of photoacoustic spectroscopy (see Section 8.2.13), but with the distinct advantage of accurate determination of absolute absorption intensity.

With a cavity 1 m long and a dye laser pumped with a nitrogen laser the 688.25 nm ($14\,530\,cm^{-1}$) band of O_2 was observed[310] and is shown in Figure 8.90. This is the $\upsilon = 1\text{–}0$ band of the $b^1\Sigma_g^+\text{–}X^3\Sigma_g^-$ electronic band system which was first observed[311] in 1948 by the absorption of oxygen in the earth's atmosphere with the sun, at low altitude, as the source of radiation.

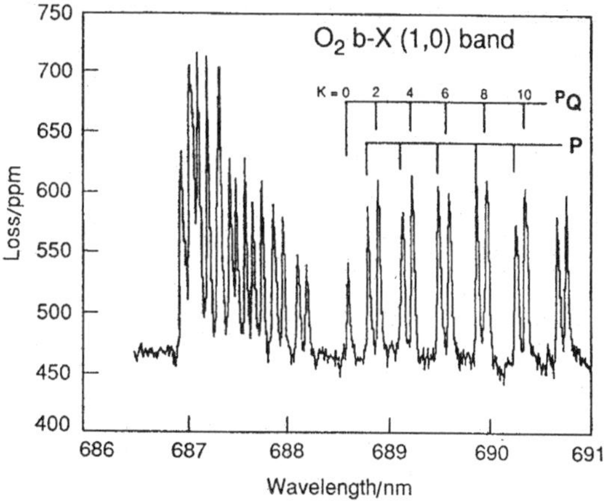

Figure 8.90 The cavity ring-down absorption spectrum of O_2 showing part of the 1–0 band of the $b^1\Sigma_g^+\text{–}X^3\Sigma_g^-$ system
[Reproduced, with permission, from O'Keefe, A. and Deacon, D. A. G. (1988). *Rev. Sci. Instrum.*, **59**, 2544]

The b–X band system of oxygen is doubly forbidden. The transition is electric dipole forbidden but, as in this case, may be allowed by magnetic dipole selection rules (see Section 6.2.3.4). In addition, the b–X transition is spin forbidden, as a singlet–triplet transition, but this is partly overcome by spin–orbit coupling (see Section 6.2.3.3). For these reasons, the b–X system is extremely weak and the oscillator strength [see equation (2.26)] for the 1–0 band is only 2×10^{-11}.

Rotational selection rules[312] for the b–X transition allow a P branch, an R branch and two Q branches, ^rQ and ^pQ, where r and p refer to $\Delta N = +1$ and -1, respectively. Assignments for the P and ^PQ branches are shown; the R and ^rQ branches form the low wavelength group of lines.

As an extension of the photoacoustic spectra of HCN,[302] cavity ring-down absorption spectra have been obtained[313] with six, seven and eight quanta of various mixtures of the C—N and C—H stretching vibrations, ν_1 and ν_3, respectively. A dye laser pumped with an excimer (XeCl) laser was used together with a 1.3 m long ring-down cavity. The laser line width used was $0.18\,cm^{-1}$, so that a fairly high pressure,

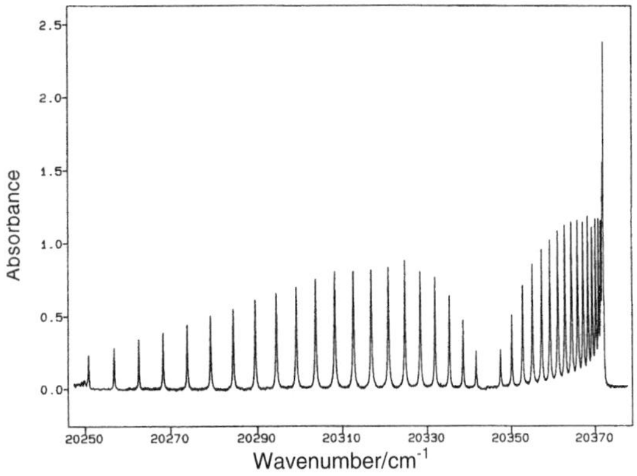

Figure 8.91 The cavity ring-down $1_0^1 3_0^6$ absorption spectrum of HCN
[Reproduced, with permission, from Romanini, D. and Lehmann, K. K. (1993). *J. Chem. Phys.*, **99**, 6287]

100 Torr, of HCN could be used without pressure broadening becoming important. The sample was at room temperature.

The $1_0^1 3_0^5$, 3_0^6, $1_0^2 3_0^5$, $1_0^1 3_0^6$, 3_0^7, $1_0^3 3_0^5$, $1_0^2 3_0^6$ and $1_0^1 3_0^7$ bands, in addition to the $1_0^1 2_1^1 3_0^6$ and $1_0^1 2_1^1 3_0^7$ hot bands, involving the bending vibration ν_2, were observed. Figure 8.91 shows the $1_0^1 3_0^6$ band with the band centre at 20 344.510 cm^{-1}. At the low wavenumber extremity, the beginning of the weak $1_0^1 2_1^1 3_0^6$ hot band can just be seen. The strong degradation of the band, with strong convergence in the *R* branch and divergence in the *P* branch is typical of a vibrational level so high in energy on the potential surface. The change in the rotational constant *B* is -75.69×10^{-3} cm^{-1} compared with 1.478 221 834 cm^{-1} in the ground state.

As with the lower energy states of HCN observed by photoacoustic spectroscopy,[302] there are no appreciable perturbations and, therefore, no evidence for IVR.

The cavity ring-down method has been extended from the visible into the infrared region[314] by using a laser system comprising an optical parametric oscillator (OPO), pumped by a pulsed Nd^{3+}:YAG laser, and a two-stage optical parametric amplifier (OPA) system (see Section 8.1.8.15). This provides tunable radiation from 1.3 to 4.5 μm (from 7700 to 2200 cm^{-1}).

One of the main problems in extending the method to the shorter visible and ultraviolet wavelengths is the decreasing reflectivity of the mirrors. Mirrors are specially coated for a limited wavelength range but, whereas the reflectivity may be as high as 99.99% in the infrared, this decreases to less than 99.7% at wavelengths lower than 450 nm. This reduces the number of traversals of the cavity by the laser pulse, before the transmitted intensity is reduced to 1/e of the original, from about 5000 to 300.

In spite of the consequent reduction in sensitivity, electronic spectra of copper silicide,[315] CuSi, and silver silicide,[316] AgSi, have been obtained in the 420–380 nm region. The molecules were generated and injected into a skimmed supersonic jet using a laser vaporization technique.[317] An excimer laser, operating at 248 nm, was focused on to a rotating and translating copper or silver rod, before the nozzle, causing vaporization of the metal. Coincident with the vaporization pulse, a pulse of mixed helium and silane gas was passed into the vaporization region. On passing through the nozzle, various cluster species were formed and these were monitored by a mass spectrometer in order to optimize the conditions for the metal silicide production. The pulses from the molecular beam pass perpendicularly into the ring-down cavity and are timed to coincide with the pulses from a dye laser entering the cavity.

The $B^2\Sigma^+$–$X^2\Sigma^+$ systems of CuSi and AgSi and the $C^2\Sigma^+$–$X^2\Sigma^+$ system of AgSi were observed in this way. Figure 8.92 shows, for example, the 0—0 band of the B—X system of AgSi observed with a resolution of 0.04 cm^{-1}. Spin splitting in both electronic states is not resolved so the bands show simple *P*- and *R*-branch structure. Typical of all three band systems of CuSi and AgSi, the band is strongly degraded to low wavenumber owing to a large increase in bond length in the upper state. For

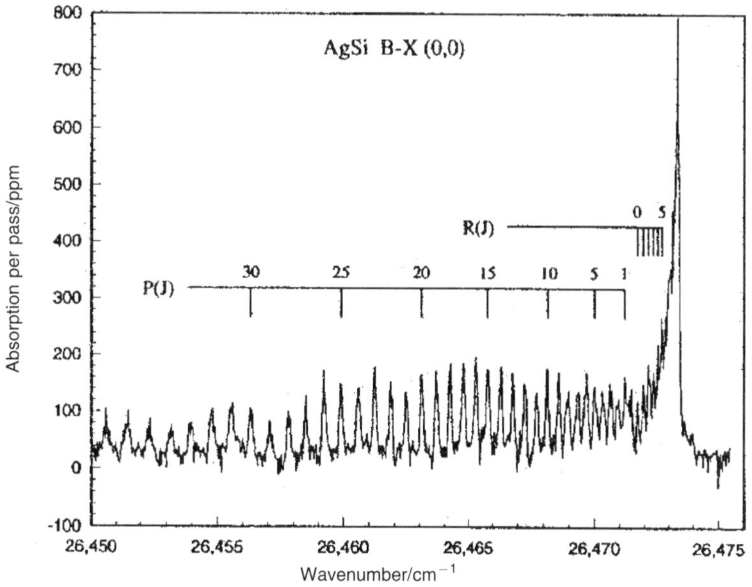

Figure 8.92 The 0–0 band of the $B^2\Sigma^+$–$X^2\Sigma^+$ system in a cavity ring-down spectrum of AgSi showing the *P*- and *R*-branch structure
[Reproduced, with permission, from Scherer, J. J., Paul, J. B., Collier, C. P. and Saykally, R. J. (1995). *J. Chem. Phys.*, **103**, 113]

example, r_e is 2.413 and 2.509 Å in the X and B states, respectively, of AgSi. Because the rotational constant *B* is very small in such heavy molecules, e.g. 0.130 23 cm^{-1} in the X state of AgSi, the spectrum is rotationally very dense owing not only to small rotational constants, but also to the appreciable population of rotational levels with high *J* values, even in the molecular beam.

A hollow cathode ring-down cavity absorption cell, 1.2 m long and with a facility for liquid nitrogen cooling, has been constructed[318] in which the spectra of molecular ions may be observed. With a pulsed discharge in a low pressure (*ca* 1 Torr) mixture of helium and nitrogen the 4–0 and 6–0 vibronic transitions, at 16 360 and 19 790 cm^{-1}, respectively, of the $A^2\Pi_u$–$X^2\Sigma_g^+$ system of N_2^+ were observed at a resolution of 0.15 cm^{-1}. A rotational temperature of about 150 K was achieved with the liquid nitrogen cooling of the discharge.

In a different approach to the cavity ring-down technique the cavity is relatively short and made to resonate in a multimode manner.[319] Mirrors, whose confocal setting would be 10 cm apart, are placed 11.5 cm apart. The effect of this is to produce such a congested mode pattern that there is transmission by the cavity in all regions of the spectrum: there are no gaps between modes. This method was employed in the ultraviolet region where, below about 450 nm, the reflectivity decreases to less than 99.7% compared to 99.99% in the infrared.

Using a dye laser operating at about 298 nm and with a 1 cm long slit-shaped Bunsen burner between the mirrors, an absorption spectrum of the OH radical was obtained with a resolution of 0.1 cm^{-1}. The OH was produced by burning a mixture of methane and air and is vibrationally and rotationally hot. Part of the $A^2\Sigma^+$–$X^2\Pi$ system, with rotational transitions in the 1–0 vibronic band, and also the 2–1 and 3–2 hot bands, is shown in Figure 8.93. The rotational structure of a $^2\Sigma^+$–$^2\Pi$ is similar to that of a $^2\Pi$–$^2\Sigma^+$ transition (see Section 6.2.5.7) in that each *P*-, *Q*- and *R*-branch line is split into two components.

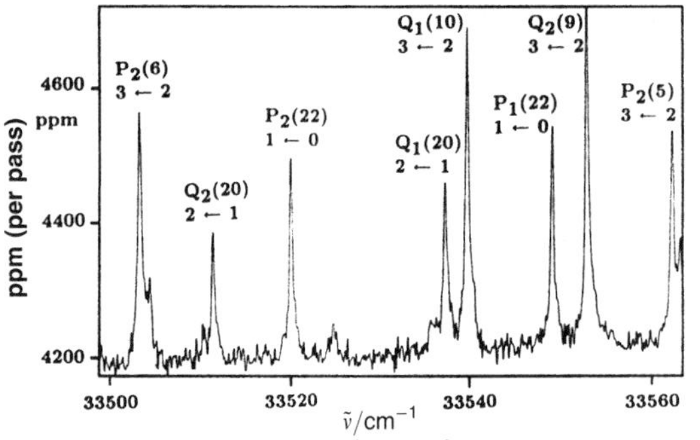

Figure 8.93 Part of the $A^2\Sigma^+$–$X^2\Pi$ spectrum of OH obtained by the coherent cavity ring-down technique [Reproduced, with permission, from Meijer, G., Boogaarts, M. G. H., Jongma, R. T., Parker, D. H. and Wodtke, A. M. (1994). *Chem. Phys. Lett.*, **217**, 112]

8.2.15 Infrared laser and microwave spectroscopy of molecules near to dissociation

Techniques such as photoacoustic and cavity ring-down spectroscopy attempt to observe high energy rovibrational states but do not closely approach the dissociation limit. This section will be devoted to spectroscopic techniques which employ infrared lasers, both carbon dioxide and diode lasers, or microwave and millimetre wave sources to obtain spectra of positive ions in regions which approach the dissociation limit very closely indeed.

An experimental arrangement which has been used[320] to obtain infrared, microwave and millimetre wave spectra in the region of dissociation is shown in Figure 8.94. Positive ions are produced by electron bombardment and accelerated out of the source to a high velocity with a potential of up to 10 kV. The large accessible range of potential, and therefore of ion velocities, allows a large wavenumber range for Doppler tuning the ions into resonance with the source radiation. It also increases the effect known as kinematic compression, or velocity bunching, which reduces the Doppler width: this is proportional to $\Delta V/v$, where ΔV is the energy spread in the ion beam and v is ion velocity.

The ion beam passes into a 55° magnetic sector which mass selects the ion of choice. This beam then passes into a 60 cm long drift tube,

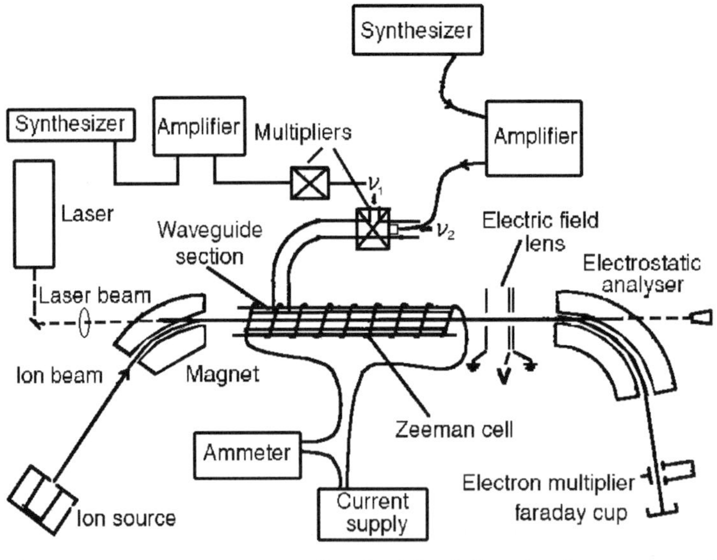

Figure 8.94 A tandem mass spectrometer system for obtaining infrared, microwave or millimetre wave spectra of positive ions. The waveguide cell is replaced by a drift tube for infrared experiments [Reproduced, with permission, from Carrington, A., Shaw, A. M. and Taylor, S. M. (1995). *J. Chem. Soc., Faraday Trans.*, **91**, 3725]

for infrared studies, or a 40 cm long waveguide cell, for microwave/millimetre wave studies. A magnetic field may be applied for observing the Zeeman effect. A product ion, resulting from a charge exchange process, an ion–molecule reaction or dissociation, is then selected by the 81.5° electrostatic analyser and detected by a Faraday cup.

For infrared investigations a $^{12}CO_2$ or $^{13}CO_2$ laser (see Section 8.1.8.10) is used. The use of a second isotopic species increases the number of laser lines available and, with Doppler tuning, a range of 872–1094 cm^{-1} is covered. Figure 8.94 shows the laser beam, which enters and exits through holes bored through the magnet sectors, travelling collinearly with, and in the same direciton as, the ion beam. However, reversal of the direction of the laser beam can be used to increase the range of Doppler tuning.

The method of Doppler tuning, and also the detection method depending on charge exchange, derive from their use in the study[321] of the $\upsilon = 1$–0 band of $^1H^2H^+$.

Doppler tuning partially overcomes the limitations of using a laser producing lines of fixed wavenumbers. The voltage applied to the ion source or, alternatively, to the drift tube is swept across a range up to 10 kV which, in turn, sweeps the ion velocities. Because of the Doppler effect ions with different velocities 'see' the laser radiation as sweeping through a range of wavenumbers. This, together with the ability to reverse the laser direction, to use many $^{12}CO_2$ laser lines and to use the lines of a $^{13}CO_2$ laser, overcomes the essential non-tunability of the CO_2 laser.

The fact that a transition has occurred in the drift tube can only be determined indirectly. One such method is that of charge exchange.[321] This involves collisions between the ions and a gas, such as H_2 or CO, which is introduced into the drift tube. Charge exchange reactions occur between the collision gas and the ions, e.g.

$$^1H^2H^+ + {}^1H_2 \rightarrow {}^1H_2^+ + {}^1H^2H \qquad (8.51)$$

If the ion being studied, $^1H^2H^+$ in this example, is excited, in this case rovibrationally, the collision cross-section is altered. This is reflected in a change in the $^1H^2H^+$ beam current. The charge exchange method has been used to study the $A^2\Pi$–$X^2\Sigma^+$ electronic transition of CO^+, using an Ar^+ laser, by monitoring the CO^+ beam current[322] and also[323] by monitoring the production of HCO^+ produced from the ion–molecule reaction

$$CO^+ + H_2 \rightarrow HCO^+ + H \qquad (8.52)$$

Photodissociation[324] and predissociation[325] methods of detection have been used. For photodissociation, two infrared photons, from the same or different lasers, are required. The first takes the molecule to a level near to dissociation and the second dissociates it into an ion, which is detected, and a neutral. For predissociation an infrared photon takes the molecule to a quasi-bound level from which it predissociates into an ion, which is detected, and a neutral. There is a restriction to this method in that the lifetime of the quasi-bound state must not be too short, otherwise the spectroscopic line observed will be very broad, and it must not be too long, otherwise predissociation will not occur before detection.

A further method of detection involves the dissociation of the ions, when they are excited to energy levels close to dissociation, by a strong electric field.[326] In Figure 8.94 this field is applied by the electric field lens. Fields of up to 40 kV cm^{-1} produce fragmentation of $^1H^2H^+$ to give $^1H^+$ or $^2H^+$ from energy levels which lie up to 5 cm^{-1} below the dissociation limit.[327] The fragment ions emerge from the lens with kinetic energies which depend on the energy levels from which they are ionized.

Microwave and millimetre wave experiments, in the ranges 2.0–26.5 and 40–170 GHz, are conducted in a similar way to that using an infrared laser except that the radiation is injected into the 40 cm long waveguide as shown in Figure 8.94. Reflection of the radiation at the open ends of the waveguide results in radiation travelling in both directions relative to the ion beam resulting in doubling of the Doppler shifted transitions.

In the ground state of $^1H_2^+$, and $^2H_2^+$, as explained in Section 6.2.1.1, the only electron is in the $\sigma_g 1s$ molecular orbital (MO) and the ground electronic state is $X^2\Sigma_g^+$. Promotion of the electron to the $\sigma_u 1s$ MO gives the first excited state, $A^2\Sigma_u^+$. (These orbitals are labelled $1s\sigma_g$ and $2p\sigma_u$, respectively, in the united atom approximation; see Section 6.2.2.) As the $\sigma_u 1s$ MO is antibonding the A state is repulsive. Since both the X and A states dissociate to give the same products, namely a proton and a ground state hydrogen atom, the potential curves should resemble those for the two lowest (X and b) states of H_2 shown in Figure 6.42(b).

In fact, there is a weak attractive force, between the proton and the dipole induced in the atom, which is proportional to r^{-4}, where r is the internuclear distance. This leads to a very shallow minimum with a well depth of only $13.3\,cm^{-1}$ and an internuclear distance of 6.6 Å.[328] The upper parts, very close to dissociation, of the potential curves for the X and A states of $^1H_2^+$ and $^2H_2^+$ are shown in Figure 8.95. The potential minimum for $^1H_2^+$ supports only the $\upsilon = 0$ level with two rotational level, with the rotational quantum number $N = 0$–2, while that for $^2H_2^+$ supports the $\upsilon = 0$ level, with $N = 0$–4, and the $\upsilon = 1$ level, with $N = 0$ and 1.

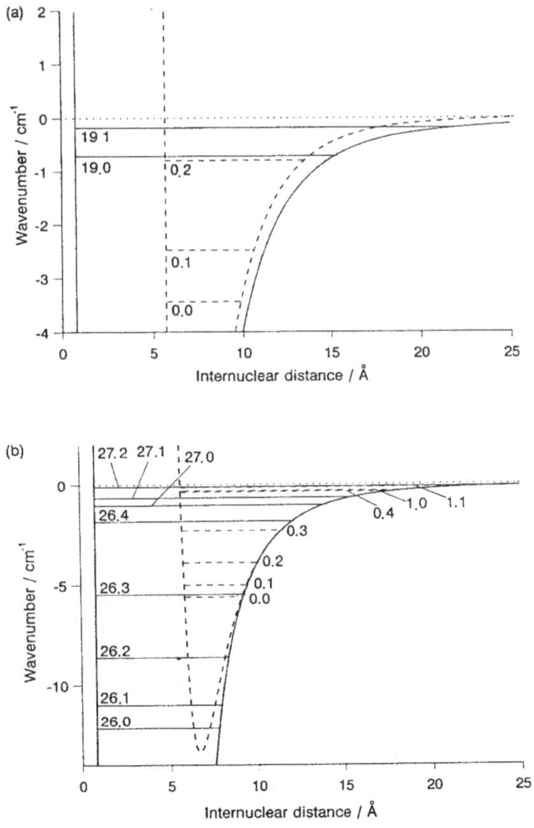

Figure 8.95 Potential curves near to dissociation for the $X^2\Sigma_g^+$ state (solid line) and the $A^2\Sigma_u^+$ state (broken line) of (a) $^1H_2^+$ and (b) $^2H_2^+$. The quantum numbers indicated are υ, N
[Reproduced, with permission, from Carrington, A., Shaw, A. M. and Taylor, S. M. (1995). *J. Chem. Soc., Faraday Trans.*, **91**, 3725]

Neither $^1H_2^+$ nor $^2H_2^+$ has a permanent dipole moment and, therefore, has no rotational microwave or infrared vibrational spectrum. However, the $A^2\Sigma_u^+$–$X^2\Sigma_g^+$ electronic transition is electric dipole allowed and Figure 8.95 indicates that vibronic transitions from high-lying vibrational levels of the X state to the few bound levels of the A state lie in the microwave/millmetre wave or infrared regions. The electron impact process used to form the ion beam produces ions in all vibrational states of the X state up to the dissociation limit. In the low pressure environment of the ion beam there is no collisonal vibrational deactivation. Population of high-lying vibrational levels is essential because it is only for transitions from these levels that Franck–Condon factors are sufficient for the transitions to be observed.

The first such transitions to be observed[329] were rovibronic transitions in the 0–21 and 1–21 vibronic bands of $^2H_2^+$. These were observed using a CO_2 laser, Doppler tuning and electric field dissociation. An $N = 3$–4 microwave transition in the 0–26 vibronic band was also observed.

Transitions in $^1H_2^+$ cannot be observed with a CO_2 laser because they lie in the wrong region of the spectrum, but many microwave and millimetre wave transitions of both $^1H_2^+$ and $^2H_2^+$ have been observed.[320]

Since H_2^+ is the simplest diatomic molecule, having only one electron, it has been an important teset of theory to calculate the spectra which have been observed. There is excellent agreement between recent theory[330,331] and the experimental results.

The disposition of the potential energy curves for the X and A states of $^1H^2H^+$ is different from that for the homonuclear species because, whereas the X state dissociates into $^1H^+ + ^2H$, the A state dissociates into $^1H + ^2H^+$. The two dissociation limits are 29.8 cm^{-1} apart, the X state limit being the lower in energy. Any A–X transitions lie outside the microwave/millimetre wave limit and are expected to be very broad owing to rapid predissociation.

However, characteristic of an ion with a large separation between the charge (at the centre of the molecule in this case) and the centre of mass (close to the deuterium nucleus), $^1H^2H^+$ is calculated to have a large dipole moment.[332] Infrared vibration–rotation transitions within the X state potential should, therefore, be observable.

Infrared spectra were obtained, initially, using a sequential two-photon technique (see Section 8.2.18) using a CO_2 laser and Doppler tuning. The first photon takes the $^1H^2H^+$ from a fairly high-lying vibrational level to one which is just below the dissociation limit and the second takes it into the A state where it dissociates to give $^2H^+$ (or $^1H^+$), which can be detected. A later, alternative, method was to use electric field dissociation[333] following the absorption of a single photon taking the ion close to dissociation. Figure 8.96 shows how much more sensitive is the second method. Transitions were also observed by infrared–microwave and infrared–radiofrequency double resonance techniques (see Section 8.2.21), and also from microwave measurements.

The observation of microwave/millimetre wave transitions, using the electric field dissociation method for fragmenting the excited

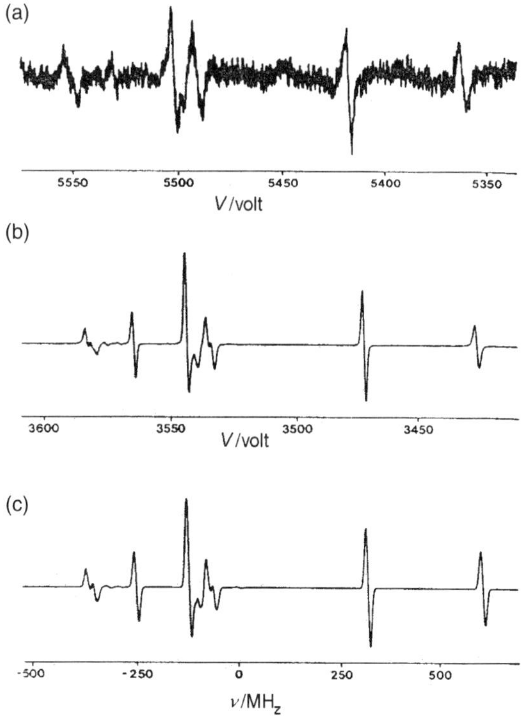

Figure 8.96 The $\upsilon = 21$–17, $N = 2$–3 transition of $^1H^2H^+$, showing hyperfine components, (a) using two photon absorption and monitoring $^2H^+$, (b) using electric field dissociation and (c) calculated using hyperfine theory[334] with a Doppler line width of 9 MHz
[Reproduced, with permission, from Carrington, A., McNab, I. R. and Montgomerie, C. A. (1988). *Chem. Phys. Lett.*, **151**, 258]

molecular ion, has been extended[320] to include the noble gas dimer ions He_2^+, $HeAr^+$, $HeKr^+$, $HeNe^+$ and $NeAr^+$.

In He_2^+ the ground MO configuration is $(\sigma_g 1s)^2(\sigma_u 1s)^1$ giving the $X^2\Sigma_u^+$ ground state. The first excited configuration is $(\sigma_g 1s)^2(\sigma_g 2s)^1$ giving the $A^2\Sigma_g^+$ state. Both states give the same dissociation products and the repulsive A state shows only a shallow minimum about 17 cm^{-1} deep. The A–X electronic transition is allowed and rovibronic transitions have been observed.

The ground states of $HeNe^+$, $HeAr^+$ and $HeKr^+$ are $X^2\Sigma^+$ but the lowest excited state,

$A^2\Pi$, is split by spin–orbit coupling into two components. The lowest one, $A_1{}^2\Pi_{\frac{3}{2}}$, is a bound state and produces the same dissociation products, $He(^1S_0) + Ne^+$, Ar^+, $Kr^+(^2P_{\frac{3}{2}})$, as the X state. Microwave rovibronic transitions have been observed near to dissociation for all three ions, but very few for $HeNe^+$. Microwave transitions observed for $NeAr^+$ have yet to be interpreted.

Microwave electric field dissociation spectra of the weakly bound complexes $He \cdots H_2^+$, which is linear, and $He \cdots N^+$ have been observed.[335,336] The upper levels of the observed transitions are near to the dissociation limit of each molecule. The dissociation of $He \cdots H_2^+$ is within the ground electronic state for which the well depth is about 1700 cm^{-1}. The well depth for $He \cdots N^+$ is about 1400 cm^{-1} but dissociation can give He in its ground state and N^+ in either of the $^3P_2$, $^3P_1$ or $^3P_0$ states corresponding to three different dissociation limits. It is thought that the electric field dissociation gives N^+ in its $^3P_1$ state.

The infrared predissociation spectra of $^1H_3^+$, $^1H_2{}^2H^+$ and $^1H^2H_2^+$ have been obtained. They are all extremely complex; that of $^1H_3^+$ shows 27 000 lines in the 872–1094 cm^{-1} region.[337] The spectra have not been assigned but theory suggests that there are relatively long-lived, so-called periodic states above the dissociation limit.[338] Some periodic states may result from, for example, large amplitude bending motion of quasi-linear $^1H_3^+$.

The infrared predissociation technique has given[339] two rovibrational transitions of $^4HeH^+$; these are 6,13–5,12 and 7,11–5,12 using the v, J notation. However, many more transitions have been observed[340,341] by generating HeH^+ in a 1 m long glow discharge tube containing a mixture of H_2 and He and observing the spectrum with a diode laser source. These transitions are indicated in Figure 8.97, which shows that four were observed in emission.

Figure 8.98 shows three transitions. The 0,24–0,23 transition is a pure rotational transition observed in absorption and is from a level

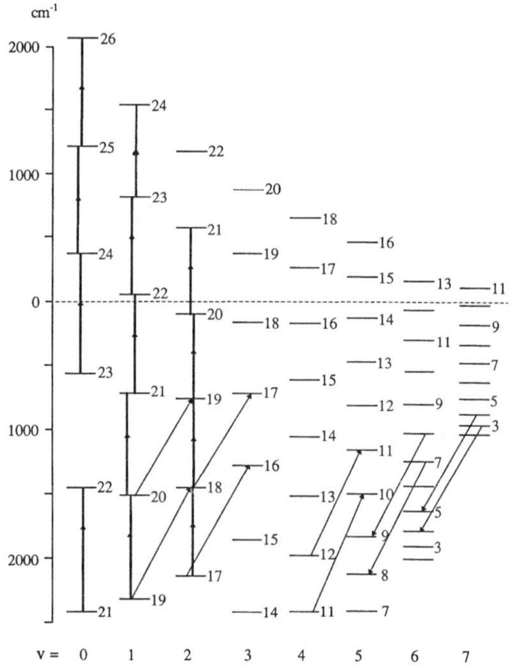

Figure 8.97 Rovibrational energy levels of HeH^+ close to the dissociation limit at 0 cm^{-1}. The stacks of levels associated with each value of the vibrational quantum number v are labelled by the rotational quantum number J. Transitions observed in absorption and emission are shown
[Reproduced, with permission, from Liu, Z. and Davies, P. B., unpublished]

below to one above the dissociation limit, a quasi-bound–bound transition, whereas the 1,23–1,22 is a quasi-bound–quasi-bound transition. The 7,4–6,5 rovibrational transition is observed in emission and appears with the opposite phase in first differential form.

Since the dissociation energy of HeH^+ is about 15 000 cm^{-1}, it is clear that HeH^+ is created in very high energy rovibrational levels in the discharge. The observation of some lines in emission is a consequence of population inversion.

The dissociation limit indicated in Figure 8.97 is that for the non-rotating molecule. When a molecule rotates it undergoes centrifugal distortion and the potential energy V_0 for

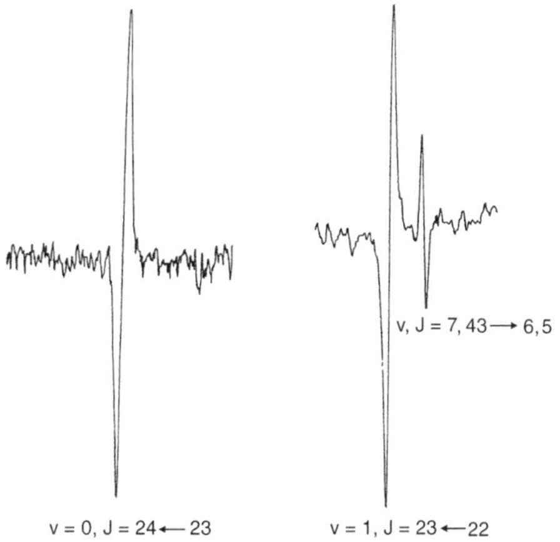

Figure 8.98 The 0,24–0,23 and 1,23–1,22 rotational and the 7,4–6,5 rovibrational transitions of HeH^+. The latter transition appears in emission and the other two in absorption
[Reproduced with permission from Liu, Z. and Davies, P. B., unpublished]

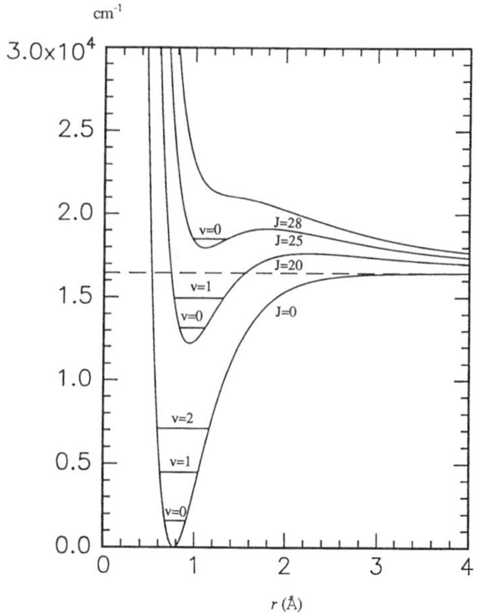

Figure 8.99 Potential curves for the ground electronic state of HeH^+ for various values of J showing quasibound states and centrifugal barriers at high values of J
[Reproduced with permission from Liu, Z. and Davies, P. B., unpublished]

the non-rotating molecule, for example in the form of the Morse function of equation (5.22), is modified, when the molecule rotates, to

$$V_J = V_0 + \frac{h}{8\pi^2 c\mu r^2} J(J+1) \qquad (8.53)$$

where the last term is the rotational term value and the equation is in dimensions of wavenumber.

Two effects of the last term are to move the potential energy curve higher in energy as J increases and also to introduce a barrier to dissociation, known as a centrifugal barrier. These effects are illustrated for HeH^+ in Figure 8.99 in which the $J = 25$ potential curve, for example, clearly shows a centrifugal barrier.

Levels above the $J = 0$ dissociation limit are said to be quasi-bound. Lifetimes of such levels depend on the ease of tunnelling through the barrier in a predissociation process and affect the observed line widths.

The effect of rotation on the potential energy is extremely large in HeH^+ so that, as shown in Figure 8.97, quasi-bound levels have been observed which are more than 2000 cm^{-1} above the $J = 0$ dissociation energy. This effect is typical of diatomic hydrides which have a very small reduced mass, making the last term in equation (8.53) very large, particularly at high values of J.

8.2.16 Rotationally resolved fluorescence excitation spectroscopy

Before tunable visible and ultraviolet lasers, notably the dye laser operating in both the fundamental and frequency doubled ranges, became available, high resolution electronic spectroscopy mostly involved a straightforward absorption experiment, as illustrated in Figure 3.2. The absorption cell might contain a stable, gas phase species or a short-lived species produced by flash photolysis, as in Figure 3.33. Emission spectra were usually obtained

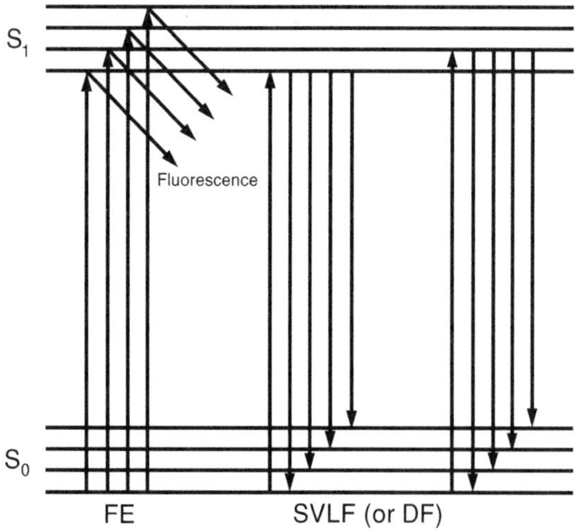

Figure 8.100 Fluorescence excitation (FE) and single vibronic level fluorescence (SVLF), or dispersed fluorescence (DF), processes

from an electrical discharge and might be in the form of fluorescence, in which there is no change of electron spin multiplicity ($\Delta S = 0$) between the upper and lower electronic states, or phosphorescence, in which there is a change of multiplicity ($\Delta S = \pm 1$).

In small molecules, particularly di-, tri- and tetraatomics, the absorption to not only the lowest but also to several higher energy electronic states might be rotationally discrete and show narrow line widths. Discrete emission is usually from various excited electronic states to the ground state but might also be between different excited states.

In this section the application of the technique of fluorescence excitation (FE) to electronic spectroscopy will be discussed. The FE process is illustrated on the left of Figure 8.100. The laser is scanned across the S_1–S_0 (in this example) absorption system while the intensity of the total, undispersed fluorescence is detected. A plot of this intensity against laser wavenumber is the FE spectrum. This spectrum is exactly the same as the corresponding absorption spectrum provided the fluorescence quantum yield [see equation (6.288)] is constant. There is often a tendency for the quantum yield to decrease with increasing wavenumber, due to the onset of competing decay processes, in which case the intensity declines, as the wavenumber increases, compared with that of the absorption spectrum.

With a laser source detection of fluorescence is more sensitive than detection of a small decrease of the laser intensity due to absorption by the species concerned. Laser absorption spectra have been obtained by directing the laser across a slit jet, thereby increasing the absorption path, and by using bolometric detection, which is so effective in the visible and infrared regions (see Section 8.2.11), but the FE technique has proved more generally useful, even when the fluorescence quantum yield is low.

The FE process is only one of several which can be classified under the general name of laser induced fluorescence (LIF). However, there is an unfortunate tendency to refer, specifically, to FE as LIF, which should not be encouraged.

In electronic spectroscopy, with a laser source, the FE technique is very commonly used, but this section will be confined to the FE spectroscopy of molecules, both large and small, which have a very congested room temperature absorption spectrum, and of short-lived and refractory molecules.

One of the first FE spectra to be obtained with laser excitation and the molecules in a supersonic jet was that of NO_2 in the 570.8–670.8 nm region.[90] The room temperature absorption spectrum in this region is due principally to the $\tilde{A}^2B_2$–$\tilde{X}^2A_1$ electronic transition (see Table 6.13), but the rovibronic transitions are so densely packed, almost uniquely so among small molecule electronic spectra, that it is extremely difficult to identify even separate vibronic bands and analysis of the spectrum proved elusive.

The FE spectrum of NO_2 in a supersonic jet at a rotational temperature of about 3 K is dramatically simplified.[90] Using an argon ion pumped dye laser, with an intracavity etalon, radiation, with a band width of 1 GHz (0.03 cm^{-1}), was scanned across the 570.8–670.8 nm region. In this 2600 cm^{-1} range 140 vibronic bands were observed and rotationally resolved.

The NO_2 molecule is a prolate near-symmetric rotor in which spin–rotation coupling splits the rotational levels into two components. At a rotational temperature of 3 K the only rotational levels in the $\tilde{X}$ state which are appreciably populated are those with $K_a = 0$ and $N = 0$, 2 and 4 (those with odd N are missing because the oxygen atoms have zero nuclear spin). Spin–rotation coupling in the $\tilde{X}$ state was not resolvable but, in the $\tilde{A}$ state, each rotational level is split into two with rotational energies given by

$$\begin{aligned} F_1(N') &= \bar{B}'N'(N'+1) + \tfrac{1}{2}\gamma' N' \\ F_2(N') &= \bar{B}'N'(N'+1) - \tfrac{1}{2}\gamma'(N'+1) \end{aligned} \qquad (8.54)$$

where, in a prolate near-symmetric rotor, the rotational constant $\bar{B}' = \frac{1}{2}(B' + C')$ and the spin–rotation constant γ is defined in equation (6.283).

All the bands are type A, polarized along the in-plane axis perpendicular to the C_2 axis. In a supersonic jet, vibrational cooling is typically to a temperature of about 100 K, much higher than the rotational temperature, although it may vary considerably from one vibration to another. Therefore, hot bands, involving vibrational excitation in the $\tilde{X}$ state, may appear in the supersonic jet FE spectrum. Of the 120 vibronic bands observed 20 were hot bands. All the 110 cold type A bands must be, vibronically, B_2–A_1.

Although large rotational perturbations were found in only a minority of the bands the band origins show no regularity at all. In addition, there are far more vibronic bands in this region than the 20 which might be expected in a vibronically unperturbed situation.

Observation of the FE spectrum has been considerably extended[342,343] using higher resolution and a lower rotational temperature. Sub-Doppler line widths of less than 150 MHz ($0.005\,cm^{-1}$) were achieved by collecting fluorescence, excited with a ring dye laser, from only the central core of an unskimmed supersonic jet. An example of a band, centred at $17\,733\,cm^{-1}$, is shown in Figure 8.101. This figure shows that each rotational line is doubled, with $J = N \pm 1$, due to spin–rotation interaction and that each line has a high resolution profile which allows an unambiguous rotational assignment to be made.

On cooling the sample to about 0.3 K, each vibronic band is dominated by only the $R(0)$ line, making identification of weaker bands much easier. In this way a total of 350 $^2B_2$ vibronic levels in the $\tilde{A}$ state have now been identified[343] in the $16\,000$–$19\,360\,cm^{-1}$ region, again showing no regularity. It is concluded that chaotic behaviour is established between these vibronic levels after a few hundred femtoseconds.

The reason for this chaotic behaviour is implied in Figure 6.66(b). This shows that the calculated potential curve for the $\tilde{X}^2A_1$ state crosses that for the $\tilde{A}^2B_2$ state near to its minimum. In a diatomic molecule this crossing would be allowed because the states are of different symmetries. However, in a polyatomic molecule, there may be a non-totally symmetric vibration which mixes the two states and results in an avoided crossing. Figure 6.66(b) shows, on the right-hand side, such an avoided crossing between the $\tilde{X}^2A_1$ and $\tilde{A}^2B_2$ states due to activity of the b_2 asymmetric stretching vibration. This has the correct symmetry to allow mixing of the vibronic levels of the $\tilde{X}$ and $\tilde{A}$ states.

The potential energy in Figure 6.66(b) is plotted against only the angle bending coordinate. If it is plotted in three dimensions, against the angle bending and asymmetric stretching coordinates, a conical intersection[344] is produced between the $\tilde{X}$ and $\tilde{A}$ states. One result is that many of the vibronic levels of the $\tilde{X}$ state, which are extremely dense in the region of the intersection, become associated with the $\tilde{A}$ state and give rise to the extraordinarily high density of vibronic bands in the $\tilde{A}$–$\tilde{X}$ spectrum.

Most diatomic molecules containing a $3d$ transition metal atom have an extremely dense

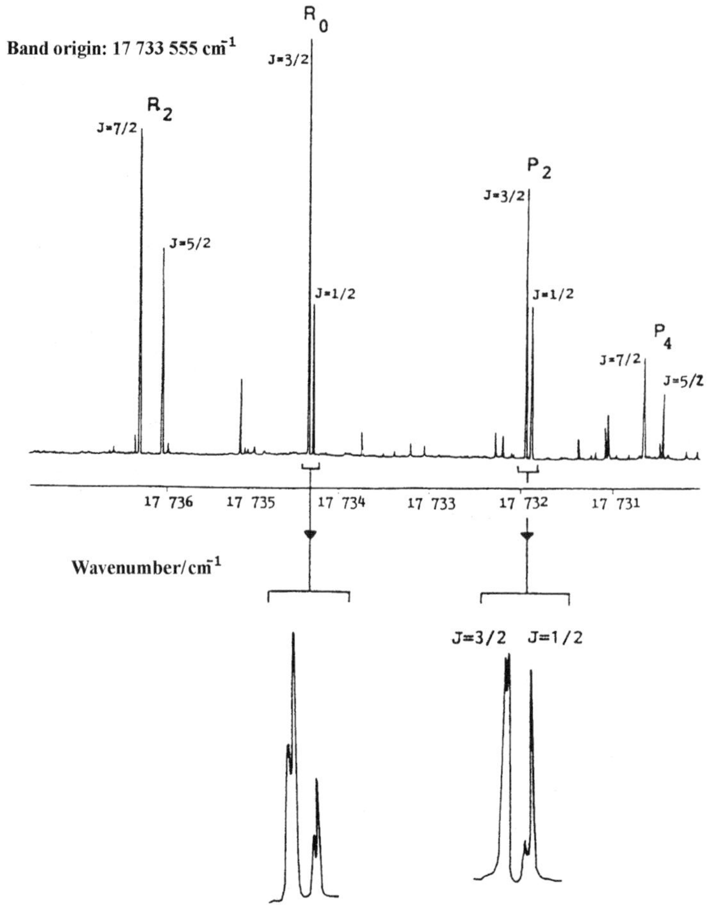

Figure 8.101 The first two *P*- and *R*-branch lines in the $K_a = 0$–0 sub-band of a type *A* band in the $\tilde{A}^2B_2$–$\tilde{X}^2A_1$ system of NO_2. Two spin components are shown for each rotational transition. The high resolution line profiles allow unambiguous rotational assignments
[Reproduced, with permission, from Delon, A., Jost, R. and Lombardi, M. (1991). *J. Chem. Phys.*, **95**, 5701]

electronic spectrum in the near infrared and visible regions owing to large numbers of low-lying electronic states with high spin multiplicity.[345] The reasons for this can be understood from the very approximate, but useful, molecular orbital (MO) diagrams in Figure 8.102 for the examples of the transition metal oxides TiO, FeO and CuO, taken to be typical of metal atoms near to the beginning, middle and end of the 3*d* transition series. This series of oxides is important in the astrophysics of the cooler stars because of the stability of the metal atom nuclei; for example, ^{56}Fe is the end product of the thermonuclear processes in these stars. The high cosmic abundance of oxygen and the high dissociation energies of these metal oxides result in their dominating the spectra of cooler stars.

The atomic orbitals (AOs) which are important in the low-lying electronic states are 4*s* and 3*d* on the metal atom and 2*p* on oxygen. As Figure 8.102 shows, their relative energies vary along the 3*d* transition series. The 3*d* and 4*s* AOs become closer and then cross over, and the 2*p* AO approaches the metal atom AOs more closely as the number of 3*d* electrons increases.

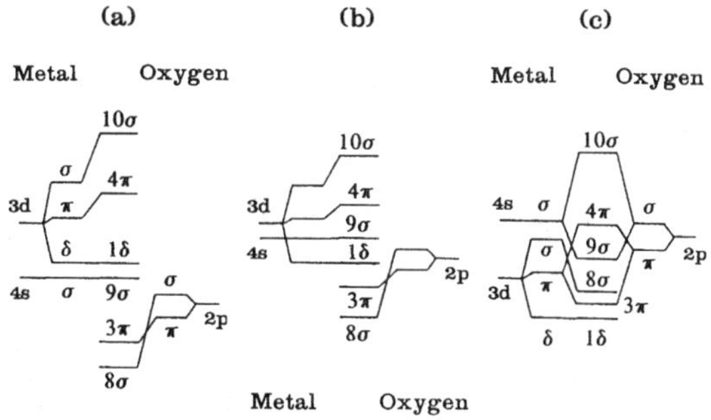

Figure 8.102 Molecular orbital diagrams for the 3*d* transition metal oxides (a) TiO, (b) FeO and (c) CuO
[Reproduced, with permission, from Merer, A. J. (1989). *Annu. Rev. Phys. Chem.*, **40**, 407]

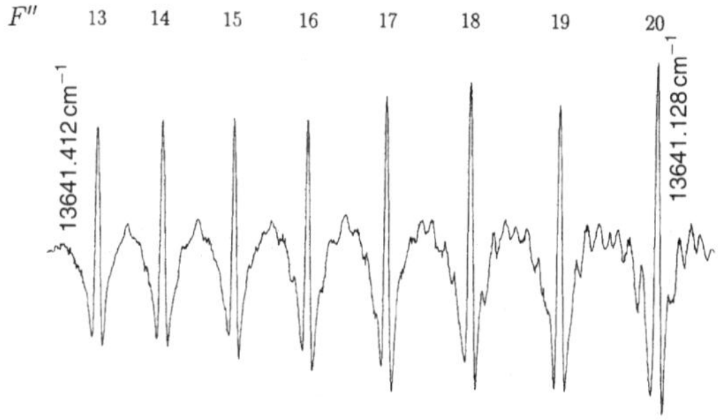

Figure 8.103 Nuclear hyperfine structure of the ${}^{r}P_{31}(15)$ line of the $\upsilon = 1$–0 band of the $B^4\Pi$–$X^4\Sigma^-$ transition of VO. The second derivative line profiles are a result of frequency modulation and lock-in detection
[Reproduced, with permission, from Huang, G., Merer, A. J. and Clouthier, D. J. (1992). *J. Mol. Spectrosc.*, **153**, 32]

Table 6.16 shows how a fivefold degenerate *d* orbital is split in a $D_{\infty h}$, and therefore in a $C_{\infty v}$, ligand field into σ, π and δ orbitals. The 4*s* AO becomes another σ orbital. The threefold degenerate 2*p* AO splits into a σ and a π orbital. Depending on how close the split AOs are in energy, those of the same symmetry may interact to form the MOs, this interaction being the greatest in CuO.

If we take VO as an example, the ground configurations of the V and O atoms are . . . $(3d)^3(4s)^2$ and . . . $(2p)^4$, respectively. These nine valence electrons feed into the MOs to give the ground configuration of VO as . . . $(8\sigma)^2(3\pi)^4(9\sigma^1)(1\delta)^2$. This high spin configuration is of lower energy than one in which there are two electrons in the 9σ MO. Of the states arising from the ground configuration the $X^4\Sigma^-$ state is the ground state. However, from the ground and other low-lying MO configurations of VO a total of 27 low-lying electronic states arise resulting in a very congested electronic spectrum. The configuration . . . $(9\sigma)^1(1\delta)^1(4\pi)^1$ gives rise to the $A^4\Pi$ state[346] and the A–X system lies in the 1050 nm region.

Observation of the FE spectrum of VO in the 790 nm region at sub-Doppler resolution, but not in a supersonic jet, has shown[347] that the upper electronic state is $B^4\Pi$. The VO was formed in a microwave discharge in flowing $VOCl_3$ and argon contained in a fluorescence cell within the cavity of a single mode, standing wave dye laser resulting in sub-Doppler resolution. The fluorescence was collected by a photomultiplier on top of the cell as the laser was scanned through the absorption.

The X and B states of VO are examples of case ($b_{\beta J}$) and case (a_β), respectively, illustrated in Figures 6.59 and 6.55. There are many rotational branches in such a transition and Figure 8.103 shows one rovibronic transition with 8 hyperfine components due to the nuclear spin $I = \frac{7}{2}$ for the ^{51}V nucleus. The hyperfine splitting is caused mostly by splitting of levels in the X state.

In case ($b_{\beta J}$) coupling, which applies to the X state, Figure 6.59 shows that $\boldsymbol{R}$ (or, in the case of a Π state, $\boldsymbol{N}$) couples to $\boldsymbol{S}$ to give $\boldsymbol{J}$. For the transition in Figure 8.103, $N = 15$ and $S = \frac{3}{2}$, giving $J = 16\frac{1}{2}$. Then $\boldsymbol{J}$ couples to $\boldsymbol{I}$ to give $\boldsymbol{F}$, resulting in $F'' = 13, 14, 15, \ldots, 20$.

Determination of the hyperfine parameters for the B state has confirmed that it arises from the . . .$(9\sigma)^0(1\delta)^2(4\pi)^1$ configuration.

The CrO molecule, with one more electron than VO, has a similarly crowded near infrared and visible spectrum owing to a large number of low-lying excited states. The ground configuration is . . .$(9\sigma)^1(1\delta)^2(4\pi)^1$ giving the high spin

$X^5\Pi$ ground state. The $A^5\Sigma^+$–$X^5\Pi$ system lies in the 1250 nm region and the $A'^5\Delta$–$X^5\Pi$ system in the 847 nm region. The CrO molecule has been prepared in a microwave discharge in a mixture of He and CrO_2Cl_2 and this latter system studied through the Doppler-limited FE spectrum[348] with the fluorescence cell external to a Ti^{3+}:sapphire laser cavity.

As was pointed out in Section 6.1.1, there is an inate stability associated with the $3d^5$, high spin, half-filled sub-shell of Mn. This stability is carried through to the hydride MnH in which there are only three excited electronic states below 20 000 cm^{-1}, all of them maintaining the half-filled $3d$ orbital with all the electrons having parallel spins. The molecule has been prepared using a hollow cathode sputtering source containing argon and hydrogen and with a copper cathode impregnated with manganese. Analysis of the Doppler limited[349] and sub-Doppler[350] $A^7\Pi$–$X^7\Sigma^+$ FE spectrum, in the 568 nm region, has shown that the electron promotion is from the 9σ to the 5π orbital which is not shown in Figure 8.102 but is derived from the $4p$ orbital of Mn. The five $3d$ electrons remain undisturbed.

Refractory molecules can be cooled in a supersonic jet provided that they are generated close to the nozzle and that the nozzle is heated to a sufficiently high temperature to prevent condensation. Such a system has been used[351] to generate $NiCl_2$ either by heating $NiCl_2$ to 1100 K or passing CCl_4 in argon over heated nickel. Even with this high temperature method of production of $NiCl_2$, close to the nozzle orifice, a rotational temperature of 12 K was achieved.

As with the $3d$ transition metal oxides and hydrides already discussed, the low-lying MOs of $NiCl_2$, involving the $3d$ AOs of nickel, also result in a large number of low-lying electronic states.

The FE spectrum in the 460 nm region has been rotationally analysed to show that the ground state is $\tilde{X}^3\Sigma_g^-$. The molecule is linear in the ground state and also in the excited state, which is probably $^3\Delta_g$, the forbidden electronic transition being vibronically induced by the bending vibration so that the observed upper vibronic state is $^3\Pi_u$. There is evidence that there are $^1\Sigma_g^+$ and $^3\Pi_g$ electronic states only a few thousand cm^{-1} above the ground state.

A variety of methods have been used for generating short-lived radicals in a supersonic jet[352] and their electronic spectra are commonly investigated by the FE process.

Figure 8.104 illustrates two methods of generation. In the photolysis method, shown in Figure 8.104(a), the supersonic jet, travelling upwards and containing the precursor molecule, is intersected by the beam from a photolysing laser, usually an excimer laser, focused into the jet close to the nozzle. The opening and closing of the nozzle orifice is synchronized with the laser pulses.

The laser ablation and photolysis method, shown in Figure 8.104(b), is particularly useful for the preparation of jet-cooled organometallic radicals. A pulsed stream containing the organic precursor is intersected by a laser beam, frequently from a Nd^{3+}:YAG laser, which is focused on to the surface of a rotating disc made from the appropriate metal. Metal atoms are ablated from the surface and enter the stream where they encounter, and combine with, organic radicals produced by photolysis of the precursor by the dual-purpose Nd^{3+}:YAG laser beam.

The upper part of Figure 8.104 shows intersection of the supersonic jet, downstream from the nozzle, by a tunable dye laser to excite fluorescence. On the right, collection of the total fluorescence gives the FE spectrum, whereas on the left, the dye laser wavenumber is fixed and the fluorescence is dispersed to give the dispersed fluorescence spectrum (see Section 8.2.17).

Other methods of generating radicals in a supersonic jet include high temperature pyrolysis of the radical precursor in the region of the nozzle. Radicals may also be produced in a corona discharge[353] in a supersonic nozzle. This is created in the gaseous mixture containing the precursor molecule by placing a tungsten wire

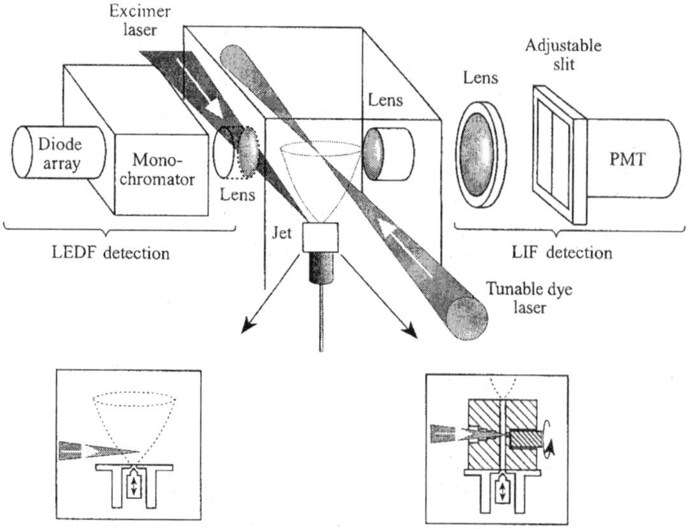

(a) Straight photolysis (b) Laser ablation and photolysis

Figure 8.104 (a) Photolysis and (b) laser ablation and photolysis methods of producing jet-cooled radicals. The upper part shows how the FE (or the dispersed fluorescence) spectra of the radicals are obtained [Reproduced, with permission, from Tan, X. Q., Wright, T. G. and Miller, T. A. (1995). *Jet Spectroscopy and Molecular Dynamics*, Hollas, J. M. and Phillips, D. (Eds.), chap. 3. Blackie, Glasgow]

tip 3–4 nozzle diameters upstream of the nozzle and applying a potential of up to 30 kV to it. As in any electrical discharge the species generated may be present in their ground or excited electronic states.

In a corona discharge an emission spectrum may be observed directly without any need for laser excitation. The first such spectrum to be observed in this way was the $A^2\Pi$–$X^2\Sigma^+$ fluorescence of OH produced from H_2O in a discharge.[354] The rotational temperature in the A state was 11 K but the vibrational temperature was very high indeed, about 3400 K, typical of a discharge with no jet cooling. This lack of vibrational cooling in the upper state causes considerable congestion in the fluorescence spectrum of a larger radical, such as CH_3O,[355] and the fluorescence intensity cannot rival that resulting from laser excitation of the radical.

In a van der Waals complex formed with one open shell moiety, the nature of the interaction between the moieties differs from that in a complex where the moieties are closed shell. The linear OH$\cdots$Ar complex provides an interesting example of such an open shell complex. This has been made[356] by the photolysis method in Figure 8.104. The OH precursor, HNO_3, in a mixture with argon, was photolysed by an ArF excimer laser. The OH was rotationally cold but vibrationally hot. In the ground state, the O—H stretching vibration was populated up to $v''_{OH} = 2$.

Figure 8.105 shows part of the FE spectrum, in the region of the $v_{OH} = 1$–0 band of the $\tilde{A}$–$\tilde{X}$ system, where the $\tilde{X}$ state derives from the $X^2\Pi_{\frac{3}{2}}$ state of OH and the ${}^1S_0$ ground state of Ar, and the $\tilde{A}$ state from the $A^2\Sigma^+$ state of OH and the ${}^1S_0$ state of Ar.

Compared with a complex between closed shell moieties, the observation of a long progression in the van der Waals stretching vibration is unusual and implies, through the Franck–Condon principle, a considerable change of the van der Waals bond length from the $\tilde{X}$ to the $\tilde{A}$ state. Rotational and

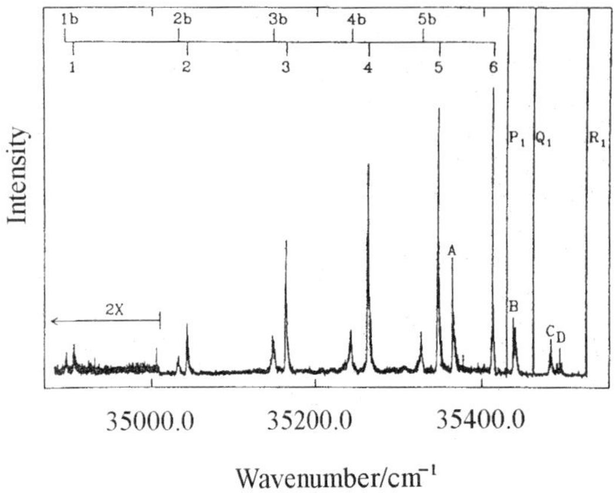

Figure 8.105 FE spectrum of OH$\cdots$Ar in the region of the $\upsilon_{OH} = 1$–0 band of the $\tilde{A}$–$\tilde{X}$ system showing a progression, labelled 1–6, in the van der Waals stretching vibration. The bands labelled 1b–5b are probably a similar progression in larger OH$\cdots$Ar$_n$ clusters. The P_1, Q_1 and R_1 lines are due to free OH
[Reproduced, with permission, from Berry, M. T., Brustein, M. R. and Lester, M. I. (1990). *J. Chem. Phys.*, **92**, 6469]

vibrational analyses are consistent with a decrease of this bond length from 3.6 Å in the $\tilde{X}$ state to the 2.9 Å in the $\tilde{A}$ state.[356] The complex is much more strongly bound in the $\tilde{A}$ state, for which $\omega_e' = 170\,cm^{-1}$ and the dissociation energy $D_0' > 742\,cm^{-1}$, than in the $\tilde{X}$ state, for which $\omega_e'' = 23$–$33\,cm^{-1}$ and $D_0'' > 93\,cm^{-1}$.

Excitation of the O—H stretching vibration, up to $\upsilon = 2$, in both the $\tilde{X}$ and $\tilde{A}$ states of OH$\cdots$Ar shows[356] that, in the $\tilde{A}$ state, there is a strongly enhanced coupling of the O—H stretching to the van der Waals stretching motion.

The CCl_2 radical has been produced, and the FE jet spectrum obtained[357] at slightly sub-Doppler resolution, by the method of pyrolysis of the precursor $(CH_3)_3SiCCl_3$. This was seeded in argon and the mixture pyrolysed at 773 K in the throat of a supersonic nozzle. Typical of any radical produced in a supersonic nozzle, there was rotational cooling to a temperature of 4–8 K but little vibrational cooling.

Whereas the ground state of CH_2 is a triplet state (see Section 6.3.1.1), those of the mono- and dihalides are singlets. Table 6.13 shows, for example, that CF_2 has an $\tilde{X}^1A_1$ ground state and the $\tilde{A}^1B_1$ excited state arises from promotion of an electron from the $6a_1$ to the $2b_1$ orbital. Similarly, CCl_2 has an $\tilde{X}^1A_1$ ground state and an $\tilde{A}^1B_1$ excited state arising from promotion of an electron from the analogous $9a_1$ to the $3b_1$ orbital.

The $\tilde{A}$–$\tilde{X}$ system of CCl_2 is dominated by progressions in the bending vibration ν_2 because of a large increase in the ClCCl angle from 109.2° in the $\tilde{X}$ state to 131.4° in the $\tilde{A}$ state. This is rationalized by the strong angle dependence of the $9a_1$ orbital energy, analogous to that of $6a_1$ in Figure 6.65(a). Rotational analysis of the type C 0_0^0 band (the molecule is a prolate near-symmetric rotor) also shows that the CCl bond length decreases from 1.716 Å in the $\tilde{X}$ to 1.652 Å in the $\tilde{A}$ state; this results in considerable activity of the symmetric C—Cl stretching vibration ν_1. These geometry changes, the lack of vibrational cooling and the presence of two dominant isotopomers lead to considerable vibrational congestion in the FE spectrum.

Much of the vibrational congestion in the fluorescence spectrum of the C_{3v} methoxy radical, CH_3O, observed in a corona discharge[355] is removed when it is made by the laser photolysis technique illustrated in Figure 8.104 and the FE supersonic jet spectrum obtained. The CH_3O was made by photolysis of CH_3NO_2 by a KrF excimer laser and the FE spectrum of the jet-cooled radical obtained by scanning a dye laser through the $\tilde{A}^2A_1$–$\tilde{X}^2E$ band system in the 316 nm region.[358]

Figure 8.106 shows the experimental and computer simulated rotational fine structure of the 0_0^0 band of the $\tilde{A}$–$\tilde{X}$ system of CH_3O, a perpendicular band of a prolate symmetric rotor. The rotational temperature is about 25 K. For the computer simulation, 12 molecular parameters for the $\tilde{X}$ and six for the $\tilde{A}$ state were required, the larger number for the $\tilde{X}$ state being a consequence of spin–orbit coupling in an electronic E state. The unusually large

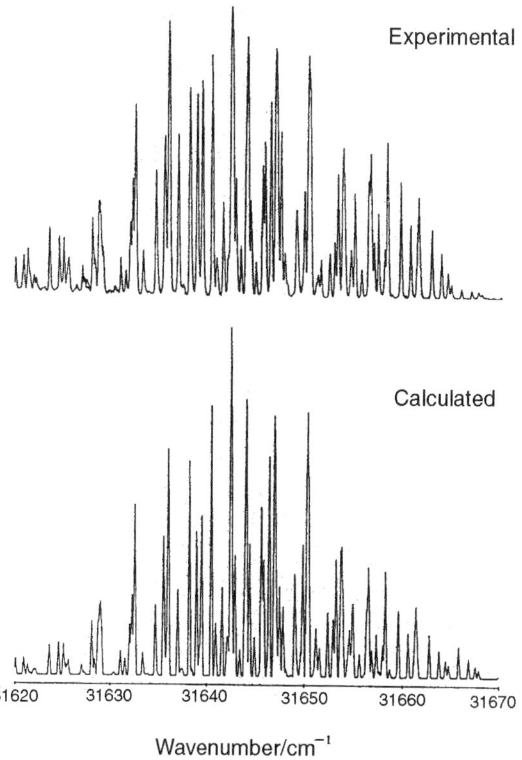

Figure 8.106 The FE spectrum of CH_3O showing the rotational fine structure of the 0_0^0 band of the $\tilde{A}^2A_1$–$\tilde{X}^2E$ system and the computer simulation [Reproduced, with permission, from Liu, X., Damo, C. P., Lin, T.-Y. D., Foster, S. C., Misra, P., Yu, L. and Miller, T. A. (1989). *J. Phys. Chem.*, **93**, 2266]

spin–rotation coupling has been explained with a Jahn–Teller type of vibronic coupling scheme.[359]

The main geometry change from the $\tilde{X}$ to the $\tilde{A}$ state is an increase in the C—O bond length from 1.36 to 1.58 Å. As a result, the spectrum is dominated by progressions in the C—O stretching vibration ν_3.[355] The *e* vibration ν_5 is weakly active in the $\tilde{X}$ state, in the $\tilde{A}$–$\tilde{X}$ fluorescence, owing to the Jahn–Teller effect (see Section 6.3.4.5).

The rotationally resolved FE spectrum of the cyclopentadienyl radical, C_5H_5, has been obtained[360] by the laser photolysis technique. Cyclopentadiene was photolysed with a KrF excimer laser and the FE spectrum obtained with a ring dye laser. Only the central core of the jet was imaged on to the photomultiplier. The consequent reduction of the Doppler broadening resulted in a line width at half intensity of < 100 MHz (0.003 cm^{-1}).

Planar C_5H_5 is an oblate symmetric rotor and belongs to the D_{5h} point group. The $\tilde{A}^2A_2''$–$\tilde{X}^2E_1''$ system lies in the 338 nm region and, since $A_2'' \times E_1'' = E_1'$ and $E_1' = \Gamma(T_x, T_y)$ (see Table A.35 in the Appendix), the transition is allowed and the 0_0^0 band is a perpendicular band. Rotational analysis of spectra obtained with rotational temperatures ranging from 0.6 to 10 K showed, unexpectedly, that there is no resolvable electron spin splitting in the $\tilde{X}$ state. A useful technique for rotational assignment is to start with the lowest temperature spectrum and work up to the highest.

There is a significant Jahn–Teller distortion in the $\tilde{X}$ state, probably in the direction of the e_2' C—C stretching vibration. Observation of the corresponding spectra of various deuterated isotopomers has led to the conclusion[361] that the average C—C bond length is 1.421 Å in the $\tilde{X}$ state with a distortion of 0.0372 Å.

Rotationally resolved FE spectra of the radicals MCH_3, where M = Zn or Cd, have been obtained[362] by the laser photolysis method using an ArF excimer laser and a ring dye laser. They are C_{3v} molecules and their $\tilde{A}^2E$–$\tilde{X}^2A_1$ systems are in the 417 and 438 nm regions, respectively. The 0_0^0 bands are perpendicular bands of a prolate symmetric rotor. Both of them show large spin–orbit splitting and it is thought that this is the reason for there being no observable Jahn–Teller interaction in the $\tilde{A}$ states.

The laser ablation and photolysis method, illustrated in Figure 8.104, has been used to prepare half sandwich radicals, such as MC_5H_5 where M = Ca, Sr, Mg, Zn and Cd. These can be regarded as sandwich complexes in which a metal atom is sandwiched between two cyclopentadienyl moieties, but one of the cyclopentadienyl moieties has been removed. In fact, the half sandwich radicals have not been made from the sandwich complexes but by laser

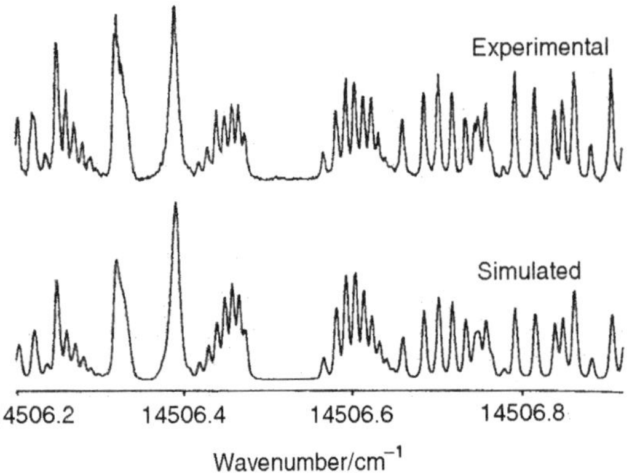

Figure 8.107 The central region of the F_1 component of the 0_0^0 band of the $\tilde{A}^2E$–$\tilde{X}^2A_1$ system of CaC_5H_5 in (above) the FE spectrum and (below) the computer simulation
[Reproduced, with permission, from Cerny, T. M., Williamson, J. M. and Miller, T. A. (1995). *J. Chem. Phys.*, **102**, 2372]

ablation of the metal which then combines, in a supersonic jet, with the cyclopentadienyl radical formed by photolysis of cyclopentadiene.

In the case of CaC_5H_5, for which the rotationally resolved FE spectrum has been obtained and analysed,[363] a KrF excimer laser ablated the Ca atoms from the surface of the metal and also photolysed the cyclopentadiene.

Figure 8.107 shows a part of the 0_0^0 band of the $\tilde{A}^2E$–$\tilde{X}^2A_1$ system of CaC_6H_5 together with the computer simulation. The 0_0^0 band is split by spin–orbit coupling into two components, F_1 and F_2 [see equation (6.270)]; Figure 8.107 shows only the F_1 component. No Jahn–Teller effect is observed in the $\tilde{A}^2E$ state. The most important geometry change is a decrease in the $\tilde{A}$ state by 0.060 Å of the distance between the Ca atom and the centre of the carbon ring.

What we call large molecules are generally understood to be those with principal moments of inertia of the order of magnitude of those of benzene, or larger. Prior to the use of lasers, electronic spectra of such molecules were studied mostly in absorption. If the molecule is closed shell the ground state is a singlet state, S_0, and there are manifolds of excited singlet and triplet states of which the lowest are S_1 and the triplet state T_1, as shown in Figure 6.115. Higher electronic states, S_2, T_2, etc., are usually rotationally very broad owing to a short lifetime process, such as intramolecular vibrational relaxation (IVR), which competes strongly with an emission process. The high resolution study of the electronic spectra of large molecules is usually confined, therefore, to the S_1–S_0 and, if spin–orbit coupling is sufficiently strong, the T_1–S_0 transitions.

At room temperature the population of rotational energy levels in the S_0 state of a large molecule extends to such high values (of the order of 100 of J and K, in a symmetric rotor, or J and K_a (or K_c), in an asymmetric rotor, that there are tens of thousands of rotational transitions accompanying an electronic or vibronic transition observed in absorption. As a result, there may be around 20 rotational transitions within a typical Doppler line width of 0.02 cm^{-1} and individual rotational transitions can never be resolved. What has been achieved for both symmetric rotors and, as was discussed in Section 6.2.5.3(i), for asymmetric rotors, is a computer simulation of the observed rotational contour resulting from the tens of thousands of unresolved rotational transitions. Matching of the computed to the observed contour results in a determination of

(a) The rotational selection rules, which depend on the direction of the electronic or vibronic transition moment. The selection rules may result in a parallel or perpendicular band in a symmetric rotor when the transition moment is, respectively, parallel or perpendicular to the top axis. In an asymmetric rotor the transition moment may be along the a, b or c inertial axis, resulting in a type A, B or C band, or, in a molecule of low symmetry, it may lie between axes, resulting in a hybrid band.

(b) The excited state rotational constants A' (or C') and B', in a symmetric rotor, or A', B' and C' in an asymmetric rotor. (In any planar molecule with zero inertial defect, the out-of-plane moment of inertia is equal to the sum of the two in-plane moments of inertia and the

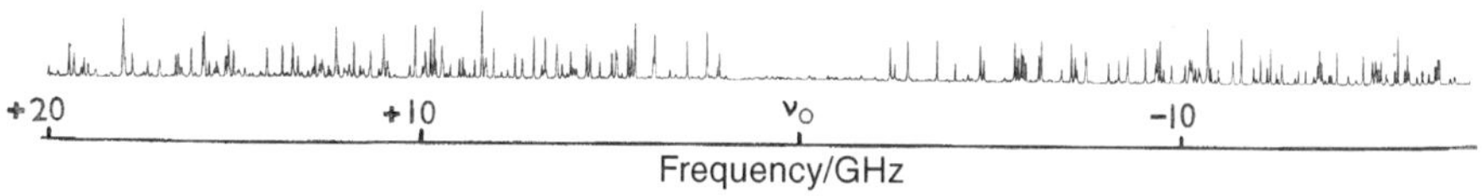

Figure 8.108 FE spectrum showing the central region of the 44^1_0 band in the $\tilde{A}^1B_{2u}$–$\tilde{X}^1A_g$ system of naphthalene
[Reproduced, with permission, from Majewski, W. and Meerts, W. L. (1981). *J. Mol. Spectrosc.*, **104**, 271]

number of independent rotational constants is reduced by one.) In such contour simulations it is necessary to know, or to have a reasonably reliable estimate of, the ground state rotational constants.

Examples of such computer simulations are shown in Figure 6.109 for difluorodiazirine and in Figure 6.110 for 1,4-difluorobenzene.

The development of the dye laser, particularly the high resolution ring dye laser with a frequency doubling facility, has revolutionized electronic spectroscopy, particularly of large molecules. Individual rotational lines can now be resolved in FE spectra of molecules even larger than, say, naphthalene.

The first high resolution, rotationally resolved, sub-Doppler FE spectrum to be obtained was that of naphthalene.[363] A quartz sample container and nozzle were heated to about 380 K to provide sufficient vapour pressure. The skimmed molecular beam was crossed by radiation from a ring dye laser, frequency doubled with a $LiIO_3$ crystal. The line width was less than 0.5 MHz (0.000 02 cm^{-1}).

Figure 8.108 shows part of the FE spectrum of the 44^1_0 vibronic band of the $\tilde{A}^1B_{2u}$–$\tilde{X}^1A_g$ system of naphthalene (see Section 6.3.1.9) at a rotational temperature of 2 K and with a sub-Doppler line width of 35 MHz (0.0012 cm^{-1}). Because the vibration ν_{44} has b_{3g} symmetry the vibronic transition is B_{1u}–A_g and is polarized along the in-plane z-axis (see Figure 6.83). This is the inertial b-axis and 44^1_0 is a type B band of an asymmetric rotor. The band is vibrationally induced and is about 10 times more intense than the electronically allowed 0^0_0 band (see Section 6.3.4.1).

Table 8.2 Changes of rotational constants from the $\tilde{X}$ to the $\tilde{A}$ state of naphthalene obtained by rotational contour simulation[364] and from the FE spectrum in a molecular beam[363]

	Ref. 364	Ref. 363
$\Delta A/cm^{-1}$	−0.002 51(2)	−0.002 487
$\Delta B/cm^{-1}$	−0.000 63(2)	−0.000 624 4
$\Delta C/cm^{-1}$	−0.000 53(1)	−0.000 550 4
$\tilde{\nu}_0/cm^{-1}$	32 453.50(4)	32 453.522(5)

The 44^1_0 band, in the room temperature absorption spectrum, had been rotationally analysed previously [364] by the method of computer simulation of the rotational contour. Ground state rotational constants had to be assumed and the inertial defect in the $\tilde{A}$ state was taken to be zero. The changes ΔA, ΔB and ΔC of the rotational constants from the $\tilde{X}$ to the $\tilde{A}$ state obtained in this way are compared in Table 8.2 with those obtained from the spectrum in Figure 8.108. The agreement is extremely good, confirming the power of the earlier contour method. However, the latter technique has several great advantages: (a) the accuracy of the rotational constants is better by a factor of 10–100; (b) the ground state rotational constants can be determined independently, and the high accuracy with which these can be determined rivals that of microwave spectroscopy and, of course, no permanent dipole moment is necessary; (c) all three rotational constants can be found independently in both states, and it is not necessary to assume zero inertial defect in a planar molecule; and (d) weakly bound species, which are not stable at room temperature, can be investigated.

The S_1–S_0, $\tilde{A}^1B_{3u}$–$\tilde{X}^1A_g$ transition of pyrazine (1,4-diazabenzene), shown in Figure 4.7(b), involves the promotion of an electron from an n to a π^* orbital. The 0_0^0 band has been observed[365] with a sub-Doppler line width of 15 MHz (0.0005 cm^{-1}). Although the band is a type C band of an oblate near-symmetric rotor, the asymmetry parameter κ is close to 1 and the band resembles very closely a parallel band of a symmetric rotor.

In such a band the $P(1)$ transition, having $J' = 0$, should be a single rotational line, having no K_c components. However, the FE spectrum in Figure 8.109 shows that, at the very high resolution obtained, there are many components (36 have been observed) of widely varying intensities spread over only about 5 GHz (0.2 cm^{-1}). The reason for this is that there is a $^3B_{3u}$ electronic state lying below the $\tilde{A}^1B_{3u}$ state and there is a very high density of triplet rovibronic states which are nearly degenerate with the $J' = 1$ level of the $\tilde{A}$ state. Spin–orbit coupling, which is appreciable in the azabenzenes, mixes the singlet with the nearby triplet levels. The singlet character imparted to the triplet levels accounts for the 36 components observed and their relative intensities.

In a smaller molecule, the sub-Doppler FE spectrum of the $\tilde{A}^1A_u$–$\tilde{X}^1\Sigma_g^+$ system of acetylene, C_2H_2, also shows rotational lines with multiplet structure due to spin–orbit interaction with levels of a nearby triplet state.[366]

The $\tilde{A}$ state of C_2H_2 is *trans* bent (see Section 6.3.1.4) and the $\tilde{A}$–$\tilde{X}$ transition from the linear ground state is dominated by progressions in the *trans* bending vibration, which becomes ν_3 in the C_{2h} point group vibrational numbering (it is ν_4 in the $\tilde{X}$ state, see Figure 5.35). Bands in the FE spectrum involving three and four quanta of ν_3 in the $\tilde{A}$ state show groups of lines where one rotational line would be expected. There is a $^3B_2$ state, which may be the lowest triplet state and in which the molecule is *cis* bent, lying below the $\tilde{A}$ state

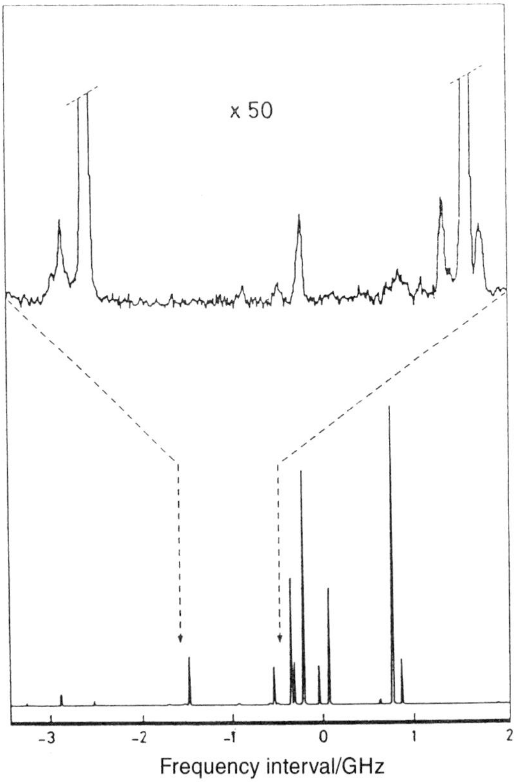

Figure 8.109 FE spectrum of pyrazine showing many components of the $P(1)$ line of the $\tilde{A}^1B_{3u}$–$\tilde{X}^1A_g$ transition
[Reproduced, with permission, from van Herpen, W. M., Meerts, W. L., Drabe, K. E. and Kommandeur, J. (1987). *J. Chem. Phys.*, **86**, 4396]

and the estimated density of triplet rovibronic states agrees with the conclusion that it is this triplet state which is interacting with the $\tilde{A}$ state. Application of a magnetic field further splits each line, confirming the triplet character resulting from spin–orbit coupling.

The examples of pyrazine and acetylene show how these sub-Doppler FE spectra can give important information about triplet state levels which are otherwise difficult to access.

Benzene is the prototype large symmetric rotor molecule for which rotational contour analysis[367] by computer simulation of bands in

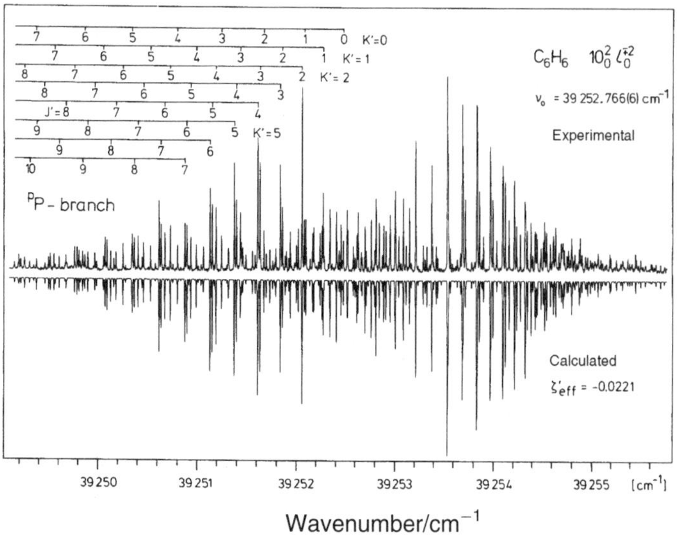

Figure 8.110 The experimental and computed rotational structure of the $10^2_0 l^{\pm2}_0$ band of the $\tilde{A}^1B_{2u}$–$\tilde{X}^1A_{1g}$ system of benzene
[Reproduced, with permission, from Riedle, E. and Pliva, J. (1991). *Chem. Phys.*, **152**, 375]

the $\tilde{A}^1B_{2u}$–$\tilde{X}^1A_{1g}$ system had to suffice before laser sources and sub-Doppler techniques became available. The system is electronically forbidden but vibrationally induced principally by the e_{2g} vibration ν_6, using the Wilson numbering.[368] However, other vibrations or their combinations or overtones may have the correct symmetry for a vibronic transition to be allowed. This is the case, for example, for the weak 10^2_0 band where ν_{10} is an e_{1g} vibration.

Table 5.10 gives the result that $(e_{1g})^2 = A_{1g} + E_{2g}$ so that the 10^2 vibrational level is triply degenerate. The symmetries of the components of the 10^2 vibronic level are obtained by multiplying the vibrational symmetries by B_{2u} giving $B_{2u} + E_{1u}$. A transition from the $^1A_{1g}$ ground state to only the E_{1u} component of the 10^2 state is allowed. Since $E_{1u} = \Gamma(T_x, T_y)$, as in Table A.36, the transition to the E_{1u} component of the 10^2_0 band should give rise to a perpendicular band.

This band has been observed in the sub-Doppler FE spectrum of benzene.[369] The method used[370] employed a frequency doubled, pulse amplified CW dye laser with a rather larger line width, of 100 MHz (0.0033 cm^{-1}), than a ring dye laser. With a skimmed jet the Doppler line width is reduced but the laser line width may make an appreciable contribution to the observed line width.

Figure 8.110 shows the experimentally observed 10^2_0 band at a rotational temperature of 6.5 K and, upside down, the computer simulation.

In an E vibronic state there is a Coriolis interaction with a coupling constant ζ. The computer simulation shows that, in the 10^2 state, $\zeta = 0.0111$, a very small value compared with 0.5788 for the 6^1 state.

For an e vibrational state we must consider the quantum number l as in equation (5.212). Since $l = v, v-2, v-4, \ldots -v$ it can be 0 and ±2 in this case. It is 0 for the unobserved B_{2u} component and ±2 for the E_{1u} component. The parameter q which splits the $l = +2$ and -2 components was found to be 3.14×10^{-4} cm^{-1}.

The 16^2_0, $6^1_0 10^2_0$ and $6^1_0 16^2_0$, where ν_{16} is an e_{2u} vibration, were also analysed,[369] the latter two showing particularly large l-type resonances.

Weakly bound complexes are stabilized in a supersonic jet and structural investigations of those involving large molecules are made

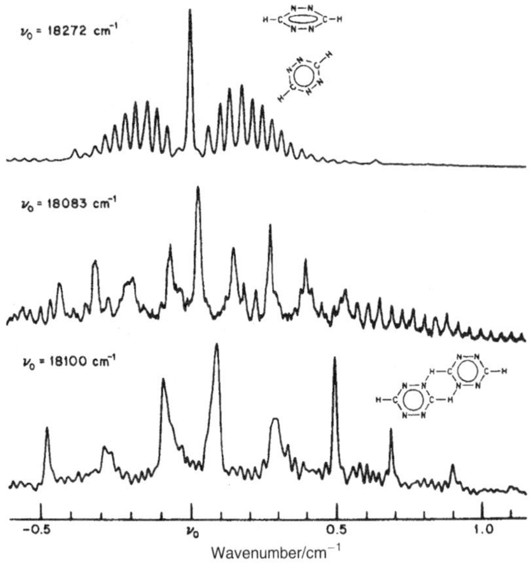

Figure 8.111 FE spectra showing 0_0^0 bands of the *s*-tetrazine dimer. Top: T-shaped dimer, with the electronic transition in the monomer unit in the top of the T. Middle: T-shaped dimer with the electronic transition in the monomer unit in the stem of the T. Bottom: planar dimer
[Reproduced, with permission, from Haynam, C. A., Brumbaugh, D. V. and Levy, D. H. (1983). *J. Chem. Phys.*, **79**, 1581]

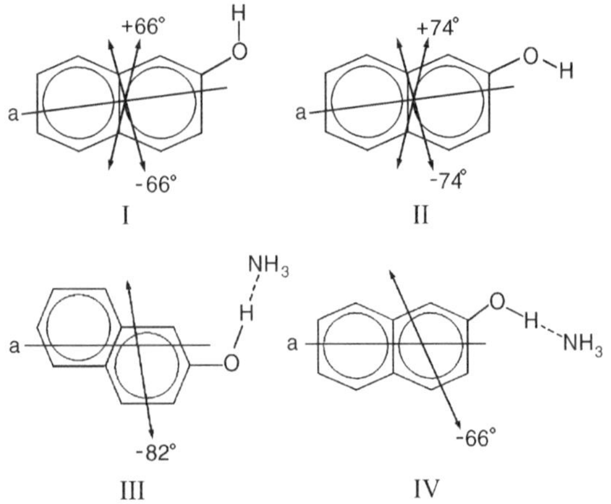

Figure 8.112 *cis*- (I) and *trans*- (II) 2-hydroxynaphthalene, and their complexes with ammonia. The alternative directions of the electronic transition moments obtained for 2-hydroxynaphthalene are shown. Only angles with the *a*-axis of −66° (*cis*) and −74° (*trans*), and −82° (*cis*) and −66° (*trans*) for the complex, are mutually consistent

possible using a high resolution laser to observe Doppler-limited or sub-Doppler rotational fine structure in their FE spectra.

The FE spectra in Figure 8.111 show the 0_0^0 bands of two isomers of the *s*-tetrazine (1,2,4,5-tetraazabenzene) dimer. They occur at about 522 nm and were observed[371] in a supersonic jet using a dye laser with a resolution of $0.006\,\mathrm{cm}^{-1}$, which corresponds to the line width in the lifetime broadened spectrum, at a rotational temperature of about 1 K.

The three bands show different rotational structure. Rotational constants were obtained by computer simulation of the band contours and are consistent with the structures shown. In the planar dimer the electronic transition moment is polarized perpendicular to the molecular plane, as in the $\tilde{A}^1B_{3u}$–$\tilde{X}^1A_g$, π^*–n transition of the monomer, resulting in a type *C* band.

The top two bands in Figure 8.111 show rotational structure which is consistent with their belonging to a T-shaped dimer in which one of the hydrogen atoms in the monomer unit in the stem of the T is attracted towards the π-electron density of the ring at the top of the T. The planes of the two rings are mutually perpendicular but the T is not symmetrical, the stem being tilted at an angle of about 50°. The electronic transition is localized in one of the rings. In the top band in Figure 8.111 the transition is in the ring at the top of the T whereas in the middle band it is in the ring forming the stem.

At higher, sub-Doppler resolution it is possible to distinguish between isomers of large molecules with very similar structures. A good example is 2-hydroxynaphthalene, which, as shown in Figure 8.112, may exist as two rotational isomers (rotamers). For convenience these are labelled *cis* and *trans* according to whether the hydroxyl hydrogen atom is directed towards or away from the central C—C bond, respectively.

Table 8.3 Rotational constants for the S_0 and S_1 states of *cis*- and *trans*-2-hydroxynaphthalene

Rotamer	Rotational constant	S_0	Change of rotational constant	S_1
cis	A''/cm^{-1}	0.095 043	$\Delta A/\mathrm{cm}^{-1}$	−0.002 68
	B''/cm^{-1}	0.027 51	$\Delta B/\mathrm{cm}^{-1}$	−0.000 097
	C''/cm^{-1}	0.021 34	$\Delta C/\mathrm{cm}^{-1}$	−0.000 19
trans	A''/cm^{-1}	0.094 902	$\Delta A/\mathrm{cm}^{-1}$	−0.002 44
	B''/cm^{-1}	0.027 53	$\Delta B/\mathrm{cm}^{-1}$	−0.000 093
	C''/cm^{-1}	0.021 35	$\Delta C/\mathrm{cm}^{-1}$	−0.000 18

The S_1–S_0 system shows two 0_0^0 bands, due to the two rotamers, separated by 317 cm^{-1}. Both rotamers are prolate asymmetric rotors. Rotational constants, given in Table 8.3, have been produced[372] for both electronic states from sub-Doppler FE spectra obtained with a ring dye laser and a skimmed supersonic jet. The A rotational constants are easily the most sensitive to the orientation of the OH group.

Determination of the rotational constants for the two rotamers of 2-hydroxynaphthalene-(O^{2}H) and the use of Kraitchman's equations (see Section 4.9) gave the coordinates of the hydroxyl hydrogen atom and allowed the assignment of the 0_0^0 bands at 30 903.348 and 30 586.277 cm^{-1} to the *cis* and *trans* rotamers, respectively.

These assignments raise the question of why the A'' rotational constant, given in Table 8.3, is larger for the *cis* than the *trans* rotamer. Figure 8.112 indicates that the opposite would be expected as the hydrogen atom is nearer to the a-axis in the *cis* than in the *trans* rotamer. However, *ab initio* calculations suggest that this is due to a tilting of the C—O bond away from the a-axis in going from the *cis* to the *trans* rotamer. The experimentally determined rotational constants cannot provide such structural detail.

2-Hydroxynaphthalene has only a plane of symmetry and the electronic transition moment in the S_1–S_0, π^*–π transition is confined to the molecular plane. However, it is not confined to either of the a or b inertial axes since neither is a symmetry axis. As a result the 0_0^0 bands are type A/B hybrid bands. The hybrid character has been found[372] to be 83/17% type A/B for the *cis* and 92/8% type A/B for the *trans* rotamer, with uncertainties of 5%. From equation (5.236) θ, the angle between the transition moment and the a-axis is $\pm 66°$ and $\pm 74°$ for the *cis* and *trans* rotamers, respectively. The alternative sign for θ is a result of the $\tan^2\theta$ term in equation (5.237) and the alternative directions are indicated in Figure 8.112.

By investigating 2-hydroxynaphthalene in this way there is no means of distinguishing these two possibilities for the transition moment direction. However, when the molecule forms a weakly bound complex with, for example, ammonia, as shown in Figure 8.112 (III and IV), the small electronic perturbation does not affect the transition moment direction in the parent molecule. What are affected are the orientations of the a and b inertial axes: the a axis is indicated in the figure.

Rotational analysis of the sub-Doppler 0_0^0 bands of the ammonia complexes of *cis*- and *trans*-2-hydroxynaphthalene[373] has shown that the transition moment is now at an angle to the a-axis of $\pm 82°$ in the *cis* and $\pm 66°$ in the *trans* complex. Assuming that the transition moment direction is unaffected by complexation we can see that only $-82°$ in the complex and $-66°$ in the parent are consistent for the *cis*, and $-66°$ in the complex and $-74°$ in the parent for the *trans* rotamer. In this way the absolute sign of θ for the parent (and also the complex) has been determined.

Figure 8.113 shows that hydroquinone (1,4-dihydroxybenzene) should exist as two planar

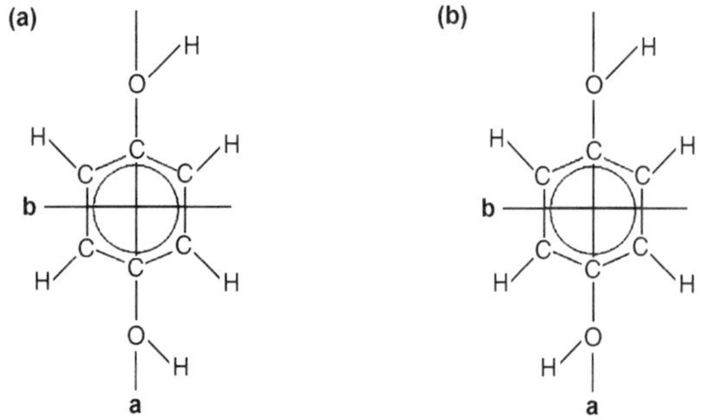

Figure 8.113 (a) *cis*- and (b) *trans*-hydroquinone. The *cis* rotamer belongs to the C_{2v} point group and the *trans* to C_{2h}

rotamers, *cis* and *trans*. The low resolution S_1–S_0, FE jet spectrum shows[374] two 0_0^0 bands separated by a 35 cm^{-1}. Rotational analysis of the sub-Doppler FE jet spectrum showed that the band at 33 534.782 cm^{-1} belongs to the *cis* and that at 33 500.054 cm^{-1} to the *trans* rotamer. Determination of the coordinates of the hydroxyl hydrogen atom, using the spectrum of the deuterated species, confirmed these assignments, but the most elegant way of doing this was by making use of the different nuclear spin statistical weights of the rotational levels of the two rotamers in S_0.

cis-Hydroquinone belongs to the C_{2v} point group and, as Figure 8.113 shows, rotation about the C_2-axis, which is the *b* inertial axis, exchanges three pairs of identical hydrogen nuclei each with the nuclear spin quantum number $I = \frac{1}{2}$. This results in the nuclear spin statistical weights shown in Table 8.4 being attached to the asymmetric rotor energy levels symbolized by K_a and K_c being even (e) or odd (o)—see also Section 4.5 for a discussion of nuclear spin statistical weights in asymmetric rotors.

trans-Hydroquinone belongs to the C_{2h} point group. It also contains three pairs of identical hydrogen nuclei but it is rotation about the *c*-axis which exchanges them. The resulting nuclear spin statistical weight factors are given in Table 8.4 and differ from those for the *cis* rotamer. It is this difference which allowed the assignment of the 0_0^0 bands to their respective rotamers using the intensities of the resolved rotational structure.

Spectra of hydroquinone-$(O^2H)_2$ show different nuclear spin statistical weight factors, given in Table 8.4. Although the differences are not so pronounced they allowed assignments of the 0_0^0 bands of the *cis* and *trans* rotamers to be made.[374]

Aniline (see Figure 4.38) readily forms a van der Waals complex with argon in a supersonic jet and the S_1–S_0 0_0^0 band is 54 cm^{-1} to low wavenumber of that of aniline. The FE spectrum of the complex has been obtained with sub-Doppler resolution using a skimmed jet and a ring dye laser.[375] Figure 8.114 shows the rotationally resolved 0_0^0 band, at a rotational temperature of 2.2 K, and compares a part of the expanded spectrum with a computer simulation.

The heavy argon atom converts aniline, which is a prolate asymmetric rotor, into an oblate asymmetric rotor. The electronic transition moment is polarized along the *b* inertial axis in aniline.[376] It is unaffected in the complex but the direction of polarization is now parallel to that of the *c*-axis of the complex resulting in a

Table 8.4 Nuclear spin statistical weight factors for *cis*- and *trans*-hydroquinone

K_aK_c	*cis*-$(O^1H)_2$	*trans*-$(O^1H)_2$	*cis*-$(O^2H)_2$	*trans*-$(O^2H)_2$
ee	7	7	13	13
oe	9	7	11	13
eo	9	9	11	11
oo	7	9	13	11

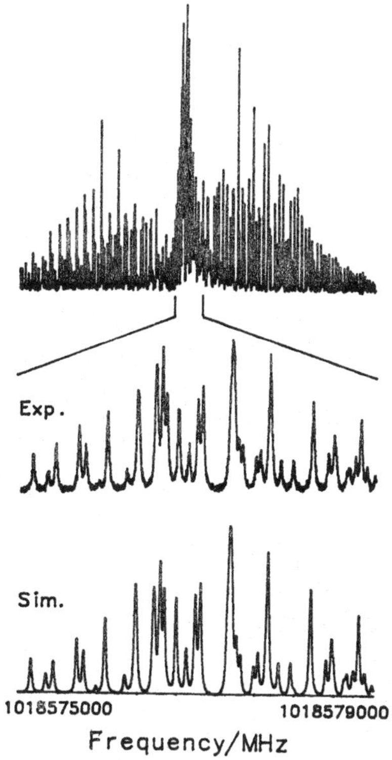

Figure 8.114 The rotationally resolved 0_0^0 band of aniline ··· Ar and a computer simulation of a part of the expanded spectrum
[Reproduced, with permission, from Sinclair, W. E. and Pratt, D. W. (1996). *J. Chem. Phys.*, **105**, 7942]

type *C* band characterized by the central intensity maximum in Figure 8.114.

Interpretation of the rotational constants of aniline-(N^1H_2 and N^2H_2)···Ar places the argon atom 3.541 and 3.472 Å above the plane of the benzene ring in S_0 and S_1, respectively. The decrease in S_1 is a result of stronger van der Waals bonding in the more polarizable upper state. Aniline is non-planar in S_0 and it retains this conformation in aniline···Ar in which the argon atom is on the opposite side of the ring to the out-of-plane hydrogen atoms of the NH_2 group. In both aniline and the complex the NH_2 group is coplanar with the benzene ring in S_1.

In aniline···Ar the argon atom is displaced slightly from above the centre of the ring towards the nitrogen atom.

8.2.17 Fluorescence excitation and single vibronic level (or dispersed) fluorescence spectroscopy with vibrational resolution

In Section 6.3.4.2(i), progressions in totally symmetric vibrations in electronic spectra of polyatomic molecules were discussed. The Frank–Condon principle tells us that a long vibrational progression indicates an appreciable change of geometry from one electronic state to the other in the direction of the corresponding normal coordinate. In a diatomic molecule there is only one vibration and the intensity distribution along a progression can be used to obtain the change in bond length, although this is usually more accurately determined from the rotational constant *B* in the two electronic states.

In polyatomic molecules it is generally much more difficult to obtain geometry changes from progression intensities. Benzene provided an early exception. The intensity distribution along the progression in the symmetrical ring-breathing vibration v_1, which, as Figure 6.92 shows, dominates the $\tilde{A}^1B_{2u}$–$\tilde{X}^1A_g$ absorption spectrum, was interpreted in terms of an increase by 0.345 Å of all the C—C bonds from the $\tilde{X}$ to the $\tilde{A}$ state.[377] This value agrees well with that from a rotational analysis of a very much higher resolution spectrum.

However, the benzene example turned out to represent something of a *cul de sac* because of its high symmetry and the fact that the geometry change is quite large and concentrated in one normal coordinate. We only have to go to, say, fluorobenzene, in which there is a similar ring expansion in the $\tilde{A}$ state, to encounter complications. The main one is that, because of the lowering of the symmetry from D_{6h} to C_{2v}, the ring breathing motion is now distributed among several normal coordinates. The $\tilde{A}$–$\tilde{X}$ spectrum is much more vibrationally complex, leading to difficulties in identifying progressions in the corresponding vibrations and measuring the intensity distributions along them.

The situation changed with the invention of the supersonic jet. Under conditions of low vibrational temperature, typically around 100 K but this may vary from one vibration to another, the electronic spectra of large molecules are greatly simplified, making assignments of progressions much more reliable. In addition, and most importantly, this allows the investigation of much larger molecules and, particularly, molecules which have large amplitude, and therefore low wavenumber, vibrations. A low wavenumber vibration has a correspondingly small force constant which is likely to change appreciably when the electronic structure of the molecule is changed as, for example, on electronic excitation. The resulting change in the vibrational potential results, as a consequence of the Franck–Condon principle, in a long progression in that vibration.

Investigation of a vibrational potential in an excited electronic state may be by an absorption experiment or, more likely, fluorescence excitation (FE). The potential in a ground electronic state can be interrogated by the single vibronic level fluorescence (SVLF) technique, also known as dispersed fluorescence (DF). In both FE and SVLF experiments the molecule is usually seeded into a supersonic jet. Because only vibrational resolution, say $< 1\,cm^{-1}$, is required, skimming of the jet is not necessary.

The principles of the FE, discussed in Section 8.2.16, and SVLF techniques are illustrated in Figure 8.100 in which the ground state, S_0, and the excited electronic state, S_1, are assumed to be singlet states. In SVLF the laser radiation, forming a beam crossing the jet, is tuned to a particular absorption band. It is a very important property of the jet that conditions are collision-free, that is, virtually no collisions occur during the fluorescence lifetime of S_1. The fluorescence is collected, sometimes by a concave mirror or just a lens, as for example in Figure 8.104, and directed into a scanning monochromator and thence to a detector.

The resulting SVLF, or DF, spectrum shows transitions which are exclusively from the vibronic level populated by the absorption. (In a small molecule it may be possible to populate a single rovibronic level and to observe dispersed fluorescence from that level. The resulting spectrum is then a single rovibronic level fluorescence, or SRLF, spectrum.)

The transitions to ground state vibrational energy levels have intensities governed largely by Franck–Condon factors. The intensity distribution along a vibrational progression may be altered by changing the laser wavenumber so that it sits on another absorption band and populates a different vibronic state in S_1. In this way ground state vibrational energy levels may be accessed which are inaccessible by any other means.

In the FE spectrum transitions are observed, mainly from the zero-point level of S_0 because of the cooling experienced in the jet, to vibrational levels of S_1. Again, the intensities are governed by the Franck–Condon factors but the flexibility of the SVLF technique, because of the ability to observe fluorescence from different vibronic levels of S_1, is missing. Some low-lying vibrational levels in S_0 may be sufficiently populated in the jet for an SVLF spectrum to be obtained with the laser sitting on a hot absorption band, but this is not generally the case.

It has already been pointed out that progressions in electronic spectra are more likely in large amplitude, low wavenumber vibrations and these generally fall into two categories, inversion and torsional vibrations.

Inversion vibrations have been discussed in Section 5.2.7.1. The classical example in a small molecule is that of NH_3. The potential for the inversion vibration v_2 is shown in Figure 5.68 and shows the typical double minimum with an energy barrier, of $2020\,cm^{-1}$ in this case, to planarity. In general an inversion potential may be represented by the gaussian-type in equation

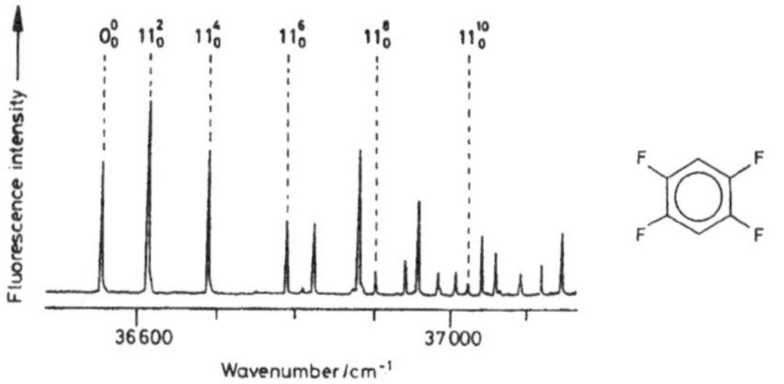

Figure 8.115 Part of the S_1–S_0 FE spectrum of 1,2,4,5-tetrafluorobenzene
[Reproduced, with permission, from Okuyama, K., Kakinuma, T., Fujii, M., Mikami, M. and Ito, M. (1986). *J. Phys. Chem.*, **90**, 3948]

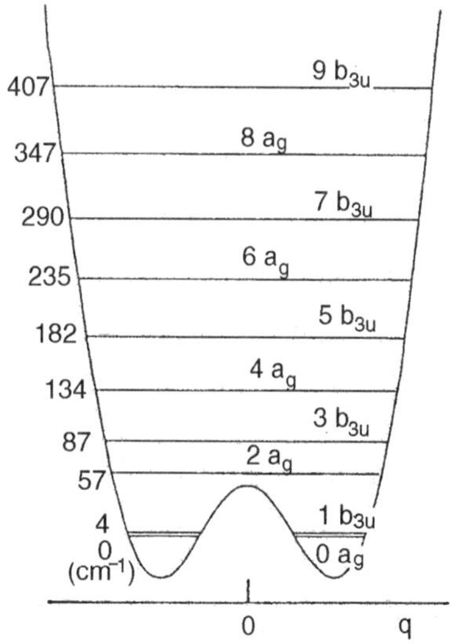

Figure 8.116 Double minimum potential for the 'butterfly' vibration of the fluorine atoms in 1,2,4,5-tetrafluorobenzene
[Reproduced, with permission, from Okuyama, K., Kakinuma, T., Fujii, M., Mikami, M. and Ito, M. (1986). *J. Phys. Chem.*, **90**, 3948]

(5.262) or the quadratic–quartic in equation (5.264).

A much-studied inversion potential in a larger molecule is that of aniline (see Figure 4.38). In the S_0 ground electronic state, $\tilde{X}^1A_1$, the hydrogen atoms of the amino group are not coplanar with the rest of the molecule. The barrier to planarity is 547 cm^{-1} and, like NH_3 in several of its excited electronic states, is planar in S_1, the lowest excited singlet state $\tilde{A}^1B_2$.[378] The inversion vibrational energy levels, from which the potentials were constructed, were obtained by various spectroscopic methods—all of them without lasers. Unlike the case of NH_3, in which the inversion vibration is totally symmetric, in aniline it is non-totally symmetric—b_1 in the C_{2v} point group. Consequently, vibronic transitions with only Δv even are allowed in the electronic spectrum.

Fluorobenzene and 1,4-difluorobenzene are planar in S_0 and S_1 and there seemed to be no reason to suspect that 1,2,4,5-tetrafluorobenzene, shown in Figure 8.115, is any different. However, the S_1–S_0, $\tilde{A}^1B_{2u}$–$\tilde{X}^1A_g$ FE spectrum in a supersonic jet, part of which is shown in the figure, indicates otherwise.[379] The region near to the 0^0_0 band shows a long progression in the b_{3u} fluorine inversion, or 'butterfly', vibration ν_{11} in which the fluorine atoms move symmetrically above and below the plane of the benzene ring. The progression, up to $v = 10$, attached to the 0^0_0 band involves only even quanta as ν_{11} is non-totally symmetric.

The length of the progression, and the irregular spacings, indicate a structural change from the planar S_0 to a non-planar S_1 in which the fluorine atoms are above or below the plane. Therefore, the potential for ν_{11} in S_1 is W-shaped, qualitatively similar to that for the ground state of NH_3. The observed energy levels have been fitted to the potential function in equation (5.262) and the result is shown in Figure 8.116. This shows that, in S_1, there is an energy barrier to planarity of 78 cm^{-1}. The C—F bonds are about 11° out-of-plane.

The FE jet spectrum in Figure 8.115 was obtained with a scanning dye laser with a line width of about 1 cm^{-1}, adequate for the vibrational resolution required.

SVLF spectra of 1,2,4,5-tetrafluorobenzene were also obtained.[379]. The laser wavenumber was tuned to match, in turn, those of the first five members of the progression. The spectra are shown in Figure 8.117 and show how the Franck–Condon factors change for the members of the progression in ν_{11} in the S_0 state as the upper vibronic state changes. For example the transition to $v''_{11} = 8$ is prominent when $v'_{11} = 6$, and $v''_{11} = 6$ is prominent when $v'_{11} = 8$. In this way many vibrational levels of S_0 can be probed by varying v'_{11}.

The low intensity of dispersed fluorescence in an SVLF spectrum, compared with that of the undispersed fluorescence in an FE spectrum, means that the resolution obtained is usually limited by the width of the entrance slit of the dispersing monochromator. For the spectra in Figure 8.117 the resolution was about 6 cm^{-1} but this is adequate for the vibrational resolution required.

The vibrational intervals in ν_{11} in S_0 are regularly spaced, consistent with the molecule being planar.

Information regarding molecular geometry derived in this way from the positions of minima in a potential energy curve refers to equilibrium, so far as the corresponding vibration is concerned, but to the zero-point levels of all the other vibrations in the molecule.

The barrier to planarity of 78 cm^{-1} is very low indeed compared with, say, a typical bond dissociation energy, but the potential in Figure 8.116 illustrates how sensitive the vibrational levels are to this barrier. There is considerable irregularity of the levels below and just above the barrier which can easily be measured from an FE spectrum at fairly low resolution.

Confirmatory information regarding the potentials in S_1 and S_0 can be derived from the intensity distributions along the inversion vibration progressions in the FE and SVLF spectra. From the potentials, the vibrational wave functions, and hence the overlap integrals [see equation (6.127)] and Franck–Condon factors (the square of the overlap integral), can be derived. If the potentials are correct the Franck–Condon factors should agree with the observed intensities.

The unexpected change of shape from S_0 to S_1 is due to the fact that, as more hydrogen

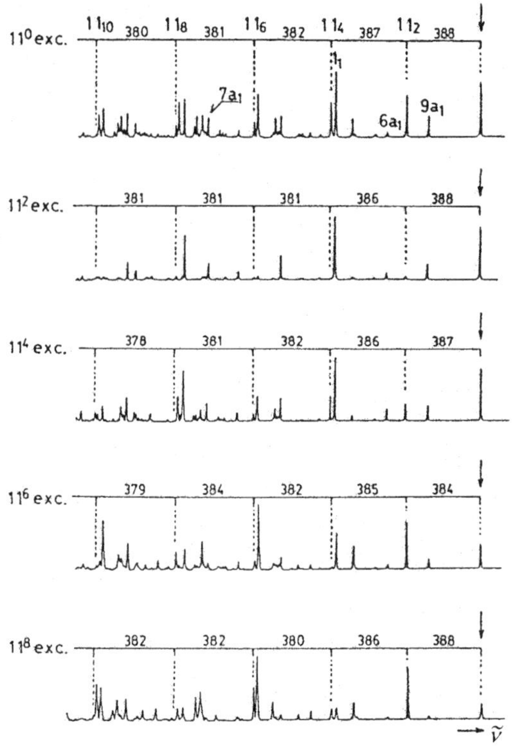

Figure 8.117 SVLF spectra of 1,2,4,5-tetrafluorobenzene obtained by dispersing the fluorescence from various levels of ν_{11} in S_1
[Reproduced, with permission, from Okuyama, K., Kakinuma, T., Fujii, M., Mikami, M. and Ito, M. (1986). *J. Phys. Chem.*, **90**, 3948]

atoms of benzene are substituted with fluorine atoms, the energy of a $\sigma^*\pi$ state is lowered. The σ^* orbital is antibonding and localized in the C—F bonds. In 1,2,4,5-tetrafluorobenzene this state is lowered in energy so much that it is involved in strong vibronic coupling with the S_1, $\pi^*\pi$ state.

Derivatives of anthracene, in which the unsaturated central carbon atoms, in the 9- and 10-positions, are replaced by saturated atoms or groups, have a low wavenumber vibration involving bending about the atoms in these positions.

Figure 8.118(a) shows 9,10-dihydroanthracene and part of the S_1–S_0 FE jet spectrum.[380] The 0^0_0 band is at 268.762 nm and a long

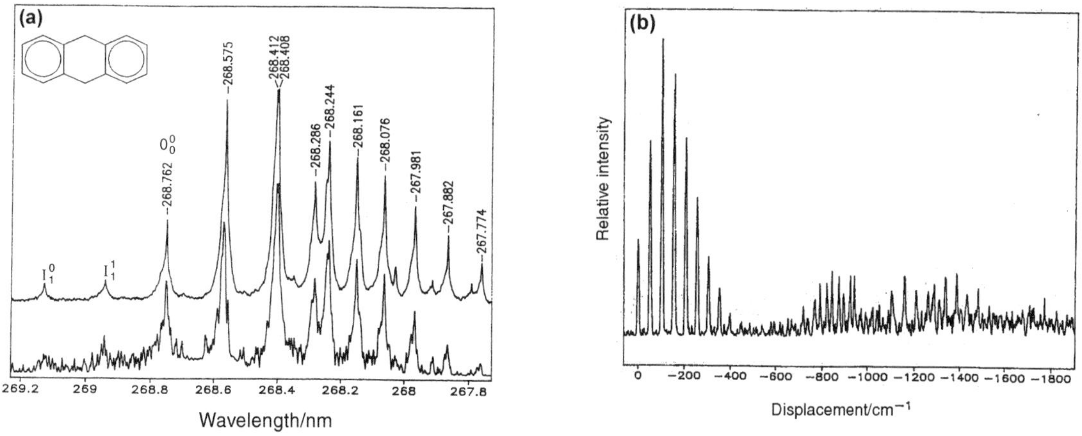

Figure 8.118 (a) Part of the FE spectrum of 9,10-dihydroanthracene. (The lower spectrum is an absorption spectrum obtained with the laser traversing a jet from a slit nozzle.) (b) An SVLF spectrum with the laser tuned to the 0_0^0 band
[Reproduced, with permission, from Shin, Y.-D., Saigusa, H., Zgierski, M. K., Zerbetto, F. and Lim, E. C. (1991). *J. Chem. Phys.*, **94**, 3511]

progression in the out-of-plane bending, or inversion, vibration about the 9- and 10-positions is built on it. The first interval is only $26\,cm^{-1}$ and there is a conspicuous doubling at 268.286/268.244 nm. SVLF spectra have been obtained with the dye laser tuned to each of these progression members in turn. Figure 8.118(b) shows the spectrum when tuned to the 0_0^0 band. The progression in the inversion vibration in S_0 shows a first interval of $51\,cm^{-1}$ and no irregularities.

Fitting the intervals in the inversion vibration in S_0 and S_1 resulted in the potentials shown in Figure 8.119. These show that the molecule is bent, in a C_{2v} conformation, in both electronic states with a decrease in the barrier to planarity from $615\,cm^{-1}$ in S_0 to $80\,cm^{-1}$ in S_1 and an increase in the angle between the planes of the benzene rings from 144.6 to 164.1°.

The barrier in S_1 is sufficiently low that there is appreciable tunnelling through it above the zero-point level. This means that the vibrational levels should be classified according to the D_{2h} point group, to which the planar molecule belongs. The inversion vibration is of b_{3u} symmetry. All levels with υ odd have this symmetry whereas those with υ even have a_g symmetry. Therefore, the zero-point level consists of an a_g/b_{3u} degenerate pair, the next, split, level consists of an a_g (lower) and b_{3u} (upper) pair, and so on. Similarly, the unsplit levels of S_0 should all be considered as a_g/b_{3u} degenerate pairs. Then the transitions which are allowed by symmetry in the inversion progressions in the FE and SVLF spectra can be derived using the selection rules relevant to the D_{2h} group. (This is analogous to using the D_{3h} point group to derive selection rules for NH_3.)

In xanthene, one of the bridging CH_2 groups is replaced by an oxygen atom. FE and SVLF spectra show long progressions in the inversion vibration. The vibrational intervals and progression intensity distributions have been interpreted[381] in terms of an S_0 state which is nonplanar, but with a barrier to planarity of only $49\,cm^{-1}$, and an S_1 state which is planar.

In tropolone, shown in Figure 4.48, the largest amplitude vibration involves the transfer of the internally hydrogen bonded hydrogen atom between the oxygen atoms. The double minimum, inversion-like potential for this motion in the S_0 state is of the type shown in

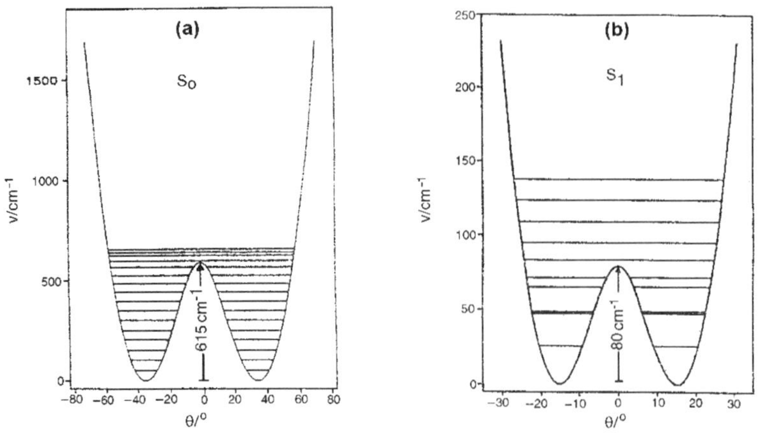

Figure 8.119 Inversion potentials for (a) the S_0 and (b) the S_1 state of 9,10-dihydroanthracene [Reproduced, with permission, from Shin, Y. -D., Saigusa, H., Zgierski, M. K., Zerbetto, F. and Lim, E. C. (1991). *J. Chem. Phys.*, **94**, 3511]

Figure 4.49(b). The splitting of the zero-point level due to tunnelling through the barrier has been obtained by microwave spectroscopy[37] and is only 0.9777 cm^{-1}. Investigation of the gas phase $\tilde{A}^1B_2$–$\tilde{X}^1A_1$, π^*–π absorption spectrum[36] at room temperature has shown that this splitting increases by 18 cm^{-1} from S_0 to S_1, and by 2 cm^{-1} in tropolone-(O^{2}H), indicating a lower barrier to proton transfer in S_1.

The laser FE jet spectra[382] of tropolone-(O^{1}H) and -(O^{1}H), parts of which are shown in Figure 8.120, confirm these results. The 0^0_0 band is split by 19 cm^{-1} (O^{1}H) and 2 cm^{-1} (O^{2}H), the selection rules, derived using C_{2v} classification, allowing the lower-to-lower and upper-to-upper transitions between the components of the split zero-point levels.

The progression with an interval of about 76 and 62 cm^{-1} in the spectra of tropolone-(O^{1}H) and -(O^{2}H), respectively, involves even quanta of a low wavenumber b_1 out-of-plane vibration.[383] The fact that the tunnelling splitting decreases appreciably along this progression shows that the tunnelling barrier increases as this b_1 vibration is increasingly excited. Coupling of the hydrogen tunnelling vibration to other vibrational modes in the molecule involving heavy atom motions, particularly those resulting from the switching of double and single bonds when tropolone goes from the structure in Figure 4.48(c)(i) to that in 4.48(c)(iii), has been discussed.[384]

The conformational problem in 1,3-benzodioxole, shown in Figure 8.121, concerns the coplanarity, or otherwise, of the skeletal atoms of the five-membered ring with the benzene ring—whether either or both of the angles ψ and τ are zero.

Figure 8.122 shows an SVLF jet spectrum of the $\tilde{A}^1A_1$–$\tilde{X}^1A_1$ system of 1,3-benzodioxole with the laser tuned to the 0^0_0 band.[385] Many of the bands involve short progressions in totally symmetric vibrations, and their combinations, but the bands labelled 20^0_2 and 20^0_4 are of particular interest. The vibration ν_{20} can be described, approximately, as the ring puckering, or inversion, b_1 vibration of the CH_2 group for which the selection rule is that Δv must be even. Levels of ν_{20} were observed up to $v_{20} = 7$ and are anharmonic: the molecule must be non-planar. The levels were fitted to a potential of the type in equation (5.264) and gave a barrier to planarity of 157 cm^{-1}.

Investigation of the microwave spectrum[386] showed that the low wavenumber vibration ν_{20} involves primarily ring puckering, in which the

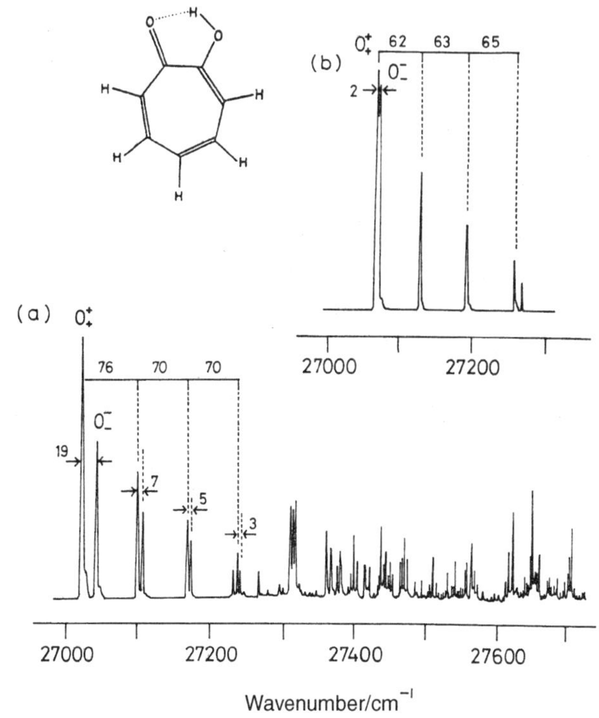

Figure 8.120 Parts of the FE jet spectra of (a) tropolone-(O^{1}H) and (b) tropolone-(O^{2}H) [Reproduced, with permission, from Tomioka, Y., Ito, M. and Mikami, N. (1983). *J. Phys. Chem.*, **87**, 4401]

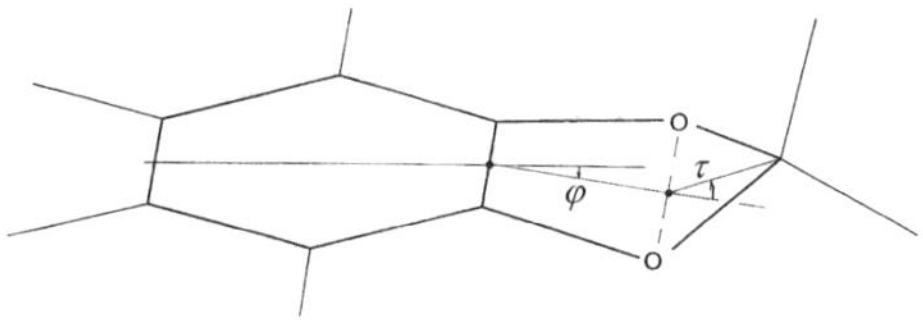

Figure 8.121 1,3-Benzodioxole showing the puckering angle τ and the 'butterfly' bending angle ψ [Reproduced, with permission, from Caminati, W., Melandri, S., Corbelli, G., Favero, B. and Meyer, R. (1993). *Mol. Phys.*, **80**, 1297]

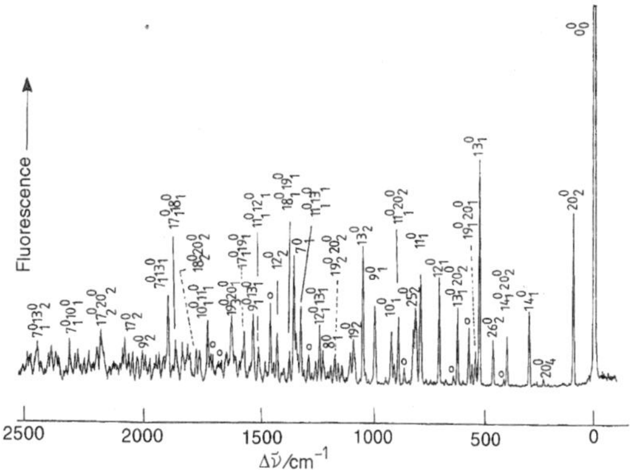

Figure 8.122 The SVLF spectrum of 1,3-benzodioxole in a supersonic jet with the laser tuned to the 0_0^0 band of the $\tilde{A}^1A_1$–$\tilde{X}^1A_1$ system [Reproduced, with permission, from Hassan, K. H. and Hollas, J. M. (1989), *Chem. Phys. Lett.*, **157**, 183]

angle τ is changing, but that it also involves a smaller change of the angle ψ. The values of τ and ψ, for the zero-point level, were found to be 26.3 and 8.3°, respectively.

The form of the potential function and the symmetry classification of the energy levels for a torsional vibration have been discussed in

Section 5.2.7.4. Investigation of these large amplitude, low wavenumber vibrations, involving identification of torsional energy levels, construction of the torsional potential and determination of torsional energy barriers, has been greatly facilitated by the use of a supersonic jet. FE and SVLF spectra have provided this information for ground and excited electronic states, respectively.

The CH_3- and SiH_3-torsional vibrations in toluene (methylbenzene) and phenylsilane (silylbenzene) have a sixfold energy barrier and, typical of a high-fold barrier, it is expected to be low. The potential involves only the V_6 term in the general expression [see equation (5.269)]

$$V(\phi) = \tfrac{1}{2}\sum_n V_n(1 - \cos n\phi) \tag{8.55}$$

The torsional energy levels are classified according to the G_{12} molecular symmetry group, which is isomorphous with the D_{3h} point group, in Figure 5.83. In toluene and phenylsilane, the only contributors to the internal rotation constant F, where, with dimensions of wavenumber,

$$F = h/8\pi^2 cI \tag{8.56}$$

and I is the reduced moment of inertia for the motion, are the three off-axis hydrogen atoms. Therefore, F is relatively large, about $5\,\text{cm}^{-1}$, and the energy levels are widely spaced.

In an electronic transition in a G_{12} molecule the selection rules for transitions between torsional levels are

$$a_1'\text{–}a_1',\ a_2'\text{–}a_2',\ e'\text{–}e',\ a_1''\text{–}a_1'',\ a_2''\text{–}a_2'',\ e''\text{–}e'',\ a_1'\text{–}a_1'',\ a_2'\text{–}a_2'',\ e'\text{–}e'' \tag{8.57}$$

If there is interaction between internal and overall rotation, which tends to occur near the top of the barrier to internal rotation, transitions obeying the selection rules

$$a_1'\text{–}a_2',\ a_1''\text{–}a_2'' \tag{8.58}$$

become weakly allowed.

Values of 4.875 and $6.22\,\text{cm}^{-1}$ have been found for V_6 in the ground electronic states of toluene[387] and phenylsilane,[388] respectively, typically very low torsional barriers. However, microwave spectroscopy cannot determine the sign of V_6 which indicates whether the methyl or silyl group is staggered (negative) or eclipsed (positive) in relation to the benzene ring.[389] It is assumed that the configuration is staggered but the sign can be obtained from an electronic spectrum if the order of, for example, the $3a_2''$ and $3a_1''$ levels can be established. In the case of a staggered configuration the $3a_1''$ level is the higher one.

It has been shown from the REMPI spectrum (similar to an FE spectrum—see Section 8.2.18) of toluene[390] that $V_6 = -25\,\text{cm}^{-1}$ in the S_1 electronic state and, from the FE spectrum of phenylsilane,[391] that $V_6 = -45\,\text{cm}^{-1}$ in S_1. Both spectra were obtained with the molecule in a supersonic jet in which cooling of the torsional levels with e symmetry is independent of that of those with a symmetry. The reason for this is that a–e transitions are nuclear spin forbidden and cannot occur, even in a collisional process. Consequently, if the lowest a and e levels of Figure 5.83 are split by tunnelling, their relative populations will not be represented by a Boltzmann distribution but by their nuclear spin statistical weights.

In both molecules, in S_1, the barrier increases slightly and the negative sign of V_6 shows that the configuration is staggered.

When there is further substitution in the toluene ring the CH_3 torsional potential may be considerably changed in both S_0 and S_1. Good examples of this are provided by 2-, 3- and 4-fluorotoluene.[392].

Figure 8.123 shows part of the FE jet spectrum of the S_1–S_0 system of 3-fluorotoluene and parts of the SVLF spectra with the laser tuned to each of the six torsional bands in the FE spectrum. The torsional transitions in the FE spectrum show the symmetry species of the lower states involved. The molecule belongs to the G_6 molecular symmetry group and the selection rules are those in equation (8.57)

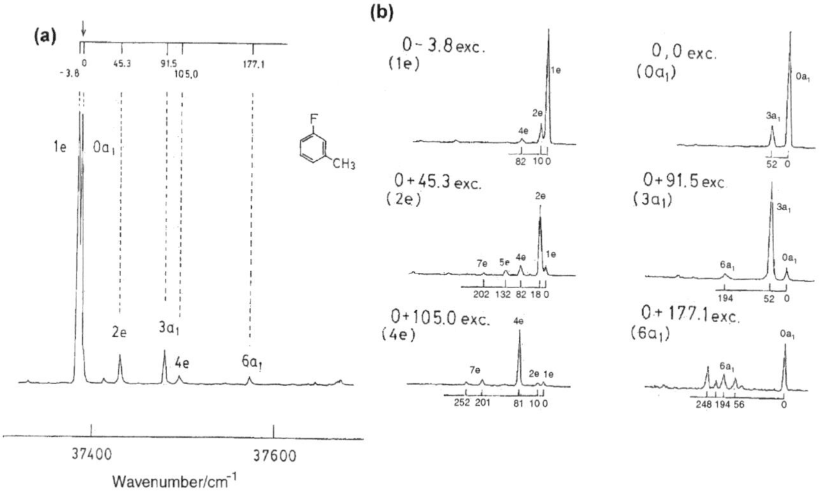

Figure 8.123 (a) Part of the FE spectrum and (b) parts of various SVLF spectra of 3-fluorotoluene [Reproduced, with permission, from Okuyama, K., Mikami, N. and Ito, M. (1985). *J. Phys. Chem.*, **89**, 5617]

with the primes and double primes removed. As Figure 8.123(b) shows, SVLF spectra from an a_1 or an e level in S_1 gives a_1 and e levels, respectively, in S_0.

Figure 8.124 shows the torsional potentials for 3-fluorotoluene, and also 2-fluorotoluene, constructed from their observed FE and SVLF spectra. In 3-fluorotoluene the $1e$–$0a_1$ splitting in the FE spectrum in Figure 8.123(a) is seen to be due largely to the tunnelling splitting of the $1e$ and $0a_1$ levels in S_0 where the torsional barrier is low.

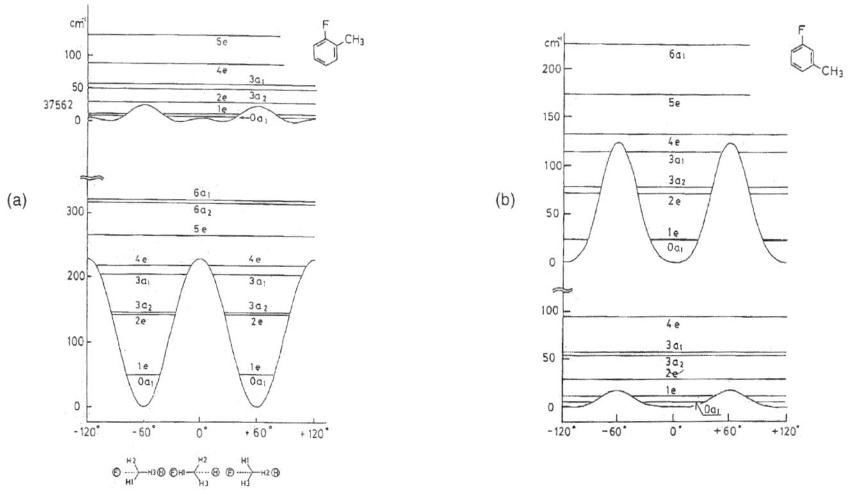

Figure 8.124 Torsional potential functions in the S_0 and S_1 states of (a) 2-fluorotoluene and (b) 3-fluorotoluene
[Reproduced, with permission, from Okuyama, K., Mikami, N. and Ito, M. (1985). *J. Phys. Chem.*, **89**, 5617]

The asymetrically substituted fluorine atom in 2- and 3-fluorotoluene introduces the possibility that a V_3 term in the potential of equation (8.52) will dominate the V_6 term present in toluene itself. However, in the symmetrically substituted 4-fluorotoluene[392] there can only be a V_6 term. Table 8.5 gives the parameters of the torsional potentials for all the monofluorotoluenes. (The cations will be discussed in Section 8.2.19.)

The parameters in Table 8.5 and the potentials in Figure 8.124 show that there is a fairly high torsional barrier of 228.3 cm^{-1} in the S_0 state of 2-fluorotoluene and of 123.7 cm^{-1} in the S_1 state of 3-fluorotoluene. These contrast with very low barriers in the S_1 state of 2-fluorotoluene and the S_0 state of 3-fluorotoluene, in which the magnitudes of V_3 and V_6 are comparable. The potentials also show that there is a change of equilibrium conformation from S_0 to S_1 in 2- but not in 3-fluorotoluene. Relative conformations in the two states are obtained from the intensity distributions observed in the torsional progressions.

The depths of the torsional potentials and the equilibrium conformations of 2- and 3-substituted toluenes have been attributed[393,394] to a balance between two possible causes, steric effects and the effect of different π-electron density in the C—C bonds of the ring, on either side of the CH_3 group. The effect of different π-electron density is exemplified by the example of propene ($CH_3CH{=}CH_2$).[395] In the ground electronic state the stable conformation is that in which one of the C—H bonds of the methyl group eclipses the C=C bond. In monosubstituted toluenes the C—H bond of a methyl group will tend to eclipse the adjacent C—C bond of the ring, which has the higher π-electron density.

In 3-fluorotoluene there can be no steric effect and the orientation of the CH_3 group must be determined entirely by effects of π-electron density. Molecular orbital considerations[393] show that the pseudo-*trans* conformation, in which the eclipsing C—H bond is *trans* to the fluorine atom, is the stable one in both S_0 (but only just) and S_1.

In 2-fluorotoluene the steric effect is clearly important. In S_0 it is overwhelming and stabilizes the pseudo-*trans* conformation whereas, in S_1, the π-electron density effect more than cancels this making the pseudo-*cis* the lower energy conformation [but with an extremely small energy barrier, as Figure 8.124(a) shows].

In 4-fluorotoluene there is a sixfold barrier and the potential in both S_0 and S_1 is similar to that of toluene.

In 2- and 3-aminotoluene[396] and 4-aminotoluene[397] (the toluidines) the torsional potentials follow a similar pattern to those of the fluorobenzenes as shown by the values of V_3 and V_6 in Table 8.6. Here they are compared with the torsional potentials for 2-, 3- and 4-aminobenzotrifluoride[398–400] in which the CH_3 group is replaced by CF_3. In general, the potentials for CF_3 behave in a similar way to

Table 8.5 Torsional potential parameters for monofluorotoluenes

Compound	Parameter	S_0	S_1	D_0 (of cation)
2-Fluorotoluene	V_3/cm^{-1}	228.3[a]	21.8[b]	334[a]
	V_6/cm^{-1}	6.6	−13.8	−3
3-Fluorotoluene	V_3/cm^{-1}	15.9[a]	123.7[a]	318[c]
	V_6/cm^{-1}	8.0	−26.4	−8
4-Fluorotoluene	V_6/cm^{-1}	−4.8[d]	−33.7[d]	−27[d]

[a]Pseudo-*trans*.
[b]Almost pseudo-*cis*.
[c]Pseudo-*cis*.
[d]Staggered.

Table 8.6 Torsional potential parameters for the monoaminotoluenes (AT) and the monoaminobenzotrifluorides (ABTF)

State	Parameter	2ABTF	2AT	3ABTF	3AT	4ABTF	4AT
S_0	V_3/cm^{-1}	450	703	9	9	–	–
	V_6/cm^{-1}	83	62	−10	−10	< 5	−5.6
S_1	V_3/cm^{-1}	240	40	155	317	–	–
	V_6/cm^{-1}	−67	−11	−40	−19	−33	−43.9

those for CH_3 except that the energy barriers tend to be lower for CF_3, the S_1 state of 2-aminobenzotrifluoride being an exception.

Although the potentials may be similar, their spectra are very different. This is mainly because of the large difference in the internal rotation constant F of equation (8.53). It is much smaller, about 0.3 cm^{-1}, in the aminobenzotrifluorides than in the aminotoluenes, in which it about 5 cm^{-1}. As a consequence, there is little tunnelling through the torsional barriers in the former, where the motion tends to behave like a torsional vibration. In the latter there is a much more effective tunnelling and the motion tends to behave more like internal rotation.

This is illustrated by the part of the FE spectrum and the torsional potential energy curves in S_0 and S_1 of 3-aminobenzotrifluoride shown in Figure 8.125. The FE spectrum shows a long progression in the torsional vibration ν_τ

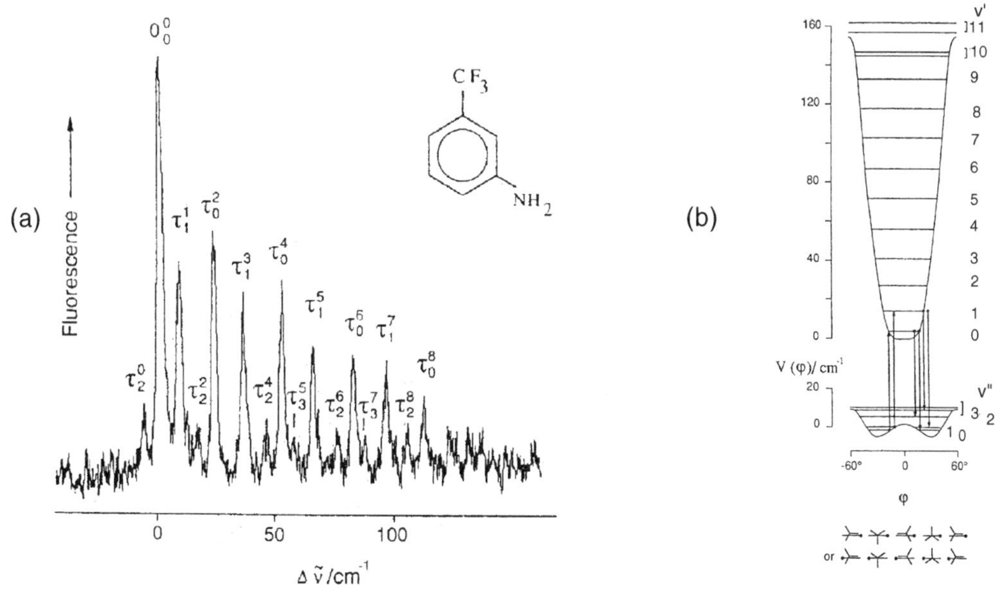

Figure 8.125 (a) Part of the FE jet spectrum of 3-aminobenzotrifluoride and (b) the CF_3-torsional potential in the S_0 (lower) and S_1 (upper) electronic states
[Reproduced, with permission, from Gordon, R. D., Hollas, J. M., Ribeiro-Claro, P. J. A. and Teixeira-Dias, J. J. C. (1991). *Chem. Phys. Lett.*, **183**, 377]

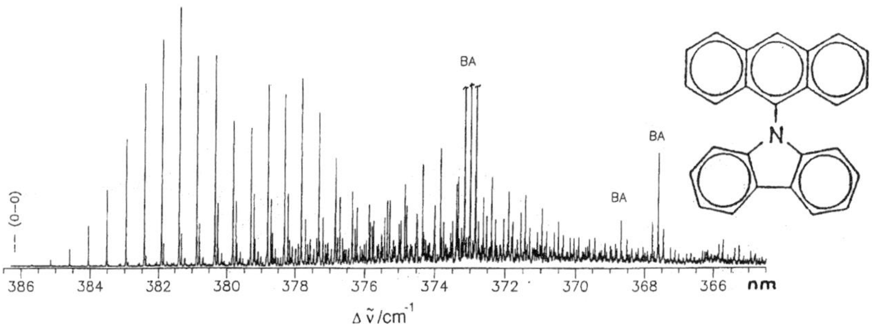

Figure 8.126 Part of the FE jet spectrum of the S_1–S_0 system of 9-(*N*-carbazolyl)anthracene (bands marked BA are due to 9, 9′-bianthryl impurity)
[Reproduced, with permission, from Monte, Ch., Roggan, A., Subaric-Leitis, A., Rettig, W. and Zimmermann, P. (1993). *J. Chem. Phys.*, **98**, 2580]

up to $v_\tau = 8$, with Δv_τ even. The potential curve for S_1 shows that the vibrational levels up to $v_\tau = 10$ are not appreciably doubled by tunnelling. As a consequence of the heavy CF_3 group, and of the V_6 term which flattens the potential, the $v_\tau = 1$–0 interval is only 11 cm^{-1}. In S_0 the energy barriers are very low but, even so, there is little tunnelling up to $v_\tau = 3$. The first three vibrational levels of S_0 are so low-lying that they are sufficiently populated in the supersonic jet for hot bands involving $v_\tau = 1$, 2 and 3 in S_0 to be observed in the FE spectrum in Figure 8.125(a).

9-(*N*-Carbazolyl)anthracene, shown in Figure 8.126, is a very large molecule with an interesting torsional vibrational problem. The anthryl and carbazolyl moieties undergo a large amplitude torsional motion about the C—N bond. The conformation in S_0 is expected to be one in which the two moieties are at about 90° to each other because of the steric hindrance between hydrogen atoms.

Figure 8.126 shows part of the FE jet spectrum[401] of the S_1–S_0 system in which there is a progression of more than 20 members in the torsional vibration. For this heavy molecule the value of F [equation (8.56)] is only 0.035 cm^{-1}. This and many SVLF spectra have been assigned and the torsional energy levels and intensity distributions along progressions fitted to the potentials in Figure 8.127. In S_0 the angle of twist from the perpendicular configuration is only 12.5° with a very small barrier of 17 cm^{-1}. In S_1 the twist angle is increased to 26° and the barrier to 1050 cm^{-1} due to the increased π-electron conjugation across the C—N bond, giving it more double bond character.

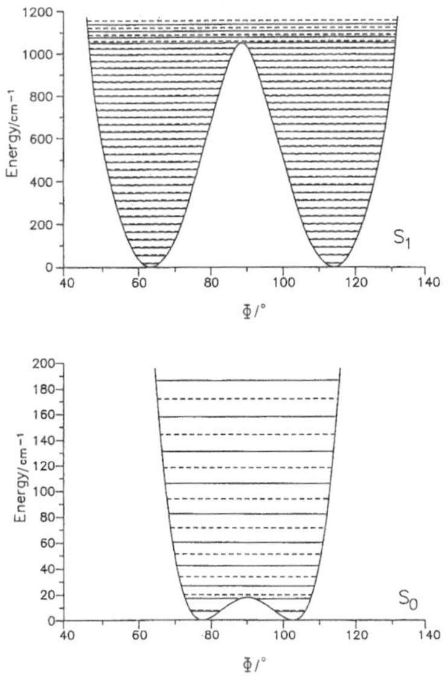

Figure 8.127 Torsional potentials for 9-(*N*-carbazolyl)anthracene in S_0 and S_1 in which Φ is the angle between the two moieties
[Reproduced, with permission, from Monte, Ch., Roggan, A., Subaric-Leitis, A., Rettig, W. and Zimmermann, P. (1993). *J. Chem. Phys.*, **98**, 2580]

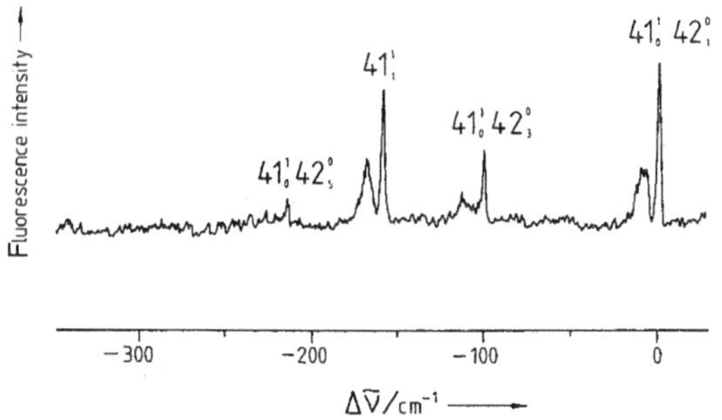

Figure 8.128 Part of an SVLF spectrum of styrene in the gas phase with the laser tuned to the $41_0^1 42_1^0$ cross-sequence band. Each band shows a resolved rotational contour
[Reproduced, with permission, from Hollas, J. M. and Ridley, T. (1980). *Chem. Phys. Lett.*, **75**, 94]

Before 1978, observation of the C(1)—C(α) torsional vibration of styrene, shown in Figure 8.129, had proved elusive. Then, identification of hot bands in the S_1–S_0, $\tilde{A}^1A'$–$\tilde{X}^1A'$ gas phase absorption spectrum showed[402] that the $v = 1$–0 separation of the torsional levels is much lower (38 cm^{-1}) than had been anticipated and that the higher levels show an unexpected strong divergence. It was also shown that, in S_1, the normal coordinates Q_{42} and Q_{41}, which correspond, in S_0, to torsion and out-of-plane bending about C(α), respectively, are heavily mixed in S_1. This was confirmed from gas phase SVLF spectra in which the gas pressure is sufficiently low that there are no collisions during the fluorescence lifetime. For example, Figure 8.128 shows part of the SVLF spectrum with the laser tuned to the $41_0^1 42_1^0$ cross-sequence band.[403] Fluorescence occurs from the exclusively populated 41^1 level to both the 41_1 and 42_1 with almost equal intensity.

The mixing of normal coordinates of vibrations belonging to the same symmetry species when a molecule goes from one electronic state to another is known as the Duschinsky effect and was discussed in Section 6.3.4.3. The effect is likely to be of greatest importance in molecules of low symmetry such as styrene, which belongs to the C_s point group and has only a' and a'' vibrations,

The spectrum in Figure 8.128 allows the identification of the $v_{42} = 1$, 3 and 5 levels in S_0. Torsional levels up to $v_{42} = 12$ have been identified from SVLF, absorption and Raman spectra in the gas phase.[404] Figure 8.129 shows a progression in v_{42}, with Δv even as ν_{42} has a'' symmetry, in an SVLF jet spectrum with excitation in the 0_0^0 band. The levels have been fitted to a potential of the type given in equation (8.52) but, because the potential must repeat every π rad, n must be even. There is an

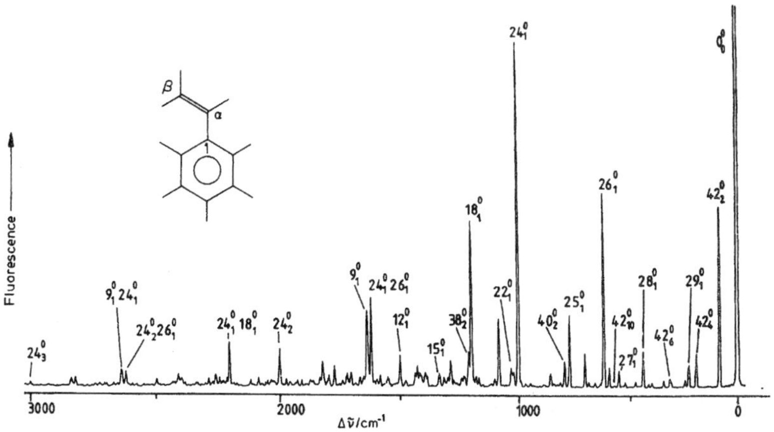

Figure 8.129 SVLF spectrum of styrene, in a supersonic jet, with excitation in the 0_0^0 band
[Reproduced, with permission, from Taday, P. F. (1991). *PhD Thesis*, University of Reading]

additional complication in a case like this where the axis about which torsion takes place does not have at least C_2 symmetry. The rotational 'constant' F, in equation (8.56), is no longer constant but varies with the torsional angle ϕ according to the equation

$$F(\phi) = \sum_n F_n \cos n\phi \tag{8.59}$$

where, in general, $n = 0, 1, 2, \ldots$, but, in this case, n must be even.

The parameters $V_2 = 1070\,\text{cm}^{-1}$ (the barrier height), $V_4 = -275\,\text{cm}^{-1}$ and $V_6 = 7.0\,\text{cm}^{-1}$ reproduced the observed torsional levels and the potential is shown in Figure 8.130.

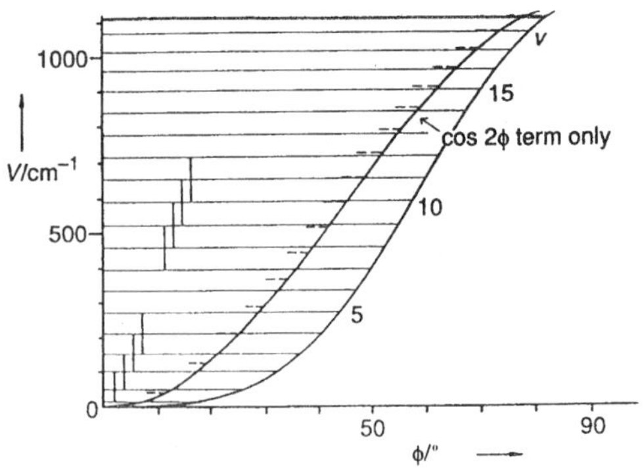

Figure 8.130 The C(1)—C(α) torsional potential for the S_0 state of styrene. The upper curve shows only the V_2 term. The lower curve shows the effect of including the V_4 term
[Reproduced, with permission, from Hollas, J. M. and Ridley, T. (1980). *Chem. Phys. Lett.*, **75**, 94]

An important, and unusual, feature of this potential is the large, negative value of V_4. Figure 8.130 shows that the inclusion of this term flattens the potential and closes up the energy levels so that the $v_{42} = 1$–0 interval is only $38.0\,\text{cm}^{-1}$ and subsequent levels diverge. For example, the $v_{42} = 2$–1 interval is $48.1\,\text{cm}^{-1}$.

The torsional potential shows that, in S_0, styrene is planar, but only just. In fact, if $4|V_4| > V_2$ the potential shows a secondary maximum at $\phi = 0°$. The flat nature of the potential causes the molecule to be very floppy about the planar conformation. This is probably due to repulsion between the π-electrons in the C=C bond and those in the adjacent ring C—C bond, which favours non-planarity, competing strongly with the effect of π-electron conjugation in the C(1)—C(α) bond, which favours planarity.

Because of the large Duschinsky effect in the S_1 state torsion becomes a two-dimensional motion—it is divided between ν_{41} and ν_{42} whereas, in S_0, it can be treated as one-dimensional. However, both vibrations show harmonic behaviour in S_1, indicating a more rigidly planar molecule, consistent with the expectation of increased conjugation in the C(1)—C(α) bond.

At the level of accuracy afforded by vibrational resolution it is probably a sufficiently good approximation to treat methyl and trifluoromethyl torsional vibration, in the ground and excited electronic states, as a one-dimensional motion. At a higher level of accuracy an *ab initio* study of internal rotation in propene ($CH_3CH{=}CH_2$) suggests[405] that methyl torsional motion is accompanied by some C—C stretching.

2-, 3- and 4-methylbenzyl radicals ($CH_3C_6H_4CH_2$) have been made by photolysing the corresponding α-chloroxylene ($CH_3C_6H_4CH_2Cl$) with an excimer laser near the tip of a supersonic nozzle.[406] FE and SVLF spectra of the radicals show methyl torsional progressions from which the torsional potentials in the D_0 (doublet) ground state and the D_1 excited state have been obtained. The parameters V_6 and V_3 are given in Table 8.7.

These parameters show some similarities to those for the fluorotoluenes, in Table 8.5, and the aminotoluenes, in Table 8.6. In the ground state 2-methylbenzyl has a high barrier whereas 3-methylbenzyl has a much lower barrier. However, the barrier in the excited state of 2-methylbenzyl remains fairly high and the barrier in the excited state of 3-methylbenzyl is almost unchanged from the ground state. The long torsional progressions observed for 3-methylbenzyl are a consequence of a change

Table 8.7 Torsional potential parameters for the methylbenzyl radicals

State	Parameter	2-Methylbenzyl	3-Methylbenzyl	4-Methylbenzyl
D_0	V_3/cm^{-1}	754	74	–
	V_6/cm^{-1}	0	28	−4
D_1	V_3/cm^{-1}	362	71.6	–
	V_6/cm^{-1}	0	22	100

of 42° in the equilibrium torsional angle from D_0 to D_1, which contrasts with no change in 3-fluoro- and 3-aminotoluene.

For molecules in which there is a strong spin–orbit interaction, the electronic transition from the ground state S_0 to the first excited triplet state T_1 may be sufficiently intense that a phosphorescence excitation (PE) spectrum may be obtained with the molecule in a supersonic jet. To obtain a PE spectrum the total, undispersed phosphorescence is monitored as the laser is scanned through the T_1–S_0 system.

Such a PE spectrum has been obtained[407] for *trans*-methylglyoxal (CH_3COCHO). [*trans*-Glyoxal is shown in Figure 5.27(g).] The torsional levels in T_1 have been fitted to a potential with $V_3 = 115\,\text{cm}^{-1}$, which is appreciably different from $V_3 = 269\,\text{cm}^{-1}$ in S_0 and $190\,\text{cm}^{-1}$ in S_1. In S_0 a C—H bond eclipses the adjacent C=O bond but, in S_1 and T_1, there is a 60° rotation of the methyl group.

8.2.18 Multiphoton absorption and ionization spectroscopy

In the discussion in Section 8.1.7 of harmonic generation of laser radiation we have seen how the high photon density produced by focusing a laser beam into a crystalline material, such as potassium dihydrogenphosphate, produces frequency doubling, or second harmonic generation, of the laser radiation. Similarly, if a laser beam of wavenumber $\tilde{\nu}_L$ is focused into a material which is known to absorb at a

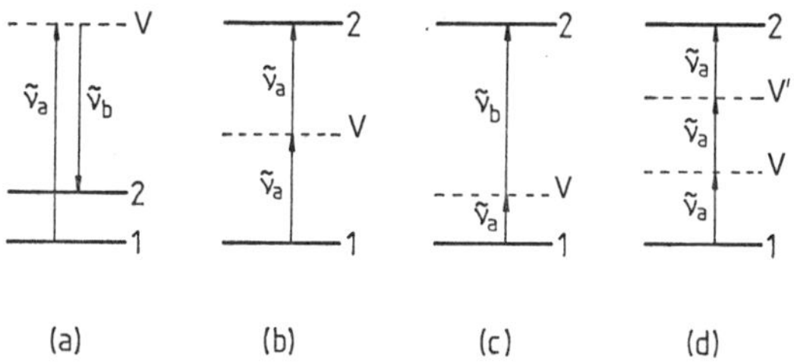

Figure 8.131 Multiphoton processes shown are (a) Raman scattering, (b) two-photon absorption via a virtual state V and where the two photons are of the same wavenumber, (c) two-photon absorption where the photons are of different wavenumber and (d) three-photon absorption where all photons are of the same wavenumber

wavenumber $2\tilde{\nu}_L$, in an ordinary one-photon process, the laser radiation may be absorbed in a two-photon process provided it is allowed by the relevant selection rules.

The similarity between two-photon absorption and harmonic generation is, however, somewhat superficial as the fluorescence which follows two-photon absorption is emitted in all directions and, owing to various processes in the sample, over a range of wavelengths.

The analogy between two-photon and Raman spectroscopy is much closer. Figure 8.131(a) shows that a Raman transition between states 1 and 2 is really a two-photon process, the first photon taking the molecule from state 1 to the virtual state V by absorption at a wavenumber ν_a. The second photon is scattered at a wavenumber $\tilde{\nu}_b$.

In a two-photon absorption process the first photon takes the molecule from an initial state

1 to a virtual state V and the second photon takes it from V to 2. As in the Raman process, the state V is not an eigenstate of the molecule. (The case where it is an eigenstate corresponds to two-step double resonance, which will be discussed in Section 8.2.21.2. The two photons absorbed may be of equal or unequal energies, as illustrated in Figure 8.131(b) and 8.131(c), respectively. It is also possible that more than two photons may be absorbed in going from state 1 to 2. Figure 8.131(d) illustrates the case of three photons, involving two virtual states V and V'.

For the case shown in Figure 8.131(b), in which both photons are identical, the two photon absorption probability P^{12} for the process follows from the original equations of Göppert-Mayer[408] and is given by

$$P^{12} \propto \sum_i E_a^2 \frac{\langle 1|\mu|i\rangle |\mu|2\rangle}{E_2 - E_1 - hc\tilde{\nu}_a} \tag{8.60}$$

where μ is the transition dipole moment, E_1 and E_2 are the energies of the initial and final states, i is an intermediate state (explained in Section 8.2.1 where the label used is r), $\tilde{\nu}_a$ is the wavenumber of the radiation inducing the transition and E_a is the electric field component of the laser photon. (The 'bra' and 'ket' notation in, for example, $\langle 1|\mu|i\rangle$ is an alternative notation for the expression $\int \psi_1 \mu\, \psi_i \mathrm{d}\tau$.)

The fact that P^{12} is proportional to the product of *two* electric dipole transition moments illustrates the point that, like a Raman scattering process, a two-photon absorption process can be regarded as two successive electric dipole transitions, to and from state i. Clearly, then, the only intermediate states which contribute to the summation in equation (8.57) are those for which the transition $2 - i$ and $i - 1$ are allowed by electric dipole selection rules. It follows, also, that the overall selection rules for two-photon absorption are the same as for Raman scattering and the reader is referred to Chapters 4 and 5 for discussions of rotational, vibrational and rovibrational selection rules.

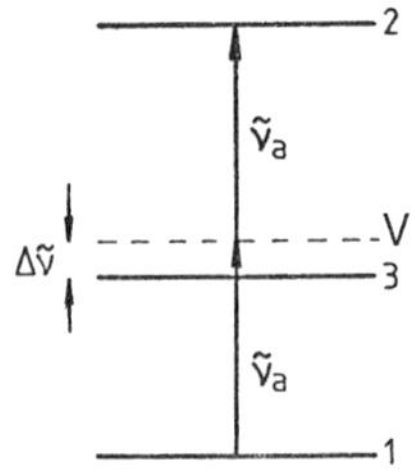

Figure 8.132 Absorption of two identical photons when an eigenstate 3 is close to the virtual state V

The summation in equation (8.60) is over all states i, provided that they fulfil the selection rule requirements. However, if there is an eigenstate 3 close to V, as shown in Figure 8.132, which fulfils these requirements its contribution to P^{12} will be overwhelming. The denominator becomes, when i is state 3,

$$E_2 - E_3 - hc\tilde{\nu}_a = hc\Delta\tilde{\nu} \tag{8.61}$$

where $hc\Delta\tilde{\nu}$ is the energy separation of states V and 3.

Examples of intermediate eigenstates making a large contribution to P^{12} are not uncommon when the two photons are provided by an intense microwave or radiofrequency source and states 1 and 2 are rotational states. Such an example, the first microwave two-photon transition to be observed,[409] is illustrated in Figure 8.133. The $J = 2$–0 transition for $K = 0$ was observed for the symmetric rotors C^2H_3CN and PF_3 in which the $J = 1$, $K = 0$ state is close to the virtual state V. A high intensity klystron of the type used in double resonance experiments (Section 8.2.21.1) is tuned and, when the wavenumber $\tilde{\nu}_M$ is such that $2\tilde{\nu}_M$ matches the $J = 2$–0 transition, there is a sudden decrease in intensity of the probed $J = 1$–0 transition at a wvenumber $\tilde{\nu}_P$. In this example the $J = 1$–0 and 2–1 transitions are electric dipole allowed and the $J = 2$–0 transition is electric dipole forbidden but two-photon (Raman) allowed.

In general, two-photon transitions with $\Delta\tilde{\nu}$ as large as 0.33 cm^{-1} (1 GHz) may be sufficiently intense to be observed.

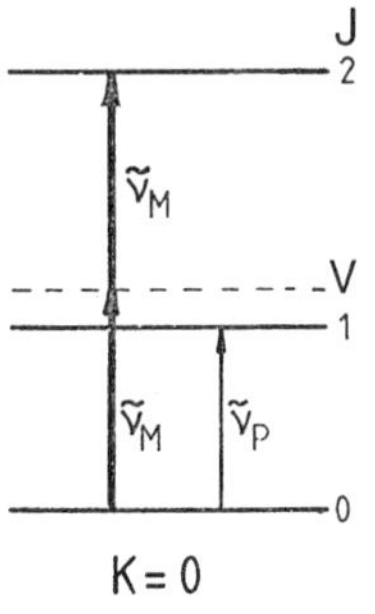

Figure 8.133 Microwave two-photon absorption process in which the $J = 1$ state is close to the virtual state V, which is midway between the $J = 2$ and $J = 0$ states

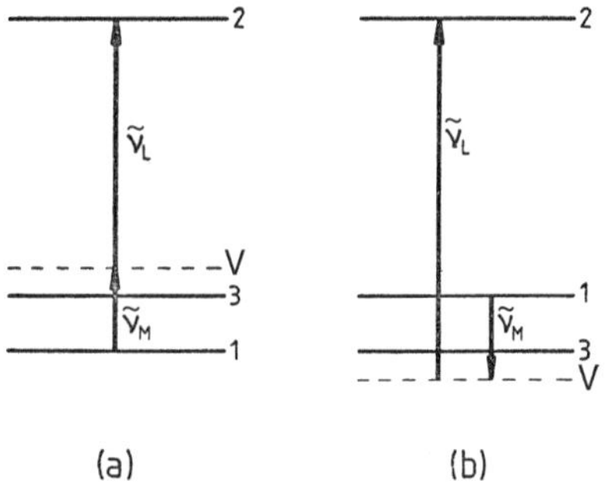

Figure 8.134 Two processes involving an infrared and a microwave photon when there is an eigenstate 3 close to the virtual state V

Microwave two-photon transitions have also been observed where there is no eigenstate between the initial and final states. These may occur in symmetric rotors with $K > 0$ for which the double degeneracy of the J levels (see Figure 5.48) causes transitions with $\Delta J = \pm 1$ to be two-photon allowed.

Whereas it is possible to achieve sufficiently high power for two-photon transitions to be observed using conventional microwave or radiofrequency sources, in the infrared, visible or ultraviolet regions laser sources are essential.

Following the development of infrared lasers two-photon absorption experiments were carried out with one microwave (or radiofrequency) photon and one infrared photon. Figure 8.134(a) shows an energy level scheme similar to that in Figure 8.131(c) but in which there is a (rotational) eigenstate 3 close to the virtual state V. The process in Figure 8.134(a) involves the sum of the infrared laser wavenumber $\tilde{\nu}_L$ and the microwave wavenumber $\tilde{\nu}_M$. On the other hand, a two-photon process involving the difference between the wavenumbers, $\tilde{\nu}_L - \tilde{\nu}_M$, may also occur; this is shown in Figure 8.134(b).

In the cases illustrated in Figure 8.134, where there is an eigenstate near to the virtual state, equations (8.60) and (8.61) give

$$\boldsymbol{P}^{12} = \frac{\langle 1|\mu|3\rangle\langle 3|\mu|2\rangle}{hc\Delta\tilde{\nu}} \tag{8.62}$$

where $\Delta\tilde{\nu}$ is the wavenumber separation of states V and 3.

The first microwave–infrared two-photon absorption experiments were carried out with the microwave absorption cell outside the cavity of the infrared (CO_2 or N_2O) laser.[410] Because the two-photon transition moment is proportional to the product of the laser and microwave field strengths [equation (8.59)], much greater sensitivity can be achieved by placing the microwave cell inside the laser cavity.[411] The microwave source is modulated and phase sensitive detection used in measuring the laser output power. When the microwave source is tuned to a wavenumber $\tilde{\nu}_M$, such that $\tilde{\nu}_L + \tilde{\nu}_M$ corresponds to a two-photon transition in the gaseous sample in the microwave cell, there is a sharp decrease in the laser power. With this intracavity technique microwave power of only about 20 mW is necessary compared to several hundred milliwatts with the cell outside the cavity. By using not only $^{12}C^{16}O_2$ laser lines but also those of $^{13}C^{16}O_2$ and $^{12}C^{18}O_2$, many two-photon transitions have been observed in NH_3.[412]

Under normal circumstances the width of the two-photon signal is limited by the Doppler width characteristic of the infrared region in which it occurs. However, in the intracavity experiments, saturation of some two-photon transitions has been achieved with the consequent observation of Lamb dips. The likelihood of this happening increases as $\Delta\tilde{\nu}$, the separation of states V and 3 in Figure 8.134(a) or (b), decreases.

When the two photons involved in a transition are both in the visible region of the spectrum the transition is likely to be an electronic transition. Typically both photons are of the same wavenumber, $\tilde{\nu}_a$, and come from a tunable dye laser. Since the transition probability depends on the square of the laser power [equation (8.57)] a pulsed laser, producing short pulses of high peak power, is suitable.

Detection of a two-photon electronic transition by direct absorption measurements has been used but is not a very sensitive method.

Two more useful methods are illustrated in Figure 8.135. That in Figure 8.135(a) is fluorescence excitation (FE). This is very similar to that for one-photon absorption described in Section 8.2.16 except that two-photon absorption has much lower probability: the absorption cross-section is lower, typically by a factor of 10^{33}, than for one-photon absorption. (It is lower by a factor of about 10^{67} for three-photon absorption.)

The experimental method of obtaining a two-photon FE spectrum is very similar to that for a one-photon spectrum except that the laser beam is focused into the sample to provide a greater photon flux. To provide a sufficiently large flux it is necessary to use a pulsed laser. A 'hard focus' by a short focal length lens is necessary for a three-photon process, but a 'soft focus' is more suitable for a two-photon process. For high resolution spectroscopy the sample may be in the gas phase, under low pressure, collision-free conditions, or seeded into a supersonic jet.

The two-photon absorption cross-section is so low that the obtaining a dispersed fluorescence spectrum with the laser fixed on an absorption transition is rarely possible. A second method of detection is that employing multiphoton ionization illustrated in Figure 8.135(b). In the case shown, two photons of wavenumber $\tilde{\nu}_a$ take the molecule into an eigenstate 2 and a third ionizes it. The process is known as resonance-enhanced multiphoton ionization (REMPI), the probability of ionization being enhanced by the fact that the first two photons are resonant with an absorption. That shown in Figure 8.135 is a one-colour $2+1$ REMPI process in which all three photons have the same 'colour', i.e. have the same wavenumber. In some cases it may require a further two photons to ionize the molecule from state 2 in a one-colour $2+2$ REMPI process. Alternatively, a second laser may provide a single ionizing photon from state 2 in a $2+1'$ two-colour REMPI process. There are other variations on $n+m$, one- or two-colour REMPI processes.

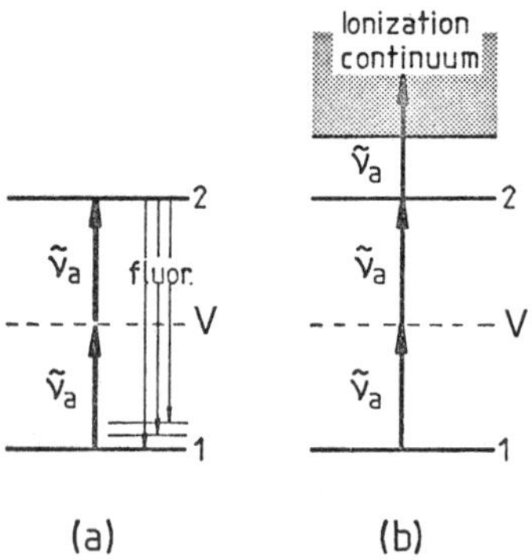

Figure 8.135 A two-photon absorption process can be detected by (a) a fluorescence excitation (FE) process or (b) a resonant multiphoton ionization (REMPI) process, in this example a $2+1$ process

The experimental technique is similar to that for obtaining a multiphoton FE spectrum except for the method of detection. In principle, the positive ions or the photoelectrons may be detected. There are advantages of sensitivity in detecting the ions and, most importantly, of selectivity. In the case, for example, of investigation of a van der Waals complex in a supersonic jet mass selection of the ions by a time-of-flight mass spectrometer allows the separate recording of the REMPI spectrum of the complex without interference from the parent species or other complexes.

An advantage of using ion rather than fluorescence detection is that molecules with a very low fluorescence quantum yield, that is, molecules which we tend to say 'do not fluoresce', may produce ions by absorption of a further photon(s) from the excited electronic state. An ionization process, under typical conditions, takes of the order of 1 ps. However, if there are efficient non-radiative processes which take place within this time a molecule with a very low fluorescence quantum yield may not give a REMPI spectrum either.

The sensitivity of the method of ion detection is such that obtaining a $1+1$, one-colour REMPI spectrum may be preferred to obtaining an FE spectrum. In principle, a $1+1$ REMPI spectrum might be obtained for a molecule which does not fluoresce but, in practice, such molecules often have such rapid competing decay processes for the excited electronic state that there is no REMPI spectrum either.

The electronic selection rules for two-photon absorption are analogous to vibrational Raman selection rules and we can rewrite equation (5.184) to give

$$\Gamma(\psi'_e) \times \Gamma(S_{ij}) \times \Gamma(\psi''_e) \supset A \tag{8.63}$$

as the requirement for an allowed transition between two electronic states. If the lower state is the ground state of a closed shell molecule, this equation simplifies to

$$\Gamma(\psi'_e) = \Gamma(S_{ij}) \tag{8.64}$$

The S_{ij} in these equations are the elements of the two-photon tensor $\boldsymbol{S}$. When the two photons are identical the tensor is symmetric and of the form

$$\boldsymbol{S} = \begin{pmatrix} S_{xx} & S_{xy} & S_{xz} \\ S_{xy} & S_{yy} & S_{yz} \\ S_{xz} & S_{yz} & S_{zz} \end{pmatrix} \tag{8.65}$$

This is analogous to the polarizability tensor in equation (4.117) and the symmetry species of elements of both kinds of tensor are the same

$$\Gamma(S_{ij}) = \Gamma(\alpha_{ij}) \tag{8.66}$$

The first continuously recorded two-photon FE spectrum was that of crystalline biphenyl ($C_6H_5C_6H_5$) at a temperature of 2 K.[413] The electronic transition observed was the electric dipole forbidden $\tilde{A}^1B_{3g}$–$\tilde{X}^1A_g$ transition.

If biphenyl is assumed to be planar it belongs to the D_{2h} point group and the $\tilde{A}$–$\tilde{X}$ transition is forbidden as a one-photon process. However, since $B_{3g} = \Gamma(\alpha_{yz})$, it follows from equations (8.63) and (8.66) that it is allowed as a two-photon process. In the two-photon FE spectrum the 0_0^0 band is observed with high intensity.[413]

The same transition has been observed in a supersonic jet by a $1+1$ REMPI process.[414] Progressions are observed in the vibration involving torsional motion of the two rings about the central C—C bond. These show that the molecule is planar in the $\tilde{A}$ state but that it is twisted with a torsional angle of about 44° in the $\tilde{X}$ state. The selection rules, based on the molecular symmetry group G_{12} to which the non-rigid molecule in the $\tilde{X}$ state belongs, allow the $\tilde{A}$–$\tilde{X}$ transition as a one-photon process and the 0_0^0 band is weakly observed.

The $A^2\Sigma^+$–$X^2\Pi_{\frac{1}{2},\frac{3}{2}}$ transition of NO is allowed as a one- or two-photon process but the two-photon FE spectrum shows[415] that the selection rules for the accompanying rotational transitions are modified from $\Delta J = 0, \pm 1$, in a one-photon spectrum, to $\Delta J = 0, \pm 1, \pm 2$ giving O, P, Q, R and S branches, just as for a Raman vibration–rotation transition [see Section 5.2.5.1(ii)].

In general, the selection rules for any molecule with an inversion centre allow $u \leftrightarrow u$ and $g \leftrightarrow g$ two-photon transitions, the opposite of the $g \leftrightarrow u$ selection rule which applies to a one- and also a three-photon transition. This means that, in homonuclear diatomic molecules with a g ground state, two-photon transitions may access g excited electronic states which could not be reached from the ground state by a one-photon process.

A particularly interesting example of this is the $E,F^1\Sigma_g^+$ excited state of H_2. As shown in Figure 8.136, the E,F state shows a double minimum in the potential. This is due to an avoided crossing between the potential curves for the two ${}^1\Sigma_g^+$, E and F, states. The E and F states are described by, predominantly, the $(\sigma_g 1s)^1(\sigma_g 2s)^1$ and the doubly excited $(\sigma_u^* 2p)^2$ configurations, respectively (see Figure 6.29). The inner minimum, due to the E state, is at 1.012 Å and the outer minimum, due to the F state, is at 2.322 Å.

The E,F state can be populated in an electrical discharge in H_2 and had been studied by subsequent fluorescence to the $B^1\Sigma_u^+$ state shown in the figure. However, it can be populated directly by a two-photon transition from the X state. This was achieved[416] using a pulsed dye laser to produce radiation which was frequency doubled and Raman shifted, the fourth anti-Stokes radiation being used. In this way, tunable radiation in the 98 800–104 000 cm^{-1} (101–96 nm) region, which covered the E,F–X absorption, was obtained. Detection was by counting H_2^+ ions resulting from the absorption of a further photon in a 2 + 1 REMPI process.

The inner potential well contains only three ($\upsilon_E = 0$–2) and the outer only five ($\upsilon_F = 0$–4) vibrational levels. As the vibrational wave functions for the $\upsilon_E = 1$ and $\upsilon_F = 2$ levels in Figure 8.136 show, there is considerable tunnelling through the barrier between the minima. Although Franck–Condon considerations demand that excitation from the X state will be initially to levels of the inner minimum, the tunnelling mechanism feeds molecules through to levels of the outer minimum. In this way $\upsilon_E = 1$–0 and $\upsilon_F = 2$–0 and 3–0 transitions were observed and rotational constants obtained from the rotational structure which is dominated by Q branches.

This same 2 + 1 REMPI process has been used to probe the H_2 product of the reaction

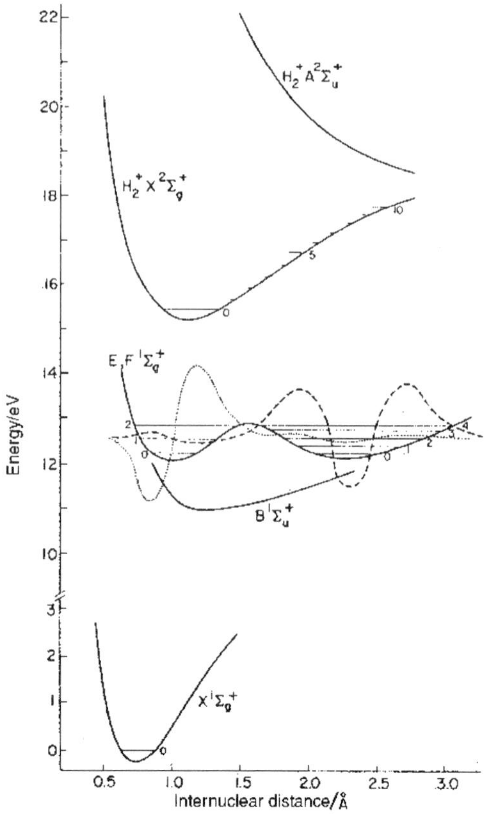

Figure 8.136 Potential energy curves for the X, B and E,F states of H_2. Calculated vibrational wave functions for $\upsilon_E = 1$ of the E state (dotted line) and $\upsilon_F = 2$ of the F state (dashed line) are also shown [Reproduced, with permission, from Marinero, E. E., Vasudev, R. and Zare, R. N. (1983). *J. Chem. Phys.*, **78**, 692]

$H + HI = H_2 + I$ in a skimmed supersonic jet.[417] The H_2 is formed predominantly in the $\upsilon = 1$, $J = 11$ level of the X state and two reaction pathways for its production, one involving cluster chemistry in the molecular beam, have been distinguished.

The example of NH provides an illustration of the power of multiphoton spectroscopy in identifying high-lying electronic states. Transitions to some of these are forbidden as one-photon processes but, even for those which are allowed, a multiphoton absorption allows

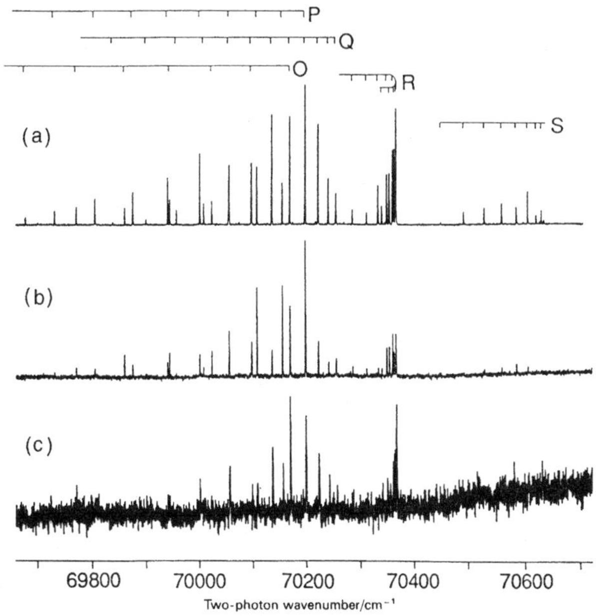

Figure 8.137 2 + 1 (or 2) REMPI spectra showing the 0–0 band of the $d^1\Sigma^+$–$a^1\Delta$ system of NH using mass selection of the ions with (a) $m/z = 15$, (b) $m/z = 14$ and (c) $m/z = 1$
[Reproduced, with permission, from Ashfold, M. N. R., Clement, S. G., Howe, J. D. and Western, C. M. (1991). *J. Chem. Soc., Faraday Trans.*, **87**, 2515]

access to states which could only be reached, in one-photon absorption, using a vacuum ultraviolet source of laser radiation which is not so readily available.

The ground configuration of NH is $\ldots 3\sigma^2 1\pi^2$, which gives rise to the $X^3\Sigma^-$ ground state and the low-lying $a^1\Delta$ and $b^1\Sigma^+$ excited states, both of which are metastable because transitions to the ground state are orbitally and spin forbidden. The lower $a^1\Delta$ state lies $12\,688\,\mathrm{cm}^{-1}$ above the X state and, when NH is prepared, for example, from HN_3, a high proportion is in the $a^1\Delta$ state.

About $70\,200\,\mathrm{cm}^{-1}$ above the a state is the $d^1\Sigma^+$ state arising, predominantly, from the $\ldots 3\sigma^0 1\pi^4$ configuration. This has been identified, for example, by observing fluorescence to the $b^1\Sigma^+$ state,[418] a transition which is one-photon allowed. However, the $d^1\Sigma^+$–$a^1\Delta$ transition is two-photon allowed by the selection rule[419]

$$\Delta\Lambda = 0, \pm 1, \pm 2 \qquad (8.67)$$

giving direct access to the d state from the abundant NH in the a state.

The d–a transition has been observed[420] using a frequency doubled dye laser providing one photon to dissociate the HN_3, two more to excite NH from the a to the d state and a further one, or more, photons for ionization. The importance of mass selection of the ions in REMPI spectroscopy, usually by a time-of-flight mass spectrometer, is illustrated by the REMPI spectra in Figure 8.137. This shows the 0–0 band of the $d^1\Sigma^+$–$a^1\Delta$ system of NH with mass selection of NH^+, N^+ and H^+ in Figure 8.137(a), (b) and (c), respectively.

The spectrum in Figure 8.137(a) shows *O*, *P*, *Q*, *R* and *S* branches resulting from the $\Delta J = 0$, ± 1, ± 2 rotational selection rule. This spectrum was obtained using plane-polarized laser radiation with the electric vector along the axis of the mass spectrometer.

The spectrum in Figure 8.137(b) shows that the yield of N^+, relative to NH^+, varies according to which rotational line is involved in the two-photon excitation: $P(3)$ to $P(5)$ and $O(4)$ to $O(7)$ are particularly productive of N^+. The high noise level in Figure 8.137(c) shows that the yield of H^+ is small. These results can be interpreted in terms of alternative mechanisms for formation of the fragment ions.[420]

Circularly polarized radiation, which can be produced from plane polarized radiation with a Fresnel rhomb, may also be used when the effect on the selection rules[421] is to suppress the Q-branch intensity. This has been used, for example, to aid the assignment[422] of the 0–0 band of the $g^1\Delta$–$a^1\Delta$ Rydberg transition of N^2H.

In considering the rotational transitions accompanying two-photon electronic or vibronic transitions we have to use spherical tensor elements, which are simple functions of the cartesian tensor elements of equation (8.62). There are nine spherical tensor elements $S_{j,k}$, where $j = 0, 1, 2$ represents the rank of the element and $k = j, j-1, \ldots, -j$. Only the tensor elements of zero and second rank, $S_{0,k}$ and $S_{2,k}$, are involved in two-photon transitions. For a diatomic molecule those of zero rank involve transitions with $\Delta J = 0$ and those of second rank $\Delta J = 0, \pm 1, \pm 2$, except that $J = 0 \leftrightarrow\!\!\!\!\!/\;\, J = 0$. With circular polarization the zero rank contribution is removed, with a consequent reduction in the Q-branch intensity.

High energy electronic states may be accessed by the absorption of three photons using laser radiation in the visible region. For example, a 3 + 1 REMPI spectrum of NO has been obtained[423] with a laser wavelength in the 460–490 nm region. The electronic states involved in resonance with the third photon are $F^2\Delta$, $H^2\Sigma^+$ and $H'^2\Pi$, which are in the range 62 000–63 000 cm^{-1} above the ground state.

Spherical tensor elements of the first and third rank, $S_{1,k}$ and $S_{3,k}$, are involved in three-photon transitions.[419] For a diatomic molecule those of first rank involve transitions with $\Delta J = 0, \pm 1$, except that $J = 0 \leftrightarrow\!\!\!\!\!/\;\, J = 0$, and those of third rank $\Delta J = 0, \pm 1, \pm 2, \pm 3$, except that $J = 0 \leftrightarrow\!\!\!\!\!/\;\, J = 0, 1, 2$ and $J = 1 \leftrightarrow\!\!\!\!\!/\;\, J = 1$. First rank contributions are suppressed with circularly polarized radiation.

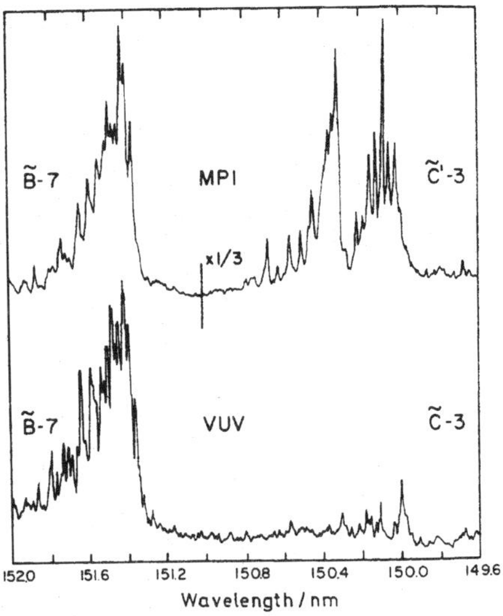

Figure 8.138 The upper spectrum is part of the 3 + 1 REMPI spectrum of NH_3 showing the 2^7_0 band of the $\tilde{B}$–$\tilde{X}$ system and the 2^3_0 band of the $\tilde{C}'$–$\tilde{X}$ system, a band which is missing from the lower, vacuum ultraviolet one-photon absorption spectrum [Reproduced, with permission, from Nieman, G. C. and Colson, S. D. (1979). *J. Chem. Phys.*, **71**, 571]

Investigation of the 3 + 1 REMPI spectrum of NH_3, in the gas phase[424] and in a supersonic jet,[425] revealed a previously unknown electronic state. Figure 8.138 shows, in the lower half, a region of the vacuum ultraviolet absorption spectrum in which the 2^7_0 band (v_2 is the inversion vibration) of the $\tilde{B}^1E''$–$\tilde{X}^1A'_1$ system appears. There is also a much weaker band which was assigned, at that time, to the 2^3_0 band of the $\tilde{C}$–$\tilde{X}$ system which was later shown[134] to be a vibronic component of the $\tilde{B}^1E''$–$\tilde{X}^1A'_1$ system (see Section 6.3.4.5). In the upper half of the figure is the 3 + 1 REMPI spectrum, obtained with plane polarized radiation, showing the 2^7_0, $\tilde{B}$–$\tilde{X}$ band and, in addition, another intense band, which is shifted from the '$\tilde{C}$–$\tilde{X}$' band in the lower spectrum. This additional band is assigned to the 2^3_0 band of a new, $\tilde{C}'^1A'_1$–$\tilde{X}^1A'_1$ system which is forbidden as a one-photon process.

The $\tilde{C}'$–$\tilde{X}$ system has also been observed[426] in an FE spectrum following two-photon absorption in the gas phase. The fluorescence monitored was not to the ground state but to the first excited singlet state $\tilde{A}^1A''_2$. Figure 8.139 shows the 0^0_0 band of the $\tilde{C}'$–$\tilde{X}$ system observed in this way with plane polarization. Also shown is the computer simulation of the rotational structure with a line width limited by the $1.8\,cm^{-1}$ laser band width.

In this case the rotational selection rules are $\Delta J = 0$, $\Delta K = 0$ for the zero rank component and $\Delta J = 0, \pm 1, \pm 2$, $\Delta K = 0$ for the second rank component. A ratio of 1:0.75 for the zero : second rank contributions was assumed.

As discussed in Section 6.3.5.4, bands observed in emission from an electrical discharge in N^1H_3 (or N^2H_3), the so-called Schuster bands, had been attributed[147] to NH_4 but were later shown to be due to $\tilde{C}'$–$\tilde{A}$ emission[148] of NH_3. In particular, the strongest band was assigned to the 2^2_2 sequence band in the $\tilde{C}'$–$\tilde{X}$ system. Dispersed $\tilde{C}'$–$\tilde{X}$ fluorescence spectra[426] of both N^1H_3 and N^2H_3, following two-photon excitation, confirm the reassignments.

Transitions from the ground state of NH_3 to each of the $\tilde{A}$, $\tilde{B}$ and $\tilde{C}$, excited states show rotational line broadening due to predissociation yielding the products $H + NH_2$. The broadening is least in the $\tilde{C}'$ state, because the NH_2 fragment is highly excited, and is less pronounced in N^2H_3 than in N^1H_3. Sub-Doppler line widths have been measured[427] in the 2^2_0 band of the $\tilde{C}'$–$\tilde{X}$ system of N^2H_3. A pulsed dye laser amplifier produced a two-photon band width of only 275 MHz ($0.0092\,cm^{-1}$). Counter-propagating beams[428] were used to obtain 2 + 1 sub-Doppler REMPI spectra by reflecting the laser beam, focused into the cell containing the NH_3

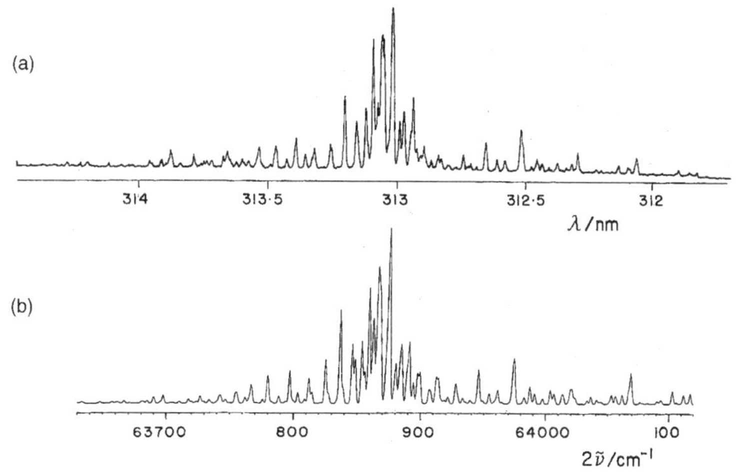

Figure 8.139 (a) Two-photon FE spectrum showing the 0^0_0 band of the $\tilde{C}'$–$\tilde{X}$ system of NH_3 detected by monitoring $\tilde{C}'$–$\tilde{A}$ fluorescence. (b) Computer simulation of the 0^0_0 band of the $\tilde{C}'$–$\tilde{X}$ system
[Reproduced, with permission, from Ashfold, M. N. R., Bennett, C. L., Dixon, R. N., Fielden, P., Rieley, H. and Stickland, R. J. (1986). *J. Mol. Spectrosc.*, **117**, 216]

gas, back through the cell with a concave mirror which re-focuses the beam to a point coincident with the first focal point.

When a molecule with a particular velocity absorbs one photon from the incident laser beam, it experiences a Doppler shift in one direction and, when it absorbs the second photon from the reflected beam, the Doppler shift is of the same magnitude but in the opposite direction. In this way Doppler broadening is removed or, in practice, considerably reduced. An additional advantage of this sub-Doppler technique is that all molecules of all velocity groups contribute to the signal, thereby increasing the sensitivity compared with techniques which involve only selected velocity groups.

The narrowest rotational line in the $\tilde{C}'$–$\tilde{X}$ system of N^2H_3, that with $J = K = 0$, has a sub-Doppler width, deconvoluted from the two-photon laser line width, of 350 MHz ($0.012\,cm^{-1}$) corresponding to a lifetime [see equation (8.45)] of 0.45 ns. The lifetime decreases with increasing rotational excitation, consistent with increasing rovibronic coupling to the $\tilde{A}$ state in which predissociation is much more pronounced.

The technique of laser vaporization has been used to obtain REMPI spectra of the unstable species Cu_3[429] and SiC_2.[430] Copper or silicon carbide is vaporized by 532 nm radiation from a Q-swtiched Nd^{3+}:YAG laser and the products combined with helium which is pumped through a supersonic nozzle. The 2 + 1 REMPI spectrum of the skimmed jet was obtained using mass selection.

Investigation of the $\tilde{A}^2E''$–$\tilde{X}^1E'$ system of Cu_3 showed that it is triangular in both states. There is a small Jahn–Teller distortion in the $\tilde{A}$ state in which one Cu—Cu bond length differs from the others by 0.03 Å. The distortion is larger than in the $\tilde{X}$ state.

The 2 + 1 REMPI spectrum of SiC_2 showed that the molecule is triangular in both the ground state, $\tilde{X}^1A_1$, and first excited singlet state, $\tilde{A}^1B_2$, contrary to previous studies in which it was assumed to be linear

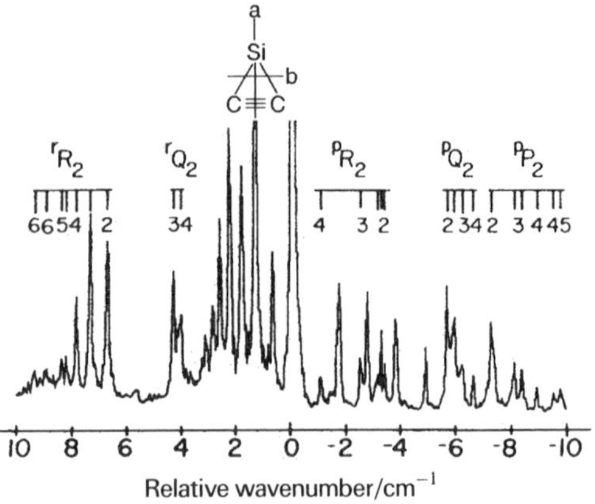

Figure 8.140 The 0_0^0 band of the $\tilde{A}^1B_2$–$\tilde{X}^1A_1$, 2 + 1 REMPI spectrum of SiC_2
[Reproduced, with permission, from Michalopoulos, D. L., Geusic, M. E., Langridge-Smith, P. R. R. and Smalley, R. E. (1984). *J. Chem. Phys.*, **80**, 3556]

in both states, analogous to C_3 [see Section 6.3.5.1(i)].

Figure 8.140 shows the 0_0^0 band of the $\tilde{A}^1B_2$–$\tilde{X}^1A_1$ system at a rotational temperature of 15 K. The molecule is a prolate asymmetric rotor belonging to the C_{2v} point group and the electronic transition moment is polarized along the b inertial axis (the y cartesian axis). Since $B_2 = \Gamma(\alpha_{yz})$ the transition is two-photon allowed. The rotational transitions are dominated by $\Delta J = 0, \pm 1$ and $\Delta K_a = 0$. Since rotation about the a-axis exchanges identical carbon atoms with zero nuclear spin, levels in the $\tilde{X}$ state with K_a odd are missing. The spectrum in Figure 8.140 shows transitions with $K_a = 0$ and 2, only. The C—Si—C angle is 41–42° in both states, the C—C bond length increases from 1.25 Å in the $\tilde{X}$ to 1.30 Å in the $\tilde{A}$ state, consistent with the π^*–π nature of the transition, and the Si—C bond length increases from 1.81 Å in the $\tilde{X}$ to 1.88 Å in the $\tilde{A}$ state.

The $\tilde{A}^1B_1$–$\tilde{X}^1A_1$, two-photon FE spectrum of the cyclic C_{2v} molecule difluorodiazirine, CF_2N_2, a highly asymmetric rotor ($\kappa = -0.08$), has been obtained in the gas phase.[431] In the 0_0^0 band the transition moment is polarized along the c-axis, normal to the

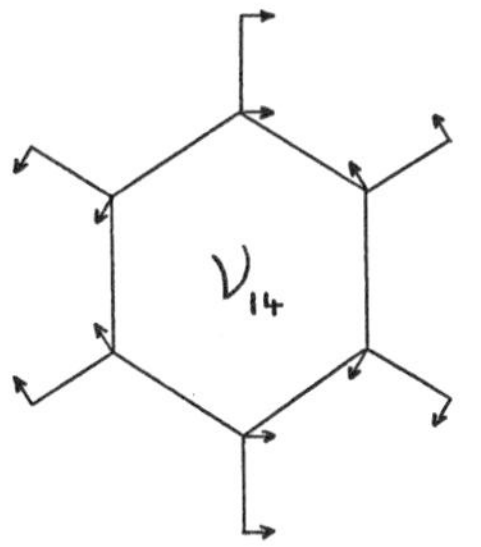

Molecule	Symmetry species of ν_{14}	$\tilde{\nu}''_{14}/cm^{-1}$	$\tilde{\nu}'_{14}/cm^{-1}$
C_6H_6	b_{2u}	1309	1570
$C_6H_5CH_3$	b_2	1286	1562
C_6H_5F	b_2	1299	1592
1,4-$C_6H_4F_2$	b_{2u}	1285	1591
1,2-$C_6H_4F_2$	a_1	1292	1603
1,3-$C_6H_4F_2$	b_2	1260	1609

Figure 8.141 The ν_{14} vibration of benzene and its wavenumber $\tilde{\nu}''_{14}$ and $\tilde{\nu}'_{14}$ in the $\tilde{X}$ and $\tilde{A}$ states, respectively, of benzene and some substituted benzenes

plane of the CN_2 three-membered ring and the rotational transitions include $\Delta J = 0, \pm 1, \pm 2$ with plane polarized radiation. In the two-photon spectrum, vibronic transitions of all symmetries are allowed. This fact, together with the use of circularly polarized radiation, allowed all the vibrations in the $\tilde{A}$ state to be identified.[432]

Analogous to equation (8.63), the requirement for a two-photon vibronic transition to be allowed is that

$$\Gamma(\psi'_{ev}) \times \Gamma(S_{ij}) \times \Gamma(\psi''_{ev}) \supset A \tag{8.68}$$

where $\Gamma(\psi_{ev}) = \Gamma(\psi_e) \times \Gamma(\psi_v)$. In the C_{2v} point group, to which difluorodiazirine belongs, each of the symmetry species is that of an element of the two-photon (or polarizability) tensor and, therefore, all two-photon vibronic transitions are allowed.

The $\tilde{A}^1B_{2u}$–$\tilde{X}^1A_{1g}$ system of benzene is forbidden as a one-photon process (see Section 6.3.1.8) and gains most of its intensity by Herzberg–Teller intensity stealing through the e_{2g} vibration ν_6 [see Section 6.3.4.1 and equation (6.230)]. Since $B_{2u} \neq \Gamma(\alpha_{ij})$ the $\tilde{A}$–$\tilde{X}$ transition is forbidden as a two-photon process also. However, vibrations of b_{2u}, e_{2u} or e_{1u} symmetry may be involved in intensity stealing[433] since

$$\begin{aligned} B^e_{2u} \times b^v_{2u} &= A^{ev}_{1g} = \Gamma(\alpha_{xx} + \alpha_{yy}, \alpha_{zz}) \\ B^e_{2u} \times e^v_{2u} &= E^{ev}_{1g} = \Gamma(\alpha_{xz}, \alpha_{yz}) \\ B^e_{2u} \times e^v_{1u} &= E^{ev}_{2g} = \Gamma(\alpha_{xx-yy}, \alpha_{xy}) \end{aligned} \tag{8.69}$$

In fact, using the Wilson vibrational numbering,[368] ν_{14} and ν_{15} (b_{2u}), ν_{16} and ν_{17} (e_{2u}), and ν_{18} (e_{1u}) are all active in single quanta. The vibration ν_{14} gives rise to most of the intensity while ν_{18} and ν_{17} are the next most important.[434] The ν_{14} normal mode has been called the 'Kekule vibration'; Figure 8.141 shows that it involves lengthening and shortening of alternate C—C bonds. It has wavenumbers of 1309 and 1570 cm^{-1} in the $\tilde{X}$ and $\tilde{A}$ states, respectively.

Figure 8.142 shows the 14^1_0 (A_{1g}–A_{1g}) band in the two-photon FE $\tilde{A}$–$\tilde{X}$ spectrum of benzene[435] in the gas phase. For an A_{1g}–A_{1g} vibronic transition the rotational selection rules for benzene, an oblate symmetric rotor are

$$\Delta K = 0, \Delta J = 0, \pm 1, \pm 2 \tag{8.70}$$

giving O, P, Q, R and S branches which show different intensity behaviour for plane and circularly polarized radiation.[421] Figure 8.142 shows how the ^qQ branch intensities are reduced, with circularly polarized radiation, relative to those of the ^qS and ^qR branches.

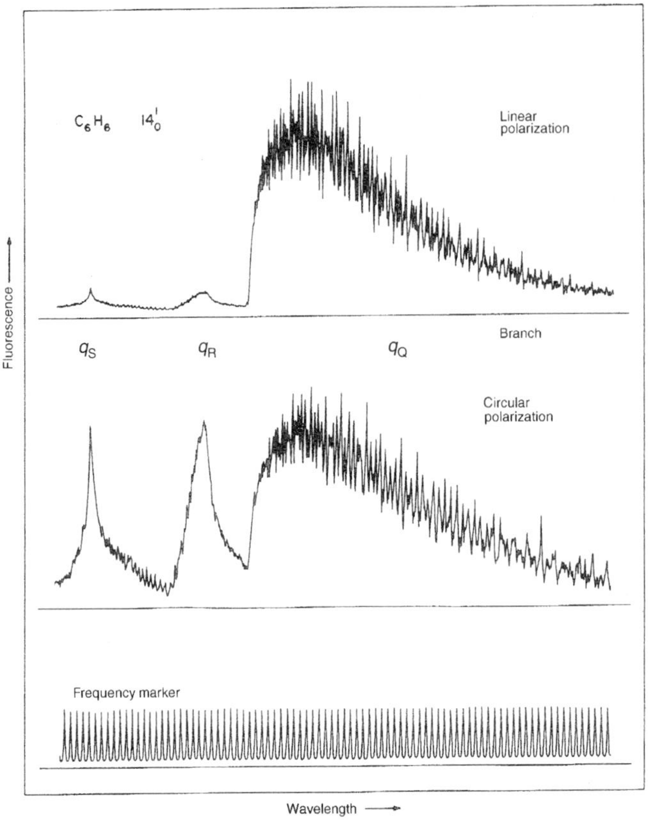

Figure 8.142 The 14^1_0 band in the two-photon FE $\tilde{A}^1B_{2u}$–$\tilde{X}^1A_{1g}$ spectrum of benzene with plane (linear) and circular polarization
[Reproduced, with permission, from Lombardi, J. R., Wallenstein, R., Hänsch, T. W. and Friedrich, D. M. (1976). *J. Chem. Phys.*, **65**, 2357]

The ΔK selection rule is changed to ± 1 for an E_{1g}–A_{1g} and to ± 2 for an E_{2g}–A_{1g} vibronic transition.

Figure 8.143 shows part of the 14^1_0 band in the two-photon FE spectrum obtained,[436] in the gas phase, at higher resolution than in Figure 8.141. This sub-Doppler resolution was achieved by using two counter-propagating beams. Computer simulation of the rotational structure gave a value of 0.181 330(45) cm^{-1} for the rotational constant B in the 14^1 level of the $\tilde{A}$ state. This compares with a value of 0.181 721 cm^{-1} for the zero-point level of the $\tilde{A}$ state obtained from a 1 + 1 REMPI spectrum of benzene in a skimmed supersonic jet using narrow-band (90 MHz) stimulated Raman shifted laser radiation.[437] The C—C bond length increases by 0.035 Å from the $\tilde{A}$ state.

The $\tilde{A}^1B_{2u}$–$\tilde{X}^1A_g$ system of 1,4-difluorobenzene, belonging to the D_{2h} point group, is the analogue of the $\tilde{A}$–$\tilde{X}$ system of benzene and is allowed as a one-photon but forbidden as a two-photon transition. In the two-photon FE spectrum,[438] single quanta of a_u, b_{1u}, b_{2u} and b_{3u}

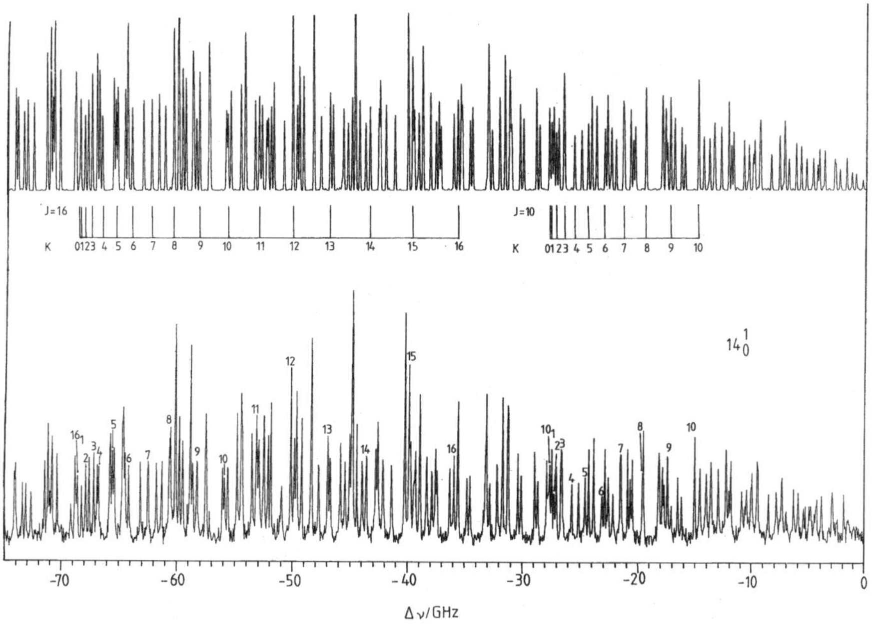

Figure 8.143 Part of the observed (above) and computed (below) 14^1_0 band in the two-photon $\tilde{A}$–$\tilde{X}$, FE spectrum of benzene. The observed spectrum was obtained with sub-Doppler resolution using two counter-propagating beams
[Reproduced, with permission, from Riedle, E., Neusser, H. J. and Schlag, E. W. (1981) *J. Chem. Phys.*, **75**, 4231]

vibrations are all involved in vibronically induced bands but the b_{2u} 'Kekule' vibration ν_{14} is much the most important giving a very intense 14^1_0, A_g–A_g band. As Figure 8.141 shows, the wavenumber of ν_{14} is very insensitive to mono- and di-substitution in benzene.

Benzene forms weakly bound complexes with the noble gases and these are stabilized in a supersonic expansion. The $\tilde{A}$–$\tilde{X}$ $1 + 1'$ REMPI spectra of the complexes C_6H_6Ar and $C_6H_6Ar_2$ have been obtained[439–441] using a two-colour method. The two laser beams are directed collinearly into a skimmed supersonic molecular beam of benzene seeded into argon. The wavenumber of the first photon excites the complex to a chosen vibronic level of the $\tilde{A}$ state whereas that of the second ionizes it, but with only a few hundred cm^{-1} of excess energy in order to avoid any further $\tilde{A}$–$\tilde{X}$ excitation. Both laser intensities are attenuated to avoid saturation of the rovibronic transitions observed.

Figure 8.144 shows a low resolution spectrum in the region of the 6^1_0 bands of C_6H_6Ar and $C_6H_6Ar_2$ where ν_6 is the vibration in the complex corresponding to the e_{2g} vibration ν_6 in benzene. Although this spectrum was obtained by monitoring $C_6H_6Ar^+$ ions, it also contains bands due to $C_6H_6Ar_2$ because of the fast dissociation of $C_6H_6Ar_2$ to give C_6H_6Ar.

Benzene is an oblate symmetric rotor with the c-axis perpendicular to the ring. In C_6H_6Ar the argon atom is on this axis 3.582 and 3.523 Å above the plane in the $\tilde{X}$ and $\tilde{A}$ state, respectively. The molecule is a prolate symmetric rotor belonging to the C_{6v} point group. In $C_6H_6Ar_2$ the argon atoms are symmetrically positioned on either side of the ring at a distance of 3.58 and 3.52 Å from the ring in the $\tilde{X}$ and $\tilde{A}$ state, respectively, showing that

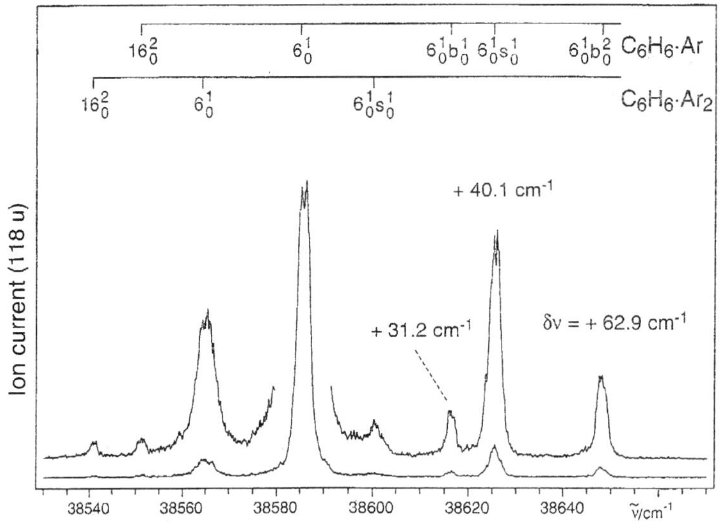

Figure 8.144 Low resolution, two-colour, 1 + 1′ REMPI spectrum of benzene–argon complexes in the region of the 6^1_0 band of the Ã–X̃ system showing bands involving van der Waals stretching (s) and bending (b) vibrations
[Reproduced, with permission, from Riedle, E., Sussmann, R., Weber, Th. and Neusser, H. J. (1996). *J. Chem. Phys.*, **104**, 865]

the addition of a second argon atom does not affect the position of the first. The molecule belongs to the D_{6h} point group.

In C_6H_6Ar or $C_6H_6Ar_2$, υ_6 is an e_2 or e_{2g} vibration, respectively, the Ã electronic state is B_2 or B_{2u} and the 6^1_0 transition is E_1^{ev}–A_1^e or E_{1u}^{ev}–A_{1g}^e, giving a perpendicular band in both cases. Analysis of these bands has given values[439] of the Coriolis constant ζ' [see equation (5.212)] for ν_6' of −0.5869 and −0.5851 for C_6H_6Ar and $C_6H_6Ar_2$, respectively, very similar to the value of −0.5785 for benzene.

In C_6H_6Ar there are, in comparison with benzene, two additional (van der Waals) vibrations, the Ar-ring stretching, ν_s, and bending, ν_b, vibrations, of a_1 and e_1 symmetry, respectively. The band labelled $6^1_0s^1_0$ has been assigned[440] on the basis of the small separation of $+40.1\,cm^{-1}$ from 6^1_0 and its being a perpendicular band. The band labelled $6^1_0b^2_0$, $+62.9\,cm^{-1}$ from 6^1_0, is also a perpendicular band. This is because the symmetry species of the 6^1b^2 state is given by $E_1 \times (e_1)^2$, which gives (see Tables 5.9 and 5.10) $E_1 + B_1 + B_2 + E_1$, and only the E_1–A_1 component is allowed.

The band labelled $6^1_0b^1_0$, with a separation of $+31.2\,cm^{-1}$ from the 6^1_0 band, should be a parallel band because the symmetry of the 6^1b^1 state is given by (see Table 5.9) $E_1 \times e_1 = A_1 + A_2 + E_2$ and only the transition to the A_1 component is allowed. The 1 + 1, two-colour REMPI spectrum in Figure 8.145 shows this band at sub-Doppler resolution. There is a complication in this parallel band due to the fact there is an additional Coriolis interaction due to the e_1 bending vibration. Jahn's rule, given by equation (5.194), shows that the A_1 and A_2 components of the 6^1b^1 state are connected by a rotation about the z-axis. The effective Coriolis constant ζ' in this state is given by $\zeta'_{ev} - \zeta'_b$, the difference between the 6^1 state vibronic contribution and that from the

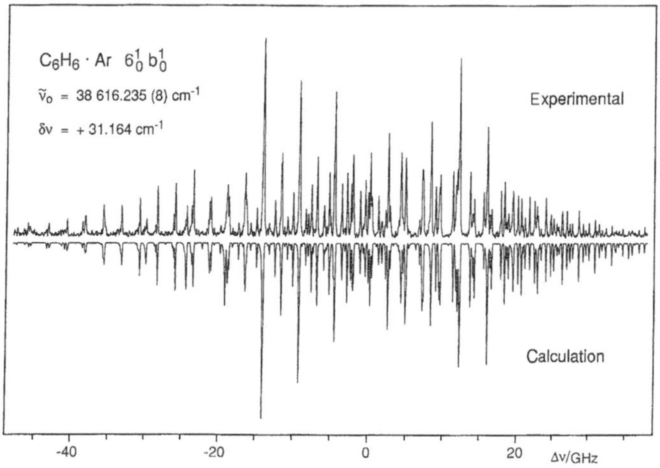

Figure 8.145 The experimental (above) and calculated (below) sub-Doppler rotational structure of the $6_0^1b_0^1$ band of C_6H_6Ar
[Reproduced, with permission, from Riedle, E. and van der Avoird, A. (1996). *J. Chem. Phys.*, **104**, 882]

van der Waals bending vibration. The value of ζ' was found[441] to be -0.6367 from the computed spectrum in Figure 8.145.

The intensity distributions along the progressions in ν_s and ν_b, shown in Figure 8.141, do not follow normal expectations from the Franck–Condon principle. There is a Fermi resonance between the 6^1s^1 and 6^1b^2 levels and the $6_0^1b_0^1$ band gains intensity, not through Herzberg–Teller vibronic coupling, but through libration of the vibronic transition moment in the benzene moiety.

The band separated from the 6_0^1 band of $C_6H_6Ar_2$ by $+35.7\,cm^{-1}$ has been assigned to $6_0^1s_0^1$ involving the symmetric van der Waals stretching vibration. It is a perpendicular band very similar in appearance to the 6_0^1 band.

In a complex between a monosubstituted (C_{2v}) benzene and argon there are three van der Waals vibrations, a stretching vibration (s_z) and two bending vibrations, one parallel (b_x) to and the other perpendicular (b_y) to the C_2 axis of the monomer. These have been observed with vibrational resolution for C_6H_5XAr where $X = NH_2$, OH, F and Cl in one-colour 1 + 1 REMPI spectra[442] with mass selection. As in C_6H_6Ar the intensities do not follow Franck–Condon expectations. There are Fermi resonances between one quantum of s_z and two quanta of both b_x and b_y.

The one-colour 2 + 1 REMPI spectrum of catechol (1,2-dihydroxybenzene) shown in Figure 8.146, obtained with vibrational resolution, has shown[443] that it exists predominantly as only one of three possible planar rotational isomers.

The part of the REMPI spectrum shown is dominated by three bands which had been assigned previously as the 0_0^0 bands of the three rotamers, the one shown in the figure, one in which both hydroxyl hydrogens are directed towards each other and one in which they are directed away from each other.

The hole-burning (HB) spectrum, also shown in Figure 8.146, shows that this interpretation is incorrect. For the HB spectrum a second laser was tuned to the band at $35\,762\,cm^{-1}$ and crossed the supersonic jet before the exciting and ionizing laser. The HB laser depletes the population of the zero-point level of the S_0 state of the rotamer responsible for this band. Therefore, the HB spectrum will show a

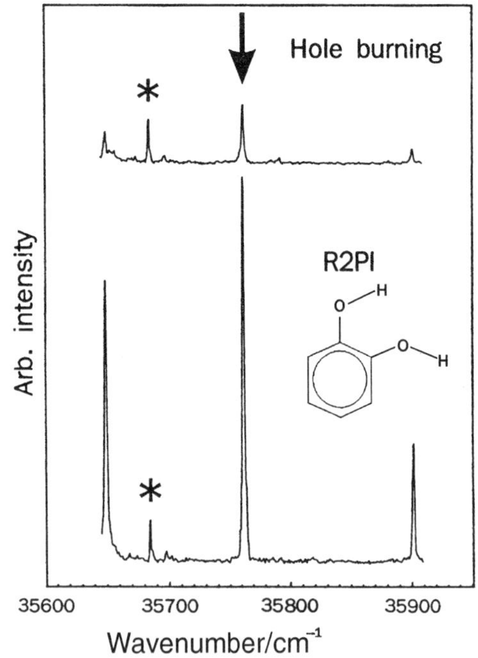

Figure 8.146 Part of the 2 + 1 REMPI spectrum and the hole-burning spectrum of catechol. The band marked with an asterisk is due to catechol.H_2O [Reproduced, with permission, from Burgi, T. and Leutwyler, S. (1994). *J. Chem. Phys.*, **101**, 8418]

reduction of the signal for all bands whose lower level is the S_0 zero-point level of the same rotamer. The HB spectrum shows that the signal is reduced equally for all three bands and that they must all be due to the same rotamer. This ability of the HB spectrum to single out molecular species is illustrated by the fact that, in the HB spectrum, the signal due to the van der Waals complex catechol.H_2O is not reduced.

The 35 649 cm^{-1} band is the 0^0_0 band of the only detected rotamer and the other two bands form a progression, with Δv even, in an OH-torsional vibration.

Ab initio calculations predict that the rotamer of minimum energy is that shown in Figure 8.146.

The sub-Doppler two-photon technique using counter-propagating beams has been used to measure the Lamb shift in the $n = 1$, $^2S_{\frac{1}{2}}$ state of the hydrogen atom, mentioned in Section 6.1.4 and shown in Figure 6.16. In this state, unlike the $n = 2$ states, the Lamb shift can be measured only as a small shift in a very large transition wavenumber, such as that of the $n = 2$–$n = 1$ member of the Lyman series. This was first achieved by using a tunable dye laser at about 486 nm with a frequency-doubling crystal to convert to 243 nm which was passed through a cell containing hydrogen atoms. Two-photon absorption occurs in the $2^2S_{\frac{1}{2}}$–$1^2S_{\frac{1}{2}}$ transition, forbidden as a one-photon process. The $2^2S_{\frac{1}{2}}$–$1^2S_{\frac{1}{2}}$ fluorescence is forbidden but collisions easily promote transfer from $2^2S_{\frac{1}{2}}$ to $2^2P_{\frac{1}{2}}$ from which fluorescence is allowed and is monitored. The various processes are shown in Figure 8.147.

To overcome the difficulty of measuring, with sufficient accuracy, the laser wavelength at which the two-photon process occurs part of the dye laser fundamental, at 486 nm, was

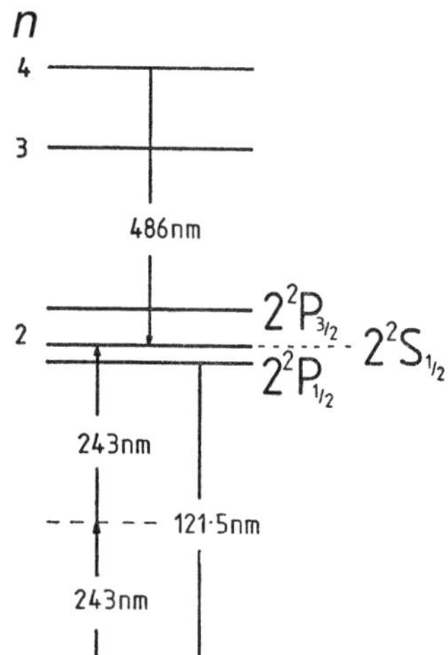

Figure 8.147 Energy level scheme for the hydrogen atom showing how the Doppler-free two-photon absorption technique was used to measure the Lamb shift in the $1^2S_{\frac{1}{2}}$ state

directed into the positive column of a hydrogen discharge in order to observe, simultaneously, the $n = 4$–2 transition of atomic hydrogen. According to the Bohr theory, the $n = 2$–1 separation is exactly four times the $n = 4$–2 separation. When the more accurate quantum electrodynamics is applied this is no longer exactly true. Nevertheless, they are sufficiently close for the calibration to be achieved by observing, simultaneously and with the same laser but not frequency doubled, the fine structure components of the $n = 4$–2 transition (second line of the Balmer series).

The most accurate value[445] of the Lamb shift of the $1^2S_{\frac{1}{2}}$ state is 8.161 ± 0.029 GHz ($0.2722\,\text{cm}^{-1}$) and was found using polarization spectroscopy (Section 8.2.23) to measure the $n = 4$–2 fine structure components more accurately.

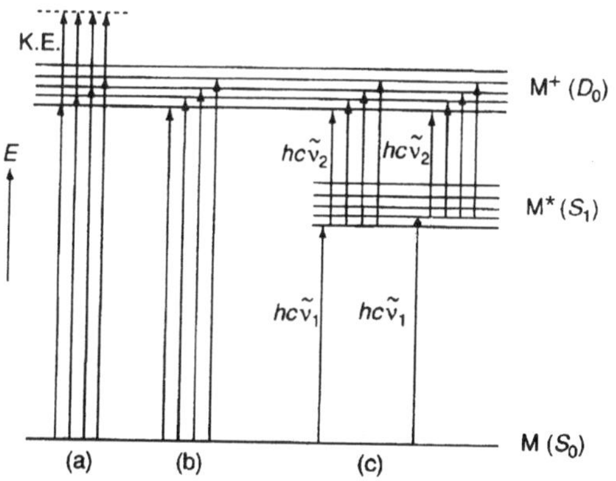

Figure 8.148 Processes involved in obtaining (a) a UPS spectrum, (b) a ZEKE-PE spectrum by a one-photon process and (c) a ZEKE-PE spectrum by a two-photon process in which the first photon is resonant with an excited electronic or vibronic state of the molecule

8.2.19 Zero kinetic energy photoelectron spectroscopy

Ultraviolet photoelectron spectroscopy (UPS) has been discussed in Section 7.5.1. It is a low resolution technique, about $32\,\text{cm}^{-1}$ at best, compared with that of most other spectroscopies. This means that resolution of rotational fine structure in molecular spectra is not possible, except for molecules, such as H_2, with a large rotational constant. In larger molecules, even vibrational structure is not sufficiently well resolved for many purposes.

The factor limiting the resolution in UPS is the inability to measure the kinetic energy of the photoelectrons with sufficient accuracy. The source of the problem pointed to a possible solution. If the photoelectrons could be produced with zero kinetic energy, this cause of the lack of resolution would be largely removed. This is the basis of zero kinetic energy photoelectron (ZEKE-PE) spectroscopy.

Figure 8.148(a) illustrates the ionization process in a UPS experiment in which a photon of the incident radiation always has much more energy than is necessary to ionize the molecule M into either the zero-point level or a vibrationally excited level of M^+. The excess energy is then removed as kinetic energy of the photoelectron.

In Figure 8.148(b) is shown a process in which a monochromatic, tunable source of high energy (vacuum ultraviolet) radiation is scanned across the M to M^+ band system while detecting the photoelectrons which have zero kinetic energy. The resulting ZEKE-PE spectrum shows progressions involving the vibrational levels of the ion in the same way as a UPS spectrum. The main differences from a UPS experiment are that the ions have zero kinetic energy and that the ionizing radiation is usually a laser resulting in a much higher resolution of $1\,\text{cm}^{-1}$ or better. In addition, a ZEKE-PE spectrum is usually obtained with the sample molecules in a supersonic jet, which reduces substantially the initial states populated in the molecule M, thereby simplifying the spectrum.

Instead of using a laser operating in the vacuum ultraviolet region, a laser operating at half the wavenumber may be used. Then the ionization process in Figure 8.148(b) involves the absorption of two photons to go from M to

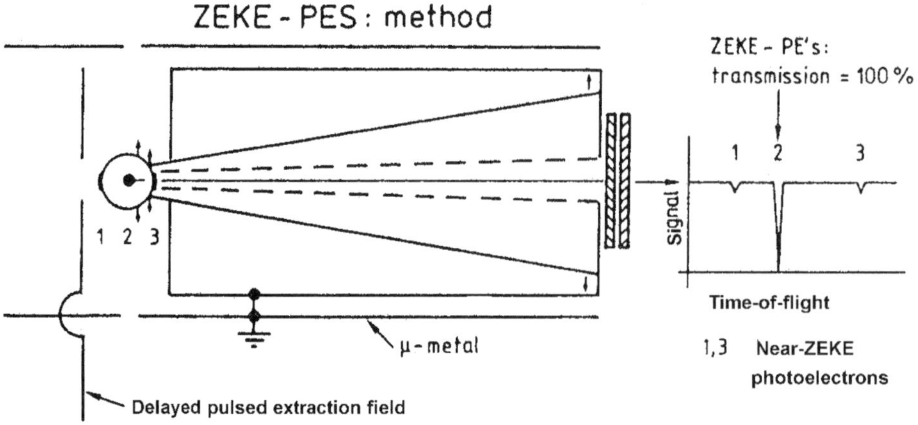

Figure 8.149 Method of distinguishing between ZEKE (2) and near-ZEKE (1 and 3) photoelectrons [Reproduced, with permission, from Müller-Dethlefs, K. and Schlag, E. W. (1991). *Annu. Rev. Phys. Chem.*, **42**, 109]

M^+, similar to the two-photon absorption process in Figure 8.131(b).

More commonly, the resonant two-photon process in Figure 8.148(c) is employed. This is a two-colour experiment requiring two lasers, one at a fixed wavenumber $\tilde{\nu}_1$ and the other at a wavenumber $\tilde{\nu}_2$ which is tunable. The first photon takes the molecule, which, again, is usually in a supersonic jet, to the zero-point level of an excited electronic state. This excitation may also be achieved by a two- or three-photon absorption. If M is a closed shell molecule, the ground state is a singlet state, S_0, and the lowest excited singlet state is S_1. Excitation may be to S_1 or a higher electronic state. The wavenumber of the second photon is scanned across the S_1 to D_0 band system, where D_0 is the ground (doublet) state of M^+.

As shown on the right of Figure 8.148(c), the first photon (or photons) may take the molecule to a vibrationally excited level in S_1. Then the intensity distribution along the progression in the corresponding vibration in D_0 will be changed, thereby giving access to more vibrational levels.

This general method of attempting to detect electrons with zero kinetic energy, known as threshold photoelectron spectroscopy, suffers from interference by electrons which are produced with a small amount of kinetic energy. These electrons result from an autoionization process in which an electron is removed from the molecule when the latter is in a high-lying Rydberg state just below the ionization limit. The fact that the photoelectrons have a small spread of kinetic energies reduces the resolution of which ZEKE-PE is capable.

Considerable progress was made in realizing this high resolution capability by employing delayed pulsed field extraction. The photoelectrons with zero kinetic energy are separated from those with kinetic energy by waiting for about 1 μs after they have been produced by the pulsed laser(s). In this time interval the two sets of electrons have drifted apart and application of a pulsed voltage accelerates them differently and they can be distinguished by their different times of flight to the electron detector. Figure 8.149 shows how photoelectrons such as 1 and 3, resulting from autoionization and which have drifted away from the region of the supersonic jet in which they were produced, have different times of flight from photoelectron 2 which was produced with zero kinetic energy. Gated boxcar detection allows the collection of the

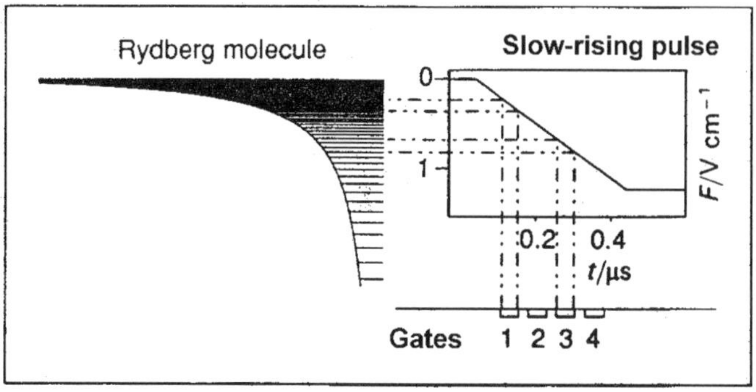

Figure 8.150 Pulsed field ionization of long-lived Rydberg states using a slow-rising extraction pulse to remove the electrons
[Reproduced, with permission, from Fischer, I., Lindner, R. and Müller-Dethlefs, K. (1994). *J. Chem. Soc., Faraday Trans.*, **90**, 2425]

signal due only to the ZEKE electrons. This technique is known as ZEKE spectroscopy and resolution is *ca* 1 cm^{-1}, a considerable improvement on threshold photoelectron (TPE) spectroscopy.

A further improvement in resolution has been achieved by using pulsed field ionization, in ZEKE-PFI spectroscopy.[447]

Figure 8.150 shows a typical stack of Rydberg states, typically with $n > 150$, lying up to about 3–4 cm^{-1} below the ionization limit to which they converge. Such states are long-lived with lifetimes of up to tens of microseconds. Following the laser(s) pulse exciting the molecule into these Rydberg states, a pulsed voltage is applied, rising slowly in the example shown from 0 to 1 V cm^{-1} in about 0.4 μs. This pulse removes photoelectrons from Rydberg states which are further below the ionization limit the higher is the applied voltage. As the figure shows, gated detection of the electrons, with a narrow gate, selects electrons from a very narrow range of Rydberg states. In this way the ZEKE-PFI technique allows the resolution of the ZEKE technique to be increased to about 0.2 cm^{-1}, thereby allowing resolution of rotational fine structure.

The rotational selection rules for a ZEKE ionization process in diatomic[448,449] and polyatomic[449] molecules are less restrictive than for transitions between bound states. For example, for a diatomic or linear molecule, the selection rule for the angular momentum of the core is

$$\Delta J = |J^+ - J''| < l + \tfrac{3}{2} \qquad (8.71)$$

where J^+ refers to the ion and J'' to either the ground state of the neutral molecule, in a one-photon process in Figure 8.148(b), or an excited electronic state, in a $1 + 1'$ two-photon process as in Figure 8.148(c). The quantum number l refers to the orbital angular momentum of the photoelectron and is not, in general, restricted. For a homonuclear diatomic molecule the parity selection rule is

$$+ \leftrightarrow + \text{ and } - \leftrightarrow - \qquad (8.72)$$

The one-photon ZEKE-PFI spectrum of N_2, in a molecular beam but with little rotational cooling, has been obtained[450] using extreme ultraviolet (XUV) radiation tunable in the range 125 000–128 500 cm^{-1}. This radiation is produced by sum-frequency mixing of two laser beams in a pulsed atomic beam of krypton. The first laser beam is of fixed wavenumber $\tilde{\nu}_1 = 47\,046.8\ cm^{-1}$ and is in two-photon resonance with a transition from the ground state of Kr. The wavenumber of the second laser is

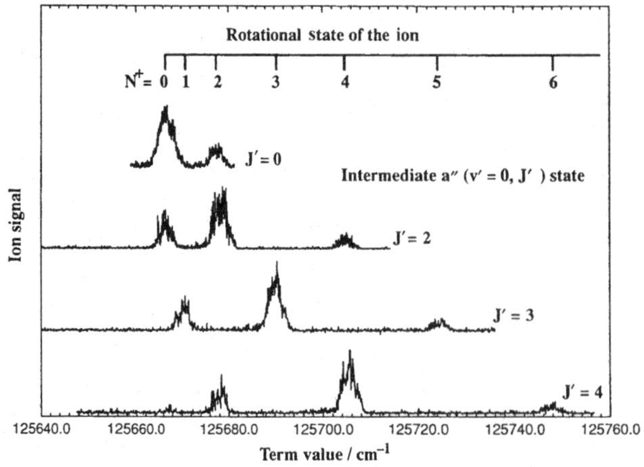

Figure 8.151 ZEKE-PFI, $2+1'$ spectra of N_2 $X^1\Sigma_g^+$ ($v''=0$, $J''=0$) via different intermediate $a''^1\Sigma_g^+$ ($v'=0$, J') states
[Reproduced, with permission, from Mackenzie, S. R., Merkt, F., Halse, E. J. and Softley, T. P. (1995). *Mol. Phys.*, **86**, 1283]

scanned through the region $\tilde{\nu}_2 = 31\,000$–$34\,500\,\mathrm{cm}^{-1}$, thereby producing XUV radiation of wavenumber $2\tilde{\nu}_1 + \tilde{\nu}_2 = 125\,000$–$128\,500\,\mathrm{cm}^{-1}$.

The ionization process observed was N_2^+ $X^2\Sigma_g^+(v^+=0, 1)$–N_2 $X^1\Sigma_g^+(v=0)$, where v^+ is the vibrational quantum number in the ion. The best-resolved rotational structure in the corresponding conventional photoelectron spectrum (the bands in the 15.5–16.0 eV region in Figure 7.17) shows strong Q branches, with $N^+ - J'' = 0$, where N^+ is the quantum number in the ion referring to the total angular momentum excluding electron spin, and weaker S ($N^+ - J'' = 2$) and O ($N^+ - J'' = -2$) branches. However, the rotational line intensities are different in the ZEKE-PFI spectrum. The 0–0 band shows an O branch about 10 times stronger than the S branch, whereas the 1–0 band shows a strong U branch, with $N^+ - J'' = 4$, in addition to the O, Q and S branches. These distorted intensity distributions are a consequence of complex interactions affecting the high-lying Rydberg states accessed prior to pulsed field ionization.

Ion–molecule reactions are extremely common throughout the universe and it is important to be able to study them in the laboratory with ions in a selected rovibronic state. Selection of such ions is made possible by a ZEKE-PFI process and is illustrated by the production of N_2^+, in the $X^2\Sigma_g^+(v^+=0)$ vibronic state with particular values of the rotational quantum number N.

This has been achieved[451] by a $2+1'$ two-colour excitation. The N_2 $X^1\Sigma_g^+(v=0)$ is excited, in a two-photon process, to a particular rotational level, J', of the $a''^1\Sigma_g^+$ electronic state. A third photon takes the molecule to a high-lying Rydberg state and this is followed by pulsed field ionization. In this case the ions, rather than the photoelectrons, are detected and selection is by a time-of-flight technique.

Figure 8.151 shows that ions with N^+ predominantly 0, 2, 3 and 4 can be produced with excitation to levels with $J = 0$, 2, 3 and 4, respectively, in the a'' state. Use of this $2+1'$ multiphoton technique, producing a simplified spectrum, is clearly advantageous, in comparison with the one-photon technique,[450] in enhancing ionic state selectivity.

A $2+1'$ ZEKE-PFI spectrum of NH_3 has been obtained[452] with the two-photon absorption taking the molecule to the $\tilde{B}^1E''$ state in which it is planar. The conventional UPS spectrum [see Section 7.5.1.4(i)] shows a long progression in the inversion vibration ν_2 in going to the ground state, $\tilde{X}^2A_2''$, of the ion indicating planarity in this state also. Therefore, the Franck–Condon principle dictates that the ionization process NH_3^+ $\tilde{X}^2A_2''$–NH_3 $\tilde{B}^1E''$ will be dominated by transitions with $\Delta v_2 = 0$.

Rotationally resolved spectra with $v_2 = 2$ in the $\tilde{B}$ state of NH_3 and $v_2^+ = 1$ and 2 in NH_3^+ have been analysed[452] and the selection rules and intensities interpreted.[453]

ZEKE-PFI spectra of benzene have been obtained[449] in a supersonic jet by a $1 + 1'$ process in which the first photon takes the molecule from the ground state to the $v_6 = 1$ level of the $\tilde{A}^1B_{2u}$ state, where ν_6 is an e_{2g} vibration [see equation (6.230)]. The $\tilde{X}^2E_{1g}$–$\tilde{A}^1E_{1u}^{ev}(v_6 = 1)$ band, in which the second photon forms the cation in the zero-point level of the ground electronic state, has been rotationally analysed and D_{6h} symmetry confirmed. If the cation was distorted by a static Jahn–Teller effect in the degenerate E_{1g} state rotational transitions which are forbidden with D_{6h} symmetry would be allowed—but these are not observed.

The $\tilde{X}$ state of the cation is doubly degenerate and is subject to a dynamic Jahn–Teller effect. Excitation of ν_6 in this state splits the 6^1 level into two components with $J = \pm\frac{1}{2}$ and $\pm\frac{3}{2}$ [see equation (6.248)] and two bands, split by $320\,\text{cm}^{-1}$, result.

ZEKE-PFI supersonic jet spectra of the asymmetric rotors H_2O and H_2S have been obtained with rotational resolution by a one-photon process[454] and, of H_2S, by a $1 + 1'$ two-photon process[455] in which the first photon is resonant with the $\tilde{A}$ state. The rotational selection rules for these asymmetric rotors have been discussed.[449]

We have seen in Section 8.2.17 how fluorescence excitation and single vibronic level fluorescence spectra of the monofluorotoluenes, with vibrational resolution, show progressions in the methyl torsional vibration (see, for example, Figure 8.123). The spacings and intensity distributions along these progressions can be interpreted to give the torsional potentials in the S_0 and S_1 electronic states (see Table 8.5).

$1 + 1'$ ZEKE-PFI spectra of the monofluorotoluenes have been obtained[456] in a supersonic jet. The first photon can be tuned to populate the zero-point level or excited methyl torsional levels in the S_1 state. Figure 8.152 shows, for example, ZEKE-PFI spectra of 3-fluorotoluene with the first laser tuned to various torsional bands in the S_1–S_0 spectrum in Figure 8.123. These result in the exclusive populations of the $0a_1$, $1e$, $2e$, $3a_1$ or $4e$ levels in S_1. Transitions in the ZEKE-PFI spectra from the $1e$, $2e$ and $4e$ levels give the e levels in the ground doublet state, D_0, of the cation and spectra from the $0a_1$ and $3a_1$ levels give the a_1 and, weakly, the a_2 levels in D_0. The Franck–Condon principle results in varied intensity distributions along the progressions in the ZEKE-PFI spectra giving access to a wide range of torsional levels in D_0 which can be used to construct the torsional potential.

Figure 8.153 shows the resulting torsional potential for the D_0 state of the cation compared with those for the S_0 and S_1 states. These show that there is not only the highest barrier of $318\,\text{cm}^{-1}$ in D_0 but that there is also a 60° conformational change from either state of the molecule to the ground state of the cation.

The V_3/V_6 parameters obtained experimentally[456] for the monofluorotoluene cations are given in Table 8.5, which also indicates whether the lowest energy conformation is pseudo-*cis* or pseudo-*trans*.

Modest *ab initio* calculations[393] of the torsional potentials in the S_0 states of the molecules and the D_0 states of the cations reproduce the experimentally determined values very well. In the cations of 2- and 3-fluorotoluene the removal of a π-electron creates a pattern of long and short bonds in the ring which is mostly responsible for the relatively high torsional barriers.

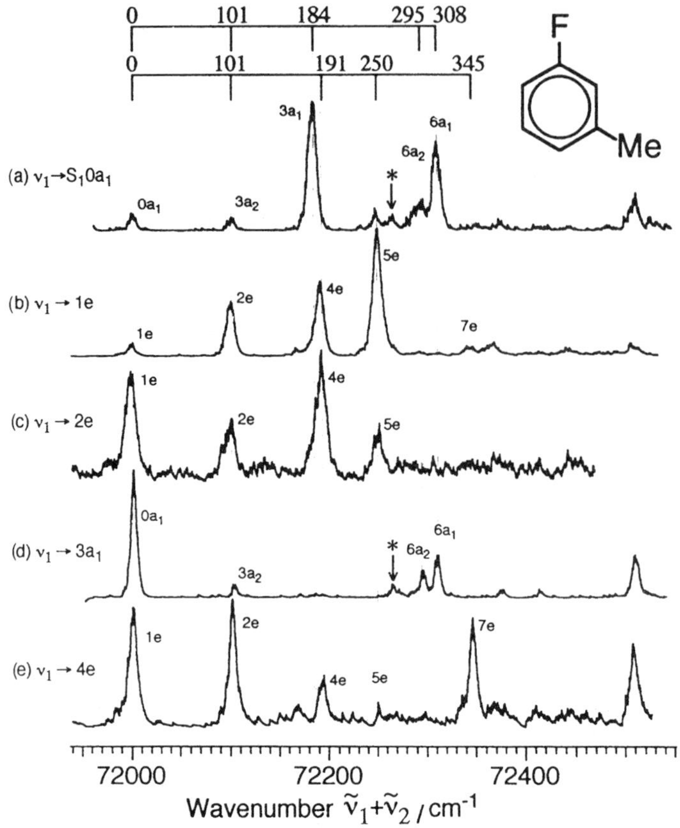

Figure 8.152 ZEKE-PFI spectra of jet-cooled 3-fluorotoluene following excitation of the (a) $0a_1$, (b) $1e$, (c) $2e$, (d) $3a_1$ and (e) $4e$ levels of S_1
[Reproduced, with permission, from Takazawa, K., Fujii, M. and Ito, M. (1993). *J. Chem. Phys.*, **99**, 3205]

ZEKE-PFI spectra of the complexes aniline–Ar ($C_6H_5NH_2$–Ar) and aniline–Ar_2, in a skimmed supersonic jet have shown that there is a significant change in the van der Waals bonding on going from the S_1 state to the ground state D_0 of the cation.[457] Whereas 1 + 1 REMPI spectra of these complexes show only short progressions in their S_1–S_0 transitions,[442] their $1 + 1'$ ZEKE-PFI spectra, with the first photon exciting to the zero-point level of S_1, show longer progressions, as shown in Figure 8.154.

The spectrum due to An^+ shows no vibronic structure close to the 0_0^0 band but those of An–Ar^+ and An–Ar_2^+ show harmonic progressions with intervals of 15 and 11 cm^{-1}, respectively. These have been assigned, by comparison with the REMPI spectra,[442] to the totally symmetric bending vibration. In An–Ar this involves bending along the direction of the C_2 axis of aniline. In An–Ar_2, in which one argon atom is above and the other below the planar An^+, it involves in-phase bending of the two argon atoms. Bending vibration wavenumbers are reduced in both cations compared with their values of 22 and 15 cm^{-1} in the S_1 states of An–Ar and An–Ar_2, respectively, even though the van der Waals bond strength increases.

The length of the progressions in the bending vibration implies an appreciable change of

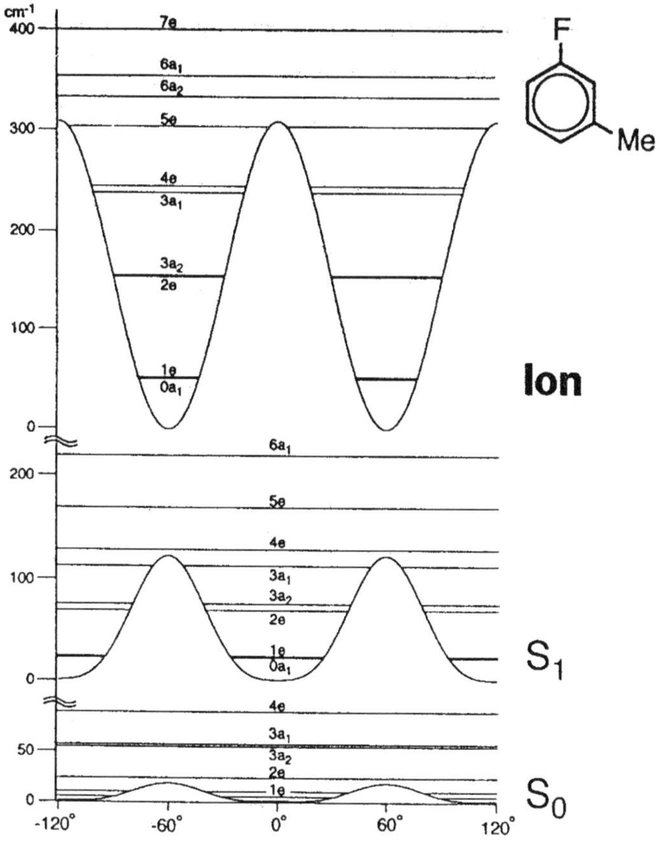

Figure 8.153 Methyl torsional potential for the D_0 state of the 3-fluorotoluene cation compared with those for the S_0 and S_1 states of the neutral molecule
[Reproduced, with permission, from Takazawa, K., Fujii, M. and Ito, M. (1993). *J. Chem. Phys.*, **99**, 3205]

geometry from S_1 to D_0 in this coordinate. It seems reasonable to suppose that the positive charge is located close to the nitrogen atom and this attracts the argon atom towards it.

The phenol–methanol (C_6H_5OH–$HOCH_3$) complex is stabilized by hydrogen bonding between the two OH groups. The $1+1'$ ZEKE-PFI spectrum[458] of the complex, formed in a skimmed supersonic jet, with excitation to the S_1 zero-point level, is shown in Figure 8.155.

The spectrum is dominated by two progressions. One of these is very harmonic, showing about 10 members with a vibrational interval of 34 cm^{-1}. This vibration has been assigned as an intermolecular bending vibration with a wavenumber tentatively assigned as 27 and 22 cm^{-1} in the S_1 and S_0 states, respectively. This progression is repeatedly associated with each member of a second progression with an initial spacing of 278 cm^{-1} and negative anharmonicity. This is assigned to the intermolecular stretching vibration and its wavenumber compares with 176 and 162 cm^{-1} in the S_1 and S_0 states, respectively.

The binding energy of the complex increases by 5421 cm^{-1} from S_0 to D_0 owing to the additional charge–dipole interaction in the cation. This leads to a shorter hydrogen bond and the consequent long progression in the

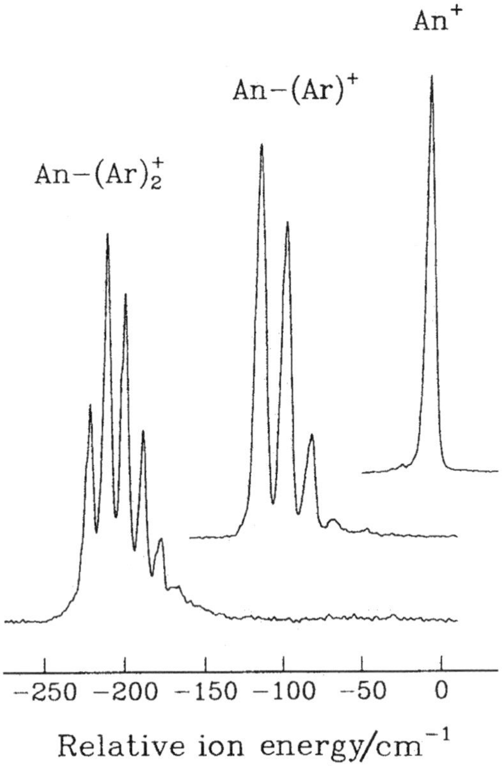

Figure 8.154 ZEKE-PFI spectra of aniline, aniline–Ar and aniline–Ar_2 with excitation close to the zero-point level of the S_1 state
[Reproduced, with permission, from Zhang, X., Smith, J. M. and Knee, J. L. (1992). *J. Chem. Phys.*, **97**, 2843]

stretching vibration and its large increase in wavenumber from 162 to 278 cm^{-1}. As in the examples of aniline–Ar and aniline–Ar_2, ionization has a much larger effect on the intermolecular potential than electronic excitation—a fairly general observation.

The ZEKE-PFI spectrum of a radical, CH_3, has been obtained[459] by a one-photon process. The radical was produced by flash (pulsed) pyrolysis of azomethane, CH_3NNCH_3, in a supersonic jet. The source of tunable vacuum ultraviolet radiation at about 126 nm (79 000 cm^{-1}) was produced by difference frequency mixing in a similar way to that used for generating XUV radiation used for the one-photon ZEKE-PFI spectrum of N_2.[450] Laser beams of wavenumbers $\tilde{\nu}_1$ and $\tilde{\nu}_2$, with wavelengths of 216.67 nm (fixed) and about 770 nm (tunable), respectively, are focused into an atomic beam of krypton where difference frequency mixing occurs, via a two-photon transition in krypton, to give a wavenumber of $2\tilde{\nu}_1 - \tilde{\nu}_2$.

The VUV radiation excites the CH_3 to high Rydberg states close to the ionization limit and a pulsed field ionizes it. The resulting spectrum, with a resolution of 1.5 cm^{-1}, is shown in the upper part of Figure 8.156.

The CH_3 radical is a planar, oblate symmetric rotor in the ground state, X^2A_2'', and the CH_3^+

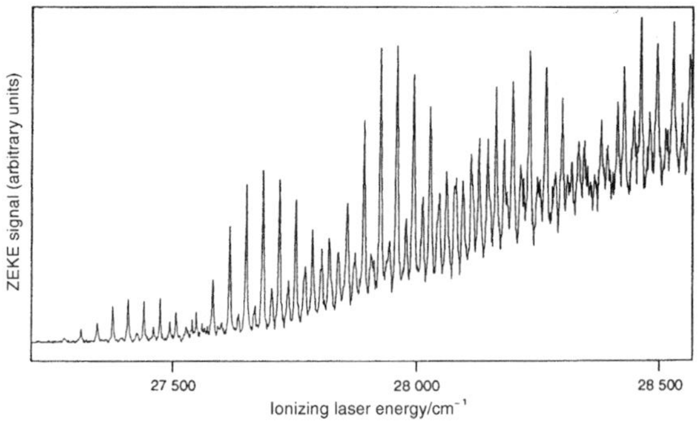

Figure 8.155 ZEKE-PFI spectrum of the phenol–methanol complex with initial excitation to the zero-point level of the S_1 state
[Reproduced, with permission, from Wright, T. G., Cordes, E., Dopfer, O. and Müller-Dethlefs, K. (1993). *J. Chem. Soc., Faraday Trans.*, **89**, 1609]

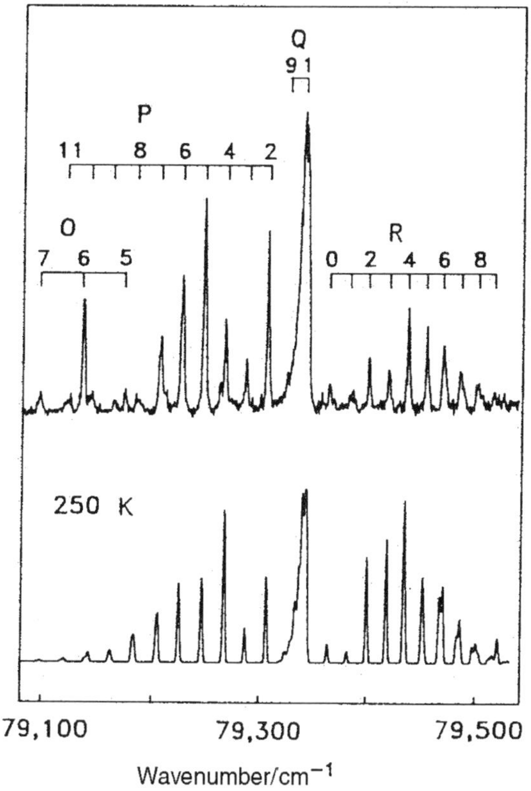

Figure 8.156 Experimental (top) and calculated (bottom) one-photon ZEKE-PFI spectrum of CH_3. For the calculated spectrum a rotational temperature of 250 K was assumed and selection rules $\Delta K = 0$ and $J^+ - N = 0, \pm 1$
[Reproduced, with permission, from Blush, J. A., Chen, P., Wiedmann, R. T. and White, M. G. (1993). *J. Chem. Phys.*, **98**, 3557]

cation is also planar in its X^1A_1' ground state. The rotational selection rules depend on the nature of the wave function of the photoelectron but the calculated spectrum in Figure 8.156 shows that they are dominated by $\Delta K = 0$, with almost coincident sub-bands, and $J^+ - N = 0, \pm 1$, giving *P*, *Q* and *R* branches. In addition, there is an *O* branch which is weak apart from the line with $N = 6$; this high intensity is attributed to a near resonant autoionization.

8.2.20 Laser photoelectron or photodetachment spectroscopy of negative ions

Until relatively recently, spectroscopy of negative ions remained largely elusive with the exception of C_2^-, whose $\mathrm{B}^2\Sigma_u^+ - \mathrm{X}^2\Sigma_g^+$ transition was observed[460] in a flash discharge in CH_4. Negative ions can be formed in electrical discharges, for example, but they tend to be very short-lived because of the ease with which they lose an electron.

The situation began to change with the development of laser photoelectron (LPE) spectroscopy.[461] The anions, produced from the parent molecule in a discharge, are extracted from it by mass selection. Beyond this stage the experiment resembles that used to obtain an ultraviolet photoelectron spectrum (see Section 7.2.1). The main difference is that the ionization energy of an anion is typically small, of the order of 1 eV, compared with that of a neutral species, which is of the order of 10 eV. For this reason, the vacuum ultraviolet, monochromatic source, such as HeI (21.22 eV) which is commonly used in photoelectron spectroscopy, can be replaced by a visible laser such as an argon ion or dye laser. The laser beam crosses the ion beam and the ejected photoelectrons are collected and velocity analysed with, typically, a hemispherical analyser. The resolution, typical of photoelectron spectroscopy, is low, of the order of 10 meV (80 cm^{-1}).

In this way, laser photoelectron spectra of cations such as OH^-, CH^-, SiH^-, SiH_2^- and C_2^- were obtained.

The technique has been called, alternatively, laser photodetachment spectroscopy. Of course, all photoelectron spectroscopy involves the photodetachment of an electron, but it may be that the much easier photodetachment of an electron from a cation has led to the alternative name.

The LPE spectrum of CH_2^- has generated considerable interest as it is capable of giving, directly, the energy difference between the ground state, $\tilde{\mathrm{X}}^3B_1$, and the lowest excited state, $\tilde{a}^1A_1$, of CH_2 [see equations (6.151) and (6.152)].

An early LPE spectrum[462] of CH_2^- is shown in Figure 8.157(a). The cations were produced in a discharge in CH_4 or diazomethane (CH_2N_2) and the ionization of CH_2^- achieved with an argon ion laser operating at 488 nm (2.540 eV). The spectrum shows two band systems, one dominated by the band G and the other showing a long progression which was presumed to be in the bending vibration and starting with band A.

The ground molecular orbital configuration of CH_2^- is (see Section 6.3.1.1) ... $(2a_1)^2(1b_2)^2(3a_1)^2(1b_1)^1$, giving the $\tilde{X}^2B_1$ ground state. It is known that the ground state geometry of a cation is very similar to that of the isoelectronic neutral species, in this case NH_2 for which $r_{NH} = 1.024$ Å and $\angle HNH = 103.4°$. Removal of an electron from CH_2^- can be either from the $3a_1$ orbital, to give the $\tilde{X}^3B_1$ ground state of CH_2 [equation (6.152)], or from the $1b_1$ orbital to give the low-lying excited state $\tilde{a}^1A_1$ [equation (6.151)]. However, the geometry of CH_2 in the $\tilde{a}^1A_1$ state, in which $r_{CH} = 1.11$ Å and $\angle HCH = 102.4°$, is very similar to the assumed geometry of CH_2^- in the $\tilde{X}^2B_1$ state, whereas that of CH_2 in the $\tilde{X}^3B_1$ state is not.

From these deductions, and the observation, or otherwise, of a progression in the bending vibration in the two band systems in Figure 8.157(a), it follows that band G and, apparently, band A correspond to ionization of CH_2^- ($\tilde{X}^2B_1$) to give CH_2 in the zero-point level of the $\tilde{a}^1A_1$ and $\tilde{X}^3B_1$ states, respectively. The separation of bands A and G of 0.845 ± 0.03 eV ($6815\,cm^{-1}$) was then assumed to give the separation of the ã and $\tilde{X}$ states of CH_2, a value far outside the range of $2800 \pm 700 cm^{-1}$ expected at that time.

A later LPE spectrum of CH_2^- was obtained[463] in a flowing afterglow ion source and cooled by collisions with helium buffer gas at room temperature and is shown in Figure 8.157(b). This shows that bands A, B and C in the earlier spectrum are hot bands (involving a progression in the bending vibration) and that band D is the 0_0^0 band of the $\tilde{X}^3B_1$–$\tilde{X}^2B_1$ ionization band system. This gives a revised splitting between the $\tilde{X}$ and ã states of CH_2 of about 0.38 eV ($3100\,cm^{-1}$). This value agrees with the more accurate value of 0.3902 eV ($3147\,cm^{-1}$) for the separation of the zero-point levels obtained[72] from a fitting of all experimental rotation–vibration data for CH_2 in the $\tilde{X}^3B_1$ state, some of which result from perturbations by the $\tilde{a}^1A_1$ state. The corresponding separation resulting from a high-level *ab initio* calculation[464] is 0.3914 eV ($3157\,cm^{-1}$).

The ground MO configuration of HO_2 is (see Table 6.14) ... $(6a')^2(7a')^2(2a'')^1$, giving the $\tilde{X}^2A''$ ground state. The ground configuration of HO_2^- has a second electron in the $2a''$ MO and the ground state is $\tilde{X}^1A'$. The first band system in the LPE spectrum[464] has been obtained with 488 nm argon ion laser radiation and the cation formed in a discharge in a mixture of oxygen and ethyl nitrite (CH_3CH_2ONO). The system is dominated by a progression in the O—O stretching vibration. The intensity distribution along the progression is consistent with an increase in the O—O bond length from 1.364 Å in the ground state of HO_2 to 1.50 Å in the ground state of HO_2^-.

Anions of transition metal clusters can be formed[465] in a flowing afterglow ion source by sputtering from the cathode made from the appropriate metal, or mixture of metals in an alloy, in a d.c. discharge. In this way the LPE spectra of $NiCu^-$, $NiAg^-$, $NiAg_2^-$, and Ni_2Ag^- were obtained[466] using 351.1 nm (3.531 eV) argon ion laser radiation. Figure 8.158(b) shows, for example, the spectrum of $NiCu^-$, comparing it with those of Cu_2^- and Ni_2^- in Figure 8.158(a) and (c).

The ground electronic configurations of Cu_2^- and Ni_2^- are ... $d^{20}(\sigma_g)^2(\sigma_u^*)^1$ and ... $d^{18}(\sigma_g)^2(\sigma_u^*)^1$, respectively, in which two electrons are in a bonding and one in an antibonding σ orbital and the 20 or 18 d electrons remain in atomic orbitals on the metal atoms. Similarly, the ground configuration of $NiCu^-$ is ... $d_{Ni}^9 d_{Cu}^{10}(\sigma)^2(\sigma^*)^1$. All three anions have a bond order of $\frac{1}{2}$.

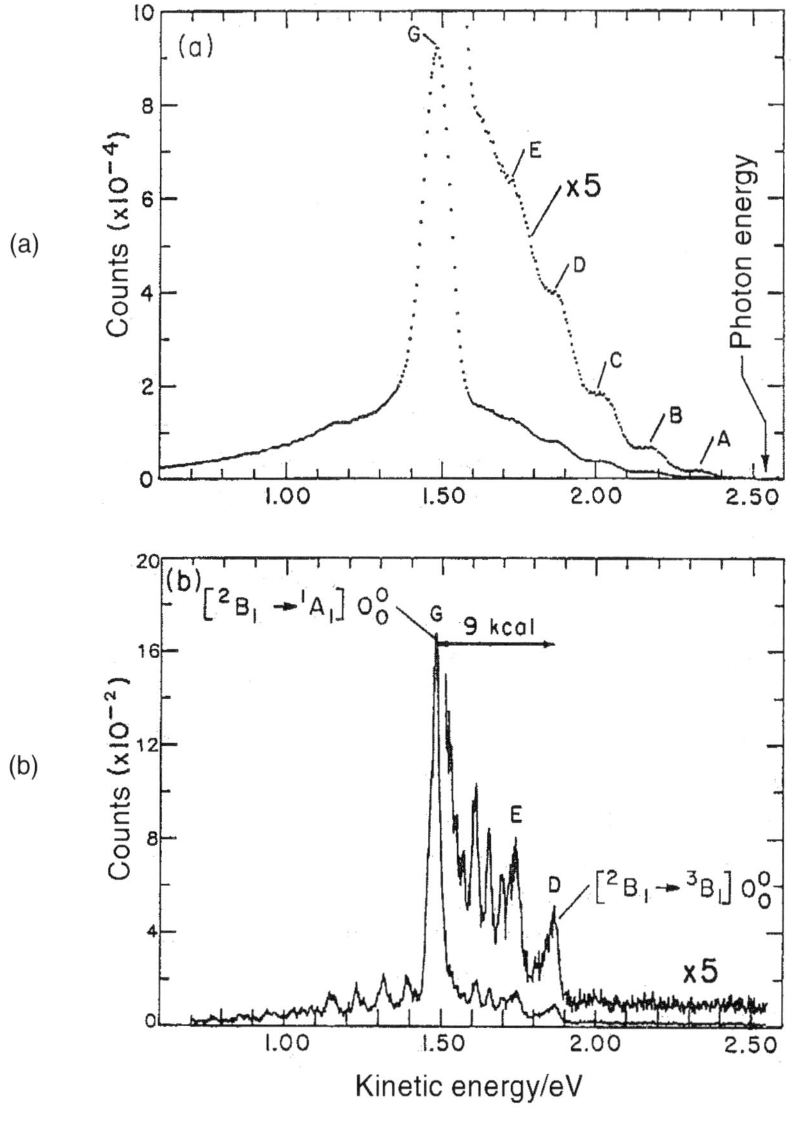

Figure 8.157 Laser photoelectron spectra, obtained with 488 nm excitation, of CH_2^- produced in (a) a gas discharge ion source, resulting in vibrationally hot CH_2^-, and (b) a flowing afterglow ion source, with collisional cooling
[Reproduced, with permission, from Leopold, D. G., Murray, K. K. and Lineberger, W. C. (1984). *J. Chem. Phys.*, **81**, 1048]

In the ground configurations of Cu_2, Ni_2 and NiCu the electron has been ejected from the anti-bonding σ orbital to give ground states $X^1\Sigma_g^+$, $X[0_g^+$ or $0_u^-]$ and $X^2\Delta_{\frac{5}{2}}$, respectively, with Hund's case (c) coupling of angular momenta in the case of NiCu. The corresponding band system in each LPE spectrum shows a long vibrational progression. That for NiCu is shown in Figure 8.159. Simulation of the progression intensities shows that ω_e is

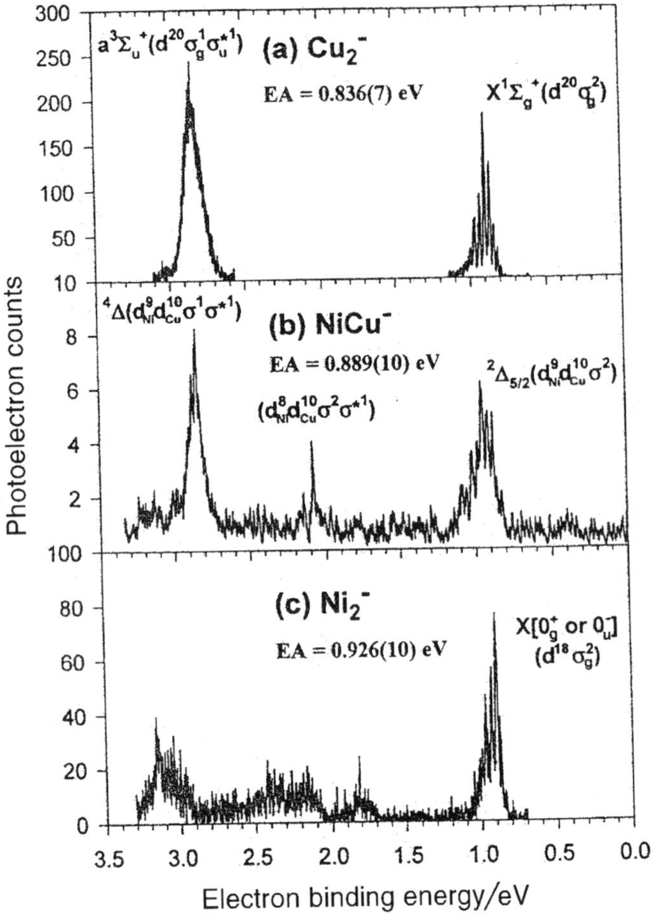

Figure 8.158 Laser photoelectron spectra, with 351.1 nm radiation, of (a) Cu_2^-, (b) $NiCu^-$ and (c) Ni_2^- [Reproduced, with permission, from Dixon-Warren, St. J., Gunion, R. F. and Lineberger, W. C. (1996). *J. Chem. Phys.*, **104**, 4902]

$235 \pm 25\,cm^{-1}$, the vibrational temperature is 500 ± 150 K and the bond length in the anion is 2.36 Å compared with the previously known value of 2.2346 Å in NiCu. The decrease in bond length in NiCu is consistent with an increase in bond order from $\frac{1}{2}$ to 1. There are similar bond length decreases in Cu_2 and Ni_2.

The band systems at about 2.8 eV in the spectra of Cu_2^- and $NiCu^-$ involve states of the neutrals in which an electron in the anion has been ejected from the bonding orbital. This, presumably, leads to a decrease in bond length.

The band system at about 2.1 eV in the $NiCu^-$ spectrum is probably due to ejection of a non-bonding *d* electron. The complex 1.7–3.2 eV region of the Ni_2^- spectrum may also involve *d*-electron ejection.

The infrared laser diode spectrum[254] of Si_2^-, created in an electrical discharge in a mixture of hydrogen and silane, was discussed in Section 8.2.8. The spectrum is part of the $A^2\Pi_u$–$X^2\Sigma_g^+$ electronic band system which falls in the infrared region. The $A^2\Pi_u$ state is inverted with the $^2\Pi_{u(\frac{3}{2})}$ component below $^2\Pi_{u(\frac{1}{2})}$. The

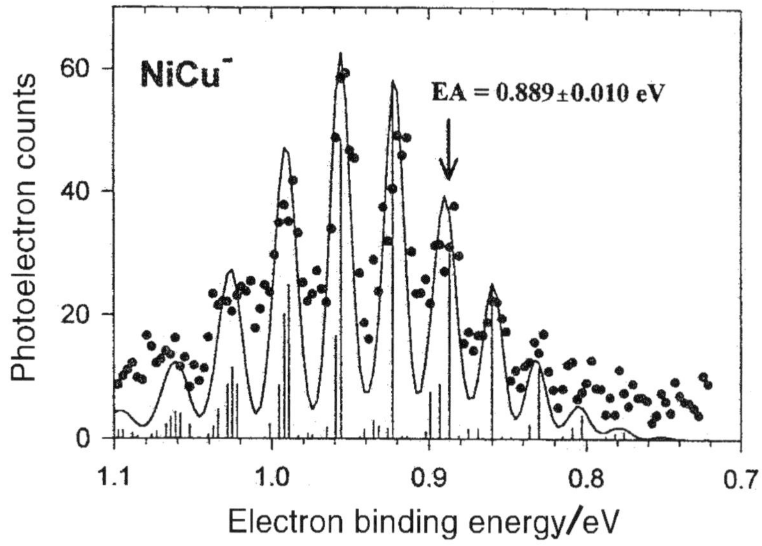

Figure 8.159 Franck–Condon simulation of the progression in the 0.976 eV band system in the laser photoelectron spectrum of $NiCu^-$. The 0–0 band is at 0.976 eV and there are a few hot bands due to the high vibrational temperature of 500 ± 150 K
[Reproduced, with permission, from Dixon-Warren, St. J., Gunion, R. F. and Lineberger, W. C. (1996). *J. Chem. Phys.*, **104**, 4902]

potential curves for the X and A states are shown in figure 8.64.

The LPE and ZEKE-PE spectra of Si_2^- shed further light on the excited electronic states.[467] The Si_2^- was made in a plasma resulting from vaporization from the surface of a rotating and translating silicon rod with an XeCl laser operating at 308 nm. A Nd^{3+}:YAG laser, operating at 355 nm (third harmonic), was the source for the LPE spectrum.

For the ZEKE-PE spectrum (see Section 8.2.19) a pulsed dye laser was used with wavelength tuning so that the photoelectrons produced have zero kinetic energy. However, anions do not have a series of Rydberg states converging to the energy limit for photodetachment. The reason for this is that photodetachment leads to a charged particle (the electron) and a neutral particle. Converging Rydberg states result only when the resulting particles have opposite charges and the interparticle force is coulombic. One result of this is that the pulsed field ionization technique cannot be applied when the ZEKE-PE spectroscopic method (also known as the threshold photodetachment spectroscopic method) is used for anions.

Although the LPE spectrum is of lower resolution ($64\,cm^{-1}$) than the ZEKE-PE spectrum ($15\,cm^{-1}$), it has the advantage of showing some of the allowed photodetachment transitions which do not appear in the ZEKE-PE spectrum.

The separation between the average of the two components of the $A^2\Pi_u$ state and the $X^2\Sigma_g^+$ state of Si_2^- is only $216.5\,cm^{-1}$ and the splitting of the components of the $A^2\Pi_u$ state is $115.292\,cm^{-1}$.[254] Therefore, the X and A states will have similar populations when the Si_2^- is prepared. Figure 8.160 shows the ZEKE-PE spectrum of Si_2^- which contains band systems with both X and A lower states. The five

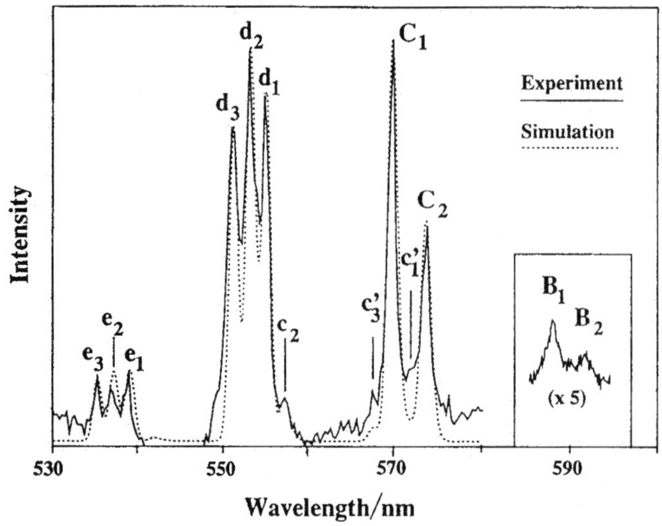

Figure 8.160 The observed and calculated ZEKE-PE spectrum of Si_2^-
[Reproduced, with permission, from Kitsopoulos, T. N., Chick, C. J., Zhao, Y. and Neumark, D. M. (1991). *J. Chem. Phys.*, **95**, 1441]

strongest bands are 0–0 bands of the following band systems:

$$C_2\ X^3\Sigma_g^- - A^2\Pi_{u(\frac{1}{2})};\ C_1\ X^3\Sigma_g^- - A^2\Pi_{u(\frac{3}{2})};$$
$$d_1\ D^3\Pi_{u(2)} - X^2\Sigma_g^+;\ d_2\ D^3\Pi_{u(1)} - X^2\Sigma_g^+;$$
$$d_3\ D^3\Pi_{u(0)} - X^2\Sigma_g^+$$

The remaining bands involve vibrational excitation in either the upper or lower electronic state. The D state is an example of Hund's case (a) coupling with large splitting between the three components. At higher energies the $a^1\Delta_g$, $b^1\Pi_u$ and $c^1\Sigma_g^+$ states of Si_2 have been observed, the last one detected only in the LPE spectrum.

8.2.21 Double resonance spectroscopy

Double resonance techniques involve the interaction of an atom or molecule with two sources of monochromatic radiation in a period of time which is short compared with the relaxation rate of the processes which are being studied. The sources are of different wavenumbers (or frequencies†) $\tilde{\nu}_1$ and $\tilde{\nu}_2$ and the associated energies $hc\tilde{\nu}_1$ and $hc\tilde{\nu}_2$ correspond to the separation of two pairs of eigenstates which have one state in common. This represents what is called a three-level double resonance and Figure 8.161 shows four possible arrangements of the three energy levels. A further requirement for a double resonance experiment is that one of the sources of radiation, $\tilde{\nu}_1$ in Figure 8.161, is sufficiently intense to saturate the corresponding transition.

The double resonance technique is used routinely in nuclear magnetic resonance spectroscopy, in which it has proved invaluable in the assignment of complex spectra.

In 1959–60 the technique was first applied to microwave spectroscopy.[468,469] Since the most

†In this section the use of microwave, infrared, visible and ultraviolet radiation, and combinations of them, will be discussed but, for the purposes of uniformity, I shall refer to 'wavenumber' throughout.

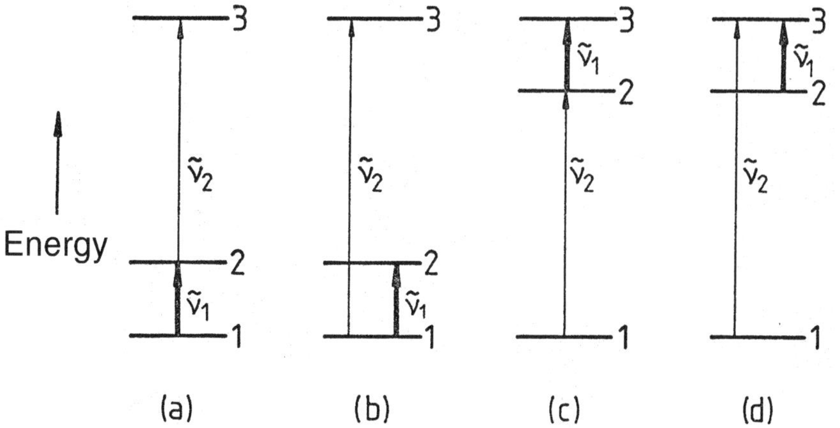

Figure 8.161 Four possible arrangements of energy levels involved in three-level double resonance. The radiation $\tilde{\nu}_1$ is sufficiently intense to saturate the corresponding transition

important processes promoting rotational relaxation are collisional, the pressure of the sample must be sufficiently low that interaction with two photons can take place between collisions, or it must be in a supersonic jet.

We consider molecules initially in rotational states 1 and 2 in Figure 8.161(a) and assume that the populations N_1 and N_2 of these states are equilibrium Boltzmann populations. If the states are closely spaced N_1 is only slightly greater than N_2. Under the influence of intense radiation of wavenumber $\tilde{\nu}_1$, transitions are induced in both directions between the two levels so that, eventually, $N_1 = N_2$ and no further absorption of $\tilde{\nu}_1$ radiation can take place. In other words, saturation has occurred. With power of the order of $100\,\mathrm{mW\,cm^{-2}}$ at $\lambda = 1\,\mathrm{cm}$, obtainable from a typical klystron, saturation is easily obtained. The intensity of absorption of the second rotational transition 3–2 at a wavenumber $\tilde{\nu}_2$ is monitored, using a second klystron or radiofrequency oscillator, while the first is tuned through $\tilde{\nu}_1$. The intensity of the 3–2 absorption goes through a maximum corresponding to resonance between the radiation from the first klystron and the 2–1 transition.

As described in Section 3.3, the relative insensitivity of microwave spectroscopy is overcome, to some extent, by Stark modulation. However, in a double resonance experiment it is preferable to replace Stark modulation with frequency modulation of the pumping klystron.[470] The frequency oscillates between two close, fixed values and it is only when one of these values corresponds to $\tilde{\nu}_1$ that a signal is detected.

In the three-level double resonance illustrated in Figure 8.161(b), saturation of the 2–1 transition *decreases* the intensity of absorption at $\tilde{\nu}_2$ because the population of level 1 is decreased. For the cases illustrated in Figure 8.161(c) and 8.161(d) the absorption intensity at $\tilde{\nu}_2$ increases and decreases, respectively, when transition 3–2 is saturated.

Microwave–microwave double resonance is used to aid the assignment of microwave transitions. It is also a powerful tool for the study of the transfer of rotational energy by inelastic collisions. Both the time between collisions and for rotational energy transfer can be studied.[471]

In principle the kind of double resonance experiments illustrated in Figure 8.161 may be carried out for two photons in any region of the spectrum. However, it was not until the development of lasers in the so-called 'optical' regions of the spectrum (infrared, visible and ultraviolet) that the pump radiation $\tilde{\nu}_1$, could be any other than microwave or radiofrequency radiation.

Double resonance experiments in which one photon is microwave or radiofrequency and the other is optical are abbreviated to MODR (*m*icrowave–*o*ptical *d*ouble *r*esonance) or RODR.

In experiments where both photons are optical we refer to OODR (*o*ptical–*o*ptical *d*ouble *r*esonance), although the term infrared–optical double resonance may be used to distinguish the case where one photon is in the infrared region.

8.2.21.1 Microwave–optical double resonance spectroscopy

In this type of double resonance spectroscopy, any of the four types of experiment illustrated in Figure 8.161 may be carried out. If the radiation $\tilde{\nu}_2$ is from an infrared laser, level 3 [Figure 8.161(a) and (b)] or levels 2 and 3 [Figure 8.161(c) and (d)] will normally be in an excited vibrational state whereas, if it is from a visible or ultraviolet laser, they will normally be in an excited electronic state. In either case the microwave (or radiofrequency) radiation $\tilde{\nu}_1$ disturbs the Boltzmann populations of the rotational levels 1 and 2 (or 2 and 3) provided the corresponding rotational transitions are allowed. Under conditions of saturation the populations become equal. As in microwave–microwave double resonance the intensity of absorption of the radiation $\tilde{\nu}_2$ reflects the population change.

It may be helpful to point out here that there is a close analogy between MODR and level anticrossing spectroscopy[472] (Section 8.2.25). In both, two close-lying levels are connected by a perturbation. In level anticrossing the perturbation is in the molecule itself, whereas in MODR it is imposed externally in the form of microwave radiation.

In MODR experiments the main aim is to obtain rotational constants with high accuracy in excited vibrational or electronic states. Line widths in pure microwave spectroscopy are typically about 0.1 MHz ($3 \times 10^{-6}\,\mathrm{cm}^{-1}$ but, in MODR, line widths are typically 10–50 MHz (3×10^{-4}–$2 \times 10^{-3}\,\mathrm{cm}^{-1}$). Consequently, although rotational constants can often be obtained more accurately in MODR than from the absorption or emission spectrum their accuracy is not as high as from a microwave spectrum.

One of the reasons for greater line widths in MODR is that the natural line width is greater. For example, for an excited electronic state with a radiative lifetime of 10 ns, the natural line width [equation (8.42)] is 16 MHz. Pressure, laser power and microwave power may also contribute to a greater line width in MODR.

Many MODR experiments in which the optical radiation is in the visible region have been carried out with a fixed wavenumber laser, such as an argon ion or krypton ion laser, but the use of a tunable dye laser introduces much greater flexibility. This method was first used to obtain rotational constants in the ground and excited electronic states of the $A^1\Sigma^+$–$X^1\Sigma^+$ transition of BaO.[473]

Figure 8.162(a) and (b) illustrate how the X and A states, respectively, were investigated. In principle, the laser radiation may or may not saturate the transition 3–1. If it is not saturated the Beer–Lambert law holds and the optical pumping is said to be linear; if the transition is saturated the Beer–Lambert law does not hold [see equation (8.38)] and the pumping is non-linear.

In the case of linear optical pumping and microwave saturation of a rotational transition in the ground electronic state the population N_1 of level 1 [Figure 8.162(a)] is not affected appreciably by the optical pumping but only by the microwave saturation. For example, in the BaO experiment, where levels 1 and 2 were the $J'' = 1$ and 2 rotational levels, saturation

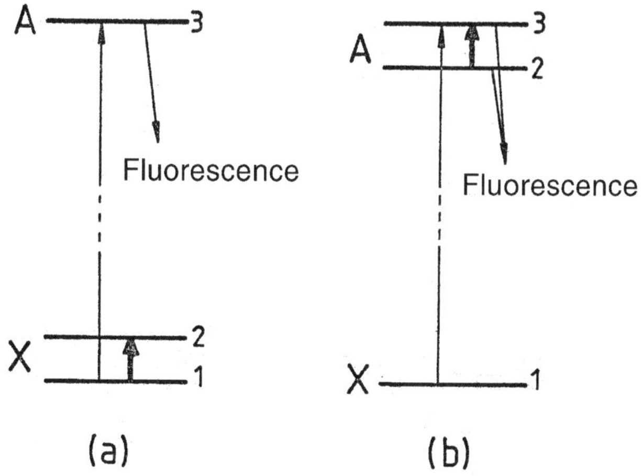

Figure 8.162 Method of MODR investigation to obtain rotational constants in (a) the $X^1\Sigma^+$ state, and (b) the $A^1\Sigma^+$ state of BaO

decreased the population N_1 by 0.2%. The population change was monitored by observing the fluorescence from level 3, the intensity being proportional to N_1.

For linear optical pumping and saturation of an excited electronic state rotational transition [Figure 8.162(b)] the effect on the fluorescence from levels 2 and 3 is more subtle. Saturation of the 3–2 transition hardly affects N_3 relative to N_1 because $N_3 \ll N_1$. However, if the *total* fluorescence from levels 3 and 2 is monitored a change of the polarization characteristics occurs when the microwave radiation saturates the 3–2 transition. Alternatively, if the fluorescence is resolved, that from levels 2 and 3 have different wavelengths; in addition, fluorescence intensity changes can be observed when microwave saturation occurs.

In the case of nonlinear optical pumping, obtainable with a 100 mW laser, the population of level 3 in Figure 8.162(a) and (b) is appreciable compared with that of level 1. Microwave saturation in the ground state, as shown in Figure 8.162(a), replenishes the population of level 1 and therefore decreases the fluorescence intensity. Microwave saturation in the excited state, illustrated in Figure 8.162(b), transfers molecules from level 3 to 2 and affects the polarization of the total fluorescence from these levels.

Figure 8.163 shows an example of MODR in which the optical radiation is provided by an infrared laser. The $R(16)$ line of a CO_2 laser at 9.4 μm is coincident with the $J' = 8$, $K' = 2$–$J'' = 7$, $K'' = 2$ rotational line of the 1_0^1 vibrational transition of the symmetric rotor CF_3I, where ν_1 is the CF_3 symmetric stretching vibration. The microwave spectrum in Figure 8.163(a), shows that, without laser irradiation (upper trace), there is no $J' = 9$–8 microwave spectrum but, with irradiation (lower trace), the $J' = 9$–8 transition is observed in absorption due to the increased population of the $J' = 8$ level. Conversely, Figure 8.163(b) shows that the $J' = 8$–7 transition is observed in emission because of the high population of the $J' = 8$ relative to the $J' = 7$ level.

Similarly, in the ground vibrational state laser irradiation depletes the population of the $J'' = 7$ level resulting in the $J'' = 8$–7 and 7–6 transitions being observed in emission and absorption, respectively, as shown in Figure 8.163(c) and (d).

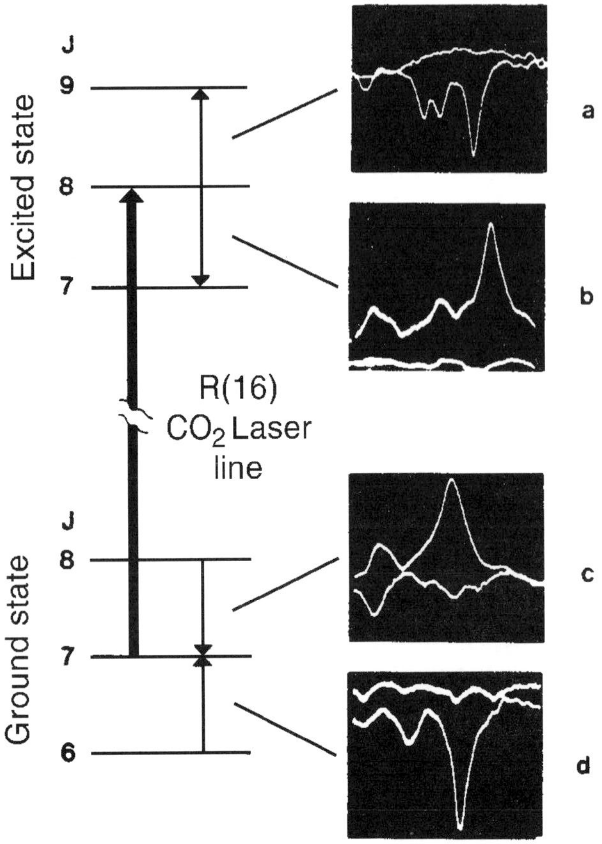

Figure 8.163 MODR spectra of the CF_3I molecule. The optical radiation is from a CO_2 infrared laser and four microwave transitions, two in the $v_1 = 1$ state and two in the ground vibrational state, are observed in absorption, shown in (a) and (d), or emission, shown in (b) and (c), when the laser pumps the $J' = 8, K' = 2$–$J'' = 7, K'' = 2$ transition
[Reproduced, with permission, from Jones, H. (1978). *Comments At. Mol. Phys.*, **8**, 51]

The fact that each microwave signal in Figure 8.163(a)–(d) shows several components is due to nuclear hyperfine splitting.

8.2.21.2 Optical–optical double resonance spectroscopy

In this type of double resonance spectroscopy two photons, of wavenumbers $\tilde{\nu}_1$ and $\tilde{\nu}_2$, undergo absorption in a three-level system of the type illustrated in Figure 8.164(a), (b) or (c). In Figure 8.164(a) and (b) levels 1 and 2 are close together and the Doppler profiles of the transitions to or from level 3 may overlap. These energy level diagrams have much in common with those illustrating MODR in Figure 8.161 but, in optical–optical double resonance (OODR), either both photons are infrared, visible or ultraviolet or each is from a different one of these spectral regions. Most OODR experiments have been carried out with infrared and visible lasers and the type of

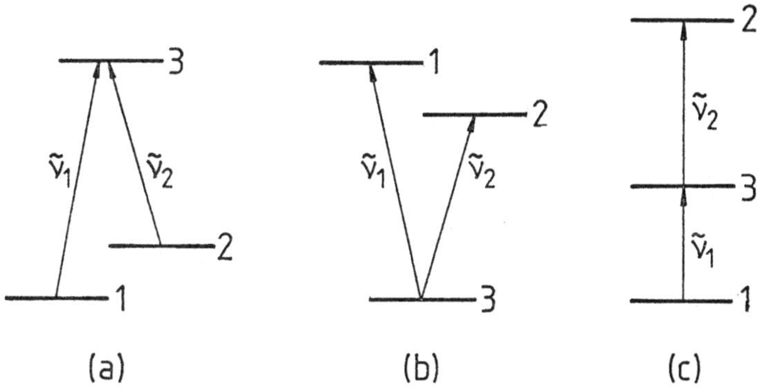

Figure 8.164 Three possible arrangements of energy levels involved in three-level optical–optical double resonance

experiment in which one of each is used is sometimes referred to as infrared–optical double resonance.

Another important difference from MODR is that transitions are broadened predominantly by the Doppler effect (inhomogeneous broadening), whereas MODR line widths are due to the natural line width and pressure effects (homogeneous broadening). As a result, OODR spectra may be observed with sub-Doppler line widths if the laser line widths are small compared with the Doppler line width.

In general, there are two kinds of experimental arrangement, one in which the two laser beams are travelling in the same direction and the other in which they travel in opposite directions—counterpropagating beams.

For beams in the same direction the transition schemes illustrated by Figure 8.164(a) and (b) result in very sharp, sub-Doppler double resonance signals because each beam interacts with molecules having the same range of velocities.[474] A double resonance signal will be observed if the laser wavenumbers $\tilde{\nu}_1^L$ and $\tilde{\nu}_2^L$ are related to the wavenumbers $\tilde{\nu}_1$ and $\tilde{\nu}_2$ of the centres of the Doppler profiles of the corresponding transitions by

$$\tilde{\nu}_1^L - \tilde{\nu}_2^L = \tilde{\nu}_1 - \tilde{\nu}_2 \qquad (8.73)$$

This resonance requirement is illustrated by the two holes burnt in the Doppler profiles shown in Figure 8.165(a). When $\tilde{\nu}_1 \approx \tilde{\nu}_2$ i.e. levels 1 and 2 are close together, and saturation broadening can be ignored, the width γ_N of the narrow OODR signal is given by

$$\gamma_N = \gamma_1 + \gamma_2 \qquad (8.74)$$

where γ_1 and γ_2 are the natural widths of levels 1 and 2.

In the case where the two laser beams are in opposite directions, we assume that the laser radiation $\tilde{\nu}_2^L$ lies within the Doppler profile of the 3–2 (or 2–3) transition, as shown in Figure 8.165(b). Since $\tilde{\nu}_2^L > \tilde{\nu}_2$, the group of molecules which interact with the radiation have a velocity component in the opposite direction to that of propagation of the radiation. Conversely, the same molecules have a velocity component in the same direction as the laser radiation $\tilde{\nu}_1$ and the corresponding hole in the Doppler profile of the 3–1 (or 1–3) transition is shifted to low wavenumber from $\tilde{\nu}_1$, as Figure 8.165(b) shows. If the OODR signal is being detected through the effect on the radiation $\tilde{\nu}_1^L$, the hole in the $\tilde{\nu}_1$ profile is relatively broad; the width, γ_B, is given by

$$\gamma_B = \gamma_1 + \gamma_2 + 2\gamma_3 \qquad (8.75)$$

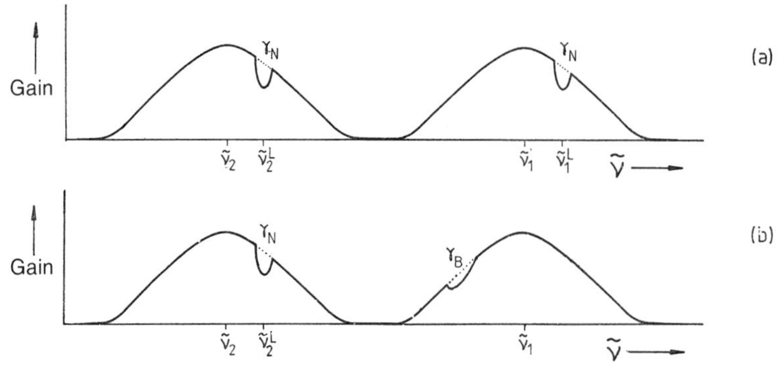

Figure 8.165 (a) Holes burnt in the Doppler profiles of two close-lying transitions when, for two laser beams $\tilde{\nu}_1^L$ and $\tilde{\nu}_2^L$ in the same direction in the sample, $\tilde{\nu}_1^L - \tilde{\nu}_2^L = \tilde{\nu}_1 - \tilde{\nu}_2$. (b) Holes burnt in the Doppler profiles when the two laser beams are in opposite directions

In the situations illustrated in Figure 8.164(a) and (b) levels 1 and 2 may be very close in the absence of any field; on the other hand, they may be Zeeman- or Stark-tuned into resonance. An example of a Stark-tuned OODR experiment, employing two CO_2 lasers in infrared–infrared double resonance, is the accurate determination of the dipole moment of the symmetric rotor CH_3F in various rotational and vibrational states.[475] The molecule exhibits a first-order Stark effect [equation (4.72)]. The two laser beams travel in the same direction and are polarized parallel to each other but perpendicular to the electric field. The selection rule is $\Delta M_J = \pm 1$ and the resonance condition for the Stark-tuned levels is

$$\tilde{\nu}_1^L - \tilde{\nu}_2^L = 2\mu\mathscr{E}K/hcJ(J+1) \qquad (8.76)$$

where $\mathscr{E}$ is the applied electric field, for non-linear absorption of the laser radiation to be detected. Line widths at half intensity of 200 kHz (7×10^{-6} cm^{-1}) were observed and dipole moments accurate to ± 0.001 D were obtained.

Sub-Doppler OODR spectra have been obtained, involving infrared transitions of the types in Figure 8.164(a) and (b) using two beams of radiation originating from the same CO_2 laser.[476] The CO_2 radiation, $\tilde{\nu}_L$, is plane polarized and falls on a germanium acousto-optic modulator. Radiofrequency radiation, $\tilde{\nu}_R$, in the range 50–70 MHz is directed on to the modulator normal to the laser beam. The modulator acts as a three-wave mixing device and the emergent beam contains not only $\tilde{\nu}_L$ but also $\tilde{\nu}_L - \tilde{\nu}_R$. The $\tilde{\nu}_L - \tilde{\nu}_R$ radiation is slightly displaced and can be spatially filtered from the $\tilde{\nu}_L$ radiation. The two beams, $\tilde{\nu}_L$ and $\tilde{\nu}_L - \tilde{\nu}_R$, are directed through a sample cell in either the same or opposite directions. The radiofrequency radiation is tuned so that the separation of levels 1 and 2 in Figure 8.164(a) and (b) is equal to $\tilde{\nu}_R$. In addition, the two levels can be Stark-tuned into resonance with $\tilde{\nu}_R$. In this way, sub-Doppler OODR spectra of CH_3OH, CF_3I and the short-lived molecules H_2CS and CH_2NH were obtained.

OODR experiments of the type illustrated in Figure 8.164(a) and (b) have also been performed using two dye lasers to observe transitions where levels 1 and 2 are in a different electronic state from level 3. An alternative is to use a single dye laser and an acousto-optic modulator, as described above, to provide radiation which differs in wavenumber from the laser radiation by $\pm\tilde{\nu}_R$, the wavenumber of the tunable radiofrequency radiation. This technique has been used for investigating the $\tilde{A}^1A''$–$\tilde{X}^1A'$ system of the short-lived molecule HNO.[477]

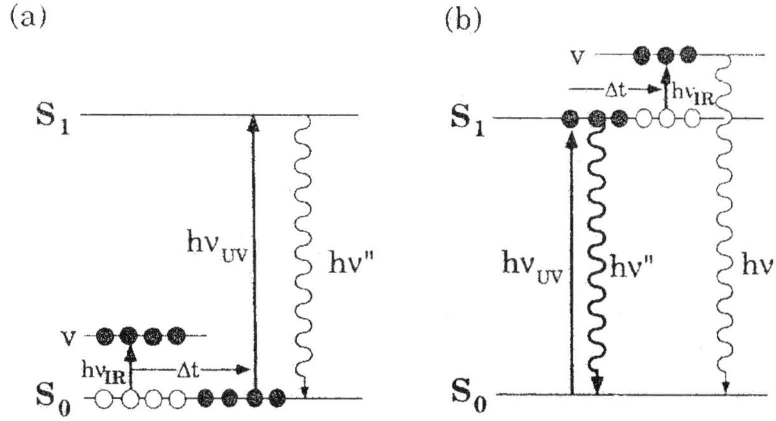

Figure 8.166 Scheme for fluorescence depletion spectroscopy used to obtain infrared spectra in (a) the ground electronic state S_0 and (b) an excited state S_1
[Reproduced, with permission, from Ebata, T., Mizuochi, N., Watanabe, T. and Mikami, N. (1996). *J. Phys. Chem.*, **100**, 546]

Like so many techniques of laser spectroscopy, that of observing OODR signals using modulated exciting radiation was first used in atomic spectroscopy.[478]

The OODR process in Figure 8.164(c) has been called a cascade OODR process or, by analogy with two-photon absorption (Section 8.2.18), two-step two-photon absorption. For this type of OODR process the double resonance signal is very sharp for counter-propagating laser beams and broader, but still sub-Doppler, for beams in the same direction.

Cascade-type OODR experiments have been carried out on BaO using a fixed frequency argon ion laser ($\tilde{\nu}_1$) and a tunable dye laser ($\tilde{\nu}_2$),[479] and a more flexible system of two tunable dye lasers[480] propagating in the same direction. The line width was sub-Doppler but limited to about 15 MHz (0.0005 cm^{-1}) by the laser line width.

When $\tilde{\nu}_1^L$ is resonant with the 3–1 transition, resonance of $\tilde{\nu}_2^L$ with the 2–3 transition can be detected either by a decrease in fluorescence from level 3 or an increase in fluorescence from level 2. There is a possibility that two photons from the same laser, i.e. $\tilde{\nu}_1^L = \tilde{\nu}_2^L$, will take part in a cascade process and such OODR signals have also been observed.

In an OODR experiment, in which one laser was an infrared and the other a laser in the visible region, an argon ion laser was used to excite NO_2 molecules electronically and a chopped CW CO_2 laser to induce vibrational excitation in the excited electronic state.[481] The double resonance was detected by observing the amplitude-modulated fluorescence.

Spectroscopy involving a double resonance experiment in which transitions are detected through the diminution of a fluorescence signal is called fluorescence depletion (FD) or, sometimes, fluorescence dip spectroscopy. Figure 8.166 shows how FD spectroscopy can be used to obtain infrared spectra of molecules in the ground electronic state (labelled S_0 if it is a singlet state), in Figure 8.166(a), and in an excited electronic state (labelled S_1 if it is the lowest excited singlet state), in Figure 8.166(b).

In the scheme in Figure 8.166(a), which is an example of the general scheme in 8.164(b), a pulse from a tunable infrared laser of frequency ν_{IR} reaches the sample before the UV pulse which excites the molecule, in the

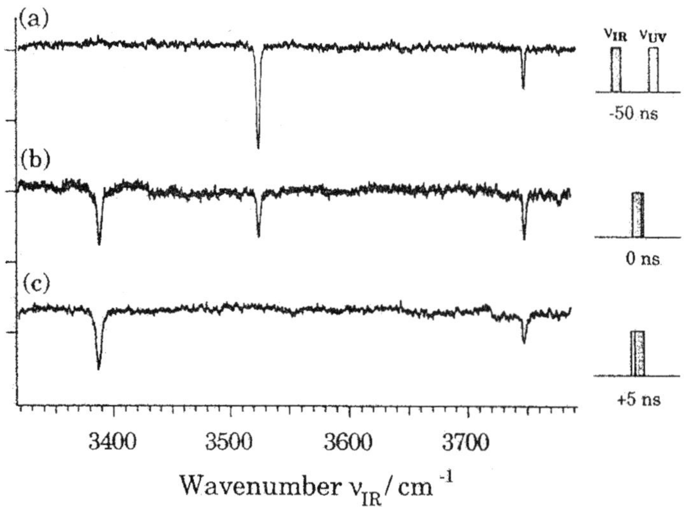

Figure 8.167 Fluorescence depletion spectra showing infrared vibrational transitions in phenol–H_2O (a) in S_0, with a delay of −50 ns between the IR and UV pulses, (b) in S_0 and S_1, with no delay, and (c) in S_1, with a delay of +5 ns. Fluorescence intensity is plotted on the vertical axis
[Reproduced, with permission, from Ebata, T., Mizuochi, N., Watanabe, T. and Mikami, N. (1996). *J. Phys. Chem.*, **100**, 546]

0_0^0 transition, to S_1. The total, undispersed fluorescence from S_1 is monitored as the infrared laser is scanned across the region of infrared-active vibrational transitions in S_0. When an infrared absorption is encountered there is a decrease, or depletion, of the fluorescence signal due to a decrease in the population of the zero-point level of S_0.

In the scheme in Figure 8.166(b), which is an example of the general scheme in Figure 8.164(c), the pulse from a tunable infrared laser reaches the sample after the UV pulse taking the molecule into the zero-point level of S_1. Then an infrared absorption in S_1 results in a decrease in the population of the zero-point level of S_1 and depletion of the fluorescence signal.

Figure 8.167 shows an example in which an infrared spectrum of the hydrogen-bonded complex phenol–H_2O ($C_6H_5OH\cdots OH_2$), formed in a supersoic jet, was obtained[482] in both S_0 and S_1 by varying the delay time, Δt, between the IR and UV pulses. The decay times for vibrational states are much longer than for the S_1 electronic state, which is about 10 ns.

Figure 8.167(a) shows the FD spectrum with the IR pulse arriving at the sample, and exciting vibrational transitions in S_0, 50 ns before the UV pulse. The spectrum shows infrared bands in S_0, in the region of the OH-stretching vibration of phenol which appears at 3524 cm^{-1}. The band at 3748 cm^{-1} is the asymmetric OH-stretching vibration in the H_2O moiety.

For the FD spectrum in Figure 8.167(c) the IR pulse arrived 5 ns after the UV pulse, thereby exciting infrared vibrational transitions in S_1. The phenolic OH-stretching vibration is seen to shift to 3388 cm^{-1} in S_1 whereas the shift of the asymmetric OH-stretching vibration in the H_2O moiety is very small.

The spectrum in Figure 8.167(b), with no delay between the pulses, shows vibrational transitions in both S_0 and S_1.

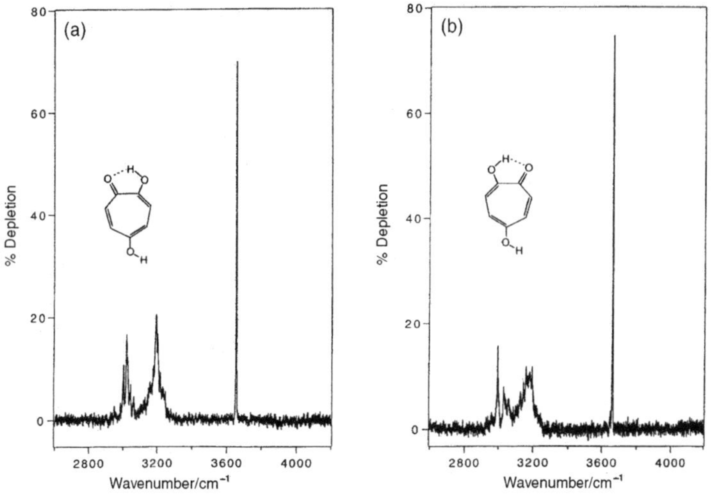

Figure 8.168 Infrared fluorescence depletion spectra of the (a) *syn* and (b) *anti* rotamers of 5-hydroxytropolone in the region of the OH- and CH-stretching vibrations
[Reproduced, with permission, from Frost, R. K., Hagemeister, F., Schleppenbach, D., Laurence, G. and Zwier, T. S. (1996). *J. Phys. Chem.*, **100**, 16835]

The UV radiation was generated from a frequency doubled dye laser and the tunable IR radiation by a difference frequency laser system (see Section 8.2.9) comprising the second harmonic of a Nd^{3+} : YAG laser and a dye laser.

This is a very powerful method of obtaining vibration wavenumbers in S_1 where they may not be accessible by other means. In this case, neither of the two OH-stretching vibrations in S_1 (or S_0) can be obtained from, say, an FE or SVLF spectrum because, although activity of the vibrations is symmetry allowed, there is no appreciable Franck–Condon intensity in the OH-stretching progressions. In any case, they would be difficult to identify in a region which would be dominated by other fundamentals and combinations.

The reduction by 136 cm^{-1} in the phenolic OH-stretching vibration wavenumber from S_0 to S_1 shows that the O—H bond is weaker, and that the phenol moiety is more acidic, in S_1. This is what is found in the bare moelcule but, in the complex, the wavenumber is 133 cm^{-1} less in S_0 and 193 cm^{-1} less in S_1.

The fact that the wavenumber of the asymmetric OH-stretching vibration, 3748 cm^{-1}, is unchanged from S_0 to S_1, and is also little changed from its value of 3758 cm^{-1} in free H_2O, shows that the H_2O moiety in the complex behaves almost as a free molecule, unaffected by the acidity of the phenol moiety.

Using the scheme shown in Figure 8.166(a), with tunable infrared radiation from a Nd^{3+}:YAG pumped OPO (see Section 8.1.8.15), ground state infrared FD spectra of the two planar rotamers (*syn* and *anti*) of 5-hydroxytropolone, shown in Figure 8.168, have been obtained.[483] The rotamers can be excited separately because their S_1–S_0, 0_0^0 bands are separated by 486 cm^{-1}. The wavenumber of the OH-stretching vibration (in the 5-position) is shown to be slightly different in the two rotamers, being 3654 and 3664 cm^{-1} in *syn* and *anti*, respectively.

These bands are shown in Figure 8.168, which also shows, in the 3200 cm^{-1} region, the OH-stretching region of the internally hydrogen-bonded hydrogen. For each rotamer the region shows a broad, structured set of bands typical of the infrared spectrum of a hydrogen-bonded OH-stretching vibration. However, the shift in the centre of gravity of the set of bands and the change in their structure indicates a surprisingly large effect of the orientation of the distant OH group. The bands below 3100 cm^{-1} are due to CH-stretching vibrations.

A novel method of observing 'dark' states using FD spectroscopy has been devised and applied to the $\tilde{A}^2A_1$ state of the C_{3v} radical CF_3S,[484] dark states being those to which absorption occurs but from which fluorescence is too weak to be detected.

The CF_3S radicals were produced by laser photolysis of $(CF_3S)_2$ and entrained in helium in a supersonic jet expansion. The FE spectrum was obtained in the usual way and shows several progressions, one of the strongest being in ν_3, the CS-stretching vibration.

Two lasers, a probe and a dump laser, were used. The wavenumber of the probe laser was fixed, sitting on the intensely fluorescing 3_0^3 band and nearly saturating the transition, although any intensely fluorescing band with the zero-point level as the lower state could be used. The dump laser was then scanned across the whole of the $\tilde{A}^2A_1$–$\tilde{X}^2E$ band system while monitoring the total, undispersed fluorescence, including that resulting from the probe laser. Both lasers were pulsed and there was a time delay of 50 ns between the dump laser (first) and the probe laser. This delay is short compared to the fluorescence lifetimes, of a few microseconds, of strongly fluorescing states. The fluorescence was collected with a CCD detector (see Section 3.8.1) using a 2 μs gate following the probe laser pulse.

If the upper level of a state reached by the dump laser is a dark state which does not fluoresce, there is a decrease in the intensity of the fluorescence produced by the pump laser because the population of the zero-point level of the ground state has been reduced. On the other hand, if the dump laser reaches an upper level which fluoresces strongly, the total fluorescence intensity produced by both lasers will increase. An upper level with competitive radiative and non-radiative pathways may give an increase or decrease in the signal depending on the relative rates of the two processes.

Figure 8.169 shows part of the resulting $\tilde{A}^2A_1$–$\tilde{X}^2E$ fluorescence depletion spectrum where it is compared with the corresponding part of the FE spectrum. A particularly interesting feature of these spectra is the behaviour of bands forming a progression in the CS-stretching vibration ν_3. A similar progression is observed in the corresponding spectrum of CH_3O (see Section 8.2.16) and is associated with the promotion of a C—O or, in this case C—S, bonding electron to a non-bonding π orbital on O or S.

In the lower energy part of the FE spectrum, not shown in Figure 8.169, the intensities of the early members of the ν_3 progression suggest a much longer progression than is observed but the member with $v_3' = 4$ is weak and subsequent members are missing. However, the FD spectrum in Figure 8.169 clearly shows those members with $v_3' = 6$–8, with a negative signal, which do not appear in the FE spectrum.

The reason for this is that, with increasing excitation of ν_3, there is increasing photofragmentation of CF_3S to give $CF_3 + S$. Excitation to these higher levels of ν_3 depletes the population of the ground electronic state but does not produce any fluorescence—hence the negative signal in the FD spectrum.

Comparison of the FD and FE spectra shows that photofragmentation is mode-selective. For example, vibrational levels involving ν_1 and ν_2 do not show photofragmentation except at the highest energies investigated.

Ion-pair states of diatomic molecules have been discussed in Section 6.2.4.7. Unless the potential for such a state involves an avoided crossing, as for LiF illustrated in Figure 6.44(b), the potential energy is given by equation (6.132b) and there are an infinite number of vibrational levels below the dissociation limit.

The fact that ion-pair states tend to have large equilibrium bond lengths compared to the

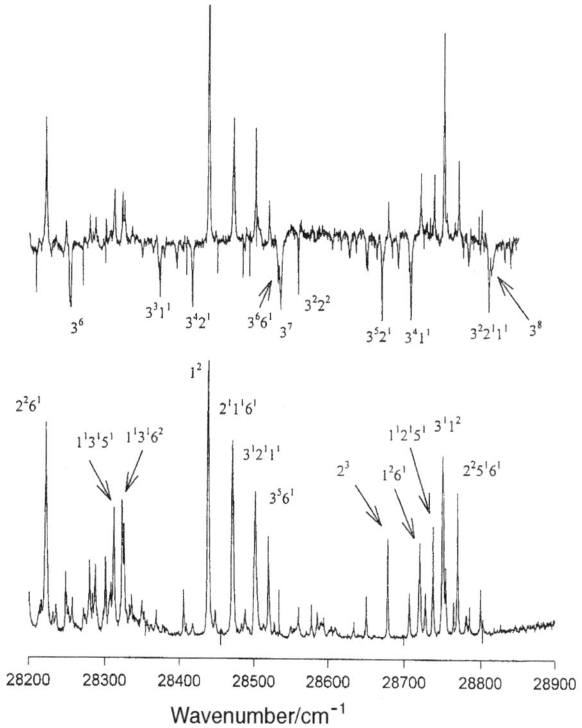

Figure 8.169 Part of the $\tilde{A}^2 A_1$–$\tilde{X}^2 E$ fluorescence excitation spectrum (bottom) and the fluorescence depletion spectrum (top) of CF_3S
[Reproduced, with permission, from Powers, D. E., Pushkarsky, M. and Miller, T. A. (1995), *Chem. Phys. Lett.*, **247**, 548]

ground state means that they are difficult to access because of the low Franck–Condon probability. This has been overcome for ion-pair states of, for example, I_2[485,486] and Cl_2[487,488] by OODR techniques based on the cascade process in Figure 8.164(c).

In one of these OODR techniques, applied to Cl_2[487] in a molecular beam, the first photon, provided by the pump laser, takes $^{35}Cl_2$ molecules from the $X^1\Sigma_g^+$ ground state to the $B^3\Pi_{0+u}$ state in the $R(30)$ rotational transition of the $\upsilon = 8$–0 vibronic transition. As in the corresponding B state of I_2 there is a large increase in bond length from the X state: in Cl_2 it increases from 1.988 Å in the X to 2.435 Å in the B state. There is a further increase to 2.876 Å in the E 0_g^+ ion-pair state. This state correlates with $Cl^-(^1S) + Cl^+(^3P_2)$ and is best described by Hund's case (c). It lies 57 945 cm^{-1} above the X state but the transition probability for the second photon to take the molecule from the $\upsilon = 8$ level of the B state to vibrational levels of the ion-pair state is appreciable. The two-colour OODR spectrum was obtained by scanning the second laser across the E–B absorption system while monitoring the undispersed E–B fluorescence. The time delay between the two laser pulses must be sufficiently short, of the order of 10 ns, that no relaxation occurs in the B state before the second photon is absorbed.

The resulting OODR spectrum is shown in Figure 8.170. The vibrational progression gives the first six vibrational levels in the E state. The pairs of lines which form each progression member are due to the $\Delta J = \pm 1$ selection rule

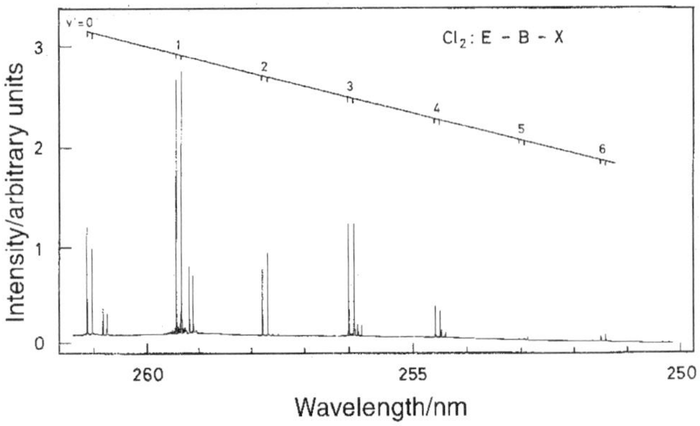

Figure 8.170 Part of the optical–optical double resonance spectrum of $^{35}Cl_2$ with the pump laser tuned to the 8–0, *R*(30) transition of the B–X system. The labelled progression is in the E 0_g^+–$B^3\Pi_{0^+u}$ system of $^{35}Cl_2$ and the unlabelled progression is due to $^{35}Cl^{37}Cl$
[Reproduced, with permission, from Shinzawa, T., Tokunaga, A., Ishiwata, T. and Tanaka, I. (1985). *J. Chem. Phys.*, **83**, 5407]

governing transitions from the $J = 31$, $v = 8$ level of the B state. The unlabelled progression is due to $^{35}Cl^{37}Cl$.

In this way, 23 vibrational levels in the E 0_g^+ state were identified. The absolute numbering was found from single vibronic level (dispersed) fluorescence E 0_g^+–$B^3\Pi_{0^+u}$ spectra following the $1 + 1'$ excitation to the E state. The intensity distribution along the vibrational progression in the B state is characteristic of the value of v in the E state. Figure 8.171 shows, for example, the progressions resulting from excitation to the $v = 0$ and 1 levels of the E state with a typical single or double intensity maximum, respectively.

An alternative OODR method of accessing ion-pair states is to use an unbound intermediate level which may be in a repulsive electronic state or above the dissociation limit of a bound electronic state. This has been used in one- and two-colour experiments[485,486] to identify ion-pair states in I_2.

Figure 8.172 shows potential energy curves for the ground state, $X^1\Sigma_g^+$, the $B^3\Pi_{0^+u}$ state and the E 0_g^+ and f 0_g^+ ion-pair states. There is also a repulsive $B''^1\Pi_{1u}$ state falling in the region just above the dissociation limit of the B state, but it is not shown here. The difference between the close-lying E and f states is that they dissociate to give I^+ in a $^3P_2$ and a $^3P_0$ state, respectively.

To observe an OODR process via an unbound state it is necessary that the pump photon ($h\nu_1$) lies near the maximum of the absorption continuum, in this case the absorption to the continuum of levels above the dissociation limit of the B state. The probe photon ($h\nu_2$) should lie close to the low wavenumber limit of the fluorescence from the final state to the B state, and the two sequential transitions should have large cross sections. The use of two colours[486] rather than one[485] is desirable to achieve these conditions but, when the intermediate state is an unbound state, the second photon must arrive within 10^{-13}–10^{-14} s of the first.

The OODR spectrum of I_2 in Figure 8.173 was obtained[486] by the two-colour process illustrated in Figure 8.172 with the I_2 cooled in a supersonic jet. The spectrum was produced not by monitoring fluorescence from the E and f states but by counting the I^+ ions resulting from the absorption of a further photon of frequency ν_2.

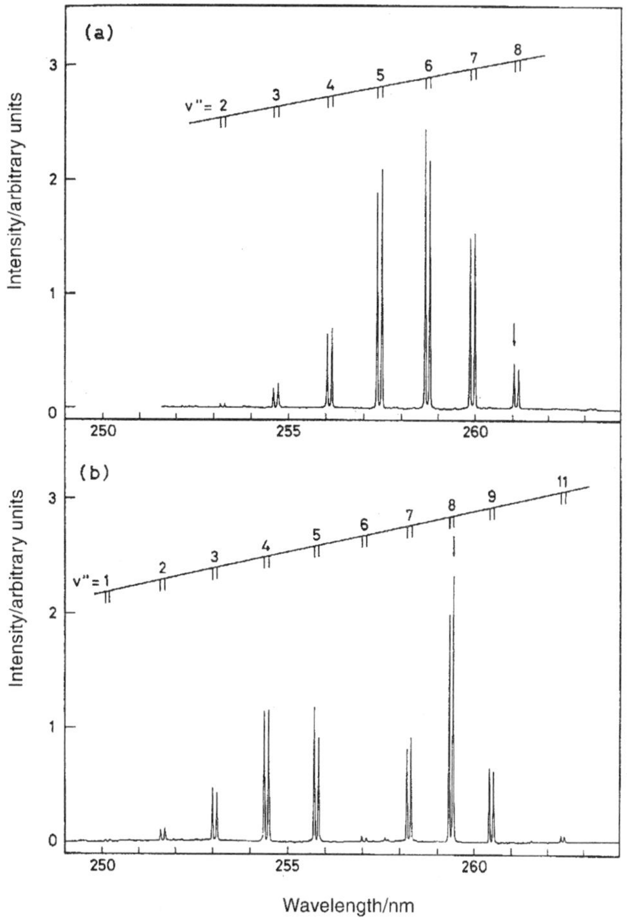

Figure 8.171 Single vibronic level (dispersed) fluorescence E 0_g^+–$B^3\Pi_{0^+u}$ spectra originating in the (a) $v' = 0$, $J' = 32$ and (b) $v' = 1$, $J' = 30$ levels of the E state of $^{35}Cl_2$
[Reproduced, with permission, from Shinzawa, T., Tokunaga, A., Ishiwata, T. and Tanaka, I. (1985). *J. Chem. Phys.*, **83**, 5407]

For a two-colour OODR process either of the laser wavenumbers can be fixed. In this case it was the probe laser in order to avoid three probe laser photons taking part in a competitive 2 + 1 REMPI process.

The spectrum in Figure 8.173 shows overlapping progressions in the E and f states. The most probable transitions are those of the type shown in Figure 8.172 which conserve the classical momentum. There is also a superimposed general decrease in intensity from 500 to 427 nm due to the decrease in the B–X absorption intensity in this range.

The rotational selection rule for a 0_g^+–0_g^+ two-photon transition is $\Delta J = 0, \pm 2$. Rotational analysis[486] of the 106–0 band of the E–X system and the 30–0 band of the f–X system, observed with a resolution of $0.04\,\mathrm{cm}^{-1}$, showed that the Q branches are dominant. The intensity ratio of the Q to the O and S branches[489] showed that the $B''^1\Pi_{1u}$ continuum hardly contributes to the intermediate state.

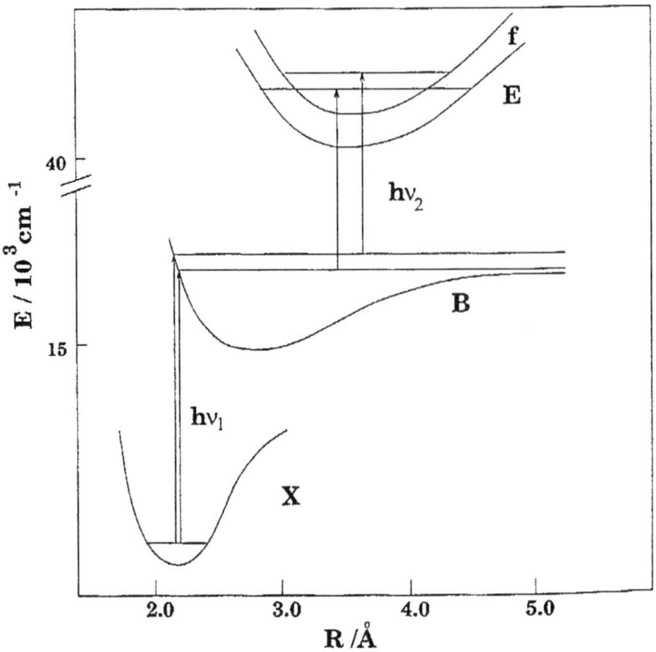

Figure 8.172 Schematic potential energy curves for the X and B states of I_2, together with the E and f ion-pair states, showing how a two-colour OODR experiment can access the ion-pair states
[Reproduced, with permission, from Donovan, R. J., Lawley, K. P., Min, Z., Ridley, T. and Yarwood, A. J. (1994). *Chem. Phys. Lett.*, **226**, 525]

A similar two-colour OODR experiment has been carried out with Cl_2 via repulsive intermediate states and a third photon, from the probe laser, producing Cl^+ detected in a time-of-flight mass spectrometer. In this way many vibrational levels of the E 0_g^+, β 1_g, f 0_g^+ and G 1_g ion-pair states have been identified.[488]

8.2.22 Stimulated emission pumping spectroscopy

The stimulated emission pumping (SEP) process, illustrated in Figure 8.174, is similar to a two-colour OODR process of the cascade type, illustrated in Figure 8.164(c) except that the second photon stimulates a downward, rather than an upward transition.

In Figure 8.174(a) the pump laser takes the molecule from the zero-point level of the ground electronic state $\tilde{X}$ to the zero-point level of an excited electronic state, the $\tilde{A}$ state in this example. In a small molecule, the transition may be to an individual rovibronic state; in a larger molecule, where these may not be resolved, this will be an individual vibronic state. The dump laser is tuned to stimulate emission from the state populated by the pump laser to a particular rovibrational (or vibrational) level of the $\tilde{X}$ state. Scanning the dump laser across the $\tilde{A}$–$\tilde{X}$ emission band system produces a decrease in the population of the upper level whenever the dump laser coincides with an allowed transition.

Monitoring of this population decrease, giving the SEP spectrum, may be achieved in

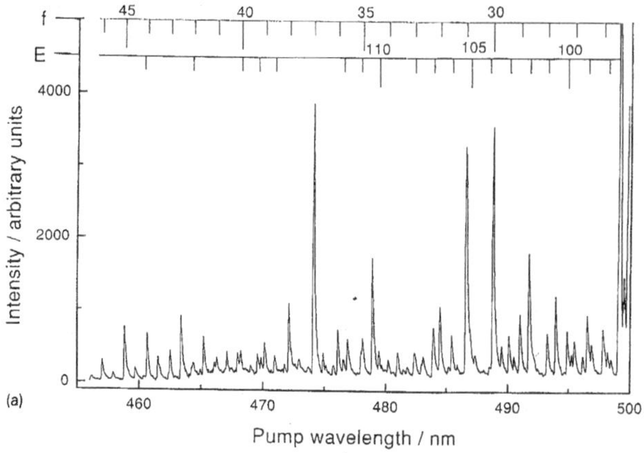

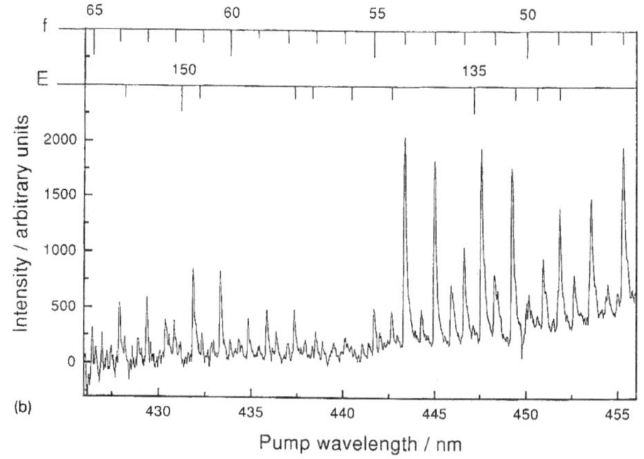

Figure 8.173 Two-colour OODR, E, f 0_g^+–$X^1\Sigma_g^+$, spectrum of I_2 obtained via an unbound intermediate $B^3\Pi_{0^+u}$ state and with detection of absorption by ionization resulting from absorption of a third photon [Reproduced, with permission, from Donovan, R. J., Lawley, K. P., Min, Z., Ridley, T. and Yarwood, A. J. (1994). *Chem. Phys. Lett.*, **226**, 525]

various ways. A commonly used method is to measure the intensity of the undispersed fluorescence following excitation by the pump laser. Then depopulation by the dump laser results in a depletion of the fluorescence signal. Both lasers are pulsed and the dump laser pulse is delayed by, typically, 20 ns to allow time for population of the $\tilde{A}$ state to build up. The emission stimulated by the dump laser is propagated collinearly with the dump beam while the spontaneous fluorescence is generated in all directions so that they can be readily separated.

As for the cascade OODR process in Figure 8.164(c) sub-Doppler resolution may be achieved if the laser line widths are small compared with the Doppler width.

Other methods of detecting a depletion of the population of the excited electronic state may be used such as ionization of the molecule in this state with a further photon.[490] Then stimulated emission results in a dip in the signal

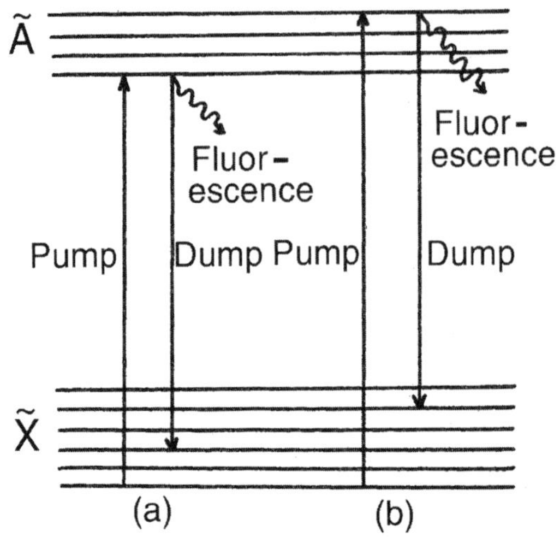

Figure 8.174 Illustration of the stimulated emission pumping process in which the pump laser takes the molecule to (a) the zero-point level or (b) an excited vibrational level of an excited electronic state

resulting from counting the ions. Polarization labelling[491] may also be used in which one laser induces birefringence or dichroism in the sample which results in changes in the transmission of the second, polarization blocked laser.

In addition to being related to OODR the SEP process is also related to stimulated Raman scattering discussed in Section 8.2.3 and illustrated in Figure 8.38. The main differences are that the upper state in stimulated Raman scattering is a virtual state and there is no need for a dump laser: the forward stimulated Raman scattering radiation takes its place.

The SEP process results in a considerable transfer of up to 30% of the population from the zero-point level of the X̃ state to excited rovibrational levels within the X̃ state using dye lasers with pulse lengths of a few nanoseconds. With single mode lasers and coherent excitation, up to 100% population transfer is possible. Total population transfer can also be achieved by the method of stimulated Raman scattering by delayed pulses (STIRAP).[492] This is a similar process to SEP except that the dump pulse precedes the pump pulse although there must be some temporal overlap. The STIRAP method is particularly adapted to the sample being in a molecular beam. For maximum population transfer the dump laser beam crosses the molecular beam a distance Δ before the pump beam where $\Delta/r_L \approx 1.4$ and r_L is 1/e of the radius of the intensity profile of the laser beam.

Appreciable population transfer presents the possibility of studying the spectroscopy and, particularly, the dynamics of molecules in a single rovibrational or vibrational state in the ground electronic state.

For spectroscopic studies of the ground electronic state SEP has been the preferred technique.[493,494]

We have seen in Sections 8.2.13 and 8.2.14 how photoacoustic and cavity ring-down spectroscopies have provided access to high-lying vibrational levels of the ground electronic state. These spectroscopies are confined almost entirely to hydrogen-stretching vibrations. These extremely sensitive techniques are necessary because of the very low probabilities for transitions to high overtone vibrational levels, these probabilities being much higher than for other vibrations because of the high wavenumber, typically about 3000 cm^{-1}, and the unusually large anharmonicity of hydrogen-stretching vibrations.

Transitions observed by SEP obtain their intensity in an entirely different way. Transition probabilities depend on the Franck–Condon factors which, in turn, depend on the geometry changes from the ground to the excited electronic state concerned. For example, acetylene (HC≡CH) changes from a linear $\tilde{X}$ state to a *trans* bent $\tilde{A}$ state (see Section 6.3.1.4) so that an SEP spectrum[495] with pumping to the zero-point level of the $\tilde{A}$ state shows a long progression in the *trans* bending vibration ν_4. If the Franck–Condon factor is too small a transition cannot be observed by SEP.

The intensity of an SEP spectrum is far higher than that of a single vibronic level, or dispersed, fluorescence spectrum because of the stimulated nature of the transitions and the fact that all the intensity is concentrated in a particular transition at any particular time.

Another important advantage of SEP is that the pump laser may be tuned to populate other vibronic levels, as in Figure 8.174(b), provided that the Franck–Condon factor is sufficiently high. Then the Franck–Condon factors for transitions to levels of the ground state are changed providing access to still more vibrational levels. When the sample is in a static gas cell the pump laser may be tuned to a hot band which may result in access to further vibrational levels.

The SEP method, then, is complementary to the photoacoustic and cavity ring-down methods as it can access vibrational levels of the ground state which the other methods cannot reach. It also provides essential data to be compared with increasingly accurate calculations of rovibronic levels of small molecules and information regarding the participation of a variety of vibrations in IVR[300] and their approach to chaotic behaviour.

To complement the photoacoustic spectrum of HCN, observed[302] in the 15 000–18 500 cm^{-1} region, and the cavity ring-down spectrum observed[313] in the 17 500–23 000 cm^{-1} region, the SEP spectrum,[496] using fluorescence depletion detection, has revealed 67 vibrational states in the 8900–18 900 cm^{-1} region.

The pump laser was tuned to a rovibronic transition in the $\tilde{A}^1A''$–$\tilde{X}^1\Sigma^+$ system in which the major geometry changes from the $\tilde{X}$ to the $\tilde{A}$ state are a decrease in the ∠HCN from 180 to 125° and an increase in the CN bond length from 1.156 to 1.297 Å.[497] Pumping, in the gas phase at room temperature, was in the Q branch of the hot band $2^{1(K=1)}_{2,0}$ where, ν_2 is the bending vibration and the subscript 2,0 implies $v_2 = 2$ and $\ell = 0$ in the $\tilde{X}$ state. Resulting from the geometry changes and the corresponding Franck–Condon factors the bands in the SEP spectrum involve excitation of ν_2, up to 14 quanta, and the CN-stretching vibration ν_1, up to seven quanta, but not the CH-stretching vibration ν_3, which is active exclusively in the photoacoustic and cavity ring-down spectra.

Observation of so many vibrational levels in the $\tilde{X}$ state of HCN has proved an excellent test of theoretical methods. Using a variational technique,[498] in which the potential energy surface is refined by a least-squares fit to observed data, agreement with observed stretching vibrational states up to 18 000 cm^{-1} is within ±0.8 cm^{-1} and, with their B rotational constants, is within ±0.001 cm^{-1}. Bending vibrational states are less well reproduced but are sufficiently accurate for vibrational assignments to be made.[499]

As with the high energy CH-stretching vibrational states observed in the photoacoustic and cavity ring-down spectra of HCN, there is no evidence of IVR in the high energy bending vibrational states because the density of states is low, being only about 6 per 100 cm^{-1} at 16 000 cm^{-1}. Neither is there any evidence of perturbations which might result from approaching the barrier to isomerization of HCN to the known species HNC. Increasing amplitude of the bending motion should result, eventually, in this rearrangement. The barrier height is not known but is predicted to be at least 12 000 cm^{-1}.

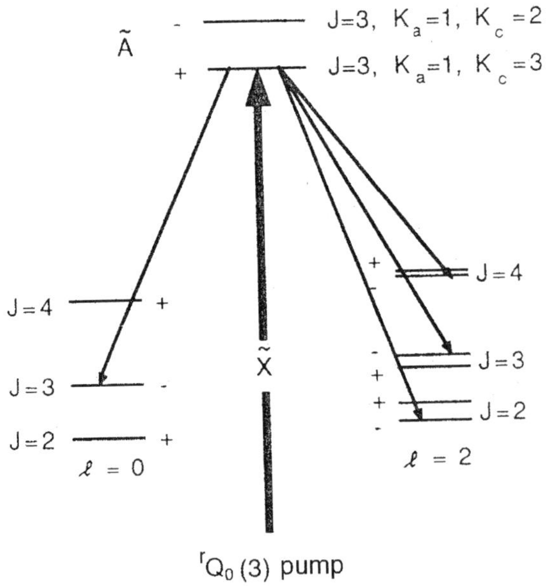

Figure 8.175 Illustration of SEP dump transitions in acetylene when the pump laser is tuned to the $^rQ_0(3)$ transition of the Ã–X̃ system
[Reproduced, with permission, from Chen, Y., Halle, S., Jonas, D. M., Kinsey, J. L. and Field, R. W. (1990). *J. Opt. Soc. Am. B.*, **7**, 1805]

Acetylene (HC≡CH) is *trans* bent in the $\tilde{A}^1A_u$ excited state[500] with ∠HCC = 120°. The large change of angle from the linear $\tilde{X}^1\Sigma_g^+$ ground state results in a long progression in the *trans* bending vibration ν_3. (The *trans* bending vibration is ν_4 in the linear lower state and ν_3 in the bent upper state.) An appreciable increase in the CC bond length from 1.208 Å in the X̃ state to 1.39 Å in the Ã state causes progressions, in the Ã–X̃ absorption spectrum, in the CC-stretching vibration ν_2 to be combined with that in ν_3.

The Franck–Condon principle, therefore, allows access, in SEP spectra, to high-lying vibrational levels of ν_2 and ν_4 in the X̃ state which have been investigated[501] in the range 11 400–15 700 cm^{-1} with fluorescence depletion detection. This information complements that from the photoacoustic spectrum,[304] discussed in Section 8.2.13, in which levels of the symmetric CH-stretching vibration, ν_1, were investigated in the range 14 900–18 500 cm^{-1}.

The SEP spectrum of the room temperature gas, unlike that of HCN,[496] shows regions of both regularity and severe perturbations. This behaviour is relevant to the onset of chaotic dynamic motion. In the transition region between regular and chaotic motion there is expected to be coexistence of both types.

Figure 8.175 illustrates, for the SEP spectrum of acetylene, how the rotational selectivity greatly simplifies the rotational structure associated with each vibronic transition. The Ã–X̃ transition is polarized perpendicular to the plane of the molecule in the C_{2h} excited state. The pump laser, tuned to the $^rQ_0(3)$ line of the 3_0^2 (or 3_0^3) band, populates the $J = 3$, $K_a = 1$ level in the Ã state and the dump laser stimulates transitions to only the $J = 3$ level, of the stack with $\ell = 0$, and the $J = 2$, 3 and 4

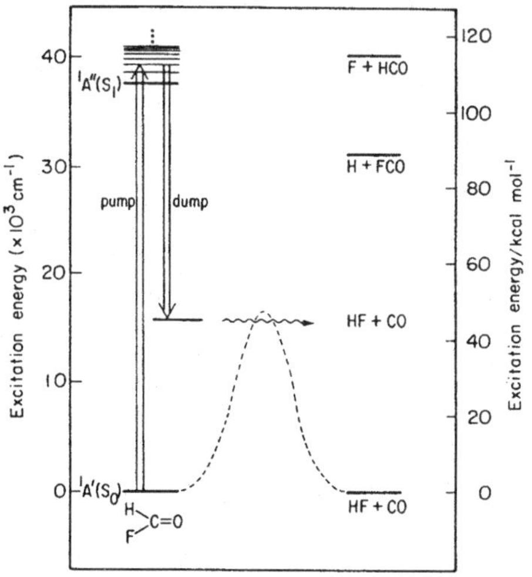

Figure 8.176 Energy level diagram for HFCO showing the SEP process used to investigate the dissociation to HF + CO
[Reproduced, with permission, from Choi, Y. S. and Moore, C. B. (1991). *J. Chem. Phys.*, **94**, 5414]

levels with (−) parity, of the stack with $\ell = 2$. Levels with the opposite parity were accessed by pumping the $^rR_0(2)$ transition.

Analysis of the SEP spectra showed that, for vibrational levels with $l = 0$, the rotational term values are those of an unperturbed rigid rotor but those with $l = 2$ show considerable rotational perturbation. Analysis of the vibrational level spacings shows that they tend to follow Poisson statistics rather than showing evidence of classical chaos.

There is an interesting comparison between the SEP spectra and a single rovibronic level, dispersed fluorescence spectrum of acetylene.[502] Although the resolution in the latter is much lower, 7–30 cm^{-1}, because of the reduced intensity of emission necessitating a large monochromator slit width, a spectrum could be obtained, in a pulsed supersonic jet, with excitation in the 0^0_0 band. This spectrum has two advantages over the SEP spectra. First, it avoids the nodes in the Franck–Condon intensity envelope which are inevitable when the upper state involves excitation of ν_3 and, second, it allows access to vibrational levels with no lower wavenumber limit. Such a limit is imposed on SEP spectra by the range covered by the dump laser. The dispersed fluorescence spectrum identifies IVR pathways for the excitation of the *trans* bending vibration ν_4 up to $v_4 = 14$.

The Rice–Ramsperger–Kassel–Marcus (RRKM) theory is generally successful when applied to unimolecular reactions. This theory assumes that there is complete redistribution of vibrational energy immediately prior to reaction whatever initial vibrational excitation took place. However, there is experimental and theoretical evidence that, in many cases, this intramolecular vibrational redistribution (IVR) is incomplete and that some reactions may be mode specific.

SEP is a powerful technique for investigating IVR in high-lying vibrational states and for determining whether it is complete prior to a dissociation process. Such a process is an example of a unimolecular reaction and SEP can give information on the mode specific character, or otherwise, of the dissociation.

Fluoroformaldehyde (HFCO), shown in Figure 8.176, is an ideal molecule for study because, as the figure shows, the dissociation

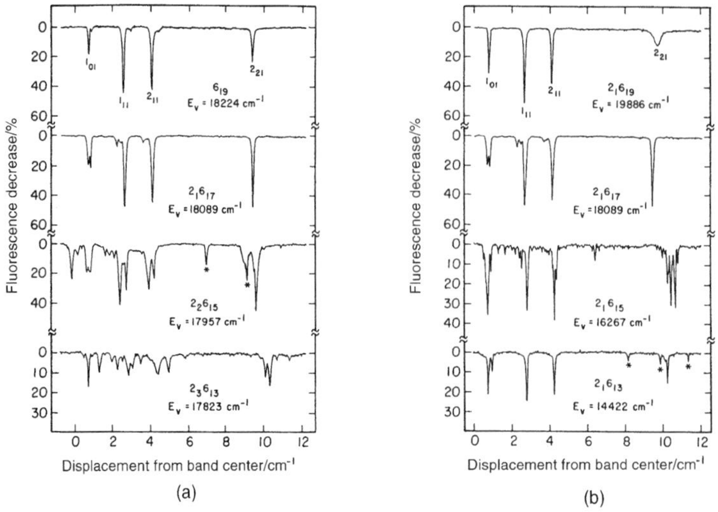

Figure 8.177 SEP spectra of HFCO (a) in the region of 18 000 cm^{-1} of vibrational energy in the $\tilde{X}$ state involving various combinations of ν_2 (CO stretch) and ν_6 (out-of-plane bend) and (b) in various regions with one quantum of ν_2 and increasing quanta of ν_6. Lines marked with asterisks belong to different vibrational states
[Reproduced, with permission, from Choi, Y. S. and Moore, C. B. (1991) *J. Chem. Phys.*, **94**, 5414]

energy for the process HFCO → HF + CO is relatively low: the *ab initio* calculated value[503] is about 16 500 cm^{-1}.

The dispersed fluorescence spectrum, in a pulsed supersonic jet, has been obtained[504] with excitation in the 6_0^2 band of the $\tilde{A}^1A$–$\tilde{X}^1A'$ system, where the molecule is planar in the $\tilde{X}$ state and, like formaldehyde, non-planar, with an out-of-plane hydrogen atom, in the $\tilde{A}$ state.[505] The vibration ν_6 is the out-of-plane bending of the hydrogen atom. This, together with ν_2, the CO-stretching vibration, is strongly excited in the $\tilde{A}$–$\tilde{X}$ system because of the change to non-planarity and an increase in the CO bond length by 0.18 Å in the $\tilde{A}$ state. Although the resolution of the spectrum was only about 40 cm^{-1}, it shows sharp, resolved bands above the dissociation threshold up to at least 22 000 cm^{-1}

Much higher resolution (0.05 cm^{-1}) SEP spectra have been recorded,[504] with fluorescence depletion detection and HFCO in a pulsed supersonic jet, and vibrational levels in the $\tilde{X}$ state assigned with energy in the 13 000–23 000 cm^{-1} region. Variation of the relative polarization properties of the pump and dump lasers was used to aid these assignments. The rotational transitions obey different ΔM_J selection rules according to whether the pump and dump lasers are polarized parallel or perpendicular to each other.

With the pump laser tuned to the 1_{10}–0_{00} rotational transition (the symbolism is $J_{K_aK_c}$) in the intense 6_0^2 band SEP spectra, in the region of a vibrational energy of 18 000 cm^{-1} in the $\tilde{X}$ state, were obtained and are shown in Figure 8.177(a). Figure 177(b) shows four more SEP bands but with a wide range of vibrational energies.

All the bands in Figure 8.177(a) and (b) show sharp rotational structure including six which involve vibrational energies well above the dissociation limit. Closer inspection reveals two further important features. First, although it is not obvious on the scale of these spectra, there are many line widths which are greater than the laser band width of 0.05 cm^{-1}. This homogeneous broadening is due to a reduction of the lifetimes of the lower states of these transitions caused by dissociation to HF + CO.

The second feature is that there are many more lines in some of the bands than the four that might be expected and which appear in the upper spectrum of Figure 8.177(a). The additional lines are due to coupling of the observed bright states to the nearby pseudo-continuum of dark states. However, the degree of coupling clearly depends on the nature of the bright states. The spectra in Figure 8.177 show that, at vibrational energies around 18 000 cm^{-1}, the 6^0_{19} band shows no coupling while increasing involvement of ν_2, particularly with more than one quantum, results in increased coupling.

Figure 8.177(b) shows that, with constant involvement of one quantum of ν_2 and increasing vibrational energy through increasing v_6, coupling increases from an energy of 14 422 to 16 267 cm^{-1}, which is in the region of the dissociation energy barrier, and then decreases again. The band with 19 886 cm^{-1} of vibrational energy shows no coupling at all.

These results show that IVR is mode selective. For a particular ground state vibrational energy those states with the most quanta of ν_6 show the least coupling. Excitation of one quantum of ν_2 increases the coupling only slightly but further quanta increase it dramatically. For increasing quanta of ν_6, and constant excitation of ν_2 (with v_2 not greater than one), coupling goes through a maximum near to the dissociation limit and then decreases at higher energies. This is strong evidence for the regular states, or quasi-periodic trajectories, which theoretical studies have predicted are embedded in a continuum of background states.

The HCO radical has an unusually weak C—H bond leading to a dissociation energy, for the process HCO → H + CO, of only about 5000 cm^{-1}. SEP spectra have been obtained[506] by the processes of both fluorescence depletion and two-colour resonant four-wave mixing, also known as laser-induced grating spectroscopy (LIGS).

LIGS is a variant of degenerate four-wave mixing.[507] In the latter, three laser beams of identical wavenumber are incident on a nonlinear medium which produces a fourth, signal, beam of the same wavenumber which is coherent, highly collimated and intense. Two of the incident, pump, beams are coaxial, but counterpropagating, and the third incident, probe, beam crosses the other two at an angle of, typically, 1–4°. The probe beam interferes with the pump beams, giving a fringe pattern resembling a diffraction grating. This interference results in scattering of the pump beams producing the signal beam. The signal beam intensity is increased whenever the wavenumber of the pump beams and the probe beam corresponds to an absorption.

In the case of LIGS[508] the probe (or dump) laser is of a different wavenumber from the pump laser. The two pump laser beams cross at a small angle and intensity modulation forms a grating pattern. If the pump laser wavenumber corresponds to an absorption in the sample, the grating is in the form of a spatial modulation in the populations of the ground and excited states connected by the transition. If the dump laser wavenumber corresponds to a transition from either of the levels connected by the pump laser, the dump beam detects a spatially modulated absorption and is scattered by the ground or excited state population grating.

The two possible schemes for LIGS involving three input laser wavenumbers of $\tilde{\nu}_1$, $\tilde{\nu}_2$ and $\tilde{\nu}_3$ and an output wavenumber $\tilde{\nu}_4$ are illustrated in Figure 8.178. In Figure 8.178(a), $\tilde{\nu}_1 = \tilde{\nu}_2$ (the pumping wavenumber) and $\tilde{\nu}_3 = \tilde{\nu}_4$ (the dumpling wavenumber) and, for the other, illustrated in Figure 8.178(b), $\tilde{\nu}_1 = \tilde{\nu}_4$ (pump) and $\tilde{\nu}_2 = \tilde{\nu}_3$ (dump). For SEP spectra the scheme in Figure 8.178(a) is appropriate. An advantage of this technique is that saturation broadening, which is a problem with fluorescence depletion, is avoided and more accurate line widths are obtained.

The HCO was prepared by laser photolysis of acetaldehyde, CH_3CHO, and the pump laser was tuned to rotational lines in either the 0^0_0 or 1^1_0 band of the $\tilde{B}^2A'$–$\tilde{X}^2A'$ system, where ν_1 is the CH-stretching vibration. The molecule is

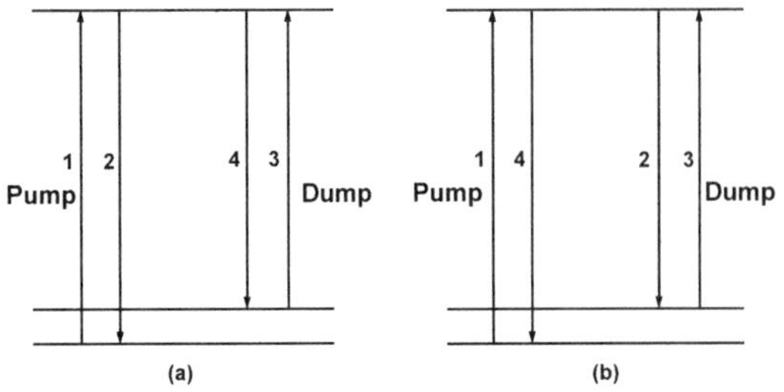

Figure 8.178 Two schemes for two-colour resonant four-wave mixing, or laser grating, spectroscopy using three input beams, 1, 2 and 3, and a signal beam 4

bent in both states with an increased C—O bond length and decreased bond angle in the $\tilde{B}$ state resulting in progressions in ν_2, the CO-stretching vibration and ν_3, the bending vibration. A combination of dispersed fluorescence and SEP spectra allowed identification[506] of vibrational energy levels in the $\tilde{X}$ state in the range 2000–21 000 cm^{-1}. These included states with $\upsilon_2 = 1$–12 and $\upsilon_3 = 0$–5. The $\upsilon_1 = 1$ level was observed only by pumping the 1_0^1 transition.

As in the case of HFCO, the unimolecular dissociation of HCO does not follow the RRKM theory. For HCO there is a much lower density of states above the dissociation threshold, so low that the states that lie above the threshold can be regarded as bound states embedded in a pseudo-continuum consisting of the states of the H and CO products. Consequently, there is a very slow rate of IVR.

The distributions among the vibrational and rotational states of the CO product, following state selection of HCO by SEP, have been studied.[509] The vibrational and rotational populations were obtained from fluorescence excitation spectra of the CO product using VUV laser radiation. It was found that, for states in the 10 800–16 000 cm^{-1} range involving υ_2 in the range 5–9 and υ_3 in the range 0–2, CO is produced, predominantly, with $\upsilon = 2$, some being produced with $\upsilon = 3$ for the higher energy states of HCO. The rotational populations favour low J and tend to be non-statistical.

An *ab initio* potential energy surface[510] reproduces very well the vibrational levels of HCO up to 16 000 cm^{-1} above the dissociation threshold.

A system of bands at 405 nm, observed originally in the emission spectra of comets[511] and much more recently in the spectrum of a carbon star,[512] has been assigned[513] to the $\tilde{A}^1\Pi_u$–$\tilde{X}^1\Sigma_g^+$ system of C_3 observed in the laboratory. The molecule is linear in both states although there may be a small barrier of 16 cm^{-1} or less in the $\tilde{X}$ state. Considerable interest attaches to the bending vibration ν_2, which has a very low wavenumber of 61.94 cm^{-1} in this state, concerning whether it is harmonic and whether the bending potential is affected by the excitation of the symmetric or antisymmetric stretching vibration, ν_1 and ν_3, respectively. A prediction[514] that the cyclic form of C_3 lies about 10 500 cm^{-1} above the linear form provides a further incentive to investigate high-lying vibrational levels of the $\tilde{X}$ state, particularly those involving combinations of bend and antisymmetric stretch because the reaction coordinate for the isomerization resembles such combinations.

In the $\tilde{A}$ state there is a large increase in the bending vibration wavenumber to 308 from 62 cm^{-1} in the $\tilde{X}$ state leading to progressions in ν_2 in the $\tilde{A}$–$\tilde{X}$ system thereby allowing access to high-lying levels of ν_2 in SEP spectra with $\tilde{A}$ as the intermediate state.

With C_3 generated by laser vaporization of graphite SEP supersonic jet spectra, with fluorescence depletion detection, have been obtained[515] with dump transitions which terminate in $2_\upsilon 3_2$ levels, with $\upsilon = 0$–8, and $2_\upsilon 3_4$ levels with $\upsilon = 0$–14. The pump laser was tuned to either the 0_0^0 or 2_0^2 band. The resulting vibrational term values show that there is a significant increase in the barrier to linearity as ν_3 is excited.

In complementary work, with C_3 generated by laser photolysis of furan, C_4H_4O, SEP spectra have been obtained[516] with lower levels 2_υ, with $\upsilon = 2$–34, and also $1_1 2_\upsilon$, with $\upsilon = 0$–8, and $1_3 2_\upsilon$ levels with $\upsilon = 4$–8. In order to

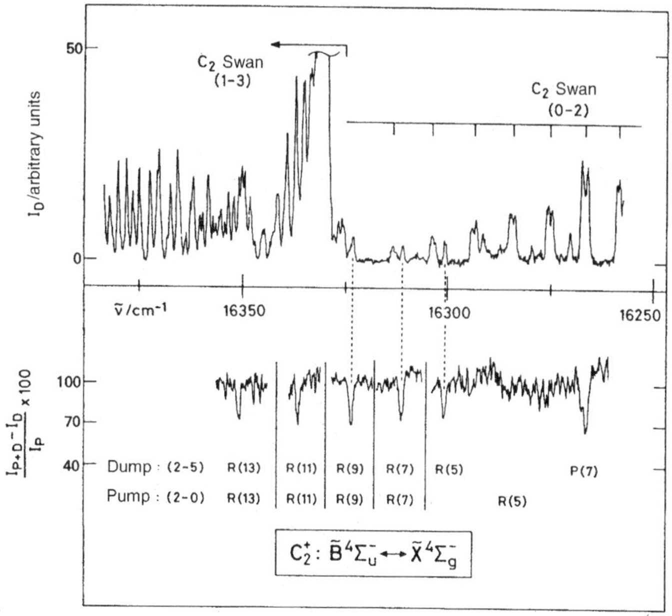

Figure 8.179 The lower part shows the SEP spectrum of C_2^+, the $\upsilon = 2$–5 transition of the B–X system, extracted from the spectrum in the upper part which includes, also, fluorescence from C_2
[Reproduced, with permission, from Celii, F. G. and Maier, J. P. (1990). *Chem. Phys. Lett.*, **166**, 517]

observe a large number of vibrational levels in the $\tilde{X}$ state the pump laser was tuned to a variety of absorption bands such that the Franck–Condon overlap between the upper state of the pump laser and the lower state of interest was maximized. Unlike the antisymmetric stretching vibration, the symmetric stretching vibration does not produce non-linearity of the molecule. The energy levels for ν_2 diverge strongly with increasing υ_2 resulting from a steep-sided potential energy curve.

Positive ions can be generated in a Penning ionization source cooled with liquid nitrogen. For example, C_2^+ is formed by reacting metastable He, in the 2^1S_0 or 2^3S_1 state, created in a d.c. discharge, with acetylene. SEP spectra of the resulting C_2^+ have been obtained[517] by pumping the $\upsilon = 2$–0 transition of the $B^4\Sigma_u^-$–$X^4\Sigma_g^-$ system resulting in characterization of the $\upsilon = 4$–6 levels of the X state.

Two problems were encountered which are typical of such an experiment. First, there may be other species present in the discharge which fluoresce with pump or dump laser radiation. In this case, C_2 is produced which fluoresces in the $d^3\Pi_g$–$a^3\Pi_u$, Swan band system excited by the dump laser. This is shown in the upper part of Figure 8.179 severely overlapping the C_2^+ spectrum. A modulation technique was used to extract the C_2^+ spectrum shown in the lower part.

The second problem arose because the C_2^+ is not as cold as it would be in a supersonic jet. As a result, the $\upsilon = 5$ level of the X state, to which the dump laser is stimulating transitions in the SEP spectrum in Figure 8.179, is populated under the conditions of production of C_2^+ so that the intensity measured at the wavelength of the dump laser contains contributions from B–X, $\upsilon = 2$–5 fluorescence excitation. The

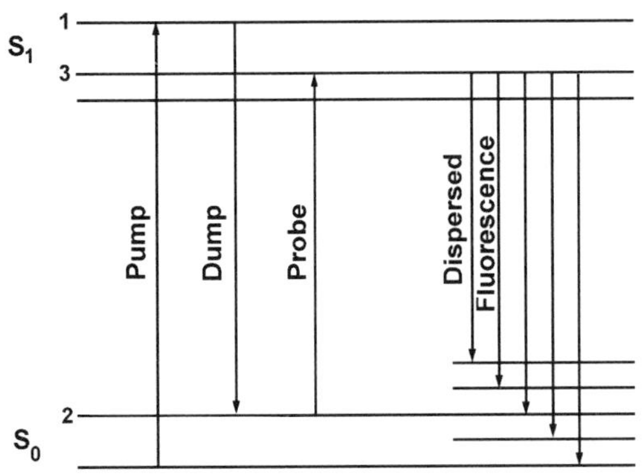

Figure 8.180 Scheme for probing, by dispersed fluorescence, the population of a vibrational level in the ground electronic state achieved, initially, by SEP

modulation scheme used discriminates against these signals also. This is confirmed by the observation that, when pumping the $R(5)$, $\upsilon = 2$–0 transition of the B–X system, only $J = 6$ is populated in the B state and the stimulated emission is to $J = 5$ and 7 in the $\upsilon = 5$ level of the X state.

A similar technique has been used[518] to obtain SEP spectra of $C_4H_2^+$, the diacetylene cation, and $BrCN^+$

The preparation of a large proportion of molecules in a particular excited vibrational state within the ground electronic state by SEP allows a study of rates of collisional deactivation of this state by probing the decrease in population with time.

This method has been used[519] to determine collision cross-sections for 1,4-difluorobenzene, in the gas phase at room temperature, in the vibrational state $5_2 30_2$. The vibrational motions involved are not important here. What is important is that the level is 2036 cm^{-1} above the zero-point level of the ground electronic state S_0 and that the estimated density of vibrational states at this level is 6–8 per cm^{-1}.

Figure 8.180 illustrates the experimental scheme. The pump laser overlies the $5^1_0 30^2_2$ hot band and populates the $5^1 30^2$ level (1) in the excited singlet state S_1. The dump laser stimulates emission in the $5^1_2 30^2_2$ transition thereby transferring population to the $5_2 30_2$ level (2). The decrease of this population with time, due to collisions with a wide range of partners ranging from small, such as He, to large, such as pentan-1-ol, was monitored. The method used was to tune a probe laser, with various delay times after the dump laser, to the $5^0_2 30^2_2$ transition. The intensity of the single vibronic level, dispersed fluorescence from the $5^0 30^2$ level (3) was then a measure of the population of the $5_2 30_2$ level.

Figure 8.181 shows a small part of the single vibronic level, dispersed fluorescence spectrum in which the decline of the intensity of the $3^0_1 30^2_2$ band with time is a measure of the collisional decay rate for the $5_2 30_2$ population. In this case the collision partner is 1,4-difluorobenzene itself at a pressure of 110 mTorr. (The weaker $5^0_1 6^0_1 30^2_2$ band is a less accurate measure, and the $5^1_2 6^0_1 30^2_2$ 'rogue' band is part of the fluorescence produced by the pump laser.)

It was found that the collision cross-sections, derived from the rate constants for vibrational relaxation, vary from 0.25 to 2.7 times the hard sphere value for the collision partners He and $(C_2H_5)_2O$, respectively, similar to values obtained previously for excited electronic states of aromatic molecules.

When the fluorescence dip method is used for detection, following an SEP process, the depth of the dip can be used to determine rates of IVR. However, some unavoidable spatial mismatching of the pump and dump lasers leads to some unreliability of the depth. This problem can be alleviated by using the intensity of ionization from the excited electronic state to monitor its changing population; then, any spatial mismatch is unimportant.

This method has been applied to a study of the vibration wavenumbers in the ground electronic state S_0 of the phenol–water complex,[520] $C_6H_5OH \cdots OH_2$, which is readily formed in a supersonic jet. The dump laser was also used for ionization from S_1. The

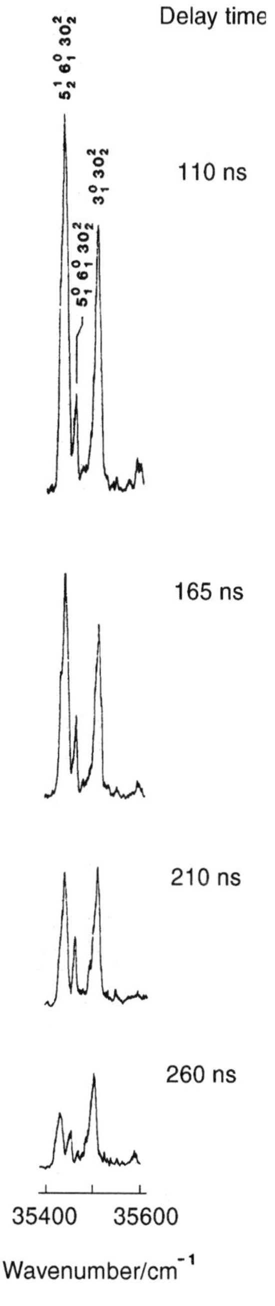

Figure 8.181 Part of the dispersed fluorescence spectrum of 1,4-difluorobenzene, obtained at different delay times following the dump pulse which populates the $5_2 30_2$ level by stimulated emission pumping [Reproduced, with permission, from Lawrance, W. D. and Knight, A. E. W. (1983). *J. Chem. Phys.*, **79**, 6030]

intensity of the pump laser was kept to a minimum to avoid ionization by this laser alone.

Figure 8.182 makes an interesting comparison between the single vibronic level, dispersed fluorescence spectrum, with excitation in the 0^0_0 band, and the ionization dip SEP spectrum. The latter is much sharper and more extensive and the resolution is higher, that of the fluorescence spectrum being limited by the slit width of the monochromator necessary to collect the weak signal. There are also important intensity differences.

The great complexity of the SEP spectrum compared with that of bare phenol is due to the low wavenumber modes introduced by the hydrogen bonding.

Rates of IVR were estimated from the ionization dip intensities and it was shown that there is a large increase in the rate of IVR on complexation and also as the density of vibrational states in the complex increases.

8.2.23 Saturation, intermodulated fluorescence and polarization spectroscopy

Any branch of spectroscopy which involves the saturation of transitions and the consequent observation of Lamb dips would seem to qualify for inclusion under the general heading of saturation spectroscopy. Nevertheless, it seems to be general practice to refer separately to techniques such as laser magnetic resonance and laser Stark spectroscopy and reserve saturation spectroscopy for the kind of experiment illustrated in Figure 8.183.

In this experiment, radiation from a tunable, single mode laser falls on a beam splitter which reflects a small proportion of the beam on to the mirror M_1. From M_1 this beam is directed into the sample cell and falls on to the detector. The other, more intense, beam is periodically interrupted by a chopper and passes, via mirror M_2, through the sample cell. The figure exaggerates the angle at which the weak beam, the probe beam, and the intense beam, the

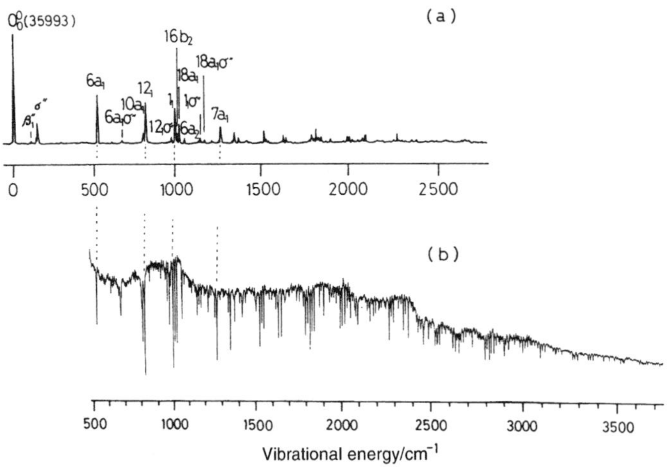

Figure 8.182 (a) Single vibronic level, dispersed fluorescence spectrum of the phenol–water complex with excitation in the 0_0^0 band. (b) The SEP spectrum, detected by the ionization dip method, with the pump laser tuned to the 0_0^0 band
[Reproduced, with permission, from Ebata, T., Furukawa, M., Suzuki, T. and Ito, M. (1990). *J. Opt. Soc. Am. B.*, **7**, 1890]

saturation beam, intersect; in practice it is only a few milliradians.

The two counterpropagating beams produce an effect in the atoms or molecules of the sample which is the same as that produced by using a single source beam and a plane mirror, normal to the beam, which reflects it back through the sample, as illustrated in Figure 3.8. As explained in Section 3.4, the only sample particles which can interact with both the saturation and probe beams are those with zero velocity component along the direction of beam propagation. The saturating beam is sufficiently intense to reduce appreciably the population of the lower state of the transition to which the laser is tuned and to cause the probe

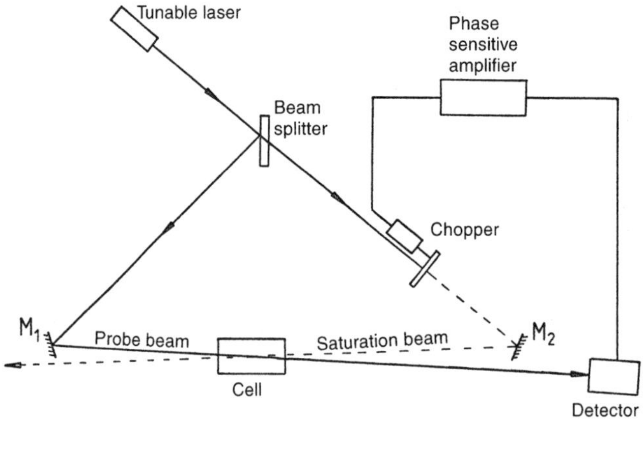

Figure 8.183 Experimental arrangement for saturation spectroscopy using counterpropagating saturation and probe beams

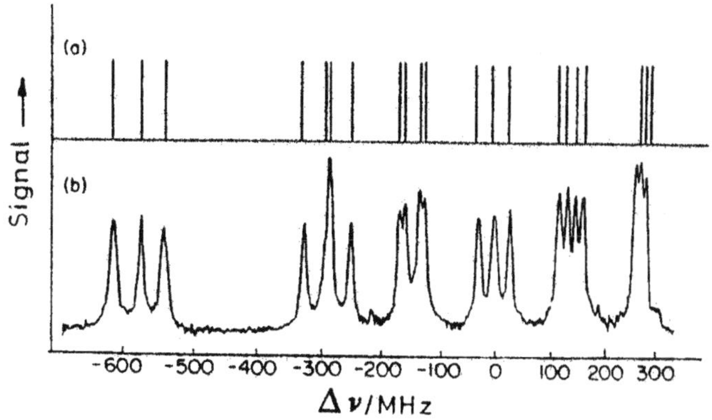

Figure 8.184 The (a) calculated, and (b) observed 21 hyperfine components of the $P(117)$ line of the $\upsilon = 21$–1 band of the $B^3\Pi_{0^+u}$–$X^1\Sigma_g^+$ system of the I_2 molecule
[Reproduced, with permission, from Hänsch, T. W., Levenson, M. D. and Schawlow, A. L. (1971). *Phys. Rev. Lett.*, **26**, 946]

beam to undergo less absorption, by the particles with zero velocity component, than it would otherwise do. Therefore, as the laser is tuned across the Doppler profile of an absorption line the absorption experienced by the probe beam shows a sharp minimum, a Lamb dip, at the centre. The width of the dip is the natural line width of the transition. As Figure 8.183 shows, phase-sensitive detection at the chopping frequency is used for observing the signal. The technique was developed[521,522] in 1970–71.

One of the early applications resulted in the observation[523] of resolved hypefine structure in the $B^3\Pi_{0^+u}$–$X^1\Sigma_g^+$ transition of I_2. This experiment was performed before tunable dye lasers had been developed and used the 556.2 nm line of a krypton ion laser which could be tuned over a 5500 MHz range with a scanning interferometer. The $P(117)$ line of the 21–1 band of the B–X system of I_2 lies within this range and is normally Doppler broadened to a line width at half intensity of *ca* 1400 MHz. As the saturation spectrum in Figure 8.184 shows, there is almost complete resolution of the 21 hyperfine components.

The hyperfine splitting is due to two effects. The first is small and due to an interaction between the magnetic field of the electrons and the magnetic moment of the nuclei and is characterized by a quantity C. The second is much larger and is due to the interaction between the electric quadrupole moment of the nuclei ($I = \frac{5}{2}$ for each) and the electric field gradient of the molecule (see Section 4.2.5) and is characterized by $\Delta(eqQ)$, the change in eqQ from the lower to the upper state. Interpretation of the spectrum gives $\Delta(eqQ) = -940 \pm 30$ MHz and $C = 53 \pm 5$ kHz.

The technique of saturation spectroscopy with counterpropagating beams has been extended into the infrared.[524] Two separate CO_2 waveguide lasers are used. An important property of these lasers is that each vibration–rotation laser line can be tuned over several hundred megahertz. The counterpropagating beams are directed independently into a multiple reflection absorption cell in which 'cat's eyes' retroflectors set up independent standing waves between pairs of reflecting mirrors. One of the lasers is used as a standard and the frequency of the second laser is measured by the beat frequency of the two together. Resolution of a few kilohertz has been obtained, permitting hyperfine splitting of vibration–rotation lines in the infrared spectra of such large molecules as $SF_6(\nu_3)$ and $OsO_4(\nu_3)$ to be resolved and interpreted.

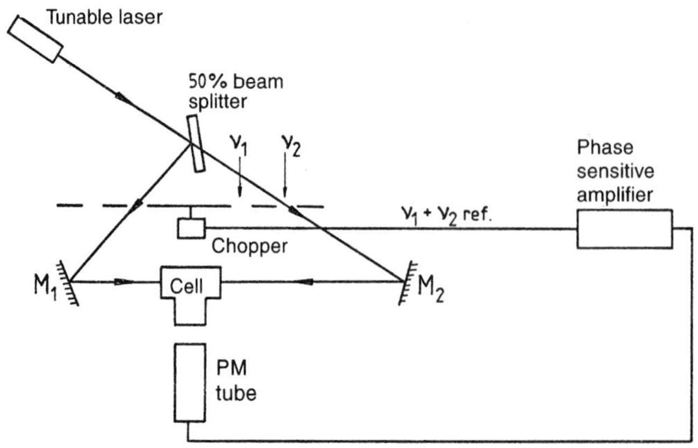

Figure 8.185 Saturation spectroscopy by the technique of inter-modulated fluorescence

A more sensitive technique has been developed involving saturation of electronic transitions and monitoring of fluorescence rather than absorption. The technique is called intermodulated fluorescence[525] and is illustrated in Figure 8.185.

Radiation from a tunable laser is split into equal parts by a beam splitter and both beams are chopped, but at different frequencies, ν_1 and ν_2. The fluorescence from the sample is monitored by a photomultiplier with phase sensitive detection at the frequency $\nu_1 + \nu_2$.

Just as the intensity of absorption is decreased by saturation, so is the intensity of fluorescence. However, the use of intermodulated fluorescence greatly increases the sensitivity compared to monitoring fluorescence with modulation of only one of the counterpropagating beams.

An example of the sensitivity of the technique is the observation[526,527] of hyperfine structure in a rovibrational line in the $\tilde{A}^2\Pi_u$–$\tilde{X}^2\Pi_g$ electronic transition of the short-lived molecule $^{11}BO_2$.

Another technique related to saturation spectroscopy is polarization spectroscopy.[491] The experimental method is very similar to that illustrated in Figure 8.183 employing an intense, saturating pump beam and a weaker probe beam. Before passing into the sample cell the pump beam is circularly polarized and the probe beam is plane polarized. The detector monitors the probe beam but, by introducing a polarizing filter in front, it detects only that radiation which is plane polarized at 90° to the plane of polarization of the probe beam incident on the sample. It follows that only radiation from the probe beam whose plane of polarization has been rotated can reach the detector. A rotation of this kind can be introduced by the saturating beam, because of its circular polarization, but only if the probe and saturating beams interact with the same atoms or molecules of the sample: these are the ones with zero velocity component in the line of the beams. For this reason, polarization spectroscopy is a form of sub-Doppler spectroscopy and, like intermodulated fluorescence spectroscopy, has high sensitivity.

The $C^1\Pi_u$–$X^1\Sigma_g^+$ system of Cs_2 has been observed, with sub-Doppler resolution, by polarization spectroscopy.[528] The rotational structure shows a *P*, *Q* and *R* branch but is very crowded. One reason for this is that the molecule is heavy, with a *B* value of about $0.0127\,cm^{-1}$, and a second reason is that the molecule is produced by heating caesium to 220 °C resulting in levels with very high values of *J* being appreciably populated. More than 800 rotational lines in the 0–0 band lie within about $22\,cm^{-1}$.

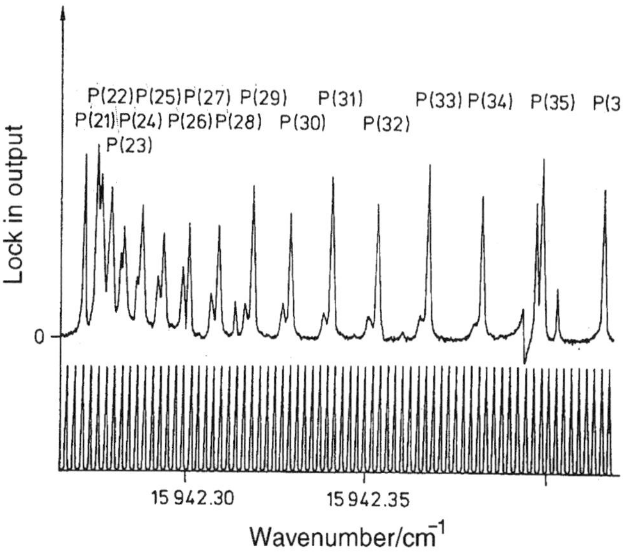

Figure 8.186 Part of the P branch of the 0–0 band of the $C^1\Pi_u$–$X^1\Sigma_g^+$ system of Cs_2 observed by sub-Doppler polarization spectroscopy
[Reproduced, with permission, from Raab, M., Höning, G., Castell, R. and Demtröder, W. (1979). *Chem. Phys. Lett.*, **66**, 307]

Simplification of the spectrum was achieved not only by sub-Doppler resolution but by separating the Q branch from the P and R branches. The Q branch was observed with plane polarization of the pump laser and the planes of polarization of the polarizer and analyser for the probe laser at an angle of about 2°. Under these conditions, the P and R branches are suppressed. Alternatively, the Q branch can be suppressed by using circular pump polarization with the polarizer and analyser completely crossed. Part of the P branch, observed in this way, is shown in Figure 8.186. The intensity alternation of 7 : 9 for J even : odd can be clearly seen and is a consequence (see equation 4.84) of the nuclear spin quantum number $I = \frac{7}{9}$ for Cs.

8.2.24 Level crossing spectroscopy, including the Hanle effect

The development of level crossing spectroscopy has its origins in the Hanle effect discovered in

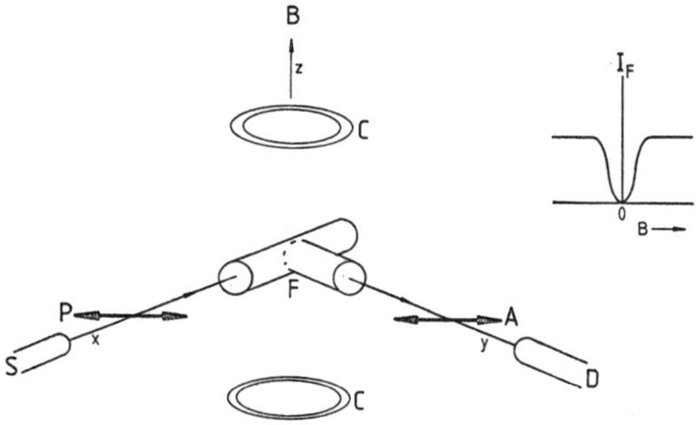

Figure 8.187 An experiment to demonstrate the Hanle effect

1924.[529] The original experiment is illustrated in Figure 8.187.

The source lamp S is a mecury vapour discharge lamp. The radiation from this includes the intense 253.7 nm, $^3P_1$–$^1S_0$ (. . . $6s^16p^1$– . . . $6s^2$), radiation and is plane-polarized by a polarizer P. The radiation is directed in the x direction into a fluorescence cell F containing mercury vapour in which resonance

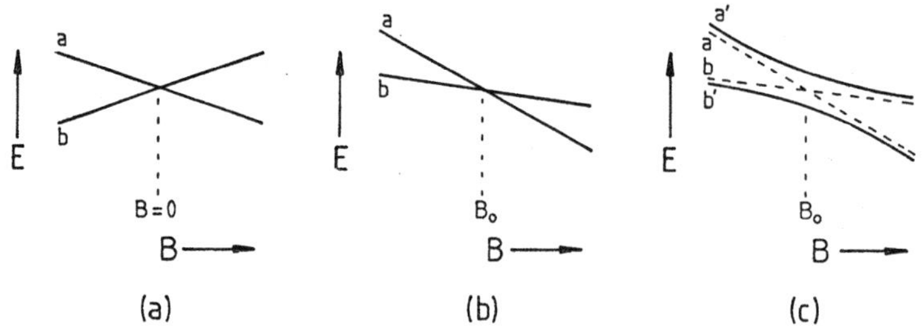

Figure 8.188 (a) Zero-field level crossing (the Hanle effect), (b) level crossing at $B = B_0 \neq 0$ and (c) level anticrossing at $B = B_0$

fluorescence occurs, excited by the incident radiation. The fluorescence in the y direction passes through the analyser A oriented so that the detector D receives only fluorescence which is polarized in the same plane as the exciting radiation. A magnetic field of strength B is applied in the z direction by Helmholtz coils C and the plane of polarization of incident and detected radiation is perpendicular to the field. As B is varied in magnitude and sign the detected fluorescence intensity I_F passes through a minimum at $B = 0$ as shown in Figure 8.187. A similar experiment may be carried out with an electric rather than a magnetic field.

Figure 8.188(a) illustrates how the separation of two Zeeman fine-structure levels a and b varies with B. The splitting is zero at $B = 0$ where the levels cross. The Hanle effect is a special case of the general phenomenon of level crossing and has been given the alternative name of zero field level crossing.

Figure 8.188(b) illustrates the more general case of level crossing which occurs accidentally for two levels a and b when $B = B_0$. If, in an experiment analogous to that in which the Hanle effect is observed, resonance fluorescence from the levels a and b is monitored with changing magnetic (or electric) field, a level crossing signal analogous to that in Figure 8.187 is observed at $B = B_0$.

An extremely important feature of the Hanle effect and level crossing spectroscopy is that the width of the signal is governed only by the natural line width and not the Doppler width. The precision of level crossing spectroscopy contrasts with that of a conventional Zeeman effect experiment (see Section 6.3.5.5) which is Doppler limited, leading to serious overlap of Zeeman components of a transition.

The change of polarized fluorescence intensity in a Hanle effect or level crossing experiment is due, not to a change in the total fluorescence intensity from the levels concerned, but to a redistribution in space: the fluorescence becomes anisotropic. This is due to interference effects between the radiation from levels a and b.[530]

Figure 8.189 shows an example where interference might occur. The atom or molecule is initially in the Zeeman state indicated by the quantum numbers J, M_J. The exciting radiation, plane-polarized perpendicular to the magnetic field, takes it into the levels (J', $M_J + 1$) and (J', $M_J - 1$). The emission processes to the J'' Zeeman components obey the selection rule $\Delta M_J = 0, \pm 1$. The figure shows that there are two distinct ways of the overall transition (J'', M_J)–(J, M_J) occurring, via either the (J', $M_J + 1$) or the (J', $M_J - 1$) levels. For this reason, these two levels may be said to share the same photon and the consequence is interference between the radiation from them. Zare[530] drew the analogy with the double slit interference experiment, carried out by Young, in which monochromatic light emerges from a single slit and falls on two symmetrically disposed slits. Interference occurs

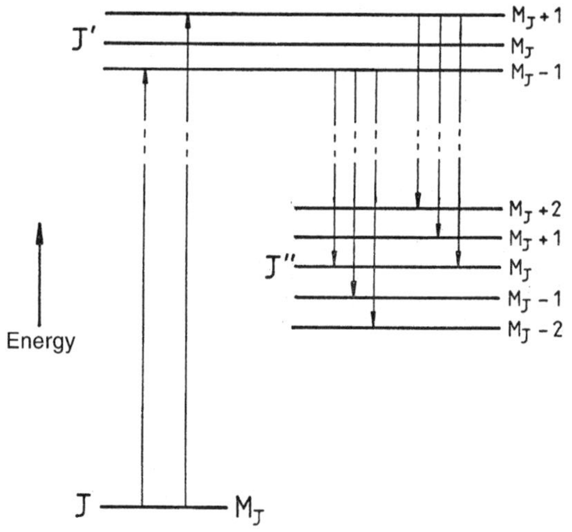

Figure 8.189 In a fluorescence experiment the levels (J', M_J+1) and (J', M_J-1) are populated and fluorescence to various components of J'' occurs. The transition (J'', M_J)–(J, M_J) may occur by two routes and this results in interference between the radiation from the two upper levels

between the light coming through these two slits and fringes can be observed on a screen. The separation of the fringes is small when the slits are close and increases when the slits are moved apart. The interference can be attributed to the two slits sharing a photon.

For the fluorescence system in Figure 8.189 the levels (J', M_J+1) and (J', M_J-1) are the analogues of the two identical slits in Young's experiment and, therefore, the closer the levels are the greater is the interference, which reaches a maximum when the levels are degenerate. The lifetimes of the states corresponding to the two levels are analogous to the slit widths in Young's experiment. This last analogy makes it apparent that the width of the level crossing signal is dependent on the radiative lifetime of the emitting levels. In the case of the Hanle effect the width and the lifetime are related through the uncertainty principles as stated in equation (1.29). In the present case ΔE is given by

$$\Delta E = \mu_0 g H_{\frac{1}{2}}[(M_J+1)-(M_J-1)] \qquad (8.77)$$

which is to be compared with equation (6.71) for the anomalous Zeeman effect in atoms. In equation (8.74), μ_0 is the electronic Bohr magneton, g is the Landé g-factor and $H_{\frac{1}{2}}$ is the half width at half intensity of the Hanle signal. The uncertainty Δt is simply the radiative lifetime τ of the zero-field state so that, from equations (1.29) and (8.77),

$$H_{\frac{1}{2}} = \hbar/2\mu_0 g\tau \qquad (8.78)$$

Therefore, by measuring $H_{\frac{1}{2}}$, the product $g\tau$ can be obtained. Since the lifetime τ is dependent on the pressure, P, it is usual to measure $H_{\frac{1}{2}}$ as a function of P and extrapolate to $P=0$ when τ is the radiative lifetime.

In order to obtain a value for τ an estimate for g must be made.

When a high power laser source is being used it is possible that the emission observed in a level crossing experiment may be stimulated rather than spontaneous. In this case a level crossing signal is observed as a sharp decrease in fluorescence due to the transitions being saturated. The decrease is proportional to the second and higher powers of the laser intensity, unlike the spontaneous level crossing signal, which is linear.

One of the first observations of the Hanle effect in a molecule was in the $\upsilon=10$, $J=12$ level of the $B^1\Pi_u$ state of Na_2, the exciting radiation being the 476.5 nm line of an argon ion laser.[531] The g-value was calculated assuming Hund's case (a) coupling (Section 6.2.5.3) between the orbital and nuclear rotational angular momenta and ignoring nuclear spin angular momenta; this is valid provided that the hyperfine splittings are much less than the natural line width. The resulting value for τ is 6.41 ± 0.38 ns.

Pumping the $R(7)$ line of the 0–0 band of the $A^1\Pi$–$X^1\Sigma^+$ transition of CS with an Mn^+ line at 257.61 nm and the observation[537] of the Hanle effect gave $\tau = 200 \pm 40$ ns for the $\upsilon=0$, $J=8$ level of the $A^1\Pi$ state. In addition, the corresponding effect, employing an electric field, gave an approximate value of the electic dipole moment μ of 0.72 D; this compares with 1.958 D in the $X^1\Sigma^+$ state.

The Hanle effect has been measured in several rotational levels of the $v_2 = 10$, Π vibronic level of the $\tilde{A}^2A_1$ state of NH_2 using a tunable dye laser.[533] Calculation of g was difficult in this case because there are angular momentum contributions due to nuclear spin, electron spin and orbital motions, and there may also be a contribution from vibrational angular momentum. An approximate value for g gave $\tau \approx 350$ ns.

The observation of a level crossing signal at $B \neq 0$, as in Figure 8.188(b), requires a coherence in the excitation process which results from the levels a and b being populated by transitions from the same initial state, for example the state (J, M_J) in Figure 8.189, and also a coherence in the emission process which results only if the transitions are from a and b to the same final state, for example (J'', M_J) in Figure 8.189.

Level crossing signals at $B \neq 0$ are observed with an experimental arrangement similar to that in Figure 8.187. As with Hanle effect measurements, the source of exciting radiation, particularly where atoms are being studied, may be an atomic emission lamp emitting resonance radiation which then excites the sample consisting of the same atoms as the source. For molecular samples methods of excitation include molecular resonance lamps, accidental atomic emission line coincidences, white light sources and tunable or fixed frequency lasers, the last relying on accidental coincidences with molecular transitions.

It might appear from the behaviour in Figure 8.188(b) of levels a and b that excitation of resonance fluorescence with a monochromatic source, particularly a laser source, would be possible only for a particular value of B and, at other values, the exciting radiation would be off-resonance. However, Series has pointed out[534] that the perturbations involved in level crossing result in a modified form of signal being observed when the atoms or molecules of the sample are static and, more importantly, the motion of the particles results in absorption of the radiation occurring anywhere within the Doppler line profile which is usually broad compared with the natural line width; the result of this is that the level crossing signal is observed just as for broad band excitation.

The first observation of level crossing signals at $B \neq 0$ was in the 2^3P–2^3S_1 transition of He using a He resonance lamp for excitation.[535] Level crossings observed involve Zeeman sublevels of the $^3P_1$ and $^3P_2$ components of the 2^3P term (see Figure 6.18, which illustrates the closeness of $^3P_1$ and $^3P_2$ in zero field).

Level crossings have also been observed[536] in the 2^2P term of ^{7}Li (and ^{6}Li) using a Li resonance source lamp to excite 2^2P–2^2S transitions. The 2^2P term gives rise to two states, $^2P_{\frac{1}{2}}$ and $^2P_{\frac{3}{2}}$, forming a normal multiplet. For $J = \frac{1}{2}$, $M_J = \frac{1}{2}, -\frac{1}{2}$ and, for $J = \frac{3}{2}$, $M_J = \frac{3}{2}, \frac{1}{2}, -\frac{1}{2}, -\frac{3}{2}$ and there is a crossing between the $^2P_{\frac{3}{2}}$, $M_J = -\frac{3}{2}$ and the $^2P_{\frac{1}{2}}$, $M_J = \frac{1}{2}$ levels at $B \approx 0.32$ T (3.2 kG). Figure 8.190 shows that the crossing is complicated by the splitting of the $^2P_{\frac{1}{2}}$, $M_J = \frac{1}{2}$ level into four components due to the nuclear spin $(I = \frac{3}{2})$ of ^{7}Li. In the presence of a magnetic field there is a nuclear Paschen–Back effect (see Section 6.1.8.2 and Figure 6.23) and the four components have $M_I = \frac{3}{2}, \frac{1}{2}, -\frac{1}{2}, -\frac{3}{2}$. The analogous splitting of the $^2P_{\frac{3}{2}}$, $M_J = -\frac{3}{2}$ level

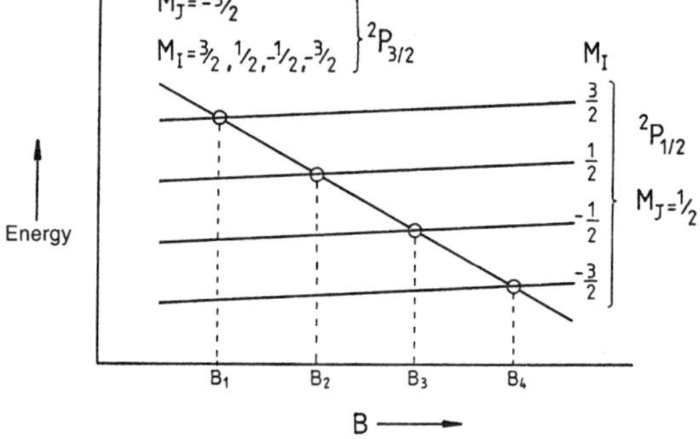

Figure 8.190 Level crossings in the ^{7}Li atom

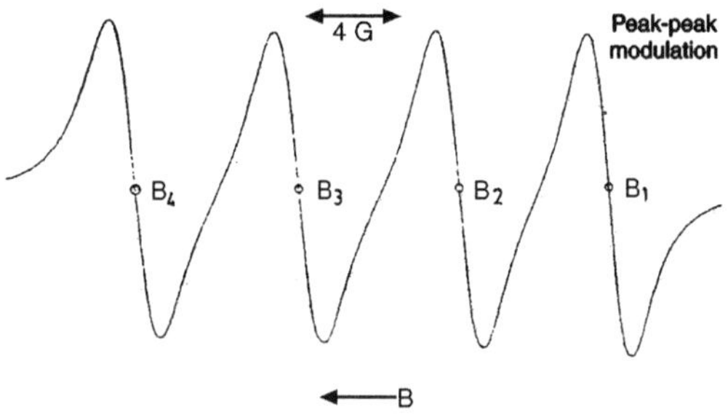

Figure 8.191 Level crossing spectrum of the ^{7}Li atom showing, in first derivative form, the four level crossings, at magnetic fields B_1, B_2, B_3 and B_4, marked in Figure 8.190 ($1\ \mathrm{G} = 10^{-4}\ \mathrm{T}$) [Reproduced, with permission, from Brog, K. C., Eck, T. G. and Wieder, H. (1967). *Phys. Rev.*, **153**, 91]

into four components is too small for it to be observed in the level crossing spectrum so that a total of four level crossings, marked with circles in Figure 8.190, are observed. Figure 8.191 shows the spectrum which, because of field modulation and phase sensitive detection, is in first derivative form.

From level crossing signals at $B \neq 0$ two kinds of information can be obtained. As in the Hanle effect, the radiative lifetime τ, taken here to be the same for the two Zeeman sublevels, of the 2^2P term can be found from the width ΔB of the level crossing signal since

$$\Delta B = \frac{1}{3^{\frac{1}{2}}\pi\tau}\left(\frac{\partial \varDelta}{\partial B}\right)^{-1} \tag{8.79}$$

where $\delta\varDelta/\delta B$ is the rate of change of separation $\varDelta$ of the crossing levels with magnetic field. For the ^2P term of ^{7}Li it was found that $\tau = 27.8 \pm 0.4\,\mathrm{ns}$.

The second kind of information from level crossing signals relates to hyperfine splitting constants in the states involved in the crossings. For example, from the spectrum of ^{7}Li, the Fermi contact and dipole–dipole contributions to magnetic hyperfine interaction in the $2^2P_{\frac{3}{2}}$ state and the nuclear quadrupole moment of ^{7}Li were obtained.

An example of molecular level crossings is an observation[537] of crossings between nuclear hyperfine components of a number of levels of the $v_2 = 9$ and $v_2 = 10$ vibrational states within the $\tilde{A}^2A_1$ excited electronic state of NH_2. Level crossings were observed involving only the *para* rotational levels. Since $I = \frac{1}{2}$ for both the hydrogen nuclei, rotational levels in the $\tilde{A}$ state with the rotational quantum number N odd are *ortho* levels, with parallel nuclear spins (compare Figure 4.22 for H_2), and those with N even are *para* levels.

One set of level crossings observed was between the hyperfine components of an $N = 2$ level for which, since $S = \frac{1}{2}$ in a doublet state, $J = \frac{5}{2}$. For a *para* level the only contributor to nuclear spin is the nitrogen atom, for which $I = 1$. Consequently [see equation (6.40)], $F = \frac{7}{2}, \frac{5}{2}, \frac{3}{2}$ and these levels are split at zero magnetic field. However, the splitting is only of the order of 100 MHz (0.003 cm^{-1}). For this reason it requires a sub-Doppler technique, such as level crossing, for the hyperfine splitting to be observed.

In a magnetic field each nuclear hyperfine component is split into $(2F + 1)$ components and, with the incident laser radiation polarized perpendicular to the field, those crossings with $\Delta M_F = 2$ are observed. These can be interpreted to give the ratio of the zero-field hyperfine splitting to the magnetic moment of NH_2.

Level crossing signals have been observed[538] in a laser Stark spectrum of POF_3 using CO_2 laser lines in the 10 µm region to investigate the 4_0^1 band, where ν_4 is the degenerate PF-stretching e vibration of the C_{3v} molecule. In addition to many Lamb dips observed in the laser Stark spectrum (see Section 8.2.7) in the presence of an electric field, 15 additional sub-Doppler features were observed with the same sign as the Lamb dips and assigned to level crossings in the ground vibrational state. In addition, many more signals were detected, with the opposite sign to the Lamb dips, and are due to level anticrossings (see Section 8.2.25).

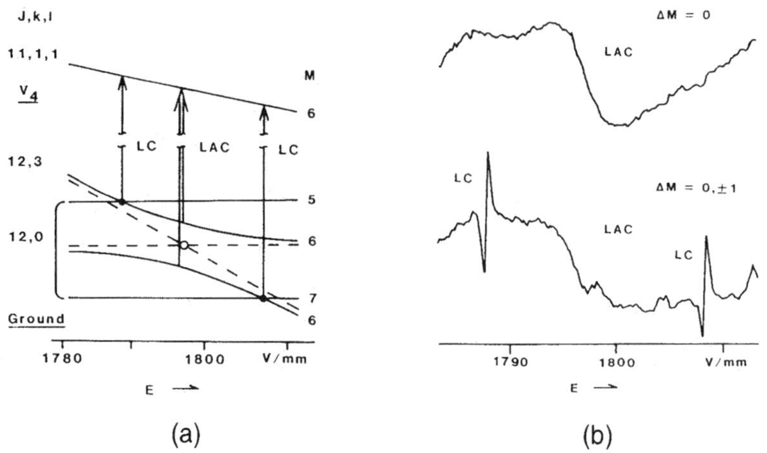

Figure 8.192 (a) Two level crossings and one anticrossing in the ground vibrational state of POF_3. (b) Laser Stark spectrum showing (above) only the level anticrossing signal, LAC, and (below) with two level crossing signals, LC, superimposed
[Reproduced, with permission, from Tanaka, K., Someya, K. and Tanaka, T. (1991). *Chem. Phys.*, **152**, 229]

The level crossing, and anticrossing, features are a consequence of POF_3 being a near-spherical rotor. As a result, rotational levels with the same J but different K are very close in zero field. When a Stark field is applied those levels which cannot interact with each other can cross. Such a situation is illustrated in Figure 8.192(a) for the pairs of levels 12,0,7 and 12,0,6 and also 12,0,5 and 12,3,6, where the symbolism is J,K,M_J, in the ground state.

The two sub-Doppler level crossing signals are shown in the lower part of Figure 8.192(b), superimposed on the broader level anticrossing signal. The spectra were obtained with the polarization of the laser such as to restrict the selection rule to $\Delta M_J = 0$ in the upper spectrum and $\Delta M_J = 0, \pm 1$ in the lower spectrum.

8.2.25 Level anticrossing spectroscopy

Figure 8.188(c) illustrates what happens when two close-lying levels a' and b' approach each other when the magnetic field is increased but, because of interaction between them, they repel each other; there is an avoided crossing or anticrossing. In the absence of the interaction the levels would cross, as shown by a and b.

Because there is *always* some interaction, however small, between two levels which may be usefully described as 'crossing', all level crossings are really anticrossings but clearly the converse is not true.

The experimental detection of a level anticrossing depends, as in the Hanle effect and level crossings, on observing a change in the fluorescence signal, at the point of closest approach of the two levels, as the magnetic (or electric) field is tuned through the anticrossing. At this point the wave function of each state is a 50:50 mixture of the wave functions of a and b before interaction. There is an important difference, however, from a level crossing experiment in that the observation of an anticrossing signal does not require coherence in the excitation and detection processes. For this reason, the detection of a signal in the absence of coherence can be used to distinguish a level anticrossing from a level crossing signal.

A second difference from a level crossing experiment is that fluorescence must involve only one of the states a or b. This discrimination may be natural or by the conditions of detection. If a and b are equally populated and their radiative lifetimes are equal, the anticrossing signal disappears. On the other hand, if one of the states does not radiate ($\tau = \infty$) there is no anticrossing (or, for levels which cross, any crossing) signal.

The first observation of level anticrossing signals was in the fluorescence spectrum of ^{7}Li.[539,540] Hyperfine levels of the $2^2P_{\frac{3}{2}}$, $M_J = -\frac{3}{2}$ state show anticrossings with those of the $2^2P_{\frac{1}{2}}$, $M_J = -\frac{1}{2}$ state. Since $I = \frac{3}{2}$, $M_I = \frac{3}{2}, \frac{1}{2}, -\frac{1}{2}, -\frac{3}{2}$ giving values of M_F ($= M_J + M_I$, as in Figure 6.23) of 0, -1, -2, -3 for $M_J = -\frac{3}{2}$ and 1, 0, -1, -2 for $M_J = -\frac{1}{2}$. Anticrossings occur between pairs of levels with the same M_F but different M_I so that there are three anticrossings, with $M_F = 0, -1, -2$. As in the case of level crossings, the interpretation and measurement of anticrossings result in values for fine and hyperfine structure constants in excited electronic states.

The line width of an anticrossing decreases as $\partial\Delta/\partial B$ increases and τ_a or τ_b increases, as for level crossing [equation (8.76)], and also as the minimum level separation decreases.

Level anticrossing spectroscopy is a powerful tool for deriving accurate information about electronic states which cannot be observed directly. This is the case, for example, with the $a^4\Sigma^+$ state of CN.

The CN molecule is short-lived but very readily produced, for example in flames, arcs and in interstellar space. The ground state is $X^2\Sigma^+$ and several band systems have been observed in emission involving this as the lower state. In addition, several other systems appear in emission but all involve only doublet states; no quartet states have been observed by conventional spectroscopic techniques.

However, it turns out that the $a^4\Sigma^+$ state is close to the $B^2\Sigma^+$ state and there are many level anticrossings between fine structure components of both states.[541] Anticrossings result from the spin-forbidden nature of the interactions. The radiative lifetime of the $B^2\Sigma^+$ state is short compared to that of $a^4\Sigma^+$ because radiation from the latter to the $X^2\Sigma^+$ state is spin forbidden.

Figure 8.193 shows the Zeeman sub-levels of the $v = 11$, $N = 20$ level of the $B^2\Sigma^+$ state and of an $N = 20$ level of an unknown vibrational state of the $a^4\Sigma^+$ state. Also in the figure is a calculated anticrossing spectrum.

The experimental spectrum was obtained by monitoring the $B^2\Sigma^+$–$X^2\Sigma^+$ emission from CN produced by reaction of metastable Ar atoms with BrCN and polarized either parallel or perpendicular to the direction of the magnetic field. Only a contour of the many unresolvable anticrossings could be observed. From the contour, the spin–rotation coupling constant γ was found for the $a^4\Sigma^+$ state, for which it was shown to be zero, and also for the $B^2\Sigma^+$ state. In addition, the spin–spin coupling constant γ for the $a^4\Sigma^+$ state, the separation of the rotational levels at zero field, and the width at half intensity of the anticrossing signals were all determined.

Observations of anticrossing signals have made a very important contribution to our knowledge of the triplet states of H_2. As in the He atom, spin–orbit coupling is so small in H_2 that no triplet–singlet transitions have been observed, only singlet–singlet and triplet–triplet. This makes it difficult to relate, in energy terms, the singlet and triplet state manifolds although energy separations within each are mostly well known from conventional spectroscopy.

If rotational levels of a singlet and a nearby triplet electronic state can be made to approach each other by introducing a magnetic field, anticrossings may be observed. Figure 8.194 shows a simple anticrossing involving levels of the $B'^1\Sigma_u^+$ and $f^3\Sigma_u^+$ electronic states of H_2. The B' state corresponds to the electron configuration $(\sigma_g 1s)^1(\sigma_u{*}2s)^1$ and the f state to the configuration $(\sigma_g 1s)^1(\sigma_u{*}2p)^1$, using the separated atom orbital nomenclature (see Figure 6.29).

The $v = 0$, $N = 0$ level of the f state and the $v = 3$, $N = 0$ level of the B' state are only 4.277 cm^{-1} apart at zero magnetic field. When

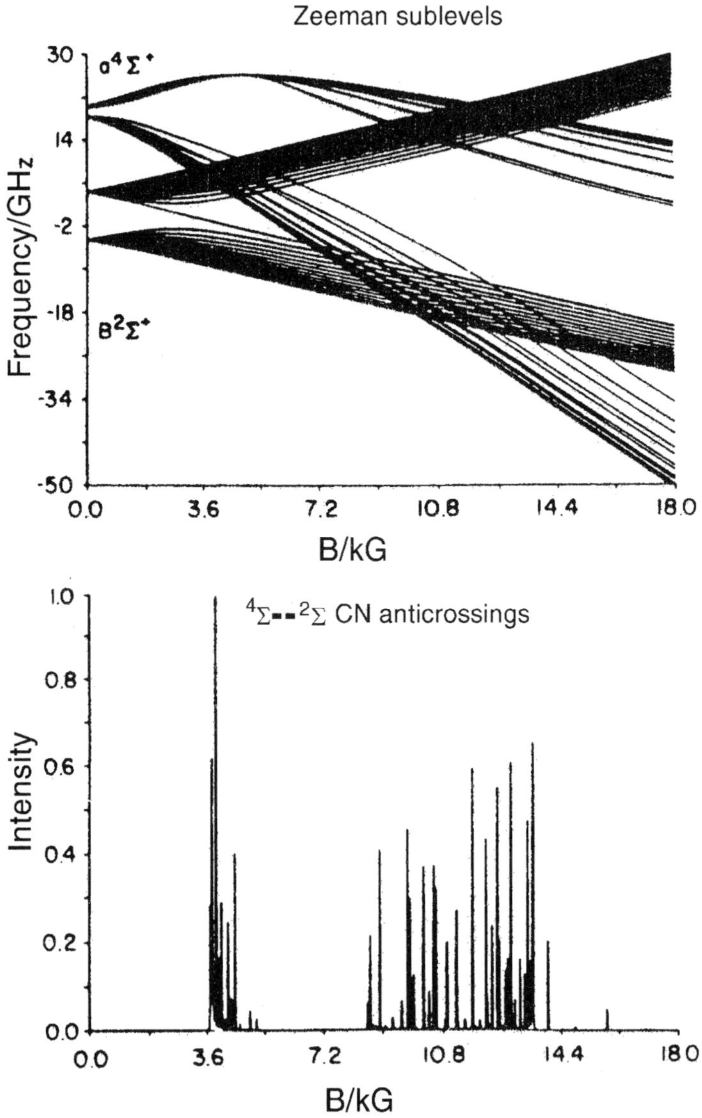

Figure 8.193 Zeeman sub-levels of the $v = 11$, $N = 20$ level of the $B^2\Sigma^+$ and of an $N = 20$ level of an unknown vibrational state of the $a^4\Sigma^+$ state of the CN molecule together with the calculated level anticrossing spectrum ($1G = 10^{-4}T$)
[Reproduced, with permission, from Cook, J. M., Zegarski, B. R. and Miller, T. A. (1979). *J. Chem. Phys.*, **70**, 3739]

a magnetic field is introduced the $M_J = 1$ component of the level of the f state would, if it were allowed, cross the single component of the level of the B state. In fact, as shown in the inset to Figure 8.194, the $M_J = 1$ component of the level of the f state is split into three sub-levels due to the nuclear spin I. Since, for both nuclei together, $I = 1$ it follows that $M_I = 1, 0, -1$ and $M_F (= M_I + M_J) = 2, 1, 0$. As the levels concerned are *ortho* levels the

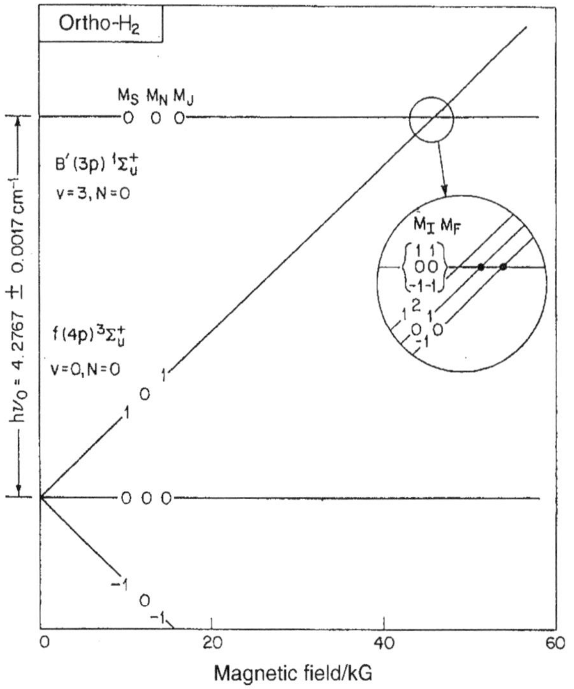

Figure 8.194 Anticrossings of Zeeman sub-levels of the $v = 0$, $N = 0$ level of the f state and the $v = 3$, $N = 0$ level of the B′ state of the H_2 molecule ($1 G = 10^{-4}$ T)
[Reproduced, with permission, from Miller, T. A. and Freund, R. S. (1976), *J. Mol. Spectrosc.*, **63**, 193]

three components of the B′ level, with $M_I = M_F = 1, 0, -1$, are not split. Because the anticrossings obey the selection rule[542] $\Delta M_F = 0$, there are only two of them.

Experimentally, the anticrossings were observed as an increase in fluorescence intensity in the 0–0, $P(1)$ line of the $f^3\Sigma_u^+ – a^3\Sigma_g^+$ transition and a decrease of fluorescence intensity in the 3–1, $P(1)$ line of the $B'^1\Sigma_u^+ – E^1\Sigma_g^+$ transition.

The observation of these anticrossings results in an accurate value of the zero-field splitting, given in Figure 8.194, of the levels of the singlet and triplet states, the Fermi contact interaction in the f state and the difference in the orbital angular momentum g values for the two electronic states.

Anticrossings, created by the presence of an electric field, have been observed[538] in the 4_0^1 laser Stark vibrational spectrum of POF_3, in addition to the level crossings discussed in Section 8.2.24. Figure 8.192 shows one such example.

In Figure 8.192(a) the anticrossing between the interacting rotational levels 12,0,6 and 12,3,6, where the symbolism is J,K,M_J, is shown. The interaction is a consequence of both levels having the same value of M_J. The upper spectrum in Figure 8.192(b) shows the resulting level anticrossing observed through the ${}^rP_0(12)$ $M_J = 6–6$ line of the 4_0^1 transition. The first differential form shows that the sign of the level anticrossing signal is opposite to that of the level crossing signals in the lower spectrum. Although both types of signal are sub-Doppler, the level anticrossing signal is clearly much broader due to the interaction width, this being the energy separation of the interacting levels.

Information derived from assignments of level crossing and anticrossing signals, together with infrared–microwave double resonance spectra of POF_3, includes an accurate value of 1.868 513(14) D for the dipole moment in the zero-point level of the ground electronic state.[538]

8.2.26 Quantum beat spectroscopy

Interference effects in the radiation emitted from two close-lying levels following coherent excitation from the same initial level was discussed in the context of level crossing spectroscopy in Section 8.2.24.

Apart from the effect on fluorescence anisotropy in a level crossing experiment, interference can be made use of more directly in two other types of experiment. In the first type,[543] the continuously emitted exciting radiation is modulated. This results in a corresponding modulation of the fluorescence signal. Changes in the amplitude of this modulation occur when the modulation frequency corresponds to the interval between two close-lying energy levels, such as fine or hyperfine Zeeman components separated in a magnetic field.

The second kind of experiment is more direct and employs pulsed exciting radiation.[544,545] with a pulse length which is short compared with the radiative lifetime of the states involved. The result is that the fluorescence is modulated at a frequency corresponding to the separation of the two states which are excited coherently and the damping of the modulation as a function of time gives the radiative lifetime of the excited states. This kind of experiment constitutes quantum beat spectroscopy, the beats resulting from the interference between the radiation emitted from two states excited coherently.

An ideal source for observation of quantum beats is a pulsed dye laser with a pulse length of the order of 10 ps, in order to resolve closely spaced beats, and a band width of a few cm^{-1}, in order to populate coherently states which lie within this range.

Figure 8.195(a) illustrates an early experiment in which the quantum beats were observed[546] involving nuclear hyperfine components in states of ^{133}Cs, for which $I = \frac{7}{2}$. The ground state, $6^2S_{\frac{1}{2}}$, has two hyperfine components with $F = 3, 4$ while the excited state, $7^2P_{\frac{3}{2}}$, has four components with $F = 2, 3, 4, 5$. The exciting radiation was a nitrogen-pumped dye laser with a line width of only about 1 GHz ($0.03\,cm^{-1}$), sufficiently small to isolate the groups of transitions a and b in Figure 8.195(a). The duration of the laser pulse was about 2 ns. Since

$$\tau^{-1} < (\nu_{54} + \nu_{43} + \nu_{32}) \tag{8.80}$$

where ν_{ij} is the separation of the $F = i$ and j components, the laser radiation prepares the ^{133}Cs atoms in a coherent superposition of $7^2P_{\frac{3}{2}}$ states with $F = 2, 3, 4, 5$.

The groups of transitions a and b obey the usual selection rule $\Delta F = 0, \pm 1$. The laser radiation was plane polarized and an analyser could be rotated so that the detected fluorescence I was polarized parallel (I_π) or perpendicular (I_σ) to the exciting radiation.

Figure 8.195(b) shows the quantum beat spectrum, recorded as $I_\pi - I_\sigma$, for the a group of transitions. The reciprocal of the damping rate of the beats gives a radiative lifetime of about 135 ns for the $7^2P_{\frac{3}{2}}$ state.

Early quantum beat experiments with diatomic molecules involved fluorescence from individual rotational levels, split in a magnetic field by the Zeeman effect, in the $B^3\Pi_{0^+u}$ state of I_2[547] and the $B^3\Sigma_u^+$ state of Se_2.[548] From these observations, accurate values of the Landé factors determining the splittings were obtained.

Quantum beats were observed for the first time in a polyatomic molecule in the fluorescence of biacetyl[549] ($CH_3COCOCH_3$) cooled to a rotational temperature of 1 K in a supersonic jet. Fluorescence from single rotational levels of the zero-point level of the $\tilde{A}^1A_u$ excited electronic state was excited by a pulsed dye laser and the decay of the fluorescence monitored as a function of time. Beats observed in

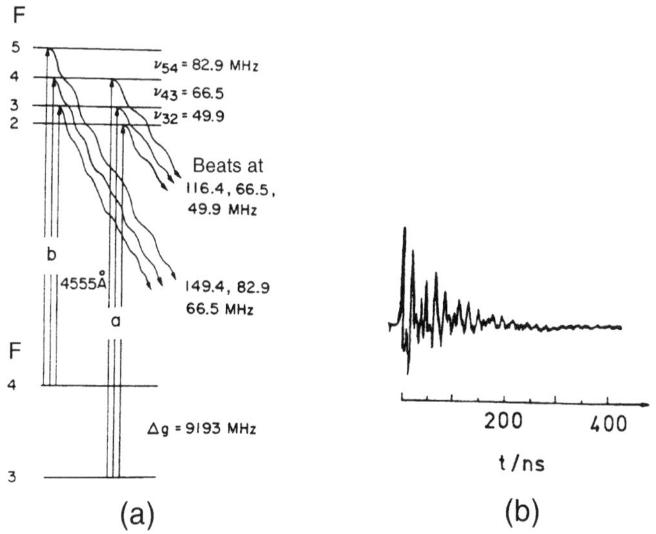

Figure 8.195 (a) Formation of quantum beats involving nuclear hyperfine components of the $6^2S_{\frac{1}{2}}$ ground state and the $7^2P_{\frac{3}{2}}$ excited state of ^{133}Cs. (b) The quantum beat spectrum following excitation from the $F = 3$ component of the ground state to three components of the excited state
[Reproduced, with permission, from Haroche, S., Paisner, J. A. and Schawlow, A. L. (1973). *Phys. Rev. Lett.*, **30**, 948]

the decay patterns are due to several (up to 12) rovibrational levels of the lower lying $\tilde{a}^3B_g$ state which are sufficiently close to be coupled to the rotational levels of the $\tilde{A}^1A_u$ state and, therefore, excited simultaneously.

For large molecules, quantum beat spectroscopy is a powerful technique for investigating intramolecular vibrational redistribution (IVR) in excited electronic states, discussed, in the context of the ground electronic state, in Section 8.2.9. A very well studied example is the $\tilde{A}$ state of anthracene, shown in Figure 8.196(b), through the observation of quantum beats[550] in the $\tilde{A}^1B_{1u}$–$\tilde{X}^1A_g$ fluorescence using a picosecond pulsed laser and the sample cooled in a supersonic jet.

The efficiency of IVR relating to a bright state depends on the number, N, of nearby bath states to which it is coupled. IVR has been usefully divided into three regimes,[551] a low energy regime with $N = 1$, an intermediate energy regime with $N = 2$–*ca* 10 and a high energy regime with $N > 10$. In the low energy regime, there is no IVR. In the $\tilde{A}$ state of anthracene this regime extends up to a vibrational energy of about 1200 cm^{-1}. For example, the 10^2_0 band, with 766 cm^{-1} of vibrational energy, shows no quantum beats and a fluorescence lifetime of 18 ns.

The 6^1_0 band, with 1380 cm^{-1} of vibrational energy, shows prominent quantum beats which can be seen in Figure 8.196(a). The 6^1 level lies in the intermediate energy regime in which there is restricted IVR in which the vibrational energy oscillates between a small number of coupled levels. The figure shows that the depth of the beat pattern varies if the fluorescence which is detected is not the total, undispersed fluorescence but only that to a particular vibrational level in the $\tilde{X}$ state. In this case, the depth is greatest with detection of fluorescence to the level in the $\tilde{X}$ state with vibrational energy of 1460 cm^{-1}. The phase of the beats may also vary with the detected fluorescence. This

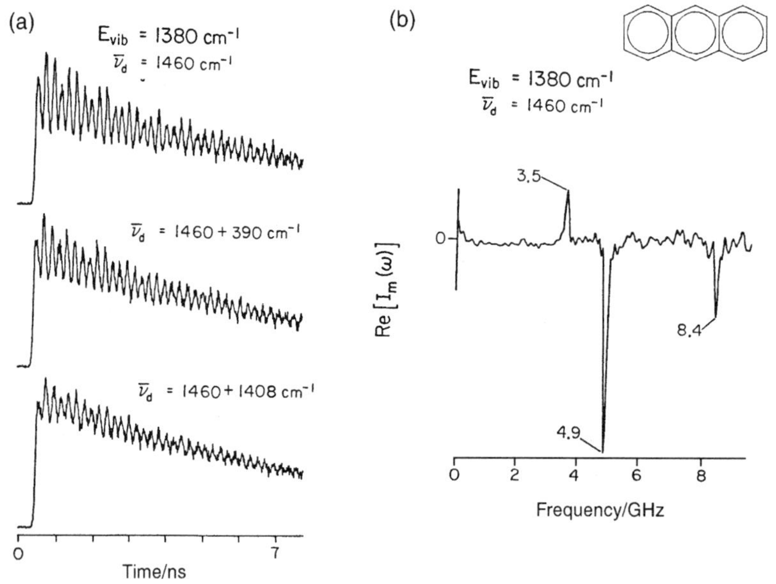

Figure 8.196 (a) Quantum beat pattern observed for the 6^1_0 band of anthracene with three different detection wavenumbers and (b) Fourier transform of the upper beat pattern in (a)
[Reproduced, with permission, from Felker, P. M. and Zewail, A. H. (1995). In *Jet Spectroscopy and Molecular Dynamics*, Hollas, J. M. and Phillips, D. (Eds), chap. 7. Blackie, Glasgow]

happens, for example, in the beat pattern for fluorescence from the 5^1 level, with a vibrational energy of $1420\,\mathrm{cm}^{-1}$. The phase of the beats is reversed for detection of fluorescence to vibrational levels in the $\tilde{X}$ state with energies of 390 and $1750\,\mathrm{cm}^{-1}$.

Fourier transformation of the beat pattern, which is in the time domain, gives the spectrum in the frequency domain. This is analogous to the Fourier transformation required in the Fourier transform infrared (FTIR) technique, discussed in Section 3.5.4. As in FTIR, the resolution increases with the extent of the beat pattern which can be observed. In the case of Fourier transformation of a quantum beat pattern, the resolution increases with the fluorescence lifetime.

The fluorescence excitation (FE) spectrum in the frequency domain could also be obtained directly with a high resolution laser as, for example, for the spectrum of pyrazine shown in Figure 8.109. A quantum beat spectrum has the disadvantage of the difficulty of obtaining absolute frequencies but the advantage of not requiring a high resolution laser.

Figure 8.196(b) shows the Fourier transform of the upper quantum beat pattern in Figure 8.196(a) indicating three coupled vibrational levels within a range of 4.9 GHz ($0.16\,\mathrm{cm}^{-1}$). The beat frequencies of about 5 GHz show that the time-scale of restricted IVR in this region is of the order of 200 ps.

In the region of restricted IVR, in which the 6^1 level lies, the density of vibrational states is estimated to be 25–40 per cm^{-1} However, those involved in coupling must have the same symmetry and, in the D_{2h} point group to which anthracene belongs, only about one in eight of these satisfies the symmetry restriction.

Fluorescence from vibrational levels in the high energy region of dissipative IVR shows a more complex and rapidly declining beat

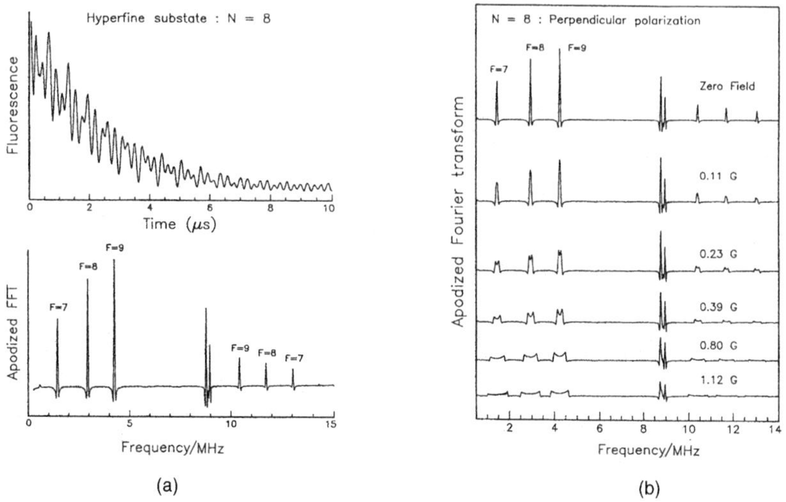

Figure 8.197 (a) Quantum beats and Fourier transformation following excitation into the 4_0^1 band of the Ã–X̃ system of cyanogen and (b) the effect of a magnetic field on the Fourier transformed signal [Reproduced, with permission, from Hemmi, N. and Cool, T. A. (1996). *J. Chem. Phys.*, **104**, 5721]

pattern. That obtained with excitation to a level with 1792 cm^{-1} of vibrational energy indicates a lifetime for IVR of about 20 ps.

Quantum beat patterns in the Ã–X̃ system of perylene (two naphthalene moieties joined at the 1- and 8-positions) show[552] qualitatively similar behaviour to those of anthracene. Absorptions in vibronic transitions with vibrational energies in the Ã state of 0, 353, 705 and 1057 cm^{-1} do not give quantum beats and lie in the low energy regime where there is no IVR. An exception in this regime is the fluorescence time profile with only 900 cm^{-1} of vibrational energy which does show quantum beats; this is due to a Fermi resonance with a nearby level, of the same symmetry, with 886 cm^{-1} of vibrational energy.

Fluorescence from a level with 1292 cm^{-1} of vibrational energy, which lies in the intermediate energy regime where there is restricted IVR, shows a pronounced beat pattern.

Propynal, HC≡CCHO, is a much smaller molecule, with C_s symmetry, and a quantum beat fluorescence profile can be obtained with excitation to a single rovibronic level in the $\tilde{A}^1A''$ electronic state. For example, with the molecule in a supersonic jet, time domain, undispersed fluorescence from a 0_{00} $(J_{K_aK_c})$ rotational level in the $\tilde{A}^1A''$ state, with a vibrational energy of 2945 cm^{-1}, shows pronounced quantum beats.[553] Fourier transformation shows that these result from two very close-lying levels only 610 kHz (0.000 02 cm^{-1}) apart. Further splitting into Zeeman sub-levels by a magnetic field shows that the coupling of the 0_{00} level is to a rotational level of the lower lying $\tilde{a}^3A''$ state.

Quantum beats have been observed[554] in the time domain $\tilde{A}^1\Sigma_u^-$–$\tilde{X}^1\Sigma_g^+$ fluorescence of cyanogen, N≡C—C≡N, in a supersonic jet. Figure 8.197(a) shows the quantum beat pattern observed following excitation into the N' (or J') = 8 rotational level via the 4_0^1 band of the Ã–X̃ system, where ν_4' has a wavenumber of 274 cm^{-1}.

The Fourier transform shows that there are eight transitions contributing to the beat pattern. Figure 8.197(b) shows Fourier transformed spectra in which a magnetic field has been applied. The fact that the two central lines

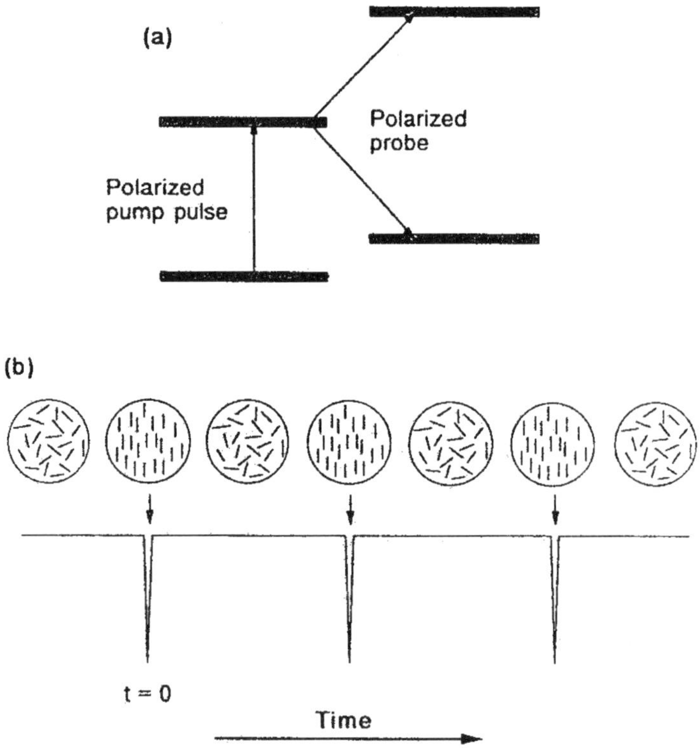

Figure 8.198 (a) Schematic diagram of a polarized pump–probe rotational coherence experiment in which the probe laser takes the molecule to a higher or lower energy manifold of rovibronic states. (b) At $t = 0$ the plane-polarized pump laser aligns those molecules whose transition moments have a large component in the polarization direction of the excitation field. Owing to their rotation, they dephase and then rephase after a time which is dependent on their rotational frequencies and, therefore, their rotational constants [Reproduced, with permission, from Felker, P. M. (1992). *J. Phys. Chem.*, **96**, 7844]

are not affected shows that the splitting of lines is due to interaction with another singlet state, probably a high-lying vibrational level of the ground electronic state. Each of the other six lines is split by the magnetic field, showing that their presence is due to spin–orbit interaction with levels of a nearby triplet state, $\tilde{a}^3\Sigma_u^+$. The splitting within each set of three lines, on either side of the central doublet, is due to nuclear hyperfine splitting. Spin–orbit coupling between hyperfine levels obeys the selection rule $\Delta F = 0$ and, when the hyperfine splitting is much smaller than the triplet splitting, $\Delta J = 0$. Therefore, only the $J = N$ triplet state is significantly coupled and quantum beats are observed for hyperfine levels with $F = N - 1$, N and $N + 1$, in this case 7, 8 and 9.

8.2.27 Rotational coherence spectroscopy

The principles of rotational coherence spectroscopy are illustrated in Figure 8.198. A plane polarized laser pulse, with a pulse width of about 30 ps and a band width of a few cm^{-1}, pumps the molecular sample at a rotational temperature of a few kelvin in a supersonic jet to a vibronic level of an excited electronic state. The technique is suited to molecules which are sufficiently large that the band width of the laser overlies many rotational transitions within the vibronic transition so that many rotational levels within the excited vibronic state are excited coherently. As shown in Figure 8.198(b), a polarized laser pulse at time $t = 0$ aligns those molecules whose transition moments have large components along the polarization direction of the excitation field. Because the molecules are rotating, this alignment lasts only for about 10 ps before dephasing takes place and the orientations are randomized. After a much longer period of time, owing to the quantized nature of the rotation, the molecules become aligned, or rephase, once more. This repetitive dephasing and rephasing can be observed for a period of time which depends on the fluorescence lifetime of the excited vibronic state.

Various methods of detection of this effect have been employed.[555] In the first reported use of rotational coherence spectroscopy[556,557] (RCS) the dephasing and rephasing were probed by simply monitoring the fluorescence from the excited vibronic level as a function of time with the plane of polarization of the detected fluorescence (a) parallel, (b) perpendicular and (c) at the magic angle of 54.7° to the plane of polarization of the pump radiation. The method of fluorescence detection was that of time-resolved single photon counting (TRSPC). The time resolution in such an experiment is typically 50 ps.

The jet-cooled molecule was *trans*-stilbene ($C_6H_5CH{=}CHC_6H_5$) and Figure 8.199 shows the rotational coherence (RC) signals superimposed on the fluorescence decay signals with these three relative orientations of the polarization planes. The RC signals disappear at the magic angle. In the lower part of the figure the RC signals are isolated by plotting $r(t)$ against time, where

$$r(t) = \frac{I_{\parallel}(t) - I_{\perp}(t)}{I_{\parallel}(t) + 2I_{\perp}(t)} \tag{8.81}$$

trans-Stilbene is a prolate near-symmetric rotor and the transition moment of the $\tilde{A}^1B_u$–$\tilde{X}^1A_g$ system lies in the molecular plane (the two phenyl groups are slightly tilted out-of-plane, but that need not concern us here). The 0_0^0 band is, therefore, a hybrid band but the type A component is dominant. In a prolate symmetric rotor recurrences in the RC signals, or transients, appear at intervals of $1/2B$ when the transition moment is parallel to the a-axis and, in a prolate near-symmetric rotor, they appear at intervals of $1/(B + C)$. The figure shows that the transients appear with alternating polarity. The lower RC spectra in the figure were calculated with rotational constants which give the best agreement with the observed. The assumed rotational temperature was 5 K. There is very little difference between the spectra calculated assuming the molecule is a symmetric or near-symmetric rotor.

The rotational constants $A = 0.089\,27\,\text{cm}^{-1}$, $B = 0.008\,73\,\text{cm}^{-1}$ and $C = 0.008\,33\,\text{cm}^{-1}$ used in Figure 8.199 to simulate the observed RC recurrences are those of the $\tilde{A}$ state. The fact that they are not ground state constants, which would be preferred for structural determination, does not matter very much because their accuracy is not sufficient for the typical differences between rotational constants in the $\tilde{A}$ and $\tilde{X}$ states to be very significant, particularly when rotational levels with only small quantum numbers are populated in the supersonic jet.

The technique of RCS is one in which Doppler broadening has no effect because it is the differences between transition energies that are determined. However, the accuracy with which the rotational constants can be obtained does not approach that of sub-Doppler fluorescence excitation (see Section 8.2.16). On the other hand, a great advantage of RCS is that the spectra are very much less crowded. Consequently, the method can be applied to considerably larger molecules and complexes.

Which rotational coherence signals (transients) appear in a particular spectrum depends on (a) whether the molecule is a prolate or oblate symmetric, or near-symmetric, rotor and (b) the directions of the transition moments for

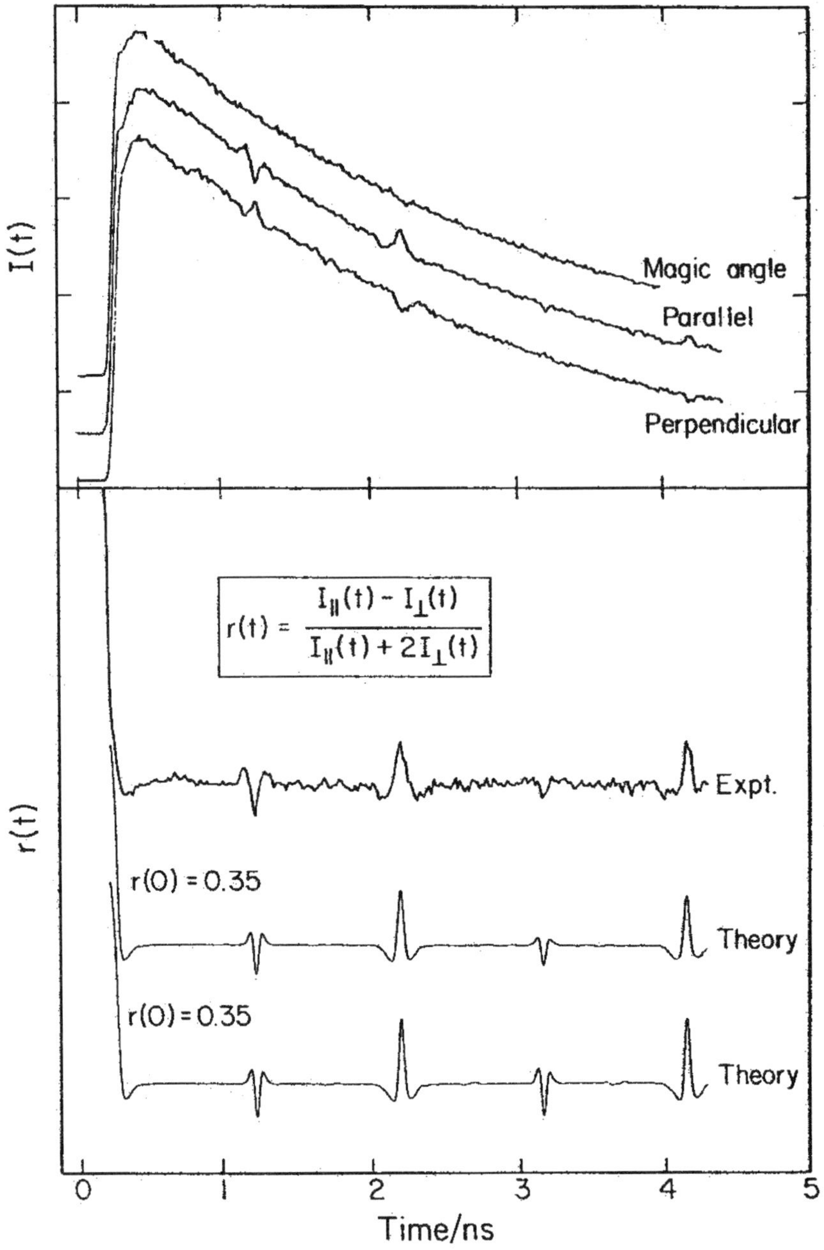

Figure 8.199 The upper part shows rotational coherence signals from the $\tilde{A}$–$\tilde{X}$, 0_0^0 band of *trans*-stilbene superimposed on the fluorescence decay with the planes of polarization of the pump radiation and the fluorescence parallel, perpendicular and at the magic angle to each other. The lower part shows the isolated experimental rotational coherence signals and those calculated for an asymmetric rotor (upper), with $A = 0.089\,27\,\text{cm}^{-1}$, $B = 0.008\,73\,\text{cm}^{-1}$, $C = 0.008\,33\,\text{cm}^{-1}$, and for a symmetric rotor (lower)
[Reproduced, with permission, from Felker, P. M., Baskin, J. S. and Zewail, A. H. (1986). *J. Phys. Chem.*, **90**, 724]

Table 8.8 Transients in rotational coherence spectra[a]

Transient	Position[b]	Polarity[c]	Contributing coherences[d]	Pump, probe dipole directions[e]
J-type	P: $t \approx n/2(B+C)$	Alternating	$\|\Delta J\| = 1,2$	All
	O: $t \approx n/2(A+B)$	$(-, +, -, \ldots)$	$\|\Delta K\| = 0$	
K-type	P: $t \approx n/(4A-2B-2C)$	Non-alternating	$\|\Delta J\| = 0$	$(\perp, \perp), (\perp, \mathrm{H})$
	O: $t \approx n/\|4C-2A-2B\|$	$(-, -, -, \ldots)$	$\|\Delta K\| = 2$	(H, H)
Hybrid	P: $t \approx n/(2A-B-C)$	Alternating	$\|\Delta J\| = 0$	(H, H)
	O: $t \approx n/\|2C-A-B\|$	$(-, +, -, \ldots)$	$\|\Delta K\| = 1$	
C-type	$t \approx n/4C$	Undefined	P: $\Delta J = 2$, $\Delta K = 0$	P: all
			O: $\Delta J = 2$, $\Delta K = 2$	O: $(\perp, \perp), (\perp, \mathrm{H})$
A-type	$t \approx n/4A$	Undefined	P: $\Delta J = 2$, $\Delta K = 2$	P: $(\perp, \perp), (\perp, \mathrm{H})$
			O: $\Delta J = 2$, $\Delta K = 0$	O: all

[a]P and O indicate a prolate or oblate symmetric, or near-symmetric, rotor, respectively. In a prolate symmetric rotor $B = C$ and, in an oblate symmetric rotor, $A = B$.
[b]n is an integer.
[c]Specific polarities given, e.g. $(-, +, -, \ldots)$, refer only to time-resolved fluorescence depletion experiments in which the pump and probe lasers have the same plane polarization.
[d]J and K refer to the symmetric rotor limit.
[e] $\|$, $\perp$ and H refer to parallel, perpendicular and hybrid transition moment directions in the symmetric rotor limit. The transition moment direction for the pump transition is given first and that for the probe transition second. Wherever $(\perp, \mathrm{H})$ is given, $(\mathrm{H}, \perp)$ also applies.

the pump and probe processes. These requirements have been derived[555,558,559] and are summarized in Table 8.8, which is taken from ref. 555.

The fact that C-type or A-type transients appear in the RC spectrum of an asymmetric rotor is a consequence of a prolate asymmetric rotor behaving like an oblate symmetric rotor when $J \gg K_a$ or an oblate asymmetric rotor behaving like a prolate symmetric rotor when $J \gg K_c$, as was discussed in Section 6.3.5.3(i)—see equations (6.280) and (6.281).

A second, and more frequently used, method of obtaining a RC spectrum employs a pump–probe method in which the probe beam is provided by the same pulsed laser as the pump beam and they are both of the same wavenumber. In Figure 8.198(a) this corresponds to the probe laser taking the molecule to a lower state which is the same as the initial ground state. The two beams come from the same picosecond laser via a 50% beam splitter, one of the beams being delayed by an optical delay line. The total fluorescence intensity is then monitored as the time delay between the two beams is smoothly varied from zero to such a time when the fluorescence intensity has declined to zero. Rotational coherences then result in fluorescence depletion and the technique is called time-resolved fluorescence depletion (TRFD).[560] The time resolution, typically about 25 ps, is better by about a factor of two than that for the method of TRSPC using direct fluorescence detection following excitation with the pump laser.

The TRFD method produces transients resulting from both ground and excited state coherences. If the rotational constants in these two states are sufficiently different the transients will be resolved but, if they are not, the transients will occur at positions reflecting the average of the ground and excited state rotational constants. The fact that the molecules are cooled to a rotational temperature of about 5 K means that any changes of rotational constants will be more difficult to detect because of the low values of the rotational quantum numbers which are involved.

4-Cyclohexylaniline, shown in Figure 8.200, is a prolate near-symmetric rotor with the a-axis almost parallel to the C—N bond. The $\tilde{A}$–$\tilde{X}$ transition is localized in the benzene ring and

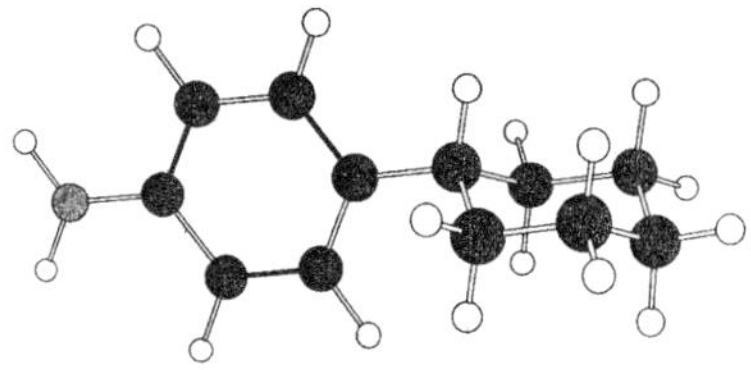

Figure 8.200 The lowest energy conformer of 4-cyclohexylaniline in which the cyclohexyl moiety is in the chair form and the substituent is in the equatorial position
[Reproduced, with permission, from Smith, P. G. and McDonald, J. D. (1990). *J. Chem. Phys.*, **92**, 1004]

the transition moment is polarized in a direction almost perpendicular to the *a*-axis, as in aniline itself.

Rotational coherences, detected by TFRD, in the 0_0^0 band of the $\tilde{A}$–$\tilde{X}$ system[561] of 4-cyclohexylaniline are shown in Figure 8.201. The top, middle and bottom scans were obtained with the two laser pulses plane-polarized with the plane of polarization at 90°, 54.7° (magic angle) and 0^0, respectively. The bottom scan shows *J*-type transients at 1383 ps $[1/(B+C)]$ and 696.5 ps $[1/2(B+C)]$, with opposite polarity, and a series, indicated by the lower set of lines shown in the figure, of 12 *K*-type transients with a spacing of 122.0 ps $[1/4(A-B)]$. From these transients the rotational constants $A = 0.0804$, $B = 0.0121$ and $C = 0.0121\ \mathrm{cm}^{-1}$ were obtained.

The second series of transients, indicated by the upper set of lines shown in the figure, has a spacing of 125.8 ps but the source was not identified.[561]

By using the time-resolved single photon counting (TRSPC) method, which gives transients related only to the excited state rotational

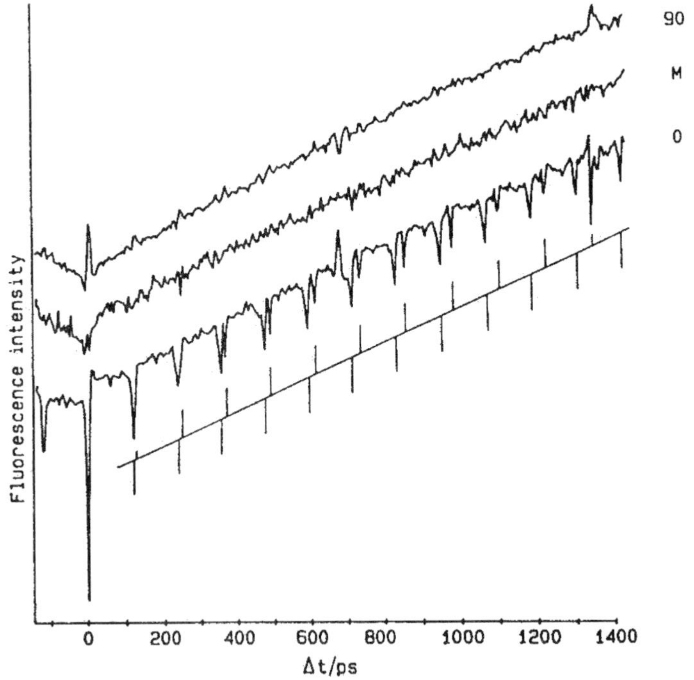

Figure 8.201 Rotational coherence transients in time-resolved fluorescence depletion scans of 4-cycolhexylaniline with the polarization planes of the pump and probe laser pulses at 90, 54.7 and 0° to each other
[Reproduced, with permission, from Smith, P. G. and McDonald, J. D. (1990). *J. Chem. Phys.*, **92**, 3991]

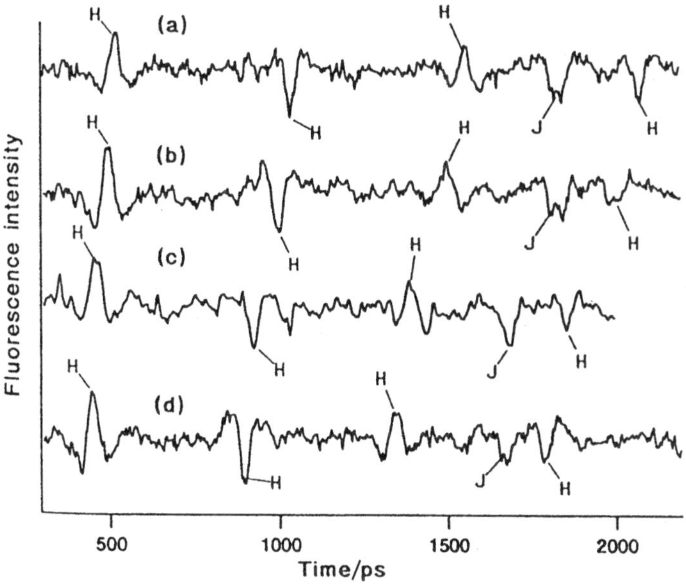

Figure 8.202 Rotational coherence transients in the spectra of four isotopomers of the phenol dimer: (a) $^2H_6$–$^2H_6$, (b) $^2H_5$–$^2H_5$, (c) $^2H_1$–$^2H_1$ and (d) ^2H_O–^2H_O
[Reproduced, with permission, from Connell, L. L., Ohline, S. M., Joireman, P. W., Corcoran, T. C. and Felker, P. M. (1992). *J. Chem. Phys.*, **96**, 2585]

constants, it has been shown[562] that the transients with a spacing of 122.0 ps in the TRFD scan do not appear in the TRSPC scan and are due to 4-cyclohexylaniline in the ground state. Therefore, those with an interval of 125.4 ps, in the later work, relate to the excited state. Combination of the TRSPC and TRFD results gives $A' = 0.0785 \pm 0.0005\,\text{cm}^{-1}$, $B' = 0.0122 \pm 0.0001\,\text{cm}^{-1}$ and $C' = 0.0120 \pm 0.0001\,\text{cm}^{-1}$ for the excited state and $A'' = 0.08033\,\text{cm}^{-1}$ and $B'' + C'' = 0.0242\,\text{cm}^{-1}$ for the ground state. These constants are consistent with the conformation in Figure 8.200.

A second example of a molecule whose conformation can be determined by RCS, but which is too large for high resolution, sub-Doppler fluorescence excitation spectroscopy in a skimmed supersonic jet, is the phenol dimer [$(C_5H_5OH)_2$].

Figure 8.202 shows the transients in the RC spectra[563] resulting from excitation in the 0^0_0 band of the $\tilde{A}$–$\tilde{X}$ system of four isotopomers of the phenol dimer formed in a supersonic jet and at a rotational temperature of about 5 K. The dimer is a prolate near-symmetric rotor and the spectra show hybrid (H) and *J*-type transients. As Table 8.8 shows, the hybrid transients show separations of $1/(2A - B - C)$ and have alternating polarities. Of the *J*-type transients, which should show alternating polarity, those with negative polarity are expected to be much stronger when TRFD detection is used. These are the only ones which are observed and show a spacing of $1/(B + C)$.

With approximate values of $2A - B - C$ and $B + C$ obtained from the positions of the transients RC spectra were calculated with various values of $B - C$ to determine the possible range of these values and also to determine the direction of the transition moment. Then a non-linear least-squares method was used to optimize the values of $2A - B - C$, $B + C$ and $B - C$, and hence the values of A, B and C, with the direction of the transition moment fixed and using the

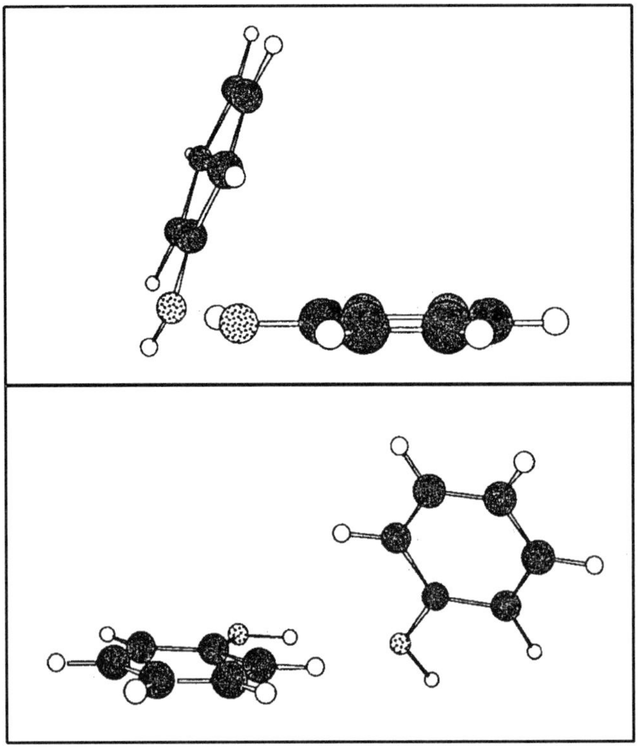

Figure 8.203 Two views of the phenol dimer
[Reproduced, with permission, from Connell, L. L., Ohline, S. M., Joireman, P. W., Corcoran, T. C. and Felker, P. M. (1992). *J. Chem. Phys.*, **96**, 2585]

additional information regarding the relative intensities of the transients. In the case of the phenol dimer the rotational constants are the same, within their experimental uncertainties, in both the ground and excited electronic states.

In order to obtain information about the geometry of the dimer, it was assumed that, as is usually the case, the structure of each of the monomer moieties is unchanged from the free molecule. Then it was found that the structure shown in Figure 8.203 is consistent with the rotational constants. The fact that the dimer is non-planar, in contrast to some earlier conclusions, is clear from the large negative inertial defect $\Delta I = I_c - I_a - I_b$, which is of the order of $-200\,\mathrm{u\,\AA^2}$ for each of the isotopomers. The O···O distance across the single hydrogen bond is 3.05 Å.

The positions of one or more argon atoms in relation to aromatic rings, in complexes of aromatic molecules with several condensed aromatic rings, are of particular interest. This problem has been tackled[564] by RCS for anthracene–$(\mathrm{Ar})_n$, with $n = 1$–3, and tetracene–Ar. The complexes were formed in a supersonic jet and the TRFD method of detection used with the pump laser tuned to the 0_0^0 band of the $\tilde{A}$–$\tilde{X}$ system.

In anthracene–Ar (anthracene is shown in Figure 8.196) the rotational constants $2A - B - C$, $B + C$ and $B - C$ were obtained from the pattern of J and K transients, shown in Figure 8.204(a), by the same procedure as that

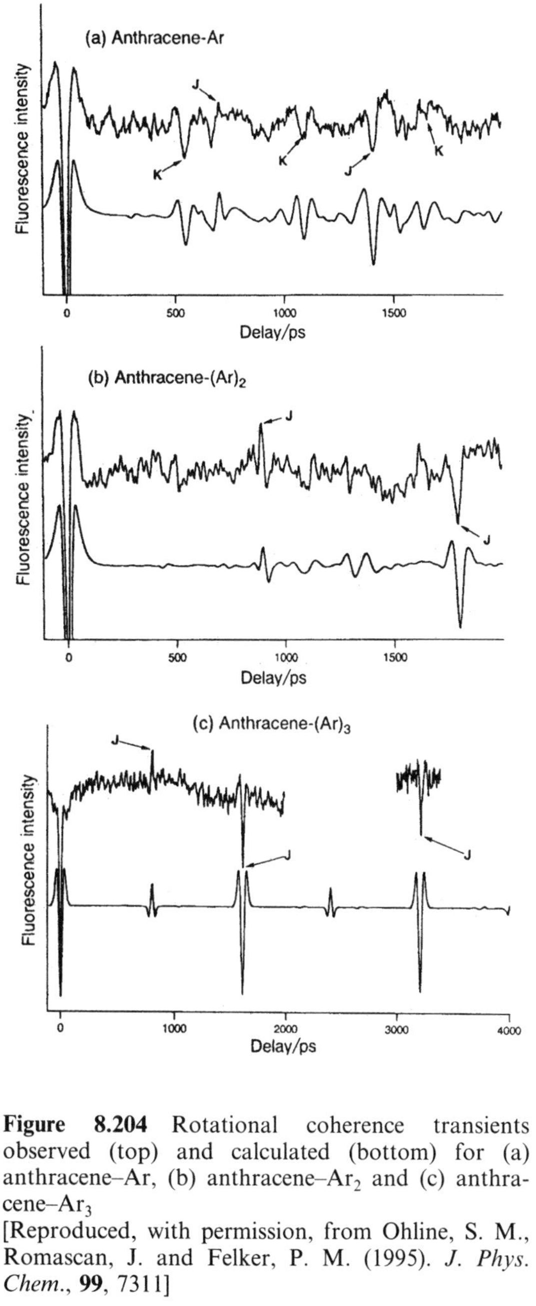

Figure 8.204 Rotational coherence transients observed (top) and calculated (bottom) for (a) anthracene–Ar, (b) anthracene–Ar_2 and (c) anthracene–Ar_3
[Reproduced, with permission, from Ohline, S. M., Romascan, J. and Felker, P. M. (1995). *J. Phys. Chem.*, **99**, 7311]

used for the phenol dimer.[563] The complex is a prolate asymmetric rotor with the argon atom sitting 3.43 ± 0.03 Å above the centre of the central aromatic ring.

The pattern of transients shown in Figure 8.204(b) for anthracene–Ar_2 shows only *J*-type transients. However, at this stage it was not known whether the complex is a prolate or an oblate asymmetric rotor. Therefore, the separation of the *J*-type transients could be $1/(B + C)$ or $1/(A + B)$. Detailed simulation of the transients, and those of the perdeuterated isotopomer, showed that it is a prolate asymmetric rotor with the argon atoms on the same side of the anthracene moiety 3.43 Å above each of the outer rings, not above the centres of the rings but 1.88 Å to either side of the centre of the central ring.

Figure 8.204(c) shows the *J*-type transients obtained for anthracene–Ar_3. Again, depending on whether the complex is a prolate or an oblate asymmetric rotor, their separations give a value for either $B + C$ or $A + B$. Simulation of the transient pattern shows that it is a prolate asymmetric rotor with two argon atoms on the same side of the anthracene moiety, with a similar geometry to that of anthracene–Ar_2, and the third on the opposite side above the centre of the central ring. All the argon atoms are at a similar vertical distance from the anthracene moiety to that in anthracene–Ar_2.

Simulation of the rotational coherence transients of tetracene–Ar (tetracene is a linear condensed aromatic ring system with one more ring than anthracene) has shown[564] that the argon atom is about 3.43 Å above the tetracene moiety and close to being above the centre of one of the inner rings.

Rotational depletion coherence spectroscopy (RDCS) is a modification of RCS and is a technique which allows the determination of only ground state rotational constants. These are determined from rotational coherences detected in a set of molecules which remain after substantial depletion of the original sample.

The first application of RDCS was to obtaining the structure of Ar_3^+ in the ground electronic state.[565]

Cold Ar_3^+ was generated by sputtering of solid argon with argon atoms having 8 keV of energy. The cold ions were irradiated with 20 ps pulses of 532 nm Nd^{3+}:YAG, frequency doubled, plane polarized laser radiation, with the electric field along the z-axis. This radiation is absorbed by the linear Ar_3^+ in an electronic transition with the transition moment along the internuclear axis, but the only ions which absorb are those whose internuclear axes are tilted towards the z-axis. The ion readily dissociates in the excited state and the laser intensity was adjusted so that about 50% of the ions absorb and dissociate. The experiment was conducted in a triple quadrupole mass spectrometer and the irradiation took place in the central quadrupole. The mass spectrometer subsequently contains only half the original parent ions; these are in the ground electronic state and have their transition moments tilted towards the xy-plane at time t_0. These ions continue to rotate but periodically realign their transition moments in the xy-plane. This rotational coherence occurs, in general for a linear molecule, at time intervals of $1/4B$, where B is the rotational constant.

The alignment of the parent ions at $t > t_0$ was probed by a second picosecond pulsed laser and the ions detected in the mass spectrometer. What was measured was the ratio of the number of ions $I_\perp$, dissociated by the second pulse when it was polarized along the y-axis, to the number $I_\parallel$ dissociated when it was polarized along the z-axis. Immediately after t_0 this ratio is large because ions with their transition moments tilted towards the z-axis are severely depleted. As the ions rotate the ratio decreases but increases again after a period of $1/8B$. This is half the time interval of $1/4B$ expected for a linear molecule because the argon atom has zero nuclear spin and alternate rotational energy levels are missing.

Figure 8.205 shows the recurrences, at 144, 295, 421, and 553 ps, in the RDC spectrum of $^{40}Ar_3^+$. The average interval gives $B = 0.030 \pm 0.001\ \mathrm{cm}^{-1}$ and an Ar–Ar bond length of 2.65 ± 0.03 Å. Similar results were obtained for $^{36}Ar_3^+$, selected by the mass spectrometer.

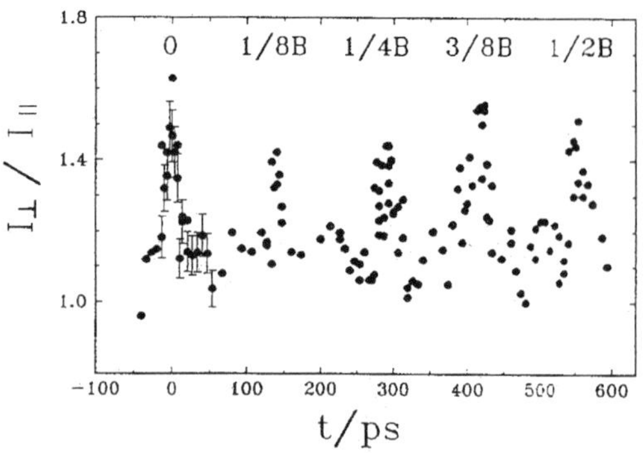

Figure 8.205 Rotational recurrences in the rotational depletion coherence spectrum of $^{40}Ar_3^+$ [Reproduced, with permission, from Magnera, T. F., Sammond, D. M. and Michl, J. (1993). *Chem. Phys. Lett.*, **211**, 378]

8.2.28 Hydrogen atom photofragment translational spectroscopy

The dynamics of photodissociation, the dissociation by a photon of a molecule ABC into photofragments AB and C, which may be atomic or molecular,

$$\mathrm{AB{-}C} \xrightarrow{h\nu} \mathrm{AB} + \mathrm{C} \tag{8.82}$$

have been investigated in a variety of ways[566–568] by analysis of the characteristic motions of the products. For a molecular (often diatomic) product AB this includes the directions of the velocity vectors relative to the electric vector of the laser photon, and therefore to the transition moment direction in AB, the distribution of the rotational $\boldsymbol{J}$ vectors of the fragments relative to the electric vector of the laser photon, and the mutual correlation between the velocity vectors and $\boldsymbol{J}$ vectors of the fragments.

Techniques which have been used for interrogating the internal energy (electronic, vibrational and rotational) include fluorescence excitation (FE) spectroscopy and resonance enhanced multiphoton ionization (REMPI) spectroscopy. However, many fragments are

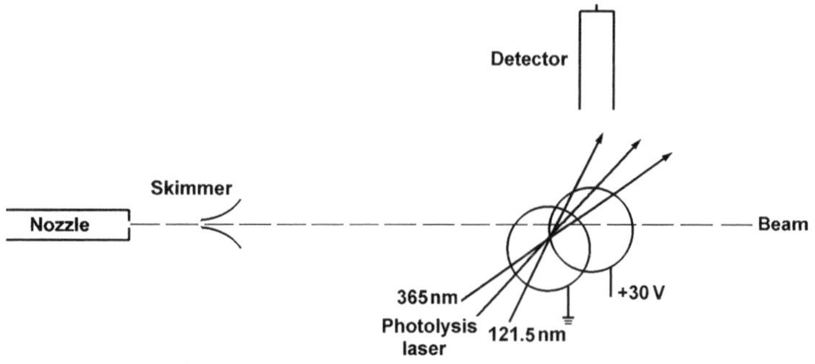

Figure 8.206 The experimental method of obtaining hydrogen atom photofragment translational spectra

not suited to these techniques. A more generally applicable technique is that of time-of-flight (TOF) spectroscopy, in which the translational energy of a mass-selected photofragment is measured from its time of flight between the region of photodissociation and the detector. The best resolution can be achieved when the photofragment is light, preferably a hydrogen (or deuterium) atom.

The internal energy of a photofragment AB resulting from photodissociation of the hydride ABH in the process

$$\text{AB—H} \xrightarrow{h\nu} \text{AB} + \text{H} \tag{8.83}$$

can be obtained by measuring the translational energy of the hydrogen atom. Since this has no internal energy, that of the molecular fragment, $E_{\text{int}}(\text{AB})$, is given by

$$E_{\text{int}}(\text{AB}) = h\nu - D_0(\text{AB—H}) - E_{\text{trans}} \tag{8.84}$$

where $h\nu$ is the photon energy, $D_0(\text{AB—H})$ is the dissociation energy of the B—H bond and E_{trans} is the total translational energy of the two fragments. The advantages of the hydrogen atom as a photofragment is that it carries most of the translational energy. For example, in the photodissociation of H_2O to give H + OH, conservation of translational energy requires that

$$E_{\text{trans}}(\text{total}) = \tfrac{18}{17} E_{\text{trans}}(\text{H}) \tag{8.85}$$

Photofragment translational spectroscopy, where C in equation (8.82) is a fragment other than a hydrogen or deuterium atom, is possible but the relative masses of AB and C should be such that C does not have too large a mass and carries most of the translational energy. The resolution and signal to noise ratio achieved by earlier TOF methods[568] was improved upon using an experimental method illustrated in Figure 8.206.

A skimmed supersonic jet, seeded with the sample, is crossed by three laser beams focused into a small volume of the molecular beam. One laser beam has a wavelength suitable for photolysis of the sample. A second laser beam, at 121.5 nm, a wavelength obtained by frequency tripling, in krypton gas, the output at 364.6 nm from an excimer pumped dye laser, excites the hydrogen atoms from the ground, $n = 1$, state to the excited $n = 2$ state.

In an early version of the experiment,[569] the third laser beam came from the same dye laser as the second beam. Absorption of a further 364.6 nm photon from the dye laser takes the hydrogen atom from the $n = 2$ level to the ionization threshold and the detector counts the hydrogen ions as a function of their time of flight from the photolysis volume. Higher resolution than had been obtained by previous methods resulted from a much better defined

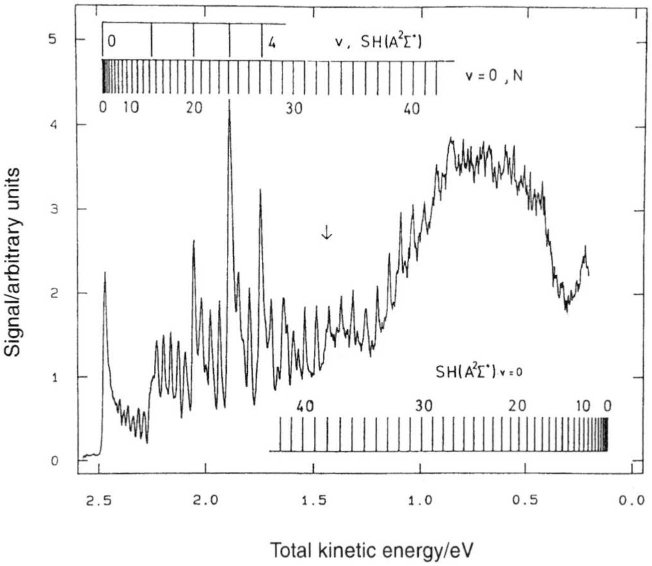

Figure 8.207 Total kinetic energy spectrum of the fragments resulting from photolysis of jet-cooled H_2S, in the $\tilde{B}^1A_1$ state, at 121.5 nm
[Reproduced, with permission, from Schneider, L., Meier, W., Welge, K. H., Ashfold, M. N. R. and Western, C. M. (1990). *J. Chem. Phys.*, **92**, 7027]

flight distance from the photolysis volume to the detector.

Unfortunately, coulombic repulsion between the ions occurs in the region of their production thereby disturbing the distribution of translational energies and limiting the resolution.

This problem has been overcome[570] by the third laser beam coming from a second dye laser tuned to a wavelength of 364.886 nm, slightly longer than in the earlier method, so that the hydrogen atoms are not ionized but excited to a high-lying state with $n = 90$, just below the ionization limit. Any ions which may be formed are removed by the electric field of about $2\,\mathrm{V\,mm^{-1}}$ applied to electrodes on either side of the photolysis region. The highly excited hydrogen atoms travel a distance of 419.8 mm to the detector, the lifetime of the $n = 90$ state being sufficiently long that they do not decay before reaching it. Immediately before detection the atoms in the $n = 90$ state are ionized by the application of a small field. In this way a kinetic energy resolving power, $E/\Delta E$, of about 330 was achieved. For example, at an energy of 2 eV, this corresponds to a resolution, ΔE, of 0.003 eV, or a $\Delta\tilde{\nu}$ of $50\,\mathrm{cm^{-1}}$. This resolution is not very high by general standards of high resolution spectroscopy but is much better, by an order of magnitude, than in earlier methods.[568]

Photodissociation of H_2S in the $\tilde{B}^1A_1$ electronic state might proceed by various energetically allowed channels:

$$
\begin{aligned}
H_2S &\rightarrow H + SH(X^2\Pi) &\quad (I)\\
&\rightarrow H + SH(A^2\Sigma^+) &\quad (II)\\
&\rightarrow H + H + S(^3P) &\quad (III)\\
&\rightarrow H + H + S(^1D) &\quad (IV)\\
&\rightarrow H_2 + S(^3P) &\quad (V)\\
&\rightarrow H_2 + S(^1D) &\quad (VI)\\
&\rightarrow H_2 + S(^1S) &\quad (VII)
\end{aligned}
\qquad (8.86)
$$

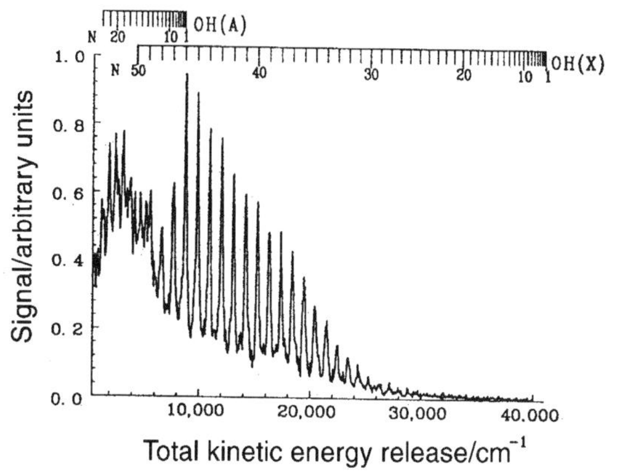

Figure 8.208 Total kinetic energy spectrum of the fragments resulting from photolysis of jet-cooled H_2O, in the $\tilde{B}^1A_1$ state, at 121.5 nm
[Reproduced, with permission, from Mordaunt, D. H., Ashfold, M. N. R. and Dixon, R. N. (1994). *J. Chem. Phys.*, **100**, 7360]

and hydrogen atom photofragment translational spectroscopy gives information on the branching ratios and the internal energy distribution in the SH fragment.[570] In this case the information obtained by interrogation of SH using the alternative method of fluorescence excitation was restricted owing to predissociation in the $A^2\Sigma^+$ state.

Figure 8.207 shows the total kinetic energy spectrum resulting from photolysis of jet-cooled H_2S with 121.5 nm radiation which excites it to the $\tilde{B}^1A_1$ state. The sharp onset at about 2.46 eV and the subsequent separations of rotational energy levels, which are characteristic of the $v = 0$–4 vibrational levels of the $A^2\Sigma^+$ state of SH, show that channel II in equation (8.83) is very effective with $v = 0$ dominant. Vibrationally excited SH($A^2\Pi$) tends to be produced in the lower rotational states. There is no evidence for any SH($X^2\Pi$) fragment which would travel faster and appear at lower kinetic energy: any fragmentation by channel I is negligible.

Observation of previously unobserved rotational levels of SH in the $A^1\Sigma^+$ state allowed an improvement of the potential energy function in this state giving a well depth of $9280 \pm 600\,\text{cm}^{-1}$ and, consequently, a bond dissociation energy of 3.71 ± 0.07 eV in the $X^2\Pi$ state.

Hydrogen atoms are also produced as a result of predissociation of the SH($A^2\Sigma^+$) produced by channel II. In principle, these should produce a spectrum which is a mirror image of that due to channel II. The position of this mirror image spectrum is indicated by the comb at the bottom of Figure 8.207 but this spectrum is blurred and contributes only to the underlying continuum.

The broad maximum at about 0.7 eV is attributed to fragmentation by the channel IV three-body mechanism.

Photofragmentation in the analogous $\tilde{B}^1A_1$ state of H_2O, following excitation at 121.5 nm, may proceed by the following channels:

$$\begin{aligned} H_2O &\rightarrow H + OH(X^2\Pi) && \text{(I)} \\ &\rightarrow H + OH(A^2\Sigma^+) && \text{(II)} \\ &\rightarrow H + H + O(^3P) && \text{(III)} \\ &\rightarrow H_2 + O(^1D) && \text{(IV)} \end{aligned} \qquad (8.87)$$

The total kinetic energy spectrum[571] in Figure 8.208 can provide information on channels I–III. The spectrum is dominated by OH($X^2\Pi$) produced by channel I, unlike the corresponding photofragmentation of H_2S, with high values (mostly >30) of the rotational quantum number N but with $v = 0$. These fragments are probably produced as a result of radiationless transitions from the $\tilde{B}$ to the $\tilde{X}$ state of H_2S.

Some OH fragments are also produced in the $A^2\Sigma^+$ state, with $v = 0$, by fragmentation in the $\tilde{B}$ state of H_2O.

Simulation of the spectrum in Figure 8.208 indicates that the branching ratios for channels I, II, III and IV are 0.64, 0.14, 0.11 and 0.11, respectively.

Photolysis of acetylene (H—C≡C—H) by 211.53 nm laser radiation takes place in the $\tilde{A}^1A_u$ *trans* bent excited state and produces, predominantly, the photofragments $H + C_2H$. The total kinetic energy spectrum of the fragments, is shown in Figure 8.209 and was

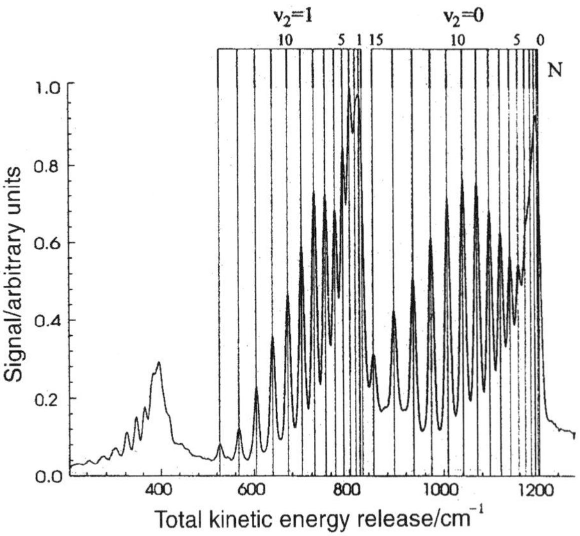

Figure 8.209 Total kinetic energy spectrum of the fragments resulting from photolysis of jet-cooled acetylene, in the $\tilde{A}^1A_u$ state, at 211.53 nm. The values of the v_2, where ν_2 is the bending vibration, and N quantum numbers refer to C_2H in the $X^2\Sigma^+$ state
[Reproduced, with permission, from Mordaunt, D. H. and Ashfold, M. N. R. (1994). *J. Chem. Phys.*, **101**, 2630]

obtained[572] by measuring the translational energy of the hydrogen atoms.

The spectrum is dominated by rotational levels of C_2H with $v_2 = 0$ and 1 in the ground electronic state $\tilde{X}^2\Sigma^+$, where ν_2 is the bending vibration. That they belong to the $\tilde{X}$ and not the $\tilde{A}$ state is confirmed by comparison of these levels with those previously determined for the $\tilde{X}$ state.[573,574] It is likely that the mechanism for fragmentation to ground state C_2H involves predissociation of the $\tilde{A}^1A_u$ state of acetylene to nearly resonant levels of a lower lying triplet state T_1.

The spectrum in Figure 8.209 shows that there is a pronounced peak in the rotational energy distribution in the $\tilde{X}$ state of C_2H at about $n = 10$ and 8 for $v_2 = 0$ and 1, respectively.

BIBLIOGRAPHY

Lasers

BEESLEY, M. J. (1976). *Lasers and their Applications*. Taylor and Francis, London

LENGYEL, B. A. (1971). *Lasers*, 2nd edn. Wiley-Interscience, New York

SIEGMAN, A. E. (1971). *An Introduction to Lasers and Masers*. McGraw-Hill, New York

SVELTO, O. (1976). *Principles of Lasers* (transl. by D. C. Hanna). Heyden, London

Laser spectroscopy

ALVES, A. C. P., BROWN, J. M. and HOLLAS, J. M. (Eds) (1988). *Frontiers of Laser Spectroscopy of Gases*. Kluwer, Dordrecht

ANDREWS, D. L. (Ed.) (1992). *Applied Laser Spectroscopy*. VCH, Weinheim

ANDREWS, D. L. (1997). *Lasers in Chemistry*. Springer, Berlin

AMBARTZUMIAN, R. V. and LETOKHOV, V. S. (1977). 'Selective dissociation of polyatomic molecules by intense infrared laser fields', *Acc. Chem. Res.*, **10**, 61

ANDERSON, H. C. and HUDSON, B. S. (1978). 'Coherent anti-Stokes Raman scattering', chap. 4 in *Molecular Spectroscopy*, vol. 5, Specialist Periodical Report. Chemical Society, London

ASHFOLD, M. N. R. and BAGGOTT, J. E. (Eds) (1987). *Molecular Photodissociation*. Royal Society of Chemistry, London

BALIAN, R., HAROCHE, S. and LIBERMAN, S. (Eds) (1977). *Frontiers in Laser Spectroscopy*, vols 1–2. North-Holland, Amsterdam

BEHRINGER, J. (1975). 'Experimental resonance Raman spectroscopy', chap. 3 in *Molecular Spectroscopy*, vol. 3, Specialist Periodical Report. Chemical Society, London.

BLOEMBERGEN, N. and YABLONOVITCH, E. (1978). 'Infrared laser induced unimolecular reactions'. *Phys. Today*, p 23 (May)

BREWER, R. G. (1972). 'Nonlinear spectroscopy', *Science, NY*, **178**, 247

BREWER, R. G. and MOORADIAN, A. (Eds) (1974). *Laser Spectroscopy*. Plenum, New York

CORNEY, A. (1979). *Atomic and Laser Spectroscopy*. Oxford University Press, Oxford

DEMTRÖDER, W. (1981). *Laser Spectroscopy*. Springer, Berlin

DODD, J. N. and SERIES, G. W. (1978). 'Time-resolved fluorescence spectroscopy', chap. 14 in *Progress in Atomic Spectroscopy*, pt. A. Plenum, New York

ERMAN, P. (1979). 'Time resolved spectroscopy of small molecules', chap. 5 in *Molecular Spectroscopy*, vol. 6, Specialist Periodical Report. Chemical Society, London

FAUSTO, R. (Ed.) (1993). *Recent Experimental and Computational Advances in Molecular Spectroscopy*. Kluwer, Dordrecht

FAUSTO, R. (Ed.) (1996). *Low Temperature Molecular Spectroscopy*. Kluwer, Dordrecht

FELD, M. S., JAVAN, A. and KURNIT, N. A. (Eds) (1973). 'Fundamental and applied laser physics', in *Proc. Esfahan Symp.* (Aug. 29 to Sept. 5, 1971). Wiley-Interscience, New York

FELD, M. S. and LETOKHOV, V. S. (1973). 'Laser spectroscopy'. *Sci. Am.*, **229**, 69

FRENCH, M. J. and LONG, D. A. (1976). 'Non-linear Raman effects': pt 1, chap. 6 in *Molecular Spectroscopy*, vol. 4, Specialist Periodical Report. Chemical Society, London

GARETZ, B. A. and LOMBARDI, J. R. (Eds) (1986). *Advances in Laser Spectroscopy*, vol. 3. Wiley, New York

GRYNBERG, G. and CAGNAC, B. (1977). 'Doppler-free multiphotonic spectroscopy', *Rep. Prog. Phys.*, **40**, 791

HALL, J. L. and CARLSTEIN, J. L. (1977). *Laser Spectroscopy III*. Springer, Berlin

HAROCHE, S., PEBAY-PEYRONLA, J. C., HÄNSCH, T. W. and Harris, S. E. (Eds) (1975). *Laser Spectroscopy II*. Springer, Berlin

HOLLAS, J. M. and PHILLIPS, D. (Eds) (1975). *Jet Spectroscopy and Molecular Dynamics*. Blackie, Glasgow

KAMPA, K. L. and SMITH, S. D. (Eds) (1979). *Laser-Induced Process in Molecules*. Springer, Berlin

KIMEL, S. and SPEISER, S. (1977). 'Lasers in chemistry', *Chem. Rev.*, **77**, 437

KLIGER, D. S. (Ed.) (1983). *Ultrasensitive Laser Spectroscopy*. Academic Press, New York

LANGE, W., LUTHER, J. and STEUDEL, A. (1974). 'Dye lasers in atomic spectroscopy', chap. 4 in *Advances in Atomic and Molecular Physics*, vol. 10. Academic Press, New York

LETOKHOV, V. S. (1975). 'Nonlinear high resolution laser spectroscopy', *Science, NY*, **190**, 344

LETOKHOV, V. S. (1979). 'Laser isotope separation', *Nature (London)*, **277**, 605

McCLAIN, W. M. and HARRIS, R. A. (1977). 'Two-photon molecular spectroscopy in liquids and gases', chap. 1 in *Excited States*, vol. 3. Academic Press, New York

MOORE, C. B. (Ed.) (1974–79). *Chemical and Biochemical Applications of Lasers*, vols. 1–4. Academic Press, New York

POWIS, I., BAER, T. and NG, C.-Y. (Eds) (1995). *High Resolution Laser Photoionization and Photoelectron Studies*. Wiley, Chichester

RHODES, C. K. (Ed.) (1979). 'Excimer lasers', in *Topics in Applied Physics*, vol. 30. Springer, Berlin

SCHÄFER, F. P. (Ed.) (1973). 'Dye lasers', in *Topics in Applied Physics*, vol. 1. Springer, Berlin

SHIMODA, K. (Ed.) (1976). 'High resolution laser spectroscopy', in *Topics in Applied Physics*, vol. 13. Springer, Berlin

SMITH, S. D. (1978). 'Infrared spectroscopy with the spin–flip laser', chap. 5 in *Molecular Spectroscopy*, vol. 5, Specialist Periodical Report. Chemical Society, London

STEINFELD, J. I. (Ed.) (1978). *Laser and Coherence Spectroscopy*. Plenum, New York

WALTHER, H. (Ed.) (1976). 'Laser spectroscopy of atoms and molecules', in *Topics in Applied Physics*, vol. 2. Springer, Berlin

WALTHER, H. and ROTHE, K. W. (Eds) (1979). *Laser Spectroscopy IV*. Springer, Berlin

WEST, M. A. (Ed.) (1977). *Lasers in Chemistry*. Elsevier, Amsterdam

ZARE, R. N. (1977). 'Laser separation of isotopes', *Sci. Am.*, **236**, 86

APPENDIX: CHARACTER TABLES

INDEX TO CHARACTER TABLES

Table No.	Point group	Table No.	Point group
A.1	C_s	A.24	C_{4h}
A.2	C_i	A.25	C_{5h}
A.3	C_1	A.26	C_{6h}
A.4	C_2	A.27	D_{2d}
A.5	C_3	A.28	D_{3d}
A.6	C_4	A.29	D_{4d}
A.7	C_5	A.30	D_{5d}
A.8	C_6	A.31	D_{6d}
A.9	C_7	A.32	D_{2h}
A.10	C_8	A.33	D_{3h}
A.11	C_{2v}	A.34	D_{4h}
A.12	C_{3v}	A.35	D_{5h}
A.13	C_{4v}	A.36	D_{6h}
A.14	C_{5v}	A.37	$D_{\infty h}$
A.15	C_{6v}	A.38	S_4
A.16	$C_{\infty v}$	A.39	S_6
A.17	D_2	A.40	S_8
A.18	D_3	A.41	T_d
A.19	D_4	A.42	T
A.20	D_5	A.43	O_h
A.21	D_6	A.44	O
A.22	C_{2h}	A.45	K_h
A.23	C_{3h}	A.46	I_h

CHARACTER TABLES

Table A.1

C_s	I	σ		
A'	1	1	T_x, T_y, R_z	$\alpha_{xx}, \alpha_{yy}, \alpha_{zz}, \alpha_{xy}$
A''	1	−1	T_z, R_x, R_y	α_{yz}, α_{xz}

Table A.2

C_i	I	i		
A_g	1	1	R_x, R_y, R_z	$\alpha_{xx}, \alpha_{yy}, \alpha_{zz}, \alpha_{xy}, \alpha_{xz}, \alpha_{yz}$
A_u	1	−1	T_x, T_y, T_z	

Table A.3

C_1	I	
A	1	All R, T, α

Table A.4

C_2	I	C_2		
A	1	1	T_z, R_z	α_{xx}, α_{yy}, α_{zz}, α_{xy}
B	1	−1	T_x, T_y, R_x, R_y	α_{yz}, α_{xz}

Table A.5

C_3	I	C_3	C_3^2		
A	1	1	1	T_z, R_z	$\alpha_{xx}+\alpha_{yy}$, α_{zz}
E	{1 1	ε ε^*	ε^* ε}	(T_x, T_y), (R_x, R_y)	$(\alpha_{xx}-\alpha_{yy}, \alpha_{xy})$ $(\alpha_{yz}, \alpha_{xz})$

$\varepsilon = \exp(2\pi i/3)$; $\varepsilon^* = \exp(-2\pi i/3)$.

Table A.6

C_4	I	C_4	C_2	C_4^3		
A	1	1	1	1	T_z, R_z	$\alpha_{xx}+\alpha_{yy}$, α_{zz}
B	1	−1	1	−1		$\alpha_{xx}-\alpha_{yy}$, α_{xy}
E	{1 1	i $-i$	−1 −1	$-i$ i}	(T_x, T_y), (R_x, R_y)	$(\alpha_{yz}, \alpha_{xz})$

Table A.7

C_5	I	C_5	C_5^2	C_5^3	C_5^4		
A	1	1	1	1	1	T_z, R_z	$\alpha_{xx}+\alpha_{yy}$, α_{zz}
E_1	{1 1	ε ε^*	ε^2 ε^{2*}	ε^{2*} ε^2	ε^* ε}	(T_x, T_y), (R_x, R_y)	$(\alpha_{yz}, \alpha_{yz})$
E_2	{1 1	ε^2 ε^{2*}	ε^* ε	ε ε^*	ε^{2*} ε^2}		$(\alpha_{xx}-\alpha_{yy}, \alpha_{xy})$

$\varepsilon = \exp(2\pi i/5)$; $\varepsilon^* = \exp(-2\pi i/5)$.

Table A.8

C_6	I	C_6	C_3	C_2	C_3^2	C_6^5		
A	1	1	1	1	1	1	T_z, R_z	$\alpha_{xx}+\alpha_{yy}, \alpha_{zz}$
B	1	−1	1	−1	1	−1		
E_1	1	ε	$-\varepsilon^*$	−1	$-\varepsilon$	ε^*	$(T_x, T_y), (R_x, R_y)$	$(\alpha_{xz}, \alpha_{yz})$
	1	ε^*	$-\varepsilon$	−1	$-\varepsilon^*$	ε		
E_2	1	$-\varepsilon^*$	$-\varepsilon$	1	$-\varepsilon^*$	$-\varepsilon$		$(\alpha_{xx}-\alpha_{yy}, \alpha_{xy})$
	1	$-\varepsilon$	$-\varepsilon^*$	1	$-\varepsilon$	$-\varepsilon^*$		

$\varepsilon = \exp(2\pi i/6)$; $\varepsilon^* = \exp(-2\pi i/6)$.

Table A.9

C_7	I	C_7	C_7^2	C_7^3	C_7^4	C_7^5	C_7^6		
A	1	1	1	1	1	1	1	T_z, R_z	$\alpha_{xx}+\alpha_{yy}, \alpha_{zz}$
E_1	1	ε	ε^2	ε^3	ε^{3*}	ε^{2*}	ε^*	$(T_x, T_y), (R_x, R_y)$	$(\alpha_{xz}, \alpha_{yz})$
	1	ε^*	ε^{2*}	ε^{3*}	ε^3	ε^2	ε		
E_2	1	ε^2	ε^{3*}	ε^*	ε	ε^3	ε^{2*}		$(\alpha_{xx}-\alpha_{yy}, \alpha_{xy})$
	1	ε^{2*}	ε^3	ε	ε^*	ε^{3*}	ε^2		
E_3	1	ε^3	ε^*	ε^2	ε^{2*}	ε	ε^{3*}		
	1	ε^{3*}	ε	ε^{2*}	ε^2	ε^*	ε^3		

$\varepsilon = \exp(2\pi i/7)$; $\varepsilon^* = \exp(-2\pi i/7)$.

Table A.10

C_8	I	C_8	C_4	C_8^3	C_2	C_8^5	C_4^3	C_8^7		
A	1	1	1	1	1	1	1	1	T_z, R_z	$\alpha_{xx}+\alpha_{yy}, \alpha_{zz}$
B	1	−1	1	−1	1	−1	1	−1		
E_1	1	ε	i	$-\varepsilon^*$	−1	$-\varepsilon$	$-i$	ε^*	$(T_x, T_y), (R_x, R_y)$	$(\alpha_{xz}, \alpha_{yz})$
	1	ε^*	$-i$	$-\varepsilon$	−1	$-\varepsilon^*$	i	ε		
E_2	1	i	−1	$-i$	1	i	−1	$-i$		$(\alpha_{xx}-\alpha_{yy}, \alpha_{xy})$
	1	$-i$	−1	i	1	$-i$	−1	i		
E_3	1	$-\varepsilon^*$	$-i$	ε	−1	ε^*	i	$-\varepsilon$		
	1	$-\varepsilon$	i	ε^*	−1	ε	$-i$	$-\varepsilon^*$		

$\varepsilon = \exp(2\pi i/8)$; $\varepsilon^* = \exp(-2\pi i/8)$.

Table A.11

C_{2v}	I	C_2	$\sigma_v(xz)$	$\sigma_v'(yz)$		
A_1	1	1	1	1	T_z	$\alpha_{xx}, \alpha_{yy}, \alpha_{zz}$
A_2	1	1	−1	−1	R_z	α_{xy}
B_1	1	−1	1	−1	T_x, R_y	α_{xz}
B_2	1	−1	−1	1	T_y, R_x	α_{yz}

Table A.12

C_{3v}	I	$2C_3$	$3\sigma_v$		
A_1	1	1	1	T_z	$\alpha_{xx}+\alpha_{yy}, \alpha_{zz}$
A_2	1	1	−1	R_z	
E	2	−1	0	$(T_x, T_y), (R_x, R_y)$	$(\alpha_{xx}-\alpha_{yy}, \alpha_{xy}), (\alpha_{xz}, \alpha_{yz})$

Table A.13

C_{4v}	I	$2C_4$	C_2	$2\sigma_v$	$2\sigma_d$		
A_1	1	1	1	1	1	T_z	$\alpha_{xx}+\alpha_{yy}, \alpha_{zz}$
A_2	1	1	1	−1	−1	R_z	
B_1	1	−1	1	1	−1		$\alpha_{xx}-\alpha_{yy}$
B_2	1	−1	1	−1	1		α_{xy}
E	2	0	−2	0	0	$(T_x, T_y), (R_x, R_y)$	$(\alpha_{xz}, \alpha_{yz})$

Table A.14

C_{5v}	I	$2C_5$	$2C_5{}^2$	$5\sigma_v$		
A_1	1	1	1	1	T_z	$\alpha_{xx}+\alpha_{yy}, \alpha_{zz}$
A_2	1	1	1	−1	R_z	
E_1	2	2 cos 72°	2 cos 144°	0	$(T_x, T_y), (R_x, R_y)$	$(\alpha_{xz}, \alpha_{yz})$
E_2	2	2 cos 144°	2 cos 72°	0		$(\alpha_{xx}-\alpha_{yy}, \alpha_{xy})$

Table A.15

C_{6v}	I	$2C_6$	$2C_3$	C_2	$3\sigma_v$	$3\sigma_d$		
A_1	1	1	1	1	1	1	T_z	$\alpha_{xx}+\alpha_{yy}, \alpha_{zz}$
A_2	1	1	1	1	−1	−1	R_z	
B_1	1	−1	1	−1	1	−1		
B_2	1	−1	1	−1	−1	1		
E_1	2	1	−1	−2	0	0	$(T_x, T_y), (R_x, R_y)$	$(\alpha_{xz}, \alpha_{yz})$
E_2	2	−1	−1	2	0	0		$(\alpha_{xx}-\alpha_{yy}, \alpha_{xy})$

Table A.16

$C_{\infty v}$	I	$2C_\infty{}^\phi$	...	$\infty\sigma_v$		
$A_1 \equiv \Sigma^+$	1	1	...	1	T_z	$\alpha_{xx}+\alpha_{yy}, \alpha_{zz}$
$A_2 \equiv \Sigma^-$	1	1	...	−1	R_z	
$E_1 \equiv \Pi$	2	$2\cos\phi$	...	0	$(T_x, T_y), (R_x, R_y)$	$(\alpha_{xz}, \alpha_{yz})$
$E_2 \equiv \Delta$	2	$2\cos 2\phi$	...	0		$(\alpha_{xx}-\alpha_{yy}, \alpha_{xy})$
$E_3 \equiv \Phi$	2	$2\cos 3\phi$	...	0		
⋮	⋮	⋮	...	⋮		

Table A.17

D_2	I	$C_2(z)$	$C_2(y)$	$C_2(x)$		
A	1	1	1	1		$\alpha_{xx}, \alpha_{yy}, \alpha_{zz}$
B_1	1	1	−1	−1	T_z, R_z	α_{xy}
B_2	1	−1	1	−1	T_y, R_y	α_{xz}
B_3	1	−1	−1	1	T_x, R_x	α_{yz}

Table A.18

D_3	I	$2C_3$	$2C_2$		
A_1	1	1	1		$\alpha_{xx}+\alpha_{yy}, \alpha_{zz}$
A_2	1	1	−1	T_z, R_z	
E	2	−1	0	$(T_x, T_y), (R_x, R_y)$	$(\alpha_{xx}-\alpha_{yy}, \alpha_{xy}), (\alpha_{xz}, \alpha_{yz})$

Table A.19

D_4	I	$2C_4$	$C_2\ (=C_4^2)$	$2C_2'$	$2C_2''$		
A_1	1	1	1	1	1		$\alpha_{xx}+\alpha_{yy}, \alpha_{zz}$
A_2	1	1	1	−1	−1	T_z, R_z	
B_1	1	−1	1	1	−1		$\alpha_{xx}-\alpha_{yy}$
B_2	1	−1	1	−1	1		α_{xy}
E	2	0	−2	0	0	$(T_x, T_y), (R_x, R_y)$	$(\alpha_{xz}, \alpha_{yz})$

Table A.20

D_5	I	$2C_5$	$2C_5^2$	$5C_2$		
A_1	1	1	1	1		$\alpha_{xx}+\alpha_{yy}, \alpha_{zz}$
A_2	1	1	1	−1	T_z, R_z	
E_1	2	2 cos 72°	2 cos 144°	0	$(T_x, T_y), (R_x, R_y)$	$(\alpha_{xz}, \alpha_{yz})$
E_2	2	2 cos 144°	2 cos 72°	0		$(\alpha_{xx}-\alpha_{yy}, \alpha_{xy})$

Table A.21

D_6	I	$2C_6$	$2C_3$	C_2	$3C_2'$	$3C_2''$		
A_1	1	1	1	1	1	1		$\alpha_{xx}+\alpha_{yy}, \alpha_{zz}$
A_2	1	1	1	1	−1	−1	T_z, R_z	
B_1	1	−1	1	−1	1	−1		
B_2	1	−1	1	−1	−1	1		
E_1	2	1	−1	−2	0	0	$(T_x, T_y), (R_x, R_y)$	$(\alpha_{xz}, \alpha_{yz})$
E_2	2	−1	−1	2	0	0		$(\alpha_{xx}-\alpha_{yy}, \alpha_{xy})$

Table A.22

C_{2h}	I	C_2	i	σ_h		
A_g	1	1	1	1	R_z	$\alpha_{xx}, \alpha_{yy}, \alpha_{zz}, \alpha_{xy}$
B_g	1	−1	1	−1	R_x, R_y	α_{xz}, α_{yz}
A_u	1	1	−1	−1	T_z	
B_u	1	−1	−1	1	T_x, T_y	

Table A.23

C_{3h}	I	C_3	$C_3^{\,2}$	σ_h	S_3	$S_3^{\,5}$		
A'	1	1	1	1	1	1	R_z	$\alpha_{xx}+\alpha_{yy}, \alpha_{zz}$
A''	1	1	1	−1	−1	−1	T_z	
E'	1	ε	ε^*	1	ε	ε^*	(T_x, T_y)	$(\alpha_{xx}-\alpha_{yy}, \alpha_{xy})$
	1	ε^*	ε	1	ε^*	ε		
E''	1	ε	ε^*	−1	$-\varepsilon$	$-\varepsilon^*$	(R_x, R_y)	$(\alpha_{xz}, \alpha_{yz})$
	1	ε^*	ε	−1	$-\varepsilon^*$	$-\varepsilon$		

$\varepsilon = \exp(2\pi i/3)$; $\varepsilon^* = \exp(-2\pi i/3)$.

Table A.24

C_{4h}	I	C_4	C_2	$C_4^{\,3}$	i	$S_4^{\,3}$	σ_h	S_4		
A_g	1	1	1	1	1	1	1	1	R_z	$\alpha_{xx}+\alpha_{yy}, \alpha_{zz}$
B_g	1	−1	1	−1	1	−1	1	−1		$\alpha_{xx}-\alpha_{yy}, \alpha_{xy}$
E_g	1	i	−1	$-i$	1	i	−1	$-i$	(R_x, R_y)	$(\alpha_{xz}, \alpha_{yz})$
	1	$-i$	−1	i	1	$-i$	−1	i		
A_u	1	1	1	1	−1	−1	−1	−1	T_z	
B_u	1	−1	1	−1	−1	1	−1	1		
E_u	1	i	−1	$-i$	−1	$-i$	1	i	(T_x, T_y)	
	1	$-i$	−1	i	−1	i	1	$-i$		

Table A.25

C_{5h}	I	C_5	$C_5^{\,2}$	$C_5^{\,3}$	$C_5^{\,4}$	σ_h	S_5	$S_5^{\,7}$	$S_5^{\,3}$	$S_5^{\,9}$		
A'	1	1	1	1	1	1	1	1	1	1	R_z	$\alpha_{xx}+\alpha_{yy}, \alpha_{zz}$
$E_1{}'$	1	ε	ε^2	ε^{2*}	ε^*	1	ε	ε^2	ε^{2*}	ε^*	(T_x, T_y)	
	1	ε^*	ε^{2*}	ε^2	ε	1	ε^*	ε^{2*}	ε^2	ε		
$E_2{}'$	1	ε^2	ε^*	ε	ε^{2*}	1	ε^2	ε^*	ε	ε^{2*}		$(\alpha_{xx}-\alpha_{yy}, \alpha_{xy})$
	1	ε^{2*}	ε	ε^*	ε^2	1	ε^{2*}	ε	ε^*	ε^2		
A''	1	1	1	1	1	−1	−1	−1	−1	−1	T_z	
$E_1{}''$	1	ε	ε^2	ε^{2*}	ε^*	−1	$-\varepsilon$	$-\varepsilon^2$	$-\varepsilon^{2*}$	$-\varepsilon^*$	(R_x, R_y)	$(\alpha_{xz}, \alpha_{yz})$
	1	ε^*	ε^{2*}	ε^2	ε	−1	$-\varepsilon^*$	$-\varepsilon^{2*}$	$-\varepsilon^2$	$-\varepsilon$		
$E_2{}''$	1	ε^2	ε^*	ε	ε^{2*}	−1	$-\varepsilon^2$	$-\varepsilon^*$	$-\varepsilon$	$-\varepsilon^{2*}$		
	1	ε^{2*}	ε	ε^*	ε^2	−1	$-\varepsilon^{2*}$	$-\varepsilon$	$-\varepsilon^*$	$-\varepsilon^2$		

$\varepsilon = \exp(2\pi i/5)$; $\varepsilon^* = \exp(-2\pi i/5)$.

Table A.26

C_{6h}	I	C_6	C_3	C_2	C_3^2	C_6^5	i	S_3^5	S_6^5	σ_h	S_6	S_3		
A_g	1	1	1	1	1	1	1	1	1	1	1	1	R_z	$\alpha_{xx}+\alpha_{yy}$, α_{zz}
B_g	1	−1	1	−1	1	−1	1	−1	1	−1	1	−1		
E_{1g}	1	ε	$-\varepsilon^*$	−1	$-\varepsilon$	ε^*	1	ε	$-\varepsilon^*$	−1	$-\varepsilon$	ε^*	(R_x, R_y)	$(\alpha_{xz}, \alpha_{yz})$
	1	ε^*	$-\varepsilon$	−1	$-\varepsilon^*$	ε	1	ε^*	$-\varepsilon$	−1	$-\varepsilon^*$	ε		
E_{2g}	1	$-\varepsilon^*$	$-\varepsilon$	1	$-\varepsilon^*$	$-\varepsilon$	1	$-\varepsilon^*$	$-\varepsilon$	1	$-\varepsilon^*$	$-\varepsilon$		$(\alpha_{xx}-\alpha_{yy}, \alpha_{xy})$
	1	$-\varepsilon$	$-\varepsilon^*$	1	$-\varepsilon$	$-\varepsilon^*$	1	$-\varepsilon$	$-\varepsilon^*$	1	$-\varepsilon$	$-\varepsilon^*$		
A_u	1	1	1	1	1	1	−1	−1	−1	−1	−1	−1	T_z	
B_u	1	−1	1	−1	1	−1	−1	1	−1	1	−1	1		
E_{1u}	1	ε	$-\varepsilon^*$	−1	$-\varepsilon$	ε^*	−1	$-\varepsilon$	ε^*	1	ε	$-\varepsilon^*$	(T_x, T_y)	
	1	ε^*	$-\varepsilon$	−1	$-\varepsilon^*$	ε	−1	$-\varepsilon^*$	ε	1	ε^*	$-\varepsilon$		
E_{2u}	1	$-\varepsilon^*$	$-\varepsilon$	1	$-\varepsilon^*$	$-\varepsilon$	−1	ε^*	ε	−1	ε^*	ε		
	1	$-\varepsilon$	$-\varepsilon^*$	1	$-\varepsilon$	$-\varepsilon^*$	−1	ε	ε^*	−1	ε	ε^*		

$\varepsilon = \exp(2\pi i/6)$; $\varepsilon^* = \exp(-2\pi i/6)$.

Table A.27

D_{2d}	I	$2S_4$	C_2	$2C_2'$	$2\sigma_d$		
A_1	1	1	1	1	1		$\alpha_{xx}+\alpha_{yy}$, α_{zz}
A_2	1	1	1	−1	−1	R_z	
B_1	1	−1	1	1	−1		$\alpha_{xx}-\alpha_{yy}$
B_2	1	−1	1	−1	1	T_z	α_{xy}
E	2	0	−2	0	0	(T_x, T_y), (R_x, R_y)	$(\alpha_{xz}, \alpha_{yz})$

Table A.28

D_{3d}	I	$2C_3$	$3C_2$	i	$2S_6$	$3\sigma_d$		
A_{1g}	1	1	1	1	1	1		$\alpha_{xx}+\alpha_{yy}$, α_{zz}
A_{2g}	1	1	−1	1	1	−1	R_z	
E_g	2	−1	0	2	−1	0	(R_x, R_y)	$(\alpha_{xx}-\alpha_{yy}, \alpha_{xy})$, $(\alpha_{xz}, \alpha_{yz})$
A_{1u}	1	1	1	−1	−1	−1		
A_{2u}	1	1	−1	−1	−1	1	T_z	
E_u	2	−1	0	−2	1	0	(T_x, T_y)	

Table A.29

D_{4d}	I	$2S_8$	$2C_4$	$2S_8^3$	C_2	$4C_2'$	$4\sigma_d$		
A_1	1	1	1	1	1	1	1		$\alpha_{xx}+\alpha_{yy}$, α_{zz}
A_2	1	1	1	1	1	−1	−1	R_z	
B_1	1	−1	1	−1	1	1	−1		
B_2	1	−1	1	−1	1	−1	1	T_z	
E_1	2	$\sqrt{2}$	0	$-\sqrt{2}$	−2	0	0	(T_x, T_y)	
E_2	2	0	−2	0	2	0	0		$(\alpha_{xx}-\alpha_{yy}, \alpha_{xy})$
E_3	2	$-\sqrt{2}$	0	$\sqrt{2}$	−2	0	0	(R_x, R_y)	$(\alpha_{xz}, \alpha_{yz})$

Table A.30

D_{5d}	I	$2C_5$	$2C_5{}^2$	$5C_2$	i	$2S_{10}{}^3$	$2S_{10}$	$5\sigma_d$		
A_{1g}	1	1	1	1	1	1	1	1		$\alpha_{xx}+\alpha_{yy}$, α_{zz}
A_{2g}	1	1	1	−1	1	1	1	−1	R_z	
E_{1g}	2	2 cos 72°	2 cos 144°	0	2	2 cos 72°	2 cos 144°	0	(R_x, R_y)	$(\alpha_{xz}, \alpha_{yz})$
E_{2g}	2	2 cos 144°	2 cos 72°	0	2	2 cos 144°	2 cos 72°	0		$(\alpha_{xx}-\alpha_{yy}, \alpha_{xy})$
A_{1u}	1	1	1	1	−1	−1	−1	−1		
A_{2u}	1	1	1	−1	−1	−1	−1	1	T_z	
E_{1u}	2	2 cos 72°	2 cos 144°	0	−2	−2 cos 72°	−2 cos 144°	0	(T_x, T_y)	
E_{2u}	2	2 cos 144°	2 cos 72°	0	−2	−2 cos 144°	−2 cos 72°	0		

Table A.31

D_{6d}	I	$2S_{12}$	$2C_6$	$2S_4$	$2C_3$	$2S_{12}{}^5$	C_2	$6C_2'$	$6\sigma_d$		
A_1	1	1	1	1	1	1	1	1	1		$\alpha_{xx}+\alpha_{yy}$, α_{zz}
A_2	1	1	1	1	1	1	1	−1	−1	R_z	
B_1	1	−1	1	−1	1	−1	1	1	−1		
B_2	1	−1	1	−1	1	−1	1	−1	1	T_z	
E_1	2	$\sqrt{3}$	1	0	−1	$-\sqrt{3}$	−2	0	0	(T_z, T_y)	
E_2	2	1	−1	−2	−1	1	2	0	0		$(\alpha_{xx}-\alpha_{yy}, \alpha_{xy})$
E_3	2	0	−2	0	2	0	−2	0	0		
E_4	2	−1	−1	2	−1	−1	2	0	0		
E_5	2	$-\sqrt{3}$	1	0	−1	$\sqrt{3}$	−2	0	0	(R_x, R_y)	$(\alpha_{xz}, \alpha_{yz})$

Table A.32

D_{2h}	I	$C_2(z)$	$C_2(y)$	$C_2(x)$	i	$\sigma(yz)$	$\sigma(xz)$	$\sigma(yz)$		
A_g	1	1	1	1	1	1	1	1		α_{xx}, α_{yy}, α_{zz}
B_{1g}	1	1	−1	−1	1	1	−1	−1	R_z	α_{xy}
B_{2g}	1	−1	1	−1	1	−1	1	−1	R_y	α_{xz}
B_{3g}	1	−1	−1	1	1	−1	−1	1	R_x	α_{yz}
A_u	1	1	1	1	−1	−1	−1	−1		
B_{1u}	1	1	−1	−1	−1	−1	1	1	T_z	
B_{2u}	1	−1	1	−1	−1	1	−1	1	T_y	
B_{3u}	1	−1	−1	1	−1	1	1	−1	T_x	

Table A.33

D_{3h}	I	$2C_3$	$3C_2$	σ_h	$2S_3$	$3\sigma_v$		
A_1'	1	1	1	1	1	1		$\alpha_{xx}+\alpha_{yy}$, α_{zz}
A_2'	1	1	−1	1	1	−1	R_z	
E'	2	−1	0	2	−1	0	(T_x, T_y)	$(\alpha_{xx}-\alpha_{yy}, \alpha_{xy})$
A_1''	1	1	1	−1	−1	−1		
A_2''	1	1	−1	−1	−1	1	T_z	
E''	2	−1	0	−2	1	0	(R_x, R_y)	$(\alpha_{xz}, \alpha_{yz})$

Table A.34

D_{4h}	I	$2C_4$	C_2	$2C_2'$	$2C_2''$	i	$2S_4$	σ_h	$2\sigma_v$	$2\sigma_d$		
A_{1g}	1	1	1	1	1	1	1	1	1	1		$\alpha_{xx}+\alpha_{yy}$, α_{zz}
A_{2g}	1	1	1	−1	−1	1	1	1	−1	−1	R_z	
B_{1g}	1	−1	1	1	−1	1	−1	1	1	−1		$\alpha_{xx}-\alpha_{yy}$
B_{2g}	1	−1	1	−1	1	1	−1	1	−1	1		α_{xy}
E_g	2	0	−2	0	0	2	0	−2	0	0	(R_x, R_y)	$(\alpha_{x}z, \alpha_{yz})$
A_{1u}	1	1	1	1	1	−1	−1	−1	−1	−1		
A_{2u}	1	1	1	−1	−1	−1	−1	−1	1	1	T_z	
B_{1u}	1	−1	1	1	−1	−1	1	−1	−1	1		
B_{2u}	1	−1	1	−1	1	−1	1	−1	1	−1		
E_u	2	0	−2	0	0	−2	0	2	0	0	(T_x, T_y)	

Table A.35

D_{5h}	I	$2C_5$	$2C_5^2$	$5C_2$	σ_h	$2S_5$	$2S_5^3$	$5\sigma_v$		
A_1'	1	1	1	1	1	1	1	1		$\alpha_{xx}+\alpha_{yy}$, α_{zz}
A_2'	1	1	1	−1	1	1	1	−1	R_z	
E_1'	2	2 cos 72°	2 cos 144°	0	2	2 cos 72°	2 cos 144°	0	(T_x, T_y)	
E_2'	2	2 cos 144°	2 cos 72°	0	2	2 cos 144°	2 cos 72°	0		$(\alpha_{xx}-\alpha_{yy}, \alpha_{xy})$
A_1''	1	1	1	1	−1	−1	−1	−1		
A_2''	1	1	1	−1	−1	−1	−1	1	T_z	
E_1''	2	2 cos 72°	2 cos 144°	0	−2	−2 cos 72°	−2 cos 144°	0	(R_x, R_y)	$(\alpha_{xz}, \alpha_{yz})$
E_2''	2	2 cos 144°	2 cos 72°	0	−2	−2 cos 144°	−2 cos 72°	0		

Table A.36

D_{6h}	I	$2C_6$	$2C_3$	C_2	$3C_2'$	$3C_2''$	i	$2S_3$	$2S_6$	σ_h	$3\sigma_d$	$3\sigma_v$		
A_{1g}	1	1	1	1	1	1	1	1	1	1	1	1		$\alpha_{xx}+\alpha_{yy}$, α_{zz}
A_{2g}	1	1	1	1	−1	−1	1	1	1	1	−1	−1	R_z	
B_{1g}	1	−1	1	−1	1	−1	1	−1	1	−1	1	−1		
B_{2g}	1	−1	1	−1	−1	1	1	−1	1	−1	−1	1		
E_{1g}	2	1	−1	−2	0	0	2	1	−1	−2	0	0	(R_x, R_y)	$(\alpha_{xz}, \alpha_{yz})$
E_{2g}	2	−1	−1	2	0	0	2	−1	−1	2	0	0		$(\alpha_{xx}-\alpha_{yy}, \alpha_{xy})$
A_{1u}	1	1	1	1	1	1	−1	−1	−1	−1	−1	−1		
A_{2u}	1	1	1	1	−1	−1	−1	−1	−1	−1	1	1	T_z	
B_{1u}	1	−1	1	−1	1	−1	−1	1	−1	1	−1	1		
B_{2u}	1	−1	1	−1	−1	1	−1	1	−1	1	1	−1		
E_{1u}	2	1	−1	−2	0	0	−2	−1	1	2	0	0	(T_x, T_y)	
E_{2u}	2	−1	−1	2	0	0	−2	1	1	−2	0	0		

Table A.37

$D_{\infty h}$	I	$2C_{\infty}^{\phi}$	...	$\infty\sigma_v$	i	$2S_{\infty}^{\phi}$	...	∞C_2		
$A_{1g} \equiv \Sigma_g^+$	1	1	...	1	1	1	...	1		$\alpha_{xx} + \alpha_{yy}, \alpha_{zz}$
$A_{2g} \equiv \Sigma_g^-$	1	1	...	−1	1	1	...	−1	R_z	
$E_{1g} \equiv \Pi_g$	2	$2\cos\phi$	...	0	2	$-2\cos\phi$	...	0	(R_x, R_y)	$(\alpha_{xz}, \alpha_{yz})$
$E_{2g} \equiv \Delta_g$	2	$2\cos 2\phi$	...	0	2	$2\cos 2\phi$	...	0		$(\alpha_{xx} - \alpha_{yy}, \alpha_{xy})$
$E_{3g} \equiv \Phi_g$	2	$2\cos 3\phi$	...	0	2	$-2\cos 3\phi$	...	0		
⋮	⋮	⋮		⋮	⋮	⋮		⋮		
$A_{1u} \equiv \Sigma_u^+$	1	1	...	1	−1	−1	...	−1	T_z	
$A_{2u} \equiv \Sigma_u^-$	1	1	...	−1	−1	−1	...	1		
$E_{1u} \equiv \Pi_u$	2	$2\cos\phi$	...	0	−2	$2\cos\phi$	...	0	(T_x, T_y)	
$E_{2u} \equiv \Delta_u$	2	$2\cos 2\phi$	...	0	−2	$-2\cos 2\phi$	...	0		
$E_{3u} \equiv \Phi_u$	2	$2\cos 3\phi$	...	0	−2	$2\cos 3\phi$	...	0		
⋮	⋮	⋮		⋮	⋮	⋮		⋮		

Table A.38

S_4	I	S_4	C_2	S_4^3		
A	1	1	1	1	R_z	$\alpha_{xx} + \alpha_{yy}, \alpha_{zz}$
B	1	−1	1	−1	T_z	$\alpha_{xx} - \alpha_{yy}, \alpha_{xy}$
E	1	i	−1	$-i$	$(T_x, T_y), (R_x, R_y)$	$(\alpha_{xz}, \alpha_{yz})$
	1	$-i$	−1	i		

Table A.39

S_6	I	C_3	C_3^2	i	S_6^5	S_6		
A_g	1	1	1	1	1	1	R_z	$\alpha_{xx} + \alpha_{yy}, \alpha_{zz}$
E_g	1	ε	ε^*	1	ε	ε^*	(R_x, R_y)	$(\alpha_{xx} - \alpha_{yy}, \alpha_{xy}), (\alpha_{xz}, \alpha_{yz})$
	1	ε^*	ε	1	ε^*	ε		
A_u	1	1	1	−1	−1	−1	T_z	
E_u	1	ε	ε^*	−1	$-\varepsilon$	$-\varepsilon^*$	(T_x, T_y)	
	1	ε^*	ε	−1	$-\varepsilon^*$	$-\varepsilon$		

$\varepsilon = \exp(2\pi i/3)$; $\varepsilon^* = \exp(-2\pi i/3)$.

Table A.40

S_8	I	S_8	C_4	S_8^3	C_2	S_8^5	C_4^3	S_8^7		
A	1	1	1	1	1	1	1	1	R_z	$\alpha_{xx} + \alpha_{yy}, \alpha_{zz}$
B	1	−1	1	−1	1	−1	1	−1	T_z	
E_1	1	ε	i	$-\varepsilon^*$	−1	$-\varepsilon$	$-i$	ε^*	$(T_x, T_y), (R_x, R_y)$	
	1	ε^*	$-i$	$-\varepsilon$	−1	$-\varepsilon^*$	i	ε		
E_2	1	i	−1	$-i$	1	i	−1	$-i$		$(\alpha_{xx} - \alpha_{yy}, \alpha_{xy})$
	1	$-i$	−1	i	1	$-i$	−1	i		
E_3	1	$-\varepsilon^*$	$-i$	ε	−1	ε^*	i	$-\varepsilon$		$(\alpha_{xz}, \alpha_{yz})$
	1	$-\varepsilon$	i	ε^*	−1	ε	$-i$	$-\varepsilon^*$		

$\varepsilon = \exp(2\pi i/8)$; $\varepsilon^* = \exp(-2\pi i/8)$

Table A.41

T_d	I	$8C_3$	$3C_2$	$6S_4$	$6\sigma_d$		
A_1	1	1	1	1	1		$\alpha_{xx}+\alpha_{yy}+\alpha_{zz}$
A_2	1	1	1	−1	−1		
E	2	−1	2	0	0		$(\alpha_{xx}+\alpha_{yy}-2\alpha_{zz}, \alpha_{xx}-\alpha_{yy})$
$T_1 \equiv F_1$	3	0	−1	1	−1	(R_x, R_y, R_z)	
$T_2 \equiv F_2$	3	0	−1	−1	1	(T_x, T_y, T_z)	$(\alpha_{xy}, \alpha_{xz}, \alpha_{yz})$

Table A.42

T	I	$4C_3$	$4C_3^2$	$3C_2$		
A	1	1	1	1		$\alpha_{xx}+\alpha_{yy}+\alpha_{zz}$
E	$\begin{cases}1\\1\end{cases}$	ε ε^*	ε^* ε	1 1		$(\alpha_{xx}+\alpha_{yy}-2\alpha_{zz}, \alpha_{xx}-\alpha_{yy})$
$T \equiv F$	3	0	0	−1	$(T_x, T_y, T_z), (R_x, R_y, R_z)$	$(\alpha_{xy}, \alpha_{xz}, \alpha_{yz})$

$\varepsilon = \exp(2\pi i/3)$; $\varepsilon^* = \exp(-2\pi i/3)$.

Table A.43

O_h	I	$8C_3$	$6C_2$	$6C_4$	$3C_2'$ $(=3C_4^2)$	i	$6S_4$	$8S_6$	$3\sigma_h$	$6\sigma_d$		
A_{1g}	1	1	1	1	1	1	1	1	1	1		$\alpha_{xx}+\alpha_{yy}+\alpha_{zz}$
A_{2g}	1	1	−1	−1	1	1	−1	1	1	−1		
E_g	2	−1	0	0	2	2	0	−1	2	0		$(\alpha_{xx}+\alpha_{yy}-2\alpha_{zz}, \alpha_{xx}-\alpha_{yy})$
$T_{1g} \equiv F_{1g}$	3	0	−1	1	−1	3	1	0	−1	−1	(R_x, R_y, R_z)	
$T_{2g} \equiv F_{2g}$	3	0	1	−1	−1	3	−1	0	−1	1		$(\alpha_{xz}, \alpha_{yz}, \alpha_{xy})$
A_{1u}	1	1	1	1	1	−1	−1	−1	−1	−1		
A_{2u}	1	1	−1	−1	1	−1	1	−1	−1	1		
E_u	2	−1	0	0	2	−2	0	1	−2	0		
$T_{1u} \equiv F_{1u}$	3	0	−1	1	−1	−3	−1	0	1	1	(T_x, T_y, T_z)	
$T_{2u} \equiv F_{2u}$	3	0	1	−1	−1	−3	1	0	1	−1		

Table A.44

O	I	$8C_3$	$6C_2$	$6C_4$	$3C_2'(=3C_4^2)$		
A_1	1	1	1	1	1		$\alpha_{xx}+\alpha_{yy}+\alpha_{zz}$
A_2	1	1	−1	−1	1		
E	2	−1	0	0	2		$(\alpha_{xx}+\alpha_{yy}-2\alpha_{zz}, \alpha_{xx}-\alpha_{yy})$
$T_1 \equiv F_1$	3	0	−1	1	−1	$(T_x, T_y, T_z), (R_x, R_y, R_z)$	
$T_2 \equiv F_2$	3	0	1	−1	−1		$(\alpha_{xy}, \alpha_{xz}, \alpha_{yz})$

Table A.45

K_h	I	$\infty C_\infty^\phi \ldots$	$\infty S_\infty^\phi \ldots$	i		
S_g	1	1	1	1		$\alpha_{xx}+\alpha_{yy}+\alpha_{zz}$
S_u	1	1	-1	-1		
P_g	3	$1+2\cos\phi$	$1-2\cos\phi$	1	$(R_x,\ R_y,\ R_z)$	
P_u	3	$1+2\cos\phi$	$-1+2\cos\phi$	-1	$(T_x,\ T_y,\ T_z)$	
D_g	5	$1+2\cos\phi+2\cos 2\phi$	$1-2\cos\phi+2\cos 2\phi$	1		$(\alpha_{xx}+\alpha_{yy}-2\alpha_{zz},\ \alpha_{xx}-\alpha_{yy},\ \alpha_{xy},\ \alpha_{xz},\ \alpha_{yz})$
D_u	5	$1+2\cos\phi+2\cos 2\phi$	$-1+2\cos\phi-2\cos 2\phi$	-1		
F_g	7	$1+2\cos\phi+2\cos 2\phi+2\cos 3\phi$	$1-2\cos\phi+2\cos 2\phi-2\cos 3\phi$	1		
F_u	7	$1+2\cos\phi+2\cos 2\phi+2\cos 3\phi$	$-1+2\cos\phi-2\cos 2\phi+2\cos 3\phi$	-1		
$\vdots$	$\vdots$	$\vdots$	$\vdots$	$\vdots$		

Table A.46

I_h	E	$12C_5$	$12C_5^{\ 2}$	$20C_3$	$15C_2$	i	$12S_{10}$	$12S_{10}^{\ 3}$	$20S_6$	15σ		
A_g	1	1	1	1	1	1	1	1	1	1		$\alpha_{xx}+\alpha_{yy}+\alpha_{zz}$
$T_{1g} \equiv F_{1g}$	3	$\frac{1}{2}(1+\sqrt{5})$	$\frac{1}{2}(1-\sqrt{5})$	0	−1	3	$\frac{1}{2}(1-\sqrt{5})$	$\frac{1}{2}(1+\sqrt{5})$	0	−1	(R_z, R_y, R_z)	
$T_{2g} \equiv F_{2g}$	3	$\frac{1}{2}(1-\sqrt{5})$	$\frac{1}{2}(1+\sqrt{5})$	0	−1	3	$\frac{1}{2}(1+\sqrt{5})$	$\frac{1}{2}(1-\sqrt{5})$	0	−1		
G_g	4	−1	−1	1	0	4	−1	−1	1	0		
H_g	5	0	0	−1	1	5	0	0	−1	1		$(2\alpha_{zz}-\alpha_{xx}-\alpha_{yy}$, $\alpha_{xx}-\alpha_{yy}$, $\alpha_{xy}, \alpha_{yz}, \alpha_{xz})$
A_u	1	1	1	1	1	−1	−1	−1	−1	−1		
$T_{1u} \equiv F_{1u}$	3	$\frac{1}{2}(1+\sqrt{5})$	$\frac{1}{2}(1-\sqrt{5})$	0	−1	−3	$-\frac{1}{2}(1-\sqrt{5})$	$-\frac{1}{2}(1+\sqrt{5})$	0	1	(T_x, T_y, T_z)	
$T_{2u} \equiv F_{2u}$	3	$\frac{1}{2}(1-\sqrt{5})$	$\frac{1}{2}(1+\sqrt{5})$	0	−1	−3	$-\frac{1}{2}(1+\sqrt{5})$	$-\frac{1}{2}(1-\sqrt{5})$	0	1		
G_u	4	−1	−1	1	0	−4	1	1	−1	0		
H_u	5	0	0	−1	1	−5	0	0	1	−1		

REFERENCES

1. BORN, M. and OPPENHEIMER, J. R. (1927). *Ann. Phys.*, **84**, 457
2. VAN VLECK, J. H. and WEISSKOPF, V. F. (1945). *Rev. Mod. Phys.*, **17**, 227
3. MITCHELL, A. C. G. and ZEMANSKY, M. W. (1961). *Resonance Radiation and Excited Atoms*. Cambridge University Press, Cambridge
4. RADFORD, H. E. (1966). *Rev. Sci. Instrum.*, **37**, 790
5. BERNSTEIN, H. J. and HERZBERG, G. (1948). *J. Chem. Phys.*, **16**, 30
6. CONNES, J. and CONNES, P. (1966). *J. Opt. Soc. Am.*, **56**, 896
7. NORRISH, R. G. W. and PORTER, G. (1949). *Nature (London)*, **164**, 658
8. KING, G. W. (1958). *J. Sci. Instrum.*, **35**, 11
9. COWAN, M. and GORDY, W. (1957). *Bull. Am. Phys. Soc.*, **2**, 212
10. JONES, G. and GORDY, W. (1964). *Phys. Rev.*, **135**, A295
11. FLEMING, J. W. (1976). *J. Quant. Spectrosc. Radiat. Transfer*, **16**, 63
12. MAES, S. (1960). *Cah. Phys.*, **14**, 125
13. OTAKE, M., MATSUMARA, C. and MORINO, Y. (1968). *J. Mol. Spectrosc.*, **28**, 316
14. WATSON, J. K. G. (1977). In *Vibrational Spectra and Structure* (Ed. J. R. Durig), Vol. 6, pp. 1–89. Elsevier, New York
15. BOGEY, M., BOLVIN, H., CORDONNIER, M., DEMUYNCK, C., DESTOMBES, J. L. and CSÁSZÁR, A. G. (1994). *J. Chem. Phys.*, **100**, 8614
16. HECHT, K. T. (1960). *J. Mol. Spectrosc.*, **5**, 355, 390
17. BELL, M. P., FELDMAN, P. A., KWOK, S. and MATTHEWS, H. E. (1982). *Nature (London)*, **295**, 389
18. BUHL, D. and SNYDER, L. E. (1970). *Nature (London)*, **228**, 267
19. KLEMPERER, W. (1970). *Nature (London)*, **227**, 1230
20. WOODS, R. C., DIXON, T. A., SAYKALLY, R. J. and SZANTO, P. G. (1975). *Phys. Rev. Lett.*, **35**, 1269
21. RAMAN, C. V. and KRISHNAN, K. S. (1928). *Nature (London)*, **121**, 501
22. BUTCHER, R. J. and JONES, W. J. (1974). *J. Chem. Soc., Faraday Trans. II*, **70**, 560
23. WELSH, H. L., CUMMING, C. and STANSBURY, E. J. (1951). *J. Opt. Soc. Am.*, **41**, 712
24. GRIGGS, J. L., JR, RAO, K. N., JONES, L. H. and POTTER, R. M. (1968). *J. Mol. Spectrosc.*, **25**, 34
25. KRAITCHMAN, J. (1953). *Am. J. Phys.*, **21**, 17
26. LISTER, D. G., TYLER, J. K., HØG, J. H. and WESSEL LARSEN, N. (1974). *J. Mol. Struct.*, **23**, 253
27. BENNEWITZ, H. G., PAUL, W. and SCHLIER, C. (1955). *Z. Phys.*, **141**, 6
28. WHARTON, L., BERG, R. A. and KLEMPERER, W. (1963). *J. Chem. Phys.*, **39**, 2023
29. NOVICK, S. E., DAVIES, P. B., DYKE, T. R. and KLEMPERER, W. (1973). *J. Am. Chem. Soc.*, **95**, 8547
30. NOVICK, S. E., DAVIES, P. B., HARRIS, S. J. and KLEMPERER, W. (1973). *J. Chem. Phys.*, **59**, 2273
31. HUTSON, J. M. and HOWARD, B. J. (1982). *Mol. Phys.*, **45**, 769
32. COSTAIN, C. C. and SRIVASTAVA, G. P. (1964). *J. Chem. Phys.*, **41**, 1620
33. MARTINACHE, L., KRESA, W., WEGENER, M., VONMONT, U. and BAUDER, A. (1990). *Chem. Phys.*, **148**, 129
34. PICKETT, H. M. (1973). *J. Am. Chem. Soc.*, **95**, 1770
35. ROWE, W. F., JR, DUERST, R. W. and WILSON, E. B. (1976). *J. Am. Chem. Soc.*, **98**, 4021
36. ALVES, A. C. P. and HOLLAS, J. M. (1972). *Mol. Phys.*, **23**, 927
37. TANAKA, T., unpublished results
38. DYKE, T. R., HOWARD, B. J. and KLEMPERER, W. (1972). *J. Chem. Phys.*, **56**, 2442
39. BALLE, T. J., CAMPBELL, E. J., KEENEN, M. R. and FLYGARE, W. H. (1979). *J. Chem. Phys.*, **71**, 2723
40. BRUPBACHER, TH., MAKAREWICZ, J. and BAUDER, A. (1994). *J. Chem. Phys.*, **101**, 9736
41. BRUPBACHER, TH. and BAUDER, A. (1990). *Chem. Phys. Lett.*, **173**, 435
42. KLOTS, T. D., EMILSSON, T. and GUTOWSKY, H. S. (1992). *J. Chem. Phys.*, **97**, 5335
43. BETTENS, F. L., BETTENS, R. P. A. and BAUDER, A. (1995). 'Rotational spectroscopy of weakly bound complexes', in *Jet Spectroscopy and Molecular Dynamics* (Eds J. M. Hollas and D. Phillips), chap. 1. Blackie, Glasgow
44. LEGON, A. C. and REGO, C. A. (1993). *J. Chem. Phys.*, **99**, 1463, and references cited therein

45. XU, Y., JÄGER, W., DJAUHARI, J. and GERRY, M. C. L. (1995). *J. Chem. Phys.*, **103**, 2827, and references cited therein
46. MORSE, P. M. (1929). *Phys. Rev.*, **34**, 57
47. HULBERT, H. M. and HIRSCHFELDER, J. O. (1941). *J. Chem. Phys.*, **9**, 61
48. MULLIKEN, R. S. (1955). *J. Chem. Phys.*, **23**, 1997
49. KROTO, H. W., HEATH, J., O'BRIEN, S. C., CURL, R. F. and SMALLEY, R. E. (1985). *Nature (London)*, **318**, 312
50. TAYLOR, R., HARE, J. P., ABDUL-SADA, A. K. and KROTO, H. W. (1990). *J. Chem. Soc., Chem. Commun.*, 1423
51. KRATSCHMER, W., LAMB, L. D., FOSTIROPOULOS, K. and HUFFMAN, D. R. (1990). *Nature (London)*, **347**, 354
52. BETHUNE, D. S., MEIJER, G., TANG, W. C., ROSEN, H. J., GOLDEN, W. G., SEKI, H., BROWN, C. A. and DE VRIES, M. S. (1991). *Chem. Phys. Lett.*, **179**, 181
53. DI LAURO, C. and MILLS, I. M. (1966). *J. Mol. Spectrosc.*, **21**, 386
54. GORA, E. K. (1965). *J. Mol. Spectrosc.*, **16**, 378
55. FERMI, E. (1931). *Z. Phys.*, **71**, 250
56. AMAT, G. and PIMBERT, M. (1965). *J. Mol. Spectrosc.*, **16**, 278
57. DARLING, B. T. and DENNISON, D. M. (1940). *Phys. Rev.*, **57**, 128
58. SIEBRAND, W. and WILLIAMS, D. F. (1968). *J. Chem. Phys.*, **49**, 1860
59. CHILD, M. S. and LAWTON, R. T. (1981). *Faraday Discuss. Chem. Soc.*, **71**, 273
60. SWALLEN, J. D. and IBERS, J. A. (1962). *J. Chem. Phys.*, **36**, 1914
61. BELL, R. P. (1945). *Proc. R. Soc. London*, **A183**, 328
62. DURIG, J. R. and CARRIERA, L. A. (1972). *J. Chem. Phys.*, **56**, 4966
63. DIXON, R. N. (1964). *Trans. Faraday. Soc.*, **60**, 1363
64. LONGUET-HIGGINS, H. C. (1963). *Mol. Phys.*, **6**, 445
65. PITZER, K. S. (1946). *J. Chem. Phys.*, **14**, 239
66. BUTZ, K. W., KRAJNOVICH, D. J. and PARMENTER, C. S. (1990). *J. Chem. Phys.*, **93**, 1557
67. RUDOLPH, H. D. and TRIMKAUS, A. (1968). *Z. Naturforsch., Teil A*, **23**, 68
68. LAMB, W. E. and RETHERFORD, R. C. (1947). *Phys. Rev.*, **72**, 241; see also **79**, 549 (1950) and **81**, 222 (1952)
69. STEINFELD, J. I., ZARE, R. N., JONES, L., LESK, M. and KLEMPERER, W. (1965). *J. Chem. Phys.*, **42**, 25
70. HERZBERG, G. (1970). *J. Mol. Spectrosc.*, **33**, 147
71. CARRINGTON, A., LEACH, C. A., MARR, A. J., SHAW, A. M., VIANT, M. R., HUTSON, J. M. and LAW, M. M. (1995). *J Chem. Phys.*, **102**, 2379
72. JENSEN, P. and BUNKER, P. R. (1988). *J. Chem. Phys.*, **89**, 1327
73. KABBADJ, Y., HUET, T. R., ULY, D. and OKA, T. (1996). *J. Mol. Spectrosc.*, **175**, 277
74. JENSEN, P., BUNKER, P. R. and McCLEAN, A. D. (1987). *Chem. Phys. Lett.*, **141**, 53
75. GIBSON, S. T., GREENE, P. J. and BERKOWITZ, J. (1985). *J. Chem. Phys.*, **83**, 4319
76. DUBOIS, I. (1968). *Can. J. Phys.*, **46**, 2485
77. CRAMER, C. J., DULLES, F. J., STORER, J. W. and WORTHINGTON, S. E. (1994). *Chem. Phys. Lett.*, **218**, 387
78. SAITO, K. and OBI, K. (1993). *Chem. Phys. Lett.*, **215**, 193
79. KAROLCZAK, J., HARPER, W. W., GREV, R. S. and CLOUTHIER, D. J. (1995). *J. Chem. Phys.*, **103**, 2839
80. JUNGEN, CH., HALLIN, K.-E. J. and MERER, A. J. (1980). *Mol. Phys.*, **40**, 25
81. HERZBERG, G. and LEW, H. (1974). *Astron. Astrophys.*, **31**, 123
82. DUXBURY, G., HORANI, M. and ROSTAS, J. (1972). *Proc. R. Soc. London*, **A331**, 109
83. DIXON, R. N., DUXBURY, G. and LAMBERTON, H. M. (1968). *Proc. R. Soc. London*, **A305**, 271
84. OKA, T. (1980). *Phys. Rev. Lett.*, **45**, 531
85. DROISSART, P., MAILLARD, J.-P., CALDWELL, J., KIM, S. J., WATSON, J. K. G., MAJEWSKI, W. A., TENNYSON, J., MILLER, S., ATREYA, S. K., CLARKE, J. T., WAITE, J. H., JR and WAGENER, R. (1989). *Nature (London)*, **340**, 539
86. TRAFTON, L. M., GEBALLE, T. R., MILLER, S., TENNYSON, J. and BALLESTER, G. E. (1993). *Astrophys. J.*, **405**, 761
87. GABALLE, T. R., JAGOD, M.-F. and OKA, T. (1993). *Astrophys. J.*, **408**, L109
88. HERZBERG, G., LEW, H., SLOAN, J. J. and WATSON, J. K. G. (1981). *Can. J. Phys.*, **59**, 428
89. HALLER, E., KOPPEL, H. and CEDERBAUM, L. S. (1985). *J. Mol. Spectrosc.*, **111**, 377
90. SMALLEY, R. E., WHARTON, L. and LEVY, D. H. (1975). *J. Chem. Phys.*, **63**, 4977
91. TUCKER, K. D., KUTNER, M. L. and THADDEUS, P. (1974). *Astrophys. J.*, **193**, L115
92. OAKES, J. M., HARDING, L. B. and ELLISON, G. B. (1985). *J. Chem. Phys.*, **83**, 5400
93. SNYDER, L. E. and HOLLIS, J. M. (1978). *Astrophys. J.*, **204**, L139
94. THADDEUS, P. and TURNER, B. E. (1975). *Astrophys. J.*, **201**, L25
95. KLEMPERER, W. (1970). *Nature (London)*, **227**, 1230
96. CRESSWELL, R. A., PEARSON, E. F., WINNEWISSER, M. and WINNEWISSER, G. (1976). *Z. Naturforsch., Teil A*, **31**, 221
97. SAYKALLY, R. J., DIXON, T. A., ANDERSON, T. G., SZANTO, P. G. and WOODS, R. C. (1976). *Astrophys. J.*, **205**, L101

98. WOODS, R. C., DIXON, T. A., SAYKALLY, R. J. and SZANTO, P. G. (1975). *Phys. Rev. Lett.*, **35**, 1269
99. DIXON, R. N. (1969). *Trans. Faraday Soc.*, **65**, 3141
100. TUCKETT, R. P., FREEDMAN, P. A. and JONES, W. J. (1979). *Mol. Phys.*, **37**, 379
101. WOODMAN, C. M. (1970). *J. Mol. Spectrosc.*, **33**, 311
102. RICHARDS, W. G. (1979). *Nature (London)*, **278**, 507
103. BACK, R. A., WILLIS, C. and RAMSAY, D. A. (1978). *Can. J. Phys.*, **56**, 1575
104. CALLOMON, J. H. and INNES, K. K. (1963). *J. Mol. Spectrosc.*, **10**, 166
105. RAYNES, W. T. (1966). *J. Chem. Phys.*, **44**, 2755
106. CURRIE, G. N. and RAMSAY, D. A. (1971). *Can. J. Phys.*, **49**, 317
107. ALVES, A. C. P., CHRISTOFFERSEN, J. and HOLLAS, J. M. (1971). *Mol. Phys.*, **20**, 625
108. DOERING, J. P. (1977). *J. Chem. Phys.*, **67**, 4065
109. HERZBERG, G. and TELLER, E. (1933). *Z. Phys. Chem.*, **B21**, 410
110. SHARF, B. (1971). *J. Chem. Phys.*, **55**, 1379
111. ORLANDI, G. and SIEBRAND, W. (1972). *Chem. Phys. Lett.*, **15**, 465
112. ATKINSON, G. H. and PARMENTER, C. S. (1978). *J. Mol. Spectrosc.*, **73**, 52
113. TER HORST, G. and KOMMANDEUR, J. (1979). *Chem. Phys.*, **44**, 287
114. CALLOMON, J. H. and INNES, K. K. (1963). *J. Mol. Spectrosc.*, **10**, 166
115. CRAIG, D. P. and GORDON, R. D. (1965). *Proc. R. Soc. London*, **A288**, 69
116. SMITH, W. L. (1966). *Proc. Phys. Soc.*, **89**, 1021
117. HOLLAS, J. M., HOWSON, M. R., RIDLEY, T. and HALONEN, L. (1983). *Chem. Phys. Lett.*, **98**, 611
118. BRAND, J. C. D., WILLIAMS, D. R. and COOK, T. J. (1966). *J. Mol. Spectrosc.*, **20**, 359
119. ROSS, I. G., HOLLAS, J. M. and INNES, K. K. (1966). *J. Mol. Spectrosc.*, **20**, 312
120. DUSCHINSKY, F. (1937). *Acta Phys.-Chim. URSS*, **7**, 551
121. HOLLAS, J. M. and RIDLEY, T. (1981). *J. Mol. Spectrosc.*, **89**, 232
122. BIST, H. D., BRAND, J. C. D. and WILLIAMS, D. R. (1967). *J. Mol. Spectrosc.*, **24**, 413
123. HOLLAS, J. M. and KHALILIPOUR, E. (1977). *J. Mol. Spectrosc.*, **66**, 452
124. ATKINSON, G. H. and PARMENTER, C. S. (1978). *J. Mol. Spectrosc.*, **73**, 52
125. HERZBERG, G. and TELLER, E. (1933). *Z. Phys. Chem.*, **B21**, 410
126. RENNER, R. (1934). *Z. Phys.*, **92**, 172
127. GAUSSET, L., HERZBERG, G., LAGERQUIST, A. and ROSEN, B. (1963). *Discuss. Faraday Soc.*, **35**, 113
128. POPLE, J. A. and LONGUET-HIGGINS, H. C. (1958). *Mol. Phys.*, **1**, 372
129. JUNGEN, Ch. and MERER, A. J. (1980). *Mol. Phys.*, **40**, 1
130. GREEN, W. H., JR, HANDY, N. C., KNOWLES, P. J. and CARTER, S. (1991). *J. Chem. Phys.*, **94**, 118
131. JAHN, H. A. and TELLER, E. (1937). *Proc. R. Soc. London*, **A161**, 220
132. CHILD, M. S. (1963). *J. Mol. Spectrosc.*, **10**, 357
133. DOUGLAS, A. E. and HOLLAS, J. M. (1961). *Can. J. Phys.*, **39**, 479
134. GLOWNIA, J. H., RILEY, S. J., COLSON, S. D. and NIEMAN, G. C. (1980). *J. Chem. Phys.*, **73**, 4296
135. SEARS, T. J., MILLER, T. A. and BONDYBEY, V. E. (1980). *J. Chem. Phys.*, **72**, 6070
136. SEARS, T. J., MILLER, T. A. and BONDYBEY, V. E. (1981). *J. Chem. Phys.*, **74**, 3240
137. WEINSTOCK, B. and GOODMAN, G. L. (1965). *Adv. Chem. Phys.*, **9**, 169
138. DOUGLAS, A. E. (1963). *Discuss. Faraday Soc.*, **35**, 158
139. YAMADA, C., HIROTA, E. and KAWAGUCHI, K. (1981). *J. Chem. Phys.*, **75**, 5256
140. ASHFOLD, M. N. R., DIXON, R. N., LITTLE, N., STICKLAND, R. J. and WESTERN, C. M. (1988). *J. Chem. Phys.*, **89**, 1754
141. HOUGEN, J. T. (1964). *Can. J. Phys.*, **42**, 433
142. JOHNS, J. W. C. (1963). *Can. J. Phys.*, **41**, 209
143. POLO, S. R. (1957). *Can. J. Phys.*, **35**, 880
144. BROWN, J. M. (1969). *J. Mol. Spectrosc.*, **31**, 118
145. DIXON, R. N., DUXBURY, G. and RAMSAY, D. A. (1967). *Proc. R. Soc. London*, **A296**, 137
146. RAYNES, W. T. (1964). *J. Chem. Phys.*, **41**, 3020
147. HERZBERG, G. (1981). *Faraday Discuss. Chem. Soc.*, **71**, 165
148. WATSON, J. K. G., MAJEWSKI, W. A. and GLOWNIA, J. H. (1986). *J. Mol. Spectrosc.*, **115**, 82
149. DOUGLAS, A. E. and MILTON, E. R. V. (1964). *J. Chem. Phys.*, **41**, 357
150. EBERHARDT, W. H. and RENNER, H. (1961). *J. Mol. Spectrosc.*, **6**, 483
151. BROWN, J. M., BUCKINGHAM, A. D. and RAMSAY, D. A. (1976). *Can. J. Phys.*, **54**, 895
152. BIRSS, F. W., RAMSAY, D. A. and TILL, S. M. (1978). *Chem. Phys. Lett.*, **53**, 14
153. FREEMAN, D. E. and KLEMPERER, W. (1966). *J. Chem. Phys.*, **45**, 52
154. BUCKINGHAM, A. D., RAMSAY, D. A. and TYRRELL, J. (1970). *Can. J. Phys.*, **48**, 1242
155. LOMBARDI, J. R. (1970). *J. Am. Chem. Soc.*, **92**, 1831
156. FREEMAN, D. E. and KLEMPERER, W. (1964). *J. Chem. Phys.*, **40**, 604
157. BANCROFT, J. L., HOLLAS, J. M. and RAMSAY, D. A. (1962). *Can. J. Phys.*, **40**, 322

158. DIXON, R. N., JONES, K. B., NOBLE, M. and CARTER, S. (1981). *Mol. Phys.*, **42**, 455
159. BYRNE, J. P. and ROSS, I. G. (1965). *Can. J. Chem.*, **43**, 3253
160. CALLOMON, J. H., PARKIN, J. E. and LOPEZ-DELGADO, R. (1972). *Chem. Phys. Lett.*, **13**, 125
161. WUNSCH, L., NEUSSER, H. J. and SCHLAG, E. W. (1981). *Z. Naturforsch., Teil A*, **36**, 1340
162. SCHUBERT, U., RIEDLE, E., NEUSSER, H. J. and SCHLAG, E. W. (1986). *J. Chem. Phys.*, **84**, 6182
163. AL-JOBOURY, M. I. and TURNER, D. W. (1962). *J. Chem. Phys.*, **37**, 3007
164. VILESOV, F. I., KURBATOV, B. L. and TERENIN, A. A. (1961). *Sov. Phys. Dokl.*, **6**, 490
165. PRICE, W. C. (1968). In *Molecular Spectroscopy* (Ed. P. Hepple), p. 221. Institute of Petroleum, London
166. FROST, D. C., McDOWELL, C. A. and VROOM, D. A. (1967). *Proc. R. Soc. London*, **A296**, 566
167. KOOPMANS, T. (1933). *Physica, Eindhoven*, **1**, 104
168. SICHEL, J. M. (1970). *Mol. Phys.*, **18**, 95
169. CADE, P. E., SALES, K. D. and WAHL, A. C. (1966). *J. Chem. Phys.*, **44**, 1973
170. KRUPENIE, P. H. (1972). *J. Phys. Chem. Ref. Data*, **1**, 423
171. DIXON, R. N. and HULL, S. E. (1969). *Chem. Phys. Lett.*, **3**, 367
172. JONATHAN, N., MORRIS, A., OKUDA, M., ROSS, K. J. and SMITH, D. J. (1974). *J. Chem. Soc., Faraday Trans. II*, **70**, 1810
173. DYKE, J. M., GOLOB, L., JONATHAN, N. and MORRIS, A. (1975). *J. Chem. Soc., Faraday Trans. II*, **71**, 1026
174. JONATHAN, N., SMITH, D. J. and ROSS, K. J. (1971). *Chem. Phys. Lett.*, **9**, 217
175. EDQVIST, O., ÅSBRINK, L. and LINDHOLM, E. (1971). *Z. Naturforsch., Teil A*, **26**, 1407
176. HOLLAS, J. M. and SUTHERLEY, T. A. (1971). *Mol. Phys.*, **22**, 213
177. SO, S. P. and RICHARDS, W. G. (1975). *J. Chem. Soc., Faraday Trans. II*, **71**, 62
178. CEDERBAUM, L. S., DOMCKE, W., KÖPPEL, H. and VON NIESSEN, W. (1977). *Chem. Phys.*, **26**, 169
179. BOTTER, R. and ROSENSTOCK, H. M. (1969). *J. Res. Natl. Bur. Stand.*, **73A**, 313
180. ÅSBRINK, L. and RABALAIS, J. W. (1971). *Chem. Phys. Lett.*, **12**, 182
181. DIXON, R. N., DUXBURY, G., HORANI, M. and ROSTAS, J. (1971). *Mol. Phys.*, **22**, 977
182. RABALAIS, J. W., KARLSSON, L., WERME, L. O., BERGMARK, T. and SIEGBAHN, K. (1973). *J. Chem. Phys.*, **58**, 3370
183. DYKE, J., JONATHAN, N., LEE, E. and MORRIS, A. (1976). *J. Chem. Soc., Faraday Trans. II*, **72**, 1385
184. MERER, A. J. and MULLIKEN, R. S. (1969). *Chem. Rev.*, **69**, 639
185. POTTS, A. W. and PRICE, W. C. (1972). *Proc. R. Soc. London*, **A326**, 165
186. DIXON, R. N. (1971). *Mol. Phys.*, **20**, 113
187. SCHULMAN, J. M. and MOSKOWITZ, J. W. (1967). *J. Chem. Phys.*, **47**, 3491
188. VON NIESSEN, W., CEDERBAUM, L. S. and KRAEMER, W. P. (1976). *J. Chem. Phys.*, **65**, 1378
189. SHIRLEY, D. A. (1974). *J. Electron Spectrosc.*, **5**, 135
190. PAULING, L. (1960). *Nature of the Chemical Bond.* Cornell University Press, Ithaca, NY
191. GORDON, J. P., ZEIGER, H. J. and TOWNES, C. H. (1954). *Phys. Rev.*, **95**, 282; see also **99**, 1264 (1955)
192. MAIMAN, T. H. (1960). *Nature (London)*, **187**, 493
193. MOORADIAN, A. (1979). *Rep. Prog. Phys.*, **42**, 1533
194. AMAT, G. and PIMBERT, M. (1965). *J. Mol. Spectrosc.*, **16**, 278
195. BENEDICT, W. S., POLLACK, M. A. and TOMLINSON, W. J., III, (1969). *IEEE J. Quantum Electron.*, **QE-5**, 108
196. LIDE, D. R., JR and MAKI, A. G. (1967). *Appl. Phys. Lett.*, **11**, 62
197. MAKI, A. G. (1968). *Appl. Phys. Lett.*, **12**, 122
198. EVENSON, K. M., JENNINGS, D. A., PETERSEN, F. R., MUCHA, J. A., JIMÉNEZ, J. J., CHARLTON, R. M. and HOWARD, C. J. (1977). *IEEE J. Quantum Electron.*, **QE-13**, 442
199. LEE, W., KIM, D., MALK, E. and LEAP, J. (1979). *IEEE J. Quantum Electron.*, **QE-15**, 838
200. HÄNSCH, T. W. (1972). *Appl. Opt.*, **11**, 895
201. ALBRECHT, A. C. (1961). *J. Chem. Phys.*, **34**, 1476
202. HOLZER, W., MURPHY, W. F. and BERNSTEIN, H. J. (1970). *J. Chem. Phys.*, **52**, 399
203. SENSION, R. J., BRUDZYNSKI, R. J., LI, S., HUDSON, B. S., ZERBETTO, F. and ZGIERSKI, M. K. (1992). *J Chem. Phys.*, **96**, 2617
204. KUMBLE, R., LOPPNOW, G. R., HU, S., MUKHERJEE, A., THOMPSON, M. A. and SPIRO, T. G. (1995). *J. Phys. Chem.*, **99**, 5809
205. JAS, G. S., WAN, C., KUCZERA, K. and JOHNSON, C. K. (1996). *J. Chem. Phys.*, **100**, 11857
206. CYVIN, S. J., RAUCH, J. E. and DECIUS, J. C. (1965). *J. Chem. Phys.*, **43**, 4083
207. STANTON, L. (1973). *J. Raman Spectrosc.*, **1**, 53
208. BONANG, C. C. and CAMERON, S. M. (1992). *J. Chem. Phys.*, **97**, 5377
209. ECKHARDT, G., HELLWARTH, R. W., McCLUNG, F. C., SCHWARZ, S. E. and WEINER, D. (1962). *Phys. Rev. Lett.*, **9**, 455

210. MINCK, R. W., TERHUNE, R. W. and RADO, W. G. (1963). *Appl. Phys. Lett.*, **3**, 181
211. OWYOUNG, A. and JONES, E. D. (1977). *Opt. Lett.*, **1**, 152
212. HARTLAND, G. V., HENSON, B. F., VENTURO, V. A., HERTZ, R. A. and FELKER, P. M. (1990). *J. Opt. Soc. Am. B*, **7**, 1950
213. ESHERICK, P., OWYOUNG, A. and PLIVA, J. (1985). *J. Chem. Phys.*, **83**, 3311
214. KING, D. A., HAINES, R., ISENOR, N. R. and ORR, B. J. (1983). *Opt. Lett.*, **8**, 629
215. JONES, W. J. and STOICHEFF, B. P. (1964). *Phys. Rev. Lett.*, **13**, 657
216. McLAREN, R. A. and STOICHEFF, B. P. (1970). *Appl. Phys. Lett.*, **16**, 140
217. ALFANO, R. R. and SHAPIRO, S. L. (1971). *Chem. Phys. Lett.*, **8**, 631
218. VALENTINI, J. J., ESHERICK, P. and OWYOUNG, A. (1980). *Chem. Phys. Lett.*, **75**, 590
219. HUBER-WÄLCHLI, P., GUTHALS, D. M. and NIBLER, J. W. (1979). *Chem. Phys. Lett.*, **67**, 233
220. DUNCAN, M. D., ÖSTERLIN, P. and BYER, R. L. (1981). *Opt. Lett.*, **6**, 90
221. BOZLEE, B. J. and NIBLER, J. W. (1986). *J. Chem. Phys.*, **84**, 3798
222. BECAN, T. S., JONUSCHEIT, J., LEHNER, U. and SCHRÖTTER, H. W. (1995). *J. Raman Spectrosc.*, **26**, 699
223. LAMB, W. E. (1964). *Phys. Rev.*, **134**, A1429
224. WELLS, J. S. and EVENSON, K. M. (1970). *Rev. Sci. Instrum.*, **41**, 226
225. DALE, R. M., JOHNS, J. W. C., McKELLAR, A. R. W. and RIGGIN, M. (1977). *J. Mol. Spectrosc.*, **67**, 440
226. EVENSON, K. M., BROIDA, H. P., WELLS, J. S., MAHLER, R. J. and MIZUSHIMA, M. (1968). *Phys. Rev. Lett.*, **21**, 1038
227. EVENSON, K. M., WELLS, J. S. and RADFORD, H. E. (1970). *Phys. Rev. Lett.*, **25**, 199
228. RADFORD, H. E., EVENSON, K. M. and HOWARD, C. J. (1974). *J. Chem. Phys.*, **60**, 3178
229. HOUGEN, J. T., RADFORD, H. E., EVENSON, K. M. and HOWARD, C. J. (1975). *J. Mol. Spectrosc.*, **56**, 210
230. TUCKETT, R. P., FREEDMAN, P. A. and JONES, W. J. (1979). *Mol. Phys.*, **37**, 379
231. RADFORD, H. E. and RUSSELL, D. K. (1977). *J. Chem. Phys.*, **66**, 2222
232. McKELLAR, A. R. W. (1979). *Can. J. Phys.*, **57**, 2106
233. BEATON, S. P., EVENSON, K. M. and BROWN, J. M. (1994). *J. Mol. Spectrosc.*, **164**, 395
234. BROWN, J. M., EVENSON, K. M., ZINK, L. R. and VASCONCELLOS, E. C. C. (1996). *Astrophys. J.*, **464**, L203
235. KÖRSGEN, H., EVENSON, K. M. and BROWN, J. M. (1997). *J. Chem. Phys.*, **107**, 1025
236. BREWER, R. G., KELLY, M. J. and JAVAN, A. (1969). *Phys. Rev. Lett.*, **23**, 559
237. SHIMIZU, F. (1970). *J. Chem. Phys.*, **52**, 3572
238. DUXBURY, G. and JONES, R. G. (1971). *Mol. Phys.*, **20**, 721
239. JOHNS, J. W. C. and McKELLAR, A. R. W. (1977). *J. Chem. Phys.*, **66**, 1217
240. JOHNS, J. W. C. and McKELLAR, A. R. W. (1975). *J. Chem. Phys.*, **63**, 1682
241. TAKAMI, M., JONES, H. and OKA, T. (1979). *J. Chem. Phys.*, **70**, 3557
242. SCHLOSSBERG, H. R. and JAVAN, A. (1966). *Phys. Rev.*, **150**, 267
243. DAVIES, P. B., MARTIN, N. A., NUNES, M. D. and RUSSELL, D. K. (1990). *J. Chem. Phys.*, **93**, 1576
244. McILROY, A., LASCOLA, R., LOVEJOY, C. M. and NESBITT, D. J. (1991). *J. Phys. Chem.*, **95**, 2636
245. GUTOWSKY, H. S., CHUANG, C., KLOTS, T. D., EMILSSON, T., RUOFF, R. S. and KRAUSE, K. R. (1988). *J. Chem. Phys.*, **88**, 2919
246. WALSH, M. A., ENGLAND, T. H., DYKE, T. R. and HOWARD, B. J. (1987). *Chem. Phys. Lett.*, **142**, 265
247. WEIDA, M. J., SPERHAC, J. M. and NESBITT, D. J. (1995). *J. Chem. Phys.*, **103**, 7685
248. XU, Y., CIVIS, S., McKELLAR, A. R. W., KONIG, S., HAVERLAG, M., HILPERT, G. and HAVENITH, M. (1996). *Mol. Phys.*, **87**, 1071
249. YAMADA, C., NAGAI, K. and HIROTA, E. (1981). *J. Mol. Spectrosc.*, **85**, 416
250. YAMADA, C., HIROTA, E. and KAWAGUCHI, K. (1981). *J. Chem. Phys.*, **75**, 5256
251. YAMADA, C. and HIROTA, E. (1986). *Phys. Rev. Lett.*, **56**, 923
252. GUDEMAN, C. S., BEGEMANN, M. H., PFAFF, J. and SAYKALLY, R. J. (1983). *Phys. Rev. Lett.*, **50**, 727
253. OWRUTSKY, J., ROSENBAUM, N., TACK, L., GRUEBELE, M., POLAK, M. and SAYKALLY, R. J. (1987). *Philos. Trans. R. Soc. London*, **A234**, 97
254. LIU, Z. and DAVIES, P. B. (1996). *J. Chem. Phys.*, **105**, 3443
255. XU, Y., FUKUSHIMA, M., AMANO, T. and McKELLAR, A. R. W. (1995). *Chem. Phys. Lett.*, **242**, 126
256. PINE, A. S. (1974). *J. Opt. Soc. Am.*, **64**, 1683
257. PINE, A. S. (1976). *J. Opt. Soc. Am.*, **66**, 97
258. HERZBERG, G. (1949). *Nature* (London), **163**, 170
259. McKELLAR, A. R. W. and OKA, T. (1978). *Can. J. Phys.*, **56**, 1315
260. OKA, T. (1980). *Phys. Rev. Lett.*, **45**, 531
261. SUHM, M. A., FARRELL, J. T., JR, ASHWORTH, S. H. and NESBITT, D. J. (1993). *J. Chem. Phys.*, **98**, 5985

262. LEHMANN, K. K., SCOLES, G. and PATE, B. H. (1994). *Annu. Rev. Phys. Chem.*, **45**, 241
263. ANDERSON, D. T., DAVIS, S., ZWIER, T. S. and NESBITT, D. J. (1996). *Chem. Phys. Lett.*, **258**, 207
264. McILROY, A. and NESBITT, D. J. (1991. *Chem. Phys. Lett.*, **187**, 215
265. FARRELL, J. T., JR, SNEH, O., McILROY, A., KNIGHT, A. E. W. and NESBITT, D. J. (1992). *J. Chem. Phys.*, **97**, 7967
266. HUTSON, J. M. (1992). *J. Chem. Phys.*, **96**, 6752
267. FARHOOMAND, J., BLAKE, G. A., FRERKING, M. A. and PICKETT, H. M. (1985). *J. Appl. Phys.*, **57**, 1763
268. BUSAROW, K. L., BLAKE, G. A., LAUGHLIN, K. B., COHEN, R. C., LEE, Y. T. and SAYKALLY, R. J. (1988). *J. Chem. Phys.*, **89**, 1268
269. COUDERT, L. H., LOVAS, F. J., SUENRAM, R. D. and HOUGEN, J. T. (1987). *J. Chem. Phys.*, **87**, 6290
270. ODUTOLA, J. A., HU, T. A., PRINSLOW, D., O'DELL, S. E. and DYKE, T. R. (1988). *J. Chem. Phys.*, **88**, 5352
271. BUSAROW, K. L., COHEN, R. C., BLAKE, G. A., LAUGHLIN, K. B., LEE, Y. T. and SAYKALLY, R. J. (1989). *J. Chem. Phys.*, **90**, 3937
272. COUDERT, L. H. and HOUGEN, J. T. (1988). *J. Mol. Spectrosc.*, **130**, 86
273. GOUGH, T. E., MILLER, R. E. and SCOLES, G. (1977). *Appl. Phys. Lett.*, **30**, 338
274. PATE, B. H., LEHMANN, K. K. and SCOLES, G. (1991). *J. Chem. Phys.*, **95**, 3891
275. McILROY, A., NESBITT, D. J., KERSTEL, E. R. TH., PATE, B. H., LEHMANN, K. K. and SCOLES, G. (1994). *J. Chem. Phys.*, **100**, 2596
276. BRUMMEL, C. L., MORK, S. W. and PHILIPS, L. A. (1991). *J. Chem. Phys.*, **95**, 7041
277. PINE, A. S., FRASER, G. T. and PLIVA, J. M. (1988). *J. Chem. Phys.*, **89**, 2720
278. PINE, A. S., LAFFERTY, W. J. and HOWARD, B. J. (1984). *J. Chem. Phys.*, **81**, 2939
279. HUANG, Z. S., JUCKS, K. W. and MILLER, R. E. (1986). *J. Chem. Phys.*, **85**, 3338
280. JUCKS, K. W. and MILLER, R. E. (1988). *J. Chem. Phys.*, **89**, 1262
281. KERSTEL, E. R. TH., SCOLES, G. and YANG, X. (1993). *J. Chem. Phys.*, **99**, 876
282. YANG, X., KERSTEL, E. R. TH., SCOLES, G., BEMISH, R. J. and MILLER, R. E. (1995). *J. Chem. Phys.*, **103**, 8828
283. SCOTONI, M., BOSCHETTI, A., OBERHOFER, N. and BASSI, D. (1991). *J. Chem. Phys.*, **94**, 971
284. SCOTONI, M., LEONARDI, C. and BASSI, D. (1991). *J. Chem. Phys.*, **95**, 8655
285. FRASER, G. T., SUENRAM, R. D. and COUDERT, L. H. (1989). *J. Chem. Phys.*, **90**, 6077
286. FRASER, G. T. and PINE, A. S. (1989). *J. Chem. Phys.*, **91**, 637
287. FRASER, G. T., PINE, A. S., DOMENECH, J. L. and PATE, B. H. (1993). *J. Chem. Phys.*, **99**, 2396
288. OKUMURA, M., YEH, L. I., MYERS, J. D. and LEE, Y. T. (1988). *J. Chem. Phys.*, **88**, 79
289. BIESKE, E. J., NIZKORODOV, S. A., BENNETT, F. R. and MAIER, J. P. (1995). *J. Chem. Phys.*, **102**, 5152
290. CHANG, H.-C. and KLEMPERER, W. (1993). *J. Chem. Phys.*, **98**, 2497
291. CHANG, H.-C. and KLEMPERER, W. (1993). *J. Chem. Phys.*, **98**, 9266
292. CHANG, H.-C. and KLEMPERER, W. (1994). *J. Chem. Phys.*, **100**, 1
293. BELL, A. G. (1880). *Proc. Am. Assoc. Adv. Sci.*, **29**, 115
294. KERR, E. L. and ATWOOD, J. G. (1968). *Appl. Opt.*, **7**, 915
295. DAVIDSSON, J., GUTOW, J. H. and ZARE, R. N. (1990). *J. Phys. Chem.*, **94**, 4069
296. BRAY, R. G. and BERRY, M. J. (1979). *J Chem. Phys.*, **71**, 4909
297. CROFTON, M. W., STEVENS, C. G., KLENERMAN, D., GUTOW, J. H. and ZARE, R. N. (1988). *J. Chem. Phys.*, **89**, 7100
298. REDDY, K. V., HELLER, D. F. and BERRY, M. J. (1982). *J. Chem. Phys.*, **76**, 2814
299. SEGALL, J., ZARE, R. N., DUBAL, H. R., LEWERENZ, M. and QUACK, M. (1987). *J. Chem. Phys.*, **86**, 634
300. NESBITT, D. J. and FIELD, R. W. (1996). *J. Phys. Chem.*, **100**, 12735
301. SIBERT, E. L., III, REINHARDT, W. P. and HYNES, J. T. (1984). *J. Chem. Phys.*, **81**, 1115
302. LEHMANN, K. K., SCHERER, G. J. and KLEMPERER, W. (1982). *J. Chem. Phys.*, **77**, 2853
303. CARTER, S., MILLS, I. M. and MURRELL, J. N. (1980). *J. Mol. Spectrosc.*, **81**, 110
304. SCHERER, G. J., LEHMANN, K. K. and KLEMPERER, W. (1983). *J. Chem. Phys.*, **78**, 2817
305. HENRY, B. R. (1977). *Acc. Chem. Res.*, **10**, 207
306. ZHAN, X., VAITTINEN, O., KAUPPI, E. and HALONEN, L. (1991). *Chem. Phys. Lett.*, **180**, 310
307. ZHAN, X., KAUPPI, E. and HALONEN, L. (1992). *Rev. Sci. Instrum.*, **63**, 5546
308. NEWNHAM, D., ZHAN, X., VAITTINEN, O., KAUPPI, E. and HALONEN, L. (1992). *Chem. Phys. Lett.*, **189**, 205
309. O'KEEFE, A. and DEACON, D. A. G. (1988). *Rev. Sci. Instrum.*, **59**, 2544
310. LEHMANN, K. K. and ROMANINI, D. (1996). *J. Chem. Phys.*, **105**, 10263
311. BABCOCK, H. D. and HERZBERG, L. (1948). *Astrophys. J.*, **108**, 167

312. HERZBERG, G. (1950). *Spectra of Diatomic Molecules*. Van Nostrand, New York
313. ROMANINI, D. and LEHMANN, K. K. (1993). *J. Chem. Phys.*, **99**, 6287
314. SCHERER, J. J., VOELKEL, D., RAKESTRAW, D. J., PAUL, J. B., COLLIER, C. P., SAYKALLY, R. J. and O'KEEFE, A. (1995). *Chem. Phys. Lett.*, **245**, 273
315. SCHERER, J. J., PAUL, J. B., COLLIER, C. P. and SAYKALLY, R. J. (1995). *J. Chem. Phys.*, **102**, 5190
316. SCHERER, J. J., PAUL, J. B., COLLIER, C. P. and SAYKALLY, R. J. (1995). *J. Chem. Phys.*, **103**, 113
317. POWERS, D. E., HANSEN, S. G., GEUSIC, M. E., MICHALOPOULOS, D. L. and SMALLEY, R. E. (1982). *J. Phys. Chem.*, **86**, 2556
318. KOTTERER, M., CONCEICAO, J. and MAIER, J. P. (1996). *Chem. Phys. Lett.*, **259**, 233
319. MEIJER, G., BOOGAARTS, M. G. H., JONGMA, R. T., PARKER, D. H. and WODTKE, A. M. (1994). *Chem. Phys. Lett.*, **217**, 112
320. CARRINGTON, A., SHAW, A. M. and TAYLOR, S. M. (1995). *J. Chem. Soc. Faraday Trans.*, **91**, 3725
321. WING, W. H., RUFF, G. A., LAMB, W. E. and SPEZESKI, J. J. (1976). *Phys. Rev. Lett.*, **36**, 1488
322. CARRINGTON, A. and SARRE, P. J. (1977). *Mol. Phys.*, **33**, 1495
323. CARRINGTON, A., MILVERTON, D. R. J. and SARRE, P. J. (1978). *Mol. Phys.*, **35**, 1505
324. CARRINGTON, A., BUTTENSHAW, J. and ROBERTS, P. G. (1979). *Mol. Phys.*, **38**, 1711
325. CARRINGTON, A., ROBERTS, P. G. and SARRE, P. J. (1978). *Mol. Phys.*, **35**, 1523
326. BJERRE, N. and KEIDING, S. R. (1986). *Phys. Rev. Lett.*, **56**, 1459
327. CARRINGTON, A., McNAB, I. R. and MONTGOMERIE, C. A. (1988). *Chem. Phys. Lett.*, **151**, 258
328. PEEK, J. M. (1969). *J. Chem. Phys.*, **50**, 4595
329. CARRINGTON, A., McNAB, I. R., MONTGOMERIE, C. A. and KENNEDY, R. A. (1989). *Mol. Phys.*, **67**, 711
330. MOSS, R. E. (1993). *J. Chem. Soc., Faraday Trans.*, **89**, 3851
331. MOSS, R. E. (1993). *Chem. Phys. Lett.*, **206**, 83
332. BUNKER, P. R. (1974). *Chem. Phys. Lett.*, **27**, 322
333. CARRINGTON, A., McNAB, I. R. and MONTGOMERIE, C. A. (1988). *Chem. Phys. Lett.*, **151**, 258
334. CARRINGTON, A. and KENNEDY, R. A. (1984). *Mol. Phys.*, **56**, 935
335. CARRINGTON, A., GAMMIE, D. I., SHAW, A. M., TAYLOR, S. M. and HUTSON, J. M. (1996). *Chem. Phys. Lett.*, **260**, 395
336. CARRINGTON, A., GAMMIE, D. I., SHAW, A. M. and TAYLOR, S. M. (1996). *Chem. Phys. Lett.*, **262**, 598
337. CARRINGTON, A. and KENNEDY, R. A. (1984). *J. Chem. Phys.*, **81**, 91
338. McNAB, I. R. (1995). *Adv. Chem. Phys.*, **89**, 91
339. CARRINGTON, A., BUTTENSHAW, J., KENNEDY, R. A. and SOFTLEY, T. P. (1981). *Mol. Phys.*, **44**, 1233
340. LIU, Z. and DAVIES, P. B. (1997). *J. Chem. Phys.*, **107**, 337
341. LIU, Z. and DAVIES, P. B. (1997). *Phys. Rev. Lett.*, **79**, 2779
342. DELON, A., JOST, R. and LOMBARDI, M. (1991). *J. Chem. Phys.*, **95**, 5701
343. GEORGES, R., DELON, A. and JOST, R. (1995). *J. Chem. Phys.*, **103**, 1732
344. KÖPPEL, H., DOMCKE, W. and CEDERBAUM, L. S. (1984). *Adv. Chem. Phys.*, **57**, 59
345. MERER, A. J. (1989). *Annu. Rev. Phys. Chem.*, **40**, 407
346. CHEUNG, A. S.-C., TAYLOR, A. W. and MERER, A. J. (1982). *J. Mol. Spectrosc.*, **92**, 391
347. HUANG, G., MERER, A. J. and CLOUTHIER, D. J. (1992). *J. Mol. Spectrosc.*, **153**, 32
348. BARNES, M., HAJIGEORGIOU, P. G. and MERER, A. J. (1993). *J. Mol. Spectrosc.*, **160**, 289
349. VARBERG, T. D., GRAY, J. A., FIELD, R. W. and MERER, A. J. (1992). *J. Mol. Spectrosc.*, **156**, 296
350. VARBERG, T. D., FIELD, R. W. and MERER, A. J. (1991). *J. Chem. Phys.*, **95**, 1563
351. ASHWORTH, S. H., GRIEMAN, F. J. and BROWN, J. M. (1996). *J. Chem. Phys.*, **104**, 48
352. TAN, X.-Q., WRIGHT, T. G. and MILLER, T. A. (1995). In *Jet Spectroscopy and Molecular Dynamics* (Eds J. M. Hollas and D. Phillips), chap. 3. Blackie, Glasgow
353. ENGELKING, P. C. (1986). *Rev. Sci. Instrum.*, **57**, 2274
354. DROEGE, A. T. and ENGELKING, P. C. (1983). *Chem. Phys. Lett.*, **96**, 316
355. BROSSARD, S. D., CARRICK, P. G., CHAPPELL, E. L., HULEGAARD, S. C. and ENGELKING, P. C. (1986). *J. Chem. Phys.*, **84**, 2459
356. BERRY, M. T., BRUSTEIN, M. R. and LESTER, M. I. (1990). *J. Chem. Phys.*, **92**, 6469
357. CLOUTHIER, D. J. and KAROLCZAK, J. (1991). *J. Chem. Phys.*, **94**, 1
358. LIU, X., DAMO, C. P., LIN, T.-Y., FOSTER, S. C., MISRA, P., YU, L. and MILLER, T. A. (1989). *J. Phys. Chem.*, **93**, 2266
359. LIU, X., FOSTER, S. C., WILLIAMSON, J. M., YU, L. and MILLER, T. A. (1990). *Mol. Phys.*, **69**, 357
360. YU, L., FOSTER, S. C., WILLIAMSON, J. M., HEAVEN, M. C. and MILLER, T. A. (1988). *J. Phys. Chem.*, **92**, 4263
361. YU, L., CULLIN, D. W., WILLIAMSON, J. M. and MILLER, T. A. (1993). *J. Chem. Phys.*, **98**, 2682

362. CERNY, T. M., TAN, X.-Q., WILLIAMSON, J. M., ROBLES, E. S. J., ELLIS, A. M. and MILLER, T. A. (1993). *J. Chem. Phys.*, **99**, 9376
363. MAJEWSKI, W. and MEERTS, W. L. (1981). *J. Mol. Spectrosc.*, **104**, 271
364. HOLLAS, J. M. and THAKUR, S. N. (1971). *Mol. Phys.*, **22**, 203
365. VAN HERPEN, W. M., MEERTS, W. L., DRABE, K. E. and KOMMANDEUR, J. (1987). *J. Chem. Phys.*, **86**, 4396
366. DRABBELS, M., HEINZE, J. and MEERTS, W. L. (1994). *J. Chem. Phys.*, **100**, 165
367. CALLOMON, J. H., DUNN, T. M. and MILLS, I. M. (1966). *Philos. Trans. R. Soc. London*, **259A**, 499
368. WILSON, E. B. (1934). *Phys. Rev.*, **45**, 706
369. RIEDLE, E. and PLIVA, J. (1991). *Chem. Phys.*, **152**, 375
370. RIEDLE, E., KNITTEL, TH., WEBER, TH. and NEUSSER, H. J. (1989). *J. Chem. Phys.*, **91**, 4555
371. HAYNAM, C. A., BRUMBAUGH, D. V. and LEVY, D. H. (1983). *J. Chem. Phys.*, **79**, 1581
372. JOHNSON, J. R., JORDAN, K. D., PLUSQUELLIC, D. F. and PRATT, D. W. (1990). *J. Chem. Phys.*, **93**, 2258
373. PLUSQUELLIC, D. F., TAN, X.-Q. and PRATT, D. W. (1992). *J. Chem. Phys.*, **96**, 8026
374. HUMPHREY, S. J. and PRATT, D. W. (1993). *J. Chem. Phys.*, **99**, 5078
375. SINCLAIR, W. E. and PRATT, D. W. (1996). *J. Chem. Phys.*, **105**, 7942
376. CHRISTOFFERSEN, J., HOLLAS, J. M. and KIRBY, G. H. (1969). *Mol. Phys.*, **16**, 441
377. CRAIG, D. P. (1950). *J. Chem. Soc.*, 2146
378. HOLLAS, J. M., HOWSON, M. R. and RIDLEY, T. (1983). *Chem. Phys. Lett.*, **98**, 611
379. OKUYAMA, K., KAKINUMA, T., FUJII, M., MIKAMI, M. and ITO, M. (1986). *J. Phys. Chem.*, **90**, 3948
380. SHIN, Y.-D., SAIGUSA, H., ZGIERSKI, M. K., ZERBETTO, F. and LIM, E. C. (1991). *J. Chem. Phys.*, **94**, 3511
381. CHAKRABORTY, T. and LIM, E. C. (1993). *J. Chem. Phys.*, **98**, 836
382. TOMIOKA, Y., ITO, M. and MIKAMI, N. (1983). *J. Phys. Chem.*, **87**, 4401
383. ALVES, A. C. P., HOLLAS, J. M., MUSA, H. and RIDLEY, T. (1985). *J. Mol. Spectrosc.*, **109**, 99
384. REDINGTON, R. L. (1996). In *Low Temperature Spectroscopy* (Ed. R. Fausto), p. 227. Kluwer, Dordrecht
385. HASSAN, K. H. and HOLLAS, J. M. (1989). *Chem. Phys. Lett.*, **157**, 183
386. CAMINATI, W., MELANDRI, S., CORBELLI, G., FAVERO, L. B. and MEYER, R. (1993). *Mol. Phys.*, **80**, 1297
387. RUDOLPH, H. D., DREIZLER, H., JAESCKE, A. and WENDLING, P. (1967). *Z Naturforsch. Teil A*, **22**, 940
388. CAMINATI, W., CAZZOLI, G. and MIRRI, A. M. (1975). *Chem. Phys. Lett.*, **35**, 475
389. GORDON, R. D. and HOLLAS, J. M. (1989). *Chem. Phys. Lett.*, **164**, 255
390. BREEN, P. J., WARREN, J. A., BERNSTEIN, E. R. and SEEMAN, J. I. (1987). *J. Chem. Phys.*, **87**, 1917
391. ISHIKAWA, H., KAJIMOTO, O. and KATO, S. (1993). *J. Chem. Phys.*, **99**, 800
392. OKUYAMA, K., MIKAMI, N. and ITO, M. (1985). *J. Phys. Chem.*, **89**, 5617
393. LU, K.-T., WEINHOLD, F. and WEISSHAAR, J. C. (1995). *J. Chem. Phys.*, **102**, 6787
394. SPANGLER, L. H. and PRATT, D. W. (1995). In *Jet Spectroscopy and Molecular Dynamics* (Eds J. M. Hollas and D. Phillips), chap. 9. Blackie, Glasgow
395. HOFFMANN, R. (1971). *Acc. Chem. Res.*, **4**, 1
396. OKUYAMA, K., MIKAMI, N. and ITO, M. (1987). *Laser Chem.*, **7**, 197
397. TAN, X.-Q. and PRATT, D. W. (1994). *J. Chem. Phys.*, **100**, 7061
398. GORDON, R. D., HOLLAS, J. M., RIBEIRO-CLARO, P. J. A. and TEIXEIRA-DIAS, J. J. C. (1993). *Chem. Phys. Lett.*, **211**, 392
399. GORDON, R. D., HOLLAS, J. M., RIBEIRO-CLARO, P. J. A. and TEIXEIRA-DIAS, J. J. C. (1991). *Chem. Phys. Lett.*, **183**, 377
400. GORDON, R. D., HOLLAS, J. M., RIBEIRO-CLARO, P. J. A. and TEIXEIRA-DIAS, J. J. C. (1991). *Chem. Phys. Lett.*, **182**, 649
401. MONTE, CH., ROGGAN, A., SUBARIC-LEITIS, A., RETTIG, W. and ZIMMERMANN, P. (1993). *J. Chem. Phys.*, **98**, 2580
402. HOLLAS, J. M., KHALILIPOUR, E. and THAKUR, S. N. (1978). *J. Mol. Spectrosc.*, **73**, 240
403. HOLLAS, J. M. and RIDLEY, T. (1980). *Chem. Phys. Lett.*, **75**, 94
404. HOLLAS, J. M., MUSA, H., RIDLEY, T., TURNER, P. H., WEISSENBERGER, K. H. and FAWCETT, V. (1982). *J. Mol. Spectrosc.*, **94**, 437
405. KUNDU, T., GOODMAN, L. and LESZCZYNSKI, J. (1995). *J. Chem. Phys.*, **103**, 1523
406. LIN, T.-Y. D., and MILLER, T. A. (1990). *J. Phys. Chem.*, **94**, 3554
407. SPANGLER, L. H. and PRATT, D. W. (1986). *J. Chem. Phys.*, **84**, 4789
408. GÖPPERT-MAYER, M. (1931). *Ann. Phys.*, **9**, 273
409. OKA, T. and SHIMIZU, T. (1970). *Phys. Rev. A*, **2**, 587
410. OKA, T. and SHIMIZU, T. (1971). *Appl. Phys. Lett.*, **19**, 88
411. JONES, H. (1977). *Appl. Phys.*, **14**, 169

412. JONES, H. (1978). *J. Mol. Spectrosc.*, **70**, 279
413. HOCHSTRASSER, R. M., SUNG, H.-N. and WESSEL, J. E. (1973). *J. Chem. Phys.*, **58**, 4694
414. IM, H.-S. and BERNSTEIN, E. R. (1988). *J. Chem. Phys.*, **88**, 7337
415. BRAY, R. G., HOCHSTRASSER, R. M. and WESSEL, J. E. (1974). *Chem. Phys. Lett.*, **27**, 167
416. MARINERO, E. E., VASUDEV, R. and ZARE, R. N. (1983). *J. Chem. Phys.*, **78**, 692
417. BUNTINE, M. A., BALDWIN, D. P., ZARE, R. N. and CHANDLER, D. W. (1991). *J. Chem. Phys.*, **94**, 4672
418. GRAHAM, W. R. M. and LEW, H. (1978). *Can. J. Phys.*, **56**, 85
419. ASHFOLD, M. N. R. and HOWE, J. D. (1994). *Annu. Rev. Phys. Chem.*, **45**, 57
420. ASHFOLD, M. N. R., CLEMENTS, S. G., HOWE, J. D. and WESTERN, C. M. (1991). *J. Chem. Soc., Faraday Trans.*, **87**, 2515
421. McCLAIN, W. M. and HARRIS, R. A. (1977). In *Excited States* (Ed. E. C. Lim), vol. 3, chap. 1. Academic Press, New York
422. CLEMENT, S. G., ASHFOLD, M. N. R. and WESTERN, C. M. (1992). *J. Chem. Soc., Faraday Trans.*, **88**, 3121
423. JOHNSON, P. M., BERMAN, M. R. and ZAKHEIM, D. (1975). *J. Chem. Phys.*, **62**, 2500
424. NIEMAN, G. C. and COLSON, S. D. (1979). *J. Chem. Phys.*, **71**, 571
425. GLOWNIA, J. H., RILEY, S. J. and COLSON, S. D. (1980). *J. Chem. Phys.*, **72**, 5998
426. ASHFOLD, M. N. R., BENNETT, C. L., DIXON, R. N., FIELDEN, P., RIELEY, H. and STICKLAND, R. J. (1986). *J. Mol. Spectrosc.*, **117**, 216
427. ASHFOLD, M. N. R., DIXON, R. N., ROSSER, K. N., STICKLAND, R. J. and WESTERN, C. M. (1986). *Chem. Phys.*, **101**, 467
428. BIRABEN, F., CAGNAC, B. and GRYNBERG, G. (1974). *Phys. Rev. Lett.*, **32**, 643
429. MORSE, M. D., HOPKINS, J. B., LANGRIDGE-SMITH, P. R. R. and SMALLEY, R. E. (1983). *J. Chem. Phys.*, **79**, 5316
430. MICHALOPOULOS, D. L., GEUSIC, M. E., LANGRIDGE-SMITH, P. R. R. and SMALLEY, R. E. (1984). *J. Chem. Phys.*, **80**, 3556
431. SIEBER, H., RIEDLE, E. and NEUSSER, H. J. (1990). *Chem. Phys. Lett* **169**, 191
432. SIEBER, H., BRUNO, A. E. and NEUSSER, H. J. (1990). *J. Phys. Chem.*, **94**, 203
433. HOCHSTRASSER, R. M. and WESSEL, J. E. (1974). *Chem. Phys. Lett.*, **24**, 1
434. WUNSCH, L., METZ, F., NEUSSER, H. J. and SCHLAG, E. W. (1977). *J. Chem. Phys.*, **66**, 386
435. LOMBARDI, J. R., WALLENSTEIN, R., HÄNSCH, T. W. and FRIEDRICH, D. M. (1976). *J. Chem. Phys.*, **65**, 2357
436. RIEDLE, E., NEUSSER, H. J. and SCHLAG, E. W. (1981). *J. Chem. Phys.*, **75**, 4231
437. SUSSMAN, R., WEBER, TH., RIEDLE, E. and NEUSSER, H. J. (1992). *Opt. Commun.*, **88**, 408
438. ROBEY, M. J. and SCHLAG, E. W. (1978). *Chem. Phys.*, **30**, 9
439. WEBER, TH. and NEUSSER, H. J. (1991). *J. Chem. Phys.*, **94**, 7689
440. RIEDLE, E., SUSSMAN, R., WEBER, TH. and NEUSSER, H. J. (1996). *J. Chem. Phys.*, **104**, 865
441. RIEDLE, E, and VAN DER AVOIRD, A. (1996). *J. Chem. Phys.*, **104**, 882
442. BIESKE, E. J., RAINBIRD, M. W., ATKINSON, I. M. and KNIGHT, A. E. W. (1989). *J. Chem. Phys.*, **91**, 752
443. BÜRGHI, T. and LEUTWYLER, S. (1994). *J. Chem. Phys.*, **101**, 8418
444. HÄNSCH, T. W., LEE, S. A., WALLENSTEIN, R. and WIEMAN, C. (1975). *Phys. Rev. Lett.*, **34**, 307
445. HÄNSCH, T. W., SCHAWLOW, A. L. and SERIES, G. W. (1979). *Sci. Am.*, **240**, 72
446. MÜLLER-DETHLEFS, K., SANDER, M. and SCHLAG, E. W. (1984). *Chem. Phys. Lett.*, **112**, 291
447. REISER, G., HABENICHT, W., MÜLLER-DETHLEFS, K. and SCHLAG, E. W. (1988). *Chem. Phys. Lett.*, **152**, 119
448. XIE, J. and ZARE, R. N. (1990). *J. Chem. Phys.*, **93**, 3033
449. FISCHER, I., LINDNER, R. and MÜLLER-DETHLEFS, K., (1994). *J. Chem. Soc., Faraday Trans.*, **90**, 2425
450. MERKT, F. and SOFTLEY, T. P. (1992). *Phys. Rev. A*, **46**, 302
451. MACKENZIE, S. R., MERKT, F., HALSE, E. J. and SOFTLEY, T. P. (1995). *Mol. Phys.*, **86**, 1283
452. HABENICHT, W., REISER, G. and MÜLLER-DETHLEFS, K. (1991). *J. Chem. Phys.*, **95**, 4809
453. MULLER-DETHLEFS, K. (1991). *J. Chem. Phys.*, **95**, 4821
454. WHITE, M. G., and WIEDEMANN, R. T. (1992). *Proc. SPIE–Int. Soc. Opt. Eng.*, **1638**, 273
455. FISCHER, I., LOCHSCHMIDT, A., STROBEL, A., NIEDNER-SCHATTEBURG, G., MÜLLER-DETHLEFS, K. and BONDYBEY, V. E. (1993). *J. Chem. Phys.*, **98**, 3592
456. TAKAZAWA, K., FUJII, M. and ITO, M. (1993). *J. Chem. Phys.*, **99**, 3205
457. ZHANG, X., SMITH, J. M. and KNEE, J. L. (1992). *J. Chem. Phys.*, **97**, 2843
458. WRIGHT, T. G., CORDES, E., DOPFER, O. and MÜLLER-DETHLEFS, K. (1993). *J. Chem. Soc., Faraday Trans.*, **89**, 1609
459. BLUSH, J. A., CHEN, P., WIEDEMANN, R. T. and WHITE, M. G. (1993). *J. Chem. Phys.*, **98**, 3557

460. HERZBERG, G, and LAGERQVIST, A. (1968). *Can. J. Phys.*, **46**, 2363
461. LINEBERGER, W. C. and WOODWARD, B. W. (1970). *Phys. Rev. Lett.*, **25**, 424
462. ZITTEL, P. F., ELLISON, G. B., O'NEIL, S. W., HERBST, E., LINEBERGER, W. C. and REINHARDT, W. P. (1976). *J. Am. Chem. Soc.*, **98**, 3731
463. LEOPOLD, D. G., MURRAY, K. K. and LINEBERGER, W. C. (1984). *J. Chem. Phys.*, **81**, 1048
464. OAKES, J. M., HARDING, L. B. and ELLISON, G. B. (1985). *J. Chem. Phys.*, **83**, 5400
465. HO, J., ERVIN, K. M. and LINEBERGER, W. C. (1990). *J. Chem. Phys.*, **93**, 6987
466. DIXON-WARREN, ST. J., GUNION, R. F. and LINEBERGER, W. C. (1996). *J. Chem. Phys.*, **104**, 4902
467. KITSOPOULOS, T. N., CHICK, C. J., ZHAO, Y. and NEUMARK, D. M. (1991). *J. Chem. Phys.*, **95**, 1441
468. BATTAGLIA, A., GOZZINI, A. and POLACCO, E. (1959). *Nuovo Cimento*, **14**, 1076
469. YAJIMA, T. and SHIMODA, K. (1960). *J. Phys. Soc. Jpn.*, **15**, 1668
470. WOODS III, R. C., RONN, A. M. and WILSON, E. B., JR (1966). *Rev. Sci. Instrum.*, **37**, 927
471. OKA, T. (1967). *J. Chem. Phys.*, **47**, 13
472. SERIES, G. W. (1963). *Phys. Rev. Lett.*, **11**. 13
473. FIELD, R. W., ENGLISH, A. D., TANAKA, T., HARRIS, D. O. and JENNINGS, D. A. (1973). *J. Chem. Phys.*, **59**, 2191
474. FELD, M. S. and JAVAN, A. (1969). *Phys. Rev.*, **177**, 540
475. BREWER, R. G. (1970). *Phys. Rev. Lett.*, **25**, 1639
476. BEDWELL, D. J. and DUXBURY, G. (1979). *Chem. Phys.*, **37**, 445
477. DIXON, R. N. and NOBLE, M. (1980). *Chem. Phys.*, **50**, 331
478. CORNEY, A. and SERIES, G. W. (1964). *Proc. Phys. Soc.*, **83**, 213
479. FIELD, R. W., CAPELLE, G. A., and REVELLI, M. A. (1975). *J. Chem. Phys.*, **63**, 3228
480. GOTTSCHO, R. A., KOFFEND, J. B., FIELD, R. W. and LOMBARDI, J. R. (1978). *J. Chem. Phys.*, **68**, 4110
481. HERMAN, I. P., JAVAN, A. and FIELD, R. W. (1978). *J. Chem. Phys.*, **68**, 2398
482. EBATA, T., MIZUOCHI, N., WATANABE, T. and MIKAMI, N. (1996). *J. Phys. Chem.*, **100**, 546
483. FROST, R. K., HAGEMEISTER, R. F., SCHLEPPENBACH, D., LAURENCE, G. and ZWIER, T. S. (1996). *J. Phys. Chem.*, **100**, 16835
484. POWERS, D. E., PUSHKARSKY, M. and MILLER, T. A. (1995). *Chem. Phys. Lett.*, **247**, 548
485. AL-KAHALI, M. S. N., DONOVAN, R. J., LAWLEY, K. P. and RIDLEY, T. (1994). *Chem. Phys. Lett.*, **220**, 225
486. DONOVAN, R. J., LAWLEY, K. P., MIN, Z., RIDLEY, T. and YARWOOD, A. J. (1994). *Chem. Phys. Lett.*, **226**, 525
487. SHINZAWA, T., TOKUNAGA, A., ISHIWATA, T. and TANAKA, I. (1985). *J. Chem. Phys.*, **83**, 5407
488. AL-KAHALI, M. S. N., DONOVAN, R. J., LAWLEY, K. P., MIN, Z. and RIDLEY, T. (1996). *J. Chem. Phys.*, **104**, 1825
489. DALBY, F. W., PETTY-SIL, G., PRYCE, M. H. L. and TAI, C. (1977). *Can. J. Phys.*, **55**, 1033
490. KLIMCAK, C. M. and WESSEL, J. E. (1980). *J. Opt. Soc. Am.*, **70**, 659
491. WIEMAN, C. and HÄNSCH, T. W. (1976). *Phys. Rev. Lett.*, **36**, 1170
492. HE, G.-Z., KUHN, A., SCHIEMANN, S. and BERGMANN, K. (1990). *J. Opt. Soc. Am. B*, **7**, 1960
493. HAMILTON, C. E., KINSEY, J. L. and FIELD, R. W, (1986). *Annu. Rev. Phys. Chem.*, **37**, 493
494. NORTHRUP, F. J. and SEARS, T. J. (1992). *Annu. Rev. Phys. Chem.*, **43**, 127
495. CHEN, Y., HALLE, S., JONAS, D. M., KINSEY, J. L. and FIELD, R. W. (1990). *J. Opt. Soc. Am. B*, **7**, 1805
496. YANG, X., ROGASKI, C. A. and WODTKE, A M. (1990). *J. Opt. Soc. Am. B*, **7**, 1835
497. HERZEBERG, G. and INNES, K. K. (1957). *Can. J. Phys.*, **35**, 842
498. CARTER, S., HANDY, N. C, and MILLS, I. M. (1990). *Philos. Trans. R. Soc. London*, **A332**, 309
499. CARTER, S., MILLS, I. M. and HANDY, N. C. (1990). *J. Chem. Phys.*, **93**, 3722
500. INNES, K. K. (1954). *J. Chem. Phys.*, **22**, 863
501. CHEN, Y., HALLE, S., JONAS, D. M., KINSEY, J. L. and FIELD, R. W. (1990). *J. Opt. Soc. Am. B*, **7**, 1805
502. SOLINA, S. A. B., O'BRIEN, J. P., FIELD, R. W. and POLIK, W. F. (1996). *J. Phys. Chem.*, **100**, 7797
503. GODDARD, J. D. and SCHAEFER, H. F., III (1990). *J. Chem. Phys.*, **93**, 4907
504. CHOI, Y. S. and MOORE, C. B. (1991). *J. Chem. Phys.*, **94**, 5414
505. PARKIN, J. E. and INNES, K. K. (1965). *J. Mol. Spectrosc.*, **16**, 93
506. TOBIASON, J. D., DUNLOP, J. R. and ROHLFING, E. A. (1995). *J. Chem. Phys.*, **103**, 1448
507. FARROW, R. L. and RAKESTRAW, D. J. (1992). *Science*, **257**, 1894
508. BUNTINE, M. A., CHANDLER, D. W. and HAYDEN, C. C. (1992). *J. Chem. Phys.*, **97**, 707
509. NEYER, D. W., LUO, X., BURAK, I. and HOUSTON, P. L. (1995). *J. Chem. Phys.*, **102**, 1645
510. KELLER, H.-M., FLOETHMANN, H., DOBBYN, A. J., SCHINKE, R., WERNER, H.-J., BAUER, C. and ROSMUS, P. (1996). *J. Chem. Phys.*, **105**, 4983
511. HUGGINS, W. (1882). *Proc. R. Soc. London*, **33**, 1

512. HINKLE, K. H., KEADY, J. J. and BERNATH, P. F. (1988). *Science*, **241**, 1319
513. DOUGLAS, A. E. (1951). *Astrophys. J.*, **114**, 466
514. WHITESIDE, R. A., KRISHNAN, R., FRISCH, M. J., POPLE, J. A. and SCHLEYER, P. VON R. (1981). *Chem. Phys. Lett.*, **80**, 547
515. ROHLFING, E. A. and GOLDSMITH, J. E. M. (1990). *J. Opt. Soc. Am. B*, **7**, 1915
516. NORTHRUP, F. J. and SEARS, T. J. (1990). *J. Opt. Soc. Am. B*, **7**, 1924
517. CELII, F. G. and MAIER, J. P. (1990). *Chem. Phys. Lett.*, **166**, 517
518. CELII, F. G. and MAIER, J. P. (1990). *J. Opt. Soc. Am. B*, **7**, 1944
519. LAWRANCE, W. D. and KNIGHT, A. E. W. (1983). *J. Chem. Phys.*, **79**, 6030
520. EBATA, T., FURUKAWA, M., SUZUKI, T. and ITO, M. (1990). *J. Opt. Soc. Am. B*, **7**, 1890
521. BORDÉ, C. J. (1970). *C. R. Acad. Sci.*, **271**B, 371
522. SMITH, P. W. and HÄNSCH, T. W. (1971). *Phys. Rev. Lett.*, **26**, 29
523. HÄNSCH, T. W., LEVENSON, M. D. and SCHAWLOW, A. L. (1971). *Phys. Rev. Lett.*, **26**, 946
524. BORDÉ, C. J., OUHAYOUN, M., VAN LERBERGHE, A., SALOMON, C., AVRILLIER, S., CANTRELL, C. D. and BORDÉ, J. (1980). In *Laser Spectroscopy IV*. Springer, Berlin
525. SOREM, M. S. and SCHAWLOW, A. L. (1972). *Opt. Commun.*, **5**, 148
526. MUIRHEAD, A., SASTRY, K. V. L. N., CURL, R. F., COOK, J. and TITTEL, F. K. (1974). *Chem. Phys. Lett.*, **24**, 208
527. LOWE, R. S., GERHARDT, H., DILLENSCHNEIDER, W., CURL, R. F. and TITTEL, F. K. (1979). *J. Chem. Phys.*, **70**, 42
528. RAAB, M., HÖNING, G., CASTELL, R. and DEMTRÖDER, W. (1979). *Chem. Phys. Lett.*, **66**, 307
529. HANLE, W. (1924). *Z. Phys.*, **30**, 93
530. ZARE, R. N. (1971). *Acc. Chem. Res.*, **4**, 361
531. McCLINTOCK, M., DEMTRÖDER, W. and ZARE, R. N. (1969). *J. Chem. Phys.*, **51**, 5509
532. SILVERS, S. J., BERGEMAN, T. H. and KLEMPERER, W. (1970). *J. Chem. Phys.*, **52**, 4385
533. KROLL, M. (1975). *J. Chem. Phys.* **63**, 1803
534. SERIES, G. W. (1966). *Proc. Phys. Soc.*, **89**, 1017
535. COLEGROVE, F. D., FRANKEN, P. A., LEWIS, R. R. and SANDS, R. H. (1959). *Phys. Rev. Lett.*, **3**, 420
536. BROG, K. C., ECK, T. G. and WIEDER, H. (1967). *Phys. Rev.*, **153**, 91
537. DIXON, R. N. and FIELD, D. (1977), *Mol. Phys.*, **34**, 1563
538. TANAKA, K., SOMEYA, K. and TANAKA, T. (1991). *Chem. Phys.*, **152**, 229
539. ECK, T. G., FOLDY, L. L. and WIEDER, H. (1963). *Phys. Rev. Lett.*, **10**, 239
540. WIEDER, H. and ECK, T. G. (1967). *Phys. Rev.*, **153**, 103
541. COOK, J. M., ZEGARSKI, B. R. and MILLER, T. A. (1979). *J. Chem. Phys.*, **70**, 3739
542. MILLER, T. A. and FREUND, R. S. (1974). *J. Chem. Phys.*, **61**, 2160
543. CORNEY. A. and SERIES, G. W. (1964). *Proc. Phys. Soc.*, **83, 207**
544. DODD, J. N., KAUL, R. D. and WARRINGTON, D. M. (1964). *Proc. Phys. Soc.*, **84**, 176
545. ALEKSANDROV, E. B. (1964). *Opt. Spectrosc.*, **17**, 522
546. HAROCHE, S., PAISNER, J. A. and SCHAWLOW, A. L. (1973). *Phys. Rev. Lett.*, **30**, 948
547. WALLENSTEIN, R., PAISNER, J. A. and SCHAWLOW, A. L. (1974). *Phys. Rev. Lett.*, **32**, 1333
548. GOUEDARD, G. and LEHMANN, J. C. (1979). *J. Phys. Lett.*, **40**, L119
549. CHAIKEN, J., BENSON, T., GURNICK, M. and McDONALD, J. D. (1979) *Chem. Phys. Lett.*, **61**, 195
550. FELKER, P. M. and ZEWAIL, A. H. (1985). *J. Chem. Phys.*, **82**, 2975
551. FELKER, P. M. and ZEWAIL, A. H. (1995). In Jet Spectroscopy and Molecular Dynamics (Eds J. M. Hollas and D. Phillips), chap. 7. Blackie, Glasgow
552. TOPP, M. R. (1995). In *Jet Spectroscopy and Molecular Dynamics* (Eds J. M. Hollas and D. Phillips), chap. 8. Blackie, Glasgow
553. WILLMOTT, P. R., BITTO, H. and HUBER, J. R. (1992). *Chem. Phys. Lett.*, **188**, 369
554. HEMMI, N. and COOL, T. A. (1996). *J. Chem. Phys.*, **104**, 5721
555. FELKER, P. M. (1992). *J. Phys. Chem.*, **96**, 7844
556. BASKIN, J. S., FELKER, P. M. and ZEWAIL, A. H. (1986). *J. Chem. Phys.*, **84**, 4708
557. FELKER P. M., BASKIN, J. S. and ZEWAIL, A. H. (1986). *J. Phys. Chem.*, **90**, 724
558. FELKER, P. M. and ZEWAIL, A. H. (1995) In *Jet Spectroscopy and Molecular Dynamics* (Eds J. M. Hollas and D. Phillips), chap. 6. Blackie, Glasgow
559. JOIREMAN, P. W., CONNELL, L. L., OHLINE, S. M. and FELKER, P. M. (1992). *J. Chem. Phys.*, **96**, 4118
560. KAUFFMAN, J. F., CÔTÉ, M. J., SMITH., P. G. and McDONALD, J. D. (1989). *J. Chem. Phys.*, **90**, 2874
561. SMITH, P. G. and McDONALD, J. D. (1990). *J. Chem. Phys.*, **92**, 3991
562. SMITH, P. G., TROXLER, T. and TOPP, M. R. (1993). *J. Phys. Chem.*, **97**, 6983
563. CONNELL, L. L., OHLINE, S. M., JOIREMAN, P. W., CORCORAN, T. C. and FELKER, P. M. (1992). *J. Chem. Phys.*, **96**, 2585
564. OHLINE, S. M., ROMASCAN, J. and FELKER, P. M. (1995). *J. Phys. Chem.*, **99**, 7311

565. MAGNERA, T. F., SAMMOND, D. M. and MICHL, J. (1993). *Chem. Phys. Lett.*, **211**, 378
566. SIMONS, J. P. (1984). *J. Phys. Chem.*, **88**, 1287
567. ANDRESEN, P. (1987) In *Frontiers of Laser Spectroscopy of Gases* (Eds A. C. P. Alves, J. M. Brown and J. M. Hollas), p. 379. Kluwer, Dordrecht
568. WODTKE, A. M. and LEE, Y. T. (1987). In *Molecular Photodissociation Dynamics* (Eds M. N. R. Ashfold and J. E. Baggott), p. 31. Royal Society of Chemistry, London
569. KRAUTWALD, H. J., SCHNIEDER, L., WELGE, K. H. and ASHFOLD, M. N. R. (1986). *Faraday Discuss. Chem. Soc.*, **82**, 99
570. SCHNIEDER, L., MEIER, W., WELGE, K. H., ASHFOLD, M. N. R. and WESTERN, C. M. (1990). *J. Chem. Phys.*, **92**, 7027
571. MORDAUNT, D. H., ASHFOLD, M. N. R. and DIXON, R. N. (1994). *J. Chem. Phys.*, **100**, 7360
572. MORDAUNT, D. H. and ASHFOLD, M. N. R. (1994). *J. Chem. Phys.*, **101**, 2630
573. SASTRY, K. V. L. N., HELMINGER, P., CHARO, A., HERBST, E. and DELUCIA, F. C. (1981). *Astrophys. J.*, **251**, L119
574. KANAMORI, H. and HIROTA, E. (1988). *J. Chem. Phys.*, **89**, 3962

INDEX OF ATOMS AND MOLECULES

The system of indexing molecules is, first, according to the number of atoms in the molecule. Then, with chemical formulae written in what seems a natural way, they are arranged in alphabetical order of the atoms as they appear in the formulae: for example $C_4H_4N_2$ comes before C_3H_6O. For two molecules containing the same atoms, but with different numbers, that with the smallest number of the first (or second, or third . . .) atom takes precedence, so that CH_4 comes before C_4H.

The system of labelling isotopically substituted molecules is the same as that used in the text. Except for a few cases where it seems useful to indicate the isotopic species of all the nuclei, the only nuclei which are labelled are those which are *not* the abundant species.

In the case of van der Waals and hydrogen bonded complexes the smaller moiety comes first in the formula. For example, the formulae for benzene–argon and 2-hydroxynaphthalene–ammonia are written as ArC_6H_6, rather than C_6H_6Ar, and $NH_3C_{10}H_7OH$, rather than $C_{10}H_7OHNH_3$, respectively.

One Atom

Ag
  electron diffraction by foil, 9
  in Stern–Gerlach experiment, 274
Ar
  photoelectron spectrum (UPS), 437 ff
Ar^+
  from photoelectron spectrum (UPS) of Ar, 437 ff
  laser, 490
^{133}Cs
  quantum beat spectrum, 668
Cu
  conductor, 483 ff
F
  dissociation of SF_6, 496
Fe
  LMR spectrum, 528
H
  atomic spectrum, 1, 5 ff, 287 ff
  Bohr theory, 3 ff
  in interstellar medium, 127, 287, 289
  in translational spectroscopy, 680 ff
  in stars, 287
  Lamb shift
    in $1^2S_{1/2}$ state, 616 ff
    in $2^2S_{1/2}$ state, 288 ff
He
  atomic spectrum, 289 ff
  in CO_2 laser, 494
  in He–Ne laser, 487 ff
  level crossing spectrum, 662
Hg
  photoelectron spectrum (UPS), 438
Kr
  photoelectron spectrum (UPS), 438
Kr^+
  laser, 490
    use, 657
Li
  atomic spectrum, 283 ff
  level crossing spectrum, 662 ff
^7Li
  level anticrossing spectrum, 665
N^+
  LMR spectrum, 529
Na
  atomic spectrum, 285 ff
  quantum defects, 285
Nd^{3+}
  laser, 481 ff
Ne
  in He–Ne laser, 487 ff
  shakeup and shakeoff in XPS, 465 ff
O
  LMR spectrum, 528
P^+
  LMR spectrum, 529
S
  LMR spectrum, 528
Se
  LMR spectrum, 528
Si
  LMR spectrum, 528
  semiconductor, 484 ff

Sn
LMR spectrum, 528
U
photoelectron spectrum (UPS), 438
Xe
photoelectron spectrum (UPS), 438

Two Atoms

AgSi
cavity ring-down spectrum, 564 ff
AlCl
in interstellar medium, 129
AlF
in interstellar medium, 129
AlH
$A^1\Pi$–$X^1\Sigma^+$ system, 336 ff
Λ-doubling in $A^1\Pi$ state, 401
production, 79
Ar_2
in supersonic jet, 145
ArBr
laser, 492
ArCl
laser, 492
ArF
laser, 492
ArH^+
difference frequency laser spectrum, 543
ArKr
microwave spectrum, 156 ff
ArNa
in supersonic jet, 146
ArXe
microwave spectrum, 156 ff
AsH
LMR spectrum, 528
AsO
molecular orbitals, 308
BaO
MODR spectrum, 632
OODR spectrum, 637
BeO
plasma tube, 490
BiO
$A^2\Pi_{1/2}$–$X_1\ ^2\Pi_{1/2}$ system, 343
$B^4\Sigma^-$–$X_1\ ^2\Pi_{1/2}$ system, 343
$C^2\Delta_{3/2}$–$X_1\ ^2\Pi_{1/2}$ system, 343
C_2
$D^1\Sigma_u^+$–$X^1\Sigma_g^+$ system, 321
avoided crossing in $C^1\Pi_g$ state, 328
electronic states and transitions, 316 ff
production, 79
C_2^+
SEP spectrum, 653 ff
C_2^-
in flash discharge, 625
laser photoelectron spectrum, 625
CdH
$A^2\Pi$–$X^2\Sigma^+$ system, 341
CF
LMR spectrum, 528
CH
in interstellar medium, 129
LMR spectrum, 527 ff
production, 527
CH^+
in interstellar medium, 129
CH^-
laser photoelectron spectrum, 625
Cl_2
ion-pair E state, 641 ff
CN
$A^2\Pi$–$X^2\Sigma^+$ system, 339
$a^4\Sigma^+$, $B^2\Sigma^+$ anticrossing, 665 ff
in interstellar medium, 129
production, 79
CO
$A^1\Pi$–$X^1\Sigma^+$ system, 324 ff
$a^3\Pi$–$X^1\Sigma^+$ system, 313, 341
centrifugal distortion, 102
dipole moment, 101 ff, 103
in interstellar medium, 129
laser, 495
uses, 524, 527, 531
photoelectron spectrum (XPS), 459
rotational IR spectrum, 97 ff
intensities, 101
vibration–rotation Raman spectrum, 170 ff
CO^+
$A^2\Pi$–$X^2\Sigma^+$ system, 567
in interstellar medium, 129
CoH
LMR spectrum, 528 ff
production, 528
CP
in interstellar medium, 129
CrH
LMR spectrum, 528
CrO
fluorescence excitation spectrum, 576
molecular orbitals, 575 ff
CS
$A^1\Pi$–$X^1\Sigma^+$ Hanle effect, 661 ff
in interstellar medium, 129
Cs_2
polarization spectrum, 658 ff
Cu_2^-
laser photoelectron spectrum, 626 ff

CuH
  $A^1\Sigma^+$–$X^1\Sigma^+$ system, 332 ff
  production, 79
CuO
  molecular orbitals, 574 ff
CuSi
  cavity ring-down spectrum, 564 ff
F_2
  nuclear spin statistical weights, 116, 170
FeH
  LMR spectrum, 528
FeO
  molecular orbitals, 574 ff
FO
  LMR spectrum, 528
GaAs
  laser, for far IR sideband laser, 545
  semiconductor laser, 485 ff
GaP
  semiconductor laser, 486
GeH
  LMR spectrum, 528
H_2
  $a^3\Sigma_g^+$–$b^3\Sigma_u^+$ system, 328
  $B^1\Sigma_u^+$–$X^1\Sigma_g^+$ system, 381
  $B'^1\Sigma_u^+$–$E^1\Sigma_g^+$ system, 667
  E, F state double minimum, 606
  $f^3\Sigma_u^+$–$a^3\Sigma_g^+$ system, 667
  $f^3\Sigma_u^+$, $B'^1\Sigma_u^+$ anticrossings, 667
  dissociation energy, 327
  molecular orbitals, 299 ff
  nuclear spin statistical weights, 115 ff, 170
  photoelectron spectrum (UPS), 439 ff
  quadrupole IR spectrum, 541
  stimulated Raman spectrum, 514
$^1H^2H$
  dipole moment, 95
$^2H_2$
  nuclear spin statistical weights, 115 ff, 170
  stimulated Raman spectrum, 514
H_2^+
  from photoelectron spectrum (UPS) of H_2, 439 ff
  spectroscopy near dissociation, 567 ff
$^1H^2H^+$
  spectroscopy near dissociation, 567 ff
$^2H_2^+$
  spectroscopy near dissociation, 567 ff
HBr
  photoelectron spectrum (UPS), 449 ff
HBr^+
  from photoelectron spectrum (UPS) of HBr, 449 ff
  LMR spectrum, 529
HCl
  anharmonicity, 166
  in chemical laser, 497
  in interstellar medium, 129
  molecular orbitals, 308
  photoelectron spectrum (UPS), 449
$^1H^{35}Cl$
  vibration–rotation spectrum, 167 ff
$^1H^{37}Cl$
  vibration–rotation spectrum, 167 ff
He_2
  $X^1\Sigma_g^+$ and $A^1\Sigma_u^+$ states, 327 ff, 492
He_2^+
  spectroscopy near dissociation, 569
$HeAr^+$
  Hund's case (e) coupling, 340
  spectroscopy near dissociation, 569 ff
HeH^+
  IR diode laser spectrum, 570 ff
  IR predissociation spectrum, 570
HeN^+
  spectroscopy near dissociation, 570
$HeNe^+$
  spectroscopy near dissociation, 569 ff
$HeKr^+$
  spectroscopy near dissociation, 569 ff
HF
  dipole moment, 95
  in chemical laser, 496 ff
  vibration, 15
^2HF
  in chemical laser, 497
HgH^+
  $A^2\Pi_{1/2}$–$X^2\Sigma^+$ system, 341, 343
HI
  nuclear quadrupole coupling constant, 156
  photoelectron spectrum (UPS), 450
I_2
  $B^3\Pi_{0^+u}$–$X^1\Sigma_g^+$ system
    dissociation energies, 326 ff
    emission, 324
    Franck–Condon principle, 322
    in chemical laser, 497
    ion-pair states, E, F, 642 ff
    quantum beat spectrum, 668
    resonance Raman spectrum, 506
    saturation spectrum, 657
    vibrational numbering, 323
    vibrational progressions, 320
  case (c) coupling, 313, 340
InSb
  semiconductor, 487
KCl
  F centres, 487
  in interstellar medium, 129
KF
  F centres, 487

KrBr
laser, 492
KrCl
laser, 492
KrF
laser, 492 ff
KrI
laser, 492
KrXe
microwave spectrum, 156 ff
LiF
dissociation, 330
MgH
$A^2\Pi$–$X^2\Sigma^+$ system, 341
MnH
fluorescence excitation spectrum, 576
molecular orbitals, 576
N_2
$A^3\Sigma_u^+$–$X^1\Sigma_g^+$ system, 315, 341
$a^1\Pi_g$–$X^1\Sigma_g^+$ system, 315
$B^3\Pi_g$–$A^3\Sigma_u^+$ system, 491
$C^3\Pi_u$–$B^3\Pi_g$ system, 320, 491 ff
bond length in a, B and X states, 321
electrical discharge, 79
excited configurations, 305
ground configuration, 304
in CO laser, 495
in CO_2 laser, 494
laser, 491 ff
nuclear spin statistical weights, 116 ff, 171
photoelectron spectrum (UPS), 441 ff
ZEKE-PE spectrum, 619 ff
$^{14}N^{15}N$
rotational Raman spectrum, 134 ff
$^{15}N_2$
rotational Raman spectrum, 134 ff
N_2^+
cavity ring-down spectrum, 565
from photoelectron spectrum (UPS) of N_2, 442 ff
Na_2
Hanle effect, 661
NaCl
in interstellar medium, 129
NaI
nuclear quadrupole coupling constant, 156
NbN (or NbO^+)
$^3\Phi$–$^3\Delta$ system, 341 ff
NbO
$A^4\Sigma^-$–$X^4\Sigma^-$ system, 343
NeAr
microwave spectrum, 156 ff
$NeAr^+$
spectroscopy near dissociation, 569 ff
NeF
laser, 492
NeKr
microwave spectrum, 156 ff
NeXe
microwave spectrum, 156 ff
NH
$d^1\Sigma^+$–$a^1\Delta$ REMPI spectrum, 606 ff
in interstellar medium, 129
LMR spectrum, 528
Ni_2^-
laser photoelectron spectrum, 626 ff
$NiAg^-$
laser photoelectron spectrum, 626
$NiCu^-$
laser photoelectron spectrum, 626 ff
NiH
LMR spectrum, 528
NO
$A^2\Sigma^+$–$X^2\Pi$ two-photon absorption, 605
excited configurations, 306 ff
ground configuration, 306
in interstellar medium, 129
LMR spectrum, 527 ff
photoelectron spectrum (UPS), 448 ff
REMPI spectrum, 608
NO^+
from photoelectron spectrum (UPS) of NO, 448 ff
NS
in interstellar medium, 129
NSe
LMR spectrum, 528
O_2
$a^1\Delta_g$–$X^3\Sigma_g^-$ system, 316
$B^3\Sigma_u^-$–$X^3\Sigma_g^-$ system, 381
$b^1\Sigma_g^+$–$X^3\Sigma_g^-$ system, 316, 563
ground configuration, 305 ff
in chemical laser, 497
nuclear spin statistical weights, 117, 171
photoelectron spectrum (UPS), 443 ff
Zeeman effect, 525
O_2^+
from photoelectron spectrum (UPS) of O_2, 443 ff
OH
$A^2\Sigma^+$–$X^2\Pi$ cavity ring-down spectrum, 565
difference frequency laser spectrum, 543
in interstellar medium, 127 ff
LMR spectrum, 525 ff, 527
photofragment of H_2O, 683
production, 525
OH^+
LMR spectrum, 529
OH^-
colour centre laser spectrum, 538
laser photoelectron spectrum, 625

PO
molecular orbitals, 308
PN
in interstellar medium, 129
S_2
photoelectron spectrum (UPS), 447
S_2^+
from photoelectron spectrum (UPS) of S_2, 447
ScO
$A^2\Pi$–$X^2\Sigma^+$ system, 343
$B^2\Sigma^+$–$X^2\Sigma^+$ system, 343
Se_2
$B^3\Sigma_u^-$–$X^3\Sigma_g^-$ system
quantum beats, 668
SeH
LMR spectrum, 528
SH
LMR spectrum, 528
photofragment of H_2S, 682 ff
SH^-
IR absorption spectrum, 538
Si_2^-
diode laser spectrum, 538 ff
laser photoelectron spectrum, 629 ff
SiC
in interstellar medium, 129
SiCl
in interstellar medium, 129
SiF
$a^4\Sigma^-$–$X^2\Pi$ system, 313
SiH^-
laser photoelectron spectrum, 625
SiN
in interstellar medium, 129
SiO
in interstellar medium, 129
SiS
in interstellar medium, 129
SO
in interstellar medium, 129
molecular orbitals, 308
photoelectron spectrum (UPS), 447 ff
SO^+
from photoelectron spectrum (UPS) of SO, 447 ff
in interstellar medium, 129
TeH
LMR spectrum, 528
TiO
molecular orbitals, 574 ff
VO
$B^4\Pi$–$X^4\Sigma^-$ system, 343, 575
$C^4\Sigma^-$–$X^4\Sigma^-$ system, 343
molecular orbitals, 575
Xe_2
laser, 492
XeBr
laser, 492
XeCl
laser, 492
XeF
laser, 492 ff
XeI
laser, 492
ZnSe
near IR window material, 525

Three Atoms

AlH_2
$\tilde{A}^2B_1$–$\tilde{X}^2A_1$ system, 348
molecular orbitals, 348
production, 348
AlGaAs
semiconductor laser, 486
Ar_3^+
rotational depletion coherence spectrum, 679 ff
ArCO
diode laser spectrum, 536
ArHCl
far IR spectrum, 544
rotational spectrum, 145 ff
ArHF
difference frequency laser spectrum, 544
diode laser spectrum, 534 ff
IR laser induced fluorescence spectrum, 556 ff
ArHO
fluorescence excitation spectrum, 577 ff
ArNO
in supersonic jet, 145
AsH_2
$\tilde{A}^2A_1$–$\tilde{X}^2B_1$ system
by flash photolysis, 350
progression in ν_2, 390
Renner–Teller effect, 394
molecular orbitals, 350 ff
$BaBr_2$
dipole moment, 144
$BaCl_2$
dipole moment, 144
BaF_2
dipole moment, 144
BaI_2
dipole moment, 144
BeH_2
molecular orbitals, 347
BH_2
$\tilde{A}^2\Pi_u$–$\tilde{X}^2A_1$ system
by flash photolysis, 348
Renner–Teller effect, 394

BH_2 (*cont.*)
molecular orbitals, 347 ff
production, 348
BO_2
molecular orbitals, 354
production, 355
$^{11}B^{16}O_2$
intermodulated fluorescence spectrum, 658
BrCN
photoelectron spectrum (UPS), 450
$BrCN^+$
from photoelectron spectrum (UPS) of BrCN, 450
SEP spectra, 654
C_3
$\tilde{A}^1\Pi_u$–$\tilde{X}^1\Sigma_g^+$ system, 401
Renner–Teller effect, 394, 396, 401
rotational fine structure, 401
SEP spectra, 652 ff
discovery, 355
in interstellar medium, 129
molecular orbitals, 354
$CaBr_2$
dipole moment, 144
$CaCl_2$
dipole moment, 144
CaF_2
dipole moment, 144
CCl_2
fluorescence excitation spectrum, 578
CCN
LMR spectrum, 528
molecular orbitals, 356
production, 356
CCO
molecular orbitals, 356
production, 356
CF_2
molecular orbitals, 354
production, 355
CH_2
$\tilde{B}^3A_2$–$\tilde{X}^3B_1$ system, 349
$\tilde{b}^1B_1$–$\tilde{a}^1A_1$ system
by flash photolysis, 349
progression in v_2, 390
Renner–Teller effect, 394, 397
$\tilde{c}^1\Sigma_g^+$–$\tilde{a}^1A_1$ system, 349
in interstellar medium, 129
LMR spectrum, 528
molecular orbitals, 348 ff
production, 349
CH_2^-
laser photoelectron spectrum, 625 ff
C_2H
in interstellar medium, 129
LMR spectrum, 528
molecular orbitals, 357
photofragment of C_2H_2, 683 ff
C_2H^-
infrared absorption spectrum, 538
ClCN
photoelectron spectrum (UPS), 450
$^{35}Cl^{12}C^{15}N$
$J = 3$–2 microwave transition, 104 ff
$ClCN^+$
from photoelectron spectrum (UPS) of ClCN, 450
CNC
molecular orbitals, 354
production, 354
CO_2
Fermi resonance, 248 ff
in ^2HF-CO_2 laser, 497
laser, 493 ff
uses, 524, 530, 545, 567, 603, 633
molecular orbitals, 354
photoelectron spectrum (XPS), 459
potential energy surface, 244 ff
vibrations, 255, 493 ff
waveguide laser, 657
$^{13}C^{16}O_2$
laser, 603
$^{12}C^{18}O_2$
laser, 603
CO_2^+
molecular orbitals, 354
production, 355
CO_2^-
molecular orbitals, 354
C_2O
in interstellar medium, 129
CS_2
$\tilde{a}^3A_2$–$\tilde{X}^1\Sigma_g^+$ system, 419
molecular orbitals, 356
CS_2^+
molecular orbitals, 356
C_2S
in interstellar medium, 129
Cu_3
REMPI spectrum, 610
FeH_2
LMR spectrum, 528, 529
FNO
laser Stark spectrum, 530 ff
FOH
force constants, 203
GeH_2
molecular orbitals, 350
H_3
$H_2 + H$ intermediate, 245
molecular orbitals, 351 ff
transitions between Rydberg states, 352

H_3^+
difference frequency laser spectrum, 543
stability in $\tilde{X}^1A_1'$ state, 352
infrared predissociation spectrum, 570
$^1H^2H_2^+$
infrared predissociation spectrum, 570
$^1H_2{}^2H^+$
infrared predissociation spectrum, 570
$^2H_3^+$
diode laser spectrum, 539
HCF
molecular orbitals, 357
production, 358
HCN
$\tilde{A}^1A''-\tilde{X}^1\Sigma^+$ system, 413
C_∞ axis, 183
$C_{\infty v}$ point group, 189, 192 ff
cavity ring-down spectrum, 563 ff
Coriolis interaction, 108 ff
in interstellar medium, 129
laser, 496
uses, 525, 530
molecular orbitals, 357
photoacoustic spectrum, 560
photoelectron spectrum (UPS), 450 ff
principal axes, 90
SEP spectra, 647 ff
stretching vibrations, 206
vibrational numbering, 213
vibration–rotation spectrum (IR), 220 ff, 223
vibrations, 107 ff
$^2H^{12}C^{14}N$
laser, 496
HCN^+
from photoelectron spectrum (UPS) of HCN, 451
HCO
in interstellar medium, 129
LMR spectrum, 528
molecular orbitals, 357
production, 357 ff
SEP spectra, 651 ff
HCO^+
colour centre laser spectrum, 538
in interstellar medium, 129 ff
molecular orbitals, 357
production, 358
HCP
molecular orbitals, 357
HCS^+
in interstellar medium, 129
HeH_2^+
spectroscopy near dissociation, 570
HNC
in interstellar medium, 129
molecular orbitals, 357
production, 357
HNF
molecular orbitals, 357
production, 358
HNO
$\tilde{A}^1A''-\tilde{X}^1A'$ system
OODR spectrum, 636
predissociation, 421 ff
rotational fine structure, 412
in interstellar medium, 129
molecular orbitals, 357
production, 358
HO_2
$\tilde{A}^2A'-\tilde{X}^2A''$ system, 358, 528
LMR spectrum, 528
molecular orbitals, 357
production, 358, 528
HO_2^-
laser photoelectron spectrum, 626
molecular orbitals, 357
H_2O
$\tilde{C}^1B_1-\tilde{X}^1A_1$ system, 351
rotational fine structure, 412
barrier to linearity, 259
bending vibration potential, 259 ff
C_2 axis, 183
C_{2v} point group, 188, 193
force field, 202 ff
in interstellar medium, 129
IR vibrational selection rules, 210, 211 ff
laser, 495 ff
uses, 525 ff
molecular orbitals, 346 ff, 351
nuclear spin statistical weights, 124
photoelectron spectrum (UPS), 451 ff
Raman vibrational selection rules, 217
rotations, 195
symmetry coordinates, 208 ff
symmetry elements, 187
translational spectroscopy, 683
translations, 194
vibrational numbering, 213
vibrations, 171 ff, 195
ZEKE-PE spectrum, 621
$^1H^2HO$
far IR spectrum, 545
$^2H_2{}^{16}O$
laser, 496
nuclear spin statistical weights, 124
$^1H_2{}^{18}O$
laser, 496
H_2O^+
$\tilde{A}^2A_1-\tilde{X}^2B_1$ system, 350
from photoelectron spectrum (UPS) of H_2O, 451 ff

H_2O^+ (*cont.*)
molecular orbitals, 348
HS_2
LMR spectrum, 528
H_2S
in interstellar medium, 129
molecular orbitals, 351
photoelectron spectrum (UPS), 452 ff
translational spectroscopy, 682 ff
ZEKE-PE spectrum, 621
H_2S^+
$\tilde{A}^2A_1$–$\tilde{X}^2B_1$ system, 350
from photoelectron spectrum (UPS) of H_2S, 452 ff
molecular orbitals, 350
H_2Se
molecular orbitals, 351
HSiCl
$\tilde{A}^1A''$–$\tilde{X}^1A'$ system, 413
H_2Te
molecular orbitals, 351
ICN
photoelectron spectrum (UPS), 450
ICN^+
from photoelectron spectrum (UPS) of ICN, 450
InGaN
semiconductor laser, 486
KrClF
rotational spectrum, 146
MgH_2
molecular orbitals, 347
MgNC
in interstellar medium, 129
N_3
$\tilde{B}^2\Sigma_u^+$–$\tilde{X}^2\Pi_{g(1/2)}$ system
nuclear spin statistical weights, 401
rotational fine structure, 403
LMR spectrum, 528
molecular orbitals, 354
production, 355
N_3^-
IR absorption spectrum, 538
NaCN
in interstellar medium, 129
NCN
$\tilde{A}^3\Pi_u$–$\tilde{X}^3\Sigma_g^-$ system, 403 ff
LMR spectrum, 528
molecular orbitals, 354
production, 355
NCO
molecular orbitals, 356
production, 356
NCO^-
IR absorption spectrum, 538
NeHCl
in supersonic jet, 145
Ne^2HCl
in supersonic jet, 145
NH_2
$\tilde{A}^2A_1$–$\tilde{X}^2B_1$ system
geometry of states, 350
progression in ν_2, 390
Renner–Teller effect, 394, 396
Hanle effect, 662
in interstellar medium, 129
level crossing spectrum, 663
LMR spectrum, 528
molecular orbitals, 350 ff
production, 79 ff, 350
NH_2^+
molecular orbitals, 349
NH_2^-
IR absorption spectrum, 538
N_2H^+
difference frequency laser spectrum, 543
in interstellar medium, 129
molecular orbitals, 357
production, 357
$NiAg_2^-$
laser photoelectron spectrum, 626
$NiCl_2$
fluorescence excitation spectrum, 576
Ni_2Ag^-
laser photoelectron spectrum, 626
NO_2
fluorescence excitation spectrum, 572 ff
electronic transitions, 355 ff
molecular orbitals, 354
OODR spectrum, 637
N_2O
in interstellar medium, 129
laser, uses, 530, 603
structure determination, 140
N_2O^+
molecular orbitals, 356
production, 356
O_3
molecular orbitals, 354
OCS
in interstellar medium, 129
$J = 3$–2 Stark effect, 103
$J = 1$–0 and 42–41 transitions, 97
rotational levels, 94
PH_2
$\tilde{A}^2A_1$–$\tilde{X}^2B_1$ system
geometry of states, 350
progression in ν_2, 390
Renner–Teller effect, 394
rotational fine structure, 418
molecular orbitals, 350
production, 350

SiC_2
in interstellar medium, 129
REMPI spectrum, 610
SiCCl
in interstellar medium, 129
SiF_2
molecular orbitals, 356
SiH_2
$\tilde{A}^1B_1$–$\tilde{X}^1A_1$ system, 350
molecular orbitals, 350
production, 350
SiH_2^-
laser photoelectron spectrum, 625
SO_2
in interstellar medium, 129
$SrBr_2$
dipole moment, 144
SrF_2
dipole moment, 144
SrI_2
dipole moment, 144
XeHCl
rotational spectrum, 145
ZnCdSe
semiconductor laser, 486
ZnTeSe
semiconductor laser, 486

Four Atoms

$AgGaS_2$
in optical parametric oscillator, 502
$AgGaSe_2$
frequency doubling, 477
in optical parametric oscillator, 502
AlInGaP
semiconductor laser, 486
$ArCO_2$
rotational spectrum, 145
Ar_2HF
diode laser spectrum, 534 ff
AsH_3
inversion vibration, 255
molecular orbitals, 360
BF_3
D_{3h} point group, 190
nuclear spin statistical weights, 117
rotational spectrum, 127
symmetry elements, 185 ff, 187, 192
BH_3
molecular orbitals, 360
CH_3
$\tilde{B}^2A_1'$–$\tilde{X}^2A_2''$ system, 404 ff, 421
diode laser spectrum, 536 ff
electronic transitions, 359
molecular orbitals, 359
photoelectron spectrum (UPS), 453
planarity, 405 ff
production, 359
ZEKE-PE spectrum, 624 ff
C^2H_3
$\tilde{B}^2A_1'$–$\tilde{X}^2A_2''$ system
0_0^0 parallel band, 405
nuclear spin statistical weights, 405 ff
planarity, 405 ff
C_2H_2 (acetylene)
$\tilde{A}^1A_u$–$\tilde{X}^1\Sigma_g^+$ system
axis tilting, 413
fluorescence excitation spectrum, 582
geometry of states, 361
progression in ν_3, 390
$\tilde{b}^3B_2$–$\tilde{a}^3A_2$ system, 361
$\tilde{G}^1\Pi_u$–$\tilde{X}^1\Sigma_g^+$ system, 401
$D_{\infty h}$ point group, 190
inversion centre, 97
IR vibrational selection rules, 215
IR vibration–rotation spectrum, 221 ff
molecular orbitals, 360 ff
photoacoustic spectrum, 561 ff
Raman vibrational selection rules, 217
Renner–Teller effect, 394
rotational levels, 94
SEP spectra, 647, 648 ff
translational spectroscopy, 683 ff
vibrations, 210 ff
$C_2{}^1H^2H$ (acetylene)
dipole moment, 95
C_2N_2 (cyanogen)
$\tilde{a}^3\Sigma_u^+$–$\tilde{X}^1\Sigma_g^+$ system, 404
quantum beat spectrum, 671 ff
C_3H (cyclic)
in interstellar medium, 129
C_3H (linear)
in interstellar medium, 129
C_3N
in interstellar medium, 129
COH_2 (carbon monoxide–hydrogen)
in supersonic jet, 145
C_3O
in interstellar medium, 129
C_3S
in interstellar medium, 129
HCCN
in interstellar medium, 129
$(HCl)_2$
in supersonic jet, 145
$HCNH^+$
in interstellar medium, 129
HC_2F (fluoroacetylene)
photoacoustic spectrum, 561 ff
H_2CN
in interstellar medium, 129

HCNO (fulminic acid)
ν_4, ν_5 Coriolis interaction, 110
HClHF
optothermal spectra, 553 ff
rotational spectrum, 146
H_2CO (formaldehyde)
$\tilde{A}^1A_2$–$\tilde{X}^1A_1$ system
f-value, 381
Herzberg–Teller intensity stealing, 386 ff
Kerr effect spectrum, 421
magnetic dipole transition, 364
nuclear spin statistical weights, 124
progression in ν_2, 389
progression in ν_4, 389
$\tilde{a}^3A_2$–$\tilde{X}^1A_1$ system
f-value, 381
magnetic rotation spectrum, 420
spin-orbit coupling, 364, 418
asymmetric rotor, 92
dipole moment in $\tilde{A}$ state, 421
dipole moment in $\tilde{a}$ state, 421
far IR spectrum, 545
in interstellar medium, 129
inversion vibration, 255, 389 ff
molecular orbitals, 363
nuclear spin statistical weights, 124
structure determination, 139
symmetry elements, 38 ff
HCOOH (formic acid)
saturable absorber, 476
H_2CS (thioformaldehyde)
$\tilde{A}^1A_2$–$\tilde{X}^1A_1$ system, 364
$\tilde{a}^3A_2$–$\tilde{X}^1A_1$ system, 364
in interstellar medium, 129
molecular orbitals, 364
OODR spectrum, 636
$(HF)_2$
IR laser induced fluorescence spectrum, 557
optothermal spectrum, 550 ff
HFCO (fluoroformaldehyde)
dipole moment in $\tilde{A}$ state, 421
SEP spectra, 649 ff
$^1HF^2HF$
rotational spectrum, 146
HNCO (isocyanic acid)
in interstellar medium, 129
HNCS (isothiocyanic acid)
in interstellar medium, 129
$HOCO^+$
in interstellar medium, 129
H_2O_2 (hydrogen peroxide)
C_2 point group, 188
molecular orbitals, 362
H_3O^+
colour centre laser spectrum, 538
difference frequency laser spectrum, 543
in interstellar medium, 129
H_2S_2 (hydrogen persulphide)
molecular orbitals, 362
MgNCO (magnesium isocyanate)
in interstellar medium, 129
NF_3 (nitrogen trifluoride)
α_i^B and B_e values, 113 ff
equilibrium structure, 140
$^{15}NF_3$
α_i^B and B_e values, 113 ff
equilibrium structure, 140
NH_3 (ammonia)
$\tilde{A}^1A_2''$–$\tilde{X}^1A_1'$ system, 390 ff
$\tilde{B}^1E''$–$\tilde{X}^1A_1'$ system
Jahn–Teller effect, 399 ff, 409
j-type doubling, 409
Zeeman effect, 419
$\tilde{C}'^1A_1'$–$\tilde{X}^1A_1'$ system, 609 ff
C_{3v} point group, 189
dipole moment, 255
in interstellar medium, 129
inversion barrier, 252 ff
inversion time, 255
inversion vibration, 224, 252 ff, 256
IR vibrational selection rules, 214 ff
laser Stark spectrum, 530, 531 ff
maser, 478 ff
molecular orbitals, 359 ff
nuclear spin statistical weights, 408
photoelectron spectrum (UPS), 453
Raman vibrational selection rules, 217
REMPI spectra, 608 ff
Rydberg transitions, 360, 390
Schuster bands, 418 ff
symmetry coordinates, 207
two-photon transitions, 603
vibrations, 196 ff
ZEKE-PE spectrum, 620 ff
N^2H_3
$\tilde{A}^1A_2''$–$\tilde{X}^1A_1'$ system
parallel bands, 407
inversion time, 255
nuclear spin statistical weights, 408
$^{15}NH_3$
inversion time, 255
NH_3^+
from photoelectron spectrum (UPS) of NH_3, 453
N_2H_2 (diimide)
$\tilde{A}^1B_g$–$\tilde{X}^1A_g$ system, 361 ff
molecular orbitals, 361 ff
$(NO)_2$
in supersonic jet, 145
N_2O_2
in supersonic jet, 145

PF_3 (phosphorus trifluoride)
microwave two-photon transition, 602
PH_3 (phosphine)
molecular orbitals, 360
photoelectron spectrum (UPS), 453
SbH_3 (stibine)
molecular orbitals, 360
SiH_3
diode laser spectrum, 536 ff
Si_2H_2 (disilyne)
millimetre wave spectrum, 124 ff
$ZnGeP_2$
in optical parametric oscillator, 502

Five Atoms

$ArBF_3$ (boron trifluoride–argon)
in supersonic jet, 145
ArC_2H_2 (acetylene–argon)
in supersonic jet, 146
Ar_3HF
diode laser spectrum, 534 ff
$CaCH_3$ (calcium methyl)
fluorescence excitation spectrum, 580
CF_3I
$\tilde{C}^1E$–$\tilde{X}^1A_1$ system, 399
MODR spectrum, 633 ff
OODR spectrum, 636
CF_2N_2 (difluorodiazirine)
$\tilde{A}^1B_1$–$\tilde{X}^1A_1$ system, 413 ff
two-photon spectrum, 610 ff
CF_3S
fluorescence depletion spectrum, 640
fluorescence excitation spectrum, 640
CH_4 (methane)
CARS spectrum, 520 ff
difference-frequency IR spectrum, 541
electronic spectrum, 418
hybridization, 174
inverse Raman spectrum, 520
IR vibration–rotation spectrum, 233
nuclear spin statistical weights, 233
photoelectron spectrum (UPS), 454 ff
photoelectron spectrum (XPS), 464
Raman gain spectrum, 515 ff
rotational IR spectrum, 126
rotational Raman spectrum, 134
spherical rotor, 93
stimulated Raman spectrum, 514
T_d point group, 190
C^2H_4
nuclear spin statistical weights, 233
Raman vibration–rotation spectrum, 234
CH_4^+
from photoelectron spectrum (UPS) of CH_4, 454 ff
from photoelectron spectrum (XPS) of CH_4, 464
C_3H_2 (cyclic)
in interstellar medium, 129
C_3H_2 (linear)
in interstellar medium, 129
C_4H
in interstellar medium, 129
CHBrClF (bromochlorofluoromethane)
asymmetric rotor, 92
C_1 point group, 188
symmetry elements, 39
CH_2CN
in interstellar medium, 129
CH_2CO
in interstellar medium, 129
CHF_3 (fluoroform)
FTIR spectrum, 560
optothermal spectrum, 549 ff
photoacoustic spectrum, 560
CH_2F_2 (difluoromethane)
axis labelling, 195
C_{2v} point group, 188
far IR laser, 545
symmetry elements, 186
CH_3F (methyl fluoride)
C_3 axis, 183
C_{3v} point group, 189
Coriolis interaction, 229 ff
dipole moment, 113, 636
inversion, 224
nuclear spin statistical weights, 226
OODR spectrum, 636
C^2H_3F
1_0^1 IR parallel band, 226
nuclear spin statistical weights, 226
CH_3I (methyl iodide)
prolate symmetric rotor, 90
CH_2N_2 (diazirine)
$\tilde{A}$–$\tilde{X}$ system, 389 ff
CH_2NH (methanimine)
OODR spectrum, 636
CH_3O (methoxy)
fluorescence excitation spectrum, 578 ff
LMR spectrum, 528
C_4Si
in interstellar medium, 129
HC_2CN (cyanoacetylene)
in interstellar medium, 129
HC_2NC (isocyanoacetylene)
in interstellar medium, 129
HCNHF (hydrogen cyanide–hydrogen fluoride)
microwave spectrum, 146
HCOOH (formic acid)
far IR laser, 545
in interstellar medium, 129

H_2HCO^+ (hydrogen–HCO^+)
vibrational predissociation spectrum, 554 ff
HNC_3
in interstellar medium, 129
H_2OHF (water–hydrogen fluoride)
hydrogen bond vibrations, 149
microwave spectrum, 146
$H_2{}^{32}SH^{35}Cl$ (hydrogen sulphide–hydrogen chloride)
microwave spectrum, 153
$KrBF_3$ (boron trifluoride–krypton)
in supersonic jet, 145
NH_4
possible electronic spectrum, 418 ff
NH_2CN (cyanamide)
in interstellar medium, 129
N_2HCN (nitrogen–hydrogen cyanide)
optothermal spectrum, 551 ff
OCHCN (carbon monoxide–hydrogen cyanide)
optothermal spectrum, 551 ff
OsO_4 (osmium tetroxide)
saturation spectrum, 657
POF_3
level anticrossing spectrum, 663 ff, 667
level crossing spectrum, 663 ff
$[PtCl_4]^{2-}$
D_{4h} point group, 190
oblate symmetric rotor, 91
symmetry elements, 185
SiH_4 (silane)
IR rotational spectrum, 126 ff
T_d point group, 190
SiH_3F (silyl fluoride)
6_0^1 IR perpendicular band, 229
$ZnCH_3$ (zinc methyl)
fluorescence excitation spectrum, 579

Six Atoms

Ar_4HF
diode laser spectrum, 534 ff
BF_3CO (boron trifluoride–carbon monoxide)
in supersonic jet, 145
BF_3NO (boron trifluoride–nitric oxide)
in supersonic jet, 145
CF_3NO
CARS spectrum of photofragments, 521 ff
C_2H_4 (ethylene)
$\tilde{A}^1B_{1u}$–$\tilde{X}^1A_g$ system, 365
$\tilde{a}^3B_{1u}$–$\tilde{X}^1A_g$ system, 365
axis labelling, 243
D_{2h} point group, 189 ff
electronic transition moment, 36
elements of symmetry, 182
hybridization, 173
7_0^1 IR type C band, 241 ff
9_0^1 IR type B band, 240 ff
11_0^1 IR type A band, 239
molecular orbitals, 36, 364 ff
mutual exclusion rule, 181
nuclear spin statistical weights, 124
photoelectron spectrum (UPS), 454
torsional barrier, 260 ff
torsional vibration, 215, 217, 260
vibrations, 175, 239 ff
$C_2{}^2H_4$
nuclear spin statistical weights, 124
$^{13}C_2H_4$
nuclear spin statistical weights, 124
$C_2H_4^+$
from photoelectron spectrum (UPS) of C_2H_4, 454
$C_4H_2^+$ (diacetylene cation)
SEP spectra, 654
C_5H
in interstellar medium, 129
C_2H_2BrF (1-bromo-1-fluoroethylene)
hybrid IR band, 237
CH_3CN (methyl cyanide)
in interstellar medium, 129
C^2H_3CN
microwave two-photon transition, 602
$C_2H_2F_2$ (1,1-difluoroethylene)
C_{2v} point group, 188
nuclear spin statistical weights, 124
structure determination, 139
symmetry elements, 182
$C_2H_2F_2$ (*cis*-1,2-difluoroethylene)
C_{2v} point group, 188
symmetry elements, 182, 185
torsional barrier, 264
$C_2H_2F_2$ (*trans*-1,2-difluoroethylene)
C_{2h} point group, 189
symmetry elements, 182, 185
torsional barrier, 264
C_2H_3F (fluoroethylene)
C_s point group, 188
symmetry elements, 182
vibrations, 175
CH_3NC (methyl isocyanide)
in interstellar medium, 129
$(CHO)_2$ (s-*cis*-glyoxal)
s-*cis*-s-*trans*-energy difference, 264, 369
torsional vibration, 263 ff
$(CHO)_2$ (s-*trans*-glyoxal)
$\tilde{A}^1A_u$–$\tilde{X}^1A_g$ system, 369
progression in CO stretch, 389
$\tilde{a}^3A_u$–$\tilde{X}^1A_g$ system, 369
magnetic rotation spectrum, 420
C_{2h} point group, 189
molecular orbitals, 368 ff

torsional barrier, 264
torsional vibration, 263 ff
CH_3OH (methyl alcohol)
far IR laser, 645
in interstellar medium, 129
OODR spectrum, 636
saturable absorber, 476
torsional barrier, 261
torsional vibration, 260
vibrations, 173
CH_3SH (methanthiol)
in interstellar medium, 129
$(CO_2)_2$
diode laser spectrum, 535
in supersonic jet, 145
HC_2CHO (propynal)
$\tilde{A}^1A''$–$\tilde{X}^1A'$ system, 389
dipole moment in $\tilde{A}$ state, 421
in interstellar medium, 129
quantum beat spectra, 571
$(HCN)_2$
microwave spectrum, 148
HC_3NH^+
in interstellar medium, 129
$(HF)_3$
difference frequency laser spectrum, 542 ff
$(H_2O)_2$
far IR spectrum, 546 ff
in supersonic jet, 146
microwave spectrum, 553
N_2H_4 (hydrazine)
molecular orbitals, 365
NH_2CHO (formamide)
in interstellar medium, 129
inversion barrier height, 255
NH_4Cl (ammonium chloride)
microwave spectrum, 153 ff
NH_3HI (ammonia–hydrogen iodide)
nuclear quadrupole coupling constant, 156
$(NO)_3$
in supersonic jet, 145
PH_3HI (phosphine–hydrogen iodide)
nuclear quadrupole coupling constant, 156
$XeOF_4$
C_{4v} point group, 189
symmetry elements, 183

Seven Atoms

β-BaB_2O_4 (β-barium borate or BBO)
frequency doubling, 477
in optical parametric oscillator, 502
$CH_2{=}C{=}CH_2$ (allene)
D_{2d} point group, 189
rotational spectrum, 127
symmetry elements, 91 ff, 185, 186
CH_3C_2H (methylacetylene)
in interstellar medium, 129
CH_2CHCN (cyanoethylene)
in interstellar medium, 129
CH_3CHO (acetaldehyde)
$\tilde{A}^1A''$–$\tilde{X}^1A'$ system, 389
in interstellar medium, 129
CH_3NH_2 (methylamine)
in interstellar medium, 129
CH_3NO_2 (nitromethane)
torsional barrier, 261
C_6H
in interstellar medium, 129
$[CoF_6]^{3-}$
molecular orbitals, 378
HC_2CF_3 (1,1,1-trifluoropropyne)
optothermal spectrum, 548 ff
HC_2CH_3 (propyne)
optothermal spectrum, 549
HC_4CN (cyanodiacetylene)
in interstellar medium, 128 ff
microwave spectrum, 98
IrF_6 (iridium hexafluoride)
electronic spectrum, 418
MoF_6 (molybdenum hexafluoride)
electronic spectrum, 418
OsF_6 (osmium hexafluoride)
Jahn–Teller effect, 400
ReF_6 (rhenium hexafluoride)
electronic spectrum, 418
Jahn–Teller effect, 400
RuF_6 (ruthenium hexafluoride)
Jahn–Teller effect, 400
SF_6 (sulphur hexafluoride)
electronic spectrum, 418
O_h point group, 191
possible rotation spectrum, 127, 134
saturation spectrum, 657
source of F atoms, 497
symmetry elements, 94, 185
SiH_3NCS (silyl isothiocyanate)
$J = 8$–7 transition, 112
nuclear spin statistical weights, 114, 117
TcF_6 (technetium hexafluoride)
Jahn–Teller effect, 400

Eight Atoms

$(BF_3)_2$
in supersonic jet, 145
CF_3CH_3 (1,1,1-trifluoroethane)
optothermal spectrum, 554
C_2H_6 (ethane)
CARS spectrum, 522
D_{3d} point group, 189

C_2H_6 (ethane) (*cont.*)
hyper Raman spectrum, 511 ff
torsional barrier, 261
torsional vibration, 260
$CH_2{=}C{=}C{=}CH_2$ (butatriene)
photoelectron spectrum (UPS), 451
$CH_2{=}CHC{\equiv}CH$ (vinyl acetylene)
vibrations, 174
$CH_2{=}CHCHO$ (s-*cis*-acrolein)
s-*cis*-s-*trans* energy difference, 264, 369
torsional vibration, 263 ff
$CH_2{=}CHCHO$ (s-*trans*-acrolein)
$\tilde{A}^1A''$–$\tilde{X}^1A'$ system, 369
f-value, 381
progression in CO stretch, 389
rotational line width, 423
$\tilde{a}^3A''$–$\tilde{X}^1A'$ system, 369
near-symmetric rotor, 92
torsional vibration, 263 ff
$(CHClF)_2$ (1,2-dichloro-1,2-difluoroethane)
C_i point group, 188
CH_3C_2CN (methyl cyanoacetylene)
in interstellar medium, 129
CH_3CNHF (methyl cyanide–hydrogen fluoride)
microwave spectrum, 148 ff
$C_2H_2N_4$ (s-(or sym-)tetrazine)
molecular orbitals, 374
$HCOOCH_3$ (methyl formate)
in interstellar medium, 129
KH_2PO_4 (potassium dihydrogen phosphate or KDP)
frequency doubling, 477
Pockels effect, 475
$K^2H_2PO_4$ (potassium dideuterium phosphate or KD*P)
frequency doubling, 477
Pockels effect, 475
K_3NbO_4 (potassium niobate or KN)
frequency doubling, 477
in optical parametric oscillator, 502
$KTiOAsO_4$ (potassium titanyl arsenate or KTA)
frequency doubling 477
$KTiOPO_4$ (potassium titanyl phosphate or KTP)
frequency doubling, 477
in optical parametric oscillator, 502
Li_3NbO_4 (lithium niobate or LN)
frequency doubling, 477
frequency mixing, 540
in optical parametric oscillator, 502
$(NO)_4$
in supersonic jet, 145

Nine Atoms

$CH_3CH{=}CH_2$ (propene)
conformation, 596
methyl torsion, 600
CH_3CH_2CN (ethyl cyanide)
in interstellar medium, 129
CH_3C_4H (methyl diacetylene)
in interstellar medium, 129
CH_3COCHO (*trans*-methylglyoxal)
torsional vibration, 601
CH_2FCH_2OH (2-fluoroethanol)
optothermal spectrum, 549
$(CH_3)_2O$ (dimethylether)
in interstellar medium, 129
C_4H_4O (furan)
photoelectron spectrum (XPS), 462 ff
$(CHO)CH{=}CHOH$ (malonaldehyde)
intramolecular hydrogen bonding, 150 ff
C_2H_5OH (ethyl alcohol)
in interstellar medium, 129
C_4H_4S (thiophene)
photoelectron spectrum (XPS), 462 ff
$(CO_2)_3$
diode laser spectrum, 535 ff
HC_6CN (cyanotriacetylene)
in interstellar medium, 129
LiB_3O_5 (lithium triborate)
frequency doubling, 477
$Ni(CO)_4$ (nickel tetracarbonyl)
diode laser spectrum, 534
molecular orbitals, 374 ff
nuclear spin statistical weights, 232
T_d point group, 190

Ten Atoms

$CH_2{=}CHCH{=}CH_2$ (s-*cis*-buta-1,3-diene)
s-*cis*-s-*trans* energy difference, 264
torsional vibration, 263 ff
$CH_2{=}CHCH{=}CH_2$ (s-*trans*-buta-1,3-diene)
$\tilde{A}^1B_u$–$\tilde{X}^1A_g$ system, 368
C_{2h} point group, 189
molecular orbitals, 367 ff
symmetry elements, 185
torsional barrier, 261
torsional vibration, 263 ff
$C_4H_4N_2$ (pyridazine)
molecular orbitals, 373
$C_4H_4N_2$ (pyrimidine)
molecular orbitals, 373
$C_4H_4N_2$ (pyrazine)
$\tilde{A}^1B_{3u}$–$\tilde{X}^1A_g$ system
fluorescence excitation spectrum, 582
pre-resonance Raman spectrum, 505 ff
progression in b_{1g} vibration, 391
molecular orbitals, 373
symmetry elements, 93
C_4H_4NH (pyrrole)
photoelectron spectrum (XPS), 462 ff

C_3H_6O (oxetane)
ring-puckering vibration, 258
C_5H_5 (cyclopentadienyl)
$\tilde{A}^2A_2''-\tilde{X}^2E_1''$ fluorescence excitation spectrum, 579
Jahn–Teller effect in $\tilde{X}$ state, 579
$Cr_2O_3:Al_2O_3$ (ruby)
ruby laser, 479 ff
uses, 513, 519
$(HC_2CN)_2$
optothermal spectrum, 552

Eleven Atoms

CaC_5H_5 (calcium cyclopentadienyl)
$\tilde{A}^2E-\tilde{X}^2A_1$ fluorescence excitation spectrum, 579 ff
production, 579
CdC_5H_5 (cadmium cyclopentadienyl)
production, 579
$CH_3CH{=}CHCHO$ (crotonaldehyde)
IR and Raman vibrational spectra, 176 ff
C_5H_5N (pyridine)
$\tilde{A}^1B_1-\tilde{X}^1A_1$ system, 373
f-value, 381
hydrogen bonding with phenol, 142
molecular orbitals, 373
HC_8CN (cyanotetraacetylene)
in interstellar medium, 129
MgC_5H_5 (magnesium cyclopentadienyl)
production, 579
SrC_5H_5 (strontium cyclopentadienyl)
production, 579
ZnC_5H_5 (zinc cyclopentadienyl)
production, 579

Twelve Atoms

$(BF_3)_3$
in supersonic jet, 145
$C_6F_6^+$ (hexafluorobenzene positive ion)
Jahn–Teller effect, 400
C_4H_8 (cyclobutane)
ring-puckering vibration, 256 ff
C_6H_6 (benzene)
$\tilde{A}^1B_{2u}-\tilde{X}^1A_{1g}$ system, 370 ff
band contour analysis, 409 ff
channel three, 424 ff
decay of $\tilde{A}$ state, 424 ff
fluorescence excitation spectrum, 582 ff
Herzberg–Teller intensity stealing, 385 ff
progression in ring-breathing vibration, 388, 587
sequence in e_{2u} vibration, 393
SVLF spectrum, 389
two-photon absorption, 611 ff
$\tilde{a}^3B_{1u}-\tilde{X}^1A_{1g}$ system, 371
$\tilde{B}^1B_{1u}-\tilde{X}^1A_{1g}$ system, 370 ff
$\tilde{C}^1E_{1u}-\tilde{X}^1A_{1g}$ system, 385
D_{6h} point group, 190
inverse Raman spectrum, 519
ionization gain stimulated Raman spectrum, 517
molecular orbitals, 369 ff
oblate symmetric rotor, 91
optothermal spectrum, 552 ff
photoacoustic spectrum, 559 ff
photoelectron spectrum (UPS), 456 ff, 459
Raman gain spectrum, 514
REMPI spectrum, 611 ff
resonance Raman spectrum, 507
stimulated Raman spectrum, 513 ff
symmetry coordinates, 207 ff
symmetry elements, 183, 185
ZEKE-PE spectrum, 621
$C_6H_6^+$
from photoelectron spectrum (UPS) of C_6H_6, 456 ff
Jahn–Teller effect, 621
$CH_3CH{=}CHCOOH$ (s-*cis* and s-*trans*-crotonic acid)
microwave spectrum, 119 ff
$CH_3COCOCH_3$ (biacetyl)
$\tilde{A}^1A_u-\tilde{X}^1A_g$ system
quantum beats, 668 ff
$C_6H_3Cl_3^+$ (1,3,5-trichlorobenzene positive ion)
Jahn–Teller effect, 400
C_6H_4ClF (1-chloro-2-fluorobenzene)
vibrations, 172, 207 ff
$C_6H_2F_4$ (1,2,4,5-tetrafluorobenzene)
fluorescence excitation spectrum, 589 ff
inversion (butterfly) vibration, 589 ff
$C_6H_3F_3^+$ (1,3,5-trifluorobenzene positive ion)
Jahn–Teller effect, 400
$C_6H_4F_2$ (1,2-difluorobenzene)
two-photon absorption, 611
$C_6H_4F_2$ (1,3-difluorobenzene)
two-photon absorption, 611
$C_6H_4F_2$ (1,4-difluorobenzene)
$\tilde{A}^1B_{2u}-\tilde{X}^1A_{1g}$ system, 416 ff
two-photon absorption, 611, 612 ff
SEP spectra, 654
C_6H_5F (fluorobenzene)
molecular orbitals, 371
vibrational numbering, 213
$C_6H_4O_2$ (*p*-benzoquinone)
$\tilde{A}^1B_{1g}-\tilde{X}^1A_g$ system
Herzberg–Teller intensity stealing, 386
magnetic dipole transition, 386
positive-running sequences, 391
unusual sequence intensities, 392

$\tilde{a}^3A_u-\tilde{X}^1A_g$ system
magnetic rotation spectrum, 420
unusual sequence intensities, 392
$(NH_4)H_2PO_4$ (ammonium dihydrogen phosphate—ADP)
frequency doubling, 477
Pockels effect, 475

Thirteen Atoms

ArC_6H_6 (benzene–argon)
ionization gain stimulated Raman spectrum, 517
microwave spectrum, 153
REMPI spectrum, 613 ff
ArC_6H_5Cl (chlorobenzene–argon)
REMPI spectrum, 615
ArC_6H_5F (fluorobenzene–argon)
REMPI spectrum, 615
$CF_3COOHHCOOH$ (trifluoroacetic acid–formic acid)
microwave spectrum, 147 ff
C_5H_8 (cyclopentene)
ring-puckering vibration, 257 ff
C_6H_5OH (phenol)
$\tilde{A}^1B_2-\tilde{X}^1A_1$ system
cross sequences, 393
G_4 permutation-inversion group, 261 ff
hydrogen bonding with diethyl ether, 176
hydrogen bonding with pyridine, 142
torsional barrier, 261
torsional vibration, 260 ff
vibrations, 173, 213
$[Fe(CN)_6]^{3-}$
spherical rotor, 94
$[Fe(CN)_6]^{4-}$
$\tilde{A}^1T_{1g}-\tilde{X}^1A_{1g}$ system, 379
$\tilde{B}^1T_{2g}-\tilde{X}^1A_{1g}$ system, 379
low spin complex, 378 ff
molecular orbitals, 374 ff, 380
$HC_{10}CN$ (cyanopentaacetylene)
in interstellar medium, 129
$^{84}KrC_6H_6$ (benzene–krypton)
microwave spectrum, 153
NeC_6H_6 (benzene–neon)
microwave spectrum, 153
$^{129}XeC_6H_6$ (benzene–xenon)
microwave spectrum, 153

Fourteen Atoms

$Ar_2C_6H_6$ (benzene–(argon)$_2$)
REMPI spectrum, 613 ff
ArC_6H_5OH (phenol–argon)
REMPI spectrum, 615
B_5H_9 (pentaborane-9)
photoelectron spectrum (XPS), 463
$(CH_3)_3HI$
nuclear quadrupole coupling constant, 156
$CF_3COOCH_2CH_3$ (ethyltrifluoroacetate)
photoelectron spectrum (XPS), 460 ff
$C_6H_5NH_2$ (aniline)
$\tilde{A}^1B_2-\tilde{X}^1A_1$ system
inversion vibration, 392, 589
C_s point group, 188
inversion barrier, 255, 261
r_s structure, 140 ff
$C_6H_5N^2H_2$
$\tilde{A}^1B_2-\tilde{X}^1A_1$ system
inversion vibration, 392
$C_6H_5NO_2$ (nitrobenzene)
Kerr effect, 475
$C_6H_4(OH)_2$ (catechol)
hole-burning spectrum, 615 ff
REMPI spectrum, 615 ff
$C_6H_4(OH)_2$ (*cis*- and *trans*-hydroquinone)
fluorescence excitation spectrum, 585 ff
KB_5O_8 (potassium pentaborate)
frequency doubling, 477

Fifteen Atoms

$ArC_6H_5NH_2$ (aniline–argon)
fluorescence excitation spectrum, 586 ff
REMPI spectrum, 615
ZEKE-PE spectrum, 622 ff
C_5H_{10} (cyclopentane)
pseudorotation, 258 ff
ring-puckering vibration, 258 ff
$C_6H_5CH_3$ (toluene)
$\tilde{A}^1B_2-\tilde{X}^1A_1$ system
two-photon absorption, 611
inverse Raman spectrum, 519
torsional vibration, 260 ff, 594
$C_5H_3(CHO)CHOH$ (6-hydroxy-2-formylfulvene)
intramolecular hydrogen bonding, 150 ff
$C_6H_4FCH_3$ (2-fluorotoluene)
torsional vibration, 265, 594 ff
ZEKE-PE spectrum, 621
$C_6H_4FCH_3$ (3-fluorotoluene)
torsional vibration, 260, 264, 594 ff
ZEKE-PE spectrum, 621
$C_6H_4FCH_3$ (4-fluorotoluene)
torsional vibration, 594 ff
ZEKE-PE spectrum, 621
$(CH_3)_3NHI$ (trimethylammonium iodide)
microwave spectrum, 155 ff
nuclear quadrupole coupling constant, 156
$(C_2H_5)_2O$ (diethyl ether)
hydrogen bonding with phenol, 176

$C_6H_3(OH)_3$ (1,3,5-trihydroxybenzene)
  C_{3h} point group, 189
$C_6H_4(OH)CHO$ (2-hydroxybenzaldehyde)
  intramolecular hydrogen bonding, 149
$C_7H_5O(OH)$ (tropolone)
  $\tilde{A}^1B_2$–$\tilde{X}^1A_1$ spectrum, 591 ff
  intramolecular hydrogen bonding, 150 ff
$C_7H_6O_2$ (1,3-benzodioxole)
  $\tilde{A}^1A_1$–$\tilde{X}^1A_1$ spectrum, 592 ff
  microwave spectrum, 592 ff
$C_4H_{10}Si$ (silacyclopentane)
  pseudorotation, 259
$C_6H_5SiH_3$ (phenylsilane)
  torsional vibration, 594
$CO_2C_6H_6$ (benzene–carbon dioxide)
  ionization gain stimulated Raman spectrum, 517
$(HC_2CN)_3$
  optothermal spectrum, 552

Sixteen Atoms

$Ar_2C_6H_5NH_2$ (aniline–(argon)$_2$)
  ZEKE-PE spectrum, 622 ff
$(BF_3)_4$
  in supersonic jet, 145
$CF_3COOHCH_3COOH$ (trifluoroacetic acid–acetic acid)
  microwave spectrum, 146 ff
$CF_3COOHCH_2FCOOH$ (trifluoroacetic acid–fluoroacetic acid)
  microwave spectrum, 147
$C_6H_5CH{=}CH_2$ (styrene)
  $\tilde{A}^1A'$–$\tilde{X}^1A'$ system
    cross sequence, 393, 599 ff
  torsional barrier, 600
  torsional vibration, 599 ff
$(CH_3)_3CNO_2$ (trimethylnitromethane)
  torsional barrier, 261, 263
$(C_2H_2N_4)_2$ (s-tetrazine dimer)
  fluorescence excitation spectrum, 584
$C_7H_4O(OH)_2$ (5-hydroxytropolone)
  fluorescence depletion spectrum, 639 ff
$H_2OC_6H_5OH$ (phenol–water)
  fluorescence depletion spectrum, 638 ff
  SEP spectrum, 654 ff

Seventeen Atoms

$CH_4C_6H_6$ (benzene–methane)
  ionization gain stimulated Raman spectrum, 517
$C_6H_4(CH_3)CH_2$ (2-methylbenzyl)
  torsional vibration, 600 ff
$C_6H_4(CH_3)CH_2$ (3-methylbenzyl)
  torsional vibration, 600 ff
$C_6H_4(CH_3)CH_2$ (4-methylbenzyl)
  torsional vibration, 600 ff
$C_6H_4(NH_2)CF_3$ (2-aminobenzotrifluoride)
  torsional vibration, 596 ff
$C_6H_4(NH_2)CF_3$ (3-aminobenzotrifluoride)
  torsional vibration, 596 ff
$C_6H_4(NH_2)CF_3$ (4-aminobenzotrifluoride)
  torsional vibration, 596 ff
$C_6H_4(NH_2)CH_3$ (2-aminotoluene)
  torsional vibration, 596 ff
$C_6H_4(NH_2)CH_3$ (3-aminotoluene)
  torsional vibration, 596 ff
$C_6H_4(NH_2)CH_3$ (4-aminotoluene)
  torsional vibration, 596 ff
$HF(CH_3)_3CCN$ (t-butyl cyanide–hydrogen fluoride)
  microwave spectrum, 148

Eighteen Atoms

$C_{10}H_8$ (naphthalene)
  $\tilde{A}^1B_{2u}$–$\tilde{X}^1A_g$ system, 372 ff
    fluorescence excitation spectrum, 581 ff
    f-value, 381
    Herzberg–Teller intensity stealing, 387
    rotational line width, 423
    sequences, 391
  $\tilde{B}^1B_{1u}$–$\tilde{X}^1A_g$ system, 372 ff
    f-value, 381
  $\tilde{C}^1B_{2u}$–$\tilde{X}^1A_g$ system, 372 ff
    f-value, 381
  configuration interaction, 372 ff
  D_{2h} point group, 189
  molecular orbitals, 371 ff
  resonance hyper Raman spectrum, 512
  symmetry elements, 185
$C_{10}{}^2H_8$
  type *A* IR band, 239
  type *B* IR band, 241
  type *C* IR band, 242

Nineteen Atoms

C_9H_{10} (indane)
  $\tilde{A}^1A_1$–$\tilde{X}^1A_1$ system
    sequences, 393
$CH_3OHC_6H_5OH$ (phenol–methanol)
  ZEKE-PE spectrum, 623 ff
$C_{10}H_7OH$ (*cis*- and *trans*-2-hydroxynaphthalene)
  fluorescence excitation spectrum, 584 ff
$[Co(H_2O)_6]^{2+}$
  $\tilde{A}^4T_{2g}$–$\tilde{X}^4T_{1g}$ system, 379
  $\tilde{B}^4A_{2g}$–$\tilde{X}^4T_{1g}$ system, 379
  $\tilde{C}^4T_{1g}$–$\tilde{X}^4T_{1g}$ system, 379
  molecular orbitals, 379

$[Cr(H_2O)_6]^{2+}$
$\tilde{A}^5T_{2g}$–$\tilde{X}^5E_g$ system, 378
molecular orbitals, 377 ff
$[Cr(H_2O)_6]^{3+}$
$\tilde{A}^4T_{2g}$–$\tilde{X}^4A_{2g}$ system, 377
$\tilde{B}^4T_{1g}$–$\tilde{X}^4A_{2g}$ system, 377
molecular orbitals, 377
$[Cu(H_2O)_6]^{2+}$
$\tilde{A}^2T_{2g}$–$\tilde{X}^2E_g$ system, 376 ff
molecular orbitals, 376 ff
$[Fe(H_2O)_6]^{2+}$
$\tilde{A}^5E_g$–$\tilde{X}^5T_{2g}$ system, 378
molecular orbitals, 378
$[Mn(H_2O)_6]^{2+}$
molecular orbitals, 378
$[Ni(H_2O)_6]^{2+}$
$\tilde{A}^3T_{2g}$–$\tilde{X}^3A_{2g}$ system, 377
$\tilde{B}^3T_{1g}$–$\tilde{X}^3A_{2g}$ system, 377
molecular orbitals, 377
$[Ti(H_2O)_6]^{3+}$
$\tilde{A}^2E_g$–$\tilde{X}^2T_{2g}$ system, 376
molecular orbitals, 375 ff
$[V(H_2O)_6]^{3+}$
$\tilde{A}^3T_{2g}$–$\tilde{X}^3T_{1g}$ system, 377
$\tilde{B}^3T_{1g}$–$\tilde{X}^3T_{1g}$ system, 377
molecular orbitals, 377

Twenty or More Atoms

$ArC_{14}H_{10}$ (anthracene–argon)
rotational coherence spectrum, 678 ff
$ArC_{18}H_{12}$ (tetracene–argon)
rotational coherence spectrum, 679
$Ar_2C_{14}H_{10}$ (anthracene–(argon)$_2$)
rotational coherence spectrum, 678 ff
$Ar_3C_{14}H_{10}$ (anthracene–(argon)$_3$)
rotational coherence spectrum, 678 ff
$B_{12}H_{12}^-$
icosahedron, 191
C_{60} (buckminsterfullerene)
infrared spectrum, 217 ff
Raman spectrum, 217 ff
truncated icosahedron, 191
$C_{14}H_{10}$ (anthracene)
$\tilde{A}^1B_{1u}$–$\tilde{X}^1A_g$ system
spectral congestion, 422
quantum beat spectrum, 669 ff
time-resolved resonance Raman spectrum, 510
$C_{14}H_{10}$ (phenanthrene)
$\tilde{A}^1A_1$–$\tilde{X}^1A_1$ system
Herzberg–Teller intensity stealing, 387
C_{2v} point group, 188
$C_{14}H_{12}$ (9,10-dihydroanthracene)
inversion vibration, 590 ff
$C_{20}H_{12}$ (perylene)
quantum beat spectrum, 671
$C_{20}H_{20}$ (dodecahedrane)
dodecahedron, 191
$C_6H_5CH{=}CHC_6H_5$ (*trans*-stilbene)
rotational coherence spectrum, 673
C_6H_5–C_6H_5 (biphenyl)
$\tilde{A}^1B_{3g}$–$\tilde{X}^1A_g$ system
two-photon absorption, 605
$C_{14}H_9$–$C_{12}H_8N$ (9-(*N*-carbazolyl) anthracene)
torsional vibration, 598
$C_6H_4(NH_2)C_6H_{11}$ (4-cyclohexylaniline)
rotational coherence spectrum, 675 ff
$(C_6H_5OH)_2$ (phenol dimer)
rotational coherence spectrum, 677 ff
$C_{13}H_{10}O$ (xanthene)
inversion vibration, 591
$[Co(H_2NCH_2CH_2NH_2)_3]^{3+}$
D_3 point group, 189
$[Co(NH_3)_6]^{3+}$
$\tilde{A}^1T_{1g}$–$\tilde{X}^1A_{1g}$ system, 379
$\tilde{B}^1T_{2g}$–$\tilde{X}^1A_{1g}$ system, 379
low spin complex, 378
molecular orbitals, 378 ff
$NH_3C_{10}H_7OH$ (*cis*- and *trans*-2-hydroxy-naphthalene-ammonia)
fluorescence excitation spectrum, 585
Cryptocyanine dye
for *Q*-switch, 476
Rhodamine B dye
in dye laser, 498
Rhodamine 6G dye
in dye laser, 500
Rhodamine 6G-cresyl violet perchlorate dye
in dye laser, 500
Vanadium phthalocyanine
for *Q*-switch, 476
Zinc octaethylporphyrin
time-resolved resonance Raman spectrum, 508 ff

SUBJECT INDEX

Abelian group, 192
Absorbance, 36 ff
Absorption cell, 48
Absorption coefficient, 42
  molar, 37, 51
Absorption process, 33 ff
Acousto-optic modulator, 636
Actinide configurations, 273
Action, 12
Active medium, of laser, 468
Alexandrite laser, 480 ff
Alkali metals
  configurations, 269
  spectra, 283 ff
Alkaline earth metals
  configurations, 269
  spectra, 289 ff
Amplification, in laser, 468 ff
Amplitude of radiation, 32
Angular distribution of photoelectrons, 457 ff
Angular momentum, 3 ff
  coupling in atoms, 275 ff
  electron spin, 14, 22 ff, 26 ff, 30, 273 ff
  nuclear spin, 14, 22 ff, 26, 30, 275
  orbital, 13 ff, 22, 30, 273 ff
  rotational, 14 ff, 22 ff, 30, 110 ff
  total, 25, 30, 273, 278, 331
  vibrational, 107 ff, 260, 398 ff, 409
  vibronic, 395, 398 ff
Angular wave function, 16 ff
Anharmonic constants, 163, 166, 247
Anharmonic oscillator, 102, 142, 162 ff, 244 ff
Anharmonicity
  electrical, 162 ff
  mechanical, 162 ff
Anisotropy parameter β, 458 ff
Anti-Stokes Raman scattering, 133 ff
Antisymmetric part, of direct product, 198, 306
Arc source, 84
Argon ion laser, 490 ff
  in difference-frequency laser, 540
  in Hanle effect measurement, 661
  in OODR spectroscopy, 637
  in photodetachment spectroscopy, 625
  in Raman gain spectroscopy, 515
  in resonance Raman spectroscopy, 506
Associated Laguerre functions, 17
Associated Legendre polynomials, 17
Asymmetric rotor, 15, 92 ff
Asymmetry doubling, 412
Asymmetry parameter
  b, 118 ff
  κ, 121
Atomic absorption spectroscopy, 78, 82 ff
Atomic beam, 42
Aufbau principle, 269, 300
Auger electron spectroscopy, 427
Autoionization, 446 ff, 618
Avogadro constant, 3
Axial mode, of laser cavity, 473
Axis of symmetry, C_n, 38 ff, 91, 183
Axis switching 412
Axis tilting, 412 ff
Axis-labelling convention, 195
Azimuthal quantum number, 5, 16

Background radiation, in universe, 130
Backward scattering, 180
Backward wave oscillator, 50 ff
Balmer series of H atom, 1, 4 ff
Band, definition, 160
Band centre, 167
Band contour, 239 ff
Band contour analysis, 409, 413 ff
Band gap, 484
Band head, 325, 333
Barrier
  in hydrogen bonding, 150 ff
  to inversion, 252 ff
  to linearity, 259 ff
  to ring-puckering, 256 ff
  to torsion, 261 ff
Bath states, 543
Beam expander, in dye laser, 501
Beamsplitter, 66 ff, 68, 78
Beat frequency, 153
Beer–Lambert law, 37, 562
  breakdown due to saturation, 523, 632
Bending vibration, in triatomic molecules, 259 ff
Binding energy, 435
Birefringence, 475, 477
Birge–Sponer extrapolation, 163 ff, 325 ff
Black body radiation, 1, 35
Blaze angle, 64
Blazing of grating, 63 ff

Bohr magneton, 275
Bohr radius, 17
Bohr theory of H atom, 3 ff
Bolometer, 73, 78
  in optothermal spectroscopy, 548 ff
Boltzmann distribution law, 34, 44, 100 ff, 148, 319
Boolean symbol ⊂, 211
Born–Oppenheimer approximation, 24 ff, 322, 323
  breakdown, 385, 395, 398
Bose–Einstein statistics, 115
Bosons, 115
Boxcar, 618 ff
Bremsstrahlung, 433, 464
Brewster angle windows, 489, 495, 524, 526
Brewster's angle, 489
Bright state, 424, 542
Brightness, of laser, 470

C_i character table, 686
C_n axis (*see* Axis of symmetry)
C_n point groups, 187 ff
C_{nh} point groups, 189
C_{nv} point groups, 188 ff
C_s character table, 686
C_1 character table, 687
C_2 character table, 687
C_3 character table, 687
C_4 character table, 687
C_5 character table, 687
C_6 character table, 688
C_7 character table, 688
C_8 character table, 688
C_{2h} character table, 691
C_{3h} character table, 691
C_{4h} character table, 691
C_{5h} character table, 691
C_{6h} character table, 692
C_{2v} character table, 193 ff, 688
C_{3v} character table, 196 ff, 689
C_{4v} character table, 689
C_{5v} character table, 689
C_{6v} character table, 689
$C_{\infty v}$ character table, 198 ff, 689
Carbon dioxide laser, 493 ff, 497, 657
  in laser Stark spectroscopy, 530
  in LMR spectroscopy, 524
  in MODR spectroscopy, 633
  in OODR spectroscopy, 636
  in two-photon spectroscopy, 603
  pumping far IR lasers, 496
  pumping spin–flip lasers, 487
Carbon monoxide laser, 495
  in laser Stark spectroscopy, 531
  in LMR spectroscopy, 524, 527
  pumping spin–flip lasers, 487
CARS spectroscopy, 520 ff
Cascade OODR process, 637
Cavity, of laser, 469
Cavity modes, of laser, 472 ff
  of diode laser, 486
Cavity ring-down absorption spectroscopy, 562 ff
Centre burst, 69
Centre dip, 531 ff
Centre of inversion *i*, 95
Centre of symmetry *i*, 95, 182, 185
Centrifugal barrier, 571
Centrifugal distortion
  asymmetric rotor, 125
  linear molecule, 102, 108, 168 ff
  spherical rotor, 126, 134
  symmetric rotor, 111 ff, 114
Chaotic behaviour
  in $\tilde{A}$–$\tilde{X}$ system of NO_2, 573
Character, of symmetry species, 193, 195 ff
Character tables, 193 ff, 686 ff
Characteristic temperature, 3
Charge-coupled device (CCD), 82
Charge exchange reactions, 567
Charge-transfer transitions, 381
Chemical lasers, 496 ff
Chemical shift, in XPS, 459 ff
Chopping, 74
Chromophore, 381 ff
Circular vibration frequency, 28
Circularly polarized radiation, 179 ff, 608
Class, of symmetry elements, 192, 196
Clebsch–Gordan series, 104, 274
Coherence, of laser radiation
  spatial, 470
  temporal, 470 ff
Coherent anti-Stokes Raman scattering (CARS), 520 ff
Coherent Stokes Raman scattering (CSRS), 520 ff
Collision diameter, 389
Collisions, 40, 42, 43 ff, 48, 533
Colour centre, 487
Colour centre lasers, 487
Colour centre laser spectroscopy, 538 ff, 548 ff
Combination differences
  in vibration–rotation bands, 168, 221 ff
  in vibronic bands, 333, 336
Combination states, symmetry species of, 195 ff
Comet
  Ikeya, 350
  Kohoutek, 350
Commutation, of symmetry elements, 187, 192
Complex conjugate, of wave function, 300
Concentric resonator, 469
Conduction band, 73, 483 ff
Configuration (electron), 269, 278 ff
Configuration interaction, 344 ff, 372 ff

Confocal resonator, 469
Congestion, in electronic spectra, 422
Continuous electronic spectra
  diatomic molecules, 327 ff
  polyatomic molecules, 421 ff
Conventions
  for anharmonic constants, 163
  for labelling cartesian axes in a molecule, 195
  for labelling electronic states in a molecule, 312
  for numbering vibrations, 213
  for vibrational transitions, 213
  for vibronic transitions
    diatomic molecules, 320
    polyatomic molecules, 385
Coriolis force, 108
Coriolis interaction
  electronic spectra
    linear molecules, 401, 413
    symmetric rotors, 408
  in IVR, 543
  Jahn's rule, 221, 232
  vibrational spectra
    linear molecules, 108 ff, 219 ff, 496
    spherical rotors, 231 ff
    symmetric rotors, 113 ff, 223 ff
Corona discharge, 539, 577, 578
Correlation, of electrons, 344, 434, 436
Correspondence principle, 27, 30
Coulomb integral, 268, 301
  in Hückel MO's, 366
Counterpropagating beams, 609, 612, 635, 656
Critical inversion, 470
Cross sequence, 393
Crossover resonance, 531 ff
Cryptocyanine dye, 476
Crystal conversion noise, 51
Crystal diode detector, 51
Crystal field MO's, 374 ff
CSRS spectroscopy, 520 ff
CW laser, 472
Czerny–Turner grating mounting, 71 ff, 77, 80

D-line of Na, 286, 296
d-orbitals, 19 ff
  splitting by ligands, 374 ff
D_n point groups, 189
D_{nd} point groups, 189
D_{nh} point groups, 189 ff
D_2 character table, 690
D_3 character table, 690
D_4 character table, 690
D_5 character table, 690
D_6 character table, 690
D_{2d} character table, 692
D_{3d} character table, 692
D_{4d} character table, 692
D_{5d} character table, 693
D_{6d} character table, 693
D_{2h} character table, 693
D_{3h} character table, 693
D_{4h} character table, 694
D_{5h} character table, 694
D_{6h} character table, 694
$D_{\infty h}$ character table, 695
Damping rate, of quantum beats, 668
Dark state, 424, 542
Darling–Dennison resonance, 249
  in H_2O, 249
de Broglie relation, 7 ff
Deflection analysers, in photoelectron spectroscopy, 430 ff
Degeneracy, 18, 21, 34
Degenerate four-wave mixing, 651
Degenerate point group, 192
Degradation, of a band, 332 ff, 336, 408 ff
Degree of circularity, 180
Dephasing, of rotational alignment, 673
Deplorization ratio, 179 ff
Derivative polarizability, 215 ff
Deslandres table, 324 ff
  for $A^1\Pi$–$X^1\Sigma^+$ system of CO, 325
Detector, 49
Diatomic molecules
  electronic spectra, 298 ff
  rotational spectra, 94 ff
  vibrational spectra, 159 ff
Difference frequency laser, 540 ff
Difference frequency laser spectroscopy, 540 ff, 639
Diffraction, of electrons, 9
Diffuse electronic spectra
  diatomic molecules, 327 ff
  polyatomic molecules, 421 ff
Dihedral axes, 185
Diode lasers, 483 ff
  pumping Nd^{3+} : YAG laser, 483
Diode laser spectroscopy, 533 ff
Dipole–dipole force, 141
Dipole–induced-dipole force, 141
Dipole moment (*see* Electric dipole moment)
Direct product
  degenerate species, 197
  non-degenerate species, 195
Directionality, of laser beam, 470
Discharge, high voltage, 79
  for UV sources, 84 ff
Dispersed fluorescence, 587 ff
Dispersing element, 48, 58 ff
Dispersion, 58 ff, 61 ff
Dispersion force, 141

Dissociation, 162, 249 ff, 329 ff
by electric field, 569
energy, 163 ff, 316, 325 ff
Dodecahedron, 191
Doping, of semiconductor, 484 ff
Doppler broadening, 41 ff, 51, 54 ff, 57, 409, 421, 473 ff
Doppler effect, 41, 54 ff, 128
Doppler tuning, of laser, 566 ff
Double beam operation, 74, 76, 79
Double resonance
four-level, 533
three-level, 532, 630 ff
Double resonance spectroscopy, 630 ff
microwave–microwave, 630 ff
microwave–optical (MODR), 632 ff
optical–optical (OODR), 634 ff
Drift tube, 566 ff
Duschinsky effect, 393, 599
Dye laser, 497 ff
in CARS spectroscopy, 520 ff
in difference frequency laser, 540 ff
in Hanle effect measurement, 662
in MODR spectroscopy, 632
in OODR spectroscopy, 636
in photodetachment spectroscopy, 625
in quantum beat spectroscopy, 668
in Raman gain spectroscopy, 515
in translational spectroscopy, 681
Dyes, for lasers, 497 ff
Dynamic Jahn–Teller effect, 398

Eagle grating mounting, 86
Ebert grating mounting, 81 ff
Ebert–Fastie grating mounting, 71 ff, 77
Echelle grating, 63
Efficiency, of lasers, 471
Effusive beam, 53 ff
Eigenfunction, 13
Eigenstate, 13
Eigenvalue, 13
Einstein coefficients, 35 ff, 40 ff
Electric dipole moment
definition, 36
excited electronic states, 420 ff
laser Stark spectroscopy, 529 ff
of FNO, 531
of NH_3, 531
rotational spectra, 95 ff
spherical rotor, 126 ff
Stark effect, 52 ff, 102, 113, 126, 144
variation with r in diatomic, 160
Electric dipole selection rules, 37 ff, 210 ff, 380, 383
Electric field strength, 32
Electromagnetic radiation, 32 ff, 46 ff
Electromagnetic spectrum, 46 ff
Electron, wave nature, 7 ff
Electron affinity, 329
Electron bombardment analyser, 144
Electron correlation, 344, 434, 436
Electron impact spectroscopy, 427 ff
Electron multiplier detector, 431
Electron reorganization, 344, 436, 460
Electron repulsions, 21
Electron spin, 14 ff, 26 ff
Electronegativity, in XPS, 460 ff
Electronic selection rules
in atoms, 284 ff
in diatomic molecules, 312 ff, 332
in polyatomic molecules, 380 ff, 383
Elements of symmetry, 181 ff
Elliptically polarized radiation, 178
Emission processes, 33 ff
Energy loss spectroscopy, 466 ff
Equivalent electrons, 276
ESCA (electron spectroscopy for chemical analysis), 427
Etalon, 65 ff
in dye laser, 501
in Nd^{3+} : YAG laser, 483
Étendu advantage, 70
Exchange integral, 268
Excimer, 492
Excimer lasers, 492 ff
Exciplex, 492
Exciplex lasers, 492 ff

2f detection method, 530
Fabry-Perot interferometer, 64 ff, 81, 153
False origin, 385
Far infrared region, 47 ff
Far infrared sideband laser, 544 ff
Far ultraviolet region, 47 ff
Faraday cup detector, 567
Faraday effect, 501
F-centre, 487
F-centre lasers, 487
Feasible operation, 224, 406
Feedback, in laser, 469
Fellgett advantage, 70, 153
Femtosecond laser
optical parametric oscillator, 502 ff
Ti^{3+} : sapphire, 481
Fermi diad, 248
Fermi energy, 484 ff
Fermi resonance, 181, 247 ff
in IVR, 543
Fermi-Dirac statistics, 115
Fermions, 115
FG matrix method, 209

Fine structure, in atomic spectrum, 283
Fingerprint region, of IR, 175
First-order decay process, 40
Flash photolysis, 79 ff
Flashlamp pumping of laser, 472, 479 ff, 481, 500
Flop in mode in Rabi spectrometer, 144
Flop out mode in Rabi spectrometer, 144
Fluorescence, 320, 388, 422
  of dyes, 498 ff
  of ions, 435
Fluorescence depletion
  detection in SEP spectroscopy, 647 ff
Fluorescence depletion spectroscopy, 637 ff
Fluorescence dip spectroscopy, 637 ff
Fluorescence excitation spectroscopy
  infrared, 555 ff
  multiphoton, 604 ff
  rotationally resolved, 571 ff
  vibrationally resolved, 587 ff
Fluorescence lifetime, 423
Force constant, 27, 159
  bending, definition, 203
  typical bending, 174
  typical stretching, 173 ff
Force field
  central, 203
  general, 203
  valence, 203
    for H_2O, 208
Fore-grating, 62
Fore-prism, 62 ff
Fortrat curves, 410
Forward scattering, 180, 487, 512
Fountain design, of Nd^{3+} : YAG laser, 483
Four group, 236
Fourier transform microwave spectroscopy, 152 ff
Fourier transform infrared (FTIR) spectroscopy, 64 ff
Fourier transform Raman spectroscopy, 138 ff
Fourier transformation, 68, 78
  of quantum beat spectra, 670 ff
Four-level system, for laser, 471 ff
  Nd^{3+} : YAG laser, 482
Frame, of internal rotor, 265
Franck–Condon approximation, breakdown, 383 ff, 385
Franck–Condon factors, 322
  in photoelectron spectra, 450, 451, 464
  in resonance Raman spectra, 507
  in SEP spectroscopy, 647 ff
Franck–Condon principle
  in dye fluorescence, 499
  in electronic spectra, 318 ff, 320 ff, 387 ff, 587 ff
  in photoelectron spectra, 439 ff, 464
Free molecule, 182
Free radical, 79
Frequency, 46
Frequency doubling, 477, 481, 500, 501, 510
Frequency mixing, 500, 515
Frequency modulation, 631
Frequency tripling
  of dye laser, 681
  of Nd^{3+} : YAG laser, 481

g-value
  of electron, 274
  of nucleus, 275
G_4 group, 262
G_6 group, 265 ff
G_{12} group, 266
γ-ray region, 47
Γ, symbol for representation, 195
Gain, of laser, 470
  in *P*-branches, 494
Galaxy, 127 ff
Gas dynamic (CO_2) laser, 495
Gauche rotamer, 264
Gaussian laser beam, 483
Gaussian line shape, 42, 551
Generation, of symmetry elements, 185 ff, 187
Giant *l*-type doubling, 408
Giant pulse, 474 ff
Globar, 75
Golay cell, 72 ff, 78
Grating ghosts, 64
Grating, 61 ff
Grazing incidence, 86
Grotrian diagram, 284
  for He, 291
  for Li, 284
Group vibration wavenumber (frequency), 174 ff
Group vibrations, 171 ff
Groups, properties of, 191 ff

Half width at half maximum (HWHM), 39
Half-life, of excited state, 40
Hamiltonian, 12 ff
  for H atom, 16 ff, 267
  for He atom, 267
  for other atoms, 268
Hanle effect, 659 ff
Harmonic generation, 477
  in Nd^{3+} : YAG laser, 483
  microwave, 51
Harmonic oscillator, 27 ff, 159 ff
  definition, 163
Hartree–Fock SCF method, 268 ff, 436, 438, 462
Heat capacity, 2, 3
Heiles' cloud 2, 128
Heisenberg uncertainty principle, 10 ff, 40 ff
Helium discharge lamp, 428 ff
Helium–neon laser, 487 ff

Hemiconfocal resonator, 469
Hemispherical resonator, 469
Hermite polynomials, 29 ff
Herzberg–Teller intensity stealing, 376 ff, 382 ff, 451, 505, 507
  in two-photon absorption, 611 ff
Heterogeneous perturbation, 328
Heterojunction diode, 486
High spin configuration, 377
Hole burning, 55, 490, 501, 615 ff, 635
Hollow cathode source, 83
Holgraphic grating, 64
Holography, 470
Homogeneous line broadening, 40, 41
  in OODR spectroscopy, 635
Homogeneous perturbation, 328
Hönl–London factors, 226, 229, 408
Hooke's law, 27, 162
Hot band, definition, 160, 214
Hückel MO method, 366 ff
Hund's coupling case
  case (a), 312, 313, 317, 337 ff, 401, 403, 404
  case (a_α), 341
  case (a_β), 341
  case (b), 338 ff, 401, 403, 404
  case ($b_{\beta J}$), 343
  case ($b_{\beta R}$), 342
  case ($b_{\beta S}$), 342
  case (c), 312, 313, 339 ff
  case (c_β), 343
  case (d), 340
  case (e), 340
Hund's rules, 280, 306, 316, 377
Hybrid band
  infrared, 237
  Raman 242 ff
Hybrid laser, 497
Hybridization, 105
  effect on force constant, 173 ff
Hydrocarbon flame bands, 358
Hydrogen bonded complexes, 58, 141 ff
Hydrogen bonding
  bond lengths, 149
  intermolecular, 142, 146 ff
  intramolecular, 142, 149 ff
Hydrogen cyanide laser, 496
  in laser Stark spectroscopy, 530
  in LMR spectroscopy, 524 ff
Hydrogen discharge source, 79
Hyper Rayleigh scattering, 510 ff
Hyper Raman spectroscopy, 510 ff
  rotational selection rules, 512
  vibrational selection rules, 510 ff
Hyperfine structure, in atomic spectrum
  isotopic, 283
  nuclear, 283
Hyperfine structure, in molecular spectra, 341 ff
Hyperpolarizability, 477, 510 ff

I symmetry element (*see* Identity element)
i symmetry element (*see* Centre of symmetry)
I_h character table, 698
I_h point group, 191
Icosahedron, 191
Identity element, *I*, 186, 192
Idler frequency, in optical parametric oscillator, 501 ff
Ikeya comet, 350
Improper rotation axis, 186
Incoherent radiation, 470
Induced emission, 34 ff, 100, 469 ff
Induced-dipole–induced-dipole force, 141
Inert gas configurations, 269
Inertial defect, 409, 414
Infrared laser spectroscopy, near dissociation, 566 ff
Inhomogeneous line broadening, 40, 42
  in MODR spectroscopy, 635
Insulator, 484
Intensity
  electronic transition, 380 ff
  rotational transition, 100 ff, 117 ff
  rovibronic transition, 333
  vibration–rotation transition, 169 ff, 226
Intensity stealing
  by spin-forbidden transitions, 314 ff
  by vibronic transitions (*see* Herzberg–Teller intensity stealing)
Interference effects, in fluorescence, 660 ff, 668 ff
Interferogram, 69
Interferometer, 64 ff
Intermediate states
  in Raman scattering, 505
  in two-photon transitions, 602 ff
Intermodulated fluorescence, 658
Internal coordinates, for stretching in linear molecule, 204
Internal rotation, 264 ff
Interstellar dust, 127 ff
Interstellar molecules, 128 ff
Intersystem crossing, 422
Intramolecular vibrational redistribution (IVR), 424 ff, 542 ff, 549, 559 ff, 649 ff, 655, 669 ff
Intrinsic semiconductor, 484
Inverse Lamb dip, 524
Inverse, of a symmetry element, 192
Inverse Raman spectroscopy, 518 ff
Inversion operation (*see* Centre of symmetry)
Inversion vibration, 175, 252 ff, 588 ff
Inverted multiplet, 281

Ion beam, 566 ff
Ion-pair states, 329 ff, 640 ff
Ionization dip spectroscopy, 654 ff
Ionization energy, 426, 435
adiabatic, 441
vertical, 441
Ionization gain stimulated Raman spectroscopy, 516 ff
Ionization loss stimulated Raman spectroscopy, 516 ff
Irreducible representation, 195
Isomerism, s-*cis*-s-*trans*
acrolein, 263 ff
buta-1,3-diene, 263 ff
crotonic acid, 119 ff
glyoxal, 263 ff
Isomorphous groups, 193
Isotopes
effect on dissociation energy, 164
in structure determination, 139 ff

jj-coupling, 276, 281 ff
j-type doubling, 408
Jacquinot advantage, 70
Jahn's rule, for Coriolis interaction, 221, 232
Jahn–Teller effect, 397 ff
in CF_3I, 399
in CH_4^+, 455 ff
in C_5H_5, 579
in C_6H_6, 508
in $C_6H_6^+$, 457
in $C_6H_3Cl_3^+$, 400
in $C_6H_3F_3^+$, 400
in $C_6F_6^+$, 400
in NH_3, 399 ff, 408, 419 ff
in NH_3^+, 453
in OsF_6, 400
in ReF_6, 400
in RuF_6, 400
in TcF_6, 400
Johnson noise, 51

K_h character table, 697
K_h point group, 191, 293 ff
K-type doubling, 401
KBr disk, 76, 176
Kerr cell, 475
Kerr effect, 475, 481
Kerr effect spectrum, 421
Kinematic compression, 566
Kinetic energy, of harmonic oscillator, 202
Kinetic theory of gases, 388 ff
Klystron, 50
Kohoutek comet, 350
Koopmans' theorem, 436
breakdown in N_2, 436, 443
Kraitchman equations, 140, 585
Krypton ion laser, 490
in resonance Raman spectroscopy, 506
in saturation spectroscopy, 657

Λ-type doubling, 336 ff, 401, 527, 529
l-type doubling
linear molecule, 107 ff, 221 ff
symmetric rotor, 113 ff, 227
Lagrange equation, 201 ff
Laguerre functions, 17
Lamb dip
in He–Ne laser, 489 ff
in laser Stark spectroscopy, 530 ff
in LMR spectroscopy, 523 ff
Lamb dip spectroscopy, 42, 51, 54 ff
Lamb peak, 525
Lamb shift, in H atom, 288, 616 ff
Landé *g*-factor, 295 ff, 661
Lanthanide configurations, 273
Laplacian, 12
Laporte rule, 292
Laser ablation, 576
Laser cavity, 469
Laser electric resonance spectroscopy, 529 ff
Laser induced fluorescence, 572
infrared, 555 ff
Laser-induced grating spectroscopy, 651
Laser magnetic resonance (LMR) spectroscopy, 522 ff
Laser Stark spectroscopy, 529 ff
Laser Zeeman spectroscopy, 522 ff
Lasers, 34, 468 ff
LCAO method, 300 ff
in solids, 483
Legendre polynomials, 17
Lennard–Jones 12-6 potential, 142
Level anticrossing spectroscopy, 664 ff
comparison with MODR, 632
Level crossing spectroscopy, 659 ff
Libration, of transition moment, 615
Lifetime, relation to line width, 551
Lifetimes of vibrational states, 514
Ligand, 374
Ligand field MO's, 374, 379 ff
Line, definition, 160
Line width, 39 ff
Linear (napierian) absorption coefficient, 562
Linear pumping, 632 ff
Lissajous motion, 171
Littrow mounting, 77, 80
LMR spectroscopy, 522 ff
Local (vibrational) modes, 249 ff
in C_6H_6, 249 ff

Local (vibrational) modes (*cont.*)
in CO_2, 251 ff
in H_2O, 250 ff
Longitudinal mode
of laser cavity, 473 ff
of ring-down cavity, 562 ff
Lorentzian line shape, 42, 551
Loss, of laser, 474
Low spin configuration, 377
Lummer–Gehrke plate, 475
Lyman far UV source, 85

Mach number, in supersonic jet, 57
Magic angle, in rotational coherence spectroscopy, 673
Magnetic dipole selection rules, 38, 315
Magnetic dipole transitions
diatomic molecules, 315 ff
formaldehyde, 364, 386 ff, 390 ff
p-benzoquinone, 386
Magnetic field strength, 32
Magnetic moment
Bohr magneton, 275
coupling in atoms, 275 ff
electron spin, 274 ff
nuclear spin, 275, 341
orbital, 273 ff
Magnetic quantum number, 7, 16 ff, 53
Magnetic rotation spectrum, 420
Magnetogyric ratio
of electron, 274
of nucleus, 275
Maser, 468 ff
ammonia, 478 ff
Mass spectrometer
tandem, 566 ff
time-of-flight, 604 ff, 620, 644, 681
Matrix inversion, 205
Matrix mechanics, 13
Maxwell velocity distribution, 41, 55, 56
Mean free path, for collisions, 54, 56
Mean polarizability, 161
Mercury arc, 70 ff
Metastable states, of He atom, 290, 488
Michelson interferometer, 66 ff, 77 ff, 126
Microphone, in photoacoustic spectroscopy, 558 ff
Microwave bands, 49
Microwave region, 47
Microwave spectrometer, 49 ff
Microwave spectroscopy, near dissociation, 566 ff
Mid infrared region, 47
Milky Way, 127, 289
Millimetre wave region, 47
Millimetre wave spectroscopy
advantages, 97, 106
experimental methods, 49 ff
near to dissociation, 566 ff
Minor, of a matrix, 205
Mode locking, 476 ff
He–Ne laser, 489
ruby laser, 480
Modulation, of LMR signal, 522 ff
Modulation broadening, 44
Modulation frequency, 44
Molar absorption coefficient, 37, 51, 562
Molar absorptivity, 37
Molecular beam, 42, 55 ff, 143
Molecular beam electric resonance spectroscopy, 143 ff
with optothermal detection, 553 ff
Molecular orbital (MO) method, 299 ff
Molecular orbitals
AB_2 molecules, 353 ff
AH_2 molecules, 344 ff
AH_3 molecules, 358 ff
HAAH molecules, 360 ff
H_2AAH_2 molecules, 364 ff
HAB molecules, 353 ff
H_2AB molecules, 362 ff
XHCCHY molecules, 367 ff
aza-aromatics, 373 ff
benzene and substituted benzenes, 369 ff
naphthalene, 371 ff
transition metal complexes, 374 ff
Momental ellipsoid, 89 ff
Moments of inertia, 89 ff
Monochromatic character, of laser, 470
Monochromator
for synchrotron radiation, 430
for X-ray radiation, 433 ff
Monopole transitions, 465 ff
Morse potential function, 165 ff, 250
Multichannel analyser, 82, 510
Multichannel electron multiplier, 431
Multilayer mirror coating, 470
Multimode operation, of laser, 474
Multiphoton absorption, 601 ff
Multiphoton ionization, 604 ff
Multiple reflection cell, 75 ff, 79, 80, 137, 534
Multiplex advantage, 70, 82, 153
Multiplication of symmetry species, 195 ff
Multiplication table, of group, 193
Multiplicity, 277
Mutual exclusion, of Raman and IR vibrations, 182, 217
Mylar, windows in far IR, 524

Natural lifetime, of excited state, 41
Natural line width, 41, 55, 421
Near infrared region, 47
Near ultraviolet region, 47
Near-symmetric rotor, 92 ff

Nebula, 127 ff
Negative ions
  photoelectron spectra, 625 ff
Neodymium–YAG laser, 481 ff
  for FT Raman spectroscopy, 138
  pumping a dye laser, 500 ff
  pumping a titanium–sapphire laser, 481
  pumping an optical parametric oscillator, 502
Neodymium–YLF laser, 481
Neon discharge lamp, 429
Nernst filament, 75
Newton's rings, 64
Nitrogen laser, 491 ff
  pumping a dye laser, 500
Nitrous oxide laser
  in laser Stark spectroscopy, 530
  in two-photon spectroscopy, 603
Noble gases, configurations, 269
Node, 21
Non-crossing rule, 299, 328
Non-degenerate point group, 193
Non-equivalent electrons, 276
Non-feasible operation, 224
Nonlinear effects, 477
Nonlinear pumping, 632 ff
Nonlinear Raman effects, 503
Nonradiative decay, 421 ff
Non-rigid molecules, 252
Nontotally symmetric species, 194
Nontotally symmetric vibration, 180
Normal modes of vibration, 171
  determination of, 201 ff
Normal multiplet, 280 ff
Nuclear hyperfine splitting
  electric resonance spectra, 145
  electronic spectra of diatomics, 341 ff
  rotational spectra, 104 ff, 128, 153
Nuclear magneton, 275
Nuclear spin, 14 ff, 26 ff, 104 ff
Nuclear spin statistical weights, 114 ff
  C_3, 401
  CF_2N_2, 413
  C^1H_3, 405 ff
  C^2H_3, 405 ff
  C^1H_4, 233
  C^2H_4, 233
  $^{12}C_2{}^1H_4$, 124
  $^{12}C_2{}^2H_4$, 124
  $^{13}C_2{}^1H_4$, 124
  C^1H_3F, 226
  C^2H_3F, 226
  $^{12}C_2{}^1H_2{}^{19}F_2$, 124
  1,4-$C_6H_4F_2$, 416
  1,4-$C_6H_4(OH)_2$, 586
  F_2, 116 ff, 171
  $^1H_2$, 115 ff, 171, 334
  $^2H_2$, 116 ff, 171
  $^1H_2{}^{12}C^{16}O$, 124
  $^1H_2{}^{16}O$, 124
  $^2H_2{}^{16}O$, 124
  6-hydroxy-2-formylfulvene, 150 ff
  malonaldehyde, 150 ff
  $^{14}N_2$, 116 ff, 171
  $^{14}N^{15}N$, 134
  $^{15}N_2$, 134
  N_3, 401
  N^1H_3, 255
  N^2H_3, 405
  O_2, 116 ff, 171
  Si_2H_2, 124 ff
  SiH_3F, 229
  SiH_3NCS, 114, 117
  tropolone, 150 ff
Nuclear quadrupole coupling constant, 156
Nujol mull, 76
Nutation, 110

O character table, 696
O_h character table, 696
O_h point group, 190 ff
Oblate symmetric rotor, 91 ff
Operator, 13
Optical null method, 74
Optical parametric amplifier (OPA), 502, 564
Optical parametric oscillator (OPO), 501 ff, 564, 639
Optothermal spectroscopy, 548 ff
Order of a group, 192
Order of diffraction, 62
Ortho-hydrogen (H_2), 116
Oscillator–amplifier dye laser, 501
Oscillator strength, 37, 380 ff
Overlap, of atomic orbitals, 300 ff
Overlap integral, 301
  in Hückel MO's, 366
Overlapping orders, of grating, 62 ff
Overtone states, symmetry species of, 195 ff, 197

p-orbitals, 18 ff
Para-hydrogen (H_2), 115 ff
Parallel band, 220, 404
Parity of rotational levels
  asymmetric rotor, 122 ff
  diatomic molecule, 331 ff
  linear molecule
    +, − and *e*, *f* symbols, 109, 220
Partition function, 34
  electronic, 100
  rotational, 100, 118, 124
  vibrational, 100

Paschen–Back effect, 297 ff, 528, 662
Pauli principle, 115, 198, 269, 279, 290, 306, 483
Penetration, 285
Penning ionization spectroscopy, 427 ff
Periodic states, 570
Periodic table, 267 ff
Permutation of nuclei, 261 ff
Perpendicular band, 220, 408
Phase of radiation, 32
Phosphorescence, 320, 388, 422
Photoacoustic spectroscopy, 557 ff
Photochemical reaction, 40
Photoconductive detector, 73 ff, 78
Photodetachment spectroscopy, 625
Photodiode detector, 82
Photodissociation, 567, 680 ff
Photo-double ionization, 434
Photoelectric effect, 2, 426
Photoelectron spectrometer, 428 ff
Photoelectron spectroscopy, 426 ff
  laser, 625
  ultraviolet (UPS), 427, 437 ff
  X-ray (XPS), 427, 459 ff
Photoelectron, 2, 426
Photofragment translational spectroscopy, 680 ff
Photofragmentation
  of CF_3S, 640
Photographic plate, 81
Photoionization cross section, 438, 441, 444, 445, 458
Photomultiplier, 82, 86
Photon, 32
Photon counting, 510
Piezoelectric translator, 495, 524
Pixel, 82
Planck constant, 2 ff
Plane of symmetry σ, 39, 183 ff
Plane-polarized radiation, 32, 53
Pockels cell, 475
Pockels effect, 475
Point groups, 187 ff
Polar diagram, 19 ff
Polarizability, 131, 141
  variation with vibration, 161, 216
Polarizability ellipsoid, 131
Polarization labelling, 646
Polarization of transition moment, 38 ff
Polarization spectroscopy, 658 ff
Polyethylene, windows in far IR, 524
Population inversion, 34, 468, 471 ff
Populations of energy levels
  rotational, 100 ff, 118, 169
  vibrational, 105 ff, 148, 160, 214
Population transfer, by SEP, 644 ff, 646
Potassium bromide disk, 76
Potential energy, of harmonic oscillator, 27, 202 ff
Potential energy curve, 24, 27 ff, 316 ff
  for ion-pair state, 329 ff
Power saturation broadening, 44, 51
Precession, 23, 275 ff, 283, 297 ff, 302
Predispersion, 62 ff
Predissociation, 327 ff, 421 ff, 450, 567
  vibrational, 550, 551 ff
Pre-resonance Raman scattering, 505 ff
Pressure broadening, 42 ff, 51
Principal moments of inertia, 15, 89 ff
Principal quantum number, 16, 267
Prism, 58 ff
  infrared, 76 ff
  visible/UV, 80
Probability, 11 ff
Products of inertia, 89
Progression, 318 ff, 382 ff
Prolate symmetric rotor, 90 ff
Proper rotation axis, 186
Pseudo- (or quasi-) continuum of levels, 423
Pseudo-Jahn–Teller distortion, 507
Pseudorotation, 258 ff
Pulsed field extraction, 618 ff
Pulsed field ionization, 619 ff
Pulsed laser, 472
Pumping, of a transition
  linear, 632 ff
  nonlinear, 632 ff
Pumping, of laser, 468
  electrical, 472
  optical, 472

Q, of laser, 474 ff
Q-switching, 474 ff, 510, 514
  ruby laser, 480, 513
Quadrupole field, 143 ff, 478
Quadrupole moment, 104 ff
Quantum, 2
Quantization, 13 ff
Quantum beats, 543
Quantum beat spectroscopy, 668 ff
Quantum defect, 285, 310
Quantum electrodynamics, 288 ff
Quantum yield, of fluorescence, 423 ff
Quasibound–bound transition
  in HeH^+, 570 ff
Quasibound–quasibound transition
  in HeH^+, 570 ff
Quasibound state, 567, 570 ff
Quasi- (or pseudo-) continuum of levels, 423
Quasi-planar, 258
Quenching of rotation, 48

r_0 structure, 140

r_e structure, 139 ff
r_s structure, 140
Rabi spectrometer, 143 ff
Radial charge density, 17 ff, 21
Radial probability distribution, 17
Radial wave function, 17 ff
Radiation density, 2
Radiation trapping, 488 ff
Radiative decay, 421 ff
Radical (*see* Short-lived species)
Radio wave region, 46 ff
Radiotelescope, 127 ff
Raman gain spectroscopy, 514 ff
Raman loss spectroscopy, 514 ff
Raman scattering, 131 ff
  depolarization, 179 ff
  depolarized, 180
  intensity, 504
  polarized, 180
  two-photon process, 601
Raman shifting, of laser wavelength, 507, 508 ff, 514
Raman spectra (*see* Rotational Raman or Vibrational Raman spectra)
Ratio recording, 74
Rayleigh criterion, 59 ff
Rayleigh scattering, 133
  depolarization, 179 ff
Reducible representation, 198
Reduction of reducible representation, 198
Rephasing, of rotational alignment, 573
Relativity, 5, 12, 436
Relaxation time, of excited state, 40
Renner parameter, 396
Renner–Teller effect, 393 ff
  in AsH_2, 394
  in C_3, 394, 396, 401
  in CH_2, 394, 397
  in C_2H_2, 394
  in HCO, 357
  in H_2O^+, 452
  in H_2S^+, 452 ff
  in NH_2, 394, 396
  in PH_2, 394
Reorganization, of orbitals, 344, 436, 460
Repulsive states, 327 ff
Resolution, 59, 70, 86 ff, 432, 464
Resolving power, 59
  diffraction grating, 62
  photoelectron spectrometer, 431 ff
  prism, 60
Resonance, 33
Resonance-enhanced multiphoton ionization (REMPI), 604 ff
Resonance integral, 301 ff
  in Hückel MO's, 366
Resonance Raman spectroscopy, 504 ff
Resonant two-photon process, 618
Retardation, 67
Retarding grid analyser, in photoelectron spectroscopy, 430
Reversal factor, 180
Rhodamine B dye, 498 ff
Rhodamine 6G dye, 500
Rhodamine 6G-cresyl violet perchlorate dye, 500
Rice–Rampsberger–Kassel–Marcus (RRKM) theory, 649
Rigid rotor approximation, 25
Ring-down time, 563
Ring dye laser, 501, 561, 581 ff
Ring-breathing vibration, 175
Ring-puckering vibration, 256 ff
Rocking vibration, 175
Rotamers, 263 ff
Rotating mirror Q-switch, 475
Rotation axis, C_n, 38 ff
Rotation, of diatomic molecule, 14, 25 ff, 94 ff, 166 ff, 330 ff
Rotational coherence spectroscopy, 672 ff
Rotational depletion coherence spectroscopy, 679 ff
Rotational electric dipole spectra
  asymmetric rotor, 118 ff
  diatomic molecule, 94 ff
  linear polyatomic molecule, 94 ff
  spherical rotor, 126 ff
  symmetric rotor, 110 ff
Rotational fine structure, in electronic spectra
  asymmetric rotor, 411 ff
    doublet states, 417 ff
    hybrid bands, 411
    triplet–singlet transitions, 418
    type A band, 411, 415 ff
    type B band, 411, 413 ff, 415 ff
    type C band, 411, 412, 415 ff
  diatomic molecules
    $^1\Delta$–$^1\Delta$ transition, 337
    $^1\Delta$–$^1\Pi$ transition, 337
    $^1\Pi$–$^1\Sigma$ transition, 334 ff
    $^1\Pi$–$^1\Pi$ transition, 337
    $^2\Pi$–$^2\Sigma$ transition, 340 ff
    $^1\Sigma$–$^1\Sigma$ transition, 330 ff
    $^2\Sigma$–$^2\Sigma$ transition, 340
    $^3\Sigma$–$^3\Sigma$ transition, 340
  linear polyatomics
    $^1\Pi$–$^1\Sigma$ transition, 401
    $^3\Pi$(a)–$^1\Sigma$ transition, 404
    $^3\Pi$(b)–$^1\Sigma$ transition, 404
    $^3\Pi$(a,b)–$^3\Sigma$ transition, 403 ff
    $^1\Sigma$–$^1\Sigma$ transition, 400
    $^2\Sigma$–$^2\Pi$ transition, 403
    $^2\Sigma$–$^2\Sigma$ transition, 401

Rotational fine structure (*cont.*)
linear polyatomics (*cont.*)
$^3\Sigma$–$^1\Sigma$ transition, 404
$^3\Sigma$–$^3\Sigma$ transition, 401
spherical rotor, 418 ff
symmetric rotor, 404 ff
doublet states, 411
triplet states, 411
triplet–singlet transitions, 411
Rotational fine structure, in photoelectron spectra, 441, 452
in ZEKE-PE spectra, 619 ff
Rotational fine structure, in vibrational spectra
infrared
asymmetric rotor, 234 ff
diatomic molecule, 166 ff
linear polyatomic molecule, 219 ff
spherical rotor, 231 ff
symmetric rotor, 223 ff
Raman
asymmetric rotor, 242 ff
diatomic molecule, 170 ff
linear polyatomic molecule, 219, 223
spherical rotor, 233 ff
symmetric rotor, 230 ff
Rotational Raman spectra
asymmetric rotor, 136
diatomic molecule, 130 ff
experimental method, 135 ff
linear polyatomic molecule, 130 ff
spherical rotor, 134
symmetric rotor, 136
Rotational temperature, 333
in supersonic jet, 58
Rotation–reflection axis, S_n, 92, 185 ff
Rotations, symmetry species of, 194 ff
Rowland circle, 86
Ruby laser, 479 ff
in inverse Raman spectroscopy, 519
in stimulated Raman spectroscopy, 513
Russell–Saunders coupling, 276 ff
equivalent electrons, 279 ff
non-equivalent electrons, 276 ff
Rydberg constant, 1, 6, 285, 308
Rydberg orbitals, 308 ff
Rydberg transitions, 380
Rydberg–Klein–Rees (RKR) method, 164 ff

S_n axis (*see* Rotation–reflection axis)
S_n point groups, 188
s, *p*, *d*, *f* symbolism, 284
s-orbitals, 19, 21
σ plane (*see* Plane of symmetry)
S_4 character table, 695
S_6 character table, 695
S_8 character table, 695
Sagittarius B2, 128
Saturable absorber, 476
Saturable dye, 476
Saturation, 44, 55, 471, 523 ff, 631, 655 ff
Saturation spectroscopy, 655 ff
Scattering, 33, 130 ff
Scattering plane, 178 ff
Schrödinger equation, 11 ff
time-dependent, 12, 34
time-independent, 12
Schumann photographic plate, 86
Schuster bands, of NH_3, 609
Scissoring vibration, 175
s-*cis*-s-*trans* rotamers, 263 ff
Second hyperpolarizability, 477
Second-order perturbation
asymmetric rotor levels, 415
Secular determinant
factorization, 207
MO theory, 301, 366 ff
stretching vibrations of linear molecule, 206
Selection rules, 37 ff
breakdown in liquid and solid phase, 215
for photoionization, 434
Self-consistent field (SCF) method, 268 ff, 436
Self-termination, of laser, 491
Semiconductor, 71, 78, 483 ff
n-type, 484 ff
p-type, 485
Semiconductor detector, 71, 78
Semiconductor diode, 485
Semiconductor laser, 483 ff
Separated atom approximation, 299
Septum, 52
Sequence
diatomic molecule, 318 ff
polyatomic molecule, 391 ff
Shakeoff satellites, in XPS, 464 ff
Shakeup satellites, in XPS, 464 ff
Shell, of electrons, 269
Shielding, 285
Short-lived species
flash photolysis, 79
microwave spectra, 53
Sideband laser, 544 ff
Signal frequency, in optical parametric oscillator, 501 ff
Single mode operation, of laser, 474
Single photon counting, 673
Single rovibronic level fluorescence (SRLF), 588
Single vibronic level fluorescence (SVLF), 320, 388 ff, 424, 505, 572, 587 ff
Skeletal vibrations, 175
Skimmer, 57
Slit jet, 534

Slit nozzle, for supersonic jet, 534
  in infrared laser induced fluorescence, 555 ff
Solar blind photomultiplier, 86
Sommerfeld theory of H atom, 5 ff
Source modulation, 53
Space groups, 187
Space quantization, 7, 14, 16 ff, 23, 26, 102 ff
Spark gap laser trigger, 491
Spark source, 84
Spectrograph, 53, 58
Spectrometer, 53, 58
Spectroscope, 53
Spherical harmonics, 16 ff
Spherical rotor, 15, 93 ff
Spin–flip Raman laser, 486 ff
Spin–orbit coupling
  atom, 276, 278 ff
  diatomic molecule, 314 ff, 337 ff
  formaldehyde, 364
  polyatomic molecule, 401 ff, 411, 417 ff
Spin–orbit coupling constant
  linear molecule, 401 ff
Spin–orbital, 269
Spin–rotation coupling constant
  asymmetric rotor, 417 ff
  diatomic molecule, 338 ff
  symmetric rotor, 411
Spin–spin coupling constant
  asymmetric rotor, 417
  diatomic molecule, 339
  symmetric rotor, 411
Spontaneous emission, 35 ff, 40 ff
Standing wave, 9 ff
Stark effect, 51 ff, 144 ff
  asymmetric rotor, 126
  atom, 298
  laser Stark spectra, 529 ff
  linear molecule, 102 ff
  molecular electronic spectra, 419 ff
  OODR spectra, 636
  symmetric rotor, 112 ff
Stark modulation, 44, 51 ff
State, as opposed to configuration, 269, 278 ff
Static Jahn–Teller effect, 398
Stationary state, 11, 13, 40 ff
Statistical weights (*see* Nuclear spin)
Stern–Gerlach experiment, 274
Stimulated anti-Stokes Raman scattering, 514
Stimulated emission, 35 ff
Stimulated emission pumping (SEP), 644 ff
Stimulated Raman scattering, 487
Stimulated Raman spectroscopy, 512 ff
  by delayed pulses (STIRAP), 646
  relation to stimulated emission pumping, 646
Stokes Raman scattering, 133 ff
Storage ring, 429 ff
Structure, from rotational constants, 139 ff
Sub-band, 225
Sub-Doppler spectra, 41, 57, 145, 534, 546, 548, 576, 582 ff, 610, 612, 635, 657, 658, 673
Sub-Rydberg transitions, 381
Sub-shell, 269
Superradiance, 472 ff, 491, 493
Supersonic jet, 55 ff, 143 ff
Symmetric part, of direct product, 198, 306
Symmetric rotor, 15, 90 ff
Symmetry, 181 ff
Symmetry coordinates, 206 ff
Symmetry operation, 183 ff
Symmetry orbitals, 345
Synchrotron source, 85, 430

T character table, 696
T_d character table, 696
T_d point group, 190
TEA laser (CO_2), 494 ff
TEM modes, of laser cavity, 473
Tandem mass spectrometer, 566 ff
Temperature
  rotational, 58
  translational, 57
  vibrational, 58
Term value, 94
Term, of an atom, 277, 278 ff
Thermocouple, 73, 78
Third channel, for decay in benzene, 424 ff
Three-level system, for laser, 471 ff
  ruby laser, 479 ff
Threshold condition, of laser, 473 ff
Threshold frequency, 2, 426
Threshold photoelectron spectroscopy, 618 ff
Thyratron laser trigger, 491
Time-resolved fluorescence depletion spectroscopy, 675 ff
Time-resolved resonance Raman (TR^3) spectroscopy, 508 ff
Time-resolved single photon counting, 673, 676
Titanium–sapphire laser, 481 ff
  femtosecond pulses, 481
  pumping an optical parametric oscillator, 502
  ring configuration, 555 ff, 561
Top, of internal rotor, 266
Toronto arc, 137
Torsional vibration, 175, 260 ff, 593 ff
Totally symmetric species, 194
Totally symmetric vibration, 180
Trace of a matrix, 197
Transient species, 79
Transition metal atoms, configurations, 272 ff
Transition metal complexes
  molecular orbitals, 374 ff

Transition moment, 36 ff, 48
  infrared vibrational, 159 ff
  Raman vibrational, 161
  rotational, 95
Translational spectroscopy, 680 ff
Translations, symmetry species of, 194
Transverse mode, of laser cavity, 473
Tungsten source, 78
Tunnelling through barrier, 252 ff, 571
  donor–acceptor barrier in $(H_2O)_2$, 547 ff
Twisting vibration, 175
Two-level system, for laser, 471
  ammonia maser, 478
  diode laser, 485
  excimer laser, 493
Two-photon absorption, 601 ff
  selection rules, 602, 605
  transition tensor, 605
Two-step, two-photon absorption, 637
Type *A* band
  infrared, 238 ff
  Raman, 242 ff
Type *B* infrared band, 240 ff
Type B_a Raman band, 242 ff
Type B_b Raman band, 242 ff
Type B_c Raman band, 242 ff
Type *C* infrared band, 241 ff

Ultraviolet photoelectron spectroscopy (UPS), 427, 437 ff
Ultraviolet region, 47
Umbrella vibration, 175
Uncertainty principle, 10 ff, 477
United atom approximation, 299
Unstable laser cavity, 483

V (*a*, *b*, *c*) group, 236
Valence band, 73, 484
Valence bond method, 299 ff
Valence force field, 173 ff
Van der Waals complex, 58, 141 ff
Van der Waals force, 141 ff
Vanadium phthalocyanine dye, 476
Variation principle, 301
Velocity bunching, 566
Velocity modulation, 538 ff
Vibration
  diatomic molecule, 15, 159 ff
  polyatomic molecule, 15
Vibrational combination tone, 172
Vibrational numbering
  Mulliken convention, 213
Vibrational overlap integral, 322
Vibrational overtone, 172, 548 ff, 555 ff, 557 ff
Vibrational predissociation spectroscopy, 554 ff
Vibrational quantum number, 28
Vibrational Raman spectra
  diatomic molecule, 160 ff
  polyatomic molecule, 172
Vibrational satellites, 98
  diatomic molecule, 105 ff
  hydrogen bonded complex, 148 ff
  intermolecular hydrogen bonding, 151
  linear polyatomic molecule, 107 ff
Vibrational selection rules
  diatomic molecule
    infrared, 160, 163
    Raman, 161, 163
  polyatomic molecule
    infrared, 210 ff
    Raman, 215 ff
Vibrational temperature, in supersonic jet, 58
Vibrational term values
  diatomic molecule
    anharmonic oscillator, 163
    harmonic oscillator, 159
  polyatomic molecule
    anharmonic oscillator, 246 ff
    harmonic oscillator, 172
Vibrational wave function, 29 ff
Vibration–rotation interaction constant, 106, 113, 125, 140, 169, 230, 247
Vibronic interaction (*see* Herzberg–Teller intensity stealing)
Vibronic laser, 480, 481
Vibronic selection rules, 383
Vibronic transition, 319
Vibronic wave function, 321 ff
Virtual state, 133, 504, 601 ff
Visible region, 46 ff
Voigt line shape, 45, 551

Wagging vibration, 175
Wall collision broadening, 43 ff
Walsh MO diagrams, 344 ff
Water laser, 496
  in LMR spectroscopy, 525
Wave mixing, 515, 520, 636, 651
Wave packet, picture of electron, 10
Waveguide, 49
Wavelength, 46
Wavenumber, 46
Welsh multiple reflection mirrors, 137
White multiple reflection mirrors, 75
Work function, 2, 426

Xenon arc source, 78 ff
X-ogen, 129 ff

X-ray photoelectron spectroscopy (XPS), 427, 459 ff
X-ray region, 46 ff
X-ray sources, in photoelectron spectroscopy, 432 ff

Zeeman effect, 53, 294 ff
  anomalous, 296 ff
  in level crossing spectroscopy, 660 ff
  in LMR spectroscopy, 522 ff
  in molecular electronic spectra, 419 ff
  in OODR spectra, 636
  in spin–flip laser, 487
Zeeman modulation, 53
Zero field level crossing, 660
Zero gap, 167
Zero kinetic energy photoelectron (ZEKE-PE) spectroscopy, 432, 617 ff